HANDBOOK OF GAME THEORY
with Economic Applications
VOLUME 3

HANDBOOKS
IN
ECONOMICS

11

Series Editors

KENNETH J. ARROW
MICHAEL D. INTRILIGATOR

ELSEVIER

AMSTERDAM · BOSTON · LONDON · NEW YORK · OXFORD · PARIS
SAN DIEGO · SAN FRANCISCO · SINGAPORE · SYDNEY · TOKYO

HANDBOOK OF GAME THEORY
with Economic Applications

VOLUME 3

Edited by

ROBERT J. AUMANN
The Hebrew University of Jerusalem

and

SERGIU HART
The Hebrew University of Jerusalem

N·H

2002

ELSEVIER

AMSTERDAM · BOSTON · LONDON · NEW YORK · OXFORD · PARIS
SAN DIEGO · SAN FRANCISCO · SINGAPORE · SYDNEY · TOKYO

ELSEVIER SCIENCE B.V.
Sara Burgerhartstraat 25
P.O. Box 211, 1000 AE Amsterdam, The Netherlands

First edition 2002

Library of Congress Cataloging in Publication Data
A catalog record from the Library of Congress has been applied for.

ISBN: 0-444-89428-4
ISSN: 0169-7218 (Handbooks in Economics Series)

⊗ The paper used in this publication meets the requirements of ANSI/NISO Z39.48-1992 (Permanence of Paper).
Printed in The Netherlands.

INTRODUCTION TO THE SERIES

The aim of the *Handbooks in Economics* series is to produce Handbooks for various branches of economics, each of which is a definitive source, reference, and teaching supplement for use by professional researchers and advanced graduate students. Each Handbook provides self-contained surveys of the current state of a branch of economics in the form of chapters prepared by leading specialists on various aspects of this branch of economics. These surveys summarize not only received results but also newer developments, from recent journal articles and discussion papers. Some original material is also included, but the main goal is to provide comprehensive and accessible surveys. The Handbooks are intended to provide not only useful reference volumes for professional collections but also possible supplementary readings for advanced courses for graduate students in economics.

KENNETH J. ARROW and MICHAEL D. INTRILIGATOR

PUBLISHER'S NOTE

For a complete overview of the Handbooks in Economics Series, please refer to the listing on the last two pages of this volume.

CONTENTS OF THE HANDBOOK

VOLUME 1

PREFACE

This is the third and last volume of the *HANDBOOK OF GAME THEORY with Economic Applications*. For an introduction to the entire *Handbook*, please see the Preface to the first volume. Here we provide an overview of the organization of this third volume. As before, the space devoted in the Preface to the various chapters is no indication of their relative importance.

We follow the rough division into "noncooperative", "cooperative", and "general" adopted in the previous volumes. Chapters 41 through 52 are mainly noncooperative; 53 through 58, cooperative; 59 through 62, general. This division should not be taken too seriously; chapters may well contain aspects of both approaches. Indeed, we hope that the *Handbook* will help demonstrate that noncooperative and cooperative game theory are two sides of the same coin, which complement each other well.

The noncooperative part of the volume starts with three chapters on the basic concepts of the noncooperative approach. Chapter 41 discusses strategic equilibrium — "Nash", its refinements, and its extensions — without doubt the most used solution of game theory. The conceptual foundations and significance of these notions are by no means straightforward; Chapter 42 delves into these matters from various perspectives. The following chapter, 43, deals with incomplete information in multi-person interactive contexts, with special reference to "Bayesian" games.

Chapters 44 through 48 survey three important classes of games. The case of two players is of particular significance: A two-person interaction is the simplest and most basic there is; there are no non-trivial group structures. The theoretical and computational implications of this in the general (non-zero-sum) case are studied in Chapters 44 and 45 (two-person zero-sum games are covered in Chapter 20 in Volume 2). The other extreme — when there are many individually insignificant players (a continuum or "ocean") — is the subject of Chapter 46. The next two chapters are devoted to stochastic games: multi-stage games in which current actions physically affect future opportunities (unlike repeated games, surveyed in Volume 1, which reflect mainly the informational side of ongoing interactions). Chapter 47 covers the period up to 1995; Chapter 48, the period since then.

Considerable attention is paid in this *Handbook* to applications, economic and otherwise; in this volume, Chapters 49 through 52 and 57 through 61. Among the "hottest" applications is Industrial Organization, studied in Chapter 49, which discusses matters like collusion, entry and entry deterrence, predation, price wars, subsidies, strategic international trade, and "clearance" sales. The next chapter, 50, continues the discussion of noncooperative models of bargaining (Chapter 7 in Volume 1), with special reference to the implications of incomplete information about the other player (his reservation prices, red lines, true preferences, and so on). Here, the participants must credibly convey their position; a special section is devoted to empirical data from strikes. Inspection games — which may be viewed as statistical inference when data can be strategically manipulated — are covered in Chapter 51. This area became popular in the mid-sixties, at the height of the "cold war", when arms control and disarmament treaties were being negotiated; more recently, these techniques have been applied to auditing, environmental control, material accountancy, and so on. The last "noncooperative" chapter, 52, shows how the area of economic history benefits from the discipline of game theory.

The cooperative part of this volume centers on the Shapley value, its extensions and applications. Perhaps the most "successful" cooperative solution concept, the value is universally applicable (unlike the core), and leads to significant insights in widely varying contexts. The basic definitions and results are presented in Chapter 53. Relaxing and generalizing the axioms defining the value leads to the extensions discussed in Chapter 54. The previous two chapters study the basic model with transferable utility; in many contexts (especially economic ones) non-transferable utilities, covered in Chapter 55, are more realistic. Chapter 56 treats values of games with many individually insignificant players (a continuum or "ocean"). Inter alia, this case is basic to understanding perfectly competitive economies, whose value theory is reviewed in Chapter 57. Other applications of the value in continuum games — taxation, public goods and fixed prices — are dealt with in Chapter 58.

The *Handbook* closes with four "mixed" chapters, each combining cooperative and noncooperative tools and approaches in four very different areas of application. Chapter 59 treats political systems from the "micro", strategic viewpoint (sophisticated voting equilibria, manipulation of agendas and voting rules, and so on). This is to be distinguished from the more "macro"-oriented power considerations discussed in Chapter 32 in Volume 2. Chapter 60 treats a relatively new area of game-theoretic application: Law. It has to do both with the "game" implicit in a given legal system, and with the design of "optimal" legal systems. This brings us to the general mechanism design problem, or "implementation", the subject of Chapter 61: how to design the "rules of the game" to achieve certain desired results (as in auctions, matching markets, and final offer arbitration). The last chapter, 62, discusses some of the interplay between theory and experiments in games, and also the role that game theory plays in the design of interactive simulations (as in business games and war games).

This concludes our summary of the contents of the *Handbook*. Unfortunately, certain topics that we planned to cover were in the end omitted for one reason or another. They include adaptive dynamic learning, social psychology, macroeconomics, and the history of game theory. Needless to say, game theory is constantly expanding its horizons and increasing its depth (as attested to by the large and varied participation in the First Congress of the Game Theory Society in 2000 in Bilbao). Thus, any coverage in a *Handbook* is necessarily incomplete.

We would like to thank heartily all the many people who were involved in this project: the contributors, referees, series editors, and all those who helped us with their advice and support. Finally, we are grateful to our editorial assistant Mike Borns, without whom this volume would never have been completed.

ROBERT J. AUMANN and SERGIU HART

CONTENTS OF VOLUME 3

Chapter 46
Non-Cooperative Games with Many Players

Chapter 56
Values of Games with Infinitely Many Players
ABRAHAM NEYMAN

Chapter 41

STRATEGIC EQUILIBRIUM*

ERIC VAN DAMME

CentER for Economic Research, Tilburg University, Tilburg, The Netherlands

Contents

*This paper was written in 1994, and no attempt has been made to provide a survey of the developments since then. The author thanks two anonymous referees and the editors for their comments.

Handbook of Game Theory, Volume 3, Edited by R.J. Aumann and S. Hart

Abstract

This chapter of the Handbook of Game Theory (Vol. 3) provides an overview of the theory of Nash equilibrium and its refinements. The starting-point is the rationalistic approach to games and the question whether there exists a convincing, self-enforcing theory of rational behavior in non-cooperative games. Given the assumption of independent behavior of the players, it follows that a self-enforcing theory has to prescribe a Nash equilibrium, i.e., a strategy profile such that no player can gain by a unilateral deviation. Nash equilibria exist and for generic (finite) games there is a finite number of Nash equilibrium outcomes. The chapter first describes some general properties of Nash equilibria. Next it reviews the arguments why not all Nash equilibria can be considered self-enforcing. For example, some equilibria do not satisfy a backward induction property: as soon as a certain subgame is reached, a player has an incentive to deviate. The concepts of subgame perfect equilibria, perfect equilibria and sequential equilibria are introduced to solve this problem. The chapter defines these concepts, derives properties of these concepts and relates them to other refinements such as proper equilibria and persistent equilibria. It turns out that none of these concepts is fully satisfactory as the outcomes that are implied by any of these concepts are not invariant w.r.t. inessential changes in the game. In addition, these concepts do not satisfy a forward induction requirement. The chapter continues with formalizing these notions and it describes concepts of stable equilibria that do satisfy these properties. This set-valued concept is then related to the other refinements. In the final section of the chapter, the theory of equilibrium selection that was proposed by Harsanyi and Selten is described and applied to several examples. This theory selects a unique equilibrium for every game. Some drawbacks of this theory are noted and avenues for future research are indicated.

Keywords

non-cooperative games, Nash equilibrium, equilibrium refinements, stability, equilibrium selection

JEL classification: C70, C72

1. Introduction

It has been said that "the basic task of game theory is to tell us what strategies rational players will follow and what expectations they can rationally entertain about other rational players' strategies" [Harsanyi and Selten (1988, p. 342)]. To construct such a theory of rational behavior for interactive decision situations, game theorists proceed in an indirect, roundabout way, as suggested in von Neumann and Morgenstern (1944, §17.3). The analyst assumes that a satisfactory theory of rational behavior exists and tries to deduce which outcomes are consistent with such a theory. A fundamental requirement is that the theory should not be self-defeating, i.e., players who know the theory should have no incentive to deviate from the behavior that the theory recommends. For noncooperative games, i.e., games in which there is no external mechanism available for the enforcement of agreements or commitments, this requirement implies that the recommendation has to be self-enforcing. Hence, if the participants act independently and if the theory recommends a unique strategy for each player, the profile of recommendations has to be a Nash equilibrium: the strategy that is assigned to a player must be optimal for this player when the other players follow the strategies that are assigned to them. As Nash writes

> "By using the principles that a rational prediction should be unique, that the players should be able to make use of it, and that such knowledge on the part of each player of what to expect the others to do should not lead him to act out of conformity with the prediction, one is led to the concept" [Nash (1950a)].

Hence, a satisfactory normative theory that advises people how to play games necessarily must prescribe a Nash equilibrium in each game. Consequently, one wants to know whether Nash equilibria exist and what properties they have. These questions are addressed in the next section of this paper. In that section we also discuss the concept of rationalizability, which imposes necessary requirements for a satisfactory set-valued theory of rationality. A second immediate question is whether a satisfactory theory can prescribe just any Nash equilibrium, i.e., whether all Nash equilibria are self-enforcing. Simple examples of extensive form games have shown that the answer to this question is no: some equilibria are sustained only by incredible threats and, hence, are not viable as the expectation that a rational player will carry out an irrational (nonmaximizing) action is irrational. This observation has stimulated the search for more refined equilibrium notions that aim to formalize additional necessary conditions for self-enforcingness. A major part of this paper is devoted to a survey of the most important of these so-called refinements of the Nash equilibrium concept. (See Chapter 62 in this Handbook for a general critique of this refinement program.)

In Section 3 the emphasis is on extensive form solution concepts that aim to capture the idea of backward induction, i.e., the idea that rational players should be assumed to be forward-looking and to be motivated to reach their goals in the future, no matter what happened in the past. The concepts of subgame perfect, sequential, perfect and proper equilibria that are discussed in Section 3 can all be viewed as formalizations of this basic idea. Backward induction, however, is only one aspect of self-enforcingness, and it turns out that it is not sufficient to guarantee the latter. Therefore, in Section 4 we turn to another aspect of self-enforcingness, that of forward induction. We will discuss stability concepts that aim at formalizing this idea, i.e., that actions taken by rational actors in the past should be interpreted, whenever possible, as being part of a grand plan that is globally optimal. As these concepts are related to the notion of persistent equilibrium, we will have an opportunity to discuss this latter concept as well. Furthermore, as these ideas are most easily discussed in the normal form of the game, we take a normal-form perspective in Section 4. As the concepts discussed in this section are set-valued solution concepts, we will also discuss the extent to which set-valuedness contradicts the uniqueness of the rational prediction as postulated by Nash in the above quotation.

The fact that many games have multiple equilibria poses a serious problem for the "theory" rationale of Nash equilibrium discussed above. It seems that, for Nash's argument to make sense, the theory has to select a unique equilibrium in each game. However, how can a rational prediction be unique if the game has multiple equilibria? How can one rationally select an equilibrium? A general approach to this latter problem has been proposed in Harsanyi and Selten (1988), and Section 5 is devoted to an overview of that theory as well as a more detailed discussion of some of its main elements, such as the tracing procedure and the notion of risk-dominance. We also discuss some related theories of equilibrium selection in that section and show that the various elements of self-enforcingness that are identified in the various sections may easily be in conflict; hence, the search for a universal solution concept for non-cooperative games may continue in the future.

I conclude this introduction with some remarks concerning the (limited) scope of this chapter. As the Handbook contains an entire chapter on the conceptual foundations of strategic equilibrium (Chapter 42 in this Handbook), there are few remarks on this topic in the present chapter. I do not discuss the epistemic conditions needed to justify Nash equilibrium [see Aumann and Brandenburger (1995)], nor how an equilibrium can be reached. I'll focus on the formal definitions and mathematical properties of the concepts. Throughout, attention will be restricted to finite games, i.e., games in which the number of players as well as the action set of each of these players is finite. It should also be stressed that several other completely different rationales have been advanced for Nash equilibria, and that these are not discussed at all in this chapter. Nash (1950a) already discussed the "mass-action" interpretation of equilibria, i.e., that equilibria can result when the game is repeatedly played by myopic players who learn over time. I refer to Fudenberg and Levine (1998), and the papers cited therein for a discussion of the contexts in which learning processes can be expected to converge to Nash equilibria. Maynard Smith and Price (1973) showed that Nash equilibria can result as outcomes of evolutionary processes that wipe out less fit strategies through time. I refer to Hammerstein and Selten (1994) and Van Damme (1994) for a discussion of the role of Nash equilibrium in the biological branch of game theory, and to Samuelson (1997), Vega-Redondo (1996) and Weibull (1995) for more general discussions on evolutionary processes in games.

2. Nash equilibria in normal form games

2.1. Generalities

A (finite) *game in normal form* is a tuple $g = (A, u)$ where $A = A_1 \times \cdots \times A_I$ is a Cartesian product of finite sets and $u = (u_1, \ldots, u_I)$ is an I-tuple of functions $u_i : A \to \mathbb{R}$. The set $I = \{1, \ldots, I\}$ is the set of players, A_i is the set of pure strategies of player i and u_i is this player's payoff function. Such a game is played as follows: simultaneously and independently players choose strategies; if the combination $a \in A$ results, then each player i receives $u_i(a)$. A *mixed strategy* of player i is a probability distribution s_i on A_i and we write S_i for the set of such mixed strategies, hence

$$S_i = \left\{ s_i : A_i \to \mathbb{R}_+, \sum_{a_i \in A_i} s_i(a_i) = 1 \right\}. \tag{2.1}$$

(Generally, if C is any finite set, $\Delta(C)$ denotes the set of probability distributions on C, hence, $S_i = \Delta(A_i)$.) A mixed strategy may be interpreted as an act of deliberate randomization of player i or as a probability assessment of some player $j \neq i$ about how i is going to play. We return to these different interpretations below. We identify $a_i \in A_i$ with the mixed strategy that assigns probability 1 to a_i. We will write S for the set of mixed strategy profiles, $S = S_1 \times \cdots \times S_I$, with s denoting a generic element of S. Note that when strategies are interpreted as beliefs, taking strategy profiles as the primitive concept entails the implicit assumption that any two opponents j, k of player i have a common belief s_i about which pure action i will take. Alternatively, interpreting s as a profile of deliberate acts of randomization, the expected payoff to i when $s \in S$ is played, is written $u_i(s)$, hence

$$u_i(s) = \sum_{a \in A} \prod_{j \in I} s_j(a_j) u_i(a). \tag{2.2}$$

If $s \in S$ and $s_i' \in S_i$, then $s \backslash s_i'$ denotes the strategy profile in which each $j \neq i$ plays s_j while i plays s_i'. Occasionally we also write $s \backslash s_i' = (s_{-i}, s_i')$, hence, s_{-i} denotes the strategy vector used by the opponents of player i. We also write $S_{-i} = \prod_{j \neq i} S_j$ and $A_{-i} = \prod_{j \neq i} A_j$. We say that s_i' is a *best reply* against s in g if

$$u_i(s \backslash s_i') = \max_{s_i'' \in S_i} u_i(s \backslash s_i'') \tag{2.3}$$

and the set of all such best replies is denoted as $\mathcal{B}_i(s)$. Obviously, $\mathcal{B}_i(s)$ only depends on s_{-i}, hence, we can also view $\mathcal{B}_i$ as a correspondence from S_{-i} to S_i. If we write $B_i(s)$ for the set of pure best replies against s, hence $B_i(s) = \mathcal{B}_i(s) \cap A_i$, then obviously $\mathcal{B}_i(s)$ is the convex hull of $B_i(s)$. We write $\mathcal{B}(s) = \mathcal{B}_1(s) \times \cdots \times \mathcal{B}_I(s)$ and refer to $\mathcal{B} : S \to S$ as the *best-reply correspondence* associated with g. The pure best-reply correspondence is denoted by B, hence $B = \mathcal{B} \cap A$.

2.2. Self-enforcing theories of rationality

We now turn to solution concepts that try to capture the idea of a theory of rational behavior being self-enforcing. We assume that it is common knowledge that players are rational in the Bayesian sense, i.e., whenever a player faces uncertainty, he constructs subjective beliefs representing that uncertainty and chooses an action that maximizes his subjective expected payoffs. We proceed in the indirect way outlined in von Neumann and Morgenstern (1944, §17.3). We assume that a self-enforcing theory of rationality exists and investigate its consequences, i.e., we try to determine the theory from its necessary implications. The first idea for a solution of the game g is a definite strategy recommendation for each player, i.e., some $a \in A$. Already in simple examples like matching pennies, however, no such simple theory can be self-enforcing: there is no $a \in A$ that satisfies $a \in B(a)$, hence, there is always at least one player who has an incentive to deviate from the strategy that the theory recommends for him. Hence, a general theory of rationality, if one exists, must be more complicated.

Let us now investigate the possibilities for a theory that may recommend more than one action for each player. Let $C_i \subset A_i$ be the nonempty set of actions that the theory recommends for player i in the game g and assume that the theory, i.e., the set $C = X_i C_i$, is common knowledge among the players. If $|C_j| > 1$, then player i faces uncertainty about player j's action, hence, he will have beliefs $s_j^i \in S_j$ about what j will do. Assuming beliefs associated with different opponents to be independent, we can represent player i's beliefs by a mixed strategy vector $s^i \in S_{-i}$. (Below we also discuss the case of correlated beliefs; a referee remarked that he considered that to be the more relevant case.) The crucial question now is which beliefs can player i rationally entertain about an opponent j. If the theory C is self-enforcing, then no player j has an incentive to choose an action that is not recommended, hence, player i should assign zero possibility to any $a_j \in A_j \backslash C_j$. Writing $C_j(s_j)$ for the support of $s_j \in S_j$,

$$C_j(s_j) = \{a_j \in A_j \colon s_j(a_j) > 0\}, \tag{2.4}$$

we can write this requirement as

$$C_j(s_j^i) \subset C_j \quad \text{for all } i, j. \tag{2.5}$$

The remaining question is whether all beliefs s_j^i satisfying (2.5) should be allowed, i.e., whether i's beliefs about j can be represented by the set $\Delta(C_j)$. One might argue yes: if the opponents of j had an argument to exclude some $a_j \in C_j$, our theory would not be very convincing; the players would have a better theory available (simply replace C_j by $C_j \backslash \{a_j\}$). Hence, let us insist that all beliefs s_j^i satisfying (2.5) are allowed. Being Bayesian rational, player i will choose a best response against his beliefs s^i. His opponents, although not necessarily knowing his beliefs, know that he behaves in this way, hence, they know that he will choose an action in the set

$$B_i(C) = \bigcup \{B_i(s^i): s_j^i \in \Delta(C_j) \text{ for all } j\}. \tag{2.6}$$

Write $B(C) = X_i B_i(C)$. A necessary requirement for C to be self-enforcing now is that

$$C \subset B(C). \tag{2.7}$$

For, if there exists some $i \in I$ and some $a_i \in A_i$ with $a_i \in C_i \backslash B_i(C)$, then the opponents know that player i will not play a_i, but then they should assign probability zero to a_i, contradicting the assumption made just below (2.5). Write 2^A for the collection of subsets of A. Obviously, 2^A is a finite, complete lattice and the mapping $B : 2^A \to 2^A$ (defined by (2.6) and $B(\emptyset) = \emptyset$) is monotonic. Hence, it follows from Tarski's fixed point theorem [Tarski (1955)], or by direct verification that

   (i) there exists a nonempty set C satisfying (2.7),

  (ii) the set of all sets satisfying (2.7) is again a complete lattice, and

 (iii) the union of all sets C satisfying (2.7), to be denoted R, is a fixed point of B, i.e., $R = B(R)$, hence, R is the largest fixed point.

The set R is known as the set of pure *rationalizable strategy profiles* in g [Bernheim (1984), Pearce (1984)]. It follows by the above arguments that any self-enforcing set-valued theory of rationality has to be a subset of R and that R itself is such a theory. The reader can also easily check that R can be found by repeatedly eliminating the non-best responses from g, hence

$$\text{if } C^0 = A \text{ and } C^{t+1} = B(C^t), \quad \text{then } R = \bigcap_t C^t. \tag{2.8}$$

It is tempting to argue that, for C to be self-enforcing, it is not only necessary that (2.7) holds, but also that conversely

$$B(C) \subset C; \tag{2.9}$$

hence, that C actually must be a fixed point of B. The argument would be that, if (2.9) did not hold and if $a_i \in B_i(C) \backslash C_i$, player i could conceivably play a_i, hence, his opponents should assign positive probability to a_i. This argument, however, relies on the assumption that a rational player can play any best response. Since not all best responses might be equally good (some might be dominated, inadmissible, inferior or non-robust (terms that are defined below)), it is not completely convincing. We note that sets with the property (2.9) have been introduced in Basu and Weibull (1991) under the name of *curb sets*. (Curb is mnemonic for closed under rational behavior.) The set of all sets satisfying (2.9) is a complete lattice, i.e., there are minimal nonempty elements and such minimal elements are fixed points. (Fixed points are called tight curb sets in Basu and Weibull (1991).) We will encounter this concept again in Section 4.

Above we allowed two different opponents i and k to have different beliefs about player j, hence $s_j^i \neq s_j^k$. In such situations one should actually discuss the beliefs that i has about k's beliefs. To avoid discussing such higher-order beliefs, let us assume that players' beliefs are summarized by one strategy vector $s \in S$, hence we are discussing a theory that recommends a unique mixed strategy vector. For such a theory s to be self-enforcing, we obtain, arguing exactly as above, as a necessary requirement

$$C(s) \subset B(s), \tag{2.10}$$

where $C(s) = X_i C_i(s_i)$, hence, each player believes that each opponent will play a best response against his beliefs. A condition equivalent to (2.10) is

$$s \in \mathcal{B}(s) \tag{2.11}$$

or

$$u_i(s) = \max_{s_i' \in S_i} u_i\left(s \backslash s_i'\right) \quad \text{for all } i \in I. \tag{2.12}$$

A strategy vector s satisfying these conditions is called a *Nash equilibrium* [Nash (1950b, 1951)]. A standard application of Kakutani's fixed point theorem yields:

THEOREM 1 [Nash (1950b, 1951)]. *Every (finite) normal form game has at least one Nash equilibrium.*

We note that Nash (1951) provides an elegant proof that relies directly on Brouwer's fixed point theorem. We have already seen that some games only admit equilibria in mixed strategies. Dresher (1970) has computed that a large game with randomly drawn payoffs has a pure equilibrium with probability $1 - 1/e$. More recently, Stanford (1995) has derived a formula for the probability that a randomly selected game has exactly k pure equilibria. Gul et al. (1993) have shown that, for generic games, if there are $k \geqslant 1$ pure equilibria, then the number of mixed equilibria is at least $2k - 1$, a result to which we return below. An important class of games that admit pure equilibria are *potential games* [Monderer and Shapley (1996)]. A function $P : A \to \mathbb{R}$ is said to be an *ordinal potential* of $g = \langle A, u \rangle$ if for every $a \in A, i \in I$ and $a_i' \in A_i$

$$u_i(a) - u_i\left(a \backslash a_i'\right) > 0 \quad \text{iff} \quad P(a) - P\left(a \backslash a_i'\right) > 0. \tag{2.13}$$

Hence, if (2.13) holds, then g is ordinally equivalent to a game with common pay-offs and any maximizer of the potential P is a pure equilibrium of g. Consequently, a game g that has an ordinal potential, has a pure equilibrium. Note that g may have pure equilibria that do not maximize P and that there may be mixed equilibria as well. The function P is said to be an *exact potential* for g if

$$u_i(a) - u_i(a \backslash a_i') = P(a) - P(a \backslash a_i') \tag{2.14}$$

and Monderer and Shapley (1996) show that such an exact potential, when it exists, is unique up to an additive constant. Hence, the set of all maximizers of the potential is a well-defined refinement. Neyman (1997) shows that if the multilinear extension of P from A to S (as in (2.2)) is concave and continuously differentiable, every equilibrium of g is pure and is a maximizer of the potential. Another class of games, with important applications in economics, that admit pure strategy equilibria are *games with strategic complementaries* [Topkis (1979), Vives (1990), Milgrom and Roberts (1990, 1991), Milgrom and Shannon (1994)]. These are games in which each A_i can be ordered so that it forms a complete lattice and in which each player's best-response correspondence is monotonically nondecreasing in the opponents' strategy combination. The latter is guaranteed if each u_i is supermodular in a_i (i.e., $u_i(a_i, a_{-i}) + u_i(a_i', a_{-i}) \leqslant u_i(a_i \wedge a_i', a_{-i}) + u_i(a_i \vee a_i', a_{-i})$) and has increasing differences in (a_i', a_{-i}) (i.e., if $a_{-i} \geqslant a_{-i}'$, then $u_i(a_i, a_{-i}) - u_i(a_i, a_{-i}')$ is increasing in a_i). Topkis (1979) shows that such a game has at least one pure equilibrium and that there exists a largest and a smallest equilibrium, $\bar{a}$ and $\underline{a}$ respectively. Milgrom and Roberts (1990, 1991) show that $\bar{a}_i$ (resp. $\underline{a}_i$) is the largest (resp. smallest) serially undominated action of each player i, hence, by iterative elimination of strictly dominated strategies, the game can be reduced to the interval $[\underline{a}, \bar{a}]$. It follows that, if a game with strategic complementarities has a unique equilibrium, it is dominance-solvable, hence, that only the unique equilibrium strategies are rationalizable.

An equilibrium s^* is called *strict* if it is the unique best reply against itself, hence $\{s^*\} = B(s^*)$. Obviously, strict equilibria are necessarily in pure strategies, consequently they need not exist. An equilibrium s^* is called *quasi-strict* if all pure best replies are chosen with positive probability in s^*, that is, if $a_i \in B_i(s^*)$, then $s_i^*(a_i) > 0$. Also, quasi-strict equilibria need not exist: Van Damme (1987a, p. 56) gives a 3-player example. Norde (1999) has shown, however, that quasi-strict equilibria do exist in 2-person games.

An axiomatization of the Nash concept, using the notion of *consistency*, has been provided in Peleg and Tijs (1996). Given a name g, a strategy profile s and a coalition of players C, define the reduced game $g^{C,s}$ as the game that results from g if the players in $I \backslash C$ are committed to play strategies as prescribed by s. A family of games Γ is called closed if all possible reduced games, of games in Γ, again belong to Γ. A solution concept on Γ is a map φ that associates to each g in Γ a nonempty set of strategy profiles in g. φ is said to satisfy one-person rationality (OPR) if in every one-person game it selects all payoff maximizing actions. On a closed set of games Γ, φ is said to be consistent (CONS) if, for every g in Γ and s and C: if $s \in \varphi(g)$, then $s_C \in \varphi(g^{C,s})$, in other words, if some players are committed to play a solution, the remaining players find that the solution prescribed to them is a solution for their reduced game. Finally, a solution concept φ on a closed set Γ is said to satisfy converse consistency (COCONS) if, whenever s is such that $s_C \in \varphi(g^{C,s})$ for all $C \neq \phi$, then also $s \in \varphi(g)$; in other words, if the profile is a solution in all reduced games, then it is also a solution in the overall game. Peleg and Tijs (1996, Theorem 2.12) show that, on any closed family of games, the Nash equilibrium correspondence is characterized by the axioms OPR, CONS and COCONS.

Next, let us briefly turn to the assumption that strategy sets are finite. We note, first of all, that Theorem 1 can be extended to games in which the strategy sets A_i are nonempty, compact subsets of some finite-dimensional Euclidean space and the payoff functions u_i are continuous [Glicksberg (1952)]. If, in addition, A_i is convex and u_i is quasi-concave in a_i, there exists a pure equilibrium. Existence theorems for discontinuous games have been given in Dasgupta and Maskin (1986) and Simon and Zame (1990). In the latter paper it is pointed out that discontinuities typically arise from indeterminacies in the underlying (economic) problem and that these may be resolved by formulating an endogenous sharing rule. In this paper, emphasis will be on finite games. All games will be assumed finite, unless explicitly stated otherwise.

To conclude this subsection, we briefly return to the independence assumption that underlies the above discussion, i.e., the assumption that player i represents his uncertainty about his opponents by a mixed strategy vector $s^i \in S_{-i}$. A similar development is possible if we allow for correlation. In that case, (2.8) will be replaced by the procedure of iterative elimination of strictly dominated strategies, and the analogous concept to (2.9) is that of formations [Harsanyi and Selten (1988), see also Section 5]. The concept that corresponds to the parallel version of (2.12) is that of correlated equilibrium [Aumann (1974)]. Formally, if σ is a correlated strategy profile (i.e., σ is a probability distribution on A, $\sigma \in \Delta(A)$), then σ is a *correlated equilibrium* if for each player i and each $a_i \in A_i$

$$\text{if } \sigma_i(a_i) > 0 \quad \text{then} \sum_{a_{-i}} \sigma_{-i}(a_{-i}|a_i)u_i(a_{-i}, a_i) \geqslant \sum_{a_{-i}} \sigma_{-i}(a_{-i}|a_i)u_i\left(a_{-i}, a_i'\right)$$

$$\text{for all } a_i' \in A_i,$$

where $\sigma_i(a_i)$ denotes the marginal probability of a_i and where $\sigma_{-i}(a_{-i}|a_i)$ is the conditional probability of a_{-i} given a_i. One interpretation is as follows. Assume that an impartial mediator (a person or machine through which the players communicate) selects an outcome (a recommendation) $a \in A$ according to σ and then informs each player i privately about this player's personal recommendation a_i. If the above conditions hold, then, assuming that the opponents will always follow their recommendations, no player has any incentive to deviate from his recommendation, no matter what σ may recommend to him, hence, the recommendation σ is self-enforcing. Note that correlated equilibrium allows for private communication between the mediator and each player i: after hearing his recommendation a_i, player i does not necessarily know what action has been recommended to j, and two players i and k may have different posterior beliefs about what j will do. Aumann (1974) shows that a correlated equilibrium is nothing but a Nash equilibrium of an extended game in which the possibilities for communicating and correlating have been explicitly modeled, so in a certain sense there is nothing new here, but, of course, working with a reduced form solution concept may have its advantages. More importantly, Aumann (1985) argues that correlated beliefs arise naturally and he shows that, if it is common knowledge that each player is rational (in the Bayesian sense) and if players analyse the game by using a common prior, then the resulting distribution over outcomes must be a correlated equilibrium. Obviously, each Nash equilibrium is a correlated equilibrium, so that existence is guaranteed. An elementary proof of existence, which uses the fact that the set of correlated equilibria is a polyhedral set, has been given in Hart and Schmeidler (1989). Moulin and Vial (1978) gives an example of a correlated equilibrium with a payoff that is outside the convex hull of the Nash equilibrium payoffs, thus showing that players may benefit from communication with the mediator not being public. Myerson (1986) shows that, in extensive games, the timing of communication becomes of utmost importance. For more extensive discussion on communication and correlation in games, we refer to Myerson's Chapter 24 in this Handbook.

2.3. Structure, regularity and generic finiteness

For a game g we write $E(g)$ for the set of its Nash equilibria. It follows from (2.10) that $E(g)$ can be described by a finite number of polynomial inequalities, hence, $E(g)$ is a semi-algebraic set. Consequently, $E(g)$ has a finite triangulation, hence

THEOREM 2 [Kohlberg and Mertens (1986, Proposition 1)]. *The set of Nash equilibria of a game consists of finitely many connected components.*

Two equilibria s, s' of g are said to be *interchangeable* if, for each $i \in I$, also $s \backslash s'_i$ and $s' \backslash s_i$ are equilibria of g. Nash (1951) defined a *subsolution* as a maximal set of interchangeable equilibria and he called a game solvable if all its equilibria are interchangeable. Nash proved that each subsolution is a closed and convex set, in fact, that it is a product of polyhedral sets. Subsolutions need not be disjoint and a game may have uncountably many subsolutions [Chin et al. (1974)]. In the 2-person case, however, there are only finitely many subsolutions [Jansen (1981)]. A special class of solvable games is the 2-person *zero-sum games*, i.e., $u_1 + u_2 = 0$. For such games, all equilibria yield the same payoff, the so-called value of the game, and a strategy is an equilibrium strategy if and only if it is a minmax strategy. The reader is referred to Chapter 20 in this Handbook for a more extensive discussion of zero-sum 2-person games.

Let us now take a global perspective. Write $\Gamma = \Gamma_A$ for the set of all normal form games g with strategy space $A = A_1 \times \cdots \times A_I$. Obviously, $\Gamma = \mathbb{R}^{I \times A}$, a finite-dimensional linear space. Write E for the graph of the equilibrium correspondence, hence, $E = \{(g, s) \in \Gamma \times S : s \in E(g)\}$. Kohlberg and Mertens have shown that this graph E is itself a relatively simple object as it is homeomorphic to the space of games Γ. Kohlberg and Mertens show that the graph E (when compactified by adding a point ∞) looks like a deformation of a rubber sphere around the (similarly compactified) sphere of games. Hence, the graph is "simple", it just has folds, there are no holes, gaps or knots. Formally

THEOREM 3 [Kohlberg and Mertens (1986, Theorem 1)]. *Let π be the projection from E to Γ. Then there exists a homeomorphism φ from Γ to E such that $\pi \circ \varphi$ is homotopic to the identity on Γ under a homotopy that extends from Γ to its one-point compactification $\overline{\Gamma}$.*

Kohlberg and Mertens use Theorem 3 to show that each game has at least one component of equilibria that does not vanish entirely when the payoffs of the game are slightly perturbed, a result that we will further discuss in Section 4. We now move on to show that the graph E is really simple as generically (i.e., except on a closed set of games with measure zero) the equilibrium correspondence consists of a finite (odd number) of differentiable functions. We proceed in the spirit of Harsanyi (1973a), but follow the more elegant elaboration of Ritzberger (1994). At the end of the subsection, we briefly discuss some related recent work that provides a more general perspective.

Obviously, if s is a Nash equilibrium of g, then s is a solution to the following system of equations

$$s_i(a_i)\left[u_i(s \backslash a_i) - u_i(s)\right] = 0 \quad \text{all } i \in I, \ a_i \in A_i. \tag{2.15}$$

(The system (2.15) also admits solutions that are not equilibria – for example, any pure strategy vector is a solution – but this fact need not bother us at present.) For each player i, one equation in (2.15) is redundant; it is automatically satisfied if the others are. If we select, for each player i, one strategy $\bar{a}_i \in A_i$ and delete the corresponding equation, we are left with $m = \sum_i |A_i| - I$ equations. Similarly we can delete the variable $s_i(\bar{a}_i)$ for each i as it can be recovered from the constraint that probabilities add up to one. Hence, (2.15) reduces to a system of m equations with m unknowns. Taking each pair (i, a) with $i \in I$ and $a \in A_i \backslash \{\bar{a}_i\}$ as a coordinate, we can view S as a subset of $\mathbb{R}^m$ and the left-hand side of (2.15) as a mapping from S to $\mathbb{R}^m$, hence

$$f_{ia_i}(s) = s_i(a_i)\left[u_i(s \backslash a_i) - u_i(s)\right], \quad i \in I, \ a_i \in A_i \backslash \{\bar{a}_i\}. \tag{2.16}$$

Write $\partial f(s)$ for the Jacobian matrix of partial derivates of f evaluated at s and $|\partial f(s)|$ for its determinant. We say that s is a *regular equilibrium* of g if $|\partial f(s)| \neq 0$, hence, if the Jacobian is nonsingular. The reader easily checks that for all $i \in I$ and $a_i \in A_i$, if $s_i(a_i) = 0$, then $u_i(s \backslash a_i) - u_i(s)$ is an eigenvalue of $\partial f(s)$, hence, it follows that a regular equilibrium is necessarily quasi-strict. Furthermore, if s is a strict equilibrium, the above observation identifies m (hence, all) eigenvalues, so that any strict equilibrium is regular. A straightforward application of the implicit function theorem yields that, if s^* is a regular equilibrium of a game g^*, there exist neighborhoods U of g^* in Γ and

V of s^* in S and a continuous map $s : U \to V$ with $s(g^*) = s^*$ and $\{s(g)\} = E(g) \cap V$ for all $g \in U$. Hence, if s^* is a regular equilibrium of g^*, then around (g^*, s^*) the equilibrium graph E looks like a continuous curve. By using Sard's theorem (in the manner initiated in Debreu (1970)) Harsanyi showed that for almost all normal form games all equilibria are regular. Formally, the proof proceeds by constructing a subspace $\tilde{\Gamma}$ of Γ and a polynomial map $\varphi : \tilde{\Gamma} \times S \to \Gamma$ with the following properties (where $\tilde{g}$ denotes the projection of g in $\tilde{\Gamma}$):

(1) $\varphi(\tilde{g}, s) = g$ if $s \in E(g)$;
(2) $|\partial\varphi(\tilde{g}, s)| = 0$ if and only if $|\partial f(s)| = 0$.

Hence, if s is an irregular equilibrium of g, then g is a critical value of φ and Sard's theorem guarantees that the set of such critical values has measure zero. (For further details we refer to Harsanyi (1973a) and Van Damme (1987a).) We summarize the above discussion in the following theorem.

THEOREM 4 [Harsanyi (1973a)]. *Almost all normal form games are regular, that is, they have only regular equilibria. Around a regular game, the equilibrium correspondence consists of a finite number of continuous functions. Any strict equilibrium is regular and any regular equilibrium is quasi-strict.*

Note that Theorem 4 may be of limited value for games given originally in extensive form. Any such nontrivial extensive form gives rise to a strategic form that is not in general position, hence, that is not regular. We will return to generic properties associated with extensive form games in Section 4. We will now show that the finiteness mentioned in Theorem 4 can be strengthened to oddness. Again we trace the footsteps of Harsanyi (1973a) with minor modifications as suggested by Ritzberger (1994), a paper that in turn builds on Dierker (1972).

Consider a regular game g and add to it a logarithmic penalty term so that the payoff to i resulting from s becomes

$$u_i^\varepsilon(s) = u_i(s) + \varepsilon \sum_{a_i \in A_i} \ln s_i(a_i) \quad (i \in I, \ s \in S). \tag{2.17}$$

Obviously, an equilibrium of this game has to be in completely mixed strategies. (Since the payoff function is not multilinear, (2.10) and (2.12) are no longer equivalent; by an equilibrium we mean a strategy vector satisfying (2.12) with u_i replaced by u_i^ε. It follows easily from Kakutani's theorem that an equilibrium exists.) Hence, the necessary and sufficient conditions for equilibrium are given by the first order conditions:

$$f_{ia_i}^\varepsilon(s) = f_{ia_i}(s) + \varepsilon\left(1 - |A_i| s_i(a_i)\right) = 0, \quad i \in I, \ a_i \in A_i \backslash \{\bar{a}_i\}. \tag{2.18}$$

Because of the regularity of g, g has finitely many equilibria, say $s^1, \ldots, s^K$. The implicit function theorem tells us that for small ε, system (2.18) has at least K solutions $\{s^k(\varepsilon)\}_{k=1}^K$ with $s^k(\varepsilon) \to s^k$ as $\varepsilon \to 0$. In fact there must be exactly K solutions for small ε: because of regularity there cannot be two solution curves converging to the same s^k, and if a solution curve remained bounded away from the set $\{s^1, \ldots, s^K\}$, then it would have a cluster point and this would be an equilibrium of g. However, the latter is impossible since we have assumed g to be regular. Hence, if ε is small, f^ε has exactly as many zero's as g has equilibria. An application of the Poincaré–Hopf theorem for manifolds with boundary shows that each f^ε has an odd number of zero's, hence, g has an odd number of equilibria. (To apply the Poincaré–Hopf theorem, take a smooth approximation to the boundary of S, for example,

$$S(\delta) = \left\{ s \in S; \prod_{a_i \in A_i} s_i(a_i) \geqslant \delta \text{ all } i \right\}. \tag{2.19}$$

Then the Euler characteristic of $S(\delta)$ is equal to 1 and, for fixed ε, if δ is sufficiently small, f^ε points outward at the boundary of $S(\delta)$.) To summarize, we have shown:

THEOREM 5 [Harsanyi (1973a), Wilson (1971), Rosenmüller (1971)]. *Generic strategic form games have an odd number of equilibria.*

Ritzberger notes that actually we can say a little more. Recall that the index of a zero s of f is defined as the sign of the determinant $|\partial f(s)|$. By the Poincaré–Hopf theorem and the continuity of the determinant

$$\sum_{s \in E(g)} \text{sgn} |\partial f(s)| = 1. \tag{2.20}$$

It is easily seen that the index of a pure equilibrium is $+1$. Hence, if there are l pure equilibria, there must be at least $l - 1$ equilibria with index -1, and these must be mixed. This latter result was also established in Gul et al. (1993). In this paper, the authors construct a map g from the space of mixed strategies S into itself such that s is a fixed point of g if and only if s is a Nash equilibrium. They define an equilibrium s to be regular if it is quasi-strict and if $\det(I - g'(s)) \neq 0$. Using the result that the sum of the Lefschetz indices of the fixed points of a Lefschetz function is $+1$ and the observation that a pure equilibrium has index $+1$, they obtain their result that a regular game that has k pure equilibria must have at least $k - 1$ mixed ones. The authors also show that almost all games have only regular equilibria.

Recall that already Nash (1951) worked with a function f of which the fixed points correspond with the equilibria of the game. (See also the remark immediately below Theorem 1.) Nash's function is, however, different from that of Gul et al. (1993), and different from the function that we worked with in (2.15). This raises the question of whether the choice of the function matters. In recent work, Govindan and Wilson (2000) show that the answer is no. These authors define a Nash map as a continuous function $f : \Gamma \times S \to S$ that has the property that for each fixed game g the induced map $f_g : S \to S$ has as its fixed points the set of Nash equilibria of g. Given such a Nash map, the index ind(C, f) of a component C of Nash equilibria of g is defined in the usual way [see Dold (1972)]. The main result of Govindan and Wilson (2000) states that for any two Nash maps f, f' and any component C we have ind$(C, f) = $ ind(C, f'). Furthermore, if the degree of a component, deg(C), is defined as the local degree of the projection map from the graph E of the equilibrium correspondence to the space of games (cf. Theorem 3), then ind$(C, f) = $ deg(C) [see Govindan and Wilson (1997)].

2.4. Computation of equilibria: The 2-person case

The papers of Rosenmüller and Wilson mentioned in the previous theorem proved the generic oddness of the number of equilibria of a strategic form game in a completely different way than we did. These papers generalized the Lemke and Howson (1964) algorithm for the computation of equilibria in bimatrix games to n-person games. Lemke and Howson had already established the generic oddness of the number of equilibria for bimatrix games and the only difference between the 2-person case and the n-person case is that in the latter the pivotal steps involve nonlinear computations rather than the linear ones in the 2-person case. In this subsection we restrict ourselves to 2-person games and briefly outline the Lemke–Howson algorithm, thereby establishing another proof for Theorem 5 in the 2-person case. The discussion will be based upon Shapley (1974).

Let $g = \langle A, u \rangle$ be a 2-person game. The nondegeneracy condition that we will use to guarantee that the game is regular is

$$|C(s)| \geqslant |B(s)| \quad \text{for all } s \in S. \tag{2.21}$$

This condition is clearly satisfied for almost all bimatrix games and indeed ensures that all equilibria are regular. We write $L(s_i)$ for the set of "labels" associated with $s_i \in S_i$

$$L(s_i) = A_i \backslash C_i(s_i) \cup B_j(s_i). \tag{2.22}$$

If $m_i = |A_i|$, then, by (2.21), the number of labels if s_i is at most m_i. We will be interested in the set N_i of those s_i that have exactly m_i labels. This set is finite: the regularity condition (2.21) guarantees that for each set $L \subset A_1 \cup A_2$ with $|L| = m_i$ there is at most one $s_i \in S_i$ such that $L(s_i) = L$. Hence, the labelling identifies the strategy, so that the word label is appropriate. If $s_i \in N_i \backslash A_i$, then for each $a_i \in L(s_i)$ there exists (because of (2.21)) a unique ray in S_i emanating at s_i of points s_i' with $L(s_i') = L(s_i) \backslash \{a_i\}$, and moving in the direction of this ray we find a new point $s_i'' \in N_i$ after a finite distance. A similar remark applies to $s_i \in N_i \cap A_i$, except that in that case we cannot eliminate the label corresponding to $B_j(s_i)$. Consequently, we can construct a graph T_i with node set N_i that has m_i edges (of points s_i' with $|L(s_i')| = m_i - 1$) originating from each node in $N_i \backslash A_i$ and that has $m_i - 1$ edges originating from each node in $N_i \cap A_i$. We say that two nodes are adjacent if they are connected by an edge, hence, if they differ by one label.

Now consider the "product graph" T in the product set S: the set of nodes is $N = N_1 \times N_2$ and two nodes s, s' are adjacent if for some i, $s_i = s_i'$ while for $j \neq i$ we have that s_j and s_j' are adjacent in N_j. For $s \in S$, write $L(s) = L(s_1) \cup L(s_2)$. Obviously, we have that $L(s) = A_1 \cup A_2$ if and only if s is a Nash equilibrium of g. Hence, equilibria correspond to fully labelled strategy vectors and the set of such vectors will be denoted by E. The regularity assumption (2.21) implies that $E \subset N$, hence, E is a finite set. For $a \in A_1 \cup A_2$ write N^a for the set of $s \in N$ that miss at most the label a. The observations made above imply the following fundamental lemma:

LEMMA 1.
  (i) *If $s \in E$, $s_i = a$, then s is adjacent to no node in N^a.*
  (ii) *If $s \in E$, $s_i \neq a$, then s is adjacent to exactly one node in N^a.*
  (iii) *If $s \in N^a \backslash E$, $s_i = a$, then s is adjacent to exactly one node in N^a.*
  (iv) *If $s \in N^a \backslash E$, $s_i \neq a$, then s is adjacent to exactly two nodes in N^a.*

PROOF.
  (i) In this case s is a pure and strict equilibrium, hence, any move away from s eliminates labels other than a.
  (ii) If s is a pure equilibrium, then the only move that eliminates only the label a is to increase the probability of a in T_i. If s_i is mixed, then (2.21) implies that s_j is mixed as well. We either have $s_i(a) = 0$ or $a \in B_i(s_j)$. In the first case the only move that eliminates only label a is one in T_i (increase the probability of a), in the second case it is the unique move in T_j away from the region where a is a best response.
  (iii) The only possibility that this case allows is $s = (a, b)$ with b being the unique best response to a. Hence, if a' is the unique best response against b, the a' is the unique action that is labelled twice. The only possible move to an adjacent point in N^a is to increase the probability of a' in T_i.
  (iv) Let b be the unique action that is labelled by both s_1 and s_2, hence $\{b\} = L(s_1) \cap L(s_2)$. Note that s_i is mixed. If s_j is mixed as well, then we can either drop b from $L(s_1)$ in T_i or drop b from $L(s_2)$ in T_j. This yields two different possibilities and these are the only ones. If s_j is pure, then $b \in A_i$ and the same argument applies. □

The lemma now implies that an equilibrium can be found by tracing a path of almost completely labelled strategy vectors in N^a, i.e., vectors that miss at most a. Start at the pure strategy pair (a, b) where b is the best response to a. If a is also the best response to b, we are done. If not, then we are in case (iii) of the lemma and we can follow a unique edge in N^a starting at (a, b). The next node s we encounter is one satisfying either condition (ii) of the lemma (and then we are done) or condition (iv). In the latter case, there are two edges of N^a at s. We came in via one route, hence there is only one way to continue. Proceeding in similar fashion, we encounter distinct nodes of type (iv) until we finally hit upon a node of type (ii). The latter must eventually happen since N^a has finitely many nodes.

The lemma also implies that the number of equilibria is odd. Consider an equilibrium s' different from the one found by the above construction. Condition (ii) from the lemma guarantees that this equilibrium is connected to exactly one node in N^a as in condition (iv) of the lemma. We can now repeat the above constructive process until we end up at yet another equilibrium s''. Hence, all equilibria, except the distinguished one constructed above, appear in pairs: the total number of equilibria is odd.

Note that the algorithm described in this subsection offers no guarantee to find more than one equilibrium, let alone to find all equilibria. Shapley (1981) discusses a way of transforming the paths so as to get access to some of the previously inaccessible equilibria.

2.5. Purification of mixed strategy equilibria

In Section 2.1 we noted that mixed strategies can be interpreted both as acts of deliberate randomization as well as representations of players' beliefs. The former interpretation seems intuitively somewhat problematic; it may be hard to accept the idea of making an important decision on the basis of a toss of a coin. Mixed strategy equilibria also seem unstable: to optimize his payoff a player does not need to randomize; any pure strategy in the support is equally as good as the equilibrium strategy itself. The only reason a player randomizes is to keep the other players in equilibrium, but why would a player want to do this? Hence, equilibria in mixed strategies seem difficult to interpret [Aumann and Maschler (1972), Rubinstein (1991)].

Harsanyi (1973a) was the first to discuss the more convincing alternative interpretation of a mixed strategy of player i as a representation of the ignorance of the opponents as to what player i is actually going to do. Even though player i may follow a deterministic rule, the opponents may not be able to predict i's actions exactly, since i's decision might depend on information that the opponents can only assess probabilistically. Harsanyi argues that each player always has a tiny bit of private information about his own payoffs and he modifies the game accordingly. Such a slightly perturbed game admits equilibria in pure strategies and the (regular) mixed equilibria of the original unperturbed game may be interpreted as the limiting beliefs associated with these pure equilibria of the perturbed games. In this subsection we give Harsanyi's construction and state and illustrate his main result.

Let $g = \langle A, u \rangle$ be an I-person normal form game and, for each $i \in I$, let X_i be a random vector taking values in $\mathbb{R}^A$. Let $X = (X_i)_{i \in I}$ and assume that different components of X are stochastically independent. Let F_i be the distribution function of X_i and assume that F_i admits a continuously differentiable density f_i that is strictly positive on some ball Θ_i around zero in $\mathbb{R}^A$ (and 0 outside that ball). For $\varepsilon > 0$, write $g^\varepsilon(X)$ for the game described by the following rules:

    (i) nature draws x from X,
    (ii) each player i is informed about his component x_i,

(iii) simultaneously and independently each player i selects an action $a_i \in A_i$,

(iv) each player i receives the payoff $u_i(a) + \varepsilon x_i(a)$, where a is the action combination resulting from (iii).

Note that, if ε is small, a player's payoff is close to the payoff from g with probability approximately 1. What a player will do in $g^\varepsilon(X)$ depends on his observation and on his beliefs about what the opponents will do. Note that these beliefs are independent of his observation and that, no matter what the beliefs might be, the player will be indifferent between two pure actions with probability zero. Hence, we may assume that each player i restricts himself to a pure strategy in $g^\varepsilon(X)$, i.e., to a map $\sigma_i : \Theta_i \to A_i$. (If a player is indifferent, he himself does not care what he does and his opponents do not care since they attach probability zero to this event.) Given a strategy vector σ^ε in $g^\varepsilon(X)$ and $a_i \in A_i$ write $\Theta_i^{a_i}(\sigma^\varepsilon)$ for the set of observations where σ_i^ε prescribes to play a_i. If a player $j \neq i$ believes i is playing σ_i^ε, then the probability that j assigns to i choosing a_i is

$$s_i^\varepsilon(a_i) = \int_{\Theta_i^{a_i}(\sigma^\varepsilon)} dF_i. \tag{2.23}$$

The mixed strategy vector $s^\varepsilon \in S$ determined by (2.23) will be called the vector of beliefs associated with the strategy vector σ^ε. Note that all opponents j of i have the same beliefs about player i since they base themselves on the same information. The strategy combination σ^ε is an equilibrium of $g^\varepsilon(X)$ if, for each player i, it assigns an optimal action at each observation, hence

$$\text{if } x_i \in \Theta_i^{a_i}(\sigma^\varepsilon), \quad \text{then } a_i \in \arg\max\left[u_i\left(s^\varepsilon \backslash a_i\right) + \varepsilon x_i\left(s^\varepsilon \backslash a_i\right)\right]. \tag{2.24}$$

We can now state Harsanyi's theorem

THEOREM 6 [Harsanyi (1973b)]. *Let g be a regular normal form game and let the equilibria be $s^1, \ldots, s^K$. Then, for sufficiently small ε, the game $g^\varepsilon(X)$ has exactly K equilibrium belief vectors, say $s^1(\varepsilon), \ldots, s^K(\varepsilon)$, and these are such that $\lim_{\varepsilon \to 0} s^k(\varepsilon) = s^k$ for all k. Furthermore, the equilibrium $\sigma^k(\varepsilon)$ underlying the belief vector $s^k(\varepsilon)$ can be taken to be pure.*

We will illustrate this theorem by means of a simple example, the game from Figure 1. (The "t" stands for "tough", the "w" for "weak", the game is a variation of the battle of the sexes.) For analytical simplicity, we will perturb only one payoff for each player, as indicated in Figure 1.

The unperturbed game g ($\varepsilon = 0$ in Figure 1) has 3 equilibria, (t_1, w_2), (w_1, t_2) and a mixed equilibrium in which each player i chooses t_i with probability $s_i = 1 - u_j$ ($i \neq j$). The pure equilibria are strict, hence, it is easily seen that they can be approximated by equilibrium beliefs of the perturbed games in which the players have private information: if ε is small, then (t_i, w_j) is a strict equilibrium of $g^\varepsilon(x_1, x_2)$ for a set of

	w_2	t_2
t_1	$1, u_2 + \varepsilon x_2$	$0, 0$
w_1	$u_1 + \varepsilon x_1, u_2 + \varepsilon x_2$	$u_1 + \varepsilon x_1, 1$

Figure 1. A perturbed game $g^\varepsilon(x_1, x_2)$ $(0 < u_1, u_2 < 1)$.

(x_1, x_2)-values with large probability. Let us show how the mixed equilibrium of g can be approximated. If player i assigns probability s_j^ε to j playing t_j, then he prefers to play t_i if and only if

$$1 - s_j^\varepsilon > u_i + \varepsilon x_i. \tag{2.25}$$

Writing F_i for the distribution of X_i we have that the probability that j assigns to the event (2.25) is $F_i((s_j^\varepsilon - u_i)/\varepsilon)$, hence, to have an equilibrium of the perturbed game we must have

$$s_i^\varepsilon = F_i\big((1 - s_j^\varepsilon - u_i)/\varepsilon\big), \quad i, j \in \{1, 2\}, \ i \neq j. \tag{2.26}$$

Writing G_i for the inverse of F_i, we obtain the equivalent conditions

$$1 - s_j^\varepsilon - u_i - \varepsilon G_i(s_i^\varepsilon) = 0, \quad i, j \in \{1, 2\}, \ i \neq j. \tag{2.27}$$

For $\varepsilon = 0$, the system of equations has the regular, completely mixed equilibrium of g as a solution, hence, the implicit function theorem implies that, for ε sufficiently small, there is exactly one solution $(s_1^\varepsilon, s_2^\varepsilon)$ of (2.27) with $s_i^\varepsilon \to 1 - u_j$ as $\varepsilon \to 0$. These beliefs are the ones mentioned in Theorem 6. A corresponding pure equilibrium strategy for each player i is: play w_i if $x_i \leqslant (1 - s_j^\varepsilon - u_i)/\varepsilon$ and play b_i otherwise.

For more results on purification of mixed strategy equilibria, we refer to Aumann et al. (1983), Milgrom and Weber (1985) and Radner and Rosenthal (1982). These papers consider the case where the private signals that players receive do not influence the payoffs and they address the question of how much randomness there should be in the environment in order to enable purification. In Section 5 we will show that completely different results are obtained if players make common noisy observations on the entire game: in this case even some strict equilibria cannot be approximated.

3. Backward induction equilibria in extensive form games

Selten (1965) pointed out that, in extensive form games, not every Nash equilibrium can be considered self-enforcing. Selten's basic example is similar to the game g from Figure 2, which has (l_1, l_2) and (r_1, r_2) as its two pure Nash equilibria. The equilibrium (l_1, l_2) is not self-enforcing. Since the game is noncooperative, player 2 has no ability

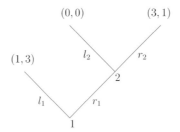

Figure 2. A Nash equilibrium that is not self-enforcing.

to commit himself to l_2. If he is actually called upon to move, player 2 strictly prefers to play r_2, hence, being rational, he will indeed play r_2 in that case. Player 1 can foresee that player 2 will deviate to r_2 if he himself deviates to r_1, hence, it is in the interest of player 1 to deviate from an agreement on (l_1, l_2). Only an agreement on (r_1, r_2) is self-enforcing.

Being a Nash equilibrium, (l_1, l_2) has the property that no player has an incentive to deviate from it if he expects the opponent to stick to this strategy pair. The example, however, shows that player 1's expectation that player 2 will abide by an agreement on (l_1, l_2) is nonsensical. For a self-enforcing agreement we should not only require that no player can profitably deviate if nobody else deviates, we should also require that the expectation that nobody deviates be rational. In this section we discuss several solution concepts, refinements of Nash equilibrium, that have been proposed as formalizations of this requirement. In particular, attention is focussed on sequential equilibria [Kreps and Wilson (1982a)] and on perfect equilibria [Selten (1975)]. Along the way we will also discuss Myerson's (1978) notion of proper equilibrium. First, however, we introduce some basic concepts and notation related to extensive form games.

3.1. Extensive form and related normal forms

Throughout, attention will be confined to *finite extensive form games with perfect recall*. Such a game g is given by
  (i) a collection I of players,
  (ii) a game tree K specifying the physical order of play,
  (iii) for each player i a collection H_i of information sets specifying the information a player has when he has to move. Hence H_i is a partition of the set of decision points of player i in the game and if two nodes x and y are in the same element h of the partition H_i, then i cannot distinguish between x and y,
  (iv) for each information set h, a specification of the set of choices C_h that are feasible at that set,
  (v) a specification of the probabilities associated with chance moves, and
  (vi) for each end point z of the tree and each player i a payoff $u_i(z)$ that player i receives when z is reached.

For formal definitions, we refer to Selten (1975), Kreps and Wilson (1982a) or Hart (1992). For an extensive form game g we write $g = (\Gamma, u)$ where Γ specifies the structural characteristics of the game and u gives the payoffs. Γ is called a *game form*. The set of all games with game form Γ can be identified with an $|I| \times |Z|$ Euclidean space, where I is the player set and Z the set of end points. The assumption of *perfect recall*, saying that no player ever forgets what he has known or what he has done, implies that each H_i is a partially ordered set.

A *local strategy* s_{ih} of player i at $h \in H_i$ is a probability distribution on the set C_h of choices at this information set h. It is interpreted as a plan for what i will do at h or as the beliefs of the opponents of what i will do at that information set. Note that the latter interpretation assumes that different players hold the same beliefs about what i will do at h and that these beliefs do not change throughout the game. A *behavior strategy* s_i of player i assigns a local strategy s_{ih} to each $h \in H_i$. We write S_{ih} for the set of local strategies at h and S_i for the set of all behavior strategies of player i. A behavior strategy s_i is called pure if it associates a pure action at each $h \in H_i$ and the set of all these strategies is denoted A_i.

A *behavior strategy combination* $s = (s_1, \ldots, s_I)$ specifies a behavior strategy for each player i. The probability distribution p^s that s induces on Z is called the *outcome* of s. Two strategies s_i' and s_i'' of player i are said to be *realization equivalent* if $p^{s \setminus s_i'} = p^{s \setminus s_i''}$ for each strategy combination s, i.e., if they induce the same outcomes against any strategy profile of the opponents. Player i's *expected payoff* associated with s is $u_i(s) = \sum_z p^s(z) u_i(z)$. If x is a node of the game tree, then p_x^s denotes the probability distribution that results on Z when the game is started at x with strategies s and $u_{ix}(s)$ denotes the associated expectation of u_i. If every information set h of g that contains a node y after x actually has all its nodes after x, then that part of the tree of g that comes after x is a game of its own. It is called the *subgame* of g starting at x.

The *normal form* associated with g is the normal form game $\langle A, u \rangle$ which has the same player set, the same sets of pure strategies and the same payoff functions as g has. A mixed strategy from the normal form induces a behavioral strategy in the extensive form and Kuhn's (1953) theorem for games with perfect recall guarantees that, conversely, for every mixed strategy, there exists a behavior strategy that is realization equivalent to it. [See Hart (1992) for more details.] Note that the normal form frequently contains many realization equivalent pure strategies for each player: if the information set $h \in H_i$ is excluded by player i's own strategy, then it is "irrelevant" what the strategy prescribes at h. The game that results from the normal form if we replace each equivalence class (of realization equivalent) pure strategies by a representative from that class, will be called the *semi-reduced normal form*. Working with the semi-reduced normal form implies that we do not specify player j's beliefs about what i will do at an information set $h \in H_i$ that is excluded by i's own strategy.

The *agent normal* form associated with g is the normal form game $\langle C, u \rangle$ that has a player ih associated with every information set h of each player i in g. This player ih has the set C_h of feasible actions as his pure strategy set and his payoff function is the payoff of the player i to whom he belongs. Hence, if $c_{ih} \in C_h$ for each $h \in \bigcup_i H_i$, then $s = (c_{ih})_{ih}$ is a (pure) strategy combination in g and we define $u_{ih}(s) = u_i(s)$ for $h \in H_i$. The agent normal form was first introduced in Selten (1975). It provides a local perspective, it decentralizes the strategy decision of player i into a number of local decisions. When planning his decision for h, the player does not necessarily assume that he is in full control of the decision at an information set $h' \in H_i$ that comes after h, but he is sure that the player/agent making the decision at that stage has the same objectives as he has. Hence, a player is replaced by a team of identically motivated agents.

Note that a pure strategy combination is a Nash equilibrium of the agent normal form if and only if it is a Nash equilibrium of the normal form. Because of perfect recall, a similar remark applies to equilibria that involve randomization, provided that we identify strategies that are realization equivalent. Hence, we may define a Nash equilibrium of the extensive form as a Nash equilibrium of the associated (agent) normal form and obtain (2.12) as the defining equations for such an equilibrium. It follows from Theorem 1 that each extensive form game has at least one Nash equilibrium. Theorems 2 and 3 give information about the structure of the set of Nash equilibria of extensive form games. Kreps and Wilson proved a partial generalization of Theorem 4:

THEOREM 7 [Kreps and Wilson (1982a)]. *Let Γ be any game form. Then, for almost all u, the extensive form game $\langle \Gamma, u \rangle$ has finitely many Nash equilibrium outcomes (i.e., the set $\{p^s(u): s$ is a Nash equilibrium of $\langle \Gamma, u \rangle\}$ is finite) and these outcomes depend continuously on u.*

Note that in this theorem, finiteness cannot be strengthened to oddness: any extensive form game with the same structure as in Figure 2 and with payoffs close to those in Figure 2 has l_1 and (r_1, r_2) as Nash equilibrium outcomes. Hence, Theorem 5 does not hold for extensive form games. Little is known about whether Theorem 6 can be extended to classes of extensive form games. However, see Fudenberg et al. (1988) for results concerning various forms of payoff uncertainty in extensive form games.

Before moving on to discuss some refinements in the next subsections, we briefly mention some coarsenings of the Nash concept that have been proposed for extensive form games. Pearce (1984), Battigalli (1997) and Börgers (1991) propose concepts of extensive form rationalizability. Some of these also aim to capture some aspects of forward induction (see Section 4). Fudenberg and Levine (1993a, 1993b) and Rubinstein and Wolinsky (1994) introduce, respectively, the concepts of "self-confirming equilibria" and of "rationalizable conjectural equilibria" that impose restrictions that are in between those of Nash equilibrium and rationalizability. These concepts require players to hold identical and correct beliefs about actions taken at information sets that are on the equilibrium path, but allow players to have different beliefs about opponents' play at information sets that are not reached. Hence, in such an equilibrium, if players only observe outcomes, no player will observe play that contradicts his predictions.

3.2. Subgame perfect equilibria

The Nash equilibrium condition (2.12) requires that each player's strategy be optimal from the ex ante point of view. Ex ante optimality implies that the strategy is also ex post optimal at each information set that is reached with positive probability in equilibrium, but, as the game of Figure 2 illustrates, such ex post optimality need not hold at the unreached information sets. The example suggests imposing ex post optimality as a necessary requirement for self-enforcingness but, of course, this requirement is meaningful only when conditional expected payoffs are well-defined, i.e., when the information set is a singleton. In particular, the suggestion is feasible for *games with perfect information*, i.e., games in which all information sets are singletons, and in this case one may require as a condition for s^* to be self-enforcing that it satisfies

$$u_{ih}(s^*) \geqslant u_{ih}(s^* \backslash s_i) \quad \text{for all } i, \text{ all } s_i \in S_i, \text{ all } h \in H_i. \tag{3.1}$$

Condition (3.1) states that at no decision point h can a player gain by deviating from s^* if after h no other player deviates from s^*. Obviously, equilibria satisfying (3.1) can be found by rolling back the game tree in a dynamic programming fashion, a procedure already employed in Zermelo (1912). It is, however, also worthwhile to remark that already in von Neumann and Morgenstern (1944) it was argued that this backward induction procedure was not necessarily justified as it incorporates a very strong assumption of "persistent" rationality. Recently, Hart (1999) has shown that the procedure may be justified in an evolutionary setting. Adopting Zermelo's procedure one sees that, for perfect information games, there exists at least one Nash equilibrium satisfying (3.1) and that, for generic perfect information games, (3.1) selects exactly one equilibrium. Furthermore, in the latter case, the outcome of this equilibrium is the unique outcome that survives iterated elimination of weakly dominated strategies in the normal form of the game. (Each elimination order leaves at least this outcome and there exists a sequence of eliminations that leaves nothing but this outcome; cf. Moulin (1979).)

Selten (1978) was the first paper to show that the solution determined by (3.1) may be hard to accept as a guide to practical behavior. (Of course, it was already known for a long time that in some games, such as chess, playing as (3.1) dictates may be infeasible since the solution s^* cannot be computed.) Selten considered the finite repetition of the game from Figure 2, with one player 2 playing the game against a sequence of different players in each round and with players always being perfectly informed about the outcomes in previous rounds. In the story that Selten associates with this game, player 2 is the owner of a chain store who is threatened by entry in each of finitely many towns. When entry takes place (r_1 is chosen), the chain store owner either acquiesces (chooses r_2) or fights entry (chooses l_2). The backward induction solution has players play (r_1, r_2) in each round, but intuitively, we expect player 2 to behave aggressively (choose l_2) at the beginning of the game with the aim of inducing later entrants to stay out. The *chain store paradox* is the paradox that even people who accept the logical validity of the backward induction reasoning somehow remain unconvinced by it and do not act in the manner that it prescribes, but rather act according to the intuitive solution. Hence, there is an inconsistency between plausible human behavior and game-theoretic reasoning. Selten's conclusion from the paradox is that a theory of perfect rationality may be of limited relevance for actual human behavior and he proposes a theory of limited rationality to resolve the paradox. Other researchers have argued that the paradox may be caused more by the inadequacy of the model than by the solution concept that is applied to it. Our intuition for the chain store game may derive from a richer game in which the deterrence equilibrium indeed is a rational solution. Such richer models have been constructed in Kreps and Wilson (1982b), Milgrom and Roberts (1982) and Aumann (1992). These papers change the game by allowing a tiny probability that player 2 may actually find it optimal to fight entry, which has the consequence that, when the game still lasts for a long time, player 2 will always play as if it were optimal to fight entry which forces player 1 to stay out.

The cause of the chain store paradox is the assumption of *persistent rationality* that underlies (3.1), i.e., players are forced to believe that even at information sets h that can be reached only by many deviations from s^*, behavior will be in accordance with s^*. This assumption that forces a player to believe that an opponent is rational even after he has seen the opponent make irrational moves has been extensively discussed and criticized in the literature, with many contributions being critical [see, for example, Basu (1988, 1990), Ben-Porath (1993), Binmore (1987), Reny (1992a, 1992b, 1993) and Rosenthal (1981)]. Binmore argues that human rationality may differ in systematic ways from the perfect rationality that game theory assumes, and he urges theorists to build richer models that incorporate explicit human thinking processes and that take these systematic deviations into account. Reny argues that (3.1) assumes that there is common knowledge of rationality throughout the game, but that this assumption is self-contradicting: once a player has "shown" that he is irrational (for example, by playing a strictly dominated move), rationality can no longer be common knowledge and solution concepts that build on this assumption are no longer appropriate. Aumann and Brandenburger (1995) however argue that Nash equilibrium does not build on this common knowledge assumption. Reny (1993), on the other hand, concludes from the above that a theory of rational behavior cannot be developed in a context that does not allow for irrational behavior, a conclusion similar to the one also reached in Selten (1975) and Aumann (1987). Aumann (1995), however, disagrees with the view that the assumption of common knowledge of rationality is impossible to maintain in extensive form games with perfect information. As he writes, "The aim of this paper is to present a coherent formulation and proof of the principle that in PI games, common knowledge of rationality implies backward induction" (p. 7) (see also Aumann (1998) for an application to Rosenthal's centipede game; the references in that paper provide further information, also on other points of view).

We now leave this discussion on backward induction in games with perfect information and move on to discuss more general games. Selten (1965) notes that the argument leading to (3.1) can be extended beyond the class of games with perfect information. If the game g admits a subgame γ, then the expected payoffs of s^* in γ depend only on what s^* prescribes in γ. Denote this restriction of s^* to γ by s_γ^*. Once the subgame γ is reached, all other parts of the game have become strategically irrelevant, hence, Selten argues that, for s^* to be self-enforcing, it is necessary that s_γ^* be self-enforcing for every subgame γ. Selten defined a *subgame perfect equilibrium* as an equilibrium s^* of g that induces a Nash equilibrium s_γ^* in each subgame γ of g and he proposed subgame perfection as a necessary requirement for self-enforcingness. Since every equilibrium of a subgame of a finite game can be "extended" to an equilibrium of the overall game, it follows that every finite extensive form game has at least one subgame perfect equilibrium.

Existence is, however, not as easily established for games in which the strategy spaces are continuous. In that case, not every subgame equilibrium is part of an overall equilibrium: players moving later in the game may be forced to break ties in a certain way, in order to guarantee that players who moved earlier indeed played optimally. (As a simple example, let player 1 first choose $x \in [0, 1]$ and let then player 2, knowing x, choose $y \in [0, 1]$. Payoffs are give by $u_1(x, y) = xy$ and $u_2(x, y) = (1 - x)y$. In the unique subgame perfect equilibrium both players choose 1 even though player 2 is completely indifferent when player 1 chooses $x = 1$.) Indeed, well-behaved continuous extensive form games need not have a subgame perfect equilibrium, as Harris et al. (1995) have shown. However, these authors also show that, for games with almost perfect information ("stage" games), existence can be restored if players can observe a common random signal before each new stage of the game which allows them to correlate their actions. For the special case where information is perfect, i.e., information sets are singletons, Harris (1985) shows that a subgame perfect equilibrium does exist even when correlation is not possible [see also Hellwig et al. (1990)].

Other chapters of this Handbook contain ample illustrations of the concept of subgame perfect equilibrium, hence, we will not give further examples. It suffices to remark here that subgame perfection is not sufficient for self-enforcingness, as is illustrated by the game from Figure 3.

The left-hand side of Figure 3 illustrates a game where player 1 first chooses whether or not to play a 2×2 game. If player 1 chooses r_1, both players are informed that r_1 has been chosen and that they have to play the 2×2 game. This 2×2 game is a subgame of the overall game and it has (t, l_2) as its unique equilibrium. Consequently,

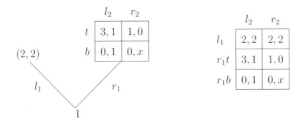

Figure 3. Not all subgame perfect equilibria are self-enforcing.

$(r_1 t, l_2)$ is the unique subgame perfect equilibrium. The game on the right is the (semi-reduced) normal form of the game on the left. The only difference between the games is that, in the normal form, player 1 chooses simultaneously between l_1, $r_1 t$ and $r_1 b$ and that player 2 does not get to hear that player 1 has not chosen l_1. However, these changes appear inessential since player 2 is indifferent between l_2 and r_2 when player 1 chooses l_1. Hence, it would appear that an equilibrium is self-enforcing in one game only if it is self-enforcing in the other. However, the sets of subgame perfect equilibria of these games differ. The game on the right does not admit any proper subgames so that the Nash equilibrium (l_1, r_2) is trivially subgame perfect.

3.3. Perfect equilibria

We have seen that Nash equilibria may prescribe irrational, non-maximizing behavior at unreached information sets. Selten (1975) proposes to eliminate such non-self-enforcing equilibria by eliminating the possibility of unreached information sets. He proposes to look at complete rationality as a limiting case of incomplete rationality, i.e., to assume that players make mistakes with small vanishing probability and to restrict attention to the limits of the corresponding equilibria. Such equilibria are called (trembling hand) perfect equilibria.

Formally, for an extensive form game g, Selten (1975) assumes that at each information set $h \in H_i$ player i will, with a small probability $\varepsilon_h > 0$, suffer from "momentary insanity" and make a mistake. Note that ε_h is assumed not to depend on the intended action at h. If such a mistake occurs, player i's behavior is assumed to be governed by some unspecified psychological mechanism which results in each choice c at h occurring with a strictly positive probability $\sigma_h(c)$. Selten assumes each of these probabilities ε_h and $\sigma_h(c)$ ($h \in H_i$, $c \in C_h$) to be independent of each other and also to be independent of the corresponding probabilities of the other players. As a consequence of these assumptions, if a player i intends to play the behavior strategy s_i, he will actually play the behavior strategy $s_i^{\varepsilon,\sigma}$ given by

$$s_i^{\varepsilon,\sigma}(c) = (1 - \varepsilon_h)s_{ih}(c) + \varepsilon_h \sigma_h(c) \quad (c \in C_h, \ h \in H_i). \tag{3.2}$$

Obviously, given these mistakes all information sets are reached with positive probability. Furthermore, if players intend to play $\bar{s}$, then, given the mistake technology specified by (ε, σ), each player i will at each information set h intend to choose a local strategy s_{ih} that satisfies

$$u_i\left(\bar{s}^{\varepsilon,\sigma} \backslash s_{ih}\right) \geqslant u_i\left(\bar{s}^{\varepsilon,\sigma} \backslash s_{ih}'\right) \quad \text{all } s_{ih}' \in S_{ih}. \tag{3.3}$$

If (3.3) is satisfied by $s_{ih} = \bar{s}_{ih}$ at each $h \in \bigcup_i H_i$ (i.e., if the intended action optimizes the payoff taking the constraints into account), then $\bar{s}$ is said to be an equilibrium of the perturbed game $g^{\varepsilon,\sigma}$. Hence, (3.3) incorporates the assumption of persistent rationality. Players try to maximize whenever they have to move, but each time they fall short of the ideal. Note that the definitions have been chosen to guarantee that $\bar{s}$ is an equilibrium of $g^{\varepsilon,\sigma}$ if and only if $\bar{s}$ is an equilibrium of the corresponding perturbation of the agent normal form of g. A straightforward application of Kakutani's fixed point theorem yields that each perturbed game has at least one equilibrium. Selten (1975) then defines $\bar{s}$ to be a *perfect equilibrium* of g if there exist sequences ε^k, σ^k of mistake probabilities ($\varepsilon^k > 0$, $\varepsilon^k \to 0$) and mistake vectors $\sigma_{ih}^k(c) > 0$ and an associated sequence s^k with s^k being an equilibrium of the perturbed game $g^{\varepsilon^k,\sigma^k}$ such that $s^k \to \bar{s}$ as $k \to \infty$. Since the set of strategy vectors is compact, it follows that each game has at least one perfect equilibrium. It may also be verified that $\bar{s}$ is a perfect equilibrium of g if and only if there exists a sequence s^k of completely mixed behavior strategies ($s_{ih}^k(c) > 0$ for all i, h, c, k) that converges to $\bar{s}$ as $k \to \infty$, such that $\bar{s}_{ih}$ is a local best reply against any element in the sequence, i.e.,

$$u_i\left(s^k \backslash \bar{s}_{ih}\right) = \max_{s_{ih} \in S_{ih}} u_i\left(s^k \backslash s_{ih}\right) \quad \text{(all } i,\ h,\ k\text{)}. \tag{3.4}$$

Note that for $\bar{s}$ to be perfect, it is sufficient that $\bar{s}$ can be rationalized by some sequence of vanishing trembles, it is not necessary that $\bar{s}$ be robust against all possible trembles. In the next section we will discuss concepts that insist on such stronger stability. We will also encounter concepts that require robustness with respect to specific sequences of trembles. For example, Harsanyi and Selten's (1988) concept of uniformly perfect equilibria is based on the assumption that all mistakes are equally likely. In contrast, Myerson's (1978) properness concept builds on the assumption that mistakes that are more costly are much less likely.

It is easily verified that each perfect equilibrium is subgame perfect. The converse is not true: in the game on the right of Figure 3 with $x \leqslant 1$, player 2 strictly prefers to play l_2 if player 1 chooses $r_1 t$ and $r_1 b$ by mistake, hence, only $(r_1 t, l_2)$ is perfect. However, since there are no subgames, (l_1, r_2) is subgame perfect.

By definition, the perfect equilibria of the extensive form game g are the perfect equilibria of the agent normal form of g. However, they need not coincide with the perfect equilibria of the associated normal form. Applying the above definitions to the normal form shows that $\bar{s}$ is a perfect equilibrium of a normal form game $g = \langle A, u \rangle$ if there exists a sequence of completely mixed strategy profiles s^k with $s^k \to \bar{s}$ such that $\bar{s} \in B(s^k)$ for all k, i.e.,

$$u_i\left(s^k \backslash \bar{s}_i\right) = \max_{s_i \in S_i} u_i\left(s^k \backslash s_i\right) \quad \text{(all } i,\ k\text{)}. \tag{3.5}$$

Hence, we claim that the global conditions (3.5) may determine a different set of solutions than the local conditions (3.4). As a first example, consider the game from Figure 4. In the extensive form, player 1 is justified to choose L if he expects himself, at his second decision node, to make mistakes with a larger probability than player 2 does. Hence, the outcome $(1, 2)$ is perfect in the extensive form. In the normal form, however, Rl_1 is a strategy that guarantees player 1 the payoff 1. This strategy dominates all

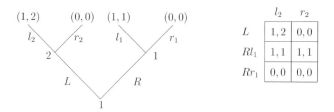

Figure 4. A perfect equilibrium of the extensive form need not be perfect in the normal form.

others, so that perfectness forces player 1 to play it, hence, only the outcome $(1, 1)$ is perfect in the normal form. Motivated by the consideration that a player may be more concerned with mistakes of others than with his own, Van Damme (1984) introduces the concept of a quasi-perfect equilibrium. Here each player follows a strategy that at each node specifies an action that is optimal against mistakes of other players, keeping the player's own strategy fixed throughout the game. Mertens (1992) has argued that this concept of "quasi-perfect equilibria" is to be preferred above "extensive form perfect equilibria". (We will return to the concept below.)

Conversely, we have that a perfect equilibrium of the normal form need not even be subgame perfect in the extensive form. The game from Figure 3 with $x > 1$ provides an example. Only the outcome $(3,1)$ is subgame perfect in the extensive form. In the normal form, player 2 is justified in playing r_2 if he expects that player 1 is (much) more likely to make the mistake $r_1 b$ than to make the mistake $r_1 t$. Hence, (l_1, r_2) is a perfect equilibrium in the normal form. Note that in both examples there is at least one equilibrium that is perfect in both the extensive and the normal form. Mertens (1992) discusses an example in which the sets of perfect equilibria of these game forms are disjoint: the normal form game has a dominant strategy equilibrium, but this equilibrium is not perfect in the extensive form of the game.

It follows from (3.5) that a perfect equilibrium strategy of a normal form game cannot be weakly dominated. (Strategy s_i' is said to be *weakly dominated* by s_i'' if $u_i(s \backslash s_i'') \geqslant u_i(s \backslash s_i')$ for all s and $u_i(s \backslash s_i'') > u_i(s \backslash s_i')$ for some s.) Equilibria in undominated strategies are not necessarily perfect, but an application of the separating hyperplane theorem shows that the two concepts coincide in the 2-person case [Van Damme (1983)]. (In the general case a strategy s_i is not weakly dominated if and only if it is a best reply against a completely mixed correlated strategy of the opponents.)

Before summarizing the discussion from this section in a theorem we note that games in which the strategy spaces are continua and payoffs are continuous need not have equilibria in undominated strategies. Consider the 2-player game in which each player i chooses x_i from $[0, 1/2]$ and in which $u_i(x) = x_i$ if $x_i \leqslant x_j/2$ and $u_i(x) = x_j(1 - x_i)/2 - x_j$ otherwise. Then the unique equilibrium is $x = 0$, but this is in dominated strategies. We refer to Simon and Stinchcombe (1995) for definitions of perfectness concepts for continuous games.

THEOREM 8 [Selten (1975)]. *Every game has at least one perfect equilibrium. Every extensive form perfect equilibrium is a subgame perfect equilibrium, hence, a Nash equilibrium. An equilibrium of an extensive form game is perfect if and only if it is perfect in the associated agent normal form. A perfect equilibrium of the normal form need not be perfect in the extensive form and also the converse need not be true. Every perfect equilibrium of a strategic form game is in undominated strategies and, in 2-person normal form games, every undominated equilibrium is perfect.*

3.4. Sequential equilibria

Kreps and Wilson (1982a) propose to eliminate irrational behavior at unreached information sets in a somewhat different way than Selten does. They propose to extend the applicability of (3.1) by explicitly specifying beliefs (i.e., conditional probabilities) at each information set so that posterior expected payoffs can always be computed. Hence, whenever a player reaches an information set, he should, in conformity with Bayesian decision theory, be able to produce a probability distribution on the nodes in that set that represents his uncertainty. Of course, players' beliefs should be consistent with the strategies actually played (i.e., beliefs should be computed from Bayes' rule whenever possible) and they should respect the structure of the game (i.e., if a player has essentially the same information at h as at h', his beliefs at these sets should coincide). Kreps and Wilson ensure that these two conditions are satisfied by deriving the beliefs from a sequence of completely mixed strategies that converges to the strategy profile in question.

Formally, a *system of beliefs* μ is defined as a map that assigns to each information set $h \in \bigcup_i H_i$ a probability distribution μ_h on the nodes in that set. The interpretation is that, when $h \in H_i$ is reached, player i assigns a probability $\mu_h(x)$ to each node x in h. The system of beliefs μ is said to be *consistent* with the strategy profile s if there exists a sequence s^k of completely mixed behavior strategies ($s^k_{ih}(c) > 0$ for all i, h, k, c) with $s^k \to s$ as $k \to \infty$ such that

$$\mu_h(x) = \lim_{k \to \infty} p^{s^k}(x|h) \quad \text{for all } h, x, \tag{3.6}$$

where $p^{s^k}(x|h)$ denotes the (well-defined) conditional probability that x is reached given that h is reached and s^k is played. Write $u^\mu_{ih}(s)$ for player i's expected payoff at h associated with s and μ, hence $u^\mu_{ih}(s) = \sum_{x \in h} \mu_h(x) u_{ix}(s)$, where u_{ix} is as defined in Section 3.1. The profile s is said to be *sequentially rational* given μ if

$$u^\mu_{ih}(s) \geqslant u^\mu_{ih}(s \backslash s'_i) \quad \text{all } i, h, s'_i. \tag{3.7}$$

An assessment (s, μ) is said to be a *sequential equilibrium* if μ is consistent with s and if s is sequentially rational given μ. Hence, the difference between perfect equilibria and sequential equilibria is that the former concept requires ex post optimality approaching the limit, while the latter requires this only at the limit. Roughly speaking, perfectness amounts to sequentiality plus admissibility (i.e., the prescribed actions are not locally dominated). Hence, if s is perfect, then there exists some μ such that (s, μ) is a sequential equilibrium, but the converse does not hold: in a normal form game every Nash equilibrium is sequential, but not every Nash equilibrium is perfect. The difference between the concepts is only marginal: for almost all games the concepts yield the same outcomes. The main innovation of the concept of sequential equilibrium is the explicit incorporation of the system of beliefs sustaining the strategies as part of the definition of equilibrium. In this, it provides a language for discussing the relative plausibility of various systems of beliefs and the associated equilibria sustained by them. This language has proved very effective in the discussion of equilibrium refinements in games with incomplete information [see, for example, Kreps and Sobel (1994)]. We summarize the above remarks in the following theorem. (In it, we abuse the language somewhat: $s \in S$ is said to be a sequential equilibrium if there exists some μ such that (s, μ) is sequential.)

THEOREM 9 [Kreps and Wilson (1982a), Blume and Zame (1994)]. *Every perfect equilibrium is sequential and every sequential equilibrium is subgame perfect. For any game structure Γ we have that for almost all games $\langle \Gamma, u \rangle$ with that structure the sets of perfect and sequential equilibria coincide. For such generic payoffs u, the set of perfect equilibria depends continuously on u.*

Let us note that, if the action spaces are continua, and payoffs are continuous, a sequential equilibrium need not exist. A simple example is the following signalling game [Van Damme (1987b)]. Nature first selects the type t of player 1, $t \in \{0, 2\}$ with both possibilities being equally likely. Next, player 1 chooses $x \in [0, 2]$ and thereafter player 2, knowing x but not knowing t, chooses $y \in [0, 2]$. Payoffs are $u_1(t, x, y) = (x - t)(y - t)$ and $u_2(t, x, y) = (1 - x)y$. If player 2 does not choose $y = 2 - t$ at $x = 1$, then type t of player 1 does not have a best response. Hence, there is at least one type that does not have a best response, and a sequential equilibrium does not exist.

In the literature one finds a variety of solution concepts that are related to the sequential equilibrium notion. In applications it might be difficult to construct an approximating sequence as in (3.6), hence, one may want to work with a more liberal concept that incorporates just the requirement that beliefs are consistent with s whenever possible, hence $\mu_h(x) = p^s(s|h)$ whenever $p^s(h) > 0$. Combining this condition with the sequential rationality requirement (3.7) we obtain the concept of *perfect Bayesian equilibrium* which has frequently been applied in dynamic games with incomplete information. Some authors have argued that in the context of an incomplete information game, one should impose a support restriction on the beliefs: once a certain type of a player is assigned probability zero, the probability of this type should remain at zero for the remainder of the game. Obviously, this restriction comes in handy when doing backward induction. However, the restriction is not compelling and there may exist no Nash equilibria satisfying it [see Madrigal et al. (1987), Neyman and Van Damme (1990)]. For further discussions on variations of the concept of perfect Bayesian equilibrium, the reader is referred to Fudenberg and Tirole (1991).

Since the sequential rationality requirement (3.7) has already been discussed extensively in Section 3.2, there is no need to go into detail here. Rather let us focus on the consistency requirement (3.6). When motivating this requirement, Kreps and Wilson refer to the intuitive idea that when a player reaches an information set h with $p^s(h) = 0$, he reassesses the game, comes up with an alternative hypothesis s' (with $p^{s'}(h) > 0$) about how the game is played and then constructs his beliefs at h from s'. A system of beliefs is called structurally consistent if it can be constructed in this way. Kreps and Wilson claimed that consistency, as in (3.6), implies structural consistency, but this claim was shown to be incorrect in Kreps and Ramey (1987): there may not exist an equilibrium that can be sustained by beliefs that are both consistent and structurally consistent. At first sight this appears to be a serious blow to the concept of sequential equilibrium, or at least to its motivation. However, the problem may be seen to lie in the idea of reassessing the game, which is not intuitive at all. First of all, it goes counter to the idea of rational players who can foresee the play in advance: they would have to reassess at the start. Secondly, interpreting strategy vectors as beliefs about how the game will be played implies there is no reassessment: all agents have the same beliefs about the behavior of each agent. Thirdly, the combination of structural consistency with the sequential rationality requirement (3.7) is problematic: if player i believes at h that s' is played, shouldn't he then optimize against s' rather then against s? Of course, rejecting structural consistency leaves us with the question of whether an alternative justification for (3.6) can be given. Kohlberg and Reny (1997) provide such a natural interpretation of consistency by relying on the idea of consistent probability systems.

3.5. Proper equilibria

In Section 3.1 we have seen that perfectness in the normal form is not sufficient to guarantee (subgame) perfectness in the extensive form. This observation raises the question of whether backward induction equilibria (say sequential equilibria) from the extensive form can already be detected in the normal form of the game. This question is important since it might be argued that, since a game is nothing but a collection of simultaneous individual decision problems, all information that is needed to solve these problems is already contained in the normal form of the game. The criteria for self-enforcingness in the normal form are no different from those in the extensive form: if the opponents of player i stick to s, then the essential information for i's decision problem is contained in this normal form: if i decides to deviate from s at a certain information set h, he can already plan that deviation beforehand, hence, he can deviate in the normal form. It turns out that the answer to the opening question is yes: an equilibrium that is proper in the normal form induces a sequential equilibrium outcome in every extensive form with that normal form.

Proper equilibria were introduced in Myerson (1978) with the aim of eliminating certain deficiencies in Selten's perfectness concept. One such deficiency is that adding strictly dominated strategies may enlarge the set of perfect equilibria. As an example, consider the game from the right-hand side of Figure 3 with the strategy $r_1 b$ eliminated. In this 2×2 game only $(r_1 t, b)$ is perfect. If we then add the strictly dominated strategy $r_1 b$, the equilibrium (l_1, r_2) becomes perfect. But, of course, strictly dominated strategies should be irrelevant; they cannot determine whether or not an outcome is self-enforcing. Myerson argues that, in Figure 3, player 2 should not believe that the mistake $r_1 b$ is more likely than $r_1 t$. On the contrary, since $r_1 t$ dominates $r_1 b$, the mistake $r_1 b$ is more severe than the mistake $r_1 t$; player 1 may be expected to spend more effort at preventing it and as a consequence it will occur with smaller probability. In fact, Myerson's concept of proper equilibrium assumes such a more costly mistake to occur with a probability that is of smaller order.

Formally, for a normal form game $\langle A, u \rangle$ and some $\varepsilon > 0$, a strategy vector $s^\varepsilon \in S$ is said to be an ε-proper equilibrium if it is completely mixed (i.e., $s_i^\varepsilon(a_i) > 0$ for all i, all $a_i \in A_i$) and satisfies

$$\text{if } u_i\big(s^\varepsilon \backslash a_i\big) < u_i\big(s^\varepsilon \backslash b_i\big) \quad \text{then } s_i^\varepsilon(a_i) \leqslant \varepsilon s_i^\varepsilon(b_i) \quad (\text{all } i, \ a_i, \ b_i). \tag{3.8}$$

A strategy vector $s \in S$ is a *proper equilibrium* if it is a limit, as $\varepsilon \to 0$, of a sequence s^ε of ε-proper equilibria.

Myerson (1978) shows that each strategic form game has at least one proper equilibrium and it is easily seen that any such equilibrium is perfect. Now, let g be an extensive form game with semi-reduced normal form $n(g)$ and, for $\varepsilon \to 0$, let s^ε be an ε-proper equilibrium of $n(g)$ with $s^\varepsilon \to s$ as $\varepsilon \to 0$. Since s^ε is completely mixed, it induces a completely mixed behavior strategy $\bar{s}^\varepsilon$ in g. Let $\bar{s} = \lim_{\varepsilon \to 0} \bar{s}^\varepsilon$. Then $\bar{s}$ is a behavior strategy vector that induces the same outcome as s does, $p^{\bar{s}} = p^s$. (Note that s need not induce a full behavior strategy vector; as s was defined in the semi-reduced normal form, it does not necessarily specify a unique action at information sets that are excluded by the players themselves.) Condition (3.8) now implies that at each information set h, $\bar{s}_i$ assigns positive probability only to the pure actions at h that maximize the local payoff at h against $\bar{s}^\varepsilon$. Namely, if c is a best response at h and c' is not, then for each pure strategy in the normal form that prescribes to play c' there exists a pure strategy that prescribes to play c and that performs strictly better against s^ε. (Take strategies that differ only at h.) Condition (3.8) then implies that in the normal form the total probability of the set of strategies choosing c' is of smaller order than the total probability of choosing c, hence, the limiting behavior strategy assigns probability 0 to c'. Hence, we have shown that each player always maximizes his local payoff, taking the mistakes of opponents into account. In other words, using the terminology of Van Damme (1984), the profile $\bar{s}$ is a quasi-perfect equilibrium. By the same argument, $\bar{s}$ is a sequential equilibrium. Formally, let μ^ε be the system of beliefs associated with $\bar{s}^\varepsilon$ and let $\mu = \lim_{\varepsilon \to 0} \mu^\varepsilon$. Then the assessment $(\bar{s}, \mu)$ satisfies (3.6) and (3.7), hence, it is a sequential equilibrium of g. The following theorem summarizes the above discussion.

THEOREM 10.
(i) [Myerson (1978)]. *Every strategic form game has at least one proper equilib-rium. Every proper equilibrium is perfect.*
(ii) [Van Damme (1984), Kohlberg and Mertens (1986)]. *Let g be an extensive form game with semi-reduced normal form $n(g)$. If s is a proper equilibrium of $n(g)$, then p^s is a quasi-perfect and a sequential equilibrium outcome in g.*

Mailath et al. (1997) have shown that sorts of converses to Theorem 10(ii) hold as well. Let $\{s^\varepsilon\}$ be a converging sequence of completely mixed strategies in a semi-reduced normal form game $n(g)$. This sequence induces a quasi-perfect equilibrium in every extensive form game with semi-reduced normal form $n(g)$ if and only if the limit of $\{s^\varepsilon\}$ is a proper equilibrium that is supported by the sequence. It is important that the same sequence be used: Hillas (1996) gives an example of a strategy profile that is not proper and yet is quasi-perfect in every associated extensive form. Secondly, Mailath et al. (1997) define a concept of normal form sequential equilibrium and they show that an equilibrium is normal form sequential if and only if it is sequential in every extensive form game with that semi-reduced normal form.

Theorem 10(ii) appears to be the main application of proper equilibrium. One other application deserves to be mentioned: in 2-person zero-sum games, there is essentially one proper equilibrium and it is found by the procedure of cautious exploitation of the mistakes of the opponent that was proposed by Dresher (1961) [see Van Damme (1983, Section 3.5)].

4. Forward induction and stable sets of equilibria

Unfortunately, as the game of Figure 5 (a modification of a game discussed by Kohlberg (1981)) shows, none of the concepts discussed thus far provides sufficient conditions for self-enforcingness. In this game player 1 first chooses between taking up an outside option that yields him 2 (and the opponent 0) and playing a battle-of-the-sexes game. Player 2 only has to move when player 1 chooses to play the subgame. In this game player 1 taking up his option and players continuing with (w_1, s_2) in the subgame con-stitutes a subgame perfect equilibrium. The equilibrium is even perfect: player 2 can

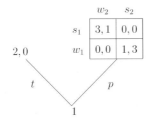

Figure 5. Battle of the sexes with an outside option.

argue that player 1 must have suffered from a sudden stroke of irrationality at his first move, but that he (player 1) will come back to his senses before his second move and continue with the plan (i.e., play w_1) as if nothing had happened. In fact, the equilibrium (t, s_2) is even proper in the normal form of the game: properness allows player 2 to conclude that the mistake pw_1 is more likely than the mistake ps_1 since pw_1 is better than ps_1 when player 2 plays s_2.

However, the outcome where player 1 takes up his option does not seem self-enforcing. If player 1 deviates and decides to play the battle-of-the-sexes game, player 2 should not rush to conclude that player 1 must have made a mistake; rather he might first investigate whether he can give a rational interpretation of this deviation. In the case at hand, such an explanation can indeed be given. For a rational player 1 it does not make sense to play w_1 in the subgame since the plan pw_1 is strictly dominated by the outside option. Hence, combining the rationality of player 1 with the fact that this player chose to play the subgame, player 2 should come to the conclusion that player 1 intends to play s_1 in the subgame, i.e., that player 1 bets on getting more than his option and that player 2 is sufficiently intelligent to understand this. Consequently, player 2 should respond by w_2, a move that makes the deviation of player 1 profitable, hence, the equilibrium is not self-enforcing.

Essentially what is involved here is an argument of forward induction: players' deductions about other players should be consistent with the assumption that these players are pursuing strategies that constitute rational plans for the overall game. The backward induction requirements discussed before were local requirements only taking into account rational behavior in the future. Forward induction requires that players' deductions be based on overall rational behavior whenever possible and forces players to take a global perspective. Hence, one is led to an analysis by means of the normal form. In this section we take such a normal form perspective and ask how forward induction can be formulated. The discussion will be based on the seminal work of Elon Kohlberg and Jean-François Mertens [Kohlberg and Mertens (1986), Kohlberg (1989), Mertens (1987, 1989a, 1989b, 1991)]. At this stage the reader may wonder whether there is no loss of information in moving to the normal form, i.e., whether the concepts that were discussed before can be recovered in the normal form. Theorem 10(ii) already provides part of the answer as it shows that sequential equilibria can be recovered. Mailath et al. (1993) discuss the question in detail and they show that also subgames and subgame perfect equilibria can be recovered in the normal form.

4.1. Set-valuedness as a consequence of desirable properties

Kohlberg and Mertens (1986) contains a first and partial axiomatic approach to the problem of what constitutes a self-enforcing agreement. (It should, however, be noted that the authors stress that their requirements should not be viewed as axioms since some of them are phrased in terms that are outside of decision theory.) Kohlberg and Mertens argue that a solution of a game should:
    (i) always exist,

(ii) be consistent with standard one-person decision theory,

(iii) be independent of irrelevant alternatives, and

(iv) be consistent with backward induction.

(The third requirement states that strategies which certainly will not be used by rational players can have no influence on whether a solution is self-enforcing; it is the formalization of the forward induction requirement that was informally discussed above; it will be given a more precise meaning below.) In this subsection we will discuss these requirements (except for (iv), which was extensively discussed in the previous section), and show that they imply that a solution cannot just be a single strategy profile but rather has to be a set of strategy profiles. In the next subsection, we give formalized versions of these basic requirements.

The existence requirement is fundamental and need not be discussed further. It guarantees that, if our necessary conditions for self-enforcingness leave only one candidate solution, that solution is indeed self-enforcing. Without having an existence theorem, we would run the risk of working with requirements that are incompatible, hence, of proving vacuous theorems.

The second requirement from the above list follows from the observation that a game is nothing but a simultaneous collection of one-person decision problems. In particular, it implies that the solution of a game can depend only on the normal form of that game. As a matter of fact, Kohlberg and Mertens argue that even less information than is contained in the normal form should be sufficient to decide on self-enforcingness. Namely, they take mixed strategies seriously as actions, and argue that a player is always able to add strategies that are just mixtures of strategies that are already explicitly given to them. Hence, they conclude that adding or deleting such strategies can have no influence on self-enforcingness. Formally, define the *reduced normal form* of a game as the game that results when all pure strategies that are equivalent to mixtures of other pure strategies have been deleted. (Hence, strategy $a_i \in A_i$ is deleted if there exists $s'_i \in S_i$ with $s'_i(a_i) = 0$ such that $u_j(s \backslash s'_i) = u_j(s \backslash a_i)$ for all j. The reader may ask whether the reduced normal form is well-defined. We return to this issue in the next subsection.) As a first consequence of consistency with one-person decision theory, Kohlberg and Mertens insist that two games with the same reduced normal form be considered equivalent and, hence, as having the same solutions.

Kohlberg and Mertens accept as a basic postulate from standard decision theory that a rational agent will only choose undominated strategies, i.e., that he will not choose a strategy that is weakly dominated. Hence, a second consequence of (ii) is that game solutions should be undominated (admissible) as well. Furthermore, if players do not choose undominated strategies, such strategies are actually irrelevant alternatives, hence, (iii) requires that they can be deleted without changing the self-enforcingness of the solution. Hence, the combination of (ii) and (iii) implies that self-enforcing solutions should survive iterated elimination of weakly dominated strategies. Note that the requirement of independence of dominated strategies is a "global" requirement that is applicable independent of the specific game solution that is considered. Once one has a specific candidate solution, one can argue that, if the solution is self-enforcing, no player will use a strategy that is not a best response against the solution, and, hence, that such inferior strategies should be irrelevant for the study of the self-enforcingness of the solution. Consequently, Kohlberg and Mertens require as part of (iii) that a self-enforcing solution remain self-enforcing when a strategy that is not a best response to this solution is eliminated.

Note that "axioms" (ii) and (iii) force the conclusion that only (3,1) can be self-enforcing in the game of Figure 5: only this outcome survives iterated elimination of weakly dominated strategies. The same conclusion can also be obtained without using such iterative elimination: it follows from backward induction together with the requirement that the solution should depend only on the reduced normal form. Namely, add to the normal form of the game of Figure 5 the mixed strategy $m = \lambda t + (1 - \lambda)s_1$ with $1/2 < \lambda < 1$ as an explicit pure strategy. The resulting game can be viewed as the normal form associated with the extensive form game in which first player 1 decides between the outside option t and playing a subgame with strategy sets $\{s_1, w_1, m\}$ and $\{s_2, w_2\}$. This extensive form game is equivalent to the extensive form from Figure 5, hence, they should have the same solutions. However, the newly constructed game only has (3,1) as a subgame perfect equilibrium outcome. (In the subgame w_1 is strictly dominated by m, hence player 2 is forced to play w_2.)

We will now show why (subsets of) the above "axioms" can only be satisfied by set-valued solution concepts. Consider the trivial game from Figure 6. Obviously, player 1 choosing his outside option is self-enforcing. The question is whether a solution should contain a unique recommendation for player 2. Note that the unique subgame perfect equilibrium of the game requires player 2 to choose $\frac{1}{2}l_2 + \frac{1}{2}r_2$, hence, according to (i) and (iv) this strategy should be the solution for player 2. However, according to (i), (ii), and (iii), the solution should be l_2 (eliminate the strategies in the order pl_1, r_2), while according to these same axioms the solution should also be r_2 (take the elimination order pr_1, l_2). Hence, we see that, to guarantee existence, we have to allow for set-valued solution concepts. Furthermore, we see that, even with set-valued concepts, only weak versions of the axioms – that just require set inclusion – can be satisfied. Actually, these weak versions of the axioms imply that in this game all equilibria should belong to the solution. Namely, add to the normal form of the game of Figure 6 the mixed strategy $\lambda t + (1 - \lambda)pl_1$ with $0 < \lambda < 1/2$ as a pure strategy. Then the resulting

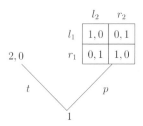

Figure 6. All Nash equilibria are equally good.

game is the normal form of an extensive form game that has $\mu l_2 + (1 - \mu) r_2$ with $\mu = (1 - 2\lambda)/(2 - \lambda)$ as the unique subgame perfect equilibrium strategy for player 2. Hence, as λ moves through $(0, 1/2)$, we trace half of the equilibrium set of player 2, viz. the set $\{\mu l_2 + (1 - \mu) r_2 \colon 0 < \mu < 1/2\}$. The other half can be traced by adding the mixed strategy $\lambda t + (1 - \lambda) p r_1$ in a similar way. Hence, axioms (i), (ii) and (iv) imply that all equilibrium strategies of player 2 belong to the solution.

The game of Figure 6 suggests that we should broaden our concept of a solution, that we should not always aim to give a unique strategy recommendation for a player at an information set that can be reached only by irrational moves of other players. In the game of Figure 6, it is unnecessary to give a specific recommendation to player 2 and any such recommendation is somewhat artificial. Player 2 is dependent upon player 1 so that his optimal choice seems to depend on the exact way in which player 1 is irrational. However, our analysis has assumed rational players, and since no model of irrationality has been provided, the theorist could be content to remain silent. Hence, a self-enforcing solution should not necessarily pin down completely the behavior of players at unreached points of the game tree. We may be satisfied if we can recommend what players do in those circumstances that are consistent with players being rational, i.e., as long as the play is according to the self-enforcing solution.

Note that by extending our solution notion to allow for multiple beliefs and actions after irrational moves we can also get rid of the unattractive assumption of persistent rationality that was discussed in Section 3.2 and that corresponds to a narrow reading of axiom (iv). We might just insist that a solution contains a backward induction equilibrium, not that it consists exclusively of backward induction equilibria. We should not fully exclude the possibility that a player just made a one-time mistake and will continue to optimize, but we should not force this assumption. In fact, the axioms imply that the solution of a perfect information game frequently cannot just consist of the subgame perfect equilibrium. Namely, consider the game TOL(3) represented in Figure 7, which is a variation of a game discussed in Reny (1993). (TOL(n) stands for "Take it or leave it with n rounds".) The game starts with \$1 on the table in round 1 and each time the

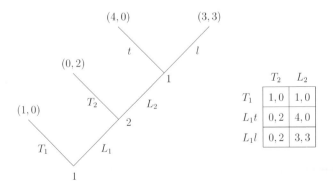

Figure 7. The game TOL(3).

game moves to a next round, the amount of money doubles. In round t, the player i with $i(\bmod 2) = t(\bmod 2)$ has to move. The game ends as soon as a player takes the money; if the game continues till the end, each player receives $\$2^{n-1} - 1$. (In the unique backwards induction equilibrium, player 1 takes the first dollar.)

The unique subgame perfect equilibrium of TOL(3) is $(T_1 t, T_2)$, which corresponds to (T_1, T_2) in the semi-reduced normal form. If the solution of the game were just (T_1, T_2), then $L_1 t$ would not be a best reply against the solution and according to "axiom" (iii), (T_1, T_2) should remain a solution when $L_1 t$ is eliminated. However, in the resulting 2×2 game, the unique perfect equilibrium is $(L_1 l, L_2)$ so that the axioms force this outcome to be the solution. Hence, the axioms imply that the strategy $\frac{1}{4}L_2 + \frac{3}{4}T_2$ of player 2 has to be part of the solution (in order to make $L_1 t$ a best response against the solution): player 1 cannot believe that, after player 2 has seen player 1 making the move L_1, player 2 believes player 1 to be rational. Intuitively, stable sets have to be large since they must incorporate the possibility of irrational play. Once we start eliminating dominated and/or inferior strategies, we attribute more rationality to players, make them more predictable and hence can make do with smaller stable sets. In formalizations of iterated elimination, we naturally have set inclusion.

The question remains of what type of mathematical objects are candidates for solutions of games now that we know that single strategy profiles do not qualify. In the above examples, the set of all equilibria was suggested, but the examples were special since there was only one connected component of Nash equilibria. More generally, one might consider connected components as solution candidates; however, this might be too coarse. For example, if, in Figure 6, we were to change player 2's payoffs in the subgame in such a way as to make l_2 strictly dominant, we would certainly recommend player 2 to play l_2 even if player 1 has made the irrational move. Hence, the answer to the question appears unclear. Motivated by constructions like the above, and by the interpretation of stable sets as patterns in which equilibria vary smoothly with beliefs or presentation effects, Kohlberg and Mertens suggest connected subsets of equilibria as solution candidates. Hence, a solution is a subset of a component of the equilibrium set (cf. Theorem 2). Note that since a generic game has only finitely many Nash equilibrium outcomes (Theorem 7), all equilibria in the same connected component yield the same outcome (since outcomes depend continuously on strategies); hence, for generic games each Kohlberg and Mertens solution indeed generates a unique outcome. (See also Section 4.3.)

4.2. Desirable properties for strategic stability

In this subsection we rephrase and formalize (some consequences of) the requirements (i)–(iv) from the previous subsection, taking the discussion from that subsection into account.

Let Γ be the set of all finite games. A *solution concept* is a map S that assigns to each game $g \in \Gamma$ a collection of non-empty subsets of mixed strategy profiles for the game. A *solution* T of g is a subset T of the set of mixed strategies of (the normal form of) g with $T \in \mathcal{S}(g)$, hence, it is a set of profiles that $\mathcal{S}$ allows. The first fundamental requirement that we encountered in the previous subsection was:

(E) *Existence*: $\mathcal{S}(g) \neq \emptyset$.

(We adopt the convention that, whenever a quantifier is missing, it should be read as "for all", hence (E) requires existence of at least one solution for each game.) Secondly, we will accept Nash equilibrium as a necessary requirement for self-enforcingness:

(NE) *Equilibrium*: If $T \in \mathcal{S}(g)$, then $T \subset E(g)$.

A third requirement discussed above was:

(C) *Connectedness*: If $T \in \mathcal{S}(g)$, then T is connected.

As discussed in the previous subsection, Kohlberg and Mertens insist that rational players only play admissible strategies. One formalization of admissibility is the restriction to undominated strategies, i.e., strategies that are best responses to correlated strategies of the opponents with full support. If players make their choices independently, a stronger admissibility suggests itself, viz. each player chooses a best response against a completely mixed strategy combination of the opponents. Formally, say that s_i' is an *admissible best reply* against s if there exists a sequence s^k of completely mixed strategy vectors converging to s such that s_i' is a best response against any element in the sequence. Write $\mathcal{B}_i^a(s)$ for the set of all such admissible best replies, $B_i^a(s) = \mathcal{B}_i^a(s) \cap A_i$, and let $\mathcal{B}^a(s) = X_i \mathcal{B}_i^a(s)$. For any subset S' of S write $\mathcal{B}^a(S') = \bigcup_{s \in S'} \mathcal{B}^a(s)$. We can now write the admissibility requirement as:

(A) *Admissibility*: If $T \in \mathcal{S}(g)$, then $T \subset \mathcal{B}^a(S)$.

Note that the combination of (NE) and (A) is almost equivalent to requiring perfection. The difference is that, as (3.5) shows, perfectness requires the approximating sequence s^k to be the same for each player. Accepting that players only use admissible best responses implies that a strategy that is not an admissible best response against the solution is certain not to be played and, hence, can be eliminated. Consequently, we can write the independence of irrelevant alternatives requirement as:

(IIA) *Independence of irrelevant alternatives*: If $a \notin B_i^a(T)$, then T contains a solution of the game in which a has been eliminated.

Note that $B_i^a(T) \subset B_i(T) \cap B_i^a(S)$, hence, (IIA) implies the requirements that strategies that are not best responses against the solution can be eliminated and that strategies that are not admissible can be eliminated.

It is also a fundamental requirement that irrelevant players should have no influence on the solutions. Formally, following Mertens (1990), say that a subset J of the player set constitutes a *small world* if their payoffs do not depend on the actions of the players not in J, i.e.,

$$\text{if } s_j = s_j' \text{ for all } j \in J, \quad \text{then } u_j(s) = u_j(s') \text{ for all } j \in J. \tag{4.1}$$

A solution has the small worlds property if the players outside the small world have no influence on the solutions inside the small world. Formally, if we write g_J for the game played by the insiders, then

> (SMW) *Small worlds property*: If J is a small world in g, then T_J is a solution in g_J if and only if it is a projection of a solution T in g.

Closely related to the small worlds property is the decomposition property: if two disjoint player sets play different games in different rooms, it does not matter whether one analyzes the games separately or jointly. Formally, say that g decomposes at J if both J and $\bar{J} = I \setminus J$ are small worlds in g.

> (D) *Decomposition*: If g decomposes at J, then $T \in \mathcal{S}(g)$ if and only if $T = T_J \times T_{\bar{J}}$ with $T_k \in \mathcal{S}(g_k)$ $(k \in \{J, \bar{J}\})$.

We now discuss the "player splitting property" which deals with another form of decomposition. Suppose g is an extensive form game and assume that there exists a partition P_i of H_i (the set of information sets of player i) such that, if h, h' belong to different elements of P_i, there is no path in the tree that cuts both h and h'. In such a case, the player can plan his actions at h without having to take into consideration his plans at h'. More generally, plans at one element of the partition can be made independently of plans at the other part and we do not limit the freedom of action of player i if we replace this player by a collection of agents, one agent for each element of P_i. Consequently, we should require the two games to have the same self-enforcing solutions:

> (PSP) *Player splitting property*: If g' is obtained from g by splitting some player i into a collection of independent agents, then $\mathcal{S}(g) = \mathcal{S}(g')$.

Note that for a solution concept having this property it does not matter whether a signalling game [Kreps and Sobel (1994)] is analyzed in normal form (also called the Harsanyi-form in this case) or in agent normal form (also called the Selten-form). Also note that in (PSP) the restriction to independent agents is essential: in the agent normal form of the game from Figure 5, the first agent of player 1 taking up his outside option is a perfectly sensible outcome: once the decisions are decoupled, the first action cannot signal anything about the second action. We will return to this in Section 5.

We will now formalize the requirement that the solution of a game depends only on those aspects of the problem that are relevant for the players' individual decision problems, i.e., that the solution is ordinal [cf. Mertens (1987)]. As already discussed above, Mertens argues that rational players will only play admissible best responses. A natural invariance requirement thus is that the solutions depend only on the admissible best-reply correspondence, formally

> (BRI) *Best reply invariance*: If $\mathcal{B}_g^a = \mathcal{B}_{g'}^a$, then $\mathcal{S}(g) = \mathcal{S}(g')$.

Note that the application of (BRI) is restricted to games with the same player sets and the same strategy spaces, hence, this requirement should be supplemented with requirements that the names of the players and the strategies do not matter, etc.

In the previous subsection we also argued that games with the same reduced normal form should be considered equivalent. In order to be able to properly formalize this invariance requirement it turns out to be necessary to extend the domain of games somewhat: after one has eliminated all equivalent strategies of a player, this player's strategy set need no longer be a full simplex. To deal with such possibilities, define an I-person *strategic form game* as a tuple $\langle S, u \rangle$ where $S = X_i S_i$ is a product of compact polyhedral sets and u is a multilinear map on S. Note that each such strategic form game has at least one equilibrium, and that the equilibrium set consists of finitely many connected components. Furthermore, all the requirements introduced above are meaningful for strategic form games. Say that an I-person strategic form game $g' = \langle S', u' \rangle$ is a *reduction* of the I-person normal form game $g = \langle A, u \rangle$ if there exists a map $f = (f_i)_{i \in I}$ with $f_i : S_i \to S_i'$ being linear and surjective, such that $u = u' \circ f$, hence, f preserves payoffs. Call such a map f an isomorphism from g onto g'. The requirement that the solution depends only on the reduced normal form may now be formalized as:

(I) *Invariance*: If f is an isomorphism from g onto g', then $S(g') = \{f(T) : T \in S(g)\}$ and $f^{-1}(T') = \bigcup \{T \in S(g) : f(T) = T'\}$ for all $T' \in S(g')$.

It should be stressed here that in Mertens (1987) the requirements (BRI) and (I) are derived from more abstract requirements of ordinality.

The final requirement that was discussed in the previous subsection was the backwards induction requirement, which, in view of Theorem 10, can be formalized as:

(BI) *Backwards induction*: If $T \in S(g)$, then T contains a proper equilibrium of g.

4.3. Stable sets of equilibria

In Kohlberg and Mertens (1986), three set-valued solution concepts are introduced that aim to capture self-enforcingness. Unfortunately, each of these fails to satisfy at least one of the above requirements so that that seminal paper does not come up with a definite answer as to what constitutes a self-enforcing outcome. The definitions of these concepts build on Theorem 3 that describes the structure of the Nash equilibrium correspondence. The idea is to look at components of Nash equilibria that are robust to slight perturbations in the data of the game. The structure theorem implies that at least one such component exists. By varying the class of perturbations that are allowed, different concepts are obtained. Formally define

(i) T is a *stable set of equilibria* of g if it is minimal among all the closed sets of equilibria T' that have the property that each perturbed game $g^{\varepsilon, \sigma}$ with ε close to zero has an equilibrium close to T'.

(ii) T is a *fully stable set of equilibria* of g if it is minimal among all the closed sets of equilibria T' that have the property that each game $\langle S', u \rangle$ with S_i' a polyhedral set in the interior of S_i (for each i) that is close to g has an equilibrium close to T'.

(iii) T is a *hyperstable set of equilibria* of g if it is minimal among all the closed sets of equilibria T' that have the property that for each game $g' = \langle A', u' \rangle$ that is equivalent to g and for each small payoff perturbation $\langle A', u^\varepsilon \rangle$ of g' there exists an equilibrium close to T'.

Kohlberg and Mertens (1986) show that every hyperstable set contains a set that is fully stable and that every fully stable set contains a stable set. Furthermore, from Theorem 3 they show that every game has a hyperstable set that is contained in a single connected component of Nash equilibria and, hence, that the same property holds for fully stable sets and stable sets. They, however, reject the (preliminary) concepts of hyperstability and full stability because these do not satisfy the admissibility requirement. Kohlberg and Mertens write that stability seems to be the "right" concept but they are forced to reject it since it violates (C) and (BI). (This concept does satisfy (E), (NE), (A), (IIA), (BRI), and (I).) Kohlberg and Mertens conclude with "we hope that in the future some appropriately modified definition of stability will, in addition, imply connectedness and backwards induction". Mertens (1989a, 1991) gives such a modification. We will consider it below.

An example of a game in which every fully stable set contains an inadmissible equilibrium (and hence in which every hyperstable set contains such an equilibrium) is obtained by changing the payoff vector $(0, 2)$ in TOL(3) (Figure 7) to $(5, 5)$. The unique admissible equilibrium then is $(L_1 t, T_2)$ but every fully stable set has to contain the strategy $(L_1 l)$ of player 1. Namely, if (in the normal form) player 1 trembles with a larger probability to T_1 when playing $L_1 t$ than when playing $L_1 l$, we obtain a perturbed game in which only $(L_1 l, T_2)$ is an equilibrium.

We now describe a 3-person game (attributed to Faruk Gul in Kohlberg and Mertens (1986)) that shows that stable sets may contain elements from different equilibrium components and need not contain a subgame perfect equilibrium. Player 3 starts the game by choosing between an outside option T (which yields payoffs $(0, 0, 2)$) or playing a simultaneous move subgame with players 1 and 2 in which each of the three players has strategy set $\{a, b\}$ and in which the payoffs are as in the matrix from the left-hand side of Figure 8 ($x, y \in \{a, b\}, x \neq y$). Hence, players 1 and 2 have identical payoffs and they want to make the same choice as player 3. Player 3 prefers these play-

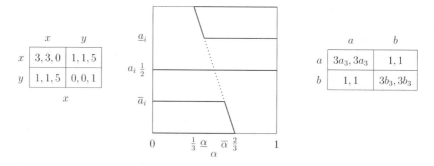

Figure 8. Stable sets need not contain a subgame perfect equilibrium.

ers to make different choices, but, if they make the same choice, he wants his choice to be different from theirs.

The game g described in the above story has a unique subgame perfect equilibrium: player 3 chooses to play the subgame and each player chooses $\frac{1}{2}a + \frac{1}{2}b$ in this subgame. This strategy vector constitutes a singleton component of the set of Nash equilibria. In addition, there are two components in which player 3 takes up his option T. Writing a_i (resp. b_i) for the probability with which player i ($i = 1, 2$) chooses a (resp. b), the strategies of players 1 and 2 in this component are the solutions to the pair of inequalities

$$4(a_1 + a_2) - 9a_1a_2 \leqslant 1 \quad \text{and} \quad 4(b_1 + b_2) - 9b_1b_2 \leqslant 1. \tag{4.2}$$

Note that the solution set of (4.2) indeed consists of two connected components, one around (a, a) (i.e., $a_1 = a_2 = 1$) and one around (b, b). Now, let us look at perturbations of (the normal form of) g. If player 3 chooses to play the subgame with positive stability ε and if, conditional on such a mistake, he chooses a (resp. b) with probability a_3 (resp. b_3), players 1 and 2 face the game from the right-hand side of Figure 8. The equilibria of this game are given by

$$a_i = \begin{cases} 0 & \text{if } a_3 < 1/3, \\ 0, 2 - 3a_3, 1 & \text{if } 1/3 < a_3 < 2/3, \\ 1 & \text{if } 2/3 < a_3, \end{cases}$$

hence, restricted to players 1 and 2, each perturbed game has (a, a) or (b, b) (or both) as a strict equilibrium. If players 1 and 2 coordinate on any of these strict equilibria, player 3 strictly prefers to play T, hence, $\{(a_1, a_2, T), (b_1, b_2, T)\}$ is a stable set of g. Obviously, this set does not contain the subgame perfect equilibrium, and even yields a different outcome.

A closer investigation may reveal the source of the difficulty and suggest a resolution of the problem. Since problematic zero-probability events arise only from player 3 choosing T, let us insist that he chooses to play the subgame with probability ε but, for simplicity, let us not perturb the strategies of players 1 and 2. Formally, consider a perturbed game $g^{\varepsilon,\sigma}$ with $\varepsilon_1 = \varepsilon_2 = 0$, $\varepsilon_3 = \varepsilon > 0$ and $\sigma_3 = (0, \alpha, 1 - \alpha)$, hence α is the probability that player 3 chooses a if he makes a mistake. The middle panel in Figure 8 displays, for any small $\varepsilon > 0$, the equilibrium correspondence as a function of α. (The horizontal axis corresponds to α, the vertical one to a_i.) Each perturbed game has an equilibrium close to the subgame perfect equilibrium of g. This equilibrium is represented by the horizontal line at $a_i = 1/2$. The inverted z-shaped figure corresponds to the solutions of (4.3). If players 1 and 2 play such a solution that is sufficiently close to a pure strategy, then T is the unique best response of player 3, hence, in that case we have an equilibrium of the perturbed game with $a_3 = \alpha$. If players 1 and 2 play a solution of (4.3) that is sufficiently close to $a_i = 1/2$ (i.e., they choose $a_i \in (\bar{a}_i, \underline{a}_i)$ corresponding to the dashed part of the z-curve), then we do not have an equilibrium unless $a_1 = a_2 = a_3 = 1/2$. (If $a_i > 1/2$, then the unique best response of player 3 is to play b, hence $a_3 = \varepsilon\alpha < 1/3$ so that by (4.3) we should have $a_i = 0$.) The points $\underline{\alpha}$ and $\bar{\alpha}$ where the solid z-curve changes into the dashed z-curve are somewhat special. Writing $\underline{a}_i = 2 - 3\underline{\alpha}$ we have that if each player i ($i = 1, 2$) chooses a with probability $\underline{a}_i$, then player 3's best responses are T and b. Consequently, by playing b voluntarily with the appropriate probability, player 3 can enforce any $a_3 \in (\varepsilon\alpha, \alpha)$, hence, if ε is sufficiently small and $\alpha > \underline{\alpha}$, player 3 can enforce $a_3 = \underline{\alpha}$. We see that for each $\alpha \geqslant \underline{\alpha}$, the perturbed game has an equilibrium with $a_i = \underline{a}_i$. In the diagram, this branch is represented by the horizontal line at $\underline{a}_i$. Of course, there is a similar branch at $\bar{a}_i$. Since the above search was exhaustive, the middle panel in Figure 8 contains a complete description of the equilibrium graph, or at least of its projection on the (α, a_i)-space.

The critical difference between the "middle" branch of the equilibrium correspondence and each of the other two branches is that in the latter cases it is possible to continuously deform the graph, leaving the part over the extreme perturbations ($\alpha \in \{0, 1\}$) intact, in such a way that the interior is no longer covered, i.e., such that there are no longer "equilibria" above the positive perturbations. Hence, although the projection from the union of the top and bottom branches to the perturbations is surjective (as required by stability), this projection is homologically trivial, i.e., it is homologous to the identity map of the boundary of the space of perturbations. Building on this observation, and on the topological structure of the equilibrium correspondence more generally, Mertens (1989a, 1991) proposes a refinement of stability (to be called M-stability) that essentially requires that the projection from a neighborhood of the set to a neighborhood of the game should be homologically nontrivial. As the formal definition is somewhat involved we will not give it here but confine ourselves to stating its main properties. Let us, however, note that Mertens does not insist on minimality; he shows that this conflicts with the ordinality requirement (cf. Section 4.5).

THEOREM 11 [Mertens (1989a, 1990, 1991)]. *M-stable sets are closed sets of normal form perfect equilibria that satisfy all properties listed in the previous subsection.*

We close this subsection with a remark and with some references to recent literature. First of all, we note that also in Hillas (1990) a concept of stability is defined that satisfies all properties from the list of the previous subsection. (We will refer to this concept as H-stability. To avoid confusion, we will refer to the stability concept that was defined in Kohlberg and Mertens as KM-stability.) T is an *H-stable set of equilibria* of g if it is minimal among all the closed sets of equilibria T' that have the following property: each upper-hemicontinuous compact convex-valued correspondence that is pointwise close to the best-reply correspondence of a game that is equivalent to g has a fixed point close to T'. The solution concept of H-stable sets satisfies the requirements (E), (NE), (C), (A), (IIA), (BRI), (I) and (BI), but it does not satisfy the other requirement from Section 4.2. (The minimality requirement forces H-stable sets to be connected, hence, in the game of Figure 8 only the subgame perfect equilibrium outcome is H-stable.) In Hillas et al. (1999) it is shown that each M-stable set contains an H-stable set. That paper discusses a couple of other related concepts as well.

I conclude this section by referring to some other recent work. Wilson (1997) discusses the role of admissibility in identifying self-enforcing outcomes. He argues that admissibility criteria should be deleted when selecting among equilibrium components, but that they may be used in selecting equilibria from a component, hence, Wilson argues in favor of perfect equilibria in essential components, i.e., components for which the degree (cf. Section 2.3) is non-zero. Govindan and Wilson (1999) show that, in 2-player games, maximal M-stable sets are connected components of perfect equilibria, hence, such sets are relatively easy to compute and their number is finite. (On finiteness, see Hillas et al. (1997).) The result implies that an essential component contains a stable set; however, as Govindan and Wilson illustrate by means of several examples, inessential components may contain stable sets as well.

4.4. Applications of stability criteria

Concepts related to strategic stability have been frequently used to narrow down the number of equilibrium outcomes in games arising in economic contexts. (Recall that in generic extensive games all equilibria in the same component have the same outcome so that we can speak of stable and unstable outcomes.) Especially in the context of signalling games many refinements have been proposed that were inspired by stability or by its properties [cf. Cho and Kreps (1987), Banks and Sobel (1987) and Cho and Sobel (1990)]. As this literature is surveyed in the chapter by Kreps and Sobel (1994), there is no need to discuss these applications here [see Van Damme (1992)]. I shall confine myself here to some easy applications and to some remarks on examples where the fine details of the definitions make the difference.

It is frequently argued that the Folk theorem, i.e., the fact that repeated games have a plethora of equilibrium outcomes (see Chapter 4 in this Handbook) shows a fundamental weakness of game theory. However, in a repeated game only few outcomes may actually be strategically stable. (General results, however, are not yet available.) To illustrate, consider the twice-repeated battle-of-the-sexes game, where the stage game payoffs are as in (the subgame occurring in) Figure 5 and that is played according to the standard information conditions. The path $\langle (s_1, w_2), (s_1, w_2) \rangle$ in which player 1's most preferred stage equilibrium is played twice is not stable. Namely, the strategy $s_2 w_2$ (i.e., deviate to s_2 and then play w_2) is not a best response against any equilibrium that supports this path, hence, if the path were stable, then according to (IIA) it should be possible to delete this strategy. However, the resulting game does not have an admissible equilibrium with payoff $(6, 2)$ so that the path cannot be stable. (Admissibility forces player 1 to respond with w_1 after 2 has played s_2; hence, the deviation $s_2 s_2$ is profitable for player 2.) For further results on stability in repeated games, the reader is referred to Balkenborg (1993), Osborne (1990), Ponssard (1991) and Van Damme (1989a).

Stability implies that the possibility to inflict damage on oneself confers power. Suppose that before playing the one-shot battle-of-the-sexes game, player 1 has the opportunity to burn 1 unit of utility in a way that is observable to player 2. Then the only stable outcome is the one in which player 1 does not burn utility and players play (s_1, w_1), hence, player 1 gets his most preferred outcome. The argument is simply that the game can be reduced to this outcome by using (IIA). If both players can throw away utility, then stability forces utility to be thrown away with positive probability: any other outcome can be upset by (IIA). [See Van Damme (1989a) for further details and Ben-Porath and Dekel (1992), Bagwell and Ramey (1996), Glazer and Weiss (1990) for applications.]

Most applications of stability in economics use the requirements from Section 4.2 to limit the set of solution candidates to one and they then rely on the existence theorem to conclude that the remaining solution must be stable. Direct verification of stability may be difficult; one may have to enumerate all perturbed games and investigate how the equilibrium graph hangs together [see Mertens (1987, 1989a, 1989b, 1991) for various illustrations of this procedure and for arguments as to why certain shortcuts may not work]. Recently, Wilson (1992) has constructed an algorithm to compute a simply stable component of equilibria in bimatrix games. Simply stable sets are robust against a restricted set of perturbations, viz. one perturbs only one strategy (either its probability or its payoff). Wilson amends the Lemke and Howson algorithm from Section 2.3 to make it applicable to nongeneric bimatrices and he adds a second stage to it to ensure that it can only terminate at a simply stable set. Whenever the Lemke and Howson algorithm terminates with an equilibrium that is not strict, Wilson uses a perturbation to transit onto another path. The algorithm terminates only when all perturbations have been covered by some vertex in the same component. Unfortunately, Wilson cannot guarantee that a simply stable component is actually stable.

In Van Damme (1989a) it was argued that stable sets (as originally defined by Kohlberg and Mertens) may not fully capture the logic of forward induction. Following an idea originally discussed in McLennan (1985) it was argued that if an information set $h \in H_i$ can be reached only by one equilibrium s^*, and if s^* is self-enforcing, player i should indeed believe that s^* is played if h is reached and, hence, only s^*_{ih} should be allowed at h. A 2-person example in Van Damme (1989a) showed that stable equilibria need not satisfy this forward-induction requirement. (Actually Gul's example (Figure 8) already shows this.) Hauk and Hurkens (1999) have recently shown that this forward-induction property is satisfied by none of the stability concepts discussed above. On the other hand they show that this property is satisfied by some evolutionary equilibrium concepts that are related to those discussed in Section 4.5 below.

Gul and Pearce (1996) argue that forward induction loses much of its power when public randomization is allowed; however, Govindan and Robson (1998) show that the Gul and Pearce argument depends essentially on the use of inadmissible strategies.

Mertens (1992) describes a game in which each player has a unique dominant strategy, yet the pair of these dominant strategies is not perfect in the agent normal form. Hence, the M-stable sets of the normal form and those of the agent normal form may be disjoint. That same paper also contains an example of a nongeneric perfect information game (where ties are not noticed when doing the backwards induction where the unique M-stable set contains other outcomes besides the backwards induction outcome). [See also Van Damme (1987b, pp. 32, 33).]

Govindan (1995) has applied the concept of M-stability to the Kreps and Wilson (1982b) chain store game with incomplete information. He shows that only the outcome that was already identified in Kreps and Wilson (1982b) as the unique "reasonable" one, is indeed the unique M-stable outcome. Govindan's approach is to be preferred to Kreps and Wilson's since it does not rely on ad hoc methods. It is worth remarking that Govindan is able to reach his conclusion just by using the properties of M-stable equilibria (as mentioned in Theorem 11) and that the connectedness requirement plays an important role in the proof.

4.5. Robustness and persistent equilibria

Many game theorists are not convinced that equilibria in mixed strategies should be treated on an equal footing with pure, strict equilibria; they express a clear preference for pure equilibria. For example, Harsanyi and Selten (1988, p. 198) write: "Games that arise in the context of economic theory often have many strict equilibrium points. Obviously in such cases it is more natural to select a strict equilibrium point rather than a weak one. Of course, strict equilibrium points are not always available (...) but it is still possible to look for a principle that helps us to avoid those weak equilibrium points that are especially unstable." (They use the term "strong" where I write "strict".) In this subsection we discuss such principles.

Harsanyi and Selten discuss two forms of instability associated with mixed strategy equilibria. The first, weak form of instability results from the fact that even though a player might have no incentive to deviate from a mixed equilibrium, he has no positive incentive to play the equilibrium strategy either: any pure strategy that is used with positive probability is equally good. As we have seen in Section 2.5, the reinterpretation of mixed equilibria as equilibria in beliefs provides an adequate response to the criticism that is based on this form of instability. The second, strong form of instability is more serious and cannot be countered so easily. This form of instability results from the fact that, in a mixed equilibrium, if a player's beliefs differ even slightly from the equilibrium beliefs, optimizing behavior will typically force the player to deviate from the mixed equilibrium strategy. In contrast, if an equilibrium is strict, a player is forced to play his equilibrium strategy as long as he assigns a sufficiently high probability to the opponents playing this equilibrium. For example, in the battle-of-the-sexes game (that occurs as the subgame in Figure 5), each player is willing to follow the recommendation to play a pure equilibrium as long as he believes that the opponent follows the recommendation with a probability of at least $2/3$. In contrast, player i is indifferent between s_i and w_i only if he assigns a probability of exactly $1/3$ to the opponent playing w_j. Hence, it seems that strict equilibria possess a type of robustness property that the mixed equilibrium lacks. However, this difference is not picked up by any of the stability concepts that have been discussed above: the mixed strategy equilibrium of the battle-of-the-sexes game constitutes a singleton stable set according to each of the above stability definitions. In this subsection, we will discuss some set-valued generalizations of strict equilibria that do pick up the difference. They all aim at capturing the idea that equilibria should be robust to small trembles in the equilibrium beliefs, hence, they address the question of what outcome an outsider would predict who is quite sure, but not completely sure, about the players' beliefs. The discussion that follows is inspired by Balkenborg (1992).

If s is a strict equilibrium of $g = \langle A, u \rangle$, then s is the unique best response against s, hence $\{s\} = B(s)$. We have already encountered a set-valued analogue of this uniqueness requirement in Section 2.2, viz. the concept of a minimal curb set. Recall that $C \subset A$ is a curb set of g if

$$B(C) \subset C, \tag{4.3}$$

i.e., if every best reply against beliefs that are concentrated on C again belongs to C. Obviously, a singleton set C satisfies (4.4) only if it is a strict equilibrium. Nonsingleton curb sets may be very large (for example, the set A of all strategy profiles trivially satisfies (4.4)), hence in order to obtain more definite predictions, one can investigate minimal sets with the property (4.4). In Section 2.2 we showed that such minimal curb sets exist, that they are tight, i.e., $B(C) = C$, and that distinct minimal curb sets are disjoint. Furthermore, curb sets possess the same neighborhood stability property as strict equilibria, viz. if C satisfies (4.4), then there exists a neighborhood U of $X_i \Delta(C_i)$ in S such that

$$B(U) \subset C. \tag{4.4}$$

Despite all these nice properties, minimal curb sets do not seem to be the appropriate generalization of strict equilibria. First, if a player i has payoff equivalent strategies, then (4.4) requires all of these to be present as soon as one is present in the set, but optimizing behavior certainly does not force this conclusion: it is sufficient to have at least one member of the equivalence class in the curb set. (Formally, define the strategies s_i' and s_i'' of player i to be *i-equivalent* if $u_i(s \backslash s_i') = u_i(s \backslash s_i'')$ for all $s \in S$, and write $s_i' \sim_i s_i''$ if s_i' and s_i'' are i-equivalent.) Secondly, requirement (4.4) does not differentiate among best responses; it might be preferable to work with the narrower set of admissible best responses. As a consequence of these two observations, curb sets may include too many strategies and minimal curb sets do not provide a useful generalization of the strict equilibrium concept.

Kalai and Samet's (1984) concept of persistent retracts does not suffer from the two drawbacks mentioned above. Roughly, this concept results when requirement (4.5) is weakened to "$B(s) \cap C \neq \emptyset$ for any $s \in U$". Formally, define a *retract* R as a Cartesian product $R = X_i R_i$ where each R_i is a nonempty, closed, convex subset of S_i. A retract is said to be *absorbing* if

$$\mathcal{B}(s) \cap R \neq \emptyset \quad \text{for all } s \text{ in a neighborhood } U \text{ of } R, \tag{4.5}$$

that is, if against any small perturbation of strategy profile in R there exists a best response that is in R. A retract is defined to be *persistent* if it is a minimal absorbing retract. Zorn's lemma implies that persistent retracts exist; an elementary proof is indicated below. Kakutani's fixed point theorem implies that each absorbing retract contains a Nash equilibrium. A Nash equilibrium that belongs to a persistent retract is called a *persistent equilibrium*. A slight modification of Myerson's proof for the existence of proper equilibrium actually shows that each absorbing retract contains a proper equilibrium. Hence, each game has an equilibrium that is both proper and persistent. Below we give examples to show that a proper equilibrium need not be persistent and that a persistent equilibrium need not be proper.

Note that each strict equilibrium is a singleton persistent retract. The reader can easily verify that in the battle-of-the-sexes game only the pure equilibria are persistent and that (in the normal form of) the overall game in Figure 5 only the equilibrium (ps_1, w_2) is persistent, hence, in this example, persistency selects the forward induction outcome. As a side remark, note that $\bar{s}$ is a Nash equilibrium if and only if $\{\bar{s}\} = R$ is a minimal retract with the property "$\mathcal{B}(s) \cap R \neq \emptyset$ for all $s \in R$", hence, persistency corresponds to adding neighborhood robustness to the Nash equilibrium requirement.

Kalai and Samet (1984) show that persistent retracts have a very simple structure, viz. they contain at most one representative from each i-equivalence class of strategies for each player i. To establish this result, Kalai and Samet first note that two strategies s_i' and s_i'' are i-equivalent if and only if there exists an open set U in S such that, against any strategy in U, s_i' and s_i'' are equally good. Hence, it follows that, up to equivalence, the best response of a player is unique (and pure) on an open and dense subset of S. Note that, to a certain extent, a strategy that is not a best response against an open set of beliefs is superfluous, i.e., a player always has a best response that is also a best response to an open set in the neighborhood. Let us call s_i' a *robust best response* against s if there exists an open set $U \in S$ with s in its closure such that s_i' is a best response against all elements in U. (Balkenborg (1992) uses the term semi-robust best response, in order to avoid confusion with Okada's (1983) concept.) Write $\mathcal{B}_i^r(s)$ for all robust best responses of player i against s and $\mathcal{B}^r(s) = X_i \mathcal{B}_i^r(s)$. Note that $\mathcal{B}^r(s) \subset \mathcal{B}^a(s) \subset \mathcal{B}(s)$ for all s. Also note that a mixed strategy is a robust best response only if it is a mixture of equivalent pure robust best responses. Hence, up to equivalence, robustness restricts players to using pure strategies. Finally, note that an outside observer, who is somewhat uncertain about the players' beliefs and who represents this uncertainty by continuous distributions on S, will assign positive probability only to players playing robust best responses.

The reader can easily verify that (4.6) is equivalent to

$$\text{if } s \in R \text{ and } a \in \mathcal{B}_i^r(s) \quad \text{then } a \sim_i s_i' \text{ for some } s_i' \in R_i \quad (\text{all } i, s). \tag{4.6}$$

Hence, up to equivalence, all robust best responses against the retract must belong to the absorbing retract. Minimality thus implies that a persistent retract contains at most one representative from each equivalence class of robust best responses. From this observation it follows that there exists an absorbing retract that is spanned by pure strategies and that there exists at least one persistent retract. (Consider the set of all retracts that are spanned by pure strategies. The set is finite, partially ordered and the maximal element $(R = S)$ is absorbing, hence, there exists a minimal element.) Of course, for generic strategic form games, no two pure strategies are equivalent and any pure best response is a robust best response. For such games it thus follows that R is a persistent retract if and only if there exists a minimal curb set C such that $R_i = \Delta(C_i)$ for each player i.

We will now investigate which properties from Section 4.2 are satisfied by persistent retracts. We have already seen that persistent retracts exist; they are connected and contain a proper equilibrium. Hence, the properties (E), (C), and (BI) hold. Also (IIA) is satisfied, as follows easily from (4.7) and the fact that $B^r(s) \subset B^a(s)$. Also (BRI) follows easily from (4.7). However, persistent retracts do not satisfy (NE). For example, in the matching pennies game the entire set of strategies is the unique persistent retract. Of course, persistency satisfies a weak form of (NE): any persistent retract contains a Nash equilibrium. In fact, it can be shown that each persistent retract contains a stable set of equilibria. (This is easily seen for stability as defined by Kohlberg and Mertens, Mertens (1990) proves it for M-stability and Balkenborg (1992) proves the property for H-stable sets.) Similarly, persistency satisfies a weak form of (A): (4.7) implies that if R is a persistent retract and s_i is an extreme point of R_i, then s_i is a robust best response, hence, s_i is admissible. Consequently, property (A) holds for the extreme points of R, and each element in R only assigns positive probability to admissible pure strategies. This, however, does not imply that the elements of R are themselves admissible. For example, in the game of Figure 9, the only persistent retract is the entire game, but the strategy $(\frac{1}{2}, \frac{1}{2}, 0)$ of player 1 is dominated. In particular the equilibrium $\langle(\frac{1}{2}, \frac{1}{2}, 0), (0, 0, 1)\rangle$ is persistent but not perfect.

Persistent retracts are not invariant. In Figure 9, replace the payoff "2" by "$\frac{3}{2}$" so that the third strategy becomes a duplicate of the mixture $(\frac{1}{2}, \frac{1}{2}, 0)$. The unique persistent retract contains the mixed strategy $(\frac{1}{2}, \frac{1}{2}, 0)$, but it does not contain the equivalent strategy $(0, 0, 1)$. Hence, the invariance requirement (I) is violated. Balkenborg (1992), however, shows that the extreme points of a persistent retract satisfy (I). He also shows that this set of extreme points satisfies the small worlds property (SWP) and the decomposition property (D).

A serious drawback of persistency is that it does not satisfy the player splitting property: the agent normal form and the normal form of an incomplete information game can have different persistent retracts. The reason is that the normal form forces different types to have the same beliefs about the opponent, whereas the Selten form (i.e., the agent normal form) allows different types to have different conjectures. (Cf. our

3, 0	0, 3	0, 2
0, 3	3, 0	0, 2
2, 0	2, 0	0, 0

Figure 9. A persistent equilibrium need not be perfect.

discussion in Section 2.2.) Perhaps it is even more serious that also other completely inessential changes in the game may induce changes in the persistent retracts and may make equilibria persistent that were not persistent before. As an example, consider the game from Figure 5 in which only the outcome (3, 1) is persistent. Now change the game such that, when (pw_1, w_2) is played, the players do not receive zero right away, but are rather forced to play a matching pennies game. Assume players simultaneously choose "heads" or "tails", that player 1 receives 4 units from player 2 if choices match and that he has to pay 4 units if choices differ. The change is completely inessential (the game that was added has unique optimal strategies and value zero), but it has the consequence that in the normal form, only the entire strategy space is persistent. In particular, player 1 taking up his outside option is a persistent and proper equilibrium outcome of the modified game.

For applications of persistent equilibria the reader is referred to Kalai and Samet (1985), Hurkens (1996), Van Damme and Hurkens (1996), Blume (1994, 1996), and Balkenborg (1993). Kalai and Samet consider "repeated" unanimity games. In each of finitely many periods, players simultaneously announce an outcome. The game stops as soon as players announce that same outcome, and then that outcome is implemented. Kalai and Samet show that if there are at least as many rounds as there are outcomes, players will agree on an efficient outcome in a (symmetric) persistent equilibrium. Hurkens (1996) analyzes situations in which some players can publicly burn utility before the play of a game. He shows that if the players who have this option have common interests [Aumann and Sorin (1989)], then only the outcome that these players prefer most is persistent. Van Damme and Hurkens (1996) study games in which players have common interests and in which the timing of the moves is endogenous. They show that persistency forces players to coordinate on the efficient equilibrium. Blume (1994, 1996) applies persistency to a class of signalling games and he also obtains that persistent equilibria have to be efficient. Balkenborg (1993) studies finitely repeated common interest games. He shows that persistent equilibria are almost efficient.

The picture that emerges from these applications (as well as from some theoretical considerations not discussed here, see Van Damme (1992)) is that persistency might be more relevant in an evolutionary and/or learning context, rather than in the pure deductive context we have assumed in this chapter. Indeed, Hurkens (1994) discusses an explicit learning model in which play eventually settles down in a persistent retract.

The following proposition summarizes the main elements from the discussion in this section:

THEOREM 12.
(i) [Kalai and Samet (1985)]. *Every game has a persistent retract. Each persistent retract contains a proper equilibrium. Each strategy in a persistent retract assigns positive probability only to robust best replies.*
(ii) [Balkenborg (1992)]. *For generic strategic form games, persistent retracts correspond to minimal curb sets.*

(iii) [Balkenborg (1992)]. *Persistent retracts satisfy the properties* (E), (C), (IIA), (BRI) *and* (BI) *from Section* 4.2, *but violate the other properties. The set of extreme points of persistent retracts satisfies* (SWP), (D) *and* (I).

(iv) [Mertens (1990), Balkenborg (1992)]. *Each persistent retract contains an M-stable set. It also contains an H-stable set as well as a KM-stable set.*

5. Equilibrium selection

Up to now this paper has been concerned just with the first and basic question of non-cooperative game theory: which outcomes are self-enforcing? The starting point of our investigations was that being a Nash equilibrium is necessary but not sufficient for self-enforcingness, and we have reviewed several other necessary requirements that have been proposed. We have seen that frequently even the most stringent refinements of the Nash concept allow multiple outcomes. For example, many games admit multiple strict equilibria and any such equilibrium passes every test of self-enforcingness that has been proposed up to now. In the introduction, however, we already argued that the "theory" rationale of Nash equilibrium relies essentially on the assumption that players can co-ordinate on a single outcome. Hence, we have to address the questions of when, why and how players can reach such a coordinated outcome. One way in which such coordi-nation might be achieved is if there exists a convincing theory of rationality that selects a unique outcome in every game and if this theory is common knowledge among the players. One such theory of equilibrium selection has been proposed in Harsanyi and Selten (1988). In this section we will review the main building blocks of that theory.

The theory from Harsanyi and Selten may be seen as derived from three basic postu-lates, viz. that a theory of rationality should make a recommendation that is (i) a unique strategy profile, (ii) self-enforcing, and (iii) universally applicable. The latter require-ment says that no matter the context in which the game arises, the theory should apply. It is a strong form of history-independence. Harsanyi and Selten (1988, pp. 342, 343) re-fer to it as the assumption of endogenous expectations: the solution of the game should depend only on the mathematical structure of the game itself, no matter the context in which this structure arises. The combination of these postulates is very powerful; for ex-ample, one implication is that the solution of a symmetric game should be symmetric. The postulates also force an agent normal form perspective: once a subgame is reached, only the structure of the subgame is relevant, hence, the solution of a game has to project onto the solution of the subgame. Harsanyi and Selten refer to this requirement as "sub-game consistency". It is a strong form of the requirement of "persistent rationality" that was extensively discussed in Section 3. Of course, subgame consistency is naturally ac-companied by the axiom of truncation consistency: to find the overall solution of the game it should be possible to replace a subgame by its solution. Indeed, Harsanyi and Selten insist on truncation consistency as well. It should now be obvious that the require-ments that Harsanyi and Selten impose are very different from the requirements that we discussed in Section 4.2. Indeed the requirements are incompatible. For example, the

Harsanyi and Selten requirements imply that the solution of the game from Figure 5 is (tm_1, m_2) where $m_i = \frac{1}{4}s_i + \frac{3}{4}w_i$. Symmetry requires the solution of the subgame to be (m_1, m_2) and the axioms of subgame and truncation consistency prevent player 1 from signalling anything. If one accepts the Harsanyi and Selten postulates, then it is common knowledge that the battle-of-the-sexes subgame has to be played according to the mixed equilibrium, hence, if he has to play, player 2 *must* conclude that player 1 has made a mistake. Note that uniqueness of the solution is already incompatible with the pair (I), (BI) from Section 4.2. We showed that (I) and (BI) leave only the payoff $(3, 1)$ in the game of Figure 5, hence, uniqueness forces $(3, 1)$ as the unique solution of the "battle of the sexes". However, if we would have given the outside option to player 2 rather than to player 1, we would have obtained $(1, 3)$ as the unique solution. Hence, to guarantee existence, the approach from Section 4 must give up uniqueness, i.e., it has to allow multiple solutions. Both $(3,1)$ and $(1,3)$ have to be admitted as solutions of the battle-of-the-sexes game, in order to allow the context in which the game is played to determine which of these equilibria will be selected. The approach to be discussed in this section, which requires context independence, is in sharp conflict with that from the previous section. However, let us note that, although the two approaches are incompatible, each of the approaches corresponds to a coherent point of view. We confine ourselves to presenting both points of view, to allow the reader to make up his own mind.

5.1. Overview of the Harsanyi and Selten solution procedure

The procedure proposed by Harsanyi and Selten to find the solution of a given game generates a number of "smaller" games which have to be solved by the same procedure. The process of reduction and elimination should continue until finally a basic game is reached which cannot be scaled down any further. The solution of such a basic game can be determined by applying the tracing procedure to which we will return below. Hence, the theory consists of a process of reducing a game to a collection of basic games, a rule for solving each basic game, and a procedure for aggregating these basic solutions to a solution of the overall game. The solution process may be said to consist of five main steps, viz. (i) initialization, (ii) decomposition, (iii) reduction, (iv) formation splitting, and (v) solution using dominance criteria. To describe these steps in somewhat greater detail, we first introduce some terminology.

The Harsanyi and Selten theory makes use of the so-called standard form of a game, a form that is in between the extensive form and the normal form. Formally, the standard form consists of the agent normal form together with information about which agents belong to the same player. Write I for the set of players in the game and for each $i \in I$, let $H_i = \{ij : j \in J_i\}$ be the set of agents of player i. Writing $H = \bigcup_i H_i$ for the set of all agents in the game, a *game in standard form* is a tuple $g = \langle A, u \rangle_H$ where $A = X_{ij} A_{ij}$ with A_{ij} being the action set of agent ij, and $u_i : A \to \mathbb{R}$ for each player i. Harsanyi and Selten work with this form since on the one hand they want to guarantee perfectness in the extensive form, while on the other hand they want different agents of the same player to have the same expectations about the opponents.

Given a game in extensive form, the Harsanyi and Selten theory should not be directly applied to its associated standard form g; rather, for each $\varepsilon > 0$ that is sufficiently small, the theory should be applied to the uniform ε-perturbation g^ε of the game. The solution of g is obtained by taking the limit, as ε tends to zero, of the solution of g^ε. The question of whether the limit exists is not treated in Harsanyi and Selten (1988); the authors refer to the unpublished working paper Harsanyi and Selten (1977) in which it is suggested that there should be no difficulties. Formally, g^ε is defined as follows. For each agent ij let σ_{ij} be the *centroid* of A_{ij}, i.e., the strategy that chooses all pure actions in A_{ij} with probability $|A_{ij}|^{-1}$. For $\varepsilon > 0$ sufficiently small, write $\varepsilon_{ij} = \varepsilon|A_{ij}|$ and let $\tilde{\varepsilon} = (\varepsilon_{ij})_{ij \in H}$. Recall, from Equation (3.2) that $s^{\tilde{\varepsilon},\sigma}$ denotes the strategy vector that results when each player intends to play s, but players make mistakes with probabilities determined by $\tilde{\varepsilon}$ and mistakes are given by σ. The *uniformly perturbed game* g^ε is the standard form game $\langle A, u^\varepsilon \rangle_H$ where the payoff function u^ε is defined by $u_i^\varepsilon(s) = u_i(s^{\tilde{\varepsilon},\sigma})$. Hence, in g^ε each agent ij mistakenly chooses each action with probability ε and the total probability that agent ij makes a mistake is $|A_{ij}|\varepsilon$.

Let C be a collection of agents in a standard form game g and denote the complement of C by $\overline{C}$. Given a strategy vector t for the agents in $\overline{C}$, write $g_C^t = \langle A, u^t \rangle_C$ for the *reduced game* faced by the agents in C when the agents in $\overline{C}$ play t, hence $u_{ij}^t(s) = u_{ij}(s,t)$ for $ij \in C$. Write $g_C^t = g_C$ and $u^t = u^C$ in the special case where t is the centroid strategy for each agent in $\overline{C}$. The set C is called *cell* in g if for each t and each player i with an agent in C there exist constants $\alpha_i(t) > 0$ and $\beta_i(t) \in \mathbb{R}$ such that

$$u_i^t(s) = \alpha_i(t)u_i^C(s) + \beta_i(t) \quad \text{(for all } s\text{)}. \tag{5.1}$$

Hence, if C is a cell, then up to positive linear transformations, the payoffs to agents in C are completely determined by the agents in C. Since the intersection of two cells is again a cell whenever this intersection is nonempty, there exist minimal cells. Such cells are called *elementary cells*. Two elementary cells have an empty intersection. Note that for the special case of a normal form game (each player has only one agent), each cell is a small world. Also note that a transformation as in (5.1) leaves the best-reply structure unchanged. Hence, if we had defined a small world as a set of players whose (admissible) best responses are not influenced by outsiders, then each small world would have been a cell. A solution concept that assigns to each standard form game g a unique strategy vector $f(g)$ is said to satisfy *cell and truncation consistency* if for each C that is a cell in g we have

$$f_{ij}(g) = \begin{cases} f_{ij}(g^C) & \text{if } ij \in C, \\ f_{ij}\big(g_{\overline{C}}^{f(g^C)}\big) & \text{if } ij \notin C. \end{cases} \tag{5.2}$$

The reader may check that a subgame of a uniformly perturbed extensive form game induces a cell in the associated perturbed standard form; hence, the axiom of cell and truncation consistency formalizes the idea that the solution is determined by backward induction in the extensive form.

If g is a standard form game and B_{ij} is a nonempty set of actions for each agent ij, then $B = X_{ij} B_{ij}$ is called a *formation* if for each agent ij, each best response against any correlated strategy that only puts probability on actions in B belongs to B_{ij}. Hence, in normal form games, formations are just like curb sets (cf. Section 2.2), the only difference being that formations allow for correlated beliefs. As the intersection of two formations is again a formation, we can speak about *primitive formations*, i.e., formations that do not contain a proper subformation.

An action a of an agent ij is said to be *inferior* if there exists another action b of this agent that is a best reply against a strictly larger set of (possibly) correlated beliefs of the agents. Hence, noninferiority corresponds to the concept of robustness that we encountered (for the case of independent beliefs) in Section 4.5. Any strategy that is weakly dominated is inferior, but the converse need not hold.

Using the concepts introduced above, we can now describe the main steps employed in the Harsanyi and Selten solution procedure:

1. *Initialization*: Form the standard normal form g of the game, and, for each $\varepsilon > 0$ that is sufficiently small, compute the uniformly perturbed game g^ε; compute the solution $f(g^\varepsilon)$ according to the steps described below and put $f(g) = \lim_{\varepsilon \downarrow 0} f(g^\varepsilon)$.

2. *Decomposition*: Decompose the game into its elementary cells; compute the solution of an indecomposable game according to the steps described below and form the solution of the overall game by using cell and truncation consistency.

3. *Reduction*: Reduce the game by using the next three operations:
   (i) Eliminate all inferior actions of all agents.
   (ii) Replace each set of equivalent actions of each agent ij (i.e., actions among which all players are indifferent) by the centroid of that set.
   (iii) Replace, for each agent ij, each set of ij-equivalent actions (i.e., actions among which ij is indifferent no matter what the others do) by the centroid strategy of that set.

   By applying these steps, an irreducible game results. The solution of such a game is by means of Step 4.

4. *Solution*:
   (i) *Initialization*: Split the game into its primitive formations and determine the solution of each basic game associated with each primitive formation by applying the tracing procedure to the centroid of that formation. The set of all these solutions constitutes the first candidate set Ω^1.
   (ii) *Candidate elimination and substitution*: Given a candidate set Ω, determine the set $M(\Omega)$ of maximally stable elements in Ω. These are those equilibria in Ω that least dominated in Ω. Dominance involves both payoff dominance and risk dominance and payoff dominance ranks more important than risk dominance. The latter is defined by means of the tracing procedure (see below) and need not be transitive. Form the chain $\Omega = \Omega^1$, $\Omega^{t+1} = M(\Omega^t)$ until $\Omega^{T+1} = \Omega^T$. If $|\Omega^T| = 1$, then Ω^T is the solution, otherwise replace

Ω^T by the trace, $t(\Omega^T)$, of its centroid and repeat the process with the new candidate set $\Omega = \Omega^{T-1}\backslash\Omega^T \cup \{t(\Omega^T)\}$.

It should be noted that it may be necessary to go through these steps repeatedly. Furthermore, the steps are hierarchically ordered, i.e., if the application of step 3(i) (i.e., the elimination of inferior actions) results in a decomposable game, one should first return to step 2. The reader is referred to the flow chart on p. 127 of Harsanyi and Selten (1988) for further details.

The next two sections of the present paper are devoted to step 4, the core of the solution procedure. We conclude this subsection with some remarks on the other steps.

We already discussed step 2, as well as the reliance on the agent normal form in the previous subsection. Deriving the solution of an unperturbed game as a limit of solutions of uniformly perturbed games has several consequences that might be considered undesirable. For one, duplicating strategies in the unperturbed game may have an effect on the outcome. Consider the normal form of the game from Figure 6. If we duplicate the strategy pl_1 of player 1, the limit solution prescribes r_2 for player 2 (since the mistake pl_1 is more likely than the mistake pr_1), but if we duplicate pr_1 then the solution prescribes player 2 to choose l_2. Hence, the Harsanyi–Selten solution does not satisfy the invariance requirement (I) from Section 4.2, nor does it satisfy (IIA). Secondly, an action that is dominated in the unperturbed game need no longer be dominated in the ε-perturbed version of the game and, consequently, it is possible to construct an example in which the Harsanyi–Selten solution is an equilibrium that uses dominated strategies [Van Damme (1990)]. Hence, the Harsanyi–Selten solution violates (A). Turning now to the reduction step, we note that the elimination procedure implies that invariance is violated. (Cf. the discussion on persistency in Section 4.5; note that any pure strategy that is a mixture of non-equivalent pure strategies is inferior.) Let us also remark that stable sets need not survive when an inferior strategy is eliminated. [See Van Damme (1987a, Figure 10.3.1) for an example.] Finally, we note that since the Harsanyi–Selten theory makes use of payoff comparisons of equilibria, the solution of that theory is not best reply invariant. We return to this below.

5.2. Risk dominance in 2 × 2 games

The core of the Harsanyi–Selten theory of equilibrium selection consists of a procedure that selects, in each situation in which it is common knowledge among the players that there are only two viable solution candidates, one of these candidates as the actual solution for that situation. A simple example of a game with two obvious solution candidates (viz. the strict equilibria (a, a) and $(\bar{a}, \bar{a})$) is the stag-hunt game of the left-hand panel of Figure 10, which is a slight modification of a game first discussed in Aumann (1990). (The only reason to discuss this variant is to be able to draw simpler pictures.)

The stag hunt from the left-hand panel is a symmetric game with common interests [Aumann and Sorin (1989)], i.e., it has (a, a) as the unique Pareto-efficient outcome. Playing a, however, is quite risky: if the opponent plays his alternative equilibrium strategy $\bar{a}$, the payoff is only zero. Playing $\bar{a}$ is much safer: one is guaranteed the equilibrium payoff and, if the opponent deviates, the payoff is even higher. Harsanyi and Selten

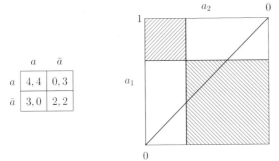

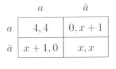

	a	$\bar{a}$
a	4, 4	0, $x+1$
$\bar{a}$	$x+1$, 0	x, x

	a	$\bar{a}$
a	4, 4	0, 3
$\bar{a}$	3, 0	2, 2

Figure 10. The stag hunt.

discuss a variant of this game extensively since it is a case where the two selection criteria that are used in their theory (viz. those of payoff dominance and risk dominance) point in opposite directions. [See Harsanyi and Selten (1988, pp. 88, 89, and 358, 359).] Obviously, if each player could trust the other to play a, he would also play a, and players clearly prefer such mutual trust to exist. The question, however, is under which conditions such trust exists and how it can be created if it does not exist. As Aumann (1990) has argued, preplay communication cannot create trust where it does not exist initially. In the end, Harsanyi and Selten decide to give precedence to the payoff dominance criterion, i.e., they assume that rational players can rely on collective rationality and they select (a, a) in the game of Figure 10. However, the arguments given are not fully convincing. We will use the game of Figure 10 to illustrate the concept of risk dominance, which is based on strictly individualistic rationality considerations.

Intuitively, the equilibrium s risk dominates the equilibrium $\bar{s}$ if, when players are in a state of mind where they think that either s or $\bar{s}$ should be played, they eventually come to the conclusion that $\bar{s}$ is too risky and, hence, they should play s. For general games, risk dominance is defined by means of the tracing procedure. For the special case of 2-player 2×2 normal form games with two strict equilibria, the concept is also given an axiomatic foundation. Before discussing this axiomatization, we first illustrate how riskiness of an equilibrium can be measured in 2×2 games.

Let $G(a, \bar{a})$ be the set of all 2-player normal form games in which each player i has the strategy set $\{a, \bar{a}\}$ available and in which (a, a) and $(\bar{a}, \bar{a})$ are strict Nash equilibria. For $g \in G(a, \bar{a})$, we identify a mixed strategy of player i with the probability a_i that this strategy assigns to a and we write $\bar{a}_i = 1 - a_i$. We also write $d_i(a)$ for the loss that player i incurs when he unilaterally deviates from (a, a) (hence, $d_1(a) = u_1(a, a) - u_1(\bar{a}, a)$) and we define $d_i(\bar{a})$ similarly. Note that when player j plays a with probability a_j^* given by

$$a_j^* = d_i(\bar{a})/\big(d_i(a) + d_i(\bar{a})\big), \tag{5.3}$$

player i is indifferent between a and $\bar{a}$. Hence, the probability a_j^* as in (5.3) represents the risk that i is willing to take at $(\bar{a}, \bar{a})$ before he finds it optimal to switch to a. In a symmetric game (such as that of Figure 10) $a_1^* = a_2^*$, hence a_1^* (resp. $\bar{a}_1^*$) is a natural measure of the riskiness of the equilibrium $(\bar{a}, \bar{a})$ (resp. (a, a)) and $(\bar{a}, \bar{a})$ is more risky if $a_1^* < \bar{a}_1^*$, that is, if $a_1^* < 1/2$. In the game of Figure 10, we have that $a_1^* = 2/3$, hence (a, a) is more risky than $(\bar{a}, \bar{a})$. More generally, let us measure the riskiness of an equilibrium as the sum of the players' risks. Formally, say that (a, a) *risk dominates* $(\bar{a}, \bar{a})$ in g (abbreviated $a \succ_g \bar{a}$) if

$$a_1^* + a_2^* < 1; \tag{5.4}$$

say that $(\bar{a}, \bar{a})$ risk dominates (a, a) (written $\bar{a} \succ_g a$) if the reverse strict inequality holds, and say that there is no dominance relationship between (a, a) and $(\bar{a}, \bar{a})$ (written $a \sim_g \bar{a}$) if (5.4) holds with equality. In the game of Figure 10, we have that $(\bar{a}, \bar{a})$ risk dominates (a, a).

To show that these definitions are not "ad hoc", we now give an axiomatization of risk-dominance. On the class $G(a, \bar{a})$, Harsanyi and Selten (1988, Section 3.9) characterize this relation by the following axioms.

1. (Asymmetry and completeness): For each g exactly one of the following holds: $a \succ_g \bar{a}$ or $\bar{a} \succ_g a$ or $a \sim_g \bar{a}$.
2. (Symmetry): If g is symmetric and player i prefers (a, a) while player j $(j \neq i)$ prefers $(\bar{a}, \bar{a})$, then $a \sim_g \bar{a}$.
3. (Best-reply invariance): If g and g' have the same best-reply correspondence, then $a \succ_g \bar{a}$, if and only if $a \succ_{g'} \bar{a}$.
4. (Payoff monotonicity): If g' results from g by making (a, a) more attractive for some player i while keeping all other payoffs the same, then $a \succ_{g'} \bar{a}$ whenever $a \succ_g \bar{a}$ or $a \sim_g \bar{a}$.

The proof is simple and follows from the observations that

 (i) games are best-reply-equivalent if and only if they have the same (a_1^*, a_2^*),
 (ii) symmetric games with conflicting interests satisfy (5.4) with equality, and
 (iii) increasing $u_i(a, a)$ decreases a_j^*.

Harsanyi and Selten also give an alternative characterization of risk-dominance. Condition (5.4) is equivalent to the (Nash) product of players' deviation losses at (a, a) being larger than the corresponding Nash product at $(\bar{a}, \bar{a})$, hence

$$d_1(a)d_2(a) > d_1(\bar{a})d_2(\bar{a}) \tag{5.5}$$

and, in fact, the original definition is by means of this inequality. Yet another equivalent characterization is that the area of the stability region of (a, a) (i.e., the set of mixed strategies against which a is a best response for each player) is larger than the area of the stability region of $(\bar{a}, \bar{a})$. (Obviously, the first area is $\bar{a}_1^* \bar{a}_2^*$, the second is $a_1^* a_2^*$.) For the stag hunt game, the stability regions have been displayed in the middle panel of Figure 10. (The diagonal represents the line $a_1 + a_2 = 1$; the upper left corner of the diagram is the point $a_1 = 1$, $a_2 = 1$, it corresponds to the upper left corner of the matrix, and similarly for other points.)

In Carlsson and Van Damme (1993a), equilibrium selection according to the risk-dominance criterion is derived from considerations related to uncertainty concerning the payoffs of the game. These authors assume that players can observe the payoffs in a game only with some noise. In contrast to Harsanyi's model that was discussed in Section 2.5, Carlsson and Van Damme assume that each player is uncertain about both players' payoffs. Because of the noise, the actual best-reply structure will not be common knowledge and as a consequence of this lack of common knowledge, players' behavior at each observation may be governed by the behavior at some remote observation [also cf. Rubinstein (1989)]. In the noisy version of the stag hunt game of Figure 8, even though players may know to a very high degree that (a, a) is the Pareto-dominant equilibrium, they might be unwilling to play it since each player i might think that j will play $\bar{a}$ since i will think that j will think... that $\bar{a}$ is a dominant action. Hence, even though this model superficially resembles that of Harsanyi (1973a), it leads to completely different results.

As a simple and concrete illustration of the model, suppose that it is common knowledge among the players that payoffs are related to actions as in the right panel $g(x)$ of Figure 10. A priori, players consider all values $x \in [-1, 4]$ to be possible and they consider all such values to be equally likely. (Carlsson and Van Damme (1993a) show that the conclusion is robust with respect to such distributional assumptions, as well as with respect to assumptions on the structure of the noise.) Note that $g(x) \in G(a, \bar{a})$ for $x \in (0, 3)$, that a is a dominant strategy if $x < 0$ and that $\bar{a}$ is dominant if $x > 3$. Suppose now that players can observe the actual value of x that prevails only with some slight noise. Specifically, assume player i observes $x_i = x + \varepsilon e_i$ where x, e_1, e_2 are independent and e_i is uniformly distributed on $[-1, 1]$. Obviously, if $x_i < -\varepsilon$ (resp. $x_i > 3 + \varepsilon$), player i will play a (resp. $\bar{a}$) since he knows that that action is dominant at each actual value of x that corresponds to such an observation. Forcing players to play their dominant actions at these observations will make a and $\bar{a}$ dominant at a larger set of observations and the process can be continued iteratively. Let $\underline{x}$ (resp. $\bar{x}$) be the supremum (resp. infimum) of the set of observations y for which each player i has a (resp. $\bar{a}$) as an iteratively dominant action for each $x_i < y$ (resp. $x_i > y$). Then there must be a player i who is indifferent between a and $\bar{a}$ when he observes $\underline{x}$ (resp. $\bar{x}$). Writing $a_j(x_i)$ for the probability that i assigns to j playing a when he observes x_i, we can write the indifference condition of player i at x_i (approximately) as

$$4a_j(x_i) = a_j(x_i) + x_i. \tag{5.6}$$

Now, at $x_i = \underline{x}$, we have that $a_j(x_i)$ is at least $1/2$ because of our symmetry assumptions and since j has a as an iteratively dominant strategy for each $x_j < \underline{x}$. Consequently, $\underline{x} \geqslant 3/2$. A symmetric argument establishes that $\bar{x} \leqslant 3/2$, hence $\underline{x} = \bar{x} = 3/2$, and each player i should choose a if he observes $x_i < 3/2$ while he should choose $\bar{a}$ if $x_i > 3/2$. Hence, in the noisy version of the game, each player should always play the risk-dominant equilibrium of the game that corresponds to his observation.

To conclude this subsection, we remark that the concept of risk dominance also plays an important role in the literature that derives Nash equilibrium as a stationary state of processes of learning or evolution. Even though each Nash equilibrium may be a stationary state of such a process, occasional experimentation or mutation may result in only the risk-dominant equilibrium surviving in the long run: this equilibrium has a larger stability region, hence, a larger basin of attraction, so that the process is more easily trapped there and mutations have more difficulty in upsetting it [see Kandori et al. (1993), Young (1993a, 1993b), Ellison (1993)].

5.3. Risk dominance and the tracing procedure

Let us now consider a more general normal form game $g = \langle A, u \rangle$ where the players are uncertain which of two equilibria, s or $\bar{s}$, should be played. Risk dominance tries to capture the idea that in this state of confusion the players enter a process of expectation formation that converges on that equilibrium which is the least risky of the two. (Note that a player i with $s_i = \bar{s}_i$ is not confused at all. Harsanyi and Selten first eliminate all such players before making risk comparisons. For the remaining players they similarly delete strategies not in the formation spanned by s and $\bar{s}$ since these are never best responses, no matter what expectations the players have. To the smaller game that results in this way, one should then first apply the decomposition and reduction steps from Section 5.2. We shall assume that all these transformations have been made and we will denote the resulting game again by g.)

Harsanyi and Selten view the rational formation of expectations as a two-stage process. In the first stage, players form preliminary expectations which are based on the structure of the game. These preliminary expectations take the form of a mixed strategy vector s^0 for the game. On the basis of s^0, players can already form plans about how to play the game. A naive plan would be for each player to play the best response against s^0, but, of course, these plans are not necessarily consistent with the preliminary expectations. The second stage of the expectation formation process then consists of a procedure that gradually adjusts plans and expectations until they are consistent and yield an equilibrium of the game g. Harsanyi and Selten actually make use of two adjustment processes, the linear tracing procedure T and the logarithmic tracing procedure $\tilde{T}$. Formally, each of these is a map that assigns to a mixed strategy vector s^0 exactly one equilibrium of g. The linear tracing procedure is easier to work with, but it is not always well-defined. The logarithmic tracing procedure is well-defined and yields the same outcome as the linear one whenever the latter is well-defined. We now first discuss these tracing procedures. Thereafter, we return to the question of how to form the preliminary expectations and how to define risk dominance for general games.

Let $g = \langle A, u \rangle$ be a normal form game and let p be a vector of mixed strategies for g, interpreted as the players' prior expectations. For $t \in [0, 1]$ define the game $g^{t,p} = \langle A, u^{t,p} \rangle$ by

$$u_i^{t,p} = tu_i(s) + (1 - t)u_i(p \backslash s_i). \tag{5.7}$$

Hence, for $t = 1$ the game coincides with g, while $g^{0,p}$ is a trivial game in which each player's payoff depends only on this player's prior expectations, not on what the opponents are actually doing. Write $\Gamma(p)$ for the graph of the equilibrium correspondence, hence

$$\Gamma(p) = \{(t, s) \in [0, 1] \times S \colon s \text{ is an equilibrium of } g^{t,p}\}. \tag{5.8}$$

In nondegenerate cases, $g^{0,p}$ will have exactly one (and strict) equilibrium $s(0, p)$ and this equilibrium will remain an equilibrium for sufficiently small t. Let us denote it by $s(t, p)$. The linear tracing procedure now consists in following the curve $s(t, p)$ until, at its endpoint $T(p) = s(1, p)$, an equilibrium of g is reached. Hence, as the tracing procedure progresses, plans and expectations are continuously adjusted until an equilibrium is reached. The parameter t may be interpreted as the degree of confidence players have in the solution $s(t, p)$. Formally, the *linear tracing procedure* with prior p is well-defined if the graph $\Gamma(p)$ contains a unique connected curve that contains endpoints both at $t = 0$ and $t = 1$. In this case, the endpoint $T(p)$ at $t = 1$ is called the *linear trace* of p. (Note the requirement that there be a unique connecting curve. Herings (2000) shows that there will always be at least one such curve, hence, the procedure is feasible in principle.)

We can illustrate the procedure by means of the stag hunt game from Figure 10. Write p_i for the prior probability that i plays a. If $p_i > 2/3$ for $i = 1, 2$, then $g^{0,p}$ has (a, a) as its unique equilibrium and this strategy pair remains an equilibrium for all t. Furthermore, for any $t \in [0, 1]$, (a, a) is disconnected in $\Gamma(p)$ from any other equilibrium of $g^{t,p}$. Hence, in this case the linear tracing procedure is well-defined and we have $T(p) = (a, a)$. Similarly, $T(p) = (\bar{a}, \bar{a})$ if $p_i < 2/3$ for $i = 1, 2$. Next, assume $p_1 < 2/3$ and $p_2 > 2/3$ so that $s(0, p) = (a, \bar{a})$. In this case the initial plans do not constitute an equilibrium of the final game so that adjustments have to take place along the path. The strategy pair $(a, \bar{a})$ remains an equilibrium of $g^{t,p}$ as long as

$$4(1 - t)p_2 \geqslant 2t + (1 - t)(2 + p_2) \tag{5.9}$$

and

$$(1 - t)(2 + p_1) + 3t \geqslant 4p_1(1 - t) + 4t. \tag{5.10}$$

Hence, provided that no player switches before t, player 1 has to switch at the value of t given by

$$\frac{t}{1 - t} = \frac{3p_2 - 2}{2} \tag{5.11}$$

while player 2 has to switch when

$$\frac{t}{1 - t} = 2 - 3p_1. \tag{5.12}$$

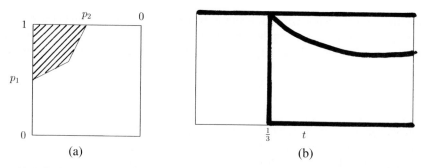

Figure 11. (a) In the interior of the shaded area $T(p) = (a, a)$. In the interior of the complement $T(p) = (\bar{a}, \bar{a})$. (b) A case where the linear tracing procedure is not well-defined.

Assume $p_1 + p_2/2 < 1$ so that the t-value determined by (5.11) is smaller than the value determined by (5.12). Hence, player 1 has to switch first and, following the branch $(a, \bar{a})$, the linear tracing procedure continues with a branch $(\bar{a}, \bar{a})$. Since $(\bar{a}, \bar{a})$ is a strict equilibrium of g, this branch continues until $t = 1$, hence $T(p) = (\bar{a}, \bar{a})$ in this case. Similarly, $T(p) = (a, a)$ if $p_1 < 2/3$, $p_2 > 2/3$ and $p_1 + p_2/2 > 1$. In the case where $p_1 > 2/3$ and $p_2 < 2/3$, the linear trace of p follows by symmetry. The results of our computations are summarized in the left-hand panel of Figure 11.

If $p_1 < 2/3$, $p_2 > 2/3$ and $p_1 + p_2/2 = 1$, then the equations (5.11)–(5.12) determine the same t-value, hence, both players want to switch at the same time $\tilde{t}$. In this case, the game $g^{\tilde{t}, p}$ is degenerate with equilibria both at (a, a) and at $(\bar{a}, \bar{a})$. Now there exists a path in Γ that connects $(a, \bar{a})$ with (a, a) as well as a path that connects $(a, \bar{a})$ with $(\bar{a}, \bar{a})$. In fact, all three equilibria of g (including the mixed one) are connected to the equilibrium of $g^{0, p}$, hence, the linear tracing procedure is not well-defined in this case. Figure 11(b) gives a graphical display of this case. (The picture is drawn for the case where $p_1 = 1/2$, $p_2 = 1$ and displays the probability of 1 choosing a.)

The logarithmic tracing procedure has been designed to resolve ambiguities such as those in Figure 11(b). For $\varepsilon \in (0, 1]$, $t \in [0, 1)$ and $p \in S$, define the game $g^{\varepsilon, t, p}$ by means of

$$u_i^{\varepsilon, t, p}(s) = u_i^{t, p}(s) + \varepsilon(1 - t)\alpha_i \sum_a \ln s_i(a), \qquad (5.13)$$

where α_i is a constant defined by

$$\alpha_i = \max_s \left[\max_{s_i'} u_i(s \backslash s_i') - \min_{s_i'} u_i(s \backslash s_i') \right]. \qquad (5.14)$$

Hence, $u_i^{\varepsilon, t, p}(s)$ results from adding a logarithmic penalty term to $u_i^{t, p}(s)$. This term ensures that all equilibria are completely mixed and that there is a unique equilibrium

$s(\varepsilon, 0, p)$ if $t = 0$. Write $\widetilde{\Gamma}(p)$ for the graph of the equilibrium correspondence

$$\widetilde{\Gamma}(p) = \{(\varepsilon, t, s) \in (0, 1] \times [0, 1) \times S : s \text{ is an equilibrium of } g^{\varepsilon, t, p}\}. \tag{5.15}$$

$\widetilde{\Gamma}(p)$ is the zero set of a polynomial and, hence, is an algebraic set. Loosely speaking, the *logarithmic tracing procedure* consists of following, for each $\varepsilon > 0$, the analytic continuation $s(\varepsilon, t, p)$ of $s(\varepsilon, 0, p)$ till $t = 1$ and then taking the limit, as $\varepsilon \to 0$, of the end points. Harsanyi and Selten (1988) and Harsanyi (1975) claim that this construction can indeed be carried out, but Schanuel et al. (1991) pointed to some difficulties in this construction: the analytic continuation need not be a curve and there is no reason for the limit to exist. Fortunately, these authors also showed that, apart from a finite set E of ε-values, the construction proposed by Harsanyi and Selten is indeed feasible. Specifically, if $\varepsilon \notin E$, then there exists a unique analytic curve in $\widetilde{\Gamma}(p)$ that contains $s(\varepsilon, 0, p)$. If we write $s(\varepsilon, t, p)$ for the strategy component of this curve, then $\widetilde{T}(p) = \lim_{\varepsilon \downarrow 0} \lim_{t \to 1} s(\varepsilon, t, p)$ exists. $\widetilde{T}(p)$ is called the *logarithmic trace* of p. Hence, the logarithmic tracing procedure is well-defined. Furthermore, Schanuel et al. (1991) show that there exists a connected curve in $\Gamma(p)$ connecting $\widetilde{T}(p)$ to an equilibrium in $g^{0, p}$ implying that $\widetilde{T}(p) = T(p)$ whenever the latter is well-defined. Hence, we have

THEOREM 13 [Harsanyi (1975), Schanuel et al. (1991)]. *The logarithmic tracing procedure $\widetilde{T}$ is well-defined. The linear tracing procedure T is well-defined for almost all priors and $\widetilde{T}(p) = T(p)$ whenever the latter is well-defined.*

The logarithmic penalty term occurring in (5.13) gives players an incentive to use completely mixed strategies. It has the consequence that in Figure 11(b) the interior mixed strategy path is approximated as $\varepsilon \to 0$. Hence, if p is on the south-east boundary of the shaded region in Figure 11(a), then $\widetilde{T}(p)$ is the mixed strategy equilibrium of the game g.

We finally come to the construction of the prior probability distribution p used in the risk dominance comparison between s and $\bar{s}$. According to Harsanyi and Selten, each player i will initially assume that his opponents already know whether s or $\bar{s}$ is the solution. Player i will assign a subjective probability z_i to the solution being s and a probability $\bar{z}_i = 1 - z_i$ to the solution being $\bar{s}$. Given his beliefs z_i player i will then choose a best response $b_i^{z_i}$ to the correlated strategy $z_i s_{-i} + \bar{z}_i \bar{s}_{-i}$ of his opponents. (In case of multiple best responses, i chooses all of them with the same probability.) An opponent j of player i is assumed not to know i's subjective probability z_i; however, j knows that i is following the above reasoning process. Applying the principle of insufficient reasoning, Harsanyi and Selten assume that j considers all values of z_i to be equally likely, hence, j considers z_i to be uniformly distributed on $[0, 1]$. Consequently, j believes that i will play $a_i \in A_i$ with a probability given by

$$p_i(a_i) = \int b_i^{z_i}(a_i) \, dz_i. \tag{5.16}$$

Equation (5.16) determines the players' *prior expectations* p to be used for risk-dominance comparison between s and $\bar{s}$. If $\tilde{T}(p) = s$ (resp. $\tilde{T}(p) = \bar{s}$) then s is said to *risk dominate* $\bar{s}$ (resp. $\bar{s}$ risk dominates s). If $\tilde{T}(p) \notin \{s, \bar{s}\}$, neither equilibrium risk dominates the other. The reader may verify that for 2×2 games this definition of risk dominance is in agreement with the one given in the previous section. For example, in the stag hunt game from Figure 8 we have that $b_i^{z_i}(a) = 1$ if $z_i > 2/3$ and $b_i^{z_i}(a) = 0$ if $z_i < 2/3$, hence $p_i(a) = 1/3$. Consequently, p lies in the non-shaded region in Figure 11(a) and $T(p) = (\bar{a}, \bar{a})$, hence, $(\bar{a}, \bar{a})$ risk dominates (a, a).

Unfortunately, for games larger than 2×2, the risk dominance relation need not be transitive [see Harsanyi and Selten (1988, Figure 3.25) for an example] and selection on the basis of this criterion need not be in agreement with selection on the basis of stability with respect to payoff perturbations [Carlsson and Van Damme (1993b)]. To illustrate the latter, consider the n-player stag hunt game in which each player i has the strategy set $\{a, \bar{a}\}$. A player choosing a gets the payoff 1 if all players choose a, and 0 otherwise. A player choosing $\bar{a}$ gets the payoff $x \in (0, 1)$ irrespective of what the others do. There are two strict Nash equilibria, viz. "all a" and "all $\bar{a}$". If player i assigns prior probability z to his opponents playing the former, then he will play a if $z > x$, hence, $p_i(a) = 1 - x$ according to (5.16). Consequently, the risk-dominant solution is "all a" if

$$(1 - x)^{n-1} > x \tag{5.17}$$

and it is "all $\bar{a}$" if the reverse strict inequality is satisfied. On the other hand, Carlsson and Van Damme (1993b) derive that, whenever there is slight payoff uncertainty, a player should play a if $1/n > x$. It is interesting to note that this n-person stag hunt game has a potential (cf. Section 2.3) and that the solution identified by Carlsson and Van Damme maximizes the potential. More generally, suppose that, when there are k players choosing a, the payoff to a player choosing a equals $f(k)$ (with $f(0) = 0$, $f(n) = 1$) and that the payoff to a player choosing $\bar{a}$ equals $x \in (0, 1)$. Then the function p that assigns to each outcome in which exactly k players cooperate the value

$$p(k) = \sum_{l=1}^{k} [f(l) - x] \tag{5.18}$$

is an exact potential for the game. "All a" maximizes the potential if and only if $\sum_{l=1}^{k} f(l)/n > x$ and this condition is identical to the one that Carlsson and Van Damme derive for a to be optimal in their model.

To conclude this subsection, we remark that, in order to derive (5.16), it was assumed that player i's uncertainty can be represented by a correlated strategy of the opponents. Güth (1985) argues that such correlated beliefs may reflect the strategic aspects rather poorly and he gives an example to show that such a correlated belief may lead to counterintuitive results. Güth suggests computing the prior as above, save by starting from the assumption that i believes $j \neq i$ to play $z_j s_j + \bar{z}_j \bar{s}_j$ with z_j uniform on $[0, 1]$ and different z's being independent.

5.4. Risk dominance and payoff dominance

We already encountered the fundamental conflict between risk dominance and payoff dominance when discussing the stag hunt game in Section 5.2 (Figure 10). In that game, the equilibrium (a, a) Pareto dominates the equilibrium $(\bar{a}, \bar{a})$, but the latter is risk dominant. In cases of such conflict, Harsanyi and Selten have given precedence to the payoff dominance criterion, but their arguments for doing so are not compelling, as they indeed admit in the postscript of their book, when they discuss Aumann's argument (also already mentioned in Section 5.2) that pre-play communication cannot make a difference in this game. After all, no matter what a player intends to play he will always attempt to induce the other to play a as he always benefits from this. Knowing this, the opponent cannot attach specific meaning to the proposal to play (a, a), communication cannot change a player's beliefs about what the opponent will do and, hence, communication can make no difference to the outcome of the game [Aumann (1990)]. As Harsanyi and Selten (1988, p. 359) write, "This shows that in general we cannot expect the players to implement payoff dominance unless, from the very beginning, payoff dominance is part of the rationality concept they are using. Free communication among the players in itself might not help. Thus if one feels that payoff dominance is an essential aspect of game-theoretic rationality, then one must explicitly incorporate it into one's concept of rationality".

Several equilibrium concepts exist that explicitly incorporate such considerations. The most demanding concept is Aumann's (1959) notion of a *strong equilibrium*: it requires that no coalition can deviate in a way that makes all its members better off. Already in simple examples such as the prisoners' dilemma, this concept generates an empty set of outcomes. (In fact, generically all Nash equilibria are inefficient [see Dubey (1986)].) Less demanding is the idea that the grand coalition not be able to renegotiate to a more attractive stable outcome. This idea underlies the concept of renegotiation-proof equilibrium from the literature on repeated games [see Bernheim and Ray (1989), Farrell and Maskin (1989) and Van Damme (1988, 1989a)]. Bernheim et al. (1987) have proposed the interesting concept of *coalition-proof Nash equilibrium* as a formalization of the requirement that no subcoalition should be able to profitably deviate to a strategy vector that is stable with respect to further renegotiation. The concept is defined for all normal form games and the formal definition is by induction on the number of players. For a one-person game any payoff-maximizing action is defined to be coalition-proof. For an I-person game, a strategy profile s is said to be weakly coalition-proof, if, for any proper subcoalition coalition C of I, the strategy profile s_C is coalition-proof in the reduced game in which the complement $\bar{C}$ is restricted to play $s_{\bar{C}}$, and s is said to be coalition-proof if there is no other weakly coalition-proof profile s' that strictly Pareto dominates it. For 2-player games, coalition-proof equilibria exist, but existence for larger games is not guaranteed. Furthermore, coalition-proof equilibria may be Pareto dominated by other equilibria.

	a	$\bar{a}$
a	$2, 2, 2$	$0, 0, 0$
$\bar{a}$	$0, 0, 0$	$3, 3, 0$

Figure 12. Renegotiation as a constraint.

The tension between "global" payoff dominance and "local" efficiency was already pointed out in Harsanyi and Selten (1988): an agreement on a Pareto-efficient equilibrium may not be self-enforcing since, with the agreement in place, and accepting the logic of the concept, a subcoalition may deviate to an even more profitable agreement. The following provides a simple example. Consider the 3-player game g in which player 3 first decides whether to take up an outside option T (which yields all players the payoff 1) or to let players 1 and 2 play a subgame in which the payoffs are as in Figure 12.

The game g from Figure 12 has two Nash equilibrium outcomes. In the first, player 3 chooses T (in the belief that 1 and 2 will choose $\bar{a}$ with sufficiently high probability); in the second, player 3 chooses p, i.e., he gives the move to players 1 and 2, who play (a, a). Both outcomes are subgame perfect (even stable) and the equilibrium (a, a, p) Pareto dominates the equilibrium T. At the beginning of the game it seems in the interest of all players to play (a, a, p). However, once player 3 has made his move, his interests have become strategically irrelevant and it is in the interest of players 1 and 2 to renegotiate to $(\bar{a}, \bar{a})$.

Although the above argument was couched in terms of the extensive form of the game, it is equally relevant for the case in which the game is given in strategic form, i.e., when players have to move simultaneously. After agreeing to play (a, a, p), players 1 and 2 could secretly get together and arrange a joint deviation to $(\bar{a}, \bar{a})$. This deviation is in their interest and it is stable since no further deviations by subgroups are profitable. Hence, the profile (a, a, p) is not coalition-proof.

The reader may argue that these "cooperative refinements" in which coalitions of players are allowed to deviate jointly have no place in the theory of strategic equilibrium, and that, as suggested in Nash (1953), it is preferable to stay squarely within the non-cooperative framework and to fully incorporate possibilities for communication and cooperation in the game rather than in the solution concept. The present author agrees with that view. The above discussion has been included to show that, while it is tempting to argue that equilibria that are Pareto-inferior should be discarded, this view encounters difficulties and may not stand up to closer scrutiny. Nevertheless, the shortcut may sometimes yield valuable insights. The interested reader is referred to Bernheim and Whinston (1987) for some applications using the shortcut of coalition-proofness.

5.5. *Applications and variations*

Nash (1953) already noted the need for a theory of equilibrium selection for the study of bargaining. He wrote: "Thus the equilibrium points do not lead us immediately to a solution of the game. But if we discriminate between them by studying their relative stabilities we can escape from this troublesome nonuniqueness" [Nash (1953, pp. 131, 132)]. Nash studied 2-person bargaining games in which the players simultaneously make payoff demands, and in which each player receives his demand if and only if the pair of demands is feasible. Since each pair that is just compatible (i.e., is Pareto optimal) is a strict equilibrium, there are multiple equilibria. Using a perturbation argument, Nash suggested taking that equilibrium in which the product of the utility gains is largest as the solution of the game. The desire to have a solution with this "Nash product property" has been an important guiding principle for Harsanyi and Selten when developing their theory (cf. (5.5)). One of the first applications of that theory was to unanimity games, i.e., games in which each player's payoff is zero unless all players simultaneously choose the same alternative. As the reader can easily verify, the Harsanyi and Selten solution of such a game is indeed the outcome in which the product of the payoffs is largest, provided that there is such a unique maximizing outcome.

Another early application of the theory was to market entry games [Selten and Güth (1982)]. In such a game there are I players who simultaneously decide whether to enter a market or not. If k players enter, the payoff to a player i that enters is $\pi(k) - c_i$, while his payoff is zero otherwise (π is a decreasing function). The Harsanyi and Selten solution prescribes entry of the players with the lowest entry costs up to the point where entry becomes unprofitable.

The Harsanyi and Selten theory has been extensively applied to bargaining problems [cf. Harsanyi and Selten (1988, Chapters 6–9), Harsanyi (1980, 1982), Leopold-Wildburger (1985), Selten and Güth (1991), Selten and Leopold (1983)]. Such problems are modelled as unanimity games, i.e., a set of possible agreements is specified, players simultaneously choose an agreement and an agreement is implemented if and only if it is chosen by all players. In case there is no agreement, trade does not take place. For example, consider bargaining between two risk-neutral players about how to divide one dollar and suppose that one of the players, say player 1, has an outside option of α. The Harsanyi–Selten solution allocates $\max(\sqrt{\alpha}, 1/2)$ to player 1 and the rest to player 2. Hence, the outside option influences the outcome only if it is sufficiently high [Harsanyi and Selten (1988, Chapter 6)]. As another example, consider bargaining between one seller and n identical buyers about the sale of an indivisible object. If the seller's value is 0 and each buyer's value is 1, the Harsanyi–Selten solution is that each player proposes a sale at the price $p(n) = (2^n - 1)/(2^n - 1 + n)$.

Harsanyi and Selten (1988, Chapters 8 and 9) apply the theory to simple bargaining games with incomplete information. Players bargain about how to divide one dollar; if there is disagreement, a player receives his conflict payoff, which may be either 0 or x (both with probability $1/2$) and which is private information. In the case of one-sided incomplete information (it is common knowledge that player 1's conflict payoff is zero), player 1 proposes that he get a share $x(\alpha)$ of the cake, where $x(\alpha)$ is some decreasing square root function of α with $x(0) = 50$. The weak type of player 2 (i.e., the one with conflict payoff 0) proposes that player 1 get $x(\alpha)$, while the strong type proposes $x(\alpha)$ if $\alpha < \alpha^*$ (≈ 81) and 0 in case $\alpha > \alpha^*$. Hence, the bargaining outcome may be ex post inefficient. Güth and Selten (1991) consider a simple version of Akerlof's lemons problem [Akerlof (1970)]. A seller and a buyer are bargaining about the price of an object of art, which may be either worth 0 to each of them (it is a forgery) or which may be worth 1 to the seller and $v > 1$ to the buyer. The seller knows whether the object is original or fake, but the buyer only knows that both possibilities have positive probability. The solution either is disagreement, or exploitation of the buyer by the seller (i.e., the price equals the buyer's expected value), or some compromise in which the buyer bears a greater part of the fake risk than the seller does. At some parameter values, the solution (the price) changes discontinuously, and Güth and Selten admit that they cannot give plausible intuitive interpretations for these jumps.

Van Damme and Güth (1991a, 1991b) apply the Harsanyi–Selten theory to signalling games. In Van Damme and Güth (1991a) the most simple version of the Spence (1973) signalling game is considered. There are two types of workers, one productive, the other unproductive, who differ in their education costs and who can use the education level to signal their type to uninformed employers who compete in prices à la Bertrand. It turns out that the Harsanyi and Selten solution coincides with the E_2-equilibrium that was proposed in Wilson (1977). Hence, the solution is the sequential equilibrium that is most preferred by the high quality worker, and this worker signals his type if and only if signalling yields higher utility than pooling with the unproductive worker does. It is worth remarking that this solution is obtained without invoking payoff dominance. Note that the solution is again discontinuous in the parameter of the problem, i.e., in the ex ante probability that the worker is productive. The discontinuity arises at points where a different element of the Harsanyi–Selten solution procedure has to be invoked. Specifically, if the probability of the worker being unproductive is small, then there is only one primitive formation and this contains only the Pareto-optimal pooling equilibrium. As soon as this probability exceeds a certain threshold, however, also the formation spanned by the Pareto-optimal separating equilibrium is primitive, and, since the separating equilibrium risk dominates the pooling equilibrium, the solution is separating in this case.

We conclude this subsection by mentioning some variations of the Harsanyi–Selten theory that have recently been proposed. Güth and Kalkofen (1989) propose the ESBORA theory, whose main difference to the Harsanyi–Selten theory is that the (intransitive) risk dominance relation is replaced by the transitive relation of resistance dominance. The latter takes the intensity of the dominance relation into account. Formally, given two equilibria s and s', define player i's resistance at s against s' as the largest probability z such that, when each player $j \neq i$ plays $(1 - z)s_j + zs'_j$, player i still prefers s_i to s'_i. Güth and Kalkofen propose ways to aggregate these individual resistances into a resistance of s against s' which can be measured by a number $r(s, s')$. The resistance against s' can then be represented by the vector $R(s') = \langle r(s, s') \rangle_s$ and Güth and Kalkofen propose to select that equilibrium s' for which the vector $R(s')$, written in nonincreasing order, is lexicographically minimal. At present the ESBORA theory is still incomplete: the individual resistances can be aggregated in various ways and the solution may depend in an essential way on which aggregation procedure is adopted, as examples in Güth and Kalkofen (1989) show [see also Güth (1992) for different aggregation procedures].

For a restricted class of games (specifically, bipolar games with linear incentives), Selten (1995) proposes a set of axioms that determine a unique rule to aggregate the players individual resistances into an overall measure of resistance (or risk) dominance. For 2×2 games, selection on the basis of this measure is in agreement with selection as in Section 5.2, but for larger games, this need no longer be true. In fact, for 2-player games with incomplete information, selection according to the measure proposed in Selten (1995) has close relations with selection according to the "generalized Nash product" as in Harsanyi and Selten (1972).

Finally, we mention that Harsanyi (1995) proposes to replace the bilateral risk comparisons between pairs of equilibria by a multilateral comparison involving all equilibria that directly identifies the least risky of all of them. He also proposes not to make use of payoff comparisons, a suggestion that brings us back the to fundamental conflict between payoff dominance and risk dominance that was discussed in Section 5.4.

5.6. Final remark

We end this section and chapter by mentioning a result from Norde et al. (1996) that puts all the attempts to select a unique equilibrium in a different perspective. Recall that in Section 2 we discussed the axiomatization of Nash equilibrium using the concept of consistency, i.e., the idea that a solution of a game should induce a solution of any reduced game in which some players are committed to playing the solution. Norde et al. (1996) show that if s is a Nash equilibrium of a game g, g can be embedded in a larger game that only has s as an equilibrium; consequently consistency is incompatible with equilibrium selection. More precisely, Norde et al. (1996) show that the only solution concept that satisfies consistency, nonemptiness and one-person rationality is the Nash concept itself, so that not only equilibrium selection, but even the attempt to refine the Nash concept is frustrated if one insists on consistency.

References

Akerlof, G. (1970), "The market for lemons", Quarterly Journal of Economics 84:488–500.

Aumann, R.J. (1959), "Acceptable points in general cooperative n-person games", in: R.D. Luce and A.W. Tucker, eds., Contributions to the Theory of Games, IV, Annals of Mathematics Studies, Vol. 40 (Princeton, NJ) 287–324.

Aumann, R.J. (1974), "Subjectivity and correlation in randomized strategies", Journal of Mathematical Economics 1:67–96.

Aumann, R.J. (1985), "What is game theory trying to accomplish?", in: K. Arrow and S. Honkapohja, eds., Frontiers of Economics (Basil Blackwell, Oxford) 28–76.

Aumann, R.J. (1987), "Game theory", in: J. Eatwell, M. Milgate and P. Newman, eds., The New Palgrave Dictionary of Economics (Macmillan, London) 460–482.

Aumann, R.J. (1990), "Nash equilibria are not self-enforcing", in: J.J. Gabszewicz, J.-F. Richard and L.A. Wolsey, eds., Economic Decision-Making: Games, Econometrics and Optimisation (Elsevier, Amsterdam) 201–206.

Aumann, R.J. (1992), "Irrationality in game theory", in: P. Dasgupta et al., Economic Analysis of Markets and Games (MIT Press, Cambridge) 214–227.

Aumann, R.J. (1995), "Backward induction and common knowledge of rationality", Games and Economic Behavior 8:6–19.

Aumann, R.J. (1998), "On the centipede game", Games and Economic Behavior 23:97–105.

Aumann, R.J., and A. Brandenburger (1995), "Epistemic conditions for Nash equilibrium", Econometrica 63:1161–1180.

Aumann, R.J., Y. Katznelson, R. Radner, R.W. Rosenthal and B. Weiss (1983), "Approximate purification of mixed strategies", Mathematics of Operations Research 8:327–341.

Aumann R.J., and M. Maschler (1972), "Some thoughts on the minimax principle", Management Science 18:54–63.

Aumann R.J., and S. Sorin (1989), "Cooperation and bounded recall", Games and Economic Behavior 1:5–39.

Bagwell, K., and G. Ramey (1996), "Capacity, entry and forward induction", Rand Journal of Economics 27:660–680.

Balkenborg, D. (1992), "The properties of persistent retracts and related concepts", Ph.D. thesis, Department of Economics, University of Bonn.

Balkenborg, D. (1993), "Strictness, evolutionary stability and repeated games with common interests", CARESS, WP 93-20, University of Pennsylvania.

Banks J.S., and J. Sobel (1987), "Equilibrium selection in signalling games", Econometrica 55:647–663.

Basu, K. (1988), "Strategic irrationality in extensive games", Mathematical Social Sciences 15:247–260.

Basu, K. (1990), "On the non-existence of rationality definition for extensive games", International Journal of Game Theory 19:33–44.

Basu K., and J. Weibull (1991), "Strategy subsets closed under rational behavior", Economics Letters 36:141–146.

Battigalli, P. (1997), "On rationalizability in extensive games", Journal of Economic Theory 74:40–61.

Ben-Porath, E. (1993), "Common belief of rationality in perfect information games", mimeo (Tel Aviv University).

Ben-Porath, E., and E. Dekel (1992), "Signalling future actions and the potential for sacrifice", Journal of Economic Theory 57:36–51.

Bernheim, B.D. (1984), "Rationalizable strategic behavior", Econometrica 52:1007–1029.

Bernheim, B.D., B. Peleg and M.D. Whinston (1987), "Coalition-proof Nash equilibria I: Concepts", Journal of Economic Theory 42:1–12.

Bernheim, B.D., and D. Ray (1989), "Collective dynamic consistency in repeated games", Games and Economic Behavior 1:295–326.

Bernheim, B.D., and M.D. Whinston (1987), "Coalition-proof Nash equilibria II: Applications", Journal of Economic Theory 42:13–29.

Binmore, K. (1987), "Modeling rational players I", Economics and Philosophy 3:179–214.

Binmore, K. (1988), "Modeling rational players II", Economics and Philosophy 4:9–55.

Blume, A. (1994), "Equilibrium refinements in sender–receiver games", Journal of Economic Theory 64:66–77.

Blume, A. (1996), "Neighborhood stability in sender–receiver games", Games and Economic Behavior 13:2–25.

Blume, L.E., and W.R. Zame (1994), "The algebraic geometry of perfect and sequential equilibrium", Econometrica 62:783–794.

Börgers, T. (1991), "On the definition of rationalizability in extensive games", DP 91-22, University College London.

Carlsson, H., and E. van Damme (1993a), "Global games and equilibrium selection", Econometrica 61:989–1018.

Carlsson, H., and E. van Damme (1993b), "Equilibrium selection in stag hunt games", in: K. Binmore, A. Kirman and P. Tani, eds., Frontiers of Game Theory (MIT Press, Cambridge) 237–254.

Chin, H.H., T. Parthasarathy and T.E.S. Raghavan (1974), "Structure of equilibria in n-person non-cooperative games", International Journal of Game Theory 3:1–19.

Cho, I.K., and D.M. Kreps (1987), "Signalling games and stable equilibria", Quarterly Journal of Economics 102:179–221.

Cho, I.K., and J. Sobel (1990), "Strategic stability and uniqueness in signaling games", Journal of Economic Theory 50:381–413.

Van Damme, E.E.C. (1983), Refinements of the Nash Equilibrium Concept, Lecture Notes in Economics and Mathematical Systems, Vol. 219 (Springer-Verlag, Berlin).

Van Damme, E.E.C. (1984), "A relation between perfect equilibria in extensive form games and proper equilibria in normal form games", International Journal of Game Theory 13:1–13.

Van Damme, E.E.C. (1987a), Stability and Perfection of Nash Equilibria (Springer-Verlag, Berlin). Second edition 1991.

Van Damme, E.E.C.(1987b), "Equilibria in non-cooperative games", in: H.J.M. Peters and O.J. Vrieze, eds., Surveys in Game Theory and Related Topics, CWI Tract, Vol. 39 (Amsterdam) 1–37.

Van Damme, E.E.C. (1988), "The impossibility of stable renegotiation", Economics Letters 26:321–324.

Van Damme, E.E.C. (1989a), "Stable equilibria and forward induction", Journal of Economic Theory 48:476–496.

Van Damme, E.E.C. (1989b), "Renegotiation-proof equilibria in repeated prisoners' dilemma", Journal of Economic Theory 47:206–217.

Van Damme, E.E.C. (1990), "On dominance solvable games and equilibrium selection theories", CentER DP 9046, Tilburg University.

Van Damme, E.E.C. (1992), "Refinement of Nash equilibrium", in: J.J. Laffont, ed., Advances in Economic Theory, 6th World Congress, Vol. 1. Econometric Society Monographs No. 20 (Cambridge University Press) 32–75.

Van Damme, E.E.C. (1994), "Evolutionary game theory", European Economic Review 38:847–858.

Van Damme, E.E.C., and W. Güth (1991a), "Equilibrium selection in the Spence signalling game", in: R. Selten, ed., Game Equilibrium Models, Vol. 2, Methods, Morals and Markets (Springer-Verlag, Berlin) 263–288.

Van Damme, E.E.C., and W. Güth (1991b), "Gorby games: A game theoretic analysis of disarmament campaigns and the defence efficiency hypothesis", in: R. Avenhaus, H. Kavkar and M. Rudniaski, eds., Defence Decision Making. Analytical Support and Crises Management (Springer-Verlag, Berlin) 215–240.

Van Damme, E.E.C., and S. Hurkens (1996), "Commitment robust equilibria and endogenous timing", Games and Economic Behavior 15:290–311.

Dasgupta, P., and E. Maskin (1986), "The existence of equilibria in discontinuous games, 1: Theory", Review of Economic Studies 53:1–27.

Debreu, G. (1970), "Economics with a finite set of equilibria", Econometrica 38:387–392.

Dierker, E. (1972), "Two remarks on the number of equilibria of an economy", Econometrica 40:951–953.

Dold, A. (1972), Lectures on Algebraic Topology (Springer-Verlag, New York).

Dresher, M. (1961), Games of Strategy (Prentice-Hall, Englewood Cliffs, NJ).

Dresher, M. (1970), "Probability of a pure equilibrium point in n-person games", Journal of Combinatorial Theory 8:134–145.

Dubey, P. (1986), "Inefficiency of Nash equilibria", Mathematics of Operations Research 11:1–8.

Ellison, G. (1993), "Learning, local interaction, and coordination", Econometrica 61:1047–1072.

Farrell, J., and M. Maskin (1989), "Renegotiation in repeated games", Games and Economic Behavior 1:327–360.

Forges, F. (1990), "Universal mechanisms", Econometrica 58:1341–1364.

Fudenberg, D., D. Kreps and D.K. Levine (1988), "On the robustness of equilibrium refinements", Journal of Economic Theory 44:354–380.

Fudenberg, D., and D.K. Levine (1993a), "Self-confirming equilibrium", Econometrica 60:523–545.

Fudenberg, D., and D.K. Levine (1993b), "Steady state learning and Nash equilibrium", Econometrica 60:547–573.

Fudenberg, D., and D.K. Levine (1998), The Theory of Learning in Games (MIT Press, Cambridge, MA).

Fudenberg, D., and J. Tirole (1991), "Perfect Bayesian equilibrium and sequential equilibrium", Journal of Economic Theory 53:236–260.

Glazer, J., and A. Weiss (1990), "Pricing and coordination: Strategically stable equilibrium", Games and Economic Behavior 2:118–128.

Glicksberg, I.L. (1952), "A further generalization of the Kakutani fixed point theorem with application to Nash equilibrium points", Proceedings of the National Academy of Sciences 38:170–174.

Govindan, S. (1995), "Stability and the chain store paradox", Journal of Economic Theory 66:536–547.

Govindan, S., and A. Robson (1998), "Forward induction, public randomization and admissibility", Journal of Economic Theory 82:451–457.

Govindan, S., and R. Wilson (1997), "Equivalence and invariance of the index and degree of Nash equilibria", Games and Economic Behavior 21:56–61.

Govindan, S., and R.B. Wilson (1999), "Maximal stable sets of two-player games", mimeo (University of Western Ontario and Stanford University).

Govindan, S., and R. Wilson (2000), "Uniqueness of the index for Nash equilibria of finite games", mimeo (University of Western Ontario and Stanford University).

Gul, F., and D. Pearce (1996), "Forward induction and public randomization", Journal of Economic Theory 70:43–64.

Gul F., D. Pearce and E. Stacchetti (1993), "A bound on the proportion of pure strategy equilibria in generic games", Mathematics of Operations Research 18:548–552.

Güth, W. (1985), "A remark on the Harsanyi–Selten theory of equilibrium selection", International Journal of Game Theory 14:31–39.

Güth, W. (1992), "Equilibrium selection by unilateral deviation stability", in: R. Selten, ed., Rational Interaction, Essays in Honor of John C. Harsanyi (Springer-Verlag, Berlin) 161–189.

Güth, W., and B. Kalkofen (1989), "Unique solutions for strategic games", Lecture Notes in Economics and Mathematical Systems (Springer-Verlag, Berlin).

Güth, W., and R. Selten (1991), "Original or fake – a bargaining game with incomplete information", in: R. Selten, ed., Game Equilibrium Models, Vol. 3, Strategic Bargaining (Springer-Verlag, Berlin) 186–229.

Hammerstein, P., and R. Selten (1994), "Game theory and evolutionary biology", in: R.J. Aumann and S. Hart, eds., Handbook of Game Theory, Vol. 2 (North-Holland, Amsterdam) Chapter 28, 929–994.

Harris, C. (1985), "Existence and characterization of perfect equilibrium in games of perfect information", Econometrica 53:613–628.

Harris, C., P. Reny and A. Robson (1995), "The existence of subgame-perfect equilibrium in continuous games with almost perfect information: A case for public randomization", Econometrica 63:507–544.

Harsanyi, J.C. (1973a), "Games with randomly disturbed payoffs: A new rationale for mixed strategy equilibrium points", International Journal of Game Theory 2:1–23.

Harsanyi, J.C. (1973b), "Oddness of the number of equilibrium points: A new proof", International Journal of Game Theory 2:235–250.

Harsanyi, J.C. (1975), "The tracing procedure: A Bayesian approach to defining a solution for n-person non-cooperative games", International Journal of Game Theory 4:61–94.

Harsanyi, J.C. (1980), "Analysis of a family of two-person bargaining games with incomplete information", International Journal of Game Theory 9:65–89.

Harsanyi, J.C. (1982), "Solutions for some bargaining games under Harsanyi–Selten solution theory, Part I: Theoretical preliminaries; Part II: Analysis of specific bargaining games", Mathematical Social Sciences 3:179–191, 259–279.

Harsanyi, J.C. (1995), "A new theory of equilibrium selection for games with complete information", Games and Economic Behavior 8:91–122.

Harsanyi, J.C., and R. Selten (1972), "A generalized Nash solution for two-person bargaining games with incomplete information", Management Science 18:80–106.

Harsanyi, J.C., and R. Selten (1977), "Simple and iterated limits of algebraic functions", WP CP-370, Center for Research in Management, University of California, Berkeley.

Harsanyi, J.C., and R. Selten (1988), A General Theory of Equilibrium Selection in Games (MIT Press, Cambridge, MA).

Hart, S. (1992), "Games in extensive and strategic forms", in: R.J. Aumann and S. Hart, eds., Handbook of Game Theory, Vol. 1 (North-Holland, Amsterdam) Chapter 2, 19–40.

Hart, S. (1999), "Evolutionary dynamics and backward induction", DP 195, Center for Rationality, Hebrew University.

Hart, S., and D. Schmeidler (1989), "Existence of correlated equilibria", Mathematics of Operations Research 14:18–25.

Hauk, E., and S. Hurkens (1999), "On forward induction and evolutionary and strategic stability", WP 408, University of Pompeu Fabra, Barcelona, Spain.

Hellwig, M., W. Leininger, P. Reny and A. Robson (1990), "Subgame-perfect equilibrium in continuous games of perfect information: An elementary approach to existence and approximation by discrete games", Journal of Economic Theory 52:406–422.

Herings, P.J.-J. (2000), "Two simple proofs of the feasibility of the linear tracing procedure", Economic Theory 15:485–490.

Hillas, J. (1990), "On the definition of the strategic stability of equilibria", Econometrica 58:1365–1390.

Hillas, J. (1996), "On the relation between perfect equilibria in extensive form games and proper equilibria in normal form games", mimeo (SUNY, Stony Brook).

Hillas, J., and E. Kohlberg (2002), "Foundations of strategic equilibrium", in: R.J. Aumann and S. Hart, eds., Handbook of Game Theory, Vol. 3 (North-Holland, Amsterdam) Chapter 42, 1597–1663.

Hillas, J., J. Potters and A.J. Vermeulen (1999), "On the relations among some definitions of strategic stability", mimeo (Maastricht University).

Hillas, J., A.J. Vermeulen and M. Jansen (1997), "On the finiteness of stable sets: Note", International Journal of Game Theory 26:275–278.

Hurkens, S. (1994), "Learning by forgetful players", Games and Economic Behavior 11:304–329.

Hurkens, S. (1996), "Multi-sided pre-play communication by burning money", Journal of Economic Theory 69:186–197.

Jansen, M.J.M. (1981), "Maximal Nash subsets for bimatrix games", Naval Research Logistics Quarterly 28:147–152.

Jansen, M.J.M., A.P. Jurg and P.E.M. Borm (1990), "On the finiteness of stable sets", mimeo (University of Nijmegen).

Kalai, E., and D. Samet (1984), "Persistent equilibria", International Journal of Game Theory 13:129–141.

Kalai, E., and D. Samet (1985), "Unanimity games and Pareto optimality", International Journal of Game Theory 14:41–50.

Kandori, M., G.J. Mailath and R. Rob (1993), "Learning, mutation and long-run equilibria in games", Econometrica 61:29–56.

Kohlberg, E. (1981), "Some problems with the concept of perfect equilibrium", NBER Conference on Theory of General Economic Equilibrium, University of California, Berkeley.

Kohlberg, E. (1989), "Refinement of Nash equilibrium: The main ideas", mimeo (Harvard University).

Kohlberg, E., and J.-F. Mertens (1986), "On the strategic stability of equilibria", Econometrica 54:1003–1037.

Kohlberg, E., and P. Reny (1997), "Independence on relative probability spaces and consistent assessments in game trees", Journal of Economic Theory 75:280–313.

Kreps, D., and G. Ramey (1987), "Structural consistency, consistency, and sequential rationality", Econometrica 55:1331–1348.

Kreps, D., and J. Sobel (1994), "Signalling", in: R.J. Aumann and S. Hart, eds., Handbook of Game Theory, Vol. 2 (North-Holland, Amsterdam) Chapter 25, 849–868.

Kreps, D., and R. Wilson (1982a), "Sequential equilibria", Econometrica 50:863–894.

Kreps, D., and R. Wilson (1982b), "Reputation and imperfect information", Journal of Economic Theory 27:253–279.

Kuhn, H.W. (1953), "Extensive games and the problem of information", Annals of Mathematics Studies 48:193–216.

Lemke, C.E., and J.T. Howson (1964), "Equilibrium points of bimatrix games", Journal of the Society for Industrial and Applied Mathematics 12:413–423.

Leopold-Wildburger, U. (1985), "Equilibrium selection in a bargaining game with transaction costs", International Journal of Game Theory 14:151–172.

Madrigal, V., T. Tan and S. Werlang (1987), "Support restrictions and sequential equilibria", Journal of Economic Theory, 43:329–334.

Mailath, G.J., L. Samuelson and J.M. Swinkels (1993), "Extensive form reasoning in normal form games", Econometrica 61:273–302.

Mailath, G.J., L. Samuelson and J.M. Swinkels (1997), "How proper is sequential equilibrium", Games and Economic Behavior 18:193–218.

Maynard Smith, J., and G. Price (1973), "The logic of animal conflict", Nature 246:15–18.

McLennan, A. (1985), "Justifiable beliefs in sequential equilibrium", Econometrica 53:889–904.

Mertens, J.-F. (1987), "Ordinality in non-cooperative games", CORE DP 8728, Université Catholique de Louvain, Louvain-la-Neuve.

Mertens, J.-F. (1989a), "Stable equilibria – a reformulation, Part I, Definition and basic properties", Mathematics of Operations Research 14:575–625.

Mertens, J.-F. (1989b), "Equilibrium and rationality: Context and history-dependence", CORE DP, October 1989, Université Catholique de Louvain, Louvain-la-Neuve.

Mertens, J.-F. (1990), "The 'Small Worlds' axiom for stable equilibria", CORE DP 9007, Université Catholique de Louvain, Louvain-la-Neuve.

Mertens, J.-F. (1991), "Stable equilibria – a reformulation, Part II, Discussion of the definition, and further results", Mathematics of Operations Research 16:694–753.

Mertens, J.-F. (1992), "Two examples of strategic equilibrium", CORE DP 9208, Université Catholique de Louvain, Louvain-la-Neuve.

Milgrom, P., and D.J. Roberts (1982), "Predation, reputation and entry deterrence", Journal of Economic Theory 27:280–312.

Milgrom, P., and D.J. Roberts (1990), "Rationalizability, learning and equilibrium in games with strategic complementarities", Econometrica 58:1255–1278.

Milgrom, P., and D.J. Roberts (1991), "Adaptive and sophisticated learning in repeated normal form games", Games and Economic Behavior 3:1255–1278.

Milgrom, P., and C. Shannon (1994), "Monotone comparative statics", Econometrica 62:157–180.

Milgrom, P., and R. Weber (1985), "Distributional strategies for games with incomplete information", Mathematics of Operations Research 10:619–632.

Monderer, D., and L.S. Shapley (1996), "Potential games", Games and Economic Behavior 14:124–143.

Moulin, H. (1979), "Dominance solvable voting games", Econometrica 47:1337–1351.

Moulin, H., and J.P. Vial (1978), "Strategically zero-sum games: The class of games whose completely mixed equilibria cannot be improved upon", International Journal of Game Theory 7:201–221.

Myerson, R. (1978), "Refinements of the Nash equilibrium concept", International Journal of Game Theory 7:73–80.

Myerson, R.B. (1986), "Multistage games with communication", Econometrica 54:323–358.

Myerson, R.B. (1994), "Communication, correlated equilibria, and incentive compatibility", in: R.J. Aumann and S. Hart, eds., Handbook of Game Theory, Vol. 2 (North-Holland, Amsterdam) Chapter 24, 827–848.

Nash, J.F. (1950a), "Non-cooperative games", Ph.D. Dissertation, Princeton University.

Nash, J.F. (1950b), "Equilibrium points in n-person games", Proceedings from the National Academy of Sciences, U.S.A. 36:48–49.

Nash, J.F. (1951), "Non-cooperative games", Annals of Mathematics 54:286–295.

Nash, J.F. (1953), "Two-person cooperative games", Econometrica 21:128–140.

Neumann, J. von, and O. Morgenstern (1947), Theory of Games and Economic Behavior (Princeton University Press, Princeton, NJ). First edition 1944.

Neyman, A. (1997), "Correlated equilibrium and potential games", International Journal of Game Theory 26:223–227.

Nöldeke, G., and E.E.C. van Damme (1990), "Switching away from probability one beliefs", DP A-304, University of Bonn.

Norde, H. (1999), "Bimatrix games have quasi-strict equilibria", Mathematical Programming 85:35–49.

Norde, H., J. Potters, H. Reijnierse and A.J. Vermeulen (1996), "Equilibrium selection and consistency", Games and Economic Behavior 12:219–225.

Okada, A. (1983), "Robustness of equilibrium points in strategic games", DP B137, Tokyo Institute of Technology.

Osborne, M. (1990), "Signaling, forward induction, and stability in finitely repeated games", Journal of Economic Theory 50:22–36.

Pearce, D. (1984), "Rationalizable strategic behavior and the problem of perfection", Econometrica 52:1029–1050.

Peleg, B., and S.H. Tijs (1996), "The consistency principle for games in strategic form", International Journal of Game Theory 25:13–34.

Ponssard, J.-P. (1991), "Forward induction and sunk costs give average cost pricing", Games and Economic Behavior 3:221–236.

Radner, R., and R.W. Rosenthal (1982), "Private information and pure strategy equilibria", Mathematics of Operations Research 7:401–409.

Raghavan, T.E.S. (1994), "Zero-sum two-person games", in: R.J. Aumann and S. Hart, eds., Handbook of Game Theory, Vol. 2 (North-Holland, Amsterdam) Chapter 20, 735–768.

Reny, P. (1992a), "Backward induction, normal form perfection and explicable equilibria", Econometrica 60:627–649.

Reny, P.J. (1992b), "Rationality in extensive form games", Journal of Economic Perspectives 6:103–118.

Reny, P.J. (1993), "Common belief and the theory of games with perfect information", Journal of Economic Theory 59:257–274.

Ritzberger, K. (1994), "The theory of normal form games from the differentiable viewpoint", International Journal of Game Theory 23:207–236.

Rosenmüller, J. (1971), "On a generalization of the Lemke–Howson algorithm to noncooperative n-person games", SIAM Journal of Applied Mathematics 21:73–79.

Rosenthal, R. (1981), "Games of perfect information, predatory pricing and the chain store paradox", Journal of Economic Theory 25:92–100.

Rubinstein, A. (1989), "The electronic mail game: Strategic behavior under 'almost common knowledge' ", American Economic Review 79:385–391.

Rubinstein, A. (1991), "Comments on the interpretation of game theory", Econometrica 59:909–924.

Rubinstein, A., and A. Wolinsky (1994), "Rationalizable conjectural equilibrium: Between Nash and rationalizability", Games and Economic Behavior 6:299–311.

Samuelson, L. (1997), Evolutionary Games and Equilibrium Selection (MIT Press, Cambridge, MA).

Schanuel, S.H., L.K. Simon and W.R. Zame (1991), "The algebraic geometry of games and the tracing procedure", in: R. Selten, ed., Game Equilibrium Models, Vol. 2: Methods, Morals and Markets (Springer-Verlag, Berlin) 9–43.

Selten, R. (1965), "Spieltheoretische Behandlung eines Oligopolmodels mit Nachfragetragheit", Zeitschrift für die Gesamte Staatswissenschaft 12:301–324, 667–689.

Selten, R. (1975), "Re-examination of the perfectness concept for equilibrium points in extensive games", International Journal of Game Theory 4:25–55.

Selten, R. (1978), "The chain store paradox", Theory and Decision 9:127–159.

Selten, R. (1995), "An axiomatic theory of a risk dominance measure for bipolar games with linear incentives", Games and Economic Behavior 8:213–263.

Selten, R., and W. Güth (1982), "Equilibrium point selection in a class of market entry games", in: M. Deistler, E. Fürst and G. Schwödiauer, eds., Games, Economic Dynamics, and Time Series Analysis – A Symposium in Memoriam of Oskar Morgenstern (Physica-Verlag, Würzburg) 101–116.

Selten, R., and U. Leopold (1983), "Equilibrium point selection in a bargaining situation with opportunity costs", Economie Appliqueé 36:611–648.

Shapley, L.S. (1974), "A note on the Lemke–Howson algorithm", Mathematical Programming Study 1:175–189.

Shapley, L.S. (1981), "On the accessibility of fixed points", in: O. Moeschlin and D. Pallaschke, eds., Game Theory and Mathematical Economics (North-Holland, Amsterdam) 367–377.

Shubik, M. (2002), "Game theory and experimental gaming", in: R.J. Aumann and S. Hart, eds., Handbook of Game Theory, Vol. 3 (North-Holland, Amsterdam) Chapter 62, 2327–2351.

Simon, L.K., and M.B. Stinchcombe (1995), "Equilibrium refinement for infinite normal-form games", Econometrica 63:1421–1443.

Simon, L.K., and W.R. Zame (1990), "Discontinuous games and endogenous sharing rules", Econometrica 58:861–872.

Sorin, S. (1992), "Repeated games with complete information", in: R.J. Aumann and S. Hart, eds., Handbook of Game Theory, Vol. 1 (North-Holland, Amsterdam) Chapter 4, 71–108.

Spence, M. (1973), "Job market signalling", Quarterly Journal of Economics 87:355–374.

Stanford, W. (1995), "A note on the probability of k pure Nash equilibria in matrix games", Games and Economic Behavior 9:238–246.

Tarski, A. (1955), "A lattice theoretical fixed point theorem and its applications", Pacific Journal of Mathematics 5:285–308.

Topkis, D. (1979), "Equilibrium points in nonzero-sum n-person submodular games", SIAM Journal of Control and Optimization 17:773–787.

Vega-Redondo, F. (1996), Evolution, Games and Economic Behavior (Oxford University Press, Oxford, UK).

Vives, X. (1990), "Nash equilibrium with strategic complementarities", Journal of Mathematical Economics 19:305–321.

Weibull, J. (1995), Evolutionary Game Theory (MIT Press, Cambridge, MA).

Wilson, C. (1977), "A model of insurance markets with incomplete information", Journal of Economic Theory 16:167–207.

Wilson, R.B. (1971), "Computing equilibria of n-person games", SIAM Journal of Applied Mathematics 21:80–87.

Wilson, R.B. (1992), "Computing simply stable equilibria", Econometrica 60:1039–1070.

Wilson, R.B. (1997), "Admissibility and stability", in: W. Albers et al., eds., Understanding Strategic Interaction; Essays in Honor of Reinhard Selten (Springer-Verlag, Berlin) 85–99.

Young, H.P. (1993a), "The evolution of conventions", Econometrica 61:57–84.

Young, H.P. (1993b), "An evolutionary model of bargaining", Journal of Economic Theory 59:145–168.

Zermelo, E. (1912), "Über eine Anwendung der Mengenlehre auf die Theorie des Schachspiels", in: E.W. Hobson and A.E.H. Love, eds., Proceedings of the Fifth International Congress of Mathematicians, Vol. 2 (Cambridge University Press) 501–504.

Chapter 42

FOUNDATIONS OF STRATEGIC EQUILIBRIUM

JOHN HILLAS

Department of Economics, University of Auckland, Auckland, New Zealand

ELON KOHLBERG

Harvard Business School, Boston, MA, USA

Contents

Handbook of Game Theory, Volume 3, Edited by R.J. Aumann and S. Hart
© 2002 Elsevier Science B.V. All rights reserved

Abstract

This chapter examines the conceptual foundations of the concept of strategic equilibrium and its various variants and refinements. The emphasis is very much on the underlying ideas rather than on any technical details.

After an examination of some pre-equilibrium ideas, in particular the concept of rationalizability, the concept of strategic (or Nash) equilibrium is introduced. Various interpretations of this concept are discussed and a proof of the existence of such equilibria is sketched.

Next, the concept of correlated equilibrium is introduced. This concept can be thought of as retaining the self-enforcing aspect of the idea of equilibrium while relaxing the independence assumption.

Most of the remainder of the chapter is concerned with the ideas underlying the refinement of equilibrium: admissibility and iterated dominance; backward induction; forward induction; and ordinality and various invariances to changes in the player set. This leads to a consideration of the concept of strategic stability, a strong refinement satisfying these various ideas.

Finally there is a brief examination of the epistemic approach to equilibrium and the relation between strategic equilibrium and correlated equilibrium.

Keywords

Nash equilibrium, strategic equilibrium, correlated equilibrium, equilibrium refinement, strategic stability

JEL classification: C72

1. Introduction

The central concept of noncooperative game theory is that of the *strategic equilibrium* (or Nash equilibrium, or noncooperative equilibrium). A strategic equilibrium is a profile of strategies or plans, one for each player, such that each player's strategy is optimal for him, given the strategies of the others.

In most of the early literature the idea of equilibrium was that it said something about how players would play the game or about how a game theorist might recommend that they play the game. More recently, led by Harsanyi (1973) and Aumann (1987a), there has been a shift to thinking of equilibria as representing not recommendations to players of how to play the game but rather the expectations of the others as to how a player will play. Further, if the players all have the same expectations about the play of the other players we could as well think of an outside observer having the same information about the players as they have about each other.

While we shall at times make reference to the earlier approach we shall basically follow the approach of Harsanyi and Aumann or the approach of considering an outside observer, which seem to us to avoid some of the deficiencies and puzzles of the earlier approach. Let us consider the example of Figure 1. In this example there is a unique equilibrium. It involves Player 1 playing T with probability $1/2$ and B with probability $1/2$ and Player 2 playing L with probability $1/3$ and R with probability $2/3$.

There is some discomfort in applying the first interpretation to this example. In the equilibrium each player obtains the same expected payoff from each of his strategies. Thus the equilibrium gives the game theorist absolutely no reason to recommend a particular strategy and the player no reason to follow any recommendation the theorist might make. Moreover one often hears comments that one does not, in the real world, see players actively randomizing.

If, however, we think of the strategy of Player 1 as representing the uncertainty of Player 2 about what Player 1 will do, we have no such problem. Any assessment other than the equilibrium of the uncertainty leads to some contradiction. Moreover, if we assume that the uncertainty in the players' minds is the objective uncertainty then we also have tied down exactly the distribution on the strategy profiles, and consequently the expected payoff to each player, for example $2/3$ to Player 1.

This idea of strategic equilibrium, while formalized for games by Nash (1950, 1951), goes back at least to Cournot (1838). It is a simple, beautiful, and powerful concept. It seems to be the natural implementation of the idea that players do as well as they can, taking the behavior of others as given. Aumann (1974) pointed out that there is some-

	L	R
T	2,0	0,1
B	0,1	1,0

Figure 1.

thing more involved in the definition given by Nash, namely the independence of the strategies, and showed that it is possible to define an equilibrium concept that retains the idea that players do as well as they can, taking the behavior of others as given while dropping the independence assumption. He called such a concept *correlated equilibrium*. Any strategic equilibrium "is" a correlated equilibrium, but for many games there are correlated equilibria which are quite different from any of the strategic equilibria.

In another sense the requirements for strategic equilibrium have been seen to be too weak. Selten (1965, 1975) pointed out that irrational behavior by each of two different players might make the behavior of the other look rational, and proposed additional requirements, beyond those defining strategic equilibrium, to eliminate such cases. In doing so Selten initiated a large literature on the refinement of equilibrium. Since then many more requirements have been proposed. The question naturally arises as to whether it is possible to simultaneously satisfy all, or even a large subset of such requirements. The program to define strategically stable equilibria, initiated by Kohlberg and Mertens (1986) and brought to fruition by Mertens (1987, 1989, 1991b, 1992) and Govindan and Mertens (1993), answers this question in the affirmative.

This chapter is rather informal. Not everything is defined precisely and there is little use, except in examples, of symbols. We hope that this will not give our readers any problem. Readers who want formal definitions of the concepts we discuss here could consult the chapters by Hart (1992) and van Damme (2002) in this Handbook.

2. Pre-equilibrium ideas

Before discussing the idea of equilibrium in any detail we shall look at some weaker conditions. We might think of these conditions as necessary implications of assuming that the game and the rationality of the players are common knowledge, in the sense of Aumann (1976).

2.1. Iterated dominance and rationalizability

Consider the problem of a player in some game. Except in the most trivial cases the set of strategies that he will be prepared to play will depend on his assessment of what the other players will do. However it *is* possible to say a little. If some strategy was strictly preferred by him to another strategy s whatever he thought the other players would do, then he surely would not play s. And this remains true if it was some lottery over his strategies that was strictly preferred to s. We call a strategy such as s a *strictly dominated strategy*.

Perhaps we could say a little more. A strategy s would surely not be played unless there was some assessment of the manner in which the others might play that would lead s to be (one of) the best. This is clearly at least as restrictive as the first requirement. (If s is best for some assessment it cannot be strictly worse than some other strategy for all assessments.) In fact, if the set of assessments of what the others might do is convex (as a set of probabilities on the profiles of pure strategies of the others) then the two requirements are equivalent. This will be true if there is only one other player, or if a player's assessment of what the others might do permits correlation. However, the set of product distributions over the product of two or more players' pure strategy sets is not convex.

Thus we have two cases: one in which we eliminate strategies that are strictly dominated, or equivalently never best against some distribution on the vectors of pure strategies of the others; and one in which we eliminate strategies that are never best against some product of distributions on the pure strategy sets of the others. In either case we have identified a set of strategies that we argue a rational player would not play. But since everything about the game, including the rationality of the players, is assumed to be common knowledge no player should put positive weight, in his assessment of what the other players might do, on such a strategy. And we can again ask: are there any strategies that are strictly dominated when we restrict attention to the assessments that put weight only on those strategies of the others that are not strictly dominated? If so, a rational player who knew the rationality of the others would surely not play such a strategy. And similarly for strategies that were not best responses against some assessment putting weight only on those strategies of the others that are best responses against some assessment by the others.

And we can continue for an arbitrary number of rounds. If there is ever a round in which we don't find any new strategies that will not be played by rational players commonly knowing the rationality of the others, we would never again "eliminate" a strategy. Thus, since we start with a finite number of strategies, the process must eventually terminate. We call the strategies that remain *iteratively undominated* or *correlatedly rationalizable* in the first case; and *rationalizable* in the second case. The term rationalizable strategy and the concept were introduced by Bernheim (1984) and Pearce (1984). The term correlatedly rationalizable strategy and the concept were explicitly introduced by Brandenburger and Dekel (1987), who also show the equivalence of this concept to what we are calling iteratively undominated strategies, though both the concept and this equivalence are alluded to by Pearce (1984).

The issue of whether or not the assessments of one player of the strategies that will be used by the others should permit correlation has been the topic of some discussion in the literature. Aumann (1987a) argues strongly that they should. Others have argued that there is at least a case to be made for requiring the assessments to exhibit independence. For example, Bernheim (1986) argues as follows.

> Aumann has disputed this view [that assessments should exhibit independence]. He argues that there is no *a priori* basis to exclude any probabilistic beliefs. Correlation between opponents' strategies may make perfect sense for a variety of reasons. For example, two players who attended the same "school" may have similar dispositions. More generally, while each player knows that his decision does not directly affect the choices of others, the substantive information which leads him to make one choice rather than another also affects his beliefs about other players' choices.

Yet Aumann's argument is not entirely satisfactory, since it appears to make our theory of rationality depend upon some ill-defined "dispositions" which are, at best, extra-rational. What is the "substantive information" which disposes an individual towards a particular choice? In a pure strategic environment, the only available substantive information consists of the features of the game itself. This information is the same, regardless of whether one assumes the role of an outside observer, or the role of a player with a particular "disposition". Other information, such as the "school" which a player attended, is simply extraneous. Such information could only matter if, for example, different schools taught different things. A "school" may indeed teach not only the information embodied in the game itself, but also "something else"; however, differences in schools would then be substantive only if this "something else" was substantive. Likewise, any apparently concrete source of differences or similarities in dispositions can be traced to an amorphous "something else", which does not arise directly from considerations of rationality.

This addresses Aumann's ideas in the context of Aumann's arguments concerning correlated equilibrium. Without taking a position here on those arguments, it does seem that in the context of a discussion of rationalizability the argument for independence is not valid. In particular, even if one accepts that one's opponents actually choose their strategies independently and that there is nothing substantive that they have in common outside their rationality and the roles they might have in the game, another player's assessment of what they are likely to play could exhibit correlation.

Let us go into this in a bit more detail. Consider a player, say Player 3, making some assessment of how Players 1 and 2 will play. Suppose also that Players 1 and 2 act in similar situations, that they have no way to coordinate their choices, and that Player 3 knows nothing that would allow him to distinguish between them. Now we assume that Player 3 forms a probabilistic assessment as to how each of the other players will act. What can we say about such an assessment? Let us go a bit further and think of another player, say Player 4, also forming an assessment of how the two players will play. It is a hallmark of rationalizability that we do *not* assume that Players 3 and 4 will form the same assessment. (In the definition of rationalizability each strategy can have its own justification and there is no assumption that there is any consistency in the justifications.) Thus we do not assume that they somehow know the true probability that Players 1 and 2 will play a certain way.

Further, since we allow them to differ in their assessments it makes sense to also allow them to be uncertain not only about how Players 1 and 2 will play, but indeed about the probability that a rational player will play a certain way. This is, in fact, exactly analogous to the classic problem discussed in probability theory of repeated tosses of a possibly biased coin. We assume that the coin has a certain fixed probability p of coming up heads. Now if an observer is uncertain about p then the results of the coin tosses will not, conditional on his information, be statistically independent. For example, if his assessment was that with probability one half $p = 1/4$ and with probability one

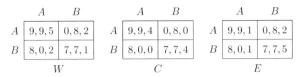

Figure 2.

half $p = 3/4$ then after seeing heads for the first three tosses his assessment that heads will come up on the next toss will be higher than if he had seen tails on the first three tosses.

Let us be even more concrete and consider the three player game of Figure 2. Here Players 1 and 2 play a symmetric game having two pure strategy equilibria. This game has been much discussed in the literature as the "Stag Hunt" game. [See Aumann (1990), for example.] For our purposes here all that is relevant is that it is not clear how Players 1 and 2 will play and that how they will play has something to do with how rational players in general would think about playing a game. If players in general tend to "play safe" then the outcome (B, B) seems likely, while if they tend to coordinate on efficient outcomes then (A, A) seems likely. Player 3 has a choice that does not affect the payoffs of Players 1 and 2, but whose value to him does depend on the choices of 1 and 2. If Players 1 and 2 play (B, B) then Player 3 does best by choosing E, while if they play (A, A) then Player 3 does best by choosing W. Against any product distribution on the strategies of Players 1 and 2 the better of E or W is better than C for Player 3.

Now suppose that Player 3 knows that the other players were independently randomly chosen to play the game and that they have no further information about each other and that they choose their strategies independently. As we argued above, if he doesn't know the distribution then it seems natural to allow him to have a nondegenerate distribution over the distributions of what rational players commonly knowing the rationality of the others do in such a game. The action taken by Player 1 will, in general, give Player 3 some information on which he will update his distribution over the distributions of what rational players do. And this will lead to correlation in his assessment of what Players 1 and 2 will do in the game. Indeed, in this setting, requiring independence essentially amounts to requiring that players be certain about things about which we are explicitly allowing them to be wrong.

We indicated earlier that the set of product distributions over two or more players' strategy sets was not convex. This is correct, but somewhat incomplete. It is possible to put a linear structure on this set that would make it convex. In fact, that is exactly what we do in the proof of the existence of equilibrium below. What we mean is that if we think of the product distributions as a subset of the set of all probability distributions on the profiles of pure strategies and use the linear structure that is natural for that latter set then the set of product distributions is not convex. If instead we use the product of the linear structures on the spaces of distributions on the individual strategy spaces, then the set of product distributions will indeed be convex. The nonconvexity however reappears in the fact that with this linear structure the expected payoff function is no longer linear – or even quasi-concave.

2.2. *Strengthening rationalizability*

There have been a number of suggestions as to how to strengthen the notion of rationalizability. Many of these involve some form of the iterated deletion of (weakly) dominated strategies, that is, strategies such that some other (mixed) strategy does at least as well whatever the other players do, and strictly better for some choices of the others. The difficulty with such a procedure is that the order in which weakly dominated strategies are eliminated can affect the outcome at which one arrives. It is certainly possible to give a definition that unambiguously determines the order, but such a definition implicitly rests on the assumption that the other players will view a strategy eliminated in a later round as infinitely more likely than one eliminated earlier.

Having discussed in the previous section the reasons for rejecting the requirement that a player's beliefs over the choices of two of the other players be independent we shall not again discuss the issue but shall allow correlated beliefs in all of the definitions we discuss. This has two implications. The first is that our statement below of Pearce's notion of cautiously rationalizable strategies will not be his original definition but rather the suitably modified one. The second is that we shall be able to simplify the description by referring simply to rounds of deletions of dominated strategies rather than the somewhat more complicated notions of rationalizability.

Even before the definition of rationalizability, Moulin (1979) suggested using as a solution an arbitrarily large number of rounds of the elimination of all weakly dominated strategies for all players. Moulin actually proposed this as a solution only when it led to a set of strategies for each player such that whichever of the allowable strategies the others were playing the player would be indifferent among his own allowable strategies.

A somewhat more sophisticated notion is that of *cautiously rationalizable strategies* defined by Pearce (1984). The set of such strategies is the set obtained by the following procedure. One first eliminates all strictly dominated strategies, and does this for an arbitrarily large number of rounds until one reaches a game in which there are no strictly dominated strategies. One then has a single round in which all weakly dominated strategies are eliminated. One then starts again with another (arbitrarily long) sequence of rounds of elimination of strictly dominated strategies, and again follows this with a single round in which all weakly dominated strategies are removed, and so on. For a finite game such a process ends after a finite number of rounds.

Each of these definitions has a certain apparent plausibility. Nevertheless they are not well motivated. Each depends on an implicit assumption that a strategy eliminated at a later round is much more likely than a strategy eliminated earlier. And this in turn

	X	Y	Z
A	1,1	1,0	1,0
B	0,1	1,0	2,0
C	1,0	1,1	0,0
D	0,0	0,0	1,1

Figure 3.

depends on an implicit assumption that in some sense the strategies deleted at one round are equally likely. For suppose we could split one of the rounds of the elimination of weakly dominated strategies and eliminate only part of the set. This could completely change the entire process that follows.

Consider, for example, the game of Figure 3. The only cautiously rationalizable strategies are A for Player 1 and X for Player 2. In the first round strategies C and D are (weakly) dominated for Player 1. After these are eliminated strategies Y and Z are strictly dominated for Player 2. And after these are eliminated strategy B is strictly dominated for Player 1. However, if (A, X) is indeed the likely outcome then perhaps strategy D is, in fact, much less likely than strategy C, since, given that 2 plays X, strategy C is one of Player 1's best responses, while D is not. Suppose that we start by eliminating just D. Now, in the second round only Z is strictly dominated. Once Z is eliminated, we eliminate B for Player 1, but nothing else. We are left with A and C for Player 1 and X and Y for Player 2.

There seems to us one slight strengthening of rationalizability that is well motivated. It is one round of elimination of weakly dominated strategies followed by an arbitrarily large number of rounds of elimination of strictly dominated strategies. This solution is obtained by Dekel and Fudenberg (1990), under the assumption that there is some small uncertainty about the payoffs, by Börgers (1994), under the assumption that rationality was "almost" common knowledge, and by Ben-Porath (1997) for the class of generic extensive form games with perfect information.

The papers of Dekel and Fudenberg and of Börgers use some approximation to the common knowledge of the game and the rationality of the players in order to derive, simultaneously, admissibility and some form of the iterated elimination of strategies. Ben-Porath obtains the result in extensive form games because in that setting a natural definition of rationality implies more than simple ex ante expected utility maximization. An alternative justification is possible. Instead of deriving admissibility, we include it in what we mean by rationality. A choice s is admissibly rational against some conjecture c about the strategies of the others if there is some sequence of conjectures putting positive weight on all possibilities and converging to c such that s is maximizing against each conjecture in the sequence. Now common knowledge of the game and of the admissible rationality of the players gives precisely the set we described. The argument is essentially the same as the argument that common knowledge of rationality implies correlated rationalizability.

3. The idea of equilibrium

In the previous section we examined the extent to which it is possible to make predictions about players' behavior in situations of strategic interaction based solely on the common knowledge of the game, including in the description of the game the players' rationality. The results are rather weak. In some games these assumptions do indeed restrict our predictions, but in many they imply few, if any, restrictions.

To say something more, a somewhat different point of view is productive. Rather than starting from only knowledge of the game and the rationality of the players and asking what implications can be drawn, one starts from the supposition that there is some established way in which the game will be played and asks what properties this manner of playing the game must satisfy in order not to be self-defeating, that is, so that a rational player, knowing that the game will be played in this manner, does not have an incentive to behave in a different manner. This is the essential idea of a strategic equilibrium, first defined by John Nash (1950, 1951).

There are a number of more detailed stories to go along with this. The first was suggested by von Neumann and Morgenstern (1944), even before the first definition of equilibrium by Nash. It is that players in a game are advised by game theorists on how to play. In each instance the game theorist, knowing the player's situation, tells the player what the theory recommends. The theorist does offer a (single) recommendation in each situation and all theorists offer the same advice. One might well allow these recommendations to depend on various "real-life" features of the situation that are not normally included in our models. One would ask what properties the theory should have in order for players to be prepared to go along with its recommendations. This idea is discussed in a little more detail in the introduction of the chapter on "Strategic equilibrium" in this Handbook [van Damme (2002)].

Alternatively, one could think of a situation in which the players have no information beyond the rules of the game. We'll call such a game a *Tabula-Rasa game*. A player's optimal choice may well depend on the actions chosen by the others. Since the player doesn't know those actions we might argue that he will form some probabilistic assessment of them. We might go on to argue that since the players have precisely the same information they will form the same assessments about how choices will be made. Again, one could ask what properties this common assessment should have in order not to be self-defeating. The first to make the argument that players having the same information should form the same assessment was Harsanyi (1967–1968, Part III). Aumann (1974, p. 92) labeled this view the Harsanyi doctrine.

Yet another approach is to think of the game being preceded by some stage of "pre-play negotiation" during which the players may reach a non-binding agreement as to how each should play the game. One might ask what properties this agreement should have in order for all players to believe that everyone will act according to the agreement. One needs to be a little careful about exactly what kind of communication is available to the players if one wants to avoid introducing correlation. Bárány (1992) and Forges (1990) show that with at least four players and a communication structure that allows for private messages any correlated equilibrium "is" a Nash equilibrium of the game augmented with the communication stage. There are also other difficulties with the idea of justifying equilibria by pre-play negotiation. See Aumann (1990).

Rather than thinking of the game as being played by a fixed set of players one might think of each player as being drawn from a population of rational individuals who find themselves in similar roles. The specific interactions take place between randomly selected members of these populations, who are aware of the (distribution of) choices that had been made in previous interactions. Here one might ask what distributions are self-enforcing, in the sense that if players took the past distributions as a guide to what the others' choices were likely to be, the resulting optimal choices would (could) lead to a similar distribution in the current round. One already finds this approach in Nash (1950).

A somewhat different approach sees each player as representing a whole population of individuals, each of whom is "programmed" (for example, through his genes) to play a certain strategy. The players themselves are not viewed as rational, but they are assumed to be subject to "natural selection", that is, to the weeding out of all but the payoff-maximizing programs. Evolutionary approaches to game theory were introduced by Maynard Smith and Price (1973). For the rest of this chapter we shall consider only interpretations that involve rational players.

3.1. Self-enforcing plans

One interpretation of equilibrium sees the focus of the analysis as being the actual strategies chosen by the players, that is, their plans in the game. An equilibrium is defined to be a self-enforcing profile of plans. At least a necessary condition for a profile of plans to be self-enforcing is that each player, given the plans of the others, should not have an alternate plan that he strictly prefers.

This is the essence of the definition of an equilibrium. As we shall soon see, in order to guarantee that such a self-enforcing profile of plans exists we must consider not only deterministic plans, but also random plans. That is, as well as being permitted to plan what to do in any eventuality in which he might find himself, a player is explicitly thought of as planning to use some lottery to choose between such deterministic plans.

Such randomizations have been found by many to be somewhat troubling. Arguments are found in the literature that such "mixed strategies" are less stable than pure strategies. (There is, admittedly, a precise sense in which this is true.) And in the early game theory literature there is discussion as to what precisely it means for a player to choose a mixed strategy and why players may choose to use such strategies. See, for example, the discussion in Luce and Raiffa (1957, pp. 74–76).

Harsanyi (1973) provides an interpretation that avoids this apparent instability and, in the process, provides a link to the interpretation of Section 3.2. Harsanyi considers a model in which there is some small uncertainty about the players' payoffs. This uncertainty is independent across the players, but each player knows his own payoff. The uncertainty is assumed to be represented by a probability distribution with a continuously differentiable density. If for each player and each vector of pure actions (that is, strategies in the game without uncertainty) the probability that the payoff is close to some particular value is high, then we might consider the game close to the game in which the payoff is exactly that value. Conversely, we might consider the game in which the payoffs are known exactly to be well approximated by the game with small uncertainty about the payoffs.

Harsanyi shows that in a game with such uncertainty about payoffs all equilibria are essentially pure, that is, each player plays a pure strategy with probability 1. Moreover, with probability 1, each player is playing his unique best response to the strategies of the other players; and the expected mixed actions of the players will be close to an equilibrium of the game without uncertainty. Harsanyi also shows, modulo a small technical error later corrected by van Damme (1991), that any regular equilibrium can be approximated in this way by pure equilibria of a game with small uncertainty about the payoffs. We shall not define a regular equilibrium here. Nor shall we give any of the technical details of the construction, or any of the proofs. The reader should instead consult Harsanyi (1973), van Damme (1991), or van Damme (2002).

3.2. Self-enforcing assessments

Let us consider again Harsanyi's construction described in the previous section. In an equilibrium of the game with uncertainty no player consciously randomizes. Given what the others are doing the player has a strict preference for one of his available choices. However, the player does not know what the others are doing. He knows that they are not randomizing – like him they have a strict preference for one of their available choices – but he does not know precisely their payoffs. Thus, if the optimal actions of the others differ as their payoffs differ, the player will have some probabilistic assessment of the actions that the others will take. And, since we assumed that the randomness in the payoffs was independent across players, this probabilistic assessment will also be independent, that is, it will be a vector of mixed strategies of the others.

The mixed strategy of a player does not represent a conscious randomization on the part of that player, but rather the uncertainty in the minds of the others as to how that player will act. We see that even without the construction involving uncertainty about the payoffs, we could adopt this interpretation of a mixed strategy. This interpretation has been suggested and promoted by Robert Aumann for some time [for example, Aumann (1987a), Aumann and Brandenburger (1995)] and is, perhaps, becoming the preferred interpretation among game theorists. There is nothing in this interpretation that compels us to assume that the assessments over what the other players will do should exhibit independence. The independence of the assessments involves an additional assumption.

Thus the focus of the analysis becomes, not the choices of the players, but the assessments of the players about the choices of the others. The basic consistency condition that we impose on the players' assessments is this: a player reasoning through the conclusions that others would draw from their assessments should not be led to revise his own assessment.

More formally, Aumann and Brandenburger (1995, p. 1177) show that if each player's assessment of the choices of the others is independent across the other players, if any two players have the same assessment as to the actions of a third, and if these assessments, the game, and the rationality of the players are all mutually known, then the assessments constitute a strategic equilibrium. (A fact is *mutually known* if each player knows the fact and knows that the others know it.) We discuss this and related results in more detail in Section 15.

4. The mixed extension of a game

Before discussing exactly how rational players assess their opponents' choices, we must reflect on the manner in which the payoffs represent the outcomes. If the players are presumed to quantify their uncertainties about their opponents' choices, then in choosing among their own strategies they must, in effect, compare different lotteries over the outcomes. Thus for the description of the game it no longer suffices to ascribe a payoff to each outcome, but it is also necessary to ascribe a payoff to each lottery over outcomes. Such a description would be unwieldy unless it could be condensed to a compact form.

One of the major achievements of von Neumann and Morgenstern (1944) was the development of such a compact representation ("cardinal utility"). They showed that if a player's ranking of lotteries over outcomes satisfied some basic conditions of consistency, then it was possible to represent that ranking by assigning numerical "payoffs" just to the outcomes themselves, and by ranking lotteries according to their expected payoffs. See the chapter by Fishburn (1994) in this Handbook for details.

Assuming such a scaling of the payoffs, one can expand the set of strategies available to each player to include not only definite ("pure") choices but also probabilistic ("mixed") choices, and extend the definition of the payoff functions by taking the appropriate expectations. The strategic form obtained in this manner is called the *mixed extension* of the game.

Recall from Section 3.2 that we consider a situation in which each player's assessment of the strategies of the others can be represented by a product of probability distributions on the others' pure strategy sets, that is, by a mixed strategy for each of the others. And that any two players have the same assessment about the choices of a third.

Denoting the (identical) assessments of the others as a probability distribution (mixed strategy) over a player's (pure) choices, we may describe the consistency condition as follows: each player's mixed strategy must place positive probability only on those pure strategies that maximize the player's payoff given the others' mixed strategies. Thus a profile of consistent assessments may be viewed as a strategic equilibrium in the mixed extension of the game.

Let us consider again the example of Figure 1 that we looked at in the introduction. Player 1 chooses the row and Player 2 (simultaneously) chooses the column. The resulting (cardinal) payoffs are indicated in the appropriate box of the matrix, with Player 1's payoff appearing first.

What probabilities could characterize a self-enforcing assessment? A (mixed) strategy for Player 1 (that is, an assessment by 2 of how 1 might play) is a vector $(x, 1 - x)$, where x lies between 0 and 1 and denotes the probability of playing T. Similarly, a strategy for 2 is a vector $(y, 1 - y)$. Now, given x, the payoff-maximizing value of y is indicated in Figure 4(a), and given y the payoff-maximizing value of x is indicated in Figure 4(b). When the figures are combined as in Figure 4(c), it is evident that the game possesses a single equilibrium, namely $x = 1/2$, $y = 1/3$. Thus in a self-enforcing assessment Player 1 must assign a probability of $1/3$ to 2's playing L, and Player 2 must assign a probability of $1/2$ to Player 1's playing T.

Note that these assessments do not imply a recommendation for action. For example, they give Player 1 no clue as to whether he should play T or B (because the expected payoff to either strategy is the same). But this is as it should be: it is impossible to expect rational deductions to lead to definite choices in a game like Figure 1, because whatever those choices would be they would be inconsistent with their own implications. (Figure 1 admits no pure-strategy equilibrium.) Still, the assessments do provide the players with an a priori evaluation of what the play of the game is worth to them, in this case $2/3$ to Player 1 and $1/2$ to Player 2.

The game of Figure 1 is an instance in which our consistency condition completely pins down the assessments that are self-enforcing. In general, we cannot expect such a sharp conclusion.

Consider, for example, the game of Figure 5. There are three equilibrium outcomes: $(8, 5)$, $(7, 6)$ and $(6, 3)$ (for the latter, the probability of T must lie between 0.5 and 0.6). Thus, all we can say is that there are three different consistent ways in which the players could view this game. Player 1's assessment would be: either that Player 2 was (defi-

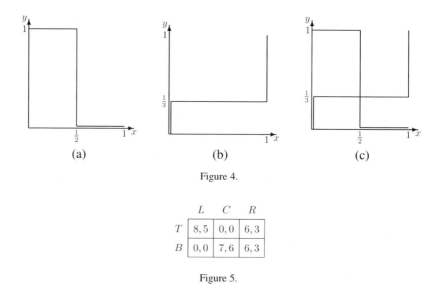

(a)   (b)   (c)

Figure 4.

	L	C	R
T	8, 5	0, 0	6, 3
B	0, 0	7, 6	6, 3

Figure 5.

nitely) going to play L, or that Player 2 was going to play C, or that Player 2 was going to play R. Which one of these assessments he would, in fact, hold is not revealed to us by means of equilibrium analysis.

5. The existence of equilibrium

We have seen that, in the game of Figure 1, for example, there may be no pure self-enforcing plans. But what of mixed plans, or self-enforcing assessments? Could they too be refuted by an example? The main result of non-cooperative game theory states that no such example can be found.

THEOREM 1 [Nash (1950, 1951)]. *The mixed extension of every finite game has at least one strategic equilibrium.*

(A game is *finite* if the player set as well as the set of strategies available to each player is finite.)

SKETCH OF PROOF. The proof may be sketched as follows. (It is a multi-dimensional version of Figure 4(c).) Consider the set-valued mapping (or correspondence) that maps each strategy profile, x, to all strategy profiles in which each player's component strategy is a *best response* to x (that is, maximizes the player's payoff given that the others are adopting their components of x). If a strategy profile is contained in the set to which it is mapped (is a *fixed point*) then it is an equilibrium. This is so because a strategic equilibrium is, in effect, defined as a profile that is a best response to itself.

Thus the proof of existence of equilibrium amounts to a demonstration that the "best response correspondence" has a fixed point. The fixed-point theorem of Kakutani (1941) asserts the existence of a fixed point for every correspondence from a convex and compact subset of Euclidean space into itself, provided two conditions hold. One, the image of every point must be convex. And two, the graph of the correspondence (the set of pairs (x, y) where y is in the image of x) must be closed.

Now, in the mixed extension of a finite game, the strategy set of each player consists of all vectors (with as many components as there are pure strategies) of non-negative numbers that sum to 1; that is, it is a simplex. Thus the set of all strategy profiles is a product of simplices. In particular, it is a convex and compact subset of Euclidean space.

Given a particular choice of strategies by the other players, a player's best responses consist of all (mixed) strategies that put positive weight only on those pure strategies that yield the highest expected payoff among all the pure strategies. Thus the set of best responses is a subsimplex. In particular, it is convex.

Finally, note that the conditions that must be met for a given strategy to be a best response to a given profile are all weak polynomial inequalities, so the graph of the best response correspondence is closed.

Thus all the conditions of Kakutani's theorem hold, and this completes the proof of Nash's theorem.   □

Nash's theorem has been generalized in many directions. Here we mention two.

THEOREM 2 [Fan (1952), Glicksberg (1952)]. *Consider a strategic form game with finitely many players, whose strategy sets are compact subsets of a metric space, and whose payoff functions are continuous. Then the mixed extension has at least one strategic equilibrium.*

(Here "mixed strategies" are understood as Borel probability measures over the given subsets of pure strategies.)

THEOREM 3 [Debreu (1952)]. *Consider a strategic form game with finitely many players, whose strategy sets are convex compact subsets of a Euclidean space, and whose payoff functions are continuous. If, moreover, each payoff function is quasi-concave in the player's own strategy, then the game has at least one strategic equilibrium.*

(A real-valued function on a Euclidean space is *quasi-concave* if, for each number a, the set of points at which the value of the function is at least a is convex.)

Theorem 2 may be thought of as identifying conditions on the strategy sets and payoff functions so that the game is like a finite game, that is, can be well approximated by finite games. Theorem 3 may be thought of as identifying conditions under which the strategy spaces are like the mixed strategy spaces for the finite games and the payoff functions are like expected utility.

6. Correlated equilibrium

We have argued that a self-enforcing assessment of the players' choices must constitute an equilibrium of the mixed extension of the game. But our argument has been incomplete: we have not explained why it is sufficient to assess each player's choice separately.

Of course, the implicit reasoning was that, since the players' choices are made in ignorance of one another, the assessments of those choices ought to be independent. In fact, this idea is subsumed in the definition of the mixed extension, where the expected payoff to a player is defined for a product distribution over the others' choices.

Let us now make the reasoning explicit. We shall argue here in the context of a Tabula-Rasa game, as we outlined in Section 3. Let us call the common assessment of the players implied by the Harsanyi doctrine the *rational assessment*. Consider the assessment over the pure-strategy profiles by a rational observer who knows as much about the players as they know about each other. We claim that observation of some player's choice, say Player 1's choice, should not affect the observer's assessment of the other players' choices. This is so because, as regards the other players, the observer and Player 1 have identical information and – by the Harsanyi doctrine – also identical analyses of that information, so there is nothing that the observer can learn from the player.

Thus, for any player, the conditional probability, given that player's choice, over the others' choices is the same as the unconditional probability. It follows [Aumann and Brandenburger (1995, Lemma 4.6, p. 1169) or Kohlberg and Reny (1997, Lemma 2.5, p. 285)] that the observer's assessment of the choices of all the players must be the product of his assessments of their individual choices. In making this argument we have taken for granted that the strategic form encompasses all the information available to the players in a game. This assumption, of "completeness", ensured that a player had no more information than was available to an outside observer.

Aumann (1974, 1987a) has argued against the completeness assumption. His position may be described as follows: it is impractical to insist that every piece of information available to some player be incorporated into the strategic form. This is so because the players are bound to be in possession of all sorts of information about random variables that are strategically irrelevant (that is, that cannot affect the outcome of the game). Thus he proposes to view the strategic form as an incomplete description of the game, indicating the available "actions" and their consequences; and to take account of the possibility that the actual choice of actions may be preceded by some unspecified observations by the players.

Having discarded the completeness assumption (and hence the symmetry in information between player and observer), we can no longer expect the rational assessment over the pure-strategy profiles to be a product distribution. But what can we say about it? That is, what are the implications of the rational assessment hypothesis itself? Aumann (1987a) has provided the answer. He showed that a distribution on the pure-strategy profiles is consistent with the rational assessment hypothesis if and only if it constitutes a correlated equilibrium.

Before going into the details of Aumann's argument, let us comment on the significance of this result. At first blush, it might have appeared hopeless to expect a direct method for determining whether a given distribution on the pure-strategy profiles was consistent with the hypothesis: after all, there are endless possibilities for the players' additional observations, and it would seem that each one of them would have to be tried out. And yet, the definition of correlated equilibrium requires nothing but the verification of a finite number of linear inequalities.

Specifically, a distribution over the pure-strategy profiles constitutes a *correlated equilibrium* if it imputes positive marginal probability only to such pure strategies, s, as are best responses against the distribution on the others' pure strategies obtained by conditioning on s. Multiplying throughout by the marginal probability of s, one obtains linear inequalities (if s has zero marginal probability, the inequalities are vacuous).

Consider, for example, the game of Figure 1. Denoting the probability over the ijth entry of the matrix by p_{ij}, the conditions for correlated equilibrium are as follows:

$$2p_{11} \geqslant p_{12}, \quad p_{22} \geqslant 2p_{21}, \quad p_{21} \geqslant p_{11}, \quad \text{and} \quad p_{12} \geqslant p_{22}.$$

There is a unique solution: $p_{11} = p_{21} = 1/6$, $p_{12} = p_{22} = 1/3$. So in this case, the correlated equilibria and the Nash equilibria coincide.

For an example of a correlated equilibrium that is not a Nash equilibrium, consider the distribution $1/2(T, L) + 1/2(B, R)$ in Figure 6. The distribution over II's choices obtained by conditioning on T is L with probability 1, and so T is a best response. Similarly for the other pure strategies and the other player.

For a more interesting example, consider the distribution that assigns weight $1/6$ to each non-zero entry of the matrix in Figure 7. [This example is due to Moulin and Vial (1978).] The distribution over Player 2's choices obtained by conditioning on T is C with probability $1/2$ and R with probability $1/2$, and so T is a best response (it yields 1.5 while M yields 0.5 and B yields 1). Similarly for the other pure strategies of Player 1, and for Player 2.

It is easy to see that the correlated equilibria of a game contain the convex hull of its Nash equilibria. What is less obvious is that the containment may be strict [Aumann (1974)], even in payoff space. The game of Figure 7 illustrates this: the unique Nash equilibrium assigns equal weight, $1/3$, to every pure strategy, and hence gives rise to the expected payoffs $(1, 1)$; whereas the correlated equilibrium described above gives rise to the payoffs $(1.5, 1.5)$.

Let us now sketch the proof of Aumann's result. By the rational assessment hypothesis, a rational observer can assess in advance the probability of each possible list of observations by the players. Furthermore, he knows that the players also have the same assessment, and that each player would form a conditional probability by restricting attention to those lists that contain his actual observations. Finally, we might as well assume that the player's strategic choice is a function of his observations (that is, that if

	L	R
T	3, 1	0, 0
B	0, 0	1, 3

Figure 6.

	L	C	R
T	0, 0	1, 2	2, 1
M	2, 1	0, 0	1, 2
B	1, 2	2, 1	0, 0

Figure 7.

the player must still resort to a random device in order to decide between several payoff-maximizing alternatives, then the observer has already included the various outcomes of that random device in the lists of the possible observations).

Now, given a candidate for the rational assessment of the game, that is, a distribution over the matrix, what conditions must it satisfy if it is to be consistent with some such assessment over lists of observations? Our basic condition remains as in the case of Nash equilibria: by reasoning through the conclusions that the players would reach from their assessments, the observer should not be led to revise his own assessment. That is, conditional on any possible observation of a player, the pure strategy chosen must maximize the player's expected payoff.

As stated, the condition is useless for us, because we are not privy to the rational observer's assessment of the probabilities over all the possible observations. However, by lumping together all the observations inducing the same choice, s (and by noting that, if s maximized the expected payoff over a number of disjoint events then it would also maximize the expected payoff over their union), we obtain a condition that we can check: that the choice of s maximizes the player's payoff against the conditional distribution given s. But this is precisely the condition for correlated equilibrium.

To complete the proof, note that the basic consistency condition has no implications beyond correlated equilibria: any correlated equilibrium satisfies this condition relative to the following assessment of the players' additional observations: each player's additional observation is the name of one of his pure strategies, and the probability distribution over the possible lists of observations (that is, over the entries of the matrix) is precisely the distribution of the given correlated equilibrium.

For the remainder of this chapter we shall restrict our attention to uncorrelated strategies. The issues and concepts we discuss concerning the refinement of equilibrium are not as well developed for correlated equilibrium as they are in the setting where stochastic independence of the solution is assumed.

7. The extensive form

The strategic form is a convenient device for defining strategic equilibria: it enables us to think of the players as making single, simultaneous choices. However, actually to describe "the rules of the game", it is more convenient to present the game in the form of a tree.

The *extensive form* [see Hart (1992)] is a formal representation of the rules of the game. It consists of a rooted tree whose nodes represent decision points (an appropriate label identifies the relevant player), whose branches represent moves and whose endpoints represent outcomes. Each player's decision nodes are partitioned into *information sets* indicating the player's state of knowledge at the time he must make his move: the player can distinguish between points lying in different information sets but cannot distinguish between points lying in the same information set. Of course, the actions available at each node of an information set must be the same, or else the player could distinguish between the nodes according to the actions that were available. This means that the number of moves must be the same and that the labels associated with moves must be the same.

Random events are represented as nodes (usually denoted by open circles) at which the choices are made by Nature, with the probabilities of the alternative branches included in the description of the tree.

The information partition is said to have *perfect recall* [Kuhn (1953)] if the players remember whatever they knew previously, including their past choices of moves. In other words, all paths leading from the root of the tree to points in a single information set, say Player i's, must intersect the same information sets of Player i and must display the same choices by Player i.

The extensive form is "finite" if there are finitely many players, each with finitely many choices at finitely many decision nodes. Obviously, the corresponding strategic form is also finite (there are only finitely many alternative "books of instructions"). Therefore, by Nash's theorem, there exists a mixed-strategy equilibrium.

But a mixed strategy might seem a cumbersome way to represent an assessment of a player's behavior in an extensive game. It specifies a probability distribution over complete plans of action, each specifying a definite choice at each of the player's information sets. It may seem more natural to specify an independent probability distribution over the player's moves at each of his information sets. Such a specification is called a *behavioral strategy*.

Is nothing lost in the restriction to behavioral strategies? Perhaps, for whatever reason, rational players do assess their opponents' behavior by assigning probabilities to complete plans of action, and perhaps some of those assessments cannot be reproduced by assigning independent probabilities to the moves?

Kuhn's theorem (1953) guarantees that, in a game with perfect recall, nothing, in fact, is lost. It says that in such a game every mixed strategy of a player in a tree is equivalent to some behavioral strategy, in the sense that both give the same distribution on the endpoints, whatever the strategies of the opponents.

For example, in the (skeletal) extensive form of Figure 8, while it is impossible to reproduce by means of a behavioral strategy the correlations embodied in the mixed strategy $0.1TLW + 0.1TRY + 0.5BLZ + 0.1BLW + 0.2BRX$, nevertheless it is possible to construct an equivalent behavioral strategy, namely $((0.2, 0.8), (0.5, 0, 0.5), (0.125, 0.25, 0, 0.625))$.

To see the general validity of the theorem, note that the distribution over the endpoints is unaffected by correlation of choices that anyway cannot occur at the same "play" (that is, on the same path from the root to an endpoint). Yet this is precisely the type of correlation that is possible in a mixed strategy but not in a behavioral strategy. (Correlation among choices lying on the same path is possible also in a behavioral strategy. Indeed, this possibility is already built into the structure of the tree: if two plays differ in a certain move, then (because of perfect recall) they also differ in the information set at which any later move is made and so the assessment of the later move can be made dependent on the earlier move.)

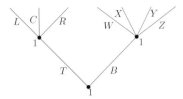

Figure 8.

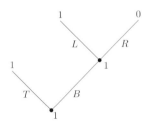

Figure 9.

Kuhn's theorem allows us to identify each equivalence class of mixed strategies with a behavioral strategy – or sometimes, on the boundary of the space of behavioral strategies, with an equivalence class of behavioral strategies. Thus, in games with perfect recall, for the strategic choices of payoff-maximizing players there is no difference between the mixed extension of the game and the "behavioral extension" (where the players are restricted to their behavioral strategies). In particular, the equilibria of the mixed and of the behavioral extension are equivalent, so either may be taken as the set of candidates for the rational assessment of the game.

Of course, the equivalence of the mixed and the behavioral extensions implies the existence of equilibrium in the behavioral extension of any finite game with perfect recall. It is interesting to note that this result does not follow directly from either Nash's theorem or from Debreu's theorem. The difficulty is that the convex structure on the behavioral strategies does not reflect the convex structure on the mixed strategies, and therefore the best-reply correspondence need not be convex. For example, in Figure 9, the set of optimal strategies contains (T, R) and (B, L) but not their behavioral mixture $(\frac{1}{2}T + \frac{1}{2}B, \frac{1}{2}L + \frac{1}{2}R)$. This corresponds to the difference between the two linear structures on the space of product distributions on strategy vectors that we discussed at the end of Section 2.1.

8. Refinement of equilibrium

Let us review where we stand: assuming that in any game there is one particular assessment of the players' strategic choices that is common to all rational decision makers ("the rational assessment hypothesis"), we can deduce that that assessment must con-

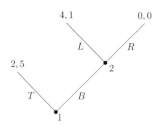

Figure 10.

stitute a strategic equilibrium (which can be expressed as a profile of either mixed or behavioral strategies).

The natural question is: can we go any further? That is, when there are multiple equilibria, can any of them be ruled out as candidates for the self-enforcing assessment of the game? At first blush, the answer seems to be negative. Indeed, if an assessment stands the test of individual payoff maximization, then what else can rule out its candidacy?

And yet, it turns out that it *is* possible to rule out some equilibria. The key insight was provided by Selten (1965). It is that irrational assessments by two different players might each make the other look rational (that is, payoff-maximizing).

A typical example is that of Figure 10. The assessment (T, R) certainly does not appear to be self-enforcing. Indeed, it seems clear that Player 1 would play B rather than T (because he can bank on the fact that Player 2 – who is interested only in his own payoff – will consequently play L). And yet (T, R) constitutes a strategic equilibrium: Player 1's belief that Player 2 would play R makes his choice of T payoff-maximizing, while Player 2's belief that Player 1 would play T makes his choice of R (irrelevant hence) payoff-maximizing.

Thus Figure 10 provides an example of an equilibrium that can be ruled out as a self-enforcing assessment of the game. (In this particular example there remains only a single candidate, namely (B, L).) By showing that it is sometimes possible to narrow down the self-enforcing assessments beyond the set of strategic equilibria, Selten opened up a whole field of research: the *refinement* of equilibrium.

8.1. The problem of perfection

We have described our project as identifying the self-enforcing assessments. Thus we interpret Selten's insight as being that not all strategic equilibria are, in fact, self-enforcing. We should note that this is not precisely how Selten described the problem. Selten explicitly sees the problem as the prescription of disequilibrium behavior in "unreached parts of the game" [Selten (1975, p. 25)]. Harsanyi and Selten (1988, p. 17) describe the problem of imperfect equilibria in much the same way.

Kreps and Wilson (1982) are a little less explicit about the nature of the problem but their description of their solution suggests that they agree. "A sequential equilibrium provides at each juncture an equilibrium in the subgame (of incomplete information) induced by restarting the game at that point" [Kreps and Wilson (1982, p. 864)].

$$\begin{array}{c c} & \begin{array}{c c} L & \quad R \end{array} \\ \begin{array}{c} T \\ \\ B \end{array} & \begin{array}{|c|c|} \hline 10,10 & 0,0 \\ \hline 0,0 & 1,1 \\ \hline \end{array} \end{array}$$

Figure 11.

Also Myerson seems to view his definition of proper equilibrium as addressing the same problem as addressed by Selten. However he discusses examples and defines a solution explicitly in the context of normal form games where there are no unreached parts of the game. He describes the program as eliminating those equilibria that "may be inconsistent with our intuitive notions about what should be the outcome of a game" [Myerson (1978, p. 73)].

The description of the problem of the refinement of strategic equilibrium as looking for a set of necessary and sufficient conditions for self-enforcing behavior does nothing without specific interpretations of what this means. Nevertheless it seems to us to tend to point us in the right direction. Moreover it does seem to delineate the problem to some extent. Thus in the game of Figure 11 we would be quite prepared to concede that the equilibrium (B, R) was unintuitive while at the same time claiming that it was quite self-enforcing.

8.2. Equilibrium refinement versus equilibrium selection

There is a question separate from but related to that of equilibrium refinement. That is the question of equilibrium selection. Equilibrium refinement is concerned with establishing necessary conditions for reasonable play, or perhaps necessary and sufficient conditions for "self-enforcing". Equilibrium selection is concerned with narrowing the prediction, indeed to a single equilibrium point. One sees a problem with some of the equilibrium points, the other with the *multiplicity* of equilibrium points.

The central work on equilibrium selection is the book of Harsanyi and Selten (1988). They take a number of positions in that work with which we have explicitly disagreed (or will disagree in what follows): the necessity of incorporating mistakes; the necessity of working with the extensive form; the rejection of forward-induction-type reasoning; the insistence on subgame consistency. We are, however, somewhat sympathetic to the basic enterprise. Whatever the answer to the question we address in this chapter there will remain in many games a multiplicity of equilibria, and thus some scope for selecting among them. And the work of Harsanyi and Selten will be a starting point for those who undertake this enterprise.

9. Admissibility and iterated dominance

It is one thing to point to a specific equilibrium, like (T, R) in Figure 10, and claim that "clearly" it cannot be a self-enforcing assessment of the game; it is quite another matter to enunciate a principle that would capture the underlying intuition.

One principle that immediately comes to mind is *admissibility*, namely that rational players never choose dominated strategies. (As we discussed in the context of rational-izability in Section 2.2 a strategy is dominated if there exists another strategy yielding at least as high a payoff against any choice of the opponents and yielding a higher pay-off against some such choice.) Indeed, admissibility rules out the equilibrium (T, R) of Figure 10 (because R is dominated by L).

Furthermore, the admissibility principle immediately suggests an extension, *iterated dominance*: if dominated strategies are never chosen, and if all players know this, all know *this*, and so on, then a self-enforcing assessment of the game should be unaffected by the (iterative) elimination of dominated strategies. Thus, for example, the equilibrium (T, L) of Figure 12(a) can be ruled out even though both T and L are admissible. (See Figure 12(b).)

At this point we might think we have nailed down the underlying principle separating self-enforcing equilibria from ones that are not self-enforcing. (Namely, that rational equilibria are unaffected by deletions of dominated strategies.) However, nothing could be further from the truth: first, the principle cannot possibly be a general property of self-enforcing assessments, for the simple reason that it is self-contradictory; and second, the principle fails to weed out *all* the equilibria that appear not to be self-enforcing.

On reflection, one realizes that admissibility and iterated dominance have somewhat inconsistent motivations. Admissibility says that whatever the assessment of how the game will be played, the strategies that receive zero weight in this assessment neverthe-less remain relevant, at least when it comes to breaking ties. Iterated dominance, on the other hand, says that some such strategies, those that receive zero weight because they are inadmissible, are irrelevant and may be deleted.

To see that this inconsistency in motivation actually leads to an inconsistency in the concepts, consider the game of Figure 13. If a self-enforcing assessment were unaffected by the elimination of dominated strategies then Player 1's assessment of Player 2's choice would have to be L (delete B and then R) but it would also have to be R

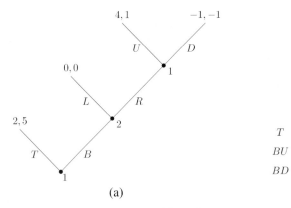

(a)   (b)

Figure 12.

	L	R
T	2, 0	1, 0
M	0, 1	0, 0
B	0, 0	0, 1

Figure 13.

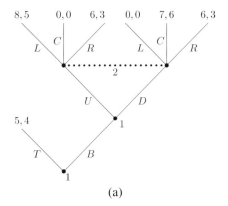

	L	C	R
T	5, 4	5, 4	5, 4
BU	8, 5	0, 0	6, 3
BD	0, 0	7, 6	6, 3

(a)　　　　　　　(b)

Figure 14.

(delete M and then L). Thus the assessment of the outcome would have to be both $(2, 0)$ and $(1, 0)$.

To see the second point, consider the game of Figure 14(a). As is evident from the normal form of Figure 14(b), there are no dominance relationships among the strategies, so all the equilibria satisfy our principle, and in particular those in which Player 1 plays T (for example, $(T, 0.5L + 0.5C)$). And yet those equilibria appear not to be self-enforcing: indeed, if Player 1 played B then he would be faced with the game of Figure 5, which he assesses as being worth more than 5 (recall that any self-enforcing assessment of the game of Figure 5 has an outcome of either $(8, 5)$ or $(7, 6)$ or $(6, 3)$); thus Player 1 must be expected to play B rather than T.

In the next section, we shall concentrate on the second problem, namely how to capture the intuition ruling out the outcome $(5, 4)$ in the game of Figure 14(a).

10. Backward induction

10.1. The idea of backward induction

Selten (1965, 1975) proposed several ideas that may be summarized as the following *principle of backward induction*:

A self-enforcing assessment of the players' choices in a game tree must be consistent with a self-enforcing assessment of the choices from any node (or, more generally, information set) in the tree onwards.

This is a multi-person analog of "the principle of dynamic programming" [Bellman (1957)], namely that an optimal strategy in a one-person decision tree must induce an optimal strategy from any point onward.

The force of the backward induction condition is that it requires the players' assessments to be self-enforcing even in those parts of the tree that are ruled out by their own assessment of earlier moves. (As we have seen, the equilibrium condition by itself does not do this: one can take the "wrong" move at a node whose assessed probability is zero and still maximize one's expected payoff.)

The principle of backward induction indeed eliminates the equilibria of the games of Figure 10 and Figure 12(a) that do not appear to be self-enforcing. For example, in the game of Figure 12(a) a self-enforcing assessment of the play starting at Player 1's second decision node must be that Player 1 would play U, therefore the assessment of the play starting at Player 2's decision node must be BU, and hence the assessment of the play of the full game must be BRU, that is, it is the equilibrium (BU, R).

Backward induction also eliminates the outcome $(5, 4)$ in the game of Figure 14(a). Indeed, any self-enforcing assessment of the play starting at Player 1's second decision node must impute to Player 1 a payoff greater than 5, so the assessment of Player 1's first move must be B.

And it eliminates the equilibrium (T, R, D) in the game of Figure 15 (which is taken from Selten (1975)). Indeed, whatever the self-enforcing assessment of the play starting at Player 2's decision node, it certainly is not (R, D) (because, if Player 2 expected Player 3 to choose D, then he would maximize his own payoff by choosing L rather than R).

There have been a number of attacks [Basu (1988, 1990), Ben-Porath (1997), Reny (1992a, 1993)] on the idea of backward induction along the following lines. The requirement that the assessment be self-enforcing implicitly rests on the assumption that the

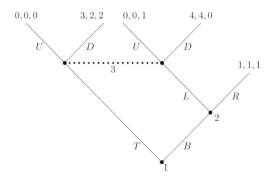

Figure 15.

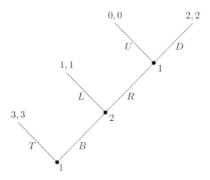

Figure 16.

players are rational and that the other players know that they are rational, and indeed, on higher levels of knowledge. Also, the requirement that a self-enforcing assessment be consistent with a self-enforcing assessment of the choices from any information set in the tree onwards seems to require that the assumption be maintained at that information set and onwards. And yet, that information set might only be reached if some player has taken an irrational action. In such a case the assumption that the players are rational and that their rationality is known to the others should not be assumed to hold in that part of the tree. For example, in the game of Figure 16 there seems to be no compelling reason why Player 2, if called on to move, should be assumed to know of Player 1's rational-ity. Indeed, since he has observed something that contradicts Player 1 being rational, perhaps Player 2 *must* believe that Player 1 is not rational.

The example however does suggest the following observation: the part of the tree following an irrational move is anyway irrelevant (because a rational player is sure not to take such a move), so whether or not rational players can assess what would happen there has no bearing on their assessment of how the game would actually be played (for example, in the game of Figure 16 the rational assessment of the outcome is (3, 3), regardless of what Player 2's second choice might be). While this line of reasoning is quite convincing in a situation like that of Figure 16, where the irrationality of the move *B* is self-evident, it is less convincing in, say, Figure 12(a). There, the rationality or irrationality of the move *B* becomes evident only after consideration of what would happen if *B* were taken, that is, only after consideration of what in retrospect appears "counterfactual" [Binmore (1990)].

One approach to this is to consider what results when we no longer assume that the players are known to be rational following a deviation by one (or more) from a can-didate self-enforcing assessment. Reny (1992a) and, though it is not the interpretation they give, Fudenberg, Kreps and Levine (1988) show that such a program leads to few restrictions beyond the requirements of strategic equilibrium. We shall discuss this a little more at the end of Section 10.4.

And yet, perhaps one can make some argument for the idea of backward induction. The argument of Reny (1992a), for example, allows a deviation from the candidate equilibrium to be an indication to the other players that the assumption that all of the players were rational is not valid. In other words the players are no more sure about the nature of the game than they are about the equilibrium being played. We may recover some of the force of the idea of backward induction by requiring the equilibrium to be robust against a little strategic uncertainty.

Thus we argue that the requirement of backward induction results from a series of tests. If indeed a rational player, in a situation in which the rationality of all players is common knowledge, would not take the action that leads to a certain information set being reached then it matters little what the assessment prescribes at that information set. To check whether the hypothesis that a rational player, in a situation in which the rationality of all players is common knowledge, would not take that action we suppose that he would and see what could arise.

If all self-enforcing assessments of the situation following a deviation by a particular player would lead to him deviating then we reject the hypothesis that such a deviation contradicts the rationality of the players. And so, of course, we reject the candidate assessment as self-enforcing.

If however our analysis of the situation confirms that there is a self-enforcing assessment in which the player, if rational, would not have taken the action, then our assessment of him not taking the action is confirmed. In such a case we have no reason to insist on the results of our analysis following the deviation. Moreover, since we assume that the players are rational and our analysis leads us to conclude that rational players will not play in this part of the game we are forced to be a little imprecise about what our assessment says in that part of the game. This relates to our discussion of sets of equilibria as solutions of the game in Section 13.2.

This modification of the notion of backward induction concedes that there may be conceivable circumstances in which the common knowledge of rationality of the players would, of necessity, be violated. It argues, however, that if the players are sure enough of the nature of the game, including the rationality of the other players, that they abandon this belief only in the face of truly compelling evidence, then the behavior in such circumstances is essentially irrelevant.

The principle of backward induction is completely dependent on the extensive form of the game. For example, while it excludes the equilibrium (T, L) in the game of Figure 12(a), it does not exclude the same equilibrium in the game of Figure 12(b) (that is, in an extensive form where the players simultaneously choose their strategies).

Thus one might see an inconsistency between the principle of backward induction and von Neumann and Morgenstern's reduction of the extensive form to the strategic form. We would put a somewhat different interpretation on the situation. The claim that the strategic form contains sufficient information for strategic analysis is not a denial that some games have an extensive structure. Nor is it a denial that valid arguments, such as backward induction arguments, can be made in terms of that structure. Rather the point is that, were a player, instead of choosing through the game, required to decide in advance what he will do, he could consider in advance any of the issues that would lead him to choose one way or the other during the game. And further, these issues will affect his incentives in precisely the same way when he considers them before playing as they would had he considered them during the play of the game.

In fact, we shall see in Section 13.6 that the sufficiency of the normal form substantially strengthens the implications of backward induction arguments. We put off that discussion for now. We do note however that others have taken a different position. Selten's position, as well as the position of a number of others, is that the reduction to the strategic form is unwarranted, because it involves loss of information. Thus Figures 14(a) and 14(b) represent fundamentally different games, and (5, 4) is indeed not self-enforcing in the former but possibly is self-enforcing in the latter. (Recall that this equilibrium cannot be excluded by the strategic-form arguments we have given to date, such as deletions of dominated strategies, but can be excluded by backward induction in the tree.)

10.2. Subgame perfection

We now return to give a first pass at giving a formal expression of the idea of backward induction. The simplest case to consider is of a node such that the part of the tree from the node onwards can be viewed as a separate game (a "subgame"), that is, it contains every information set which it intersects. (In particular, the node itself must be an information set.)

Because the rational assessment of any game must constitute an equilibrium, we have the following implication of backward induction [*subgame perfection*, Selten (1965)]:

The equilibrium of the full game must induce an equilibrium on every subgame.

The subgame-perfect equilibria of a game can be determined by working from the ends of the tree to its root, each time replacing a subgame by (the expected payoff of) one of its equilibria. We must show that indeed a profile of strategies obtained by means of step-by-step replacement of subgames with equilibria constitutes a subgame-perfect equilibrium. If not, then there is a smallest subgame in which some player's strategy fails to maximize his payoff (given the strategies of the others). But this is impossible, because the player has maximized his payoff given his own choices in the subgames of the subgame, and those he is presumed to have chosen optimally.

For example, in the game of Figure 12(a), the subgame whose root is at Player 1's second decision node can be replaced by (4, 1), so the subgame whose root is at Player 2's decision node can also be replaced by this outcome, and similarly for the whole tree.

Or in the game of Figure 14(a), the subgame (of Figure 5) can be replaced by one of its equilibria, namely (8, 5), (7, 6) or (6, 3). Since any of them give Player 1 more than 5, Player 1's first move must be B. Thus all three outcomes are subgame perfect, but the additional equilibrium outcome, (5, 4), is not.

Because the process of step-by-step replacement of subgames by their equilibria will always yield at least one profile of strategies, we have the following result.

THEOREM 4 [Selten (1965)]. *Every game tree has at least one subgame-perfect equilibrium.*

Subgame perfection captures only one aspect of the principle of backward induction. We shall consider other aspects of the principle in Sections 10.3 and 10.4.

10.3. Sequential equilibrium

To see that subgame perfection does not capture all that is implied by the idea of backward induction it suffices to consider quite simple games. While subgame perfection clearly isolates the self-enforcing outcome in the game of Figure 10 it does not do so in the game of Figure 17, in which the issues seem largely the same. And we could even modify the game a little further so that it becomes difficult to give a presentation of the game in which subgame perfection has any bite. (Say, by having Nature first decide whether Player 1 obtains a payoff of 5 after M or after B and informing Player 1, but not Player 2.)

One way of capturing more of the idea of backward induction is by explicitly requiring players to respond optimally at all information sets. The problem is, of course, that, while in the game of Figure 17 it is clear what it means for Player 2 to respond optimally, this is not generally the case. In general, the optimal choice for a player will depend on his assessment of which node of his information set has been reached. And, at an out of equilibrium information set this may not be determined by the strategies being played.

The concept of sequential equilibrium recognizes this by defining an equilibrium to be a pair consisting of a behavioral strategy and a system of beliefs. A *system of beliefs* gives, for each information set, a probability distribution over the nodes of that information set. The behavioral strategy is said to be *sequentially rational* with respect to the system of beliefs if, at every information set at which a player moves, it maximizes the conditional payoff of the player, given his beliefs at that information set and the strategies of the other players. A system of beliefs is said to be *consistent* with a behavioral strategy if it is the limit of a sequence of beliefs each being the actual conditional distribution on nodes of the various information sets induced by a sequence of completely mixed behavioral strategies converging to the given behavioral strategy. A *sequential equilibrium* is a pair such that the strategy is sequentially rational with respect to the beliefs and the beliefs are consistent with the strategy.

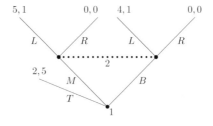

Figure 17.

The idea of a strategy being sequentially rational appears quite straightforward and intuitive. However the concept of consistency is somewhat less natural. Kreps and Wilson (1982) attempted to provide a more primitive justification for the concept, but, as was shown by Kreps and Ramey (1987) this justification was fatally flawed. Kreps and Ramey suggest that this throws doubt on the notion of consistency. (They also suggest that the same analysis casts doubt on the requirement of sequential rationality. At an unreached information set there is some question of whether a player should believe that the future play will correspond to the equilibrium strategy. We shall not discuss this further.)

Recent work has suggested that the notion of consistency is a good deal more natural than Kreps and Ramey suggest. In particular Kohlberg and Reny (1997) show that it follows quite naturally from the idea that the players' assessments of the way the game will be played reflects certainty or stationarity in the sense that it would not be affected by the actual realizations observed in an identical situation. Related ideas are explored by Battigalli (1996a) and Swinkels (1994). We shall not go into any detail here.

The concept of sequential equilibrium is a strengthening of the concept of subgame perfection. Any sequential equilibrium is necessarily subgame perfect, while the converse is not the case. For example, it is easy to verify that in the game of Figure 17 the unique sequential equilibrium involves I choosing M. And a similar re-

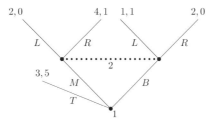

Figure 18.

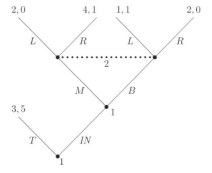

Figure 19.

sult holds for the modification of that game involving a move of Nature discussed above.

Notice also that the concept of sequential equilibrium, like that of subgame perfection, is quite sensitive to the details of the extensive form. For example in the extensive form game of Figure 18 there is a sequential equilibrium in which Player 1 plays T and Player 2 plays L. However in a very similar situation (we shall later argue strategically identical) – that of Figure 19 – there is no sequential equilibrium in which Player 1 plays T. The concepts of extensive form perfect equilibrium and quasi-perfect equilibrium that we discuss in the following section also feature this sensitivity to the details of the extensive form. In these games they coincide with the sequential equilibria.

10.4. Perfect equilibrium

The sequential equilibrium concept is closely related to a similar concept defined earlier by Selten (1975), the perfect equilibrium. Another closely related concept, which we shall argue is in some ways preferable, was defined by van Damme (1984) and called the quasi-perfect equilibrium. Both of these concepts, like sequential equilibrium, are defined explicitly on the extensive form, and depend essentially on details of the extensive form. (They can, of course, be defined for a simultaneous move extensive form game, and to this extent can be thought of as defined for normal form games.) When defining these concepts we shall assume that the extensive form games satisfy perfect recall.

Myerson (1978) defined a normal form concept that he called proper equilibrium. This is a refinement of the concepts of Selten and van Damme when those concepts are applied to normal form games. Moreover there is a remarkable relation between the normal form concept of proper equilibrium and the extensive form concept quasi-perfect equilibrium.

Let us start by describing the definition of perfect equilibrium. The original idea of Selten was that however close to rational players were they would never be perfectly rational. There would always be some chance that a player would make a mistake. This idea may be implemented by approximating a candidate equilibrium strategy profile by a nearby completely mixed strategy profile and requiring that any of the deliberately chosen actions, that is, those given positive probability in the candidate strategy profile, be optimal, not only against the candidate strategy profile, but also against the nearby mixed strategy profile. If we are defining extensive form perfect equilibrium, a strategy is interpreted to mean a behavioral strategy and an action to mean an action at some information set. More formally, a profile of behavioral strategies b is a perfect equilibrium if there is a sequence of completely mixed behavioral strategy profiles $\{b^t\}$ such that at each information set and for each b^t, the behavior of b at the information set is optimal against b^t, that is, is optimal when behavior at all other information sets is given by b^t. If the definition is applied instead to the normal form of the game the resulting equilibrium is called a *normal form perfect equilibrium*.

Like sequential equilibrium, (extensive form) perfect equilibrium is an attempt to express the idea of backward induction. Any perfect equilibrium is a sequential equilibrium (and so is a subgame perfect equilibrium). Moreover the following result tells us that, except for exceptional games, the converse is also true.

THEOREM 5 [Kreps and Wilson (1982), Blume and Zame (1994)]. *For any extensive form, except for a closed set of payoffs of lower dimension than the set of all possible payoffs, the sets of sequential equilibrium strategy profiles and perfect equilibrium strategy profiles coincide.*

The concept of normal form perfect equilibrium, on the other hand, can be thought of as a strong form of admissibility. In fact for two-player games the sets of normal form perfect and admissible equilibria coincide. In games with more players the sets may differ. However there is a sense in which even in these games normal form perfection seems to be a reasonable expression of admissibility. Mertens (1987) gives a definition of the admissible best reply correspondence that would lead to fixed points of this correspondence being normal form perfect equilibria, and argues that this definition corresponds "to the intuitive idea that would be expected from a concept of 'admissible best reply' in a framework of independent priors" [Mertens (1987, p. 15)].

Mertens (1995) offers the following example in which the set of extensive form perfect equilibria and the set of admissible equilibria have an empty intersection. The game may be thought of in the following way. Two players agree about how a certain social decision should be made. They have to decide who should make the decision and they do this by voting. If they agree on who should make the decision that player decides. If they each vote for the other then the good decision is taken automatically. If each votes for himself then a fair coin is tossed to decide who makes the decision. A player who makes the social decision is not told if this is so because the other player voted for him, or because the coin toss chose him. The extensive form of this game is given in Figure 20. The payoffs are such that each player prefers the good outcome to the bad outcome. (In Mertens (1995) there is an added complication to the game. Each player does slightly worse if he chooses the bad outcome than if the other chooses it. However this additional complication is, as Mertens pointed out to us, totally unnecessary for the results.)

In this game the only admissible equilibrium has both players voting for themselves and taking the right choice if they make the social decision. However, any perfect equilibrium must involve at least one of the players voting for the other with certainty. At least one of the players must be at least as likely as the other to make a mistake in the second stage. And such a player, against such mistakes, does better to vote for the other.

10.5. Perfect equilibrium and proper equilibrium

The definition of perfect equilibrium *may* be thought of as corresponding to the idea that players really do make mistakes, and that in fact it is not possible to think coherently about games in which there is no possibility of the players making mistakes. On the

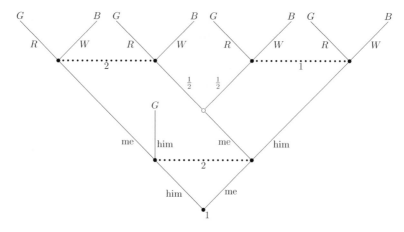

Figure 20.

other hand one might think of the perturbations as instead encompassing the idea that the players should have a little strategic uncertainty, that is, they should not be completely confident as to what the other players are going to do. In such a case a player should not be thought of as being uncertain about his own actions or planned actions. This is (one interpretation of) the idea behind van Damme's definition of quasi-perfect equilibrium.

Recall why we used perturbations in the definition of perfect equilibrium. We wanted to require that players act optimally at all information sets. Since the perturbed strategies are completely mixed all information sets are reached and so the conditional distribution on the nodes of the information set is well defined. In games with perfect recall however we do not need that all the strategies be completely mixed. Indeed the player himself may affect whether one of his information sets is reached or not, but cannot affect what will be the distribution over the nodes of that information set if it is reached – that depends only on the strategies of the others.

The definition of quasi-perfect equilibrium is largely the same as the definition of perfect equilibrium. The definitions differ only in that, instead of the limit strategy b being optimal at each information set against behavior given by b^t at *all* other information sets, it is required that b be optimal at all information sets against behavior at other information sets given by b for information sets that are owned by the same player who owns the information set in question, and by b^t for other information sets. That is, the player does not take account of his own "mistakes", except to the extent that they may make one of his information sets reached that otherwise would not be. As we explained above, the assumption of perfect recall guarantees that the conditional distribution on each information set is uniquely defined for each t and so the requirement of optimality is well defined.

This change in the definition leads to some attractive properties. Like perfect equilibria, quasi-perfect equilibria are sequential equilibrium strategies. But, unlike perfect equilibria, quasi-perfect equilibria are always normal form perfect, and thus admissible. Mertens (1995) argues that quasi-perfect equilibrium is precisely the right mixture of admissibility and backward induction.

Also, as we remarked earlier, there is a relation between quasi-perfect equilibria and proper equilibria. A *proper equilibrium* [Myerson (1978)] is defined to be a limit of ε-proper equilibria. An ε-*proper equilibrium* is a completely mixed strategy vector such that for each player if, given the strategies of the others, one strategy is strictly worse than another the first strategy is played with probability at most ε times the probability with which the second is played. In other words, more costly mistakes are made with lower frequency. Van Damme (1984) proved the following result. (Kohlberg and Mertens (1982, 1986) independently proved a slightly weaker result, replacing quasi-perfect with sequential.)

THEOREM 6. *A proper equilibrium of a normal form game is quasi-perfect in any extensive form game having that normal form.*

In van Damme's paper the theorem is actually stated a little differently referring simply to a pair of games, one an extensive form game and the other the corresponding normal form. (It is also more explicit about the sense in which a quasi-perfect equilibrium, a behavioral strategy profile, *is* a proper equilibrium, a mixed strategy profile.) Thus van Damme correctly states that the converse of his theorem is not true. There are such pairs of games and quasi-perfect equilibria of the extensive form that are in no sense equivalent to a proper equilibrium of the normal form. Kohlberg and Mertens (1986) state their theorem in the same form as we do, but refer to sequential equilibria rather than quasi-perfect equilibria. They too correctly state that the converse is not true. For any normal form game one could introduce dummy players, one for each profile of strategies having payoff one at that profile of strategies and zero otherwise. In any extensive form having that normal form the set of sequential equilibrium strategy profiles would be the same as the set of equilibrium strategy profiles originally.

However it is not immediately clear that the converse of the theorem as we have stated it is not true. Certainly we know of no example in the previous literature that shows it to be false. For example, van Damme (1991) adduces the game given in extensive form in Figure 21(a) and in normal form in Figure 21(b) to show that a quasi-perfect equilibrium may not be proper. The strategy (BD, R) is quasi-perfect, but not proper. Nevertheless there is a game – that of Figure 22 – having the same normal form, up to duplication of strategies, in which that strategy is not quasi-perfect. Thus one might be tempted to conjecture, as we did in an earlier version of this chapter, that given a normal form game, any strategy vector that is quasi-perfect in any extensive form game having that normal form would be a proper equilibrium of the normal form game.

A fairly weak version of this conjecture is true. If we fix not only the equilibrium under consideration but also the sequence of completely mixed strategies converging to it then if in every equivalent extensive form game the sequence supports the equilibrium

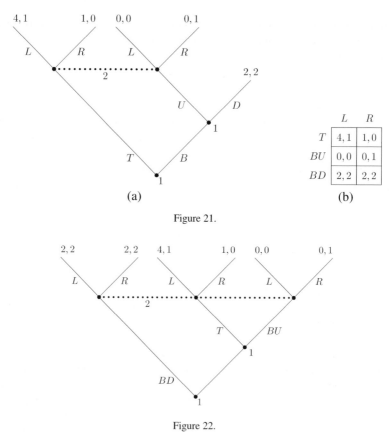

Figure 21.

(a)   (b)

Figure 22.

as a quasi-perfect equilibrium then the equilibrium is proper. We state this a little more formally in the following theorem.

THEOREM 7. *An equilibrium σ of a normal form game G is supported as a proper equilibrium by a sequence of completely mixed strategies $\{\sigma^k\}$ with limit σ if and only if $\{\sigma^k\}$ induces a quasi-perfect equilibrium in any extensive form game having the normal form G.*

SKETCH OF PROOF. This is proved in Hillas (1998c) and in Mailath, Samuelson and Swinkels (1997). The if direction is implied by van Damme's proof. (Though not quite by the result as he states it, since that leaves open the possibility that different extensive forms may require different supporting sequences.)

The other direction is quite straightforward. One first takes a subsequence such that the conditional probability on any subset of a player's strategy space converges. (This conditional probability is well defined since the strategy vectors in the sequence are assumed to be completely mixed.) Thus $\{\sigma^k\}$ defines, for any subset of a player's strategy space, a conditional probability on that subset. And so the sequence $\{\sigma^k\}$ partitions the strategy space S_n of each player into the sets $S_n^0, S_n^1, \ldots, S_n^J$ where S_n^j is the set of those strategies that receive positive probability conditional on one of the strategies in $S_n \setminus (\bigcup_{i<j} S_n^i)$ being played.

Now consider the following extensive form. The players move in order of their names, but without observing anything about the choices of those who chose earlier. (That is, they move essentially simultaneously.) Each Player n first decides whether to play one of the strategies in S_n^0 or not. If he decides to do so then he chooses one of those strategies. If he decides not to then he decides whether to play one of the strategies in S_n^1, and so on.

The only behavioral strategy in such a game consistent with (the limiting behavior of) $\{\sigma^k\}$ has each player choosing at each opportunity to play one of the strategies in S_n^j rather than continuing with the process, and then choosing among those strategies according to the (strictly positive) limiting conditional (on S_n^j) probability distribution.

Now, if such a vector of strategies is a quasi-perfect equilibrium, then for sufficiently large k, for each player, every strategy in S_n^j is at least as good as any strategy in $S_n^{j'}$ for any $j' > j$ and is assigned probability arbitrarily greater than any such strategy. Thus $\{\sigma^k\}$ supports σ as a proper equilibrium. $\qquad\square$

However, the conjecture as we originally stated it – that is, without any reference to the supporting sequences of completely mixed strategies – is not true. Consider the game of Figure 23. The equilibrium (A, V) is not proper. To see this we argue as follows. Given that Player 1 plays A, Player 2 strictly prefers W to Y and X to Y. Thus in any ε-proper equilibrium Y is played with at most ε times the probability of W, and also at most ε times the probability of X. The fact that Y is less likely than W implies that Player 1 strictly prefers B to C, while the fact that Y is less likely than X implies that Player 1 strictly prefers B to D. Thus in an ε-proper equilibrium C and D are both played with at most ε times the probability of B. This in turn implies that Player 2 strictly prefers Z to V, and so there can be no ε-proper equilibrium in which V is played with probability close to 1. Thus (A, V) is not proper.

Nevertheless, in any extensive form game having this normal form there are perfect – and quasi-perfect – equilibria equivalent to (A, V). The idea is straightforward and not

	V	W	X	Y	Z
A	1,1	3,1	3,1	0,0	1,1
B	1,1	2,0	2,0	0,0	1,2
C	1,1	1,0	2,1	1,0	1,0
D	1,1	2,1	1,0	1,0	1,0

Figure 23.

at all delicate. In order to argue that (A, V) was not proper we needed to deduce, from the fact that Player 2 strictly prefers W to Y and X to Y, that Y is played with much smaller probability than W, and much smaller probability than X. However there is no extensive representation of this game in which quasi-perfection has this implication. For example, consider an extensive form in which Player 2 first decides whether to play W and, if he decides not to, then decides between X and Y. Now if all we may assume is that Player 2 strictly prefers W to Y and X to Y then we cannot rule out that Player 2 strictly prefers X to W. And in that case it is consistent with the requirements of either quasi-perfectness or perfectness that the action W is taken with probability ε^2 and the action Y with probability ε. This results in the strategy Y being played with substantially greater probability than the strategy W. Something similar results for any other way of structuring Player 2's choice. See Hillas (1998c) for greater detail.

10.6. Uncertainties about the game

Having now defined normal form perfect equilibrium we are in a position to be a little more explicit about the work of Fudenberg, Kreps and Levine (1988) and of Reny (1992a), which we mentioned in Section 10.1. These papers, though quite different in style, both show that if an out of equilibrium action can be taken to indicate that the player taking that action is not rational, or equivalently, that his payoffs are not as specified in the game, then any normal form perfect equilibrium is self-enforcing. Fudenberg, Kreps and Levine also show that if an out of equilibrium action can be taken to indicate that the game was not as originally described – so that others' payoffs may differ as well – then *any* strategic equilibrium is self-enforcing.

11. Forward induction

In the previous section we examined the idea of backward induction, and also combinations of backward induction and admissibility. If we suppose that we have captured the implications of this idea, are there any further considerations that would further narrow the set of self-enforcing assessments?

Consider the game of Figure 24(a) [Kohlberg and Mertens (1982, 1986)]. Here, every node can be viewed as a root of a subgame, so there is no further implication of "backward induction" beyond "subgame perfection". Since the outcome $(2, 5)$ (that is, the equilibrium where Player 1 plays T and the play of the subgame is (D, R)) is subgame perfect, it follows that backward induction cannot exclude it.

On the other hand, this outcome cannot possibly represent a self-enforcing assessment of the game. Indeed, in considering the contingency of Player 1's having played B, Player 2 must take it for granted that Player 1's subsequent move was U and not D (because a rational Player 1 will not plan to play B and then D, which can yield a maximum of 1, when he can play T for a sure 2). Thus the assessment of Player 1's second move must be U, and hence the assessment of Player 2's move must be L, which implies that the assessment of the whole game must be $(4, 1)$.

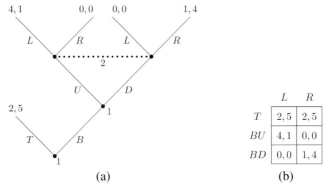

Figure 24.

Notice that this argument does not depend on the order in which Players 1 and 2 move in the subgame. If the order was reversed so that Player 2 moved first in the subgame the argument could be written almost exactly as it has been requiring only that "was U and not D" be changed to "will be U and not D".

Thus a self-enforcing assessment of the game must not only be consistent with deductions based on the opponents' rational behavior in the future (backward induction) but it must also be consistent with deductions based on the opponents' rational behavior in the past (forward induction).

A formal definition of forward induction has proved a little elusive. One aspect of the idea of forward induction is captured by the idea that a solution should not be sensitive to the deletion of a dominated strategy, as we discussed in Section 9. Other aspects depend explicitly on the idea of a solution as a set of equilibria. Kohlberg and Mertens (1986) give such a definition; and Cho and Kreps (1987) give a number of definitions in the context of signaling games, as do Banks and Sobel (1987). We shall leave this question for the moment and return to it in Sections 13.3, 13.5, and 13.6.

12. Ordinality and other invariances

In the following section we shall look again at identifying the self-enforcing solutions. In this section we examine in detail part of what will be involved in providing an answer, namely, the appropriate domain for the solution correspondence. That is, if we are asking simply what behavior (or assessment of the others' behavior) is self-enforcing, which aspects of the game are relevant and which are not?

We divide the discussion into two parts, the first takes the set of players (and indeed their names) as given, while the second is concerned with changes to the player set. The material in this section and the next is a restatement of some of the discussion in and of some of the results from Kohlberg and Mertens (1982, 1986) and Mertens (1987, 1989, 1991b, 1991a, 1992, 1995).

Recall that we are asking only a rather narrow question: what solutions are self-enforcing? Or, equivalently, what outcomes are consistent with the individual incentives of the players (and the interaction of those incentives)? Thus if some change in the game does not change the individual incentives then it should not change the set of solutions that are self-enforcing. The following two sections give some formal content to this informal intuition.

Before that, it is well to emphasize that we make no claim that a solution in any other sense, or the strategies that will actually be played, should satisfy the same invariances. This is a point that has been made in a number of places in the papers mentioned above,[1] but is worth repeating.

It is also worth reminding the reader at this point that we are taking the "traditional" point of view, that all uncertainty about the actual game being played has been explicitly included in the description of the game. Other analyses take different approaches. Harsanyi and Selten (1988) assume that with any game there are associated error probabilities for every move, and focus on the case in which all errors have the same small probability. Implicit, at least, in their analysis is a claim that one cannot think coherently, or at least cannot analyze, a game in which there is no possibility of mistakes. Others are more explicit. Aumann (1987b) argues that

> to arrive at a strong form of rationality (one must assume that) irrationality cannot be ruled out, that the players ascribe irrationality to each other with small probability. True rationality requires 'noise'; it cannot grow on sterile ground, it cannot feed on itself only.

We differ. Moreover, if we are correct and it is possible to analyze sensibly games in which there are no possibilities of mistakes, then such an analysis is, in theory, more general. One can apply the analysis to a game in which all probabilities of error (or irrationality) are explicitly given.

12.1. Ordinality

Let us now consider a candidate solution of a game. This candidate solution would not be self-enforcing if some player, when he came to make his choices, would choose not to play as the solution recommends. That is, if given his beliefs about the choices of the others, the choice recommended for him was not rational. For fixed beliefs about the choices of the others this is essentially a single-agent decision problem, and the concept of what is rational is relatively straightforward. There remains the question of whether all payoff-maximizing choices, or just admissible ones, are rational, but for the moment we don't need to take a position on that issue.

Now consider two games in which, for some player, for any beliefs on the choices of the others the set of that player's rational choices is the same. Clearly if he had the

---

[1] Perhaps most starkly in Mertens (1987, p. 6) where he states "There is here thus no argument whatsoever according to which the choice of equilibrium should depend only on this or that aspect of the game".

same beliefs in both games then those recommendations that would be self-enforcing (rational) for him in one game would also be self-enforcing in the other.

Now suppose that this is true for all the players. If the beliefs of the players are consistent with the individual incentives in one of the games they will also be in the other, since given those beliefs the individual decision problems have, by assumption, the same solutions in the two games. Thus a combination of choices and beliefs about the others' choices will be self-enforcing in one game if and only if it is self-enforcing in the other. Moreover, the same argument means that if we were to relabel the strategies the conclusion would continue to hold. A solution would be self-enforcing in the one game if and only if its image under the relabeling was self-enforcing in the other.

Let us flesh this out and make it a little more formal. Previously we have been mainly thinking of the mixed strategy of a player as the beliefs of the other players about the pure action that the player will take. On the other hand there it is difficult to think how we might prevent a player from randomizing if he wanted to. (It is not even completely clear to us what this might mean.) Thus the choices available to a player include, not only the pure strategies that we explicitly give him, but also all randomizations that he might decide to use, that is, his mixed strategy space. A player's beliefs about the choices of the others will now be represented by a probability distribution over the product of the (mixed) strategy spaces of the other players. Since we are not using any of the structure of the mixed strategy spaces, we consider only probability distributions with finite support.

Let us consider an example to make things a little more concrete. Consider the games of Figures 25(a) and 25(b). The game of Figure 25(a) is the Stag Hunt game of Figure 2 that we discussed in Section 2.1. The game of Figure 25(b) is obtained from the game of Figure 25(a) by subtracting 7 from Player 1's payoff when Player 2 plays L and subtracting 7 from Player 2's payoff when Player 1 plays T. This clearly does not, for any assessment of what the other player will do, change the rational responses of either player. Thus the games are ordinally equivalent, and the self-enforcing behaviors should be the same.

There remains the question of whether we should restrict the players' beliefs to be independent across the other players. We argued in Section 2 that in defining rationalizability the assumption of independence was not justified. That argument is perhaps not as compelling in the analysis of equilibrium. And, whatever the arguments, the assumption of independence seems essential to the spirit of Nash equilibrium. In any case the analysis can proceed with or without the assumption.

	L	R
T	9, 9	0, 7
B	7, 0	8, 8

(a)

	L	R
T	2, 2	0, 0
B	0, 0	8, 8

(b)

Figure 25.

For any such probability distribution the set of rational actions of the player is defined. Here there are essentially two choices: the set of that player's choices (that is, mixed strategies) that maximize his expected payoff, given that distribution; or the set of admissible choices that maximize his payoff. And if one assumes independence there is some issue of how exactly to define admissibility. Again, the analysis can go ahead, with some differences, for any of the assumptions.

For the rest of this section we focus on the case of independence and admissibility. Further, we assume a relatively strong form of admissibility. To simplify the statement of the requirement, the term belief will always mean an independent belief with finite support.

We say a player's choice is an *admissible best reply* to a certain belief p about the choices of the others if, for any other belief q of that player about the others, there are beliefs $\tilde{q}$ arbitrarily close to p and having support containing the support of q, such that the player's choice is expected payoff-maximizing against $\tilde{q}$.

The statement of this requirement is a little complicated. The idea is that the choice should be optimal against some completely mixed strategies of the other players close to the actual strategy of the others. However, because of the ordinality requirement we do not wish to use the structure of the strategy spaces, so we make no distinction between completely mixed strategies and those that are not completely mixed. Nor do we retain any notion of one mixed strategy being close to another, except information that may be obtained simply from the preferences of the players. Thus the notion of "close" in the definition is that the sum over all the mixed strategies in the support of either p or $\tilde{q}$ of the difference between p and $\tilde{q}$ is small.

One now requires that if two games are such that a relabeling of the mixed strategy space of one gives the mixed strategy space of the other and that this relabeling transforms the best reply correspondence of the first game into the best reply correspondence of the second then the games are equivalent in terms of the individual incentives and the (self-enforcing) solutions of the one are the same as the (appropriately relabeled) solutions of the other.

Mertens (1987) shows that this rather abstract requirement has a very concrete implementation. Two games exhibit the same self-enforcing behavior if they have the same pure strategy sets and the same best replies for every vector of completely mixed strategies; this remains true for any relabeling of the pure strategies, and also for any addition or deletion of strategies that are equivalent to existing mixed strategies. That is, the question of what behavior is self-enforcing is captured completely by the reduced normal form, and by the best reply structure on that form. And it is only the best reply structure on the interior of the strategy space that is relevant.

If one dropped the assumption that beliefs should be independent one would require a little more, that the best replies to any correlated strategy of the others should be the same. And if one thought of the best replies as capturing rationality rather than admissible best replies, one would need equivalence of the best replies on the boundary of the strategy space as well.

12.2. Changes in the player set

We turn now to questions of changes in the player set. We start with the most obvious. While, as Mertens (1991a) points out, the names of the players may well be relevant for what equilibrium is actually played, it is clearly not relevant for the prior question of what behavior is self-enforcing. That is, the self-enforcing outcomes are anonymous.

Also it is sometimes the case that splitting a player into two agents does not matter. In particular the set of self-enforcing behaviors does not change if a player is split into two subsets of his agents, provided that in *any* play, whatever the strategies of the players at most one of the subsets of agents moves.

This requirement too follows simply from the idea of self-enforcing behavior. Given the player's beliefs about the choices of the others, if his choice is not rational then it must be not rational for at least one of the subsets, and conversely. Moreover the admissibility considerations are the same in the two cases. It is the actions of Nature and the other players that determines which of the agents actually plays.

Now, at least in the independent case, the beliefs of the others over the choices of the player that are consistent with the notion of self-enforcing behavior and the beliefs over the choices of the subsets of the agents that are consistent with the notion of self-enforcing behavior may differ. But only in the following way. In the case of the agents it is assumed that the beliefs of the others are the independent product of beliefs over each subset of the agents, while when the subsets of the agents are considered a single player no such assumption is made. And yet, this difference has no impact on what is rational behavior for the others. Which of the agents plays in the game is determined by the actions of the others. And so, what is rational for one of the other players depends on his beliefs about the choices of the rest of the other players and his beliefs about the choices of each agent of the player in question conditional on that agent actually playing in the game. And the restrictions on these conditional beliefs will be the same in the two cases.

Notice that this argument breaks down if both subsets of the agents might move in some play. In that case dependence in the beliefs may indeed have an impact. Further, the argument that the decision of the player is decomposable into the individual decisions of his agents in not correct.

We now come to the final two changes, which concern situations in which subsets of the players can be treated separately. The first, called the decomposition property, says that if the players of a game can be divided into two subsets and that the players in each subset play distinct games with no interaction with the players in the other subset – that is, the payoffs of the one subset do not depend on the actions of the players in the other subset – then the self-enforcing behaviors in the game are simply the products of the self-enforcing behaviors in smaller games. Mertens (1989, p. 577) puts this more succinctly as "if disjoint player sets play different games in different rooms, one can as well consider the compound game as the separate games".

Finally, we have the small worlds axiom, introduced in Mertens (1991b) and proved for stable sets in Mertens (1992). This says that if the payoffs of some subset of the players do not depend on the choices of the players outside this subset, then the self-enforcing behavior of the players in the subset is the same whether one considers the whole game or only the game between the "insiders".

These various invariance requirements can be thought of as either restrictions on the permissible solutions or as a definition of what constitutes a solution. That is, they define the domain of the solution correspondence. Also, the requirements become stronger in the presence of the other requirements. For example, the player splitting requirement means that the restriction to nonoverlapping player sets in the decomposition requirement can be dispensed with. [For details see Mertens (1989, p. 578).] And, ordinality means that in the decomposition and small worlds requirements we can replace the condition that the payoffs of the subset of players not change with the condition that their best replies not change, or even that their admissible best replies not change.

13. Strategic stability

We come now to the concept of strategic stability introduced by Kohlberg and Mertens (1986) and reformulated by Mertens (1987, 1989, 1991a, 1991b, 1992) and Govindan and Mertens (1993). This work is essentially aimed at answering the same question as the concepts of Nash equilibrium and perfect equilibrium, namely, what is self-enforcing behavior in a game?

We argued in the previous section that the answer to such a question should depend only on a limited part of the information about the game. Namely, that the answer should be ordinal, and satisfy other invariances. To this extent, as long as one does not require admissibility, Nash equilibrium is a good answer. Nash equilibria are, in some sense, self-enforcing. However, as was pointed out by Selten (1965, 1975), this answer entails only a very weak form of rationality. Certainly, players choose payoff-maximizing strategies. But these strategies may prescribe choices of strategies that are (weakly) dominated and actions at unreached information sets that seem clearly irrational.

Normal form perfect equilibria are also invariant in all the senses we require, and moreover are also clearly admissible, even in the rather strong sense we require. However, they do not always satisfy the backward induction properties we discussed in Section 9. Other solutions that satisfy the backward induction property do not satisfy the invariance properties. Sequential equilibrium is not invariant. Even changing a choice between three alternatives into two binary choices can change the set of sequential equilibria. Moreover, sequential equilibria may be inadmissible. Extensive form perfect equilibria too are not invariant. Moreover they too may be inadmissible. Recall that in Section 10.4 (Figure 20) we examined a game of Mertens (1995) in which *all* of the extensive form perfect equilibria are inadmissible. Quasi-perfect equilibria, defined by van Damme (1984), and proper equilibria, defined by Myerson (1978), are both admissible and satisfy backward induction. However, neither is ordinal – quasi-perfect equilibrium depends on the particular extensive form and proper equilibrium may change with the addition of mixtures as new strategies – and proper equilibrium does not satisfy the player splitting requirement.

We shall see in the next section that it is impossible to satisfy the backward induction property for any ordinal single-valued solution concept. This leads us to consider set-valued solutions and, in particular, to strategically stable sets of equilibria.

13.1. The requirements for strategic stability

We are seeking an answer to the question: What are the self-enforcing behaviors in a game? As we indicated, the answer to this question should satisfy the various invariances we discussed in Section 12. We also require that the solution satisfy stronger forms of rationality than Nash equilibrium and normal form perfect equilibrium, the two "non-correlated" equilibrium concepts that do satisfy those invariances. In particular, we want our solution concept to satisfy the admissibility condition we discussed in Section 9, and some form of the iterated dominance condition we also discussed in Section 9, the backward induction condition we discussed in Section 10, and the forward induction condition we discussed in Section 11. We also want our solution to give some answer for all games.

As we indicated above, it is impossible for a single-valued solution concept to satisfy these conditions. In fact, two separate subsets of the conditions are inconsistent for such solutions. Admissibility and iterated dominance are inconsistent, as are backward induction and invariance.

To see the inconsistency of admissibility and iterated dominance consider the game in Figure 26 [taken from Kohlberg and Mertens (1986, p. 1015)]. Strategy B is dominated (in fact, strictly dominated) so by the iterated dominance condition the solution should not change if B is deleted. But in the resulting game admissibility implies that (T, L) is the unique solution. Similarly, M is dominated so we can delete M and then (T, R) is the unique solution.

To see the inconsistency of ordinality and backward induction consider the game of Figure 27(a) [taken from Kohlberg and Mertens (1986, p. 1018)]. Nature moves at the node denoted by the circle going left with probability α and up with probability $1 - \alpha$. Whatever the value of α the game has the reduced normal form of Figure 27(b). (Notice that strategy Y is just a mixture of T and M.) Since the reduced normal form does not depend on α, ordinality implies that the solution of this game should not depend on α. And yet, in the extensive game the unique sequential equilibrium has Player 2 playing L with probability $(4 - 3\alpha)/(8 - 4\alpha)$.

	L	R
T	3, 2	2, 2
M	1, 1	0, 0
B	0, 0	1, 1

Figure 26.

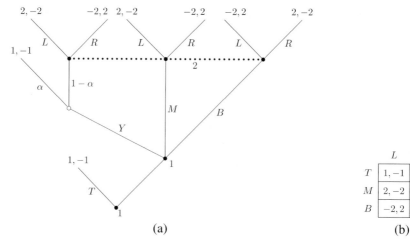

Figure 27.

Thus it may be that elements of a solution satisfying the requirements we have discussed would be sets. However we would not want these sets to be too large. We are still thinking of each element of the solution as, in some sense, a single pattern of behavior. In generic extensive form games we might think of a single pattern of behavior as being associated with a single equilibrium outcome, while not specifying exactly the out of equilibrium behavior. One way to accomplish this is to consider only connected sets of equilibria. In the definition of Mertens discussed below the connectedness requirement is strengthened in a way that corresponds, informally, to the idea that the particular equilibrium should depend continuously on the "beliefs" of the players. Without a better understanding of exactly what it means for a set of equilibria to be the solution we cannot say much more. However some form of connectedness seems to be required.

In Section 13.4 we shall discuss a number of definitions of strategic stability that have appeared in the literature. All of these definitions are motivated by the attempt to find a solution satisfying the requirements we have discussed. As we have just seen such a solution must of necessity be set-valued, and we have claimed that such sets should be connected. For the moment, without discussing the details of the definition we shall use the term strategically stable sets of equilibria (or simply stable sets) to mean one of the members of such a solution.

13.2. Comments on sets of equilibria as solutions to non-cooperative games

We saw in the previous section that single-valued solutions could not satisfy the requirements that were implied by a relatively strong form of the notion of self-enforcing. There are two potential responses to this observation. One is to abandon the strong conditions and view the concept of Nash equilibrium or normal form perfect equilibrium as the appropriate concept. The other is to abandon the notion that solutions to noncoop-erative games should be single-valued.

It may seem that the first is more natural, that perhaps the results of Section 13.1 tell us that there is some inconsistency between the invariances we have argued for and the backward induction requirements as we have defined them. We take the oppo-site view. Our view is that these results tell us that there is something unwarranted in insisting on single-valued solutions. Our own understanding of the issue is very incom-plete and sketchy and this is reflected in the argument. Nevertheless the issue should be addressed and we feel it is worthwhile to state here what we *do* know, however incom-plete.

Consider again the game of Figure 16. In this game the unique stable set contains all the strategy profiles in which Player 1 chooses T. The unique subgame perfect equilib-rium has Player 2 choosing R. Recall our interpretation of mixed strategies as the as-sessment of others of how a player will play. In order to know his best choices Player 1 does not need to know anything about what Player 2 will choose, so there is no need to make Player 1's assessment precise. Moreover from Player 2's perspective one could argue for either L or R equally well on the basis of the best reply structure of the game. They do equally well against T while L is better against BU and R is better against BD. Neither BU nor BD are ever best replies for Player 1, so we cannot distinguish between them on the basis of the best reply correspondence.

What then of the backward induction argument? We have said earlier that backward induction relies on an assumption that a player is rational whatever he has done at pre-vious information sets. This is perhaps a bit inaccurate. The conclusion here is that Player 1, being rational, will not choose to play B. The "assumption" that Player 2 at his decision node will treat Player 1 as rational is simply a test of the assumption that Player 1 will not play B. Once this conclusion is confirmed one cannot continue to maintain the assumption that following a choice of B by Player 1, Player 2 will continue to assume that Player 1 is rational. Indeed he cannot.

In this example Player 1's choices in the equilibrium tell us nothing about his as-sessment of Player 2's choices. If there is not something else – such as admissibility considerations, for example – to tie down his assessment then there is no basis to do so.

Consider the slight modification of this game of Figure 28(a). In this game it is not so immediately obvious that Player 1 will not choose B. However if we assume that Player 1 does choose B, believes that Player 2 is rational, and that Player 2 still be-lieves that Player 1 is rational we quickly reach a contradiction. Thus it must not be that Player 1, being rational and believing in the rationality of the other, chooses B. Again we can no longer maintain the assumption that Player 1 will be regarded as rational in the event that Player 2's information set is reached. Again both L and R are admissible for Player 2. In this case there is a difference between BU and BD for Player 1 – BD is ad-missible while BU is not. However, having argued that a rational Player 1 will choose T there is no compelling reason to make any particular assumption concerning Player 2's assessment of the relative likelihoods of BU and BD. The remaining information we have to tie down Player 1's assessment of Player 2's choices is Player 1's equilibrium

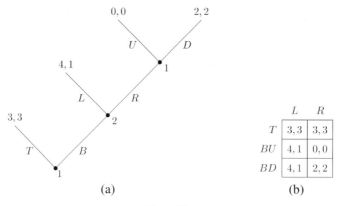

Figure 28.

choice. Since Player 1 chooses T, it must be that his assessment of Player 2's choices made this a best response. And this is true for any assessment that puts weight no greater than a half on L. Again this set of strategies is precisely the unique strategically stable set.

These examples point to the fact that there may be something unwarranted in specifying single strategy vectors as "equilibria". We have emphasized the interpretation of mixed strategy vectors as representing the uncertainty in one player's mind about what another player will do. In many games there is no need for a player to form a very precise assessment of what another player will do. Put another way, given the restrictions our ordinality arguments put on the "questions we can ask a player" we cannot elicit his subjective probability on what the others will do. Rather we must be satisfied with less precise information.

As we said, this is at best suggestive. In fact, if there are more than two players there are real problems. In either of the examples discussed we could add a player with a choice between two strategies, his optimal choice varying in the stable set and not affecting the payoffs of the others. Here too we see a sense in which the set-valued solution does represent the strategic uncertainty in the interaction among the players. However we do not see any way in which to make this formal, particularly since the notion of a "strategic dummy" is so elusive. For example, we could again amend the game by adding another game in which the third player interacts in an essential way with the other two and, say, having nature choose which of the two games will be played.

Finally, it is worth pointing out that in an important class of games, the sets of equilibria we consider are small. More specifically, we have the following result.

THEOREM 8 [Kohlberg and Mertens (1986)]. *For generic extensive games, every connected set of equilibria is associated with a single outcome (or distribution over endpoints) in the tree.*

This follows from a result of Kreps and Wilson (1982) and Kohlberg and Mertens (1986) that the set of equilibrium probability distributions on endpoints is finite. If there are a finite number of equilibrium distributions then, since the map from strategies to distributions on endpoints is continuous, a connected set of equilibria must all lead to the same distribution on endpoints.

13.3. Forward induction

We discussed in general terms in Section 11 the idea of forward induction; and discussed one aspect of this idea in Section 9. We are now in a position to be a little more explicit about another aspect of the idea.

The idea that the solution should not be sensitive to the (iterated) deletion of dominated strategies was that since it was known that players are never willing to choose such strategies it should not make a difference if such strategies were removed from the game. This suggests another aspect of forward induction. If, in the solution in question, there is some strategy of a player that the player would never be willing to choose then it should not make a difference if such a strategy were removed from the game.

In fact, like the deletion of a dominated strategy, it is clear that the deletion of a strategy that a player is never willing to play in the solution must change the situation. (It can, for example, change undominated strategies into dominated ones, and vice versa.) So our intuition is that it should not matter much. This is reflected in the fact that various versions of this requirement have been stated by requiring, for example, that a stable set *contain* a stable set of the game obtained by deleting such a strategy, or by looking at equilibrium outcomes in generic extensive form games, without being explicit about what out of equilibrium behavior is associated with the solution.

There is one aspect of this requirement that has aroused some comment in the literature. It is that in deleting dominated strategies we delete weakly dominated strategies. In deleting strategies that are inferior in the solution we delete only strategies that are strictly dominated at all points of the solution. In the way we have motivated the definition this difference appears quite naturally. The weakly dominated strategy is deleted because the player would not in any case be willing to choose it. In order to be sure that the player would not anywhere in the solution be willing to choose the strategy it is necessary that the strategy be strictly dominated at all points of the solution. Another way of saying almost the same thing is that the idea that players should not play dominated strategies might be thought of as motivated by a requirement that the solution should not depend on the player being absolutely certain that the others play according to the solution. Again, the same requirement implies that only strategies that are dominated in some neighborhood of the solution be deleted.

13.4. The definition of strategic stability

The basic ideas of backward induction and forward induction seem at first sight to require in an essential way the extensive form; and implicit in much of the literature following Selten (1965) was that essential insights about what was or was not reasonable in a game could not be seen in the normal form. However we have argued in Section 12 that only the normal form of the game *should* matter. This argument had been made originally by von Neumann and Morgenstern (1944) and was forcefully reintroduced by Kohlberg and Mertens (1986).

In fact, we saw in discussing, for example, the sequential equilibrium concept that even looking explicitly at the extensive form the notion of sequential equilibrium seems sensitive to details of the extensive form that appear irrelevant. Dalkey (1953) and Thompson (1952), and more recently Elmes and Reny (1994), show that there is a list of inessential transformations of one extensive form into another such that any two extensive form games having the same normal form can be transformed into each other by the application of these transformations. Thus even if we wish to work in the extensive form we are more or less forced to look for solutions that depend only on the normal form.

Kohlberg and Mertens reconciled the apparent dependence of the concepts of backward and forward induction by defining solutions that depended only on the reduced normal form of the game and that satisfied simultaneously rather strong forms of the backward and forward induction requirements, and showing that all games had such solutions. The solutions they defined consisted of sets of equilibria rather than single equilibria, as they showed must be the case. Nevertheless, in generic extensive form games all equilibria in such a set correspond to a single outcome.

Kohlberg and Mertens (1986) define three different solution concepts which they call hyper-stable, fully stable, and stable. Hyper-stable sets are defined in terms of perturbations to the payoffs in the normal form, fully stable sets in terms of convex polyhedral restrictions of the mixed strategy space, and stable sets in terms of Selten-type perturbations to the normal form of a game. A closed set of equilibria is called strategically stable in one of these senses if it is minimal with respect to the property that all small perturbations have equilibria close to the set. Kohlberg and Mertens explicitly viewed these definitions as a first pass and were well aware of their shortcomings. In particular hyper-stable and fully stable sets may contain inadmissible equilibria, while stable sets might fail to satisfy quite weak versions of backward induction. They give an example, due to Faruk Gul, of a game with such a stable set that does not yield the same outcome as the unique subgame perfect equilibrium. Moreover such sets are often not connected, though there always exist stable sets contained within a connected component of equilibria.

In spite of the problems with these definitions this work already shows that the program of defining solutions in the normal form that satisfy apparently extensive form requirements is practical; and the solutions defined there go most of the way to solving the problems. A minor modification of this definition somewhat expanding the notion of a perturbation remedies part of the deficiencies of the original definition of strategic stability. Such a definition essentially expands the definition of a perturbation enough so that backward induction is satisfied while not so much as to violate admissibility, as the definition of fully stable sets by Kohlberg and Mertens does. This modification was known to Kohlberg and Mertens and was suggested in unpublished work of Philip

Reny. It was independently discovered and called, following a suggestion of Mertens, fully stable sets of equilibria in Hillas (1990). While remedying some of the deficiencies of the original definition of Kohlberg and Mertens this definition does not satisfy at least one of the requirements given above. The definition defines different sets depending on whether two agents who never both move in a single play of a game are considered as one player or two.

Hillas (1990) also gave a different modification to the definition of strategic stability in terms of perturbations to the best reply correspondence. That definition satisfied the requirements originally proposed by Kohlberg and Mertens, though there is some problem with invariance and an error in the proof of the forward induction properties. This is discussed in Vermeulen, Potters and Jansen (1997) and Hillas et al. (2001a). Related definitions are given in Vermeulen, Potters and Jansen (1997) and Hillas et al. (2001b). This definition also fails to satisfy the small worlds axiom.

In a series of papers Mertens (1987, 1989, 1991b, 1992) gave and developed another definition – or, more accurately, a family of definitions – of strategic stability that satisfies all of the requirements we discussed above. The definitions we referred to in the previous paragraph are, in our opinion, best considered as rough approximations to the definitions of Mertens.

Stable sets are defined by Mertens in the following way. One again takes as the space of perturbations the Selten-type perturbations to the normal form. The stable sets are the limits at zero of some connected part of the graph of the equilibrium correspondence above the interior of a small neighborhood of zero in the space of perturbations. Apart from the minimality in the definition of Kohlberg and Mertens and the assumption that the part of the graph was connected in the definition of Mertens, the definition of Kohlberg and Mertens could be stated in exactly this way. One would simply require that the projection map from the part of the equilibrium correspondence to the space of perturbations be onto.

Mertens strengthens this requirement in a very natural way, requiring that the projection map not only be onto, but also be nontrivial in a stronger sense. The simplest form of this requirement is that the projection map not be homotopic to a map that is not onto, under a homotopy that left the projection map above the boundary unchanged. However such a definition would not satisfy the ordinality we discussed in Section 12.1. Thus Mertens gives a number of variants of the definition all involving coarser notions of maps being equivalent. That is, more maps are trivial (equivalent to a map that is not onto) and so fewer sets are stable. These definitions require that the projection map be nontrivial either in homology or in cohomology, with varying coefficient modules. Mertens shows that, with some restrictions on the coefficient modules, such definitions are ordinal.

One also sees in this formulation the similarities of the two aspects of forward induction we discussed in the previous section. In Mertens's work they appear as special cases of one property of stable sets. A stable set contains a stable set of the game obtained by deleting a strategy that is nowhere in the relevant part of the graph of the equilibrium correspondence played with more than the minimum required probability.

Govindan and Mertens (1993) give a definition of stability directly in terms of the best reply correspondence. One can think of the set of equilibria as the intersection of the graph of the best reply correspondence with the diagonal. To define stable sets one looks at the graph of the best reply correspondence in a neighborhood of the diagonal. It is required that a map that takes points consisting of a strategy and a best reply to that strategy to points in the strategy space in the following way be essential. The points on the diagonal are projected straight onto the strategy space. Other points are taken to the best reply and then shifted by some constant times the difference between the original strategy vector and the best reply. As long as this constant is large enough the boundary of the neighborhood will be taken outside the strategy space (at which point the function simply continues to take the boundary value). The form of the essentiality requirement in this paper is that the map be essential in Čech cohomology. Loosely, the requirement says that the intersection of the best reply correspondence with the diagonal should be essential. Govindan and Mertens show that this definition gives the same stable sets as the original definition of Mertens in terms of the graph of the equilibrium correspondence with the same form of essentiality.

The following theorem applies to the definitions of Mertens and to the definition of Govindan and Mertens.

THEOREM 9. *Stable sets are ordinal, satisfy the player splitting, the decomposition property, and the small worlds axiom. Every game has at least one stable set. Any stable set is a connected set of normal form perfect equilibria and contains a proper (and hence quasi-perfect and hence sequential) equilibrium. Any stable set contains a stable set of a game obtained by deleting a dominated strategy or by deleting a strategy that is strictly inferior everywhere in the stable set.*

13.5. Strengthening forward induction

In this section we discuss an apparently plausible strengthening of the forward induction requirement that we discussed in Section 11 that is not satisfied by the definitions of stability we have given. Van Damme (1989) suggests that the previous definitions of forward induction do not capture all that the intuition suggests.

Van Damme does not, in the published version of the paper, give a formal definition of forward induction but rather gives

> a (weak) property which in my opinion should be satisfied by any concept that is consistent with forward induction... The proposed requirement is that in generic 2-person games in which player i chooses between an outside option or to play a game Γ of which a unique (viable) equilibrium e^* yields this player more than the outside option, only the outcome in which i chooses Γ and e^* is played in Γ is plausible.

He gives the example in Figure 29(a) to show that strategic stability does not satisfy this requirement. In this game there are three components of normal form perfect equi-

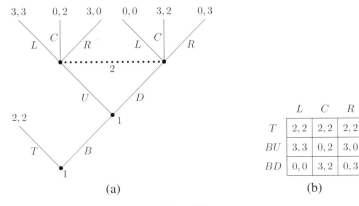

Figure 29.

libria, (BU, L), $\{(T, (q, 1 - q, 0)) \mid 1/3 \leqslant q \leqslant 2/3\}$, and $\{(T, (0, q, 1 - q)) \mid 1/3 \leqslant q \leqslant 2/3\}$. (Player 2's mixed strategy gives the probabilities of L, C, and R, in that order.) All of these components are stable (or contain stable sets) in any of the senses we have discussed. Since (U, L) is the unique equilibrium of the subgame that gives Player 1 more than his outside option the components in which Player 1 plays T clearly do not satisfy van Damme's requirement.

It is not clear if van Damme intends to address the question we ask, namely, what outcomes are self-enforcing? (Recall that he doesn't claim that outcomes not satisfying his condition are not self-enforcing, but rather that they are not plausible.) Nevertheless let us address that question. There does not seem, to us, to be a good argument that the outcome in which Player 1 plays T is not self-enforcing. There are two separate "patterns of behavior" by Player 2 that lead Player 1 to play T. To make matters concrete consider the patterns in which Player 2 plays L and C. Let us examine the hypothesis that Player 1 plays T because he is uncertain whether Player 2 will play L or C. If he is uncertain (in the right way) this is indeed the best he can do. Further at the boundary of the behavior of Player 2 that makes T the best for Player 1 there is an assessment of Player 2's choices that make BU best for Player 1 and another assessment of Player 2's choices that make BD best for Player 1. If Player 2 can be uncertain as to whether Player 1 has deviated to BU or to BD then indeed both strategies L and C can be rationalized for Player 2, as can mixtures between them.

To be very concrete, let us focus on the equilibrium $(T, (\frac{1}{3}, \frac{2}{3}, 0))$. In this equilibrium Player 1 is indifferent between T and BD and prefers both to BU. It seems quite consistent with the notion of self-enforcing that Player 2, seeing a deviation by Player 1, should not be convinced that Player 1 had played BU.

In a sense we have done nothing more than state that the stable sets in which Player 1 plays T satisfy the original statement of forward induction given by Kohlberg and Mertens (1986). The argument does however suggest to us that there is something missing in a claim that the stronger requirement of van Damme is a necessary implication of the idea of self-enforcing. To be much more explicit would require a better developed notion of what it means to have set-valued solutions, as we discussed in Section 13.2.

Before leaving this example, we shall address one additional point. Ritzberger (1994) defines a vector field, which he calls the Nash field, on the strategy space. This definition loosely amounts to moving in (one of) the direction(s) of increasing payoff. Using this structure he defines an index for isolated zeros of the field (that is, equilibria). He extends this to components of equilibria by taking the sum of indices of regular perturbations of the Nash field in a small neighborhood of the component. He points out that in this game the equilibrium (BU, L) has index $+1$, and so since the sum of the indices is $+1$ the index of the set associated with Player 1 playing T must have index zero. However, as we have seen there are two separate stable sets in which Player 1 plays T. Now these two sets are connected by a further set of (nonperfect) equilibria. The methods used by Ritzberger seem to require us to define and index for this component in its entirety. As we said above both of the connected subsets of normal form perfect equilibria are stable in the sense of Mertens's reformulation. And it would seem that there might be some way to define an index so that these sets, considered separately, would have nonzero index, $+1$ and -1. Ritzberger comments that "the reformulation of Stable Sets [Mertens (1989)] eliminates the component with zero index" and speculates that "as a referee suggested, it seems likely that sets which are stable in the sense of Mertens will all be contained in components whose indices are non-zero". However his claim that neither of the sets are stable is in error and his conjecture is incorrect. It seems to us that the methods used by Ritzberger force us to treat two patterns of behavior, that are separately quite well behaved, together and that this leads, in some imprecise sense, to their indices canceling out. There seems something unwarranted in rejecting an outcome on this basis.

13.6. Forward induction and backward induction

The ideas behind forward induction and backward induction are closely related. As we have seen backward induction involves an assumption that players assume, even if they see something unexpected, that the other players will choose rationally in the future. Forward induction involves an assumption that players assume, even if they see something unexpected, that the other players chose rationally in the past. However, a distinction between a choice made in the past whose outcome the player has not observed, and one to be made in the future goes very much against the spirit of the invariances that we argued for in Section 12.

In fact, in a much more formal sense there appears to be a relationship between backward and forward induction. In many examples – in fact, in all of the examples we have examined – a combination of the invariances we have discussed and backward induction gives the results of forward induction arguments and, in fact, the full strength of stability. Consider again the game of Figure 24(a) that we used to motivate the idea of forward induction. The game given in extensive form in Figure 30(a) and in normal form in Figure 30(b) has the same reduced normal form as the game of Figure 24(a). In

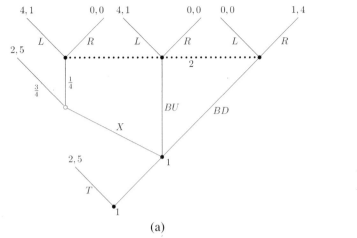

(a)   (b)

Figure 30.

that game the unique equilibrium of the subgame is (BU, L). Thus the unique subgame perfect equilibrium of the game is (BU, L). Recall that this is the same result that we argued for in terms of forward induction in Section 11.

This is perhaps not too surprising. After all this was a game originally presented to illustrate in the most compelling way possible the arguments in favor of forward induction. Thus the issues in this example are quite straightforward. We obtain the same result from two rounds of the deletion of weakly dominated strategies. What is a little surprising is that the same result occurs in cases in which the arguments for forward induction are less clear cut. A number of examples are examined in Hillas (1998b) with the same result, including the examples presented by Cho and Kreps (1987) to argue against the strength of some definitions of forward induction and strategic stability.

We shall illustrate the point with one further example. We choose the particular example both because it is interesting in its own right and because in the example the identification of the stable sets seems to be driven mostly by the definition of stability and not by an obvious forward induction logic. The example is from Mertens (1995). It is a game with perfect information and a unique subgame perfect equilibrium – though the game is not generic. Moreover there are completely normal form arguments that support the unique subgame perfect equilibrium. In spite of this the unique strategically stable set of equilibria consists of all the admissible equilibria, a set containing much besides the subgame perfect equilibrium. The extensive form of the game is given here in Figure 31(a) and the normal form in Figure 31(b).

The unique subgame perfect equilibrium of this game is (T, R). Thus this is also the unique proper equilibrium – though the uniqueness of the proper equilibrium does involve an identification of duplicate strategies that is not required for the uniqueness of the subgame perfect equilibrium – and so (T, R) is a sequential equilibrium of any extensive form game having this normal form. Nevertheless there are games having

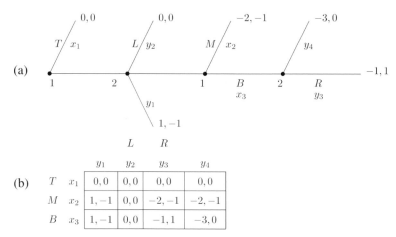

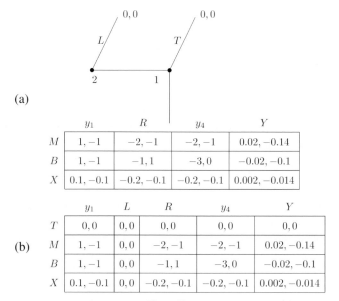

Figure 31.

Figure 32.

the same *reduced* normal form as this game for which (T, R) is not sequential, and moreover the outcome of any sequential equilibrium of the game is not the same as the outcome from (T, R).

Suppose that the mixtures $X = 0.9T + 0.1M$ and $Y = 0.86L + 0.1y_1 + 0.04y_4$ are added as new pure strategies. This results in the normal form game of Figure 32(b). The extensive form game of Figure 32(a) has this normal form. It is straightforward to see that the unique equilibrium of the simultaneous move subgame is (M, Y) in which the payoff to Player 1 is strictly positive (0.02) and so in any subgame perfect equilibrium Player 1 chooses to play "in" – that is, M, or B, or X – at his first move. Moreover in the equilibrium of the subgame the payoff to Player 2 is strictly negative (-0.14) and so in any subgame perfect equilibrium Player 2 chooses to play L at his first move. Thus the unique subgame perfect equilibrium of this game is (M, L).

This equilibrium is far from the unique backward induction equilibrium of the original game. Moreover its outcome is also different. Thus, if one accepts the arguments that the solution(s) of a game should depend only on the reduced normal form of the game and that any solution to a game should contain a "backward induction" equilibrium, then one is forced to the conclusion that in this game the solution must contain much besides the subgame perfect equilibrium of the original presentation of the game.

14. An assessment of the solutions

We offer here a few remarks on how the various solutions we have examined stand up. As we indicated in Section 12 we do not believe that the kinds of invariance we discussed there should be controversial. (We are aware that for some reason in some circles they are, but have every confidence that as the issues become better understood this will change.)

On the other hand we have less attachment to or confidence in our other maintained assumption, that of stochastic independence. We briefly discussed in the introduction our reasons for adopting that assumption in this chapter. Many of the issues we have discussed here in the context of strategic equilibrium have their analogs for correlated equilibrium. However the literature is less well developed – and our own understanding is even more incomplete.

The most straightforward, and in some ways weakest, requirement beyond that of equilibrium is admissibility. If one is willing to do without admissibility the unrefined strategic equilibrium seems a perfectly good self-enforcing outcome. Another way of thinking about this is given in the work of Fudenberg, Kreps and Levine (1988). If the players are permitted to have an uncertainty about the game that is of the same order as their strategic uncertainty – that is, their uncertainty about the choices of the others given the actual game – then any strategic equilibrium appears self-enforcing in an otherwise strong manner.

There are a number of different ways that we could define admissibility. The standard way leads to the standard definition of admissible equilibrium, that is, a strategic equilibrium in undominated strategies. A stronger notion takes account of both the strategy being played and the assumption of stochastic independence. In Section 12.1 we gave a definition of an admissible best reply correspondence such that a fixed point of such a correspondence is a normal form perfect equilibrium. Again Fudenberg, Kreps and Levine give us another way to think about this. If the players have no uncertainty about their own payoffs but have an uncertainty about the payoffs (or rationality) of the others that is of the same order as their strategic uncertainty then any normal form perfect equilibrium is self-enforcing. Reny's (1992a) work makes much the same point.

To go further and get some version of the backward induction properties that we discussed in Section 10 one needs to make stronger informational assumptions. We claimed in Section 10.1 that the idea of an equilibrium being robust to a little strategic uncertainty in the sense that the strategic uncertainty should be when possible of a higher order than the uncertainty about the game led us towards backward induction. At best this was suggestive. However as we saw in Section 13.6 it takes us much further. In combination with the invariances of Section 12 it seems to imply something close to the full strength of strategic stability. As yet we have seen this only in examples, but the examples include those that were constructed precisely to show the unreasonableness of the strength of strategic stability. We regard these examples to be, at worst, very suggestive.

Mertens (1989, pp. 582, 583) described his attitude to the project that led to his work on strategic stability as follows:

> I should say from the outset that I started this whole series of investigations with substantial scepticism. I had (and still have) some instinctive liking for the bruto Nash equilibrium, or a modest refinement like admissible equilibria, or normal-form perfect equilibria.

It is not entirely coincidental that we arrive at a similar set of preferred solutions. Our approaches have much in common and our work has been much influenced by Mertens.

15. Epistemic conditions for equilibrium

There have been a number of papers relating the concept of strategic equilibrium to the beliefs or knowledge of the players. One of the most complete and successful is by Aumann and Brandenburger (1995). In the context of a formal theory of *interactive belief systems* they find sets of sufficient conditions on the beliefs of the players for strategic equilibrium. In keeping with the style of the rest of this paper we shall not develop the formal theory but shall rather state the results and sketch some of the proofs in an informal style. Of course, we can do this with any confidence only because we know the formal analysis of Aumann and Brandenburger.

There are two classes of results. One concerns the actions or the plans of the players and the other the conjectures of the players concerning the actions or plans of the other players. These correspond to the two interpretations of strategic equilibrium that we discussed in Section 3. The first class consists of a single result. It is quite simple; its intuition is clear; and its proof is, as Aumann and Brandenburger say, immediate.

THEOREM 10. *Suppose that each player is rational and knows his own payoff function and the vector of strategies chosen by the others. Then these strategies form a strategic equilibrium.*

The results involving conjectures are somewhat more involved. The reader is warned that, while the results we are presenting are precisely Aumann and Brandenburger's, we are presenting them with a quite different emphasis. In fact, the central result we present next appears in their paper as a comment well after the central arguments. They give central place to the two results that we give as corollaries (Theorems A and B in their paper).

THEOREM 11.   *Suppose that each player knows the game being played, knows that all the players are rational, and knows the conjectures of the other players. Suppose also that*
  (a) *the conjectures of any two players about the choices of a third are the same, and*
  (b) *the conjectures of any player about the choices of two other players are independent.*
Then the vector of conjectures about each player's choice forms a strategic equilibrium.

The proof is, again, fairly immediate. Suppositions (a) and (b) give immediately that the conjectures "are" a vector of mixed strategies, σ. Now consider the conjecture of Player n' about Player n. Player n' knows Player n's conjecture, and by (a) and (b) it is σ. Also Player n' knows that Player n is rational. Thus his conjecture about Player n's choices, that is, σ_n, should put weight only on those choices of Player n that maximize n's payoff given n's conjecture, that is, given σ. Thus σ is a strategic equilibrium.

COROLLARY 12.   *Suppose that there are only two players, that each player knows the game being played, knows that both players are rational, and knows the conjecture of the other player. Then the pair of conjectures about each player's choice forms a strategic equilibrium.*

The proof is simply to observe that for games with only two players the conditions (a) and (b) are vacuous.

COROLLARY 13.   *Suppose that the players have a common prior that puts positive weight on the event that a particular game g is being played, that each player knows the game and the rationality of the others, and that it is common knowledge among the players that the vector of conjectures takes on a particular value φ. Then the conjectures of any two players about the choices of a third are the same, and the vector consisting of, for each Player n, the conjectures of the others about Player n's choice forms a strategic equilibrium.*

We shall not prove this result. It is clear that once it is proved that the conjectures of any two players about the choices of a third are the same and that the conjectures of any player about the choices of two other players are independent the rest of the result follows from Theorem 11. The proof that the conjectures of any two players about the choices of a third are the same is essentially an application of Aumann's (1976) "agreeing to disagree" result. The proof that the conjectures of any player about the choices of two other players are independent involves further work, for which we refer the reader to Aumann and Brandenburger's paper.

Aumann and Brandenburger point to the absence of the common knowledge assumption in the two-player version of the result and to the limited role that common knowledge plays in the general result. They cite a long list of prominent game theorists (including Aumann and Brandenburger) who have written various statements connecting strategic equilibrium with common knowledge.

And yet, as we have seen earlier common knowledge of the game and the rationality of the players does characterize (correlated) rationalizability, and common knowledge of the game and the rationality of the players and a common prior does characterize correlated equilibrium. Since strategic equilibrium is a refinement of correlated equilibrium it seems a little paradoxical that common knowledge seems to play such a central role in providing an epistemic characterization of correlated equilibrium and such a limited role in providing a characterization of strategic equilibrium.

The assumption that the conjectures of the players are known (or with more than two players are commonly known) is sufficiently strong that it renders unnecessary the common knowledge assumptions that are used in the characterization of correlated equilibrium. It will be interesting to see what *additional* conditions are needed beyond the conditions that characterize correlated equilibrium to obtain strategic equilibrium, even if taken together these conditions are stronger than the conditions proposed by Aumann and Brandenburger.

Let us return to an argument that we made briefly in Section 6. There we assumed that the description of the game completely described the information that the players have, that is, the players have no information beyond the rules of the game. We also assumed that the game, the rationality of the players, and this completeness are common knowledge, and, following Harsanyi, argued that players having the same information should form the same assessments.

We think of an outside observer having the same information as the players. The observation of some player's choice, say Player 1's choice, will not affect the observer's assessment of the other players' choices because the observer and Player 1 have identical information concerning the other players.

Thus, for any player, the conditional probability, given that player's choice, of the others' choices is the same as the unconditional probability. It follows that the observer's assessment of the choices of all the players is the product of his assessments of their individual choices.

Of course, if we need the completeness assumption to justify Nash equilibrium then we may not be very comfortable in applying this concept in our models. However we do *not* need the full strength of the completeness assumption. All that was required for our argument was that we could think of an outside observer having, for any player, the *same* information as that player about the choices of the others.

We see that the players may indeed have private information as long as it says nothing about how the others will play. And they may have information about how the others will play that is not contained in the description of the game, as long as such information is common to all the players.

If the information beyond the rules of the game held by the players satisfies these restrictions then we say that the players have a *common understanding of the game*.

THEOREM 14. *If the players have a common understanding of the game then at any state their common assessment of how the other players will play constitutes a strategic equilibrium.*

Notice that we do *not* claim that there is anything compelling about the assumption that the players have a common understanding of the game. In many circumstances correlated equilibria that are not independent may be perfectly reasonable. Rather, in an epistemic setting, common understanding of the game seems to us to be what distinguishes strategic equilibrium from correlated equilibrium.

References

Aumann, R.J. (1974), "Subjectivity and correlation in randomized strategies", Journal of Mathematical Economics 1:67–96.

Aumann, R.J. (1976), "Agreeing to disagree", Annals of Statistics 4:1236–1239.

Aumann, R.J. (1985), "What is game theory trying to accomplish", in: K.J. Arrow and S. Honkapohja, eds., Frontiers of Economics (Basil Blackwell, Oxford) 28–76.

Aumann, R.J. (1987a), "Correlated equilibrium as an expression of Bayesian rationality", Econometrica 55:1–18.

Aumann, R.J. (1987b), "Game theory", in: J. Eatwell, M. Milgate and P. Newman, eds., The New Palgrave Dictionary of Economics (Macmillan, London) 460–482.

Aumann, R.J. (1990), "Nash equilibria are not self-enforcing", in: J.J. Gabszewicz, J.-F. Richard and L. Wolsey, eds., Economic Decision-Making: Games, Econometrics, and Optimisation (Elsevier, Amsterdam) 201–206.

Aumann, R.J. (1992), "Irrationality in game theory", in: P. Dasgupta, D. Gale, O. Hart and E. Maskin, eds., Economic Analysis of Markets and Games (MIT Press, Cambridge, MA) 214–227.

Aumann, R.J. (1995), "Backward induction and common knowledge of rationality", Games and Economic Behavior 8:6–19.

Aumann, R.J. (1997), "Rationality and bounded rationality", Games and Economic Behavior 21:2–14.

Aumann, R.J. (1998), "Common priors: A reply to Gul", Econometrica 66:929–938.

Aumann, R.J. (1999a), "Interactive epistemology I: Knowledge", International Journal of Game Theory 28:263–300.

Aumann, R.J. (1999b), "Interactive epistemology II: Probability", International Journal of Game Theory 28:301–314.

Aumann, R.J., and A. Brandenburger (1995), "Epistemic conditions for Nash equilibrium", Econometrica 63:1161–1180.

Aumann, R.J., and M. Maschler (1972), "Some thoughts on the minimax principle", Management Science 18:54–63.

Aumann, R.J., and S. Sorin (1989), "Cooperation and bounded recall", Games and Economic Behavior 1:5–39.

Banks, J., and J. Sobel (1987), "Equilibrium selection in signaling games", Econometrica 55:647–661.

Bárány, I. (1992), "Fair distribution protocols or how the players replace fortune", Mathematics of Operations Research 17:327–340.

Basu, K. (1988), "Strategic irrationality in extensive games", Mathematical Social Sciences 15:247–260.

Basu, K. (1990), "On the non-existence of rationality definition for extensive games", International Journal of Game Theory 19:33–44.

Battigalli, P. (1996a), "Strategic independence and perfect Bayesian equilibria", Journal of Economic Theory 70:201–234.

Battigalli, P. (1996b), "Strategic rationality orderings and the best rationalization principle", Games and Economic Behavior 13:178–200.

Battigalli, P. (1997a), "Dynamic consistency and imperfect recall", Games and Economic Behavior 20:31–50.

Battigalli, P. (1997b), "On rationalizability in extensive games", Journal of Economic Theory 74:40–61.

Battigalli, P., and P. Veronesi (1996), "A note on stochastic independence without Savage-null events", Journal of Economic Theory 70:235–248.

Bellman, R. (1957), Dynamic Programming (Princeton University Press, Princeton, NJ).

Ben-Porath, E. (1997), "Rationality, Nash equilibrium and backwards induction in perfect information games", Review of Economic Studies 64:23–46.

Ben-Porath, E., and E. Dekel (1992), "Signaling future actions and the potential for sacrifice", Journal of Economic Theory 57:36–51.

Bernheim, B.D. (1984), "Rationalizable strategic behavior", Econometrica 52:1007–1028.

Bernheim, B.D. (1986), "Axiomatic characterizations of rational choice in strategic environments", Scandinavian Journal of Economics 88(3):473–488.

Binmore, K. (1987), "Modeling rational players, Part I", Journal of Economics and Philosophy 3:179–214.

Binmore, K. (1988), "Modeling rational players, Part II", Journal of Economics and Philosophy 4:9–55.

Binmore, K. (1990), Essays on the Foundations of Game Theory (Basil Blackwell, Cambridge, MA).

Blume, L., A. Brandenburger and E. Dekel (1991a), "Lexicographic probabilities and choice under uncertainty", Econometrica 59:61–79.

Blume, L., A. Brandenburger and E. Dekel (1991b), "Lexicographic probabilities and equilibrium refinements", Econometrica 59:81–98.

Blume, L.E., and W.R. Zame (1994), "The algebraic geometry of perfect and sequential equilibrium", Econometrica 62:783–794.

Bonnano, G. (1992), "Player's information in extensive games", Mathematical Social Sciences 24:35–48.

Börgers, T. (1989), "Bayesian optimization and dominance in normal form games", unpublished (University of Basel).

Börgers, T. (1991), "On the definition of rationalizability in extensive games", Discussion Paper 91–22, University College, London.

Börgers, T. (1994), "Weak dominance and approximate common knowledge", Journal of Economic Theory 64:265–276.

Brandenburger, A. (1992a), "Knowledge and equilibrium in games", Journal of Economic Perspectives 6(4):83–102.

Brandenburger, A. (1992b), "Lexicographic probabilities and iterated admissibility", in: P. Dasgupta, D. Gale, O. Hart and E. Maskin, eds., Economic Analysis of Markets and Games (MIT Press, Cambridge, MA).

Brandenburger, A., and E. Dekel (1987), "Rationalizability and correlated equilibrium", Econometrica 55:1391–1402.

Cho, I.-K. (1987), "A refinement of sequential equilibrium", Econometrica 55:1367–1389.

Cho, I.-K., and D.M. Kreps (1987), "Signaling games and stable equilibria", Quarterly Journal of Economics 102:179–221.

Cho, I.-K., and J. Sobel (1990), "Strategic stability and uniqueness in signaling games", Journal of Economic Theory 50:381–413.

Cournot, A.A. (1838), Recherches sur les principes mathématiques de la théorie des richesse (M. Riviere, Paris). Translated in: Researches into the Mathematical Principles of Wealth (A.M. Kelly, New York) 1960.

Dalkey, N. (1953), "Equivalence of information patterns and essentially determinate games", in: H.W. Kuhn and A.W. Tucker, eds., Contributions to the Theory of Games, Vol. 2 (Princeton University Press, Princeton, NJ) 127–143.

Debreu, G. (1952), "A social equilibrium existence theorem", Proceedings of the National Academy of Sciences 38:886–893.

Dekel, E., and D. Fudenberg (1990), "Rational behavior with payoff uncertainty", Journal of Economic Theory 52:243–267.

Dhillon, A., and J.-F. Mertens (1996), "Perfect correlated equilibria", Journal of Economic Theory 68:279–302.

Elmes, S., and P.J. Reny (1994), "On the strategic equivalence of extensive form games", Journal of Economic Theory 62:1–23.

Fan, K. (1952), "Fixed-point and minimax theorems in locally convex linear spaces", Proceedings of the National Academy of Sciences 38:121–126.

Fishburn, P.C. (1994), "Utility and subjective probability", in: R.J. Aumann and S. Hart, eds., Handbook of Game Theory, Vol. 2 (North-Holland, Amsterdam) Chapter 39, 1397–1435.

Forges, F. (1990), "Universal mechanisms", Econometrica 58:1341–1364.

Fudenberg, D., and D.M. Kreps (1991), "Learning and equilibrium in games, Part I: Strategic-form games", unpublished.

Fudenberg, D., and D.M. Kreps (1993), "Learning mixed equilibria", Games and Economic Behavior 5:320–367.

Fudenberg, D., and D.M. Kreps (1995), "Learning in extensive-form games. 1. Self-confirming equilibria", Games and Economic Behavior 8:20–55.

Fudenberg, D., D.M. Kreps and D.K. Levine (1988), "On the robustness of equilibrium refinements", Journal of Economic Theory 44:351–380.

Fudenberg, D., and D.K. Levine (1993a), "Self-confirming equilibrium", Econometrica 61:523–545.

Fudenberg, D., and D.K. Levine (1993b), "Steady state learning and Nash equilibrium", Econometrica 61:547–573.

Fudenberg, D., and J. Tirole (1991), "Perfect Bayesian equilibrium and sequential equilibrium", Journal of Economic Theory 53:236–260.

Geanakoplos, J. (1992), "Common knowledge", Journal of Economic Perspectives 6(4):53–82.

Geanakoplos, J. (1994), "Common knowledge", in: R.J. Aumann and S. Hart, eds., Handbook of Game Theory, Vol. 2 (North-Holland, Amsterdam) Chapter 40, 1437–1496.

Glicksberg, I.L. (1952), "A further generalization of the Kakutani fixed-point theorem with applications to Nash equilibrium points", Proceedings of the American Mathematical Society 3:170–174.

Govindan, S. (1995a), "Every stable set contains a fully stable set", Econometrica 63:191–193.

Govindan, S. (1995b), "Stability and the chain store paradox", Journal of Economic Theory 66:536–547.

Govindan, S. (1996), "A subgame property of stable equilibria", Mathematics of Operations Research 21:991–999.

Govindan, S., and A. McLennan (1995), "A game form with infinitely many equilibria on an open set of payoffs", unpublished.

Govindan, S., and J.-F. Mertens (1993), "An equivalent definition of stable equilibria", unpublished.

Govindan, S., and A.J. Robson (1997), "Forward induction, public randomization, and admissibility", Journal of Economic Theory 82:451–457.

Govindan, S., and R. Wilson (1997a), "Equivalence and invariance of the index and degree of Nash equilibria", Games and Economic Behavior 21:56–61.

Govindan, S., and R. Wilson (1997b), "Uniqueness of the index for Nash equilibria of two-player games", Economic Theory 10:541–549.

Grossman, S., and M. Perry (1986), "Perfect sequential equilibrium", Journal of Economic Theory 39:97–119.

Gul, F. (1998), "A comment on Aumann's Bayesian view", Econometrica 66:923–927.

Hammerstein, P., and R. Selten (1994), "Game theory and evolutionary biology", in: R.J. Aumann and S. Hart, eds., Handbook of Game Theory, Vol. 2 (North-Holland, Amsterdam) Chapter 28, 929–993.

Hammond, P.J. (1993), "Aspects of rationalizable behavior", in: K. Binmore, A. Kirman and P. Tani, eds., Frontiers of Game Theory (MIT Press, Cambridge, MA).

Harsanyi, J.C. (1967–1968), "Games with incomplete information played by 'Bayesian' players, I–III", Management Science 14:159–182, 320–334, 486–502.

Harsanyi, J.C. (1973), "Games with randomly distributed payoffs: A new rationale for mixed strategy equilibrium points", International Journal of Game Theory 2:1–23.

Harsanyi, J.C. (1977), Rational Behavior and Bargaining Equilibrium in Games and Social Situations (Cambridge University Press, Cambridge).

Harsanyi, J.C., and R. Selten (1988), A General Theory of Equilibrium Selection in Games (MIT Press, Cambridge, MA).

Hart, S. (1992), "Games in extensive and strategic form", in: R.J. Aumann and S. Hart, eds., Handbook of Game Theory, Vol. 1 (North-Holland, Amsterdam) Chapter 2, 19–40.

Hart, S., and D. Schmeidler (1989), "Existence of correlated equilibrium", Mathematics of Operations Research 14:18–25.

Herings, P.J.-J., and V.J. Vannetelbosch (1999), "Refinements of rationalizability for normal-form games", International Journal of Game Theory 28:53–68.

Hillas, J. (1990), "On the definition of the strategic stability of equilibria", Econometrica 58:1365–1390.

Hillas, J. (1998a), "A game illustrating some features of the definition of strategic stability", unpublished.

Hillas, J. (1998b), "How much of 'forward induction' is implied by 'backward induction' and 'ordinality'?", unpublished.

Hillas, J. (1998c), "On the relation between perfect equilibria in extensive form games and proper equilibria in normal form games", unpublished.

Hillas, J., M. Jansen, J. Potters and A.J. Vermeulen (2001a), "Best reply stable equilibria and forward induction: A direct proof", unpublished.

Hillas, J., M. Jansen, J. Potters and A.J. Vermeulen (2001b), "On the relation among some definitions of strategic stability", Mathematics of Operations Research, forthcoming.

Hillas, J., and E. Kohlberg (2000), "On the epistemic conditions for Nash equilibrium", unpublished.

Hillas, J., A.J. Vermeulen and M. Jansen (1997), "On the finiteness of stable sets", International Journal of Game Theory 26:275–278.

Jansen, M.J.M., A.P. Jurg and P.E.M. Borm (1994), "On strictly perfect sets", Games and Economic Behavior 6:400–415.

Jansen, M., and A.J. Vermeulen (2001), "On the computation of stable sets and strictly perfect equilibria", Economic Theory 17:325–344.

Kakutani, S. (1941), "A generalization of Brouwer's fixed-point theorem", Duke Mathematics Journal 8:457–459.

Kalai, E., and D. Samet (1984), "Persistent equilibria in strategic games", International Journal of Game Theory 13:129–144.

Kohlberg, E. (1990), "Refinement of Nash equilibrium: The main ideas", in: T. Ichiishi, A. Neyman and Y. Tauman, eds., Game Theory and Applications (Academic Press, San Diego, CA).

Kohlberg, E., and J.-F. Mertens (1982), "On the strategic stability of equilibria", CORE Discussion Paper 8248, Université Catholique de Louvain, Louvain-la-Neuve, Belgium.

Kohlberg, E., and J.-F. Mertens (1986), "On the strategic stability of equilibria", Econometrica 54:1003–1038.

Kohlberg, E., and P. Reny (1992), "On the rationale for perfect equilibrium", Harvard Business School Working Paper 92–011.

Kohlberg, E., and P. Reny (1997), "Independence on relative probability spaces and consistent assessments in game trees", Journal of Economic Theory 75:280–313.

Kreps, D. (1986), "Out of equilibrium beliefs and out of equilibrium behavior", unpublished.

Kreps, D.M. (1987), "Nash equilibrium", in: J. Eatwell, M. Milgate and P. Newman, eds., The New Palgrave Dictionary of Economics (Norton, New York) 460–482.

Kreps, D.M. (1990), Game Theory and Economic Modeling (Oxford University Press, New York, NY).

Kreps, D.M., P. Milgrom, J. Roberts and R. Wilson (1982), "Rational cooperation in the finitely repeated prisoner's dilemma", Journal of Economic Theory 27:245–252.

Kreps, D.M., and G. Ramey (1987), "Structural consistency, consistency, and sequential rationality", Econometrica 55:1331–1348.

Kreps, D.M., and R. Wilson (1982), "Sequential equilibria", Econometrica 50:863–894.

Kuhn, H.W. (1953), "Extensive games and the problem of information", in: Contributions to the Theory of Games, Vol. 2 (Princeton University Press, Princeton, NJ) 193–216.

Luce, R.D., and H. Raiffa (1957), Games and Decisions: Introduction and Critical Survey (John Wiley & Sons, New York).

Mailath, G.J., L. Samuelson and J.M. Swinkels (1993), "Extensive form reasoning in normal form games", Econometrica 61:273–302.

Mailath, G.J., L. Samuelson and J.M. Swinkels (1994), "Normal form structures in extensive form games", Journal of Economic Theory 64:325–371.

Mailath, G.J., L. Samuelson and J.M. Swinkels (1997), "How proper is sequential equilibrium?", Games and Economic Behavior 18:193–218.

Marx, L.M., and J.M. Swinkels (1997), "Order independence for iterated weak dominance", Games and Economic Behavior 18:219–245.

Marx, L.M., and J.M. Swinkels (2000), "Order independence for iterated weak dominance", Games and Economic Behavior 31:324–329.

Maynard Smith, J., and G.R. Price (1973), "The logic of animal conflict", Nature 246:15–18.

McLennan, A. (1985a), "Justifiable beliefs in sequential equilibrium", Econometrica 53:889–904.

McLennan, A. (1985b), "Subform perfection", unpublished (Cornell University).

McLennan, A. (1989a), "Consistent conditional systems in noncooperative game theory", International Journal of Game Theory 18:141–174.

McLennan, A. (1989b), "Fixed points of contractible valued correspondences", International Journal of Game Theory 18:175–184.

McLennan, A. (1995), "Invariance of essential sets of Nash equilibria", unpublished.

Mertens, J.-F. (1987), "Ordinality in noncooperative games", CORE Discussion Paper 8728, Université Catholique de Louvain, Louvain-la-Neuve, Belgium.

Mertens, J.-F. (1989), "Stable equilibria – a reformulation, Part I: Definition and basic properties", Mathematics of Operations Research 14:575–624.

Mertens, J.-F. (1991a), "Equilibrium and rationality: Context and history dependence", in: K.J. Arrow, ed., Issues in Contemporary Economics, Vol. 1: Markets and Welfare, Proceedings of the Ninth World Congress of the International Economic Association, I.E.A. Conference, Vol. 98, (MacMillan, New York) 198–211.

Mertens, J.-F. (1991b), "Stable equilibria – a reformulation, Part II: Discussion of the definition and further results", Mathematics of Operations Research 16:694–753.

Mertens, J.-F. (1992), "The small worlds axiom for stable equilibria", Games and Economic Behavior 4:553–564.

Mertens, J.-F. (1995), "Two examples of strategic equilibria", Games and Economic Behavior 8:378–388.

Mertens, J.-F., and S. Zamir (1985), "Formulation of Bayesian analysis for games with incomplete information", International Journal of Game Theory 14:1–29.

Moulin, H. (1979), "Dominance solvable voting schemes", Econometrica 47:1337–1351.

Moulin, H., and J.-P. Vial (1978), "Strategically zero-sum games: The class of games whose completely mixed equilibria cannot be improved upon", International Journal of Game Theory 7:1337–1351.

Myerson, R. (1978), "Refinement of the Nash equilibrium concept", International Journal of Game Theory 7:73–80.

Myerson, R. (1986a), "Acceptable and predominant correlated equilibria", International Journal of Game Theory 15:133–154.

Myerson, R. (1986b), "Multistage games with communication", Econometrica 54:323–358.

Myerson, R. (1994), "Communication, correlated equilibria, and incentive compatibility", in: R.J. Aumann and S. Hart, eds., Handbook of Game Theory, Vol. 2 (North-Holland, Amsterdam) Chapter 24, 827–847.

Nash, J. (1950), "Equilibrium points in N-person games", Proceedings of the National Academy of Sciences 36:48–49.

Nash, J. (1951), "Non-cooperative games", Annals of Mathematics 54:286–295.

Nau, R.F., and K.F. McCardle 1990), "Coherent behavior in noncooperative games", Journal of Economic Theory 50:424–444.

Neyman, A. (1999), "Cooperation in repeated games when the number of stages is not commonly known", Econometrica 60:45–64.

Nöldeke, G., and L. Samuelson (1993), "An evolutionary analysis of backward and forward induction", Games and Economic Behavior 5:425–454.

Okada, A. (1987), "Complete inflation and perfect recall in extensive games", International Journal of Game Theory 16:85–91.

Osborne, M. (1990), "Signaling, forward induction, and stability in finitely repeated games", Journal of Economic Theory 50:22–36.

Pearce, D.G. (1984), "Rationalizable strategic behavior and the problem of perfection", Econometrica 52:1029–1050.

Perea y Monsuwé, A., M. Jansen and A.J. Vermeulen (2000), "Player splitting in extensive form games", International Journal of Game Theory 29:433–450.

Polak, B. (1999), "Epistemic conditions for Nash equilibrium, and common knowledge of rationality", Econometrica 67:673–676.

Reny, P.J. (1992a), "Backward induction, normal form perfection and explicable equilibria", Econometrica 60:627–649.

Reny, P.J. (1992b), "Rationality in extensive form games", Journal of Economic Perspectives 6:103–118.

Reny, P.J. (1993), "Common belief and the theory of games with perfect information", Journal of Economic Theory 59:257–274.

Ritzberger, K. (1994), "The theory of normal form games from the differentiable viewpoint", International Journal of Game Theory 23:207–236.

Ritzberger, K. (1999), "Recall in extensive form games", International Journal of Game Theory 28:69–87.

Rosenthal, R. (1981), "Games of perfect information, predatory pricing, and the chain store paradox", Journal of Economic Theory 25:92–100.

Rubinstein, A. (1989), "The electronic mail game: Strategic behavior under 'almost common knowledge' ", American Economic Review 79:385–391.

Rubinstein, A. (1991), "Comments on the interpretation of game theory", Econometrica 59:909–924.

Samet, D. (1996), "Hypothetical knowledge and games with perfect information", Games and Economic Behavior 17:230–251.

Samuelson, L. (1992), "Dominated strategies and common knowledge", Games and Economic Behavior 4:284–313.

Selten, R. (1965), "Spieltheoretische Behandlung eines Oligopolmodells mit NachFraGetragheit", Zeitschrift fur die gesamte Staatswissenschaft 121:301–324, 667–689.

Selten, R. (1975), "Reexamination of the perfectness concept for equilibrium points in extensive games", International Journal of Game Theory 4:25–55.

Selten, R. (1978), "The chain-store paradox", Theory and Decision 9:127–159.

Sorin, S. (1992), "Information and rationality: Some comments", Annales d'Economie et de Statistique 25/26:315–325.

Swinkels, J.M. (1992a), "Evolutionary stability with equilibrium entrants", Journal of Economic Theory 57:306–332.

Swinkels, J.M. (1992b), "Stability and evolutionary stability: From Maynard Smith to Kohlberg and Mertens", Journal of Economic Theory 57:333–342.

Swinkels, J.M. (1993), "Adjustment dynamics and rational play in games", Games and Economic Behavior 5:455–484.

Swinkels, J.M. (1994), "Independence for conditional probability systems", unpublished (Northwestern University).

Tan, T.C.-C., and S.R. da Costa Werlang (1988), "The Bayesian foundations of solution concepts of games", Journal of Economic Theory 45:370–391.

Thompson, F.B. (1952), "Equivalence of games in extensive form", RM 759, The Rand Corporation.

van Damme, E. (1984), "A relation between perfect equilibria in extensive form games and proper equilibria in normal form games", International Journal of Game Theory 13:1–13.

van Damme, E. (1989), "Stable equilibria and forward induction", Journal of Economic Theory 48:476–496.

van Damme, E. (1991), Stability and Perfection of Nash Equilibria, 2nd edn. (Springer-Verlag, Berlin).

van Damme, E. (2002), "Strategic equilibrium", R.J. Aumann and S. Hart, eds., Handbook of Game Theory, Vol. 3 (North-Holland, Amsterdam) Chapter 41, 1521–1596.

Vermeulen, A.J. (1995), Stability in non-cooperative game theory, Ph.D. thesis, Department of Mathematics, University of Nijmegen.

Vermeulen, A.J., and M.J.M. Jansen (1996), "Are strictly perfect equilibria proper? A counterexample", Journal of Optimization Theory and Applications 90:225–230.

Vermeulen, A.J., and M.J.M. Jansen (1997a), "Extending invariant solutions", Games and Economic Behavior 21:135–147.

Vermeulen, A.J., and M.J.M. Jansen (1997b), "On the invariance of solutions of finite games", Mathematical Social Sciences 33:251–267.

Vermeulen, A.J., and M.J.M. Jansen (1998), "The reduced form of a game", European Journal of Operational Research 106:204–211.

Vermeulen, A.J., and M.J.M. Jansen (2000), "Ordinality of solutions of noncooperative games", Journal of Mathematical Economics 33:13–34.

Vermeulen, A.J., M.J.M. Jansen and J.A.M. Potters (1994), "On a method to make solution concepts invariant", unpublished.

Vermeulen, A.J., J.A.M. Potters and M.J.M. Jansen (1996), "On quasi-stable sets", International Journal of Game Theory 25:34–39.

Vermeulen, A.J., J.A.M. Potters and M.J.M. Jansen (1997), "On stable sets of equilibria", in: T. Parthasarathy, B. Dutta and J.A.M. Potters, eds., Game Theoretical Applications to Economics and Operations Research (Kluwer Academic, The Netherlands) 133–148.

von Neumann, J., and O. Morgenstern (1944), Theory of Games and Economic Behavior (Princeton University Press, Princeton, NJ).

Wilson, R. (1992), "Computing simply stable equilibria", Econometrica 60:1039–1070.

Wilson, R. (1995), "Admissibility and stability", unpublished (Stanford University).

Wen-Tsün, W., and J. Jia-He (1962), "Essential equilibrium points of n-person non-cooperative games", Scientia Sinica 10:1307–1322.

Chapter 43

INCOMPLETE INFORMATION*

ROBERT J. AUMANN[†]

Center for Rationality and Institute of Mathematics, The Hebrew University of Jerusalem,
Center for Game Theory, State University of New York at Stony Brook, and Stanford University

AVIAD HEIFETZ

Tel Aviv University and California Institute of Technology

Contents

*Important input from Sergiu Hart, Martin Meier, and Dubi Samet is gratefully acknowledged.
[†]Research partially supported under NSF grant SES-9730205.

Handbook of Game Theory, Volume 3, Edited by R.J. Aumann and S. Hart

Abstract

In interactive contexts such as games and economies, it is important to take account not only of what the players believe about substantive matters (such as payoffs), but also of what they believe about the beliefs of other players. Two different but equivalent ways of dealing with this matter, the semantic and the syntactic, are set forth. Canonical and universal semantic systems are then defined and constructed, and the concepts of common knowledge and common priors formulated and characterized. The last two sections discuss relations with Bayesian games of incomplete information and their applications, and with interactive epistemology – the theory of multi-agent knowledge and belief as formulated in mathematical logic.

Keywords

incomplete or differential information, interactive epistemology, semantic belief systems, syntactic belief systems, common priors

JEL classification: D82, C70

MSC: Primary 03B42, Secondary 91A10; 91A40; 91A80; 91B44

1. Introduction

In interactive contexts such as games and economies, it is important to take account not only of what the players believe about substantive matters (such as payoffs), but also of what they believe about the beliefs of other players. This chapter sets forth several ways of dealing with this matter, all essentially equivalent.

There are two basic approaches, the syntactic and the semantic; while the former is conceptually more straightforward, the latter is more prevalent, especially in game and economic contexts. Each appears in the literature in several variations.

In the *syntactic* approach, the beliefs are set forth explicitly: One specifies what each player believes about the substantive matters in question, about the beliefs of the others about these substantive matters, about the beliefs of the others about the beliefs of the others about the substantive matters, and so on ad infinitum. This sounds – and is – cumbersome and unwieldy, and from the beginning of research into the area, a more compact, manageable way was sought to represent interactive beliefs.

Such a way was found in the *semantic* approach, which is "leaner" and less elaborate, but also less transparent. It consists of a set of *states of the world* (or simply *states*), and for each player and each state, a probability distribution on the set of all states. We will see that this provides the same information as the syntactic approach.

The subject of *common priors* is discussed in Section 9; applications to Game Theory, in Section 10; and finally, Section 11 contains a brief discussion of the theory from the viewpoint of mathematical logic.

2. The semantic approach

Given a set N of players, a (*finite*) *semantic belief system* consists of
  (i) a finite set Ω, the *state space,* whose elements are called *states of the world,* or simply *states,* and whose subsets are called *events*; and
  (ii) for each player i and state ω, a probability distribution $\pi_i(\cdot; \omega)$ on Ω; if E is an event, $\pi_i(E; \omega)$ is called *i's probability for E in ω.*

Conceptually, a "state of the world" is meant to encompass all aspects of reality that are relevant to the matter under consideration, including the beliefs of all players in that state. As in probability theory, an event is a set of states; thus the event "it will snow tomorrow" is represented by a set of states – those in which it snows tomorrow.

We assume that

$$\text{if } \pi_i(\{v\}; \omega) > 0, \text{ then } \pi_i(E; v) = \pi_i(E; \omega) \text{ for all } E. \tag{2.1}$$

In words: If in state ω, player i considers state v possible (in the sense of assigning to it positive probability), then his probabilities for all events are the same in state v as in state ω. That is, he does not seriously entertain the possibility that his own probability for some event E is different from what it actually is; he is sure[1] of his own probabilities.

The restriction to finite systems in this section is for simplicity only; for a general treatment, see Section 6.

3. An example

Here $N = \{Ann, Bob\}$, $\Omega = \{\alpha, \beta, \gamma\}$, and the probabilities $\pi_i(\{v\}; \omega)$ are as follows:

$\omega =$	$v =$	α	β	γ
	α	1/2	1/2	0
	β	1/2	1/2	0
	γ	0	0	1

π_{Ann}

$\omega =$	$v =$	α	β	γ
	α	1	0	0
	β	0	1/2	1/2
	γ	0	1/2	1/2

π_{Bob}

In each of the states α and β, Ann attributes probability $1/2$ to each of α and β, whereas in γ, she knows that γ is the state; and Bob, in each of the states β and γ, attributes probability $1/2$ to each of β and γ, whereas in α, he knows that α is the state.

Now let $E = \{\alpha, \gamma\}$, and let us consider the probabilities of the players in state β. To start with, each player assigns probability $1/2$ to E. At the next level, each player assigns probability $1/2$ to the other assigning probability $1/2$ to E, and probability $1/2$ to the other being sure[2] of E. At the next level, Ann assigns probability $1/2$ to Bob being sure that she assigns probability $1/2$ to E; and probability $1/2$ to Bob's assigning probability $1/2$ to her being sure of E, and probability $1/2$ to her assigning probability $1/2$ to E. The same holds if Ann and Bob are interchanged. That's already quite a mouthful; we refrain from describing the subsequent levels explicitly.

The previous paragraph, with its explicit description of each player's probabilities for the other player's probabilities, is typical of the syntactic approach. The complexity of the description increases exponentially with the "depth" of the level, and there are infinitely many levels. Because this example is particularly symmetric, the description is relatively simple; in general, things are even more complex, by far. Note, moreover, that the syntactic treatment corresponds to just *one* of the states in the semantic model.

By contrast, the semantic model encapsulates the whole mess, simultaneously for *all* the states, in just two compact (3×3) tables.

---

[1] By "sure" we mean "assigns probability 1 to" (probabilists use "almost sure" for this, but here it would make the text unnecessarily cumbersome and opaque). Actually, in these models players "know" their own probabilities; i.e., they do not admit *any* possibility, even with probability 0, of having a different probability (see Section 7). Indeed, a person's own probability judgments would appear to be among the few things of which he can justifiably be absolutely certain.

[2] Indeed, Ann assigns $1/2$–$1/2$ probabilities to the states α and β. In β, Bob assigns probability $1/2$ to E, and in α, Bob is sure of E. This demonstrates our assertion about Ann's probabilities for Bob's probabilities. The calculation of Bob's probabilities for Ann's probabilities is similar.

4. The syntactic approach

The syntactic approach may be formalized by "belief hierarchies". One starts with an exhaustive set of mutually exclusive "states of nature" (like "snow", "rain", "cloudy", or "clear" at noon tomorrow). The first level of the hierarchy specifies, for each player, a probability distribution on the states of nature. The second level specifies, for each player, a joint probability distribution on the states of nature and the others' probability distributions. The third level specifies, for each player, a joint probability distribution on the states of nature and the probability distributions of the others at the first and second levels. And so on. Certain consistency conditions are required [Mertens and Zamir (1985); see Section 8 below].

Since there is a continuum of possibilities for the probability distributions at the first level, the probability distributions at the second and higher levels may well[3] be continuous (i.e., have nonatomic components). So one needs some kind of additional structure – a topology or a measurable structure (σ-field) – on the space of probability distributions at each level. With finitely many states of nature, this offers no difficulty at the first and second levels. But already the third level consists of a probability distribution on the space of probability distributions on a continuum, which requires a topology or measurable structure on the space of probability distributions on a continuum – a nontrivial matter. Of course, this applies also to higher levels.

An alternative syntactic formalism, which avoids these complications, works with sentences[4] or assertions rather than with probability distributions.[5] Start with a set $\{x, y, \ldots\}$ of "natural sentences" (like "rain", "warm", and "humid" at noon tomorrow), which need be neither exhaustive nor mutually exclusive. One may operate on these in two ways: by logical operators and connectives, like "not" ($\neg$), "or" ($\vee$), and "and" ($\wedge$), and by belief operators p_i^α, whose interpretation is "player i attributes probability at least α to ...". The operations may be concatenated in any way one wishes; for example, $p_{Ann}^{1/2}(x \vee p_{Bob}^{3/4}(\neg p_{Ann}^{1/4} y))$ means that Ann ascribes probability at least $1/2$ to the contingency that either x or that Bob ascribes probability at least $3/4$ to her ascribing probability less than $1/4$ to y. A *syntactic belief system* is a collection of such sentences that satisfies certain natural completeness, coherence, and consistency conditions (Section 8).

It may be seen that a syntactic belief system contains precisely the same substantive information as a belief hierarchy, without the topological or measure-theoretic complications.

---

[3] A discrete distribution would mean, e.g., that Ann *knows* that Bob assigns probability precisely 0, 1/2 or 1 to snow. While possible, this seems unlikely. When modelling beliefs about the beliefs of others, some fuzziness seems natural.

[4] Usually called *formulas* in the literature.

[5] Some workers reserve the term "syntactic" for this kind of system only – i.e., one using sentences and syntactic rules. We prefer a more substantive terminology, in which *syntactic* refers to any system that directly describes the actual beliefs of the players, whereas *semantic* refers to a (states-of-the-world) model from which these beliefs can be derived.

5. From semantics to syntax

We now indicate how the syntactic formalism is derived from the semantic one in a general framework.

Suppose given a semantic belief system with state space Ω, and a family of "distinguished" events $X, Y, \ldots$ (corresponding[6] to the above "natural sentences" $x, y, \ldots$); such a pair is called an *augmented* semantic belief system. Let $\pi_i(E; \omega)$ be the probability of player i for event E in state ω. For numbers α between 0 and 1, let $P_i^\alpha E$ be the event that i's probability for E is at least α; i.e., the set of all states ω at which $\pi_i(E; \omega) \geqslant \alpha$. Thus the operator P_i^α in the semantic model corresponds to the operator p_i^α in the syntactic model. As usual, the set operations $\cup$ (union) and $\Omega \backslash$ (complementation) correspond to the logical operations $\vee$ (disjunction) and $\neg$ (negation). Thus each sentence in the syntactic formalism corresponds to an event in the semantic formalism. For example, the sentence $p_{Ann}^{1/2}(x \vee p_{Bob}^{3/4}(\neg p_{Ann}^{1/4} y))$ discussed above corresponds to $P_{Ann}^{1/2}(X \cup P_{Bob}^{3/4}(\Omega \backslash P_{Ann}^{1/4} Y))$, namely the event that Ann ascribes probability at least $1/2$ to the event that either X or that Bob ascribes probability at least $3/4$ to her ascribing probability less than $1/4$ to Y.

Now fix a state ω. If e is a sentence and E the corresponding event, say that e *holds in* ω if ω is in E. If we think of ω as the "true" state of the world, then the "true" sentences are precisely those that hold in ω. These sentences constitute a syntactic belief system in the sense of the previous section.

To summarize: Starting from a pair consisting of an augmented semantic belief system and a state ω, we have constructed a syntactic belief system $\mathfrak{L}$. We call this pair (or just the state ω) a *model* for $\mathfrak{L}$.

6. Removing the finiteness restriction

Given a set N of players, a *(general) semantic belief system* consists of
  (ia) a set Ω, the *state space,* whose elements are the *states*;
  (ib) a σ-field $\mathcal{F}$ of subsets of Ω – the *events*; and
  (ii) for each player i and state ω, a probability distribution $\pi_i(\cdot; \omega)$ on $\mathcal{F}$.
We assume that

$$\pi_i(E; \omega) \text{ is } \mathcal{F}\text{-measurable in } \omega \text{ for each fixed event } E, \tag{6.1}$$

and

$$\pi_i\big(\{v: \pi_i(E; v) \neq \pi_i(E; \omega)\}; \omega\big) = 0 \text{ for each event } E \text{ and state } \omega. \tag{6.2}$$

The interpretation is as in Section 2.

---

[6] I.e., with the same conceptual content. Thus if in the syntactic treatment, x is the sentence "it will snow tomorrow", then in the semantic treatment, X is the event "it will snow tomorrow" (the set of all states in which it snows tomorrow).

7. Knowledge and common knowledge

In the formalism of Sections 2 and 6, the concept of knowledge – in the sense of absolute certainty[7] rather than probability 1 – plays no explicit role. However, this concept can be *derived* from that formalism. Indeed, for each state ω and player i, let $\mathbf{I}_i(\omega)$ be the set of all states in which i's probabilities (for all events) are the same as in ω; call it *i's information set in ω*. The information sets of i form a partition $\mathcal{P}_i$ of Ω, called *i's information partition*. Say that *i knows* an event E in a state ω if E includes $\mathbf{I}_i(\omega)$, and denote by $K_i E$ the set of all states ω in which i knows E. Conceptually, this formulation of knowledge presupposes that players know their own probabilities with absolute certainty, which is not unreasonable. It may be seen that

$$K_i E \subset P_i^1 E, \tag{7.1}$$

i.e., that players ascribe probability 1 to whatever they know, and that

$$P_i^\alpha E \subset K_i P_i^\alpha E, \tag{7.2}$$

which formally expresses the principle, enunciated above, that players know their own probabilities.

Let $K E$ be the set $\bigcap_i K_i E$ of all states in which all players know E, and set

$$K^\infty E := K E \cap K K E \cap K K K E \cap \cdots;$$

thus $K^\infty E$ is the set of all states in which all players know E, all players know that, all players know *that,* and so on ad infinitum. We say that E is *commonly known in ω* if $\omega \in K^\infty E$ [Lewis (1969)]; for a comprehensive survey of common knowledge, see Geanakoplos, this *Handbook,* Volume 2, Chapter 40. If $\mathcal{P}^\infty$ is the meet (finest common coarsening) of the information partitions $\mathcal{P}_i$ of all the players i, then it may be seen [Aumann (1976, 1999a)] that $K^\infty E$ is the union of all the atoms of $\mathcal{P}^\infty$ that are included in E. The atoms of $\mathcal{P}^\infty$ are called *common knowledge components* (or simply components) of Ω. In a sense, one can always restrict the discussion to such a component: The "true" state is always in some component, and then it is commonly known that it is, and all considerations relating to other components become irrelevant.

Some interactive probability formalisms [like Aumann (1999b)] use a separate, exogenous, concept of knowledge, and *assume* analogues of (7.1) and (7.2) (see (7.3) and (7.4) below). In the semantic framework, such a concept is redundant: Surprisingly, *knowledge is implicit in probability.*

---

[7] I.e., with error impossible: If i "knows" an event E at a state ω, then ω *must* be in E.

To see this, let $\mathcal{P}_i$, $\mathbf{I}_i$, and K_i, as above, be the knowledge concepts derived from the probabilities $\pi_i(\cdot; \omega)$, and let $\mathcal{Q}_i$, $\mathbf{J}_i$, and L_i be the corresponding exogenous concepts. Assume

$$L_i E \subset P_i^1 E \text{ (players ascribe probability 1 to what they know}^8\text{),} \tag{7.3}$$

and

$$P_i^\alpha E \subset L_i P_i^\alpha E \text{ (players know their own probabilities).} \tag{7.4}$$

It suffices to show that $\mathbf{J}_i(\omega) = \mathbf{I}_i(\omega)$ for all ω. This follows from (7.3) and (7.4); we argue from the verbal formulations. If $\nu \in \mathbf{J}_i(\omega)$, then by (7.4), i's probabilities for all events must be the same in ν and ω, so $\nu \in \mathbf{I}_i(\omega)$. Conversely, i knows in ν that he is in $\mathbf{J}_i(\nu)$, so by (7.3), his probability for $\mathbf{J}_i(\nu)$ is 1. So if $\nu \notin \mathbf{J}_i(\omega)$, then $\mathbf{J}_i(\omega)$ and $\mathbf{J}_i(\nu)$ are disjoint, so in ν his probability for $\mathbf{J}_i(\omega)$ is 0, while in ω it is 1. So his probabilities are different in ω and ν, so $\nu \notin \mathbf{I}_i(\omega)$. Thus $\nu \in \mathbf{J}_i(\omega)$ if and only if $\nu \in \mathbf{I}_i(\omega)$, as claimed.

8. Canonical semantic systems

One may think of a semantic belief system purely technically, simply as providing the setting for a convenient, compact representation of a syntactic belief system, as in the example of Section 3. But one may also think of it substantively, as setting forth a model of interactive epistemology that reflects reality, including those states of the world that are "actually" possible,[9] and what the players know about them. This substantive viewpoint raises several questions, foremost among them being what the players know about the model itself. Does each know the space of states? Does he know the others' probabilities (in each state)? If so, from where does this knowledge derive? If not, how can the formalism indicate what each player believes about the others' beliefs? For example, why would the event $P_{Ann}^{1/2} P_{Bob}^{3/4} E$ then signify that Ann ascribes probability at least $1/2$ to Bob ascribing probability at least $3/4$ to E?

More generally, the whole idea of "state of the world", and of probabilities that accurately reflect the players' beliefs about other players' beliefs, is not transparent. What *are* the states? Can they be explicitly described? Where do they come from? Where do the probabilities come from? What justifies positing this kind of model, and what justifies a particular array of probabilities?

One way of overcoming these problems is by means of "canonical" or "universal" semantic belief systems. Such a system comprises a standard state space with standard probabilities for each player in each state. The system does not depend on reality; it is a

---

[8] In the remainder of this section we use "know" in the exogenous sense.
[9] Philosophers like the term "possible worlds".

framework, it fits any reality, so to speak, like the frames that one buys in photo shops, which do not depend on who is in the photo – they fit any photo with any subject, as long as the size is right. In brief, there is no substantive information in the system.

There are basically two ways of doing this. Both depend on two parameters: the set (or simply number n) of players and the set (or simply number) of "natural" eventualities. The first way [Mertens and Zamir (1985)] is hierarchical. The zero'th level of the hierarchy is a (finite) set $\mathfrak{X} := \{x, y, \ldots\}$, whose members represent mutually exclusive and exhaustive "states of nature", but formally are just abstract symbols. The first level H^1 is the Cartesian product of $\mathfrak{X}$ with n copies of the $\mathfrak{X}$-simplex (the set of all probability distributions on $\mathfrak{X}$); a point in this Cartesian product describes the "true" state of nature, and also the probabilities of each player for each state of nature.

A point h^2 at the second level H^2 consists of a first-level point h^1 together with an n-tuple of probability distributions on H^1, one for each player; this describes the "true" state of nature, the probabilities of each player for the state of nature, and what each player believes about the other players' beliefs about the state of nature. There are two "consistency" conditions: First, the distribution on H^1 that h^2 assigns to i must attribute[10] probability 1 to the distribution on $\mathfrak{X}$ that h^1 assigns to i; that is, i must know what he himself believes. Second, the distribution[11] on $\mathfrak{X}$ that h^2 assigns to i must coincide with the one that h^1 assigns to i.

A third-level point consists of a second-level point h^2 together with an n-tuple of probability distributions[12] on H^2, one for each player; again, i's distribution on H^2 must assign probability 1 to the distribution on H^1 that h^2 assigns to i, and its marginal on H^1 must coincide with the distribution on H^1 that h^1 assigns to i.

And so on. Thus an $(m + 1)$th-level point h^{m+1} is a pair consisting of an mth-level point h^m and a distribution on H^m; we say that h^{m+1} *elaborates* h^m. Define a state h in the canonical semantic belief system H as a sequence $h^1, h^2, \ldots$ in which each term elaborates the previous one. To define the probabilities, proceed as follows: For each subset S^m of H^m, set $\widehat{S^m} = \{h \in H: h^m \in S^m\}$; let $\mathcal{F}$ be the σ-field generated by all the $\widehat{S^m}$, where S^m is measurable; and let $\pi_i(\widehat{S^m}; h)$ be the probability that i's component of h^{m+1} assigns to S^m. In words, a state consists of a sequence of probability distributions of ever-increasing depth; to get the probability of a set of such sequences in a given state, one simply reads off the probabilities specified by that state. That this indeed yields a probability on $\mathcal{F}_\sigma$ follows from the Kolmogorov extension theorem, which requires some topological assumptions.[13]

The second construction [Heifetz and Samet (1998); Aumann (1999b)] of a canonical belief system is based on sentences. It uses an "alphabet" $x, y, \ldots$, as well as more

---

[10] Technically, h^2 assigns to i a distribution on $H^1 = \mathfrak{X} \times \Delta(\mathfrak{X})$, where Δ denotes "the set of all distributions on ...". The first consistency condition says that the marginal of this distribution on the second factor (i.e., on $\Delta(\mathfrak{X})$) is a unit mass concentrated on the distribution on $\mathfrak{X}$ that h^1 assigns to i.

[11] Technically, the marginal on $\mathfrak{X}$ of the distribution on $H^1 = \mathfrak{X} \times \Delta(\mathfrak{X})$ that h^2 assigns to i. See the previous footnote.

[12] As indicated in Section 5, this involves topological issues, which we will not discuss.

[13] Without which things might not work [Heifetz and Samet (1998)].

elaborate sentences constructed from the alphabet by means of logical operations and probability operators p_i^α (as in Section 4). The letters of the alphabet represent "natural sentences", not necessarily mutually exclusive or exhaustive; but formally, as above, they are just abstract symbols. In this construction, the states γ in the canonical space Γ are simply lists of sentences; specifically, lists that are complete, coherent, and closed in a sense presently to be specified. The point is that a state is determined – or better, *defined* – by the sentences that hold there; conceptually, the state *is* simply what happens.

More precisely: A list is *complete* if for each sentence f, it contains either f itself or its negation $\neg f$; *coherent,* if for each f, it does not contain both f and $\neg f$; and *closed,* if it contains every logical consequence[14] of the sentences in it.

This defines the states γ in the canonical semantic belief system; to complete the definition of the system, we must define the σ-field $\mathcal{F}$ of events and the probabilities $\pi_i(\cdot; \gamma)$. For any sentence f, let $E_f := \{\delta \in \Gamma: f \in \delta\}$; that is, E_f is the set of all states[15] in the canonical system that contain f. We define $\mathcal{F}$ as the σ-field generated by all the events E_f for all sentences f. Then to define i's probabilities on $\mathcal{F}$ in the state γ, it suffices to define his probabilities for each E_f. Since the state is simply the list of all sentences that "hold in that state", it follows that E_f is the set of all states in which f holds – in brief, the event that f holds. The probability that i assigns to E_f in the state γ is then implicit in the specification of γ; namely, it is the supremum of all the rational α for which $p_i^\alpha f$ is in γ – the supremum α such that in γ, player i assigns probability at least α to f. This defines $\pi_i(E_f; \gamma)$ for each i, γ, and f; and since the E_f generate $\mathcal{F}$, it follows that it defines[16] $\pi_i(\cdot; \gamma)$. This completes the second construction, which has the advantage of requiring no topological machinery.

The canonical semantic belief system Γ is *universal* in a sense that we now describe. Let $x, y, \ldots$ be letters in an alphabet, and Ω, Ω' semantic belief systems with the same players, augmented by "distinguished" events $X, Y, \ldots$ and $X', Y', \ldots$, corresponding to the letters $x, y, \ldots$ (simply "systems" for short). A mapping M from Ω into Ω' is called a *belief morphism* if $M^{-1}(X') = X$, $M^{-1}(Y') = Y, \ldots$, and $\pi_i(M^{-1}(E'); \omega) = \pi_i'(E'; M(\omega))$ for all players i, events E' in Ω', and states ω in Ω; i.e., if it preserves the relevant aspects of the system (as usual for morphisms). A system Υ is called *universal* if for each system Ω, there is a belief morphism from Ω into Υ.

---

[14] As used here, the term "logical consequence" is *defined* purely syntactically, by axioms and inference rules [Meier (2001)]. But it may be *characterized* semantically: A sentence g is a logical consequence of sentences $f_1, f_2, \ldots$ if and only if g holds at any state of any augmented semantic belief system at which $f_1, f_2, \ldots$ hold; i.e., if any model for $f_1, f_2, \ldots$ is also a model for g [Meier (2001)]. The axiomatic system is infinitary, as indeed it must be: For example, $p_i^1 f$ is a logical consequence of $p_i^{1/2} f, p_i^{3/4} f, p_i^{7/8} f, \ldots$, but of no finite subset thereof. Meier's beautiful, path-breaking paper is not yet published.

[15] Recall that a state is a list of sentences.

[16] That is, since the E_f generate $\mathcal{F}$, there cannot be two different probability measures on $\mathcal{F}$ with the given values on the E_f. That there is *one* such measure follows from a theorem of Caratheodory on extending measures from a field to the σ-field it generates.

To see that Γ is universal, let Ω be a system. Then for each state ω in Ω, the family $D(\omega)$ of sentences that hold in ω (see Section 5) is a state in the canonical space Γ, and the mapping $D : \Omega \to \Gamma$ is a belief morphism – indeed the only one [Heifetz and Samet (1998)]. The construction of Mertens and Zamir (1985) also leads to a universal system, and the two canonical systems are isomorphic.[17]

At the end of the previous section, we noted that a separate, exogenous notion of knowledge is semantically redundant. Syntactically, it is not. To understand why, consider two two-player semantic belief systems, Ω and Ω'. The first has just one state ω, to which, of course, both players assign probability 1. The second has two states, ω and ω'. In ω, both players assign probability 1 to ω; in ω', Ann assigns probability 1 to ω, whereas Bob assigns probability 1 to ω'. Now augment both systems by assigning x to ω. Then the syntactic belief system derived from ω in Ω – the family $D(\omega)$ of all sentences that hold there – is identical to that derived from ω in Ω'. So syntactically, the two situations are indistinguishable. But semantically, they are different: In Ω, Ann knows *for sure* in ω that Bob assigns probability 1 to x; in Ω', she does not. There is no way to capture this syntactically without explicitly introducing knowledge operators k_i for the various players i [as in Aumann (1999b); see also (1998)]. In particular, the canonical semantic system Γ has just one component[18] – the whole space.

9. Common priors

Call a semantic belief system *regular*[19] if it is finite, has a single common knowledge component, and in each state, each player assigns positive probability to that state. A *common prior* on a regular semantic belief system Ω is defined as a probability distribution π on Ω such that

$$\pi\big(E \cap \mathbf{I}_i(\omega)\big) = \pi_i(E; \omega)\pi\big(\mathbf{I}_i(\omega)\big) \tag{9.1}$$

for each event E, player i, and state ω; that is, the personal probabilities π_i in a state ω are obtained from the common prior π by conditioning on each player's private information in that state. In Section 3, for example, the distribution that assigns probability $1/3$ to each of the three states is the unique common prior. Common priors can easily fail to exist; for example, if Ω has just two states, in each of which Ann assigns $1/2$–$1/2$ probabilities to the two states, and Bob, $1/3$–$2/3$ probabilities. Conceptually, existence of a common prior signifies that differences in probability assessments are due to differences in information only [Aumann (1987, 1998)]; that people who have always been fed precisely the same information will not differ in their probability assessments.

---

[17] I.e., there is a one-one belief morphism from the one onto the other.

[18] Except when there is only one player. This is because the probability syntax can only express beliefs of the players, not that a player knows anything about another one *for sure*.

[19] Regularity enables uncluttered statements of the definitions and results. No real loss of generality is involved.

With a common prior, players cannot "agree to disagree" about probabilities. That is, if it is commonly known in some state that for some specific α, β, and E, Ann and Bob assign probabilities α and β respectively to the event E, then $\alpha = \beta$. This is the "Agreement Theorem" [Aumann (1976)].

The question arises [Gul (1998)] whether the existence of a common prior can be characterized syntactically. That is, can one characterize those syntactic belief systems $\mathfrak{L}$ that are associated with common priors?[20] By a *syntactic* characterization, we mean one couched directly in terms of the sentences in $\mathfrak{L}$.

The answer is "yes". Indeed, there are two totally different characterizations. The first is based on the following, which is both a generalization of and a converse to the agreement theorem.

PROPOSITION 9.2 [Morris (1994); Samet[21] (1998b); Feinberg (2000)].[22] *Let Ω be a regular semantic belief system, ω a state in Ω. Suppose first that there are just two players, Ann and Bob. Then there is no common prior on Ω if and only if there is a random variable[23] $\mathbf{x}$ for which it is commonly known in ω that Ann's expectation of $\mathbf{x}$ is positive and Bob's is negative.[24]*

With n players, there is no *common prior on Ω if and only if there are n random variables $\mathbf{x}_i$ identically[25] summing to 0 for whom it is commonly known in ω that each player i's expectation of $\mathbf{x}_i$ is positive.*

In the two-person case, one may think of $\mathbf{x}$ as the amount of a bet between Ann and Bob, possibly with odds that depend on the outcome. Thus the proposition says that with a common prior, it is impossible for both players to expect to gain from such a bet. The interpretation in the n-person case is similar; one may think of pari-mutuel betting in horse-racing, the track being one of the players.

In brief, common priors exist if and only if it is commonly known that you can't make something out of nothing: I.e., in a situation that is objectively zero-sum, it cannot be commonly known that all sides expect to gain.

The importance of this proposition lies in that it is formulated in terms of one single state ω only. We are talking about what Ann expects *in that state,* what she thinks *in that*

---

[20] More precisely, those $\mathfrak{L}$ for which there is a state ω in some finite augmented semantic belief system with a common prior such that $\mathfrak{L}$ is the collection of sentences holding at ω.

[21] Samet's proof, which uses the elementary theory of convex polyhedra, is the briefest and most elegant.

[22] For an early related result, see Nau and McCardle (1990).

[23] A real function on Ω. Although the concept of random variable may seem essentially semantic, it is not; in an augmented semantic system, random variables can in general be expressed in terms of sentences [Feinberg (2000); Heifetz (2001)].

[24] Explicitly, the commonly known event in question is

$$\left\{ v: \sum_{\xi \in \Omega} \pi_{Ann}(\{\xi\}; v)\mathbf{x}(\xi) > 0 > \sum_{\xi \in \Omega} \pi_{Bob}(\{\xi\}; v)\mathbf{x}(\xi) \right\}.$$

[25] At each state in Ω.

state that Bob may expect, what she thinks *in that state* that Bob thinks that she may expect, and so on; and similarly for Bob. That is precisely what is needed for a syntactic characterization.[26]

Actually to derive a fully syntactic characterization from this proposition is, however, a different matter. To do so, one must provide syntactic formulations of (i) common knowledge, (ii) random variables and their expectations, and (iii) finiteness of the state space. The first is not difficult; setting $k := \bigwedge_i k_i$, call a sentence e *syntactically commonly known* if $e, ke, kke, kkke, \dots$ all obtain. But (ii) and (iii), though doable – indeed elegantly [Feinberg (2000); Heifetz (2001)] – are more involved; we will not discuss the matter further here.

Proposition 9.2 holds also for infinite state spaces that are "compact" in a natural sense. As before, this enables a syntactic characterization in the compact case; to do this properly one must characterize compactness syntactically. Without compactness, the proposition fails. Feinberg (2000) discusses these matters fully.

The second syntactic characterization of common priors is in terms of iterated expectations. Again, consider first the two-person case. If v is a state and $\mathbf{y}$ a random variable, then Ann's and Bob's expectations of $\mathbf{y}$ in v are

$$(\mathbf{E}_{Ann}\mathbf{y})(v) := \sum_{\xi \in \Omega} \pi_{Ann}(\{\xi\}; v)\mathbf{y}(\xi) \quad \text{and}$$

$$(\mathbf{E}_{Bob}\mathbf{y})(v) := \sum_{\xi \in \Omega} \pi_{Bob}(\{\xi\}; v)\mathbf{y}(\xi).$$

Both $\mathbf{E}_{Ann}\mathbf{y}$ and $\mathbf{E}_{Bob}\mathbf{y}$ are themselves random variables, as they are functions of the state v. So one may form the iterated expectations

$$\mathbf{E}_{Ann}\mathbf{y}, \, \mathbf{E}_{Bob}\mathbf{E}_{Ann}\mathbf{y}, \, \mathbf{E}_{Ann}\mathbf{E}_{Bob}\mathbf{E}_{Ann}\mathbf{y}, \, \mathbf{E}_{Bob}\mathbf{E}_{Ann}\mathbf{E}_{Bob}\mathbf{E}_{Ann}\mathbf{y}, \dots \tag{9.3}$$

and

$$\mathbf{E}_{Bob}\mathbf{y}, \, \mathbf{E}_{Ann}\mathbf{E}_{Bob}\mathbf{y}, \, \mathbf{E}_{Bob}\mathbf{E}_{Ann}\mathbf{E}_{Bob}\mathbf{y}, \, \mathbf{E}_{Ann}\mathbf{E}_{Bob}\mathbf{E}_{Ann}\mathbf{E}_{Bob}\mathbf{y}, \dots. \tag{9.4}$$

PROPOSITION 9.5 [Samet (1998a)]. *Each of these two sequences* (9.3) *and* (9.4) *of random variables converges to a constant (independent of the state); the system has a common prior if and only if for each random variable* $\mathbf{y}$*, these two constants coincide; and in that case, the common value of the two constants is the expectation of* $\mathbf{y}$ *over* Ω *w.r.t. the common prior* π.

The proof uses finite state Markov chains.

---

[26] Recall that we use the term *syntactic* rather broadly, as referring to anything that "actually obtains", such as the beliefs – and hence expectations – of the players. See Footnote 5.

Samet illustrates the proposition with a story about two stock analysts. Ann has an expectation for the price $\mathbf{y}$ of IBM in one month from today. Bob does not know Ann's expectation, but has some idea of what it might be; his expectation of her expectation is $\mathbf{E}_{Bob}\mathbf{E}_{Ann}\mathbf{y}$. And so on.

Like the previous characterization of common priors (Proposition 9.2), this one appears semantic, but in fact is syntactic: All the iterated expectations can be read off from a single syntactic belief system. As before, actually deriving a fully syntactic characterization requires formulating finiteness of the state space syntactically, which we do not do here. Unlike the previous characterization, this one has not been established for compact state spaces.

With n players, one considers arbitrary infinite sequences $\mathbf{E}_{i_1}\mathbf{y}$, $\mathbf{E}_{i_2}\mathbf{E}_{i_1}\mathbf{y}$, $\mathbf{E}_{i_3}\mathbf{E}_{i_2}\mathbf{E}_{i_1}\mathbf{y}$, $\mathbf{E}_{i_4}\mathbf{E}_{i_3}\mathbf{E}_{i_2}\mathbf{E}_{i_1}\mathbf{y}, \ldots$, where $i_1, i_2, \ldots$ is any sequence of players (with repetitions, of course), the only requirement being that each player appears in the sequence infinitely often. The result says that each such sequence converges to a constant; the system has a common prior if and only if all the constants coincide; and in that case, the common value of the constants is the expectation of $\mathbf{y}$ over Ω w.r.t. the common prior π.

10. Incomplete information games

The theory of incomplete information, whose general, abstract form is outlined in the foregoing sections of this chapter, has historical roots in a concrete application: games. Until the mid-sixties, game theorists did not think carefully about the informational underpinnings of their analyses. Luce and Raiffa (1957) did express some malaise on this score, but left the matter at that. The primary problem was each player's uncertainty about the payoff (or utility) functions of the others; to a lesser extent, there was also concern about the players' uncertainty about the strategies available to others, and about their own payoffs.

In path-breaking research in the mid-sixties, for which he got the Nobel Prize some thirty years later, John Harsanyi (1967-8) succeeded both in formulating the problem precisely, and in solving it. In brief, the formulation is the syntactic approach, whereas the solution is the semantic approach. Let us elaborate.

Harsanyi started by noting that though usually the players do not know the others' payoff functions, nevertheless, as good Bayesians, each has a (subjective) probability distribution over the possible payoff functions of the others. But that is not enough to analyze the situation. Each player must also take into account what the others think that he thinks about them. Even there it does not end; he must also take into account what the others think that he thinks that they think about him. And so on, ad infinitum. Harsanyi saw this infinite regress as a Gordian knot, not given to coherent, useful analysis.

To cut the knot, Harsanyi invented the notion of "type". Consider the case of two players, Ann and Bob. Each player, he said, may be one of several types. The type of a player determines his payoff function, and also a probability distribution on the other player's possible types. Since Bob's type determines his payoff function, Ann's probability distribution on his types induces a probability distribution on his payoff functions. But it also induces a probability distribution on his probability distributions on Ann's types, and so on Ann's payoff functions. And so on. Thus a "type structure" yields the whole infinite regress of payoff functions, distributions on payoff functions, distributions on distributions on payoff functions, and so on.

The reader will realize that the "infinite regress" is a syntactic belief hierarchy in the sense of Section 6, the "states of nature" being n-tuples of payoff functions; whereas a "type structure" is a semantic belief system, the "states of the world" being n-tuples of types. In modern terms, Harsanyi's insight was that a semantic system yields a syntactic system. Though this may seem obvious today (see Section 3), it was far from obvious at the time; indeed it was a major conceptual breakthrough, which enabled extending many of the fundamental concepts of game theory to the incomplete information case, and led to the opening of entirely new areas of research (see below).

The converse – that *every* syntactic system can be encapsulated in a semantic system – was not proved by Harsanyi. Harsanyi argued for the type structure on intuitive grounds. Roughly speaking, he reasoned that every player's brain must be configured in some way, and that this configuration should determine both the player's utility or payoff function and his probability distribution on the configuration of other players' brains.

The *proof* that indeed every syntactic system can be encapsulated in a semantic system is outlined in Section 8 above; the semantic system in question is simply the canonical one. This proof was developed over a number of years by several workers [Armbruster and Böge (1979); Böge and Eisele (1979)], culminating in the work of Mertens and Zamir (1985) cited in Section 8 above.

Also the assumption of common priors (Section 9) was introduced by Harsanyi (1967-8), who called this the *consistent* case. He pointed out that in this case – and *only* in this case – an n-person game of incomplete information in strategic (i.e., normal) form can be represented by an n-person game of complete information in extensive form, as follows: First, "chance" chooses an n-tuple of types (i.e., a state of the world), using the common prior π for probabilities. Then, each player is informed of his type, but not of the others' types. Then the n players simultaneously choose strategies. Finally, payoffs are made in accordance with the types chosen by nature and the strategies chosen by the players.

The concept of strategic ("Nash") equilibrium generalizes to incomplete information games in a natural way. Such an equilibrium – often called a *Bayesian Nash* equilibrium – assigns to each type t_i of each player i a (mixed or pure) strategy of i that is optimal for i given t_i's assessment of the probabilities of the others' types and the strategies that the equilibrium assigns to those types. In the consistent (common prior) case, the Bayesian Nash equilibria of the incomplete information strategic game are precisely the same as the ordinary Nash equilibria of the associated complete information extensive game (described in the previous paragraph).

Since Harsanyi's seminal work, the theory of incomplete information games has been widely developed and applied. Several areas of application are of particular interest.

Repeated games of incomplete information deal with situations where the same game is played again and again, but the players have only partial information as to what it is. This is delicate because by taking advantage of his private information, a player may implicitly reveal it, possibly to his detriment. For surveys up to 1991, see this *Handbook*, Volume I, Chapter 5 (Zamir) and Chapter 6 (Forges). Since 1992, the literature on this subject has continued to grow; see Aumann and Maschler (1995), whose bibliography is fairly complete up to 1994. A more complete and modern treatment,[27] unfortunately as yet unpublished, is Mertens, Sorin, and Zamir (1994).

Other important areas of application include auctions (see Wilson, this *Handbook*, Volume I, Chapter 8), bargaining with incomplete information (Binmore, Osborne, and Rubinstein I,7 and Ausubel, Cramton, and Deneckere III,50), principal-agent problems (Dutta and Radner II,26), inspection (Avenhaus, von Stengel, and Zamir III,51), communication and signalling (Myerson II,24 and Kreps and Sobel II,25), and entry deterrence (Wilson I,10).

Not surveyed in this *Handbook* are coalitional (cooperative) games of incomplete information. Initiated by Wilson (1978) and Myerson (1984), this area is to this day fraught with unresolved conceptual difficulties. Allen (1997) and Forges, Minelli and Vohra (2001) are surveys.

Finally, games of incomplete information are useful in understanding games of *complete* information – ordinary, garden-variety games. Here the applications are of two kinds. In one, the given complete information game is "perturbed" by adding some small element of incomplete information. For example, Harsanyi (1973) uses this technique to address the question of the significance of mixed strategies: Why would a player wish to randomize, in view of the fact that whenever a mixed strategy μ is optimal, it is also optimal to use any pure strategy in the support of μ? His answer is that indeed players never actually use mixed strategies. Rather, even in a complete information game, the payoffs should be thought of as commonly known only approximately. In fact, there are small variations in the payoff to each player that are known only to that player himself; these small variations determine which pure strategy s in the support of μ he actually plays. It turns out that the probability with which a given pure strategy s is actually played in this scenario approximates the coefficient of s in μ. Thus, a mixed strategy of a player i appears not as a deliberate randomization on i's part, but as representing the estimate of *other* players as to what i will do.

Another application of this kind comes under the heading of reputational effects. Seminal in this genre was the work of the "gang of four" [Kreps, Milgrom, Roberts, and Wilson (1982)], in which a repeated prisoner's dilemma is perturbed by assuming that with some arbitrarily small exogenous probability, the players are "irrational" automata who always play "tit-for-tat". It turns out that in equilibrium, the irrationality "takes over" in some sense: Almost until the end of the game, the rational players themselves play tit-for-tat. Fudenberg and Maskin (1986) generalize this to "irrational"

---

[27] Including also the general theory of incomplete information covered in this chapter.

strategies other than tit-for-tat; as before, equilibrium play mimics the irrational perturbation, even when (unlike tit-for-tat) it is inefficient. Aumann and Sorin (1989) allow any perturbation with "bounded recall"; in this case equilibrium play "automatically" selects an efficient equilibrium.

The second kind of application of incomplete information technology to complete information games is where the object of incomplete information is not the payoffs of the players but the actual strategies they use. For example, rather than perturbing payoffs a la Harsanyi (1973), see above, one can say that even without perturbed payoffs, players other than i simply do not know what pure strategy i will play; i's mixed strategy represents the probabilities of the other players as to what i will do. Works of this kind include Aumann (1987), in which correlated equilibrium in a complete information game is characterized in terms of common priors and common knowledge of rationality; and Aumann and Brandenburger (1995), in which Nash equilibrium in a complete information game is characterized in terms of mutual knowledge of rationality and of the strategies being played, and when there are more than two players, also of common priors and common knowledge of the strategies being played (for a more detailed account, see Hillas and Kohlberg, this *Handbook,* Volume III, Chapter 42).

The key to *all* these applications is Harsanyi's "type" definition – the semantic representation – without which building a workable model for applications would be hopeless.

11. Interactive epistemology

Historically, the theory of incomplete information outlined in this chapter has two parents: (i) incomplete information games, discussed in the previous section, and (ii) that part of mathematical logic, sometimes called *modal* logic, that treats knowledge and other epistemological issues, which we now discuss. In turn, (ii) itself has various ancestors. One is probability theory, in which the "sample space" (or "probability space") and its measurable subsets (there, like here, "events") play a role much like that of the space Ω of states of the world; whereas subfields of measurable sets, as in stochastic processes, play a role much like that of our information partitions $\mathcal{P}_i$ (Section 7). Another ancestor is the theory of extensive games, originated by von Neumann and Morgenstern (1944), whose "information sets" are closely related to our information sets $\mathbf{I}_i(\omega)$ (Section 7).

Formal epistemology, in the tradition of mathematical logic, began some forty years ago, with the work of Kripke (1959) and Hintikka (1962); these works were set in a single-person context. Lewis (1969) was the first to define common knowledge, which of course is a multi-person, interactive concept; though verbal, his treatment was entirely rigorous. Computer scientists became interested in the area in the mid-eighties; see Fagin et al. (1995).

Most work in formal epistemology concerns knowledge (Section 7) rather than probability. Though the two have much in common, knowledge is more elementary, in a sense that will be explained presently. Both interactive knowledge and interactive probability can be formalized in two ways: semantically, by a states-of-the-world model, or syntactically, either by a hierarchic model or by a formal language with sentences and logical inference governed by axioms and inference rules. The difference lies in the nature of logical inference. In the case of knowledge, sentences, axioms, and inference rules are all finitary [Fagin et al. (1995); Aumann (1999a)]. But probability is essentially infinitary (Footnote 14); there is no finitary syntactic model for probability.

Nevertheless, probability *does* have a finitary aspect. Heifetz and Mongin (2001) show that there is a finitary system $\mathcal{A}$ of axioms and inference rules such that if f and g are *finitary* probability sentences then g is a logical consequence[28] of f if and only if it is derivable from f via the *finitary* system $\mathcal{A}$. That is, though in general the notion of "logical consequence" is infinitary, for finitary sentences it *can* be embodied in a finitary framework.

Finally, we mention the matter of backward induction in perfect information games, which has been the subject of intense epistemic study, some of it based on probability 1 belief. This is a separate area, which should have been covered elsewhere in the *Handbook*, but is not.

Appendix: Limitations of the syntactic approach (by Aviad Heifetz)

The interpretation of the universal model of Section 8 as "canonical", with the association of beliefs to states "self-evident" by virtue of the states' inner structure, relies nevertheless on the non-trivial assumption that the players' beliefs cannot be but σ-additive. This is highlighted by the following example.

In the Mertens and Zamir (1985) construction mentioned in Section 8, we shall focus our attention on a subspace Ω whose states ω can be represented in the form

$$\omega = (s, \, i_1 i_2 i_3 \ldots, \, j_1 j_2 j_3 \ldots)$$

where each digit s, i_k, j_k may assume the value 0 or 1. The state of nature is s. The sequences $i_1 i_2 i_3 \ldots, \, j_1 j_2 j_3 \ldots$ encode the beliefs of the players i and j, respectively, in the following way. If the sequence starts with 1, the player assigns probability 1 to the true state of nature s in ω. Otherwise, if her sequence starts with 0, she assigns probabilities half-half to the two possible states of nature. Inductively, suppose we have already described the beliefs of the players up to level n. If $i_{n+1} = 1$, i believes with probability 1 that the nth digit of j equals the actual value of j_n in ω, and otherwise – if $i_{n+1} = 0$ – she assigns equal probabilities to each of the possible values of this digit, 0 or 1. This belief of i is *independent* of i's lower-level beliefs. In addition, i assigns probability 1 to her own lower-level beliefs. The $(n + 1)$th level of belief of individual j is defined symmetrically.

---

[28] In the infinitary sense of Meier (2001) discussed in Section 8 (Footnote 14).

Notice that up to every finite level n there are finitely many events to consider, so the beliefs are trivially σ-additive. Therefore, by the Kolmogorov extension theorem, for the hierarchy of beliefs of every sequence, there is a unique σ-additive coherent extension to a probability measure[29] over Ω. When $i_n = 0$ for all n, the strong law of large numbers asserts that this limit extension assigns probability 1 to the sequences of j where

$$\lim_{n \to \infty} \frac{j_1 + j_2 + \cdots + j_n}{n} = \frac{1}{2}.$$

However, there are finitely additive coherent extensions to i's finite level beliefs that are concentrated on the disjoint set of j's sequences where

$$\liminf_{n \to \infty} \frac{j_1 + j_2 + \cdots + j_n}{n} > \frac{1}{2} \tag{A.1}$$

and in fact there are finitely additive coherent extensions concentrated on any tail event of sequences $j_1 j_2 j_3 \ldots$, i.e., an event where every possible initial segment $j_1 \ldots j_n$ appears in some sequence.[30]

---

[29] With the σ-field generated by the events that depend on finitely many digits.

[30] To see this, observe that to any event defined by k out of j's digits, the sequence

$$i_1 i_2 i_3 \ldots = 000 \ldots \tag{A.2}$$

assigns the probability 2^{-k}. Therefore, the integral I w.r.t. this belief is well defined over the vector space $\mathcal{V}$ of real-valued functions which are each measurable w.r.t. finitely many of j's digits. Now, for every real-valued function g on j's sequences, define the functional

$$\overline{I}(g) = \inf \{ I(f) \colon f \in \mathcal{V}, \ g \leqslant f \}.$$

Then $\overline{I}$ is clearly sub-additive, $\overline{I}(\alpha g) = \alpha \overline{I}(g)$ for $\alpha \geqslant 0$, and $\overline{I} = I$ on $\mathcal{V}$. Therefore, by the Hahn–Banach theorem, there is an extension (in fact, many extensions) of I as a positive linear functional to the limits of sequences of functions in $\mathcal{V}$, and further to all the real-valued functions on Ω, satisfying

$$I(g) \leqslant \overline{I}(g).$$

Restricting our attention to characteristic functions g, we thus get a finitely additive coherent extension of (A.2) over Ω.

The proof of the Hahn–Banach theorem proceeds by consecutively considering functions g to which I is not yet extended, and defining

$$I(g) = \overline{I}(g).$$

If the first function g_1 to which I is extended is a characteristic function of a tail event (like A.1), the smallest $f \in \mathcal{F}$ majorizing g_1 is the constant function $f \equiv \mathbf{1}$, so

$$I(g) = I(\mathbf{1}) = 1,$$

and the resulting coherent extension of (A.2) assigns probability 1 to the chosen tail event.

Thus, though the finite-level beliefs single out unique σ-additive limit beliefs over Ω, nothing in them can specify that these limit beliefs must be σ-additive. If finitely additive beliefs are not ruled out in the players' minds, we cannot assume that the inner structure of the states $\omega \in \Omega$ specifies uniquely the beliefs of the players.

A similar problem presents itself if we restrict our attention to knowledge. In the syntactic formalism of Section 4, replace the belief operators p_i^α with knowledge operators k_i. When associating sentences to states of an augmented semantic belief system in Section 5, say that the sentence $k_i e$ holds in state ω if $\omega \in K_i E$, [31] where E is the event that corresponds to the sentence e. The canonical *knowledge* system Γ will now consist of those lists of sentences that hold in some state of some augmented semantic belief system. The information set $\mathbf{I}_i(\gamma)$ of player i at the list $\gamma \in \Gamma$ will consist of all the lists $\gamma' \in \Gamma$ that contain exactly the same sentences of the form $k_i f$ as in γ. It now follows that

$$k_i f \in \gamma \;\Leftrightarrow\; \gamma \in K_i E_f \tag{A.3}$$

where $E_f \subseteq \Gamma$ is the event that f holds,[32] and the knowledge operator K_i is as in Section 7: $K_i E = \{\gamma \in \Gamma: \mathbf{I}_i(\gamma) \subseteq E\}$.

However, there are many alternative definitions for the information sets $\mathbf{I}_i(\gamma)$ for which (A.3) would still obtain.[33] Thus, by no means can we say that the information sets $\mathbf{I}_i(\gamma)$ are "self-evident" from the inner structure of the lists γ which constitute the canonical knowledge system.

To see this, consider the following example [similar to that in Fagin et al. (1991)]. Ana, Bjorn and Christina participate in a computer forum over the web. At some point Ana invites Bjorn to meet the next evening. At that stage they leave the forum to continue the chat in private. If they eventually exchange n messages back and forth regarding the meeting, there is mutual knowledge of level n between them that they will meet, but not common knowledge, which they could attain, say, by eventually talking over the phone.

Christina doesn't know how many messages were eventually exchanged between Ana and Bjorn, so she does not exclude any finite level of mutual knowledge between them about the meeting. Nevertheless, Christina could still rule out the possibility of common knowledge (say, if she could peep into Bjorn's room and see that he was glued to his computer the whole day and spoke with nobody). But if this situation is formalized by a list of sentences γ, Christina does *not* exclude the possibility of common knowledge between Ana and Bjorn *with the above definition of* $I_{Christina}(\gamma)$. This is because there exists an augmented belief system Ω with a state ω' in which there is common knowledge between Ana and Bjorn, while Christina does not exclude any finite level

---

[31] The semantic knowledge operator $K_i E$ is defined in Section 7.

[32] As in Section 8.

[33] In fact, as many as there are subsets of Γ! [Heifetz (1999)].

of mutual knowledge between them, exactly as in the situation above. We then have $\gamma' \in \mathbf{I}_{Christina}(\gamma)$, where γ' is the list of sentences that hold in ω'. Thus, if we redefine $\mathbf{I}_{Christina}(\gamma)$ by omitting γ' from it, (A.3) would still obtain, because (A.3) refers only to sentences or events that describe finite levels of mutual knowledge.

At first it may seem that this problem can be mitigated by enriching the syntax with a common knowledge operator (for every subgroup of two or more players). Such an operator would be shorthand for the infinite conjunction "everybody (in the subgroup) knows, and everybody knows that everybody knows, and...". This would settle things in the above example, but create numerous new, analogous problems. *The discontinuity of knowledge is essential*: The same phenomena persist (i.e., (A.3) does not pin down the information sets $\mathbf{I}_i(\gamma)$) even if the syntax explicitly allows for infinite conjunctions and disjunctions of sentences of whatever chosen cardinality [Heifetz (1994)].

References

Allen, B. (1997), "Cooperative theory with incomplete information", in: S. Hart and A. Mas-Colell, eds., Cooperation: Game-Theoretic Approaches (Springer, Berlin) 51–65.

Armbruster, W., and W. Böge (1979), "Bayesian game theory", in: O. Moeschlin and D. Pallaschke, eds., Game Theory and Related Topics (North Holland, Amsterdam) 17–28.

Aumann, R. (1976), "Agreeing to Disagree", Annals of Statistics 4:1236–1239.

Aumann, R. (1987), "Correlated equilibrium as an expression of Bayesian rationality", Econometrica 55:1–18.

Aumann, R. (1998), "Common priors: A reply to Gul", Econometrica 66:929–938.

Aumann, R. (1999a), "Interactive epistemology I: Knowledge", International Journal of Game Theory 28:263–300.

Aumann, R. (1999b), "Interactive epistemology II: Probability", International Journal of Game Theory 28:301–314.

Aumann, R., and A. Brandenburger (1995), "Epistemic conditions for Nash equilibrium", Econometrica 63:1161–1180.

Aumann, R., and M. Maschler (1995), Repeated Games of Incomplete Information (MIT Press, Cambridge).

Aumann, R., and S. Sorin (1989), "Cooperation and bounded recall", Games and Economic Behavior 1:5–39.

Böge, W., and T. Eisele (1979), "On solutions of Bayesian games", International Journal of Game Theory 8:193–215.

Fagin, R., J.Y. Halpern and M.Y. Vardi (1991), "A model-theoretic analysis of knowledge", Journal of the Association for Computing Machinery (ACM) 91:382–428.

Fagin, R., J.Y. Halpern, M. Moses and M.Y. Vardi (1995), Reasoning about Knowledge (MIT Press, Cambridge).

Feinberg, Y. (2000), "Characterizing common priors in the form of posteriors", Journal of Economic Theory 91:127–179.

Forges, F., E. Minelli and R. Vohra (2001), "Incentives and the core of an exchange economy: A survey", Journal of Mathematical Economics, forthcoming.

Fudenberg, D., and E. Maskin (1986), "The Folk theorem in repeated games with discounting and incomplete information", Econometrica 54:533–554.

Gul, F. (1998), "A comment on Aumann's Bayesian view", Econometrica 66:923–927.

Harsanyi, J. (1967-8), "Games with incomplete information played by 'Bayesian' players", Parts I–III, Management Science 8:159–182, 320–334, 486–502.

Harsanyi, J. (1973), "Games with randomly disturbed payoffs: A new rationale for mixed strategy equilibrium points", International Journal of Game Theory 2:1–23.

Heifetz, A. (1994), "Infinitary epistemic logic", in: Proceedings of the Fifth Conference on Theoretical Aspects of Reasoning about Knowledge (TARK 5) (Morgan-Kaufmann, Los Altos, CA) 95–107. Extended version in: Mathematical Logic Quarterly 43 (1997):333–342.

Heifetz, A. (1999), "How canonical is the canonical model? A comment on Aumann's interactive epistemology", International Journal of Game Theory 28:435–442.

Heifetz, A. (2001), "The positive foundation of the common prior assumption", California Institute of Technology, mimeo.

Heifetz, A., and P. Mongin (2001), "Probability logic for type spaces", Games and Economic Behavior 35:31–53.

Heifetz, A., and D. Samet (1998), "Topology-free typology of beliefs", Journal of Economic Theory 82:324–341.

Hintikka, J. (1962), Knowledge and Belief (Cornell University Press, Ithaca).

Kreps, D., P. Milgrom, J. Roberts and R. Wilson (1982), "Rational cooperation in the finitely repeated Prisoners' Dilemma", Journal of Economic Theory 27:245–252.

Kripke, S. (1959), "A completeness theorem in modal logic", Journal of Symbolic Logic 24:1–14.

Lewis, D. (1969), Convention (Harvard University Press, Cambridge).

Luce, R.D., and H. Raiffa (1957), Games and Decisions, (Wiley, New York).

Meier, M. (2001), "An infinitary probability logic for type spaces", Bielefeld and Caen, mimeo.

Mertens, J.-F., S. Sorin and S. Zamir (1994), "Repeated games", Discussion Papers 9420, 9421, 9422 (Center for Operations Research and Econometrics, Université Catholique de Louvain).

Mertens, J.-F., and S. Zamir (1985), "Formulation of Bayesian analysis for games with incomplete information", International Journal of Game Theory 14:1–29.

Morris, S. (1994), "Trade with heterogeneous prior beliefs and asymmetric information", Econometrica 62:1327–1347.

Myerson, R.B. (1984), "Cooperative games with incomplete information", International Journal of Game Theory 13:69–96.

Nau, R.F., and K.F. McCardle (1990),"Coherent behavior in noncooperative games", Journal of Economic Theory 50:424–444.

Samet, D. (1990), "Ignoring ignorance and agreeing to disagree", Journal of Economic Theory 52:190–207.

Samet, D. (1998a), "Iterative expectations and common priors", Games and Economic Behavior 24:131–141.

Samet, D. (1998b), "Common priors and separation of convex sets", Games and Economic Behavior 24:172–174.

Von Neumann, J., and O. Morgenstern (1944), Theory of Games and Economic Behavior (Princeton University Press, Princeton).

Wilson, R. (1978), "Information, efficiency, and the core of an economy", Econometrica 46:807–816.

Chapter 44

NON-ZERO-SUM TWO-PERSON GAMES

T.E.S. RAGHAVAN

Department of Mathematics, Statistics & Computer Science, University of Illinois at Chicago, Chicago, IL, USA

Contents

Handbook of Game Theory, Volume 3, Edited by R.J. Aumann and S. Hart

Abstract

This chapter is devoted to the study of Nash equilibria, and correlated equilibria in both finite and infinite games. We restrict our discussions to only those properties that are somewhat special to the case of two-person games. Many of these properties fail to extend even to three-person games. The existence of quasi-strict equilibria, and the uniqueness of Nash equilibrium in completely mixed games, are very special to two-person games. The Lemke–Howson algorithm and Rosenmüller's algorithm which locate a Nash equilibrium in finite arithmetic steps are not extendable to general n-person games. The enumerability of extreme Nash equilibrium points and their inclusion among extreme correlated equilibrium points fail to extend beyond bimatrix games. Fictitious play, which works in zero-sum two-person matrix games, fails to extend even to the case of bimatrix games. Other algorithms that would locate certain refinements of Nash equilibria are also discussed. The chapter also deals with the structure of Nash and correlated equilibria in infinite games.

Keywords

bimatrix games, Nash and correlated equilibria, computing equilibria, refinements, strategically zero-sum games

JEL classification: C720, C610

1. Introduction

Ever since von Neumann (1928) proved the minimax theorem for zero-sum two-person finite games, considerable effort in game theory has been devoted to understanding the structure and characterization of optimal strategies [Karlin (1959a); Dresher (1961)], developing algorithms to compute values and optimal strategies, extending the theory to special classes of infinite games, and, more generally, studying general minimax theorems [Parthasarathy and Raghavan (1971)].

However, major applications and models in social sciences are usually non-zero-sum. For example, models of interaction between husband and wife, between employer and employee, and between landlord and tenant are not always antagonistic. Problems of communication gaps, variations in perception, inherent personality traits, taste differences, and many other factors influence the decision-making of rational players. In the strategic form, such games are called *bimatrix games*.

Let player I select an $i \in I = \{1, \ldots, m\}$ secretly, and let player II select a $j \in J = \{1, \ldots, n\}$ secretly. Let player I receive a payoff a_{ij}, and let player II receive a payoff b_{ij}. This game is represented by an $m \times n$ matrix with vector entries (a_{ij}, b_{ij}). Any extensive game with two players and finitely many moves and actions at each such move is reducible to a bimatrix game with payoffs in von Neumann and Morgenstern (1944) utilities of the two players (see Chapter 2 in this Handbook). Given a bimatrix game $G = (A, B)_{m \times n}$, a pair of actions (i^*, j^*) is a *Nash equilibrium in pure strategies* or a *pure equilibrium* if $a_{i^*j^*} \geqslant a_{ij^*}$ for all i and $b_{i^*j^*} \geqslant b_{i^*j}$ for all j. This means that i^* is best against j^* and j^* is best against i^*, so neither player has any incentive to deviate unilaterally from this strategy.

In a bimatrix game, a pure equilibrium may not exist. Even if it does, there may be several equilibria, giving different payoffs. In the absence of communication among players, it is not clear which one is to be chosen. In one of the examples of bimatrix games below, one may wish to model the game as a *repeated game* (see Chapter 4 in this Handbook) to recover certain types of tacit cooperation as equilibrium behavior. In such an *infinite repetition* of a game, players want to maximize their long-run average payoff per play, and tend to choose an action that takes into account the past history of actions by the two players.

Consider the following bimatrix games:

$$G_1 = \begin{bmatrix} (5,1) & (0,0) \\ (0,0) & (1,5) \end{bmatrix}, \qquad G_2 = \begin{bmatrix} (0,0) & (4,-4) \\ (-4,4) & (2,2) \end{bmatrix},$$

$$G_3 = \begin{bmatrix} (1,50) & (50,1) \\ (1,-50) & (1,-49) \end{bmatrix}, \qquad G_4 = \begin{bmatrix} (2,2) & (4,1) \\ (4,1) & (3,3) \end{bmatrix}.$$

The game G_1 is called the *Battle of Sexes*. The two payoffs $(1, 5)$ and $(5, 1)$ are Nash equilibrium payoffs in pure strategies. In zero-sum games where $A = -B$, each pure equilibrium will correspond to a saddle point for A, and any two saddle-point payoffs are the same. This is no longer true for the non-zero-sum game G_1. The game G_2 is called the *Prisoner's Dilemma*. Here $(0, 0)$ is an equilibrium payoff, while the payoff $(2, 2)$, which is certainly more desirable to both players, fails to be an equilibrium payoff. In the case of G_3, one may wish to consider the corresponding *repeated game*. The threat to punish player II for choosing column 1 in earlier plays cannot be captured by modeling the game as an ordinary bimatrix game and using its Nash equilibrium. The game G_4 has no pure equilibrium. However, $((\frac{2}{3}, \frac{1}{3}), (\frac{1}{3}, \frac{2}{3}))$ is its unique Nash equilibrium and yields an expected payoff of $\frac{10}{3}$ to player I and $\frac{5}{3}$ to player II. These are also the best payoffs that the players can guarantee for themselves. The corresponding strategies are called *maxmin* strategies. For a discussion of this type of game, see Aumann and Maschler (1972).

For a long time, the Nash equilibrium was the only solution concept for noncooperative games, and in particular for bimatrix games. It held exclusive sway in the field until the introduction of *correlated equilibrium* by Aumann (1974, 1987). As a generalization of the minimax concept, the Nash equilibrium theorem for bimatrix games carries a lot of intuitive import in many economic problems. An added attraction is its avoidance of interpersonal comparison of utilities.

The initial thrust to noncooperative games was given by the following fundamental existence theorem, which is stated for the two-person case below.

For convenience, we will write mixed strategies as row tuples. When we need to manipulate with matrix multiplications or dot products or expectations we will assume all vectors to be column vectors.

THEOREM [Nash (1950)]. *Let* $(A, B)_{m \times n}$ *be the payoffs of a bimatrix game. Then there exists a mixed strategy* $\mathbf{x}^* = (x_1^*, x_2^*, \ldots, x_m^*)$ *for player I and a mixed strategy* $\mathbf{y}^* = (y_1^*, y_2^*, \ldots, y_n^*)$ *for player II such that for any mixed strategy* $\mathbf{x} = (x_1, x_2, \ldots, x_m)$ *for player I and for any mixed strategy* $\mathbf{y} = (y_1, y_2, \ldots, y_n)$ *for player II,*

$$\langle \mathbf{x}^*, A\mathbf{y}^* \rangle = \sum_{i=1}^{m} \sum_{j=1}^{n} a_{ij} x_i^* y_j^* \geqslant \sum_{i=1}^{m} \sum_{j=1}^{n} a_{ij} x_i y_j^* = \langle \mathbf{x}, A\mathbf{y}^* \rangle, \tag{1}$$

and

$$\langle \mathbf{x}^*, B\mathbf{y}^* \rangle = \sum_{i=1}^{m} \sum_{j=1}^{n} b_{ij} x_i^* y_j^* \geqslant \sum_{i=1}^{m} \sum_{j=1}^{n} b_{ij} x_i^* y_j = \langle \mathbf{x}^*, B\mathbf{y} \rangle. \tag{2}$$

For a proof, see Chapter 42 in this Handbook. Intuitively, if players I and II are somehow convinced that the opponents are using the respective mixed strategies $\mathbf{y}^*$ and $\mathbf{x}^*$ in their decision-making, then neither player can unilaterally deviate and strictly increase his expected gain. Equivalently, substituting the unit vectors $\mathbf{f}_i \in \mathbf{R}^m$ and $\mathbf{e}_j \in \mathbf{R}^n$ for $\mathbf{x}$, $\mathbf{y}$, we have

$$v_1 = \langle \mathbf{x}^*, A\mathbf{y}^* \rangle \geqslant \langle \mathbf{f}_i, A\mathbf{y}^* \rangle = \sum_{j=1}^{n} a_{ij} y_j^* \quad \text{for } i = 1, \ldots, m, \tag{3}$$

and

$$v_2 = \langle \mathbf{x}^*, B\mathbf{y}^* \rangle \geqslant \langle \mathbf{x}^*, B\mathbf{e}_j \rangle = \sum_{i=1}^{m} b_{ij} x_i^* \quad \text{for } j = 1, \ldots, n. \tag{4}$$

Here, v_1, v_2 are the expected payoffs for players I and II at the equilibrium $(\mathbf{x}^*, \mathbf{y}^*)$.

Multiplying both sides of (3) by x_i^* and then summing both sides gives the following equality:

$$x_i^* v_1 = x_i^* \sum_{j=1}^{n} a_{ij} y_j^* \quad \text{for } i = 1, \ldots, m.$$

Thus, $\sum_j a_{ij} y_j^* = v_1$ when $x_i^* > 0$, or equivalently, $x_i^* = 0$ when $\sum_j a_{ij} y_j^* < v_1$. Similarly, $\sum_i b_{ij} x_i^* = v_2$ when $y_j^* > 0$, and $y_j^* = 0$ when $\sum_i b_{ij} x_i^* < v_2$.

The set $C(\mathbf{x}) = \{i \colon x_i > 0\}$ is called the *carrier* of $\mathbf{x}$. The set

$$B_I(\mathbf{y}) = \left\{ i \colon \sum_j a_{ij} y_j = \max_k \sum_j a_{kj} y_j \right\}$$

is called the *set of best pure replies* of player I against the mixed strategy $\mathbf{y}$ of player II. The carriers $C(\mathbf{y})$ and $B_{II}(\mathbf{x})$ are similarly defined. Thus, $(\mathbf{x}^*, \mathbf{y}^*)$ is an equilibrium of the bimatrix game if and only if $C(\mathbf{x}^*) \subseteq B_I(\mathbf{y}^*)$ and $C(\mathbf{y}^*) \subseteq B_{II}(\mathbf{x}^*)$. If the game is zero-sum, then $A + B = 0$ and the inequalities (3) and (4) reduce to

$$\sum_{j=1}^{n} a_{ij} y_j^* \leqslant v_1, \quad \text{for } i = 1, \ldots, m,$$

$$\sum_{i=1}^{m} a_{ij} x_i^* \geqslant v_1, \quad \text{for } j = 1, \ldots, n.$$

The minimax theorem of von Neumann follows from the above inequalities. (See Chapter 21 in this Handbook.)

2. Equilibrium refinement for bimatrix games

A serious conceptual problem for many games is to find a way to discard unsatisfactory equilibria. This approach of weeding out unwanted equilibria is known as *equilibrium selection* in the literature [see van Damme (1983)].

The first of such pruning procedures was suggested by Selten (1975). It is based on the idea of stability under mild irrationality of the players. Selten's procedure aims at removing those equilibria that prescribe irrational behavior at unreached information sets in extensive games. For the case of bimatrix games, Selten's definition of *perfect equilibrium* reduces to the following:

For $k = 1, 2, \ldots$, let $\mathbf{e}^k$ and $\mathbf{d}^k$ be sequences of strictly positive vectors in $\mathbf{R}^m$ and $\mathbf{R}^n$ respectively with $\mathbf{e}^k \to 0$ and $\mathbf{d}^k \to 0$. Let $(\mathbf{x}^k, \mathbf{y}^k)$ be a sequence of Nash equilibria for the bimatrix game (A, B) on the restricted strategy spaces

$$X^k = \{\mathbf{x}: \mathbf{x} \geqslant \mathbf{e}^k > 0\}, \qquad Y^k = \{\mathbf{y}: \mathbf{y} \geqslant \mathbf{d}^k > 0\} \tag{5}$$

for players I and II, respectively. The above inequalities are to be understood coordinate-wise.

If $(\mathbf{x}^*, \mathbf{y}^*)$ is a Nash equilibrium of (A, B) *and* a limit point of the above sequence $(\mathbf{x}^k, \mathbf{y}^k)$, then $(\mathbf{x}^*, \mathbf{y}^*)$ is called a *perfect* equilibrium of the bimatrix game (A, B).

Note that the requirement is not for all sequences $0 < \mathbf{e}^k \to 0$ and $0 < \mathbf{d}^k \to 0$, but only for *some* such sequence. Alternatively, one can define perfect equilibrium as follows: Let (A, B) be a bimatrix game of order $m \times n$. Let $0 < \varepsilon^k \to 0$. Then $(\mathbf{x}^*, \mathbf{y}^*)$ is a perfect equilibrium if and only if there are completely mixed strategies $\mathbf{x}^k, \mathbf{y}^k$ converging to $(\mathbf{x}^*, \mathbf{y}^*)$ such that for any i

$$\left(A\mathbf{y}^k\right)_i < \max_t \left(A\mathbf{y}^k\right)_t \Rightarrow x_i^k < \varepsilon^k \tag{6}$$

and for any j

$$\left(B^{\mathrm{T}}\mathbf{x}^k\right)_j < \max_s \left(B^{\mathrm{T}}\mathbf{x}^k\right)_s \Rightarrow y_j^k < \varepsilon^k. \tag{7}$$

Perfection weeds out some unwanted equilibria, but not all. For example, in the bimatrix game

$$\begin{bmatrix} (1, 1) & (100, 0) \\ (0, 100) & (100, 100) \end{bmatrix}$$

$(1, 1)$ is the unique perfect equilibrium, although players will prefer $(100, 100)$, which is also an equilibrium payoff. Here, the intuition behind the notion of perfect equilibrium is that player I, suspecting that player II might choose column 1, will hesitate to choose row 2, and this leads to the rejection of the equilibrium payoff $(100, 100)$.

Suppose that in the above 2×2 game, each player can use a third action with payoffs inferior to the original payoffs. Consider the 3×3 bimatrix game,

$$\begin{bmatrix} (1, 1) & (100, 0) & (-1, -2) \\ (0, 100) & (100, 100) & (0, -2) \\ (-2, -1) & (-2, 0) & (-2, -2) \end{bmatrix}.$$

The same strategy $i = 2$, $j = 2$ with payoff $(100, 100)$ is now a perfect equilibrium strategy for the above game even though it was not a perfect equilibrium for the original game! For bimatrix games, the notion of domination is a useful tool in finding perfect equilibria.

Let (A, B) be a bimatrix game. We will denote by $M = \{1, \ldots, m\}$, $N = \{1, \ldots, n\}$, the set of pure strategies of players I and II. We will denote by Δ_M^I, Δ_N^{II}, the set of mixed strategies for players I and II respectively. A mixed strategy $\mathbf{x}$ *dominates* the mixed strategy $\mathbf{x}'$ if $\langle \mathbf{x}, A\mathbf{y} \rangle \geq \langle \mathbf{x}', A\mathbf{y} \rangle$ for all mixed strategies $\mathbf{y} \in \Delta_N^{II}$ with strict inequality for some $\mathbf{y}$. Equivalently, $\mathbf{x}$ *dominates* $\mathbf{x}'$ if $A^T\mathbf{x} \geq A^T\mathbf{x}'$ and $A^T\mathbf{x} \neq A^T\mathbf{x}'$. One can similarly define domination for player II.

THEOREM [van Damme (1983)]. *A Nash equilibrium* $(\mathbf{x}^*, \mathbf{y}^*)$ *for a bimatrix game* (A, B) *is perfect if and only if* $\mathbf{x}^*$ *and* $\mathbf{y}^*$ *are undominated for the respective players.*

PROOF. Let $\mathbf{x}^*$ and $\mathbf{y}^*$ be a Nash equilibrium such that $\mathbf{x}^*$ and $\mathbf{y}^*$ are undominated for the respective players. We will show that the Nash equilibrium is perfect.

Let $\mathbf{x}^*$ be undominated. Consider the zero-sum matrix game $C = (c_{ij})$, where $c_{ij} = a_{ij} - \sum_k a_{kj} x_k^*$. One can easily check that since $\mathbf{x}^*$ is undominated, the game has value $v(C) = 0$ and, further, $\mathbf{x}^*$ is optimal for player I. If $\mathbf{q}$ is any other optimal strategy for player I, then, since $\mathbf{x}^*$ is undominated,

$$\sum_i c_{ij} q_i = 0, \quad j = 1, \ldots, n.$$

Thus, all columns of player II are equalizers. In that case, we have an optimal $\mathbf{y}^\circ > 0$ for player II. Also, $\mathbf{x}^*$ is a best reply for $\mathbf{y}^\circ$ in the matrix game C and a best reply for $\mathbf{y}^*$ in the bimatrix game (A, B). Hence, it is a best reply for $\varepsilon\mathbf{y}^\circ + (1 - \varepsilon)\mathbf{y}^*$, which converges to $\mathbf{y}^*$ as $\varepsilon \to 0$. Similarly, for some $\mathbf{x}^\circ > 0$ we can prove $\mathbf{y}^*$ is a best reply for $\varepsilon\mathbf{x}^\circ + (1 - \varepsilon)\mathbf{x}^*$, which converges to $\mathbf{x}^*$. Thus, $(\mathbf{x}^*, \mathbf{y}^*)$ is a perfect equilibrium point.

The converse, which we will not prove, is true even for any general n-person game in normal form. The problem of testing whether a strategy $\mathbf{x}^*$ is undominated can be reduced to a linear programming problem. $\square$

3. Quasi-strict equilibria

An equilibrium $(\mathbf{x}^*, \mathbf{y}^*)$ of a bimatrix game (A, B) is *quasi-strict* if and only if $B_{II}(\mathbf{y}^*) = C(\mathbf{x}^*)$ and $B_I(\mathbf{x}^*) = C(\mathbf{y}^*)$; that is, the set of best pure replies of player I against the strategy $\mathbf{y}^*$ of player II is the carrier of the strategy $\mathbf{x}^*$ of player I and conversely. This definition can be generalized to any n-person game in normal form. For example, for a 3-person normal-form game with payoffs $(A, B, C)_{ijk}$, $(\mathbf{x}^*, \mathbf{y}^*, \mathbf{z}^*)$ is quasi-strict if and only if

$$\sum_{jk} a_{ijk} y_j^* z_k^* = \max_\alpha \sum_{jk} a_{\alpha jk} y_j^* z_k^* \Leftrightarrow i \in C(\mathbf{x}^*),$$

$$\sum_{ik} b_{ijk} x_i^* z_k^* = \max_\beta \sum_{ik} a_{i\beta k} x_i^* z_k^* \Leftrightarrow j \in C(\mathbf{y}^*),$$

$$\sum_{ij} c_{ijk} x_i^* y_j^* = \max_\gamma \sum_{ij} a_{ij\gamma} x_i^* y_j^* \Leftrightarrow j \in C(\mathbf{z}^*).$$

Harsanyi (1973b) first introduced these equilibria and proved that *almost all* non-zero-sum n-person games have only quasi-strict equilibria. The following example shows that, in general, quasi-strict equilibria may not exist.

EXAMPLE. Consider a 3-person game in normal form where player I chooses a row, player II chooses a column, and player III chooses a matrix. Here the first matrix corresponds to pure strategy 1 of player III and the second matrix corresponds to pure strategy 2 of player III:

$$\begin{bmatrix} (0,0,0) & (2,0,0) \\ (0,0,2) & (0,2,0) \end{bmatrix} \quad \begin{bmatrix} (0,1,0) & (0,0,1) \\ (1,0,0) & (0,0,0) \end{bmatrix}.$$

The game has exactly two Nash equilibria corresponding to the pure strategy triples $(i, j, k) = (1, 1, 1)$ and $(2, 2, 2)$. Neither one is quasi-strict, for when players II and III use action 1, the best-reply set for player I is to choose either action 1 or 2. This is not the carrier of $(1, 1, 1)$ for player I. Similarly, one can show that $(2, 2, 2)$ is not quasi-strict.

However, in the case of bimatrix games, Norde (1998) proved the following theorem:

THEOREM. *Every bimatrix game has at least one quasi-strict Nash equilibrium in mixed strategies.*

We cannot apply Harsanyi's (1973b) generic theorem on quasi-strict equilibria and use limiting arguments to exhibit a quasi-strict equilibrium for general bimatrix games. For example, the bimatrix game

$$\left(A^\varepsilon, B^\varepsilon \right) = \begin{bmatrix} (1,0) & (0,\varepsilon) \\ (0,1) & (\varepsilon,0) \end{bmatrix}$$

has a unique Nash equilibrium for each ε, and the limit as $\varepsilon \to 0$ is the mixed strategy $((1, 0), (0, 1))$. This is not quasi-strict for the limiting bimatrix game (A^0, B^0).

The proof of the above theorem involves a judicious application of Brouwer's fixed-point theorem. Given any subset T of pure strategies for player II, let K_T^{II} be a closed polyhedron contained in the interior of the probability simplex Δ_T^{II}. A suitable continuous map $f_T : K_T^{II} \to \Delta_M^I$ (here M is the set of all pure strategies of player I) is constructed to act like the approximate best response of player I to the strategies of player II in K_T^{II}. Depending on the payoff B for player II, there exists a sequence of positive integers $k_i \to \infty$ such that approximate best responses $g_T^{k_i} : \Delta_M^I \to K_T^{II}$ for player II can be constructed. These are taken as best responses when the player is restricted to choosing strategies in T. Any fixed point $\mathbf{p}^k$ of the continuous map $f_T \circ g_T^k : \Delta_M^I \to \Delta_M^I$ has a limit point $\mathbf{p}$ which, with a suitable complementary $\mathbf{q}$ of Δ_N^{II} (here N is the set of all pure strategies of player II), becomes a quasi-strict equilibrium.

Although quasi-strictness is a refinement, it could involve mixing dominated strategies.

EXAMPLE. Consider the bimatrix game

$$
\begin{bmatrix} (1,1) & (1,0) \\ (1,1) & (0,0) \end{bmatrix}.
$$

Notice that row 1 weakly dominates row 2 and yet $\mathbf{x}^* = (0.5, 0.5)$, $\mathbf{y}^* = (1,0)$ is a quasi-strict equilibrium.

In trying to prove the generic finiteness of equilibria for n-person games, Harsanyi (1973b) introduced the notion of *regular equilibria*. The following is a slight modification of the same [van Damme (1983)].

4. Regular and stable equilibria

For a given bimatrix game (A, B), let $(\mathbf{x}^*, \mathbf{y}^*)$ be a mixed strategy Nash equilibrium with their pth and qth coordinates, x_p^*, y_q^* positive. Consider the functions $\mathbf{f} = (f^{\alpha p})$, $\alpha = 1, \ldots, m$, $\mathbf{g} = (g^{\beta q})$, $\beta = 1, \ldots, n$, defined on $\Delta_M^I \times \Delta_N^{II}$ by

$$
f^{\alpha p}(\mathbf{x}, \mathbf{y}) = x_\alpha \big((A\mathbf{y})_\alpha - (A\mathbf{y})_p \big), \quad \alpha \neq p;
$$

$$
f^{pp}(\mathbf{x}, \mathbf{y}) = \sum_i x_i - 1;
$$

$$
g^{\beta q}(\mathbf{x}, \mathbf{y}) = y_\beta \big((B^{\mathsf{T}}\mathbf{x})_\beta - (B^{\mathsf{T}}\mathbf{x})_q \big), \quad \beta \neq q;
$$

$$
g^{qq}(\mathbf{x}, \mathbf{y}) = \sum_j y_j - 1.
$$

The map $\phi_{pq} = (\mathbf{f}(\mathbf{x}, \mathbf{y}), \mathbf{g}(\mathbf{x}, \mathbf{y}))$ is called *regular* at $(\mathbf{x}', \mathbf{y}')$ if the Jacobian $J(\phi_{pq})$ is nonsingular at $(\mathbf{x}', \mathbf{y}')$. Motivated by this we have the following

DEFINITION. An equilibrium $(\mathbf{x}^*, \mathbf{y}^*)$ is called *regular* if and only if for each pure strategy pair (p, q) in the carrier of $(\mathbf{x}^*, \mathbf{y}^*)$, the Jacobian $J(\phi_{pq})$ is nonsingular.

REMARK. Since for any pure strategy pair (r, s) in the carrier of $(\mathbf{x}^*, \mathbf{y}^*)$, the Jacobian $J(\phi_{rs})(\mathbf{x}^*, \mathbf{y}^*)$ is obtained from $J(\phi_{pq})(\mathbf{x}^*, \mathbf{y}^*)$ by elementary transformations, it is enough to verify the conditions at one pair (p, q) in the carrier. The map τ defined on $\mathbf{R}^{2mn} \times \mathbf{R}^m \times \mathbf{R}^n$ given by $\tau : (A, B, \mathbf{x}, \mathbf{y}) \to (A, B, \mathbf{f}, \mathbf{g})$ as defined above has the same Jacobian as ϕ_{pq} and hence is nonsingular at $(A, B, \mathbf{x}^*, \mathbf{y}^*)$.

THEOREM. *For bimatrix games, if all equilibrium points are regular, then the equilibrium set is finite with an odd number of elements.*

PROOF. Let $(\mathbf{x}^*, \mathbf{y}^*)$ be a mixed strategy Nash equilibrium with positive pth and qth coordinates, $\mathbf{x}_p^*, \mathbf{y}_q^*$. From the above remark, the Jacobian of the map $\tau : (A, B, \mathbf{x}, \mathbf{y}) \to (A, B, \mathbf{f}, \mathbf{g})$ is nonsingular at $(A, B, \mathbf{x}^*, \mathbf{y}^*)$. Also $\mathbf{f}(\mathbf{x}^*, \mathbf{y}^*) = 0$, $\mathbf{g}(\mathbf{x}^*, \mathbf{y}^*) = 0$. By the implicit function theorem, there exist open sets $U \ni (A, B)$ and $V \ni (\mathbf{x}^*, \mathbf{y}^*)$ such that
  (i) there exists a differential map $s : U \to V$,
  (ii) the set $\{(A, B, \mathbf{x}, \mathbf{y}) \in U \times V : \mathbf{f}(\mathbf{x}, \mathbf{y}) = 0, \mathbf{g}(\mathbf{x}, \mathbf{y}) = 0\} = \{((A, B), s(A, B)) : (A, B) \in U\}$.

Further, one can show for U, V sufficiently small $s(A, B) \in \mathcal{E}(A, B)$, the set of Nash equilibrium points when $(A, B) \in U$. By (ii) we have $\mathcal{E}(A, B) \cap V = \{(\mathbf{x}^*, \mathbf{y}^*)\}$. Thus each equilibrium is isolated. Since $\mathcal{E}(A, B)$ is compact, it is a finite set. Let $(A^k, B^k) \to (A, B)$ where (A^k, B^k) are nondegenerate in the sense of Lemke and Howson (1964) with an odd number of equilibria (see Section 9). Suppose $(\mathbf{x}^*, \mathbf{y}^*) = \lim(\mathbf{x}^k, \mathbf{y}^k) = \lim(\tilde{\mathbf{x}}^k, \tilde{\mathbf{y}}^k)$, where $(\mathbf{x}^k, \mathbf{y}^k) \in s(A^k, B^k)$ and $(\tilde{\mathbf{x}}^k, \tilde{\mathbf{y}}^k) \in \mathcal{E}(A^k, B^k)$. Since $(\mathbf{x}^*, \mathbf{y}^*)$ is regular, the solutions $\tilde{x}^k, \tilde{y}^k$, for all large k, are determined by the same nonsingular square subsystem corresponding to their common carrier. Thus, for all large k, $\mathcal{E}(A, B)$ and $\mathcal{E}(A^k, B^k)$ have the same cardinality. Hence $\mathcal{E}(A, B)$ is finite and odd. [See also Rosenmüller (1971); Wilson (1971); Harsanyi (1975); Jansen (1981a).]

An equilibrium $(\mathbf{x}^*, \mathbf{y}^*)$ of the bimatrix game (A, B) is called *stable* if all bimatrix games in the neighborhood of (A, B) have equilibria close to $(\mathbf{x}^*, \mathbf{y}^*)$ [see Wu and Jia-He (1962)]. $\square$

EXAMPLE. In the bimatrix game

$$(A, B) = \begin{bmatrix} (6, 0) & (4, 2) \\ (4, 2) & (4, 4) \\ (4, 2) & (6, 4) \\ (2, 4) & (6, 6) \end{bmatrix}$$

the pure equilibrium at row 3, column 2 is perfect because it is undominated. Though the pure equilibrium at row 4, column 2 is not perfect, it is the only equilibrium which is stable.

Let $\varepsilon_1, \varepsilon_2$ be small positive quantities. Given the bimatrix game (A, B), the following bimatrix game is in its neighborhood for $\varepsilon_1, \varepsilon_2$ sufficiently small.

$$\begin{bmatrix} (6, 0) & (4, 2) \\ (4 - 2\varepsilon_1 + 2\varepsilon_2, 2 + 2\varepsilon_1 - 2\varepsilon_2) & (4 + 2\varepsilon_1, 4 + 2\varepsilon_1 - 2\varepsilon_2) \\ (4 + 2\varepsilon_1 - 2\varepsilon_2, 2 - 2\varepsilon_1 + 2\varepsilon_2) & (6 - 2\varepsilon_1, 4 - 2\varepsilon_1 + 2\varepsilon_2) \\ (2, 4) & (6, 6) \end{bmatrix}.$$

This game has the unique equilibrium value $(6, 6)$ at row 4, column 2 for all small $\varepsilon_1, \varepsilon_2$.

THEOREM. *Let $(\mathbf{x}^*, \mathbf{y}^*)$ be a stable equilibrium of a bimatrix game (A, B). Then $C(\mathbf{x}^*)$ and $C(\mathbf{y}^*)$ have the same cardinality. Further, the given stable equilibrium is an isolated regular equilibrium. Conversely, every isolated regular equilibrium is stable. Every bimatrix game with finitely many equilibria has at least one stable equilibrium.*

Since our interest is in bimatrix games, we will not discuss other equilibrium refinements that have nothing special to say about bimatrix games.

5. Completely mixed games

A mixed strategy for a player is called *completely mixed* if it is positive coordinatewise. A bimatrix game (A, B) is *completely mixed* if and only if every mixed strategy of each player in any Nash equilibrium is completely mixed.

THEOREM. *Let $G = (A, B)$ be a completely mixed bimatrix game. Then A and B are square matrices, and the game has a unique Nash equilibrium point.*

PROOF. Let $(\mathbf{x}^*, \mathbf{y}^*) \in \mathcal{E}(A, B)$. If $m > n$, then for some j we have $(B^{\mathrm{T}}\mathbf{x}^*)_j = v_2$, which is the expected payoff to player II. We have $B^{\mathrm{T}}\mathbf{u} = 0$ for some $\mathbf{u} \neq 0$. Since $\mathbf{x}^* > 0$, we can find a mixed strategy $\mathbf{x}'$ on the line segment joining $\mathbf{x}^*$ and $\mathbf{u}$ that is not completely mixed. It is easy to check that $(\mathbf{x}', \mathbf{y}^*)$ will form another equilibrium. Therefore, $m \leqslant n$. Similarly, it can be shown that $m \geqslant n$. Thus, the payoff matrices are square. The uniqueness is based on the argument that for square payoffs, as in zero-sum games, if, say, player I can be in equilibrium skipping a pure strategy, so can player II [see Kaplansky (1945); Raghavan (1970)]. $\qquad\square$

REMARK. The above theorem is not valid in its generality for n-person games. For a counterexample in 3-person games see Chin, Parthasarathy and Raghavan (1974).

While completely mixed bimatrix games have a unique Nash equilibrium, it was shown by Kreps (1974) that a necessary and sufficient condition for a bimatrix game to have a unique Nash equilibrium is that the carriers of the equilibrium mixed strategies of the two players have the same cardinality. Millham (1972) and Heuer (1975) study this problem and its many ramifications.

6. On the Nash equilibrium set

For a general bimatrix game, the Nash equilibrium set, denoted by $\mathcal{E} = \mathcal{E}(A, B)$, may have a complicated geometric shape.

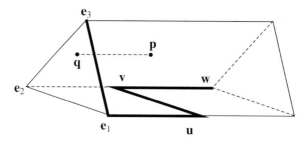

Figure 1.

EXAMPLE. Consider the bimatrix game

$$A = \begin{bmatrix} 2 & 1 & 1 \\ 2 & 1 & 0 \end{bmatrix} \quad \text{and} \quad B = \begin{bmatrix} 1 & 0 & 1 \\ 0 & 1 & 0 \end{bmatrix}.$$

We can identify each mixed strategy pair as a point $\mathbf{p}$ in a polyhedron, as in Figure 1.

Let $\mathbf{q}$ be the projection of the point $\mathbf{p}$ onto the base with vertices $\mathbf{e}_1$, $\mathbf{e}_2$, $\mathbf{e}_3$. The point $\mathbf{q}$ is the unique convex combination of the three vertices, and this gives the mixed strategy $\mathbf{y}$ for player II. If the distance from $\mathbf{p}$ to $\mathbf{q}$ is $(1 - x)$, then $(x, 1 - x)$ is a mixed strategy for player I. The shaded portion consisting of the union of the line segments $[\mathbf{e}_3\mathbf{e}_1]$, $[\mathbf{e}_1\mathbf{u}]$, $[\mathbf{uv}]$, and $[\mathbf{vw}]$ constitutes the equilibrium set (here $\mathbf{u} = ((0.5, 0.5), (1, 0, 0))$, $\mathbf{v} = ((0.5, 0.5)), (0, 1, 0))$ and $\mathbf{w} = ((0, 1), (0, 1, 0))$. The set $\mathcal{E}(A, B)$ is a simplicial complex. In this example, each line segment is a maximal convex subset of $\mathcal{E}(A, B)$, and $\mathcal{E}(A, B)$ is the union of such maximal convex sets.

7. The Vorobiev–Kuhn theorem on extreme Nash equilibria

While the set of all good strategy pairs for the players is compact and convex in a zero-sum two-person matrix game, this is not always the case with Nash equilibria in general bimatrix games. However, the Nash equilibrium set can be recovered from the extreme points of its maximal convex subsets.

Let (A, B) be an $m \times n$ bimatrix game with the Nash equilibrium set $\mathcal{E}(A, B)$. Let $M = \{1, \ldots, m\}$, $N = \{1, \ldots, n\}$. Let $\mathbf{a}_i$ and $\mathbf{b}_j$ be the ith row and the jth column of A and B, respectively. Let Δ_M^I and Δ_N^{II} be the set of mixed strategies for players I and II respectively. For any finite set $X = \{\mathbf{x}_1, \ldots, \mathbf{x}_k\} \subset \Delta_M^I$, let $S(X) = \{\mathbf{y}: (\mathbf{x}_i, \mathbf{y}) \in \mathcal{E}(A, B), \text{ for } i = 1, \ldots, k\}$. It is possible that the set $S(X)$ is empty.

LEMMA. *Let $\mathbf{y}^\circ$ be an extreme point of $S(X)$. Let $\alpha^\circ = \max_{\mathbf{x} \in X} \langle \mathbf{x}, A\mathbf{y}^0 \rangle$. Then $(\mathbf{y}^\circ, \alpha^\circ)$ is an extreme point of the polytope*

$$K = \big\{ (\mathbf{y}, \alpha): \langle \mathbf{a}_i, \mathbf{y} \rangle \leqslant \alpha, \ \langle \mathbf{x}_i, B\mathbf{y} \rangle \geqslant \langle \mathbf{x}_i, \mathbf{b}_j \rangle,$$
$$i = 1, \ldots, k; \ j = 1, \ldots, n \text{ and } \mathbf{y} \in \Delta_N^{II} \big\}.$$

PROOF. Clearly, $(\mathbf{y}^\circ, \alpha^\circ) \in K$. Suppose $(\mathbf{y}^\circ, \alpha^\circ) = \frac{1}{2}(\mathbf{y}', \alpha') + \frac{1}{2}(\mathbf{y}'', \alpha'')$, where $(\mathbf{y}', \alpha') \neq (\mathbf{y}'', \alpha'')$. Since $\max_i \langle \mathbf{a}_i, \mathbf{y}^\circ \rangle = \alpha^\circ$, we have $\max_i \langle \mathbf{a}_i, \mathbf{y}' \rangle = \alpha'$ and $\max_i \langle \mathbf{a}_i, \mathbf{y}'' \rangle = \alpha''$ and $\mathbf{y}', \mathbf{y}'' \in S(X)$ with $\mathbf{y}^\circ = \frac{1}{2}\mathbf{y}' + \frac{1}{2}\mathbf{y}''$. Clearly $\mathbf{y}', \mathbf{y}''$ are distinct and we have a contradiction. $\qquad\square$

LEMMA. *Let $(\mathbf{x}^0, \mathbf{y}^0)$ be an extreme point of some maximal convex subset of $\mathcal{E}(A, B)$. Let α^0 be the equilibrium payoff to player I. Then $(\mathbf{y}^0, \alpha^0)$ is an extreme point of the set*

$$T = \left\{ (\mathbf{y}, \alpha) : \sum_j a_{ij} y_j \leqslant \alpha, \ \forall i, \ y_j \geqslant 0, \ \forall j, \ \sum_j y_j = 1 \right\}.$$

PROOF. Suppose $(\mathbf{y}^0, \alpha^0) = \frac{1}{2}(\mathbf{y}', \alpha') + \frac{1}{2}(\mathbf{y}'', \alpha'')$, where $(\mathbf{y}', \alpha') \neq (\mathbf{y}'', \alpha'')$ and they both lie in K. By the previous lemma, $(\mathbf{y}^\circ, \alpha^\circ)$ is an extreme point of the polytope K. If $\langle \mathbf{a}_i, \mathbf{y}^\circ \rangle = \sum_j a_{ij} y_j^0 < \alpha^\circ$, then $\sum_j a_{ij} (\frac{1}{2} y_j' + \frac{1}{2} y_j'') < \alpha^0 \Rightarrow x_i^0 = 0$. When $x_i^0 > 0$ then $\sum_j a_{ij} y_j' = \sum_j a_{ij} y_j'' = \sum_j a_{ij} y_j^0 = \alpha^\circ$. Thus, $(\mathbf{x}^0, \mathbf{y}'), (\mathbf{x}^0, \mathbf{y}'') \in \mathcal{E}(A, B)$. They are both in a maximal convex subset of $\mathcal{E}(A, B)$. Since $\mathbf{y}^0 = \frac{1}{2}(\mathbf{y}' + \mathbf{y}'')$, we have a contradiction to the assumption that $(\mathbf{x}^0, \mathbf{y}^0)$ is an extreme point. $\qquad\square$

With each extreme point $(\mathbf{y}^0, \alpha^0)$ of T, we have a square submatrix A_1 of A such that $(\mathbf{y}^0, \alpha^0)$ is the unique solution to a matrix equation with a nonsingular matrix given by

$$\begin{bmatrix} A_1 & -1 \\ 1^{\mathrm{T}} & 0 \end{bmatrix} \begin{bmatrix} \mathbf{y} \\ \alpha \end{bmatrix} = \begin{bmatrix} 0 \\ 1 \end{bmatrix}.$$

A similar subsystem for player II with matrix B exists. By solving these matrix equations via Cramer's rule, we can locate all the potential extreme equilibria of maximal convex subsets of $\mathcal{E}(A, B)$. Let Y_0 and X_0 be, respectively, the finite set of such solutions for potential equilibrium components $\mathbf{y}$ of player II and $\mathbf{x}$ of player I. For any $Y \subseteq Y_0$, define $S(Y) = \{\mathbf{x} \in X_0 : (\mathbf{x}, \mathbf{y}) \text{ is a Nash equilibrium for all } \mathbf{y} \in Y\}$. ($S(Y)$ can also be empty.) $S(X)$ is similarly defined. We have the following:

THEOREM [Vorobiev (1958); Kuhn (1961)]. *Given a bimatrix game (A, B), the equilibrium set is given by*

$$\mathcal{E}(A, B) = \bigcup_{Y \subseteq Y_0} S(Y) \times \operatorname{con} Y$$

or equivalently

$$\mathcal{E}(A, B) = \bigcup_{X \subseteq X_0} S(X) \times \operatorname{con} X,$$

where for any set T, $\operatorname{con} T$ denotes the convex hull of T.

The following refinement of perfect equilibria was introduced by Myerson (1978).

DEFINITION. A Nash equilibrium $(\mathbf{x}^*, \mathbf{y}^*)$ for a bimatrix game (A, B) is called *proper* if there exists a sequence of completely mixed strategy pairs $(\mathbf{x}^n, \mathbf{y}^n) \to (\mathbf{x}^*, \mathbf{y}^*)$ such that for some sequence $0 < \varepsilon_n \to 0$,

$$\left(B^{\mathrm{T}}\mathbf{x}^n\right)_j < \left(B^{\mathrm{T}}\mathbf{x}^n\right)_k \Rightarrow \mathbf{y}_j^n \leqslant \varepsilon_n \mathbf{y}_k^n$$

and

$$\left(A\mathbf{y}^n\right)_r < \left(A\mathbf{y}^n\right)_s \Rightarrow \mathbf{x}_r^n \leqslant \varepsilon_n \mathbf{x}_s^n.$$

Intuitively, one would like more costly mistakes to occur with less probability.

REMARK. It can be shown [Borm, Jansen, Potters and Tijs (1993)] that the set of perfect equilibria for a bimatrix game (A, B) is also a union of finitely many polytopes. Further, each such polytope is a face of some maximal convex subset of $\mathcal{E}(A, B)$.

Suppose one decomposes the strategy space Δ_M^I into equivalence classes as follows. We say $\mathbf{p} \sim_1 \mathbf{p}^*$, if for all $j, l \in \{1, \ldots, n\}$

$$\left(B^{\mathrm{T}}\mathbf{p}\right)_j > \left(B^{\mathrm{T}}\mathbf{p}\right)_l \quad \text{if and only if} \quad \left(B^{\mathrm{T}}\mathbf{p}^*\right)_j > \left(B^{\mathrm{T}}\mathbf{p}^*\right)_l.$$

Using this equivalence relation, one can decompose the completely mixed strategies of player I. By a similar equivalence relation, one can decompose the set of completely mixed strategies of player II. By introducing the properness concept on these equivalence classes, one can extend the Vorobiev–Kuhn theorem to proper equilibria [Jansen (1993)]. For other extensions, see Jansen, Jurg and Borm (1994).

8. Bimatrix games and exchangeability of Nash equilibria

Two equilibria $(\mathbf{x}, \mathbf{y}), (\mathbf{x}', \mathbf{y}') \in \mathcal{E}(A, B)$ are *exchangeable* if and only if $(\mathbf{x}, \mathbf{y}') \in \mathcal{E}(A, B)$ and $(\mathbf{x}', \mathbf{y}) \in \mathcal{E}(A, B)$. A bimatrix game is *exchangeable* if and only if any two equilibria in $\mathcal{E}(A, B)$ are exchangeable. One can extend the exchangeability notion also for any non-zero-sum n-person game in strategic form. For example if (x, y, z) and (x', y', z') are any two equilibria for a 3-person game, then the equilibrium set of the game is exchangeable if and only if (p, q, r) is also an equilibrium when $p = x$ or x', $q = y$ or y', $r = z$ or z'.

One can show that a bimatrix game is exchangeable if and only if the equilibrium set is convex [Chin, Parthasarathy and Raghavan (1974)]. Huer and Millham (1976) studied maximal convex subsets of the equilibrium set. They showed that these maximal convex sets are polytopes (this can also be seen from the Vorobiev–Kuhn theorem). However, convexity is not sufficient for exchangeability for n-person games, $n \geqslant 3$ [Chin, Parthasarathy and Raghavan (1974)].

The theorems of Vorobiev (1958) and Kuhn (1961) have only been of theoretical interest and have rarely been used in practice to compute Nash equilibria. This is also the case for the algorithms of Mills (1960), Mangasarian (1964), Winkels (1979), and Mukhamediev (1978). An efficient algorithm to locate a Nash equilibrium was first given by Lemke and Howson (1964). This algorithm not only locates an equilibrium, it in fact terminates efficiently in finite arithmetic steps, showing the ordered field property for equilibria in bimatrix games. However, this algorithm will not locate all the extreme equilibria.

9. The Lemke–Howson algorithm

Let (C, D) be an $m \times n$ bimatrix game. Let E be a matrix with all entries unity. Choose $k > 0$ such that $A = kE - C > 0$, $B = kE - D > 0$. The problem of finding a Nash equilibrium pair is reduced to the following:

Given $A, B > 0$ and the vector $\mathbf{1}$ with all coordinates unity, let

$$X = \left\{\mathbf{x}: B^\mathsf{T}\mathbf{x} \geqslant \mathbf{1}, \mathbf{x} \geqslant 0\right\}, \qquad Y = \{\mathbf{y}: A\mathbf{y} \geqslant \mathbf{1}, \mathbf{y} \geqslant 0\},$$

$$\mathbf{u} = A\mathbf{y} - \mathbf{1}, \qquad \mathbf{v} = B^\mathsf{T}\mathbf{x} - \mathbf{1}. \tag{10}$$

Find $\bar{\mathbf{x}} \in X$, $\bar{\mathbf{y}} \in Y$, $\bar{\mathbf{u}} = A\bar{\mathbf{y}} - \mathbf{1}$, $\bar{\mathbf{v}} = B^\mathsf{T}\bar{\mathbf{x}} - \mathbf{1}$ such that the dot products $\bar{\mathbf{x}}.\bar{\mathbf{u}} = 0$, $\bar{\mathbf{y}}.\bar{\mathbf{v}} = 0$. Here $\bar{x}_i \bar{u}_i = 0$, $\bar{y}_j \bar{v}_j = 0$ for all coordinates i, j. By normalizing $\bar{\mathbf{x}}$ and $\bar{\mathbf{y}}$, one can get an equilibrium point for the bimatrix game (C, D).

A pair $\mathbf{x}$ of X and $\mathbf{y}$ of Y is called *almost complementary* if $x_i u_i = 0$, $y_j v_j = 0$ for all i, j except exactly for one i or one j. An extreme point $(\mathbf{x}, \mathbf{y}) \in X \times Y$ is *nondegenerate* if exactly $m + n$ of x_i, u_i, y_j and v_j are equal to zero. When an almost complementary extreme pair $(\mathbf{x}, \mathbf{y})$ fails to be a Nash equilibrium point, then for exactly one i or one j, $x_i = u_i = 0$ or $y_j = v_j = 0$. By slightly perturbing the payoffs, we can always assume that all extreme points are nondegenerate.

An *edge* of $X \times Y$ consists of all pairs $(\mathbf{x}, \mathbf{y})$ where either $\mathbf{x}$ is a fixed extreme point of X while $\mathbf{y}$ varies over a geometric edge of Y or $\mathbf{y}$ is a fixed extreme point of Y while $\mathbf{x}$ varies over a geometric edge of X. Using scalar multiples of unit vectors, we can easily initiate the algorithm at an extreme pair $(\mathbf{x}^0, \mathbf{y}^0)$, which is the end point of the unique unbounded edge, lying on the almost complementary path, where $x_1^0 u_1^0 = 0$ may possibly be violated. If it is not violated, then we are done. Suppose $x_1^0 > 0$, $u_1^0 > 0$. Then by the nondegeneracy assumption, there should be a pair $x_i^0 = 0$, $u_i^0 = 0$ or $y_j^0 = 0$, $v_j^0 = 0$ for some i or j. Suppose we have $y_2^0 = 0$, $v_2^0 = 0$. The current extreme pair is the end of precisely two edges, one violating $y_2^0 = 0$ and the other violating $v_2^0 = 0$, while the almost complementary condition still holds. Of these two edges the one violating, say, $v_2^0 = 0$, is the unbounded edge from which we started. Thus, the algorithm takes us along the other edge violating $y_2^0 = 0$ where the almost complementary condition is still satisfied. This must end in another extreme pair $(\mathbf{x}^0, \mathbf{y}^1)$. If this is still only almost complementary, it will be an end vertex of exactly two edges of $X \times Y$ lying solely on the almost complementary path. We just traveled along one edge to reach the extreme pair $(\mathbf{x}^0, \mathbf{y}^1)$. We therefore move along the other, untravelled edge.

Having reached an end vertex of an almost complementary edge, we always move along the other, untravelled edge. Since exactly one *unbounded edge* lies completely on this almost complementary path, and since we have precisely two edges meeting at each almost complementary extreme point, we have no way of reaching the unbounded edge again without retracing the traveled edges. With only finitely many extreme points present, the algorithm must terminate at an extreme point that is complementary. However, there is no warrant of success if we start at any arbitrary extreme pair lying on the almost complementary path. For, in this case, we may move along *bounded edges* of $X \times Y$ that lie on an almost complementary path that forms a cycle. Such a search is a clear waste of our efforts. Suppose we reach a complementary extreme pair in our travel. Then we can move along the other end of the initial edge lying in the almost complementary path. But this could terminate at one end of the unique unbounded edge or possibly at another complementary extreme pair.

Thus we have three types of almost complementary paths. The first consists of cycles of pure almost complementary edges. The second consists of a path terminating at two ends with complementary extreme pairs. The third consists of the unique path via the unbounded edge lying on the almost complementary path, terminating at precisely one complementary solution. Thus, the number of complementary solutions to the nondegenerate problem is odd. The algorithm establishes the following theorem.

THEOREM [Lemke and Howson (1964)]. *Any nondegenerate bimatrix game has a finite odd number of equilibria.*

A geometrically transparent approach by Rosenmüller (1981) implements the Lemke–Howson algorithm for a bimatrix game directly on the pair of mixed strategy simplices Δ_M^I, Δ_N^{II} of players I and II, respectively. We will assume that the payoff matrices A and B are nondegenerate; that is, for all square submatrices C of A and D of B, the matrices of the form

$$\begin{bmatrix} C & -\mathbf{1} \\ \mathbf{1}^T & 0 \end{bmatrix}, \quad \begin{bmatrix} D & -\mathbf{1} \\ \mathbf{1}^T & 0 \end{bmatrix}$$

are nonsingular except for the trivial 1×1 matrix $\mathbf{0}$. Under this assumption, one can restrict the search for equilibrium strategies of Player II to extreme points of polytopes

$K_1, \ldots, K_m$, where

$$K_i = \{y: \text{ the } i\text{-th row of the payoff matrix } A \text{ is the pure best reply against } y\}$$
$$\cap \Delta_N^{II}.$$

Similarly, the search for the equilibrium strategies of player I can be restricted to extreme points of polytopes L_j, $j = 1, \ldots, n$, where

$$L_j = \{x: \text{ the } j\text{-th column of the payoff matrix } B \text{ is the pure best reply against } x\}$$
$$\cap \Delta_M^{I}.$$

Clearly, $\bigcup_i K_i = \Delta_N^{II}$ and $\bigcup_j L_j = \Delta_M^{I}$. The key idea in the algorithm is to form a pair of paths, one in each simplex, which travels along one-dimensional edges of the above polytopes connecting extreme points. If the end vertices of the edges reached most recently in the two paths do not form an equilibrium, then by working on one simplex at a time a new unique edge is added to the path. We will explain Rosenmüller's algorithm through the following example.

EXAMPLE. Let

$$A = \begin{bmatrix} 3 & 5 & 1 \\ 1 & 3 & 3 \\ 5 & 1 & 2 \end{bmatrix}, \qquad B = \begin{bmatrix} 7 & 2 & 8 \\ 8 & 4 & 7 \\ 3 & 8 & 6 \end{bmatrix}.$$

The unit vectors in $\Delta_{\{1,2,3\}}^{I}$ will be denoted by $\mathbf{f}_1, \mathbf{f}_2, \mathbf{f}_3$, and the unit vectors in $\Delta_{\{1,2,3\}}^{II}$ will be denoted by $\mathbf{e}_1, \mathbf{e}_2, \mathbf{e}_3$. The simplex $\Delta_{\{1,2,3\}}^{I}$ can be represented by an equilateral triangle of altitude 1. A point P in the triangle has coordinates (x_1, x_2, x_3) where x_1, for example, is the distance of the point P from the line joining the vertices $\mathbf{e}_2$ and $\mathbf{e}_3$. In our example, say

$$K_2 = \{(y_1, y_2, y_3): y_1 + 3y_2 + 3y_3 \geqslant \max(3y_1 + 5y_2 + y_3, 5y_1 + y_2 + 2y_3)\}$$
$$\cap \Delta_{\{1,2,3\}}^{II}.$$

Similarly, say

$$L_3 = \{(x_1, x_2, x_3): 8x_1 + 7x_2 + 6x_3 \geqslant \max(7x_1 + 8x_2 + 3x_3, 2x_1 + 4x_2 + 8x_3)\}$$
$$\cap \Delta_{\{1,2,3\}}^{I}.$$

The vertices and partitions are shown in Figure 2.

The algorithm begins with a vertex $\mathbf{f}_i$ of Δ_M^{I} and a vertex $\mathbf{e}_j$ of Δ_N^{II} such that either $\mathbf{f}_i \in L_j$ or $\mathbf{e}_j \in K_i$. This is always possible. If $\mathbf{f}_i \in L_j$ *and* $\mathbf{e}_j \in K_i$, then $(\mathbf{f}_i, \mathbf{e}_j)$ is an equilibrium. In the example, we can start at $\mathbf{f}_3 \in L_2$ and then choose $\mathbf{e}_2$. Since $\mathbf{e}_2 \in K_1$,

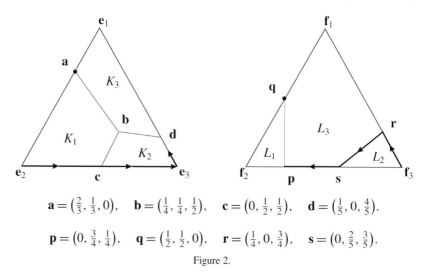

$$a = \left(\tfrac{2}{3}, \tfrac{1}{3}, 0\right), \quad b = \left(\tfrac{1}{4}, \tfrac{1}{4}, \tfrac{1}{2}\right), \quad c = \left(0, \tfrac{1}{2}, \tfrac{1}{2}\right), \quad d = \left(\tfrac{1}{5}, 0, \tfrac{4}{5}\right).$$

$$p = \left(0, \tfrac{3}{4}, \tfrac{1}{4}\right), \quad q = \left(\tfrac{1}{2}, \tfrac{1}{2}, 0\right), \quad r = \left(\tfrac{1}{4}, 0, \tfrac{3}{4}\right), \quad s = \left(0, \tfrac{2}{5}, \tfrac{3}{5}\right).$$

Figure 2.

we look for the unique edge that starts at $\mathbf{f}_3$, ends in a vertex with *first coordinate* positive, and belongs to a new polytope. This is the unique edge joining $\mathbf{f}_3$ with $\mathbf{r}$. Now $\mathbf{r} \in L_3$, and so we look for the unique edge from $\mathbf{e}_2$ to a vertex with a positive third coordinate. This is the edge joining $\mathbf{e}_2$ with $\mathbf{c}$. Since $\mathbf{c} \in K_2$, we move from $\mathbf{r}$ to $\mathbf{s}$, a vertex with a positive second coordinate. The point $\mathbf{s}$ belongs to L_2 and L_3, and we have visited them both. Since the first coordinate of $\mathbf{s}$ is 0, we leave K_1 and reach $\mathbf{e}_3$ in $\Delta^{II}_{\{1,2,3\}}$. Our current pair of strategies is $(\mathbf{s}, \mathbf{e}_3)$, which is not a Nash equilibrium. The point $\mathbf{e}_3$ is in K_2, which we have visited, and its second coordinate is 0. Thus, in $\Delta^{I}_{\{1,2,3\}}$, we leave L_2 and reach $\mathbf{p} \in L_1$. We then move from $\mathbf{e}_3$ to $\mathbf{d}$, which has a positive first coordinate. Observe that $\mathbf{d} \in \Delta^{II}_{\{1,2,3\}}$ has positive coordinates in the first and third positions while $\mathbf{p} \in \Delta^{I}_{\{1,2,3\}}$ belongs to L_1 and L_3. Thus, we have reached an equilibrium $(\mathbf{p}, \mathbf{d})$.

REMARK. To illustrate the algorithm, we calculated all the extreme points of the partitioned polytopes. This is not necessary to execute the algorithm. The procedure is based on ordinary simplex pivoting rules. For detailed implementation of the algorithm, see Krohn, Moltzahn, Rosenmüller, Sudhölter and Wallmeier (1991).

10. An algorithm for locating a perfect equilibrium in bimatrix games

While the Lemke–Howson algorithm finds an equilibrium, it need not reach a perfect or proper equilibrium. An algorithm by van den Elzen and Talman (1991, 1995), closely related to Harsanyi's tracing procedure (1975), leads to a perfect equilibrium of the positively oriented type [Shapley (1974)]. We will use the example in van den Elzen and Talman (1991) to illustrate the idea behind the algorithm.

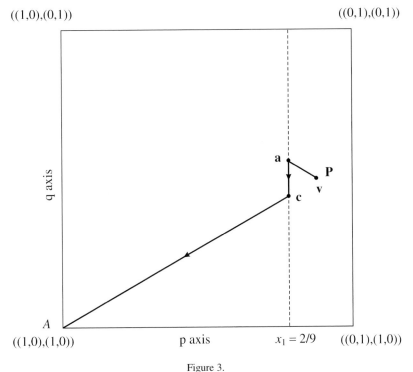

((1,0),(0,1))   ((0,1),(0,1))

q axis

a

P

v

c

A

((1,0),(1,0))   p axis $x_1 = 2/9$ ((0,1),(1,0))

Figure 3.

Consider the bimatrix game

$$A = \begin{bmatrix} 4 & 0 \\ -1 & 3 \end{bmatrix}, \qquad B = \begin{bmatrix} 4 & -3 \\ 2 & 4 \end{bmatrix}.$$

Given the unit square plotted below, let any point (p, q) in the square with $0 \leqslant p, q \leqslant 1$ denote the strategy pair $x = \begin{bmatrix} 1-p \\ p \end{bmatrix}$, $y = \begin{bmatrix} 1-q \\ q \end{bmatrix}$.

For example, the point A with (p, q) coordinates $= (0, 0)$ corresponds to the strategy pair $(1, 0)^T$, $(1, 0)^T$, namely to choosing the first row and first column in the bimatrix game. Starting from an arbitrarily chosen completely mixed strategy pair $\mathbf{x}^0$, $\mathbf{y}^0$ for the two players, the algorithm generates a piecewise linear path leading to a Nash equilibrium. For any generic point $x = (x_1, x_2)^T$, $y = (y_1, y_2)^T$ on the piecewise linear path, let $\xi_1 = (Ay)_1$, $\xi_2 = (Ay)_2$, $\eta_1 = (B^T x)_1$, $\eta_2 = (B^T x)_2$. The points on the path to be generated satisfy the following conditions:

$$
\begin{aligned}
x_i &= bx_i^0 & \text{if} \quad & \xi_i < \max_h \xi_h, & i &= 1, 2, \\
x_i &\geqslant bx_i^0 & \text{if} \quad & \xi_i = \max_h \xi_h, & i &= 1, 2, \\
y_j &= by_j^0 & \text{if} \quad & \eta_j < \max_h \eta_h, & j &= 1, 2, \\
y_j &\geqslant by_j^0 & \text{if} \quad & \eta_j = \max_h \eta_h, & j &= 1, 2,
\end{aligned}
$$

where

$$0 \leqslant b = \min_{h} \left(\frac{x_h}{x_h^0}, \frac{y_h}{y_h^0}, x_h^0, y_h^0 > 0 \right).$$

Suppose we start at the interior point P of the square with coordinates $(\frac{7}{8}, \frac{1}{2})$. This corresponds to the strategy $x^0 = (\frac{1}{8}, \frac{7}{8})^T$, $y^0 = (\frac{1}{2}, \frac{1}{2})^T$. Then $(\xi_1, \xi_2) = (2, 1)$ and $(\eta_1, \eta_2) = (\frac{9}{4}, \frac{25}{8})$. At (x^0, y^0) we have $\xi_1 > \xi_2$ and $\eta_1 < \eta_2$. So the algorithmic path leaves (x^0, y^0) in the direction of $(1, 0)^T$, $(0, 1)^T$. This way the algorithm arrives at the point $\mathbf{a} = (\frac{7}{9}, \frac{5}{9})$. This corresponds to the mixed strategy pair $\mathbf{x} = (\frac{2}{9}, \frac{7}{9})^T$, $\mathbf{y} = (\frac{4}{9}, \frac{5}{9})^T$. Along the segment $[\mathbf{p}, \mathbf{a}]$ the b value decreases from 1 to $\frac{8}{9}$. At this new pair, $\xi_1 = \frac{16}{9}$, $\xi_2 = \frac{11}{9}$. Further, $\eta_1 = \frac{22}{9}$, $\eta_2 = \frac{22}{9}$. The algorithm generates a new strategy pair keeping the value of $\eta_1 = \eta_2 = \frac{22}{9}$ fixed. This holds along the line segment $[\mathbf{a}, \mathbf{c}]$. The algorithm generates the sequence of points on the polygonal path given by $\mathbf{a} = (\frac{7}{9}, \frac{5}{9})$, $\mathbf{c} = (\frac{5}{9}, \frac{4}{9})$, and the line segment joining $\mathbf{c}$ with $((1, 0), (1, 0))$.

The algorithm is based on a certain nondegeneracy assumption (different from the nondegeneracy assumption in the Lemke–Howson algorithm) and on complementary pivoting methods. Talman and Yang (1994) and Yang (1996) also developed an iterative algorithm to approximate proper equilibria.

11. Enumerating all extreme equilibria

In their approach, Mangasarian (1964) and Winkels (1979) simply choose each vertex $\mathbf{x} \in X$, $\mathbf{y} \in Y$ (see (10) in Section 9) and check for the equilibrium conditions. Dickhaut and Kaplan (1991) and McKelvey and McLennan (1996) describe algorithms that enumerate the $(2^m - 1) \times (2^n - 1)$ possible carriers and then check for the equilibrium conditions. These are essentially enumeration methods. With an exploding number of vertices, even for low-dimensional polytopes, one faces formidable numerical problems.

To enumerate all the extreme equilibrium points, Audet et al. (1998) propose a branch and bound approach to the following pair of parametric linear programs. Let

$$\overline{X} = \left\{ (\mathbf{x}, \beta) : B^T \mathbf{x} \leqslant \beta \mathbf{1}, \ \mathbf{x} \geqslant 0, \ \sum x_i = 1 \right\},$$

$$\overline{Y} = \left\{ (\mathbf{y}, \alpha) : A\mathbf{y} \leqslant \alpha \mathbf{1}, \ \mathbf{y} \geqslant 0, \ \sum y_j = 1 \right\}.$$

Given any vertex $(\mathbf{x}, \beta) \in \overline{X}$, let

$$q(\mathbf{x}) = \max_{(\mathbf{y}, \alpha) \in \overline{Y}} \mathbf{x}^T B \mathbf{y} - \alpha.$$

Given any vertex $(\mathbf{y}, \alpha) \in \overline{Y}$, let

Problem p: $\displaystyle p(\mathbf{y}) = \max_{(\mathbf{x},\beta)\in\overline{X}} \mathbf{x}^{\mathrm{T}} A \mathbf{y} - \beta.$

Start with any vertex of either polytope $\overline{X}$ or $\overline{Y}$. Say we start with a vertex $(\mathbf{x}, \beta) \in \overline{X}$. We have an optimal solution $(\mathbf{y}, \alpha) \in \overline{Y}$ to the problem

Problem q: $\displaystyle q(\mathbf{x}) = \max_{(\mathbf{y},\alpha)\in\overline{Y}} \mathbf{x}^{\mathrm{T}} B \mathbf{y} - \alpha.$

If the pair $(\mathbf{x}, \mathbf{y})$ constitutes a Nash equilibrium, then we stop and start all over with a new vertex of $(\mathbf{x}, \beta) \in \overline{X}$. Suppose not. Then we look for

$$\max\left\{ x_i\left(\alpha - (A\mathbf{y})_i\right), \, y_j\left(\beta - \left(B^{\mathrm{T}}\mathbf{x}\right)_j\right) \right\} > 0.$$

Suppose it occurs at, say, $x_i(\alpha - (A\mathbf{y})_i)$. We introduce exactly two new subproblems p^i, u_i where p^i differs from p with an added equality constraint $x_i = 0$. Similarly, the subproblem u_i is generated from q by introducing the additional constraint $(A\mathbf{y})_i = \alpha$ on problem q. That is,

Problem p^i: $\displaystyle p^i(\mathbf{y}) = \max_{(\mathbf{x},\beta)\in\overline{X},\ x_i=0} \mathbf{x}^{\mathrm{T}} A \mathbf{y} - \beta$

and

Problem u_i: $\displaystyle u_i(\mathbf{x}) = \max_{(\mathbf{y},\alpha)\in\overline{Y},\ (A\mathbf{y})_i=\alpha} \mathbf{x}^{\mathrm{T}} B \mathbf{y} - \alpha.$

Similarly, one can define subproblems q^j, v_j. When a subproblem is infeasible, the node is discarded and we move to the other subproblems by some suitable backtracking method. If the two subproblems are feasible, either they have enough teeth (in terms of equations) to anchor the current pair to an extreme Nash equilibrium pair or we reach new nodes by branching off again.

The branching rule is to choose that index i or j with the largest complementary product $x_i(\alpha - (A\mathbf{y})_i)$ or $y_j(\beta - (B^{\mathrm{T}}\mathbf{x})_j)$.

12. Bimatrix games and fictitious play

Fictitious play was proposed by Brown (1949) as an iterative algorithm to approximate the value of zero-sum matrix games. The convergence of the fictitious play process to the value of the game was proved by Robinson (1951).

For games in normal form, fictitious play provides a learning procedure for boundedly rational players (see Chapter 4 in this Handbook for details on bounded rationality).

Here the players interpret mixed strategies as a form of beliefs about the opponent's choices. In the context of bimatrix games under repeated play, each player assumes that in each round the opponent will continue to choose actions in the same proportion as he did in the past. In each round, this gives each player a recipe to choose a pure or mixed action that is a best reply against the opponent's myopic belief.

The fictitious play is said to converge if the sequence of empirical distributions on actions as generated above for the two players is close to the Nash equilibrium set after sufficiently many stages. Miyasawa (1963) showed that by choosing an appropriate pure action, from the set of all pure best replies to the empirical distribution of the opponent's past actions, the process converges for all 2×2 bimatrix games. The proper choice of pure actions is crucial to the convergence, even in 2×2 bimatrix games [Monderer and Sela (1996)]. Shapley (1964) showed that though the bimatrix game

$$(A, B) = \begin{bmatrix} (2, 1) & (0, 0) & (1, 2) \\ (1, 2) & (2, 1) & (0, 0) \\ (0, 0) & (1, 2) & (2, 1) \end{bmatrix}$$

has a unique Nash equilibrium payoff $(1, 1)$ corresponding to the Nash equilibrium $(\mathbf{x}^*, \mathbf{y}^*) = ((\frac{1}{3}, \frac{1}{3}, \frac{1}{3}), (\frac{1}{3}, \frac{1}{3}, \frac{1}{3}))$, the fictitious play fails to converge.

To show this, suppose the play initiates with the pair of actions $(1, 1)$. After $t_1 = 1$ day, the players switch to actions $(1, 3)$ and use these actions for $t_2 = 3$ days. Then the players switch to actions $(3, 3)$ for $t_3 = 8$ days, and so on. The fictitious play process travels cyclically along the choices $(i, j) = (1, 1), (1, 3), (3, 3), (3, 2), (2, 2), (2, 1)$ and returns to $(1, 1)$. In the new round starting at $(1, 1)$, we get a new run length, say t_1', which can be shown to be at least four times the previous run length t_1. If X_i is the number of times row i is chosen by player I, neither the ratios $\frac{X_i}{X_k}$ for $i \neq k$, nor the empirical distribution functions converge. In particular, the fictitious play never converges to the unique equilibrium.

Vijay Krishna and Tomas Sjöström (1998) show that Shapley's counterexample above is simply the generic behavior of fictitious play in bimatrix games with more than two pure strategies for the two players. The following theorem makes it precise.

THEOREM. *For almost all games, if there is an open set of initial conditions and a cyclic path from the initial conditions such that the continuous version of the fictitious play converges to a Nash equilibrium, then the Nash equilibrium strategies have at most two points in their spectrum.*

PROOF. The following is the idea behind the proof. The discrete fictitious play can be replaced by a continuous fictitious play. Thus we can replace the appropriate difference equations by the differential equations:

$$\frac{dp}{dt}\bigg|_+ = \frac{i(t) - p(t)}{t}; \qquad \frac{dq}{dt}\bigg|_+ = \frac{j(t) - q(t)}{t}.$$

Here $i(t)$ is the pure best reply of player I against his myopic belief on the mixed strategy $q(t)$ of player II at time t and $j(t)$ is the pure best reply against $p(t)$. We can assume that except for a countable number of time points $t_1, t_2, \ldots, i(t)$ and $j(t)$ are unique. In an interval $t_{n-1} < t < t_n$, the pure best replies are the same, and even when there is a change, it is only for one player. Thus, the sequence of choices

$$(i_1, j_1), (i_2, j_2), \ldots, (i_K, j_K)$$

are repeated in the same order again and again. Let the run length in the r-th round be $n_k(r)$ for the strategy choice (i_k, j_k). Let $n(r) = (n_1(r), \ldots, n_K(r))^{\mathrm{T}}$. Then the run lengths satisfy a certain matrix equation $n(r+1) = Fn(r)$ for a matrix F.

The proof involves studying the eigenvalues of F. The matrix F can be shown to be singular, with the crucial observation that the product of all the nonzero eigenvalues is unity. Geometrically this means that the map F is volume-preserving in a lower dimensional invariant subspace M. For example, if $S \subset M$, then the set $F(S) = \{Fn: n \in S\}$ also has the same volume V as S. Thus there cannot exist an open set of starting positions so that the run lengths decrease from round to round. If each player uses at least three pure strategies in the cycle, then one can show by a long analysis that there will exist at least one eigenvalue $\lambda > 1$. This can be used to show that for almost all initial conditions the run lengths increase exponentially and the continuous fictitious play fails to converge. $\qquad\square$

REMARK. For the following special classes of bimatrix games, fictitious play does converge:
   (i) Strategically zero-sum games (see Section 13 for definitions).
   (ii) Bimatrix games of the form (A, A) [Monderer and Shapley (1996)].
   (iii) Strongly dominance solvable games [Milgrom and Roberts (1991)].

13. Correlated equilibrium

A Nash equilibrium $(\mathbf{x}^*, \mathbf{y}^*)$ presupposes a belief on the part of each player that the opponent will use his mixed strategy component of the Nash equilibrium in selecting his pure actions. With no communication between players, this is questionable. Aumann suggested a method to overcome this inherent deficiency by the notion of a *correlated strategy* [Aumann (1974, 1987)].

Suppose a referee advises the players to use a particular Nash equilibrium. Then neither player will gain by deviating from the referee's advice, under the assumption that the opponent will take the referee's advice. The referee's role ends just after focusing the attention of the players to a particular Nash equilibrium. A more active role by the referee is possible when he selects, based on a random device, an outcome (i, j) with probability p_{ij}. While the device and the probabilities p_{ij} are known to the two players, the referee can maintain the secrecy of the outcome by suggesting action i to player I without revealing the true outcome (i, j). Similarly, action j is suggested to player II without revealing the true outcome (i, j). If each player believes that the opponent will take the referee's advice, then the best action for, say, player I is one that maximizes the conditional expected payoff, namely, $\max_k \sum_j a_{kj} \frac{p_{ij}}{p_{i.}}$, where $p_{i.} = \sum_j p_{ij}$. Any given probability distribution $\mathbf{p} = (p_{ij})$ on the set of joint actions is called a *correlated equilibrium* if each player's conditional expected payoff is maximized at the referee's suggested action, under the assumption that the opponent will implement the referee's advice.

Equivalently, given a bimatrix game (A, B), there exists a $\mathbf{p} = (p_{ij})$ such that

$$\sum_j a_{ij} p_{ij} \geqslant \sum_j a_{kj} p_{ij} \quad \text{for all } i, k = 1, 2, \ldots, m,$$

$$\sum_i b_{ij} p_{ij} \geqslant \sum_i b_{il} p_{ij} \quad \text{for all } j, l = 1, 2, \ldots, n,$$

$$p_{ij} \geqslant 0 \quad \text{for all } i, j,$$

$$\sum_i \sum_j p_{ij} = 1.$$

Given any Nash equilibrium $(\mathbf{x}^*, \mathbf{y}^*)$, $(p_{ij}) = (x_i^* y_j^*)$ satisfies the above inequalities and is therefore a correlated equilibrium. For an alternative proof based on the minimax theorem, see Hart and Schmeidler (1989).

A correlated equilibrium may be preferred by the players because one (or both) of the players may have a strict gain over any Nash equilibrium payoff.

EXAMPLE. Consider the Nash equilibrium $(\mathbf{x}^*, \mathbf{y}^*) = ((\frac{1}{3}, \frac{1}{3}, \frac{1}{3}), (\frac{1}{3}, \frac{1}{3}, \frac{1}{3}))$ for the bimatrix game

$$(A, B) = \begin{bmatrix} (0, 0) & (1, 2) & (2, 1) \\ (2, 1) & (0, 0) & (1, 2) \\ (1, 2) & (2, 1) & (0, 0) \end{bmatrix}$$

with expected payoff $(1, 1)$. It is strictly improved for both players by the correlated equilibrium

$$p_{ij} = \begin{cases} \frac{1}{6} & \text{for all } i \neq j, \\ 0 & \text{otherwise,} \end{cases}$$

which gives the correlated payoff $(\frac{3}{2}, \frac{3}{2})$.

For zero-sum games $(B = -A)$, the correlated equilibrium inequalities reduce to

$$v = \sum_i \sum_j a_{ij} p_{ij} \geqslant \sum_j a_{kj} p_{.j} \quad \text{for all } k,$$

$$-v = \sum_i \sum_j b_{ij} p_{ij} \geqslant \sum_i b_{il} p_{i.} \quad \text{for all } l.$$

Thus, the marginals $p_{i\cdot}$ and $p_{\cdot j}$ are optimal for the game, and the value of the game coincides with the correlated equilibrium value for the two players. This implies that the correlation device gives no advantage to either player in zero-sum matrix games. In this case, we say that the game has no *good correlated equilibrium*. We call a correlated equilibrium *good* if, for any Nash equilibrium, at least one player strictly prefers the correlated equilibrium to the Nash equilibrium.

We call two bimatrix games (A, B) and (C, D) of the same order *best-response-equivalent (BRE)* if, after renumbering the rows and columns if necessary, the set of best-response mixed strategies of either player against any given mixed strategy choice of the opponent is the same for both bimatrix games. In this case, the two games have the same Nash and correlated equilibria. If the game is *BRE* to a zero-sum bimatrix game, the Nash equilibrium set will be convex and hence any two equilibria will be exchangeable (see Section 8). These games are called *strategically zero-sum games*. The following theorem characterizes algebraically bimatrix games that are strategically zero-sum.

THEOREM [Moulin and Vial (1978)]. *An $m \times n$ bimatrix game (A, B) is strategically zero-sum if and only if there exists $\lambda > 0$, $\mu > 0$ such that $\lambda a_{ij} + \mu b_{ij} = u_i + v_j$ for some u_i's and v_j's.*

THEOREM [Rosenthal (1974)]. *If a bimatrix game (A, B) is strategically zero-sum, then no correlation device is of any advantage to either player.*

PROOF. If (A, B) is *BRE* to $(C, -C)$ then the two games have the same set of correlated equilibria. For any correlated equilibrium $\mathbf{p}$, $(\mathbf{p}_{i\cdot}, \mathbf{p}_{\cdot j})$ is a Nash equilibrium in $(C, -C)$ and hence in (A, B). Given any Nash equilibrium $(\mathbf{x}^*, \mathbf{y}^*)$, the exchangeability shows that $(\mathbf{x}^*, \mathbf{p}_{\cdot j})$ is another Nash equilibrium. Since $\mathbf{x}^*$ is a best reply against $\mathbf{p}_{\cdot j}$, we have

$$\sum_j a_{kj} p_{\cdot j} \leqslant \sum_i \sum_j a_{ij} p_{ij} \quad \text{for all } k$$

with equality attained for those k for which $x_k^* > 0$. While the right-hand side of the above inequality is the correlated payoff, the left-hand side for any k with $x_k^* > 0$ is the Nash payoff to player I at the Nash equilibrium $(\mathbf{x}^*, \mathbf{y}^*)$. A similar argument shows that the Nash payoff to player II at the Nash equilibrium $(\mathbf{p}_{i\cdot}, \mathbf{y}^*)$ coincides with the correlated equilibrium payoff to player II at the correlated equilibrium $\mathbf{p}$. Thus the correlated payoffs coincide with Nash payoffs. $\square$

Best-response equivalence to zero-sum games is only a sufficient condition for games to have no good correlated equilibria. Often, games possess many properties of perfect conflict and yet possess good correlated equilibria. *Almost strictly competitive games* [Aumann (1961)] are bimatrix games (A, B) such that both (A, B) and $(-B, -A)$ share the same set of equilibrium payoffs and have at least one common equilibrium. In the second game $(-B, -A)$, when players are at an equilibrium, unilateral deviations by players have no power to inflict higher losses on the opponent. The bimatrix game

$$(A, B) = \begin{bmatrix} (6,6) & (2,7) & (-1,6.5) & (-1,1) \\ (7,2) & (0,0) & (-1,1.4) & (-1,3) \\ (6.5,-1) & (1.4,-1) & (0,0) & (0,0) \\ (1,-1) & (3,-1) & (0,0) & (0,0) \end{bmatrix}$$

is almost strictly competitive and yet has a good correlated equilibrium payoff $(5,5)$ (select only the row-column pairs $(1,1)$, $(1,2)$, $(2,1)$, each with probability $\frac{1}{3}$ [Rosenthal (1974)]).

The following theorem [Evangelista and Raghavan (1996)] shows the relationship between Nash and correlated equilibria for bimatrix games.

THEOREM. *Let $(\mathbf{x}^*, \mathbf{y}^*)$ be an extreme point of the Nash equilibrium set in the sense of Vorobiev and Kuhn. Then the correlated equilibrium $\mathbf{p}^* = (p_{ij}^*) = (x_i^* y_j^*)$ is also an extreme point of the set of correlated equilibria.*

PROOF. Let $(\mathbf{x}^*, \mathbf{y}^*)$ be an extreme Nash equilibrium of the bimatrix game (A, B). We can rearrange the rows and columns of the matrix so that

$$x_i^* > 0: \ i = 1, 2, \dots, s; \quad y_j^* > 0: \ j = 1, 2, \dots, r;$$

$$\sum_j a_{ij} y_j^* = \max_k \sum_j a_{kj} y_j^*: \ i = 1, 2, \dots, t;$$

$$\sum_i b_{ij} x_i^* = \max_l \sum_i b_{il} x_i^*: \ j = 1, 2, \dots, u.$$

Since $(\mathbf{x}^*, \mathbf{y}^*)$ is a Nash equilibrium, $s \leqslant t$ and $r \leqslant u$. If $\mathbf{p}^* = (p_{ij}^*) = (x_i^* y_j^*)$ is not an extreme point of the set of correlated equilibria, then there exist correlated equilibria $\mathbf{p}^1, \mathbf{p}^2$ such that $p_{ij}^* = \frac{1}{2} p_{ij}^1 + \frac{1}{2} p_{ij}^2$. Note that $p_{ij}^1 = p_{ij}^2 = 0$ whenever $i > s$ or $j > r$. Let $\alpha^l = \sum_i \sum_j a_{ij} p_{ij}^l$, $l = 1, 2$. Then the marginals $(p_{.1}^l, p_{.2}^l, \dots, p_{.r}^l, \alpha^l)$, $l = 1, 2$, are solutions to the system of equations

$$\sum_{j=1}^r a_{ij} z_j = \alpha, \quad i = 1, 2, \dots, t, \tag{11}$$

$$\sum_{j=1}^r z_j = 1, \quad z_j \geqslant 0, \quad j = 1, 2, \dots, r. \tag{12}$$

Since $(\mathbf{x}^*, \mathbf{y}^*)$ is an extreme Nash equilibrium, from the lemmas in Section 7 it follows that $(z_1, z_2, \dots, z_r, \alpha) = (y_1^*, y_2^*, y_r^*, \mathbf{x}^{*T} A \mathbf{y}^*)$ is the unique solution for the system of equations (11) and (12). Thus, $p^1_{\cdot j} = p^2_{\cdot j} = y^*_j$, $j = 1, 2, \ldots, r$. In a similar way, one can show that $p^1_{i.} = p^2_{i.} = x^*_i$, $i = 1, 2, \ldots, s$, where $p^l_{i.} = \sum_j p^l_{ij}$, $l = 1, 2$.

Now, consider the system of equations

$$\sum_{j=1}^{r}(a_{ij} - a_{kj})p_{ij} = 0, \quad i = 1, \ldots, s; \ k = 1, \ldots, t, \ i \neq k,$$

$$\sum_{j=1}^{r} p_{ij} = x^*_i, \quad i = 1, 2, \ldots, s.$$

We can show that the coefficient matrix of the above system has rank rs. Thus, $\mathbf{p}^* = \mathbf{p}^1 = \mathbf{p}^2$, and $\mathbf{p}^*$ is an extreme point of the correlated equilibrium set. □

REMARK. The above theorem is true only in the strategy space and not in the payoff space. In the following example [Nau and McCardle (1988)],

$$(A, B) = \begin{bmatrix} (0, 0) & (60, 30) & (30, 60) & (40, 40) \\ (30, 60) & (0, 0) & (60, 30) & (40, 40) \\ (60, 30) & (30, 36) & (0, 0) & (40, 40) \\ (40, 40) & (40, 40) & (40, 40) & (41, 41) \end{bmatrix}$$

the unique Nash equilibrium payoff $(41, 41)$ lies in the convex hull of the correlated equilibrium payoffs $(45, 45)$, $(50, 40)$, and $(40, 40\frac{10}{11})$.

14. Bayesian rationality

Another interpretation for the mixed strategy Nash equilibrium for bimatrix games can be given based on Bayesian rationality [Aumann and Brandenburger (1995)]. Suppose each player knows the true value of the opponent's payoffs and chooses a definite pure strategy called an *action*. Not knowing the choice of the opponent, each player can at best make a subjective guess at the opponent's action. This is called a *conjectural assessment* of the opponent's choice. We call a player *rational* or, more generally, *Bayesian rational* when he wants to maximize his conditional expected utility given any exogenous information about the opponent. If players I and II are *rational* and each player knows the opponent's conjectural assessment of his own choice, then the conjectural assessment must be a Nash equilibrium. The assumption about the hierarchy of knowledge is much weaker than what is normally assumed, namely *common knowledge* [Aumann (1976)].

Bayesian rationality for correlated equilibria comes from the following model. Given a bimatrix game (A, B) of order $m \times n$, let Ω be an exogenously given finite set representing possible states of the world, with generic elements ω. Let p be a given probability measure on Ω, known to both players. Let $\mathcal{P}^1$ and $\mathcal{P}^2$ be two partitions of Ω

describing the information available to players I and II respectively, about the true state of the world. Uncertainty inherent in the game can be incorporated via the notion of the state of the world. Given ω chosen according to p by a referee, the players will have to make their choices knowing only the set in their partition that contains the true state ω. Now, any pure action for player I can be thought of as a map $\Omega \to \{1, \ldots, m\}$ that is constant on any set of his partition. Let $x(\omega)$ and $y(\omega)$ be the *pure* actions chosen by players I and II respectively. We call player I *Bayes rational* if his action, given the information, maximizes his conditional expected payoff. The action pair $(x(\omega), y(\omega))$, viewed as a random vector on Ω, induces a distribution on the product space of actions. When both players are Bayes rational, this induced distribution is a correlated equilibrium [Aumann (1987)]. For an application of these ideas to market explosions and market crashes see Hart and Tauman (1996).

15. Weak correlated equilibrium

A correlated strategy $\mathbf{p}$ can be implemented by a referee choosing a pair of actions according to $\mathbf{p}$ and advising each player separately on what action to take. A variation of this situation allows a player to choose between either being bound by the referee's advice once it is given, or not being advised by the referee at all. The strategy $\mathbf{p}$ is called a *weak correlated equilibrium* [Moulin and Vial (1978)] when committing to abide by the referee's advice is best for each player if the opponent does likewise. This is equivalent to $\mathbf{p} = (p_{ij})$ satisfying the weaker inequalities

$$\sum_i \sum_j a_{ij} p_{ij} \geqslant \sum_i \sum_j a_{kj} p_{ij} \quad \text{for } k = 1, 2, \ldots, m,$$

$$\sum_i \sum_j b_{ij} p_{ij} \geqslant \sum_i \sum_j b_{il} p_{ij} \quad \text{for } l = 1, 2, \ldots, n,$$

$$p_{ij} \geqslant 0 \quad \text{for all } i, j,$$

$$\sum_{ij} p_{ij} = 1.$$

A weak correlated equilibrium can result in a strict improvement over all correlated equilibrium payoffs for both players. For example, the bimatrix game

$$(A, B) = \begin{bmatrix} (3, 3) & (1, 1) & (4, 1) \\ (1, 4) & (5, 2) & (0, 0) \\ (1, 1) & (0, 0) & (2, 5) \end{bmatrix}$$

has the unique correlated equilibrium payoff $(3, 3)$ that corresponds to the choice of row 1 and column 1 in the bimatrix game. Thus, it is also the unique Nash equilibrium. However, the strategy $\mathbf{p} = (p_{ij})$, where $p_{ij} = \frac{1}{3}$ for all $i = j$ and 0 otherwise, is a weak correlated equilibrium with payoff $(\frac{10}{3}, \frac{10}{3})$.

The following theorem isolates a class of bimatrix games with weakly correlated equilibrium payoffs that are as good as Nash equilibrium payoffs for both players and better for at least one player.

THEOREM [Moulin and Vial (1978)]. *Let* $(\bar{\mathbf{x}}, \bar{\mathbf{y}})$ *be a completely mixed equilibrium of a bimatrix game* (A, B). *If* (A, B) *is not strategically zero-sum, then the given completely mixed equilibrium payoff can be improved upon by a weak correlated equilibrium payoff.*

PROOF. Based on the previously stated algebraic characterization on strategically zero-sum games, when (A, B) is not strategically zero-sum, the set K of all $m \times n$ matrices that lie on the open line segment joining A and B is disjoint from the linear subspace $\mathcal{H}$ of matrices of the type $C = (c_{ij}) = (u_i + v_j)$. By the strong separation theorem one can find a non-null matrix $F = (f_{ij})$ such that either

$$\sum_i \sum_j a_{ij} f_{ij} \bar{x}_i \bar{y}_j \geqslant 0 \quad \text{or} \quad \sum_i \sum_j b_{ij} f_{ij} \bar{x}_i \bar{y}_j \geqslant 0$$

and at least one of them is strict. Further,

$$\sum_i \sum_j c_{ij} f_{ij} \bar{x}_i \bar{y}_j = 0 \quad \forall C \in \mathcal{H}.$$

If we define $p_{ij} = (\bar{x}_i \bar{y}_j)(1 + \frac{f_{ij}}{2\|f\|_\infty})$, then

$$\sum_i \sum_j a_{ij} p_{ij} = \sum_i \sum_j a_{ij} \left(1 + \frac{f_{ij}}{2\|f\|_\infty}\right) \bar{x}_i \bar{y}_j \geqslant \sum_i \sum_j a_{ij} \bar{x}_i \bar{y}_j,$$

$$\sum_i \sum_j b_{ij} p_{ij} = \sum_i \sum_j b_{ij} \left(1 + \frac{f_{ij}}{2\|f\|_\infty}\right) \bar{x}_i \bar{y}_j \geqslant \sum_i \sum_j b_{ij} \bar{x}_i \bar{y}_j$$

and at least one of the above two inequalities is strict. Thus $p = (p_{ij})$ has a higher expectation than the equilibrium payoff. The p above is the required weakly correlated equilibrium and the weak correlated payoff dominates the Nash payoff at $\bar{x}, \bar{y}$ for at least one player. $\square$

16. Non-zero-sum two-person infinite games

A natural generalization of bimatrix games is to consider strategy spaces that are infinite. The payoffs for each player can be defined as real functions $K_1(\mathbf{x}, \mathbf{y})$ and $K_2(\mathbf{x}, \mathbf{y})$, where $\mathbf{x}$ and $\mathbf{y}$ are in some subset of the Euclidean space. A Nash equilibrium is a pair

$(\mathbf{x}^*, \mathbf{y}^*)$ such that $K_1(\mathbf{x}^*, \mathbf{y}^*) \geqslant K_1(\mathbf{x}, \mathbf{y}^*)$ for all $\mathbf{x} \in X$ and $K_2(\mathbf{x}^*, \mathbf{y}^*) \geqslant K_2(\mathbf{x}^*, \mathbf{y})$ for all $\mathbf{y} \in Y$. In this situation, Nash's theorem can be extended as follows:

THEOREM [Nikaido and Isoda (1959)]. *Let X and Y be compact convex sets in Euclidean space. Let $K_1(\mathbf{x}, \mathbf{y})$ and $K_2(\mathbf{x}, \mathbf{y})$ be continuous real-valued functions on $X \times Y$. Let $K_1(\mathbf{x}, \mathbf{y})$ be a concave function of $\mathbf{x}$ for each $\mathbf{y}$, and let $K_2(\mathbf{x}, \mathbf{y})$ be a concave function of $\mathbf{y}$ for each $\mathbf{x}$. Then the game has a Nash equilibrium.*

PROOF. Without loss of generality, instead of the concavity conditions, we can as well assume that the functions K_1 and K_2 are strictly concave functions in the respective variables. Let $\mathbf{p} = (\mathbf{x}, \mathbf{y}) \in X \times Y$ and $\mathbf{q} = (\mathbf{x}^*, \mathbf{y}^*) \in X \times Y$. Let $f(\mathbf{p}, \mathbf{q}) = K_1(\mathbf{x}, \mathbf{y}^*) + K_2(\mathbf{x}^*, \mathbf{y})$. The map $\phi : X \times Y \to X \times Y$ given by $\mathbf{q} \to \arg\max_{\mathbf{p} \in X \times Y} f(\mathbf{p}, \mathbf{q})$ is continuous. By Brouwer's fixed-point theorem, there is a $\mathbf{q}^\circ$ such that $\max_{\mathbf{p} \in X \times Y} f(\mathbf{p}, \mathbf{q}^\circ) = f(\mathbf{q}^\circ, \mathbf{q}^\circ)$. Thus, $K_1(\mathbf{x}, \mathbf{y}^\circ) + K_2(\mathbf{x}^\circ, \mathbf{y}) \leqslant K_1(\mathbf{x}^\circ, \mathbf{y}^\circ) + K_2(\mathbf{x}^\circ, \mathbf{y}^\circ)$ for all $\mathbf{x} \in X$ and for all $\mathbf{y} \in Y$. This proves that $(\mathbf{x}^\circ, \mathbf{y}^\circ)$ is an equilibrium. $\square$

The concavity condition in the above theorem can be weakened by requiring the functions to be merely quasi-concave. We say that a function $K(\mathbf{x}, \mathbf{y})$ is *quasi-concave* in $\mathbf{y}$ if for each fixed $\mathbf{x}$ and constant c, the set $\{\mathbf{y}: K(\mathbf{x}, \mathbf{y}) > c\}$ is convex.

THEOREM [Dasgupta and Maskin (1986)]. *Let $K_1(\mathbf{x}, \mathbf{y})$ and $K_2(\mathbf{x}, \mathbf{y})$ be upper semi-continuous in $(\mathbf{x}, \mathbf{y})$. Let $K_1(., \mathbf{y})$ be a quasi-concave function of $\mathbf{x}$ for each fixed $\mathbf{y}$ and let $K_2(\mathbf{x}, .)$ be a quasi-concave function of $\mathbf{y}$ for each fixed $\mathbf{x}$. Let $\max_{\mathbf{x}} K_1(\mathbf{x}, \mathbf{y})$ be continuous in $\mathbf{y}$ and let $\max_{\mathbf{y}} K_2(\mathbf{x}, \mathbf{y})$ be continuous in $\mathbf{x}$. Then there exists a Nash equilibrium $(\mathbf{x}^*, \mathbf{y}^*)$.*

The theorems on zero-sum games on the unit square have a natural extension to non-zero-sum games on the unit square. With essentially the same technique as in the zero-sum case, the following theorems can be proved [Parthasarathy and Raghavan (1975)].

THEOREM. *Let $X = Y = [0, 1]$. Let K_1 and K_2 be polynomials in $\mathbf{x}$ and $\mathbf{y}$. Then there exists a Nash equilibrium in mixed strategies using at most finitely many points.*

THEOREM. *Let $X = Y = [0, 1]$. Let K_1 and K_2 be continuous on the unit square with K_2 concave in $\mathbf{y}$. Then there exists a Nash equilibrium of the type $(F^*, I_{\mathbf{y}^*})$, where F^* has a spectrum with at most two points and $I_{\mathbf{y}^*}$ is degenerate at $\mathbf{y}^*$. More generally, if K_1 is continuous in $(\mathbf{x}, \mathbf{y})$ and the nth partial derivative $\frac{\partial^n K_2(\mathbf{x}, \mathbf{y})}{\partial \mathbf{y}^n} > 0$ for all $(\mathbf{x}, \mathbf{y})$ for a fixed n, then there exists an equilibrium (F^*, G^*) where F^* has a finite spectrum with at most n points (0, 1 are counted as half-points) and G^* has a spectrum with at most $\frac{n}{2}$ points.*

The following is a further generalization by Radzik (1993).

THEOREM. *Let $K_1(\mathbf{x}, \mathbf{y})$ be a bounded function on the unit square and let $K_2(\mathbf{x}, \mathbf{y})$ be a function that is bounded above on the unit square. If K_1 is concave in $\mathbf{x}$ for each $\mathbf{y}$, then for any $\varepsilon > 0$, the game has an ε-Nash equilibrium, where the equilibrium strategy for player I has a spectrum consisting of at most two points that are at a distance of less than ε, and the equilibrium strategy of player II has a spectrum consisting of at most two points.*

EXAMPLE. Consider the game on the unit square with payoffs

$$K_1(x, y) = -(x - y)^2, \quad 0 \leqslant x, y \leqslant 1.$$

$$K_2(x, y) = \begin{cases} 1 & \text{if } (0 \leqslant x < \tfrac{1}{2}, y = 1) \text{ or } (\tfrac{1}{2} \leqslant x \leqslant 1, y = 0), \\ 0 & \text{otherwise.} \end{cases}$$

From the above theorems the game has an ε-equilibrium. However, the game does not have a Nash equilibrium. To see this, observe that for any equilibrium (μ, ν), $K_1(\mathbf{x}, \nu)$ is strictly concave, and μ is degenerate at a single point $\mathbf{a}$. If the game has a Nash equilibrium, then we can show, using K_2, that $\mathbf{a}$ is neither in $[0, \frac{1}{2}]$ nor in $[\frac{1}{2}, 1]$, which is a contradiction. For other extensions of ε-Nash equilibria, see Tijs (1981).

The natural models of games on the unit square with discontinuity along the diagonal are not fully explored in non-zero-sum games. Models of facility location have been formulated as non-zero-sum games on the unit square. It is known that such models fail to have pure strategy Nash equilibria [d'Aspremont et al. (1979)]. Radzik and Ravindran (1989) constructed non-zero-sum games on the unit square with upper semi-continuous quasi-concave payoff functions but with no pure Nash equilibrium.

17. Correlated equilibrium on the unit square

A correlated equilibrium on the unit square S with payoffs $K_1(x, y)$ and $K_2(x, y)$ for the two players is a probability measure μ on the square such that for any measurable function $g : [0, 1] \to [0, 1]$

$$\int_S K_1(g(x), y) \, d\mu(x, y) \leqslant \int_S K_1(x, y) \, d\mu(x, y),$$

$$\int_S K_2(x, g(y)) \, d\mu(x, y) \leqslant \int_S K_2(x, y) \, d\mu(x, y).$$

While Glicksberg's (1952) fixed-point theorem is at the heart of the existence proof for Nash equilibria on the unit square, the weakening of conditions fails to guarantee the existence not only of Nash equilibria, but even of correlated equilibria. Consider the following example of Vieille (1996).

EXAMPLE.

$$K_1(x, y) = -\left(x - \frac{y}{2}\right)^2, \quad 0 \leqslant x, y \leqslant 1.$$

$$K_2(x, y) = \begin{cases} 0 & \text{if } (x, y) = (0, 0), \\ \frac{y}{x} & \text{if } 0 \leqslant y \leqslant x \leqslant 1 \text{ and } (x, y) \neq (0, 0), \\ \frac{y}{4} - \frac{x}{2} & \text{if } 0 \leqslant 2x \leqslant y \leqslant 1 \text{ and } (x, y) \neq (0, 0), \\ \frac{2}{y}\left(x - \frac{y}{2}\right) & \text{if } 0 \leqslant y \leqslant 2x \leqslant 1, \text{ and } (x, y) \neq (0, 0). \end{cases}$$

The above payoff has the following properties:
  (i) K_1 is continuous on the square S.
  (ii) K_2 is separately continuous in each variable when the other variable is fixed,
        except at the point $(0, 0)$.
  (iii) The payoff is bounded on the square.
  The above payoff has no pure strategy Nash equilibrium. By using the same payoff K_1 and by slightly modifying the above payoff K_2, Vieille (1996) constructs another
game on the unit square. The new game admits not even one correlated equilibrium.

Acknowledgment

My sabbatical leave during 1997–98 made it possible for me to visit several research
centers in Germany, the Netherlands, Israel, Hungary, and Australia. I would like to
thank Professors Aumann and Hart for their patience with my innumerable revisions.
The referee rightly expected a lot of careful editing and rewriting. Without the substantial help of Fe Evangelista, this would have been next to impossible. I am deeply touched
by her weekend trips to Chicago to help me with this revision. Many other friends and
colleagues and students came to my rescue in polishing the language and spotting errors. I would like to thank Nero Budur, Swaminathan Sankaran, Murthy Garimella, Paul
Musial, Michael Borns, and even a professional copyeditor for their instantaneous help.
On the technical side, I would like to thank Professors Rosenmüller and Sudhölter at
Bielefeld, and Talman, Norde, and Vermeulen at Tilburg, for many interesting discussions on the algorithmic aspects of equilibria.

References

Audet, C., P. Hansen, B. Jaumard and G. Savard (1998), "Complete enumeration of equilibria for two-person
    games in strategic and sequence forms", in: 8th International Symposium on Dynamic Games and Appli-
    cations, July 5–8 (Maastricht, The Netherlands).
Aumann, R.J. (1961), "Almost strictly competitive games", Journal of the Society for Industrial and Applied
    Mathematics 9:544–551.
Aumann, R.J. (1974), "Subjectivity and correlation in randomized strategies", Journal of Mathematical Eco-
    nomics 1:67–96.
Aumann, R.J. (1976), "Agreeing to disagree", Annals of Statistics 4:1236–1239.
Aumann, R.J. (1987), "Correlated equilibrium as an extension of Bayesian Rationality", Econometrica 55:1–
    18.
Aumann, R.J., and A. Brandenburger (1995), "Epistemic conditions for Nash equilibrium", Econometrica
    63:1161–1180.
Aumann, R.J., and M. Maschler (1972), "Some thoughts on the minimax principle", Management Science
    18:54–63.

Borm, P.E.M., M.J.M. Jansen, J.A.M. Potters and S.H. Tijs (1993), "On the structure of the set of perfect equilibria in bimatrix games", OR Spektrum 15:17–20.

Brown, G.W. (1949), "Some notes on computation of game solutions", Report No. P-78 (Rand Corporation, Santa Monica, CA).

Chin, H., T. Parthasarathy and T.E.S. Raghavan (1974), "Structure of completely mixed equilibria in n-person non-cooperative games", International Journal of Game Theory 3:1–19.

Dasgupta, P., and E. Maskin (1986), "The existence of equilibrium in discontinuous economic games", Review of Economic Studies 53:1–26.

d'Aspremont, C., J. Gabszewicz and J. Thisse (1979), "On Hotelling's stability in competition", Econometrica 47:1145–1150.

Dickhaut, J., and T. Kaplan (1991), "A program for finding Nash equilibria", The Mathematica Journal 1:87–93.

Dresher, M. (1961), Games of Strategy (Prentice-Hall, Englewood Cliffs, NJ).

Evangelista, F., and T.E.S. Raghavan (1996), "A note on correlated equilibria", International Journal of Game Theory 25:35–41.

Glicksberg, I. (1952), "A further generalization of Kakutani's fixed point theorem", Proceedings of the American Mathematical Society 3:170–174.

Harsanyi, J. (1968), "Games with incomplete information played by Bayesian players", Management Science 14:159–182, 320–334, 486–502.

Harsanyi, J. (1973a), "Games with randomly disturbed payoffs: A new rationale for mixed strategy equilibrium points", International Journal of Game Theory 2:1–23.

Harsanyi, J. (1973b), "Oddness of the number of equilibrium points: A new proof", International Journal of Game Theory 2:235–250.

Harsanyi, J. (1975), "The tracing procedure: A Bayesian approach to defining a solution for n-person games", International Journal of Game Theory 4:61–94.

Hart, S., and D. Schmeidler (1989), "Existence of correlated equilibria", Mathematics of Operations Research 14:18–25.

Hart, S., and Y. Tauman (1996), "Market crashes without external shocks", Discussion Paper 124, Dec. 1996 (Center for Rationality and Interactive Decision Theory, Hebrew University of Jerusalem).

Heuer, G.A. (1975), "On completely mixed strategies in bimatrix games", Journal of London Mathematical Society 11:17–20.

Heuer, G.A. (1978), "Uniqueness of equilibrium points in bimatrix games", International Journal of Game Theory 8:13–25.

Heuer, G.A., and C.B. Millham (1976), "On Nash subsets and mobility chains in bimatrix games", Naval Research Logistics Quarterly 23:311–319.

Hillas, J., and E. Kohlberg (2002), "Foundations of strategic equilibrium", in: R.J. Aumann and S. Hart, eds., Handbook of Game Theory, Vol. 3 (North-Holland, Amsterdam) Chapter 42, 1597–1663.

Jansen, M.J.M. (1981a), "Regularity and stability of equilibrium points in bimatrix games", Mathematics of Operations Research 6:530–550.

Jansen, M.J.M. (1981b), "Equilibria and optimal threat strategies in two-person games", Ph.D. thesis (University of Nijmegen, The Netherlands).

Jansen, M.J.M. (1993), "On the set of proper equilibria of a bimatrix game", International Journal of Game Theory 22:97–106.

Jansen, M.J.M., A.P. Jurg and P.E.M. Borm (1994), "On strictly perfect sets", Games and Economic Behavior 6:400–415.

Jurg, A.P. (1993), "Some topics in the theory of bimatrix games", Ph.D. thesis (Katholic University, Nijmegen, The Netherlands).

Kaplansky, I. (1945), "A contribution to von Neumann's theory of games", Annals of Mathematics 46:474–479.

Karlin, S. (1959a), Matrix Games, Programming and Mathematical Economics, Vol. 1 (Addison-Wesley, Reading, MA).

Karlin, S. (1959b), Mathematical Methods and Theory in Games, Programming and Economics, Vol. 2 (Addison-Wesley, Reading, MA).

Kreps, V.L. (1974), "Bimatrix games with unique equilibrium points", International Journal of Game Theory 33:115–118.

Krishna, V., and T. Sjöström (1998), "On the convergence of fictitious play", Mathematics of Operations Research 23:479–511.

Krohn, I., S. Moltzahn, J. Rosenmüller, P. Sudhölter and H.-M. Wallmeier (1991), "Implementing the modified LH-algorithm", Applied Mathematics and Computation 45:35–72.

Kuhn, H.W. (1961), "An algorithm for equilibrium points in bimatrix games", Proceedings of the National Academy of Sciences U.S.A. 47:1657–1662.

Lemke, C.E., and J.T. Howson, Jr. (1964), "Equilibrium points of bimatrix games", Journal of the Society for Industrial and Applied Mathematics 12:413–423.

Mangasarian, O.L. (1964), "Equilibrium points of bimatrix games", Journal of the Society for Industrial and Applied Mathematics 12:778–780.

McKelvey, R.D., and A. McLennan (1996), "Computation of equilibria in finite games", in: H.M. Amman, D.A. Kendrick and J. Rust, eds., Handbook of Computational Economics (Elsevier, Amsterdam) 87–142.

Milgrom, P., and J. Roberts (1991), "Adaptive and sophisticated learning in normal form games", Games and Economic Behavior 3:82–100.

Millham, C.B. (1972), "Constructing bimatrix games with special properties", Naval Research Logistics Quarterly 19:709–714.

Millham, C.B. (1974), "On Nash subsets of bimatrix games", Naval Research Logistics Quarterly 21:307–317.

Mills, H. (1960), "Equilibrium points in finite games", Journal of the Society for Industrial and Applied Mathematics 8:397–402.

Miyasawa, K. (1963), "On the convergence of the learning process in a 2×2 non-zero-sum game", Princeton University Econometric Research Program, Research Memorandum No. 33.

Monderer, D., and A. Sela (1996), "A 2×2 game without the fictitious play property", Games and Economic Behavior 14:144–148.

Monderer, D., and L.S. Shapley (1996), "Fictitious play property for games with identical interests", Journal of Economic Theory 1:258–265.

Moulin, H., and J.P. Vial (1978), "Strategically zero-sum games", International Journal of Game Theory 7:201–221.

Mukhamediev, B.M. (1978), "The solution of bilinear programming problem and finding the equilibrium situations in bimatrix games", U.S.S.R. Computational Mathematics and Mathematical Physics 18:60–66.

Myerson, R. (1978), "Refinements of the Nash equilibrium concept", International Journal of Game Theory 7:73–80.

Nash, J. (1950), "Equilibrium points in n-person games", Proceedings of the National Academy of Sciences U.S.A. 36:48–49.

Nau, R., and K. McCardle (1988), "Coherent behavior in noncooperative games", Working Paper 8701 (Fuqua School of Business, Duke University).

Nikaido, H., and K. Isoda (1959), "Note on non-cooperative convex games", Pacific Journal of Mathematics 5:807–815.

Norde, H. (1998), "Bimatrix games have quasi-strict equilibria", Mathematical Programming 85:35–49.

Parthasarathy, T., and T.E.S. Raghavan (1971), Some Topics in Two-Person Games (Elsevier, New York).

Parthasarathy, T., and T.E.S. Raghavan (1975), "Equilibria of continuous two-person games", Pacific Journal of Mathematics 57:265–270.

Radzik, T. (1993), "Nash equilibria of discontinuous non-zero-sum two-person games", International Journal of Game Theory 21:429–437.

Radzik, T., and G. Ravindran (1989), "On two counterexamples of non-cooperative games without Nash equilibria", Sankhya 51:236–240.

Raghavan, T.E.S. (1970), "Completely mixed strategies in bimatrix games", Journal of the London Mathematical Society 2:709–712.

Robinson, J. (1951), "An iterative method for solving a game", Annals of Mathematics 54:296–301.

Rosenmüller, J. (1971), "On a generalization of the Lemke–Howson algorithm to non-cooperative n-person games", SIAM Journal on Applied Mathematics 21:73–79.

Rosenmüller, J. (1981), The Theory of Games and Markets (North-Holland, Amsterdam).

Rosenthal, R.W. (1973), "A class of games possessing pure strategy Nash equilibria", International Journal of Game Theory 2:65–67.

Rosenthal, R.W. (1974), "Correlated equilibria in some classes of two-person games", International Journal of Game Theory 3:119–128.

Schwarz, G. (1994), "Game theory and statistics", in: R.J. Aumann and S. Hart, eds., Handbook of Game Theory, Vol. 2 (North-Holland, Amsterdam) Chapter 21, 769–779.

Selten, R. (1975), "Re-examination of the perfectness concept for equilibrium points in extensive games", International Journal of Game Theory 4:25–55.

Shapley, L.S. (1964), "Some topics in two-person games", in: M. Dresher, L.S. Shapley and A.W. Tucker, eds., Advances in Game Theory (Princeton University Press, Princeton, NJ) 1–28.

Shapley, L.S. (1974), "A note on the Lemke–Howson algorithm", Mathematical Programming Study 1:175–189.

Sorin, S. (1992), "Repeated games with incomplete information", in: R.J. Aumann and S. Hart, eds., Handbook of Game Theory, Vol. 1 (North-Holland, Amsterdam) Chapter 4, 71–107.

Talman, A.J.J., and Z. Yang (1994), "A simplicial algorithm for computing proper Nash equilibria of finite games", CentER Discussion Paper No. 9418 (Tilburg University, Tilburg, The Netherlands).

Tijs, S.H. (1981), "Nash equilibria for non-cooperative n-person games in normal form", SIAM Review 23:225–237.

Van Damme, E.E.C. (1983), "Refinements of the Nash equilibrium concept", Ph.D. thesis (Eindhoven Technical Institute, The Netherlands).

Van Damme, E.E.C. (1991), Stability and Perfection of Nash Equilibria (Springer, Berlin).

Van den Elzen, A.H., and A.J.J. Talman (1991), "A procedure for finding Nash equilibria in bi-matrix games", ZOR Methods and Models of Operations Research 35:27–43.

Van den Elzen, A.H., and A.J.J. Talman (1995), "An algorithmic approach towards the tracing procedure of Harsanyi and Selten", CentER Discussion Paper No. 95111 (Tilburg University, Tilburg, The Netherlands).

Vermeulen, A.J., and M.J.M. Jansen (1994), "On the set of perfect equilibria of a bimatrix game", Naval Research Logistics Quarterly 41:295–302.

Vieille, N. (1996), "On equilibria on the square", International Journal of Game Theory 25:199–205.

Von Neumann, J., and O. Morgenstern (1944), Theory of Games and Economic Behavior (Princeton University Press, Princeton, NJ).

Vorobiev, N.N. (1958), "Equilibrium points in bimatrix games", Theory of Probability and its Applications 3:297–309.

Wilson, R. (1971), "Computing equilibria of n-person games", SIAM Journal on Applied Mathematics 21:80–87.

Winkels, H.-M. (1979), "An algorithm to determine all equilibrium points of a bimatrix game", in: O. Moeschler and D. Pallaschke, eds., Game Theory and Related Topics (North-Holland, Amsterdam) 137–148.

Wu, W.-T., and J. Jia-He (1962), "Essential equilibrium points of n-person non-cooperative games", Scientia Sinica 11:1307–1322.

Yang, Z. (1996), "Simplicial fixed point algorithms and applications", Ph.D. thesis (Tilburg University, Tilburg, The Netherlands).

Chapter 45

COMPUTING EQUILIBRIA FOR TWO-PERSON GAMES

BERNHARD VON STENGEL*

Mathematics Department, London School of Economics, London, UK

Contents

*This work was supported by a Heisenberg grant from the Deutsche Forschungsgemeinschaft.

Handbook of Game Theory, Volume 3, Edited by R.J. Aumann and S. Hart

Abstract

This paper is a self-contained survey of algorithms for computing Nash equilibria of two-person games. The games may be given in strategic form or extensive form. The classical Lemke–Howson algorithm finds one equilibrium of a bimatrix game, and provides an elementary proof that a Nash equilibrium exists. It can be given a strong geometric intuition using graphs that show the subdivision of the players' mixed strategy sets into best-response regions. The Lemke–Howson algorithm is presented with these graphs, as well as algebraically in terms of complementary pivoting. Degenerate games require a refinement of the algorithm based on lexicographic perturbations. Commonly used definitions of degenerate games are shown as equivalent. The enumeration of all equilibria is expressed as the problem of finding matching vertices in pairs of polytopes. Algorithms for computing simply stable equilibria and perfect equilibria are explained. The computation of equilibria for extensive games is difficult for larger games since the reduced strategic form may be exponentially large compared to the game tree. If the players have perfect recall, the sequence form of the extensive game is a strategic description that is more suitable for computation. In the sequence form, pure strategies of a player are replaced by sequences of choices along a play in the game. The sequence form has the same size as the game tree, and can be used for computing equilibria with the same methods as the strategic form. The paper concludes with remarks on theoretical and practical issues of concern to these computational approaches.

Keywords

equilibrium computation, Lemke–Howson algorithm, degenerate game, extensive game, perfect equilibrium, pivoting

JEL classification: C72, C63

1. Introduction

Finding Nash equilibria of strategic form or extensive form games can be difficult and tedious. A computer program for this task would allow greater detail of game-theoretic models, and enhance their applicability. Algorithms for solving games have been studied since the beginnings of game theory, and have proved useful for other problems in mathematical optimization, like linear complementarity problems.

This paper is a survey and exposition of *linear* methods for finding Nash equilibria. Above all, these apply to games with two players. In an equilibrium of a two-person game, the mixed strategy probabilities of one player equalize the expected payoffs for the pure strategies used by the other player. This defines an optimization problem with linear constraints. We do not consider nonlinear methods like simplicial subdivision for approximating fixed points, or systems of inequalities for higher-degree polynomials as they arise for noncooperative games with more than two players. These are surveyed in McKelvey and McLennan (1996).

First, we consider two-person games in strategic form [see also Parthasarathy and Raghavan (1971), Raghavan (1994, 2002)]. The classical algorithm by Lemke and Howson (1964) finds one equilibrium of a bimatrix game. It provides an elementary, constructive proof that such a game has an equilibrium, and shows that the number of equilibria is odd, except for degenerate cases. We follow Shapley's (1974) very intuitive geometric exposition of this algorithm. The maximization over linear payoff functions defines two *polyhedra* which provide further geometric insight. A complementary pivoting scheme describes the computation algebraically. Then we clarify the notion of *degeneracy*, which appears in the literature in various forms, most of which are equivalent. The lexicographic method extends pivoting algorithms to degenerate games. The problem of finding *all* equilibria of a bimatrix game can be phrased as a vertex enumeration problem for polytopes.

Second, we look at two methods for finding equilibria of strategic form games with additional refinement properties [see van Damme (1987, 2002), Hillas and Kohlberg (2002)]. Wilson (1992) modifies the Lemke–Howson algorithm for computing *simply stable* equilibria. These equilibria survive certain perturbations of the game that are easily represented by lexicographic methods for degeneracy resolution. Van den Elzen and Talman (1991) present a complementary pivoting method for finding a *perfect* equilibrium of a bimatrix game.

Third, we review methods for games in extensive form [see Hart (1992)]. In principle, such game trees can be solved by converting them to the reduced strategic form and then applying the appropriate algorithms. However, this typically increases the size of the game description and the computation time exponentially, and is therefore infeasible. Approaches to avoiding this problem compute with a small fraction of the pure strategies, which are generated from the game tree as needed [Wilson (1972), Koller and Megiddo (1996)]. A strategic description of an extensive game that does not increase in size is the *sequence form*. The central idea, set forth independently by Romanovskii (1962), Selten (1988), Koller and Megiddo (1992), and von Stengel (1996a), is to consider only sequences of moves instead of pure strategies, which are arbitrary combinations of moves. We will develop the problem of equilibrium computation for the strategic form in a way that can also be applied to the sequence form. In particular, the algorithm by van den Elzen and Talman (1991) for finding a perfect equilibrium carries over to the sequence form [von Stengel, van den Elzen and Talman (2002)].

The concluding section addresses issues of computational complexity, and mentions ongoing implementations of the algorithms.

2. Bimatrix games

We first introduce our notation, and recall notions from polytope theory and linear programming. Equilibria of a bimatrix game are the solutions to a linear complementarity problem. This problem is solved by the Lemke–Howson algorithm, which we explain in graph-theoretic, geometric, and algebraic terms. Then we consider degenerate games, and review enumeration methods.

2.1. Preliminaries

We use the following notation throughout. Let (A, B) be a bimatrix game, where A and B are $m \times n$ matrices of payoffs to the row player 1 and column player 2, respectively. All vectors are column vectors, so an m-vector x is treated as an $m \times 1$ matrix. A *mixed strategy* x for player 1 is a probability distribution on rows, written as an m-vector of probabilities. Similarly, a mixed strategy y for player 2 is an n-vector of probabilities for playing columns. The *support* of a mixed strategy is the set of pure strategies that have positive probability. A vector or matrix with all components zero is denoted $\mathbf{0}$. Inequalities like $x \geqslant \mathbf{0}$ between two vectors hold for all components. $B^\top$ is the matrix B transposed.

Let M be the set of the m pure strategies of player 1 and let N be the set of the n pure strategies of player 2. It is sometimes useful to assume that these sets are disjoint, as in

$$M = \{1, \ldots, m\}, \qquad N = \{m + 1, \ldots, m + n\}. \tag{2.1}$$

Then $x \in \mathbb{R}^M$ and $y \in \mathbb{R}^N$, which means, in particular, that the components of y are y_j for $j \in N$. Similarly, the payoff matrices A and B belong to $\mathbb{R}^{M \times N}$.

Denote the rows of A by a_i for $i \in M$, and the rows of $B^\top$ by b_j for $j \in N$ (so each $b_j^\top$ is a column of B). Then $a_i y$ is the expected payoff to player 1 for the pure strategy i when player 2 plays the mixed strategy y, and $b_j x$ is the expected payoff to player 2 for j when player 1 plays x.

A *best response* to the mixed strategy y of player 2 is a mixed strategy x of player 1 that maximizes his expected payoff $x^\top A y$. Similarly, a best response y of player 2 to x maximizes her expected payoff $x^\top B y$. A *Nash equilibrium* is a pair (x, y) of mixed strategies that are best responses to each other. Clearly, a mixed strategy is a best response to an opponent strategy if and only if it only plays pure strategies that are best responses with positive probability:

THEOREM 2.1 [Nash (1951)]. *The mixed strategy pair (x, y) is a Nash equilibrium of* (A, B) *if and only if for all pure strategies i in M and j in N*

$$x_i > 0 \implies a_i y = \max_{k \in M} a_k y, \tag{2.2}$$

$$y_j > 0 \implies b_j x = \max_{k \in N} b_k x. \tag{2.3}$$

We recall some notions from the theory of (convex) polytopes [see Ziegler (1995)]. An *affine combination* of points $z_1, \ldots, z_k$ in some Euclidean space is of the form $\sum_{i=1}^{k} z_i \lambda_i$ where $\lambda_1, \ldots, \lambda_k$ are reals with $\sum_{i=1}^{k} \lambda_i = 1$. It is called a *convex combination* if $\lambda_i \geqslant 0$ for all i. A set of points is *convex* if it is closed under forming convex combinations. Given points are *affinely independent* if none of these points is an affine combination of the others. A convex set has *dimension d* if and only if it has $d + 1$, but no more, affinely independent points.

A *polyhedron P* in $\mathbb{R}^d$ is a set $\{z \in \mathbb{R}^d \mid Cz \leqslant q\}$ for some matrix C and vector q. It is called *full-dimensional* if it has dimension d. It is called a *polytope* if it is bounded. A *face* of P is a set $\{z \in P \mid c^\top z = q_0\}$ for some $c \in \mathbb{R}^d$, $q_0 \in \mathbb{R}$ so that the inequality $c^\top z \leqslant q_0$ holds for all z in P. A *vertex* of P is the unique element of a 0-dimensional face of P. An *edge* of P is a one-dimensional face of P. A *facet* of a d-dimensional polyhedron P is a face of dimension $d - 1$. It can be shown that any nonempty face F of P can be obtained by turning some of the inequalities defining P into equalities, which are then called *binding* inequalities. That is, $F = \{z \in P \mid c_i z = q_i, i \in I\}$, where $c_i z \leqslant q_i$ for $i \in I$ are some of the rows in $Cz \leqslant q$. A facet is characterized by a single binding inequality which is *irredundant*, that is, the inequality cannot be omitted without changing the polyhedron [Ziegler (1995, p. 72)]. A d-dimensional polyhedron P is called *simple* if no point belongs to more than d facets of P, which is true if there are no special dependencies between the facet-defining inequalities.

A *linear program* (LP) is the problem of maximizing a linear function over some polyhedron. The following notation is independent of the considered bimatrix game. Let M and N be finite sets, $I \subseteq M$, $J \subseteq N$, $A \in \mathbb{R}^{M \times N}$, $b \in \mathbb{R}^M$, $c \in \mathbb{R}^N$. Consider the polyhedron

$$P = \left\{ x \in \mathbb{R}^N \;\middle|\; \sum_{j \in N} a_{ij} x_j = b_i, \; i \in M - I, \right.$$

$$\sum_{j \in N} a_{ij} x_j \leqslant b_i, \; i \in I,$$

$$\left. x_j \geqslant 0, \; j \in J \right\}.$$

Any x belonging to P is called *primal feasible*. The *primal LP* is the problem

$$\text{maximize} \quad c^\top x \quad \text{subject to} \quad x \in P. \tag{2.4}$$

The corresponding *dual LP* has the feasible set

$$D = \left\{ y \in \mathbb{R}^M \,\middle|\, \sum_{i \in M} y_i a_{ij} = c_j, \; j \in N - J, \right.$$

$$\sum_{i \in M} y_i a_{ij} \geqslant c_j, \; j \in J,$$

$$\left. y_i \geqslant 0, \; i \in I \right\}$$

and is the problem

$$\text{minimize} \quad y^\top b \quad \text{subject to} \quad y \in D. \tag{2.5}$$

Here the indices in I denote primal inequalities and corresponding nonnegative dual variables, whereas those in $M - I$ denote primal equality constraints and corresponding unconstrained dual variables. The sets J and $N - J$ play the same role with "primal" and "dual" interchanged. By reversing signs, the dual of the dual LP is again the primal. We recall the *duality theorem* of linear programming, which states (a) that for any primal and dual feasible solutions, the corresponding objective functions are mutual bounds, and (b) if the primal and the dual LP both have feasible solutions, then they have optimal solutions with the same value of their objective functions.

THEOREM 2.2. *Consider the primal–dual pair of LPs* (2.4), (2.5). *Then*
  (a) (Weak duality.) $c^\top x \leqslant y^\top b$ *for all* $x \in P$ *and* $y \in D$.
  (b) (Strong duality.) *If* $P \neq \emptyset$ *and* $D \neq \emptyset$ *then* $c^\top x = y^\top b$ *for some* $x \in P$ *and* $y \in D$.

For a proof see Schrijver (1986). As an introduction to linear programming we recommend Chvátal (1983).

2.2. *Linear constraints and complementarity*

Mixed strategies x and y of the two players are nonnegative vectors whose components sum up to one. These are linear constraints, which we define using

$$E = [1, \ldots, 1] \in \mathbb{R}^{1 \times M}, \quad e = 1, \quad F = [1, \ldots, 1] \in \mathbb{R}^{1 \times N}, \quad f = 1. \tag{2.6}$$

Then the sets X and Y of mixed strategies are

$$X = \left\{ x \in \mathbb{R}^M \mid Ex = e, \; x \geqslant \mathbf{0} \right\}, \qquad Y = \left\{ y \in \mathbb{R}^N \mid Fy = f, \; y \geqslant \mathbf{0} \right\}. \tag{2.7}$$

With the extra notation in (2.6), the following considerations apply also if X and Y are more general polyhedra, where $Ex = e$ and $Fy = f$ may consist of more than a single row of equations. Such polyhedrally constrained games, first studied by Charnes (1953) for the zero-sum case, are useful for finding equilibria of extensive games (see Section 4).

Given a fixed y in Y, a best response of player 1 to y is a vector x in X that maximizes the expression $x^\top(Ay)$. That is, x is a solution to the LP

$$\text{maximize} \quad x^\top(Ay) \quad \text{subject to} \quad Ex = e, \quad x \geqslant \mathbf{0}. \tag{2.8}$$

The dual of this LP with variables u (by (2.6) only a single variable) states

$$\underset{u}{\text{minimize}} \quad e^\top u \quad \text{subject to} \quad E^\top u \geqslant Ay. \tag{2.9}$$

Both LPs are feasible. By Theorem 2.2(b), they have the same optimal value.

Consider now a *zero-sum game*, where $B = -A$. Player 2, when choosing y, has to assume that her opponent plays rationally and maximizes $x^\top Ay$. This maximum payoff to player 1 is the optimal value of the LP (2.8), which is equal to the optimal value $e^\top u$ of the dual LP (2.9). Player 2 is interested in minimizing $e^\top u$ by her choice of y. The constraints of (2.9) are linear in u and y even if y is treated as a variable, which must belong to Y. So a minmax strategy y of player 2 (minimizing the maximum amount she has to pay) is a solution to the LP

$$\underset{u,y}{\text{minimize}} \quad e^\top u \quad \text{subject to} \quad Fy = f, \quad E^\top u - Ay \geqslant \mathbf{0}, \quad y \geqslant \mathbf{0}. \tag{2.10}$$

Figure 1 shows an example.

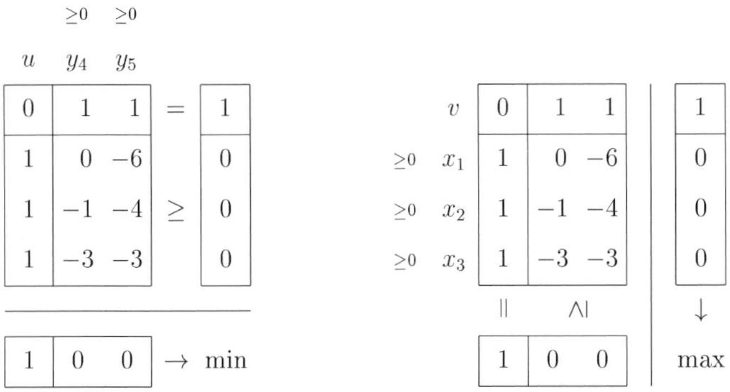

Figure 1. Left: example of the LP (2.10) for a 3×2 zero-sum game. The objective function is separated by a line, nonnegative variables are marked by "$\geqslant 0$". Right: the dual LP (2.11), to be read vertically.

The dual of the LP (2.10) has variables v and x corresponding to the primal constraints $Fy = f$ and $E^\top u - Ay \geqslant \mathbf{0}$, respectively. It has the form

$$\text{maximize} \quad f^\top v \quad \text{subject to} \quad Ex = e, \quad F^\top v - A^\top x \leqslant \mathbf{0}, \quad x \geqslant \mathbf{0}. \tag{2.11}$$

It is easy to verify that this LP describes the problem of finding a maxmin strategy x (with maxmin payoff $f^\top v$) for player 1. We have shown the following.

THEOREM 2.3. *A zero-sum game with payoff matrix A for player 1 has the equilibrium (x, y) if and only if u, y is an optimal solution to the LP (2.10) and v, x is an optimal solution to its dual LP (2.11). Thereby, $e^\top u$ is the maxmin payoff to player 1 and $f^\top v$ is the minmax payoff to player 2. Both payoffs are equal and denote the value of the game.*

Thus, the "maxmin = minmax" theorem for zero-sum games follows directly from LP duality [see also Raghavan (1994)]. This connection was noted by von Neumann and Dantzig in the late 1940s when linear programming took its shape. Conversely, linear programs can be expressed as zero-sum games [see Dantzig (1963, p. 277)]. There are standard algorithms for solving LPs, in particular Dantzig's Simplex algorithm. Usually, they compute a primal solution together with a dual solution which proves that the optimum is reached.

A best response x of player 1 against the mixed strategy y of player 2 is a solution to the LP (2.8). This is also useful for games that are not zero-sum. By strong duality, a feasible solution x is optimal if and only if there is a dual solution u fulfilling $E^\top u \geqslant Ay$ and $x^\top(Ay) = e^\top u$, that is, $x^\top(Ay) = (x^\top E^\top)u$ or equivalently

$$x^\top \left(E^\top u - Ay \right) = 0. \tag{2.12}$$

Because the vectors x and $E^\top u - Ay$ are nonnegative, (2.12) states that they have to be *complementary* in the sense that they cannot both have positive components in the same position. This characterization of an optimal primal–dual pair of feasible solutions is known as *complementary slackness* in linear programming. Since x has at least one positive component, the respective component of $E^\top u - Ay$ is zero and u is by (2.6) the maximum of the components of Ay. Any pure strategy i in M of player 1 is a best response to y if and only if the ith component of the slack vector $E^\top u - Ay$ is zero. That is, (2.12) is equivalent to (2.2).

For player 2, strategy y is a best response to x if and only if it maximizes $(x^\top B)y$ subject to $y \in Y$. The dual of this LP is the following LP analogous to (2.9): minimize $f^\top v$ subject to $F^\top v \geqslant B^\top x$. Here, a primal–dual pair y, v of feasible solutions is optimal if and only if, analogous to (2.12),

$$y^\top \left(F^\top v - B^\top x \right) = 0. \tag{2.13}$$

Considering these conditions for both players, this shows the following.

THEOREM 2.4. *The game (A, B) has the Nash equilibrium (x, y) if and only if for suitable u, v*

$$
\begin{aligned}
Ex \quad &= e, \\
Fy &= f, \\
E^\top u \qquad -Ay &\geqslant \mathbf{0}, \\
F^\top v - B^\top x \quad &\geqslant \mathbf{0}, \\
x, \quad y &\geqslant \mathbf{0}
\end{aligned}
\tag{2.14}
$$

and (2.12), (2.13) hold.

The conditions in Theorem 2.4 define a so-called mixed *linear complementarity problem* (LCP). There are various solution methods for LCPs. For a comprehensive treatment see Cottle, Pang and Stone (1992). The most important method for finding one solution of the LCP in Theorem 2.4 is the Lemke–Howson algorithm.

2.3. The Lemke–Howson algorithm

In their seminal paper, Lemke and Howson (1964) describe an algorithm for finding one equilibrium of a bimatrix game. We follow Shapley's (1974) exposition of this algorithm. It requires disjoint pure strategy sets M and N of the two players as in (2.1). Any mixed strategy x in X and y in Y is *labeled* with certain elements of $M \cup N$. These labels denote the unplayed pure strategies of the player and the pure best responses of his or her opponent. For $i \in M$ and $j \in N$, let

$$
\begin{aligned}
X(i) &= \{x \in X \mid x_i = 0\}, \\
X(j) &= \{x \in X \mid b_j x \geqslant b_k x \text{ for all } k \in N\}, \\
Y(i) &= \{y \in Y \mid a_i y \geqslant a_k y \text{ for all } k \in M\}, \\
Y(j) &= \{y \in Y \mid y_j = 0\}.
\end{aligned}
$$

Then x has label k if $x \in X(k)$ and y has label k if $y \in Y(k)$, for $k \in M \cup N$. Clearly, the best-response regions $X(j)$ for $j \in N$ are polytopes whose union is X. Similarly, Y is the union of the sets $Y(i)$ for $i \in M$. Then a Nash equilibrium is a *completely labeled* pair (x, y) since then by Theorem 2.1, any pure strategy k of a player is either a best response or played with probability zero, so it appears as a label of x or y.

THEOREM 2.5. *A mixed strategy pair (x, y) in $X \times Y$ is a Nash equilibrium of (A, B) if and only if for all $k \in M \cup N$ either $x \in X(k)$ or $y \in Y(k)$ (or both).*

For the 3×2 bimatrix game (A, B) with

$$
A = \begin{bmatrix} 0 & 6 \\ 2 & 5 \\ 3 & 3 \end{bmatrix}, \qquad
B = \begin{bmatrix} 1 & 0 \\ 0 & 2 \\ 4 & 3 \end{bmatrix},
\tag{2.15}
$$

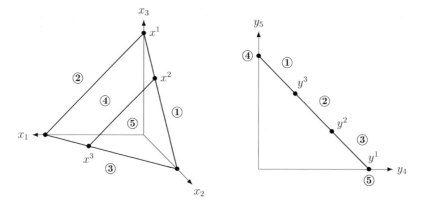

Figure 2. Mixed strategy sets X and Y of the players for the bimatrix game (A, B) in (2.15). The labels 1, 2, 3, drawn as circled numbers, are the pure strategies of player 1 and marked in X where they have probability zero, in Y where they are best responses. The pure strategies of player 2 are similar labels 4, 5. The dots mark points x and y with a maximum number of labels.

the labels of X and Y are shown in Figure 2. The equilibria are $(x^1, y^1) = ((0, 0, 1)^\top, (1, 0)^\top)$ where x^1 has the labels 1, 2, 4 (and y^1 the remaining labels 3 and 5), $(x^2, y^2) = ((0, \frac{1}{3}, \frac{2}{3})^\top, (\frac{2}{3}, \frac{1}{3})^\top)$ with labels 1, 4, 5 for x^2, and $(x^3, y^3) = ((\frac{2}{3}, \frac{1}{3}, 0)^\top, (\frac{1}{3}, \frac{2}{3})^\top)$ with labels 3, 4, 5 for x^3.

This geometric-qualitative inspection is very suitable for finding equilibria of games of up to size 3×3. It works by inspecting any point x in X with m labels and checking if there is a point y in Y having the remaining n labels. Usually, any x in X has at most m labels, and any y in Y has at most n labels. A game with this property is called *nondegenerate*, as stated in the following equivalent definition.

DEFINITION 2.6. A bimatrix game is called *nondegenerate* if the number of pure best responses to a mixed strategy never exceeds the size of its support.

A game is usually nondegenerate since every additional label introduces an equation that reduces the dimension of the set of points having these labels by one. Then only single points x in X have m given labels and single points y in Y have n given labels, and no point has more labels. Nondegeneracy is discussed in greater detail in Section 2.6 below. Until further notice, we assume that the game is nondegenerate.

THEOREM 2.7. *In a nondegenerate $m \times n$ bimatrix game (A, B), only finitely many points in X have m labels and only finitely many points in Y have n labels.*

PROOF. Let K and L be subsets of $M \cup N$ with $|K| = m$ and $|L| = n$. There are only finitely many such sets. Consider the set of points in X having the labels in K, and the set of points in Y having the labels in L. By Theorem 2.10(c) below, these sets are empty or singletons. □

The finitely many points in the preceding theorem are used to define two graphs G_1 and G_2. Let G_1 be the graph whose vertices are those points x in X that have m labels, with an additional vertex $\mathbf{0}$ in $\mathbb{R}^M$ that has all labels i in M. Any two such vertices x and x' are joined by an edge if they differ in one label, that is, if they have $m - 1$ labels in common. Similarly, let G_2 be the graph with vertices y in Y that have n labels, with the extra vertex $\mathbf{0}$ in $\mathbb{R}^N$ having all labels j in N, and edges joining those vertices that have $n - 1$ labels in common. The *product graph* $G_1 \times G_2$ of G_1 and G_2 has vertices (x, y) where x is a vertex of G_1, and y is a vertex of G_2. Its edges are given by $\{x\} \times \{y, y'\}$ for vertices x of G_1 and edges $\{y, y'\}$ of G_2, or by $\{x, x'\} \times \{y\}$ for edges $\{x, x'\}$ of G_1 and vertices y of G_2.

The Lemke–Howson algorithm can be defined combinatorially in terms of these graphs. Let $k \in M \cup N$, and call a vertex pair (x, y) of $G_1 \times G_2$ *k-almost completely labeled* if any l in $M \cup N - \{k\}$ is either a label of x or of y. Since two adjacent vertices x, x' in G_1, say, have $m - 1$ labels in common, the edge $\{x, x'\} \times \{y\}$ of $G_1 \times G_2$ is also called k-almost completely labeled if y has the remaining n labels except k. The same applies to edges $\{x\} \times \{y, y'\}$ of $G_1 \times G_2$.

Then any equilibrium (x, y) is in $G_1 \times G_2$ adjacent to exactly one vertex pair (x', y') that is k-almost completely labeled: namely, if k is the label of x, then x is joined to the vertex x' in G_1 sharing the remaining $m - 1$ labels, and $y = y'$. If k is the label of y, then y is similarly joined to y' in G_2 and $x = x'$. In the same manner, a k-almost completely labeled pair (x, y) that is completely labeled has exactly two neighbors in $G_1 \times G_2$. These are obtained by dropping the unique duplicate label that x and y have in common, joining to an adjacent vertex either in G_1 and keeping y fixed, or in G_2 and keeping x fixed. This defines a unique k-almost completely labeled path in $G_1 \times G_2$ connecting one equilibrium to another. The algorithm is started from the *artificial* equilibrium $(\mathbf{0}, \mathbf{0})$ that has all labels, follows the path where label k is missing, and terminates at a Nash equilibrium of the game.

Figure 3 demonstrates this method for the above example. Let 2 be the missing label k. The algorithm starts with $x = (0, 0, 0)^\top$ and $y = (0, 0)^\top$. Step I: y stays fixed

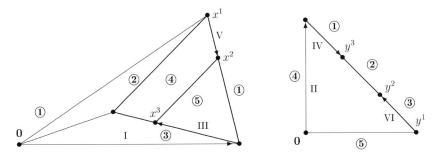

Figure 3. The graphs G_1 and G_2 for the game in (2.15). The set of 2-almost completely labeled pairs is formed by the paths with edges (in $G_1 \times G_2$) I–II–III–IV, connecting the artificial equilibrium $(\mathbf{0}, \mathbf{0})$ and (x^3, y^3), and V–VI, connecting the equilibria (x^1, y^1) and (x^2, y^2).

and x is changed in G_1 to $(0, 1, 0)^\top$, picking up label 5, which is now duplicate. Step II: dropping label 5 in G_2 changes y to $(0, 1)^\top$, picking up label 1. Step III: dropping label 1 in G_1 changes x to x^3, picking up label 4. Step IV: dropping label 4 in G_2 changes y to y^3 which has the missing label 2, terminating at the equilibrium (x^3, y^3). In a similar way, steps V and VI indicated in Figure 3 join the equilibria (x^1, y^1) and (x^2, y^2) on a 2-almost completely labeled path. In general, one can show the following.

THEOREM 2.8 [Lemke and Howson (1964), Shapley (1974)]. *Let (A, B) be a nonde-generate bimatrix game and k be a label in $M \cup N$. Then the set of k-almost completely labeled vertices and edges in $G_1 \times G_2$ consists of disjoint paths and cycles. The end-points of the paths are the equilibria of the game and the artificial equilibrium $(\mathbf{0}, \mathbf{0})$. The number of Nash equilibria of the game is odd.*

This theorem provides a constructive, elementary proof that every nondegenerate game has an equilibrium, independently of the result of Nash (1951). By different labels k that are dropped initially, it may be possible to find different equilibria. However, this does not necessarily generate all equilibria, that is, the union of the k-almost completely labeled paths in Theorem 2.8 for all $k \in M \cup N$ may be disconnected [Shapley (1974, p. 183) reports an example due to R. Wilson]. For similar observations see Aggarwal (1973), Bastian (1976), Todd (1976, 1978). Shapley (1981) discusses more general methods as a potential way to overcome this problem.

2.4. Representation by polyhedra

The vertices and edges of the graphs G_1 and G_2 used in the definition of the Lemke–Howson algorithm can be represented as vertices and edges of certain polyhedra. Let

$$H_1 = \left\{ (x, v) \in \mathbb{R}^M \times \mathbb{R} \mid x \in X,\ B^\top x \leqslant F^\top v \right\},$$
$$H_2 = \left\{ (y, u) \in \mathbb{R}^N \times \mathbb{R} \mid y \in Y,\ Ay \leqslant E^\top u \right\}. \tag{2.16}$$

The elements of $H_1 \times H_2$ represent the solutions to (2.14). Figure 4 shows H_2 for the example (2.15). The horizontal plane contains Y as a subset. The scalar u, drawn vertically, is at least the maximum of the functions $a_i y$ for the rows a_i of A and for y in Y. The maximum itself shows which strategy of player 1 is a best response to y. Consequently, projecting H_2 to Y by mapping (y, u) to y, in Figure 4 shown as $(y, 0)$, reveals the subdivision of Y into best-response regions $Y(i)$ for $i \in M$ as in Figure 2. Figure 4 shows also that the unbounded facets of H_2 project to the subsets $Y(j)$ of Y for $j \in N$. Furthermore, the maximally labeled points in Y marked by dots appear as projections of the vertices of H_2. Similarly, the facets of H_1 project to the subsets $X(k)$ of X for $k \in M \cup N$.

The graph structure of H_1 and H_2 with its vertices and edges is therefore identical to that of G_1 and G_2, except for the m unbounded edges of H_1 and the n unbounded edges of H_2 that connect to "infinity" rather than to the additional vertex $\mathbf{0}$ of G_1 and G_2, respectively.

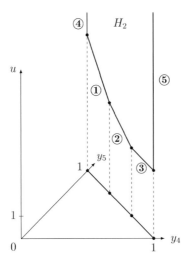

Figure 4. The polyhedron H_2 for the game in (2.15), and its projection to the set $\{(y,0) \mid (y,u) \in H_2\}$. The vertical scale is displayed shorter. The circled numbers label the facets of H_2 analogous to Figure 2.

The constraints (2.14) defining H_1 and H_2 can be simplified by eliminating the payoff variables u and v, which works if these are always positive. For that purpose, assume that

$$A \text{ and } B^\top \text{ are nonnegative and have no zero column.} \tag{2.17}$$

This assumption can be made without loss of generality since a constant can be added to all payoffs without changing the game in a material way, so that, for example, $A > \mathbf{0}$ and $B > \mathbf{0}$. For examples like (2.15), zero matrix entries are also admitted in (2.17). By (2.6), u and v are scalars and $E^\top$ and $F^\top$ are single columns with all components equal to one, which we denote by the vectors $\mathbf{1}_M$ in $\mathbb{R}^M$ and $\mathbf{1}_N$ in $\mathbb{R}^N$, respectively. Let

$$\begin{aligned}
P_1 &= \left\{ x' \in \mathbb{R}^M \mid x' \geqslant \mathbf{0},\ B^\top x' \leqslant \mathbf{1}_N \right\}, \\
P_2 &= \left\{ y' \in \mathbb{R}^N \mid Ay' \leqslant \mathbf{1}_M,\ y' \geqslant \mathbf{0} \right\}.
\end{aligned} \tag{2.18}$$

It is easy to see that (2.17) implies that P_1 and P_2 are full-dimensional polytopes, unlike H_1 and H_2.

The set H_1 is in one-to-one correspondence with $P_1 - \{\mathbf{0}\}$ with the map $(x,v) \mapsto x \cdot (1/v)$. Similarly, $(y,u) \mapsto y \cdot (1/u)$ defines a bijection $H_2 \to P_2 - \{\mathbf{0}\}$. These maps have the respective inverse functions $x' \mapsto (x,v)$ and $y' \mapsto (y,u)$ with

$$x = x' \cdot v, \quad v = 1/\mathbf{1}_M^\top x', \qquad y = y' \cdot u, \quad u = 1/\mathbf{1}_N^\top y'. \tag{2.19}$$

These bijections are not linear. However, they preserve the face incidences since a binding inequality in H_1 corresponds to a binding inequality in P_1 and vice versa. In partic-

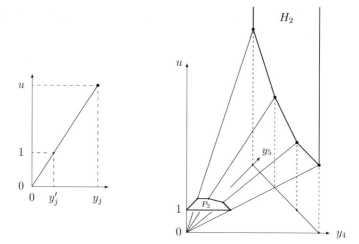

Figure 5. The map $H_2 \to P_2$, $(y, u) \mapsto y' = y \cdot (1/u)$ as a projective transformation with projection point $(\mathbf{0}, 0)$. The left-hand side shows this for a single component y_j of y, the right-hand side shows how P_2 arises in this way from H_2 in the example (2.15).

ular, vertices have the same *labels* defined by the binding inequalities, which are some of the $m + n$ inequalities defining P_1 and P_2 in (2.18).

Figure 5 shows a geometric interpretation of the bijection $(y, u) \mapsto y \cdot (1/u)$ as a *projective transformation* [see Ziegler (1995, Section 2.6)]. On the left-hand side, the pair (y_j, u) is shown as part of (y, u) in H_2 for any component y_j of y. The line connecting this pair to $(0, 0)$ contains the point $(y'_j, 1)$ with $y'_j = y_j/u$. Thus, $P_2 \times \{1\}$ is the intersection of the lines connecting any (y, u) in H_2 with $(\mathbf{0}, 0)$ in $\mathbb{R}^N \times \mathbb{R}$ with the set $\{(y', 1) \mid y' \in \mathbb{R}^N\}$. The vertices $\mathbf{0}$ of P_1 and P_2 do not arise as such projections, but correspond to H_1 and H_2 "at infinity".

2.5. Complementary pivoting

Traversing a polyhedron along its edges has a simple algebraic implementation known as *pivoting*. The constraints defining the polyhedron are thereby represented as linear equations with nonnegative variables. For $P_1 \times P_2$, these have the form

$$
\begin{aligned}
Ay' + r \quad &= \mathbf{1}_M, \\
B^\top x' \quad + s &= \mathbf{1}_N
\end{aligned}
\tag{2.20}
$$

with $x', y', r, s \geqslant \mathbf{0}$ where $r \in \mathbb{R}^M$ and $s \in \mathbb{R}^N$ are vectors of *slack* variables. The system (2.20) is of the form

$$
Cz = q
\tag{2.21}
$$

for a matrix C, right-hand side q, and a vector z of nonnegative variables. The matrix C has full rank, so that q belongs always to the space spanned by the columns C_j of C. A *basis* β is given by a basis $\{C_j \mid j \in \beta\}$ of this column space, so that the square matrix C_β formed by these columns is invertible. The corresponding *basic solution* is the unique vector $z_\beta = (z_j)_{j \in \beta}$ with $C_\beta z_\beta = q$, where the variables z_j for j in β are called *basic variables*, and $z_j = 0$ for all *nonbasic* variables z_j, $j \notin \beta$, so that (2.21) holds. If this solution fulfills also $z \geq \mathbf{0}$, then the basis β is called *feasible*. If β is a basis for (2.21), then the corresponding basic solution can be read directly from the equivalent system $C_\beta^{-1} C z = C_\beta^{-1} q$, called a *tableau*, since the columns of $C_\beta^{-1} C$ for the basic variables form the identity matrix. The tableau and thus (2.21) is equivalent to the system

$$z_\beta = C_\beta^{-1} q - \sum_{j \notin \beta} C_\beta^{-1} C_j z_j \tag{2.22}$$

which shows how the basic variables depend on the nonbasic variables.

Pivoting is a change of the basis where a nonbasic variable z_j for some j not in β *enters* and a basic variable z_i for some i in β *leaves* the set of basic variables. The pivot step is possible if and only if the coefficient of z_j in the ith row of the current tableau is nonzero, and is performed by solving the ith equation for z_j and then replacing z_j by the resulting expression in each of the remaining equations.

For a given entering variable z_j, the leaving variable is chosen to preserve feasibility of the basis. Let the components of $C_\beta^{-1} q$ be $\bar{q}_i$ and of $C_\beta^{-1} C_j$ be $\bar{c}_{ij}$, for $i \in \beta$. Then the largest value of z_j such that in (2.22), $z_\beta = C_\beta^{-1} q - C_\beta^{-1} C_j z_j$ is nonnegative is obviously given by

$$\min\{\bar{q}_i / \bar{c}_{ij} \mid i \in \beta, \ \bar{c}_{ij} > 0\}. \tag{2.23}$$

This is called a *minimum ratio test*. Except in degenerate cases (see below), the minimum in (2.23) is unique and determines the leaving variable z_i uniquely. After pivoting, the new basis is $\beta \cup \{j\} - \{i\}$.

The choice of the entering variable depends on the solution that one wants to find. The Simplex method for linear programming is defined by pivoting with an entering variable that improves the value of the objective function. In the system (2.20), one looks for a *complementary* solution where

$$x'^\top r = 0, \qquad y'^\top s = 0 \tag{2.24}$$

because it implies with (2.19) the complementarity conditions (2.12) and (2.13) so that (x, y) is a Nash equilibrium by Theorem 2.4. In a basic solution to (2.20), every nonbasic variable has value zero and represents a binding inequality, that is, a facet of the polytope. Hence, each basis defines a vertex which is labeled with the indices of the nonbasic variables. The variables of the system come in *complementary pairs* (x_i, r_i) for the indices $i \in M$ and (y_j, s_j) for $j \in N$. Recall that the Lemke–Howson algorithm follows a path of solutions that have all labels in $M \cup N$ except for a missing label k. Thus a k-almost completely labeled vertex is a basis that has exactly one basic variable from each complementary pair, except for a pair of variables (x_k, r_k), say (if $k \in M$), that are both basic. Correspondingly, there is another pair of complementary variables that are both nonbasic, representing the duplicate label. One of them is chosen as the entering variable, depending on the direction of the computed path. The two possibilities represent the two k-almost completely labeled edges incident to that vertex. The algorithm is started with all components of r and s as basic variables and nonbasic variables $(x', y') = (\mathbf{0}, \mathbf{0})$. This initial solution fulfills (2.24) and represents the artificial equilibrium.

ALGORITHM 2.9 (Complementary pivoting). For a bimatrix game (A, B) fulfilling (2.17), compute a sequence of basic feasible solutions to the system (2.20) as follows.
  (a) Initialize with basic variables $r = \mathbf{1}_M$, $s = \mathbf{1}_N$. Choose $k \in M \cup N$, and let the first entering variable be x'_k if $k \in M$ and y'_k if $k \in N$.
  (b) Pivot such as to maintain feasibility using the minimum ratio test.
  (c) If the variable z_i that has just left the basis has index k, stop. Then (2.24) holds and (x, y) defined by (2.19) is a Nash equilibrium. Otherwise, choose the complement of z_i as the next entering variable and go to (b).

We demonstrate Algorithm 2.9 for the example (2.15). The initial basic solution in the form (2.22) is given by

$$
\begin{aligned}
r_1 &= 1 && - 6y'_5, \\
r_2 &= 1 - 2y'_4 - 5y'_5, \\
r_3 &= 1 - 3y'_4 - 3y'_5
\end{aligned}
\tag{2.25}
$$

and

$$
\begin{aligned}
s_4 &= 1 - x'_1 && - 4x'_3, \\
s_5 &= 1 && - 2x'_2 - 3x'_3.
\end{aligned}
\tag{2.26}
$$

Pivoting can be performed separately for these two systems since they have no variables in common. With the missing label 2 as in Figure 3, the first entering variable is x'_2. Then the second equation of (2.26) is rewritten as $x'_2 = \frac{1}{2} - \frac{3}{2}x'_3 - \frac{1}{2}s_5$ and s_5 leaves the basis. Next, the complement y'_5 of s_5 enters the basis. The minimum ratio (2.23) in (2.25) is $1/6$, so that r_1 leaves the basis and (2.25) is replaced by the system

$$
\begin{aligned}
y'_5 &= \tfrac{1}{6} && - \tfrac{1}{6}r_1, \\
r_2 &= \tfrac{1}{6} - 2y'_4 + \tfrac{5}{6}r_1, \\
r_3 &= \tfrac{1}{2} - 3y'_4 + \tfrac{1}{2}r_1.
\end{aligned}
\tag{2.27}
$$

Then the complement x_1' of r_1 enters the basis and s_4 leaves, so that the system replacing (2.26) is now

$$
\begin{aligned}
x_1' &= 1 - 4x_3' - s_4, \\
x_2' &= \tfrac{1}{2} - \tfrac{3}{2}x_3' \quad\; - \tfrac{1}{2}s_5.
\end{aligned}
\tag{2.28}
$$

With y_4' entering, the minimum ratio (2.23) in (2.27) is $1/12$, where r_2 leaves the basis and (2.27) is replaced by

$$
\begin{aligned}
y_5' &= \tfrac{1}{6} \; - \tfrac{1}{6}r_1, \\
y_4' &= \tfrac{1}{12} + \tfrac{5}{12}r_1 - \tfrac{1}{2}r_2, \\
r_3 &= \tfrac{1}{4} \; - \tfrac{3}{4}r_1 + \tfrac{3}{2}r_2.
\end{aligned}
\tag{2.29}
$$

Then the algorithm terminates since the variable r_2, with the missing label 2 as index, has become nonbasic. The solution defined by the final systems (2.28) and (2.29), with the nonbasic variables on the right-hand side equal to zero, fulfills (2.24). Renormalizing x' and y' by (2.19) as probability vectors gives the equilibrium $(x, y) = (x^3, y^3)$ mentioned after (2.15) with payoffs 4 to player 1 and $2/3$ to player 2.

Assumption (2.17) with the simple initial basis for the system (2.20) is used by Wilson (1992). Lemke and Howson (1964) assume $A < \mathbf{0}$ and $B < \mathbf{0}$, so that P_1 and P_2 are unbounded polyhedra and the almost completely labeled path starts at the vertex at the end of an unbounded edge. To avoid the renormalization (2.19), the Lemke–Howson algorithm can also be applied to the system (2.14) represented in equality form. Then the unconstrained variables u and v have no slack variables as counterparts and are always basic, so they never leave the basis and are disregarded in the minimum ratio test. Then the computation has the following *economic interpretation* [Wilson (1992), van den Elzen (1993)]: let the missing label k belong to M. Then the basic slack variable r_k which is basic together with x_k can be interpreted as a "subsidy" payoff for the pure strategy k so that player 1 is in equilibrium. The algorithm terminates when that subsidy or the probability x_k vanishes. Player 2 is in equilibrium throughout the computation.

2.6. Degenerate games

The path computed by the Lemke–Howson algorithm is unique only if the game is nondegenerate. Like other pivoting methods, the algorithm can be extended to degenerate games by "lexicographic perturbation", as suggested by Lemke and Howson (1964). Before we explain this, we show that various definitions of nondegeneracy used in the literature are equivalent. In the following theorem, I_M denotes the identity matrix in $\mathbb{R}^{M \times M}$. Furthermore, a pure strategy i of player 1 is called *payoff equivalent* to a mixed strategy x of player 1 if it produces the same payoffs, that is, $a_i = x^\top A$. The strategy i is called *weakly dominated* by x if $a_i \leqslant x^\top A$, and *strictly dominated* by x if $a_i < x^\top A$ holds. The same applies to the strategies of player 2.

THEOREM 2.10. *Let (A, B) be an $m \times n$ bimatrix game so that (2.17) holds. Then the following are equivalent.*

(a) *The game is nondegenerate according to Definition 2.6.*

(b) *For any x in X and y in Y, the rows of $\begin{bmatrix} I_M \\ B^\top \end{bmatrix}$ for the labels of x are linearly independent, and the rows of $\begin{bmatrix} A \\ I_N \end{bmatrix}$ for the labels of y are linearly independent.*

(c) *For any x in X with set of labels K and y in Y with set of labels L, the set $\bigcap_{k \in K} X(k)$ has dimension $m - |K|$, and the set $\bigcap_{l \in L} Y(l)$ has dimension $n - |L|$.*

(d) *P_1 and P_2 in (2.18) are simple polytopes, and any pure strategy of a player that is weakly dominated by or payoff equivalent to another mixed strategy is strictly dominated by some mixed strategy.*

(e) *In any basic feasible solution to (2.20), all basic variables have positive values.*

Lemke and Howson (1964) define nondegenerate games by condition (b). Krohn et al. (1991), and, in slightly weaker form, Shapley (1974), define nondegeneracy as in (c). Van Damme (1987, p. 52) has observed the implication (b) $\Rightarrow$ (a). Some of the implications between the conditions (a)–(e) in Theorem 2.10 are easy to prove, whereas others require more work. For details of the proof see von Stengel (1996b).

The $m + n$ rows of the matrices in (b) define the inequalities for the polytopes P_1 and P_2 in (2.18), where the labels denote binding inequalities. This condition explains why a *generic* bimatrix game is nondegenerate with probability one: we call a game generic if each payoff is drawn randomly and independently from a continuous distribution, e.g., the normal distribution with small variance around an approximate value for the respective payoff. Then the rows of the matrices described in 2.10(b) are linearly independent with probability one, since a linear dependence imposes an equation on at least one payoff, which is fulfilled with probability zero. However, the strategic form of an extensive game (like Figure 8) is often degenerate since its payoff entries are not independent. A systematic treatment of degeneracy is therefore of interest.

The dimensionality condition in Theorem 2.10(c) has been explained informally before Theorem 2.7 above. The geometric interpretation of nondegeneracy in 2.10(d) consists of two parts. The polytope P_1 (and similarly P_2) is simple since a point that belongs to more than m facets of P_1 has too many labels. In the game

$$A = \begin{bmatrix} 0 & 6 \\ 2 & 5 \\ 3 & 3 \end{bmatrix}, \qquad B = \begin{bmatrix} 1 & 0 \\ 0 & 2 \\ 4 & 4 \end{bmatrix}, \tag{2.30}$$

the polytope P_1 is not simple because its vertex $(0, 0, \frac{1}{4})^\top$ belongs to four facets. This game is degenerate since the pure strategy 3 of player 1 has two best responses. Apart from this, degeneracy may result due to a redundancy of the *description* of the polytope by inequalities (for example, if A has two identical rows of payoffs to player 1). It is not hard to show that such redundant inequalities correspond to weakly dominated strategies. A binding inequality of this sort defines a face of the respective polytope. The strict dominance in (d) asserts that this face is empty if the game is nondegenerate.

Theorem 2.10(e) states that every feasible *basis* of the system is *nondegenerate*, that is, all basic variables have positive values. This condition implies that the leaving variable in step (b) of Algorithm 2.9 is unique, since otherwise, another variable that could also leave the basis but stays basic will have value zero after the pivoting step. This concludes our remarks on Theorem 2.10.

The *lexicographic method* extends the minimum ratio test in such a way that the leaving variable is always unique, even in degenerate cases. The method simulates an infinitesimal perturbation of the right-hand side of the given linear system (2.21), $z \geqslant \mathbf{0}$, and works as follows. Let Q be a matrix of full row rank with k columns. For any $\varepsilon \geqslant 0$, consider the system

$$Cz = q + Q \cdot \left(\varepsilon^1, \ldots, \varepsilon^k\right)^\top \tag{2.31}$$

which is equal to (2.21) for $\varepsilon = 0$ and which is a *perturbed* system for $\varepsilon > 0$. Let β be a basis for this system with basic solution

$$z_\beta = C_\beta^{-1} q + C_\beta^{-1} Q \cdot \left(\varepsilon^1, \ldots, \varepsilon^k\right)^\top = \bar{q} + \overline{Q} \cdot \left(\varepsilon^1, \ldots, \varepsilon^k\right)^\top \tag{2.32}$$

and $z_j = 0$ for $j \notin \beta$. It is easy to see that z_β is positive for all sufficiently small ε if and only if all rows of the matrix $[\bar{q}, \overline{Q}]$ are *lexico-positive*, that is, the first nonzero component of each row is positive. Then β is called a *lexico-feasible* basis. This holds in particular for $\bar{q} > \mathbf{0}$ when β is a nondegenerate basis for the unperturbed system. Because Q has full row rank, $\overline{Q}$ has no zero row, which implies that any feasible basis for the perturbed system is nondegenerate.

In consequence, the leaving variable for the perturbed system is always unique. It is determined by the following *lexico-minimum ratio test*. As in the minimum ratio test (2.23), let, for $i \in \beta$, the entries of the entering column $C_\beta^{-1} C_j$ be $\bar{c}_{ij}$, those of $\bar{q}$ in (2.32) be $\bar{q}_{i0}$, and those of $\overline{Q}$ be $\bar{q}_{il}$ for $1 \leqslant l \leqslant k$. Then the leaving variable is determined by the maximum choice of the entering variable z_j such that all basic variables z_i in (2.31) stay nonnegative, that is,

$$z_i = \bar{q}_{i0} + \bar{q}_{i1}\varepsilon^1 + \cdots + \bar{q}_{ik}\varepsilon^k - \bar{c}_{ij}z_j \geqslant 0$$

for all $i \in \beta$. For sufficiently small ε, the sharpest bound for z_j is obtained for that i in β with the *lexicographically smallest* row vector $1/\bar{c}_{ij} \cdot (\bar{q}_{i0}, \bar{q}_{i1}, \ldots, \bar{q}_{ik})$ where $\bar{c}_{ij} > 0$ (a vector is called lexicographically smaller than another if it is smaller in the first component where the vectors differ). No two of these row vectors are equal since $\overline{Q}$ has full row rank. Therefore, this lexico-minimum ratio test, which extends (2.23), determines the leaving variable z_i uniquely. By construction, it preserves the invariant that all computed bases are lexico-feasible, provided this holds for the initial basis like that in Algorithm 2.9(a) which is nondegenerate. Since the computed sequence of bases is unique, the computation cannot cycle and terminates as in the nondegenerate case.

The lexico-minimum ratio test can be performed without actually perturbing the system, since it only depends on the current basis β and Q in (2.32). The actual values of the basic variables are given by $\bar{q}$, which may have zero entries, so the perturbation applies as if ε is vanishing. The lexicographic method requires little extra work (and none for a nondegenerate game) since Q can be equal to C or to that part of C containing the identity matrix, so that $\overline{Q}$ in (2.32) is just the respective part of the current tableau. Wilson (1992) uses this to compute equilibria with additional stability properties, as discussed in Section 3.1 below. Eaves (1971) describes a general setup of lexicographic systems for LCPs and shows various ways (pp. 625, 629, 632) of solving bimatrix games with Lemke's algorithm [Lemke (1965)], a generalization of the Lemke–Howson method.

2.7. *Equilibrium enumeration and other methods*

For a given bimatrix game, the Lemke–Howson algorithm finds at least one equilibrium. Sometimes, one wishes to find all equilibria, for example in order to know if an equilibrium is unique. A simple approach [as used by Dickhaut and Kaplan (1991)] is to enumerate all possible equilibrium supports, solve the corresponding linear equations for mixed strategy probabilities, and check if the unplayed pure strategies have smaller payoffs. In a nondegenerate game, both players use the same number of pure strategies in equilibrium, so only supports of equal cardinality need to be examined. They can be represented as $M \cap S$ and $N - S$ for any n-element subset S of $M \cup N$ except N. There are $\binom{m+n}{n} - 1$ many possibilities for S, which is exponential in the smaller dimension m or n of the bimatrix game. Stirling's asymptotic formula $\sqrt{2\pi n}(n/e)^n$ for the factorial $n!$ shows that in a square bimatrix game where $m = n$, the binomial coefficient $\binom{2n}{n}$ is asymptotically $4^n / \sqrt{\pi n}$. The number of equal-sized supports is here not substantially smaller than the number 4^n of all possible supports.

An alternative is to inspect the vertices of $H_1 \times H_2$ defined in (2.16) if they represent equilibria. Mangasarian (1964) does this by checking if the bilinear function $x^\top(A + B)y - u - v$ has a maximum, that is, has value zero, so this is equivalent to the complementarity conditions (2.12) and (2.13). It is easier to enumerate the vertices of P_1 and P_2 in (2.18) since these are polytopes if (2.17) holds. Analogous to Theorem 2.5, a pair (x', y') in $P_1 \times P_2$, except $(\mathbf{0}, \mathbf{0})$, defines a Nash equilibrium (x, y) by (2.19) if it is completely labeled. The labels can be assigned directly to (x', y') as the binding inequalities. That is, (x', y') in $P_1 \times P_2$ has label i in M if $x'_i = 0$ or $a_i y = 1$, and label j in N if $b_j x' = 1$ or $y'_j = 0$ holds.

THEOREM 2.11. *Let (A, B) be a bimatrix game so that (2.17) holds, and let V_1 and V_2 be the sets of vertices of P_1 and P_2 in (2.18), respectively. Then if (A, B) is nonde-*

generate, (x, y) given by (2.19) is a Nash equilibrium of (A, B) if and only if (x', y') is a completely labeled vertex pair in $V_1 \times V_2 - \{(\mathbf{0}, \mathbf{0})\}$.

Thus, computing the vertex sets V_1 of P_1 and V_2 of P_2 and checking their labels finds all Nash equilibria of a nondegenerate game. This method was first suggested by Vorobiev (1958), and later simplified by Kuhn (1961). An elegant method for vertex enumeration is due to Avis and Fukuda (1992).

The number of vertices of a polytope is in general exponential in the dimension. The maximal number is described in the following theorem, where $\lfloor t \rfloor$ for a real number t denotes the largest integer not exceeding t.

THEOREM 2.12 [Upper bound theorem for polytopes, McMullen (1970)]. *The maximum number of vertices of a d-dimensional polytope with k facets is*

$$\Phi(d, k) = \binom{k - \lfloor \frac{d-1}{2} \rfloor - 1}{\lfloor \frac{d}{2} \rfloor} + \binom{k - \lfloor \frac{d}{2} \rfloor - 1}{\lfloor \frac{d-1}{2} \rfloor}.$$

For a self-contained proof of this theorem see Mulmuley (1994). This result shows that P_1 has at most $\Phi(m, n + m)$ and P_2 has at most $\Phi(n, m + n)$ vertices, including $\mathbf{0}$ which is not part of an equilibrium. In a nondegenerate game, any vertex is part of at most one equilibrium, so the smaller number of vertices of the polytope P_1 or P_2 is a bound for the number of equilibria.

COROLLARY 2.13 [Keiding (1997)]. *A nondegenerate $m \times n$ bimatrix game has at most $\min\{\Phi(m, n + m), \Phi(n, m + n)\} - 1$ equilibria.*

It is not hard to show that $m < n$ implies $\Phi(m, n + m) < \Phi(n, m + n)$. For $m = n$, Stirling's formula shows that $\Phi(n, 2n)$ is asymptotically $c \cdot (27/4)^{n/2}/\sqrt{n}$ or about $c \cdot 2.598^n/\sqrt{n}$, where the constant c is equal to $2\sqrt{2/3\pi}$ or about 0.921 if n is even, and $\sqrt{2/\pi}$ or about 0.798 if n is odd. Since 2.598^n grows less rapidly than 4^n, vertex enumeration is more efficient than support enumeration.

Although the upper bound in Corollary 2.13 is probably not tight, it is possible to construct bimatrix games that have a large number of Nash equilibria. The $n \times n$ bimatrix game where A and B are equal to the identity matrix has $2^n - 1$ Nash equilibria. Then both P_1 and P_2 are equal to the n-dimensional unit cube, where each vertex is part of a completely labeled pair. Quint and Shubik (1997) conjectured that no nondegenerate $n \times n$ bimatrix game has more equilibria. This follows from Corollary 2.13 for $n \leqslant 3$ and is shown for $n = 4$ by Keiding (1997) and McLennan and Park (1999). However, there are counterexamples for $n \geqslant 6$, with asymptotically $c \cdot (1 + \sqrt{2})^n/\sqrt{n}$ or about $c \cdot 2.414^n/\sqrt{n}$ many equilibria, where c is $2^{3/4}/\sqrt{\pi}$ or about 0.949 if n is even, and $(2^{9/4} - 2^{7/4})/\sqrt{\pi}$ or about 0.786 if n is odd [von Stengel (1999)]. These games are constructed with the help of polytopes which have the maximum number $\Phi(n, 2n)$ of vertices. This result suggests that vertex enumeration is indeed the appropriate method for finding all Nash equilibria.

For degenerate bimatrix games, Theorem 2.10(d) shows that P_1 or P_2 may be not simple. Then there may be equilibria (x, y) corresponding to completely labeled points (x', y') in $P_1 \times P_2$ where, for example, x' has more than m labels and y' has fewer than n labels and is therefore not a vertex of P_2. However, any such equilibrium is the convex combination of equilibria that are represented by vertex pairs, as shown by Mangasarian (1964). The set of Nash equilibria of an arbitrary bimatrix game is characterized as follows.

THEOREM 2.14 [Winkels (1979), Jansen (1981)]. *Let (A, B) be a bimatrix game so that (2.17) holds, let V_1 and V_2 be the sets of vertices of P_1 and P_2 in (2.18), respectively, and let R be the set of completely labeled vertex pairs in $V_1 \times V_2 - \{(\mathbf{0}, \mathbf{0})\}$. Then (x, y) given by (2.19) is a Nash equilibrium of (A, B) if and only if (x', y') belongs to the convex hull of some subset of R of the form $U_1 \times U_2$ where $U_1 \subseteq V_1$ and $U_2 \subseteq V_2$.*

PROOF. Labels are preserved under convex combinations. Hence, if the set $U_1 \times U_2$ is contained in R, then any convex combination of its elements is also a completely labeled pair (x', y') that defines a Nash equilibrium by (2.19).

Conversely, assume (x', y') in $P_1 \times P_2$ corresponds to a Nash equilibrium of the game via (2.19). Let $I = \{i \in M \mid a_i y' < 1\}$ and $J = \{j \in N \mid y'_j > 0\}$, that is, x' has at least the labels in $I \cup J$. Then the elements z in P_1 fulfilling $z_i = 0$ for $i \in I$ and $b_j z = 1$ for $j \in J$ form a face of P_1 (defined by the sum of these equations, for example) which contains x'. This face is a polytope and therefore equal to the convex hull of its vertices, which are all vertices of P_1. Hence, x' is the positive convex combination $\sum_{k \in K} x^k \lambda_k$ of certain vertices x^k of P_1, where $\lambda_k > 0$ for $k \in K$. Similarly, y' is the positive convex combination $\sum_{l \in L} y^l \mu_l$ of certain vertices y^l of P_2, where $\mu_l > 0$ for $l \in L$. This implies the convex representation

$$(x', y') = \sum_{k \in K, \, l \in L} \lambda_k \mu_l (x^k, y^l).$$

With $U_1 = \{x^k \mid k \in K\}$ and $U_2 = \{y^l \mid l \in L\}$, it remains to show $(x^k, y^l) \in G$ for all $k \in K$ and $l \in L$. Suppose otherwise that some (x^k, y^l) was not completely labeled, with some missing label, say $j \in N$, so that $b_j x^k < 1$ and $y'_j > 0$. But then $b_j x' < 1$ since $\lambda_k > 0$ and $y'_j > 0$ since $\mu_l > 0$, so label j would also be missing from (x', y') contrary to the assumption. So indeed $U_1 \times U_2 \subseteq G$.   □

The set R in Theorem 2.14 can be viewed as a bipartite graph with the completely labeled vertex pairs as edges. The subsets $U_1 \times U_2$ are *cliques* of this graph. The convex hulls of the maximal cliques of R are called *maximal Nash subsets* [Millham (1974), Heuer and Millham (1976)]. Their union is the set of all equilibria, but they are not nec-

$$A = \begin{bmatrix} 2 & 1 \\ 1 & 1 \end{bmatrix}, \qquad B = \begin{bmatrix} 1 & 1 \\ 1 & 2 \end{bmatrix}$$

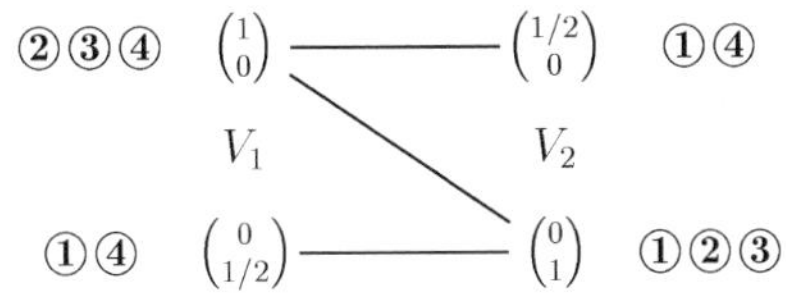

Figure 6. A game (A, B), and its set R of completely labeled vertex pairs in Theorem 2.14 as a bipartite graph. The labels denoting the binding inequalities in P_1 and P_2 are also shown for illustration.

essarily disjoint. The topological equilibrium components of the set of Nash equilibria are the unions of non-disjoint maximal Nash subsets.

An example is shown in Figure 6, where the maximal Nash subsets are, as sets of mixed strategies, $\{(1, 0)^\top\} \times Y$ and $X \times \{(0, 1)^\top\}$. This degenerate game illustrates the second part of condition 2.10(d): the polytopes P_1 and P_2 are simple but have vertices with more labels than the dimension due to weakly but not strongly dominated strategies. Dominated strategies could be iteratively eliminated, but this may not be desired here since the order of elimination matters. Knuth, Papadimitriou and Tsitsiklis (1988) study computational aspects of strategy elimination where they overlook this fact; see also Gilboa, Kalai and Zemel (1990, 1993). The interesting problem of iterated elimination of pure strategies that are *payoff equivalent* to other mixed strategies is studied in Vermeulen and Jansen (1998).

Quadratic optimization is used for computing equilibria by Mills (1960), Mangasarian and Stone (1964), and Mukhamediev (1978). Audet et al. (2001) enumerate equilibria with a search over polyhedra defined by parameterized linear programs. Bomze (1992) describes an enumeration of the *evolutionarily stable* equilibria of a symmetric bimatrix game. Yanovskaya (1968), Howson (1972), Eaves (1973), and Howson and Rosenthal (1974) apply complementary pivoting to *polymatrix games*, which are multi-player games obtained as sums of pairwise interactions of the players.

3. Equilibrium refinements

Nash equilibria of a noncooperative game are not necessarily unique. A large number of *refinement* concepts have been invented for selecting some equilibria as more "reasonable" than others. We give an exposition [with further details in von Stengel (1996b)] of two methods that find equilibria with additional refinement properties. Wilson (1992) extends the Lemke–Howson algorithm so that it computes a *simply stable* equilibrium. A complementary pivoting method that finds a *perfect* equilibrium is due to van den Elzen and Talman (1991).

3.1. Simply stable equilibria

Kohlberg and Mertens (1986) define strategic *stability* of equilibria. Basically, a set of equilibria is called stable if every game nearby has equilibria nearby [Wilson (1992)].

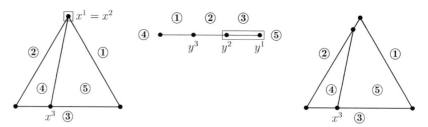

Figure 7. Left and center: mixed strategy sets X and Y for the game (A, B) in (2.30) with labels similar to Figure 2. The game has an infinite set of equilibria indicated by the pair of rectangular boxes. Right: mixed strategy set X where strategy 5 gets slightly higher payoffs, and only the equilibrium (x^3, y^3) remains.

In degenerate games, certain equilibrium sets may not be stable. In the bimatrix game (A, B) in (2.30), for example, all convex combinations of (x^1, y^1) and (x^2, y^2) are equilibria, where $x^1 = x^2 = (0, 0, 1)^\top$ and $y^1 = (0, 1)^\top$ and $y^2 = (\frac{1}{3}, \frac{2}{3})^\top$. Another, isolated equilibrium is (x^3, y^3). As shown in the right picture of Figure 7, the first of these equilibrium sets is not stable since it disappears when the payoffs to player 2 for her second strategy 5 are slightly increased.

Wilson (1992) describes an algorithm that computes a set of *simply stable* equilibria. There the game is not perturbed arbitrarily but only in certain systematic ways that are easily captured computationally. Simple stability is therefore weaker than the stability concepts of Kohlberg and Mertens (1986) and Mertens (1989, 1991). Simply stable sets may not be stable, but no such game has yet been found [Wilson (1992, p. 1065)]. However, the algorithm is more efficient and seems practically useful compared to the exhaustive method by Mertens (1989).

The perturbations considered for simple stability do not apply to single payoffs but to pure strategies, in two ways. A *primal* perturbation introduces a small *minimum probability* for playing that strategy, even if it is not optimal. A *dual* perturbation introduces a small *bonus* for that strategy, that is, its payoff can be slightly smaller than the best payoff and yet the strategy is still considered optimal. In system (2.20), the variables x', y', r, s are perturbed by corresponding vectors ξ, η, ρ, σ that have small positive components, $\xi, \rho \in \mathbb{R}^M$ and $\eta, \sigma \in \mathbb{R}^N$. That is, (2.20) is replaced by

$$A(y' + \eta) + I_M(r + \rho) = \mathbf{1}_M,$$
$$B^\top(x' + \xi) \qquad\qquad\quad + I_N(s + \sigma) = \mathbf{1}_N. \tag{3.1}$$

If (3.1) and the complementarity condition (2.24) hold, then a variable x_i or y_j that is zero is replaced by ξ_i or η_j, respectively. After the transformation (2.19), these terms denote a small positive probability for playing the pure strategy i or j, respectively. So ξ and η represent primal perturbations.

Similarly, ρ and σ stand for dual perturbations. To see that ρ_i or σ_j indeed represents a bonus for i or j, respectively, consider the second set of equations in (3.1) with $\xi = \mathbf{0}$

for the example (2.30):

$$
\begin{bmatrix} 1 & 0 & 4 \\ 0 & 2 & 4 \end{bmatrix}
\begin{pmatrix} x_1' \\ x_2' \\ x_3' \end{pmatrix}
+ \begin{pmatrix} s_4 + \sigma_4 \\ s_5 + \sigma_5 \end{pmatrix}
= \begin{pmatrix} 1 \\ 1 \end{pmatrix} .
$$

If, say, $\sigma_5 > \sigma_4$, then one solution is $x_1' = x_2' = 0$ and $x_3' = (1 - \sigma_5)/4$ with $s_5 = 0$ and $s_4 = \sigma_5 - \sigma_4 > 0$, which means that only the second strategy of player 2 is optimal, so the higher perturbation σ_5 represents a higher bonus for that strategy (as shown in the right picture in Figure 7). Dual perturbations are a generalization of primal perturbations, letting $\rho = A\eta$ and $\sigma = B^\top \xi$ in (3.1). Here, only special cases of these perturbations will be used, so it is useful to consider them both.

Denote the vector of perturbations in (3.1) by

$$
(\xi, \eta, \rho, \sigma)^\top = \delta = (\delta_1, \dots, \delta_k)^\top, \quad k = 2(m + n). \tag{3.2}
$$

For simple stability, Wilson (1992, p. 1059) considers only special cases of δ. For each $i \in \{1, \dots, k\}$, the component δ_{i+1} (or δ_1 if $i = k$) represents the largest perturbation by some $\varepsilon > 0$. The subsequent components $\delta_{i+2}, \dots, \delta_k, \delta_1, \dots, \delta_i$ are equal to smaller perturbations $\varepsilon^2, \dots, \varepsilon^k$. That is,

$$
\begin{aligned}
d_{i+j} &= \varepsilon^j \quad \text{if } i + j \leqslant k, \\
d_{i+j-k} &= \varepsilon^j \quad \text{if } i + j > k,
\end{aligned}
\qquad 1 \leqslant j \leqslant k. \tag{3.3}
$$

DEFINITION 3.1 [Wilson (1992)]. Let (A, B) be an $m \times n$ bimatrix game. Then a connected set of equilibria of (A, B) is called *simply stable* if for all $i = 1, \dots, k$, all sufficiently small $\varepsilon > 0$, and $(\xi, \eta, \rho, \sigma)$ as in (3.2), (3.3), there is a solution $r = (x', y', r, s)^\top \geqslant \mathbf{0}$ to (3.1) and (2.24) so that the corresponding strategy pair (x, y) defined by (2.19) is near that set.

Due to the perturbation, (x, y) in Definition 3.1 is only an "approximate" equilibrium. When ε vanishes, then (x, y) becomes a member of the simply stable set. A perturbation with vanishing ε is mimicked by a lexico-minimum ratio test as described in Section 2.6 that extends step (b) of Algorithm 2.9. The perturbation (3.3) is therefore easily captured computationally. With (3.2), (3.3), the perturbed system (3.1) is of the form (2.31) with

$$
z = (x', y', r, s)^\top, \quad
C = \begin{bmatrix} \mathbf{0} & A & I_M & \mathbf{0} \\ B^\top & \mathbf{0} & \mathbf{0} & I_N \end{bmatrix}, \quad
q = \begin{bmatrix} \mathbf{1}_M \\ \mathbf{1}_N \end{bmatrix} \tag{3.4}
$$

and $Q = [-C_{i+1}, \dots, -C_k, -C_1, \dots, -C_i]$ if $C_1, \dots, C_k$ are the columns of C. That is, Q is just $-C$ except for a cyclical shift of the columns, so that the lexico-minimum ratio test is easily performed using the current tableau.

The algorithm by Wilson (1992) computes a *path* of equilibria where all perturbations of the form (3.3) occur somewhere. Starting from the artificial equilibrium $(\mathbf{0}, \mathbf{0})$, the Lemke–Howson algorithm is used to compute an equilibrium with a lexicographic order shifted by some i. Having reached that equilibrium, i is increased as long as the computed basic solution is lexico-feasible with that shifted order. If this is not possible for all i (as required for simple stability), a new Lemke–Howson path is started with the missing label determined by the maximally possible lexicographic shift. This requires several variants of pivoting steps. The final piece of the computed path represents the connected set in Definition 3.1.

3.2. Perfect equilibria and the tracing procedure

An equilibrium is *perfect* [Selten (1975)] if it is robust against certain small mistakes of the players. Mistakes are represented by small positive minimum probabilities for all pure strategies. We use the following characterization [Selten (1975, p. 50, Theorem 7)] as definition.

DEFINITION 3.2 [Selten (1975)]. An equilibrium (x, y) of a bimatrix game is called *perfect* if there is a continuous function $\varepsilon \mapsto (x(\varepsilon), y(\varepsilon))$ where $(x(\varepsilon), y(\varepsilon))$ is a pair of completely mixed strategies for all $\varepsilon > 0$, $(x, y) = (x(0), y(0))$, and x is a best response to $y(\varepsilon)$ and y is a best response to $x(\varepsilon)$ for all ε.

Positive minimum probabilities for all pure strategies define a special primal perturbation as considered for simply stable equilibria. Thus, as noted by Wilson (1992, p. 1042), his modification of the Lemke–Howson algorithm can also be used for computing a perfect equilibrium. Then it is not necessary to shift the lexicographic order, so the lexico-minimum ratio test described in Section 2.6 can be used with $Q = -C$.

THEOREM 3.3. *Consider a bimatrix game (A, B) and, with (3.4), the LCP $Cz = q$, $z \geqslant \mathbf{0}$, (2.24). Then Algorithm 2.9, computing with bases β so that $C_\beta^{-1}[q, -C]$ is lexico-positive, terminates at a perfect equilibrium.*

PROOF. Consider the computed solution to the LCP, which represents an equilibrium (x, y) by (2.19). The final basis β is lexico-positive, that is, for $Q = -C$ in the perturbed system (2.32), the basic variables z_β are all positive if $\varepsilon > 0$. In (2.32), replace $(\varepsilon, \ldots, \varepsilon^k)^\top$ by

$$\delta = (\xi, \eta, \rho, \sigma)^\top = \left(\varepsilon, \ldots, \varepsilon^{m+n}, 0, \ldots, 0\right)^\top, \tag{3.5}$$

so that z_β is still nonnegative. Then z_β contains the basic variables of the solution (x', y', r, s) to (3.1), with $\rho = \mathbf{0}$, $\sigma = \mathbf{0}$ by (3.5). This solution depends on ε, so $r = r(\varepsilon)$, $s = s(\varepsilon)$, and it determines the pair $x'(\varepsilon) = x' + \xi$, $y(\varepsilon) = y' + \eta$ which represents a completely mixed strategy pair if $\varepsilon > 0$. The computed equilibrium is equal to this pair for $\varepsilon = 0$, and it is a best response to this pair since it is complementary to the slack variables $r(\varepsilon)$, $s(\varepsilon)$. Hence the equilibrium is perfect by Definition 3.2. $\qquad\square$

A different approach to computing perfect equilibria of a bimatrix game is due to van den Elzen and Talman (1991, 1999); see also van den Elzen (1993). The method uses an arbitrary *starting point* (p, q) in the product $X \times Y$ of the two strategy spaces defined in (2.7). It computes a piecewise linear path in $X \times Y$ that starts at (p, q) and terminates at an equilibrium. The pair (p, q) is used throughout the computation as a reference point. The computation uses an auxiliary variable z_0, which can be regarded as a parameter for a *homotopy* method [see Garcia and Zangwill (1981, p. 368)]. Initially, $z_0 = 1$. Then, z_0 is decreased and, after possible intermittent increases, eventually becomes zero, which terminates the algorithm.

The algorithm computes a sequence of basic solutions to the system

$$
\begin{aligned}
Ex \quad &+ \quad ez_0 = e, \\
Fy \quad &+ \quad fz_0 = f, \\
r = E^\top u \quad - Ay \quad &- (Aq)z_0 \geqslant \mathbf{0}, \\
s = \quad F^\top v - B^\top x \quad &- (B^\top p)z_0 \geqslant \mathbf{0}, \\
x, \quad y, \quad &\quad z_0 \geqslant \mathbf{0}.
\end{aligned}
\tag{3.6}
$$

These basic solutions contain at most one basic variable from each complementary pair (x_i, r_i) and (y_j, s_j) and therefore fulfill

$$
x^\top r = 0, \qquad y^\top s = 0.
\tag{3.7}
$$

The constraints (3.6), (3.7) define an *augmented* LCP which differs from (2.14) only by the additional column for the variable z_0. That column is determined by (p, q). An initial solution is $z_0 = 1$ and $x = \mathbf{0}$, $y = \mathbf{0}$. As in Algorithm 2.9, the computation proceeds by complementary pivoting. It terminates when z_0 is zero and leaves the basis. Then the solution is an equilibrium by Theorem 2.4.

As observed in von Stengel, van den Elzen and Talman (2002), the algorithm in this description is a special case of the algorithm by Lemke (1965) for solving an LCP [see also Murty (1988), Cottle et al. (1992)]. Any solution to (3.6) fulfills $0 \leqslant z_0 \leqslant 1$, and the pair

$$
(\bar{x}, \bar{y}) = (x + pz_0, y + qz_0)
\tag{3.8}
$$

belongs to $X \times Y$ since $Ep = e$ and $Fq = f$. Hence, $(\bar{x}, \bar{y})$ is a pair of mixed strategies, initially equal to the starting point (p, q). For $z_0 = 0$, it is the computed equilibrium. The set of these pairs $(\bar{x}, \bar{y})$ is the computed piecewise linear path in $X \times Y$. In particular, the computed solution is always bounded. The algorithm can therefore never encounter an unbounded ray of solutions, which in general may cause Lemke's algorithm to fail. The computed pivoting steps are unique by using lexicographic degeneracy resolution. This proves that the algorithm terminates.

In (3.8), the positive components x_i and y_j of x and y describe which pure strategies i and j, respectively, are played with higher probability than the minimum probabilities

$p_i z_0$ and $q_j z_0$ as given by (p, q) and the current value of z_0. By the complementarity condition (3.7), these are *best responses* to the current strategy pair $(\bar{x}, \bar{y})$. Therefore, any point on the computed path is an *equilibrium* of the *restricted* game where each pure strategy has at least the probability it has under $(p, q) \cdot z_0$. Considering the final line segment of the computed path, one can therefore show the following.

THEOREM 3.4 [van den Elzen and Talman (1991)]. *Lemke's complementary pivoting algorithm applied to the augmented LCP* (3.6), (3.7) *terminates at a perfect equilibrium if the starting point* (p, q) *is completely mixed.*

As shown by van den Elzen and Talman (1999), their algorithm also emulates the *linear tracing procedure* of Harsanyi and Selten (1988). The tracing procedure is an adjustment process to arrive at an equilibrium of the game when starting from a prior (p, q). It traces a pair of strategy pairs $(\bar{x}, \bar{y})$. Each such pair is an equilibrium in a parameterized game where the prior is played with probability z_0 and the currently used strategies with probability $1 - z_0$. Initially, $z_0 = 1$ and the players react against the prior. Then they simultaneously and gradually adjust their expectations and react optimally against these revised expectations, until they reach an equilibrium of the original game.

Characterizations of the sets of stable and perfect equilibria of a bimatrix game analogous to Theorem 2.14 are given in Borm et al. (1993), Jansen, Jurg and Borm (1994), Vermeulen and Jansen (1994), and Jansen and Vermeulen (2001).

4. Extensive form games

In a game in extensive form, successive moves of the players are represented by edges of a tree. The standard way to find an equilibrium of such a game has been to convert it to strategic form, where each combination of moves of a player is a strategy. However, this typically increases the description of the game exponentially. In order to reduce this complexity, Wilson (1972) and Koller and Megiddo (1996) describe computations that use mixed strategies with *small support*. A different approach uses the *sequence form* of the game where pure strategies are replaced by move sequences, which are small in number. We describe it following von Stengel (1996a), and mention similar work by Romanovskii (1962), Selten (1988), Koller and Megiddo (1992), and further developments.

4.1. Extensive form and reduced strategic form

The basic structure of an extensive game is a finite tree. The nodes of the tree represent game states. The game starts at the root (initial node) of the tree and ends at a leaf (terminal node), where each player receives a payoff. The nonterminal nodes are called *decision nodes*. The player's *moves* are assigned to the outgoing edges of the decision node. The decision nodes are partitioned into *information sets*, introduced by

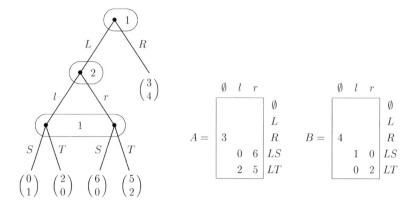

Figure 8. Left: a game in extensive form. Its reduced strategic form is (2.30). Right: the *sequence form* payoff matrices A and B. Rows and columns correspond to the sequences of the players which are marked at the side. Any sequence pair not leading to a leaf has matrix entry zero, which is left blank.

Kuhn (1953). All nodes in an information set belong to the same player, and have the same moves. The interpretation is that when a player makes a move, he only knows the information set but not the particular node he is at. Some decision nodes may belong to *chance* where the next move is made according to a known probability distribution.

We denote the set of information sets of player i by H_i, information sets by h, and the set of moves at h by C_h. In the extensive game in Figure 8, moves are marked by upper case letters for player 1 and by lower case letters for player 2. Information sets are indicated by ovals. The two information sets of player 1 have move sets $\{L, R\}$ and $\{S, T\}$, and the information set of player 2 has move set $\{l, r\}$.

Equilibria of an extensive game can be found recursively by considering *subgames* first. A subgame is a subtree of the game tree that includes all information sets containing a node of the subtree. In a game with *perfect information*, where every information set is a singleton, every node is the root of a subgame, so that an equilibrium can be found by backward induction. In games with imperfect information, equilibria of subgames are sometimes easy to find. Figure 8, for example, has a subgame starting at the decision node of player 2. It is equivalent to a 2×2 game and has a unique mixed equilibrium with probability $2/3$ for the moves S and r, respectively, and expected payoff 4 to player 1 and $2/3$ to player 2. Preceded by move L of player 1, this defines the unique *subgame perfect* equilibrium of the game.

In general, Nash equilibria of an extensive game (in particular one without subgames) are defined as equilibria of its *strategic form*. There, a *pure strategy* of player i prescribes a deterministic move at each information set, so it is an element of $\prod_{h \in H_i} C_h$. In Figure 8, the pure strategies of player 1 are the move combinations $\langle L, S \rangle$, $\langle L, T \rangle$, $\langle R, S \rangle$, and $\langle R, T \rangle$. In the *reduced strategic form*, moves at information sets that cannot be reached due to an earlier own move are identified. In Figure 8, this reduction yields the pure strategy (more precisely, equivalence class of pure strategies) $\langle R, * \rangle$, where $*$ denotes an arbitrary move. The two pure strategies of player 2 are her moves l and r.

The reduced strategic form (A, B) of this game is then as in (2.30). This game is *degenerate* even if the payoffs in the extensive game are generic, because player 2 receives payoff 4 when player 1 chooses R (the bottom row of the bimatrix game) irrespective of her own move. Furthermore, the game has an equilibrium which is not subgame perfect, where player 1 chooses R and player 2 chooses l with probability at least $2/3$.

A player may have *parallel* information sets that are not distinguished by own earlier moves. In particular, these arise when a player receives information about an earlier move by another player. Combinations of moves at parallel information sets cannot be reduced [see von Stengel (1996b) for further details]. This causes a multiplicative growth of the number of strategies even in the reduced strategic form. In general, the reduced strategic form is therefore *exponential* in the size of the game tree. Strategic form algorithms are then exceedingly slow except for very small game trees. Although extensive games are convenient modeling tools, their use has partly been limited for this reason [Lucas (1972)].

Wilson (1972) applies the Lemke–Howson algorithm to the strategic form of an extensive game while storing only those pure strategies that are actually played. That is, only the positive mixed strategy probabilities are computed explicitly. These correspond to basic variables x_i' or y_j' in Algorithm 2.9. The slack variables r_i and s_j are merely known to be nonnegative. For the pivoting step, the leaving variable is determined by a minimum ratio test which is performed *indirectly* for the tableau rows corresponding to basic slack variables. If, for example, y_k' enters the basis in step 2.9(b), then the conditions $y_j' \geqslant 0$ and $r_i \geqslant 0$ for the basic variables y_j and r_i determine the value of the entering variable by the minimum ratio test. In Wilson (1972), this test is first performed by ignoring the constraints $r_i \geqslant 0$, yielding a new mixed strategy y^0 of player 2. Against this strategy, a pure best response i of player 1 is computed from the game tree by a *subroutine*, essentially backward induction. If i has the same payoff as the currently used strategies of player 1, then $r \geqslant \mathbf{0}$ and some component of y leaves the basis. Otherwise, the payoff for i is higher and $r_i < 0$. Then at least the inequality $r_i \geqslant 0$ is violated, which is now added for a new minimum ratio test. This determines a new, smaller value for the entering variable and a corresponding mixed strategy y^1. Against this strategy, a best response is computed again. This process is repeated, computing a sequence of mixed strategies $y^0, y^1, \ldots, y^t$, until $r \geqslant \mathbf{0}$ holds and the correct leaving variable r_i is found.

Each pure strategy used in this method is stored explicitly as a tuple of moves. Their number should stay small during the computation. In the description by Wilson (1972) this is not guaranteed. However, the desired small support of the computed mixed strategies can be achieved by maintaining an additional system of linear equations for *realization weights* of the leaves of the game tree and with a *basis crashing* subroutine, as shown by Koller and Megiddo (1996).

The best response subroutine in Wilson's (1972) algorithm requires that the players have *perfect recall*, that is, all nodes in an information set of a player are preceded by the same earlier moves of that player [Kuhn (1953)]. For finding *all* equilibria, Koller and Megiddo (1996) show how to enumerate small supports in a way that can also be applied to extensive games without perfect recall.

4.2. Sequence form

The use of pure strategies can be avoided altogether by using *sequences* of moves instead. The unique path from the root to any node of the tree defines a sequence of moves for player i. We assume player i has perfect recall. That is, any two nodes in an information set h in H_i define the same sequence for that player, which we denote by σ_h. Let S_i be the set of sequences of moves for player i. Then any σ in S_i is either the empty sequence $\emptyset$ or uniquely given by its last move c at the information set h in H_i, that is, $\sigma = \sigma_h c$. Hence, $S_i = \{\emptyset\} \cup \{\sigma_h c \mid h \in H_i, \ c \in C_h\}$. So player i does not have more sequences than the tree has nodes.

The *sequence form* of the extensive game, described in detail in von Stengel (1996a), is similar to the strategic form but uses sequences instead of pure strategies, so it is a very compact description. Randomization over sequences is thereby described as follows.

A *behavior strategy* β of player i is given by probabilities $\beta(c)$ for his moves c which fulfill $\beta(c) \geqslant 0$ and $\sum_{c \in C_h} \beta(c) = 1$ for all h in H_i. This definition of β can be extended to the sequences σ in S_i by writing

$$\beta[\sigma] = \prod_{c \text{ in } \sigma} \beta(c). \tag{4.1}$$

A pure strategy π of player i can be regarded as a behavior strategy with $\pi(c) \in \{0, 1\}$ for all moves c. Thus, $\pi[\sigma] \in \{0, 1\}$ for all σ in S_i. The pure strategies π with $\pi[\sigma] = 1$ are those "agreeing" with σ by prescribing all the moves in σ, and arbitrary moves at the information sets not touched by σ.

A mixed strategy μ of player i assigns a probability $\mu(\pi)$ to every pure strategy π. In the sequence form, a randomized strategy of player i is described by the *realization probabilities* of playing the sequences σ in S_i. For a behavior strategy β, these are obviously $\beta[\sigma]$ as in (4.1). For a mixed strategy μ of player i, they are obtained by summing over all pure strategies π of player i, that is,

$$\mu[\sigma] = \sum_{\pi} \mu(\pi)\pi[\sigma]. \tag{4.2}$$

For player 1, this defines a map x from S_1 to $\mathbb{R}$ by $x(\sigma) = \mu[\sigma]$ for σ in S_1 which we call the *realization plan* of μ or a realization plan for player 1. A realization plan for player 2, similarly defined on S_2, is denoted y.

THEOREM 4.1 [Koller and Megiddo (1992), von Stengel (1996a)]. *For player 1, x is the realization plan of a mixed strategy if and only if $x(\sigma) \geqslant 0$ for all $\sigma \in S_1$ and*

$$\begin{aligned} x(\emptyset) &= 1, \\ \sum_{c \in C_h} x(\sigma_h c) &= x(\sigma_h), \quad h \in H_1. \end{aligned} \tag{4.3}$$

A realization plan y of player 2 is characterized analogously.

PROOF. Equations (4.3) hold for the realization probabilities $x(\sigma) = \beta[\sigma]$ for a behavior strategy β and thus for every pure strategy π, and therefore for their convex combinations in (4.2) with the probabilities $\mu(\pi)$.   □

To simplify notation, we write realization plans as vectors $x = (x_\sigma)_{\sigma \in S_1}$ and $y = (y_\sigma)_{\sigma \in S_2}$ with sequences as subscripts. According to Theorem 4.1, these vectors are characterized by

$$x \geqslant 0, \quad Ex = e, \quad\quad y \geqslant 0, \quad Fy = f \tag{4.4}$$

for suitable matrices E and F, and vectors e and f that are equal to $(1, 0, \ldots, 0)^\top$, where E and e have $1 + |H_1|$ rows and F and f have $1 + |H_2|$ rows. In Figure 8, the sets of sequences are $S_1 = \{\emptyset, L, R, LS, LT\}$ and $S_2 = \{\emptyset, l, r\}$, and in (4.4),

$$E = \begin{bmatrix} 1 & & & & \\ -1 & 1 & & 1 & \\ & & -1 & 1 & 1 \end{bmatrix}, \quad e = \begin{bmatrix} 1 \\ 0 \\ 0 \end{bmatrix}, \quad F = \begin{bmatrix} 1 & & \\ -1 & 1 & 1 \end{bmatrix}, \quad f = \begin{bmatrix} 1 \\ 0 \end{bmatrix}.$$

The number of information sets and therefore the number of rows of E and F is at most linear in the size of the game tree.

Mixed strategies of a player are called *realization equivalent* [Kuhn (1953)] if they define the same realization probabilities for all nodes of the tree, given any strategy of the other player. For reaching a node, only the players' sequences matter, which shows that the realization plan contains the strategically relevant information for playing a mixed strategy:

THEOREM 4.2 [Koller and Megiddo (1992), von Stengel (1996a)]. *Two mixed strategies μ and μ' of player i are realization equivalent if and only if they have the same realization plan, that is, $\mu[\sigma] = \mu'[\sigma]$ for all $\sigma \in S_i$.*

Any realization plan x of player 1 (and similarly y for player 2) naturally defines a behavior strategy β where the probability for move c is $\beta(c) = x(\sigma_h c)/x(\sigma_h)$, and arbitrary, for example, $\beta(c) = 1/|C_h|$, if $x(\sigma_h) = 0$ since then h cannot be reached.

COROLLARY 4.3 [Kuhn (1953)]. *For a player with perfect recall, any mixed strategy is realization equivalent to a behavior strategy.*

In Theorem 4.2, a mixed strategy μ is mapped to its realization plan by regarding (4.2) as a linear map with given coefficients $\pi[\sigma]$ for the pure strategies π. This maps the simplex of mixed strategies of a player to the polytope of realization plans. These polytopes are characterized by (4.4) as asserted in Theorem 4.1. They define the player's *strategy spaces* in the sequence form, which we denote by X and Y as in (2.7). The vertices of X and Y are the players' pure strategies up to realization equivalence, which is the identification of pure strategies used in the reduced strategic form. However, the dimension and the number of facets of X and Y is reduced from exponential to linear size.

Sequence form *payoffs* are defined for pairs of sequences whenever these lead to a leaf, multiplied by the probabilities of chance moves on the path to the leaf. This defines two sparse matrices A and B of dimension $|S_1| \times |S_2|$ for player 1 and player 2, respectively. For the game in Figure 1, A and B are shown in Figure 8 on the right. When the players use the realization plans x and y, the expected payoffs are $x^\top A y$ for player 1 and $x^\top B y$ for player 2. These terms represent the sum over all leaves of the payoffs at leaves multiplied by their realization probabilities.

The formalism in Section 2.2 can be applied to the sequence form without change. For zero-sum games, one obtains the analogous result to Theorem 2.3. It was first proved by Romanovskii (1962). He constructs a constrained matrix game [see Charnes (1953)] which is equivalent to the sequence form. The perfect recall assumption is weakened by Yanovskaya (1970). Until recently, these publications were overlooked in the English-speaking community.

THEOREM 4.4 [Romanovskii (1962), von Stengel (1996a)]. *The equilibria of a two-person zero-sum game in extensive form with perfect recall are the solutions of the LP (2.10) with sparse sequence form payoff matrix A and constraint matrices E and F in (4.4) defined by Theorem 4.1. The size of this LP is linear in the size of the game tree.*

Selten (1988, pp. 226, 237ff) defines sequence form strategy spaces and payoffs to exploit their linearity, but not for computational purposes. Koller and Megiddo (1992) describe the first polynomial-time algorithm for solving two-person zero-sum games in extensive form, apart from Romanovskii's result. They define the constraints (4.3) for playing sequences σ of a player with perfect recall. For the other player, they still consider pure strategies. This leads to an LP with a linear number of variables x_σ but possibly exponentially many inequalities. However, these can be evaluated as needed, similar to Wilson (1972). This solves efficiently the "separation problem" when using the ellipsoid method for linear programming.

For non-zero-sum games, the sequence form defines an LCP analogous to Theorem 2.4. Again, the point is that this LCP has the same size as the game tree. The Lemke–Howson algorithm cannot be applied to this LCP, since the missing label defines a single pure strategy, which would involve more than one sequence in the sequence form. Koller, Megiddo and von Stengel (1996) describe how to use the more general complementary pivoting algorithm by Lemke (1965) for finding a solution to the LCP derived from the sequence form. This algorithm uses an additional variable z_0 and a corresponding column to augment the LCP. However, that column is just some positive vector, which requires a very technical proof that Lemke's algorithm terminates.

In von Stengel, van den Elzen and Talman (2002), the augmented LCP (3.6), (3.7) is applied to the sequence form. The column for z_0 is derived from a starting pair (p, q) of realization plans. The computation has the interpretation described in Section 3.2.

Similar to Theorem 3.4, the computed equilibrium can be shown to be strategic-form perfect if the starting point is completely mixed.

5. Computational issues

How long does it take to find an equilibrium of a bimatrix game? The Lemke–Howson algorithm has exponential running time for some specifically constructed, even zero-sum, games. However, this does not seem to be the typical case. In practice, numerical stability is more important [Tomlin (1978), Cottle et al. (1992)]. Interior point methods that are provably polynomial as for linear programming are not known for LCPs arising from games; for other LCPs see Kojima et al. (1991). The computational complexity of finding one equilibrium is unclear. By Nash's theorem, an equilibrium exists, but the problem is to construct one. Megiddo (1988), Megiddo and Papadimitriou (1989), and Papadimitriou (1994) study the computational complexity of problems of this kind.

Gilboa and Zemel (1989) show that finding an equilibrium of a bimatrix game with maximum payoff sum is NP-hard, so for this problem no efficient algorithm is likely to exist. The same holds for other problems that amount essentially to examining all equilibria, like finding an equilibrium with maximum support. For other game-theoretic aspects of computing see Linial (1994) and Koller, Megiddo and von Stengel (1994).

The usefulness of algorithms for solving games should be tested further in practice. Many of the described methods are being implemented in the project GAMBIT, accessible by internet, and reviewed in McKelvey and McLennan (1996). The GALA system by Koller and Pfeffer (1997) allows one to generate large game trees automatically, and solves them according to Theorem 4.4. These program systems are under development and should become efficient and easily usable tools for the applied game theorist.

References

Aggarwal, V. (1973), "On the generation of all equilibrium points for bimatrix games through the Lemke–Howson algorithm", Mathematical Programming 4:233–234.

Audet, C., P. Hansen, B. Jaumard and G. Savard (2001), "Enumeration of all extreme equilibria of bimatrix games", SIAM Journal on Scientific Computing 23:323–338.

Avis, D., and K. Fukuda (1992), "A pivoting algorithm for convex hulls and vertex enumeration of arrangements and polyhedra", Discrete and Computational Geometry 8:295–313.

Bastian, M. (1976), "Another note on bimatrix games", Mathematical Programming 11:299–300.

Bomze, I.M. (1992), "Detecting all evolutionarily stable strategies", Journal of Optimization Theory and Applications 75:313–329.

Borm, P.E., M.J.M. Jansen, J.A.M. Potters and S.H. Tijs (1993), "On the structure of the set of perfect equilibria in bimatrix games", Operations Research Spektrum 15:17–20.

Charnes, A. (1953), "Constrained games and linear programming", Proceedings of the National Academy of Sciences of the U.S.A. 39:639–641.

Chvátal, V. (1983), Linear Programming (Freeman, New York).

Cottle, R.W., J.-S. Pang and R.E. Stone (1992), The Linear Complementarity Problem (Academic Press, San Diego).

Dantzig, G.B. (1963), Linear Programming and Extensions (Princeton University Press, Princeton).

Dickhaut, J., and T. Kaplan (1991), "A program for finding Nash equilibria", The Mathematica Journal 1(4):87–93.

Eaves, B.C. (1971), "The linear complementarity problem", Management Science 17:612–634.

Eaves, B.C. (1973), "Polymatrix games with joint constraints", SIAM Journal on Applied Mathematics 24:418–423.

Garcia, C.B., and W.I. Zangwill (1981), Pathways to Solutions, Fixed Points, and Equilibria (Prentice-Hall, Englewood Cliffs).

Gilboa, I., E. Kalai and E. Zemel (1990), "On the order of eliminating dominated strategies", Operations Research Letters 9:85–89.

Gilboa, I., E. Kalai and E. Zemel (1993), "The complexity of eliminating dominated strategies", Mathematics of Operations Research 18:553–565.

Gilboa, I., and E. Zemel (1989), "Nash and correlated equilibria: Some complexity considerations", Games and Economic Behavior 1:80–93.

Harsanyi, J.C., and R. Selten (1988), A General Theory of Equilibrium Selection in Games (MIT Press, Cambridge).

Hart, S. (1992), "Games in extensive and strategic forms", in: R.J. Aumann and S. Hart, eds., Handbook of Game Theory, Vol. 1 (North-Holland, Amsterdam) Chapter 2, 19–40.

Heuer, G.A., and C.B. Millham (1976), "On Nash subsets and mobility chains in bimatrix games", Naval Research Logistics Quarterly 23:311–319.

Hillas, J., and E. Kohlberg (2002), "Foundations of strategic equilibrium", in: R.J. Aumann and S. Hart, eds., Handbook of Game Theory, Vol. 3 (North-Holland, Amsterdam) Chapter 42, 1597–1663.

Howson, J.T., Jr. (1972), "Equilibria of polymatrix games", Management Science 18:312–318.

Howson, J.T., Jr., and R.W. Rosenthal (1974), "Bayesian equilibria of finite two-person games with incomplete information", Management Science 21:313–315.

Jansen, M.J.M. (1981), "Maximal Nash subsets for bimatrix games", Naval Research Logistics Quarterly 28:147–152.

Jansen, M.J.M., A.P. Jurg and P.E.M. Borm (1994), "On strictly perfect sets", Games and Economic Behavior 6:400–415.

Jansen, M.J.M., and A.J. Vermeulen (2001), "On the computation of stable sets and strictly perfect equilibria", Economic Theory 17:325–344.

Keiding, H. (1997), "On the maximal number of Nash equilibria in an $n \times n$ bimatrix game", Games and Economic Behavior 21:148–160.

Knuth, D.E., C.H. Papadimitriou and J.N. Tsitsiklis (1988), "A note on strategy elimination in bimatrix games", Operations Research Letters 7:103–107.

Kohlberg, E., and J.-F. Mertens (1986), "On the strategic stability of equilibria", Econometrica 54:1003–1037.

Kojima, M., N. Megiddo, T. Noma and A. Yoshise (1991), A Unified Approach to Interior Point Algorithms for Linear Complementarity Problems, Lecture Notes in Computer Science, Vol. 538 (Springer, Berlin).

Koller, D., and N. Megiddo (1992), "The complexity of two-person zero-sum games in extensive form", Games and Economic Behavior 4:528–552.

Koller, D., and N. Megiddo (1996), "Finding mixed strategies with small supports in extensive form games", International Journal of Game Theory 25:73–92.

Koller, D., N. Megiddo and B. von Stengel (1994), "Fast algorithms for finding randomized strategies in game trees", Proceedings of the 26th ACM Symposium on Theory of Computing, 750–759.

Koller, D., N. Megiddo and B. von Stengel (1996), "Efficient computation of equilibria for extensive two-person games", Games and Economic Behavior 14:247–259.

Koller, D., and A. Pfeffer (1997), "Representations and solutions for game-theoretic problems", Artificial Intelligence 94:167–215.

Krohn, I., S. Moltzahn, J. Rosenmüller, P. Sudhölter and H.-M. Wallmeier (1991), "Implementing the modified LH algorithm", Applied Mathematics and Computation 45:31–72.

Kuhn, H.W. (1953), "Extensive games and the problem of information", in: H.W. Kuhn and A.W. Tucker, eds., Contributions to the Theory of Games II, Annals of Mathematics Studies, Vol. 28 (Princeton Univ. Press, Princeton) 193–216.

Kuhn, H.W. (1961), "An algorithm for equilibrium points in bimatrix games", Proceedings of the National Academy of Sciences of the U.S.A. 47:1657–1662.

Lemke, C.E. (1965), "Bimatrix equilibrium points and mathematical programming", Management Science 11:681–689.

Lemke, C.E., and J.T. Howson, Jr. (1964), "Equilibrium points of bimatrix games", Journal of the Society for Industrial and Applied Mathematics 12:413–423.

Linial, N. (1994), "Game-theoretic aspects of computing", in: R.J. Aumann and S. Hart, eds., Handbook of Game Theory, Vol. 2 (North-Holland, Amsterdam) Chapter 38, 1339–1395.

Lucas, W.F. (1972), "An overview of the mathematical theory of games", Management Science 18:3–19, Appendix P.

Mangasarian, O.L. (1964), "Equilibrium points in bimatrix games", Journal of the Society for Industrial and Applied Mathematics 12:778–780.

Mangasarian, O.L., and H. Stone (1964), "Two-person nonzero-sum games and quadratic programming", Journal of Mathematical Analysis and Applications 9:348–355.

McKelvey, R.D., and A. McLennan (1996), "Computation of equilibria in finite games", in: H.M. Amman, D.A. Kendrick and J. Rust, eds., Handbook of Computational Economics, Vol. I (Elsevier, Amsterdam) 87–142.

McLennan, A., and I.-U. Park (1999), "Generic 4×4 two person games have at most 15 Nash equilibria", Games and Economic Behavior 26:111–130.

McMullen, P. (1970), "The maximum number of faces of a convex polytope", Mathematika 17:179–184.

Megiddo, N. (1988), "A note on the complexity of p-matrix LCP and computing an equilibrium", Research Report RJ 6439, IBM Almaden Research Center, San Jose, California.

Megiddo, N., and C.H. Papadimitriou (1989), "On total functions, existence theorems and computational complexity (Note)", Theoretical Computer Science 81:317–324.

Mertens, J.-F. (1989), "Stable equilibria – a reformulation, Part I", Mathematics of Operations Research 14:575–625.

Mertens, J.-F. (1991), "Stable equilibria – a reformulation, Part II", Mathematics of Operations Research 16:694–753.

Millham, C.B. (1974), "On Nash subsets of bimatrix games", Naval Research Logistics Quarterly 21:307–317.

Mills, H. (1960), "Equilibrium points in finite games", Journal of the Society for Industrial and Applied Mathematics 8:397–402.

Mukhamediev, B.M. (1978), "The solution of bilinear programming problems and finding the equilibrium situations in bimatrix games", Computational Mathematics and Mathematical Physics 18:60–66.

Mulmuley, K. (1994), Computational Geometry: An Introduction Through Randomized Algorithms (Prentice-Hall, Englewood Cliffs).

Murty, K.G. (1988), Linear Complementarity, Linear and Nonlinear Programming (Heldermann Verlag, Berlin).

Nash, J.F. (1951), "Non-cooperative games", Annals of Mathematics 54:286–295.

Papadimitriou, C.H. (1994), "On the complexity of the parity argument and other inefficient proofs of existence", Journal of Computer and System Sciences 48:498–532.

Parthasarathy, T., and T.E.S. Raghavan (1971), Some Topics in Two-Person Games (American Elsevier, New York).

Quint, T., and M. Shubik (1997), "A theorem on the number of Nash equilibria in a bimatrix game", International Journal of Game Theory 26:353–359.

Raghavan, T.E.S. (1994), "Zero-sum two-person games", in: R.J. Aumann and S. Hart, eds., Handbook of Game Theory, Vol. 2 (North-Holland, Amsterdam) Chapter 20, 735–768.

Raghavan, T.E.S. (2002), "Non-zero-sum two-person games", in: R.J. Aumann and S. Hart, eds., Handbook of Game Theory, Vol. 3 (North-Holland, Amsterdam) Chapter 44, 1687–1721.

Romanovskii, I.V. (1962), "Reduction of a game with complete memory to a matrix game", Soviet Mathematics 3:678–681.

Schrijver, A. (1986), Theory of Linear and Integer Programming (Wiley, Chichester).

Selten, R. (1975), "Reexamination of the perfectness concept for equilibrium points in extensive games", International Journal of Game Theory 4:22–55.

Selten, R. (1988), "Evolutionary stability in extensive two-person games – correction and further development", Mathematical Social Sciences 16:223–266.

Shapley, L.S. (1974), "A note on the Lemke–Howson algorithm", Mathematical Programming Study 1: Pivoting and Extensions, 175–189.

Shapley, L.S. (1981), "On the accessibility of fixed points", in: O. Moeschlin and D. Pallaschke, eds., Game Theory and Mathematical Economics (North-Holland, Amsterdam) 367–377.

Todd, M.J. (1976), "Comments on a note by Aggarwal", Mathematical Programming 10:130–133.

Todd, M.J. (1978), "Bimatrix games – an addendum", Mathematical Programming 14:112–115.

Tomlin, J.A. (1978), "Robust implementation of Lemke's method for the linear complementarity problem", Mathematical Programming Study 7: Complementarity and Fixed Point Problems, 55–60.

van Damme, E. (1987), Stability and Perfection of Nash Equilibria (Springer, Berlin).

van Damme, E. (2002), "Strategic equilibrium", in: R.J. Aumann and S. Hart, eds., Handbook of Game Theory, Vol. 3 (North-Holland, Amsterdam) Chapter 41, 1521–1596.

van den Elzen, A. (1993), Adjustment Processes for Exchange Economies and Noncooperative Games, Lecture Notes in Economics and Mathematical Systems, Vol. 402 (Springer, Berlin).

van den Elzen, A.H., and A.J.J. Talman (1991), "A procedure for finding Nash equilibria in bi-matrix games", ZOR – Methods and Models of Operations Research 35:27–43.

van den Elzen, A.H., and A.J.J. Talman (1999), "An algorithmic approach toward the tracing procedure for bi-matrix games", Games and Economic Behavior 28:130–145.

Vermeulen, A.J., and M.J.M. Jansen (1994), "On the set of (perfect) equilibria of a bimatrix game", Naval Research Logistics 41:295–302.

Vermeulen, A.J., and M.J.M. Jansen (1998), "The reduced form of a game", European Journal of Operational Research 106:204–211.

von Stengel, B. (1996a), "Efficient computation of behavior strategies", Games and Economic Behavior 14:220–246.

von Stengel, B. (1996b), "Computing equilibria for two-person games", Technical Report 253, Dept. of Computer Science, ETH Zürich.

von Stengel, B. (1999), "New maximal numbers of equilibria in bimatrix games", Discrete and Computational Geometry 21:557–568.

von Stengel, B., A.H. van den Elzen and A.J.J. Talman (2002), "Computing normal form perfect equilibria for extensive two-person games", Econometrica, to appear.

Vorobiev, N.N. (1958), "Equilibrium points in bimatrix games", Theory of Probability and its Applications 3:297–309.

Wilson, R. (1972), "Computing equilibria of two-person games from the extensive form", Management Science 18:448–460.

Wilson, R. (1992), "Computing simply stable equilibria", Econometrica 60:1039–1070.

Winkels, H.-M. (1979), "An algorithm to determine all equilibrium points of a bimatrix game", in: O. Moeschlin and D. Pallaschke, eds., Game Theory and Related Topics (North-Holland, Amsterdam) 137–148.

Yanovskaya, E.B. (1968), "Equilibrium points in polymatrix games" (in Russian), Litovskii Matematicheskii Sbornik 8:381–384 [Math. Reviews 39 #3831].

Yanovskaya, E.B. (1970), "Quasistrategies in position games", Engineering Cybernetics 1:11–19.

Ziegler, G.M. (1995), Lectures on Polytopes, Graduate Texts in Mathematics, Vol. 152 (Springer, New York).

Chapter 46

NON-COOPERATIVE GAMES WITH MANY PLAYERS*

M. ALI KHAN

Department of Economics, The Johns Hopkins University, Baltimore, MD, USA

YENENG SUN

Department of Mathematics, National University of Singapore, Singapore

Contents

The authors' first acknowledgement is to Kali Rath for collaboration and co-authorship. They also thank Graciela Chichilnisky, Duncan Foley, Peter Hammond, Andreu Mas-Colell, Lionel McKenzie, and David Schmeidler for encouragement over the years; in particular, they had access to Mas-Colell's May 1990 bibliography on the subject matter discussed herein. This work was initiated during the visit of Yeneng Sun to the Department of Economics at Johns Hopkins University in July–August 1996: the first draft was completed in September 1996 while he was at the Cowles Foundation, and parts of it were presented by Khan in a minicourse organized by Monique Florenzano at CERMSEM, Université de Paris 1, in May–June 2000. Both authors acknowledge the hospitality of their host institutions. This final version has benefited from the suggestions and careful reading of an anonymous referee, Yasar Barut, and the Editors of this Handbook.

Handbook of Game Theory, Volume 3, Edited by R.J. Aumann and S. Hart

Abstract

In this survey article, we report results on the existence of pure-strategy Nash equilibria in games with an atomless continuum of players, each with an action set that is not necessarily finite. We also discuss purification and symmetrization of mixed-strategy Nash equilibria, and settings in which private information, anonymity and idiosyncratic shocks are given particular prominence.

Keywords

pure-strategy Nash equilibria, large games, idiosyncratic shocks, Lebesgue continuum, Loeb continuum

JEL classification: G12, C60

1. Introduction

Shapiro and Shapley introduce their 1961 memorandum (published 17 years later as Shapiro and Shapley (1978)) with the remark that "institutions having a large number of competing participants are common in political and economic life", and cite as examples "markets, exchanges, corporations (from the shareholders viewpoint), Presidential nominating conventions and legislatures". They observe, however, that "game theory has not yet been able so far to produce much in the way of fundamental principles of "mass competition" that might help to explain how they operate in practice", and that it might be "worth while to spend a little effort looking at the behavior of existing n-person solution concepts, as n becomes very large". In this, they echo both von Neumann and Morgenstern (1953) and Kuhn and Tucker (1950),[1] and anticipate Mas-Colell (1998).[2]

Von Neumann and Morgenstern (1953) saw the number of participants in a game as a variable, and presented it as one determining the "total set" of variables of the problem. "Any increase in the number of variables inside a participant's partial set may complicate our problem technically, but only technically; something of a very different nature happens when the number of participants – i.e., of the partial sets of variables – is increased." After remarking that the complications arising from the "fact that every participant is influenced by the anticipated reactions of the others to his own measures" are "most strikingly the crux of the matter", the authors write:

> When the number of participants becomes really great, some hope emerges that the influence of every particular participant will become negligible, and that the above difficulties may recede and a more conventional theory become possible. Indeed, this was the starting point of much of what is best in economic theory. It is a well known phenomenon in many branches of the exact and physical sciences that very great numbers are often easier to handle than those of medium size.[3] This is of course due to the excellent possibility of applying the laws of statistics and probabilities in the first case.

Two further points are explicitly noted. First, a satisfactory treatment of such "populous games" may require "some radical theoretical innovations – a really fundamental reopening of [the] subject". Second, "only after the theory of moderate numbers has

---

[1] For the first two authors, see Section 2 in the third (1953) edition of their book (in the sequel, all quotations are from this section). For the next two, see item 11 in Kuhn and Tucker (1950, p. x) – a list of problems that Aumann (1997, p. 6) terms "remarkably prophetic".

[2] In his Nancy Schwartz Lecture, Mas-Colell (1998) observes, "I bet that [results] built on the Negligibility Hypothesis are centrally located in the trade-off frontier for the extent of coverage and the strength of results of theories. This is, however, a matter of judgement based on the conviction that mass phenomena constitute an essential part of the economic world."

[3] "An almost exact theory of a gas, containing about 10^{25} freely moving particles, is incomparably easier than that of the solar system, made up of 9 major bodies; and still more than that of a multiple star of three or four objects of about the same size."

been satisfactorily developed will it be possible to decide whether extremely great numbers of participants will simplify the situation".[4] However, an optimistic prognosis is evident.[5]

Nash (1950) contains in the space of five paragraphs a definitive formulation of the theory of non-cooperative games with an arbitrary finite number of players. This "theory, in contradistinction to that of von Neumann and Morgenstern, is based on the *absence* of coalitions in that it is assumed that each participant acts independently, without collaboration and communication from any of the others. The non-cooperative idea will be implicit, rather than explicit. The notion of an *equilibrium point* is the basic ingredient in our theory. This notion yields a generalization of the concept of a solution of a two-person zero-sum game". In a treatment that is remarkably modern, Nash presented a theorem on the existence of equilibrium in an n-person game, where n is an arbitrary finite number of participants or players. In addition to the von Neumann and Morgenstern book, the only other reference is to Kakutani's generalization of Brouwer's fixed-point theorem.[6]

With Nash's theorem in place, all that an investigation into non-cooperative games with many players requires is a mathematical framework that fruitfully articulates "many" and the attendant notions of "negligibility" and "inappreciability". This was furnished by Milnor and Shapley in 1961 in the context of cooperative game theory. They presented an idealized limit game with a "continuum of infinitesimal minor players..., an 'ocean', to emphasize the almost total absence of order or cohesion". The oceanic players were represented in measure-theoretic terms and their "voting power expressed as a measure, defined on the measurable subsets of the ocean". The authors did not devote any space to the justification of the notion of a continuum of players; they were clear about the "benefits of dealing directly with the infinite-person game, instead of with a sequence of finite approximants".[7]

With the presumption that "models with a continuum of *players* (traders in this instance) are a relative novelty,[8] [and that] the idea of a continuum of traders may seem outlandish to the reader", Aumann (1964) used such a model for a successful formalization of Edgeworth's 1881 conjecture on the relation of core and competitive allocations. Aumann's discussion proved persuasive because the framework yielded an equivalence between these two solution concepts, and thereby affected a qualitative change

---

[4] Von Neumann and Morgenstern are emphatic on this point. "There is no getting away from it: The problem must be formulated, solved and understood for small numbers of participants before anything can be proved about the changes of its character in any limiting case of large numbers such as free competition."

[5] "Let us say again: we share the hope – chiefly because of the above-mentioned analogy in other fields! – that such simplifications will indeed occur."

[6] We mention this to emphasize that Nash drew his inspiration from von Neumann and Morgenstern (1944 edition) rather than from Cournot (1838); see Nash (1950), as well as the more detailed elaboration in Nash (1951). The quotation is from the introduction to the latter paper.

[7] Again, see the reprinted version in Milnor and Shapley (1978); the original version is dated 1961.

[8] After the statement that the "references can still be counted on the fingers of one hand", Aumann lists the Shapley and Milnor–Shapley memoranda referred to above, and the papers of Davis and Peleg on von Neumann–Morgenstern solutions and their bargaining sets.

in the character of the resolution of the problem. Aumann argued that "the most natural model for this purpose contains a *continuum* of participants, similar to the continuum of points on a line or the continuum of particles in a fluid". After all, "continuous models are nothing new in economics or game theory, [even though] it is usually parameters such as price or strategy that are allowed to vary continuously". More generally, he stressed "the power and simplicity of the continuum-of-players methods in describing mass phenomena in economics and game theory", and saw his work "primarily as an illustration of this method as applied to an area where no other treatment seemed completely satisfactory". In Aumann (1964) four methodological points are made explicit.

(1) The continuum can be considered an approximation to the "true" situation in which there is a large but finite number of particles (or traders or strategies or possible prices). In economics, as in the physical sciences, the study of the ideal state has proved very fruitful, though in practice it is, at best, only approximately achieved.[9]

(2) The continuum of traders is not merely a mathematical exercise; it is the expression of an economic idea. This is underscored by the fact that the chief result holds *only* for a continuum of traders – it is false for any finite number.

(3) The purpose of adopting the continuous approximation is to make available the powerful and elegant methods of a branch of mathematics called "analysis", in a situation where treatment by finite methods would be much more difficult or hopeless.

(4) The choice of the unit interval as a model for the set of traders is of no particular significance. In technical terms, T can be any measure space *without atoms*. The condition that T have no atoms is precisely what is needed to ensure that each individual trader have no influence.[10]

In their work on the elimination of randomization (purification) in statistics and game theory, Dvoretsky, Wald and Wolfowitz (1950) had already emphasized the importance of Lyapunov's theorem,[11] and explicitly noted that the "non-atomicity hypothesis is indispensable [and that] it is this assumption that is responsible for the possibility to disregard mixed strategies in games ... opposed to the finite games originally treated by J. von Neumann".[12] With the ideas of purification and the continuum of traders in place, a natural next step was an extension of Nash's theorem to show the existence of a pure strategy equilibrium. This was accomplished in Schmeidler (1973) in the setting of an arbitrary finite number of pure strategies. Since there

---

[9] As in von Neumann and Morgenstern, a footnote refers to three ideal phenomena in the natural sciences: a freely falling body, an ideal gas, and an ideal fluid. "The individual consumer (or merchant) is as anonymous to [the policy maker in Washington] as the individual molecule is to the physicist."

[10] The quotations in this paragraph, including the four points listed above, are all taken from Aumann (1964, Section 1).

[11] We refer to this theorem at length in the sequel.

[12] See the detailed elaboration in Dvoretsky, Wald and Wolfowitz (1951a, 1951b) and in Wald and Wolfowitz (1951); also Chapter 21 of this Handbook.

does not exist such an equilibrium in general finite player games,[13] this result furnished another example of a qualitative change in the resolution of the problem. However, the analysis of situations with a continuum of actions – the continuous variation in the price or strategy variables referred to by Aumann[14] – eluded the theory. In this chapter, we sketch the shape of a general theory that encompasses, in particular, such situations. Our focus is on non-cooperative games, rather than on perfect competition, and primarily on how questions of the existence of equilibria for such games dictate, and are dictated by, the mathematical framework chosen to formalize the idea of "many" players. That being the case, we keep the methodological pointers delineated in this introduction constantly in view. The subject has a technical lure and it is important not to be unduly diverted by it. At the end of the chapter we indicate applications but leave it to the reader to delve more deeply into the relevant references. This is only because of considerations of space; of course, we subscribe to the view that there ought to be a constant interplay between the framework and the economic and game-theoretic phenomena that it aspires to address and explain.

2. Antecedent results

We motivate the need for a measure-theoretic structure on the set T of players' names by considering a model in which no restrictions are placed on the cardinality of T. For each player t in T, let the set of actions be given by A_t, and the payoff function by $u_t : A \to \mathbb{R}$, where A denotes the product $\prod_{t \in T} A_t$. Let the partial product $\prod_{t \in T, \, t \neq i} A_t$ be denoted by A_{-i}. We can present the following result.[15]

THEOREM 1. *Let $\{A_t\}_{t \in T}$ be a family of nonempty, compact convex sets of a Hausdorff topological vector space, and $\{u_t\}_{t \in T}$ be a family of real-valued continuous functions on A such that for each $t \in T$, and for any fixed $a_{-t} \in A_{-t}$, $u_t(\cdot, a_{-t})$ is a quasi-concave function on A_t. Then there exists $a^* \in A$, such that for all t in T, $u_t(a^*) \geqslant u_t(a, a^*_{-t})$ for all a in A_t.*

Nash (1950, 1951) considered games with finite action sets, and his focus on mixed strategy equilibria led him to probability measures on these action sets and to the maximization of expected utilities with respect to these measures. Theorem 1 is simply an observation that if the finiteness hypothesis is replaced by convexity and compactness,

---

[13] As is well known, there are no pure-strategy equilibria in the elementary matching pennies game.

[14] Aumann (1987) singles out Ville as the first user of continuous strategy sets in game theory. See, for example, Rauh (2001) for some very recent work using a continuous price space as the strategy space in the setting of large games.

[15] Theorem 1 in its precise form is due to Ma (1969); see also Fan (1966). The hypothesis of quasi-concavity goes back at least to Debreu (1952).

and the linearity of the payoff functions by quasi-concavity, his basic argument remains valid.[16] Once the closedness of the "one-to-many mapping of the [arbitrary] product space to itself" is established, we can invoke the full power of Tychonoff's celebrated theorem on the compactness of the product of an arbitrary set of compact spaces, and rely on a suitable extension of Kakutani's fixed-point theorem.[17] The upper semicontinuity result in Fan (1952), and the fixed-point theorem in Fan (1952) and Glicksberg (1952) furnish these technical supplements.

However, with Theorem 1 in hand, we can revert to Nash's setting and exploit the measure-theoretic structure available on each action set. For each player t, consider the measurable space $(A_t, \mathcal{B}(A_t))$, where $\mathcal{B}(A_t)$ is the Borel σ-algebra generated by the topology on A_t. Let $\mathcal{M}(A_t)$ be the set of Borel probability measures on A_t endowed with the weak* topology.[18] Without going into technical details of how to manufacture new probability spaces $(A, \mathcal{B}(A), \prod_{s \in T} \mu_s)$ and $(A_{-t}, \mathcal{B}(A_{-t}), \prod_{s \neq t} \mu_s)$ from $\{(A_t, \mathcal{B}(A_t), \mu_t)\}_{t \in T}$, and the fine points of Fubini's theorem on the interchange of integrals,[19] we can deduce[20] the following result from Theorem 1 by working with the action sets $\mathcal{M}(A_t)$ and with an explicit functional form of the payoff functions u_t.

COROLLARY 1. *Let $\{A_t\}_{t \in T}$ be a family of nonempty, compact Hausdorff spaces, and $\{v_t\}_{t \in T}$ be a family of real-valued continuous functions on A. Then there exists $\sigma^* = (\sigma_t^*: t \in T) \in \prod_{t \in T} \mathcal{M}(A_t)$ such that for all t in T,*

$$u_t(\sigma^*) = \int_A v_t(a) \, \mathrm{d} \prod_{t \in T} \sigma_t^* \geq u_t(\sigma, \sigma_{-t}^*) \quad \textit{for all } \sigma \textit{ in } \mathcal{M}(A_t).$$

---

[16] This existence proof is furnished in Nash (1950). Since the payoff function is a "polylinear form in the probabilities", the set of best responses – countering points – is convex, and the longest paragraph in the paper concerned the closedness of the graph of the "one-to-many mapping of the product space to itself". Perhaps it ought to be noted here that Nash ascribes the idea for the use of Kakutani's fixed-point theorem to David Gale. An alternative proof based on the simpler Brouwer fixed-point theorem is furnished in Nash (1951); it was to prove equally influential both for game theory and for general equilibrium theory.

[17] The setting of a Hausdorff topological vector space is a technical flourish whereby the two operations underlying the convexity hypothesis are abstracted and assumed to be continuous. The reader loses nothing of substance by thinking of each A_t as the unit interval. However, since we are no longer dealing with finite action sets, the compactness hypothesis needs to be made explicit.

[18] The use of the weak* topology in this context goes back to Glicksberg (1952); for details, see Billingsley (1968) and Parthasarathy (1967). Note also that Glicksberg did not utilize any metric hypothesis on the action sets; see Khan (1989) for dispensing with this hypothesis in another context.

[19] For measures on infinite product spaces, see, for example, Ash (1972, Sections 2.7 and 4.4) or Loéve (1977, Sections 4 and 8). For Fubini's theorem, see, in addition to these references, Rudin (1974, Chapter 7).

[20] The compactness of $\mathcal{M}(A_t)$ is a basic property of the weak* topology known as Prohorov's theorem; see, for example, Billingsley (1968) or Parthasarathy (1967). The quasi-concavity of u_t is straightforward, and its continuity follows from Proposition 1 below. One can also furnish a direct proof based on the Schauder–Tychonoff theorem along the lines of Nash (1951), as in Peleg (1969).

The question is whether any substantive meaning can be given to the continuity hypothesis on the functions v_t? The following result[21] shows that it is not merely a technical requirement but has a direct implication for the formalization of player interdependence.

DEFINITION 1. $v: A \to \mathbb{R}$ is finitely determined if for all $a, b \in A$, $v(a) = v(b)$ if there exists a finite subset F of T such that $a_t = b_t$ for all $t \notin F$. v is almost finitely determined if for every $\varepsilon > 0$, there exists a finitely determined function v_ε such that $\sup_{a \in A} |v_\varepsilon(a) - v(a)| < \varepsilon$.

PROPOSITION 1. *For a real-valued function* $v: A \to \mathbb{R}$, *the following conditions are equivalent*:
  (1) *v is almost finitely determined.*
  (2) *v is a continuous function on the space A endowed with the product topology.*
  (3) *v is integrable with respect to any $\sigma \in \mathcal{M}(A)$ and its integral is a continuous function on $\mathcal{M}(A)$.*

Thus, if we conceive of a finite set of players as a "negligible" set, the hypothesis of continuity in the product topology implies strong restrictions on how player interaction is formalized. If individual payoffs depend on the actions of a "non-negligible" set of players so that the continuity hypothesis is violated, there may not exist any Nash equilibrium in pure or in mixed strategies. The following example due to Peleg (1969) illustrates this observation.[22]

EXAMPLE 1. Consider a game in which the set of players' names T is given by the set of positive integers $\mathbb{N}$, the action set A_t by the set $\{0, 1\}$, and the individual payoffs by functions on actions that equal the action or its negative, depending on whether the sum of the actions of all the other players is respectively finite or infinite. Note that these functions are not continuous in the product topology.[23]

There is no pure strategy Nash equilibrium in this game. If the sum of all of the actions is finite in equilibrium, all players could not be playing 1, and players playing 0 would gain by playing 1. On the other hand, if the sum of all of the actions is infinite, all players could not be playing 0, and players playing 1 would gain by playing 0.

The more interesting point is that this game does not have any mixed strategy Nash equilibrium either. If $\sigma^* \in \prod_{s \in T} \mathcal{M}(\{0, 1\})$ is such an equilibrium, there must exist a

---

[21] Definition 1 and Proposition 1 are due to Peleg (1969), who should be referred to for a proof.

[22] It is worth pointing out here that there are both countably additive and non-countably additive correlated equilibria in the example below; see Hart and Schmeidler (1989) for a discussion and references.

[23] If e denotes an infinite sequence of 1's and e_n the sequence with 1 in the first n places and 0 everywhere else, the sequence $\{e_n\}_{n \in \mathbb{N}}$ converges to e, but $u_t(e_n)$ is 1 for all $n \geq 1$ while $u_t(e)$ is -1.

player t, and a mixed strategy $(1 - p, p), 0 < p < 1$, such that her payoff in equilibrium is given by

$$u_t(\sigma, \sigma_{-t}^*) = \int_{A_{-t}} \left[\sigma(\{0\}) v_t \left(0, \sum_{s \neq t} a_s \right) + \sigma(\{1\}) v_t \left(1, \sum_{s \neq t} a_s \right) \right] d\sigma_{-t}^*$$

$$= p \int_{A_{-t}} v_t \left(1, \sum_{s \neq t} a_s \right) d\sigma_{-t}^*,$$

where a_s denotes the action of player s. Since an individual player's payoff depends on whether $\sum_{s \neq t} a_s$ converges or diverges, player t obtains p or $-p$ as a consequence of the following zero-one law.[24]

PROPOSITION 2. *Let $(\Omega, \mathcal{F}, P)$ be a probability space, and X_n be a sequence of independent random variables. Then the series $\sum_{n \in \mathbb{N}} X_n(\omega)$ converges or diverges for P-almost all $\omega \in \Omega$.*

In either case, player t would gain by playing a pure strategy (1 or 0 respectively), and hence σ^* could not be a mixed strategy equilibrium. □

While the exploitation of the independence hypothesis in Example 1 is fully justified in its non-cooperative context, the fact that an individual player does not explicitly randomize on whether the sum of others' actions converges or diverges is less justifiable.[25] The important question to ask, however, is whether the above formalization of a "large" game is merely of technical interest; or does it point to something that is false for the finite case but true for the ideal, and if so, to something that we can learn about the finite case from the ideal?

3. Interactions based on distributions of individual responses

In Example 1, the set of players' names can be conceived as an infinite (but σ-finite) measure space consisting of a counting measure on the power set of $\mathbb{N}$, but it is precisely this lack of finiteness that rules out consideration of situations in which a player's payoff depends in a well-defined way on the proportion of other players taking a specific action. Such an idea admits of a precise formulation if a measure-theoretic structure on the set of players' names is explicitly brought to the fore in the form of an atomless probability

---

[24] See Ash (1972, Section 7.2) or Loéve (1977, Section 16.3).

[25] If each player is allowed to attach equal probability to the outcome under the zero-one law, there would be a mixed strategy Nash equilibrium. For the implications of introducing individual subjective mappings from the probabilities formalizing societal responses to the space of probabilities on probabilities, see Chakrabarti and Khan (1991).

space $(T, \mathcal{T}, \lambda)$, with the atomless assumption formalizing the "negligible" influence of each individual player. However, what needs to be underscored is that λ is a countably additive, rather than a finitely additive, measure.[26]

A game is now simply a random variable from T to an underlying space of characteristics, and its Nash equilibrium another random variable from T to a common action set A.[27] We shall also adopt as a working hypothesis, until Section 10, Aumann's (1964) statement that the "measurability assumption is of technical significance only and constitutes no economic restriction. Nonmeasurable sets are extremely "pathological"; it is unlikely that they would occur in the context of an economic model".

3.1. A basic result

The set of players is divided into ℓ groups or institutions,[28] with $T_1, \ldots, T_\ell$ being a partition of T with positive λ-measures $c_1, \ldots, c_\ell$. For each $1 \leqslant i \leqslant m$, let λ_i be the probability measure on T_i such that for any measurable set $B \subseteq T_i$, $\lambda_i(B) = \lambda(B)/c_i$. We assume A to be a countable compact metric space.[29] Let $\mathcal{U}_A^d$ be the space of real-valued continuous functions on $A \times \mathcal{M}(A)^\ell$, endowed with its sup-norm topology and with $\mathcal{B}(\mathcal{U}_A^d)$ its Borel σ-algebra (the superscript d denoting "distribution"). This is the space of player characteristics, with the payoff function of each player depending on her action as well as on the distribution of actions in each of the ℓ institutions. We now have all the terminology we need to present.[30]

THEOREM 2. *Let $\mathcal{G}^d$ be a measurable map from T to $\mathcal{U}_A^d$. Then there exists a measurable function $f : T \to A$ such that for λ-almost all $t \in T$,*

$$u_t\left(f(t), \lambda_1 f_1^{-1}, \ldots, \lambda_\ell f_\ell^{-1}\right) \geqslant u_t\left(a, \lambda_1 f_1^{-1}, \ldots, \lambda_\ell f_\ell^{-1}\right) \quad \text{for all } a \in A,$$

where $u_t = \mathcal{G}^d(t) \in \mathcal{U}_A^d$, f_i the restriction of f to T_i, and $\lambda_i f_i^{-1}$ the induced distribution on A.

---

[26] As discussed in the introduction, this is necessitated by the needs of mathematical analysis. For interpretive difficulties, and even absurd results, that follow from finitely additive measures, see, for example, Hart and Schmeidler (1989) and Sun (1999b).

[27] There is little doubt that an extension to different action sets can be obtained by working with the hyperspace of closed subsets of a complete separable metric space. This idea is standard in general equilibrium theory; see Hildenbrand (1974, particularly Section B.II). For the use of this hyperspace in another relevant context, see Sun (1996b, 2000).

[28] The motivation for this will become apparent when we turn to the special case of finite games with private information. One may draw a contrast here with Chakrabarti and Khan (1991) where the parameter ℓ is allowed to vary with each player deciding for herself how to conceive of societal stratification.

[29] This assumption shall also remain in force throughout the next section. However, nothing of substance is lost if the reader thinks of the set A, at first pass, as consisting of only two elements.

[30] As noted in the introduction, Schmeidler (1973) considered the case that $\ell = 1$ and A is finite.

Since Theorem 2 is phrased in terms of distributions, it stands to reason that the most relevant mathematical tools needed for its proof will revolve around the distribution of a correspondence. What is interesting is that a theory for such an object can be developed on the basis of the "marriage lemma". We turn to this.

3.2. The marriage lemma and the distribution of a correspondence

Halmos and Vaughan (1950) introduce the marriage lemma by asking for "conditions under which it is possible for each boy to marry his acquaintance if each of a (possibly infinite) set of boys is acquainted with a finite set of girls?" A general answer going beyond specific counting measures is available in the following result.[31]

PROPOSITION 3. *Let I be a countable index set, $(T_\alpha)_{\alpha \in I}$ a family of sets in $\mathcal{T}$, and $\Lambda = (\tau_\alpha)_{\alpha \in I}$ a family of non-negative numbers. There exists a family $(S_\alpha)_{\alpha \in I}$ of sets in $\mathcal{T}$ such that for all $\alpha, \beta \in I$, $\alpha \neq \beta$, one has $S_\alpha \subseteq T_\alpha$, $\lambda(S_\alpha) = \tau_\alpha$, $S_\alpha \cap S_\beta = \emptyset$ if and only if for all finite subsets I_F of I, $\lambda(\bigcup_{\alpha \in I_F} T_\alpha) \geqslant \sum_{\alpha \in I_F} \tau_\alpha$.*

We can use Proposition 3 to develop results on the non-emptiness, purification, convexity, compactness and upper semicontinuity of the distribution of a correspondence as is required for the application of fixed-point theorems.[32] However, the countability hypothesis on the range of a correspondence deserves special emphasis; all of the results reported below are false for particular correspondences from the unit Lebesgue interval to an interval,[33] the former denoted in the sequel by $([0, 1], \mathcal{B}([0, 1]), \nu)$.

A correspondence F from T to A is said to be measurable if for each $a \in A$, $F^{-1}(\{a\}) = \{t \in T: a \in F(t)\}$ is measurable. A measurable function f from $(T, \mathcal{T}, \lambda)$ to X is called a measurable selection of F if $f(t) \in F(t)$ for all $t \in T$. F is said to be closed- (compact-) valued if $F(t)$ is a closed (compact) subset of X for all $t \in T$, and its distribution is given by

$$\mathcal{D}_F = \{\lambda f^{-1}: f \text{ is a measurable selection of } F\}.$$

We can now present a simple and direct translation of Proposition 3 into a basic result on the existence of selections.

---

[31] This is in itself a special case of the result in Bollobás and Varopoulos (1974), whose paper should be referred to for a proof based on the theorems of Hall and of Krein–Milman. For the proof of the special case used here, see Rath (1996a). For the proof of the case where A is a finite set, see Hart and Kohlberg (1974, pp. 170, 171) and Hildenbrand (1974, p. 74).

[32] As discussed above, Nash (1950) is the relevant benchmark. The analogy with the theory of integration of a correspondence reported in Aumann (1965) should also be evident to the reader.

[33] See Artstein (1983), Hart and Kohlberg (1974), and Sun (1996b). For approximate results, see these papers and Hart, Hildenbrand and Kohlberg (1974) and Hildenbrand (1974). For an expositional overview, see Khan and Sun (1997) and Sun (2000).

PROPOSITION 4. *If F is measurable and $\tau \in \mathcal{M}(A)$, then $\tau \in \mathcal{D}_F$ if and only if for all finite $B \subseteq A$, $\lambda(F^{-1}(B)) \geqslant \tau(B)$.*

Proposition 3 also yields a result on purification.[34] The integral is the standard Lebesgue integral and $\{a_i: i \in \mathbb{N}\}$ is the list of all of the elements of A.

PROPOSITION 5. *Let g be a measurable function from T into $\mathcal{M}(A)$, and $\tau \in \mathcal{M}(A)$ such that for all $B \subseteq A$, $\tau(B) = \int_{t \in T} g(t)(B)\, d\lambda$. If G is a correspondence from T into A such that for all $t \in T$, $G(t) = \operatorname{supp} g(t) = \{a_i \in A: g(t)(\{a_i\}) > 0\}$, then there exists a measurable selection $\bar{g}$ of G such that $\lambda \bar{g}^{-1} = \tau$.*

After a preliminary definition, we can present basic properties of the object $\mathcal{D}_F$.

DEFINITION 2. A correspondence G from a topological space Y to another topological space Z is said to be upper semicontinuous at $y_0 \in Y$ if for any open set U which contains $G(y_0)$, there exists a neighborhood V of y_0 such that $y \in V$ implies that $G(y) \subseteq U$.

PROPOSITION 6.
   (i) *For any correspondence F, $\mathcal{D}_F$ is convex.*
   (ii) *If F is closed-valued, then $\mathcal{D}_F$ is closed, and hence compact, in the space $\mathcal{M}(A)$.*
   (iii) *If Y is a metric space, and for each fixed $y \in Y$, $G(\cdot, y)$ is a closed-valued measurable correspondence from T to A such that $G(t, \cdot)$ is upper semicontinuous on the metric space Y for each fixed $t \in T$, then $\mathcal{D}_{G(\cdot, y)}$ is upper semicontinuous on Y.*

3.3. Sketch of proofs

The convexity assertion in Proposition 6 is a simple consequence of Proposition 3. However, the other two assertions rely on what can be referred to as an analogue of Fatou's lemma, which is itself a direct consequence of Proposition 3.[35]

The proof of Theorem 2 follows Nash (1950) in its essentials; we now look for a fixed point in the product space $\mathcal{M}(A)^{\ell}$, and consider the one-to-many best-response (countering) mapping from $T \times \mathcal{M}(A)^{\ell}$ into A given by

$$(t, \mu_1, \ldots, \mu_\ell) \to F(t, \mu_1, \ldots, \mu_\ell) = \underset{a \in A}{\operatorname{Arg\,Max}}\, u_t(a, \mu_1, \ldots, \mu_\ell).$$

---

[34] The proof is an exercise, but one needs the metrizability property of the weak* topology on $\mathcal{M}(A)$; see Khan and Sun (1996b) for details. As emphasized above, this purification result is a generalization of the corresponding result of Dvoretsky, Wald, and Wolfowitz to the case of countable actions.

[35] See Lemma 1 and its proof in Khan and Sun (1995b). For full details of the proof of Proposition 6, see Khan and Sun (1995b, Section 3).

The continuity and measurability assumptions on u_t allow us to assert the upper semicontinuity of $F(t, \ldots)$ and guarantee the existence of a measurable selection from $F(\cdot, \mu_1, \ldots, \mu_\ell)$.[36] We focus on the objects $\mathcal{D}_{F^i(\cdot, \mu_1, \ldots, \mu_\ell)}$ and $G(\mu_1, \ldots, \mu_\ell) = \prod_{i=1}^{\ell} \mathcal{D}_{F^i(\cdot, \mu_1, \ldots, \mu_\ell)}$, where $F^i(t, \mu_1, \ldots, \mu_\ell) = F(t, \mu_1, \ldots, \mu_\ell)$ for each $t \in T_i$, and finish the proof by applying the Fan–Glicksberg fixed-point theorem to the one-to-many mapping $G : \mathcal{M}(A)^\ell \to \mathcal{M}(A)^\ell$.

4. Two special cases

Theorems 1 and 2 concern large non-anonymous games in that each player is identified by a particular name or index t belonging to a set T. In this section, we focus on Theorem 2 and draw out its implication for two specific contexts: one where a player is also parametrized by the information at his disposal; and another anonymous setting where a player has no identity other than his characteristics. The atomlessness assumption now formalizes "dispersed" or "diffused" characteristics rather than "numerical negligibility".

4.1. Finite games with independent private information

Building on the work of Harsanyi (1967–1968, 1973) and of Dvoretsky, Wald and Wolfowitz already referred to above, Milgrom and Weber (1981, 1985) and Radner and Rosenthal (1982) use the hypothesis of independence to present a formulation of games with incomplete information.[37] In this subsection, we show how the dependence of individual payoffs on induced distributions in this model allows us to invoke the purification and existence results furnished as Proposition 5 and Theorem 2 above.

A *game with private information* consists[38] of a finite set I of ℓ players, each of whom is endowed with an identical action set A,[39] an information space $(\Omega, \mathcal{F}, \mu)$ where $(\Omega, \mathcal{F})$ is constituted by the product space $(\prod_{i \in I}(Z_i \times X_i), \prod_{i \in I}(\mathcal{Z}_i \otimes \mathcal{X}_i))$, and a utility function $u_i : A^\ell \times X_i \to \mathbb{R}$. For any point $\omega = (z_1, x_1, \ldots, z_\ell, x_\ell) \in \Omega$, let $\zeta_i(\omega) = z_i$ and $\chi_i(\omega) = x_i$.

A *mixed strategy* for player i is a measurable function from Z_i to $\mathcal{M}(A)$. If the players play the mixed strategies $\{g_i\}_{i \in I}$, the resulting expected payoff to the ith player

---

[36] The first assertion is Berge's (1959, Section III.6) maximum theorem, and the second is its measure-theoretic version; see Castaing and Valadier (1977, Theorems III.14 and III.39) and Debreu (1967). An exposition of these results in the context of game theory is available in Khan (1986b). For the full details of the proof of Theorem 2, see Khan and Sun (1995b).

[37] A detailed elaboration of this subject is beyond the scope of this chapter; see Chapter 43 in this Handbook.

[38] We confine ourselves to settings where all the players have an identical action set, and there are no information or type variables that are common to all of the players. For these essentially notational complications, as well as for the details of the computations involved in the proofs below, see Khan and Sun (1995b).

[39] Recall that the hypothesis of a countable compact metric action set is in force.

is given by

$$U_i(g) \equiv \int_{\omega \in \Omega} \int_{a_\ell \in A} \cdots \int_{a_1 \in A} u_i\big(a_1, \ldots, a_\ell, \chi_i(\omega)\big) g_1\big(\zeta_1(\omega); \mathrm{d}a\big), \ldots,$$

$$g_\ell\big(\zeta_\ell(\omega); \mathrm{d}a\big) \mu(\mathrm{d}\omega).$$

A *pure strategy* for player i is simply a measurable function from Z_i to A. An *equilibrium in mixed strategies* is a vector of mixed strategies $\{g_i^\star\}_{i \in I}$, such that $U_i(g^\star) \geqslant U_i(g_i, g_{-i}^\star)$ for any mixed strategy g_i for player i. An equilibrium $b^\star$ in pure strategies is a *purification* of an equilibrium b in mixed strategies, if for each player i, $U_i(b) = U_i(b^\star)$.

COROLLARY 2. *If, for every player i,*
  (a) *the distribution of ζ_i is atomless, and*
  (b) *the random variables $\{\zeta_j : j \neq i\}$ together with the random variable $\xi_i \equiv (\zeta_i, \chi_i)$ form a mutually independent set,*
then every equilibrium has a purification.

PROOF. Apply the change-of-variables formula and the independence hypothesis to rewrite the individual payoff functions in a form that satisfies the hypothesis of Proposition 5. Check that the pure strategy furnished by its conclusion yields a purification of the original equilibrium.   □

COROLLARY 3. *Under the hypotheses of Corollary 2, there exists an equilibrium in pure strategies if for every player i,*
  (i) *$u_i(\cdot, \chi_i(\omega))$ is a continuous function on A^ℓ for μ-almost all $\omega \in \Omega$, and*
  (ii) *there is a real-valued integrable function h_i on $(\Omega, \mathcal{F}, \mu)$ such that μ-almost all $\omega \in \Omega$, $\|u_i(a, \chi_i(\omega))\| \leqslant h_i(\omega)$ holds for every $(a_1, \ldots, a_\ell) \in A^\ell$.*

PROOF. By an appeal to the change-of-variables formula and the independence hypothesis, rewrite the individual payoff functions in the form required in Theorem 2. Check that all of the hypotheses of this theorem are satisfied, and that the equilibrium furnished by its conclusion is also an equilibrium in pure strategies.   □

We conclude with the observation that the above results are false without the independence hypothesis or the cardinality restriction on the action set.[40]

---

[40] For the first, see Aumann et al. (1983); and for the second, Milgrom and Weber (1985, Footnote 18), Khan, Rath and Sun (1999), and Khan and Sun (1999). The possibility of a positive result without the severe cardinality restrictions is suggested in Fudenberg and Tirole (1991, Theorem 6.2, p. 236).

4.2. Large anonymous games

Once the space of characteristics has been formalized as the measurable space, $(\mathcal{U}_A^d, \mathcal{B}(\mathcal{U}_A^d))$ in Section 3 with $\ell = 1$ for example, it is natural to consider a game as simply a probability measure on such a space.[41] In this section, we show how the non-anonymous setting of Section 3 sheds light on the anonymous formulation of Mas-Colell (1984a).[42] The hypothesis of a countable compact metric action set A remains in force in this subsection.

A *large anonymous game* is a probability measure μ on the measurable space of characteristics, and it is *dispersed* if μ is atomless. A probability measure τ on the product space $(\mathcal{U}_A^d \times A)$ is a *Cournot–Nash equilibrium distribution* (*CNED*) of the large anonymous game μ if the marginal of τ on $\mathcal{U}_A^d$, $\tau_{\mathcal{U}}$, is μ, and if $\tau(B_\tau) = 1$ where

$$B_\tau = \left\{ (u, a) \in \left(\mathcal{U}_A^d \times A \right): u(a, \tau_A) \geqslant u(x, \tau_A) \text{ for all } x \in A \right\},$$

τ_A the marginal of τ on A. A CNED τ can be *symmetrized* if there exists a measurable function $f: \mathcal{U}_A^d \to A$ and another CNED τ^s such that $\tau_A = \tau_A^s$ and $\tau^s(\text{Graph}_f) = 1$, where Graph_f is simply the set $\{(u, f(u)) \in (\mathcal{U}_A^d \times A): u \in \mathcal{U}_A^d\}$. In this case, τ^s is a symmetric CNED.

We see that these reformulations[43] make heavy use of probabilistic terminology, and as in any translation, give rise to additional questions stemming from the new vocabulary. The fact that players' names are not a factor in the specification of the game, and only the statistical distribution of the types of players is given, is clear enough; what is interesting is that in the formalization of a symmetric CNED, one is asking for a "re-allocated" equilibrium in which players with identical characteristics choose identical actions. Thus, an *ad hoc* assumption common to many models can be given a rigorous basis. In any case, the simple resolution of this question is perhaps surprising.[44]

COROLLARY 4. *Every CNED of a dispersed large anonymous game can be symmetrized.*

PROOF. Let τ be a CNED of the game μ, and for each $a \in A$, let $W_a = \{u \in \mathcal{U}_A^d: (u, a) \in B_\tau\}$. $(W_a)_{a \in A}$ is a countable family of subsets of $\mathcal{B}(\mathcal{U}_A^d)$ such that for

---

[41] This idea is explicit in general equilibrium theory; see Kannai (1970, Section 7), Hart, Hildenbrand and Kohlberg (1974), and Hildenbrand (1975).

[42] Also see the formulations of Milgrom and Weber (1981, 1985) and Green (1984).

[43] This reformulation is due to Mas-Colell (1984a), and comes into non-cooperative game theory via general equilibrium theory; see Hart, Hildenbrand and Kohlberg (1974).

[44] The proof of Corollaries 4 and 6 has a somewhat tortured lineage. Corollary 4 in the case of finite A was first proved directly by Khan and Sun (1987), and the general case in Khan and Sun (1995a, 1995c). Corollary 6 in the case of finite A was proved by Mas-Colell (1984a) as a consequence of Kakutani's fixed-point theorem via results in Aumann (1965). The proof given here is due to Khan and Sun (1994).

any finite subset A_F of A,

$$
\mu\left(\bigcup_{a\in A_F} W_a\right) = \tau_{\mathcal{U}}\left(\bigcup_{a\in A_F} W_a\right) = \tau\left(\left(\bigcup_{a\in A_F} W_a\right) \times A\right) \geqslant \sum_{a\in A_F} \tau(W_a \times \{a\})
$$
$$
= \sum_{a\in A_F} \tau\left(\mathcal{U}_A^d \times \{a\}\right) = \sum_{a\in A_F} \tau_A(\{a\}).
$$

Since μ is an atomless probability measure on $(\mathcal{U}_A^d, \mathcal{B}(\mathcal{U}_A^d))$, all the hypotheses of Proposition 3 are satisfied, and there exists a family $(T_a)_{a\in A}$ of sets in $\mathcal{B}(\mathcal{U}_A^d)$ such that $T_a \subseteq W_a$, $\mu(T_a) = \tau_A(\{a\})$. Now define $h : \mathcal{U}_A^d \to A$ such that $h(t) = a$ for almost all $t \in T_a$, all $a \in A$, and note that the measure $\mu(i, h)^{-1}$, i being the identity mapping on $\mathcal{U}_A^d$, is the required symmetrization. □

This yields the interesting characterization of symmetric equilibria as the extreme points of a set of equilibria.[45]

COROLLARY 5. *Let μ be a dispersed large anonymous game. Then a CNED τ of μ is a symmetric CNED if and only if τ is an extreme point of the set $\Lambda_\tau = \{\rho \in \mathcal{M}(\mathcal{U}_A^d \times A): \rho_{\mathcal{U}} = \mu; \rho_A = \tau_A; \rho(B_\tau) = 1\}$.*

All that remains is the question of existence.

COROLLARY 6. *There exists a symmetric CNED for a dispersed large anonymous game μ.*

PROOF. In Theorem 2, use $(\mathcal{U}_A^d, \mathcal{B}(\mathcal{U}_A^d), \mu)$ as the space of players' names, and the identity mapping i as the game. If f is the equilibrium guaranteed by the theorem, the measure $\mu(i, f)^{-1}$ is a symmetric CNED. □

We conclude with the observation that these results are false without the dispersedness hypothesis.[46]

5. Non-existence of a pure strategy Nash equilibrium

In this section, we present two examples of games without Nash equilibria, in which the set of actions is a compact interval. Apart from their intrinsic methodological interest,

---

[45] For two different proofs, one based on the Douglas–Lindenstrauss theorem and the other on a direct construction of suitable measures, see Khan and Sun (1995a).

[46] See Examples 1 and 2 in Rath, Sun and Yamashige (1995). The fact that they are also false without the cardinality assumption on A will be dealt with in detail below.

these examples are useful because they anchor the abstract treatment of Section 3 to concrete specifications that one can compute and work with. Both examples are predicated on the fact that it is impossible to choose from the correspondence[47] on the Lebesgue unit interval defined by $t \to \{t, -t\}$, a measurable selection that induces a uniform distribution v^* on $[-1, 1]$.[48]

5.1. A nonatomic game with nonlinear payoffs

The following example[49] is due to Rath, Sun and Yamashige (1995) who present it in the context of Corollary 6 above.

EXAMPLE 2. Consider a game $\mathcal{G}_1$ in which the set of players $(T, \mathcal{T}, \lambda)$ is the Lebesgue unit interval $([0, 1], \mathcal{B}([0, 1]), v)$, A is the interval $[-1, 1]$, and the payoff function of any player $t \in [0, 1]$ is given by

$$u_t(a, \rho) = g(a, \beta d(v^*, \rho)) - |t - |a||,$$
$$0 < \beta < 1, \ a \in [-1, 1], \ \rho \in \mathcal{M}([-1, 1]),$$

where $d(v^*, \rho)$ is the Prohorov distance between v^* and ρ based on the natural metric on $[-1, 1]$,[50] and $g : [-1, 1] \times [0, 1] \to \mathbb{R}_+$. If $g(a, 0) \equiv 0$ for any $a \in [-1, 1]$, there is no Nash equilibrium that induces v^*. The point is that one can choose the function $g(\cdot, \cdot)$ such that the best-response function based on a distribution $\rho \neq v^*$ induces a distribution different from ρ and therefore precludes the existence of a Nash equilibrium. An example of such a function is the periodic function, with period 2ℓ, $\ell \in (0, 1]$, and defined on $[0, 2\ell]$ by

$$g(a, \ell) = \begin{cases} a/2 & \text{for } 0 \leqslant a \leqslant \ell/2, \\ (\ell - a)/2 & \text{for } \ell/2 \leqslant a \leqslant \ell, \\ -g(a - \ell, \ell) & \text{for } \ell \leqslant a \leqslant 2\ell, \end{cases}$$

with $g(a, \ell) = -g(-a, \ell)$ for $a < 0$, and extended in both directions. Indeed, this specification of $g(\cdot, \cdot)$ furnishes an equicontinuous family of payoff functions.[51]

---

[47] This correspondence has a canonical status in general equilibrium theory and goes back at least to Debreu; see Kannai (1970, Section 7).

[48] Let f be such a measurable selection, and let $E = \{t \in [0, 1]: f(t) \in [0, 1]\}$. Then $\lambda^*(E) = \lambda(f^{-1}(E)) = \lambda(E)$. Since $\lambda^* = (1/2)\lambda$, $\lambda(E) = 0$, and hence $\lambda^*([-1, 0]) = \lambda f^{-1}([-1, 0]) = \lambda(\{t \notin E\}) = 1$, a contradiction.

[49] This is Example 3 in Rath, Sun and Yamashige (1995); see also Section 2 in Khan, Rath and Sun (1997a).

[50] See Billingsley (1968, pp. 237, 238) or Hildenbrand (1974, p. 49) for a definition.

[51] This furnishes the possible interpretation that there is a "bound on the diversity of payoffs". At any rate, it shows that nonexistence has nothing to do with the measurability of the function specifying the game.

5.2. Another nonatomic game with linear payoffs

In Example 2, the distribution of societal responses enters an individual's payoff function in a non-linear way; here we present an example in which players maximize expected utilities with respect to this distribution and thereby lead to linear specification. This example is due to Khan, Rath and Sun (1997b), and unlike Example 2, does not involve the Prohorov metric.

EXAMPLE 3. Consider a game different from that in Example 2 only in that the payoff function of player t is given, for $a \in [-1, 1]$, $\rho \in \mathcal{M}([-1, 1])$, by

$$u_t(a, \rho) = \int_{-1}^{1} v_t(a, x) \, d\rho(x), \quad \text{where } v_t(a, x) = -\left| t - |a| \right| + (t - a)z(t, x),$$

and the function $z : [0, 1] \times [-1, 1] \to \mathbb{R}$ is such that for all $t \in [0, 1]$,

$$z(t, a) = \begin{cases} a & \text{if } 0 \leqslant a \leqslant t, \\ t & \text{if } t < a \leqslant 1, \\ -z(t, -a) & \text{if } a < 0. \end{cases}$$

This game does not have a Nash equilibrium. A distribution that is not uniform cannot be an equilibrium.[52] On the other hand, for a uniform distribution ρ, the value of the summary statistic $\int_A z(t, a) \, d\rho$ is zero for ν-almost all $t \in [0, 1]$, and hence requires that ρ be induced by a measurable selection from the best-response correspondence on the Lebesgue unit interval defined by $t \to \{t, -t\}$, which we have seen to be impossible. It is easy to check that $u_t(a, \rho)$ is a jointly continuous function in its three arguments (t, a, ρ), and that the family of utility payoff functions indexed by the name t is an equicontinuous family.

5.3. New games from old

Now consider another game $\mathcal{G}_2$ manufactured from the game $\mathcal{G}_1$ in Example 2. In this game the set of players $(T, \mathcal{T}, \lambda)$ is again given by the Lebesgue unit interval $([0, 1], \mathcal{B}([0, 1])\nu)$, A by $[-1, 1]$, and the payoff function of any player $t \in [0, 1]$ by $v_t : A \times \mathcal{M}([-1, 1]) \to \mathbb{R}$ where

$$v_t(\cdot, \cdot) = \begin{cases} u_{2t}(\cdot, \cdot) & \text{if } 0 \leqslant t \leqslant 1/2, \\ u_{2-2t}(\cdot, \cdot) & \text{if } 1/2 < t \leqslant 1. \end{cases}$$

It is clear that the new game $\mathcal{G}_2$ is formed from $\mathcal{G}_1$ by endowing each player with a "twin" and then by normalizing the space of players to the Lebesgue unit interval. It

---

[52] This assertion, though elementary, requires some delicate computations; see Khan, Rath and Sun (1997b) for details.

is clear that both games $\mathcal{G}_1$ and $\mathcal{G}_2$ induce identical distributions on the space of characteristics. Hence, in some essential macroscopic sense, the two games are identical. However, the point is that $\mathcal{G}_2$ has a Nash equilibrium whereas $\mathcal{G}_1$ does not! It is easy to check that $g : [0, 1] \to [-1, 1]$ is a Nash equilibrium of the modified game $\mathcal{G}_2$, where

$$
g(t) = \begin{cases} 2t & \text{if } 0 \leqslant t \leqslant 1/2, \\ 2t - 2 & \text{if } 1/2 < t \leqslant 1. \end{cases}
$$

6. Interactions based on averages of individual responses

In light of these counterexamples, the question arises as to how far one can proceed if a player's dependence on the distribution of societal responses is restricted to dependence on specific moments of the distribution. In this section, we focus on the first moment,[53] and are thereby led to integration, and a consequent linearity requirement on the action set. Since integration occurs with respect to players' names, what is important in this connection is Aumann's observation that the "chief mathematical tools are Lebesgue measure and integration [and that] Riemann integration can *not* be substituted".

6.1. A basic result

We shall follow the notation of Section 3.2 but with the difference that the common action set A is now conceived to be a nonempty compact subset of Euclidean space $\mathbb{R}^n$, and the space of player characteristics $(\mathcal{U}_A^{av}, \mathcal{B}(\mathcal{U}_A^{av}))$ is the space of real-valued continuous functions on $A \times (\mathrm{con}(A))^\ell$, endowed with its sup-norm topology and the induced Borel σ-algebra (the superscript av denoting "distribution"). We can now present[54]

THEOREM 3. *Let $\mathcal{G}^{av}$ be a measurable map from T to $\mathcal{U}_A^{av}$. Then there exists a measurable function $f : T \to A$ such that for λ-almost all $t \in T$,*

$$
u_t\left(f(t), \int_{s \in T_1} f_1(s)\, \mathrm{d}\lambda_1, \ldots, \int_{s \in T_\ell} f_\ell(s)\, \mathrm{d}\lambda_\ell \right)
$$
$$
\geqslant u_t\left(a, \int_{s \in T_1} f_1(s)\, \mathrm{d}\lambda_1, \ldots, \int_{s \in T_\ell} f_\ell(s)\, \mathrm{d}\lambda_\ell \right)
$$

for all $a \in A$, where $u_t = \mathcal{G}^{av}(t) \in \mathcal{U}_A^{av}$ and f_i is the restriction of f to T_i.

---

[53] As in Schmeidler (1973), but see Rauh (1997) for formulations involving higher moments.
[54] When ℓ equals one and the Lebesgue interval $([0, 1], \mathcal{B}([0, 1]), \nu)$ is the space of players' names, Schmeidler (1973) presents this result with A being the unit simplex in $\mathbb{R}^n$, and Rath (1992) proves the general case. Note that Remark 4 in Schmeidler (1973) does not cover this general case since a general utility function on a compact set A is usually not relevant to any quasi-concave function on the convex hull of A.

6.2. Lyapunov's theorem and the integral of a correspondence

Let the integral of a correspondence $F : T \to \mathbb{R}^n$ be defined as

$$
\mathcal{I}_F = \left\{ \int_T f(t)\, d\lambda \colon\ f \text{ is an integrable selection of } F \right\}.
$$

Since the range of F is no longer countable, we need to modify the earlier definition of measurability to require that the graph of F be an element of $\mathcal{T} \otimes \mathcal{B}(\mathbb{R}^n)$. A measurable selection theorem and Lyapunov's theorem on the range of an atomless vector measure then yield the following analogue to Proposition 6.[55]

PROPOSITION 7.
  (i) $\mathcal{I}_F$ is nonempty and convex.
 (ii) If F is integrably bounded and compact-valued, then $\mathcal{I}_F$ is compact and $\mathcal{I}_F = \mathcal{I}_{\mathrm{con}\,F}$.
(iii) If Y is a metric space, G a closed-valued correspondence, and H an integrably bounded compact-valued correspondence, both from $T \times Y$ into $\mathbb{R}^n$ such that for each fixed $t \in T$, $G(t, \cdot)$ is upper semicontinuous on Y, and for each fixed $y \in Y$, $G(t, y) \subseteq H(t)$ for λ-almost all $t \in T$, then $\mathcal{I}_{G(\cdot, y)}$ is upper semicontinuous on Y.

6.3. A sketch of the proof

With Proposition 7 in hand, one simply follows[56] all the guideposts laid out in the proof of Theorem 2, but with the Kakutani fixed-point theorem applied to a one-to-many countering or best-response mapping $G : T \times (\mathrm{con}(A))^\ell$ into A given by

$$
(t, a_1, \ldots, a_\ell) \to F(t, a_1, \ldots, a_\ell) = \operatorname*{Arg\,Max}_{a \in A} u_t(a, a_1, \ldots, a_\ell).
$$

7. An excursion into vector integration

On comparing Theorems 2 and 3, the question arises as to why cardinality restrictions on action sets are crucial when player interactions are based on distributions, but play no role when they are based on averages. This discrepancy is more apparent than real:

---

[55] For Lyapunov's theorem, see Lyapunov (1940) for the original statement; and Jamison (1974), Castaing and Valadier (1977), Diestel and Uhl (1977), and Loeb and Rashid (1987) for modern treatments. For a proof of Proposition 7, see Aumann (1965, 1976) and Hildenbrand (1974).

[56] This direct proof is due to Rath (1992). The proof in Schmeidler (1973) for the case that A is the unit simplex in $\mathbb{R}^n$ is based on a a purification argument.

the induced distribution of a random variable is also an average, in a clearly defined sense, of a related random variable taking values in an infinite-dimensional space, and the cardinality restrictions simply shift over from the sets themselves to the dimensionality of the space in which they are located. Thus, even when the primary emphasis is explicitly on distributions,[57] and the parameters of the problem do not suggest it, the relevant backdrop is still that of vector integration.[58]

To see this, return to Proposition 5 and to the discussion of purification results in Section 4, and recall that the integral of a measurable function $g : T \rightarrow \mathcal{M}(A)$ is obtained by fixing a particular element $B \in \mathcal{T}$, and then by integrating the resulting real-valued function $g(t)(B)$. However, g can also be reduced to a real-valued function by considering[59] $(x, g(t)) = \int_A x(a) g(t; da)$, where x is a particular element of $C(A)$, the space of continuous functions on the compact set A endowed with their sup norm. This procedure defines another integral, the so-called Gelfand integral $\int_T g(t) \, d\lambda(t)$, where

$$
\left(x, \int_T g(t) \, d\lambda(t) \right) = \int_T \left(x, g(t) \right) d\lambda(t)
$$
$$
= \int_T \int_A x(a) g(t; da) \, d\lambda(t) \quad \text{for all } x \in C(A).
$$

The point is that this integral is identical to the integral obtained by our first procedure, and furthermore, the Gelfand integral of $\delta_{\{g(\cdot)\}}$, the function obtained by "lifting" $g(\cdot)$ into the space of probability measures, is the same as the induced distribution λg^{-1} of g.[60]

This discussion for the specific space $C(A)$ and its dual $(C(A))^*$ can easily be transposed to the general setting of a separable Banach space[61] X and its dual X^*. We can now define the Gelfand integral[62] of any X^*-valued function $g(\cdot)$ by requiring that

$$
\left(x, \int_T g(t) \, d\lambda(t) \right) = \int_T \left(x, g(t) \right) d\lambda(t) \quad \text{for all } x \in X.
$$

In addition to its norm topology, X^* is also endowed with the weak and weak* topologies, and it is the Borel σ-algebra generated by the latter that ensures that for any $x \in X$,

---

[57] Distributions are emphasized, for example, in Milgrom and Weber (1981, 1985), Mas-Colell (1984a) and, in the context of general equilibrium theory, in Hart, Hildenbrand and Kohlberg (1974) and Hildenbrand (1975).

[58] We follow the terminology of the mathematical literature in reserving the term "vector integration" for integration of functions taking values in an infinite-dimensional space; see Diestel and Uhl (1977) and their references.

[59] Note the abuse of notation in expressing the value of the probability measure $g(t)$ at $B \in \mathcal{T}$, by $g(t; B)$.

[60] See Khan, Rath and Sun (1997a, Section 3) for details of these claims.

[61] For the basic definitions, see, for example, Royden (1968, Chapter 10) or Rudin (1973).

[62] See Diestel and Uhl (1977, Chapter II). For early applications of this integral in mathematical economics and game theory, see Bewley (1973) and Khan (1985a, 1986a).

the real-valued function $(x, g(\cdot))$ is measurable, and furnishes the relevant weak* measurability criterion for Gelfand integration.[63] However, we can also simply work with
a separable Banach space X directly. In this case, we can be guided by Lebesgue integration, and focus on functions that are pointwise limits of simple functions. This
furnishes us with the so-called strong measurability criterion, and a conventional notion
of an integral, the so-called Bochner integral, based on the convergence of the sums of
these simple functions.[64] With all of this terminology at our disposal, we can present
the following consequence of the results of Section 3.2.[65]

PROPOSITION 8. *Proposition 7 is valid for a correspondence F from T into a countably infinite subset of a separable Banach space or a dual of a separable Banach space.
In the former case, we can work with the norm or weak topologies and the Bochner
integral; and in the latter, with the weak* topology and the Gelfand integral.*

Proposition 8 is false without the cardinality restriction,[66] a fact that underlies Examples 2 and 3 above.

We conclude this section with a result that views a set of integrable functions as a
productive object in its own right. Let $L_1(\lambda, X)$ be the space of equivalence classes
of Bochner integrable functions equipped with the integral norm.[67] This is a Banach
space, and its dual space $L_\infty^w(\lambda, X^*)$ consists of equivalence classes of weak* measurable functions with essentially bounded norm functions and equipped with the essential
supremum norm.[68] We can now present analogues of results that play the same role in
the sequel that Tychonoff's theorem played in Section 2.[69]

PROPOSITION 9. *If A is a weakly compact subset of a separable Banach space X,
$L_1(\lambda, A)$ is a weakly compact subset of the Banach space $L_1(\lambda, X)$. If A is a norm
bounded, weak* closed subset of a dual Banach space X^*, $L_\infty^w(\lambda, A)$ is a weak* compact subset of the Banach space $L_\infty^w(\lambda, X^*)$.*

---

[63] For basic properties of the weak and weak* topologies, see Diestel (1984). For different notions of measurability and their interrelationships, see Talagrand (1984).

[64] Since we are ignoring the predual, the above discussion can really be phrased in terms of any separable
Banach space X. We may also point out here that there is yet another integral based on the reduction of a
vector function to a real-valued function by evaluating it with respect to elements of its dual rather than its
predual. This leads to the Pettis integral, a more elusive notion about which we shall have nothing to say here;
see Talagrand (1984) for details and further references.

[65] See Khan and Sun (1996a) for details of proof.

[66] See Sun (1997) for counterexamples, and Khan and Sun (1997) and Sun (2000) for an expositional
overview; also Rustichini (1989).

[67] For any $f \in L_1(\lambda, X)$, $\|f\|_1 = \int_T \|f(t)\| \, d\lambda$; see Diestel and Uhl (1977, Chapter II) for details.

[68] For any $f \in L_\infty^w(\lambda, X^*)$ $\|f\|_\infty = \operatorname{ess\,sup}_{t \in T} \|f(t)\|$; see Dincleanu (1973).

[69] The first statement in the proposition below is due to Diestel; see Khan (1984) for an alternative proof based
on James' theorem. The second statement is a consequence of the Banach–Alaoglu theorem; see Castaing and
Valadier (1977, Chapter V). For a leisurely discussion of the importance of these results for large games, see
Khan (1986b).

8. Interactions based on almost all individual responses

Even though in Theorem 1 payoff functions depend on the actions of each individual player rather than on some statistical summary of these actions, the hypothesis of continuity in the product topology reduces this apparent generality to dependence on the actions of an essentially finite number of players. In this section, we see how the measure-theoretic structure on the set of players' names can be exploited to yield results for situations where an individual player's payoff depends on the (equivalence class of the) entire function of individual responses.[70]

8.1. Results

Unlike earlier sections, we shall denote a game by the generic symbol $\mathcal{G}$. We continue to assume that product spaces are endowed with their product measurable or topological structures, and that the space $C(Y)$ of continuous functions on a compact set Y is equipped with its sup norm topology and the induced Borel σ-algebra. We reserve the symbol X for a separable Banach space, and except for the last result, work primarily with the weak and weak* topologies.[71]

THEOREM 4. *Let $\mathcal{G}$ be a measurable map from T to $C(A \times L_1(\lambda, A))$, where $A \subseteq X$ is convex and weakly compact, and $\mathcal{G}(t)(\cdot, g) \equiv u_t(\cdot, g)$ is quasi-concave on A for any $g \in L_1(\lambda, A)$. Then there exists $f \in L_1(\lambda, A)$ such that for λ-almost all $t \in T$, $u_t(f(t), f) \geqslant u_t(a, f)$ for all $a \in A$. This statement is valid for $L_\infty^w(\lambda, A)$ substituted for $L_1(\lambda, A)$, and with $A \subseteq X^*$ convex and weak* compact.*

It is clear in Nash (1950, 1951) that the question of the existence of pure-strategy equilibria precludes the assumptions of convexity on the action set and of quasi-concavity of the payoff function. Theorem 4 thus pertains to mixed-strategy equilibria, and one can ask whether it yields any implications for settings without these assumptions. We present three corollaries in this regard. Note that unlike the case of a finite action set, it now makes a difference whether pure strategies are conceived as extreme points of an action set or as Dirac measures on it.[72]

---

[70] Thus, in contrast to the model of Section 2, instead of an arbitrary product of the players' actions, we assume measurable "profiles" of such actions. In contrast to the models of Sections 3 and 6, the situations considered here can be conceived as one of "widespread" externalities. Note, however, that this terminology has a different meaning in the work of Hammond (1995, 1998, 1999).

[71] The reason for this is that compact sets in the norm topology sets are "small"; a norm compact set in any normed linear space is contained in the closed convex hull of a sequence converging to zero; see Diestel (1984, Theorem 5, p. 3). However, see Toussaint (1984). We may also point out here that the hypothesis of weakly compact action sets guarantees that weakly measurable profiles are strongly measurable as a consequence of the Pettis measurability theorem; see Diestel and Uhl (1977, Chapter II) and Uhl (1980). Also see Balder (1999a) for a synthetic treatment based on the so-called *feeble* topology.

[72] We shall use the notation ext(A) to denote the set of extreme points of a set A. Note that in Corollary 8 below, ext($\mathcal{M}(A)$) is the set of Dirac measures on A.

COROLLARY 7. *If A is a compact Hausdorff space, there exists a CNED for a large anonymous game.*[73]

COROLLARY 8. *Let $\mathcal{G}$ be a measurable map from T into $C(A \times L_\infty^w(\lambda, \mathcal{M}(A)))$, and A a compact metric space. Then for any $\varepsilon > 0$, there exists $f \in L_\infty^w(\lambda, \text{ext}(\mathcal{M}(A)))$ and $K_\varepsilon \in \mathcal{T}$, $\lambda(K_\varepsilon) \geqslant 1 - \varepsilon$ such that for all $t \in K_\varepsilon$,*

$$\int_A v_t(a, f) f(t; da) \geqslant \int_A v_t(a, f) \, d\rho(a) - \varepsilon$$
for all $\rho \in \mathcal{M}(A)$, where $v_t \equiv \mathcal{G}(t)$.

COROLLARY 9. *Let $\mathcal{G}$ be a measurable map from T into $C(A \times A)$, A a convex and weakly compact subset of X. Then for any $\varepsilon > 0$, there exists $f \in L_1(\lambda, \text{ext}(A))$ and $K_\varepsilon \in \mathcal{T}$, $\lambda(K_\varepsilon) \geqslant 1 - \varepsilon$ such that for all $t \in K_\varepsilon$,*

$$u_t\left(f(t), \int_T f(t) \, d\lambda(t) \right) \geqslant u_t\left(a, \int_T f(t) \, d\lambda(t) \right) - \varepsilon \quad \text{for all } a \in A,$$

where $u_t \equiv \mathcal{G}(t)$ and the integral is the Bochner integral. The above statement is also true for $L_\infty^w(\lambda, \text{ext}(A))$ where $A \subseteq X^*$ and weak* compact, with the integral the Gelfand integral.

The following result reimposes cardinality restrictions on action sets to obtain exact equilibria in the Banach setting.

THEOREM 5. *Theorem 3 is valid if A is a countable compact subset of X or of X^*, with the norm or weak topologies and the Bochner integral in the first case, and with the weak* topology and the Gelfand integral in the second.*

8.2. Uhl's theorem and the integral of a correspondence

In the discussion of his existence theorem, Aumann (1966, p. 15) noted that in "the presence of a continuum of traders, the space of assignments is no longer a subset of a finite-dimensional Euclidean space, but of an infinite-dimensional function space. This necessitates the use of completely new methods ... of functional analysis (Banach spaces) and topology". There is the additional handicap that Lyapunov's theorem fails in the infinite-dimensional setting.[74] However, it can be shown that an approximate theory of integration can be developed on the basis that the closure of the range of a vector

---

[73] See Section 4.2 above for definitions and comparisons.

[74] See Diestel and Uhl (1977, Chapter IX) for discussion and counterexamples. In particular, they observe that the examples "suggest that nonatomicity may not be a particularly strong property of vector measures, particularly from the point of view of the Lyapunov theorem in the infinite-dimensional context".

measure is convex and compact. For the weak or weak* topologies, this is a consequence of the finite-dimensional Lyapunov theorem; for the norm topology, the result is due to Uhl (1969).[75]

8.3. Sketch of proofs

Once we have access to the mathematical tools discussed above, the proofs are a technical (functional-analytical) elaboration of the basic argument of Nash (1950), with the upper semi-continuity of the best-response correspondence being the essential difficult hurdle.[76]

9. Non-existence: Two additional examples

On taking stock, we see that there exist exact pure strategy equilibria with cardinality restrictions on individual action sets (Theorem 5), and approximate pure strategy equilibria without such restrictions and even in situations where an individual player's dependence on societal responses is not limited to their distributions or averages (Corollary 8). In this section, we see that there cannot be progress on this score without additional measure-theoretic restrictions on the space of players' names.

9.1. A nonatomic game with general interdependence

The following example is due to Schmeidler (1973), and it shows that Corollary 8 cannot be improved even in the setting of two actions.

EXAMPLE 4. Consider a game in which the set of players $(T, \mathcal{T}, \lambda)$ is given by the Lebesgue unit interval $([0, 1], \mathcal{B}([0, 1]), \nu)$, A by $\{-1, 1\}$, and the payoff function of any player $t \in [0, 1]$ by

$$u_t(a, f) = \left| a - \int_0^t f(x) \, d\nu \right|, \quad a \in \{-1, 1\}, \ f \in L_1(\nu, \{-1, 1\}).$$

This game does not have a Nash equilibrium. For any equilibrium f, the value of the summary statistic $h(t) = \int_0^t f(x) \, d\nu$ must be zero for all $t \in [0, 1]$. This implies that $f(t) = 0$ for ν-almost all t, which contradicts the fact that $f(t)$ is 1 or -1.

---

[75] We leave it to the reader to develop the approximate analogues of Proposition 7; for details of the theory, see Hiai and Umegaki (1977), Artstein (1979), Khan (1985a), Khan and Majumdar (1986), Papageorgiou (1985, 1987, 1990), Yannelis (1991a).

[76] For a detailed proof of Theorem 4, see Khan (1985b, 1986b). The proof of Corollary 7 exploits the convexity hypothesis of $\mathcal{M}(A)$; see Mas-Colell (1984a) and Khan (1989) for a direct proof. For the details of proofs of Corollaries 8 and 9, see Schmeidler (1973), Khan (1986a), Pascoa (1988, 1993b). Alternative direct proofs of these corollaries based on the ideas of Rath (1992) and the approximate integration theory discussed in Section 8.2 can also be furnished. The proof of Theorem 5 is a routine consequence of Proposition 8; see Khan, Rath and Sun (1997a) for details.

9.2. A nonatomic game on the Hilbert space ℓ_2

The following example is due to Khan, Rath and Sun (1997a) and it shows that Theorem 5 cannot be improved even when action sets are norm-compact. It is based on a function $f : [0, 1] \to \ell_2$ where

$$f(t) = \left(\frac{1 - W_n(t)}{2^n} \right)_{n=1}^{\infty}, \quad W_1(t) \equiv -1, \quad W_n(t) = (-1)^{[2^{n-1}t]} \quad \text{for } n \geqslant 2,$$

where $[x]$ is the integer part of x. It can be shown[77] that the range of f is norm-compact, that it is Bochner integrable with integral $e \equiv (1, (2^{-n-1})_{n=1}^{\infty})$, and that $(e/2) \notin \int_0^1 \{0, f(t)\} \, d\nu(t)$.

EXAMPLE 5. Consider a game in which the set of players $(T, \mathcal{T}, \lambda)$ is given by the Lebesgue unit interval $([0, 1], \mathcal{B}([0, 1])\nu)$, A is a norm-compact subset of ℓ_2 containing $\{0\} \cup \{f(t) : t \in [0, 1]\}$, and the payoff function of any player $t \in [0, 1]$ is given by

$$u_t(a, b) = -h\left(t, a, f(t), \beta \left\| b - \frac{e}{2} \right\| \right) - \|a\| \cdot \|a - f(t)\|,$$

$$0 < \beta < 1, \ a \in A, \ b \in \overline{\text{con}}(A),$$

where $h : [0, 1] \times \ell_2 \times \ell_2 \times \mathbb{R}_+ \to \mathbb{R}_+$. If $h(t, a, f(t), 0) = 0$, it is easy to see that there is no Nash equilibrium. The point is that one can choose the function h such that the best-response function based on an element $b \in \overline{\text{con}}(A)$ averages to a value b' such that $\|b' - (e/2)\| \neq \|b - (e/2)\|$, and therefore precludes the existence of a Nash equilibrium. An example of such a function is given by

$$h(t, a, b, \alpha)$$
$$= \begin{cases} \alpha \left| \sin \dfrac{t}{\alpha} \pi \right| \cdot \left(\|a\| + 1 - (-1)^{[t/\alpha]} \right) \cdot \left(\|a - b\| + 1 + (-1)^{[t/\alpha]} \right) & \text{if } \alpha > 0, \\ 0 & \text{if } \alpha = 0. \end{cases}$$

Finally, we observe that the same example works for a game in which compactness and continuity are phrased in the weak rather than the norm topology on ℓ_2.

We conclude with the observation that the isomorphism between ℓ_2 and any separable infinite-dimensional L_2 space allows us to set Example 5 in the latter space. This is useful in light of the use of L_2 in models with information and uncertainty.

---

[77] For details as to these properties, see Khan, Rath and Sun (1997a). This function can be traced to Lyapunov; see Diestel and Uhl (1977, Chapter IX).

10. A richer measure-theoretic structure

In light of the counterexamples presented above, the question arises whether additional measure-theoretic structure on the set of players' names will allow the construction of a more robust and general theory. In asking this, we are guided by the emphasis of Aumann (1964) that it is not the particularity of the measure space but its atomlessness that is important from a methodological point of view.[78] As a result of a particular class of measure spaces introduced by Loeb (1975), the so-called Loeb measure spaces, a richer structure is indeed available, and it is ideally suited for studying situations where non-atomic considerations such as strategic negligibility or diffuse information are an essential and substantive issue.

10.1. Atomless Loeb measure spaces and their special properties

A Loeb space $(\overline{T}, L(\overline{\mathcal{T}}), L(\bar{\lambda}))$ is a "standardization" of a hyperfinite internal probability space $(\overline{T}, \overline{\mathcal{T}}, \bar{\lambda})$, and constructed as a simple consequence of Caratheodory's extension theorem and the countable saturation property of the nonstandard extension.[79] It bears emphasizing that a Loeb measure space, even though constituted by nonstandard entities, is a standard measure space, and in particular, a result pertaining to an abstract measure space applies to it. For applications, its specific construction can usually be ignored,[80] and one simply focuses on its special properties not shared by Lebesgue or other measure spaces.[81] We now turn to those properties.[82]

PROPOSITION 10. *If an atomless Loeb space $(\overline{T}, L(\overline{\mathcal{T}}), L(\bar{\lambda}))$ is substituted for $(T, \mathcal{T}, \lambda)$, Propositions 5, 6, and 8 are valid without any cardinality restrictions.*

---

[78] In hindsight, one sees the same emphasis in the papers of Dvoretsky, Wald and Wolfowitz. It is also worth mentioning that Aumann concludes his paper by deferring to subsequent work a discussion of "the economic significance of the Lebesgue measure of a coalition". However, Debreu (1967) did point out "the identification of economic agents with points of an analytic set seems artificial".

[79] In addition to Loeb (1975), see Cutland (1983), Hurd and Loeb (1985), Lindstrom (1988), Anderson (1991, 1994), and Khan and Sun (1997).

[80] The relevant analogy is to the situation when a user of Lebesgue measure spaces can afford to ignore the construction of a Lebesgue measure, and the Dedekind set-theoretic construction of real numbers on which it is based.

[81] The importance of Loeb measure spaces for game theory and mathematical economics is discussed, in particular, in Anderson's Chapter 14 in this Handbook; also see Rashid (1987), Anderson (1991), Khan and Sun (1997), and Sun (2000).

[82] The first proposition is due to Sun (1996b, 1997) and the second to Keisler (1988, Theorems 2.1 and 2.3); see Khan and Sun (1997) and Sun (2000) for expositional overviews. The fact that the second proposition is false for Lebesgue spaces was known to von Neumann (1932).

PROPOSITION 11. *Two real-valued random variables x and y on a Loeb counting probability space*[83] *have the same distribution iff there is an internal permutation that sends x to y.*

10.2. Results

The special properties of correspondences defined on a Loeb measure space delineated by Proposition 10 can be translated into results on exact pure-strategy equilibria for games with many players but with action sets without any cardinality restrictions.

THEOREM 6. *If an atomless Loeb space $(\overline{T}, L(\overline{T}), L(\bar{\lambda}))$ is substituted for $(T, \mathcal{T}, \lambda)$, Theorems 2 and 5 are valid without any cardinality restrictions on A.*

These results bring out the fact that the non-existence claims in Examples 2, 3 and 5 do not hold for the idealized Loeb setting. Indeed, one can show that in the finite-player versions of the games presented in these examples, approximate equilibria exist and that these approximations get finer as the number of players increases.[84] Thus, it is natural to ask what it is about the idealized Lebesgue setting that makes the exact counterparts of these equilibria disappear. Since Lebesgue measurability of a function can be represented in the Loeb setting under the condition that "infinitesimally close" points in the domain of the function have "infinitesimally close" values, the answer lies in the fact that in asking for an equilibrium that is by definition Lebesgue measurable, we have injected a cooperative requirement into an essentially non-cooperative situation. What is particularly interesting is that there may be no such equilibrium even in situations when Lebesgue measurability is fulfilled for the game itself, i.e., when players with "infinitesimally close" names have "infinitesimally close" payoff functions. Thus the use of Lebesgue measurability for the modelling of large finite game-theoretic phenomena fails at two levels: first, it restricts the types of large finite games to some special classes, and second, even with this restriction, the ideal limits of approximate equilibria cannot be modeled.

In referring to continuum-of-players methods in the context of non-cooperative game theory, Mas-Colell (1984a, p. 20) is clear that a "literal continuum of agents ... should be thought of only as a limit version of the Negligibility Hypothesis. It is an analytically useful limit because results come sharp and clean, unpolluted by ε's and δ's, but it is also the less realistic limit". Elsewhere, Mas-Colell and Vives hope that the "conclu-

---

[83] This is a specific instance of a Loeb measure space based on the hyperfinite set $N_\omega = (1, \ldots, \omega)$, where ω is an "infinitely large" integer, endowed with the counting probability measure on the set $\mathcal{N}_\omega$ of all internal subsets of N_ω. It is obtained by taking the "standard part" of the values of this finitely additive measure to obtain a countably additive measure and then its extension to the completion of the smallest σ-algebra containing $\mathcal{N}_\omega$. For details, see the references in Footnote 79 above.

[84] See Khan and Sun (1999, Section 7) for details.

sions are not too misleading when applied to realistic situations".[85] We have already emphasized the distinction between two types of idealized limiting situations, and we now observe that results based on hyperfinite Loeb measure spaces can be "asymptotically implemented" for a sequence of large but finite games as a matter of course. One only needs to control the extent to which the characteristics are allowed to vary by focussing on a "tight" sequence of mappings.[86] The following result based on a compact metric action set[87] illustrates this particular advantage of the Loeb formulation.[88]

COROLLARY 10. *For each $n \geqslant 1$, let a finite game G^n be a mapping from $T^n = \{1, 2, \ldots, n\}$ into $\mathcal{U}_A^d$, and $\{T_1^n, \ldots, T_\ell^n\}$ be a partition of T^n. Assume that the sequence of finite games is tight and that there is a positive number c such that $|T_i^n|/n > c$ for all sufficiently large n and $1 \leqslant i \leqslant \ell$. Then for any $\varepsilon > 0$, there exists $N \in \mathbb{N}$ such that for all $n \geqslant N$, there exists $g_n : T^n \to A$ such that for all $t \in T^n$, and for all $a \in A$,*

$$u_t^n\left(g_n(t), \lambda_1^n g_n^{-1}, \ldots, \lambda_\ell^n g_n^{-1}\right) \geqslant u_t^n\left(a, \lambda_1^n g_n^{-1}, \ldots, \lambda_\ell^n g_n^{-1}\right) - \varepsilon,$$

where $u_t^n \equiv G^n(t)$, and λ_i^n is the counting probability measure[89] on T_i^n, $i = 1, \ldots, \ell$.

It is also one of the strengths of a Loeb counting measure space that anomalies presented in Section 5.3 cannot arise as illustrated by the following result.

COROLLARY 11. *Let $\mathcal{G}$ and $\mathcal{F}$ be measurable maps from T to $\mathcal{U}_A^d$, such that $\overline{v}_i \mathcal{G}_i^{-1} = \overline{v}_i \mathcal{F}_i^{-1}$ for $i = 1, \ldots, \ell$, where $\mathcal{G}_i$ and $\mathcal{F}_i$ are restrictions of $\mathcal{G}$ and $\mathcal{F}$ to T_i respectively. Then there exists automorphisms $\phi_i : (T_i, \overline{v}_i) \to (T_i, \overline{v}_i)$ for each $i = 1, \ldots, \ell$ such that $\mathcal{G}_i(t) = \mathcal{F}_i(\phi_i(t))$ for almost all $t \in T_i$. Let $f : T \to A$ be a Nash equilibrium of the atomless game $\mathcal{F}$ and define $g : T \to A$ such that $g(t) = f(\phi_i(t))$ for all $t \in T_i$. Then g is a Nash equilibrium of the atomless game $\mathcal{G}$ and every Nash equilibrium of $\mathcal{G}$ is obtained in this way.*

---

[85] See Mas-Colell and Vives (1993, Paragraph 2), who begin with the statement, "We have argued elsewhere [see Mas-Colell (1984a, 1984b), Vives (1988)] that strategic games with a continuum of players constitute a useful technique in economics."

[86] For a definition, see Billingsley (1968) or Parthasarathy (1967), and in the context of economic application, Hildenbrand (1974).

[87] Thus, as in Section 3 but now without countability requirements on A, $\mathcal{U}_A^d$ is the space of real-valued continuous functions on $A \times \mathcal{M}(A)^\ell$, endowed with its sup-norm topology and with $\mathcal{B}(\mathcal{U}_A^d)$ its Borel σ-algebra.

[88] For a detailed treatment of the asymptotic theory, see Khan and Sun (1996b, 1999). Note that this theory furnishes approximate results for the large but finite case rather than for an idealized limit setting as in Milgrom and Weber (1981, 1985), Aumann et al. (1983), Khan (1986a), Housman (1987), Pascoa (1988, 1993b). Note also that these results have nothing to say about the rate of convergence problem as in Mas-Colell (1998), Kumar and Satterthwaite (1985), Gresik and Satterthwaite (1989), Satterthwaite and Williams (1989). Also see Rashid's (1983, 1985, 1987) work based on the Shapley–Folkman theorem for games with a finite number of players and a common finite action set.

[89] Note that $\lambda_i^n g_n^{-1}$ is the distribution on A induced by the restriction of g_n on T_i^n and is given for any $i = 1, \ldots, \ell$, by $(1/|T_i^n|) \sum_{s \in T_i^n} \delta_{g_n(s)}$ where for any $a \in A$, δ_a denotes the Dirac measure at a.

10.3. Sketch of proofs

Once we have access to the theory of distribution and integration of a correspondence defined on an atomless Loeb measure space, the proof of Theorem 6 is a straightforward consequence of the basic argument that we trace to Nash. Corollary 10 follows from the nonstandard extension,[90] and Corollary 11 from Proposition 11.

11. Large games with independent idiosyncratic shocks

When von Neumann and Morgenstern referred to the "excellent possibility of applying the laws of statistics and probabilities" to games with a large number of players, they did not have in mind the cancellation of individual (independent) risks through aggregation or diversification. Both "Crusoe" and a participant in a social exchange economy are "given a number of data which are 'dead'; they are the unalterable physical background of the situation [and] even when they are apparently variable, they are really governed by fixed statistical laws. Consequently [these purely statistical phenomena] can be eliminated by the known procedures of the calculus of probabilities". Instead of these individual "uncontrollable factors [that] can be described by statistical assumptions", their primary focus was on "alien" variables that are the "product of other participants' actions and volitions" and which cannot be "obviated by a mere recourse to the devices of the theory of probability".[91]

Recent and ongoing work in economic theory, however, considers economic situations in which a continuum of (albeit identical) participants are exposed to individual chance factors.[92] This literature appeals to a basic intuition underlying the theory of insurance whereby the the classical law of large numbers is used to eliminate independent (idiosyncratic or non-systematic) risks. However, the difficulty in formalizing this intuition in the usual context of a continuum of random variables was pointed out early on by Doob (1937, 1953): the assumption of independence renders the sample function of the underlying stochastic process "too irregular to be useful". What is needed is a suitable analytical framework that renders the twin assumptions of independence and joint measurability compatible with each other. In this section, we discuss the nature of the difficulty, and show how the richer measure-theoretic structure discussed in Section 10, but now on a special product space, offers a viable solution to it.

---

[90] This particular advantage of the nonstandard model, stemming from the simultaneous exploitation of finite and continuous methods, is by now well understood; see Rashid (1987), Anderson (1991), Sun (2000) and their references.

[91] See Section 2.2 titled "Robinson Crusoe" Economy and Social Exchange Economy in von Neumann and Morgenstern (1953, pp. 9–12).

[92] See, for example, the references in Feldman and Gilles (1985), Aiyagari (1994), Sun (1998a, 1999a), and Barut (2000).

11.1. On the joint measurability problem

We couple the space of players' names $(T, \mathcal{T}, \lambda)$ with another probability space $(\Omega, \mathcal{A}, P)$ to represent the sample space. Let $(T \times \Omega, \mathcal{T} \otimes \mathcal{A}, \lambda \otimes P)$ be the usual product probability space. We shall refer to the functions $f(t, \cdot)$ on Ω and $f(\cdot, \omega)$ on T respectively as the random variables and the sample functions. The random variables f_t are said to be almost surely pairwise independent if for λ-almost all $t_1 \in T$, $f(t_1, \cdot)$ is independent of $f(t_2, \cdot)$ for λ-almost all $t_2 \in T$. The following result illustrates the joint measurability problem in a particularly transparent way.[93]

PROPOSITION 12. *Let f be a jointly measurable function from the usual product space $(T \times \Omega, \mathcal{T} \otimes \mathcal{A}, \lambda \otimes P)$ to a complete separable metric space X. If the random variables f_t are almost surely pairwise independent, then for λ-almost all $t \in T$, f_t is a constant random variable.*

11.2. Law of large numbers for continua

The difficulty that is brought to light in Proposition 12 is overcome in the context of a Loeb product space $(\overline{T} \times \overline{\Omega}, L(\overline{\mathcal{T} \otimes \bar{\mathcal{A}}}), L(\bar{\lambda} \otimes \overline{P}))$ constructed as a "standardization" of the hyperfinite internal probability space $(\overline{T} \times \overline{\Omega}, \overline{\mathcal{T} \otimes \bar{\mathcal{A}}}, \bar{\lambda} \otimes \overline{P})$. This special product space extends the usual product space $(\overline{T} \times \overline{\Omega}, L(\overline{\mathcal{T}}) \otimes L(\bar{\mathcal{A}}), L(\bar{\lambda}) \otimes L(\overline{P}))$, retains the crucial Fubini property, and is rich enough to allow many hyperfinite collections of random variables with any variety of distributions.[94] For simplicity, a measurable function on $(\overline{T} \times \overline{\Omega}, L(\overline{\mathcal{T} \otimes \bar{\mathcal{A}}}), L(\bar{\lambda} \otimes \overline{P}))$ will also be called a process. We also assume that both $L(\bar{\lambda})$ and $L(\overline{P})$ are atomless. We can now present a version of the law of large numbers for a hyperfinite continuum of random variables, and refer the reader to Sun (1996a, 1998a) for details and complementary results.[95]

PROPOSITION 13. *Let f be a process[96] from $(\overline{T} \times \overline{\Omega}, L(\overline{\mathcal{T} \otimes \bar{\mathcal{A}}}), L(\bar{\lambda} \otimes \overline{P}))$ to a complete separable metric space X. Assume that the random variables $f(t, \cdot)$ are almost surely pairwise independent. Then for $L(\overline{P})$-almost all $\omega \in \overline{\Omega}$, the distribution μ_ω on X induced by the sample function $f(\cdot, \omega)$ on $\overline{T}$ is equal to the distribution μ on X induced by f viewed as a random variable on $\overline{T} \times \overline{\Omega}$.*

Since solutions to individual maximization problems are not unique, the need for a law of large numbers for set-valued processes arises in a natural way, where a set-valued

---

[93] See Sun (1998b, Proposition 1) for details of a proof based on Fubini's theorem and the uniqueness of Radon–Nikodym derivatives. Earlier versions of this result are shown in Doob (1953). For additional complementary results, as well as an expositional overview, see Sun (1999b) and Hammond and Sun (2000).

[94] The first result is in Anderson (1976), the second in Keisler (1977) and the third in Sun (1998a).

[95] For an expositional overview, see Sun (2000).

[96] In the context at hand, this implies by a version of Fubini's theorem due to Keisler (1977, 1984, 1988), the measurability of $f(\cdot, \omega)$ for $L(\overline{P})$-almost all $\omega \in \overline{\Omega}$, and of $f(t, \cdot)$ for $L(\bar{\lambda})$-almost all $t \in \overline{T}$.

process is a closed-valued measurable correspondence from a product space to X. Such a law can be derived from Proposition 13.[97] What is more difficult is the following result showing that possible widespread correlations can be removed from selections of a set-valued process.

PROPOSITION 14. *Let F be a set-valued process from $\overline{T} \times \overline{\Omega}$ to a complete separable metric space X. Assume that $F(t, \cdot)$ are almost surely pairwise independent. Let g be a selection of F with distribution μ. Then there is another selection f of F such that the distribution of f is μ, and $f(t, \cdot)$ are almost surely pairwise independent.*

11.3. A result

We can now use this substantial machinery to present a result for a large non-anonymous game in which individual agents are exposed to idiosyncratic risks, but in equilibrium the societal responses do not depend on a particular sample realization, and each agent is justified in ignoring other players' risks.

THEOREM 7. *Let $\mathcal{G}: \overline{T} \times \overline{\Omega} \to \mathcal{U}_A^d$ be a game with individual uncertainty, i.e., the random payoffs $\mathcal{G}(t, \cdot)$ are almost surely pairwise independent.[98] Then there is a process $f: \overline{T} \times \overline{\Omega} \to A$ such that f is an equilibrium of the game $\mathcal{G}$, the random strategies $f(t, \cdot)$ are almost surely pairwise independent, and for $L(\overline{P})$-almost all $\omega \in \overline{\Omega}$, $f(\cdot, \omega)$ is an equilibrium of the game $\mathcal{G}(\cdot, \omega)$ with constant societal distribution $L(\overline{\lambda} \otimes \overline{P}) f^{-1}$.*

The basic idea for the proof of Theorem 7 is straightforward. On an appeal to Theorem 6, we know that there exists a measurable function $g : T \times \Omega \to A$ such that

$$\mathcal{G}_{(t,\omega)}\big(g(t,\omega), L(\overline{\lambda} \otimes \overline{P})g^{-1}\big) \geqslant \mathcal{G}_{(t,\omega)}\big(a, L(\overline{\lambda} \otimes \overline{P})g^{-1}\big) \quad \text{for all } a \in A.$$

Now finish the proof by applying Proposition 14 to the set-valued process

$$(t, \omega) \to F(t, \omega) = \operatorname*{Arg\,Max}_{a \in A} \mathcal{G}_{(t,\omega)}\big(a, L(\overline{\lambda} \otimes \overline{P})g^{-1}\big).$$

12. Other formulations and extensions

The formulation of a large game that we have explored in this chapter hinges crucially on the formalization of players' characteristics by a metric space $\mathcal{U}$, along with its Borel

---

[97] See Sun (1999a, 1999b). The same papers contain the details pertaining to Proposition 14 below. Note that the extension of the classical law of large numbers to correspondences is well understood; see Arrow and Radner (1979), Artstein and Hart (1981) and their references.
[98] Here $\mathcal{U}_A^d$ is the space in Section 3.1 with $\ell = 1$.

σ-algebra. A large non-anonymous game is then simply a measurable function with such a range space, and its anonymous counterpart the induced measure on it. Thus a random variable and its law constitute basic elements in the relevant vocabulary, and one can exploit this observation to incorporate a variety of additional aspects into the basic framework.

Two formulations deserve special mention. The first of these considers[99] "very large" or "thick" games based on a space of characteristics given $\mathcal{U} \times [0, 1]$, where $\mathcal{U}$ is interpreted as the space of "types", and there is a continuum of each type. In such a setting, one can be explicit about the cardinality of each type through the so-called "mass revealing" function, and questions concerning symmetric equilibria in which player t of type u, where $(u, t) \in \mathcal{U} \times [0, 1]$ plays an action independent of t, can be investigated.[100] The advantage of this formulation is that a correspondence defined on such a space of characteristics has a distribution with well-behaved properties of the kind we saw in Propositions 5, 6, and 10, even when the range space is compact metric, and without having to go into the Loeb setting.[101] The second formulation concerns dynamics, and a setting in which a game is constituted by an infinite sequence of distributions over a space of actions and states.[102] By using the "value function" and other techniques from stochastic dynamic programming, questions relating to the existence and stationarity of equilibria can be investigated in such a setting.

In terms of elaborations on the basic framework, there has been a substantial amount of work that investigates games with a richer space of characteristics: different actions sets,[103] upper semicontinuous payoffs[104] and more generally, non-ordered preferences,[105] uncertainty, imperfect information, differing beliefs and imperfect observability represent a selective list.[106] Issues of existence and of continuity of equilibria

---

[99] This formulation is due to Green (1984), with further work by Housman (1987, 1988) and Pascoa (1993a, 1997). As we have seen in Sections 4.2, 8.2, and 10.3, Mas-Colell's (1984a) formulation of an anonymous game dispenses with the unit interval and focusses solely on $\mathcal{U}$.

[100] See Pascoa (1993b) for an approximate theorem in this context.

[101] The driving force behind this is the fact that any probability measure on the space $\mathcal{U} \times A$ can be represented as the induced distribution of a function $(i, f) : \mathcal{U} \times [0, 1] \to \mathcal{U} \times A$; see Hart, Hildenbrand and Kohlberg (1974, pp. 164, 165) and also Aumann (1964), Housman (1987), Rustichini (1993) and Khan and Sun (1994) for related arguments. Indeed Housman (1987) uses this fact as the basis for a definition of large games that are "thick".

[102] This formulation is due to Jovanovic and Rosenthal (1988) with further work by Massó and Rosenthal (1989), Bergin and Bernhardt (1992), Massó (1993), and Chakrabarti (2000).

[103] This is an important consideration especially in the context of applications to the existence of competitive equilibrium, and constitutes so-called "generalized games" or "abstract economies"; see Housman (1987, 1988), Khan and Sun (1990), Tian (1992a, 1992b), and Toussaint (1984); the last one is set in the context of an arbitrary index of players.

[104] See Balder (1991), Khan (1989) and Rath (1996b).

[105] For action sets in a finite-dimensional space, see Khan and Vohra (1984). For general action sets, see Khan and Papageorgiou (1987a, 1987b), Khan and Sun (1990), Kim, Prikry and Yannelis (1989), and Yannelis (1987). For difficulties of interpretation in the context of non-anonymous games, see Balder (2000).

[106] For the last four aspects, see Balder (1991, 1996), Chakrabarti and Khan (1991), Khan and Rustichini (1991, 1993), Kim and Yannelis (1997), and Shieh (1992).

have both been investigated, and this work has led both to interesting technical issues, and to changes in viewpoint. For example, in the presence of non-ordered preferences, one is led to the problem of choosing selections of functions of two variables, continuous in one and measurable in the other.[107] Even without non-ordered
preferences but with weakened continuity hypotheses on payoffs, it is fruitful to regard a player as a continuous function from societal responses to a space of preferences, and this leads to deeper topological questions.[108] Similar changes in viewpoint
have proved useful in the case of uncertainty and imperfect information where the
space of characteristics is enlarged to include a sub-σ-algebra for each player, which
leads to the formulation of a measurable structure on the space of sub-σ-algebras of
a given σ-algebra.[109] Refinements of the concept of Nash equilibrium are also considered in the setting of large games.[110] Yet another example is a focus on the space
$L(T, \mathcal{M}(A))$ of so-called Young measures, and the use of this as a unifying framework.[111] Whereas it is incontestable that this work has incorporated a variety of additional considerations into one comprehensive framework, we leave it to the reader's
judgement to determine what new game-theoretic phenomena have been brought to
light.

13. A catalogue of applications

Any discussion of applications must begin with Cournot (1838).[112] As noted by
Roberts, he was the "first to make the role of large numbers explicit in his analysis
[of] quantity-setting noncollusive oligopoly, [and his] model yields prices in excess of
marginal costs with this divergence decreasing asymptotically to zero as the number
of firms increases".[113] In addition to numerical negligibility, Cournot also raised the
question of product diversity.

> The effects of competition have reached their limit when each of the partial
> productions D_k is *inappreciable*, not only with respect to the total production

---

[107] These are the so-called "Caratheodory selections"; see Artstein and Prikry (1987), Khan and Papageorgiou
(1987a), Kim, Prikry and Yannelis (1987, 1988), and Yannelis (1991b).

[108] See Khan (1989) and Khan and Sun (1990), where the space of upper semicontinuous functions on the
action set are topologized by the hypotopology of Dolecki, Salonetti and Wets, and the space of players by
the compact-open topology.

[109] See Khan and Rustichini (1991), Chakrabarti and Khan (1991), and Balder and Rustichini (1994).

[110] See Rath (1994, 1998).

[111] See Balder (1995a, 1995b) who makes an "externality mapping" an integral part of the definition of the
game; also Valadier (1993), and Balder (1999a, 1999b).

[112] Indeed, what we have been referring to as Nash equilibria are also termed Cournot–Nash equilibria, Nash–
Cournot equilibria, or simply Cournot equilibria: Dubey et al. (1980), Allen (1994), and Novshek and Sonnenschein (1983) are respective examples.

[113] The quote is taken from the two different entries listed under Roberts (1987).

$D = F(p)$ but also with respect to the derivative $F'(p)$ so that the partial production could be subtracted from D without any appreciable variation resulting in the price of the commodity. This hypothesis is the one which is realized, in social economy, for a multitude of products, and, among them, for the most important products. It introduces a great simplification into the calculations.[114]

A particularly vigorous aspect of the research program initiated by Cournot concerns what Mas-Colell (1982b) has termed the *Cournotian foundation of perfect competition.* "Under the Negligibility Hypothesis the Cournot quantity-setting equilibrium is identical with the Walras price-taking equilibrium. Every seller has an infinitesimal upper bound on how much it can sell. Therefore no seller can influence aggregate production; hence no seller can influence the price system." This non-cooperative justification of the price-taking assumption leads naturally to the question of the minimum efficient scale of production and, more generally, to the optimality properties of Nash equilibria.[115] Indeed, once one recognizes that "price-taking behavior is, in a mass market, the natural consequence of message-taking behavior", it is a short step to the result of Hammond (1979a, 1979b) that a competitive equilibrium is "incentive compatible" in a continuum economy in the sense that no single agent can influence the terms of trade by deviations from "straightforward behavior".[116] We are thus led to the imposition of game-theoretic structures and thereby to the literature on implementation and mechanism design for the allocation of resources in a large economy.[117]

A canonical example of an anonymous mechanism is of course the Walrasian equilibrium, and the relevance of a Nash equilibrium for the existence of a Walrasian equilibrium is well understood.[118] Indeed, once one considers Walrasian equilibria in en-

---

[114] See the first paragraph of Chapter VIII titled Of Unlimited Competition in Cournot (1838).

[115] The quote is from Mas-Colell (1998); also see Jaynes et al. (1978), Mas-Colell (1980, 1983, 1984b), Novshek (1980), Novshek and Sonnenschein (1978, 1980, 1983), Postlewaite and Schmeidler (1978), and Roberts and Postlewaite (1976). Questions of the rate of convergence are explored in Gresik and Satterthwaite (1989) and Satterthwaite and Williams (1989). Price-setting competition is explored in Allen (1994), Allen and Hellwig (1986a, 1986b, 1989), Gabszewicz and Vial (1972), and Roberts (1980).

[116] An intuitive suggestion of such a result was given by Hurwicz (1972). See also Dubey et al. (1980, p. 340).

[117] See Myerson (1994). The literature stemming from Vickrey (1945) and Mirrlees (1971) is voluminous but the basic references for environments with a continuum of agents are Dasgupta and Hammond (1980), Dubey et al. (1980), Hammond (1979a, 1979b), Champsaur and Laroque (1982), Mas-Colell and Vives (1993), and Hervés-Beloso et al. (1999). Makowski and Ostroy (1987, 1992) discuss the importance of large numbers in the context of a specific mechanism; see Roberts (1976), Roberts and Postlewaite (1976), and Cordoba and Hammond (1998) for some asymptotic results. For an overview, see Sonnenschein (1998) and his references.

[118] The basic insight is of course that of Arrow and Debreu (1954) and has engendered the literature on abstract economies; see Debreu (1952), Shafer and Sonnenschein (1975). Khan and Vohra (1984), Balder and Yannelis (1991), Yannelis (1987) consider this in an environment with a continuum of agents; see Balder (2000) for a critique.

vironments with widespread externalities,[119] player interaction is no longer limited to dependence only on the mean message, and we lose the so-called *aggregation axiom* which has played a crucial role in the convergence and implementation literature. Without it, we are led to monopolistic competition.[120] As observed by Samuelson (1967), it was "Chamberlin's contention that proliferation of numbers alone need not lead to perfection of competition. [It] does not mean that the limit as $N \to \infty$ is zero-market imperfection. Instead the limit may be at an irreducible positive degree of imperfection. It is an increase, in some sense, of the *density* of numbers that everybody recognizes to be the relevant situation that needs to be appraised". There is now a substantial literature that attempts a formulation of Chamberlin–Robinson imperfect competition as a large game.[121]

The inadequacy of numerical negligibility as a sole desideratum for optimality is most transparent in the case of information, and here one has to formalize what one means by the proposition that "when agents are informationally small, the inefficiency due to asymmetric information is small".[122] However, with a viable analytical framework for discussing both types of negligibility, there is an emerging literature on the microfoundations of macroeconomics.[123] This includes economics of search,[124] and of the foundation of the firm.[125] Indeed, the importance of economic environments with many agents is ubiquitous, and limited only by the imagination and technical competence of the investigator. A selective list would certainly include applications to the stock market,[126] stochastic rationing mechanisms,[127] design of tax and subsidy schemes for

---

[119] As in McKenzie (1955), Chipman (1970), Shafer and Sonnenschein (1976), Kaneko and Wooders (1986, 1989, 1994), and Hammond (1995, 1998, 1999). For the specific form of externality stemming from public goods in the sense of Samuelson (1954), see Khan and Vohra (1985), Sonnenschein (1998), and their references.

[120] See Dubey et al. (1980, p. 346), in particular, for a defence of this axiom. However, as they observe, there are environments where "the concept of mean ... is ... irrelevant to the equilibrium problem. It may not even be defined".

[121] See Hart (1979a, 1980, 1982, 1984) and Mas-Colell (1984b) for an asymptotic setting, and Pascoa (1988, 1993a) for the continuum.

[122] We have already seen the importance of diffuse information in Section 4.1. One can alternatively consider the set-up of a large exchange economy, as in Gul and Postlewaite (1992), or of bargaining, as in Mailath and Postlewaite (1990), or of an industry, as in Rob (1987). More recent investigations of Levine and Pesendorfer (1995), and Fudenberg, Levine and Pesendorfer (1998) are also relevant here.

[123] See Jovanovic and Rosenthal (1988) for a sketch of such a research program.

[124] See Lucas and Prescott (1974), Rauh (1997, 2001), Seccia and Rauh (2001), and McMillan and Rothschild (1994).

[125] As Hart (1979b) observes, "If each firm is negligible relative to the aggregate economy – a firm's shareholders will want the firm to maximize the (net) market value of its shares." Also see Kelsey and Milne (1996), Kihlstrom and Laffont (1979), and Lucas (1978). It is of interest that Kelsey and Milne rely on results established for an abstract economy in Khan and Vohra (1984).

[126] Constantinides and Rosenthal (1984), Hart (1979a), Haller (1988b), Kihlstrom and Laffont (1982), Nti (1988).

[127] See Gale (1979), Weinrich (1984) and their references.

attaining second-best equilibria,[128] voting models,[129] evolutionary game theory,[130] and the economics of fashion and "social influences".[131]

14. Conclusion

There are two distinct motivations for the study of non-cooperative games with many players. On the one hand, they delineate qualitative changes in the resolution of particular problems,[132] and on the other, they allow formulation of questions that have resisted formal treatment. Thus, in opposition to finite games, Nash equilibria of large games without widespread externalities are efficient, their pure strategy versions exist, and in models with enough institutional features, these games are well suited for studying incentives in a variety of industrial structures, particularly monopolistic competition. However, when we take stock and evaluate where we currently stand relative to the work of Cournot and Nash, it may be worthwhile to keep in mind the distinction between technical and conceptual advances emphasized by von Neumann and Morgenstern. On a technical level, we certainly have a more sophisticated understanding of the importance of the precise form of player-interdependence and of different kinds of measurable structures, and these become especially important with infinite action sets, widespread externalities, and independent shocks. In this case, distribution and integrals assume separate and distinct identities, and the importance of geometry and Lyapunov's theorem, already explicit in Dvoretsky, Wald and Wolfowitz, and brought to the fore by Aumann, shades into probability and the law of large numbers for a continuum of random variables. On a conceptual level, however, the extent to which the mathematical theory of large games currently offers a canonical model and an array of uniform techniques for handling a variety of important applications, remains to be seen in the future.

References

Aiyagari, R. (1994), "Uninsured idiosyncratic risk and aggregate saving", Quarterly Journal of Economics 109:659–684.

---

[128] In addition to the relevant references in Footnote 117, see Guesnerie (1981, 1995) and Dierker and Haller (1990); also Mas-Colell and Vives (1993).

[129] We began this chapter with a statement of Shapiro and Shapley on the relevance, in principle, of large games to the study of voting behavior in large societies; for recent work, see Chapters 29 and 30 in this Handbook, and the relevant section in Jovanovic and Rosenthal (1988). Also see Khan (1998) for the relevance of large games to questions of a more interdisciplinary nature.

[130] See Wieczorek (1996) and Hammerstein and Selten (1994).

[131] See Karni and Schmeidler (1990), and for an application to restaurant pricing, Karni and Levin (1994).

[132] This motivation was already stressed in another context by Aumann (1964), where he is less than enthusiastic about generalizations where such changes do not obtain. Referring to the examination of Edgeworth's conjecture in the context of infinite commodities, he writes "This would serve no useful purpose. Our result holds for any number of commodities, many or few, so there is nothing to be gained by considering only the case of 'many' commodities."

Allen, B. (1994), "Randomization and the limit points of monopolistic competition", Journal of Mathematical Economics 23:205–218.

Allen, B., and M. Hellwig (1986a), "Price-setting firms and the oligopolistic foundations of perfect competition", American Economic Review 76:387–392.

Allen, B., and M. Hellwig (1986b), "Bertrand–Edgeworth oligopoly in large markets", Review of Economic Studies 53:175–204.

Allen, B., and M. Hellwig (1989), "The approximation of competitive equilibria by Bertrand–Edgeworth equilibria in large markets", Journal of Mathematical Economics 16:387–392.

Anderson, R.M. (1976), "A nonstandard representation for Brownian motion and Ito integration", Israel Journal of Mathematics 25:15–46.

Anderson, R.M. (1991), "Non-standard analysis with applications to economics", in: W. Hildenbrand and H. Sonnenschein, eds., Handbook of Mathematical Economics, Vol. 4 (North-Holland, New York).

Anderson, R.M. (1994), "The core in perfectly competitive economies", in: R.J. Aumann and S. Hart, eds., Handbook of Game Theory, Vol. 1 (North-Holland, Amsterdam) Chapter 14, 413–458.

Arrow, K.J., and G. Debreu (1954), "Existence of an equilibrium for a competitive economy", Econometrica 22:265–290.

Arrow, K.J., and R. Radner (1979), "Allocation of resources in large teams", Econometrica 47:361–392.

Artstein, Z. (1979), "A note on Fatou's lemma in several dimensions", Journal of Mathematical Economics 6:277–282.

Artstein, Z. (1983), "Distributions of random sets and random selections", Israel Journal of Mathematics 46:313–324.

Artstein, Z., and S. Hart (1981), "Law of large numbers for random sets and allocation processes", Mathematics of Operations Research 6:485–492.

Artstein, Z., and K. Prikry (1987), "Caratheodory selections and the Scorza–Dragoni property", Journal of Mathematical Analysis and Applications 127:540–547.

Ash, R.B. (1972), Real Analysis and Probability (Academic Press, New York).

Aumann, R.J. (1964), "Markets with a continuum of traders", Econometrica 32:39–50.

Aumann, R.J. (1965), "Integrals of set-valued functions", Journal of Mathematical Analysis and Applications 12:1–12.

Aumann, R.J. (1966), "Existence of competitive equilibria in markets with a continuum of traders", Econometrica 34:1–17.

Aumann, R.J. (1976), "An elementary proof that integration preserves uppersemicontinuity", Journal of Mathematical Economics 3:15–18.

Aumann, R.J. (1987), "Game theory", in: J. Eatwell, M. Milgate and P.K. Newman, eds., The New Palgrave (The Macmillan Press, London) 460–482.

Aumann, R.J. (1997), "Rationality and bounded rationality", Games and Economic Behavior 21:2–14.

Aumann, R.J., and A. Heifetz (2002), "Incomplete information", in: R.J. Aumann and S. Hart, eds., Handbook of Game Theory, Vol. 3 (North-Holland, Amsterdam) Chapter 43, 1665–1686.

Aumann, R.J., Y. Katznelson, R. Radner, R.W. Rosenthal and B. Weiss (1983), "Approximate purification of mixed strategies", Mathematics of Operations Research 8:327–341.

Balder, E.J. (1991), "On Cournot–Nash equilibrium distributions for games with differential information and discontinuous payoffs", Economic Theory 1:339–354.

Balder, E.J. (1995a), "A unifying approach to Nash equilibria", International Journal of Game Theory 24:79–94.

Balder, E.J. (1995b), "Lectures on Young measures", Cahiers de Mathématiques de la Décision, CEREMADE, to appear.

Balder, E.J. (1996), "Comments on the existence of equilibrium distributions", Journal of Mathematical Economics 25:307–323.

Balder, E.J. (1997), "Remarks on Nash equilibria for games with additively coupled payoffs", Economic Theory 9:161–167.

Balder, E.J. (1999a), "On the existence of Cournot–Nash equilibria in continuum games", Journal of Mathematical Economics 32:207–223.

Balder, E.J. (1999b), "Young measure techniques for existence of Cournot–Nash–Walras equilibria", in: M.H. Wooders, ed., Topics in Mathematical Economics and Game Theory: Essays in honor of R.J. Aumann (American Mathematical Society, Providence) 31–39.

Balder, E.J. (2000), "Incompatibility of usual conditions for equilibrium existence in continuum economies without ordered preferences", Journal of Economic Theory 93:110–117.

Balder, E.J., and A. Rustichini (1994), "An equilibrium result for games with private information and infinitely many players", Journal of Economic Theory 62:385–393.

Balder, E.J., and N.C. Yannelis (1991), "Equilibria in random and Bayesian games with a continuum of players", in: M. Ali Khan and Yannelis (1991).

Barut, Y. (2000), "Existence and computation of a stationary equilibrium in a heterogeneous agent, dynamic economic model with incomplete markets", Rice University, mimeo.

Başçi, E., and M. Sertel (1996), "Prakash and Sertel's theory of non-cooperative equilibria in social systems – twenty years later", Journal of Mathematical Economics 25:1–18.

Berge, C. (1959), Topological Spaces (Oliver and Boyd, London).

Bergin, J., and D. Bernhardt (1992), "Anonymous sequential games with aggregate uncertainty", Journal of Mathematical Economics 21:543–562.

Bewley, T.F. (1973), "The equality of the core and the set of equilibria in economies with infinitely many commodities and a continuum of agents", International Economic Review 14:383–393.

Billingsley, P. (1968), Convergence of Probability Measures (Wiley, New York).

Bollobás, B., and N.Th. Varopoulos (1974), "Representation of systems of measurable sets", Mathematical Proceedings of the Cambridge Philosophical Society 78:323–325.

Brams, S.J. (1994), "Voting procedures", in: R.J. Aumann and S. Hart, eds., Handbook of Game Theory, Vol. 2 (North-Holland, Amsterdam) Chapter 30, 1055–1090.

Castaing, C., and M. Valadier (1977), Convex Analysis and Measurable Multifunctions. Lecture Notes in Mathematics, Vol. 580 (Springer-Verlag, Berlin).

Chakrabarti, S.K. (2000), "Pure strategy Markov equilibria in stochastic games with a continuum of players", mimeo, IUPUI.

Chakrabarti, S.K., and M.A. Khan (1991), "Equilibria of large games with imperfect observability", in: C.D. Aliprantis, K.C. Border and W.A.J. Luxemburg, eds., Positive Operators, Reisz Spaces and Economics (Springer-Verlag, Berlin).

Chamberlin, E. (1933), The Theory of Monopolistic Competition (Harvard University Press, Cambridge).

Chamberlin, E. (1937), "Monopolistic or imperfect competition", Quarterly Journal of Economics LI:557.

Champsaur, P., and G. Laroque (1982), "A note on incentives in large economies", Review of Economic Studies 49:627–635.

Chipman, J.S. (1970), "External economies of scale and competitive equilibrium", Quarterly Journal of Economics 84:347–385.

Constantinides, G.M., and R.W. Rosenthal (1984), "Strategic analysis of the competitive exercise of certain financial options", Journal of Economic Theory 32:128–138.

Cordoba, J.M., and P.J. Hammond (1998), "Asymptotically Walrasian strategy-proof exchange", Mathematical Social Sciences 36:185–212.

Cournot, A. (1838), Recherches sur les Principles Mathématiques de la Théorie des Richesses (Librairie des Sciences Politiques et Sociales, Paris). Also translation by N. Bacon (1897) (Macmillan, New York).

Cutland, N. (1983), "Nonstandard measure theory and its applications", Bulletin of the London Mathematical Society 15:529–589.

Cutland, N., ed. (1988), Nonstandard Analysis and its Applications (Cambridge University Press, New York).

Dasgupta, P., and P. Hammond (1980), "Fully progressive taxation", Journal of Public Economics 13:141–154.

Debreu, G. (1952), "A social equilibrium existence theorem", Proceedings of the National Academy of Sciences U.S.A. 38:386–393.

Debreu, G. (1967), "Integration of correspondences", in: Proceedings of the Fifth Berkeley Symposium on Mathematical Statistics and Probability, Part 1, 2:351–372 (University of California Press, Berkeley).

Dierker, E., and H. Haller (1990), "Tax systems and direct mechanisms in large finite economies", Journal of Economics 52:99–116.

Diestel, J. (1984), Sequences and Series in Banach Spaces (Springer-Verlag, Berlin).

Diestel, J., and J.J. Uhl, Jr. (1977), Vector Measures (American Mathematical Society, Providence).

Dincleanu, N. (1973), "Linear operations on L^p-spaces", in: D.H. Tucker and H.B. Maynard, eds., Vector and Operator Valued Measures and Applications (Academic Press, New York).

Doob, J.L. (1937), "Stochastic processes depending on a continuous parameter", Transactions of the American Mathematical Society 42:107–140.

Doob, J.L. (1953), Stochastic Processes (Wiley, New York).

Dubey, P., A. Mas-Colell and M. Shubik (1980), "Efficiency properties of strategic market games: an axiomatic approach", Journal of Economic Theory 22:339–362.

Dubey, P., and L.S. Shapley (1977), "Noncooperative exchange with a continuum of traders", Cowles Foundation Discussion Paper No. 447.

Dubey, P., and L.S. Shapley (1994), "Noncooperative general exchange with a continuum of traders: two models", Journal of Mathematical Economics 23:253–293.

Dvoretsky, A., A. Wald and J. Wolfowitz (1950), "Elimination of randomization in certain problems of statistics and of the theory of games", Proceedings of the National Academy of Sciences U.S.A. 36:256–260.

Dvoretsky, A., A. Wald and J. Wolfowitz (1951a), "Elimination of randomization in certain statistical decision problems in certain statistical decision procedures and zero-sum two-person games", Annals of Mathematical Statistics 22:1–21.

Dvoretsky, A., A. Wald and J. Wolfowitz (1951b), "Relations among certain ranges of vector measures", Pacific Journal of Mathematics 1:59–74.

Edgeworth, F.Y. (1881), Mathematical Psychics (Kegan Paul, London).

Fan, K. (1952), "Fixed points and minimax theorems in locally convex linear spaces", Proceedings of the National Academy of Sciences U.S.A. 38:121–126.

Fan, K. (1966), "Applications of a theorem concerning sets with convex sections", Mathematische Annalen 163:189–203.

Feldman, M., and C. Gilles (1985), "An expository note on individual risk without aggregate uncertainty", Journal of Economic Theory 35:26–32.

Fudenberg, D., D.K. Levine and W. Pesendorfer (1998), "When are non-anonymous players negligible", Journal of Economic Theory 79:46–71.

Fudenberg, D., and J. Tirole (1991), Game Theory (MIT Press, Cambridge).

Gabszewicz, J.J., and J.P. Vial (1972), "Oligopoly á la Cournot in general equilibrium analysis", Journal of Economic Theory 4:381–400.

Gale, D. (1979), "Large economies with trading uncertainty", Review of Economic Studies 46:319–338. A correction in (1980) 48:363–364.

Glicksberg, I. (1952), "A further generalization of Kakutani's fixed point theorem with application to Nash equilibrium points", Proceedings of the American Mathematical Society 3:170–174.

Green, E.J. (1984), "Continuum and finite player non-cooperative models of competition", Econometrica 52:975–993.

Gresik, T., and M. Satterthwaite (1989), "The rate at which a simple market converges to efficiency as the number of traders increases: an asymptotic result for optimal trading mechanisms", Journal of Economic Theory 48:304–322.

Guesnerie, R. (1981), "On taxation and incentives: further reflections on the limits of redistribution", Discussion Paper No. 89, Sonderforschungsberech 21, University of Bonn.

Guesnerie, R. (1995), A Contribution to the Pure Theory of Taxation (Cambridge University Press, Cambridge).

Gul, F., and A. Postlewaite (1992), "Asymptotic efficiency in large exchange economies with asymmetric information", Econometrica 60:1273–1292.

Haller, H. (1988a), "Equilibria of a class of nonatomic generalized games", mimeo (Virginia Polytechnic Institute).

Haller, H. (1988b), "Competition in a stock market with small firms", Journal of Economics 3:243–261.

Halmos, P.R., and H.E. Vaughan (1950),"The marriage problem", American Journal of Mathematics 72:214–215.

Hammerstein, P., and R. Selten (1994), "Game theory and evolutionary biology", in: R.J. Aumann and S. Hart, eds., Handbook of Game Theory, Vol. 2 (North-Holland, Amsterdam) Chapter 28, 929–994.

Hammond, P.J. (1979a), "Symposium on incentive compatibility: introduction", Review of Economic Studies 46:181–184.

Hammond, P.J. (1979b), "Straightforward individual incentive compatibility in large economies", Review of Economic Studies 46:263–282.

Hammond, P.J. (1987), "Markets as constraints: multilateral incentive compatibility", Review of Economic Studies 54:399–412.

Hammond, P.J. (1995), "Four characterizations of constrained Pareto efficiency in continuum economies with widespread externalities", Japanese Economic Review 46:103–124.

Hammond, P.J. (1998), "Rights, free exchange, and widespread externalities", in: J.-F. Laslier, M. Fleurbaey, N. Gravel and A. Trannoy, eds., Freedom in Economics: New Perspectives in Normative Analysis (Routledge, London and New York) Chapter 11, 139–157.

Hammond, P.J. (1999), "History as a widespread externality in some Arrow–Debreu market games", in: G. Chichilnisky, ed., Markets, Information and Uncertainty: Essays in Economic Theory in Honor of Kenneth J. Arrow (Cambridge University Press, Cambridge) Chapter 16, 328–361.

Hammond, P.J., and Y.N. Sun (2000), "Joint measurability and the one-way Fubini property for a continuum of independent variables", Working Paper No. 00-008, Department of Economics, Stanford University.

Harsanyi, J.C. (1967–1968), "Games with incomplete information played by Bayesian players", Parts I to III, Management Science 14:159–183, 14:320–334, 14:486–502.

Harsanyi, J.C. (1973), "Games with randomly disturbed payoffs", International Journal of Game Theory 2:1–23.

Hart, O. (1979a), "Monopolistic competition in a large economy with differentiated commodities", Review of Economic Studies 46:1–30. A correction in (1982) 49:313–314.

Hart, O. (1979b), "On stockholder unanimity in large stock market economies", Econometrica 47:1057–1083.

Hart, O. (1980), "Perfect competition and optimal product differentiation", Journal of Economic Theory 22:279–312.

Hart, O. (1982), "A model of imperfect competition with Keynesian features", Quarterly Journal of Economics 97:109–138.

Hart, O. (1984), "Imperfect competition in general equilibrium: an overview of recent work", in: K.J. Arrow and S. Honkapohja, eds., Frontiers of Economics (Blackwell, Oxford).

Hart, O. (1985a), "Monopolistic competition in the spirit of Chamberlin: a general model", Review of Economic Studies 52:529–546.

Hart, O. (1985b), "Monopolistic competition in the spirit of Chamberlin: special results", Economic Journal 95:889–908.

Hart, S., W. Hildenbrand and E. Kohlberg (1974), "On equilibrium allocations as distributions on the commodity space", Journal of Mathematical Economics 1:159–166.

Hart, S., and E. Kohlberg (1974), "Equally distributed correspondences", Journal of Mathematical Economics 1:167–174.

Hart, S., and D. Schmeidler (1989), "Existence of correlated equilibria", Mathematics of Operations Research 14:18–25.

Hervés-Beloso, C., E. Moreno-Garcia and M.R. Pascoa (1999), "Manipulation-proof equilibrium in atomless economies with commodity differentiation", Economic Theory 14:545–563.

Hiai, F., and H. Umegaki (1977), "Integrals, conditional expectations and martingales of multivalued functions", Journal of Multivariate Analysis 7:167–174.

Hildenbrand, W. (1974), Core and Equilibria of a Large Economy (Princeton University Press, Princeton).

Hildenbrand, W. (1975), "Distributions of agents' characteristics", Journal of Mathematical Economics 2:129–138.

Housman, D. (1987), "Equilibria in nonatomic strategic form games", mimeo (Worcester Polytechnique Institute).

Housman, D. (1988), "Infinite player noncooperative games and the continuity of the Nash equilibrium correspondence", Mathematics of Operations Research 13:488–496.

Hurd, A.E., and P.A. Loeb (1985), An Introduction to Nonstandard Real Analysis (Academic Press, Orlando).

Hurwicz, L. (1972), "On informationally decentralized systems", in: C.B. McGuire and R. Radner, eds., Decision and Organization (North-Holland, Amsterdam).

Jamison, R.E. (1974), "A quick proof for a one-dimensional version of Liapunov's theorem", American Mathematical Monthly 81:507–508.

Jaynes, G., M. Okuno and D. Schmeidler (1978), "Efficiency in an atomless economy with fiat money", International Economic Review 19:149–157.

Jovanovic, B., and R.W. Rosenthal (1988), "Anonymous sequential games", Journal of Mathematical Economics 17:77–87.

Judd, K.L. (1985), "The law of large numbers with a continuum of iid random variables", Journal of Economic Theory 35:19–25.

Kaneko, M., and M.H. Wooders (1986), "The core of a game with a continuum of players and finite coalitions: the model and some results", Mathematical Social Sciences 12:105–137.

Kaneko, M., and M.H. Wooders (1989), "The core of a continuum economy with widespread externalities and finite coalitions: from finite to continuum economies", Journal of Economic Theory 49:135–168.

Kaneko, M., and M.H. Wooders (1994), "Widespread externalities and perfectly competitive markets: examples", in: R.P. Gilles and P.H.M. Ruys, eds., Imperfections and Behavior in Economic Organizations (Kluwer Academic, Boston).

Kannai, Y. (1970), "Continuity properties of the core of a market", Econometrica 38:791–815.

Karni, E., and D. Levin (1994), "Social attributes and strategic equilibrium: a restaurant pricing game", Journal of Political Economy 102:822–840.

Karni, E., and D. Schmeidler (1990), "Fixed preferences and changing tastes", American Economic Review 80:262–267.

Keisler, H.J. (1977), "Hyperfinite model theory", in: R.O. Gandy and J.M.E. Hyland, eds., Logic Colloquium 76 (North-Holland, Amsterdam).

Keisler, H.J. (1984), "An infinitesimal approach to stochastic analysis", Memoirs of the American Mathematical Society 48:1–184.

Keisler, H.J. (1988), "Infinitesimals in probability theory", in: Cutland (1988).

Kelsey, D., and F. Milne (1996), "The existence of equilibrium in incomplete markets and the objective function of the firm", Journal of Mathematical Economics 25:229–246.

Khan, M. Ali (1984), "An alternative proof of Diestel's theorem", Glasgow Math. Journal 25:45–46.

Khan, M. Ali (1985a), "On the integration of set-valued mappings in a non-reflexive Banach space", Simon Stevin 59:257–267.

Khan, M. Ali (1985b), "Equilibrium points of non-atomic games over a non-reflexive Banach space", Journal of Approximation Theory 43:370–376.

Khan, M. Ali (1986a), "Equilibrium points of nonatomic games over a Banach space", Transactions of the American Mathematical Society 293:737–749.

Khan, M. Ali (1986b), "On the extensions of the Cournot–Nash theorem" in: C.D. Aliprantis, O.L. Burkinshaw and N. Rothman, eds., Advances in Equilibrium Theory (Springer-Verlag, Berlin).

Khan, M. Ali (1989), "On Cournot–Nash equilibrium distributions for games with a nonmetrizable action space and upper semicontinuous payoffs", Transactions of the American Mathematical Society 315:127–146.

Khan, M. Ali (1998), "A result on large anonymous games: an elementary and self-contained exposition", in: J. Moore, R. Reizman and J. Melvin, eds., Trade, Theory, and Econometrics: Essays in Honour of John S. Chipman (Routledge, New York) 206–236.

Khan, M. Ali, and M. Majumdar (1986), "Weak sequential convergence in $L_1(\mu, X)$ and an approximate version of Fatou's lemma", Journal of Mathematical Analysis and Applications 114:569–573.

Khan, M. Ali, and N. Papageorgiou (1987a), "On Cournot–Nash equilibrium in generalized qualitative games with a continuum of players", Nonlinear Analysis 11:741–755.

Khan, M. Ali, and N. Papageorgiou (1987b), "On Cournot–Nash equilibrium in generalized qualitative games with an atomless measure space of players", Proceedings of the American Mathematical Society 100:505–510.

Khan, M. Ali, K.P. Rath and Y.N. Sun (1997a), "On the existence of pure strategy equilibria in games with a continuum of players", Journal of Economic Theory 76:13–46.

Khan, M. Ali, K.P. Rath and Y.N. Sun (1997b), "Pure-strategy Nash equilibrium points in large non-anonymous games", in: B. Dutta, T. Parthasarathy, J. Potters, T.E.S. Raghavan, D. Ray and A. Sen, eds., Game Theoretical Applications to Economics and Operations Research (Kluwer Academic, Amsterdam).

Khan, M. Ali, K.P. Rath and Y.N. Sun (1999), "On private information games without pure strategy equilibria", Journal of Mathematical Economics 31:341–359.

Khan, M. Ali, and A. Rustichini (1991), "Differential information and a reformulation of Cournot–Nash equilibria", in: M.A. Théra and J.B. Baillon, eds., Fixed Point Theory and Applications, Pitman Research Notes in Mathematics, No. 252 (Longman, London).

Khan, M. Ali, and A. Rustichini (1993), "On Cournot–Nash equilibrium distributions for games with uncertainty and imperfect information", Journal of Mathematical Economics 22:35–59.

Khan, M. Ali, and Y.N. Sun (1987), "On symmetric Cournot–Nash equilibrium distributions in a finite-action, atomless game", in: M. Ali Khan and Yannelis (1991).

Khan, M. Ali, and Y.N. Sun (1990), "On a reformulation of Cournot–Nash equilibria", Journal of Mathematical Analysis and Applications 146:442–460.

Khan, M. Ali, and Y.N. Sun (1994), "On large games with finite actions: a synthetic treatment", Mita Journal of Economics 87:73–84 (in Japanese). English original in: T. Maruyama and W. Takahashi, eds., Nonlinear and Convex Analysis in Economic Theory (Springer-Verlag, Berlin).

Khan, M. Ali, and Y.N. Sun (1995a), "Extremal structures and symmetric equilibria with countable actions", Journal of Mathematical Economics 24:239–248.

Khan, M. Ali, and Y.N. Sun (1995b), "Pure strategies in games with private information", Journal of Mathematical Economics 24:633–653.

Khan, M. Ali, and Y.N. Sun (1995c), "The marriage lemma and large anonymous games with countable actions", Mathematical Proceedings of the Cambridge Philosophical Society 117:385–387.

Khan, M. Ali, and Y.N. Sun (1996a), "On integrals of set-valued functions with a countable range", Mathematics of Operations Research 21:946–954.

Khan, M. Ali, and Y.N. Sun (1996b), "Nonatomic games on Loeb spaces", Proceedings of the National Academy of Sciences U.S.A. 93:15518–15521.

Khan, M. Ali, and Y.N. Sun (1997), "On Loeb measure spaces and their significance for non-cooperative game theory", in: M. Albers, B. Hu and J. Rosenthal, eds., Current and Future Directions in Applied Mathematics (Birkhäuser, Berlin) 183–218.

Khan, M. Ali, and Y.N. Sun (1999), "Non-cooperative games on hyperfinite Loeb spaces", Journal of Mathematical Economics 31:455–492.

Khan, M. Ali, and R. Vohra (1984), "Equilibrium in abstract economies without ordered preferences and with a measure space of agents", Journal of Mathematical Economics 13:133–142.

Khan, M. Ali, and R. Vohra (1985), "On the existence of Lindahl equilibria in economies with a measure space of non-transitive consumers", Journal of Economic Theory 36:319–332.

Khan, M. Ali, and N.C. Yannelis, eds. (1991), Equilibrium Theory in Infinite Dimensional Spaces (Springer-Verlag, Berlin).

Kihlstrom, R.E., and J.J. Laffont (1979), "A general equilibrium entrepreneural theory of firm foundation based on risk aversion", Journal of Political Economy 87:719–748.

Kihlstrom, R.E., and J.J. Laffont (1982), "A competitive entrepreneural model of a stock market", in: J.J. McCall, ed., The Economics of Information and Uncertainty (The University of Chicago Press, Chicago).

Kim, S.H. (1997), "Continuous Nash equilibria", Journal of Mathematical Economics 28:69–84.

Kim, T., K. Prikry and N.C. Yannelis (1987), "Carathéodory-type selections and random fixed point theorems", Journal of Mathematical Analysis and Applications 122:393–407.

Kim, T., K. Prikry and N.C. Yannelis (1988), "On a Carathéodory-type selection theorem", Journal of Mathematical Analysis and Applications 135:664–670.

Kim, T., K. Prikry and N.C. Yannelis (1989), "Equilibrium in abstract economies with a measure space and with an infinite dimensional strategy space", Journal of Approximation Theory 56:256–266.

Kim, T., and N.C. Yannelis (1997), "Existence of equilibrium in Bayesian games with infinitely many players", Journal of Economic Theory 77:330–353.

Kuhn, H.W., and A.W. Tucker, eds. (1950), Contributions to the Theory of Games, Vol. 1, Annals of Mathematics Studies, No. 24 (Princeton University Press, Princeton).

Kumar, K., and M. Satterthwaite (1985), "Monopolistic competition, aggregation of competitive information and the amount of product differentiation", Journal of Economic Theory 37:32–54.

Levine, D.K., and W. Pesendorfer (1995), "When are agents negligible?", American Economic Review 85:1160–1170.

Lindstrom, T. (1988), "An invitation to nonstandard analysis", in: Cutland (1988).

Loeb, P.A. (1975), "Conversion from nonstandard to standard measure spaces and applications in probability theory", Transactions of the American Mathematical Society 211:113–122.

Loeb, P.A., and S. Rashid (1987), "Lyapunov's theorem", in: J. Eatwell, M. Milgate and P.K. Newman, eds., The New Palgrave (The Macmillan Press, London).

Loéve, M. (1977), Probability Theory, Vols. I & II, 4th edn. (Springer-Verlag, New York).

Lucas, R.E., Jr. (1978), "On the size distribution of business firms", Bell Journal of Economics 9:508–523.

Lucas, R.E., Jr., and E.C. Prescott (1974), "Equilibrium search and unemployment", Journal of Economic Theory 7:188–209.

Lyapunov, A. (1940), "Sur les fonctions-vecteurs complètements additives", Bulletin Academie Sciences URSS Sér. Math. 4:465–478.

Ma, T.W. (1969), "On sets with convex sections", Journal of Mathematical Analysis and Applications 27:413–416.

Mailath, G., and A. Postlewaite (1990), "Asymmetric information bargaining problems with many agents", Review of Economic Studies 57:351–368.

Makowski, L., and J. Ostroy (1987), "The Vickrey–Clarke–Groves mechanism and perfect competition", Journal of Economic Theory 42:244–261.

Makowski, L., and J. Ostroy (1992), "The Vickrey–Clarke–Groves mechanism in continuum economies: characterization and existence", Journal of Mathematical Economics 21:1–35.

Mas-Colell, A. (1978), "Efficiency of non-cooperative equilibrium in economies with a continuum of traders", in: M. Ali Khan and N.C. Yannelis (1991).

Mas-Colell, A. (1980), "Non-cooperative approaches to the theory of perfect competition", Journal of Economic Theory 22:121–135.

Mas-Colell, A. (1982a), "Perfect competition and the core", Review of Economic Studies 69:15–30.

Mas-Colell, A. (1982b), "Cournotian foundations of Walrasian equilibrium theory", in: W. Hildenbrand, ed., Advances in Economic Theory (Cambridge University Press, Cambridge).

Mas-Colell, A. (1983), "Walrasian equilibria as limits of nonooperative equilibria, Part I: Mixed strategies", Journal of Economic Theory 30:153–170.

Mas-Colell, A. (1984a), "On a theorem of Schmeidler", Journal of Mathematical Economics 13:206–210.

Mas-Colell, A. (1984b), "The profit motive in the theory of monopolistic competition", Journal of Economics, Supplementum 4:111–116.

Mas-Colell, A. (1987), "On the second welfare theorem for anonymous net trades in exchange economies with many agents", in: T. Groves, R. Radner and S. Reiter, eds., Information, Incentives and Economic Mechanisms (University of Minnesota Press, Minneapolis).

Mas-Colell, A. (1998), "On the theory of perfect competition", in: D.P. Jacobs, E. Kalai, M.I. Kamien, eds., Frontiers of Research in Economic Theory: The Nancy L. Schwartz Memorial Lectures, 1983–1997 (Cambridge University Press, Cambridge).

Mas-Colell, A., and X. Vives (1993), "Implementation in economies with a continuum of agents", Review of Economic Studies 60:613–630.

Massó, J. (1993), "Undiscounted equilibrium payoffs of repeated games with a continuum of players", Journal of Mathematical Economics 22:243–264.

Massó, J., and R.W. Rosenthal (1989), "More on the 'anti-folk theorem'", Journal of Mathematical Economics 18:281–290.

McKenzie, L.W. (1955), "Competitive equilibrium with dependent consumer preferences", in: H.A. Antosiewicz, ed., Proceedings of the Second Symposium in Linear Programming, 277–294.

McMillan, J., and M. Rothschild (1994), "Search", in: R.J. Aumann and S. Hart, eds., Handbook of Game Theory, Vol. 2 (North-Holland, Amsterdam) Chapter 27, 905–928.

Milgrom, P.R., and R.J. Weber (1981), "Topologies on information and strategies in games with incomplete information", in: O. Moeschlin and D. Pallashke, eds., Game Theory and Mathematical Economics (North-Holland, Amsterdam).

Milgrom, P.R., and R.J. Weber (1985), "Distributional strategies for games with incomplete information", Mathematics of Operations Research 10:619–632.

Milnor, J.W., and L.S. Shapley (1978), "Values of large games, II: oceanic games", Mathematics of Operations Research 3:290–307.

Mirrlees, J.A. (1971), "An exploration in the theory of optimum income taxation", Review of Economics Studies 38:175–208.

Myerson, R.B. (1994), "Communication, correlated equilibria and incentive compatibility", in: R.J. Aumann and S. Hart, eds., Handbook of Game Theory, Vol. 2 (North-Holland, Amsterdam) Chapter 24, 827–848.

Nash, J.F. (1950), "Equilibrium points in N-person games", Proceedings of the National Academy of Sciences U.S.A. 36:48–49.

Nash, J.F. (1951), "Noncooperative games", Annals of Mathematics 54:286–295.

Novshek, W. (1980), "Cournot equilibrium and free entry", Review of Economic Studies 47:473–486.

Novshek, W., and H. Sonnenschein (1978), "Cournot and Walras equilibrium", Journal of Economic Theory 19:223–266.

Novshek, W., and H. Sonnenschein (1980), "Small efficient scale as a foundation for Walrasian equilibrium", Journal of Economic Theory 22:243–256.

Novshek, W., and H. Sonnenschein (1983), "Walrasian equilibria as limits of noncooperative equilibria, Part II: Pure strategies", Journal of Economic Theory 30:171–187.

Nti, K. (1988), "Capital asset prices in an oligopolistic market", Journal of Economics 48:35–57.

O'Neill, B. (1994), "Game theory models of peace and war", in: R.J. Aumann and S. Hart, eds., Handbook of Game Theory, Vol. 2 (North-Holland, Amsterdam) Chapter 29, 995–1054.

Palfrey, T., and S. Srivastava (1986), "Private information in large economies", Journal of Economic Theory 39:34–58.

Papageorgiou, N.S. (1985), "Representation of set-valued operators", Transactions of the American Mathematical Society 292:557–572.

Papageorgiou, N.S. (1987), "Contributions to the theory of set-valued functions and set-valued measures", Transactions of the American Mathematical Society 304:245–265.

Papageorgiou, N.S. (1990), "On a paper by Khan and Majumdar", Journal of Mathematical Analysis and Applications 150:574–578.

Parthasarathy, K.R. (1967), Probability Measures on Metric Spaces (Academic Press, New York).

Pascoa, M.R. (1988), "Approximate purification of Nash equilibrium in nonatomic games", CARESS Working Paper No. 88-21, University of Pennsylvania.

Pascoa, M.R. (1993a), "Noncooperative equilibrium and Chamberlinian monopolistic competition", Journal of Economic Theory 60:335–353.

Pascoa, M.R. (1993b), "Approximate equilibrium in pure strategies for non-atomic games", Journal of Mathematical Economics 22:223–241.

Pascoa, M.R. (1997), "Monopolistic competition and non-neighboring goods", Economic Theory 9:129–142.

Pascoa, M.R. (1998), "Nash equilibrium and the law of large numbers", International Journal of Game Theory 27:83–92.

Peleg, B. (1969), "Equilibrium points for games with infinitely many players", Journal of the London Mathematical Society 44:292–294.

Postlewaite, A.W., and D. Schmeidler (1978), "Approximate efficiency of non-Walrasian Nash equilibria", Econometrica 40:127–135.

Prakash, P., and M.R. Sertel (1996), "Existence of non-cooperative equilibria in social systems", Journal of Mathematical Economics 25:19–32.

Radner, R., and R.W. Rosenthal (1982), "Private information and pure-strategy equilibria", Mathematics of Operations Research 7:401–409.

Rashid, S. (1983), "Equilibrium points of non-atomic games: asymptotic results", Economics Letters 12:7–10.

Rashid, S. (1985), "The approximate purification of mixed strategies with finite observation sets", Economics Letters 19:133–135.

Rashid, S. (1987), Economies with Many Agents (The Johns Hopkins University Press, Baltimore).

Rashid, S. (1992), "A direct proof of purification for Schmeidler's theorem", Applied Mathematics Letters 5:23–24.

Rath, K.P. (1992), "A direct proof of the existence of pure strategy equilibria in games with a continuum of players", Economic Theory 2:427–433.

Rath, K.P. (1994), "Some refinements of Nash equilibria of large games", Games and Economic Behavior 7:92–103.

Rath, K.P. (1995), "Representation of finite action large games", International Journal of Game Theory 24:23–35.

Rath, K.P. (1996a), "On the representation of sets in finite measure spaces", Journal of Mathematical Analysis and Applications 200:506–510.

Rath, K.P. (1996b), "Existence and upper hemicontinuity of equilibrium distributions of anonymous games with discontinuous payoffs", Journal of Mathematical Economics 26:305–324.

Rath, K.P. (1998), "Perfect and proper equilibria of large games", Games and Economic Behavior 22:331–342.

Rath, K.P., Y.N. Sun and S. Yamashige (1995), "The nonexistence of symmetric equilibria in anonymous games with compact action spaces", Journal of Mathematical Economics 24:331–346.

Rauh, M.T. (1997), "A model of temporary search market equilibrium", Journal of Economic Theory 77:128–153.

Rauh, M.T. (2001), "Heterogeneous beliefs, price dispersion and welfare-improving price controls", Economic Theory 18:577–603.

Rob, R. (1987), "Entry, fixed costs and the aggregation of private information", Review of Economic Studies 54:619–630.

Roberts, D.J. (1976), "The incentives for correct revelation of preferences and the number of consumers", Journal of Public Economics 6:359–374.

Roberts, D.J. (1987), "Large economies" and "Perfectly and imperfectly competitive markets", in: J. Eatwell, M. Milgate and P.K. Newman, eds., The New Palgrave (The Macmillan Press, London).

Roberts, D.J., and A. Postlewaite (1976), "The incentives for price-taking behavior in large exchange economies", Econometrica 44:115–127.

Roberts, K. (1980), "The limit points of monopolistic competition", Journal of Economic Theory 22:256–278.

Roberts, K. (1984), "The theoretical limits to redistribution", Review of Economic Studies 51:177–186.

Robinson, J. (1933), Economics of Imperfect Competition (Macmillan, London).

Royden, H.L. (1968), Real Analysis (Macmillan, New York).

Rudin, W. (1973), Functional Analysis (McGraw-Hill, New York).

Rudin, W. (1974), Real and Complex Analysis (McGraw-Hill, New York).

Rustichini, A. (1989), "A counterexample and an exact version of Fatou's lemma in infinite dimensions", Archiv der Mathematik 52:357–362.

Rustichini, A. (1993), "Mixing on function spaces", Economic Theory 3:183–190.

Rustichini, A., M.A. Satterthwaite and S.R. Williams (1994), "Convergence to efficiency in a simple market with incomplete information", Econometrica 62:1041–1063.

Rustichini, A., and N.C. Yannelis (1991), "What is perfect competition?", in: M. Ali Khan and N.C. Yannelis (1991).

Samuelson, P.A. (1954), "The pure theory of public expenditure", Review of Economics and Statistics 36:387–389.

Samuelson, P.A. (1967), "The monopolistic competition revolution", in: R.E. Kuenne, ed., Monopolistic Competition Theory: Studies in Impact (Wiley, New York).

Satterthwaite, M., and S. Williams (1989), "The rate of convergence to efficiency in the buyer's bid double auction as the market becomes large", Review of Economic Studies 56:477–498.

Schmeidler, D. (1973), "Equilibrium points of non-atomic games", Journal of Statistical Physics 7:295–300.

Schwarz, G. (1994), "Game theory and statistics", in: R.J. Aumann and S. Hart, eds., Handbook of Game Theory, Vol. 2 (North-Holland, Amsterdam) Chapter 21, 769–779.

Seccia, G., and M. Rauh (2001), "Mean-variance analysis in temporary equilibrium", Research in Economics, 55:331–345.

Shafer, W., and H. Sonnenschein (1975), "Equilibrium in abstract economies without ordered preferences", Journal of Mathematical Economics 2:345–348.

Shafer, W., and H. Sonnenschein (1976), "Equilibrium with externalities, commodity taxation and lump-sum transfers preferences", International Economic Review 17:601–611.

Shapiro, N.Z., and L.S. Shapley (1978), "Values of large games, I: A limit theorem", Mathematics of Operations Research 3:1–9.

Shapley, L. (1976), "Noncooperative general exchange", in: S.A.Y. Lin, ed., Theory and Measurement of Economic Externalities (Academic Press, New York).

Shieh, J. (1992), "Static and dynamic large games with application to warrant pricing", Unpublished Ph.D. dissertation, The Johns Hopkins University.

Sonnenschein, H. (1998), "The economics of incentives: an introductory account", in: D.P. Jacobs, E. Kalai and M.I. Kamien, eds., Frontiers of Research in Economic Theory: The Nancy L. Schwartz Memorial Lectures, 1983–1997 (Cambridge University Press, Cambridge).

Sun, Y.N. (1996a), "Hyperfinite law of large numbers", Bulletin of Symbolic Logic 2:189–198.

Sun, Y.N. (1996b), "Distributional properties of correspondences on Loeb spaces", Journal of Functional Analysis 139:68–93.

Sun, Y.N. (1997), "Integration of correspondences on Loeb spaces", Transactions of the American Mathematical Society 349:129–153.

Sun, Y.N. (1998a), "A theory of hyperfinite processes: the complete removal of individual uncertainty via exact LLN", Journal of Mathematical Economics 29:419–503.

Sun, Y.N. (1998b), "The almost equivalence of pairwise and mutual independence and the duality with exchangeability", Probability Theory and Related Fields 112:425–456.

Sun, Y.N. (1999a), "The complete removal of individual uncertainty: multiple optimal choices and random economies", Economic Theory 14:507–544.

Sun, Y.N. (1999b), "On the sample measurability problem in modeling individual risks", Research Paper No. 99-25, Center for Financial Engineering, National University of Singapore.

Sun, Y.N. (2000), "Nonstandard analysis in mathematical economics", in: P.A. Loeb and M. Wolff, eds., Mathematics and its Applications, Vol. 510 (Kluwer Academic, Dordrecht) 261–305.

Talagrand, M. (1984), "Pettis integral and measure theory", Memoirs of the American Mathematical Society 51:307.

Tian, G. (1992a), "Equilibrium in abstract economies with a non-compact infinite dimensional strategy space, an infinite number of agents and without ordered preferences", Economics Letters 33:203–206.

Tian, G. (1992b), "On the existence of equilibria in generalized games", International Journal of Game Theory 20:247–254.

Toussaint, S. (1984), "On the existence of equilibrium with infinitely many commodities", Journal of Economic Theory 33:98–115.

Uhl, J.J., Jr. (1969), "The range of a vector-valued measure", Proceedings of the American Mathematical Society 23:158–163.

Uhl, J.J., Jr. (1980), "Pettis's measurability theorem", Contemporary Mathematics 2:135–144.

Valadier, M. (1993), "Young measures", in: A. Cellina, ed., Methods of Nonconvex Analysis, Lecture Notes in Mathematics, Vol. 1446 (Springer-Verlag, Berlin).

Vickrey, W. (1945), "Measuring marginal utility by reactions to risk", Econometrica 13:319–333.

Vives, X. (1988), "Aggregation of information in large Cournot markets", Econometrica 56:851–876.

von Neumann, J. (1932), "Einige Sätze über messbare Abbildungen", Annals of Mathematics 33:574–586.

von Neumann, J., and O. Morgenstern (1953), Theory of Games and Economic Behavior, 3rd edn. (Princeton University Press, Princeton).

Wald, A., and J. Wolfowitz (1951), "Two methods of randomization in statistics and the theory of games", Annals of Mathematics 53:581–586.

Weinrich, G. (1984), "On the theory of effective demand under stochastic rationing", Journal of Economic Theory 34:95–115.

Wieczorek, A. (1996), "Elementary large games and an application to economies with many agents", mimeo (Polish Academy of Sciences).

Yamashige, S. (1992), "Large games and large economies with incomplete information", Unpublished Ph.D. dissertation, The Johns Hopkins University.

Yannelis, N.C. (1987), "Equilibria in noncooperative models of competition", Journal of Economic Theory 41:96–111.

Yannelis, N.C. (1991a), "Integration of Banach-valued correspondences", in: M. Ali Khan and N.C. Yannelis (1991).

Yannelis, N.C. (1991b), "Set-valued functions of two variables", in: M. Ali Khan and N.C. Yannelis (1991).

Chapter 47

STOCHASTIC GAMES*

JEAN-FRANÇOIS MERTENS

CORE, Louvain-la-Neuve, Belgium

Contents

---

*Much of this chapter is an abridged version of Chapter VII in Mertens, Sorin and Zamir (1994). For developments in the theory of stochastic games since 1994, see the following chapter by Vieille.

Handbook of Game Theory, Volume 3, Edited by R.J. Aumann and S. Hart

Abstract

Stochastic games model repeated play with symmetric information. We analyze their value in the zero-sum case, and approach the study of their equilibria in the non-zero-sum case.

Keywords

games, non-cooperative, stochastic, value, dynamic programming

JEL classification: C72, C73

1. Introduction

1.1. Definition

A stochastic game is a repeated game where the state of nature may change from stage to stage, according to a lottery which, just like the current payoff, depends on current state and actions. Before choosing their actions, players are fully informed about the whole past as well as the current state. For a simple example, the reader may now turn to Section 1.2.

For most of the basic results, finiteness assumptions are not necessary. So we will work mostly in the following framework:

- the state space is a measurable space $(\Omega, \mathcal{A})$;
- the action space of player $i \in I$ (I finite) is a measurable space $(S^i, \mathcal{S}^i)$ (with $S = \prod_{i \in I} S^i$);
- P is a transition probability from $\Omega \times S$ to Ω, hence for $A \in \mathcal{A}$, $P(A|\omega, s)$ is the probability that tomorrow's state belongs to A given today's state ω and actions s;
- $g(\omega, s)$ is the (jointly measurable) current payoff vector.

The game starts at some initial state ω_1; at stage n, players are first told ω_n, then they simultaneously choose the vector $s_n = (s_n^i)_{i \in I}$ of actions, then s_n is told to the players and $g^i(\omega_n, s_n)$ is paid to each player $i \in I$, finally ω_{n+1} is selected by nature according to $P(\cdot|\omega_n, s_n)$.

In the discounted game Γ_λ with discount factor λ ($0 < \lambda \leqslant 1$), each player evaluates an income stream $(g_1, g_2, g_3, \ldots)$ as $\lambda \sum_{t=0}^{\infty} (1 - \lambda)^t g_{t+1}$. Similarly, with $\bar{g}_n = \frac{1}{n} \sum_1^n g_t$ one obtains the n-stage game Γ_n, and the "undiscounted game" Γ_∞ corresponds to the "payoff function $g_\infty = \lim \bar{g}_n$" – with appropriate precautions due to the possible non-existence of the limit.

A (behavioral) strategy for player i is thus a transition probability from histories of the form $(\omega_1, s_1, \ldots, s_{n-1}, \omega_n)$ to S^i. The strategy is Markov if, at each stage n, it depends only on the current state ω_n. It is stationary if it is a stationary – i.e., time-invariant – function of the infinite past (say continuous in the product of the discrete topologies, and accompanied by a fictitious history before time zero).[1] Hence it is stationary Markov if it is at every stage the same function of the current state. Observe that in the above we rather consider Γ_λ to be the whole family of games indexed by ω_1 – a point of view that is typical in this context.

1.2. An illustration

Consider the following simple two-player zero-sum example. There are two states: $\omega = 1$ and $\omega = 2$. The payoff matrices are

---

[1] The fictitious history is needed since, to express stationarity, we need a set of possible past histories which is independent of the current date – so, in order not to introduce unrelated Markov-type restrictions, one has to take this to be the set of all infinite past histories. The continuity requirement serves then to express that the influence of the distant past washes out – so, in particular, after some time in the game, the influence of the fictitious past becomes negligible.

$$\begin{pmatrix} 10 & -1 \\ -1 & 10 \end{pmatrix} \quad \text{in state } \omega = 1, \quad \text{and} \quad \begin{pmatrix} 0 & 6 \\ 3 & 0 \end{pmatrix} \quad \text{in state } \omega = 2.$$

The transition probabilities are

$$\begin{pmatrix} (1,0) & (\frac{1}{2},\frac{1}{2}) \\ (\frac{1}{2},\frac{1}{2}) & (1,0) \end{pmatrix} \quad \text{in state } \omega = 1, \quad \text{and} \quad \begin{pmatrix} (0,1) & (0,1) \\ (0,1) & (0,1) \end{pmatrix} \quad \text{in state } \omega = 2,$$

where the first coordinate is the probability that the state at the next period is $\omega = 1$, and the second is the probability that it is $\omega = 2$. For example, when $\omega = 1$, if the "Row" player chooses "Top" and the "Column" player chooses "Right", then "Column" gets from "Row" an immediate payoff of 1, and the next state will be $\omega = 1$ or $\omega = 2$ with equal probabilities of $1/2$ each. Note that $\omega = 2$ is an *absorbing state*: once reached, the state will never change.

Consider the λ-discounted game Γ_λ, for $\lambda = 1/2$. Let v and u denote the values starting at state $\omega = 1$ and 2, respectively. Heuristically, we have the following recursion equations (remember that the payoff in any cell is the average of today's payoff and the future payoff starting tomorrow):

$$v = \text{val}\begin{pmatrix} \frac{1}{2} \cdot 10 + \frac{1}{2} \cdot (1 \cdot v + 0 \cdot u) & \frac{1}{2} \cdot (-1) + \frac{1}{2} \cdot \left(\frac{1}{2} \cdot v + \frac{1}{2} \cdot u\right) \\ \frac{1}{2} \cdot (-1) + \frac{1}{2} \cdot \left(\frac{1}{2} \cdot v + \frac{1}{2} \cdot u\right) & \frac{1}{2} \cdot 10 + \frac{1}{2} \cdot (1 \cdot v + 0 \cdot u) \end{pmatrix},$$

$$u = \text{val}\begin{pmatrix} \frac{1}{2} \cdot 0 + \frac{1}{2} \cdot u & \frac{1}{2} \cdot 6 + \frac{1}{2} \cdot u \\ \frac{1}{2} \cdot 3 + \frac{1}{2} \cdot u & \frac{1}{2} \cdot 0 + \frac{1}{2} \cdot u \end{pmatrix},$$

where val denotes the minmax value of a matrix. We have

$$u = \frac{1}{2} \cdot u + \frac{1}{2} \cdot \text{val}\begin{pmatrix} 0 & 6 \\ 3 & 0 \end{pmatrix} = \frac{1}{2} \cdot u + \frac{1}{2} \cdot 2,$$

or $u = 2$. Substituting yields

$$v = \text{val}\begin{pmatrix} 5 + \frac{1}{2} \cdot v & \frac{1}{4} \cdot v \\ \frac{1}{4} \cdot v & 5 + \frac{1}{2} \cdot v \end{pmatrix} = \frac{5}{2} + \frac{3}{8} \cdot v,$$

so $v = 4$. Optimal strategies (which are stationary) are obtained from the above matrices, after substituting $v = 4$ and $u = 2$: "Row" should play $(\frac{1}{2}, \frac{1}{2})$ at $\omega = 1$ and $(\frac{1}{3}, \frac{2}{3})$ at $\omega = 2$, and "Column" should play $(\frac{1}{2}, \frac{1}{2})$ at $\omega = 1$ and $(\frac{2}{3}, \frac{1}{3})$ at $\omega = 2$. Shapley (1953) shows that these "heuristic" computations do indeed yield the solution; note that the recursion equations generate a "contraction" (of norm $1 - \lambda$) for the vector of values (v, u).

1.3. Relaxations

For many results, the complete information assumption made is not really necessary: for instance, when we prove there is an equilibrium in Markov strategies, this implies the same result remains true assuming only that the players are at least informed of the current state. Similarly, for the existence of a value, it suffices that the players are informed of states and payoffs – actions are not necessary.

However, if even payoffs are not known, the situation is completely different. For lack of space, we can only refer the reader to the beautiful article of Coulomb (1992) concerning this issue. In that article, he considers an arbitrary game with absorbing states, where the players get no signals whatsoever on each other's moves, and shows that the inf sup and the sup inf of the undiscounted game are equal to those computed in stationary Markov strategies.

Also, to use a stochastic game model, it is not really necessary for players to be informed about the whole past: it suffices that they are informed of anything any other player has learned (symmetric information rather than complete information). Indeed, one can always postpone for as long as possible the moves of nature in the tree, letting at each stage nature do only that part of her randomization of which players have to be currently informed, and postponing the remaining.

For example, a repeated game with incomplete information, with finite state and action sets, where at each stage all players receive the same information (thus including the I-tuple of actions), becomes in this way a stochastic game [Forges (1982)]; it has finite state space if the signals are non-random [Kohlberg and Zamir (1974)]. An example is treated in detail in Mertens (1982), which leads to Blackwell and Ferguson's (1968) Big Match – cf. Section 2 below.

1.4. Applications and motivation

The above discussion indicates that essentially any strategic economic model, with some form of stationarity, and with symmetric information, will correspond to a stochastic game, by an appropriate choice of state variables. This includes: strategic "rational expectations" models – e.g., every period a strategic market game reopens where agents can trade their current holdings (this is the state variable) of commodities and assets – [recall, e.g., Shubik and Whitt (1973)]; models where the state includes current stocks of capital or of natural resources (fish stocks) [e.g., Amir (1996), Levhari and Mirman (1980), Sobel (1982)]; and other models [e.g., Leininger (1986)].

The number of applications of stochastic games in different fields is truly enormous and there is no way even just to mention them here. The interested reader might want to have a look at some of the following (random) sample: Charnes and Schroeder (1967), Dutta and Sundaram (1993), Filar (1985), Winston (1978) and Winston and Cabot (1984).

Also, several aspects of stochastic games are insufficiently covered in this chapter – for example, algorithmic aspects; for these, the reader might profitably consult other surveys, like Parthasarathy and Stern (1976), Parthasarathy and Raghavan (1981) and Raghavan and Filar (1991).

At a more methodological level, we will show in the last section of this chapter how stochastic games (with large state and action sets, with perfect information and without moves of nature) can also be viewed as normal forms for arbitrary repeated games, even with asymmetric information, information lags, etc.

1.5. Organization of the chapter

Section 2 presents the "Big Match", as a simple introductory problem. Section 3 deals with the basic existence results in the discounted case – first zero-sum, then non-zero-sum. Section 4 analyzes the corresponding asymptotic results, when the discount rate tends to zero. Section 5 gives the existence theorem of the value for the undiscounted game, and its implications in dynamic programming and similar particular cases. Section 6 covers a number of recent results concerning equilibria of the undiscounted game; and the subject of Section 7 was mentioned above.

2. The Big Match

The game consists of repeated play of "matching pennies", until the first stage where player I plays "Heads": that stage's payoff is then repeated for all future stages. So to model it as a stochastic game, we need three states:

- In state 0 players play "matching pennies", with $\left(\begin{smallmatrix} 1 & -1 \\ -1 & 1 \end{smallmatrix}\right)$ as current payoff matrix (as usual, I is the row-player, and the maximizer).

  The transitions go back to state 0 in the bottom row, while from Top Left and Top Right they lead respectively to states 1 and 2.

- In states 1 and 2, the transitions remain at the same state, and the payoffs are respectively 1 and -1.

Such a game, where all states but one are absorbing, is called a "game with absorbing states", and can be described by giving just the payoff matrix of the single non-absorbing state, together with marks (say, stars) in some boxes to denote that those lead to an absorbing state.[2]

---

[2] Indeed, in the absorbing states one can assume that players play optimally, hence get the value of that game stage after stage; so the only thing that matters in a box leading to absorbing states is the expectation (under the transition probability) of the values of the absorbing states to which one is led; the current payoff is immaterial as it is going to be met only once, hence will not affect long-run averages: one can, for example, think of it as being equal to the above expectation. This setup has to be modified slightly only if one wants also to include the possibility of strategy vectors in the non-absorbing state leading to absorbing states with probability strictly between zero and one: in that case, the above remains applicable provided one writes in that box the conditional expectation of the absorbing payoffs, given that an absorbing state is going to be reached, and one should then add as additional information the total probability of going to an absorbing state.

Games with absorbing states occur in the literature as a natural stepping stone to the general case, since they amount to investigating what would be the solution in a single state of nature, if one already knew the solution in all the others.[3]

Thus, in this notation, our game becomes

$$\begin{pmatrix} 1^* & -1^* \\ -1 & 1 \end{pmatrix}.$$

Writing the recursion equation for v_n, the value of the n-stage game (in the non-absorbing state, of course), one gets

$$v_{n+1} = \max_{0 \leqslant x \leqslant 1} \min \left[x + \frac{1}{n+1} (nv_n - 1 + x), -x + \frac{1}{n+1} (nv_n + 1 - x) \right],$$

hence $x = \frac{1}{n+2}$ and $v_{n+1} = \frac{n}{n+1} v_n$ – so by induction $v_n = 0$ for all n. Similarly, the recursion equation for v_λ (in terms of itself) yields $v_\lambda = 0 \; \forall \lambda$ – and $x = \frac{\lambda}{\lambda+1}$. So clearly one is led to guess that v_∞ should exist and $= 0$.

Obviously, player II can guarantee himself zero – at the same time in all Γ_n, in all Γ_λ, and in Γ_∞ – by playing $(\frac{1}{2}, \frac{1}{2})$ independently at every stage.

The problem for player I is:

(1) No strategy of player I guarantees him zero: let n_0 be the first time he plays Top with positive probability; by playing Left until n_0, Right at n_0, and $(\frac{1}{2}, \frac{1}{2})$ i.i.d. afterwards, player II can get a negative payoff.

(2) No Markov strategy of player I guarantees him more than -1: if the sum of player I's probabilities of Top is infinite, player II replies by playing always Right; otherwise he plays Right until the remaining sum is $\leqslant \varepsilon$, then switches to Left.

(3) Define a (stationary) strategy with finite memory for player I as an internal finite state space, say K, and a transition probability that selects the next internal state at random as a function of the current internal state and the current move of player II, together with some transition probability from the current internal state to the set of moves (H or T) of player I. Using the ergodic theorem for finite Markov chains, it is easy to show that no such strategy guarantees more than -1.

(4) Blackwell and Ferguson showed that the strategy of player I consisting of playing Top at stage n with probability $1/f_n^2$, where f_n is his current fortune, is ε-optimal for f_0 sufficiently large.

---

[3] It is just a stepping stone: e.g., in the game with two states

$$\begin{pmatrix} 2 & \square \\ -1 & 0 \end{pmatrix} \begin{pmatrix} 0 & 1 \\ \square & -2 \end{pmatrix},$$

where the state remains the same except for the boxed entries, (v, v) for $0 \leqslant v \leqslant 1$ will clearly be a simultaneous solution: if the value of the other state were really v, the value of the current "game with absorbing states" would also be v.

3. The discounted case

The main tool when dealing with the discounted case is the following class of one-shot games:

Given a vector f $(= (f^i)_{i \in I})$ of bounded real-valued measurable functions on $(\Omega, \mathcal{A})$, define the following single-stage game $\Gamma(f)_\omega$, $\omega \in \Omega$, with action sets S^i and (vector) payoff:

$$\phi(f)_\omega(s) = g(\omega, s) + \int f(\widetilde{\omega}) P(d\widetilde{\omega}|\omega, s).$$

3.1. Zero-sum case

Here the basic technique for establishing the existence of a value is based on the "contraction mapping principle".

Let (E, d) be a complete metric space, and $f: E \to E$ such that $d(f(x), f(y)) \leqslant (1 - \varepsilon)d(x, y)$ for all (x, y). Then f has a unique fixed point $\overline{x} \in E$, and for any $x \in E$, the sequence $f^n(x)$ converges to $\overline{x}$.

The idea [going back to Shapley (1953)] behind the use of the contraction principle to prove the existence of a value of the λ-discounted game Γ_λ is that any uncertainty about tomorrow's payoff is reduced by a factor of $1 - \lambda$ when evaluated in today's terms. So if one can solve "today's" game for any given payoffs for the future, one will get a contraction mapping.

The basic requirement is thus that "today's game" have a value for any choice of a "payoff for the future" in an appropriate complete metric space – and yield a payoff in the same metric space. So for any given f in (B, d), a complete metric space of bounded measurable functions on $(\Omega, \mathcal{A})$, with d the uniform distance, our aim is to show that:

(1) for each ω in Ω the game $\Gamma(f)_\omega$ has a value, say $T(f)(\omega)$.

(2) $T(f)$ belongs to B.

(3) The games $\Gamma(f)$ have ε-optimal strategies $\forall \varepsilon > 0$ (i.e., strategies that are ε-optimal for any $\omega \in \Omega$) (this typically follows by a measurable selection theorem).

Then Γ_λ has a value λV_λ. If μ is an ε-optimal strategy in $\Gamma((1 - \lambda)V_\lambda)$, then the corresponding stationary strategy $\overline{\mu}$ is ε-optimal in Γ_λ. V_λ is the solution of $V_\lambda = T[(1 - \lambda)V_\lambda]$.

REMARK. Using V_n for the unnormalized value nv_n of the n-stage game Γ_n, one obtains in the same way $V_{n+1} = T(V_n)$, with $V_0 = 0$.

Two typical illustrations follow – merely to illustrate the method; they do not strive for the utmost generality [see, e.g., Mertens, Sorin and Zamir (1994) for the proofs]:

PROPOSITION 3.1. *The state space is a standard Borel space $(\Omega, \mathcal{A})$, the action sets are compact metric spaces S and T, the payoff function $g(\omega, s, t)$ and the transition* *probability $P(A|\omega, s, t)$ are, for each given $A \in \mathcal{A}$, measurable on $(\Omega \times S \times T)$ and are, for fixed ω, separately continuous in s and in t. Further, g is bounded. Then the discounted game has a value and Borel-measurable optimal strategies which are stationary and Markov.*

In the next proposition, we relax the very strong continuity assumption on the transition probabilities as a function of the actions, at the expense of stronger assumptions on the dependence on the state: instead of using the Banach space of all bounded measurable functions, one uses the space of continuous functions.

PROPOSITION 3.2. *Assume that the state space Ω is metrizable and the action sets S and T are compact metric, and that $g(\omega, s, t)$ and $\int f(\tilde{\omega}) P(\tilde{\omega}|\omega, s, t)$ are, for each continuous f, continuous on $\Omega \times S$ for fixed t and on $\Omega \times T$ for fixed s. Then the discounted game has a value and Borel-measurable optimal strategies which are stationary and Markov.*

Observe that in some sense Proposition 3.2 is much better than Proposition 3.1: at least, if one were to strengthen the separate continuity property in Proposition 3.1 to a joint continuity property, one could immediately construct a separable metrizable topology on Ω such that the assumptions of Proposition 3.2 would also hold (with joint continuity). So Proposition 3.2 "essentially" includes Proposition 3.1; but it allows complete flexibility in the transitions – e.g., next state is a continuous function of current state and actions – while in Proposition 3.1 one is for example constrained to a dominated set of probabilities when a player's action varies.

The "right" form of Proposition 3.1 was obtained in Nowak (1985): topological properties are required only for one player – say player I – so T becomes just a standard Borel space, and continuity as a function of I's action is weakened to upper semi-continuity. In addition, action sets can vary measurably with the state. Clearly, for player II one obtains only ε-optimal strategies, and for both players the measurability property is weakened to universal measurability.

3.2. Non-zero-sum case (finite)

We assume here that the basic spaces (S and Ω) are finite. Recall that a subgame perfect equilibrium is an I-tuple σ such that after any history h, $h = (\omega_1, s_1, \ldots, \omega_n, s_n)$, σ_h is an equilibrium in Γ_λ, where σ_h is defined by $\sigma_h(h') = \sigma(\omega_1, s_1, \ldots, \omega_n, s_n, \omega'_1, s'_1, \ldots, \omega'_m)$ if $h' = (\omega'_1, s'_1, \ldots, \omega'_m)$. Σ^i denotes the set of probability distributions on S^i, and $\Sigma = \prod_{i \in I} \Sigma^i$.

PROPOSITION 3.3. *The discounted game Γ_λ has a subgame perfect equilibrium in stationary Markov strategies.*

PROOF. With $C = \max_{i, \omega, s} |g^i(\omega, s)|$, define a correspondence ψ from $\Sigma^\Omega \times [-C, C]^{I \times \Omega} = Z$ to itself by: $\psi(\sigma, f) = \{(\tau, h) \in Z \mid \text{for each } i \text{ and each } \omega, \tau^i(\cdot|\omega) \text{ is}$ a best reply against σ in $\Gamma((1 - \lambda)f)_\omega$, yielding payoff $h^i(\omega)$ to player $i\}$. ψ is clearly u.s.c. and convex-valued, hence the Kakutani fixed-point theorem gives the existence of fixed points. Those are our equilibria.   □

REMARK. Clearly, the proof of Proposition 3.3 extends immediately to the case of compact action sets, with payoffs and transitions depending continuously on the vector of actions.

3.3. Non-zero-sum case (general)

In fact, one can similarly get the existence of subgame perfect equilibria (i.e., Borel-measurable strategies that form, for every initial state ω, a subgame perfect equilibrium), even under assumptions hardly stronger than in Proposition 3.1 [Mertens and Parthasarathy (1987)]:

- The restriction to a standard Borel space $(\Omega, \mathcal{A})$ is superfluous; an arbitrary measurable space will do.
- One can also allow for compact action sets $S^i(\omega)$ that vary measurably with ω, in the following sense: $S^i(\omega) \subseteq \overline{S}^i$, $(\overline{S}^i, \mathcal{S}^i)$ is a separable and separating measurable space, each subset $S^i(\omega)$ is endowed with some compact topology, the σ-field $\mathcal{S}^i$ is generated by the real-valued measurable functions that have a continuous restriction to each set $S^i(\omega)$, and $\{\omega \mid S^i(\omega) \cap O \neq \emptyset\}$ is measurable for each $O \in \mathcal{S}^i$ whose trace on each set $S^i(\omega)$ is open.
- The uniformly bounded payoff functions $g^i(\omega, s)$ and the transition probability $P(A|\omega, s)$ are measurable (for each $A \in \mathcal{A}$) on the graph of $S(\omega) = \prod_i S^i(\omega)$; and, for each $\omega \in \Omega$, $g^i(\omega, s)$ and $P(\cdot|\omega, s)$ are continuous functions on $S(\omega)$ – in the norm topology for P, i.e., $s_n \to s$ implies $\sup_A |P(A|\omega, s_n) - P(A|\omega, s)|$ converges to zero.

The basic idea of the proof is somewhat reminiscent of what we did in the previous section, i.e., start with a "large", compact-valued measurable correspondence from state space to payoff space, K_0 – e.g., the set of all feasible payoffs. Given a measurable map K to compact subsets, define $[T(K)]_\omega$ as the set of all Nash equilibrium payoffs for the uniform closure of all games $\Gamma((1 - \lambda)f)_\omega$, letting f vary through all measurable selections from K. Prove that $T(K)$ is a measurable map to compact subsets. Get in this way inductively a decreasing sequence of measurable maps to compact subsets $K_n = T^n(K_0)$, with $K_\infty = \bigcap_n K_n$: K_∞ is then also measurable; further $K_{n+1} = T(K_n)$ goes to the limit and yields $K_\infty = T(K_\infty)$. Observe that, at each point $s \in S(\omega)$, the set of payoffs $\Gamma[(1 - \lambda)f]_{\omega,s}$ is already closed, when f varies through all measurable selections from K_∞. Thus one can choose, first for each $(\omega, p) \in K_\infty$, a continuous payoff function $\gamma_{\omega,p}(s)$ on $S(\omega)$ which is a uniform limit of functions $\Gamma[(1 - \lambda)f]_{\omega,s}$ (f measurable selection from K_∞), and a Nash equilibrium $\sigma_{\omega,p}$ of $\gamma_{\omega,p}$ with payoff p; next for each $s \in S(\omega)$, a measurable selection $f_{\omega,p,s}(\widetilde{\omega})$ from K_∞ with $\Gamma[(1 - \lambda)f_{\omega,p,s}]_{\omega,s} = \gamma_{\omega,p}(s)$. Doing all this in a measurable way yields the strategy: if next state is $\widetilde{\omega}$, just repeat the same thing with $f_{\omega,p,s}(\widetilde{\omega})$ instead of p.

One can see the close analogy with the previous method – only in the zero-sum case the contracting aspect – i.e., the minmax theorem – insured that the correspondences K_n would decrease at a rate $1 - \lambda$, hence converge to a single point.

The proof is however technically more complex; so we refer the reader to Mertens and Parthasarathy (1987) for it. There he will also find how the above assumptions can be further relaxed – e.g., the functions g^i do not need to be uniformly bounded, and the discount factor can be allowed to depend a.o. on the player, on the stage, and on the past sequence of states.

In fact, a much simpler proof (one page) is possible under the following additional assumptions [cf. Mertens and Parthasarathy (1991)]:

- the state space $(\Omega, \mathcal{A})$ is separable;
- the action sets $S^i(\omega)$ are finite and independent of ω;
- the transition probabilities are dominated by a single measure μ on $(\Omega, \mathcal{A})$;
- the payoff function is bounded, and a fixed discount rate is used.

In the general case, the strategies obtained were neither Markov nor stationary – they only have the very weak stationarity property that strategies are stationary functions of the current state and the currently expected payoff vector for the future (the "current expectations"). In this particular case, one obtains somewhat closer to Markov: the behavioral strategies can be chosen such as to be a function only of the current and the previous state. And if in addition the transition probability is nonatomic, one can further obtain (cf. loc. cit.) stationarity: the function is the same at every period. We give the proof, since it is so simple and contains in germ already several ideas of the general case.

PROOF. Let $|g^i(\omega, s)| \leqslant C$, and $F_0 = \{ f \in [L_\infty(\mu)]^I \mid \|f_i\|_\infty \leqslant C \ \forall i \in I \}$. For $f \in F_0$, let

$$N_f(\omega) = \left\{ \text{Nash equilibrium payoffs of } G_{f,\omega} = \Gamma \big((1 - \lambda) f \big)_\omega \right\},$$

$\emptyset \neq N_f(\omega) \subseteq [-C, C]^I$. Denote by $\mathcal{N}_f$ the set of all μ-measurable selections from the convex hull of $N_f(\omega)$. $\mathcal{N}_f \neq \emptyset$ using any standard selection theorem. Observe also the correspondence $f \mapsto \mathcal{N}_f$ from F_0 to itself is convex-valued, and weak*-upper semi-continuous: if $f_n \overset{w^*}{\to} f$, then $G_{f_n,\omega} \to G_{f,\omega}$ pointwise, so $\limsup N_{f_n}(\omega) \subseteq N_f(\omega)$. Thus if $\varphi_n \in \mathcal{N}_{f_n}$ converges weak* to φ, then φ is the a.e. limit of a sequence of convex combinations of the φ_n, hence $\varphi \in \mathcal{N}_f$. It follows then from the standard fixed-point theorem that $\mathcal{N}$ has a fixed point: $f_0 \in \mathcal{N}_{f_0}$. That is, (Lyapunov) $\int f_0(\widetilde{\omega}) P(d\widetilde{\omega}|\omega)$ is a measurable selection from the graph of $\omega \mapsto \int N_{f_0}(\widetilde{\omega}) P(d\widetilde{\omega}|\omega)$. And $\omega \mapsto N_{f_0}(\omega)$ is a measurable map to compact subsets of $\mathbb{R}^I$ as the composition of the measurable map $\omega \mapsto G_{f_0,\omega}$ with the equilibrium correspondence, which is by upper semi-continuity a Borel map from games to compact sets. By the measurable choice theorem in Mertens (1987b), it thus follows that there exists a measurable selection $\psi(\omega, \widetilde{\omega}) \in N_{f_0}(\widetilde{\omega})$ such that $G_{f_0(\cdot),\omega} = G_{\psi(\omega,\cdot),\omega}$. Denote by $\sigma(p, G)$ a Borel selection of an equilibrium with payoff p of the game G. The stationary equilibrium is now to play $\sigma[\psi(\omega, \widetilde{\omega}), G_{f_0,\widetilde{\omega}}]$ at state $\widetilde{\omega}$, denoting the previous state by ω. $\qquad \square$

REMARK 1. Observe that all the trouble with respect to the Markovian character of the strategies stems from the nonatomic part of the transitions: under the same assumptions as in the general case above, if one assumes the transition probabilities to be purely atomic, one immediately obtains the existence of subgame perfect equilibria in stationary Markov strategies – e.g., by going to the limit with the result of the remark after Proposition 3.3, following an ultrafilter on the increasing net of all finite subsets of Ω. (To truncate the game to a finite subset, add, e.g., an absorbing state with payoff zero, which replaces the complement of this finite subset. The argument assumes that $\mathcal{A}$ is the class of all subsets; if $\mathcal{A}$ were separable, one could always replace it by the class of all subsets on the corresponding quotient space while preserving all assumptions; and it is shown in Mertens and Parthasarathy (1987) how to reduce the general problem to the case where $\mathcal{A}$ is separable. Measurability of the strategies is anyway almost immaterial here, since the assumptions imply that, from any initial state, only countably many states are reachable.)

REMARK 2. The above remark is well illustrated by the previous proof: one has to convexify the set of Nash equilibrium payoffs, because a weak*-limit belongs pointwise only to the convex hull of the pointwise limits (it is not because of the fixed-point argument, which is not used in the other proofs). So $f_0(\omega)$ is only a convex combination of equilibrium payoffs of $G_{f_0,\omega}$ – and one will play equilibria. So one must select the equilibria, as a function of tomorrow's state $\widetilde{\omega}$, such as to give today the same game $G_{f_0,\omega}$. Basically, this uses a measurable version of Lyapunov's theorem, because (by finiteness of action sets) at ω only finitely many measures on $\widetilde{\omega}$ have to be considered. But it is clear that the solution to such a problem depends on the vector measure, i.e., on ω: the equilibrium played tomorrow at $\widetilde{\omega}$ will depend on ω.

REMARK 3. To illustrate this in still another way, let us show that, under the same assumptions, there exist stationary Markov sunspot equilibria ("extensive form correlated equilibria" [Forges (1986)] with public signals); i.e., if one convexifies the set of Nash equilibria, there is no problem. More precisely, for any α $(0 < \alpha \leqslant (\#I + 1)^{-1})$, there exists a sequence $\sigma_n(\omega)$ of Borel-measurable strategy vectors, such that if $(n_k)_{k=1}^{\infty}$ is a sequence of i.i.d. random variables (the sunspots) with a geometric distribution with parameter (decay rate) α, and all players observe n_k just before stage k, then playing $\sigma_{n_k}(\omega_k)$ at stage k is an equilibrium.

Observe in particular that, if one takes α of the form $2^{-\ell}$, this sequence n_k can be generated by a finite automaton containing a random bit generator – so the whole result is perfectly finite and constructive.

To prove this, remember first from the above proof that $\omega \to N_{f_0}(\omega)$ is a measurable map to compact subsets of $\mathbb{R}^I$ – in particular, it has an $\mathcal{A}$-measurable selection, say $\bar{f}$. Choose now a representative $\mathcal{A}$-measurable function, say $\bar{f}_0$, from the equivalence class $f_0 \in L_\infty(\mu)$: the set where $\bar{f}_0$ does not belong to the convex hull of N_{f_0} is Borel, and negligible, so change $\bar{f}_0$ to $\bar{f}$ on that subset. The function thus obtained is an $\mathcal{A}$-measurable selection from the convex hull of N_{f_0}, and belongs to the equivalence class f_0. So call it f_0. Use now the measurable version of Caratheodory's theorem [e.g., Mertens (1987b)] to write $f_0(\omega) = \sum_{i=0}^{\#I} \lambda_i(\omega) g_i(\omega)$, where the g_i are $\mathcal{A}$-measurable selections from N_{f_0}, and $\lambda_i \geqslant 0$ are $\mathcal{A}$-measurable and satisfy $\sum_{i=0}^{\#I} \lambda_i(\omega) = 1$. Use now Blackwell's (1953) result to assign to each vector $(\lambda_i(\omega))_{i=0}^{\#I}$ a partition $(K_i(\omega))_{i=0}^{\#I}$ of the positive integers, such that the probability of $K_i(\omega)$ under our geometric distribution is exactly $\lambda_i(\omega)$. It is easy to see that Blackwell's construction is perfectly measurable, so that the sets $\{\omega \mid k \in K_i(\omega)\}$ belong to $\mathcal{A}$. Then $\sigma_n(\omega) = \sigma(g_i(\omega), G_{f_0}(\omega))$ if $n \in K_i(\omega)$ fills the bill.

Results in this direction were obtained first by Duffie, Geanakoplos, McLennan and Mas-Colell (1994), and later independently by Nowak and Raghavan (1992). The former proved in addition that, under their conditions, one could choose the sunspot equilibria so as to induce an ergodic invariant measure.

REMARK 4. The assumption of norm-continuity of the transitions as a function of the actions is quite strong. Typically, one needs some form of noise in the model to assure it. The best-behaved model where it is not satisfied, and where existence of equilibria is not known, is the following: take as action set for each player the one-point compactification of the integers, and take the Cantor set as state space. Assume the reward function, and the probability of each Borel subset of the state space, are jointly continuous in state and actions, and use standard discounting. (And assume even further that the transitions are nonatomic to fix ideas – they are then all dominated by a fixed nonatomic probability.)

4. Asymptotic analysis: The algebraic aspect

As seen above, we find the stationary equilibria by looking for fixed points V of the operator T of the previous paragraph, i.e., solutions of $V = T((1 - \lambda)V)$, or f Nash equilibrium payoff in $\Gamma((1 - \lambda)f)$.

When state and action sets are finite, this becomes a system of finitely many polynomial equations and inequalities in finitely many variables as defined by the correspondence ψ (Proposition 3.3).

The system is also polynomial (affine) in λ.

We obtain then from Tarski's elimination theorem [e.g., Jacobson (1964)], and from the existence of Puiseux-series expansions for semi-algebraic functions:

PROPOSITION 4.1. *For any (finite) stochastic game, there exist* $\lambda_0 > 0$, *a positive integer M, and Puiseux-series expansions (absolutely convergent on* $[0, \lambda_0]$*) (v denotes the normalized payoff):*

$$v_\omega^i(\lambda) = \sum_{k \geqslant 0} h_k^{i,\omega} \lambda^{k/M}, \quad and$$

$$\sigma_\omega^i(s)(\lambda) = \sum_{k \geqslant 0} \alpha_k^{i,\omega,s} \lambda^{k/M}, \quad \forall i \in I, \forall s \in S_i,$$

such that, for all $\lambda \in\,]0, \lambda_0]$, *the* $\sigma^i(\lambda)$ *form a stationary Nash equilibrium, with pay-off vector* $v(\lambda)$, *of the* λ-*discounted game. (And those functions* $\sigma^i(\lambda)$, $v(\lambda)$ *are semi-algebraic on* $[0, \lambda_0].)$

The above result is due – in the zero-sum case, and by a somewhat different method – to Bewley and Kohlberg (1976a); for the present version, cf. Mertens (1982).

The next corollary is the only result of this section that we will use later.

COROLLARY 4.2. *For such a solution and any* $\lambda_1 < \lambda_0$, $\|(\mathrm{d}v/\mathrm{d}\lambda, \mathrm{d}\sigma/\mathrm{d}\lambda)\| \leqslant A\lambda^{-(M-1)/M}$ *for some* $A > 0$ *and any* $\lambda \leqslant \lambda_1$.

PROOF. Such an absolutely convergent series can be differentiated term by term in the interior of its radius of convergence.   □

REMARKS.
   (1) Every value of M in the above results can indeed occur: consider the $n \times n$ game where below the diagonal one gets zero payoff and returns to the same state, above the diagonal one goes to an absorbing state with payoff zero, and on the diagonal to an absorbing state with payoff one: $v(\lambda) = (1 - \lambda^{1/n})/(1 - \lambda)$.
   (2) Bewley and Kohlberg (1976b) prove there is a similar expansion for the value of the n-stage game: $v_n = v_\infty + \sum_{k=1}^{M-1} a_k n^{-k/M} + n^{-1}\mathrm{O}(\log n)$.
   (3) The expansion has to stop at the term $n^{-1}\mathrm{O}(\log n)$: in $\left(\begin{smallmatrix} -1^* & 1 \\ 1 & 0 \end{smallmatrix}\right)$, one has $v_n \sim (\log n)/n$.

5. ε-optimal strategies in the undiscounted game

5.1. The theorem

We consider here two-person zero-sum games. No finiteness or other conditions are imposed, but we assume the payoffs to be uniformly bounded, and the values $v(\omega, \lambda)$ of the discounted games to exist. We will exhibit sufficient conditions for the existence of strategies that guarantee in a strong sense some function $v_\infty(\omega)$ (up to ε). The theorem does not require the moves to be announced, only the payoffs.

THEOREM 5.1. *Assume a stochastic game where*
   (a) *payoffs are uniformly bounded;*
   (b) *the values* $v_\lambda(\omega)$ *of the* λ-*discounted games exist, as well as* ε-*optimal strategies (in the sense of* (3) *of Section* 3.1);
   (c) $\forall \alpha < 1$ *there exists a sequence* λ_i $(0 < \lambda_i \leqslant 1)$ *such that*

$$\lambda_{i+1} \geqslant \alpha\lambda_i, \quad \lim_{i \to \infty} \lambda_i = 0 \quad and \quad \sum_i \|v_{\lambda_i} - v_{\lambda_{i+1}}\| < +\infty$$

(denoting by $\| \cdot \|$ *the supremum norm over the state space).*

Then the game has a value v_∞ – i.e.,

$$\forall \varepsilon > 0, \ \exists \sigma_\varepsilon, \ \exists N_0: \ \forall \omega \in \Omega, \ \forall \tau, \ \forall n = N_0, N_0 + 1, \ldots, +\infty, \quad E^\omega_{\sigma_\varepsilon \tau}(\overline{g}_n) \geqslant v_\infty(\omega) - \varepsilon$$

(where $\overline{g}_n$ denotes the average payoff up to stage n, and $\overline{g}_\infty = \liminf_{n\to\infty} \overline{g}_n$), and dually for player II.

PROOF. Cf. Mertens and Neyman (1981). □

REMARK 1. Further, those ε-optimal strategies are still ε-optimal, in the same sense, in any subgame, and they consist in playing in successive blocks k of length L_k whichever $(\frac{1}{12}\varepsilon L_k \lambda_k)$-optimal strategy in the modification $\Gamma(\lambda_k, L_k)$ of the λ_k-discounted game where, from stage L_k on, the payoff $v_{\lambda_k}(\omega_{L_k})$ is received forever. Here L_k and λ_k are given functions $L(s_k)$ and $\lambda(s_k)$, where s_k is defined as in Remark 3 below.

REMARK 2. Condition (c) of the theorem is always satisfied when v_λ is of bounded variation or when (for some function v_∞) $\|v_\lambda - v_\infty\|/\lambda$ is integrable. Indeed, the latter means the integrability of $\|v_\lambda - v_\infty\|$ as a function of $\ln \lambda$; let then λ_i denote the minimizer of $\|v_\lambda - v_\infty\|$ in $[\alpha^{(i+1)/2}, \alpha^{i/2}]$ to satisfy condition (c).

REMARK 3. Using Corollary 4.2 and the above remark, the (proof of the) theorem yields in particular the following strategy for all finite stochastic games: let $\lambda(s) = 1/[s \ln^2 s]$, $s_{n+1} = \max[M, s_n + g_n - v_\infty(\omega_{n+1}) + \varepsilon/2]$, $s_1 \geqslant M$, $\lambda_n = \lambda(s_n)$, and play at stage n an optimal strategy in the λ_n-discounted game: for M sufficiently large, this is ε-optimal.

REMARK 4. Under the assumptions of the theorem v_λ and v_n converge uniformly to v_∞. The statement itself of the theorem, with N_0 independent of the initial state, implies this immediately.

REMARK 5. Some classes of stochastic games – like Everett's recursive games[4] and Milnor and Shapley's games of survival[5] – are known to have a value, although it is not known if the variation condition (c) of the theorem applies to them.

---

[4] A recursive game is a stochastic game with finite state space – and "non-pathological" to ensure a value – where the payoff is zero in every state in which the transition probability P may lead, for some choice of actions, to a different state. Cf., e.g., Mertens, Sorin and Zamir (1994), Chapter IV, Section 5.

[5] A game of survival is a repeated two-person (and – for historical reasons? – zero-sum) game, where each player starts with a given fortune, and the losing player is the first to lose his fortune.

5.2. Applications: Particular cases (finite games, two-person zero-sum)

(1) Note first that, when the stochastic game is a normalized form of a game with perfect information, then the games $\Gamma((1-\lambda)v)_\omega$ are also normal forms of games with perfect information, and hence have pure strategy solutions, which form a closed, semi-algebraic subset of the set of all solutions: applying the theorem to these yields that, for some $\lambda_0 > 0$, there exists a pure strategy vector such that the corresponding stationary strategies are optimal in the λ-discounted game for all $\lambda \leqslant \lambda_0$. Such stationary Markov strategy vectors (pure or not) are also called "uniformly discount optimal".

REMARK. The perfect information case includes in particular the situation where one player is a dummy – i.e., Markov decision processes or dynamic programming.

(2) Whenever there exist uniformly discount optimal strategies, the expansion of v_λ is in integer powers of λ: v_λ is in fact a rational function of λ, being the solution of the linear system $v_\lambda = \lambda g + (1 - \lambda)Pv_\lambda$, where g and P are the single-stage expected payoff and the transition probability generated by the strategy pair.

(3) Whenever there exists a strategy σ in the one-shot game which is o(λ)-optimal in $\Gamma(\lambda, 1)$, then for each $\varepsilon > 0$, one ε-optimal strategy of the theorem will consist in playing this all the time: the corresponding stationary strategy is optimal (in the strong sense of the theorem) in the infinite game. (Recall Remark 1 in Subsection 5.1.)

(4) Since the value exists in such a strong sense, it follows in particular that the payoff $\underline{v}(\sigma)$ guaranteed by a stationary Markov strategy σ is also completely unambiguous – applying the theorem to the one-person case. Further, the preceding points imply the existence of a pure stationary Markov best reply, which is best for all $\lambda \leqslant \lambda_0$, and that $\underline{v}_\lambda(\sigma) \geqslant \underline{v}(\sigma) - K\lambda$. One checks similarly that $\underline{v}_n(\sigma) \geqslant \underline{v}(\sigma) - K/n$. It follows in particular that if both players have stationary Markov optimal strategies σ and τ in Γ_∞, then $\|v_\lambda - v_\infty\| \leqslant K\lambda$, $\|v_n - v_\infty\| \leqslant K/n$ (and those strategies guarantee such bounds) (and in particular $\|v_\lambda - \underline{v}_\lambda(\sigma)\| \leqslant K'\lambda$, so that the corresponding one-shot strategies are O(λ) in $\Gamma(\lambda, 1)$).

(5) It follows that (3) can be improved to: whenever there exists a stationary strategy σ which is o(1)-optimal in $\Gamma(\lambda)$ (i.e., $\|\underline{v}_\lambda(\sigma) - v_\lambda\| \to 0$ as $\lambda \to 0$), then σ is optimal in the infinite game. This is an improvement because an easy induction yields that the $\varepsilon[1 - (1 - \lambda)^L]$-optimality of σ in $\Gamma(\lambda, L)$ implies its $\varepsilon[1 - (1 - \lambda)^{KL}]$-optimality in $\Gamma(\lambda, KL)$, hence its ε-optimality in $\Gamma(\lambda)$.

(6) In particular, even in the N-person case, when applying the above for each player while considering all others together as nature, a stationary Markov strategy vector which, for all $\varepsilon > 0$, is an ε-equilibrium of $\Gamma(\lambda)$ for all sufficiently small λ, is also an equilibrium of Γ_∞.

(7) The perfect information case can always be rewritten (by extending the state space and adjusting the discount factor) as a stochastic game where in each state only one player moves.

Actually the same conclusions go through (in the two-person case) assuming only that, in each state, the transition probability depends only on one player's action ("switching control") [Vrieze et al. (1983)]: assume for example that player I controls the transitions at state ω; by Proposition 4.1, we can assume that the sets S_0 and T_0 of best replies at ω for the stationary equilibria $(\sigma_\lambda, \tau_\lambda)$ are independent of λ for $\lambda \leqslant \lambda_0$. The equilibrium conditions of Proposition 3.3 takes then the following form at ω (σ and τ being probabilities on S_0 and T_0):

$$\sum_s \sigma_{\lambda,\omega}(s)\left[g_\omega^2(s,t) - g_\omega^2(s,\tilde{t})\right] \geqslant 0 \quad \forall t \in T_0, \ \forall \tilde{t},$$

$$V_{\lambda,\omega}^2 = \sum_s \sigma_{\lambda,\omega}(s)\left[g_\omega^2(s,t) + w_{\lambda,\omega}^2(s)\right] \quad \text{for some } t \in T_0,$$

$$w_{\lambda,\omega}(s) = (1-\lambda)\sum_{\tilde{\omega}} p(\tilde{\omega}|\omega,s)V_{\lambda,\tilde{\omega}},$$

$$V_{\lambda,\omega}^1 = \sum_t \tau_{\lambda,\omega}(t)g_\omega^1(s,t) + w_{\lambda,\omega}^1(s) \quad \forall s \in S_0, \text{ with}$$

inequalities for $s \in S \setminus S_0$.

The first set of inequalities describes a polyhedron of probabilities on S_0, independent of λ, with extreme points $\sigma_\omega^1, \ldots, \sigma_\omega^k$. So $\sigma_{\lambda,\omega} = \sum_{i=1}^k \mu_\omega^\lambda(i)\sigma_\omega^i$, with $\mu_\omega^\lambda(i) \geqslant 0$. The second inequality takes then the form:

$$V_{\lambda,\omega}^2 = \sum_{i=1}^k \mu_\omega^\lambda(i)\left[G_i(\omega) + (1-\lambda)\sum_{\tilde{\omega}} q_{i,\omega}(\tilde{\omega})V_{\lambda,\tilde{\omega}}^2\right].$$

Hence, by semi-algebraicity again, there exist, for each ω, indices i_0 and i_1 such that

$$G_{i_0}(\omega) + (1-\lambda)\sum_{\tilde{\omega}} q_{i_0,\omega}(\tilde{\omega})V_{\lambda,\tilde{\omega}}^2 \leqslant V_{\lambda,\omega}^2$$

$$\leqslant G_{i_1}(\omega) + (1-\lambda)\sum_{\tilde{\omega}} q_{i_1,\omega}(\tilde{\omega})V_{\lambda,\tilde{\omega}}^2$$

for all sufficiently small λ – say $\lambda \leqslant \lambda_0$. And conversely, for any solution V_λ^2 of this pair of inequalities, one obtains a corresponding σ, by a rational computation.

Similarly, the last system of inequalities, by eliminating τ, yields a system of linear inequalities in the variables $v_{\lambda,\omega}^1 - (1-\lambda)\sum_{\tilde{\omega}} p(\tilde{\omega}|\omega,s)v_{\lambda,\tilde{\omega}}^1$.

Putting those inequalities together, for all states, yields a system of linear inequalities in V_λ^1 and a similar system in V_λ^2: those systems have a solution for all $\lambda \leqslant \lambda_0$, and any such solution can be extended by a rational computation to a solution in $\sigma_\lambda, \tau_\lambda$.

Finally, since such a system of linear inequalities has only finitely many possible bases, and each such base is valid in a semi-algebraic subset of λ's – i.e., a finite union of intervals – one such basis is valid in a whole neighborhood of zero: inverting it, one obtains solutions V^1_λ and V^2_λ in the field $K(\lambda)$ of rational fractions in λ, with coefficients in the field K generated by the data of the game: there exists a stationary Markov equilibrium (σ, τ) in $K(\lambda)$, for $0 < \lambda \leqslant \lambda_1$. Further, the above describes a finite algorithm for computing it.

Now, observe there are two cases where we can in addition assume that σ_λ is independent of λ – hence equal to its limit σ_0:

(a) In the zero-sum case, since every extreme point σ^i_ω is a best reply of player I, they all yield the same expected payoff to him, and hence – zero-sum assumption – also to player II: hence $V^2_{\lambda,\omega} = G_i(\omega) + (1-\lambda) \sum_{\widetilde{\omega}} q_{i,\omega}(\widetilde{\omega}) V^2_{\lambda,\widetilde{\omega}}$ for all i, and thus any weights $\mu^\lambda_i(\omega)$ are satisfactory – in particular constant weights.

(b) If it is in all states the same player who controls the transitions, then varying μ will only vary player II's expected payoffs – in all states – and those do not affect any other inequality in the system.

In each of those cases then, it is obvious that, if one were to replace in all states the passive player's $\tau(\lambda)$ by its limit $\tau(0)$, which differs from it (rational fractions...) by at most $K \cdot \lambda$, the expected payoffs, under whatever strategies σ of the controlling agents, would vary by at most $C \cdot K \cdot \lambda$, where C is the maximum absolute value of all payoffs: indeed, transitions are not affected at all, so the probability distribution on the sequence of states remains unaffected, and in every state, the current payoff varies by at most $C \cdot K \cdot \lambda$. Hence by (6) above, (σ_0, τ_0) is a stationary Markov equilibrium of Γ_∞.

So, in those two cases, the rational fraction solutions can be chosen constant in λ ("uniformly discount optimal") for the controlling player, and are also solutions for $\lambda = 0$ – in particular, our finite algorithm computes rationally an equilibrium of the undiscounted game.

(8) Many other cases (two-person, zero-sum) with stationary optimal strategies are known. For example, one obviously obtains it if there is some state which is reached with positive probability for every starting state and every stationary pure strategy vector. In fact, a necessary and sufficient condition, due to Vrieze [cf. Vrieze (1983)] is the following. There should exist a solution $(v_\omega, y^1_\omega, y^2_\omega)$ to the system

$$v_\omega = \operatorname*{Val}_{S \times T} \langle p^{s,t}_\omega, v \rangle \quad \text{with optimal strategy sets } \Sigma_0 \text{ and } T_0,$$

$$v_\omega + y^1_\omega = \operatorname*{Val}_{\Sigma_0 \times T} \left[g^{s,t}_\omega + \langle p^{s,t}_\omega, y^1 \rangle \right],$$

$$v_\omega + y^2_\omega = \operatorname*{Val}_{S \times T_0} \left[g^{s,t}_\omega + \langle p^{s,t}_\omega, y^2 \rangle \right].$$

6. Some results on the non-zero-sum undiscounted case

Note first that there may be no uniform equilibria (strategy vectors such that, $\forall \varepsilon > 0$, $\exists N \colon \forall n \geqslant N$ those strategy vectors form an ε-equilibrium of Γ_n), hence E_∞, the corresponding set of payoffs, may be empty. In fact, already in the zero-sum case there exist no optimal strategies (Section 2). Thus we define a set E_0 of payoffs as $\bigcap_{\varepsilon > 0} E_\varepsilon$ where

$$E_\varepsilon = \Big\{ d \in \mathbb{R}^I \mid \exists \sigma, N \colon \forall n \geqslant N, \ \max_{i \in I} \big| \bar{\gamma}_n^i(\sigma) - d^i \big| \leqslant \varepsilon$$

$$\text{and } \sigma \text{ is an } \varepsilon\text{-equilibrium of } \Gamma_n \Big\}.$$

Note that obviously $E_\infty \subset E_0$ (and they coincide for supergames, i.e., for a single state of nature).

As in the zero-sum case, one might think of using Proposition 4.1 to construct equilibria – i.e., points in E_0 – for the undiscounted game.

Sorin (1986) constructed however the following counterexample – a non-zero-sum variant of the "Big Match":

$$\begin{pmatrix} (1,0)^* & (0,2)^* \\ (0,1) & (1,0) \end{pmatrix}.$$

Here, by the theorem for the zero-sum case, the threat point v is $(\frac{1}{2}, \frac{2}{3})$; and the set of individually rational Pareto-optimal payoffs is the set $P = \{(1 - \alpha, 2\alpha) \mid 1/3 \leqslant \alpha \leqslant 1/2\}$. (The above remains true even for all Γ_λ and all Γ_n.)

Sorin showed that, for all λ and n, $\{v\}$ was the unique equilibrium payoff both of the λ-discounted and of the n-stage game, while $E_0 = P$.

Observe that the ε-equilibria in the definition of E_0 (i.e., of E_ε) are also ε-equilibria in all sufficiently long finite games, hence also in all discounted games where each player has a sufficiently small discount factor. Thus E_0 qualifies in a very strong sense as the equilibrium set of the infinite game. The example shows therefore a strong discontinuity in the equilibrium set ($v \notin P$). It thus seems, at least in this example, that E_0 may be a more appropriate concept than $\lim_{\lambda \to 0} E_\lambda$ when analyzing long games – the latter may rely too much on common knowledge of each other's exact discount factor and on exact maximization, and the corresponding equilibria are too sensitive to the exact value of λ.

Existence of equilibria for games with absorbing states (i.e., $E_0 \neq \emptyset$) was proved by Vrieze and Thuijsman (1989).

The proof makes intensive use of the previous results: it starts from Proposition 4.1, and uses the corresponding strategies and their limits in a very clever way, depending on the absorbing or non-absorbing character of the pair and on the payoff vector of the limiting pair being larger or not than the limiting payoff vector. In a folk-theorem-type fashion, the result of the zero-sum case is used to punish deviations, so the verification of the equilibrium property remains quite simple.

Most promisingly, Vieille (1993) has shown that the general existence theorem for two-person stochastic games can be reduced to the same question when the games are in addition recursive games with a specific structure (and observe that, for recursive games, all difficulties that were present in the zero-sum case disappear).

7. Stochastic games with large state and action sets as normal forms for general repeated games

By a repeated game, we mean a stochastic game-like model, but without the informational restrictions: the same lottery that selects the new state also selects a pair of signals, one for each player, and players are informed only of those signals (and recall their past signals). In addition, one can allow for information lags and other relaxations. The model thus includes in particular also all repeated games with incomplete information.

We sketch here the analysis in Mertens (1987a), in the two-person zero-sum case.

It is first shown that, by appropriate enlargement of state and action sets, one can equivalently describe every repeated game by the following purely combinatorial standard form.

A partition $\mathcal{P}$ is given of the finite state space Ω, as well as a set F of functions from Ω to itself, a winning subset W of Ω and an initial state ω_1. Players play in cyclical order, each time choosing an element of F – which determines the next state given the current one ω – before being informed of the element of $\mathcal{P}$ containing ω. The current player gets one from his opponent if $\omega \in W$ and minus one otherwise.

(Replace first each stage by #I substages, adjusting the discount factor, to let players play in cyclical order. For the payoffs, normalize payoffs so they belong to [0, 1], and view them then as lotteries on {0, 1}: we can now think of the lottery as choosing also a payoff in {0, 1} – i.e., a winner in addition to signals and new state. To get rid of the moves of nature, use jointly controlled lotteries.) This transformation preserves maxmin, minmax, $\lim_{\lambda \to 0} v_\lambda$, etc.

Next, denote by P the space of consistent probabilities on the universal beliefs space associated with Ω [Mertens and Zamir (1985)]. If the game starts with a lottery $\pi_1 \in P$ selecting a type for each player and an initial state $\omega_1 \in \Omega$, and if σ denotes a vector of mixed strategies for the players for the first stage (as a function of their types), then their after-stage-one beliefs about ω_2 will be described by some $\pi_2 = \pi(\pi_1, \sigma) \in P$. Apply this idea for instance to the above standard form: denoting by T the universal type space associated with Ω, let Σ be the set of transition probabilities from T to F, and we then obtain a map π from $P \times \Sigma$ to P. And if π_n is the consistent probability at stage n, the payoff is $\pi_n(W)$: here we have a stochastic game description – with perfect information and no moves of nature: P is the state space, players alternate choosing an action $\sigma \in \Sigma$ that leads to a new state by the map π, and the payoff at stage n is just a function $\pi_n(W)$ of the current state π_n.

It is clear that this stochastic game has the same values for the n-stage games and for the discounted games. Hence the obvious recursive formula for the stochastic game induces a recursive formula for the general repeated game we started with. It is shown in Mertens (1987a) also how the existence of a value for the undiscounted stochastic game is basic to the analysis of the original repeated game. Moreover, although Martin's theorem [cf. Martin (1985)] on Borel determinacy may help in this respect, what one would like to have, at least for those stochastic games, is an extension of the theorem in Section 5, which would require only the uniform convergence of the v_λ instead of the variation condition (cf. also Remark 5 in Section 5). Such an extension can be proved with zero players – yielding a form of ergodic theorem for Markov chains. So the first difficult case – and hopefully the "key" to the general case – is with a single player – i.e., dynamic programming: if v_λ converges uniformly, do there exist ε-optimal policies for the undiscounted case? The important case is probably that of deterministic dynamic programming, since our above stochastic games are such and since the result can be proved in the zero-player case with moves of nature. It suffices then also to consider a countable state space.

A lot of progress has recently been made on this question: Lehrer and Sorin (1992) have shown that v_λ converges uniformly iff v_n converges uniformly, and then the limits are the same; Lehrer and Monderer (1994) have shown that if v_λ converges uniformly, say to v, then v is also the value of the undiscounted dynamic programming problem, when lim sup of the expected average payoffs is taken as payoff function. They show also that when lim inf is taken as payoff function, one may get less than v. So apparently some additional ingredients will be needed in the analysis of the above-mentioned problem area for repeated games.

Finally, Maitra and Sudderth (1992, 1993) obtained a very general minmax theorem for stochastic games: assume Ω and the S^i are standard Borel – the latter can even vary measurably with the state. Assume g is a uniformly bounded and upper-analytic function on Ω, S^{II} is compact, and the transition depends continuously on S^{II}. Take $\limsup_{n\to\infty} g_n$ as payoff function for the infinite game: the game has a value.

This beautiful result is not quite comparable to the result in Theorem 5.1 above – on the one hand, the assumptions are much more general, but on the other hand, it does not imply, even for finite stochastic games, that the values computed from $\limsup_{n\to\infty} \overline{g}_n$ and from $\liminf_{n\to\infty} \overline{g}_n$ would be the same, or that they would be equal to $\lim_{\lambda\to 0} v_\lambda$, and it does not yield strategies which are ε-optimal in any sufficiently long game.

But it seems a very promising possible substitute for or complement to Martin's theorem in arguments like the above: indeed, a direct appeal to Martin's theorem transforms the game into one where in addition mixed strategies are observable – and additional, unobvious minmax arguments would be needed to show that this is in fact immaterial. The result of Maitra and Sudderth may obviate this need in a number of cases. (Also, it is intriguing whether there is any relation with the appearance of the lim sup and lim inf in the results of Lehrer and Monderer.)

References

Amir, R. (1996), "Continuous stochastic games of capital accumulation with convex transitions", Games and Economic Behavior 15:132–148.

Bewley, T., and E. Kohlberg (1976a), "The asymptotic theory of stochastic games", Mathematics of Operations Research 1:197–208.

Bewley, T., and E. Kohlberg (1976b), "The asymptotic solution of a recursion equation occurring in stochastic games", Mathematics of Operations Research 1:321–336.

Bewley, T., and E. Kohlberg (1978), "On stochastic games with stationary optimal strategies", Mathematics of Operations Research 3:104–125.

Blackwell, D. (1953), "On randomization in statistical games with k terminal actions," in: H.W. Kuhn and A.W. Tucker, eds., Contributions to the Theory of Games, Vol. II, Annals of Mathematics Studies, Vol. 24 (Princeton University Press, Princeton, NJ) 183–187.

Blackwell, D., and T.S. Ferguson (1968), "The big match", Annals of Mathematical Statistics 39:159–163.

Charnes, A., and R. Schroeder (1967), "On some tactical antisubmarine games", Naval Research Logistics Quarterly 14:291–311.

Coulomb, J.M. (1992), "Repeated games with absorbing states and no signals", International Journal of Game Theory 21:161–174.

Duffie, D., J. Geanakoplos, A. Mas-Colell and A. McLennan (1994), "Stationary Markov equilibria", Econometrica 62:745–781.

Dutta, P., and R. Sundaram (1993), "How different can strategic models be?", Journal of Economic Theory 60:42–61.

Everett, H. (1957), "Recursive games", in: M. Dresher, A.W. Tucker and P. Wolfe, eds., Contributions to the Theory of Games, Vol. III, Annals of Mathematics Studies, Vol. 39 (Princeton University Press, Princeton, NJ) 47–78.

Federgruen, A. (1978), "On N-person stochastic games with denumerable state space", Advances in Applied Probability 10:452–471.

Filar, J.A. (1985), "Player aggregation in the travelling inspector model", IEEE Transactions on Automatic Control AC-30:723–729.

Forges, F. (1982), "Infinitely repeated games of incomplete information: symmetric case with random signals", International Journal of Game Theory 11:203–213.

Forges, F. (1986), "An approach to communication equilibria", Econometrica 54:1375–1385.

Gilette, D. (1957), "Stochastic games with zero stop probabilities", in: M. Dresher, A.W. Tucker and P. Wolfe, eds., Contributions to the Theory of Games, Vol. III, Annals of Mathematics Studies, Vol. 39 (Princeton University Press, Princeton, NJ) 179–187.

Jacobson, N. (1951), Lectures on Abstract Algebra, Vol. I (Van Nostrand).

Jacobson, N. (1953), Lectures on Abstract Algebra, Vol. II (Van Nostrand).

Jacobson, N. (1964), Lectures on Abstract Algebra, Vol. III (Van Nostrand).

Kohlberg, E. (1974), "Repeated games with absorbing states", Annals of Statistics 2:724–738.

Kohlberg, E., and A. Neyman (1981), "Asymptotic behavior of nonexpansive mappings in normed linear spaces", Israel Journal of Mathematics 38:269–275.

Kohlberg, E., and S. Zamir (1974), "Repeated games of incomplete information: the symmetric case", Annals of Statistics 2:1010–1041.

Lehrer, E., and D. Monderer (1994), "Discounting versus averaging in dynamic programming", Games and Economic Behavior 6:97–113.

Lehrer, E., and S. Sorin (1992), "A uniform Tauberian theorem in dynamic programming", Mathematics of Operations Research 17:303–307.

Leininger, W. (1986), "The existence of perfect equilibrium in a model of growth with altruism between generations", Review of Economic Studies 53:349–367.

Levhari, D., and L. Mirman (1980), "The Great Fish War: An example using a dynamic Cournot–Nash solution", Bell Journal of Economics 11:322–334.

Maitra, A., and T. Parthasarathy (1970), "On stochastic games", Journal of Optimization Theory and Applications 5:289–300.

Maitra, A., and W. Sudderth (1992), "An operator solution for stochastic games", Israel Journal of Mathematics 78:33–49.

Maitra, A., and W. Sudderth (1993), "Borel stochastic games with lim sup payoff", Annals of Probability 21:861–885.

Martin, D. (1985), "A purely inductive proof of Borel determinacy", in: A. Nerode and R.A. Shore, eds., Recursion Theory, Proceedings of Symposia in Pure Mathematics, Vol. 42 (American Mathematical Society, Providence, RI) 303–308.

Mertens, J.-F. (1982), "Repeated games: an overview of the zero-sum case", in: W. Hildenbrand, ed., Advances in Economic Theory (Cambridge University Press, Cambridge, MA) 175–182.

Mertens, J.-F. (1987a), "Repeated games", in: A.M. Gleason, ed., Proceedings of the International Congress of Mathematicians, Berkeley, 1986 (American Mathematical Society, Providence, RI) 1528–1577.

Mertens, J.-F. (1987b), "A measurable 'measurable choice' theorem", CORE Discussion Paper 8749 (Université Catholique de Louvain, Louvain-la-Neuve, Belgium).

Mertens, J.-F., and A. Neyman (1981), "Stochastic games", International Journal of Game Theory 10(2):53–56.

Mertens, J.-F., and A. Neyman (1982), "Stochastic games have a value", Proceedings of the National Academy of Sciences of the USA 79:2145–2146.

Mertens, J.-F., and T. Parthasarathy (1987), "Existence and characterisation of Nash equilibria for discounted stochastic games", CORE Discussion Paper 8750 (Université Catholique de Louvain, Louvain-la-Neuve, Belgium).

Mertens, J.-F., and T. Parthasarathy (1991), "Non-zero-sum stochastic games", in: T.E.S. Raghavan et al., eds., Stochastic Games and Related Topics (Kluwer Academic, Dordrecht) 145–148.

Mertens, J.-F., S. Sorin and S. Zamir (1994), "Repeated games", CORE Discussion Papers 9420, 9421, 9422 (Université Catholique de Louvain, Louvain-la-Neuve, Belgium).

Mertens, J.-F., and S. Zamir (1985), "Formulation of Bayesian analysis for games with incomplete information", International Journal of Game Theory 14:1–29.

Milnor, J., and L.S. Shapley (1957), "On games of survival", in: M. Dresher, A.W. Tucker and P. Wolfe, eds., Contributions to the Theory of Games, Vol. III, Annals of Mathematics Studies, Vol. 39 (Princeton University Press, Princeton, NJ) 15–45.

Nowak, A.S. (1985), "Universally measurable strategies in zero-sum stochastic games", Annals of Probability 13:269–287.

Nowak, A.S., and T.E.S. Raghavan (1991), "Positive stochastic games and a theorem of Ornstein", in: T.E.S. Raghavan et al., eds., Stochastic Games and Related Topics (Kluwer Academic, Dordrecht) 127–134.

Nowak, A.S., and T.E.S. Raghavan (1992), "Existence of stationary correlated equilibria with symmetric information for discounted stochastic games", Mathematics of Operations Research 17:519–526.

Parthasarathy, T. (1973), "Discounted, positive, and noncooperative stochastic games", International Journal of Game Theory 2:25–37.

Parthasarathy, T. (1984), "Markov games II", Methods of Operations Research 51:369–376.

Parthasarathy, T., and T.E.S. Raghavan (1981), "An order field property for stochastic games when one player controls transition probabilities", Journal of Optimization Theory and Applications 33:375–392.

Parthasarathy, T., and M. Stern (1976), "Markov games: A survey", in: E.O. Roxin, P.T. Lin and R.L. Sternberg, eds., Differential Games and Control Theory III, Lecture Notes in Pure and Applied Mathematics, Vol. 30 (Marcel Dekker, New York) 1–46.

Raghavan, T.E.S., T.S. Ferguson, T. Parthasarathy and O.J. Vrieze, eds. (1991), Stochastic Games and Related Topics: In Honor of Professor L.S. Shapley (Kluwer Academic, Dordrecht).

Raghavan, T.E.S., and J.A. Filar (1991), "Algorithms for stochastic games: A survey", Zeitschrift für Operations Research (Methods and Models of Operations Research) 35:437–472.

Shapley, L.S. (1953), "Stochastic games", Proceedings of the National Academy of Sciences of the USA 39:1095–1100.

Shubik, M., and W. Whitt (1973), "Fiat money in an economy with one nondurable good and no credit: A noncooperative sequential game", in: A. Blaquière, ed., Topics in Differential Games (Elsevier, Amsterdam) 401–448.

Sobel, M.J. (1982), "Stochastic fishery games with myopic equilibria", in: L.J. Mirman and D. Spulber, eds., Essays in the Economics of Renewable Resources (Elsevier, Amsterdam).

Sorin, S. (1986), "Asymptotic properties of a non-zero-sum stochastic game", International Journal of Game Theory 15:101–107.

Thuijsman, F. (1989), "Optimality and equilibria in stochastic games", Doctoraatsproefschrift, Rijksuniversiteit Limburg te Maastricht.

Vieille, N. (1993), "Contributions à la théorie des jeux répétés", Ph.D. Thesis, Université Paris 6, Paris.

Vrieze, O.J. (1981), "Linear programming and undiscounted stochastic games in which one player controls transitions", Operations Research Spektrum 3:29–35.

Vrieze, O.J. (1983), "Stochastic games with finite state and actions spaces", Ph.D. Thesis, Catholic University of Nijmegen, The Netherlands.

Vrieze, O.J., and F. Thuijsman (1986), "Stochastic games and optimal stationary strategies", A survey. Report 8639, Department of Mathematics, University of Nijmegen, The Netherlands.

Vrieze, O.J., and F. Thuijsman (1989), "On equilibria in repeated games with absorbing states", International Journal of Game Theory 18:293–310.

Vrieze, O.J., S.H. Tijs, T.E.S. Raghavan and J.A. Filar (1983), "A finite algorithm for switching control stochastic games", Operations Research Spektrum 5:15–24.

Winston, W. (1978), "A stochastic game model of a weapons development competition", SIAM Journal of Control and Optimization 16:411–419.

Winston, W., and A.V. Cabot (1984), "A stochastic game model of football play selection", mimeo (Indiana University).

Chapter 48

STOCHASTIC GAMES: RECENT RESULTS

NICOLAS VIEILLE

HEC and École Polytechnique

Contents

Handbook of Game Theory, Volume 3, Edited by R.J. Aumann and S. Hart

Abstract

This chapter presents developments in the theory of stochastic games that have taken place in recent years. It complements the contribution by Mertens. Major emphasis is put on stochastic games with finite state and action sets. In the zero-sum case, a classical result of Mertens and Neyman states that given $\varepsilon > 0$, each player has a strategy that is ε-optimal for all discount factors close to zero. Extensions to non-zero-sum games are dealt with here. In particular, the proof of existence of uniform equilibrium payoffs for two-player games is discussed, as well as the results available for more-than-two-player games. Important open problems related to N-player games are introduced by means of a class of simple stochastic games, called quitting, or stopping, games. Finally, recent results on zero-sum games with imperfect monitoring and on zero-sum games with incomplete information are surveyed.

Keywords

games, non-cooperative, stochastic, value, equilibrium

JEL classification: C72, C73

1. Introduction

The purpose of this chapter is to complement the previous chapter of Mertens [Mertens (2002)], and to present selected recent developments of the theory that are not covered in it. We limit ourselves to some of the themes addressed there, without claiming to be exhaustive. In particular, we will not discuss topics such as algorithms or applications of stochastic games.

Our notations follow those of Mertens (2002), and we refer to Section 1.1 there for a general presentation of stochastic games. Unless otherwise specified, the state space Ω and the action sets S^i, $i \in I$, are assumed finite.

In this finite setup, any undiscounted zero-sum stochastic game has a value [Mertens and Neyman (1981); see also Theorem 5.1 in Mertens (2002)]. In recent years, this fundamental result has been extended in several directions: to non-zero-sum games (do n-player stochastic games have an equilibrium payoff?) and to zero-sum games with imperfect monitoring and/or incomplete information (do such games have a maxmin and a minmax?).

In Section 1, we report on the result that any two-player stochastic game has an equilibrium payoff. Section 2 contains the known results for games with more than two players. These two sections relate to Section 7 in Mertens (2002). Section 3 deals with the extension of the theory of zero-sum games to games with imperfect monitoring in the zero-sum case. Finally, Section 4 contains new results on games with large state spaces [see Section 7 of Mertens (2002)] and lists miscellaneous results.

To facilitate an independent reading of this chapter, we recall now the definition of equilibrium payoffs, and several other notions.

DEFINITION 1. Let Γ be an N-player game. The vector $d \in \mathbf{R}^N$ is an *ε-equilibrium payoff* of Γ if there exists a strategy vector σ, and $N_0 \in \mathbf{N}$, such that, for $n \geqslant N_0$:
(1) $\|\gamma_n(\sigma) - d\| \leqslant \varepsilon$;
(2) σ is an ε-equilibrium of Γ_n.

In this definition, Γ_n is the n-stage version of Γ. We then say that σ is an ε-equilibrium; d is an equilibrium payoff if it is an ε-equilibrium payoff, for each $\varepsilon > 0$. Given $\varepsilon' > \varepsilon > 0$, an ε-equilibrium is also an ε'-equilibrium in all discounted games, provided the discount factor is close enough to zero. We refer to Sorin (1992) for a discussion of this concept.

A strategy is *stationary* (or stationary Markov) if the distribution used to select an action depends only on the current state of the game. Thus, a stationary strategy x^i of player i can be identified with an element $(x^i_\omega)_\omega$ of $\Delta(S^i)^\Omega$, where x^i_ω is the lottery used by player i in state ω. Stationary strategies will be denoted by Latin letters, and arbitrary strategies by Greek letters. Given a profile x of (stationary) strategies, the sequence $(\omega_n)_{n \in \mathbf{N}}$ of states is a Markov chain. The mixed extensions of the stage payoff g and of the transition probability P will also be denoted by g and P. A *perturbation* of a stationary strategy $x^i = (x^i_\omega)$ is a stationary strategy $\tilde{x}^i = (\tilde{x}^i_\omega)$ such that the support of x^i_ω is a subset of the support of $\tilde{x}^i_\omega$, for each $\omega \in \Omega$.

2. Two-player non-zero-sum games

The purpose of this section is to present the main ideas of the proof of the next theorem, due to Vieille (2000a, 2000b). Several tools of potential use for subsequent studies are introduced.

THEOREM 2. *Every two-player stochastic game has an equilibrium payoff.*

2.1. An overview

W.l.o.g., we assume that the stage payoff function (g^1, g^2) of the game satisfies $g^1 < 0 < g^2$. The basic idea is to devise an ε-equilibrium profile that takes the form of a stationary-like strategy vector σ, supplemented by threats of indefinite punishment.

We give a heuristic description of σ. The profile σ essentially coincides with a stationary profile $\bar{x}$. For the Markov chain defined by $\bar{x}$, consider the partition of the state space Ω into recurrent sets and transient states. (This partition depends on $\bar{x}$, since the transitions depend on actions.) The recurrent sets are classified into *solvable* and *controlled* sets. The solvable sets are those recurrent sets C for which the average payoff induced by $\bar{x}$ starting from C is high for both players; the controlled sets are the remaining sets. In each controlled set C, σ plays a perturbation of $\bar{x}$, designed so that the play leaves C in finite time. In the other states, σ coincides with $\bar{x}$. Given σ, the play eventually reaches some solvable set (and remains within it). Whenever the play is in a controlled or solvable set, each player monitors the behavior of the other player, using statistical tests.

This description is oversimplified and inaccurate in some fairly important respects, such as the fact that we use a generalized notion of recurrent set, called a *communicating set*.

The construction of σ consists of two independent steps: first, to construct the solvable sets and some controlled sets, and reduce the existence problem to a class of recursive games; second, to deal with the class of recursive games.[1]

2.2. Some terminology

Before proceeding with these steps, we shall provide a formal definition of the notions of communicating, solvable and controlled sets.

---

[1] A game is recursive if the payoff function is identically zero outside absorbing states. Recursive games were introduced by Everett (1957).

DEFINITION 3. A subset C is *communicating* given a profile x if, given any $\omega, \omega' \in C$, there exists a perturbation $\tilde{x}$ of x for which, starting from ω, the probability of reaching ω' without leaving C equals one.

In particular, C is closed given x: $P(C|\omega, x_\omega) = 1$, for each $\omega \in C$. Note that any recurrent set for x is communicating given x.

It is convenient to have the initial state of the game vary. Given an initial state ω, we denote by $(v^1(\omega), v^2(\omega))$ the threat point of the game.[2] If, facing a stationary strategy x^2, player 1 is to play s^1 in state ω, and to be punished immediately afterwards, his best future payoff is measured by the expectation $E[v^1|\omega, s^1, x_\omega^2]$ of v^1 under $P(\cdot|\omega, s^1, x_\omega^2)$. For $C \subseteq \Omega$, and given x^2, we set

$$H^1(x^2, C) = \max_{s^1 \in S^1} \max_{\omega \in C} E[v^1|\omega, s^1, x_\omega^2],$$

which somehow measures the threat point for player 1, against x^2, and given that the play visits all states of C. The definition of $H^2(x^1, C)$ is the symmetric one. It is easily seen that $H^1(x^2, C) \geqslant \max_C v^1(\omega)$.

Let a profile x, and a recurrent set R for x be given. The (long-run) average payoff $\lim_n E_{\omega,x}[\bar{g}_n]$ exists for each ω and is independent of $\omega \in R$. We denote it by $\gamma_R(x)$.

The definition builds upon a notion first introduced by Thuijsman and Vrieze (1991).

DEFINITION 4. A set $C \subseteq \Omega$ is *solvable* if, for some profile x, the following two conditions are fulfilled:

(1) C is a communicating set given x.
(2) There exists a point $\gamma = (\gamma^1, \gamma^2) \in \mathrm{co}\{\gamma_R(x), R \text{ recurrent subset of } C\}$ such that

$$(\gamma^1, \gamma^2) \geqslant (H^1(x^2, C), H^2(x^1, C)). \tag{1}$$

This concept is motivated by the following observation. The communication requirement ensures that the players are able to visit the recurrent subsets of C cyclically by playing appropriate small perturbations of x. Given the interpretation of $(H^1(x^2, C), H^2(x^1, C))$ as a threat point, the inequality (1) may be interpreted as an individual rationality requirement. By a standard proof, one can show that γ is an equilibrium payoff of the game, provided the initial state belongs to C.

The set of equilibrium payoffs of the game does not increase when one replaces each state in a solvable set C by an absorbing state, with payoff the vector γ associated with C. Therefore, we assume throughout the chapter that all such sets coincide with absorbing states.

---

[2] By definition, $v^i(\omega)$ is the value of the zero-sum stochastic game deduced from Γ, where the other player minimizes player i's payoff.

We now describe controlled sets. A pair $(\omega, s^i) \in C \times S^i$ is a *unilateral exit* of player i from $C \subseteq \Omega$ given a strategy x^{-i} if $P(C|\omega, s^i, x_\omega^{-i}) < 1$.

A triplet $(\omega, s^1, s^2) \in C \times S^1 \times S^2$ is a *joint exit* from C given x if $P(C|\omega, s^1, s^2) < 1$, and none of the pairs (ω, s^1) and (ω, s^2) is a unilateral exit.

DEFINITION 5. Let $C \subseteq \Omega$ be a communicating set given a profile x.

The set C is *controlled by player* i if there is a unilateral exit (ω, s^i) of player i (from C given x^{-i}) such that

$$\left(E[v^1|\omega, s^i, x_\omega^{-i}], E[v^2|\omega, s^i, x_\omega^{-i}] \right) \geqslant \left(H^1(x^2, C), H^2(x^1, C) \right). \tag{2}$$

The set C is *jointly controlled* if there exists

$$\gamma \in \mathrm{co}\left\{ E[v|\omega, s^1, s^2], (\omega, s^1, s^2) \text{ joint exit from } C \text{ given } x \right\}$$

such that

$$\gamma \geqslant \left(H^1(x^2, C), H^2(x^1, C) \right).$$

The rationale behind this definition runs as follows. Let $C \subseteq \Omega$ be a set controlled by player 1, and let x, $(\omega, s^1) \in C \times S^1$ be the associated profile and exit. Assume for simplicity that $P(C|\omega, s^1, x_\omega^2) = 0$.

Assume that we are given for each $\omega' \notin C$ an equilibrium payoff $\gamma(\omega')$ for the game starting at ω'. Then $E[\gamma(\cdot)|\omega, s^1, x_\omega^2]$ is an equilibrium payoff of the game, for every initial state in C.

We give a few hints for this fact. By using appropriate perturbations of x, the players are able to come back repeatedly to ω without leaving C. If player 1 slightly perturbs x^1 by s^1 in each of these visits, the play leaves C in finite time and the exit state is distributed according to $P(\cdot|\omega, s^1, x_\omega^2)$. Given such a scenario, it takes many visits to ω before the play leaves C. Hence player 1 may check the empirical choices of player 2 in these stages. Condition (2) implies that

- player 2 prefers playing x^2 in state ω to playing any other distribution and being punished; he prefers waiting for player 1 to use the exit (ω, s^1) to using any of his own exits, since

$$E[\gamma^2(\cdot)|\omega, s^1, x_\omega^2] \geqslant E[v^2(\cdot)|\omega, s^1, x_\omega^2] \geqslant H^2(x^1, C).$$

- player 1 prefers using the exit (ω, s^1) (and getting $E[\gamma^1(\cdot)|\omega, s^1, x_\omega^2]$) to using any other exit and being punished; he prefers using the exit (ω, s^1) to using no exit at all and being punished.

A similar property holds for jointly controlled sets.

2.3. A reduction to positive recursive games

To any controlled set C, we associate in a natural way a distribution μ_C of exit, i.e., a distribution such that $\mu_C(C) < 1$. If C is controlled by player i, let $\mu_C = P(\cdot|\omega, s^i, x_\omega^{-i})$ (with the notations of Definition 5). If C is jointly controlled, let μ_C be a convex combination of the distributions $P(\cdot|\omega, s^1, s^2)$, $((\omega, s^1, s^2)$ joint exit from C given x) such that $E_{\mu_C}[v] = \gamma$.

Given a controlled set C, with its distribution μ_C of exit, define a *changed game* Γ_C by changing the transitions in each state of C to μ_C. For a collection $\mathcal{C}$ of disjoint controlled sets, the changed game $\Gamma_{\mathcal{C}}$ is obtained by applying this procedure to each element of $\mathcal{C}$.

In general, there is no inclusion between the equilibrium payoff sets of the original and the changed games Γ and $\Gamma_{\mathcal{C}}$. The goal of the next proposition, which is the main result in Vieille (2000a), is to exhibit a family $\mathcal{C}$ such that: (i) such an inclusion holds and (ii) the changed game $\Gamma_{\mathcal{C}}$ has very specific properties.

Remember that, by assumption, the solvable sets of Γ coincide with the absorbing states of Γ.

PROPOSITION 6. *There exists a family $\mathcal{C}$ of disjoint controlled sets with changed game $\Gamma_{\mathcal{C}}$ having the following property: for each strategy x^1 there exists a strategy x^2 such that*
  (i) *the play reaches an absorbing state in finite time;*
  (ii) *for each initial state ω_1, the expected termination payoff to player 2 is at least $v^2(\omega_1)$.*

Two remarks are in order. First, by (i), there must exist an absorbing state in Γ. The existence of solvable sets is therefore a corollary to the proposition. Next, the two games Γ and $\Gamma_{\mathcal{C}}$ need not have the same threat point v. The value $v^2(\omega_1)$ that appears in the statement is that of Γ.

Let $\mathcal{C}$ be given by this proposition. Let $\Gamma_{\mathcal{C}}^0$ be the game obtained from $\Gamma_{\mathcal{C}}$, after setting the payoff function to zero in each non-absorbing state.

Note that $\Gamma_{\mathcal{C}}^0$ is a recursive game such that:
(P.1) all absorbing payoffs to player 2 are positive;
(P.2) player 2 can force the play to reach an absorbing state in finite time: for any profile $x = (x^1, x^2)$ where x^2 is fully mixed, the play reaches an absorbing state in finite time, whatever the initial state.

Property (P.1) is a consequence of the assumption $g^2 > 0$; property (P.2) follows from Proposition 6(i). Recursive games that satisfy both properties (P.1) and (P.2) are called positive recursive games.

It can be shown[3] that each equilibrium payoff of $\Gamma_{\mathcal{C}}^0$ is also an equilibrium payoff of the initial game Γ. The main consequence of Proposition 6 is thus that one is led to study positive recursive games.

---

[3] This is where the assumption $g^1 < 0 < g^2$ comes into play.

2.4. Existence of equilibrium payoffs in positive recursive games

We now present some of the ideas in the proof of the result:

PROPOSITION 7. *Every (two-player) recursive game which satisfies* (P.1) *and* (P.2) *has an equilibrium payoff.*

In zero-sum recursive games, ε-optimal strategies do exist [Everett (1957)]. In non-zero-sum positive recursive games, stationary ε-equilibria need not exist. For instance, in the game[4]

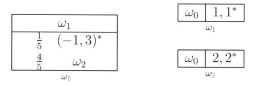

Example 1.

one can check that no stationary profile x exists that would be an ε-equilibrium for every initial state.

Throughout this section, take Γ to be a fixed positive recursive game. The basic idea of the proof is to approximate the game by a sequence of constrained games. For each $\varepsilon > 0$, let Γ_ε be the game in which player 2 is constrained to use stationary strategies that put a weight of at least ε on each single action. Player 1 is unconstrained. A crucial feature of Γ_ε is that the average payoff function, defined for stationary profiles x by $\gamma(x) = \lim_{n \to \infty} \gamma_n(x)$, is continuous.

Next, one defines B_ε as an analog of the best-reply correspondence on the space of constrained stationary profiles. This correspondence is well behaved so that: (i) it has a fixed point x_ε for each $\varepsilon > 0$, and (ii) the graph of fixed points (as a function of ε) is semialgebraic, hence there is a selection $\varepsilon \mapsto x_\varepsilon$ of fixed points such that $x_\varepsilon^{i,\omega}(s^i)$ has an expansion in Puiseux series in the neighborhood of zero [see Mertens (2002), Section 4]. This can be shown to imply that the limits $x_0 = \lim_{\varepsilon \to 0} x_\varepsilon$ and $\gamma = \lim_{\varepsilon \to 0} \gamma(x_\varepsilon)$ do exist. Finally, one proves that, for each ω, γ_ω is an equilibrium payoff for Γ starting in ω; an associated ε-equilibrium consists in playing a history-dependent perturbation of x_0, sustained by appropriate threats.

Solan (2000) proves that, by taking the usual best-reply map for B_ε, the program sketched in the previous paragraph works for games in which there are no more than two non-absorbing states, but not for more general games.

---

[4] In this example, each entry contains only the transitions. Transitions are deterministic except in state ω_0, when player 1 plays the Bottom row; the play then moves, with probability 4/5, to the state ω_2, and to an absorbing state with payoff $(-1, 3)$ otherwise.

Before defining B_ε in greater detail, we assume we have an intuitive notion of what it is. Given a fixed point x_ε of B_ε, we begin by describing the asymptotic behavior of the play, as ε goes to zero. This discussion will point out some of the requirements that a satisfactory definition of B_ε should meet.

For each $\varepsilon > 0$, given $x_\varepsilon = (x_\varepsilon^1, x_\varepsilon^2)$, the play reaches an absorbing state in finite time, since x_ε^2 is fully mixed and since Γ satisfies (P.2). As ε goes to zero, the probability of some actions may vanish, and there may exist recurrent sets for x_0 that contain non-absorbing states.

Define a binary relation $\rightarrow$ on the non-absorbing states by $\omega \rightarrow \omega'$ if and only if the probability (starting from ω, computed for x_ε) that the play visits ω' converges to one as ε goes to zero. Define an equivalence relation by

$$\omega \sim \omega' \quad \Leftrightarrow \quad (\omega \rightarrow \omega' \text{ and } \omega' \rightarrow \omega).$$

The different equivalence classes define a partition of the set of non-absorbing states. Note that a transient state (given x_0) may be included in a larger equivalence class, or constitute an equivalence class by itself. One can check that each class is either a transient state, or a set that is communicating given $x_0 = \lim_{\varepsilon \rightarrow 0} x_\varepsilon$.

Consider an equivalence class C of the latter type, and let $\varepsilon > 0$ be fixed. Since the play reaches the set of absorbing states in finite time, C is transient under x_ε. Hence, given an initial state in C, the distribution Q_C^ε of the exit state[5] from C is well defined. This distribution usually depends on the initial state in C. Since (x_ε) has a Puiseux expansion in the neighborhood of zero, it can be shown that the limit $Q_C = \lim_{\varepsilon \rightarrow 0} Q_C^\varepsilon$ exists. Moreover, it is *independent* of the initial state in C. Next, the distribution Q_C has a natural decomposition as a convex combination of the distributions

$$P(\cdot|\omega, s^i, x_0^{-i}), \quad \text{where } (\omega, s^{-i}) \text{ is a unilateral exit of } C \text{ given } x_0$$

and

$$P(\cdot|\omega, s^1, s^2), \quad \text{where } (\omega, s^1, s^2) \text{ is a joint exit from } C \text{ given } x_0.$$

It is straightforward to observe that the limit payoff vector $\gamma(\cdot) = \lim_{\varepsilon \rightarrow 0} \gamma(\cdot, x_\varepsilon)$ is such that, for $\omega \in C$, $\gamma(\omega)$ coincides with the expectation $E_{Q_C}[\gamma(\cdot)]$ of $\gamma(\cdot)$ under Q_C.

The main issue in designing the family $(B_\varepsilon)_\varepsilon$ of maps is to ensure that C is somehow controlled, in the following sense. Assuming that $\gamma(\omega')$ is an equilibrium payoff for the game starting from $\omega' \notin C$, it should be the case that $\gamma(\omega) = E_{Q_C}[\gamma(\cdot)]$ is an equilibrium payoff starting from $\omega \in C$. The main difficulty arises when the decomposition of Q_C involves two unilateral exits $(\overline{\omega}, \overline{s}^2), (\widetilde{\omega}, \widetilde{s}^2)$ of player 2, such that

---

[5] Which is defined as the actual current state, at the first stage for which the current stage does not belong to C.

$E[\gamma^2(\cdot)|\overline{\omega}, x^2_{0,\overline{\omega}}, \overline{s}^2] > E[\gamma^2(\cdot)|\widetilde{\omega}, x^2_{0,\widetilde{\omega}}, \widetilde{s}^2]$. Indeed, in such a case, player 2 is not indifferent between the two exits, and would favor using the exit $(\overline{\omega}, \overline{s}^2)$.

The approach in Vieille (2000b) is similar to proper ε-equilibrium. Given $x = (x^1, x^2)$, one measures for each pair $(\omega, s^2) \in \Omega \times B$ the *opportunity cost* of using s^2 in state ω by $\max_{s^2} E[\gamma^2(x)|\omega, x^1_\omega, \cdot] - E[\gamma^2(x)|\omega, x^2_\omega, s^2]$ (it thus compares the expected continuation payoff by playing s^2 with the maximum achievable). $B^2_\varepsilon(x)$ consists of those $\overline{x}^2$ such that whenever the pair (ω, s^2) has a higher opportunity cost than $(\overline{\omega}, \overline{s}^2)$, then the probability $\overline{x}^2_\omega(s^2)$ assigned by $\overline{x}^2$ to s^2 at state ω is quite small compared with $\overline{x}^2_{\overline{\omega}}(\overline{s}^2)$. One then sets $B_\varepsilon(x) = B^1_\varepsilon(x) \times B^2_\varepsilon(x)$, where B^1_ε is the best-reply map of player 1.

We conclude by giving a few stylized properties that show how to deal with the difficulties mentioned above. Since both exits $(\overline{\omega}, \overline{s}^2)$ and $(\widetilde{\omega}, \widetilde{s}^2)$ have a positive contribution to Q_C, it follows that $\widetilde{\omega}$ is visited (infinitely, as ε goes to zero) more often than $\overline{\omega}$, and also that, in some sense, facing x^1_0, player 2 can not reach $\overline{\omega}$ from $\widetilde{\omega}$, hence communication from $\widetilde{\omega}$ to $\overline{\omega}$ can be blocked by player 1. Thus player 1 is able to influence the relative frequency of visits in $\widetilde{\omega}$ and $\overline{\omega}$, hence the relative weight of the two exits $(\overline{\omega}, \overline{s}^2)$, $(\widetilde{\omega}, \widetilde{s}^2)$. It must therefore be the case that player 1 is indifferent between the two exits $(\widetilde{\omega}, \widetilde{s}^2)$ and $(\overline{\omega}, \overline{s}^2)$. The ε-equilibrium profile will involve a lottery performed by player 1, who chooses which of the two exits (if any) should be used to leave C.

2.5. Comments

(1) The lack of symmetry between the two players may appear somewhat unnatural. However, it is not an artifact of the proof since symmetric stochastic games need not have a symmetric ε-equilibrium. For instance, the only equilibrium payoffs of the symmetric game

0, 0	2, 1*
1, 2*	1, 1*

are $(1, 2)$ and $(2, 1)$.

(2) All the complexity of the ε-equilibrium profiles lies in the punishment phase.

(3) The main characteristics of the ε-equilibrium profile (solvable sets, controlled sets, exit distributions, stationary profiles that serve as a basis for the perturbations) are independent of ε. The value of $\varepsilon > 0$ has an influence on the statistical tests used to detect potential deviations, the size of the perturbations used to travel within a communicating set, and the specification of the punishment strategies.

(4) The above proof has many limitations. Neither of the two parts extends to games with more than two players. The ε-equilibrium profiles have no subgame perfection property. Finally, in zero-sum games, the value exists as soon as payoffs are observed (in addition to the current state). For non-zero-sum games, the tests check past choices. Whether an equilibrium exists when only the vector of current payoffs is publicly observed, is not known.

(5) These ε-equilibrium profiles involve two phases: after a solvable set is reached, players accumulate payoffs (and check for deviations); before a solvable set is reached, they care only about transitions (about which solvable set will eventually be reached). This distinction is similar to the one which appears in the proof of existence of equilibrium payoffs for games with one-sided information [Simon et al. (1995)], where a phase of information revelation is followed by payoff accumulation. This (rather vague) similarity suggests that a complete characterization of equilibrium payoffs for stochastic games would intertwine the two aspects in a complex way, by analogy with the corresponding characterization for games with incomplete information [Hart (1985)].

(6) In Example 1, the following holds: given an initial state ω, and $\varepsilon > 0$, the game starting at ω has a stationary ε-equilibrium. Whether this holds for any positive recursive game is not known.

3. Games with more than two players

It is as yet unknown whether n-player stochastic games always have an equilibrium payoff. We describe a partial result for three-player games, and explain what is specific to this number of players.

The first contribution is due to Flesch, Thuijsman and Vrieze (1997), who analyzed Example 2.

$0,0,0$	$0,1,3^*$
$1,3,0^*$	$1,0,1^*$

$3,0,1^*$	$1,1,0^*$
$0,1,1^*$	$0,0,0^*$

Example 2.

This example falls in the class of repeated games with absorbing states: there is a single non-absorbing state (in other words, the current state changes once at most during any play). We follow customary notations [see Mertens (2002)]. Players 1, 2 and 3 choose respectively a row, a column and a matrix. Starting from the non-absorbing state, the play moves immediately to an absorbing state, unless the move combination (Top, Left, Left) is played.

In this example, the set of equilibrium payoffs coincides with those convex combinations $(\gamma^1, \gamma^2, \gamma^3)$ of the three payoffs $(1, 3, 0)$, $(0, 1, 3)$, $(3, 0, 1)$ such that $(\gamma^1, \gamma^2, \gamma^3) \geqslant (1, 1, 1)$, and $\gamma^i = 1$ for at least one player i. Corresponding ε-equilibrium profiles involve cyclic perturbations of the profile of stationary (pure) strategies (Top, Left, Left). Rather than describe this example in greater detail, we discuss a class of games below that includes it.

This example gave the impetus for the study of three-player games with absorbing states [see Zamir (1992), Section 5 for some motivation concerning this class of games]. The next result is due to Solan (1999).

THEOREM 8. *Every three-player repeated game with absorbing states has an equilibrium payoff.*

SKETCH OF THE PROOF. Solan defines an auxiliary stochastic game in which the current payoff $\tilde{g}(x)$ is defined to be the (coordinatewise) minimum of the current vector payoff $g(x)$ and of the threat point.[6] He then uses Vrieze and Thuijsman's (1989) idea of analyzing the asymptotic behavior (as $\lambda \to 0$) of a family $(x_\lambda)_{\lambda > 0}$ of stationary equilibria of the auxiliary λ-discounted game.

The limits $\lim_{\lambda \to 0} x_\lambda$ and $\lim_{\lambda \to 0} \gamma_\lambda(x_\lambda)$ do exist, up to a subsequence. If it happens that $\lim_{\lambda \to 0} \gamma_\lambda(x_\lambda) = \gamma(\lim_{\lambda \to 0} x_\lambda)$, then $x = \lim_{\lambda \to 0} x_\lambda$ is a stationary equilibrium of the game. Otherwise, it must be the case that the nature of the Markov chain defined by x_λ changes at the limit: for $\lambda > 0$ close enough to zero, the non-absorbing state is transient for x_λ, whereas it is recurrent for x.

In this case, the limit payoff $\lim_{\lambda \to 0} \gamma_\lambda(x_\lambda)$ can be written as a convex combination of the non-absorbing payoff $\tilde{g}(x)$ (which by construction is dominated by the threat point) and of payoffs received in absorbing states reached when perturbing x. By using combinatorial arguments, Solan constructs an ε-equilibrium profile that coincides with cyclic perturbations of x, sustained by appropriate threats.

In order to illustrate Example 2 above and Solan's proof, we focus on the following games, called *quitting games*. Each player has two actions: *quit* and *continue*: $S^i = \{c^i, q^i\}$. The game ends as soon as at least one player chooses to quit (if no player ever quits, the payoff is zero). For simplicity, we assume that a player receives 1 if he is the only one to quit.

A stationary strategy is characterized by the probability of quitting, i.e., by a point in $[0, 1]$. Hence the space of stationary profiles is the unit cube $D = [0, 1]^3$, with $(0, 0, 0)$ being the unique non-absorbing profile.

Assume first that, for some player, say player 1, the payoff vector $\gamma(q^1, c^2, c^3)$ is of the form $(1, +, +)$, where the $+$ sign stands for "a number higher than or equal to one". Then the following stationary profile is an ε-equilibrium, provided α is small enough: player 1 quits with probability α, players 2 and 3 continue with probability 1.

We now rule out such configurations. For $\varepsilon > 0$ small, consider the constrained game where the players are restricted to stationary profiles x that satisfy $\sum_{i=1}^{3} x^i \geq \varepsilon$, i.e., the points below the triangle $T = \{x \in D, \ x^1 + x^2 + x^3 = \varepsilon\}$ are chopped off D (see Figure 1).

If it happens that at every point $x \in T$, one has $\gamma^i(x) < 1$ for some[7] i, then any stationary equilibrium of the constrained game (which exists by standard fixed-point arguments) is a stationary equilibrium of the true game.

---

[6] In particular, the current payoff is not multilinear.

[7] Player i would then rather quit than let x be played. In geometric terms, the best-reply map points inwards on T.

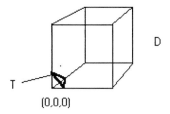

Figure 1.

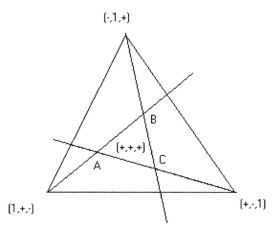

Figure 2.

It therefore remains to discuss the case where $\gamma(x_0) = (+, +, +)$ for some $x_0 \in T$. Given $x \in T$, the probability that two players quit simultaneously is of order ε^2, hence γ is close on T to the linear function

$$\frac{1}{\varepsilon}\left(x^1\gamma\left(q^1, c^{-1}\right) + x^2\gamma\left(q^2, c^{-2}\right) + x^3\gamma\left(q^3, c^{-3}\right)\right).$$

Since $\gamma^1(q^1, c^{-1}) = 1$, and $\gamma^1(x_0) \geqslant 1$, it must be that $\gamma^1(q^2, c^{-2}) \geqslant 1$ or $\gamma^1(q^3, c^{-3}) \geqslant 1$. Similar observations hold for the other two players.

If $\gamma(q^1, c^{-1})$ were of the form $(1, -, -)$, one would have $\gamma(q^2, c^{-2}) = (+, 1, +)$ or $\gamma(q^3, c^{-3}) = (+, +, 1)$, which has been ruled out. Up to a permutation of players 2 and 3, one can assume $\gamma(q^1, c^{-1}) = (1, +, -)$. The signs of $\gamma(q^2, c^{-2})$ and $\gamma(q^3, c^{-3})$ are then given by $(-, 1, +)$ and $(+, -, 1)$.

Draw the triangle T together with the straight lines $\{x, \ \gamma^i(x) = 1\}$, for $i = 1, 2, 3$.

The set of $x \in T$ for which $\gamma(x) = (+, +, +)$ is the interior of the triangle (ABC) delineated by these straight lines. We now argue that for each x on the edges of (ABC), $\gamma(x)$ is an equilibrium payoff. Consider for instance $\gamma(A)$ and let σ be the strategy profile that plays cyclically: according to the stationary profile $(\eta, 0, 0)$ during N_1 stages, then according to $(0, \eta, 0)$ and $(0, 0, \eta)$ during N_2 and N_3 stages successively. Provided N_1, N_2, N_3 are properly chosen, the payoff induced by σ coincides with $\gamma(A)$. Provided η is small enough, in the first N_1 stages (resp. next N_2, next N_3 stages), the continuation payoff[8] moves along the segment joining $\gamma(A)$ to $\gamma(B)$ (resp., $\gamma(B)$ to $\gamma(C)$, $\gamma(C)$ to $\gamma(A)$). Therefore, σ is an ε-equilibrium profile associated with $\gamma(A)$. $\square$

Clearly, this approach relies heavily upon the geometry of the three-dimensional space. Note that, for such games, there is a stationary ε-equilibrium or an equilibrium payoff in the convex hull of $\{\gamma(q^i, c^{-i}), \ i = 1, 2, 3\}$. Solan and Vieille (2001) devised a four-player quitting game for which this property does not hold. Whether or not n-player quitting games do have equilibrium payoffs remains an intriguing open problem.[9]

An important trend in the literature is to identify classes of stochastic games for which there exist ε-equilibrium profiles [see, for instance, Thuijsman and Raghavan (1997)] that exhibit a simple structure (stationary, Markovian, etc.).

To conclude, we mention that the existence of (extensive-form) correlated equilibrium payoffs is known [Solan and Vieille (2002)].

THEOREM 9. *Every stochastic game has an (autonomous) extensive-form correlated equilibrium payoff.*

The statement of the result refers to correlation devices that send (private) signals to the players at each stage. The distribution of the signals sent in stage n depends on the signal sent in stage $n - 1$, and is independent of any other information.

IDEA OF THE PROOF. The first step is to construct a "good" strategy profile, meaning a profile that yields all players a high payoff, and by which no player can profit by a unilateral deviation that is followed by an indefinite punishment. One then constructs a correlation device that imitates this profile: the device chooses for each player a recommended action according to the probability distribution given by the profile. It also reveals to all players what its recommendations were in the previous stage. In this way, a deviation is detected immediately. $\square$

4. Zero-sum games with imperfect monitoring

These are games where, at any stage, each player receives a private signal which depends, possibly randomly, on the choices of the players [see Sorin (1992), Section 5.2

---

[8] I.e., the undiscounted payoff obtained in the subgame starting at that stage.
[9] A partial existence result is given in Solan and Vieille (2001).

for the model]. In contrast to (complete information) repeated games, dropping the perfect monitoring assumption already has important implications in the zero-sum case.

It is instructive to consider first the following striking example [Coulomb (1999)]:

	L	M	R
T	100	1*	0*
B	100	0	1

Example 3.

When player 1 plays B, he receives the signal a if either the L or M column was chosen by player 2, and the signal b otherwise. The signals to player 2, and to player 1 when he plays the T row, are irrelevant for what follows.

Note that the right-hand side

1*	0*
0	1

of the game coincides (up to an affine transformation on payoffs) with the Big Match [see Mertens (2002), Section 2], which was shown to have the value $1/2$. We now show that the addition of the L column, which is apparently dominated, has the effect of bringing the max min down to zero.

Indeed, let σ be any strategy of player 1, and let y be the stationary strategy of player 2 that plays L and R with probabilities $1 - \varepsilon$ and ε, respectively. Denote by θ the absorbing stage, i.e., the first stage in which one of the two move profiles (T, M) or (T, R) is played.

If $P_{\sigma,y}(\theta < +\infty) = 1$, then $\gamma_n(\sigma, y) \to \varepsilon$ as n goes to infinity. Otherwise, choose an integer N large enough so that $P_{\sigma,y}(N \leqslant \theta < +\infty) < \varepsilon^2$. In particular,

$$P_{\sigma,y}(\theta \geqslant N, \text{ player 1 ever plays } T \text{ after stage } N) \leqslant \varepsilon.$$

Let y' be the stationary strategy of player 2 that plays M and R with probabilities $1 - \varepsilon$ and ε, respectively, and call τ the strategy that coincides with y up to stage N, and with y' afterwards. Since (B, L) and (B, M) yield the same signal to player 1, the distributions induced by (σ, y) and by (σ, τ) on sequences of signals to player 1 coincide up to the first stage after stage N, in which player 1 plays T.

Therefore, $P_{\sigma,\tau}$-almost surely,

$$\bar{g}_n \to 0 \qquad \text{if } \theta < N,$$
$$\bar{g}_n \to 1 - \varepsilon \quad \text{if } N \leqslant \theta < +\infty,$$
$$\bar{g}_n \to \varepsilon \qquad \text{if } \theta = +\infty.$$

Since $P_{\sigma, y}(N \leqslant \theta < +\infty) \leqslant \varepsilon$, $\lim_{n \to +\infty} E_{\sigma, \tau}[\bar{g}_n] \leqslant 2\varepsilon$. Thus, player 2 can defend zero.

Since player 1 clearly guarantees 0 and can defend $1/2$, the game has no value and the maxmin is equal to zero.

The following theorem is due to Coulomb (1999, 2001).

THEOREM 10. *Every zero-sum repeated game with absorbing states and partial monitoring has a* maxmin.

SKETCH OF THE PROOF. Following the steps of Kohlberg (1974), the maxmin is first shown to exist for so-called generalized Big Match games, then for all games with absorbing states.

The class of generalized Big Match games includes Example 3. Player 1 has only two actions, T and B, while the action set S^2 of player 2 is partitioned into $\overline{S}^2$ and $\widetilde{S}^2$. For $s^2 \in \overline{S}^2$, transitions and payoffs are independent of $s^1 \in \{T, B\}$. For $s^2 \in \widetilde{S}^2$, the probability of reaching an absorbing state is positive given (T, s^2) and equals zero given (B, s^2). Coulomb (1999) characterizes the maxmin for such games; as in Example 3, it depends only on the signal structure to player 1, given the action B.

As might be expected, the maxmin is quite sensitive to the signalling structure. For instance, consider again Example 3. Assume that the signal a associated with the entry (B, M) is replaced by a random device that sends the signal a with probability $1 - \eta$, and the signal a' otherwise. If $a' = b$, the maxmin is still close to zero for η small (the M column is indistinguishable from a convex combination of the L and R columns). If $a' \neq b$, the maxmin is equal to $1/2$, whatever the value of $\eta > 0$.

Let Γ be any game with absorbing states. To any pair x^1, $\tilde{x}^1$ of distributions over S^1 is associated a (fictitious) generalized Big Match $\Gamma_{x^1, \tilde{x}^1}$, in which the B and T rows correspond, respectively, to the mixed moves x^1, and x^1 slightly perturbed by $\tilde{x}^1$. It is shown by Coulomb (2001) that the maxmin of Γ is equal to the supremum over $(x^1, \tilde{x}^1)$ of the maxmin of the auxiliary game $\Gamma_{x^1, \tilde{x}^1}$. The difficult part is to show that player 2 can defend such a quantity.   □

5. Stochastic games with large state space

Consider first a stochastic game with countable state space Ω and finite actions sets S^1 and S^2. Maitra and Sudderth (1992) prove that, with $\limsup_{n \to +\infty} g_n$ as the payoff function for the infinite game,[10] the game has a value. This result was considerably extended by Martin (1998). Let Ω, S^1 and S^2 be endowed with the discrete topology, and the set H_∞ of plays be given the product topology. Let the payoff function of the infinite game be *any* Borel function f on H_∞. (Martin does not deal with stochastic

---

[10] This payoff function includes many cases of interest, including discounted stochastic games.

games, but, as argued in Maitra and Sudderth (2002), the extension to stochastic games is immediate.)

THEOREM 11. *The game with payoff function* f *has a value.*

See Martin (1998) for the proof. The proof relies on another theorem of Martin (1975) for games of perfect information. We also refer to Maitra and Sudderth (2002) for an introduction to the proof.

We conclude by citing miscellaneous results. In stochastic games with incomplete information on one side, a lottery chooses at stage 0 the stochastic game to be played, and only player 1 is informed. Such games may be analyzed through an auxiliary stochastic game in which the current posterior held by player 2 on the true game being played is part of the state variable. It is conjectured that the maxmin exists and coincides with $\lim_{n\to\infty} v_n$ and $\lim_{\lambda\to 0} v_\lambda$. The basic intuition is that the maxmin should coincide with the value of the auxiliary game, which is not known to exist [see Mertens (1987)]. Only scattered results are available so far. This has been verified by Sorin (1984) for games of the Big Match type, and by Rosenberg and Vieille (2000) when the possible stochastic games are recursive. For games with absorbing states, it is known that $\lim_{n\to\infty} v_n = \lim_{\lambda\to 0} v_\lambda$ [Rosenberg (2000)].

References

Coulomb, J.M. (1999), "Generalized big-match", Mathematics of Operations Research 24:795–816.

Coulomb, J.M. (2001), "Absorbing games with a signalling structure", Mathematics of Operations Research 26:286–303.

Everett, H. (1957), "Recursive games", in: M. Dresher, A.W. Tucker and P. Wolfe, eds., Contributions to the Theory of Games, Vol. III (Princeton University Press, Princeton, NJ) 47–78.

Flesch, J., F. Thuijsman and O.J. Vrieze (1997), "Cyclic Markov equilibrium in stochastic games", International Journal of Game Theory 26:303–314.

Hart, S. (1985), "Non-zero-sum two-person repeated games with incomplete information", Mathematics of Operations Research 10:117–153.

Kohlberg, E. (1974), "Repeated games with absorbing states", Annals of Statistics 2:724–738.

Maitra, A., and W. Sudderth (1992), "An operator solution for stochastic games", Israel Journal of Mathematics 78:33–49.

Maitra, A., and W. Sudderth (2002), "Stochastic games with Borel payoffs", in: A. Neyman and S. Sorin, eds., Proceedings of the NATO ASI on Stochastic Games, Stony Brook, 1999 (Kluwer, Dordrecht) forthcoming.

Martin, D.A. (1975), "Borel determinacy", Annals of Mathematics 102:363–371.

Martin, D.A. (1998), "The determinacy of Blackwell games", The Journal of Symbolic Logic 63:1565–1581.

Mertens, J.F. (1987), "Repeated games", in: Proceedings of the International Congress of Mathematics, Berkeley, 1986 (American Mathematical Society, Providence, RI) 1528–1577.

Mertens, J.F. (2002), "Stochastic games", in: R.J. Aumann and S. Hart, eds., Handbook of Game Theory, Vol. 3 (North-Holland, Amsterdam) Chapter 47, 1809–1832.

Mertens, J.F., and A. Neyman (1981), "Stochastic games", International Journal of Game Theory 10:53–56.

Rosenberg, D. (2000), "Zero-sum absorbing games with incomplete information on one side: Asymptotic analysis", SIAM Journal on Control and Optimization 39:208–225.

Rosenberg, D., and N. Vieille (2000), "The maxmin of recursive games with incomplete information on one side", Mathematics of Operations Research 25:23–35.

Simon, R.S., S. Spiez and H. Torunczyk (1995), "The existence of equilibria in certain games, separation for families of convex functions and a theorem of Borsuk–Ulam type", Israel Journal of Mathematics 92:1–21.

Solan, E. (1999), "Three-player absorbing games", Mathematics of Operations Research 24:669–698.

Solan, E. (2000), "Stochastic games with two non-absorbing states", Israel Journal of Mathematics 119:29–54.

Solan, E., and N. Vieille (2001), "Quitting games", Mathematics of Operations Research 26:265–285.

Solan, E., and N. Vieille (2002), "Correlated equilibrium in stochastic games", Games and Economic Behavior, forthcoming.

Sorin, S. (1984), "Big match with lack of information on one side (part 1)", International Journal of Game Theory 13:201–255.

Sorin, S. (1992), "Repeated games with complete information", in: R.J. Aumann and S. Hart, eds., Handbook of Game Theory with Economic Applications, Vol. 1 (North-Holland, Amsterdam) Chapter 4, 71–104.

Thuijsman, F., and T.E.S. Raghavan (1997), "Perfect information stochastic games and related classes", International Journal of Game Theory 26:403–408.

Thuijsman, F., and O.J. Vrieze (1991), "Easy initial states in stochastic games", in: T.E.S. Raghavan et al., eds., Stochastic Games and Related Topics (Kluwer, Dordrecht) 85–100.

Vieille, N. (2000a), "Two-player stochastic games I: A reduction", Israel Journal of Mathematics 119:55–92.

Vieille, N. (2000b), "Two-player stochastic games II: The case of recursive games", Israel Journal of Mathematics 119:93–126.

Vieille, N. (2000c), "Small perturbations and stochastic games", Israel Journal of Mathematics 119:127–142.

Vrieze, O.J., and F. Thuijsman (1989), "On equilibria in stochastic games with absorbing states", International Journal of Game Theory 18:293–310.

Zamir, S. (1992), "Repeated games of incomplete information: zero-sum", in: R.J. Aumann and S. Hart, eds., Handbook of Game Theory with Economic Applications, Vol. I (North-Holland, Amsterdam) Chapter 5, 109–154.

Chapter 49

GAME THEORY AND INDUSTRIAL ORGANIZATION*

KYLE BAGWELL

Economics Department and Graduate School of Business, Columbia University and NBER

ASHER WOLINSKY[†]

Economics Department, Northwestern University

Contents

*Financial support from the National Science Foundation is gratefully acknowledged.
[†]We thank Rob Porter and Xavier Vives for helpful discussions.

Handbook of Game Theory, Volume 3, Edited by R.J. Aumann and S. Hart

Abstract

In this article, we consider how important developments in game theory have contributed to the theory of industrial organization. Our goal is not to survey the theory of industrial organization; rather, we consider the contribution of game theory through a careful discussion of a small number of topics within the industrial organization field. We also identify some points in which developments in the theory of industrial organization have contributed to game theory. The topics that we consider are: commitment in two-stage games and the associated theories of strategic-trade policy and entry deterrence; asymmetric-information games and the associated theories of limit pricing and predation; repeated games with public moves and the associated theory of collusion in markets with public demand fluctuations; mixed-strategy equilibria and purification theory and the associated theory of sales; and repeated games with imperfect monitoring and the associated theory of collusion and price wars. We conclude with a general assessment concerning the contribution of game theory to industrial organization.

Keywords

industrial organization, game theory, entry deterrence, strategic trade, limit pricing, predation, collusion, sales

JEL classification: D4, L1

1. Introduction

Game theory has become the standard language of industrial organization: the industrial organization theory literature is now presented almost exclusively in terms of game theoretic models. But the relationship is not totally one-sided. First, the needs of industrial organization fed back and exerted a general influence on the agenda of game theory. Second, specific ideas that grew out of problems in industrial organization gained independent importance as game theoretic topics in their own right. Third, it is mostly through industrial organization that game theory was brought on a large scale into economics and achieved its current standing as a fundamental branch of economic theory.

A systematic survey of the use of game theory in industrial organization would amount in fact to a survey of industrial organization theory. This is an enormous task that has been taken up by numerous textbooks.[1] The purpose of this article is not to survey this field, but rather to illustrate through the discussion of a small selection of subjects how some important developments in game theory have been incorporated into the theory of industrial organization and to pinpoint their contribution to this theory. We will also identify some points in which industrial organization theory made a contribution to game theory. The models discussed are selected according to two criteria. First, they utilize a relatively major game theoretic idea. The second requirement is that the use of the game theoretic idea yield a relatively sharp economic insight.

Mathematical models in economics allow ideas to be expressed in a clear and precise way. In particular, they clarify the circumstances under which ideas are valid. They also facilitate the application of mathematical techniques, which sometimes yield insights that could not be obtained by simple introspection alone.[2] We will argue below that game theoretic models in industrial organization serve both of these functions.

As mentioned above, we do not intend to survey the field of industrial organization or the most important contributions to it. As a result, many important contributions and many influential contributors are not mentioned here. This should not be misinterpreted to suggest that these contributions are unimportant or that they are less important than those that were actually selected for the survey.

2. The role of commitment: An application of two-stage games

The role of commitment to future actions as a means of influencing rivals' behavior is a central theme in the analysis of oligopolistic competition. In a typical entry deterrence scenario, for example, an incumbent monopoly firm attempts to protect its market against entry of competitors by committing to post-entry behavior that would make entry unprofitable. In other scenarios, firms make partial commitments to future behavior through decisions on the adoption of technologies or through long-term contracts with

---

[1] Tirole's (1988) comprehensive text is a standard reference.

[2] For example, some dynamic models begin with simple assumptions on, say, consumption and investment behavior which then give rise to a system that displays cyclical or even erratic aggregate behavior.

their agents. The framework used in the literature for discussing these issues is that of a multi-stage game with subgame perfect equilibrium [Selten (1965)] as the solution concept.

2.1. The basic model

Two firms, 1 and 2, interact over two stages as follows. In the first stage the firms simultaneously choose the magnitudes, k_i, $i = 1, 2$. In the second, after observing the k_i's, they choose simultaneously the magnitudes x_i, $i = 1, 2$. Firm i's profit is given by the function $\pi_i(x_i, x_j; k_i)$, $i = 1, 2$, where $j \neq i$. A strategy for firm i, $[k_i, x_i(k_i, k_j)]$, prescribes a choice of k_i for stage 1 and a choice of x_i for stage 2, as a function of the k_i's chosen in the first stage. A *subgame perfect equilibrium* (SPE) is a strategy pair, $[k_i^*, x_i^*(k_i, k_j)]$, $i = 1, 2$, such that: (A) for all (k_i, k_j), $x_i^*(k_i, k_j) = \arg\max_x \pi_i[x, x_j^*(k_j, k_i); k_i]$; (B) $k_i^* = \arg\max_{k_i} \pi_i[x_i^*(k_i, k_j^*), x_j^*(k_j^*, k_i); k_i]$.

Thus, the x_i's are the direct instruments of competition in that they enter the rival's profit directly, while the k_i's have only an indirect effect. In many applications, the interpretation given is that the k_i's represent productive capacities or technologies and the x_i's describe quantities or prices of the final product. With the two-stage structure, k_i has a dual role: besides being a direct ingredient in the firm's profit, independently of the interaction, it also has a strategic role of influencing the rival's behavior in the second-stage subgame. The manner in which k_i affects x_j is credible in the sense of the SPE concept: k_i affects x_j only through shifting the second-stage equilibrium.

Perhaps the main qualitative result of this model, in its general form, is that the strategic role for k_i results in a distortion of its equilibrium level away from the level that would be optimal were x_j unaffected by k_i. When k_i is interpreted as capacity, this result means over or under investment in capacity as may be the case. The following proposition gives a precise statement of this result. Assume that π_i, $i = 1, 2$, is differentiable, $\partial \pi_i / \partial x_j \neq 0$, there exists a unique SPE equilibrium, and x_i^* is a differentiable function of (k_i, k_j).

PROPOSITION 2.1. *If* $\partial x_j^*(k_j^*, k_i^*)/\partial k_i \neq 0$, *then*

$$\partial \pi_i [x_i^*(k_i^*, k_j^*), x_j^*(k_j^*, k_i^*); k_i^*]/\partial k_i \neq 0.$$

PROOF. All the following derivatives are evaluated at the SPE point $[x_i^*(k_i^*, k_j^*), k_i^*]$, $i = 1, 2$. The first order condition for equilibrium condition (B) is $d\pi_i/dk_i = 0$, or

$$d\pi_i/dk_i \equiv (\partial \pi_i/\partial x_i)(\partial x_i^*/\partial k_i) + (\partial \pi_i/\partial x_j)(\partial x_j^*/\partial k_i) + (\partial \pi_i/\partial k_i) = 0.$$

Using the first order condition for condition (A), $\partial \pi_i/\partial x_i = 0$, we get

$$d\pi_i/dk_i = (\partial \pi_i/\partial x_j)(\partial x_j^*/\partial k_i) + (\partial \pi_i/\partial k_i) = 0$$

from which the proposition follows directly.   □

As a benchmark for comparison, consider the single-stage version of this game in which the firms choose (k_i, x_i), $i = 1, 2$, simultaneously. The Nash equilibrium [Nash (1950)] of this game is $(\overline{k}_i, \overline{x}_i)$, $i = 1, 2$, such that $(\overline{k}_i, \overline{x}_i) = \arg\max_{k,x} \pi_i(x, \overline{x}_j; k)$.

COROLLARY. $\overline{k}_i$ *is in general different from* k_i^*.

PROOF. The first order condition for the equilibrium in the one-stage game is

$$\partial \pi_i(\overline{x}_i, \overline{x}_j; \overline{k}_i)/\partial k_i = 0.$$

Since $x_i^*(\overline{k}_i, \overline{k}_j) = \overline{x}_i$, it follows that

$$\partial \pi_i[x_i^*(\overline{k}_i, \overline{k}_j), x_j^*(\overline{k}_j, \overline{k}_i); \overline{k}_i]/\partial k_i = \partial \pi_i(\overline{x}_i, \overline{x}_j; \overline{k}_i)/\partial k_i = 0.$$

Therefore, it has to be that $(\overline{k}_i, \overline{k}_j) \neq (k_i^*, k_j^*)$, or else Proposition 2.1 will be contradicted. □

From a conceptual point of view the two-stage oligopoly model is of course straightforward, and the main result of this model is an obvious implication of the SPE concept. The two-stage model does, however, provide a useful framework for discussing the role of commitment in oligopolistic competition. First, it embodies a clear notion of credibility. Second, it thereby serves to identify the features that facilitate effective commitment: durable decisions that become observable before the actual interaction takes place. Third, it has been applied to a variety of specific economic scenarios and yielded interesting economic insights. The previous literature recognized the potential importance of commitment, but it did not manage to organize and understand the central idea in the clear form that the above model does.

2.2. An application to strategic trade theory

To see the type of economic insight that this model generates, consider its application to the theory of international trade [Brander and Spencer (1985)]. Two firms, 1 and 2, are based in two different countries and export their products to a third country (the rest of the world). The product is homogeneous, production costs are 0 and the demand is given by $p = 1 - Q$, where Q is the total quantity. The interaction unfolds in two stages. In the first stage, the two governments simultaneously choose excise tax (subsidy) rates, t_i, to be levied on their home firms. In the second stage, the tax rules are observed and the firms play a Cournot duopoly game: they simultaneously choose outputs q_i and the price is determined by $p = 1 - q_1 - q_2$. The effective cost functions in the second stage are $c_i(q_i; t_i) = t_i q_i$. The objective of firm i is maximization of its after-tax profit,

$$F_i(q_i, q_j; t_i) = (1 - q_i - q_j)q_i - t_i q_i.$$

The objective of each government is maximization of its country's "true" profit: the sum of its firm's profit and the tax revenue,

$$G_i(q_i, q_j; t_i) = F_i(q_i, q_j; t_i) + t_i q_i.$$

Since the government cares only about the sum, it chooses to tax only if the tax plays a strategic role and manipulates the second-stage competition in favor of its firm.

This application may be analyzed using the two-stage model developed above, although strictly speaking this is a slightly different case. The difference is that the stage-one commitments are now made by different parties (the exporting governments) than those who interact in stage two (the firms). However, the analysis remains the same. (The function G_i and the variables q_i and t_i in this case correspond to π_i, x_i and k_i in the general model above.) Solving for the SPE of this two-stage game, we get that the governments subsidize their firms: $t_1 = t_2 = -1/5$. In comparison to the equilibrium in the absence of government intervention, outputs and firms' profits are higher but countries' profits are lower. The intuition becomes more transparent by looking at the reaction functions, $R_i(q_j; t_i) = \arg\max_{q_i} F_i[q_i, q_j; t_i]$, depicted by Figure 1. The solid curves correspond to the case with no tax or subsidy. Their intersection point gives the second-stage equilibrium in this case. The dashed R_1 curve corresponds to a subsidy for firm 1, and its intersection with the R_2 curve gives the equilibrium when firm 1 is subsidized and firm 2 is not. The subsidy makes firm 1 more aggressive in the sense that, for any expectation that it might have regarding firm 2's output, it produces more than it would with no subsidy. This induces firm 2 to contract its output in equilibrium. Notice that, for a given output of firm 2, country 1's profit is higher with no subsidization, since the subsidy induces its firm to produce "too much". But the strategic effect on the other

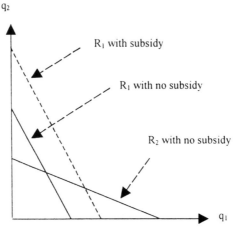

Figure 1.

firm's output makes subsidization profitable and in equilibrium both governments offer export subsidies.

This is a striking insight that provides a clear and plausible explanation for export subsidies. To believe this explanation, one need not suppose that the governments see clearly through these strategic considerations. It is enough that they somehow think that subsidization improves their firms' competitive position. Also, despite its simplicity, this insight truly requires the game theoretic framework: it cannot be obtained without rigorous consideration of the strategic consequences of export policy.

But further thought reveals that this insight is somewhat less convincing than it might have seemed at first glance. Consider an alternative version of the model [Eaton and Grossman (1986)] in which the second-stage competition is a differentiated product Bertrand game: the firms simultaneously choose prices p_i and the demands are $q_i(p_i, p_j) = 1 - p_i + \alpha p_j$, with $0 < \alpha < 1$. Now,

$$F_i(p_i, p_j; t_i) = (1 - p_i + \alpha p_j)(p_i - t_i)$$

and

$$G_i(p_i, p_j; t_i) = F_i(p_i, p_j; t_i) + t_i(1 - p_i + \alpha p_j).$$

Repeating the above analysis for this case (now, the variable p_i corresponds to x_i in the general model), the reaction functions are $R_i(p_j; t_i) = \arg\max_{p_i} F_i[p_i, p_j; t_i]$. Figure 2 depicts the reaction functions in this case. Here, too, the dashed reaction function of firm 1 corresponds to a subsidy for firm 1. As before, the subsidy makes firm 1 more aggressive, inducing it to charge a lower price for any expectation it holds. But here this change in firm 1's position induces firm 2 to choose a lower price (as seen by

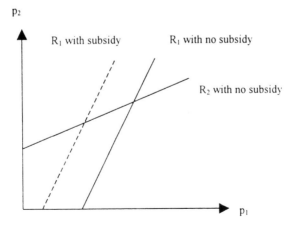

Figure 2.

comparing the two intersection points), and so an export subsidy now has a strategic cost, as it results in more aggressive behavior by the rival firm. Indeed, in the equilibrium of this scenario, the governments *tax* their firms at level $t_1 = t_2 = \alpha^2(2+\alpha)/[8 - 4\alpha^2 - \alpha^3]$, and prices and countries' profits are above their counterparts in the absence of intervention.

This, too, is a very clear insight. But, of course, the results here are almost the exact opposites of the results obtained above with the second-stage Cournot game. There are two views regarding the implications of this contrast. The more skeptical view maintains that the simple one-shot models of Cournot and Bertrand are only some sort of a parable. They are meant to capture the idea that in oligopolistic competition firms are aware that their rivals are also rational players who face similar decisions, and to point out that this sort of interaction might result in an inefficient outcome from the firms' perspective. But they are not meant to provide realistic descriptions of such competition. Thus, observations which depend on finer features of the structure of these models should not be regarded as true substantive insights. According to this view then, the only substantive insight here is that in principle there *might* be a strategic motive for the taxation/subsidization of exports. But the ambiguity of the results does not allow a useful prediction; in fact, it makes it hard to believe that this is a significant consideration in such scenarios.

A less skeptical view maintains that there is indeed a meaningful distinction between the sort of situations that are captured by the Cournot and Bertrand models. It can be argued that oligopolistic competition involves investments in production technologies (or capacities) followed by pricing decisions. The important strategic features of the Cournot model can be associated with the investment decisions, whereas the Bertrand model captures the pricing decisions.[3] The results in the Cournot case thus rationalize strategic subsidization of research and development or other investment activities aimed at reducing cost or expanding capacity in export industries. So this view attributes to this analysis further content than the general insight that export subsidization might have a strategic role. It interprets the diverse predictions of the models of Cournot and Bertrand as reflecting important differences in the environment (e.g., regarding the age of the industry), which may understandably affect the outcome.

As mentioned above, the two-stage model reviewed here has been similarly employed in a number of different economic applications (e.g., capital investment, managerial incentive schemes). The sharp distinction between the predictions of the Cournot and Bertrand models appears in many of these applications.[4] In light of this, it is useful to emphasize that the qualitative effects described in the next application arise independently of the form of oligopoly competition.

---

[3] This understanding is related to the analysis of Kreps and Scheinkman (1983) and subsequent work.

[4] Bulow, Geanakoplos and Klemperer (1985) develop a general framework to which most of these applications belong. They also coined the terms "strategic substitutes" and "strategic complements" to describe the cases of downward- and upward-sloping reaction functions, respectively. See also Fudenberg and Tirole (1984).

2.3. Application to the question of entry deterrence

We close this section by reviewing briefly the development of the theoretical literature on entry deterrence that led to the adoption of the two-stage game framework discussed above. The manner in which this literature struggled with the concept of commitment may help to illustrate the nontrivial contribution of the above described framework toward improving the quality of this discussion. We do not expand here on the economic motivations of this literature, since these issues will be discussed in the next section.

Although earlier contributions to this literature took a variety of forms, it is convenient to present the ideas in the context of the two-stage framework of this section.[5] An incumbent monopoly and a potential entrant interact over two periods. In the first (pre-entry) period, the incumbent selects a price, which is observed by the entrant. In the second (post-entry) period, the entrant decides whether or not to enter. Entry entails a fixed cost, and the incumbent's profit in the post-entry interaction with the entrant is lower than its profit as a continuing monopoly.

The earlier literature explored a particular model of this form and developed the notion of a *limit price*. In the context of this model, the incumbent limit prices when it chooses a relatively low price, typically lower than the regular monopoly price, that would render entry unprofitable, if this price were to prevail as well in the post-entry period. The potential entrant then responds by staying out. This model, however, entails the seemingly implausible assumption that the incumbent would choose to maintain its pre-entry price in the event that entry actually occurred. Furthermore, if we are unwilling to make this assumption, then it is no longer clear why the pre-entry price should affect the expected profit from entry and thus the entry decision itself. If we think, for example, that the post-entry interaction in fact takes the form of a standard duopoly game, and that the post-entry demand and cost functions are independent of the pre-entry price, then the potential entrant's expected duopoly profit should also be independent of the pre-entry price. It is therefore doubtful that the entry of rational competitors can be blocked in this manner. This suggests that limit pricing emerges as part of a credible entry deterrence strategy, only if some alternative mechanism (other than price commitment) is identified that links the incumbent's pre-entry price to the potential entrant's expected profit from entry.

Motivated by this understanding, the next step in the development of this theory [Spence (1977)] introduced the possibility that the incumbent selects a level of capacity in the pre-entry period. An investment in capacity is plausibly irreversible, and so an investment of this kind is a natural means through which an incumbent may credibly commit to be an active participant in the market. In particular, the idea was that entry would be deterred when an incumbent invested significantly in capacity, if the incumbent were to threaten that it would utilize its capacity to the fullest extent in the

---

[5] Important contributions to the early literature on limit pricing include those by Bain (1949) and Modigliani (1958).

post-entry interaction. The notion of a limit price gets here a different meaning. The pre-entry price is no longer a strategic instrument for blocking entry. But if the entry-deterring investment level reduces the incumbent's marginal cost relative to that of an unthreatened monopoly, the entry threat might have the effect of lowering the incumbent's pre-entry price. This would have the appearance of a limit price, but it is actually only a by-product of the entry-deterring investment.

While Spence's model identified capacity as a plausible pre-entry commitment variable, it still did not evaluate the credibility of the threatened utilization of the installed capacity. In fact, the threatened entry-deterring output is not always credible in the SPE sense – the equilibrium in a post-entry duopoly game does not necessarily entail utilization of the capacity installed as a threat. This shortcoming was addressed by the next step of this theory [Dixit (1980)] which introduced a formal two-stage game model, of the family discussed in this section, with SPE as the solution concept. The incumbent's threat backed by its pre-entry investment is credible in the sense that it manifests itself through its effect on the equilibrium of the post-entry duopoly game. Appropriate versions of this model thus explain excessive investment in capacity and the associated low pre-entry price as credible responses to entry threats.

From the viewpoint of pure game theory, the final model that capped this literature is rather straightforward. But the extent of its contribution, even if only to sharpen the relationship between price, capacity and deterrence, should be evident from looking at the long process of insightful research that led to that point.

3. Entry deterrence and predation: Applications of sequential equilibrium

It is widely agreed that unhindered exercise of monopoly power generally results in inefficient resource allocation and that anti-trust policy aimed at the prevention of monopolization is therefore a legitimate form of government intervention in the operation of markets. A major concern of anti-trust policy has been the identification and prevention of practices that lead to monopolization of industries. This concern has motivated a large body of theory aimed at understanding and classifying the different forms that monopolization efforts might take and their economic consequences.

Monopolization takes a variety of forms ranging from more cooperative endeavors, like merger and cartelization, to more hostile practices, like entry deterrence and predation. The last part of the previous section described some of the developments in the understanding of entry deterrence by means of pricing and preemptive investment. The related notion of predation refers to attempts to induce the exit of competitors by using similar aggressive tactics. The treatment of predation and entry deterrence raises some subtle issues, both in theory and in practice, since it is naturally difficult to distinguish between "legitimate" competitive behavior that enhances efficiency and anti-competitive behavior that ultimately reduces welfare.

In fact, it has been often argued that predation or entry deterrence through aggressive pricing behavior is not a viable strategy among rational firms. The implication is that instances of aggressive price cutting should not be interpreted along these lines. The logic of this argument was explained in the previous section: a credible predatory or entry-deterring activity must create a meaningful *commitment link* to the behavior of this firm in its future interactions. But, with rational players, an aggressive pricing policy does not in itself plausibly constitute such a commitment. This argument leads to the following conclusion: when aggressive pricing appears in instances of entry deterrence (or predation), it is a by-product of a strategic investment in capacity, rather than a strategic instrument in its own right.

The following discussion exposes a significant limitation to this conclusion. It shows that, when informational differences about cost (or other parameters that affect profit) are important, it is possible to revive the traditional view of the limit price literature that pricing policies may serve as direct instruments of monopolization. The idea is that, in the presence of such asymmetric information, prices might also transmit information and as such play a direct role in influencing the entry or exit decisions of rivals.

3.1. Limit pricing under asymmetric information

Milgrom and Roberts (1982a) consider the classic scenario of a monopoly incumbent facing a single potential entrant. The novel feature of their analysis is that the incumbent has some private information regarding its costs of production. This information is valuable to the entrant, as its post-entry profit is affected by the incumbent's level of cost: the lower is the incumbent's cost, the lower is the entrant's profit from entry. Thus, in place of the commitment link studied by the earlier literature, Milgrom and Roberts propose an *informational link* between the incumbent's pre-entry behavior and the entrant's expected post-entry profit.

The situation can be modeled as a signaling game: the entrant attempts to infer the cost information by observing the incumbent's pre-entry pricing, while the incumbent chooses its price with the understanding that this choice may affect the prospect of entry. The incumbent would like the entrant to think that its costs are low to make entry seem less profitable. As in other signaling models, the equilibrium price-signal therefore may be distorted away from its myopic level. In the present context, the price is distorted if it differs from the myopic monopoly price (i.e., the price that would prevail in the absence of the entry threat). To correspond to the original limit price conjecture, the equilibrium price has to be distorted downwards from the monopoly level. But since lower costs naturally lead to lower monopoly prices, the downwards distortion is indeed the expected result in a signaling scenario.

The details of the game are as follows.[6] The players are an incumbent monopoly firm and a potential entrant. The interaction spans two periods: the pre-entry period in which the entrant observes the incumbent's behavior and contemplates the entry decision, and

---

[6] The discussion here expands on previous presentations by Bagwell and Ramey (1988) and Fudenberg and Tirole (1986).

the post-entry period after the entry decision is resolved in either way. The incumbent is one of two possible types of firm differing in their per-unit cost of production: type $t \in \{L, H\}$ has unit cost $c(t)$, where $c(L) < c(H)$ so that L and H stand for "low" and "high", respectively. In the pre-entry period, only the incumbent knows its type. It chooses a pre-entry price, p, which the entrant observes and on the basis of which decides whether to enter. The incumbent's profit in the pre-entry period is $\Pi(p, t) = [p - c(t)]D(p)$, where D is a well-behaved downwards-sloping demand function. We abstract from the details associated with the play of firms in the post-entry period, and simply summarize the outcomes of that period. If the entrant does not enter, then its profit is 0 and the incumbent remains a monopoly and earns $\pi^m(t)$; if the entrant does enter, it learns the incumbent's cost and the resulting post-entry duopoly profits are $\pi^d(t)$ for the incumbent and $\pi^e(t)$ to the entrant. It is assumed that entry reduces the incumbent's post-entry profit, $\pi^m(t) > \pi^d(t)$, and that the entrant can recover any fixed costs associated with entry only against the high-cost incumbent, $\pi^e(H) > 0 > \pi^e(L)$. Note that $\pi^m(t)$ admits a variety of interpretations: it might be simply the discounted maximized value of $\Pi(p, t)$, or it might pertain to a different length of time and/or reflect some further future interactions.

The game theoretic model is then a simple sequential game of incomplete information. The formal description is as follows. Nature chooses the incumbent's type $t \in \{L, H\}$ with probability b_t^0, where $b_L^0 + b_H^0 = 1$. The incumbent's strategy is a pricing function $P : \{L, H\} \rightarrow [0, \infty)$. The entrant's belief function $b_t : [0, \infty) \rightarrow [0, 1]$ describes the probability it assigns to type t, given the incumbent's price p. Of course, for all p, $b_L(p) + b_H(p) = 1$. A strategy for the entrant is a function $E : [0, \infty) \rightarrow \{0, 1\}$ that describes the entry decision as a function of the incumbent's price, where "1" and "0" represent the entry and no-entry decisions, respectively. The payoffs as functions of price $p \in [0, \infty)$, entry decision $e \in \{0, 1\}$, and cost type $t \in \{L, H\}$ are: for the incumbent,

$$V(p, e, t) = \Pi(p, t) + e\pi^d(t) + (1 - e)\pi^m(t)$$

and for the entrant,

$$u(p, e, t) = e\pi^e(t).$$

It is convenient to introduce special notation for the entrant's expected payoff evaluated with its beliefs. Letting b denote the pair of functions (b_L, b_H),

$$U(p, e, b) = b_L(p)u(p, e, L) + b_H(p)u(p, e, H).$$

Observe that $U(p, 0, b) = 0$ and $U(p, 1, b) = b_L(p)\pi^e(L) + b_H(p)\pi^e(H)$.

The solution concept is *sequential equilibrium* [Kreps and Wilson (1982a)] augmented by the "intuitive criterion" refinement [Cho and Kreps (1987)]. For the present game, a sequential equilibrium is a specification of strategies and beliefs, $\{P, E, b\}$, satisfying three requirements:

(E1) Rationality for the incumbent:

$$P(t) \in \arg\max_{p} V\big(p, E(p), t\big), \quad t = L, H.$$

(E2) Rationality for the entrant:

$$E(p) \in \arg\max_{e} U\big(p, e, b(p)\big) \quad \text{for all } p \geqslant 0.$$

(E3) Bayes-consistency:

$$P(L) = P(H) \quad \text{implies} \quad b_L\big(P(L)\big) = b_L^0;$$
$$P(L) \neq P(H) \quad \text{implies} \quad b_L\big(P(L)\big) = 1, \; b_L\big(P(H)\big) = 0.$$

As (E3) indicates, there are two types of sequential equilibria. In a *pooling equilibrium* ($P(L) = P(H)$), the entrant learns nothing from the observation of the equilibrium price, and so the posterior and the prior beliefs agree; whereas in a *separating equilibrium* ($P(L) \neq P(H)$), the entrant is able to infer the incumbent's cost type upon observing the equilibrium price.

For this game, sequential equilibrium places no restrictions on the beliefs that the entrant holds when a deviant price $p \notin \{P(L), P(H)\}$ is observed. For example, the analyst may specify that the entrant is very optimistic and infers high costs upon observing a deviant price. In this event, the incumbent may be especially reluctant to deviate from a proposed equilibrium, and so it becomes possible to construct a great many sequential equilibria. The set of sequential equilibria here will be refined by imposing the following "plausibility" restriction on the entrant's beliefs following a deviant price:

(E4) Intuitive beliefs:

$$b_t(p) = 1 \quad \text{if for } t \neq t' \in \{L, H\},$$
$$V(p, 0, t) \geqslant V\big[P(t), E\big(P(t)\big), t\big] \quad \text{and}$$
$$V(p, 0, t') < V\big[P(t'), E\big(P(t')\big), t'\big].$$

The idea is that an incumbent of given type would never charge a price p that, even when followed by the most favorable response of the entrant, would be less profitable than following the equilibrium. Thus, if a deviant price is observed that could possibly improve upon the equilibrium profit only for a low-cost incumbent, then the entrant should believe that the incumbent has low costs. In what follows, we say that a triplet $\{P, E, b\}$ forms an *intuitive equilibrium* if it satisfies (E1)–(E3) and (E4).

Before turning to the equilibrium analysis, we impose some more structure on the payoffs. The first assumption is just a standard technical one, but the second injects a further meaning to the distinction between the types by ensuring that the low-cost type is also more eager to prevent entry:

(A1) The function Π is well behaved:

$\exists \, \widetilde{p} > c(H)$ such that $D(p) > 0$ iff $p < \widetilde{p}$, and Π is strictly concave and differentiable on $(0, \widetilde{p})$.

(A2) The low-cost incumbent loses more from entry:
$$\pi^m(L) - \pi^d(L) \geqslant \pi^m(H) - \pi^d(H).$$

Assumption (A2) is not obviously compelling. It is natural to assume that the low-cost incumbent fares better than the high-cost one in any case, $\pi^m(L) > \pi^m(H)$ and $\pi^d(L) > \pi^d(H)$, but this does not imply the assumed relationship. This assumption is satisfied in a number of well-behaved specifications of the post-entry duopolistic interaction, but it might be violated in other standard examples.

We observe next that a low-cost incumbent is more attracted to a low pre-entry price than is a high-cost incumbent, since the consequent increase in demand is less costly for the low-cost incumbent. Formally, for $p, p' \in (0, \widetilde{p})$ such that $p < p'$,

$$\Pi(p, L) - \Pi(p, H) \equiv \big[c(H) - c(L) \big] D(p) > \big[c(H) - c(L) \big] D(p')$$
$$\equiv \Pi(p', L) - \Pi(p', H).$$

This together with assumption (A2) immediately imply the following *single crossing property* (SCP):

For any $p < p'$ and $e \leqslant e'$, if $V(p, e, H) = V(p', e', H)$,

then $V(p, e, L) > V(p', e', L)$.

Under SCP, if a high-cost incumbent is indifferent between two price-entry pairs, then a low-cost incumbent prefers the pair with a lower price and a (weakly) lower rate of entry. In particular, to deter entry the low-cost incumbent would be willing to accept a deeper price cut than would the high-cost incumbent. As is true throughout the literature on signaling games, characterization and interpretation of the equilibria is straightforward when the preferences of the informed player, here the incumbent, satisfy an appropriate version of SCP.

Let $p_t^m = \arg\max_p \Pi(p, t)$. This is the (myopic) monopoly price of an incumbent of type t. Under our assumptions, it is easily confirmed that the low-cost monopoly price is less than that of the high-cost incumbent: $p_L^m < p_H^m$. Consider now the set of prices p such that

$$V(p, 0, H) \leqslant V\big(p_H^m, 1, H\big).$$

The concavity of Π in p assures that this inequality holds outside an interval $(\underline{p}, \overline{p}) \subset [0, \widetilde{p}]$, and its reverse holds inside the interval. Thus, $\underline{p}$ and $\overline{p}$ are the prices which give the high-cost incumbent the same payoff when it deters entry as when it selects the high-cost monopoly price and faces entry. Since entry deterrence is valuable, it follows directly that $\underline{p} < p_H^m < \overline{p}$.

Let $\widetilde{b}_L$ denote the belief that would make the entrant exactly indifferent with respect to entry. It is defined by

$$\widetilde{b}_L \pi^e(L) + (1 - \widetilde{b}_L)\pi^e(H) = 0.$$

PROPOSITION 3.1.

(i) *There exists a separating intuitive equilibrium.*

(ii) *If $p_L^m \geqslant \underline{p}$, then any separating intuitive equilibrium, $\{P, E, b\}$, satisfies $\underline{p} = P(L) < \bar{P}(H) = p_H^m$ and $E(P(L)) = 0 < 1 = E(P(H))$.*

   If $p_L^m < \underline{p}$, then any separating intuitive equilibrium, $\{P, E, b\}$, satisfies $p_L^m = P(L) < \bar{P}(H) = p_H^m$ and $E(P(L)) = 0 < 1 = E(P(H))$.

(iii) *If $p_L^m \geqslant \underline{p}$, and $b_L^0 \geqslant \tilde{b}_L$, then for every $p \in [\underline{p}, p_L^m]$, there exists an intuitive pooling equilibrium in which $P(L) = P(H) = p$.*

(iv) *In any intuitive pooling equilibrium, $P(L) = P(H) \in [\underline{p}, p_L^m]$ and $E(P(L)) = 0$.*

The proof is relegated to the appendix. The proposition establishes uniqueness of the separating equilibrium outcome. Since the high-cost incumbent faces entry in the separating equilibrium, its equilibrium price must coincide with its monopoly price p_H^m. Otherwise, it would clearly profit by deviating to p_H^m from any other price at which it anyway faces entry. The case $p_L^m \geqslant \underline{p}$ corresponds to a relatively small cost differential between the two types of incumbent. It is the more interesting case, since the separating equilibrium price quoted by the low-cost incumbent is then distorted away from its monopoly price p_L^m. The case $p_L^m < \underline{p}$ corresponds to relatively large cost differential which renders p_L^m a dominated choice for the high-cost incumbent and hence removes the tension associated with the high-cost incumbent's incentive to mimic the low-cost price.

When the prior probability of the low-cost type is sufficiently large, $b_L^0 \geqslant \tilde{b}_L$, there are also pooling equilibria. These equilibria do not exist when $b_L^0 < \tilde{b}_L$, since at a putative pooling equilibrium the entrant would choose to enter and hence the incumbent would profit from deviating to its monopoly price p_l^m.

Both the separating and the pooling equilibria exhibit limit price behavior, but these patterns are qualitatively very different. The separating equilibrium exhibits limit pricing in the sense that, for certain parameter values, $P(L) < p_L^m$. But the separating equilibrium differs from the traditional limit price theory in the important sense that equilibrium limit pricing does not deter entry, which occurs under the same conditions (namely, when the incumbent has high costs) that would generate entry in a complete-information setting. The effect of the limit price on entry is through the cost information that it credibly reveals to the entrant. Limit pricing also occurs in the pooling equilibria. The high-cost incumbent now practices limit pricing, as $P(H) < p_H^m$ in any pooling equilibrium, and the low-cost incumbent also selects a limit price, as $P(L) < p_L^m$ in all these equilibria save the one in which pooling occurs at p_L^m. In contrast with the limit pricing of the separating equilibrium, and in accordance with the traditional notion, the limit price here does deter entry. The rate of entry is lower than would occur under complete information, since the high-cost incumbent is able to deter entry when it pools its price with that of the low-cost incumbent.

Earlier literature on the traditional notion of limit pricing associated with this practice a welfare trade-off: lower prices generate immediate welfare gains but deter or reduce entry and thus lead to future welfare losses. The form of limit pricing that arises under the separating equilibrium is actually beneficial for welfare, since the low-cost incumbent signals its information with a low price and this does not come at the expense of entry. Instead, it is the pooling equilibria that exhibit the welfare trade-off that the earlier literature associated with limit pricing. While the low pre-entry prices tend to improve welfare, the reduction in entry lowers welfare in the post-entry period, as compared to the welfare that would be achieved in a complete-information setting.

The set of equilibria may be further refined with a requirement that the selected equilibrium is Pareto efficient for the low- and high-cost incumbent among the set of intuitive equilibria. When pooling equilibria exist (the conditions of part (iii) of the proposition hold), then the pooling equilibrium in which the low-cost monopoly price is selected is the efficient one for the low- and high-cost incumbent in the relevant set. This equilibrium gives the low-cost incumbent the maximum possible payoff. It also offers a higher payoff to the high-cost incumbent than occurs in the separating equilibrium, since $p_L^m \geqslant \underline{p}$ implies that

$$V\left(p_L^m, 0, H\right) \geqslant V(\underline{p}, 0, H) = V\left(p_H^m, 1, H\right).$$

3.2. Predation

Generally speaking, a firm practices predatory pricing if it charges "abnormally" low prices in an attempt to induce exit of its competitors. The ambiguity of this definition is not incidental. It reflects the inherent difficulty of drawing a clear distinction between legitimate price competition and pricing behavior that embodies predatory intent. Indeed, an important objective of the theoretical discussion of this subject is to come up with relatively simple criteria for distinguishing between legitimate price competition and predation.

Exit-inducing behavior is of course closely related to entry-deterring behavior and, indeed, a small variation on the limit-pricing model presented above can also be used to discuss predation. Consider, then, the following variation on the limit price game. In the first period, both firms (referred to as the "predator" and the "prey") are in the market and choose prices simultaneously. Then the prey, who must incur a fixed cost if it remains in the market, decides whether or not to exit. Finally, in the second period, the active firms again choose prices. If the prey exits, the predator earns monopoly profit; on the other hand, if the prey remains in the market, the two firms earn some duopoly profits. In equilibrium, the prey exits when its expected period-two profit is insufficient to cover its fixed costs. Clearly, in any SPE of the complete-information version of this game, no predation takes place, as the prey's expectation is independent of the predator's first-stage price. However, when the prey is uncertain about the predator's cost, as in the above limit-pricing model, then an informational link appears between the predator's first-period price and the prey's expected profit from remaining in the market. Recognizing that the prey will base its exit decision upon its inference of the predator's cost type, the predator may price low in order to signal that its costs are low and thus induce exit. The equilibria of this model are analogous to those described in Proposition 3.1, with exit occurring under the analogous circumstances to those under which entry was deterred. This variation provides an equilibrium foundation for the practice of predatory pricing, in which predation is identified with low prices that are selected with the intention of affecting the exit decision.

From a welfare standpoint, the predation that occurs as part of a separating equilibrium is actually beneficial. Predation brings the immediate welfare benefit of a lower price, and it induces the exit of a rival in exactly the same circumstances as would occur in a complete-information environment. When the game is expanded to include an initial entry stage [Roberts (1985)], however, a new wrinkle appears, as the rational anticipation of predatory signaling may deter the entry of the prey, resulting in a possible welfare cost.

3.3. Discussion

The notion that limit pricing can serve to deter entry has a long history in industrial organization, with a number of theoretical and empirical contributions.[7] The signaling model of entry deterrence contributed to this literature a number of new theoretical insights. First and foremost, it identified two patterns of rational behavior that may be interpreted in terms of the "anti-competitive" practices of entry deterrence and predation. One pattern, exemplified by the pooling equilibria, exhibits anti-competitive behavior in its traditional meaning of eliminating competition that would otherwise exist. The other pattern, exemplified by the separating equilibrium, takes the appearance of anti-competitive behavior but does not exhibit the traditional welfare consequences of such behavior. These observations have a believable quality to them both because the main element of this model is asymmetric information, which is surely often present in such situations, and because the pooling and separating equilibria have natural and intuitive interpretations. Furthermore, continued research has shown that the basic ideas of this theory are robust on many fronts.[8]

Even if these insights had been on some level familiar prior to the introduction of this model, and this is doubtful, they surely had not been understood as implications of a closed and internally consistent argument. In fact, it is hard to envision how these

---

[7] A recent empirical analysis is offered by Kadiyali (1996), who studies the U.S. photographic film industry and reports evidence that is consistent with the view that the incumbent (Kodak) selected a low price and a high level of advertising in the presence of a potential entrant (Fuji).

[8] There are, however, a couple of variations which alter the results in important ways. First, as Harrington (1986) shows, if the entrant's costs are positively correlated with those of the incumbent, then separating equilibria entail an upward distortion in the high-cost incumbent's price. Second, Bagwell and Ramey (1991) show that, when the industry hosts several incumbents who share private information concerning industry costs, a focal separating equilibrium exists that entails no pricing distortions whatsoever.

insights could be derived or effectively presented without the game theoretic framework. So this part of the contribution, the generation and the crisp presentation of these insights, cannot be doubted.

Still there is the question of whether these elegant insights change significantly our understanding of actual monopolization efforts. Here, it is useful to distinguish between qualitative and quantitative contributions. Certainly, as the discussion above indicates, the signaling model of entry deterrence offers a significant qualitative argument that identifies a possible role for pricing in anti-competitive behavior. It is more difficult, however, to assess the extent to which this argument will lead to a quantitative improvement in our understanding of monopolization efforts. For example, is it possible to identify with confidence industries in which behavior is substantially affected by such considerations? Can we seriously hope to estimate (even if only roughly) the quantitative implication of this theory? We do not have straightforward answers to these questions. The difficulties in measurement would make it quite hard to distinguish between the predictions of this theory and others.

Even the qualitative features of the argument may be of important use in the formulation of public policy. For example, U.S. anti-trust policy aims at curbing monopolization and specifically prohibits predatory pricing. However, the exact meaning of this prohibition as well as the manner in which it is enforced have been subject to continued review over time by the government and the courts. Two ongoing debates that influence the thinking on this matter are as follows: is predation a viable practice in rational interaction? If the possibility of predation is accepted, what is the appropriate practical definition of predation? The obvious policy implication in case predation is not deemed to be a viable practice is that government intervention is not needed. In the absence of a satisfactory framework that can supply precise answers, these policy decisions are shaped by weighing an array of incomplete arguments. Historically, some of the most influential arguments have been developed as simple applications of basic price theoretic models. Prominent among these are the "Chicago School" arguments that deny the viability of predation [McGee (1958)] and the Areeda–Turner Rule (1975) that associates the act of predation with a price that lies below marginal cost.

With this in mind, there is no doubt that the limit-pricing model enriches the arsenal of arguments in a significant way. First, it provides a theoretical framework that clearly establishes the viability of predatory behavior among rational competitors. Second, it raises some questions regarding the practical merit of cost-based definitions of predation, like the Areeda–Turner standard: it shows that predation might occur under broader circumstances than such standards admit. In a world in which a government bureaucrat or a judge has to reach a decision on the basis of imprecise impressions, arguments that rely on the logic of this theory may well have important influence.[9]

---

[9] Other game theoretic treatments of predation, like the reputation theories of Kreps and Wilson (1982b) and Milgrom and Roberts (1982b) or the war of attrition perspective of Fudenberg and Tirole (1987), also provide similar intellectual underpinning for government intervention in curbing predation.

4. Collusion: An application of repeated games

One of the main insights drawn from the basic static oligopoly models concerns the inefficiency (from the firms' viewpoint) of oligopolistic competition: industry profit is not maximized in equilibrium.[10] This inefficiency creates an obvious incentive for oligopolists to enter into a collusive agreement and thereby achieve a superior outcome that better exploits their monopoly power. Collusion, however, is difficult to sustain, since typically each of the colluding firms has incentive to opportunistically cheat on the agreement. The sustenance of collusion is further complicated by the fact that explicit collusion is often outlawed. In such cases collusive agreements cannot be enforced with reference to legally binding contracts. Instead, collusive agreements then must be "self-enforcing": they are sustained through an implicit understanding that "excessively" competitive behavior by any one firm will soon lead to similar behavior by other firms.

Collusion is an important subject in industrial organization. Its presence in oligopolistic markets tends to further distort the allocation of resources in the monopolistic direction. For this reason, public policy toward collusion is usually antagonistic. While there is both anecdotal and more systematic evidence on the existence of collusive behavior, the informal nature and often illegal status of collusion makes it difficult to evaluate its extent. But regardless of the economy-wide significance of collusion, this form of behavior is of course of great significance for certain markets.

The main framework currently used for modeling collusion is that of an infinitely repeated oligopoly game. Since the basic tension in the collusion scenario is dynamic – a colluding firm must balance the immediate gains from opportunistic behavior against the future consequences of its detection – the analysis of collusion requires a dynamic game which allows for history-dependent behavior. The repeated game is perhaps the simplest model of this type.

The earlier literature, which preceded the introduction of the repeated game model, recognized the basic tension that confronts colluding firms and the factors that affect the stability of collusive agreements [see, e.g., Stigler (1964)]. In particular, this literature contained the understanding that oligopolistic interaction sometimes results in collusion sustained by threats of future retaliation, and at other times results in non-collusive behavior of the type captured by the equilibria of the static oligopoly models. But since the formal modeling of this phenomenon requires a dynamic strategic model, which was not then available in economics, this literature lacked a coherent formal model.[11]

The main contribution of the repeated game model of collusion was the introduction of a coherent formal model. The introduction of this model offers two advantages. First,

---

[10] This result appears in an extreme form in the case of the pure Bertrand model, in which two producers of a homogeneous product who incur constant per-unit costs choose their prices simultaneously. The unique equilibrium prices are equal to marginal cost and profits are zero.

[11] As we discuss below, the earlier literature sometimes used models of the conjectural variations style, but these models were somewhat unsatisfactory or even confusing.

with such a model, it is possible to present and discuss the factors affecting collusion in a more compact and orderly manner. Second, the model enables exploration of more complex relations between the form and extent of collusive behavior and the underlying features of the market environment. A contribution of this type is illustrated below by a simple model that characterizes the behavior of collusive prices in markets with fluctuating demand [Rotemberg and Saloner (1986)]. We have selected to feature this application, since it draws a clear economic insight by utilizing closely the particular structure of the repeated game model of collusion.

4.1. Price wars during booms

Two firms play the Bertrand pricing game repeatedly in the following environment. In each period t the market demand is inelastic up to price 1 at quantity a^t,

$$Q(P) = \begin{cases} 0 & \text{if } P > 1, \\ a^t & \text{if } P \leqslant 1. \end{cases}$$

The production costs are zero. The a^t's are i.i.d. random variables which take the values H and L, where $H > L$ and $\text{Prob}\{a^t = H\} = w = 1 - \text{Prob}\{a^t = L\}$. Within each period, events unfold in the following order. First, a^t is realized and observed by the firms. Second, the firms choose prices, $p_i^t \in [0, 1]$, $i = 1, 2$, simultaneously. Third, the instantaneous profits, $\pi_i^t(p_i^t, p_j^t)$, are determined and distributed. The firm with the lower price gets the entire market, and when prices are equal the market is split equally. That is,

$$\pi_i^t(p_i^t, p_j^t) = \begin{cases} p_i^t a^t & \text{if } p_i^t < p_j^t, \\ p_i^t a^t / 2 & \text{if } p_i^t = p_j^t, \\ 0 & \text{if } p_i^t > p_j^t. \end{cases}$$

As usual, a history at t is a sequence of the form

$$\left((a^1, p_i^1, p_j^1), \ldots, (a^{t-1}, p_i^{t-1}, p_j^{t-1}), a^t\right),$$

and a strategy s_i is a sequence $(s_i^1, s_i^2, \ldots)$, where s_i^t prescribes a price after each possible history at t. A pair of strategies $s = (s_i, s_j)$ induces a probability distribution over infinite histories. Let E denote the expectation with respect to this distribution. Firm i's payoff from the profile s is

$$\Pi_i(s) = E\left[\sum_{t=1}^{\infty} \delta^{t-1} \pi_i^t(p_i^t, p_j^t)\right],$$

where $\delta \in (0, 1)$ is the discount factor. The solution concept is SPE. There are of course many SPE's. This model focuses on a symmetric SPE that maximizes the firms' payoffs over the set of all SPE's.

PROPOSITION 4.1. *There exists a symmetric SPE which maximizes the total payoff. Along its path, $p_i^t = p_j^t = p(a^t)$, where*

$$p(L) = p(H) = 1 \quad for\ \delta > \frac{H}{(1+w)H + (1-w)L},$$

$$p(L) = 1, \ p(H) = \frac{\delta(1-w)L}{H[1 - \delta(1+w)]} \quad for\ \frac{H}{(1+w)H + (1-w)L} \geqslant \delta \geqslant \frac{1}{2},$$

$$p(H) = p(L) = 0 \quad for\ \delta < 1/2.$$

The proof is relegated to the appendix. The interesting part of this result obtains for the middle range of δ's. Over this range, the equilibrium price during the high-demand state, $p(H)$, is lower than the monopoly price of 1. (For this range of δ, $p(H) = \delta(1-w)L/H[1 - \delta(1+w)] < 1$.) On the other hand, in the low-demand state, the equilibrium price achieves the monopoly price of 1. Rotemberg and Saloner refer to this result of lower prices during periods with higher demand as "price wars during booms". They argue that this is consistent with evidence on the behavior of oligopolistic industries over different phases of the business cycle. The intuition behind this result may be understood in the following general terms. A firm is willing to collude if the losses from future punishment outweigh the firm's immediate gain from deviation. In this model, owing to the independence of the shocks, the future losses are the same in "booms" and "busts". The immediate gains from defection at a given price, however, are obviously higher in booms. Therefore, when δ is not too large, it is impossible to sustain the monopoly price in a boom. To sustain collusion in a boom, it is necessary to reduce the temptation to deviate by colluding at a lower price.

4.2. Discussion

Let us highlight three points arising from this analysis. The first point concerns the substance. Rotemberg and Saloner develop the general point that the pattern of collusion and the dynamics of demand are related in a predictable way. Their analysis also uncovers the specific result that collusive prices are lower in high-demand states. This result derives to some extent from the assumption that demand shocks exhibit serial independence, and subsequent work has modified this assumption and reconsidered the relationship between demand levels and collusive prices. One finding is that collusion is easier to maintain (i.e., collusive prices tend to be higher) when future demand growth is expected to be large. The modified model can be also applied to markets with seasonal (as opposed to business-cycle) demand fluctuations, where the considerations of the oligopolists might be more transparent due to the greater predictability of the fluctuations. Indeed, recent empirical efforts offer evidence that is supportive of this hypothesis.[12] In any case, whether we consider the Rotemberg–Saloner model as is or one of its modified versions, the general approach suggested by this framework reveals considerations that plausibly influence the actual relationship between collusive prices and demand.

The second point concerns the essential role of the model. It should be noted that the repeated game model here is not incidental to the analysis. The main result of the analysis derives explicitly from the trade-off that the firm faces in a repeated game between the short-term benefit from undercutting the rival's price and the long-term cost of the consequent punishment. The result would seem somewhat counter-intuitive if one ignored the game theoretic reasoning and instead considered the situation using the standard price theoretic paradigms of monopoly and perfect competition. Of course, those who have the repeated game reasoning seated in the back of their minds can intuit through this argument easily, and may come to think that the formal model is superfluous. But then one has to have this reasoning already in the back of one's mind and to associate it with oligopolistic collusion.

The third point calls attention to one of the important strengths of this framework. The analysis illustrates clearly the flexibility with which the basic model can be adapted to incorporate alternative assumptions on the environment (in Section 6 this point will be illustrated further by another application that incorporates imperfect information into this framework). It would be difficult or even impossible to meaningfully incorporate such features into the conjectural variations paradigm (see Section 7 below).

We chose to devote much of the above presentation to a rather specific model. It would therefore be useful in closing to take a broader view and call attention to two fundamental insights of the repeated games literature that contain important lessons for oligopoly theory. The first insight is that a collusive outcome, which serves the firms better than the one-shot equilibrium, can be sustained in the interaction of fully rational competitors (in the sense formalized by the notion of SPE). The second is that repetition of the one-shot equilibrium is a robust outcome of such interaction: it is always a SPE and, in many interesting scenarios (e.g., finite horizon, high degree of impatience, short memory), it might even emerge as the unique one. The first insight provides a clear theory of collusion which can identify some key factors that facilitate collusion. The second shows the relevance of the static oligopoly models and the insights they generate: their equilibria continue to have robust and sometimes unique existence in a much richer dynamic environment. Of course, the consideration of the collusive and non-collusive

---

[12] Haltiwanger and Harrington (1991) hypothesize that demand rises and then falls as part of a deterministic cycle, and they find that collusive prices are higher when demand is rising. This model is well suited for markets that are subject to seasonal demand movements, and Borenstein and Shephard (1996) report evidence of pricing in the retail gasoline market that supports the main finding. Bagwell and Staiger (1997) hypothesize that the demand growth rate follows a Markov process, so that demand movements are both stochastic and persistent, and they find that collusive prices are higher in fast-growth (i.e., boom) phases.

outcomes predated the more recent analyses of the repeated game model. The important contribution of the repeated game framework is in establishing the validity of these as outcomes of rational and far-sighted competition that takes place over time.

5. Sales: An application of mixed-strategies equilibrium

In many retail markets prices fluctuate constantly and substantially. At any given point in time, some firms may offer a "regular" price, while other firms temporarily cut prices and offer "sales" or price promotions. The frequency and the significance of these price movements make it hard to believe that they mirror changes in the underlying demand and cost conditions.

The earlier literature had largely ignored these phenomena. Sales did not fit well into the existing price theoretic paradigm, and as a result this practice may have been viewed as reflecting irrational behavior that was better suited for psychological study than for economic analysis.

The game theoretic notion of a mixed-strategy equilibrium presents an alternative view whereby the ubiquitous phenomenon of sales can be interpreted as a stable outcome of interaction among rational players. We develop this argument with Varian's (1980) model of retail pricing.[13] This model highlights a tension that firms face between the desire to price high and profit on consumers that are poorly informed as to the available prices and the desire to price low and compete for consumers that are well informed of prices. This tension is resolved in equilibrium when firms' prices are determined by a mixed strategy. Varian's theory thus predicts that firms will offer sales on a random basis, where the event of a sale is associated with the realization of a low price from the equilibrium mixed strategy. A dynamic interpretation of this theory further implies that different firms will offer sales at different points in time.

A notable feature of this theory is that it takes the random behavior of the mixed-strategy equilibrium quite seriously and uses it to directly explain the randomness in prices observed in real markets. At the same time, the mixed-strategy approach has a well-known potential drawback, associated with the literal interpretation of the assumption that each firm selects its price in a random manner. This difficulty is often addressed with reference to Harsanyi's (1973) idea that a featured mixed-strategy equilibrium for a given game can be re-interpreted as a pure-strategy Nash equilibrium for a "nearby" game of incomplete information. However, there is of course the question of the plausibility of the nearby game for the application of interest. With these concerns in mind, we develop as well an explicit and plausible "purification" of the mixed-strategy equilibrium for Varian's pricing game. This analysis suggests that the random pattern of sales also can be understood as reflecting small private cost shocks that vary across firms and time.

---

[13] Related models are also explored by Shilony (1977) and Rosenthal (1980). Baye, Kovenock and De-Vries (1992) extend Varian's (1980) analysis, by characterizing all asymmetric equilibria.

5.1. An equilibrium theory of sales

We begin with the basic assumptions of the model. A set of $N \geqslant 2$ symmetric firms supplies a homogeneous good at unit cost c to a consumer population of unit mass. Each consumer demands one unit of the good, and the good provides a gross utility of v, where $v > c \geqslant 0$. There are two kinds of consumers. A fraction $I \in (0, 1)$ of consumers are informed about prices and hence purchase from the firm with the lowest price; if more than one low-priced firm exists, these consumers divide their purchases evenly between the low-priced firms. The complementary fraction $U = 1 - I$ of consumers are uninformed about prices and hence pick firms from which to purchase at random. Given these assumptions, we define a simultaneous-move game played by N firms as follows. A pure strategy for any firm i is a price $p_i \in [c, v]$. Letting p_{-i} denote the $(N-1)$-tuple of prices selected by firms other than firm i, the profit to firm i is defined as:

$$\Pi_i(p_i, p_{-i}) = \begin{cases} [p_i - c]U/N & \text{if } p_i > \min_{j \neq i} p_j, \\ [p_i - c](U/N + I/k) & \text{if } p_i \leqslant \min_{j \neq i} p_j \text{ and} \\ & \|\{j: p_j = p_i\}\| = k - 1. \end{cases} \tag{5.1}$$

A mixed strategy for a firm is a distribution function defined over $[c, v]$. If firm i employs the mixed strategy F_i and the strategies of its rivals are represented with the vector F_{-i}, then the expected profit to firm i is:

$$E_i(F_i, F_{-i}) = \int_c^v \cdots \int \Pi_i(p_i, p_{-i}) \, dF_1 \cdots dF_N.$$

For this game, a price vector $\{p_1, \ldots, p_N\}$ forms a Nash equilibrium in pure strategies if, for every firm i and every $p_i' \in [c, v]$, we have $\Pi_i(p_i, p_{-i}) \geqslant \Pi_i(p_i', p_{-i})$. Allowing also for mixed strategies, the distributions $(F_1, \ldots, F_N)$ form a Nash equilibrium if, for every firm i and every distribution function F_i', we have that $E_i(F_i, F_{-i}) \geqslant E_i(F_i', F_{-i})$. A symmetric Nash equilibrium is then a Nash equilibrium $(F_1, \ldots, F_N)$ satisfying $F_i = F$, for all $i = 1, \ldots, N$. Let $\overline{p}(F)$ and $\underline{p}(F)$ denote the two endpoints of the support of F; i.e., $\overline{p}(F) = \inf\{p: F(p) = 1\}$ and $\underline{p}(F) = \sup\{p: F(p) = 0\}$. We may now present the main finding:

PROPOSITION 5.1. (A) *There does not exist a pure-strategy Nash equilibrium.*
   (B) *There exists a unique symmetric Nash equilibrium F. It satisfies:*
   (i) $\overline{p}(F) = v$;
   (ii) $[\underline{p}(F) - c](U/N + I) = [v - c](U/N)$;
   (iii) $[p - c](U/N + (1 - F(p))^{N-1} I) = [v - c](U/N)$ *for every* $p \in [\underline{p}(F), \overline{p}(F)]$.

The proof is relegated to the appendix. The proposition reflects the natural economic tension that each firm faces between the incentive to cut price – offer a "sale" – in order to increase its chances of winning the informed consumers and the incentive to raise its price in order to better profit on its captive stock of uninformed consumers. This tension precludes the existence of a pure-strategy equilibrium, since the presence of informed consumers induces a firm to undercut its rivals when price exceeds marginal cost, while the presence of uninformed consumers induces a firm to raise its price when price equals marginal cost. In the mixed-strategy equilibrium, these competing incentives are resolved when firms select prices in a random manner, with some firms offering sales and other firms electing to post higher prices.

Now, while the specific predictions of the model seem to accord with casual observations and also with formal empirical studies,[14] the very literal and direct use of mixed strategies to explain price fluctuations raises some questions of interpretation. Do firms really select prices in a random manner? Correspondingly, are firms really indifferent over all prices in a certain range? And, if so, what compels a firm to draw its price from the specific equilibrium distribution? To address these questions, we develop next a purification argument that applies to this game.

5.2. Purification

We apply here Harsanyi's (1973) idea to interpret the mixed-strategy equilibrium of this game as an approximation to a pure-strategy equilibrium of a nearby game with some uncertainty over rival's costs. The uncertainty ensures that a firm is never quite sure as to the actual prices that rivals will select, and so incomplete information plays a role analogous to randomization in the mixed-strategy equilibrium presented above.

Consider now an incomplete-information version of the above game in which the firms' cost functions are private information. Firm i is of type $t_i \in [0, 1]$. A firm knows its own type but it does not know the types of the other firms. It believes that the types of the others are realizations of i.i.d random variables with uniform distribution over $[0, 1]$. The firm's type determines its cost function: firm i of type t_i has cost $c(t_i)$, where the function c is differentiable and strictly increasing and $0 < c(0) < c(1) < v$. As before, the firms simultaneously choose prices and receive the corresponding market shares and profits. Thus, this model is a standard Bayesian game with type spaces $[0, 1]$ and uniformly distributed beliefs. Notice that the uniform distribution of the beliefs is without loss of generality, since any differentiable distribution of costs can still be obtained by the appropriate choice of the cost function c. In this game, a pure strategy for any firm i is a function, $P_i(t_i)$, that maps $[0, 1]$ into $[c(0), v]$. Given a strategy profile $[P_1, \ldots, P_N]$, let P_{-i} denote the strategies of the firms other than i and let $P_{-i}(t_{-i})$ denote the vector of prices prescribed by these strategies when these firms' types are given by the $(N - 1)$-tuple t_{-i}. The profit of firm i of type t_i that charges p_i when its

---

[14] For example, Villas-Boas (1995) uses price data on coffee and saltine cracker products, and argues that the pricing patterns observed for these products are consistent with the Varian model.

rivals are of types t_{-i} is $\Pi_i(p_i, P_{-i}(t_{-i}), t_i)$ where Π_i is given by (5.1) in which c is replaced with $c(t_i)$. The profile $[P_1, \ldots, P_N]$ is a Nash equilibrium if, for all i and all t,

$$P_i(t_i) \in \arg\max_{p_i} E_{t_{-i}}\left[\Pi_i\left(p_i, P_{-i}(t_{-i}), t_i\right)\right].$$

A symmetric Nash equilibrium is such that $P_i(t_i) = P(t_i)$ for all i and t_i.

The distribution of prices induced by a strictly increasing strategy P is given by $P^{-1}(x) = \text{Prob}\{t \mid P(t) \leqslant x\}$. Let F_c be the mixed-strategy equilibrium of the complete-information game with common per-unit cost c.

PROPOSITION 5.2. *In the incomplete-information game with costs $c(\cdot)$, there exists a pure-strategy and strict Nash equilibrium, P, that satisfies the following: Given a constant $c \in (0, v)$, for any $\varepsilon > 0$, there exists $\delta > 0$ such that, if $|c(t) - c| < \delta$ for all t, then $|P^{-1}(x) - F_c(x)| < \varepsilon$, for all x.*

The proof is relegated to the appendix. In other words, the pure-strategy Nash equilibrium that arises in the incomplete-information game when the costs are near c for all t generates approximately the same distribution over prices as occurs in the mixed-strategy equilibrium of the complete-information game with common cost c. The mixed-strategy equilibrium for Varian's game can thus be interpreted as describing a pure-strategy equilibrium for a market environment in which firms acquire cost characteristics that may differ slightly and are privately known.

5.3. Discussion

The important substantive message of this theory is that price fluctuations which take the form of sales and promotions are largely explained as a consequence of straightforward price competition in the presence of buyers with varying degrees of information, rather than by some significant exogenous randomness in the basic data. When we adopt the interpretation of the purified version, the intuition can be described as follows. When there is a mix of better informed and less informed consumers, there are conflicting incentives to price high and low as explained above. Price competition arbitrates the low and high prices to the point where they are nearly equally profitable. In such a situation, relatively small differences in the firms' profit functions, such as those caused by small cost shocks, can yield large price movements. What this insight means for the empiricist is that the relevant data for understanding such markets concerns perhaps the nature of consumer information to a larger extent than it concerns technological or taste factors.

The predictions of this model appear even more compelling, when one looks at the immediate dynamic extension. As in any other static oligopoly game, the featured one-shot equilibrium corresponds to the non-collusive SPE of the repeated version of this game. We mention this obvious point specifically, since when the mixed strategy is played repeatedly, different firms can be expected to offer sales in different periods, which is a prediction that seems consistent with casual observation. In the dynamic purified version, each firm is privately informed of its current cost at the start of each period, and the cost shocks are assumed independent across time. The periodic cost shock might reflect, for example, firm-specific data like the level of the firm's inventories and the extent to which it is pressed for storage space.

Notice that the use of the game theoretic model here goes beyond a formal exposition of some natural intuition. In the absence of systematic equilibrium analysis and the concept of equilibrium in mixed strategies, it would be difficult or even impossible to come up with this explanation. Only once the result is obtained does it become possible to understand it intuitively.

6. On the contribution of industrial organization theory to game theory

Although industrial organization theory is mainly a user of concepts and ideas generated by game theory without an explicit industrial organization motivation, the relationship is not totally one-sided. There are specific ideas arising from problems in industrial organization that have gained independent importance as game theoretic topics in their own right. In what follows, we describe in detail one idea of this nature.

Repeated games with imperfect monitoring – "price wars". The development of this model was motivated by the observation that some oligopolistic industries experience spells of relatively high prices, which seem to result from implicit collusion, interrupted by spells of more aggressive price competition, referred to as "price wars".[15] A somewhat trivial theory could point out that this is consistent with the paths of certain SPE in a repeated oligopoly game. In such an equilibrium the firms coordinate for a while on collusive behavior (high prices or small quantities), then switch for a while to the one-shot equilibrium, and so on. What makes this theory rather unconvincing is that the alternation between collusion and price warfare is an artificial construct which does not reflect some more intuitive considerations. Moreover, there are simpler equilibria which Pareto dominate an equilibrium of this form. A more interesting theory for the instability of oligopolistic collusion was suggested by Rotemberg and Saloner (1986), as reviewed above, but this explanation is confined to price wars triggered by foreseen variations in demand.

The theory suggested by Green and Porter (1984) views the oligopolistic interaction as a repeated game with imperfect public information. In the repeated Cournot version of this approach (which is the version analyzed by Green and Porter), the firms simultaneously choose outputs in each period, and the price is a function of the aggregate output and some random demand shock which is unobservable to the firms. The firms

---

[15] Porter (1983) studies this pattern of behavior among firms in the U.S. railroad industry.

observe the price but cannot observe their rivals' outputs; consequently, a firm cannot tell whether a low price is the result of a bad demand shock or a high output by some rival. While the firms would like to collude on producing smaller outputs than those entailed by the static Cournot equilibrium, in this imperfect-monitoring environment it is impossible to sustain uninterrupted collusion. Intuitively, low prices cannot always go unpunished, since then firms would be induced to deviate from the collusive behavior. But this implies that collusion must sometimes break down into a "price war" along the equilibrium path. Reasoning in this way, Green and Porter constructed equilibria which exhibit on their paths spells of collusive behavior interrupted by blocks of time (following bad demand shocks) during which the firms revert to playing the Cournot equilibrium of the one-shot game.

It is somewhat easier to illustrate this point using a simple version of a repeated Bertrand duopoly game.[16] Two firms produce a homogeneous product at zero cost. The demand depends on the state of nature: with probability α there is no demand, and with probability $(1 - \alpha)$ the demand is a simple step function

$$Q(p) = \begin{cases} 2 & \text{if } p \leqslant 1, \\ 0 & \text{otherwise.} \end{cases}$$

The firms simultaneously choose prices $p_i \in [0, 1]$. If the prices are equal, the firms share the demand equally; otherwise, the low-price firm gets the entire demand. The payoffs of the firms in the high-demand state are

$$\pi_i(p_i, p_j) = \begin{cases} 2p_i & \text{if } p_i < p_j, \\ p_i & \text{if } p_i = p_j, \\ 0 & \text{if } p_i > p_j \end{cases}$$

and in the zero-demand case the payoffs to firms are zero. The firms do not observe the realization of the demand directly, but only their own shares. So, if both charged price p and the demand was high, these facts are public information. Otherwise, the only public information is that no such event occurred.

In the repeated game version, this interaction is repeated in each period $t = 1, 2, \ldots$. The firms' payoffs are the discounted sums of their profits with a common discount factor δ. In addition, assume that at the end of each period t the firms commonly observe a realization of a random variable, x^t, distributed uniformly over $[0, 1]$ and independently across periods. This variable is a mere "sunspot" which does not affect the demand or any other of the "real" magnitudes, but the possibility to condition on it enriches the set of strategies in a way that simplifies the analysis. A public history of the game at t is a sequence $h^t = (a^1, \ldots, a^{t-1})$, where either (i) $a^r = (p, x)$ which means that, in period r, demand was high, both prices were equal to $p > 0$ and the realization x was observed, or (ii) $a^r = (\sim, x)$ which means that either both prices were 0 or at least one

---

[16] The discussion here is influenced by Tirole's (1988) presentation.

of the firms sold nothing at r. A strategy for firm i prescribes a price choice for each period t after any possible public history. A sequential equilibrium (SE) is a pair of strategies (which depend on public histories) such that i's strategy is best response to j's strategy, $i \neq j = 1, 2$, after any public history.

In the one-shot game, the only equilibrium is the Bertrand Equilibrium: $p_i = 0$, $i = 1, 2$. Indefinite repetition of this equilibrium is of course a SE in the repeated game. If the demand state became observable at the end of each period, then provided that δ is sufficiently large, the repeated game would have a perfectly collusive SPE in which $p_1 = p_2 = 1$ in perpetuity. It is immediate to see that, in the present model, there is no perfectly collusive SE. If there were such a SE, in which the firms always choose $p_i = 1$ along the path, it would have to be that firm i continues to choose $p_i = 1$ after periods in which it does not get any demand. But, then it would be profitable for firm j to undercut i's price.

The interesting observation from the viewpoint of oligopoly theory is that there are equilibria which exhibit some degree of collusion and that, due to the impossibility of perfect collusion, such equilibria must involve some sort of "price warfare" on their path. Green and Porter identified a class of such equilibria that alternate along their path between a collusive phase and a punishment phase. In the present version of the model, these equilibria are described in the following manner. In the collusive phase the firms charge $p_i = 1$, and in the punishment phase they charge $p_i = 0$. The transition between the phases is then characterized by a nonnegative integer T and a number $\beta \in [0, 1]$. The punishment phase is triggered at some period t by a "bad" public observation of the form $a^{t-1} = (\sim, x)$, where $x < \beta$; the collusive phase is restarted after T periods of punishment.

To construct such a SE, let T and β be as above. Define the set of ("good") histories $G(T)$ to consist of: (i) the empty history; (ii) histories that end with $(1, x)$ for any x; (iii) histories such that since the beginning or since the last observation of the form $(1, x)$ there has been a block of exactly $k(T + 1)$ observations of the form $(\sim, x)$, where k is a natural number. Define the strategy $f_{T,\beta}$ as follows

$$f_{T,\beta}(h) = \begin{cases} 1 & \text{if } h \in G(T), \\ 1 & \text{if } h = \left(h', (\sim, x)\right) \text{ where } h' \in G(T) \text{ and } x \geqslant \beta, \\ 0 & \text{otherwise.} \end{cases}$$

Thus, the occurrence of the bad demand state does not always trigger the punishment, but only when $x < \beta$. Suppose that both firms play this strategy and let $V_{T,\beta}$ denote the expected discounted payoff for a firm calculated at the beginning of a period, after a history that belongs to $G(T)$:

$$V_{T,\beta} = 1 - \alpha + (1 - \alpha\beta)\delta V_{T,\beta} + \alpha\beta\delta^{T+1} V_{T,\beta}. \tag{6.1}$$

The RHS captures the fact that, when both follow this strategy, with probability $\alpha\beta$, there is no demand and $x < \beta$, so the interaction will switch to the punishment phase for T periods; and with probability $1 - \alpha\beta = \alpha(1 - \beta) + (1 - \alpha)$, there is either no demand and $x \geqslant \beta$ or there is high demand, in which cases the firms will continue colluding in the following period. Each firm gets a unit profit in the current period with probability $1 - \alpha$. Rearrangement of the above gives

$$V_{T,\beta} = (1 - \alpha)/\left[1 - \delta + \alpha\beta\left(\delta - \delta^{T+1}\right)\right]. \tag{6.2}$$

To verify that these strategies constitute an equilibrium, it is enough to check that there is no profitable single-period deviation after histories such that $f_{T,\beta}(h) = 1$ (since $p_1 = p_2 = 0$ is a Nash Equilibrium of the one-shot game there is clearly no incentive to deviate after other histories). Thus, the equilibrium condition is

$$2(1 - \alpha) + (1 - \beta)\delta V_{T,\beta} + \beta\delta^{T+1} V_{T,\beta} \leqslant V_{T,\beta}. \tag{6.3}$$

The LHS captures the value of a single-period deviation. The payoff 2 is the supremum over the immediate payoffs that a firm can get by undercutting its rival's price. Since the deviation yields public information of the form $(\sim, x)$, the continuation will be determined by the size of x: with probability $(1 - \beta)$, $x \geqslant \beta$ so the collusion will continue in the next period yielding the value $\delta V_{T,\beta}$; and with probability β, $x < \beta$ so the T-periods punishment phase begins yielding the value $\delta^{T+1} V_{T,\beta}$, associated with the renewed collusion after T periods. Rearrange (6.3) to get

$$V_{T,\beta} \geqslant 2(1 - \alpha)/\left[1 - \delta + \beta\left(\delta - \delta^{T+1}\right)\right]. \tag{6.4}$$

PROPOSITION 6.1.

(i) *There exists an equilibrium of this form (with possibly infinite T), iff*

$$\alpha \leqslant 1 - \frac{1}{2\delta}. \tag{6.5}$$

(ii) *For any α and δ satisfying (6.5), let $T(\alpha, \delta) = \min\{T \mid (1 - \delta)/(1 - 2\alpha)\delta(1 - \delta^T) \leqslant 1\}$.*

$$\arg\max_{T,\beta}\left[V_{T,\beta} \text{ s.t. } (6.3)\right] = \left\{(T, \beta) \mid T \geqslant T(\alpha, \delta) \text{ and } \beta = \frac{1 - \delta}{(1 - 2\alpha)\delta(1 - \delta^T)}\right\}.$$

PROOF. (i) Substitute from (6.2) to the LHS of (6.4) to get that a (T, β) equilibrium exists iff

$$(1 - \alpha)/\left[1 - \delta + \alpha\beta\left(\delta - \delta^{T+1}\right)\right] \geqslant 2(1 - \alpha)/\left[1 - \delta + \beta\left(\delta - \delta^{T+1}\right)\right]. \tag{6.6}$$

Rearrangement yields

$$\alpha \leqslant \frac{1}{2} - \frac{1 - \delta}{2\beta\left(\delta - \delta^{T+1}\right)}. \tag{6.7}$$

Since the RHS increases with T and β, if this inequality holds for some T and β, it must hold for $T = \infty$ and $\beta = 1$. Thus, there are some T and β for which (6.7) holds iff (6.5) holds.

(ii) Let (T, β) be an equilibrium configuration that maximizes $V_{T,\beta}$. For $T' \geqslant T$, define $\beta' = \beta(1 - \delta^T)/(1 - \delta^{T'})$ and observe that (T', β') is also an equilibrium configuration that maximizes $V_{T,\beta}$. To see this, note first that $\beta' \leqslant \beta \leqslant 1$. Second, since (6.6) holds for (T, β), it holds for (T', β') as the denominator remains the same and this implies that it is an equilibrium. Third, $V_{T',\beta'} = V_{T,\beta}$, since (6.2) gets the same value with (T, β) and (T', β'). Thus, in particular, an equilibrium with $T = \infty$ is always among the maximizers of $V_{T,\beta}$.

Next, observe from (6.2) that $V_{\infty,\beta}$ is decreasing in β, so that $V_{\infty,\beta}$ is maximized at the minimal β that satisfies (6.7) with $T = \infty$, which is $\beta_\infty = (1 - \delta)/(1 - 2\alpha)\delta$. Now, inspection of (6.2) implies that, for an equilibrium with $T < \infty$ to be also a maximizer of $V_{T,\beta}$, it has to be that the β of this equilibrium satisfies $\beta = \beta_\infty/(1 - \delta^T)$. This is possible only for T's such that $\beta_\infty/(1 - \delta^T) \equiv (1 - \delta)/(1 - 2\alpha)\delta(1 - \delta^T) \leqslant 1$, i.e., only for $T \geqslant T(\alpha, \delta)$. $\qquad \square$

This proposition shows that, for some range of the parameters, there are equilibria which exhibit the sought-after form of behavior: spells of collusion interrupted by price wars. Part (ii) shows that there are such equilibria among the optimal ones in the T, β-class. Moreover, a result due to Abreu, Pearce and Stacchetti (1986) implies that, in this model, the optimal equilibria in the T, β-class are also optimal among all symmetric equilibria (not just optimal in the T, β-class). So the sought-after alternation between collusion and price wars on the equilibrium path emerges, even when we insist on optimal symmetric equilibria.

This observation is somewhat qualified by the fact that the $T = \infty$ equilibrium, which does not alternate between the two regimes, is always among the optimal equilibria as well. The present model however is rather special. In particular, it has the property that the worst punishments that the parties can inflict on one another coincide with the Nash equilibrium of the one-shot game. Abreu, Pearce and Stacchetti (1986) show in a more general model that there is a joint profit-maximizing symmetric equilibrium that starts with playing the most collusive outcome and then switches into the worst sequential equilibrium. However, in models such as the repeated Cournot game, in which the one-shot equilibrium does not coincide with the worst punishment, the worst sequential equilibrium itself would involve alternation between the collusive outcome and another "punishing" outcome. So the behavior along the path of the optimal symmetric equilibrium would still have the appearance of collusion interrupted by price wars. The narrative that accompanies this equilibrium is less direct: the spells of collusion are in some sense rewards for sticking to the punishment and they are hence triggered by sufficiently bad public signals (low prices in the Cournot version) which confirm the firms' adherence to the punishment. In this sense, the behavior captured by these equilibria is not as "natural" or "straightforward" as the behavior captured by the original Green–Porter equilibrium.

The relevance of imperfect monitoring for cartel instability gets an additional twist once asymmetric equilibria are considered. Fudenberg, Levine and Maskin (1994) show that, when the public signal satisfies a certain full dimensionality property and δ is sufficiently close to 1, there are (asymmetric) equilibrium outcomes which are arbitrarily close to the Pareto efficient outcome. Thus, under such conditions, the extent of "price warfare" is insignificant along the path of the optimal equilibrium.

The imperfect-monitoring model of collusion contributes importantly to the understanding of cartel instability. In addition, while it obviously started from a clear industrial organization motivation, this model has also generated a research line in the theory of repeated games with unobservable moves that has assumed a life of its own and that continues to grow in directions that are now largely removed from the original motivation.[17]

7. An overview and assessment

Non-cooperative game theory has become the standard language and the main methodological framework of industrial organization. Before the onset of game theory, industrial organization was not without analytical methodology – the highly developed methodology of price theory served industrial organization as well. But traditional price theory addresses effectively only situations of perfect competition or pure monopoly, while industrial organization theory emphasizes the spectrum that lies between these two extremes: the study of issues like collusion and predation simply requires an oligopoly model. This gap was filled by verbal theorizing and an array of semi-formal and formal models. The formal models included game models like those of Cournot, Bertrand and Stackelberg, as well as non-game models with strategic flavor such as the conjectural variations and the contestable market models. Before proceeding with the discussion, we pause here to describe the conjectural variations model,[18] which is an important representative of the formal pre-game theoretic framework.

7.1. A pre-game theoretic model: The conjectural variations model

This model attempts to capture within an atemporal framework both the actual actions of the oligopolists and their responses to each other's choices. In the quantity-competition

---

[17] Repeated game models with imperfect monitoring had been considered somewhat earlier by Rubinstein (1979) and Radner (1981), who analyzed repeated Principal-Agent relationships. Besides the different basic game, these contributions also differ in their solution concepts (Stackelberg and Epsilon-Nash respectively) and their method of evaluating payoff streams (limit-of-the-means criterion). It seems however that, due to these differences or other reasons, the Green–Porter article has been more influential in terms of stimulating the literature.

[18] For a traditional description of this model, see Fellner (1949); for a modern view from the game theoretic perspective, see Friedman (1990).

duopoly version, two firms, 1 and 2, produce outputs, q_1 and q_2, which determine the price, $P(q_1 + q_2)$, and hence the profits, $\pi_i(q_i, q_j) = q_i P(q_1 + q_2) - c_i(q_i)$. Firm i holds a conjecture, q_i^C, regarding j's output. In the conjectural variations framework, this conjecture may in fact depend on firm i's own choice, q_i; i.e., $q_i^C = v_i(q_i)$. An equilibrium is then a pair of outputs, q_i^*, $i = 1, 2$, such that

$$q_i^* = \arg\max_{q_i} \pi_i \left[q_i, v_i(q_i) \right] \quad \text{and} \quad v_i(q_i^*) = q_j^*, \quad i = 1, 2.$$

Thus, each firm maximizes its profit under the conjecture that its rival's output will vary with its own choice, and the conjecture is not contradicted at the equilibrium.

In many applications, the v_i's were assumed linear in the q_i's (at least in the neighborhood of the solution) with slope v. Under this assumption, the parameter v indexes the equilibrium level of collusion: with $v = 1, 0$ and -1, the equilibrium outcome coincides with the joint monopoly outcome, the Cournot equilibrium outcome and the perfectly competitive outcome, respectively, and with other v's in this range the outcome falls between those mentioned.

Notice that, unlike non-cooperative game theoretic models, this model remains vague about the order of moves. In fact, if we tried to fit this model with an extensive form, it would have to be such that each firm believes that it may be moving ahead of the other firm. One possibility is that the firms hold inconsistent beliefs as to the moves of nature, who chooses the actual sequence of moves. But obviously this model was not meant to capture behavior under a special form of inconsistent beliefs. It is probably more appropriate to think of it as a reduced form of an underlying dynamic interaction that is left unmodelled.

It is important to note that, in the pre-game theoretic literature, the conjectural variations model was not viewed as different in principle from the game theoretic models. This is because the game models were viewed then somewhat differently than they are viewed now. They were not seen as specific applications of a very deep and encompassing theory, the Nash equilibrium theory, but rather as isolated specific models using a somewhat ad-hoc solution concept. In fact, the Nash equilibria of these models were often viewed as a special case of the conjectural variations model and were often referred to as the "zero-conjectural-variations" case.

Having mentioned the theoretical background against which the game theoretic models were introduced, let us try to assess briefly the contribution of this change in the theoretical framework of industrial organization. The following points discuss some aspects of this contribution both to the expositional role of the theory and to its substance.

7.2. Game theory as a language

The first contribution of game theory to industrial organization is the introduction of a language: models are described in an accurate and economical way using standard familiar formats, and basic non-cooperative solution concepts are commonly employed.

One clear benefit of this standardization is improved accessibility: a formal political theorist or a mathematician can access relatively easily the writings in modern industrial organization theory, without a long introduction to the specific culture of the field. This requires, of course, some basic familiarity with the language of game theory. But the pre-game theoretic literature also had by and large a language of its own – it was just not as universal and as efficient as game theory.

To appreciate the contribution of game theory simply as a language, one merely has to read through some of the presentations of formal models in the pre-game theoretic literature. The above description of the conjectural variations model already benefited from the game theoretic language and perhaps does not reflect appropriately the ambiguity that often surrounded the presentation of this model. The ambiguity that naturally results from the timeless unmodelled dynamics was further exacerbated in many cases by presentations that described the central assumptions of the model only in terms of derivatives (perhaps to avoid the embarrassment of describing the inconsistent beliefs explicitly).

Game theory has the additional virtue of being a more flexible language, in the sense that it allows consideration of situations involving dynamic interaction and imperfect information regarding behavior and the environment. The flexibility of the game theoretic framework in these respects derives perhaps from the rather primitive structure of non-cooperative game models, which set forth all actions and their timing. To be sure, the consideration of dynamic interaction and imperfect information complicates the models and raises additional conceptual problems (say, about the appropriate solution concept). Nevertheless, game theory does provide a systematic and formal language with which to explore these considerations – something that was not available in the pre-game theoretic literature. For example, in the conjectural variations model of collusion, it is not even clear how to begin thinking about the consequences of secret price cutting for collusive conduct. This model is not suitable for such an analysis, since it fails to describe the information that firms possess and the timing of their actions. Furthermore, there is no obvious way to add these dimensions. By contrast, such an analysis is natural in the context of the repeated game model of collusion, as discussed in the Green–Porter model reviewed in Section 6. This flexibility is an important asset of the game theoretic framework.

7.3. Game theory as a discipline

Related to its role as a language, game theory also imposes a discipline on modeling. First, a non-cooperative game model requires the analyst to specify precisely the actions available to players, the timing of the actions and the information held by players. This forces modelers to face their assumptions and hence to question them. For example, the repeated game collusion models of Sections 4 and 6 are very specific about the behavior and the information of firms. By contrast, in the conjectural variations model, no explicit assumptions on behavior are presented, so that one can judge the model only by the postulated outcome.

Second, with the game theoretic framework, results have to satisfy the requirements of known solution concepts, usually Nash equilibrium and its refinements. This forces arguments to be complete in the way dictated by these solutions. The brief review of the development of the literature on entry deterrence at the end of Section 2 illustrates this point. The argument on the role of investment and price in deterrence became complete, only after the situation was described as a simple two-stage game and analyzed with the appropriate solution concept of SPE.

The imposition of the game theoretic discipline has some drawbacks as well. First, it naturally constrains the range of ideas that can be expressed and inhibits researchers from venturing beyond its boundaries. Second, careless use of game theory may lead to substantial misconceptions. The Nash equilibrium concept is not always compelling. Standard game theoretic models often presume high degrees of rationality and knowledge on the part of the players, and the full force of such assumptions is often not acknowledged in applications. Third, there is a sense in which non-cooperative game models require the modeler to specify too much. The game theoretic collusion models specify whether the firms move simultaneously or alternately, what exactly they observe, and so on. These features are normally not known to observers and natural intuition suggests that they should not be too relevant. However, they have to be specified and even worse they often matter a great deal for the analysis. If one were very confident about the accuracy of the description and the validity of the solution concept, the important role of the fine details of the model might provide important insights. But since the models are often viewed as a rather rough sketch of what goes on, the sensitivity of predicted outcomes to modeling details is bothersome. In contrast, the conjectural variations model summarizes complicated interaction simply, without spending much effort on the specification of artificial features, so that the fine details do not interfere with the overall picture.

7.4. The substantive impact of game theory

So industrial organization has a new language/discipline and perhaps a superior one to what it had before, but has this language generated new insights into the substance of industrial organization? By taking a very broad view, one might argue that it has not. Take oligopoly theory, for example. In the pre-game theoretic era, economists clearly recognized the potential inefficiency of oligopolistic competition, the forces that work to induce and diffuse collusion and the possibility that different degrees of collusion could be sustained by threats of retaliation. In some sense, this is what is known now, too.

But a closer look reveals quite a few new specific insights. In fact, each of the previous sections described what we believe to be a new insight, and we attempted to identify the crucial role of the game theoretic framework in reaching these insights. For example, the idea that export subsidies can play a strategic role that might rationalize their use would be difficult to conceive without the game theoretic framework. In fact, it runs contrary to intuition based on standard price theory. Similarly, in the absence of this framework, it would be hard to conceive of the idea that random pricing in the form of sales is a robust phenomenon which derives from the heterogeneous price information (or search costs) that different segments of the consumer population enjoy.

But the fact that a certain relationship that exists within the model can be interpreted in terms of the underlying context, and is thus regarded as an insight, does not necessarily mean that it truly offers a better qualitative understanding of important aspects of actual market behavior. There remains the question of whether or not such an insight is more than a mere artifact of the model. For example, as discussed in Section 2, the strategic explanation of export subsidization (taxation) might be questioned in light of the sensitivity of this explanation to modeling decisions (Cournot vs. Bertrand). All things considered, however, we believe that the insights described above identify qualitative forces that plausibly play important roles in actual markets.

At the same time, we also stress that these insights should not be taken too literally. For example, the model of Section 5 tells us that sales can be a stable phenomenon through which price competition manifests itself. The insight is that this phenomenon need not reflect some important instability in the technology or pattern of demand; rather, sales emerge naturally from price competition when consumers are heterogeneously informed. Of course, this is not to say that firms know exactly some distribution and go through the precise equilibrium reasoning. The point is only that they have some rough idea that others are also pursuing these sales policies, and given this they have no clearly superior alternative than to also have sales in response to small private signals.

Let us accept then that many insights derived from the game theoretic approach offer a better qualitative understanding of important aspects of actual market behavior. We may still question the deeper significance of these insights. In particular, has the game theoretic framework delivered a new class of models that consistently facilitates better quantitative predictions than what would have been available in its absence? A serious attempt to discuss this question would take us well beyond the scope of this paper. Here, we note that important new empirical work in industrial organization makes extensive use of game theoretic models, but we also caution that there is as yet no simple basis from which to conclude that the game theoretic approach provides consistently superior quantitative predictions.

This inconclusive answer regarding the quantitative contribution of game theory does not imply that the usefulness of this framework for policy decisions is doubtful. Even if game theory has not produced a magic formula that would enable a regulator to make a definitive quantitative assessment as to the consequences of a proposed merger, this framework has enabled the regulators to think more thoroughly about the possible consequences of the merger. Likewise, it offers the regulator a deeper perspective on the issue of predation. To be sure, it does not offer a magic formula here either. But it makes it possible to have a more complete list of scenarios in which predation might be practiced and to use such arguments to justify intervention in situations that would not warrant intervention on the basis of simple price theoretic arguments.

8. Appendix

PROPOSITION 3.1.
  (i) *There exists a separating intuitive equilibrium.*
  (ii) *If $p_L^m \geqslant \underline{p}$, then any separating intuitive equilibrium, $\{P, E, b\}$, satisfies $\underline{p} = P(L) < \overline{P}(H) = p_H^m$ and $E(P(L)) = 0 < 1 = E(P(H))$.*
      If $p_L^m < \underline{p}$, then any separating intuitive equilibrium, $\{P, E, b\}$, satisfies $p_L^m = P(L) < \overline{P}(H) = p_H^m$ and $E(P(L)) = 0 < 1 = E(P(H))$.
  (iii) *If $p_L^m \geqslant \underline{p}$, and $b_L^0 \geqslant \tilde{b}_L$, then for every $p \in [\underline{p}, p_L^m]$, there exists an intuitive pooling equilibrium in which $P(L) = P(H) = p$.*
  (iv) *In any intuitive pooling equilibrium, $P(L) = P(H) \in [\underline{p}, p_L^m]$ and $E(P(L)) = 0$.*

PROOF. (i) For the case $p_L^m < \underline{p}$, define the triplet $\{P, E, b\}$ as follows: P is as in (ii) above, $E(p) = 1$ iff $p \neq p_L^m$, $b_L(p) = 0$ if $p \neq p_L^m$ and $b_L(p_L^m) = 1$. It is direct to verify that this triplet satisfies (E1)–(E4).

For the case $p_L^m \geqslant \underline{p}$, define the triplet $\{P, E, b\}$ as follows: P is as in (ii) above, $E(p) = 1$ iff $p \in (\underline{p}, \overline{p}]$, $b_L(p) = 0$ if $p \in (\underline{p}, \overline{p}]$ and $b_L(p) = 1$ otherwise. This triplet clearly satisfies (E1) when $t = H$ and when $t = L$ and $p_L^m = \underline{p}$. It also satisfies (E2)–(E4). The remaining step is to show that (E1) holds when $t = L$ and $p_L^m > \underline{p}$. For any p such that $E(p) = 0$, arguments using SCP developed in the proof of (ii) below establish that a deviation is non-improving: $V(\underline{p}, 0, L) > V(p, 0, L)$. Among p such that $E(p) = 1$, the most attractive deviation is the monopoly price, p_L^m. It is thus sufficient to confirm that $V(\underline{p}, 0, L) > V(p_L^m, 1, L)$. To this end define $p' < \underline{p}$ by $V(p', 0, H) = V(p_L^m, 1, H)$. Using the concavity of Π and thus V in p, as well as SCP, we see that this deviation is also non-improving: $V(\underline{p}, 0, L) > V(p', 0, L) > V(p_L^m, 1, L)$.

(ii) Let $\{P, E, b\}$ be a separating intuitive equilibrium. First, (E2) and (E3) imply that $E(P(L)) = 0 < 1 = E(P(H))$. Second, $P(H)$ must be equal to p_H^m, since $P(H) \neq p_H^m$ implies

$$V(P(H), 1, H) < V\left(p_H^m, 1, H\right) \leqslant V\left(p_H^m, E(p_H^m), H\right)$$

in contradiction to (E1). Third, since the H incumbent can deter entry by choosing $P(L)$, we must have

$$V\left(P(L), 0, H\right) \leqslant V\left(p_H^m, 1, H\right)$$

which implies that $P(L) \notin (\underline{p}, \overline{p})$.

Consider the case $p_L^m \geqslant \underline{p}$. The concavity of Π and hence of V in p implies that $V(\underline{p}, 0, L) > V(p, 0, L)$ for all $p < \underline{p}$ and $V(\overline{p}, 0, L) > V(p, 0, L)$ for all $p > \overline{p}$. The definition of $\underline{p}$ and $\overline{p}$ together with SCP imply $V(\underline{p}, 0, L) > V(\overline{p}, 0, L)$. Therefore, it follows that, if $P(L) \notin [\underline{p}, \overline{p})$, then there is $\varepsilon > 0$ such that

$$V(\underline{p} - \varepsilon, 0, L) > V\left(P(L), 0, L\right) \quad \text{and} \quad V(\underline{p} - \varepsilon, 0, H) < V\left(p_H^m, 1, H\right).$$

But then (E4) implies that $b_L(p - \varepsilon) = 1$. Hence $E(p - \varepsilon) = 0$ and $V(p - \varepsilon, 0, L) > V(P(L), 0, L)$ means that (E1) fails. Therefore, it must be that $P(L) \in [\underline{p}, \overline{p})$, which together with the previous conclusion that $P(L) \notin (\underline{p}, \overline{p})$ gives $P(L) = \underline{p}$.

The corresponding argument for the case $p_L^m < \underline{p}$ is that, if $P(L) \neq p_L^m$, then

$$V\left(p_L^m, 0, L\right) > V\left(P(L), 0, L\right) \quad \text{and} \quad V\left(p_L^m, 0, H\right) < V\left(p_H^m, 1, H\right).$$

Thus, by (E4), $b_L(p_L^m) = 1$ and the incumbent's deviation to p_L^m would be profitable, so $P(L) = p_L^m$.

(iii) Consider the case $p_L^m \geqslant \underline{p}$ and $b_L \geqslant \widetilde{b}_L$. Let $p' \in [\underline{p}, p_L^m]$ and define $\{P, E, b\}$ as follows: $P(L) = P(H) = p'$; $E(p) = 0$ for $p \leqslant p'$ and $E(p) = 1$ for $p > p'$; $b_L(p) = b_L^0$ for $p \leqslant p'$ and $b_L(p) = 0$ for $p > p'$. It is a routine matter to verify that $\{P, E, b\}$ satisfy (E1)–(E3). To verify that b satisfies (E4), observe that all $p < p'$ are sure to reduce profit below the equilibrium level for both types, and hence (E4) places no restriction. Next define p'' by $V(p'', 0, H) = V(p'', 0, H)$. For $p \in (p', p'']$, $V(p, 0, H) \geqslant V(p', 0, H)$, and hence $b_L(p) = 0$ satisfies (E4). For $p > p''$, observe that SCP implies $V(p'', 0, L) < V(p', 0, L)$ and then the concavity of Π in p implies $p'' > p_L^m$. Hence $V(p, 0, L) < V(p'', 0, L)$ and consequently $V(p, 0, L) < V(p', 0, L)$, so that $b_L(p) = 0$ satisfies (E4). Therefore, $\{P, E, b\}$ is a pooling intuitive equilibrium.

(iv) Let $\{P, E, b\}$ be a pooling intuitive equilibrium. Let p' denote the equilibrium price. First, $E(p')$ must be 0, since otherwise, for at least one $t \in \{L, H\}$, $p_t^m \neq p'$ and an incumbent of this type t could profitably deviate to p_t^m. Equilibrium profits are thus given by $v(L) \equiv V(p', 0, L)$ and $v(H) \equiv V(p', 0, H)$.

Clearly, $p' \geqslant \underline{p}$, since otherwise the H incumbent will deviate to p_H^m. Suppose then that $p' > p_L^m$. There are two cases to consider. First, if $p' > p_H^m$, define $p'' < p_H^m$ by $V(p'', 0, H) = v(H)$. Then SCP implies $V(p'', 0, L) > v(L)$, and so (E2) and (E4) imply $E(p'' - \varepsilon) = 0$ for small $\varepsilon > 0$. But then $V(p'' - \varepsilon, E(p'' - \varepsilon), L) > v(L)$, contradicting (E1). Second, if $p' \in (p_L^m, p_H^m]$, choose a sufficiently small $\varepsilon > 0$ such that $p' - \varepsilon > p_L^m$. Then $V(p' - \varepsilon, 0, H) < v(H)$ and $V(p' - \varepsilon, 0, L) > v(L)$, and so (E4) and (E2) imply $E(p' - \varepsilon) = 0$. Therefore, $V(p'' - \varepsilon, E(p'' - \varepsilon), L) > v(L)$, contradicting (E1) for the L incumbent. The conclusion is that $p' \leqslant p_L^m$ and hence $p' \in [\underline{p}, p_L^m]$. $\quad\square$

PROPOSITION 4.1. *There exists a symmetric SPE which maximizes the total payoff. Along its path,* $p_i^t = p_j^t = p(a^t)$, *where*

$$p(L) = p(H) = 1 \quad \text{for } \delta > \frac{H}{(1 + w)H + (1 - w)L},$$

$$p(L) = 1, \; p(H) = \frac{\delta(1 - w)L}{H[1 - \delta(1 + w)]} \quad \text{for} \quad \frac{H}{(1 + w)H + (1 - w)L} \geqslant \delta \geqslant \frac{1}{2},$$

$$p(H) = p(L) = 0 \quad \text{for } \delta < 1/2.$$

PROOF. First, let us verify that the path described in the claim is consistent with SPE. This path is the outcome of the following firms' strategies: charge p_i in state $i = L, H$, unless there has been a deviation, in which case charge 0. Obviously, these strategies are mutual best responses in any subgame following a deviation. In other subgames, there are only two relevant deviations to consider: slightly undercutting $p(H)$ in state H and slightly undercutting $p(L)$ in state L. Undercutting $p(H)$ is unprofitable if and only if

$$\{p(H)H + \delta[wp(H)H + (1-w)p(L)L]/(1-\delta)\}/2 \geqslant p(H)H$$

where the LHS captures the payoff of continuing along the path and the RHS captures the payoff associated with undercutting (a slight undercutting gives the deviant almost twice the equilibrium profit once and zero thereafter). Similarly, undercutting $p(L)$ is unprofitable if and only if

$$\{p(L)L + \delta[wp(H)H + (1-w)p(L)L]/(1-\delta)\}/2 \geqslant p(L)L.$$

Now, it can be verified that the p_i's of the proposition satisfy these two conditions in the appropriate ranges.

The following three steps show that this equilibrium maximizes the sum of the firms' payoffs, over the set of all SPE. First, for any SPE, there is a symmetric SPE in which the sum of the payoffs is the same. To see this, take a SPE in which $p_i^t \neq p_j^t$ somewhere on the path and modify it so that everywhere on the path the two prices are equal to $\min\{p_i^t, p_j^t\}$ and so that any deviation is punished by reversion to the zero prices forever. At t such that in the original SPE $p_i^t < p_j^t$, firm j still does not want to undercut, since its continuation value is at least half while its immediate gain is exactly half of the corresponding gains in the original SPE. By symmetry, this applies to i as well. At t such that in the original SPE $p_i^t = p_j^t$, for at least one of the firms, the continuation value is not smaller while the gain from undercutting is the same as in the original SPE, and by symmetry the other firm does not profit from the undercutting either. Second, let V denote the maximal sum of payoffs over the set of all SPE (since the set of SPE payoffs is compact, such maximum exists). Consider a symmetric equilibrium with sum of payoffs V (which exists by the first step). Observe that V must be the sum of payoffs in any subgame on the path that starts at the beginning of any period t before a^t was realized, i.e., after a history of the form $(a^1, p_i^1, p_j^1), \ldots, (a^{t-1}, p_i^{t-1}, p_j^{t-1})$. If it were lower for some t, then the strategies in that subgame could be changed to yield V. This would not destroy the equilibrium elsewhere, since it would only make deviations less profitable. But it will raise the sum of payoffs in the entire game, in contradiction to the maximality of V. Now, after any history along the path of this equilibrium that ends with a^t the equilibrium strategies must prescribe the price

$$\varphi(a^t) = \arg\max_p \{pa^t \text{ s.t. } (pa^t + \delta V)/2 \geqslant pa^t \text{ and } p \leqslant 1\}. \tag{8.1}$$

Otherwise, the equilibrium that prescribes these prices at t and continues according to the considered equilibrium elsewhere would have a higher sum of payoffs. Therefore, $V = [w\varphi(H)H + (1 - w)\varphi(L)L]/(1 - \delta)$. Upon substituting this for V in (8.1), a direct solution of this problem yields $\varphi(x) = p(x)$, $x = L, H$, where $p(x)$ are given in the proposition. $\qquad\qquad\qquad\qquad\qquad\qquad\qquad\qquad\qquad\qquad\qquad\qquad\qquad\qquad\qquad\square$

PROPOSITION 5.1. (A) *There does not exist a pure-strategy Nash equilibrium.*
   (B) *There exists a unique symmetric Nash equilibrium F. It satisfies:*
   (i) $\overline{p}(F) = v$;
   (ii) $[\underline{p}(F) - c](U/N + I) = [v - c](U/N)$;
   (iii) $[p - c](U/N + (1 - F(p))^{N-1}I) = [v - c](U/N)$ *for every* $p \in [\underline{p}(F), \overline{p}(F)]$.

PROOF. (A) Let k denote the number of firms selecting the lowest price, p, and begin with the possibility that $2 \leqslant k \leqslant N$. If $p > c$, then a low-priced firm would deviate from the putative equilibrium with a price just below p, since $[p - c](U/N + I) > [p - c](U/N + I/k)$. On the other hand, if $p = c$, then a low-priced firm could deviate to $p' > p$ and earn greater profit, since $(p' - c)(U/N) > 0$. Consider next the possibility that $k = 1$. Then the low-priced firm could deviate to $p + \varepsilon$, where ε is chosen so that all other firms' prices exceed $p + \varepsilon$, and earn greater profit, since $[p + \varepsilon - c](U/N + I) > [p - c](U/N + I)$.
   (B) We begin by showing that any symmetric Nash equilibrium F satisfies (i)–(iii). First, we note that, by the argument of the previous paragraph, $\underline{p}(F) > c$. We next argue that F cannot have a mass point. If p were a mass point of F, then a firm could choose a deviant strategy that is identical to the hypothesized equilibrium strategy, except that it replaces the selection of p with the selection of $p - \varepsilon$, for ε small. The firm then converts all events in which it ties for the lowest price at p with events in which it uniquely offers the lowest price at $p - \varepsilon$. Since ties at p occur with positive probability, and since $p \geqslant \underline{p}(F) > c$, the firm's expected profit would then increase if ε is small enough.
   Suppose now that $\overline{p}(F) < v$. Given that no price is selected with positive probability, ties occur with zero probability. Thus, when a firm chooses $\overline{p}(F)$, with probability one, it sells only to uninformed consumers. For ε small, the firm would increase expected profits by replacing the selection of prices in the set $[\overline{p}(F) - \varepsilon, \overline{p}(F)]$ with the selection of the price v. Thus, $\overline{p}(F) = v$. Similarly, when a firm selects the price $\underline{p}(F)$, with probability one it uniquely offers the lowest price in the market and thus sells to all informed consumers. Since expected profit must be constant throughout the support of F, it follows that $[\underline{p}(F) - c](U/N + I) = [v - c](U/N)$.
   We argue next that F is strictly increasing over $(\underline{p}(F), \overline{p}(F))$. Suppose instead that there exists an interval (p_1, p_2) such that $\underline{p}(F) < p_1$, $\overline{p}(F) > p_2$ and $F(p_1) = F(p_2)$. In this case, prices in the interval (p_1, p_2) are selected with zero probability. For ε small, a firm then would do better to replace the selection of prices in the interval $[p_1 - \varepsilon, p_1]$ with the selection of the price $p_2 - \varepsilon$. Since prices in the interval (p_1, p_2) are

selected with zero probability, the deviation would generate (approximately) the same distribution over market shares but at a higher price.

It follows that any interval of prices resting within the larger interval $[\underline{p}(F), \overline{p}(F)]$ is played with positive probability. It thus must be that all prices in the interval $[\underline{p}(F), \overline{p}(F)]$ generate the expected profit $[v - c](U/N)$. Now, the probability that a given price p is the lowest price is $[1 - F(p)]^{N-1}$. Thus, we get the iso-profit equation:

$$[p - c]\big(U/N + \big(1 - F(p)\big)^{N-1}I\big) = [v - c](U/N)$$

for all $p \in [\underline{p}(F), \overline{p}(F)]$.

Having proved that (i)–(iii) are necessary for a symmetric Nash equilibrium, we now complete the proof by confirming that there exists a unique distribution function satisfying (i)–(iii) and that it is indeed a symmetric Nash equilibrium strategy. Rewrite (iii) as $[1 - F(p)]^{N-1} = (v - p)U/N(p - c)I$ and observe from (iii) that, for $p \in (\underline{p}(F), \overline{p}(F))$, the RHS is between 0 and 1, so that there is a unique solution $F(p) \in (0, 1)$. It follows from (i)–(iii) that $F(\underline{p}(F)) = 0 < 1 = F(\overline{p}(F))$ and $F'(p) > 0$ for $p \in (\underline{p}(F), \overline{p}(F))$, confirming that F is indeed a well-defined distribution. To verify that F is a Nash equilibrium, consider any one firm and suppose that all other $N - 1$ firms adopt the strategy $F(p)$ defined by (i)–(iii). The given firm then earns a constant expected profit for any price in $[\underline{p}(F), \overline{p}(F)]$, and so it cannot improve upon F by altering the distribution over this set. Furthermore, any price below $\underline{p}(F)$ earns a lower expected profit than does the price $\underline{p}(F)$, and prices above $\overline{p}(F) = v$ are infeasible. Given that its rivals use the distribution function F, the firm can do no better than to use F as well. □

PROPOSITION 5.2. *In the incomplete-information game with costs $c(\cdot)$, there exists a pure-strategy and strict Nash equilibrium, P, that satisfies the following: Given a constant $c \in (0, v)$, for any $\varepsilon > 0$, there exists $\delta > 0$ such that, if $|c(t) - c| < \delta$ for all t, then $|P^{-1}(x) - F_c(x)| < \varepsilon$, for all x.*

PROOF. (i) Let $P : [0, 1] \to [c(0), v]$ be defined by the following differential equation and boundary condition:

$$P'(t) = \frac{[P(t) - c(t)][N - 1][1 - t]^{N-2}I}{U/N + [1 - t]^{N-1}I}, \tag{8.2}$$

$$P(1) = v. \tag{8.3}$$

Clearly, such a solution P exists and satisfies $P(t) > c(t)$ and $P'(t) > 0$ for all t.

We next show that P is a symmetric Nash equilibrium strategy. Let $\Psi(t, \tilde{t})$ denote the expected profit of a firm of type t that picks price $P(\tilde{t})$ when its rivals employ the strategy P,

$$\Psi(t, \tilde{t}) = [P(\tilde{t}) - c(t)]\{U/N + [1 - \tilde{t}]^{N-1} I\}.$$

Notice that this formula utilizes the strict monotonicity of P by letting $[1 - \tilde{t}]^{N-1}$ describe the probability that $P(\tilde{t})$ is the lowest price. To verify the optimality of $P(t)$ for a type t firm, we only have to check that $P(t)$ is more profitable than other prices in the support of P (the strict monotonicity of P implies that $P(c(0))$ is more profitable than any $p < P(c(0))$, and $P(1) = v$ is more profitable than any $p > v$). The function P thus constitutes a symmetric pure-strategy Nash equilibrium if the following incentive-compatibility condition holds:

$$\Psi(t, t) \geqslant \Psi(t, \tilde{t}) \quad \text{for all } t, \tilde{t} \in [0, 1]. \tag{8.4}$$

Observe that

$$\begin{aligned}\Psi_2(t, \tilde{t}) &= -[P(\tilde{t}) - c(t)][N - 1][1 - \tilde{t}]^{N-2} I \\ &\quad + \{U/N + [1 - \tilde{t}]^{N-1} I\} P'(\tilde{t}).\end{aligned} \tag{8.5}$$

It therefore follows from (8.2) that

$$\Psi_2(t, t) = 0 \quad \text{for all } t \in [0, 1].$$

Observe next that

$$\begin{aligned}\Psi(t, t) - \Psi(t, \tilde{t}) &= \int_{\tilde{t}}^{t} \Psi_2(t, x) \, dx = \int_{\tilde{t}}^{t} \left[\Psi_2(t, x) - \Psi_2(x, x)\right] dx \\ &= \int_{\tilde{t}}^{t} \left(\int_{x}^{t} \Psi_{12}(y, x) \, dy\right) dx \\ &= \int_{\tilde{t}}^{t} \left(\int_{x}^{t} c'(y)(N - 1)[1 - x]^{N-2} I \, dy\right) dx > 0\end{aligned}$$

where the second equality follows from $\Psi_2(x, x) = 0$ and the expression for $\Psi_{12}(y, x)$ is obtained by differentiating (8.5). Therefore, (8.4) is satisfied and this establishes that the pure strategy P defined above gives a Nash equilibrium. Notice further that a firm of type t strictly prefers the price $P(t)$ to any other.

(ii) To establish the approximation result, let $c \in (0, v)$ and let F_c denote the symmetric mixed-strategy equilibrium strategy in the complete-information game with common per-unit costs c. Define the function P_c by

$$P_c(t) = F_c^{-1}(t) \quad \text{for } t \in [0, 1].$$

This definition means that the distribution of prices induced by P_c is the same as the distribution of prices generated by the equilibrium mixed-strategy F_c of the complete-information game. Observe that P_c is the solution to (8.2)–(8.3) for $c(t) \equiv c$. First, note that $P_c(1) = F_c^{-1}(1) = v$. Next, differentiate the identity given in part B(iii) of Proposition 5.1 to get

$$1 = \frac{[p - c][N - 1][1 - F(p)]^{N-2} F'(p) I}{U/N + [1 - F(p)]^{N-1} I}. \tag{8.6}$$

Multiply both sides of (8.6) by $P_c'(t)$ and substitute there $p = P_c(t)$, $F = F_c$ and $t = F_c(P_c(t))$ to get

$$P_c'(t) = \frac{[P_c(t) - c][N - 1][1 - t]^{N-2} I}{U/N + [1 - t]^{N-1} I}.$$

So the function P_c solves (8.2).

Next observe that (8.2)–(8.3) define a continuous functional, ϕ, from the space of non-decreasing cost functions, $c: [0, 1] \rightarrow (0, v)$, into the space of price distributions on $[0, v]$. Thus, for an increasing function $c(\cdot)$, $\phi(c(\cdot))$ is the price distribution P^{-1} arising in the symmetric equilibrium P of the incomplete-information game with costs $c(\cdot)$, while for $c(\cdot) \equiv c$, $\phi(c) = F_c$. Therefore, invoking the continuity of ϕ, we conclude that, for any $\varepsilon > 0$, there exists $\delta > 0$ such that if $|c(t) - c| < \delta$, for all t, then $|P^{-1}(x) - F_c(x)| = |\phi(c(\cdot))(x) - \phi(c)(x)| < \varepsilon$, for all x. In other words, the pure-strategy Nash equilibrium that arises in the incomplete-information game generates approximately the same distribution over prices as occurs in the mixed-strategy equilibrium of the complete-information game.   □

References

Abreu, D., D. Pearce and E. Stacchetti (1986), "Optimal cartel equilibria with imperfect monitoring", Journal of Economic Theory 39:251–269.

Areeda, P., and D. Turner (1975), "Predatory pricing and related practices under Section 2 of the Sherman Act", Harvard Law Review 88:697–733.

Bagwell, K., and G. Ramey (1988), "Advertising and limit pricing", Rand Journal of Economics 19:59–71.

Bagwell, K., and G. Ramey (1991), "Oligopoly limit pricing", Rand Journal of Economics 22:155–172.

Bagwell, K., and R. Staiger (1997), "Collusion over the business cycle", Rand Journal of Economics 28:82–106.

Bain, J. (1949), "A note on pricing in monopoly and oligopoly", American Economic Review 39:448–464.

Baye, M.R., D. Kovenock and C. DeVries (1992), "It takes two to tango: Equilibria in a model of sales", Games and Economic Behavior 4:493–510.

Borenstein, S., and A. Shephard (1996), "Dynamic pricing in retail gasoline markets", Rand Journal of Economics 27:429–451.

Brander, J., and B. Spencer (1985), "Export subsidies and international market share rivalry", Journal of International Economics 18:83–100.

Bulow, J., J. Geanakoplos and P. Klemperer (1985), "Multimarket oligopoly: Strategic substitutes and complements", Journal of Political Economy 93:488–511.

Cho, I.-K., and D. Kreps (1987), "Signalling games and stable equilibria", Quarterly Journal of Economics 102:179–221.

Dixit, A. (1980), "The role of investment in entry deterrence", Economic Journal 90:95–106.

Eaton, J., and G. Grossman (1986), "Optimal trade and industrial policy under oligopoly", Quarterly Journal of Economics 101:383–406.

Fellner, W. (1949), Competition Among the Few (Knopf, New York).

Friedman, J. (1971), "A non-cooperative equilibrium in supergames", Review of Economic Studies 38:1–12.

Friedman, J. (1990), Game Theory with Applications to Economics (Oxford University Press, Oxford).

Fudenberg, D., and J. Tirole (1984), "The Fat Cat Effect, the Puppy Dog Ploy and the Lean and Hungry Look", American Economic Review 74:361–368.

Fudenberg, D., and J. Tirole (1986), Dynamic Models of Oligopoly (Harwood Academic Publishers, London).

Fudenberg, D., and J. Tirole (1987), "Understanding rent dissipation: On the use of game theory in industrial organization", American Economic Review 77:176–183.

Fudenberg, D., D. Levine and E. Maskin (1994), "The Folk theorem with imperfect public information", Econometrica 62:997–1039.

Green, E., and R. Porter (1984), "Noncooperative collusion under imperfect price information", Econometrica 52:87–100.

Haltiwanger, J., and J. Harrington (1991), "The impact of cyclical demand movements on collusive behavior", Rand Journal of Economics 22:89–106.

Harrington, J. (1986), "Limit pricing when the potential entrant is uncertain of its cost function", Econometrica 54:429–437.

Harsanyi, J. (1967–68), "Games with incomplete information played by 'Bayesian' players", Parts I, II, and III, Management Science 14:159–182, 320–324, 486–502.

Harsanyi, J. (1973), "Games with randomly disturbed payoffs: A new rationale for mixed strategy equilibrium points", International Journal of Game Theory 2:1–23.

Kadiyali, V. (1996), "Entry, its deterrence, and its accommodation: A study of the U.S. photographic film industry", Rand Journal of Economics 27:452–478.

Kreps, D., and J. Scheinkman (1983), "Quantity precommitment and Bertrand competition yield Cournot outcomes", Bell Journal of Economics 14:326–337.

Kreps, D., and R. Wilson (1982a), "Sequential equilibria", Econometrica 50:863–894.

Kreps, D., and R. Wilson (1982b), "Reputation and incomplete information", Journal of Economic Theory 27:253–279.

McGee, J. (1958), "Predatory price cutting: The Standard Oil (N.J.) case", Journal of Law and Economics 1:137–169.

Milgrom, P., and J. Roberts (1982a), "Limit pricing and entry under incomplete information: An equilibrium analysis", Econometrica 50:443–459.

Milgrom, P., and J. Roberts (1982b), "Predation, reputation and entry deterrence", Journal of Economic Theory 27:280–312.

Modigliani, F. (1958), "New developments on the oligopoly front", Journal of Political Economy 66:215–232.

Nash, J. (1950), "Equilibrium points in n-person games", Proceedings of the National Academy of Sciences 36:48–49.

Porter, R. (1983),"A study of cartel stability: The Joint Executive Committee, 1880–1886", Bell Journal of Economics 14:301–314.

Radner, R. (1981), "Monitoring cooperative agreements in a repeated principal–agent relationship", Econometrica 49:1127–1148.

Roberts, J. (1985), "A signaling model of predatory pricing", Oxford Economic Papers, Supplement 38:75–93.

Rosenthal, R. (1980), "A model in which an increase in the number of sellers leads to a higher price", Econometrica 48:1575–1580.

Rotemberg J., and G. Saloner (1986), "A supergame-theoretic model of business cycles and price wars during booms", American Economic Review 76:390–407.

Rubinstein, A. (1979), "Offenses that may have been committed by accident – an optimal policy of retribution", in: S. Brams, A. Schotter and G. Schwodiauer, eds., Applied Game Theory (Physica-Verlag, Würzburg, Vienna) 236–253.

Selten, R. (1965), "Spieltheoretische Behandlung eines Oligopolmodells mit Nachfragetragheit", Zeitschrift fur die gesamte Staatswissenschaft 12:301–324.

Shilony, Y. (1977), "Mixed pricing in oligopoly", Journal of Economic Theory 14:373–388.

Spence, M. (1977), "Entry, capacity, investment and oligopolistic pricing", Bell Journal of Economics 8:534–544.

Stigler, G. (1964), "A theory of oligopoly", Journal of Political Economy 72:44–61.

Tirole, J. (1988), The Theory of Industrial Organization (MIT Press, Cambridge).

Varian, H. (1980), "A model of sales", American Economic Review 70:651–659.

Villas-Boas, J.M. (1995), "Models of competitive price promotions: Some empirical evidence from the coffee and saltine crackers markets", Journal of Economics and Management Strategy 4:85–107.

Chapter 50

BARGAINING WITH INCOMPLETE INFORMATION

LAWRENCE M. AUSUBEL

Department of Economics, University of Maryland, College Park, MD, USA

PETER CRAMTON

Department of Economics, University of Maryland, College Park, MD, USA

RAYMOND J. DENECKERE*

Department of Economics, University of Wisconsin, Madison, WI, USA

Contents

*The authors gratefully acknowledge the support of National Science Foundation grants SBR-94-10545, SBR-94-22563, SBR-94-23104 and SBR-97-31025.

Handbook of Game Theory, Volume 3, Edited by R.J. Aumann and S. Hart

Abstract

A central question in economics is understanding the difficulties that parties have in reaching mutually beneficial agreements. Informational differences provide an appealing explanation for bargaining inefficiencies. This chapter provides an overview of the theoretical and empirical literature on bargaining with incomplete information.

The chapter begins with an analysis of bargaining within a mechanism design framework. A modern development is provided of the classic result that, given two parties with independent private valuations, ex post efficiency is attainable if and only if it is common knowledge that gains from trade exist. The classic problems of efficient trade with one-sided incomplete information but interdependent valuations, and of efficiently dissolving a partnership with two-sided incomplete information, are also reviewed using mechanism design.

The chapter then proceeds to study bargaining where the parties sequentially exchange offers. Under one-sided incomplete information, it considers sequential bargaining between a seller with a known valuation and a buyer with a private valuation. When there is a "gap" between the seller's valuation and the support of buyer valuations, the seller-offer game has essentially a unique sequential equilibrium. This equilibrium exhibits the following properties: it is stationary, trade occurs in finite time, and the price is favorable to the informed party (the Coase Conjecture). The alternating-offer game exhibits similar properties, when a refinement of sequential equilibrium is applied. However, in the case of "no gap" between the seller's valuation and the support of buyer valuations, the bargaining does *not* conclude with probability one after any finite number of periods, and it does *not* follow that sequential equilibria need be stationary. If stationarity is nevertheless assumed, then the results parallel those for the "gap" case. However, if stationarity is not assumed, then instead a folk theorem obtains, so substantial delay is possible and the uninformed party may receive substantial surplus.

The chapter also briefly sketches results for sequential bargaining with two-sided incomplete information. Finally, it reviews the empirical evidence on strategic bargaining with private information by focusing on one of the most prominent examples of bargaining: union contract negotiations.

Keywords

bargaining, sequential bargaining, incomplete information, asymmetric information, private information, Coase Conjecture

JEL classification: C78, D82

1. Introduction

A central question in economics is understanding the difficulties that parties have in reaching mutually beneficial agreements. Why do labor negotiations sometimes involve a strike by the union? Why do litigants engage in lengthy legal battles? And why does a worker with a grievance find it necessary to resort to a costly arbitration procedure? In all these cases, the parties would be better off if they could settle at the same terms without a protracted dispute. What, then, is preventing them from settling immediately? Recent theoretical work in economics has sought to answer this question.

Although the theory is still far from complete, researchers have taken promising steps in modeling bargaining disputes by focusing on the process of bargaining.[1] In the theory, costly disputes are explained by incomplete information about some aspect critical to reaching agreement, such as a party's reservation price.[2] Informational differences provide an appealing explanation for bargaining inefficiencies. If information relevant to the negotiation is privately held, the parties must learn about each other before they can identify suitable settlement terms. This learning is difficult because of incentives to misrepresent private information. Bargainers may have to engage in costly disputes to signal credibly the strength of their bargaining positions.

In this chapter, we provide an overview of the theoretical and empirical literature on bargaining under incomplete information. Since the literature on the topic is vast, it was inevitable that we had to limit the scope of our discussion. Consequently, a number of interesting and important contributions were left out. In particular, we would have liked to have had space to discuss the work on repeated bargaining [e.g., Hart and Tirole (1988), Kennan (1997), Vincent (1998)], and the extensive literature on durable goods monopoly (studying such topics as the impact of depreciation and increasing marginal cost of production, the effect of secondhand markets and transactions cost, and selling versus leasing contracts).

2. Mechanism design

We begin with an analysis of the fundamental incentives inherent in bargaining under private information. For this, we abstract from the process of bargaining. Rather than model bargaining as a sequence of offers and counteroffers, we employ mechanism design and analyze bargaining mechanisms as mappings from the parties' private information to bargaining outcomes. This allows us to identify properties shared by all Bayesian equilibria of any bargaining game.

---

[1] See Binmore, Osborne and Rubinstein (1992), Kennan and Wilson (1993), and Osborne and Rubinstein (1990) for surveys.

[2] Other motivations for disputes have been presented, such as uncertain commitments [Crawford (1982)] and multiple equilibria in the bargaining game [Fernandez and Glazer (1991), Haller and Holden (1990)].

One basic question is whether private information prevents the bargainers from reaping all possible gains from trade. Myerson and Satterthwaite (1983) find that ex post efficiency is attainable if and only if it is common knowledge that gains from trade exist; that is, uncertainty about whether gains are possible necessarily prevents full efficiency. Our development of this result follows several papers in the implementation literature [Mookherjee and Reichelstein (1992), Makowski and Mezzetti (1993), Krishna and Perry (1997), and, especially, Williams (1999)].

Consider an allocation problem with n agents. Agent i has a valuation $v_i(a, t_i)$ for the allocation $a \in A$ when its type is $t_i \in T_i$. An agent's type is private information. There is a status quo allocation, $\tilde{a}$, defining each agent's reservation utility. We normalize each v_i such that the reservation utility $v_i(\tilde{a}, t_i) = 0$. Utility for i is linear in its value and money: $u_i(a, t_i, x_i) = v_i(a, t_i) + x_i$, where x_i is the money transfer that i receives. A mechanism $\langle a, x \rangle$ determines an allocation $a(r)$ and a set of money transfers $x(r)$ based on the vector r of reported types. We wish to determine if it is possible to attain efficiency (for all t) by a mechanism that satisfies the agents' incentive and participation constraints. Let $U_i(r_i|t_i)$, $V_i(r_i|t_i)$, and $X_i(r_i)$ denote i's interim utility, valuation, and transfer when i reports r_i and the other agents honestly report t_{-i}:

$$U_i(r_i|t_i) \equiv E_{t_{-i}}\left[u_i\big(a(r_i, t_{-i}), t_i, x_i(r_i, t_{-i})\big)\right],$$

$$V_i(r_i|t_i) \equiv E_{t_{-i}}\left[v_i\big(a(r_i, t_{-i}), t_i\big)\right],$$

$$X_i(r_i) \equiv E_{t_{-i}}\left[x_i(r_i, t_{-i})\right].$$

Then $U_i(r_i|t_i) = V_i(r_i|t_i) + X_i(r_i)$. Let $U_i(t_i) \equiv U_i(t_i|t_i)$. The mechanism is incentive compatible if honest reporting is a best response: $U_i(t_i|t_i) \geqslant U_i(r_i|t_i)$ for all $t_i, r_i \in T_i$. Assume that t_i has a positive density f_i on an interval support $[\underline{t}_i, \bar{t}_i]$, and that $V_i(r_i|t_i)$ is continuously differentiable. Then from the Envelope Theorem, incentive compatibility implies for almost every $t_i \in T_i$

$$\frac{dU_i(t_i)}{dt_i} = \frac{\partial U_i(r_i = t_i|t_i)}{\partial t_i} = \frac{\partial V_i(r_i = t_i|t_i)}{\partial t_i},$$

which by the Fundamental Theorem of Calculus implies

$$U_i(t_i) = U_i(\underline{t}_i) + \int_{\underline{t}_i}^{t_i} \frac{\partial V_i(r_i = \tau_i|\tau_i)}{\partial t_i}\, d\tau_i. \qquad \text{(IC)}$$

The important implication of (IC) is that once the allocation $a(t)$ is specified, an agent's interim utility in any incentive compatible mechanism that implements $a(t)$ is uniquely determined up to a constant.

Now consider the efficient allocation $a^*(t) \in \text{argmax} \sum_i v_i(a, t_i)$, which maximizes the gains from trade. We know that the Groves mechanism implements the efficient allocation $a^*(\cdot)$ in dominant strategies. The Groves mechanism has transfers

$$x_i(t) = \sum_{j \neq i} v_j \big(a^*(t), t_j\big) - k_i(t_{-i}).$$

The second term, $k_i(t_{-i})$, is an arbitrary constant that does not distort the agent's incentives. Since the agent is concerned with its interim payoff, we can without loss of generality replace $k_i(t_{-i})$ with a single constant K_i for each agent that does not depend on the others' types. The first term provides the proper incentives. Ignoring the non-distorting constant, each agent gets the entire gains from trade. Hence, regardless of the reports of the others, honest reporting maximizes each agent's utility, since this yields the maximal gains from trade given the reports of the others. Honest reporting is a dominant strategy.

We will now develop necessary and sufficient conditions for the ex post efficient outcome to be Bayesian-implementable. Observe that a Groves mechanism automatically satisfies (IC), since it is incentive compatible. Moreover, if we vary the constants K_i, the Groves mechanisms span the set of all interim utilities that satisfy (IC) and achieve full efficiency. Thus, for any incentive-compatible and efficient mechanism, there exists a Groves mechanism that yields the same interim payoffs in dominant strategies; in checking whether efficiency can be achieved, we can simply focus on Groves mechanisms.

However, in order for efficiency to be attained in any unsubsidized mechanism where participation is voluntary, the additional requirements of (interim) individual rationality and (ex ante) budget balancing[3] must be met:

$$U_i(t_i) \geqslant 0, \quad \text{for all } i \text{ and for all } t_i \in T_i, \tag{IR}$$

$$\sum_i E_t\big[x_i(t)\big] \leqslant 0. \tag{BB}$$

That is, no type of any agent is made worse off by participating, and the sum of the expected transfers is nonnegative. Given the preceding paragraph, efficiency is attainable if and only if there exists a Groves mechanism satisfying (IR) and (BB). In the "basic" Groves mechanism with $K_i = 0$, each of the n agents needs to be awarded the

---

[3] In general, ex ante budget balancing is justified if there is a risk-neutral mediator (or other financier) who can absorb the risk of ex post budget imbalances. In the absence of such a player, one may need to impose the stronger condition of ex post budget balancing, $\sum_i x_i(t) \leqslant 0$. However, in the current context, ex ante budget balancing is equivalent to ex post budget balancing. This is because all players are risk neutral and, hence, can jointly costlessly absorb the risk associated with ex post budget imbalances [see also Cramton, Gibbons and Klemperer (1987)].

entire gains from trade, yet the gains from trade are created only once by the mechanism. Hence, the "basic" Groves mechanism generates an expected deficit, $\sum_i E_t[x_i(t)]$, equal to $(n - 1)$ times the expected gains from trade. In other words, the "basic" Groves mechanism satisfies (IR)[4] but violates (BB), whenever the expected gains from trade are positive. More general Groves mechanisms can try to finance the deficit by taxing the agents, but (IR) limits the magnitude of those taxes. Indeed, let $U_i^K(t_i)$ denote the interim utility of agent i in the Groves mechanism with taxes $K = (K_1, \ldots, K_n)$. Since $U_i^K = U_i^0 + K_i$, the tax to agent i can be no greater than $K_i = \underline{U}_i$, where

$$\underline{U}_i \equiv \inf\{U_i^0(t_i) | t_i \in T_i\}$$

is the interim utility of the worst-off type in the "basic" Groves mechanism. We therefore have:

THEOREM 1 [Williams (1999)]. *Under incentive compatibility* (IC), *individual rationality* (IR) *and budget balancing* (BB), *efficiency is attainable if and only if*:

$$(n - 1)E_t\left[\sum_i v_i(a^*(t), t_i)\right] \leqslant \sum_i \underline{U}_i. \tag{E}$$

We now apply Theorem 1 to prove the Myerson and Satterthwaite result. In this case, there are two agents, a seller S and a buyer B bargaining over the exchange of a good. Each knows its own valuation for the good, but not that of the other. The seller's valuation s is drawn from a distribution with positive density on $[\underline{s}, \bar{s}]$; the buyer's valuation b is drawn independently from a distribution with positive density on $[\underline{b}, \bar{b}]$. If $\bar{s} \leqslant \underline{b}$, it is common knowledge that gains from trade exist, and it is trivial to see that efficiency is attained by a single-price mechanism: trade for sure at a price $p \in [\bar{s}, \underline{b}]$. This is incentive compatible, since the outcome does not depend on the report, and it is individually rational, since each party receives a nonnegative payoff in every realization. We thus concentrate on the non-trivial case where $\bar{s} > \underline{b}$. The "basic" Groves mechanism has the following description: if $b > s$, trade occurs, the buyer pays s and the seller receives b, so that both get a payoff equaling $b - s$, the gains from trade; if $b \leqslant s$, then trade does not occur and both get a payoff of 0. The interim payoff to an agent is the expected gains from trade given the agent's value. Since the expected gains from trade are decreasing in the seller's value and increasing in the buyer's value, the worst-off types are seller $\bar{s}$ and buyer $\underline{b}$. Hence,

$$\underline{U}_S = E_b\big[(b - \bar{s})1_{\{b > \bar{s}\}}\big],$$
$$\underline{U}_B = E_s\big[(\underline{b} - s)1_{\{\underline{b} > s\}}\big].$$

---

[4] (IR) is satisfied, since $\sum_i v_i(a^*(t), t_i) \geqslant \sum_i v_i(\bar{a}, t_i) = 0$.

The deficit from the basic Groves mechanism is the expected gains from trade, which can be broken into four terms:

$$
\begin{aligned}
\mathrm{E}\big[(b-s)1_{\{b>s\}}\big] = {} & \mathrm{E}\big[(b-s)1_{\{b>s\}}|s>\underline{b};\, b<\bar{s}\big]\mathrm{Pr}(s>\underline{b};\, b<\bar{s}) \\
& +\mathrm{E}\big[b-s|b>\bar{s}\big]\mathrm{Pr}(b>\bar{s}) \\
& +\mathrm{E}\big[b-s|s<\underline{b}\big]\mathrm{Pr}(s<\underline{b}) \\
& -\mathrm{E}\big[b-s|s<\underline{b};\, b>\bar{s}\big]\mathrm{Pr}(s<\underline{b};\, b>\bar{s}).
\end{aligned}
$$

Since $\bar{s}>\underline{b}$, the first term is positive. Hence, (E) will be violated if the sum of the last three terms is at least as big as $\underline{U}_S+\underline{U}_B$. But

$$
\begin{aligned}
\mathrm{E}\big[b-s|b>\bar{s}\big]\mathrm{Pr}(b>\bar{s})-\underline{U}_S &= \mathrm{E}\big[\bar{s}-s|b>\bar{s}\big]\mathrm{Pr}(b>\bar{s}), \\
\mathrm{E}\big[b-s|s<\underline{b}\big]\mathrm{Pr}(s<\underline{b})-\underline{U}_B &= \mathrm{E}\big[b-\underline{b}|s<\underline{b}\big]\mathrm{Pr}(s<\underline{b}),
\end{aligned}
$$

so (E) is violated if

$$
\begin{aligned}
&\mathrm{E}\big[\bar{s}-s|b>\bar{s}\big]\mathrm{Pr}(b>\bar{s})+\mathrm{E}\big[b-\underline{b}|s<\underline{b}\big]\mathrm{Pr}(s<\underline{b}) \\
&\quad -\mathrm{E}\big[b-s|s<\underline{b};\, b>\bar{s}\big]\mathrm{Pr}(s<\underline{b};\, b>\bar{s})\geqslant 0.
\end{aligned}
$$

But this can be rewritten as

$$
\begin{aligned}
&\mathrm{E}\big[\bar{s}-s|s<\underline{b};\, b>\bar{s}\big]\mathrm{Pr}(s<\underline{b};\, b>\bar{s}) \\
&\quad +\mathrm{E}\big[b-\underline{b}|s<\underline{b};\, b>\bar{s}\big]\mathrm{Pr}(s<\underline{b};\, b>\bar{s}) \\
&\quad -\mathrm{E}\big[b-s|s<\underline{b};\, b>\bar{s}\big]\mathrm{Pr}(s<\underline{b};\, b>\bar{s}) \\
&\quad +\mathrm{E}\big[\bar{s}-s|s\geqslant\underline{b};\, b>\bar{s}\big]\mathrm{Pr}(s\geqslant\underline{b};\, b>\bar{s}) \\
&\quad +\mathrm{E}\big[b-\underline{b}|s<\underline{b};\, b\leqslant\bar{s}\big]\mathrm{Pr}(s<\underline{b};\, b\leqslant\bar{s})\geqslant 0.
\end{aligned}
$$

This follows, since the first three terms sum to $\mathrm{E}[\bar{s}-\underline{b}|s<\underline{b};\, b>\bar{s}]\mathrm{Pr}(s<\underline{b};\, b>\bar{s})\geqslant 0$, and the last two terms are both nonnegative. We thus have:

COROLLARY 1 [Myerson and Satterthwaite (1983)]. *If there is a positive probability of gains from trade (i.e., if $\bar{b}>\underline{s}$), but if it is not common knowledge that gains from trade exist (i.e., if $\bar{s}>\underline{b}$), then no incentive compatible, individually rational, budget balanced mechanism can be ex post efficient.*

Whenever there is some uncertainty about whether trade is desirable, ex post efficient trade is impossible. For this reason, private information is a compelling explanation for the frequent occurrence of bargaining breakdowns or costly delay. Inefficiencies are a necessary consequence of the strong incentives for misrepresentation between bargainers with private information.

Myerson and Satterthwaite's result depends crucially on the uncertainty being about players' valuations. For example, if players were uncertain about their respective fixed costs of delaying agreement, or about each others' discount factors, efficiency can be achieved by having players trade at a price between their (known) valuations.[5] The Myerson–Satterthwaite result also depends on independent types and risk neutrality. For example, Gresik (1991a) and McAfee and Reny (1992) show that when types are correlated efficient trade may be possible. Finally, it matters that the supports of the distributions of valuations are intervals [Matsuo (1989)].

Since ex post efficiency cannot be obtained, it is natural to ask how much of the gains from trade can be realized. Returning to the framework above of a single seller and single buyer with independent private values, then an allocation rule is simply the probability of trade as a function of the valuations: $p(s, b)$. We wish to find the allocation rule p that maximizes the expected gains from trade, subject to incentive compatibility and individual rationality. Suppose s is drawn from the distribution F with density f and b is drawn from the distribution G with density g. Myerson and Satterthwaite (1983) show that the optimal allocation rule p solves

$$\max_{p(\cdot,\cdot)} E[(b-s)p(s,b)] \quad \text{subject to}$$

$$\underline{U}_S + \underline{U}_B = E\left[\left(b - \frac{1-G(b)}{g(b)} - s - \frac{F(s)}{f(s)}\right)p(s,b)\right] \geqslant 0;$$

$$E_b[p(s,b)] \text{ decreasing;} \quad E_s[p(s,b)] \text{ increasing.}$$

The monotonicity constraints are necessary for incentive compatibility. The interim probability of trade is (weakly) decreasing in the seller's valuation and (weakly) increasing in the buyer's valuation. The first constraint is individual rationality (the worst-off types get a non-negative payoff) for a mechanism that satisfies (IC). Ignoring the monotonicity constraints, the Lagrangian is

$$\max_{p(\cdot,\cdot)} E\left[\left(d(b,\alpha) - c(s,\alpha)\right)p(s,b)\right], \quad \text{where}$$

$$c(s,\alpha) = s + \alpha\frac{F(s)}{f(s)}, \quad d(b,\alpha) = b - \alpha\frac{1-G(b)}{g(b)}.$$

Hence, by pointwise optimization the maximizing allocation rule is

$$p^\alpha(s,b) = \begin{cases} 1 & \text{if } d(b,\alpha) > c(b,\alpha), \\ 0 & \text{if } d(b,\alpha) \leqslant c(b,\alpha), \end{cases}$$

---

[5] For this reason, we will only study the outcome of dynamic trading processes when uncertainty is about players' valuations. Important contributions to extensive form bargaining when uncertainty is about players' fixed cost of bargaining include Perry (1986), Rubinstein (1985b), and Bikchandani (1992), and when uncertainty is about discount factors include Rubinstein (1985a) and Cho (1990b).

where $\alpha \in (0, 1]$ is chosen so that $\underline{U}_S + \underline{U}_B = 0$. A sufficient condition for the required monotonicity of the interim probability of trade is that $c(s, 1)$ and $d(b, 1)$ are increasing. This is the regular case.[6]

As an example, suppose both traders' valuations are drawn uniformly from $[0, 1]$. Then $\alpha = 1/3$ and the optimal allocation rule is to trade if and only if the gains from trade $b - s$ is greater than $1/4$. By surely trading when the gains from trade are largest, the mechanism reaps 84% of the possible gains from trade; there is a 16% loss due to the private information. The simultaneous-offer bargaining game studied by Chatterjee and Samuelson (1983) implements this optimal outcome. Both seller and buyer simultaneously make offers. If the seller's offer is less than the buyer's, then they trade at a price halfway between the two offers. Otherwise, they do not trade. In the ex ante efficient equilibrium, the traders use the following linear strategies: the seller offers $2s/3 + 1/4$ and the buyer offers $2b/3 + 1/12$. In choosing offers, both recognize the fundamental tradeoff between the probability of trade and the terms of trade. Whenever the probability of trade is positive, the parties have an incentive to misrepresent: the seller overstates her value and the buyer understates. The size of the misrepresentation increases with the probability of trade.

Our derivation above assumed that the seller's private information does not affect the buyer's valuation for the object, and conversely that the buyer's private information does not affect how much the seller values the object. However, as emphasized by Akerlof (1970), there are many interesting trading situations in which traders' valuations are interdependent. A seller of a used car may have information about reliability relevant to a potential buyer, and the buyer of an oil tract may have survey information relevant to its seller. While dominant strategy mechanisms no longer exist when valuations are interdependent, several authors have recently constructed generalized Groves mechanisms for which efficient trade is a Bayesian equilibrium [Ausubel (2002), Dasgupta and Maskin (2000), Jehiel and Moldovanu (2001), and Perry and Reny (1998)]. These mechanisms could be used to derive an inefficiency result analogous to Myerson and Satterthwaite's [see Gresik (1991c)]. Here we will consider the simpler environment studied by Akerlof, in which the seller's value s is private information, and the buyer's value is an increasing function of s satisfying $g(s) > s$. Note that the private values model is a special case in which $g(s)$ is constant at the level b. For this environment, Samuelson (1984) and Myerson (1985) established the following result:

THEOREM 2. *A bargaining mechanism $\{p, x\}$ is incentive compatible and individually rational if and only if $p(\cdot)$ is weakly decreasing,*

$$K \equiv \int_{\underline{s}}^{\bar{s}} \left(g(s) - s - \frac{F(s)}{f(s)} \right) f(s) p(s) \, ds \geqslant 0, \quad and$$

---

[6] Gresik (1991b) shows that we can replace interim individual rationality with the stronger ex post individual rationality without changing the set of ex ante efficient trading rules.

$$x(s) = k + sp(s) + \int_s^{\bar{s}} p(z)\,dz, \quad \text{for some } 0 \leqslant k \leqslant K.$$

Note that, since $g(s) > s$, ex post efficiency requires that $p(s) \equiv 1$. Integrating the first inequality in Theorem 2 by parts, we see that this can be a trading outcome only if $E[g(s)] \geqslant \bar{s}$, i.e., the buyer's expected value exceeds the highest seller valuation. This condition is automatically satisfied in the private values case, but is restrictive in the interdependent case. In this sense, interdependencies in valuations make trading inefficiencies more likely. For example, if $g(s) = \beta s$ and s is uniform on $[0, 1]$, ex post efficiency requires $\beta \geqslant 2$.

Akerlof went one step further, and observed that adverse selection in the above model may be so severe that no market-clearing price involving a positive level of trade can exist. This happens whenever $E[g(v) - s|v \leqslant s] < 0$ for all $s > \underline{s}$, for then any price that all seller types below s would accept yields the buyer negative expected surplus. Akerlof only considered single-price mechanisms, and it is of course conceivable that under his condition some more general trading mechanism could prove superior to competitive equilibrium. However, it is possible to use Theorem 2 to show that this cannot happen: under Akerlof's condition, the only incentive-compatible mechanism is the zero-trade mechanism. We can again illustrate this with the linear example described above; since $E[\beta v - s|v \leqslant s] = (\beta/2 - 1)s^2$, Akerlof's condition reduces to $\beta < 2$. It follows that $g(s) - s - F(s)/f(s) = (\beta - 2)s < 0$, so the incentive compatibility condition $K \geqslant 0$ can be satisfied only if $p(s) = 0$.

An important generalization of the bilateral independent values model is to multiple sellers and buyers. How does the bargaining inefficiency change as we add traders? Rustichini, Satterthwaite, and Williams (1994) consider a model with m sellers and m buyers and price is set to equate revealed demand and supply. In any equilibrium, the amount by which a trader misreports is $O(1/m)$ and the inefficiency is $O(1/m^2)$.[7] Hence, the inefficiency caused by private information quickly falls toward zero as competition increases. This provides a justification for assuming full information in competitive markets.

The mechanism design approach does not just apply to static trading procedures. Indeed, if the traders discount by the same interest rate r, then all the results above generalize to dynamic trading mechanisms, where the probability of trade $p(s, b)$ is replaced with the time of trade $t(s, b)$, where $p(s, b) = e^{-rt(s,b)}$. Hence, ex post efficiency is unobtainable as a Bayesian equilibrium in any static or dynamic bargaining game when it is uncertain whether trade is desirable.

An important feature of the ex ante efficient trading rule is that it is static. Trade either occurs immediately or not at all. Such static trading rules have been criticized, because they violate sequential rationality [Cramton (1985)]. Their implementation requires a

---

[7] See also Gresik and Satterthwaite (1989), Satterthwaite and Williams (1989), Williams (1990, 1991), and Wilson (1985).

commitment to walk away from known gains from trade. For example, in the Chatterjee–Samuelson mechanism, with probability $7/32$, the offers reveal that the gain from trade is positive, but less than $1/4$, so the parties are required not to trade, even though both know that mutually beneficial trade is possible. In addition, with probability $7/16$, at least one trader knows that both are sure to get 0 in the mechanism. This provides an incentive to propose another trading rule, even before offers are announced. An initial round of "cheap talk" may upset the equilibrium [Farrell and Gibbons (1989)].

Cramton, Gibbons, and Klemperer (CGK) (1987) generalize the Myerson and Satterthwaite (MS) problem to the case of n traders who share in the ownership of a single asset. Specifically, each trader $i \in \{1, \ldots, n\}$ owns a share $r_i \geqslant 0$ of the asset, where $r_1 + \cdots + r_n = 1$. As in MS, player i's valuation for the entire good is v_i, and the utility from owning a share r_i is $r_i v_i$, measured in monetary terms. The v_i's are independent and identically distributed according to F with positive density f on $[\underline{v}, \bar{v}]$. A partnership (r, F) is fully described by the vector of ownership rights $r = \{r_1, \ldots, r_n\}$ and the traders' beliefs F about valuations.

MS consider the case $n = 2$ and $r = \{1, 0\}$. They show that there does not exist a Bayesian equilibrium of the trading game that is individually rational and ex post efficient. In contrast, CGK show that if the ownership shares are not too unequally distributed, then it is possible to satisfy both individual rationality and ex post efficiency.

In addition to exploring the MS impossibility result, this paper considers the dissolution of partnerships, broadly construed. In a situation of joint ownership, who should buy out whom and at what price? Applications include divorce and estate fair-division problems [McAfee (1992)], and also public choice. For example, when several towns jointly need a hazardous-waste dump, which town should provide the site and how should it be compensated by the others?

In this context, ex post efficiency means giving the entire good to the partner with the highest valuation. A partnership (r, F) can be dissolved efficiently if there exists a Bayesian equilibrium of a Bayesian trading game that is individually rational and ex post efficient.

THEOREM 3. *The partnership (r, F) can be dissolved efficiently if and only if*

$$\sum_{i=1}^{n} \left[\int_{v_i^*}^{\bar{v}} [1 - F(u)] u \, dG(u) - \int_{\underline{v}}^{v_i^*} F(u) u \, dG(u) \right] \geqslant 0, \tag{D}$$

where v_i^* solves $F(v_i)^{n-1} = r_i$ and $G(u) = F(u)^{n-1}$.

Equation (D) is equivalent to (E) applied to this setting. As an example, if $n = 2$ and values are uniformly distributed on $[0, 1]$, then the partnership is dissolvable if and only if no shareholder's share is larger than 0.789. In general, the set of dissolvable partnerships is a convex, symmetric subset of the unit simplex centered at equal shares.

COROLLARY 2. *For any distribution F, the one-owner partnership* $r = \{1, 0, \ldots, 0\}$ *cannot be dissolved efficiently.*

This corollary generalizes the MS impossibility result to the case of many buyers. The one-owner partnership can be interpreted as an auction. Ex post efficiency is unattainable because the seller's reservation value v_1 is private information. The seller finds it in her best interest to set a reserve above her value v_1. The corollary also speaks to the time-honored tradition of solving complex allocation problems by resorting to lotteries: even if the winner is allowed to resell the object, such a scheme is inefficient because the one-owner partnership that results from the lottery cannot be dissolved efficiently.

CGK demonstrate that the incentives for misrepresentation depend on the ownership structure. The extreme 0–1 ownership shares in bilateral bargaining maximize the incentive for misrepresentation: sellers have a clear incentive to overstate value and buyers have a clear incentive to understate. Partial ownership introduces countervailing incentives, since the parties no longer are certain whether they are buying or selling. In the case of bilateral bargaining, the worst-off types are the highest seller type and the lowest buyer type. These trader types are unable to misrepresent (a seller cannot claim to have a value greater than $\bar{s}$ and a buyer cannot claim to have a value less than $\underline{b}$); hence, these types need not receive any information rents. With partial ownership r_i, the worst-off type is v_i^*, which solves $F(v_i)^{n-1} = r_i$. Notice that $r_i = F(v_i^*)^{n-1}$ is the probability that type v_i^* has the highest value and thus buys $1 - r_i$ of the good in the ex post efficient mechanism. Likewise, with probability $1 - r_i$, type v_i^* sells r_i. Hence, for the worst-off type, the expected purchases, $r_i(1 - r_i)$, equal the expected sales, $(1 - r_i)r_i$. In this sense, the worst-off type is the most confused about whether she is buying or selling; the incentives to overstate just balance the incentives to understate, and no bribes are required to get the trader to report the truth.

A basic insight of this analysis is that when parties have private information, bargaining efficiency depends on the assignment of property rights [see also Samuelson (1985), and Ayres and Tally (1995)]. Hence, full information is an essential ingredient in the Coase (1960) Theorem that bargaining efficiency is not affected by the assignment of property rights.

Mechanism design is a powerful theory for studying incentive problems in bargaining. We are able to characterize the set of outcomes that are attainable, recognizing each trader's voluntary participation and incentive to misrepresent private information. In addition, we are able to determine optimal trading mechanisms – mechanisms that are efficient in an ex ante (or interim) sense. Despite these virtues, mechanism design has two weaknesses. First, the mechanisms depend in complex ways on the traders' beliefs and utility functions, which are assumed to be common knowledge. Second, it allows too much commitment. In practice, bargainers use simple trading rules – such as a sequence of offers and counteroffers – that do not depend on beliefs or utility functions. And bargainers may be unable to walk away from known gains from trade. For this reason we next turn to the analysis of particular dynamic bargaining games.

3. Sequential bargaining with one-sided incomplete information: The "gap" case

In the previous section, we described bargaining as being static and mediated. Instead, we will now assume that bargaining occurs through a dynamic process of bilateral negotiation. A bargaining protocol explicitly specifies the rules that govern the negotiation process, and the bargaining outcome is described as an equilibrium of this extensive-form game.

We follow Rubinstein (1982) in requiring that only one offer can be on the bargaining table at any one time,[8] and that once an offer is rejected it becomes void (i.e., does not constrain any player's future acceptance or offer behavior). More precisely, we assume that there are an infinite number of time periods, denoted by $n = 0, 1, 2, \ldots$. In each period in which bargaining has not yet concluded, one of the players (whose identity is a function only of the time period n) can make an offer to his bargaining partner consisting of a price $p \in \mathbb{R}$ at which trade is to occur. Upon observing this offer, the partner can either accept, in which case the object is exchanged at the specified price and the bargaining ends, or reject, in which case the play moves on to the next period. Note that any terminal node of the game is uniquely identified by a pair (p, n). We assume that players are impatient, discounting surplus at the common discount factor $\delta \in [0, 1)$. Hence the payoffs assigned to terminal node (p, n) are $\delta^n (b - p)$ and $\delta^n (p - s)$, for the buyer and seller, respectively.

Three bargaining protocols of this type will be of specific interest: the *seller-offer game*, in which only the seller is allowed to make offers; the *alternating-offer game*, in which the buyer and seller alternate in making proposals; and the *buyer-offer game*, in which the buyer makes all the offers.

The private information is modeled as follows. Before the bargaining begins (i.e., prior to period 0), nature selects a signal $q \in [0, 1]$, and informs one of the two parties of its realization. The distribution of the signal is common knowledge and, without loss of generality, will be assumed to be uniform. The signal in turn determines the buyer and seller valuations through the monotone functions $v(\cdot)$ and $c(\cdot)$:

$$b = v(q), \qquad s = c(q).$$

We will say that the model has *private values*, if the uninformed party's valuation function is constant, and that the model has *interdependent values*, otherwise. We will adopt the convention that if the buyer is the informed party then the function $v(q)$ is decreasing, so that it represents an (inverse) demand function, and if the seller is the informed party then the function $c(q)$ is increasing, so that it represents an (inverse) supply function. The signal q is thus just an index indicating the rank order of the types of the

---

[8] It is well known that even in the one-shot complete-information case simultaneous offers permit any outcome. See also Sákovics (1993) for an illuminating discussion on the importance of precluding simultaneous offers.

informed party. Throughout, it will be assumed that the functions $v(\cdot)$ and $c(\cdot)$ are common knowledge.

Note that, in every period n, the information set of the offering player can be identified with a history of n rejected offers, and the information set of the receiving player can be identified with the same history concatenated with the current offer. For the offering player, a pure behavioral strategy in period n specifies the current offer as a function of this history of rejected offers. For the player receiving an offer, a pure behavioral strategy in period n specifies a decision in the set $\{A, R\}$ as a function of the n-history of rejected offers and the current offer (where A denotes acceptance and R denotes rejection of the current offer). A sequential equilibrium consists of a pair of behavioral strategies and a system of beliefs. Specifically, a sequential equilibrium associates with every node at which it is the uninformed party's turn to move a belief over the signal (rank order) of the informed party. As indicated above, the initial belief is that q is uniform on $[0, 1]$. Sequential equilibrium requires that the beliefs are "consistent", i.e., are updated from the belief in the previous period and the equilibrium strategies using Bayes' law (whenever it is applicable). Sequential equilibrium also requires that each player's strategy be optimal after any history, given the current beliefs.

Offer/counteroffer bargaining games typically have a plethora of equilibria, for two distinct reasons. First, somewhat analogous to the folk-theorem literature in repeated games, the presence of an infinite number of bargaining rounds permits history-dependent strategies that can often support a wide variety of equilibrium behavior [Ausubel and Deneckere (1989a, 1989b)]. Secondly, even if bargaining were allowed to last only a finite number of periods, there will typically still exist a multiplicity of sequential equilibria. This multiplicity arises because sequential equilibrium imposes no restrictions on players' beliefs following out-of-equilibrium moves (Bayes' law is then simply not applicable). As a consequence, an out-of-equilibrium offer by the informed party can lead to adverse inferences regarding its eagerness to conclude the transaction, resulting in poor terms of trade. In alternating-offer bargaining games, the threat of such adverse inferences can therefore often sustain a wide variety of bargaining outcomes [Fudenberg and Tirole (1983), Rubinstein (1985a, 1985b)].

In order to narrow down the range of predicted bargaining outcomes, researchers have confined attention to more restrictive equilibrium notions. One refinement that has received considerable attention is the concept of stationary equilibrium [Gul, Sonnenschein and Wilson (1986)]. Recall that a belief is a probability distribution $F(q)$ over the set of possible signals (the unit interval). We will say that a belief $G(q)$ is a *truncation* (from the left) of the belief $F(q)$ if it is the conditional probability distribution derived from $F(q)$, given that the signal exceeds some threshold level $q' > 0$. Thus $G(q) = 0$ for $q < q'$ and $G(q) = [F(q) - F(q')]/[1 - F(q')]$ for $q \geq q'$. A *stationary equilibrium* is a sequential equilibrium satisfying three additional conditions:

(1) Along the equilibrium path, the beliefs following rejection of the informed party's offer are a truncation of the beliefs entering that period;

(2) For every history such that the current beliefs are a truncation of the priors, the informed party's current acceptance behavior is a function only of the current offer; and

(3) For every history such that the current belief is the same truncation of the prior, the informed party's current offer behavior is identical.

The notion of stationarity is rather subtle, and to understand its meaning it is useful to first restrict attention to the game in which the uninformed party makes all the offers, so that only requirement (2) carries any force. Observe that, in any offer/counteroffer game, rejections by the informed party always lead to a truncation of the current beliefs:[9]

LEMMA 1 [Fudenberg, Levine and Tirole (1985)]. *Let n be a period in which it is the uninformed party's turn to make an offer, and denote the history of rejected prices entering period n by h_n. Then to every sequential equilibrium there corresponds a nonincreasing (nondecreasing) function $P(h_n, q)$ and equivalent sequential equilibrium such that if the informed party is the buyer (seller), it accepts the current offer p if and only if $p \leqslant P(h_n, q)$ (respectively, $p \geqslant P(h_n, q)$).*

PROOF. Suppose buyer type q is willing to reject the current offer p. Any buyer type $q' > q$ can always mimic the strategy of type q, and thereby secure the same expected probability of trade and expected payment from rejecting p. The single crossing property then implies that if $v(q') < v(q)$, type q' will strictly prefer rejection to acceptance. Meanwhile, if q is indifferent between accepting and rejecting, a purification argument shows that there is an equivalent sequential equilibrium and a cutoff signal level q'' with $v(q'') = v(q)$, such that all $q' < q''$ accept p and all $q' > q''$ reject p. $\square$

For the game where the uninformed party makes all the offers, Lemma 1 implies that the informed party uses a possibly history-dependent reservation price strategy, $P(h_n, q)$. Requirement (2) in the definition of stationarity requires that the acceptance functions $P(h_n, q)$ are constant over all histories h_n. It is this history independence that gives stationarity its cutting power. Stationarity is a stronger restriction than Markov-perfection [Maskin and Tirole (1994)], since the latter would only require that P be constant on histories inducing the same current beliefs. As emphasized by Gul and Sonnenschein (1988) stationarity also embodies a form of monotonicity: when the uninformed party is more optimistic (in the sense that the beliefs are truncated at lower level), the informed party must not be tougher in its acceptance behavior.

For game structures that permit the informed party to make offers, stationarity carries two additional restrictions. The informed party's offer behavior must be Markovian (re-

---

[9] Sequential equilibria of the game in which the uninformed party makes all the offers therefore have a screening structure, with higher valuation buyer types trading earlier and at higher prices than lower valuation types. Delaying agreement by rejecting the current offer credibly signals to the seller that the buyer has a lower valuation, thereby making her willing to lower price over time.

quirement (3)); and in equilibrium the beliefs following a period in which the informed party made an offer must be a truncation of the prior (requirement (1)). Thus, stationarity imposes a screening structure on the equilibrium. This assumption is very strong, since it requires the uninformed party to accept with probability zero or one following any equilibrium offer that is not made by all types, and thereby severely restricts the informed party's ability to signal its type. At the same time, however, stationarity may be insufficiently restrictive because it does not address the multiplicity of equilibria arising from "threatening with beliefs". Furthermore, refinements of sequential equilibrium designed to reduce this multiplicity are potentially at odds with the requirements of stationarity. This raises the question of whether stationary equilibria (with or without additional refinements) are always guaranteed to exist. Fortunately, as we shall see, the answer to this question is broadly positive.

In the remainder of this section, we study the trading situation in which it is common knowledge that the gains from trade are bounded away from zero, i.e., there exists $\Delta > 0$ such that $v(q) - c(q) \geqslant \Delta$ for all $q \in [0, 1]$. Section 4 studies the case where there is no such Δ, so that the gains from trade can be arbitrarily small.

3.1. Private values

To facilitate the discussion of private values model, we will henceforth assume that the informed party is the buyer (the symmetric situation in which it is the seller that is informed is treated in the subsection on interdependent values). In this case, the seller's cost is independent of the signal level and can without loss of generality be normalized to zero (by measuring buyer valuations net of cost). The model is therefore completely described by the discount factor δ and the nonincreasing buyer valuation function $v(q)$. In order to permit the existence of an equilibrium, $v(q)$ will be assumed to be left continuous (to see this is necessary, consider the seller-offer game in which $\delta = 0$).

3.1.1. The seller-offer game

Following Fudenberg, Levine and Tirole (1985) and Gul, Sonnenschein and Wilson (1986), we are interested in stationary equilibria in which the buyer's acceptance behavior depends upon previous history only to the extent it is reflected in the current price. The purification argument in the proof of Lemma 1 shows that there is no loss of generality in assuming that the buyer does not randomize in his acceptance behavior, an assumption which we will maintain henceforth. The buyer's acceptance behavior is thus completely characterized by a nonincreasing (left-continuous) acceptance function $P(q)$. Consequently, following any history the seller's belief will always be a truncation of the prior, i.e., be uniform on an interval of the form $[Q, 1]$. The lower endpoint of this interval, Q, is thus a *state variable*.

The acceptance function acts as a static demand curve for the seller, who faces a trade-off between screening more finely and delaying agreement. This tradeoff is captured by the dynamic programming equation:

$$W(Q) = \max_{Q' \geqslant Q} \left\{ P(Q') \frac{(Q' - Q)}{(1 - Q)} + \delta \frac{(1 - Q')}{(1 - Q)} W(Q') \right\}. \tag{1}$$

To understand (1), observe that if the seller brings the state to Q' (by charging the price $P(Q')$), then the buyer will accept with conditional probability $(Q' - Q)/(1 - Q)$. Rejection happens with complementary probability, moves the state to Q', and results in the seller receiving the value $W(Q')$ with a one-period delay. Letting $V(Q) = (1 - Q)W(Q)$ denote the seller's ex ante expected value from trading with buyer types in the interval $(Q, 1]$, Equation (1) can be simplified to:

$$V(Q) = \max_{Q' \geqslant Q} \left\{ P(Q')(Q' - Q) + \delta V(Q') \right\}. \tag{2}$$

Let $T(Q)$ denote the argmax correspondence in (2). By the generalized Theorem of the Maximum [Ausubel and Deneckere (1993b)], T is nonempty and compact-valued, and the value function V is continuous. A straightforward revealed preference argument also shows that T is a nondecreasing correspondence, and hence single-valued at all but at most a countable set of Q.

Define $t(Q) = \min T(Q)$, and note that $t(Q)$ is continuous at any point where $T(q)$ is single-valued. Now consider any point Q where $v(\cdot)$, $P(\cdot)$ and $t(\cdot)$ are continuous; consumer optimization then requires that:

$$P(Q) = (1 - \delta)v(Q) + \delta P\big(t(Q)\big). \tag{3}$$

Equation (3) says that when the seller charges the price $p = P(Q)$, the buyer of type $q = Q$ must be indifferent between accepting the offer p, and waiting one period to accept the next offer (which must be $P(t(Q))$). A straightforward argument establishes that the consumer indifference equation (3) must in fact hold for all $Q > 0$.[10] This fact has an important consequence: in any stationary equilibrium, *the seller will never randomize except (possibly) in the initial period*.[11] Indeed, in period zero the seller is free to randomize amongst any element of $T(0)$. However, given any such choice Q,

---

[10] Consider any of the (at most countably many) excluded states Q, and let $\{Q_n\}$ be a sequence of nonexcluded points converging from below to Q. Since (3) holds for each n, upon taking limits as $n \to \infty$, we see that (3) holds for all $Q > 0$.

[11] Gul, Sonnenschein and Wilson [(1986), Theorem 1] constructively demonstrate the absence of randomization along the equilibrium path, under the assumption that there is a gap and condition (L) of Theorem 4 (below) holds. The argument given here [drawn from Ausubel and Deneckere (1989a), Proposition 4.3] shows that it is stationarity that is the driving force behind this result.

Equation (3) requires the seller to select $t(Q)$ in the next period (even if $T(Q)$ is not single-valued). This is necessary to make the buyer's acceptance decision optimal.

The triplet $\{P(\cdot), V(\cdot), t(\cdot)\}$ completely describes a stationary equilibrium. After any history in which the seller selects a price $p = P(Q)$ for some Q, all consumer types $q \leqslant Q$ accept and all others reject; the next period the seller lowers the price to $P(t(Q))$. If the seller were ever to select a price p such that $\sup\{P(Q'): Q' > Q\} < p < P(Q)$ for some Q, then the highest consumer type to accept is again Q. However, if the gap in the range of P is due to a discontinuity in the function $t(Q)$, then to make consumer Q's acceptance rational, the seller must in the next period randomize between the offers in $P(T(Q))$ so as to make Q indifferent. Note, however, that an optimizing seller will never charge a price in this range, as she could induce exactly the same set of buyer types to accept by charging the higher price $P(Q)$. Randomization is therefore only called for if the seller made a mistake in the previous period.

Any stationary equilibrium path has the following structure. In the initial period, the seller selects (possibly randomly) a price $P(Q_0)$, for some $Q_0 \in T(0)$. Note that randomization is possible only if $T(0)$ is multiple-valued, i.e., its profit function has multiple maximizers. This should be a rare occurrence, because as a monotone correspondence, $T(Q)$ can have at most countably many points at which it is not single-valued (see the genericity statement in Theorem 4, below). The remainder of the future is then entirely deterministic, with the seller successively lowering the prices to $P(t(Q_0))$, $P(t^2(Q_0))$, $P(t^3(Q_0)), \ldots$, and corresponding buyer acceptances in $(Q_0, t(Q_0)]$, $(t(Q_0), t^2(Q_0)]$, $(t^2(Q_0), t^3(Q_0)], \ldots$.

An important question is whether the coupled pair of functional equations (2) and (3) has a solution. At the same time, the bootstrap structure of these equations suggests that there may be a severe multiplicity of stationary triplets. The pioneering work in the areas of existence and uniqueness of stationary equilibria is due to Fudenberg, Levine and Tirole (1985) and Gul, Sonnenschein and Wilson (1986). Below, we collect a number of disparate results in the literature into a single theorem:

THEOREM 4. *For any left-continuous valuation function $v(\cdot)$, there exists a stationary equilibrium of the seller-offer game. Every stationary equilibrium is supported by a stationary triplet $\{P, t, V\}$ satisfying (2) and (3). Furthermore, if there is a gap, and if the demand curve satisfies a Lipschitz condition at $q = 1$:*

There exists $L < \infty$ such that $v(q) - v(1) \leqslant L(1 - q)$, for all $q \in [0, 1]$,   (L)

then the stationary triplet is unique, every sequential equilibrium outcome coincides with a stationary equilibrium outcome, and for generic values of the state there is a unique stationary (and hence sequential) equilibrium outcome. Under these conditions, there also exists a finite integer $\widehat{N}$ such that the buyer accepts the seller's offer by period $\widehat{N}$, regardless of the discount factor δ.

PROOF. Fudenberg, Levine and Tirole [(1985), Propositions 1 and 2] prove existence and generic uniqueness of the outcome path in the case of a gap, under the assumption that the demand curve is differentiable with derivative bounded above and below. Gul, Sonnenschein and Wilson [(1986), Theorem 1] prove existence and uniqueness of a stationary triplet when there is a gap and condition (L) holds, and also demonstrate generic uniqueness of the outcome path. A general existence proof appears in Ausubel and Deneckere [(1989a), Theorem 4.2]. Deneckere (1992) proves that under condition (L) the number of bargaining rounds is uniformly bounded for fixed $v(\cdot)$. □

To make matters more concrete, and also to illustrate some of the ideas behind Theorem 4, let us work out a simple example in which the buyer's valuation can take on two possible values, $\bar{b} > \underline{b} > 0$:

$$v(q) = \begin{cases} \bar{b}, & 0 \leqslant q \leqslant \hat{q}, \\ \underline{b}, & \hat{q} < q \leqslant 1. \end{cases} \tag{4}$$

Note that this example is in the case of a "gap" and satisfies condition (L), so by Theorem 4 there exists a unique stationary triplet.

First, let us consider the case where $\hat{q}\bar{b} < \underline{b}$, i.e., the monopoly price on the static demand curve (4) equals $\underline{b}$. Observe that since the seller will never offer a price more favorable than she would if she were facing the strongest buyer type for sure, the buyer will always accept any price below $\underline{b}$ with probability one. Thus, in any sequential equilibrium, the seller's payoff must be no lower than her static monopoly profits, $\underline{b}$. Meanwhile, Stokey (1979) showed that the optimal selling policy of a dynamic monopolist with perfect commitment power consists of charging the static monopoly price, and never lowering price thereafter [see also the closely related "no-haggling" result of Riley and Zeckhauser (1983)]. Since a monopolist lacking commitment power can only do worse, the seller's equilibrium profits must also be no higher than her static equilibrium profits. We conclude that there is a unique sequential equilibrium outcome, with the seller charging the price $\underline{b}$, and all buyer types accepting. Note that this equilibrium is supported by the unique stationary triplet $V(Q) = (1 - Q)\underline{b}$, $t(Q) = 1$, and using (3), $P(q) = (1 - \delta)\bar{b} + \delta\underline{b}$ for $q \in [0, \hat{q}]$ and $P(q) = \underline{b}$ for $q \in (\hat{q}, 1]$.

When $\hat{q}\bar{b} > \underline{b}$, bargaining necessarily takes place over multiple periods, but the above argument still contains the key to uniqueness of the stationary triplet. Indeed, let us define q_1 as the lowest value of the state such that $\underline{b}$ is a monopoly price on the residual demand curve starting at q_1, i.e., $\bar{b}(\hat{q} - q_1) = \underline{b}(1 - q_1)$. A parallel argument to the one given above then establishes that once the state reaches beyond q_1, the seller will *necessarily* end the bargaining immediately, by offering the price $\underline{b}$. The role of condition (L) in Theorem 4 is to more generally guarantee the existence of a critical level $q_1 < 1$ such that whenever the state exceeds q_1 the dispersion of valuations of the remaining buyer types is such that it no longer pays the seller to price discriminate amongst them.

With the endplay tied down, backward induction on the state then completes the uniqueness argument. To see how this works, observe that there exists a $q_2 < q_1$, such that whenever the state is in $(q_2, q_1]$ the seller will select to bring the state in the interval $(q_1, 1]$. Indeed, whenever q_2 is sufficiently near q_1, any potential gain from increased price discrimination over the interval $(q_2, q_1]$ is outweighed by the loss due to delayed receipt of the profits $V(q_1)$. In our two-type model, when the state is in $(q_2, q_1]$ the monopolist will therefore offer $p_1 = (1 - \delta)\bar{b} + \delta\underline{b}$, which all buyer types in $[0, \hat{q}]$ accept. In this fashion, we can keep on recursively extending the stationary triplet to the entire interval $[0, 1]$. Buyer types in the interval $(q_i, q_{i-1}]$ will be indifferent between accepting p_i and waiting one period to receive p_{i-1}, and the state q_i is such that the monopolist is indifferent between offering p_i (with all buyer types in $(q_i, q_{i-1}]$ accepting) and offering p_{i-1} (with all buyer types in $(q_i, q_{i-2}]$ accepting). More precisely, we can compute the following explicit solution.

Let $q_{-1} = 1$, $q_0 = \hat{q}$, and inductively define the sequence $q_1 > q_2 > \cdots > q_N$ from:

$$m_n = \alpha\delta^{-(n-1)}m_{n-1} \quad (n \geqslant 2), \tag{5}$$

and the initial condition $m_1 = (\alpha - 1)m_0$, using $m_n = q_{n-1} - q_n$, $\alpha = \bar{b}/(\bar{b} - \underline{b})$, and $N = \min\{n\colon q_n \leqslant 0\}$. Also, let p_n be such that a buyer with valuation $\bar{b}$ is indifferent between accepting p_n today and waiting n periods to receive the offer $\underline{b}$:

$$p_n = (1 - \delta^n)\bar{b} + \delta^n\underline{b}. \tag{6}$$

THEOREM 5. *Let $v(q)$ be given by (4), and let $q_N \leqslant 0 < q_{N-1} < \cdots < q_0 = \hat{q}$ be defined by (5). Then with every (purified) sequential equilibrium of the seller-offer game is associated the unique stationary triplet:*

$$
\begin{aligned}
P(Q) &= p_n, & Q &\in (q_n, q_{n-1}], \\
t(Q) &= q_{n-2}, & Q &\in (q_n, q_{n-1}] \quad \text{if } n > 1, \quad \text{and} \\
& & Q &\in (q_1, 1] \quad \text{if } n = 1, \\
V(Q) &= p_{n-1}(q_{n-2} - Q) + \delta V(q_{n-2}), & Q &\in (q_n, q_{n-1}] \quad \text{if } n > 1, \quad \text{and} \\
& & Q &\in (q_1, 1] \quad \text{if } n = 1.
\end{aligned}
$$

PROOF. See Deneckere (1992). $\qquad\square$

According to Theorem 5, when $q_N < 0$ bargaining lasts for N periods. The seller starts out by offering the price $p_{N-1} = P(q_{N-2})$, which is accepted by all buyer types in the interval $[0, q_{N-2}]$. Play then continues with the seller offering p_{N-2}, which all buyer types in $(q_{N-2}, q_{N-3}]$ accept, and so on, until the state q_0 is reached at which point the seller makes the final offer p_0. When $q_N = 0$, the seller can freely randomize between charging p_N and p_{N-1}. However, given the outcome of the randomization, the remainder of the equilibrium path is uniquely determined: if the seller initially selects p_N play lasts for $(N + 1)$ periods, and if she selects p_{N-1} play lasts for N periods. Note, however, that the condition $q_N = 0$ is highly nongeneric, in two senses. First, if the initial state is slightly different from q_N the outcome is unique. Secondly, since the condition $q_N = 0$ is equivalent to $m_0 + \cdots + m_N = 1$, it follows from (5) that for generic (α, δ) the outcome path is unique.

The closed form (5) also allows us to investigate the behavior of the solution as bargaining frictions become smaller, i.e., players become more patient [see also Hart (1989), Proposition 2]. Intuitively, for fixed acceptance function P, the seller will discriminate more and more finely as she becomes more patient, approaching perfect price discrimination on the acceptance function P as δ converges to one. Counteracting this is that for fixed seller behavior, as the buyer becomes more patient, the acceptance function will become flatter and flatter, in the limit approaching the constant $\underline{b} = v(1)$ as δ converges to 1. If we fix δ_S and let δ_B converge to one, the seller loses all bargaining power. On the other hand, if we fix δ_B and let δ_S increase, the seller will gain bargaining strength [Sobel and Takahashi (1983)]. With equal discount factors, the two forces more or less balance each other out. To see this, note from (5) that m_n is decreasing, and hence that the number of bargaining rounds N is increasing in δ. However, as the limiting solution to (5) is given by $m_n = \alpha^n m_0$, we see that regardless of the discount factor, the number of bargaining rounds is bounded above by:

$$\widehat{N} = \min\{n: \alpha^n m_0 \geqslant 1\}.$$

While the number of equilibrium bargaining rounds therefore increases with δ, the existence of a uniform upper bound to the number of bargaining rounds implies that the cost of delay (as measured by the forgone surplus) vanishes as δ approaches one.

A slightly weaker, but qualitatively similar, proposition has become known in the literature as the "Coase Conjecture", after Nobel laureate Ronald Coase, who argued that a durable goods monopolist selling an infinitely durable good to a demand curve of atomistic buyers would lose its monopoly power if it could make frequent price offers [Coase (1972)]. The connection with the durable goods literature obtains because to every actual buyer type in the durable goods model, there corresponds an equivalent potential buyer type in the bargaining model. To formally state the Coase Conjecture, let us denote the length of the period between successive seller offers by z, and let r be the discount rate common to the bargaining parties, so that $\delta = e^{-rz}$. We then have:

THEOREM 6 (Coase Conjecture). *Suppose we are in the case of a gap. Then for every* $\varepsilon > 0$ *and valuation function* $v(\cdot)$, *there exists* $\bar{z} > 0$ *such that, for every time interval* $z \in (0, \bar{z})$ *between offers and for every sequential equilibrium, the initial offer in the seller-offer bargaining game is no more than* $\underline{b} + \varepsilon$ *and the buyer accepts the seller's offer with probability one by time* ε.

PROOF. Gul, Sonnenschein and Wilson [(1986), Theorem 3]. □

Note that Theorem 6 immediately follows from Theorem 4, by selecting $\bar{z} \leqslant \varepsilon/\widehat{N}$, and by noting that since the highest valuation buyer always has the option to wait until period $\widehat{N}$ to accept the price $\underline{b}$, the seller's initial price can be no more than $(1 - \delta^{\widehat{N}})v(0) + \delta^{\widehat{N}}\underline{b}$, which converges to $\underline{b}$ as z converges to zero. For empirical or experimental work, Theorem 6 has the unfortunate implication that real bargaining delays can only be explained by either exogenous limitations on the frequency with which bargaining partners can make offers, or by significant differences in the relative degree of impatience between the bargaining parties.

3.1.2. Alternating offers

When the uninformed party makes all the offers, the informed party has very limited means of communication. At any point in time, buyer types can only separate into two groups, those who accept the current offer and thereby terminate the game, and those who reject the current offer in order to trade at more favorable terms in the future. Since higher valuation buyer types stand to lose more from delaying trade, the equilibrium necessarily has a screening structure. In the alternating-offer game, screening will still occur in any seller-offer period, for exactly the same reason. During buyer-offer periods, however, the informed party has a much richer language with which to communicate, so a much richer class of outcomes becomes possible. There is now a potential for the buyer to signal his type, with higher valuation buyer types trading off higher prices for a higher probability of acceptance. But as in the literature on labor market signaling, many other types of outcomes can be sustained in sequential equilibrium, with different buyer types pooling or partially pooling on common equilibrium offers.

Researchers have long considered many of these equilibria to be implausible, because they are sustained by the threat of adverse inferences following out-of-equilibrium offers. Unfortunately, the literature on refinements has concentrated mostly on pure signaling games [Cho and Kreps (1987)], so there exist few selection criteria applicable to the more complicated extensive-form games we are considering here. In narrowing down the range of equilibrium predictions, researchers have therefore resorted to criteria which try to preserve the spirit of refinements developed for signaling games, but the necessarily ad hoc nature of those criteria has led to a variety of equilibrium predictions [Rubinstein (1985a), Cho (1990b), Bikchandani (1992)].[12]

To select plausible equilibria, Ausubel and Deneckere (1998) propose a refinement of perfect equilibrium, termed *assuredly perfect equilibrium* (APE). Assuredly perfect equilibrium requires stronger player types (e.g., lower valuation buyer types) to be infinitely more likely to tremble than weaker player types, as the tremble probabilities converge to zero. The purpose of making the strong player types much more likely to tremble is to rule out adverse inferences: following an unexpected move by the informed party, beliefs must be concentrated on the strong type, unless this action yields the weak

---

[12] One notable exception is Grossman and Perry (1986a), who develop a general selection criterion, termed perfect sequential equilibrium, and apply it to the alternating-offer bargaining game (1986). However, perfect sequential equilibria do not generally exist, and in fact fail to do so in the alternating-offer bargaining game when the discount factor is sufficiently high. This is unfortunate, as the case where bargaining frictions become small is of special importance in light of the literature on the Coase Conjecture. General existence is also a problem in Cho (1990b) and Bikchandani (1992).

type its equilibrium utility.[13] Thus beliefs are not permitted to shift to the weak type unless there is a reason why (in the equilibrium) the weak type may wish to select the deviant action. APE has the advantage of being relatively easy to apply, and is guaranteed to always exist in finite games.

Importantly, for the two-type alternating-offer bargaining model given by (4), Ausubel and Deneckere (1998) show that for generic priors there exists a unique APE.[14] We will describe this equilibrium outcome here only for the game in which the seller moves first (this facilitates comparison with the seller-offer game). For this purpose, let us define $\tilde{n} = \max\{n \in Z_+ : 1 - \delta^{2n-2} - \delta^{2n-1}\alpha^{-1} < 0\}$. The meaning of $\tilde{n}$ is that in equilibrium, regardless of the fraction of low valuation buyer types, the game always concludes in at most $2\tilde{n} + 2$ periods. This should be contrasted with the seller-offer game, where the number of bargaining rounds grows without bound as the seller becomes more and more optimistic.

The intuition behind this difference is that as the number of remaining bargaining rounds becomes larger, the seller extracts more and more surplus from the weak buyer type.[15] At the same time, there is an upper bound on how much the seller can extract, namely what he would obtain in the complete-information game against the weak buyer type. In the seller-offer game, this is all of the surplus, explaining why with this offer structure the number of effective bargaining rounds can increase without bound as the seller becomes more and more optimistic. In contrast, in the complete-information alternating-offer game the seller receives only a fraction $1/(1 + \delta)$ of the surplus (when it is his turn to move). Consequently, in the alternating-offer game the number of effective bargaining rounds must be bounded above, no matter how optimistic the seller.[16] For the sake of brevity, we will consider here only the case where $\tilde{n} > 1$ (note that this necessarily holds when δ is sufficiently high).

Qualitatively, the equilibrium has the following structure. Whenever it is the buyer's turn to make a proposal, all buyer types pool by making nonserious offers, until the seller becomes convinced he is facing the low valuation buyer. At this point, both buyer types pool by making the low valuation buyer's complete-information Rubinstein offer, $r_0 = \delta \underline{b}/(1 + \delta)$, which the seller accepts. The sequence of prices offered by the seller along the equilibrium path must keep the high valuation buyer indifferent, so we must have:

$$p_n = (1 - \delta^{2n-1})\bar{b} + \delta^{2n-1}r_0, \quad n = 1, \ldots, \tilde{n}, \tag{7}$$

---

[13] If an action yields the weak type less than its equilibrium utility, then in approximating games, the weak type must be using that action with minimum probability. As the ratio of the weak to the strong type's tremble probability converges to zero, limiting beliefs will have to be concentrated on the strong type.

[14] More precisely, they show that finite horizon versions of the alternating-offer bargaining game in which the buyer makes the last offer has a unique APE for generic values of the prior. Below, we describe the limit of this equilibrium as the horizon length approaches infinity.

[15] Formally, this is reflected in the fact that both sequences of prices (6) and (8) are increasing in n, and converge to $\bar{b}$ as n converges to infinity.

[16] Formally, $\tilde{n}$ is the largest integer such that p_n remains below $\bar{p} = \bar{b}/(1 + \delta)$, the complete information seller offer against the weak buyer type.

unless the seller is extremely optimistic, in which case the game starts out with $\bar{p} = \bar{b}/(1 + \delta)$, the seller's offer in the complete-information game against the weak buyer type.

Analogous to the seller-offer game, the sequence of cutoff levels q_n is constructed so that at q_n ($n = 1, \ldots, \tilde{n}$) the seller is indifferent between charging p_n and p_{n-1}, and at $q_{\tilde{n}+1}$ the seller is indifferent between charging $\bar{p}$ and $p_{\tilde{n}}$. Formally, let $q_{-1} = 1$, $q_0 = \hat{q}$, and inductively define the sequence of cutoff levels $q_1 > q_2 > \cdots > q_{\tilde{n}} > q_{\tilde{n}+1}$ from $m_1 = (\alpha - 1)m_0, m_2 = \beta\delta^{-1}(1 + \delta)^{-1}m_1$,

$$m_n = \beta\delta^{-(2n-3)}m_{n-1}, \quad \text{for } 3 \leqslant n \leqslant \tilde{n}, \tag{8}$$

and $m_{\tilde{n}+1} = \omega m_{\tilde{n}}$, where $\beta = \bar{b}/[\bar{b} - r_0]$ and $\omega = (1 - \delta^2)\bar{b}/[\bar{p} - p_{\tilde{n}}]$. To rule out nongeneric cases, and again analogously to the seller-offer game, let $N = \max\{n \leqslant \tilde{n} + 1: q_n \geqslant 0\}$, and suppose $q_N > 0$:

THEOREM 7 [Ausubel and Deneckere (1998)]. *Consider the alternating-offer game, and suppose that $q_N > 0$. Then in the unique APE outcome, following histories with no prior observable buyer deviations, the buyer uses a stationary acceptance strategy. If $N \leqslant \tilde{n}$, this acceptance strategy is given by:*

$$\begin{aligned}
P(q) &= p_n, && q \in (q_n, q_{n-1}], \ 0 \leqslant n \leqslant N, \\
P(q) &= \min\{p_n, \bar{p}\}, && q \in [0, q_N].
\end{aligned} \tag{9}$$

In equilibrium, the seller successively makes the offers $p_N, p_{N-1}, \ldots, p_1$, with the buyer accepting according to (9) and making nonserious counteroffers until p_1 has been rejected. The buyer then counteroffers r_0, which the seller accepts with probability one.

If $N = \tilde{n} + 1$, the buyer's acceptance strategy is given by:

$$\begin{aligned}
P(q) &= p_n, && q \in (q_n, q_{n-1}], \ 0 \leqslant n \leqslant \tilde{n}, \\
P(q) &= \bar{p}, && q \in [0, q_{\tilde{n}}].
\end{aligned} \tag{10}$$

In equilibrium, the seller starts out by offering $\bar{p}$, which all buyer types in $[0, \bar{q}_{\tilde{n}}]$ accept, and all other types reject. Following a nonserious buyer offer, the seller then randomizes between the offers $p_{\tilde{n}}$ and $p_{\tilde{n}-1}$ so as to make the weak buyer type indifferent between accepting and rejecting the previous seller offer.[17] Following the offer $p_{\tilde{n}}$ the seller continues with the offers $p_{\tilde{n}-1}, p_{\tilde{n}-2}, \ldots, p_1$, and following the offer $p_{\tilde{n}-1}$ the seller continues with the offers $p_{\tilde{n}-2}, \ldots, p_1$. In each case, the buyer accepts according to (10), and makes nonserious counteroffers until p_1 has been rejected. The game then ends with the buyer counteroffering r_0, which the seller accepts with probability one.

---

[17] In other words, denoting the weight on $p_{\tilde{n}}$ by ϕ, we have $\bar{b} - \bar{p} = \delta^2\{\phi(\bar{b} - p_{\tilde{n}}) + (1 - \phi)(\bar{b} - p_{\tilde{n}-1})\}$.

One of the main thrusts of the literature on static signaling models has been to show that refinements based on stability [Kohlberg and Mertens (1986)] tend to select signaling equilibria [Cho and Sobel (1990)]. For example, Cho and Kreps (1987) show that in the Spence labor market signaling game with two types, the Intuitive Criterion selects the Pareto efficient separating equilibrium (it is easily verified that APE would select the same outcome). In contrast, in the alternating-offer bargaining game considered above, the buyer uses only fully-pooling offers along the equilibrium path.

The intuition for why pooling obtains is that the strong buyer type *tries* to separate by making a nonserious offer and delaying trade. The only alternative for the weak buyer type is therefore to make a separating offer, which yields the worst possible (complete-information) utility level. Meanwhile, stationarity of the equilibrium acceptance strategy provides the seller with an incentive to accelerate trade, and therefore (by the usual Coase Conjecture argument) to charge a relatively low price following rejection of the nonserious offer. But then a revealing offer cannot be optimal, so the equilibrium has to be pooling (see the discussion surrounding Theorem 12 for related intuition).

In fact, from Theorem 7, we can see that the strong version of the Coase Conjecture also holds in the alternating-offer game: there exists a uniform bound M such that regardless of the discount factor δ trade occurs in at most $2M - 1$ periods. Indeed, m_n is decreasing in δ for all $n \leqslant \tilde{n}$, and $\tilde{n}$ converges to infinity as δ converges to 1, so we can find M by recursing m_n at $\delta = 1$. Note that M must be finite, because $m_2(1) = 2\theta m_1$ and $m_n(1) = \theta m_{n-1}(1)$ where $\theta = (1 + \alpha^{-1}) > 1$.

It is interesting to compare the effect of shifting bargaining power to the informed party on equilibrium bargaining delay. For this purpose, let us denote the solution to (5) by m_n^s and the solution to (8) by m_n^a. Observe that $m_0^s = m_0^a$ and $m_1^s = m_1^a$; some straightforward but tedious algebra shows that $m_n^s(\delta) < m_n^a(\delta)$, for $n \geqslant 2$. We conclude that as long as the alternating-offer game starts out with a seller offer below $\bar{p}$,[18] the alternating-offer game requires more offers, and has a lower acceptance probability than the seller-offer game. Moreover, the alternating-offer game results in additional delay because (with the exception of the final bargaining round) only seller-offer periods result in trade. Hence the traditional wisdom that bargaining becomes more efficient as the informed party gains bargaining strength proves to be incorrect.

Finally, it should be noted that when the seller is so optimistic that she starts with the highest possible offer $\bar{p}$, the equilibrium requires her to randomize with positive probability two periods later. Unlike in the seller-offer game, randomization in seller offers may thus be necessary along the equilibrium path.

3.1.3. The buyer-offer game and other extensive forms

In the game where the buyer makes all the offers, it is clearly a sequential equilibrium for the buyer to always offer the seller his cost c, and for the seller to accept any price

---

[18] As discussed above, this is necessarily the case when δ is sufficiently large.

above c with probability one. Ausubel and Deneckere [(1989b), Theorem 4] show that this is in fact the only sequential equilibrium. Intuitively, the seller can do no better than in the complete-information game where the buyer is known to have valuation $v(1)$, but since the buyer makes all the offers, he can extract all of the surplus no matter what his valuation. We conclude that the buyer-offer game always achieves an efficient outcome, regardless of whether or not there is a gap, condition (L) holds, or the magnitude of the discount factor.

More generally, we can study the impact of transferring bargaining power from the seller to the buyer by considering the (k, l)-alternating-offer bargaining game, in which the seller and buyer alternate by making k and l successive offers, respectively. The ratio k/l measures the relative frequency with which the seller gets to make offers, and hence is a measure of his bargaining strength. Note that in the complete-information case, this game yields the same outcome as the alternating-offer game in which the seller's discount factor is given by $\delta_S = \delta^l$ and the buyer's discount factor is given by $\delta_B = \delta^k$. Thus, the ratio $\rho \equiv k/l$ can also be interpreted as the relative degree of impatience between the bargaining parties. Observe now that in any sequential equilibrium, the weakest buyer type must earn at least what he would in the complete-information case, so we have $U(1) \geqslant v(1)\delta_S(1 - \delta_B)/(1 - \delta_S\delta_B)$, which converges to $v(1)/(1 + \rho)$ as δ approaches 1. Since $v(1)$ is the maximum surplus available, we conclude that when ρ is small all sequential equilibria must yield bargaining outcomes that are nearly efficient. This conclusion obtains regardless of whether or not there is a gap.

Admati and Perry (1987) consider an alternating-offer extensive form game that differs from Rubinstein's game in that the length between successive offers is chosen endogenously by the players. Thus, when a player rejects an offer, he commits unilaterally to neither make a counteroffer nor receive another offer until a length of time of his choice has elapsed. During this time period, all communications are closed off, and the commitment is irrevocable. Admati and Perry analyze the two-type model given by (4), and apply a forward-induction-like refinement. When the prior on the weak type is sufficiently high, this refinement uniquely selects a separating equilibrium.[19] The seller starts out by making the offer $\bar{p} = \bar{b}/(1 + \delta)$, which the weak buyer type accepts, and the strong buyer type rejects. The strong buyer type then delays any further negotiation until a time of length T has elapsed, at which point it makes its complete-information counteroffer $r_0 = \delta\underline{b}/(1 + \delta)$, which the seller accepts. T is chosen such that the weak buyer type is indifferent between accepting the seller's initial offer, and mimicking the low buyer type. This equilibrium has an intuitive structure strongly reminiscent of the Riley outcome in the Spence labor market signaling model, but this elegance comes at a strong price: the buyer is committed not to receive any counteroffer during the time interval of length T. Note that the seller has an incentive to make such a counteroffer, for once the buyer has chosen T, his type is revealed to be strong. In fact, both parties would

---

[19] For intermediate values of the prior, there are multiple equilibria, and for sufficiently low values of the prior the seller offers the strong buyer's complete-information price.

be better off settling immediately at the price r_0, and the buyer knows this is the case, but is committed not to reopen the lines of communication until time T [Admati and Perry (1987), Section 8.4]. If the communication channels were allowed to reopen any earlier, the signaling equilibrium would be destroyed. Indeed, in the alternating-offer game analyzed in the previous section, separation never occurs.

3.2. Interdependent values

Consider the trading situation in which a seller who is privately informed about the quality of a used car faces a potential buyer who cares about the quality of the vehicle. As we saw in Section 2, there then exists a trading mechanism that can achieve the efficient outcome if and only if the buyer's expected valuation exceeds the valuation of the owner of the highest quality car, i.e., $E[v(q)] \geqslant c(1)$. This raises two interesting questions for extensive-form bargaining. First, assuming that the above condition holds, will the same forces that operate in the private values model to produce efficient trade when bargaining frictions disappear still permit the efficient outcome to be reached when values are interdependent? Second, assuming that the above condition is violated, will the limiting trading outcome at least be ex ante efficient, in the sense that it maximizes the expected gains from trade subject to the IC and IR constraints?

So far, the literature has only studied the bargaining game in which the uninformed party (the buyer) makes all the offers [Evans (1989), Vincent (1989)]. Our discussion here is based upon Deneckere and Liang (1999). The arguments establishing existence of equilibrium for the interdependent values with a gap closely parallel those of Gul, Sonnenschein and Wilson (1986). Consequently an analogue of Theorem 6 holds, with $c(q)$ taking the role of $v(q)$, with one important difference: it is no longer the case that the number of bargaining rounds is uniformly bounded above, regardless of the discount factor (if this were the case, then as the discount factor converged to one, the efficient outcome would obtain even when $E[v(q)] < c(1)$, contradicting Theorem 2). As in the private values case, generically there is a unique equilibrium outcome, and equilibrium outcomes are sustained by a unique stationary triplet. In equilibrium, the buyer successively increases his offers over time. Low-quality seller types accept low prices, while high-quality seller types suffer delay in order to credibly prove they possess a higher-quality vehicle. The intuition for uniqueness is analogous to the one given in Section 3.1.1: under condition (L), once the buyer's beliefs cross a threshold level, he finds it no longer worthwhile to price discriminate among the remaining seller types.[20]

To illustrate consider the simple two-type model:

$$c(q) = 0, \quad v(q) = \alpha, \qquad \text{for } 0 \leqslant q \leqslant \hat{q},$$
$$c(q) = s, \quad v(q) = s + \beta, \qquad \text{for } \hat{q} < q \leqslant 1,$$

---

[20] See Samuelson (1984) for a generalization of Stokey's "no price discrimination" result to the interdependent values case.

where $\alpha > 0$ and $\beta > 0$, since we are in the case of a gap. Note that the private values case obtains when $\alpha = s + \beta$, so this is a generalization of the example studied in Section 3.1.1. See Evans (1989) for a treatment of the special case in which $\alpha = 0$.

Let $q_{-1} = 1$, $q_0 = \hat{q}$, and inductively define the sequence $q_1 > q_2 > \cdots > q_N$ from:

$$m_n = \alpha s^{-1} \delta^{-(n-1)} m_{n-1} \quad (n \geqslant 2),$$

and the initial condition $m_1 = \beta s^{-1} m_0$, using $m_n = q_{n-1} - q_n$ and the terminal condition $N = \min\{n : q_n \leqslant 0\}$. Finally, let $p_n = s \delta^n$. Then with every sequential equilibrium is associated the unique stationary triplet:

$$
\begin{aligned}
P(Q) &= p_n, & Q &\in (q_n, q_{n-1}]; \\
t(Q) &= q_{n-2}, & Q &\in (q_n, q_{n-1}] \text{ and } n > 1, \\
&= 1, & Q &\in (q_1, 1]; \\
V(Q) &= (\alpha - p_{n-1})(q_{n-2} - Q) + \delta V(q_{n-2}), & Q &\in (q_n, q_{n-1}] \text{ and } n > 1, \\
&= (\alpha - p_0)(q_0 - Q) + \beta(1 - q_0), & Q &\in (q_1, q_0], \\
&= \beta(1 - Q), & Q &\in (q_0, 1].
\end{aligned}
$$

The idea behind the above construction is as follows. Seller types in $(\hat{q}, 1]$ are held to their reservation value, because the buyer has the sole power to make offers. The last price offered will therefore be equal to s. Seller types in the interval $[0, \hat{q}]$ must be indifferent between accepting the offer p_n and waiting n periods to receive the offer $p_0 = s$, so we must have $p_n = s \delta^n$. The breakpoints q_n are constructed so that when the state is q_n the buyer is indifferent between offering p_n (and hence trading with types in $(q_n, q_{n-1}]$), and offering p_{n-1} (and hence trading with types in $(q_{n-1}, q_{n-2}]$).

Note that, in the private-values case, the sequence $\{m_1, m_2, \ldots\}$ is strictly increasing and bounded below when δ converges to 1. This is still the case here when $\alpha \geqslant s$. But when $\alpha < s$, the sequence is decreasing as long as n remains such that $\delta^n < \alpha/s$. As δ converges to 1, the range of integers for which this inequality holds increases without bound, so it is possible for the number of bargaining rounds to increase without bound as δ converges to 1. This allows us to investigate the conditions under which the Coase Conjecture will and will not hold. For this purpose, let us explicitly denote the dependence of m_i on δ by $m_i(\delta)$, and define:

$$a = \sum_{i=0}^{\infty} m_i(1) = (1 - \hat{q}) \left(1 + \frac{\beta}{s - \alpha} \right).$$

We then have:

THEOREM 8 [Deneckere and Liang (1999)]. *Consider the two-type interdependent values model defined above. Then the Coase Conjecture obtains if and only if $a \geqslant 1$. When* *a < 1, then as δ converges to* 1, *all seller types in* [0, 1 − a) *trade immediately at the price* $s\rho^2$, *and all types in* (1 − a, 1] *trade at the price s after a delay of length T discounted such that* $e^{-rT} = \rho^2$, *where* $\rho = \alpha/s$.

The condition $a \geqslant 1$ can be written in the more familiar form $E(v(q)) \geqslant c(1)$, so Theorem 8 says that when bargaining frictions disappear, inefficient delay occurs if and only if this is mandated by the basic incentive constraints presented in Theorem 2. When $E(v(q)) < c(1)$ every trading mechanism necessarily exhibits inefficiencies. However, the limiting bargaining mechanism described in Theorem 8 exhibits more delay than is necessary. To see this, observe that social welfare is increased by having all types $q \in (1 - a, \hat{q}]$ trade at the price $s\rho^2$ at time zero. In the resulting mechanism the buyer will have strictly positive surplus; this means we can increase the probability of trade on the interval $(\hat{q}, 1]$ and thereby further increase welfare.

4. Sequential bargaining with one-sided incomplete information: The "no gap" case

The case of *no gap* between the seller's valuation and the support of the buyer's valuation differs in broad qualitative fashion from the case of the *gap* which we examined in the previous section. The bargaining does not conclude with probability one after any finite number of periods. As a consequence of this fact, it is not possible to perform backward induction from a final period of trade, and it therefore does not follow that every sequential equilibrium need be stationary. If stationarity is nevertheless *assumed*, then the results parallel the results which we have already seen for the *gap* case: trade occurs with essentially no delay and the informed party receives essentially all the surplus. However, if stationarity is not assumed, then instead a folk theorem obtains, and so substantial delay in trade is possible and the uninformed party may receive a substantial share of the surplus. These qualitative conclusions hold both for the seller-offer game and alternating-offer games.

Following the same convenient notation as in Section 3, let the buyer's type be denoted by q, which is uniformly distributed on [0, 1], and let the valuation of buyer type q be given by the function $v(q)$. The seller's valuation is normalized to equal zero. The case of "no gap" is the situation where there does *not* exist $\Delta > 0$ such that it is common knowledge that the gains from trade are at least Δ. More precisely, for any $\Delta > 0$, there exists $q_\Delta \in [0, 1)$ such that $0 < v(q_\Delta) < \Delta$. Opposite the conclusion of Theorem 4 for the gap case, we have:

LEMMA 2. *In any sequential equilibrium of the infinite-horizon seller-offer game in the case of "no gap", and for any* $N < \infty$, *the probability of trade before period N is strictly less than one.*

PROOF. By Lemma 1, at the start of any period t, the set of remaining buyer types is an interval $(Q_t, 1]$. The seller never offers a negative price [Fudenberg, Levine and

Tirole (1985), Lemma 1]. Consequently, a price of $(1 - \delta)v(q) - \varepsilon$ will be accepted by all buyer types less than q, since a buyer with valuation $v(q)$ is indifferent between trading at a price of $(1 - \delta)v(q)$ in a given period and trading at a price of zero in the next period.

Suppose, contrary to the lemma, that there exists finite integer N such that $Q_N = 1$. Without loss of generality, let N be the smallest such integer, so that $Q_{N-1} < 1$. Since acceptance is individually rational, the seller must have offered a price of zero in period $N - 1$, yielding zero continuation payoff. But this was not optimal, as the seller could instead have offered $(1 - \delta)v(q) - \varepsilon$ for some $q \in (q_{N-1}, 1)$, generating a continuation payoff of at least $(q - Q_{N-1})[(1 - \delta)v(q) - \varepsilon] > 0$ (for sufficiently small ε), a contradiction. We conclude that $Q_N < 1$.   □

A result analogous to Lemma 2 also holds in the alternating-offer extensive form. However, as we have already seen in Section 3.1.3, the result for the buyer-offer game is qualitatively different: there is a unique sequential equilibrium; it has the buyer offering a price of zero in the initial period and the seller accepting with probability one.

Much of the intuition for the case of "no gap" can be developed from the example where the seller's valuation is commonly known to equal zero and the buyer's valuation is uniformly distributed on the unit interval $[0, 1]$. This example was first studied by Stokey (1981) and Sobel and Takahashi (1983). In our previous notation:

$$v(q) = 1 - q, \quad \text{for } q \in [0, 1]. \tag{11}$$

In the subsections to follow, we will see that the stationary equilibria are qualitatively similar to those for the "gap" case, but that the nonstationary equilibria may exhibit entirely different properties.

4.1. Stationary equilibria

Assuming a stationary equilibrium and given the linear specification of Equation (11), it is plausible to posit that the seller's value function $(V(Q))$ is quadratic in the measure of remaining customers, that the measure of remaining customers $(1 - t(Q))$ which the seller chooses to induce is a constant fraction of the measure of currently remaining customers, and that the seller's optimal price $(P(t(Q)))$ is linear in the measure of remaining customers. Let r denote the real interest rate and z denote the time interval between periods (so that the discount factor δ is given by $\delta \equiv e^{-rz}$). In the notation of Section 3:

$$V(Q) = \alpha_z(1 - Q)^2, \tag{12}$$
$$1 - t(Q) = \beta_z(1 - Q), \tag{13}$$
$$P(t(Q)) = \gamma_z(1 - Q), \tag{14}$$

where α_z, β_z and γ_z are constants between 0 and 1 which are parameterized by the time interval z between offers. Equations (12)–(14) can be solved simultaneously, as follows.

Since the linear-quadratic solution is differentiable and $t(Q)$ is defined to be the arg max of Equation (2), we have:

$$\frac{\partial}{\partial Q'}\left[P(Q') \cdot (Q' - Q) + \delta V(Q')\right]_{Q'=t(Q)} = 0. \tag{15}$$

Furthermore, with $t(Q)$ substituted into the right-hand side of Equation (2), the maximum must be attained:

$$V(Q) = P\left(t(Q)\right) \cdot \left(t(Q) - Q\right) + \delta V\left(t(Q)\right). \tag{16}$$

Substituting Equations (12), (13) and (14) into Equations (3), (15) and (16) yields three simultaneous equations in α_z, β_z and γ_z, which have a unique solution. In particular, the solution has $\alpha_z = 1/2\gamma_z$ and:

$$\gamma_z = 1 - \delta^{-1} + \delta^{-1}\sqrt{1 - \delta} \tag{17}$$

[Stokey (1981), Theorem 4, and Gul, Sonnenschein and Wilson (1986), pp. 163–164].

Qualitatively, the reader should observe that in the limit as the time interval z between offers approaches zero (i.e., as $\delta \to 1$), γ_z converges to zero. From Equation (14), observe that γ_z is the seller's price when the state is $Q = 0$. This means that the *initial* price in this equilibrium may be made arbitrarily close to zero (i.e., the Coase Conjecture holds). Moreover, since $\alpha_z = 1/2\gamma_z$, the seller's expected profits in this equilibrium may be made arbitrarily close to zero. According to (17), the convergence is relatively slow, but for realistic parameter values, the seller loses most of her bargaining power. For example, with a real interest rate of 10% per year and weekly offers, the seller's initial price is 4.2% of the highest buyer valuation; this diminishes to 1.63% with daily offers.

Further observe that, since the linear-quadratic equilibrium is expressed as a triplet $\{P(\cdot), V(\cdot), t(\cdot)\}$, this sequential equilibrium is stationary. However, this model is also known to have a continuum of other stationary equilibria; see Gul, Sonnenschein and Wilson [(1986), Examples 2 and 3]. Unlike the other known stationary sequential equilibria, the linear-quadratic equilibrium has the property that it does not require randomization off the equilibrium path. In the literature, stationary sequential equilibria possessing this arguably desirable property are referred to as *strong-Markov* equilibria; while stationary sequential equilibria not necessarily possessing this property are often referred to as *weak-Markov* equilibria.

The linear-quadratic equilibrium of the linear example is emblematic of all stationary sequential equilibria for the case of "no gap", as the following theorem shows:

THEOREM 9 (Coase Conjecture). *For every $v(\cdot)$ in the case of "no gap" and for every $\varepsilon > 0$, there exists $\bar{z} > 0$ such that, for every time interval $z \in (0, \bar{z})$ between offers and for every stationary sequential equilibrium, the initial price charged in the seller-offer game is less than ε.*

PROOF. Gul, Sonnenschein and Wilson (1986), Theorem 3.   □

If extremely mild additional assumptions are placed on the valuation function of buyer types, then a stronger version of the Coase Conjecture can be proven. The standard Coase Conjecture may be viewed as establishing an upper bound on the ratio between the seller's offer and the highest buyer valuation in the *initial period*; the *uniform* Coase Conjecture further bounds the ratio between the seller's offer and the highest-remaining buyer valuation in *all periods* of the game. For L, M and α such that $0 < M \leqslant 1 \leqslant L < \infty$ and $0 < \alpha < \infty$, let:

$$\mathcal{F}_{L,M,\alpha} = \left\{ v(\cdot): v(0) = 1, v(1) = 0 \text{ and } M(1-q)^\alpha \leqslant v(q) \leqslant L(1-q)^\alpha \right.$$
$$\left. \text{for all } q \in [0, 1] \right\}. \tag{18}$$

The family $\mathcal{F}_{L,M,\alpha}$ has the property that if $v \in \mathcal{F}_{L,M,\alpha}$, then *every* truncation (from above) of the probability distribution of buyer valuations (renormalized so that the valuation at the truncation point equals one) is guaranteed to also be an element of $\mathcal{F}_{L/M,M/L,\alpha}$. If a uniform $\bar{z}$ (of Theorem 9) can be found which holds for all $v \in \mathcal{F}_{L/M,M/L,\alpha}$, then the ratio between the seller's offer and the highest-remaining buyer valuation is bounded by ε in all periods of the game. We have:

THEOREM 10 (Uniform Coase Conjecture). *For every* $0 < M \leqslant 1 \leqslant L < \infty$, $0 < \alpha < \infty$, *and* $\varepsilon > 0$, *there exists* $\bar{z} > 0$ *such that for every time interval* $z \in (0, \bar{z})$ *between offers, for every* $v \in \mathcal{F}_{L,M,\alpha}$ *and for every stationary sequential equilibrium, the initial price charged in the seller-offer game is less than* ε.

PROOF. Ausubel and Deneckere (1989a), Theorem 5.4.   □

The same qualitative results hold in alternating-offer extensive forms for the case of no gap. Some additional assumptions above and beyond stationarity are made in the literature, but the stationarity assumption appears to be the driving force behind the results. Gul and Sonnenschein (1988), in analyzing the gap case, and Ausubel and Deneckere (1992a) assume stationarity.[21] They also assume that the seller's offer and acceptance rules are in pure strategies,[22] and that there is "no free screening" in the sense that any two buyer offers which each have zero probability of acceptance are required to induce the same beliefs. Similar to the seller-offer game, these imply:

---

[21] To be more precise, they assume requirement (3) and a slightly weaker version of requirement (2) in the definition of stationarity from Section 3. Their assumptions of pure strategies and no free screening imply requirement (1).

[22] In light of Section 3.1.2, the pure strategy assumption on seller acceptances may be inconsistent with refinements of sequential equilibrium; but a similar result likely holds under weaker assumptions.

THEOREM 11 (Uniform Coase Conjecture). *For every $0 < M \leqslant 1 \leqslant L < \infty, 0 < \alpha < \infty$, and $\varepsilon > 0$, there exists $\bar{z} > 0$ such that for every time interval $z \in (0, \bar{z})$ between offers, for every $v \in \mathcal{F}_{L,M,\alpha}$ and for every stationary sequential equilibrium, the initial serious (seller or buyer) offer in the alternating-offer game is less than ε.*

PROOF. Ausubel and Deneckere (1992a), Theorem 3.2. □

For the no gap case, the Coase Conjecture is equivalent to the notion of "No Delay" which Gul and Sonnenschein (1988) prove for the gap case: for sufficiently short time interval between offers, the probability that trade will occur within ε time is at least $1 - \varepsilon$. This equivalence holds in the seller-offer as well as in the alternating-offer game.

There is an especially enlightening explanation for the fact that stationary sequential equilibria of the alternating-offer game closely resemble those of the seller-offer game. In a sense which may be made precise, stationary equilibria of the alternating-offer game are *as if* the extensive form permitted offers only by the uninformed party: exogenously, both traders are permitted to make offers; endogenously, equilibrium counteroffers by the informed party degenerate to null moves.

To see this, observe that the stationarity, pure-strategy and no-free-screening restrictions on sequential equilibrium mandate that, at each time when it is the informed agent's turn to make an offer, the informed agent partitions the interval of remaining types into two subintervals (one possibly degenerate): a high subinterval (who "speak" by making a serious offer) and a low subinterval (who effectively "remain silent" by making a nonserious offer). Choosing to speak reveals a high valuation, which is information that the uninformed agent can exploit in the ensuing negotiations. Remaining silent signals a low valuation. Let $\underline{b}$ denote the lowest buyer valuation in the speaking subinterval – as well as the highest buyer valuation in the silence subinterval. Following speaking, the seller captures a price of at least $\underline{b}\delta/(1 + \delta)$, à la Rubinstein (1982), which as the time between offers shrinks toward zero, converges to $1/2\underline{b}$. Meanwhile, also as the time between offers shrinks toward zero, the terms of trade for the silence interval become increasingly favorable: à la the Uniform Coase Conjecture, the ratio between the next price and $\underline{b}$ converges to zero. Thus, silence becomes increasingly attractive relative to speaking and, for sufficiently short time intervals, delay becomes preferable to revealing the damaging information for *all* types of the informed party. In other words, you recognize that "anything you say can and will be used against you". Therefore, regardless of valuation, you decline to speak, since "you have the right to remain silent". More formally:

THEOREM 12 (Silence Theorem). *Let v belong to $\mathcal{F}_{L,M,\alpha}$ and let r be any positive interest rate. Then there exists $\bar{z} > 0$ such that, for every time interval $z \in (0, \bar{z})$ between offers and for every stationary sequential equilibrium satisfying the pure-strategy and no-free-screening restrictions, the informed party never makes any serious offers in the alternating-offer bargaining game, both along the equilibrium path and after all histories in which no prior buyer deviations have occurred.*

PROOF. Ausubel and Deneckere (1992a), Theorem 3.3.   □

Thus, stationary equilibria of the alternating-offer bargaining game with a time interval z between offers closely resemble stationary equilibria of the seller-offer bargaining game with a time interval $2z$ between offers, for sufficiently small z. Moreover, for many distributions of valuations, "sufficiently" small does not require "especially" small: for the model with linear $v(\cdot)$, the silence theorem holds whenever $\delta > 0.83929$ [Ausubel and Deneckere (1992a), Table I]; with a real interest rate r of 10% per year, this holds for all $z < 21$ months, not requiring a very quick response time between offers at all.

4.2. Nonstationary equilibria

In the case of no gap, stationarity is merely an assumption, not an implication of sequential equilibrium. As we saw in the last subsection, the stationary equilibria converge (as the time interval between offers approaches zero) in outcome to the static mechanism which maximizes the *informed* party's expected surplus. The contrast between stationary and nonstationary equilibria is most sharply highlighted by constructing nonstationary equilibria which converge in outcome to the static mechanism which maximizes the *uninformed* party's expected surplus.

Again, consider the example where the seller's valuation is commonly known to equal zero and the buyer's valuation is uniformly distributed on the unit interval [0, 1]. The static mechanism which maximizes the seller's expected surplus is given by:

$$p(q) = \begin{cases} 1, & \text{if } q \leqslant 1/2, \\ 0, & \text{if } q > 1/2, \end{cases} \qquad x(q) = \begin{cases} 1/2, & \text{if } q \leqslant 1/2, \\ 0, & \text{if } q > 1/2. \end{cases} \tag{19}$$

In terms of a sequential bargaining game, this means that, although it is possible to intertemporally price discriminate, the seller finds it optimal to merely select the static monopoly price of $1/2$ and adhere to it forever [Stokey (1979)]. The intuition for this result – in terms of the durable goods monopoly interpretation of the model – is that the sales price for a durable good equals the discounted sum of the period-by-period rental prices, and the optimal rental price for the seller in each period is always the same monopoly rental price.

A seller who lacks commitment powers will be unable to follow precisely this price path [Coase (1972)]. If the seller were believed to be charging prices of $p_n = 1/2$, for $n = 0, 1, 2, \dots$, the unique optimal buyer response would be for all $q \in [0, 1/2)$ to purchase in period 0 and for all $q \in (1/2, 1]$ to never purchase (corresponding exactly to the static mechanism of Equation (19)). But, then, the seller's continuation payoff evaluated in any period $n = 1, 2, 3, \dots$ equals zero literally. Following the same logic as in the proof of Lemma 2, there exists a deviation which yields the seller a strictly positive payoff, establishing that the constant price path is inconsistent with sequential equilibrium.

However, while the static mechanism of Equation (19) cannot literally be implemented in equilibria with constant price paths, Ausubel and Deneckere (1989a) show that the seller's optimum can nevertheless be arbitrarily closely approximated in equilibria with slowly descending price paths. The key to their construction is as follows. For any $\eta > 0$, and in the game with time interval $z > 0$ between offers, define a *main equilibrium path* by:

$$p_n = p_0 \, e^{-\eta n z}, \quad \text{for } n = 0, 1, 2, \ldots. \tag{20}$$

Also consider the (linear-quadratic) stationary equilibrium which was specified in Equations (12)–(14) and in which γ_z was solved in Equation (17). Define a *reputational price strategy* by the following seller strategy:

Offer p_m in period m, if p_n was offered in all periods $n = 0, 1, \ldots, m - 1$, (21)

Offer prices according to the stationary equilibrium, otherwise,

with the corresponding buyer strategy defined to optimize against the seller strategy (21).

It is straightforward to see that, for sufficiently short time intervals between offers, the reputational price strategy yields a (nonstationary) sequential equilibrium. This is the case for all $p_0 \in (0, 1)$; and for $p_0 = 1/2$, the sequential equilibrium converges in outcome (as $\eta \to 0$ and $z \to 0$) to the static mechanism (19) which maximizes the seller's expected payoff. A heuristic argument proceeds as follows. First, observe that the price path $\{p_n\}_{n=0}^{\infty}$ yields a relatively large measure of sales in period 0 and then a relatively slow trickle of sales thereafter. Hence, if the main equilibrium path is self-enforcing for the seller in periods $n = 1, 2, \ldots$, it will automatically be self-enforcing in period $n = 0$. Second, let us consider the seller's continuation payoff along the main equilibrium path, evaluated in any period $n = 1, 2, \ldots$. Let q denote the state at the start of period n. Given the linear distribution of types and the exponential rate of descent in price, it is easy to see that the seller's expected continuation payoff, π, is a stationary function of the state:

$$\pi(q) = \lambda_z (1 - q)^2, \tag{22}$$

where λ_z depends on η and is parameterized by z. Moreover, for every $\eta > 0$:

$$\lambda \equiv \lim_{z \to 0} \lambda_z > 0. \tag{23}$$

Meanwhile, we already saw in Equation (12) that the seller's payoff from optimally deviating from the main equilibrium path is given by $V(q) = \alpha_z (1 - q)^2$, where $\alpha_z \to 0$ as $z \to 0$. Thus, for any $\eta > 0$, there exists $\bar{z} > 0$ such that, whenever the time interval between offers satisfies $0 < z < \bar{z}$, we have $\lambda_z > \alpha_z$, and so the seller's expected payoff along the main equilibrium path exceeds the expected payoff from optimally deviating. We then conclude that the reputational price strategy yields a sequential equilibrium.

This construction generalizes to all valuation functions $v \in \mathcal{F}_{L,M,\alpha}$ and to all bargaining mechanisms. It is appropriate to restrict attention here to incentive-compatible bargaining mechanisms that are *ex post* individually rational, since the buyer will never accept a price above his valuation in any sequential equilibrium and the seller will never offer a price below her valuation. Continuing the logic developed in Theorem 2,[23] we have the following complete characterization:

LEMMA 3. *For any continuous valuation function $v(\cdot)$, the one-dimensional bargaining mechanism $\{p, x\}$ is incentive compatible and* ex post *individually rational if and only if $p : [0, 1] \to [0, 1]$ is (weakly) decreasing and x is given by the Stieltjes integral: $x(q) = -\int_q^1 v(r) \, dp(r)$.*

PROOF. Ausubel and Deneckere (1989b), Theorem 1. □

Moreover, we can translate the outcome path of any sequential equilibrium of the bargaining game into an incentive-compatible bargaining mechanism, as follows. For buyer type q, let $n(q)$ denote the period of trade for type q in the sequential equilibrium and let $\phi(q)$ denote the payment by type q. Define: $\bar{p}(q) = e^{-rn(q)z}$ and $\bar{x}(q) = \phi(q) e^{-rn(q)z}$. Then $\{\bar{p}, \bar{x}\}$ thus defined can be reinterpreted as a direct mechanism – and the fact that it derives from a sequential equilibrium immediately implies that $\{\bar{p}, \bar{x}\}$ is incentive compatible and individually rational. We will say that $\{p, x\}$ is *implemented* by sequential equilibria of the bargaining game if, for every $\varepsilon > 0$, there exists a sequential equilibrium inducing static mechanism $\{\bar{p}, \bar{x}\}$ with the property that $\{\bar{p}, \bar{x}\}$ is uniformly close to $\{p, x\}$ (except possibly in a neighborhood of $q = 1$):

$$\left| \bar{p}(q) - p(q) \right| < \varepsilon, \ \forall q \in [0, 1 - \varepsilon), \quad \text{and} \quad \left| \bar{x}(q) - x(q) \right| < \varepsilon, \ \forall q \in [0, 1]. \quad (24)$$

The reasoning described above for the static mechanism which maximizes the seller's expected payoff extends to *every* incentive-compatible bargaining mechanism. In place of the exponentially descending price path $\{p_n\}_{n=0}^{\infty}$, we substitute a general specification which approximates the incentive-compatible bargaining mechanism, for $q \in [0, 1 - \varepsilon]$ and induces an exponential evolution of the state, for $q \in (1 - \varepsilon, 1]$. In place of the linear-quadratic equilibrium following deviations, we substitute a stationary equilibrium, which is guaranteed to exist (Theorem 4) and to satisfy the Uniform Coase Conjecture (Theorem 10). We have:

---

[23] The analogue to Lemma 3 for the case where the seller is informed and the buyer uninformed follows directly from Theorem 2, as follows. With private values, i.e., $g(s) = b$ for all s, the first inequality in Theorem 2 is automatically satisfied. Since we are in the case of no gap, seller type $\bar{s}$ cannot profitably trade with the buyer, so ex post individual rationality requires $x(\bar{s}) - \bar{s}p(\bar{s}) = 0$. Consequently, it follows from Theorem 2 that:

$$x(s) = sp(s) + \int_s^{\bar{s}} p(z) \, dz = \bar{s}p(\bar{s}) - \int_s^{\bar{s}} z \, dp(z).$$

THEOREM 13 (Folk Theorem). *Let the valuation function $v(\cdot)$ belong to $\mathcal{F}_{L,M,\alpha}$. Then every incentive-compatible,* ex post *individually rational bargaining mechanism $\{p, x\}$ is implementable by sequential equilibria of the seller-offer bargaining game.*

PROOF. Ausubel and Deneckere (1989b), Theorem 2. □

A folk-theorem-like result also holds in the alternating-offer game, since each of a continuum of sequential equilibria from the seller-offer game can be embedded as equilibria in the alternating-offer game. At the same time, there is an upper bound on the price at which the buyer can be expected to trade. Suppose that the seller holds the most "optimistic" beliefs: the buyer's type equals 0 and so the buyer's valuation equals $v(0)$. Then, even in the complete-information game, if the seller offers any price greater than $(1/(1 + \delta))v(0)$, the buyer is sure to turn around and reject [Rubinstein (1982)]. In the limit as the time interval approaches zero, the seller can extract no more than one-half the surplus from the highest-valuation buyer. Thus, we have:

THEOREM 14. *Let the valuation function $v(\cdot)$ belong to $\mathcal{F}_{L,M,\alpha}$. Then an incentive-compatible,* ex post *individually rational bargaining mechanism $\{p, x\}$ is implementable by sequential equilibria of the alternating-offer bargaining game if and only if: $p(0)v(0) - x(0) \geq 1/2v(0)$.*

PROOF. Ausubel and Deneckere (1989b), Theorem 3. □

4.3. Discussion of the stationarity assumption

One useful way to understand the effect of the stationarity assumption is to see its impact on the set of equilibria of a standard supergame. Consider, for example, the infinitely repeated prisoners' dilemma – or any infinite supergame in which the stage game has a unique Nash equilibrium. Since (unlike the bargaining game) this is literally a repeated game and the play of one period has no effect on the possibilities in the next, there is no state variable at all. Stationarity restricts attention to equilibria in which the play in any period is history-independent; in other words, trigger-strategy equilibria are ruled out by assumption. (Equivalently, as in the bargaining game with one-sided incomplete information, stationarity restricts attention to equilibria of the infinite-horizon game which are limits of equilibria of finite-horizon versions of the same game.) The unique stationary equilibrium is the static Nash equilibrium played over and over.

This analogy strongly suggests that it is wrong to assume away the nonstationary equilibria. While it is interesting to know the implications of stationarity, a restriction to stationarity excludes many of the interesting effects which led economists to analyze dynamic games in the first place. Of course, stationarity is essential to the analysis of the "gap" case, since it is implied (not assumed). But to the extent that "no gap" is the appropriate condition on primitives, nonstationary equilibria and their qualitative properties are an essential part of the analysis.

5. Sequential bargaining with two-sided incomplete information

With two-sided incomplete information, incentive compatibility and individual rationality are incompatible with ex post efficiency. As we saw in Corollary 1 of Section 2, so long as the supports of the seller and buyer valuations overlapped, the static bargaining mechanism necessarily entails situations where the buyer's valuation exceeds the seller's valuation yet trade occurs with probability strictly less than one. Furthermore, as we saw in the fourth paragraph following Theorem 2, since any sequential equilibrium of the dynamic bargaining game can be expressed as a static mechanism, this immediately implies that the search for ex post efficient sequential equilibria is fruitless. The more interesting starting-point is to ask: Can the ex ante efficient static bargaining mechanism be replicated in a dynamic offer/counteroffer bargaining game, or does the dynamic game necessarily entail greater inefficiency than the static constrained optimum?

Ausubel and Deneckere (1993a) establish that, for distribution functions exhibiting monotonic hazard rates, the ex ante efficient static bargaining mechanism can essentially be replicated in very simple dynamic bargaining games:

THEOREM 15. *If $F_1(s)/f_1(s)$ and $[F_2(b) - 1]/f_2(b)$ are strictly increasing functions, then*:
  (i) *there exists $\lambda_s \in (0, 1)$ such that, for every $\lambda \in [\lambda_s, 1]$, the ex ante efficient mechanism which places weight λ on the seller is implementable in the seller-offer game; and*
  (ii) *there exists $\lambda_b \in (0, 1)$ such that, for every $\lambda \in [0, \lambda_b]$, the ex ante efficient mechanism which places weight λ on the seller is implementable in the buyer-offer game.*

PROOF. Ausubel and Deneckere (1993a), Theorem 3.1.   □

The flavor of this result is most easily seen in the standard example where the seller and buyer valuations are each uniformly distributed on the unit interval. For this special case of the theorem, calculations reveal that $\lambda_s = 1/2 = \lambda_b$. This means that, for the case of equal weighting ($\lambda = 1/2$) focused on by Chatterjee and Samuelson (1983) and Myerson and Satterthwaite (1983), we can come arbitrarily close to replicating the constrained optimum both in the seller-offer game and the buyer-offer game. Moreover, since equilibria of the seller- and buyer-offer games can be embedded in sequential equilibria of the alternating-offer game, this means that the entire ex ante Pareto frontier is implementable in the alternating-offer bargaining game. There need not be any additional inefficiency arising from the dynamic nature of the game, above and beyond the inefficiency already introduced by the two-sided incomplete information.

While the (upper) boundary of the set of all sequential equilibria is thus known, little exists in the way of results refining the set of sequential equilibrium outcomes. Cramton (1984) posited sequential equilibria of the infinite-horizon seller-offer bargaining game with the additional properties that: (a) the seller fully reveals her type in the course of making offers; and (b) in the continuation game following the seller's revelation, players adopt the strategies from a stationary equilibrium of the game of one-sided incomplete information. The seller thus uses delay to credibly signal her strength: low-valuation seller types make revealing offers early in the game, while high-valuation seller types initially make nonserious offers until revealing later in the game. Cho (1990a) posited equilibria of finite-horizon seller-offer bargaining games with the properties that: (a) the seller's pricing rule is a separating strategy after every history; (b) equilibria satisfy a continuity property resembling trembling-hand perfection; (c) equilibria satisfy a monotonicity restriction on beliefs; and (d) equilibria are stationary.

However, both the Cramton (1984) and Cho (1990a) constructions ultimately exhibit an unfortunate property, when the seller and buyer distributions have the same supports and for short time intervals between offers. By the stationarity assumption, the lowest seller type is subject to the Coase Conjecture, earning a payoff arbitrarily close to zero. Meanwhile, higher seller types offer prices which always exceed their respective types. The lowest seller type thus faces a very strong incentive to mimic a higher seller type, breaking the equilibrium unless essentially all higher types encounter extremely long delays before trading. Thus, a No Trade Theorem holds: In the limit as the time interval between offers decreases toward zero, the ex ante expected probability of trade in these equilibria converges to zero [Ausubel and Deneckere (1992b), Theorem 1].

Two other articles present plausible outcomes of dynamic bargaining games with two-sided incomplete information in which trade occurs to a substantial degree but which are inefficient compared to the constrained static optimum.[24] Cramton (1992) extends and analyzes the Admati and Perry (1987) extensive-form game to an environment with a continuum of types and two-sided incomplete information. The game begins with effectively a war of attrition between the seller and the buyer: there is a seller type $s(t)$ and a buyer type $b(t)$ who are each supposed to reveal themselves by making serious offers at time t. Thus, as the game unfolds without serious offers getting made, each party becomes more pessimistic about his counterpart's valuation. A serious offer – once made – fully reveals the offeror's type. The other player then either accepts the serious offer or further delays trade so as to credibly convey his own type and, when trade

---

[24] Perry (1986) analyzes an alternating-offer game with two-sided incomplete information about valuations, but where the cost of bargaining takes the form of a fixed cost per period rather than discounting. He establishes the existence of a unique sequential equilibrium when the players' fixed costs are unequal. When it is the turn of the player with the lower bargaining cost to make an offer, this player proposes essentially its monopoly price, which the other player accepts if it yields nonnegative utility. When it is the turn of the player with the higher bargaining cost to make an offer, this player leaves the game without making an offer. Thus, trade – if it occurs at all – occurs in the initial period. However, inefficiently little trade occurs compared to the constrained static optimum. Perry's game illustrates the principle that there is no possibility for signaling through delay when the incomplete information is about valuations but the bargaining cost is a fixed cost each period. Signaling requires the presence of an action which is relatively less costly for one type than another; in this game, the cost of delay is equal across all types.

occurs following both players' full revelation, it occurs at the complete-information price. Ausubel and Deneckere (1992b) consider the seller-offer bargaining game and construct a continuum of equilibria, all with the property that the seller's first serious offer reveals essentially all the information which she will ever reveal. One interesting equilibrium in this class is the "monopoly equilibrium": the seller fully reveals her type in the initial period by offering essentially the monopoly price relative to her valuation; and then follows a slowly descending price path thereafter. This equilibrium is also ex ante efficient – provided that all of the weight is placed on the seller.

6. Empirical evidence

Bargaining is pervasive in our economy. Thus, it is not surprising that there is a substantial empirical literature. However, only recently has this work sought to examine the data in light of strategic bargaining theories with private information.

Bargaining models with private information are especially well suited for empirical work, since a main feature of the data is the occurrence of costly disputes. These disputes arise naturally in models with incomplete information. However, private information models involve several challenges for empirical work. First, the models are often complex, making estimation difficult. Second, the results tend to be sensitive to the particular bargaining procedure, the source of private information, and the form of delay costs. In most empirical settings, the bargaining rules and the preferences of the parties cannot be fully identified. The researcher then may have too much freedom in selecting assumptions that "explain" particular facts. Finally, the theory predicts how ex post outcomes depend on realizations of private information, yet the researcher typically is unable to observe private information variables, even ex post.

We focus on one of the most prominent examples of bargaining – union contract negotiations – in understanding bargaining disputes.

Kennan and Wilson (1989) analyze attrition, screening, and signaling models, and contrast the theoretical predictions of these models with the main empirical features of strike data. They emphasize five empirical findings:

- Strikes are unusual, occurring in 10 to 20 percent of contract negotiations.
- The relationship between strike duration and wages is ambiguous. McConnell (1989) found that wages declined 3% per 100 days of strike in the U.S., but Card (1990) found no significant relationship between strike duration and wages.
- Strikes are more frequent in good times [Vroman (1989), Gunderson, Kervin, and Reid (1986)], yet strike duration decreases in good times [Kennan (1985), Harrison and Stewart (1989)].
- Strike activity varies across industries.
- Settlement rates tend to decline with strike duration [Kennan (1985), Harrison and Stewart (1989), Gunderson and Melino (1990a)].

In all of the models, strikes (or their absence) convey private information in a credible way. A key feature of attrition models is winner-take-all outcomes. In an attrition model, each side attempts to convince the other that it can last longer, so the other should concede the entire pie under negotiation. One side clearly wins at the expense of the other. In contrast, wage bargaining typically involves compromise. For this reason, we focus on screening and signaling models.

The standard setting assumes that the union is uncertain about the firm's willingness to pay. In this case, under either screening or signaling, the duration of the strike conveys information to the union about the firm's willingness to pay. A firm with a greater willingness to pay settles early at a high wage; whereas, a firm with a low willingness to pay endures a strike in order to convince the union to accept a low wage. A documentary film, Final Offer, of the 1984 negotiations between GM Canada and the UAW provides anecdotal evidence for this explanation for strikes. Early in the strike the union leaders are discussing whether they should accept GM's last offer. One says, "You might convince me that that's all there is after a month, but not after five days". Another says, "If they think it will take a short strike to convince workers to accept, they're wrong".

Screening and signaling models share several features: (1) strike incidence and strike duration increase with uncertainty over private information variables, and (2) wages fall with strike duration. However, there are important differences in wages and strike activity.

The standard screening model assumes that the union makes a sequence of declining wage demands, with each demand chosen optimally given beliefs about the firm's willingness to pay and the firm's acceptance and offer strategy. A critical assumption is that the firm employs a stationary acceptance strategy. At every point in the negotiation, a firm with value v accepts any wage demand below $w(v)$. Most importantly, this assumption means that the firm's acceptance rule cannot depend on the rate of concession by the union. This greatly limits the equilibrium set, assuring that all equilibria satisfy the Coase (1972) conjecture. As the time between offers shrinks, the union loses its bargaining power and makes offers that are close to the Rubinstein wage between the union and the lowest-value firm. Strike duration falls to zero and strike incidence increases to one, but the convergence is slow. Screening then has the property that wages, strike incidence, and strike duration all depend critically on the period over which the union can commit to a wage demand. Kennan and Wilson (1989) argue that the Coase conjecture may explain why in boom times strikes are more frequent but shorter. This would follow if the union has a shorter commitment period in boom times; however, it is not clear why the time between offers would vary with the business cycle.

One potential difficulty with the screening model is that, because of the Coase property, strike durations must be short when the commitment period is short. In the U.S., mean strike durations are about 40 days. If offers can be made every day, then the standard screening model may predict strikes that are too short given plausible interest rates and levels of uncertainty. Hart (1989) provides an explanation. If bargaining costs are low initially, but then increase at some point during the strike, say when inventories run out, then strikes can be much longer. Another explanation is given by Vincent (1989). If the parties' valuations are interdependent, then strikes of significant duration can occur even as the time between offers goes to zero.

The signaling model arises when the time between offers is endogenous [Admati and Perry (1987)]. Then the informed party (the firm) has an incentive to delay making an offer until after a sufficient time has passed to credibly reveal its private information. The critical assumption here is that the uninformed party (the union) is unable to make a counteroffer while it is waiting for the firm to make an offer. Aside from the union's initial demand, all settlements are ex post fair, in that the wage is the full-information Rubinstein (1982) wage. The union's initial demand is chosen to balance the cost of delay and the terms of settlement. This initial demand is accepted by the firm if its willingness to pay is sufficiently high. Otherwise the firm makes a counteroffer after waiting long enough to make the Rubinstein wage credible. Signaling and screening can be compared along a number of dimensions:

- Screening outcomes depend critically on the minimum time between offers; signaling outcomes are insensitive to the minimum time between offers.
- Screening outcomes strongly favor the informed party (the firm); signaling outcomes are roughly ex post fair. Hence, wages are higher under signaling and are more sensitive to the firm's private information.
- Dispute incidence and dispute durations are higher under signaling. Indeed, dispute incidence is always greater than 50% in the standard signaling model. However, introducing a fixed cost of initiating a strike can lead to any level of strike incidence.

Cramton and Tracy (1992) emphasize that the union has multiple threats. The union can strike or the union can hold out, putting pressure on the firm while continuing to work. Holdouts take the form of a slowdown, work-to-rule, sick-out, or other in-plant action. From the union's point of view, holdouts have two advantages: (1) workers are paid according to the expired contract, and (2) workers cannot be replaced. The union selects the threat, strike or holdout, that gives it the highest payoff. Since the desirability of each threat depends on observable factors, modeling this threat choice is important to understanding key features of the data. When striking is the only threat, then strike incidence depends essentially on the degree of uncertainty; whereas, with multiple threats strike incidence can vary as the composition of disputes changes with the attractiveness of each threat. For example, holdouts are more desirable when the current wage is high, and strikes are more desirable when unemployment is low and the workers have better outside options.

In Cramton and Tracy (1992), a union and a firm are bargaining over the wage to be paid over the next contract period. The union's reservation wage is common knowledge. The firm's value of the labor force is private information.

Bargaining begins with the union selecting a threat, either holdout or strike, which applies until a settlement is reached. In the holdout threat, the union is paid the current wage under the expired contract. There is some inefficiency associated with holdout. An outcome of the bargaining specifies the time of agreement, the contract wage at the time of agreement, and the threat before agreement. Following the union's threat choice, the union and firm alternate wage offers, with the union making the initial offer. The time between offers is endogenous.

The equilibrium takes a simple form. If the current wage is sufficiently low, the union decides to strike; otherwise, the union holds out. A second indifference level is determined by the union's initial offer. The firm accepts the union's initial offer if its valuation is above the indifference level, and otherwise rejects the offer and makes a counteroffer after sufficient time has passed to credibly signal the firm's value.

A primary result is that dispute activity increases with uncertainty about private information. Tracy (1986, 1987) tests this basic result by using stock price volatility as a proxy for the amount of uncertainty in contract negotiations. With U.S. data, he finds that strike incidence and strike duration increase with greater relative volatility.

Cramton and Tracy (1994a) fit the parameters of the model to match the main features of the U.S. data from 1970 to 1989. They also estimate dispute incidence and dispute composition. Consistent with the theory, strike incidence increases as the strike threat becomes more attractive, because of low unemployment or a real wage drop over the previous contract. However, the model performs less well in the 1980s than in the 1970s, suggesting a structural change in the post-1981 period. One explanation for a shift is an increase in the use of replacement workers following President Reagan's firing of striking air traffic controllers. Indeed, there was a shift away from strikes and towards holdouts in the 1980s.

Cramton and Tracy (1998) investigate the extent to which the hiring of replacement workers can account for these changes. They build a model in which a firm considers the replacement option because it improves the firm's strike payoff relative to the union's, resulting in a lower wage. However, a firm must balance this improvement in the terms of trade with the cost of replacement. A firm only uses replacements if its cost of replacement is sufficiently low. The union, anticipating the possibility of replacement, lowers its wage demand in the strike threat in order to reduce the probability of replacement. This risk of replacement, then, reduces the attractiveness of the strike threat, making it more likely that the union adopts the holdout threat at the outset of negotiations. For all large U.S. strikes in the 1980s, the likelihood of replacement is estimated. Consistent with the model, the composition of disputes shifts away from strikes as the predicted risk of replacement increases. Hence, a ban on the use of replacement workers should increase strike activity. Moreover, a ban on replacement increases uncertainty, since replacement effectively truncates the firm's distribution of willingness to pay [Kennan and Wilson (1989)].

The Canadian data provide an opportunity to test this theory. Quebec instituted a ban on replacements in 1977, and British Columbia and Ontario introduced a similar ban in 1993. Gunderson, Kervin, and Reid (1989) find that strike incidence does increase with a ban on replacements, and Gunderson and Melino (1990b) find strikes are longer after a ban. Budd (1996) and Cramton, Gunderson, and Tracy (1999) examine the effect of a ban on replacement workers on wages and strike activity. Budd does not find significant effects from the ban using a sample of single province contracts in manufacturing from 1965–1985. In contrast, with a larger sample of contract negotiations from 1967–1993, Cramton, Gunderson, and Tracy find that prohibiting the use of replacement workers during strikes is associated with significantly higher wages, and more frequent and longer strikes.

Predictions of the bargaining models are sensitive to how threat payoffs change over time. Hart (1989) shows that strike durations are much longer in a screening model when strike costs increase sharply when a crunch point is reached (say inventories run out). Cramton and Tracy (1994b) consider time-varying threats within a signaling model. Strike payoffs change as replacement workers are hired, as strikers find temporary jobs, and as inventories or strike funds run out. The settlement wage is largely determined from the long-run threat, rather than the short-run threat. As a result, if dispute costs increase in the long run, then dispute durations are longer and wages decline more slowly during the short run. Allowing time-varying threats helps explain empirical results. Settlement rates are lower during periods of eligibility for unemployment insurance [Kennan (1980)]. Strike durations are longer during business downturns [Kennan (1985), Harrison and Stewart (1989)]. Wages might not decrease with strike durations [Card (1990)]. Moreover, the theory can help explain the costly actions firms and unions take to influence threat payoffs.

An important feature of union contract negotiations is that they do not occur in isolation. Information from one contract negotiation may be linked with other contract negotiations within the same industry. Kuhn and Gu (1996) interpret holdouts in this way. In their theory, holdouts are used as a delaying tactic to get information about other bargaining outcomes in the same industry. When private information is correlated among bargaining pairs, there is an incentive to hold out, since one bargaining pair benefits from information revealed in the negotiation of another pair. Three predictions stem from this theory: (1) holdouts should increase when more bargaining pairs negotiate concurrently, (2) there should be a clustering of holdout durations within an industry, and (3) holdouts ending later are less apt to end in strikes. A panel of Canadian manufacturing contract negotiations from 1965 to 1988 support these predictions. A further implication of the linked information is that strike incidence can be reduced to the extent that private information is revealed in related contract negotiations. Kuhn and Gu (1995) find support for this hypothesis.

In addition to within-industry links, contracts are linked over time. Today's negotiation is just one in a sequence of negotiations between the union and the firm. The current negotiation affects the next negotiation in two ways: a wage linkage and an information linkage. The wage linkage is as in Cramton and Tracy (1992). The current wage is the starting point for negotiations and determines the attractiveness of striking versus holding out. An information linkage arises when the private information between contracts is correlated. Kennan (1995) studies a screening model of repeated negotiations where the firm's willingness to pay follows a Markov process. One implication of this model is a rachet effect. A firm is more hesitant to give in today, knowing that doing so will worsen its position in the next negotiation. More importantly, Kennan's model of repeated negotiation can explain some of the observed links between prior and current contract negotiations. For example, Card (1988, 1990) finds that strike incidence is higher after a short strike in the prior negotiation, and lower after either no strike or a long strike in the prior negotiation.

7. Experimental evidence

Strategic theories of bargaining with private information only recently have been evaluated in the experimental laboratory. The advantage of an experimental test of the theory, compared with an empirical test, is that the experimenter is able to observe the distribution and realizations of private information. The power of empirical tests is limited because the parties' degree of uncertainty must be estimated indirectly from the data, under the assumption that the theory is true. This has led most researchers to test other empirical implications of the model, such as the slope of the concession function. The experimenter, on the other hand, can construct an environment that conforms much more closely to the theoretical setting. In this way, less ambiguous tests of the theory can be performed. Unfortunately, even in tightly controlled experiments, some ambiguity will remain, since the subjects may have relevant private information about their preferences that the experimenter is not privy to.[25]

Most of the experimental work on strategic bargaining has focused on testing dynamic models with full information[26] or static models with private information.[27] Much could be learned by considering dynamic bargaining with private information. By introducing private information into a dynamic bargaining environment, we are able to observe how uncertainty influences the incidence and duration of disputes. This has been the focus of much of the theoretical and empirical work, and yet few experimental tests have been done.

References

Admati, A., and M. Perry (1987), "Strategic delay in bargaining", Review of Economic Studies 54:345–364.

Akerlof, G.A. (1970), "The market for 'lemons': Quality uncertainty and the market mechanism", Quarterly Journal of Economics 84:488–500.

Ausubel, L.M. (2002), "A mechanism generalizing the Vickrey auction", Econometrica, forthcoming.

Ausubel, L.M., and R.J. Deneckere (1989a), "Reputation in bargaining and durable goods monopoly", Econometrica 57:511–532.

Ausubel, L.M., and R.J. Deneckere (1989b), "A direct mechanism characterization of sequential bargaining with one-sided incomplete information", Journal of Economic Theory 48:18–46.

Ausubel, L.M., and R.J. Deneckere (1992a), "Bargaining and the right to remain silent", Econometrica 60:597–626.

---

[25] This point is emphasized by Forsythe, Kennan and Sopher (1991) and Ochs and Roth (1989).

[26] Binmore, Shaked and Sutton (1985, 1988, 1989), Neelin, Sonnenschein and Spiegel (1988), and Ochs and Roth (1989).

[27] Forsythe, Kennan and Sopher (1991) and Radner and Schotter (1989).

Ausubel, L.M., and R.J. Deneckere (1992b), "Durable goods monopoly with incomplete information", Review of Economic Studies 59:795–812.

Ausubel, L.M., and R.J. Deneckere (1993a), "Efficient sequential bargaining", Review of Economic Studies 60:435–462.

Ausubel, L.M., and R.J. Deneckere (1993b), "A generalized theorem of the maximum", Economic Theory 3:99–107.

Ausubel, L.M., and R.J. Deneckere (1994), "Separation and delay in bargaining", mimeo (University of Maryland and University of Wisconsin-Madison).

Ausubel, L.M., and R.J. Deneckere (1998), "Bargaining and forward induction", mimeo (University of Maryland and University of Wisconsin-Madison).

Ayres, I., and E. Talley (1995), "Solomonic bargaining: Dividing a legal entitlement to facilitate Coasean trade", Yale Law Journal 104:1027–1117.

Bikhchandani, S. (1992), "A bargaining model with incomplete information", Review of Economic Studies 59:187–204.

Binmore, K., M.J. Osborne and A. Rubinstein (1992), "Noncooperative models of bargaining", in: R.J. Aumann and S. Hart, eds., Handbook of Game Theory, Vol. 1 (North Holland, Amsterdam), 179–225.

Binmore, K., A. Shaked and J. Sutton (1985), "Testing noncooperative bargaining theory: A preliminary study", American Economic Review 75:1178–1180.

Binmore, K., A. Shaked and J. Sutton (1988), "A further test of noncooperative bargaining theory: Reply", American Economic Review 78:837–840.

Binmore, K., A. Shaked and J. Sutton (1989), "An outside option experiment", Quarterly Journal of Economics 104:753–770.

Budd, J.W. (1996), "Canadian strike replacement legislation and collective bargaining: Lessons for the United States", Industrial Relations 35:245–260.

Card, D. (1988), "Longitudinal analysis of strike activity", Journal of Labor Economics 6:147–176.

Card, D. (1990), "Strikes and wages: A test of a signalling model", Quarterly Journal of Economics 105:625–660.

Chatterjee, K., and W. Samuelson (1983), "Bargaining under incomplete information", Operations Research 31:835–851.

Cho, I.-K. (1990a), "Uncertainty and delay in bargaining", Review of Economic Studies 57:575–596.

Cho, I.-K. (1990b), "Characterization of equilibria in bargaining models with incomplete information", mimeo (University of Chicago).

Cho, I.-K., and D.M. Kreps (1987), "Signalling games and stable equilibria", Quarterly Journal of Economics 102:179–222.

Cho, I.-K., and J. Sobel (1990), "Strategic stability and uniqueness in signaling games", Journal of Economic Theory 50:381–413.

Coase, R.H. (1960), "The problem of social cost", Journal of Law and Economics 3:1–44.

Coase, R.H. (1972), "Durability and monopoly", Journal of Law and Economics 15:143–149.

Cramton, P. (1984), "Bargaining with incomplete information: An infinite-horizon model with continuous uncertainty", Review of Economic Studies 51:579–593.

Cramton, P. (1985), "Sequential bargaining mechanisms", in: A. Roth, ed., Game Theoretic Models of Bargaining (Cambridge University Press, Cambridge, England).

Cramton, P. (1992), "Strategic delay in bargaining with two-sided uncertainty", Review of Economic Studies 59:205–225.

Cramton, P., R. Gibbons and P. Klemperer (1987), "Dissolving a partnership efficiently", Econometrica 55:615–632.

Cramton, P., M. Gunderson and J. Tracy (1999), "The effect of collective bargaining legislation on strikes and wages", Review of Economics and Statistics 81:475–489.

Cramton, P., and J. Tracy (1992), "Strikes and holdouts in wage bargaining: Theory and data", American Economic Review 82:100–121.

Cramton, P., and J. Tracy (1994a), "Wage bargaining with time-varying threats", Journal of Labor Economics 12:594–617.

Cramton, P., and J. Tracy (1994b), "The determinants of U.S. labor disputes", Journal of Labor Economics 12:180–209.

Cramton, P., and J. Tracy (1998), "The use of strike replacements in union contract negotiations: The U.S. experience, 1980–1989", Journal of Labor Economics 16:667–701.

Crawford, V.P. (1982), "A theory of disagreement in bargaining", Econometrica 50:607–637.

Cremer, J., and R. McLean (1988), "Full extraction of the surplus in Bayesian and dominant strategy auctions", Econometrica 56:1247–1258.

Dasgupta, P., and E. Maskin (2000), "Efficient auctions", Quarterly Journal of Economics 115:341–388.

Deneckere, R.J. (1992), "A simple proof of the Coase conjecture", mimeo (Northwestern University).

Deneckere, R.J., and M.-Y. Liang (1999), "Bargaining with interdependent values", mimeo (University of Wisconsin-Madison).

Evans, R. (1989), "Sequential bargaining with correlated values", Review of Economic Studies 56:499–510.

Farrell, J., and R. Gibbons (1989), "Cheap talk can matter in bargaining", Journal of Economic Theory 48:221–237.

Fernandez, R., and J. Glazer (1991), "Striking for a bargain between two completely informed agents", American Economic Review 81:240–252.

Forsythe, R., J. Kennan and B. Sopher (1991), "An experimental analysis of strikes in bargaining games with one-sided private information", American Economic Review 81:253–278.

Fudenberg, D., D.K. Levine and J. Tirole (1985), "Infinite-horizon models of bargaining with one-sided incomplete information", in: A. Roth, ed., Game Theoretic Models of Bargaining (Cambridge University Press, Cambridge, England).

Fudenberg, D., and J. Tirole (1983), "Sequential bargaining with incomplete information", Review of Economic Studies 50:221–247.

Gresik, T.A. (1991a), "Efficient bilateral trade with statistically dependent beliefs", Journal of Economic Theory 53:199–205.

Gresik, T.A. (1991b), "Ex ante efficient, ex post individually rational trade", Journal of Economic Theory 53:131–145.

Gresik, T.A. (1991c), "Ex ante incentive efficient trading mechanisms without the private valuation restriction", Journal of Economic Theory 55:41–63.

Gresik, T.A., and M.A. Satterthwaite (1989), "The rate at which a simple market converges to efficiency as the number of traders increases: An asymptotic result for optimal trading mechanisms", Journal of Economic Theory 48:304–332.

Grossman, S.J., and M. Perry (1986a), "Perfect sequential equilibrium", Journal of Economic Theory 39:97–119.

Grossman, S.J., and M. Perry (1986b), "Sequential bargaining under asymmetric information", Journal of Economic Theory 39:120–154.

Gul, F., and H. Sonnenschein (1988), "On delay in bargaining with one-sided uncertainty", Econometrica 56:601–612.

Gul, F., H. Sonnenschein and R. Wilson (1986), "Foundations of dynamic monopoly and the Coase conjecture", Journal of Economic Theory 39:155–190.

Gunderson, M., J. Kervin and F. Reid (1986), "Logit estimates of strike incidence from Canadian contract data", Journal of Labor Economics 4:257–276.

Gunderson, M., J. Kervin and F. Reid (1989), "The effect of labour relations legislation on strike incidence", Canadian Journal of Economics 22:779–794.

Gunderson, M., and A. Melino (1990a), "Estimating strike effects in a general model of prices and quantities", Journal of Labor Economics 5:1–19.

Gunderson, M., and A. Melino (1990b), "The effects of public policy on strike duration", Journal of Labor Economics 8:295–316.

Haller, H., and S. Holden (1990), "A letter to the editor on wage bargaining", Journal of Economic Theory 52:232–236.

Harrison, A., and M. Stewart (1989), "Cyclical fluctuations in strike durations", American Economic Review 79:827–841.

Hart, O.D. (1989), "Bargaining and strikes", Quarterly Journal of Economics 104:25–44.

Hart, O.D., and J. Tirole (1988), "Contract renegotiation and Coasian dynamics", Review of Economic Studies 55:509–540.

Jehiel, P., and B. Moldovanu (2001), "Efficient design with interdependent valuations", Econometrica 69:1237–1260.

Kennan, J. (1980), "The effect of unemployment insurance payments on strike duration", Unemployment Compensation: Studies and Research 2:467–483.

Kennan, J. (1985), "The duration of contract strikes in US manufacturing", Journal of Econometrics 28:5–28.

Kennan, J. (1995), "Repeated contract negotiations with private information", Japan and the World Economy 7:447–472.

Kennan, J. (1997), "Repeated bargaining with persistent private information", mimeo (Department of Economics, University of Wisconsin-Madison).

Kennan, J., and R. Wilson (1989), "Strategic bargaining models and interpretation of strike data", Journal of Applied Econometrics 4:S87–S130.

Kennan, J., and R. Wilson (1993), "Bargaining with private information", Journal of Economic Literature 31:45–104.

Kohlberg, E., and J.-F. Mertens (1986), "On the strategic stability of equilibria", Econometrica 54:1003–1037.

Krishna, V., and M. Perry (1997), "Efficient mechanism design", Working Paper (Penn State University).

Kuhn, P., and W. Gu (1995), "The economics of relative rewards: Sequential wage bargaining", Working Paper (McMaster University).

Kuhn, P., and W. Gu (1996), "A theory of holdouts in wage bargaining", Working Paper (McMaster University).

Makowski, L., and C. Mezzetti (1993), "The possibility of efficient mechanisms for trading an indivisible object", Journal of Economic Theory 59:451–465.

Maskin, E., and J. Tirole (1994), "Markov perfect equilibrium", mimeo (Harvard University).

Matsuo, T. (1989), "On incentive compatible, individually rational, and ex-post efficient mechanisms for bilateral trading", Journal of Economic Theory 49:189–194.

McAfee, R.P. (1992), "Amicable divorce: Dissolving a partnership with simple mechanisms", Journal of Economic Theory 56:266–293.

McAfee, R.P., and P. Reny (1992), "Correlated information and mechanism design", Econometrica 60:395–422.

McConnell, S. (1989), "Strikes, wages, and private information", American Economic Review 79:801–815.

Mirrlees, J.A. (1971), "An exploration in the theory of optimal taxation", Review of Economic Studies 38:175–208.

Mookherjee, D., and S. Reichelstein (1992), "Dominant strategy implementation of Bayesian incentive compatible allocation rules", Journal of Economic Theory 56:378–399.

Myerson, R.B. (1981), "Optimal auction design", Mathematics of Operations Research 6:58–73.

Myerson, R.B. (1985), "Analysis of two bargaining problems with incomplete information", in: A. Roth, ed., Game Theoretic Models of Bargaining (Cambridge University Press, Cambridge), 115–147.

Myerson, R.B., and M.A. Satterthwaite (1983), "Efficient mechanisms for bilateral trading", Journal of Economic Theory 28:265–281.

Neelin, J., H. Sonnenschein and M. Spiegel (1988), "A further test of noncooperative bargaining theory", American Economic Review 78:824–836.

Ochs, J., and A.E. Roth (1989), "An experimental study of sequential bargaining", American Economic Review 89:355–384.

Osborne, M.J., and A. Rubinstein (1990), Bargaining and Markets (Academic Press, Boston).

Perry, M. (1986), "An example of price formation in bilateral situations: A bargaining model with incomplete information", Econometrica 54:313–321.

Perry, M., and P.J. Reny (1993), "A non-cooperative bargaining model with strategically timed offers", Journal of Economic Theory 59:50–77.

Perry, M., and P.J. Reny (1998), "An ex-post efficient auction", mimeo (University of Pittsburgh).

Radner, R., and A. Schotter (1989), "The sealed-bid mechanism: An experimental study", Journal of Economic Theory 48:179–220.

Riley, J., and R. Zeckhauser (1983), "Optimal selling strategies: When to haggle, when to hold firm", Quarterly Journal of Economics 98:267–289.

Rubinstein, A. (1982), "Perfect equilibrium in a bargaining model", Econometrica 50:97–109.

Rubinstein, A. (1985a), "A bargaining model with incomplete information about time preferences", Econometrica 53:1151–1172.

Rubinstein, A. (1985b), "Choice of conjectures in a bargaining game with incomplete information", in: A. Roth, ed., Game Theoretic Models of Bargaining (Cambridge University Press, Cambridge), 99–114.

Rubinstein, A. (1987), "A sequential strategic theory of bargaining", in: T. Bewley, ed., Advances in Economic Theory (Cambridge University Press, London), 197–224.

Rustichini, A., M.A. Satterthwaite and S.R. Williams (1994), "Convergence to efficiency in a simple market with incomplete information", Econometrica 62:1041–1064.

Sákovics, J. (1993), "Delay in bargaining games with complete information", Journal of Economic Theory 59:78-95.

Samuelson, W. (1984), "Bargaining under asymmetric information", Econometrica 52:995–1005.

Samuelson, W. (1985), "A comment on the Coase theorem", in: Alvin R., ed., Game Theoretic Models of Bargaining (Cambridge University Press, Cambridge, England).

Satterthwaite, M.A., and S.R. Williams (1989), "The rate of convergence to efficiency in the buyer's bid double auction as the market becomes large", Review of Economic Studies 56:477–498.

Sobel, J., and I. Takahashi (1983), "A multi-stage model of bargaining", Review of Economic Studies 50:411–426.

Stokey, N.L. (1979), "Intertemporal price discrimination", Quarterly Journal of Economics 93:355–371.

Stokey, N.L. (1981), "Rational expectations and durable goods pricing", Bell Journal of Economics 12:112–128.

Tracy, J.S. (1986), "An investigation into the determinants of U.S. strike activity", American Economic Review 76:423–436.

Tracy, J.S. (1987), "An empirical test of an asymmetric information model of strikes", Journal of Labor Economics 5:149–173.

Vincent, D.R. (1989), "Bargaining with common values", Journal of Economic Theory 48:47–62.

Vincent, D.R. (1998), "Repeated signalling games and dynamic trading relationships", International Economic Review 39:275–293.

Vroman, S.B. (1989), "A longitudinal analysis of strike activity in U.S. manufacturing: 1957–1984", American Economic Review 79:816–826.

Williams, S.R. (1990), "The transition from bargaining to a competitive market", American Economic Review 80:227–231.

Williams, S.R. (1991), "Existence and convergence of equilibria in the buyer's bid double auction", Review of Economic Studies 58:351–374.

Williams, S.R. (1999), "A characterization of efficient, Bayesian incentive compatible mechanisms", Economic Theory 14:155–180.

Wilson, R. (1985), "Incentive efficiency of double auctions", Econometrica 53:1101–1116.

Chapter 51

INSPECTION GAMES*

RUDOLF AVENHAUS

Universität der Bundeswehr München

BERNHARD VON STENGEL

London School of Economics

SHMUEL ZAMIR

The Hebrew University of Jerusalem

Contents

*This work was supported by the German–Israeli Foundation (G.I.F.), by the Volkswagen Foundation, and by a Heisenberg grant from the Deutsche Forschungsgemeinschaft.

Handbook of Game Theory, Volume 3, Edited by R.J. Aumann and S. Hart

Abstract

Starting with the analysis of arms control and disarmament problems in the sixties, inspection games have evolved into a special area of game theory with specific theoretical aspects, and, equally important, practical applications in various fields of human activity where inspection is mandatory. In this contribution, a survey of applications is given first. These include arms control and disarmament, theoretical approaches to auditing and accounting, for example in insurance, and problems of environmental surveillance. Then, the general problem of inspection is presented in a game-theoretic framework that extends a statistical hypothesis testing problem. This defines a game since the data can be strategically manipulated by an inspectee who wants to conceal illegal actions. Using this framework, two models are solved, which are practically significant and technically interesting: material accountancy and data verification. A second important aspect of inspection games is the fact that inspection resources are limited and have to be used strategically. This is demonstrated in the context of sequential inspection games, where many mathematically challenging models have been studied. Finally, the important concept of leadership, where the inspector becomes a leader by announcing and committing himself to his strategy, is shown to apply naturally to inspection games.

Keywords

inspection game, arms control, hypothesis testing, recursive game, leadership

JEL classification: C72, C12

1. Introduction

Inspection games form an applied field of game theory. An inspection game is a mathematical model of a situation where an *inspector* verifies that another party, called *inspectee*, adheres to certain legal rules. This legal behavior may be defined, for example, by an arms control treaty, and the inspectee has a potential interest in violating these rules. Typically, the inspector's resources are limited so that verification can only be partial. A mathematical analysis should help in designing an optimal inspection scheme, where it must be assumed that an illegal action is executed strategically. This defines a game-theoretic problem, usually with two players, inspector and inspectee. In some cases, several inspectees are considered as individual players.

Game theory is not only adequate to describe an inspection situation, it also produces results which may be used in practical applications. The first serious attempts were made in the early 1960's where game-theoretic studies of arms control inspections were commissioned by the United States Arms Control and Disarmament Agency (ACDA). Furthermore, the International Atomic Energy Agency (IAEA) performs inspections under the Nuclear Non-Proliferation Treaty. The decision rules of IAEA inspectors for detecting a deviation of nuclear material can be interpreted as equilibrium strategies in a zero-sum game with the detection probability as payoff function. In these applications, game theory has proved itself useful as a technical tool in line with statistics and other methods of operations research. Since the underlying models should be accessible to practitioners, traditional concepts of game theory are used, like zero-sum games or games in extensive form. Nevertheless, as we want to demonstrate, the solution of these games is mathematically challenging, and leads to interesting conceptual questions as well.

We will not consider inspection problems that are exclusively statistical, since our emphasis is on games. We also exclude industrial inspections for quality control and maintenance, except for an interesting worst-case analysis of timely inspections. Similarly, we do not consider *search games* [Gal (1983); O'Neill (1994)] modeling pursuit and evasion, for example in war. Search games are distinguished from inspection games by a symmetry between the two players whose actions are usually equally legitimate. In contrast, inspection games as we understand them are fundamentally asymmetrical: their salient feature is that the inspector tries to prevent the inspectee from behaving illegally in terms of an agreement.

In Section 2, we survey applications of inspection games to arms control, auditing and accounting and economics, and other areas like environmental regulatory enforcement or crime control.

In Section 3, we provide a general game-theoretic framework to inspections which extends the statistical approach used in practice. In these statistical hypothesis testing problems, the distribution of the observed random variable is strategically manipulated by the inspectee. The equilibrium of the general non-zero-sum game is found using an auxiliary zero-sum game in which the inspectee chooses a violation procedure and the inspector chooses a statistical test with a given false alarm probability. We illustrate this by two specific important models, namely material accountancy and data verification.

Inspections over time, so far primarily of methodological interest, are the subject of Section 4. In these games, the information of the players about the actions of their respective opponent is very important and is best understood if the game is represented in extensive form. If the payoffs have a simple structure, then the games can sometimes be represented recursively and solved analytically. In another timeliness game, the optimal inspection times, continuously chosen from an interval, are determined by differential equations.

In Section 5 we discuss the inspector leadership principle. It states that the inspector may commit himself to his inspection strategy in advance, and thereby gain an advantage compared to the symmetrical situation where both players choose their actions simultaneously. Obviously, this concept is particularly applicable to inspection games.

2. Applications

The majority of publications on inspection games concerns arms control and disarmament, usually relating to one or the other arms control treaty that has been signed. We survey these models first. Some of them are described in mathematical detail in later sections. Inspection games have also been applied to problems in economics, particularly in accountancy and auditing, in enforcement of environmental regulations, and in crime control and related areas. Some papers treat inspection games in an abstract setting, rather than modeling a particular application.

2.1. Arms control and disarmament

Inspection games have been applied to arms control and disarmament in three phases. Studies in the first phase, from about 1961 to 1967, analyze inspections for a nuclear test ban treaty, which was then under negotiation. The second phase, about 1968 to 1985, comprises work stimulated by the Non-Proliferation Treaty for Nuclear Weapons. Under that treaty, proper use of nuclear material is verified by the International Atomic Energy Agency (IAEA) in Vienna. The third phase lasts from about 1986 until today. The end of the Cold War brought about new disarmament treaties. Verification of these treaties has also been analyzed using game theory.

In the 1960's, a *test ban treaty* was under negotiation between the United States and the Soviet Union, and verification procedures were discussed. Tests of nuclear weapons above ground can be detected by satellites. Underground tests can be sensed by seismic methods, but in order to discriminate them safely from earthquakes, on-site inspections are necessary. The two sides never agreed on the number of such inspections that they would allow to the other side, so eventually they decided not to ban underground tests in the treaty.

While the test ban treaty was being discussed, the problem arose as to how an inspector should use a certain limited number of on-site inspections, as provided by the treaty, for verifying a much larger number of suspicious seismic events. Dresher (1962) modeled this problem as a recursive game. The estimated number of events and the number of inspections that can be used are fixed parameters. We explain this game in Section 4.1. It formed the basis for much subsequent work.

With political backing for scientific disarmament studies, the United States Arms Control and Disarmament Agency (ACDA) commissioned game-theoretic analyses of inspections to the Mathematica company in the years 1963 to 1968. Game-theoretic researchers involved in this or related work were Aumann, Dresher, Harsanyi, Kuhn, Maschler, and Selten, among others. Publications on inspection games of that group are Kuhn (1963) and, in general, the reports to ACDA edited by Anscombe et al. (1963, 1965), as well as Maschler (1966, 1967). In a related study, Saaty (1968) presents some of the existing developments in this area. Many of these papers extend Dresher's game in various ways, for example by generalizing the payoffs, or by assuming an uncertainty in the signals given by detectors.

The *Non-Proliferation Treaty* (NPT) for Nuclear Weapons was inaugurated in 1968. This treaty divided the world into weapons states, who promised to reduce or even eliminate their nuclear arsenals, and non-weapon states who promised never to acquire such weapons. All states which became parties to the treaty agreed that the IAEA in Vienna verifies the nuclear material contained in the peaceful nuclear fuel cycles of *all* states.

The verification principle of the IAEA is *material accountancy*, that is, the comparison of book and physical inventories for a given material balance area at the end of an inventory period. The plant operators report their balance data via their national organizations to the IAEA, whose inspectors verify these reported data with the help of independent measurements on a random sampling basis.

Two kinds of sampling procedures are considered in all situations, depending on the nature of the problem. *Attribute sampling* is used to test or to estimate the percentage of items in the population containing some characteristic or attribute of interest. In inspections, the attribute of interest is usually if a safeguards measure has been violated. This may be a broken seal of a container, or an unquestioned decision that a datum has been falsified. The inspector uses the rate of tampered items in the sample to estimate the population rate or to test a hypothesis.

The second kind of procedure is *variable sampling*. This is designed to provide an estimate of or a test on an average or total value of material. Each observation, instead of being counted as falling in a given category, provides a value which is totaled or averaged for the sample. This is described by a certain test statistic, like the total of Material Unaccounted For (MUF). Based on this statistic, the inspector has to decide if nuclear material has been diverted or if the result is due to measurement errors. This decision depends on the probability of a false alarm chosen by the inspector.

Game-theoretic work in this area was started by Bierlein (1968, 1969), who emphasized that payoffs should be expressed by detection probabilities only. In contrast, inspection costs are parameters that are fixed externally. This is an adequate model for the IAEA which has a certain budget limiting its overall inspection effort. The agency has no intent to minimize that effort further, but instead wants to use it most efficiently.

Since 1969, international conferences on nuclear material safeguards have regularly been held by the following institutions. The IAEA organizes about every five years conferences on Nuclear Safeguards Technology and publishes proceedings under that title. The European Safeguards Research and Development Association (ESARDA) as well as the Institute for Nuclear Material Management (INMM) in the United States meet every year and publish proceedings on that subject as well. Here, most publications concern practical matters, for example measurement technology, data processing, and safety. However, decision theoretical approaches, including game-theoretic methods, were presented throughout the years. Monographs which emphasize the theoretical aspects are Jaech (1973), Avenhaus (1986), Bowen and Bennett (1988), and Avenhaus and Canty (1996).

Some studies on nuclear material safeguards are not related to the NPT. The U.S. Nuclear Regulatory Commission (NUREG) is in charge of safeguarding nuclear plants against theft and sabotage to guarantee their safe operation, in fulfillment of domestic regulations. In a study for that commission, Goldman (1984) investigates the possible use of game theory and its potential role in safeguards.

In the mid-eighties, when the Cold War ended, new disarmament treaties were signed, like the treaty on Intermediate Nuclear Forces in 1987, or the treaty on Conventional Forces in Europe in 1990 [see Altmann et al. (1992) on verification issues]. The verification of these new treaties was investigated in game-theoretic terms by Brams and Davis (1987), by Brams and Kilgour (1988), and by Kilgour (1992). Variants of recursive games are described by Ruckle (1992). In part, these papers extend the work done before, in particular Dresher (1962).

2.2. Accounting and auditing in economics

In economic theory, inspection games have been studied for the auditing of accounts. In insurance, inspections are used against the 'moral hazard' that the client may abuse his insurance by fraud or negligence. Our summary of these topics is based on Borch (1990). We will also consider models of tax inspections.

Accounting is usually understood as a system for keeping track of the circulation of money. The monetary transactions are recorded in accounts. These may be checked in full detail by an inspector, which is called an *audit*. Auditing of accounts is often based on sampling inspection, simply because it is unnecessarily costly to check every voucher and entry in the books. The possibility that any transaction may be checked in full detail is believed to have a deterring effect likely to prevent irregularities.

A theoretical analysis of problems in this field naturally leads to a search for suitable sampling methods [for an outline see Kaplan (1973)]. The concepts of attribute and variable sampling described above for material accountancy apply to the inspection of monetary accounts as well. In particular, there are tests to validate the reasonableness of account balances, without classifying a particular observation as correct or falsified.

These may be considered as variable measurement tests and are sometimes termed 'dollar value' samples in the literature.

These theoretical investigations include the use of game-theoretic methods. An early contribution that employs both noncooperative and cooperative game theory is given by Klages (1968), who describes quite detailed models of the practical problems in accountancy and discusses the merits of a game-theoretic approach.

Borch (1982) formulates a zero-sum inspection game between an accountant and his employer. The accountant, who is the inspectee, may either record transactions faithfully or cheat to embezzle some of the company's profits, and the employer may either trust the accountant or audit his accounts. If the inspectee is honest, he receives payoff zero irrespective of the actions of the inspector. In that case, the inspector gets payoff zero if he trusts and has a certain cost if he audits. If the accountant steals while being trusted, he has a gain that is the employer's loss. If the inspector catches an illegal action, he has the same auditing cost as before but the inspectee must pay a penalty. Borch interprets mixed strategies as 'fractions of opportunities' in a repeated situation, but does not formalize this further.

The employer may buy 'fidelity guarantee insurance' to cover losses caused by dishonest accountants. The insurance may require a strict auditing system that is costly to the employer. Borch (1982) considers a three-person game where the insurance company inspects or alternatively trusts the employer if he audits properly. The employer is both inspectee, of the insurance company, and inspector, of his accountant. No interaction is assumed between insurance company and accountant, so this game is fully described by two bimatrix games. For general 'polymatrix' games of this kind see Howson (1972).

Borch (1990) sees many potential applications of games to the economics of insurance: in any situation where the insured has undertaken to take special measures to prevent accidents and reduce losses, there is a risk – usually called *moral hazard* – that he may neglect his obligations. The insurance company reserves the right to inspect that the insured really carries out the safety measures foreseen in the insurance contract. This inspection costs money, which the insured must pay for by an addition to the premium.

Moral hazard has its price. Therefore, inspections of economic transactions raise the question of the most efficient design of such a system, for example so that surveillance is minimized or even unnecessary. These problems are closely related to a variety of economic models known as *principal–agent problems*. They have been extensively studied in the economic literature, where we refer to the surveys by Baiman (1982), Kanodia (1985), and Dye (1986). Agency theory focuses on the optimal contracted relationships between two individuals whose roles are asymmetric. One, the principal, delegates work or responsibility to the other, the agent. The principal chooses, based on his own interest, the payment schedule that best exploits the agent's self-interested behavior. The agent chooses an optimal level of action contingent on the fee schedule proposed by the principal. One important issue in agency theory is the asymmetry of information available to the principal and the agent. In inspection games, a similar asymmetry exists with respect to defining the rules of the game; see Section 5.

We conclude this section with applications to tax inspections. Schleicher (1971) describes an interesting recursive game for detecting tax law violations, which extends Maschler (1966). Rubinstein (1979) analyzes the problem that it may be unjust to penalize illegal actions too hard since the inspectee might have committed them unintentionally. In a one-shot game, there is no alternative to the inspector but to use a high penalty, although its potential injustice has a disutility. Rubinstein shows that if this game is repeated, a more lenient policy also induces the inspectee to legal behavior.

Another model of tax inspections, also using repeated games, is discussed by Greenberg (1984). A tax function defines the tax to be paid by an individual with a certain income. An audited individual who did not report its income properly must pay a penalty, and the tax authorities can audit only a limited percentage of individuals. Under reasonably weak assumptions about these functions and the individuals' utility functions on income, Greenberg proposes an auditing scheme that achieves an arbitrary small percentage of tax evaders. In that scheme, the individuals are partitioned into three groups that are audited with different probabilities, and individuals are moved among these groups after an audit depending on whether they cheated or not. A similar scheme of auditing individuals with different probabilities depending on their compliance history is proposed by Landsberger and Meilijson (1982). However, their analysis does not use game theory explicitly. Reinganum and Wilde (1986) describe a model of tax compliance where they apply the sequential equilibrium concept. In that model, the income reporting process is considered explicitly as a signaling round. The tax inspector is only aware of the overall income distribution and reacts to the reported income.

2.3. Environmental control

Environmental control problems call for a game-theoretic treatment. One player is a firm which produces some pollution of air, water or ground, and which can save abatement costs by illegal emission beyond some agreed level. The other player is a monitoring agent whose responsibility is to detect or better to prevent such illegal pollution. Both agents are assumed to act strategically. Various problems of this kind have been analyzed. However, contrary to arms control and disarmament, these papers do not yet address specific practical cases.

Several papers present game-theoretic analyses of pollution problems but deal only marginally with monitoring problems. Bird and Kortanek (1974) explore various concepts in order to aid the formulation of regulations of sources of pollutant in the atmosphere related to given least cost solutions. Höpfinger (1979) models the problem of how to determine and adapt global emission standards for carbon dioxide as an infinite stage game with three players: regulator, producer, and population. Kilgour, Okada, and Nishikori (1988) describe the load control system for regulating chemical oxygen demand in water bodies. They formulate a cost sharing game and solve it in some illustrative cases.

In the last years, pollution control problems have been analyzed with game-theoretic methods. Russell (1990) characterizes current enforcement of U.S. environmental laws as very likely inadequate, while admitting that proving this proposition would be extremely difficult, exactly because there is so little information about the actual behavior of regulated firms and government activities. As a remedy, a one-stage game between a polluter and an environmental protection agency is used as a benchmark for discussing a multiple-stage game in which the source's past record of discovered violations determines its future probabilities of being monitored. It is shown that this approach can yield significant savings in limiting the extent of violations to a particular frequency in the population of polluters.

Weissing and Ostrom (1991) examine how irrigation institutions affect equilibrium rates of stealing and enforcement. Irrigators come periodically into the position of turntakers. A turntaker chooses between taking a legal amount of water and taking more water than authorized. The other irrigators are turnwaiters who must decide whether to expand resources for monitoring the behavior of the turntaker or not. For no combination of parameters the rate of stealing by the turntaker drops to zero, so in equilibrium some stealing is always going on.

Güth and Pethig (1992) consider a polluting firm that can save abatement costs by illegal waste emission, and a monitoring agent whose job it is to prevent such pollution. When deciding on whether to dispose of its waste legally or illegally the firm does not know for sure whether the controller is sufficiently qualified or motivated to detect the firm's illegal releases of pollutant. The firm has the option of undertaking a small-scale deliberate 'exploratory pollution accident' to get a hint about the controller's qualification before deciding on how to dispose of its waste. The controller may or may not respond to that 'accident' by a thorough investigation, thus perhaps revealing his type to the firm. This sequential decision process along with the asymmetric distribution of information constitutes a signaling game whose equilibrium points may signal the type of the controller to the firm.

Avenhaus (1994) considers a decision theoretic problem. The management of an industrial plant may be authorized to release some amount of pollutant per unit time into the environment. An environmental agency may decide with the help of randomly sampled measurements whether or not the real releases are larger than the permitted ones. The 'best' inspection procedure can be determined by the use of the Neyman–Pearson lemma; see Section 3 below.

2.4. Miscellaneous models

A number of papers do not belong to the above categories. Some of these deal – at least theoretically – with smuggling or crime control, other papers treat inspection games in an abstract setting.

Thomas and Nisgav (1976) consider a game where a smuggler tries to cross a strait in one of M nights. The inspecting border police has a speedboat for patrolling on k of these nights. On a patrolled night, the smuggler runs some risk of being caught. The game is described recursively in the same way as the game by Dresher (1962). The only difference is that the smuggler *must* traverse the strait even if there is a patrol every night, or else receive the same worst payoff as he would if he were caught. The resulting recurrence equation for the game has a very simple solution where the game value is a linear function of k/M.

Baston and Bostock (1991) generalize the paper by Thomas and Nisgav (1976). They clarify some implicit assumptions made by these authors, in particular the full information of the players about past events, and the detection probability associated with a night patrol. Then, they study the case of two boats and derive explicit solutions that depend on the detection probabilities if one boat or both are on patrol. The case of three boats is solved by Garnaev (1994). Sequential games with three parameters, namely number of nights, patrols, and smuggling acts, are solved by Sakaguchi (1994) and Ferguson and Melolidakis (1998). Among other things, Baston and Bostok (1991) and Sakaguchi (1994) reproduce the result by Dresher (1962), which they are not aware of. Ferguson and Melolidakis (2000) present an interesting unifying approach to these results, based on Gale (1957).

Goldman and Pearl (1976) study inspections in an abstract setting, where the inspector has to select among several sites where an inspectee can cheat, and only a limited number of sites can be inspected in total. A simple model is successively refined to study the effects of penalty levels and inspection resources.

Feichtinger (1983) considers a differential game with a suggested application to crime control. The dynamic control variables of police and thief are the 'rate of law enforcement' and the 'pilfering rate', respectively. Avenhaus (1997) analyzes inspections in local public transportation systems and shows that inspection rates used in practice coincide with the game-theoretic equilibrium.

Filar (1985) applies the theory of stochastic games to a generic 'traveling inspector model'. There are a number of sites, each with an inspectee that acts as an individual player. Each site is a state of the stochastic game. That state is deterministically controlled by the inspector who chooses the site to be inspected at the next time period. The inspector has unspecified costs associated with inspection levels, travel, and if he fails to detect a violation. The players' payoffs are either sums up to a finite stage, or limiting averages. In this model, all inspectees can be aggregated into a single player without changing equilibria. For finite horizon payoffs, the game has equilibria in Markov strategies, which depend on the time and the state but not on the history of the game. For limiting average payoffs, stationary strategies suffice, which only depend on the state.

3. Statistical decisions

Many practical inspection problems are based on random sampling procedures. Furthermore, measurement techniques are often used which inevitably produce random and systematic errors. Then, it is appropriate to think of the inspection problem as being an extension of a statistical decision problem. The extension is game-theoretic since the inspector has to decide whether the inspectee has behaved illegally, and such an action is strategic and not random.

In this section, we first present a general framework where we extend a classical statistical testing problem to an inspection game. The 'illegal part' of that game, where the inspectee has decided to violate, is equivalent to a two-person zero-sum game with the non-detection probability as payoff to the violator. If that game has a value, then the Neyman–Pearson lemma can be used to determine an optimal inspection strategy. In this framework, we then discuss two important inspection models that have emerged from statistics: material accountancy and data verification.

3.1. General game and analysis

In a classical hypothesis testing problem, the statistician has to decide, based on an observation of a random variable, between two alternatives (H_0 or H_1) regarding the distribution of the random variable. To make this an inspection game, assume that the distribution of the random variable is strategically controlled by another 'player' called the *inspectee*. More specifically, the inspectee can behave either *legally* or *illegally*. If he behaves legally, the distribution is according to the null hypothesis H_0. If he chooses to act illegally, he also decides on a *violation procedure* which we denote by ω. Thus, the distribution of the random variable Z under the alternative hypothesis H_1 depends on the procedure ω which is also a strategic variable of the inspectee. The statistician, called the *inspector*, has to decide between two actions, based on the observation z of the random variable Z: calling an *alarm* (rejecting H_0) or *no alarm* (accepting H_0). The random variable Z can be a vector, for instance in a multi-stage inspection. This rather general inspection game is described in Figure 1.

The pairs of payoffs to inspector and inspectee, respectively, are also shown in Figure 1. The status quo is legal action and no alarm, represented by the payoff 0 to both players. The payoffs for undetected illegal behavior are -1 for the inspector and 1 for the inspectee. In case of a detected violation, the inspector receives $-a$ and the inspectee $-b$. Finally, if a false alarm is raised, the inspector gets $-e$ and the inspectee $-h$. These parameters are subject to the restrictions

$$0 < e < 1, \qquad 0 < a < 1, \qquad 0 < h < b, \tag{3.1}$$

since the worst event for the inspector is an undetected violation, an alarm is undesirable for everyone, and the worst event for a violator is to be caught. Sometimes it is also assumed that $e < a$, which means that for the inspector a detected violation, representing a 'failure of safeguards', is worse than the inconvenience of a false alarm.

A pure strategy of the inspectee consists of a choice between legal and illegal behavior, and a violation procedure ω. A mixed strategy is a probability distribution on these pure strategies. Since the game under consideration is of perfect recall, such a mixed strategy is equivalent to a behavior strategy given by the probability q for acting illegally and a probability distribution on violation procedures ω given that the inspectee acts illegally. Since the set Ω of violation procedures may be infinite, we assume that it includes, if necessary, randomized violations as well, which are therefore also denoted by ω. That is, a behavior strategy of the inspectee is represented by a pair (q, ω).

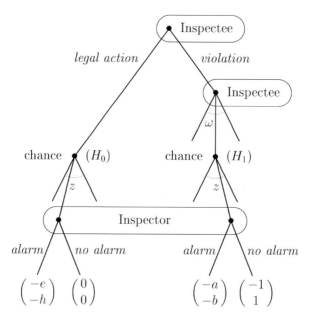

Figure 1. Inspection game extending the statistical decision problem of the inspector, who has to decide between the null hypothesis H_0 and alternative hypothesis H_1 about the distribution of the random variable Z. H_0 means legal action and H_1 means illegal action of the inspectee with violation procedure ω. The inspector is informed about the observation z of Z, but does not know if H_0 or H_1 is true. There is a separate information set for each z. The inspector receives the top payoffs, the inspectee the bottom payoffs. The parameters a, b, e, h are subject to (3.1).

A pure strategy of the inspector is an *alarm set*, that is, a subset of the range of Z, with the interpretation that the inspector calls an alarm if and only if the observation z is in that set. A mixed strategy of the inspector is a probability distribution on pure strategies. A strategy of the inspector (pure or mixed) is also called a *statistical test* and will be denoted by δ. The alarm set or sets used in such a test are usually determined by first considering the *error probabilities* of falsely rejecting or accepting the null hypothesis, as follows.

A statistical test δ and a violation strategy ω determine two conditional probabilities. The probability of an error of the first kind, that is, of a *false alarm*, is the probability $\alpha(\delta)$ of an alarm given that the inspectee acts legally, which is independent of ω. The probability of an error of the second kind, that is, of *non-detection*, is the probability $\beta(\delta, \omega)$ that no alarm is raised given that the inspectee acts illegally.

The inspection game has then the following normal form. The set of strategies δ of the inspector is Δ. The set of (behavior) strategies (q, ω) of the inspectee is given by $[0, 1] \times \Omega$. The payoffs to inspector and inspectee are denoted $I(\delta, (q, \omega))$ and $V(\delta, (q, \omega))$, respectively, where the letter 'V' indicates that the inspectee may potentially violate. In terms of the payoffs in Figure 1, these payoff functions are

$$I\big(\delta,(q,\omega)\big) = (1-q)\big(-e\alpha(\delta)\big) + q\big(-a - (1-a)\beta(\delta,\omega)\big),$$
$$V\big(\delta,(q,\omega)\big) = (1-q)\big(-h\alpha(\delta)\big) + q\big(-b + (1+b)\beta(\delta,\omega)\big). \tag{3.2}$$

We are looking for an *equilibrium* of this noncooperative game. This is a strategy pair $\delta^*, (q^*, \omega^*)$ so that

$$I\big(\delta^*,(q^*,\omega^*)\big) \geqslant I\big(\delta,(q^*,\omega^*)\big) \quad \text{for all } \delta \in \Delta,$$
$$V\big(\delta^*,(q^*,\omega^*)\big) \geqslant V\big(\delta^*,(q,\omega)\big) \quad \text{for all } q \in [0, 1], \ \omega \in \Omega. \tag{3.3}$$

Usually, there is no equilibrium in which the inspectee acts with certainty legally ($q^* = 0$) or illegally ($q^* = 1$). Namely, if $q^* = 0$, then by (3.2), the inspector would choose a test δ^* with $\alpha(\delta^*) = 0$ excluding a false alarm. However, then the equilibrium condition for the inspectee in (3.3) requires $-b + (1 + b)\beta(\delta^*, \omega^*) \leqslant 0$, which means that the non-detection probability $\beta(\delta^*, \omega^*)$ has to be sufficiently low. This is usually not possible with a test δ^* that has false alarm probability zero. On the contrary, such a test usually has a non-detection probability of one. Similarly, if $q^* = 1$, then the inspector could always raise an alarm irrespective of his observation to maximize his payoff, so that $\alpha(\delta^*) = 1$ and $\beta(\delta^*, \omega) = 0$. However, then the equilibrium choice of the inspectee would not be $q^* = 1$ since $h < b$ by (3.1). Thus, in equilibrium,

$$0 < q^* < 1, \tag{3.4}$$

and the inspectee is indifferent between legal and illegal behavior:

$$-h\alpha\big(\delta^*\big) = -b + (1+b)\beta\big(\delta^*,\omega^*\big). \tag{3.5}$$

With respect to the non-detection probability β, the payoff function in (3.2) is monotonically decreasing for the inspector and increasing for the inspectee. Thus, given any test δ, the inspectee will choose a violation mechanism ω so as to maximize $\beta(\delta, \omega)$. In equilibrium, the inspectee will therefore determine ω^* (which depends on δ) so that $\beta(\delta, \omega^*) = \max_{\omega \in \Omega} \beta(\delta, \omega)$. Typically, β is continuous and ω ranges in a compact domain, so the maximum is achieved.

On the side of the inspector, it is useful to choose first a fixed false alarm probability α and then consider only those tests that result in this error probability. Denote this set of tests $\{\delta \in \Delta \mid \alpha(\delta) = \alpha\}$ by Δ_α. For these tests, the inspector's payoff in (3.2) depends only on the non-detection probability. Thus, he will choose that test δ^* in Δ_α that minimizes the worst-case non-detection probability $\beta(\delta^*, \omega^*)$. It follows that in equilibrium, the non-detection probability is represented by

$$\beta(\alpha) := \min_{\delta \in \Delta_\alpha} \max_{\omega \in \Omega} \beta(\delta, \omega). \tag{3.6}$$

That is, the inspector's strategy is a minmax strategy with respect to the non-detection probability. This suggests that, given a false alarm probability α, the equilibrium non-detection probability $\beta(\alpha)$ is the minmax value of a certain auxiliary *zero-sum* game G_α which we call the 'non-detection' game:

DEFINITION. Given the inspection game in Figure 1 and $\alpha \in [0, 1]$, the *non-detection game* G_α is the zero-sum game $(\Delta_\alpha, \Omega, \beta)$, where: The set of strategies δ of the inspector is Δ_α, the set of all statistical tests with false alarm probability α. The set of strategies Ω of the inspectee consists of all violation strategies ω. The payoff to the inspectee is the non-detection probability $\beta(\delta, \omega)$, which he tries to maximize and the inspector tries to minimize.

In other words, the non-detection game G_α captures one part of our original game, namely the situation resulting after the inspectee has decided to behave illegally and the inspector has decided to use tests with false alarm probability α. Thus, G_α does not depend directly on the parameters a, b, e, h of the original game, but rather on the fact that these parameters determine the value of α in equilibrium.

In order to find an equilibrium as in (3.3), we assume that the minmax value $\beta(\alpha)$ in (3.6) of the game G_α is a function $\beta : [0, 1] \to [0, 1]$ that fulfills

$$\beta(0) = 1, \quad \beta(1) = 0, \quad \beta \text{ is convex, and continuous at } 0. \tag{3.7}$$

These conditions imply that β is continuous and monotonically decreasing. An extreme example is the function $\beta(\alpha) = 1 - \alpha$ which applies if the inspector ignores his observation z and calls an alarm with probability α, independently of the data. The properties (3.7) are standard for statistical tests. In particular, β is convex since the inspector may randomize.

THEOREM 3.1. *Assume that for any false alarm probability α, the non-detection game G_α has a value $\beta(\alpha)$ fulfilling (3.7), where the optimal violation procedure ω^* in G_α does not depend on α. Then the inspection game in Figure 1 has the equilibrium $(\delta^*, (q^*, \omega^*))$, where: The false alarm probability $\alpha^* = \alpha(\delta^*)$ is the solution of*

$$-h\alpha^* = -b + (1 + b)\beta(\alpha^*). \tag{3.8}$$

The probability q^* of illegal behavior is given by

$$q^* = \frac{e}{e - (1 - a)\beta'(\alpha^*)} \tag{3.9}$$

where $\beta'(\alpha^*)$ is a sub-derivative of the function β at α^*. The inspection strategy δ^* and the violation procedure ω^* are optimal strategies in G_{α^*}.

PROOF. Equation (3.8) is implied by (3.5). The probability q^* of illegal behavior is determined as follows. By (3.7), the inspector's payoff in (3.2) is a convex combination of two concave functions of α, one decreasing, the other increasing. For $q = q^*$, the resulting concave function must have its maximum at α^* (see also Figure 10 in Section 5 below). This implies (3.9), if the derivative $\beta'(\alpha^*)$ of the function β at α^* exists. Otherwise, one may take any sub-derivative $\beta'(\alpha^*)$, defined as the derivative of any tangent supporting the convex function β at its argument α^*. Since the zero-sum game G_{α^*} has a value, its solution δ^*, ω^* is part of an equilibrium, as argued above. Even if the inspector changes both α^* and δ^*, he cannot improve his payoff by the assumption [due to Wölling (2000)] on ω^*. $\qquad\square$

Usually, this equilibrium is *unique*. It is easy to see that by (3.7), Equation (3.8) has a unique solution α^* in $(0, 1)$. If β is differentiable at α^*, then q^* is also a unique solution to (3.9). The optimal strategies in the game G_{α^*} are usually unique as well. At any rate, they are equivalent since the value $\beta(\alpha^*)$ of the game is unique.

Theorem 3.1 shows that the inspection game is in effect 'decomposed' into two games which can be treated separately, such that the solution of one game becomes a parameter of the second game. Technically, this decomposition works as follows: solve the non-zero-sum game in which the strategies are $\alpha \in [0, 1]$ (false alarm probability) for the inspector, $q \in [0, 1]$ (probability for acting illegally) for the operator and in which the payoff functions are given by (3.2). The equilibrium (α^*, q^*) is determined by Equations (3.8) and (3.9) and α^* becomes a parameter of the zero-sum game G_{α^*}. The solution of G_{α^*} is given by the optimal test (of significance α^*) and the optimal diversion strategy.

The non-zero-sum game which determines the false alarm and violation probabilities is of a 'political' nature. The significance of the numerical values of these probabilities, thus, may be debated by practitioners since the latter depend on payoffs which are highly questionable: they are hard to measure since they are supposed to quantify political effects. The zero-sum game G_{α^*}, which we may refer to as the non-detection game, is of a technical nature, devoid of political considerations: given that the operator has decided to act illegally and the inspector has decided to restrict himself to false alarm probability α^*, what is the optimal statistical test for the inspector and what is the optimal illegal behavior (for the operator)? This game is less debatable 'politically' but usually more challenging mathematically. In fact, many inspection games in the literature deal only with zero-sum models of this kind. Operationally, the most important tool for solving G_α is the *Neyman–Pearson lemma* [see, e.g., Lehmann (1959)]. We *assume* first that the game G_α has a value, which can therefore be written as

$$\beta(\alpha) = \max_{\omega \in \Omega} \min_{\delta \in \Delta_\alpha} \beta(\delta, \omega).$$

Hereby, the non-detection probability $\beta(\delta, \omega)$ for a given false alarm probability α is minimized, assuming that ω (and thus the alternative hypothesis) is fully known. The corresponding *best test* δ^* is described by the Neyman–Pearson lemma. This lemma provides a test procedure to decide between two *simple hypotheses*, that is, hypotheses which determine the respective probability distributions completely. With δ^* thus found, it is often possible to describe an optimal violation procedure ω^* against this test, which usually does not depend on α. Then, it remains to show that these strategies form an equilibrium of the game G_α. In the following, we will demonstrate the use of the Neyman–Pearson lemma with two important applications.

3.2. Material accountancy

Material accountancy procedures are designed to control rare, dangerous, or precious materials. In particular, the material balance concept is fundamental to IAEA safeguards, where the inspector watches for a possible diversion of nuclear material and the inspectee is the operator of a nuclear plant. We consider the case of periodic inventory measurements. Further details of this model are described in Avenhaus (1986, Section 3.3).

Consider a certain physical material balance area and a sequence of n *inventory periods*. At the beginning of the first period, the amount I_0 of material under control is measured in the area. During the ith period, $1 \leqslant i \leqslant n$, the known net amount S_i of material enters the area, and at the end of that period, the amount I_i is measured in the area. The quantity

$$Z_i = I_{i-1} + S_i - I_i, \quad 1 \leqslant i \leqslant n,$$

is called the material balance test statistic for the ith inventory period. Under H_0, its expected value is zero (because of the preservation of matter),

$$E_0(Z_i) = 0, \quad 1 \leqslant i \leqslant n. \tag{3.10}$$

Under H_1, its expected value is μ_i,

$$E_1(Z_i) = \mu_i, \quad 1 \leqslant i \leqslant n, \quad \sum_i \mu_i = \mu, \tag{3.11}$$

where μ_i is the amount of material diverted in the ith period, and where μ is the *total* amount of material to be diverted throughout all periods.

Given normally distributed measurement errors, the random (column) vector $Z = (Z_1, \ldots, Z_n)^\top$ is multivariate normally distributed with covariance matrix Σ, which is the same for H_0 and H_1. The elements of this matrix are given by

$$\mathrm{cov}(Z_i, Z_j) = \begin{cases} \mathrm{var}(Z_i) & \text{for } i = j, \\ -\mathrm{var}(I_i) & \text{for } i = j - 1, \\ -\mathrm{var}(I_j) & \text{for } j = i - 1, \\ 0 & \text{otherwise.} \end{cases}$$

The two hypotheses to be tested by the inspector are defined by the expectation vectors $E_0(\mathbf{Z})$ and $E_1(\mathbf{Z})$ as given by (3.10) and (3.11).

According to our general model, the 'best' test procedure is an equilibrium strategy of the inspector. By Theorem 3.1, this is an optimal strategy of the non-detection game $G_\alpha = (\Delta_\alpha, \Omega, \beta)$. In this zero-sum game, the inspector's set of strategies Δ_α is, as above, the set of test procedures with a fixed false alarm probability α. The inspectee's set of strategies Ω is the set of all diversion vectors $\boldsymbol{\mu} = (\mu_1, \ldots, \mu_n)^\top$ that have fixed total diversion μ, that is, fulfill $\mathbf{1}^\top \boldsymbol{\mu} = \mu$ with $\mathbf{1} = (1, \ldots, 1)^\top$. We assume – and will later show – that this game has an equilibrium. Furthermore, this equilibrium involves only pure strategies of the players.

For fixed diversion $\boldsymbol{\mu}$, the alarm set of the best test (the set of observations where the null hypothesis is rejected) is described by the Neyman–Pearson lemma. It is given by the set of observations $z = (z_1, \ldots, z_n)^\top$

$$\left\{ z \,\middle|\, \frac{f_1(z)}{f_0(z)} > \lambda \right\}, \tag{3.12}$$

where f_1 and f_0 are the densities of the random vector $\mathbf{Z} = (Z_1, \ldots, Z_n)^\top$ under H_1 respectively H_0. These are

$$f_1(z) = \frac{1}{\sqrt{(2\pi)^n \cdot |\boldsymbol{\Sigma}|}} \exp\left(-\frac{1}{2} \cdot (z - \boldsymbol{\mu})^\top \boldsymbol{\Sigma}^{-1} (z - \boldsymbol{\mu}) \right),$$

and the density f_0 has the same form with $\boldsymbol{\mu}$ replaced by the zero vector. Finally, λ is a parameter that is determined according to the false alarm probability α. This alarm set is equivalent to the set

$$\left\{ z \,\middle|\, \boldsymbol{\mu}^\top \boldsymbol{\Sigma}^{-1} z > \lambda' \right\} \tag{3.13}$$

for some λ', which means that the test statistic is the linear expression $\boldsymbol{\mu}^\top \boldsymbol{\Sigma}^{-1} \mathbf{Z}$.

Since $\mathbf{Z}$ is multivariate normally distributed, a linear combination $\boldsymbol{a}^\top \mathbf{Z}$ of its components, for any n-vector $\boldsymbol{a}$, is univariate normally distributed with expectation 0 under H_0 and $\boldsymbol{a}^\top \boldsymbol{\mu}$ under H_1, and variance $\boldsymbol{a}^\top \boldsymbol{\Sigma} \boldsymbol{a}$. Thus, the expected value of $\boldsymbol{\mu}^\top \boldsymbol{\Sigma}^{-1} \mathbf{Z}$ is

$$E\left(\boldsymbol{\mu}^\top \boldsymbol{\Sigma}^{-1} \mathbf{Z} \right) = \begin{cases} 0 & \text{under } H_0, \\ \boldsymbol{\mu}^\top \boldsymbol{\Sigma}^{-1} \boldsymbol{\mu} & \text{under } H_1; \end{cases}$$

and its variance

$$\mathrm{var}\left(\boldsymbol{\mu}^\top \boldsymbol{\Sigma}^{-1} \mathbf{Z} \right) = \boldsymbol{\mu}^\top \boldsymbol{\Sigma}^{-1} \boldsymbol{\mu}.$$

Surprisingly, mean and variance are equal under H_1. Therefore, the probability of detection as a function of the false alarm probability is given by

$$1 - \beta = \Phi\left(\sqrt{\mu^\top \Sigma^{-1}\mu} - q_{1-\alpha}\right) \tag{3.14}$$

where Φ denotes the Gaussian or normal distribution function and $q_{1-\alpha}$ the corresponding quantile (given by the inverse of Φ).

THEOREM 3.2. *The optimal strategy of the inspectee in the non-detection game G_α for material accountancy is*

$$\mu^* = \Sigma 1 \cdot \frac{\mu}{1^\top \Sigma 1}. \tag{3.15}$$

The value β^* of the game is given by the equation

$$1 - \beta^* = \Phi\left(\frac{\mu}{\sqrt{1^\top \Sigma 1}} - q_{1-\alpha}\right), \tag{3.16}$$

which describes the probability of detection in equilibrium. The optimal strategy δ^* of the inspector is based on the test statistic $1^\top Z$ with the alarm set $\{z \mid 1^\top z > \sqrt{1^\top \Sigma 1} \cdot q_{1-\alpha}\}$. The strategy pair δ^*, μ^* is an equilibrium of G_α.

PROOF. By (3.14), the optimal diversion vector μ^* minimizes $\mu^\top \Sigma^{-1}\mu$ for all μ subject to $1^\top \mu = \mu$. Since Σ^{-1} is symmetric, the gradient of this objective function is $2\mu^\top \Sigma^{-1}$, whereas the gradient of the constraint equation is $1^\top$. For the optimum $\mu = \mu^*$, these gradients are collinear with a suitable Lagrange multiplier, which means that μ^* is a multiple of $\Sigma 1$ as in (3.15). With $\mu = \mu^*$ in (3.14) one obtains (3.16). The test statistic used in the alarm set (3.13), suitably scaled, is $1^\top Z$. Since it has variance $1^\top \Sigma 1$, the alarm set corresponding to the false alarm probability α is $\{z \mid 1^\top z > \sqrt{1^\top \Sigma 1} \cdot q_{1-\alpha}\}$.

Finally, we show that these optimal strategies fulfill the equilibrium criterion

$$\beta(\delta^*, \mu) \leqslant \beta(\delta^*, \mu^*) \leqslant \beta(\delta, \mu^*) \quad \text{for all } \delta \in \Delta_\alpha, \ \mu \in \Omega.$$

Since we have constructed the Neyman–Pearson test for any diversion strategy μ, the right inequality is just the Neyman–Pearson lemma applied to μ^*. The left inequality holds for all μ as equality since the optimal test δ^* is based on the statistic $1^\top Z$ which is normally distributed with mean $1^\top \mu$ and variance $1^\top \Sigma 1$, that is, this distribution, and hence the detection probability, does not depend on the diversion vector μ, as long as $1^\top \mu = \mu$. $\qquad\square$

This solution deserves several remarks. Both players use *pure* strategies. In particular, the inspectee does not randomize over diversion plans. By proving the equilibrium property with the Neyman–Pearson lemma, we did not have to consider all possible tests δ of the inspector explicitly. Even defining the set of these strategies would be complicated. The optimal test procedure δ^* does not depend on the total diversion μ. This adds to the appeal of the model since the total diversion μ might not be known. It is a parameter of the non-detection game which is only relevant for the violation strategy in (3.15). The optimal test statistic of the inspector is

$$\mathbf{1}^\top \mathbf{Z} = \sum_{i=1}^n Z_i = I_0 + \sum_{i=1}^n S_i - I_n$$

which is the *overall material balance for all periods*. This means that the intermediate inventories are *ignored*. In fact, the same result is obtained if one considers the problem of subdividing a plant into several material balance areas: for the general payoff structure used so far, the equilibrium strategy of the inspector does *not* require subdividing the plant.

This example demonstrates the power of the Neyman–Pearson lemma of statistics in connection with the game-theoretic concept of an equilibrium. We should emphasize that, as in many game-theoretic problems, finding a pair of equilibrium strategies can be rather difficult, while it is usually easy to verify its properties.

3.3. Data verification

As already mentioned, IAEA safeguards are based on material accountancy, which derives from the physical principle of the preservation of matter. A second, operational principle of the IAEA is *data verification*. The inspector has to compare the material balance data reported by the plant operators (via their national authorities) with his own findings in order to verify that these data are not falsified for the purpose of concealing diversion of nuclear material. Since both sets of data, those of the operators and of the inspector, are based on measurements, statistical errors cannot be avoided. Furthermore, since one has to assume that the operators will – if at all – falsify their reported data strategically, decision problems arise which are very similar to those analyzed before. Again, the Neyman–Pearson lemma has to be applied, and again very interesting solutions, both from a theoretical and practical point of view, are obtained. We present an analytical model taken from Avenhaus (1986). It has a number of variations which have been solved in the literature or which are a topic of current research, as outlined below.

Assume that a nuclear plant is shut down and that the plant operator, the inspectee, has collected the nuclear material in K classes with N_i batches in the ith class, $i = 1, \ldots, K$. Typically, each class represents a particular type of measurement method. The operator has measured all contents of all batches and has reported the results to an inspector of the IAEA. The inspector, in turn, measures some of the material contents in order to compare them with the reported ones. His decision problem will be how to allocate his limited inspection effort. Before we can describe this, we have to make assumptions about the measurement statistics, and, as we will see, about the possible falsification of the data.

Let v_{ij} be the true value of the content of the jth batch of the ith class, and let e_{ij} be the operator's random measurement error. The measurement result x_{ij} can therefore be written as

$$x_{ij} = v_{ij} + e_{ij} \quad \text{for } i = 1, \dots, K, \ j = 1, \dots, N_i. \tag{3.17}$$

We assume that the random variables E_{ij} associated with the errors are independent and normally distributed with zero expectation and variances $\sigma_{O_i}^2$ that are specific to the class i and to the operator's instruments, written as

$$E_{ij} \sim N\left(0, \sigma_{O_i}^2\right) \quad \text{for } i = 1, \dots, K, \ j = 1, \dots, N_i.$$

Under the null hypothesis H_0, no data are falsified. Then, the data reported by the operator are given by (3.17). Under the alternative hypothesis H_1, the operator is assumed to falsify all batch data of one class by a class-specific positive amount μ_i, so that each batch in that class is reported to contain μ_i more material than it actually does. In that case, the reported data are given by

$$x_{ij} = v_{ij} + \mu_i + e_{ij} \quad \text{for } i = 1, \dots, K, \ j = 1, \dots, N_i.$$

Of course, this is a very restrictive assumption about the falsification that takes place. It may be justified with practical arguments, for example manipulation of the calibration of the measuring instruments. More general assumptions are very difficult to handle. We will mention approaches in that direction below.

The inspector verifies n_i randomly chosen items in the ith class with the help of his own measurements. Without loss of generality, let these be the first n_i ones. Then, the inspector's findings y_{ij} are

$$y_{ij} = v_{ij} + d_{ij} \quad \text{for } i = 1, \dots, K, \ j = 1, \dots, n_i.$$

The random variables D_{ij} associated to the inspector's errors d_{ij} are independent and normally distributed with zero expectation and variances $\sigma_{I_i}^2$,

$$D_{ij} \sim N\left(0, \sigma_{I_i}^2\right) \quad \text{for } i = 1, \dots, K, \ j = 1, \dots, n_i.$$

The inspector is *not* interested in estimating the true values v_{ij} but only in verifying the reported data. Therefore, he uses the differences

$$z_{ij} = x_{ij} - y_{ij} \quad \text{for } i = 1, \dots, K, \ j = 1, \dots, n_i$$

for constructing his verification test. According to our assumptions, we have

$$Z_{ij} \sim N(\mu_{it}, \sigma_i^2) \quad \text{for } i = 1, \ldots, K, \; j = 1, \ldots, n_i, \; t = 0, 1,$$

where $\sigma_i^2 = \sigma_{O_i}^2 + \sigma_{I_i}^2$, and where

$$\mu_{it} = \begin{cases} 0 & \text{for } t = 0 \; (H_0), \\ \mu_i & \text{for } t = 1 \; (H_1). \end{cases}$$

The inspector uses the Neyman–Pearson test. As above in (3.12), its alarm set is with $z = (z_{11}, \ldots, z_{Kn_K})$ given by

$$\left\{ z \; \middle| \; \frac{f_1(z)}{f_0(z)} > \lambda \right\},$$

where for $t = 0, 1$

$$f_t(z) = \left(\prod_{i=1}^{K} \frac{1}{\left(\sqrt{2\pi} \sigma_i \right)^{n_i}} \right) \cdot \exp\left(-\frac{1}{2} \sum_{i=1}^{K} \sum_{j=1}^{n_i} \frac{(z_{ij} - \mu_{it})^2}{\sigma_i^2} \right).$$

Therefore, the alarm set is explicitly given by

$$\left\{ z \; \middle| \; \sum_{i=1}^{K} \frac{\mu_i}{\sigma_i^2} \sum_{j=1}^{n_i} z_{ij} > \lambda' \right\}.$$

Now we have under H_t, $t = 0, 1$,

$$\sum_{i=1}^{K} \frac{\mu_i}{\sigma_i^2} \sum_{j=1}^{n_i} Z_{ij} \sim N\left(\sum_{i=1}^{K} n_i \frac{\mu_{it}^2}{\sigma_i^2}, \; \sum_{i=1}^{K} n_i \frac{\mu_i^2}{\sigma_i^2} \right).$$

Thus, the probabilities of error of the first and second kind are given by

$$\alpha = 1 - \Phi\left(\frac{\lambda'}{\sqrt{\sum_{i=1}^{K} n_i \frac{\mu_i^2}{\sigma_i^2}}} \right) \quad \text{and} \quad \beta = \Phi\left(\frac{\lambda' - \sum_{i=1}^{K} n_i \frac{\mu_i^2}{\sigma_i^2}}{\sqrt{\sum_{i=1}^{K} n_i \frac{\mu_i^2}{\sigma_i^2}}} \right).$$

If we eliminate λ' with the help of α, then we get the non-detection probability

$$\beta = 1 - \Phi\left(\sqrt{\sum_{i=1}^{K} n_i \frac{\mu_i^2}{\sigma_i^2}} - q_{1-\alpha} \right). \tag{3.18}$$

In order to describe the strategy sets of the players, let us assume that the operator wants to falsify all data by the fixed total amount μ. Furthermore, let one measurement by the inspector in the ith class, $i = 1, \ldots, K$, cost the effort ε_i, where the total effort available, measured in monetary terms or inspection hours, is ε. This means, according to Theorem 3.1, that we consider the following non-detection game G_α.

THEOREM 3.3. *Let G_α be the zero-sum game in which the sets of strategies for inspector and inspectee are, respectively,*

$$\left\{ \boldsymbol{n} = (n_1, \ldots, n_K) \,\middle|\, \sum_{i=1}^{K} \varepsilon_i n_i = \varepsilon \right\} \quad \text{and} \quad \left\{ \boldsymbol{\mu} = (\mu_1, \ldots, \mu_K) \,\middle|\, \sum_{i=1}^{K} N_i \mu_i = \mu \right\}.$$

The payoff to the inspectee is $\beta(\boldsymbol{n}, \boldsymbol{\mu})$ defined by (3.18). Considering $\boldsymbol{n}$ as a vector of continuous variables, this game has an equilibrium $\boldsymbol{n}^*, \boldsymbol{\mu}^*$, given by

$$n_i^* = \frac{\varepsilon}{\sum_{j=1}^{K} N_j \sigma_j \sqrt{\varepsilon_j}} \cdot \frac{N_i \sigma_i}{\sqrt{\varepsilon_i}},$$

$$\qquad\qquad\qquad\qquad\qquad\qquad\qquad\qquad \text{for } i = 1, \ldots, K. \qquad\qquad (3.19)$$

$$\mu_i^* = \frac{\mu}{\sum_{j=1}^{K} N_j \sigma_j \sqrt{\varepsilon_j}} \cdot \sigma_i \sqrt{\varepsilon_i},$$

The value of the game is

$$\beta(\boldsymbol{n}^*, \boldsymbol{\mu}^*) = 1 - \Phi\left(\frac{\mu \sqrt{\varepsilon}}{\sum_{j=1}^{K} N_j \sigma_j \sqrt{\varepsilon_j}} - q_{1-\alpha} \right).$$

PROOF. The equilibrium conditions say that for all $\boldsymbol{n}, \boldsymbol{\mu}$ in the strategy sets

$$\beta(\boldsymbol{n}^*, \boldsymbol{\mu}) \leqslant \beta(\boldsymbol{n}^*, \boldsymbol{\mu}^*) \leqslant \beta(\boldsymbol{n}, \boldsymbol{\mu}^*)$$

holds. By (3.18), this is equivalent to

$$\sum_{i=1}^{K} n_i^* \frac{\mu_i^2}{\sigma_i^2} \geqslant \sum_{i=1}^{K} n_i^* \frac{\mu_i^{*2}}{\sigma_i^2} \geqslant \sum_{i=1}^{K} n_i \frac{\mu_i^{*2}}{\sigma_i^2} \quad \text{for all } \boldsymbol{n}, \boldsymbol{\mu}.$$

The right inequality is identically fulfilled. Since $\sum_{i=1}^{K} N_i \mu_i = \mu$, the left inequality is just a special case of the Cauchy–Schwarz inequality $(\sum a_i^2)(\sum b_i^2) \geqslant (\sum a_i b_i)^2$ with $a_i^2 = N_i \mu_i^2 / \sigma_i \sqrt{\varepsilon_i}$ and $b_i^2 = N_i \sigma_i \sqrt{\varepsilon_i}$. $\square$

The sampling design defined by (3.19) is known in the statistical literature as *Neyman–Chuprov* sampling [see, e.g., Cochran (1963)]. The above model provides a game-theoretic justification for this sampling design.

This data verification problem has a lot of ramifications if one assumes more general falsification strategies. It may be unrealistic to assume that all batches in class i are falsified by the same amount μ_i. In particular, this is the case for a high total diversion μ that can no longer be hidden in measurement errors for the individual batches. In that case, it turns out that the inspectee will concentrate his falsification on as few batches as possible, in the hope that these will not be measured by the inspector. A theoretical analysis of the general case is very difficult. Some approaches in this direction have been made for unstratified problems, that is, only a single class of batches; see Avenhaus, Battenberg and Falkowski (1991), Mitzrotsky (1993), Avenhaus and Piehlmeier (1994), Piehlmeier (1996), and Battenberg and Falkowski (1998).

4. Sequential inspections

Certain inspections evolve over a number of stages. The number of inspections may be limited, and statistical errors may occur. The models presented in this section vary with respect to the information of the players about the course of the game, the rules for detecting a violation, and the resulting payoffs.

4.1. Recursive inspection games

We consider sequential inspections with a finite number of inspection stages. Their models can be defined recursively and their solution is given by a recurrence equation, which in some cases is solved explicitly. The recursive description implies that each player learns about the action of the other player after each stage, which is not always justified. This is not problematic if the inspectee can violate at most once, so we consider this case first.

The inspection game introduced by Dresher (1962) has n stages. At each stage, the inspector may or may not use an inspection, with a total of m inspections permitted for all stages. The inspectee may decide at a stage to act legally or illegally, and will not perform more than one illegal act throughout the game. Illegal action is detected if and only if there is an inspection at the same stage.

Dresher modeled this game recursively, assuming zero-sum payoffs. For the inspector, this payoff is one unit for a detected violation, zero for legal action throughout the game, and a loss of one unit for an undetected violation. The value of the game, the equilibrium payoff to the inspector, is denoted by $I(n, m)$ for the parameters $0 \leqslant m \leqslant n$. For $m = n$, the inspector will inspect at every stage and the inspectee will act legally, and similarly the decision is unique for $m = 0$ where the inspectee can safely violate, so that

$$I(n, n) = 0 \quad \text{for } n \geqslant 0 \qquad \text{and} \qquad I(n, 0) = -1 \quad \text{for } n > 0. \tag{4.1}$$

For $0 < m < n$, the game is represented by the recursive payoff matrix as shown in Figure 2. The rows denote the possible actions at the first stage for the inspector and the

Inspector \ Inspectee	legal action	violation
inspection	$I(n-1, m-1)$	1
no inspection	$I(n-1, m)$	-1

Figure 2. The Dresher game, showing the decisions at the first of n stages, with at most one intended violation and m inspections, for $0 < m < n$. The game has value $I(n, m)$. The recursively defined entries denote the payoffs to the inspector.

columns those for the inspectee. If the inspectee violates, then he is either caught, where the game terminates and the inspector receives 1, or not, after which he will act legally throughout, so that the game eventually terminates with payoff -1 to the inspector. After a legal action of the inspectee, the game continues as before, with $n - 1$ instead of n stages and $m - 1$ or m inspections left.

In this game, it is reasonable to assume a circular structure of the players' preferences. That is, the inspector prefers to use his inspection if and only if the inspectee violates, who in turn prefers to violate if and only if the inspector does not inspect. This means that the game has a unique mixed equilibrium point where both players choose both of their actions with positive probability. Hence, if the inspector's probability for inspecting at the first stage is p, then the inspectee must be indifferent between legal action and violation, that is,

$$p \cdot I(n-1, m-1) + (1-p) \cdot I(n-1, m) = p + (1-p) \cdot (-1). \tag{4.2}$$

Both sides of this equation denote the game value $I(n, m)$. Solving for p and substituting yields

$$I(n, m) = \frac{I(n-1, m) + I(n-1, m-1)}{I(n-1, m) + 2 - I(n-1, m-1)}. \tag{4.3}$$

With this recurrence equation for $0 < m < n$ and the initial conditions (4.1), the game value is determined for all parameters. Dresher showed that Equations (4.1) and (4.3) have an explicit solution, namely

$$I(n, m) = -\binom{n-1}{m} \Big/ \sum_{i=0}^{m} \binom{n}{i}.$$

The *payoffs* in Dresher's model have been generalized by Höpfinger (1971). He assumed that the inspectee's gain for a successful violation need not equal his loss if he is caught, and also solved the resulting recurrence equation explicitly. Furthermore, zero-sum payoffs are not fully adequate since a caught violation, compared to legal action

Inspectee / Inspector	legal action	violation
inspection	$V(n-1, m-1)$ $I(n-1, m-1)$	$-b$ $-a$
no inspection	$V(n-1, m)$ $I(n-1, m)$	1 -1

Figure 3. Non-zero-sum game at the first of n stages with m inspections, $0 < m < n$, and equilibrium payoff $I(n, m)$ to the inspector and $V(n, m)$ to the inspectee. Legal action throughout the game has reference payoff zero to both players. A caught violation yields negative payoffs to both, where $0 < a < 1$ and $0 < b$.

throughout, is usually undesirable for *both* players since for the inspector this demonstrates a failure of his surveillance system. A non-zero-sum game that takes account of this is shown in Figure 3. There, $I(n, m)$ and $V(n, m)$ denote the equilibrium payoff to the inspector and inspectee, respectively ('V' indicating a potential violator), and a and b are positive parameters denoting the losses to both players for a caught violation, where $a < 1$. Analogous to (4.1), the cases $m = n$ and $m = 0$ are described by

$$
\begin{aligned}
I(n, n) &= 0, \\
V(n, n) &= 0,
\end{aligned}
\quad \text{for } n \geqslant 0 \quad \text{and} \quad
\begin{aligned}
I(n, 0) &= -1, \\
V(n, 0) &= 1,
\end{aligned}
\quad \text{for } n > 0. \tag{4.4}
$$

THEOREM 4.1. *The recursive inspection game in Figure 3 with initial conditions* (4.4) *has a unique equilibrium. The inspectee's equilibrium payoff is given by*

$$
V(n, m) = \binom{n-1}{m} \bigg/ \sum_{i=0}^{m} \binom{n}{i} b^{m-i} \tag{4.5}
$$

and the inspector's payoff by

$$
I(n, m) = -\binom{n-1}{m} \bigg/ \sum_{i=0}^{m} \binom{n}{i} (-a)^{m-i}. \tag{4.6}
$$

The equilibrium strategies are determined inductively, using (4.5) *and* (4.6), *by solving the game in Figure* 3.

PROOF. As an inductive hypothesis, assume that the player's payoffs in Figure 3 are such that the game has a unique mixed equilibrium; this is true for $n = 2$, $m = 1$

by (4.4). The inspectee is then indifferent between legal action and violation. As in (4.2) and (4.3), this gives the recurrence equation

$$V(n,m) = \frac{b \cdot V(n-1,m) + V(n-1,m-1)}{V(n-1,m-1) + b + 1 - V(n-1,m)}$$

for $V(n,m)$, and a similar expression for $I(n,m)$. The verification of (4.5) and (4.6) is shown by Avenhaus and von Stengel (1992), which generalizes and simplifies the derivations by Dresher (1962) and Höpfinger (1971). The assumed inequalities about the mixed equilibrium can be proved using the explicit representations. $\square$

The same payoffs, although with a different normalization, have already been considered by Maschler (1966). He derived an expression for the payoff to the inspector that is equivalent to (4.6), considering the appropriate linear transformations for the payoffs. Beyond this, Maschler introduced the *leadership* concept where the inspector announces and commits himself to his strategy in advance. Interestingly, this improves his payoff in the non-zero-sum game. We will discuss this in detail in Section 5.

The recursive description in Figures 2 and 3 assumes that all players know about the game state after each stage. This is usually not true for the inspector after a stage where he did not inspect, since then he will not learn about a violation. The constant payoffs after a successful violation indicate that the game terminates. In fact, the game continues, but any further action of the inspector is irrelevant since then the inspectee will only behave legally.

As soon as the the players' information is such that the recursive structure is not valid, the problem becomes difficult, as already noted by Kuhn (1963). Therefore, even natural and simple-looking extensions of Dresher's model are still open problems. Von Stengel (1991) shows how the recursive approach is still possible with special assumptions. Similar sequential games with three parameters, where the game continues even after a detected violation, are described by Sakaguchi (1977, 1994) and Ferguson and Melolidakis (1998). The inspector's full information after each stage is not questioned in these models.

In the situation studied by von Stengel (1991) the inspectee intends at most k violations and can violate at most once per stage. The inspectee collects one unit for each successful violation. As before, n and m denote the number of stages and inspections. Like Dresher's game, the game is zero-sum, with value $I(n,m,k)$ denoting the payoff to the inspector. That value is determined if all remaining stages either will or will not be inspected, or if the inspectee does not intend to violate:

$$I(n,n,k) = 0, \qquad I(n,0,k) = -\min\{n,k\}, \qquad I(n,m,0) = 0. \tag{4.7}$$

If the inspector detects a violation, the game terminates and the inspectee has to pay a positive penalty b to the inspector for being caught, but the inspectee does not have to return the payoff he got for previous successful violations. This means that at a stage the

Inspector \ Inspectee	legal action	violation
inspection	$I(n-1, m-1, k)$	b
no inspection	$I(n-1, m, k)$	$I(n-1, m, k-1) - 1$

Figure 4. First stage of a zero-sum game with n stages, m inspections, and up to k intended violations, for $0 < m < n$. The inspectee collects one unit for each successful violation, which he can keep even if he is caught later and has to pay the nonnegative amount b. The inspector is informed after each stage, even with no inspection, if there was a violation or not.

inspector does not inspect, a violation reduces k to $k - 1$ and a payoff of 1 is 'credited' to the inspectee, and legal action leaves k unchanged. Furthermore, it is *assumed* that even after a stage with no inspection, it becomes common knowledge whether there was a violation or not. This leads to a game with recursive structure, shown in Figure 4.

THEOREM 4.2. *The recursive inspection game in Figure 4 with initial conditions* (4.7) *has value*

$$I(n, m, k) = \frac{\binom{n-k}{m+1} - \binom{n}{m+1}}{s(n, m)} \quad \text{where } s(n, m) = \sum_{i=0}^{m} \binom{n}{i} b^{m-i}.$$

The probability p of inspection at the first stage is $p = s(n-1, m-1)/s(n, m)$.

PROOF. Analogously to (4.3), one obtains a recurrence equation for $I(n, m, k)$, which has this solution shown by von Stengel (1991). □

In this game, the inspector's equilibrium payoff $I(n, m, k)$ decreases with the number k of intended violations. However, that payoff is constant for all $k \geqslant n - m$ since then $\binom{n-k}{m+1} = 0$. Indeed, the inspectee will not violate more than $n - m$ times since otherwise he would be caught with certainty.

Theorem 4.2 asserts that the the probability p of inspection at the first stage, determined analogously to (4.2), is given by $p = s(n-1, m-1)/s(n, m)$ and thus *does not depend on k*. This means that the inspector *need not know* the value of k in order to play optimally. Hence, the assumption in this game about the knowledge of the inspector after a period without inspection can be removed: the solution of the recursive game is also the solution of the game *without* recursive structure, where – more realistically – the inspector does not know what happened at a stage without inspection. This is analyzed by von Stengel (1991) using the extensive form of the games.

Some sequential models take into account statistical errors of the first and second kind [Maschler (1967); Thomas and Nisgav (1976); Baston and Bostock (1991); Rinderle (1996)]. Explicit solutions have been found only for special cases.

4.2. Timeliness games

In certain situations, the inspector uses a number of inspections during some time interval in order to detect a single violation. This violation is detected at the earliest *subsequent* inspection, and the *time* that has elapsed in between is the payoff to the inspectee that the inspector wants to minimize. The inspectee may or may not observe the inspections.

In reliability theory, variants of this game have been studied by Derman (1961) and Diamond (1982). An operating unit may fail which creates a cost that increases with the time until the failure is detected. The overall time interval represents the time between normal replacements of the unit. A minmax analysis leads to a zero-sum game with the operating unit as inspectee which cannot observe any inspections. A more common approach in reliability theory, which is not our topic, is to assume some knowledge about the distribution of the failure time.

Another application is interim inspections of direct-use nuclear material [see, e.g., Avenhaus, Canty and von Stengel (1991)]. Nuclear plants are regularly inspected, say at the end of the year. If nuclear material is diverted, one may wish to discover this not after a year but earlier, which is the purpose of interim inspections. One may assume either that the inspectee can violate just after he has observed an interim inspection, or that such observations or alterations of his diversion plan are not possible.

The players choose their actions from the unit time interval. The inspectee violates at some time s in $[0, 1)$. The inspector has m inspections ($m \geqslant 1$) that he uses at times $t_1, \ldots, t_m$, where his last inspection must take place at the end of the interval, $t_m = 1$. He can choose the other inspection times freely, with $0 \leqslant t_1 \leqslant \cdots \leqslant t_{m-1} \leqslant t_m = 1$. The violation is discovered at the earliest inspection following the violation time s. Then, the time elapsed is the payoff to the inspectee given by the function

$$V(s, t_1, \ldots, t_{m-1}) = t_k - s \quad \text{for } t_{k-1} \leqslant s < t_k, \ k = 1, \ldots, m, \tag{4.8}$$

where $t_0 = 0$ and $t_m = 1$. It is useful to assume that if the times of inspection and violation coincide, $t_{k-1} = s$, then the violation is not detected until the next inspection at time t_k. The inspector's payoff is the negative of the payoff to the inspectee.

If the inspector deterministically inspects at times $t_i = i/m$ for $i = 1, \ldots, m$, a violation is discovered at most after time $1/m$ has elapsed. This is indeed an optimal inspection strategy if the inspectee can *observe* the inspections. In that case, the inspectee will violate *immediately* after the kth inspection assuming he can react without delay, where k is uniformly chosen from $\{0, \ldots, m-1\}$. That is, the violation time s is uniformly chosen from $\{t_0, t_1, \ldots, t_{m-1}\}$. Note that s is conditional upon the inspector's action. Then by (4.8) the expected time after the violation is discovered is $\frac{1}{m}(t_1 - t_0) + \frac{1}{m}(t_2 - t_1) + \cdots + \frac{1}{m}(t_m - t_{n-1}) = \frac{1}{m}(t_m - t_0) = 1/m$.

A more involved model of such observable inspections is studied by Derman (1961). Each interim inspection detects a violation only with a certain fixed probability. In addition to the time until detection, there is a cost to be paid for each inspection. Derman describes a deterministic minmax schedule for the inspection times where t_i is not proportional to i as before with $t_i = i/m$, but a quadratic function of i, so that the inspections accumulate towards the end of the time interval because of their cost.

If the inspections *cannot be observed*, then the violation time s and the free inspection times $t_1, \ldots, t_{m-1}$ are chosen simultaneously. This is an appropriate assumption for inspecting an unreliable unit if the inspections do not interfere with the operation of the system. This game is completely solved by Diamond (1982). We will demonstrate the solution of the first nontrivial case $m = 2$.

For $m = 2$, the inspector has one free inspection at time t_1, for short t, chosen from $[0, 1]$. The inspectee may select his violation at time $s \in [0, 1)$. This is a zero-sum game over the unit square with payoff $V(s, t)$ to the inspectee where, by (4.8),

$$V(s, t) = \begin{cases} t - s & \text{if } s < t, \\ 1 - s & \text{if } s \geqslant t. \end{cases} \tag{4.9}$$

This describes a kind of *duel* with time reversed where both players have an incentive to act early but after the other. By (4.9), the inspectee's payoff is too small if he violates too late, so that he will select s with a certain probability distribution from an interval $[0, b]$ where $b < 1$. Consequently, the inspector will not inspect later than b.

THEOREM 4.3. *The zero-sum game over the unit square with payoff $V(s, t)$ in (4.9) has the following solution. The inspector chooses the inspection time t from $[0, b]$ according to the density function $p(t) = 1/(1 - t)$, where $b = 1 - 1/e$. The inspectee chooses his violation time s according to the distribution function $Q(s) = 1/e(1 - s)$ for $s \in [0, b]$, and $Q(s) = 1$ for $s > b$. The value of the game is $1/e$.*

PROOF. The inspector chooses $t \in [0, b]$ according to a density function p, where

$$\int_0^b p(t) \, dt = 1. \tag{4.10}$$

The *expected* payoff $V(s)$ to the inspectee for $s \in [0, b]$ is then given as follows, taking into account that the inspection may take place before or after the violation:

$$\begin{aligned} V(s) &= \int_0^s (1 - s) p(t) \, dt + \int_s^b (t - s) p(t) \, dt \\ &= \int_0^s p(t) \, dt + \int_s^b t p(t) \, dt - s \int_0^b p(t) \, dt. \end{aligned}$$

If the inspectee randomizes as described, then this payoff must be constant [see Karlin (1959, Lemma 2.2.1, p. 27)]. That is, its derivative with respect to s, which by (4.10) is given by $p(s) - s p(s) - 1$, should be zero. The density for the inspection time is therefore given by

$$p(t) = \frac{1}{1 - t},$$

with b in (4.10) given by $b = 1 - 1/e$. The constant expected payoff to the violator is then $V(s) = 1/e$. For $s > b$, the inspectee's payoff is $1 - s$ which is smaller than $1/e$ so that he will indeed not violate that late.

The optimal distribution of the violation time has an atom at 0 and a density q on the remaining interval $(0, b]$. We consider its distribution function $Q(s)$ denoting the probability that a violation takes place at time s or earlier. The mentioned atom is $Q(0)$, the derivative of Q is q. The resulting expected payoff $-V(t)$ for $t \in [0, b]$ to the inspector is given by

$$
\begin{aligned}
V(t) &= t Q(0) + \int_0^t (t - s) q(s) \, ds + \int_t^b (1 - s) q(s) \, ds \\
&= t Q(0) + t \int_0^t q(s) \, ds + \int_t^b q(s) \, ds - \int_0^b s q(s) \, ds \\
&= t Q(t) + Q(b) - Q(t) - \int_0^b s q(s) \, ds.
\end{aligned}
$$

Again, the inspector randomizes only if this is a constant function of t, which means that $(t - 1) \cdot Q(t)$ is a constant function of t. Thus, because $Q(b) = Q(1 - 1/e) = 1$, the distribution function of the violation time s is for $s \in [0, b]$ given by

$$
Q(s) = \frac{1}{e} \cdot \frac{1}{1 - s},
$$

and for $s > b$ by $Q(s) = 1$. The nonzero atom is given by $Q(0) = 1/e$. $\qquad \square$

Unsurprisingly, the value $1/e$ of the game with unobservable inspections is better for the inspector than the value $1/2$ of the game with observable inspections.

The solution of the game for $m > 2$ is considerably more complex and has the following features. As before, the inspectee violates with a certain probability distribution on an interval $[0, b]$, leaving out the interval $(b, 1)$ at the end. The inspections take place in *disjoint intervals*. The inspection times are random, but it is a randomization over a *one-parameter family* of pure strategies where the inspections are fully correlated. The optimal violation scheme of the inspectee is given by a distribution function which is piecewise defined on the $m - 1$ time intervals and has an atom at the beginning of the game. For large m, the inspection times are rather uniformly distributed in their intervals and the value of the game rapidly approaches $1/(2m - \frac{4}{3})$ from above. Based on this complete analytic solution, Diamond (1982) also demonstrates computational solution methods for non-linear loss functions.

5. Inspector leadership

The leadership principle says that it can be advantageous in a competitive situation to be the first player to select and stay with a strategy. It was suggested first by von

Stackelberg (1934) for pricing policies. Maschler (1966) applied this idea to sequential inspections. The notion of leadership consists of two elements: the ability of the player first to *announce* his strategy and make it known to the other player, and second to *commit* himself to playing it. In a sense, it would be more appropriate to use the term *commitment power*. This concept is particularly suitable for inspection games since an inspector can credibly announce his strategy and stick to it, whereas the inspectee cannot do so if he intends to act illegally. Therefore, it is reasonable to assume that the inspector will take advantage of his leadership role.

5.1. Definition and introductory examples

A leadership game is constructed from a simultaneous game which is a game in normal form. Let the players be I and II with strategy sets Δ and Ω, respectively. For simplicity, we assume that Δ is closed under the formation of mixed strategies. In particular, Δ could be the set of mixed strategies over a finite set. The two players select simultaneously their respective strategies δ and ω and receive the corresponding payoffs, defined as expected payoffs if δ originally represents a mixed strategy.

In the leadership version of the game, one of the players, say player I, is called the *leader*. Player II is called the *follower*. The leader chooses a strategy δ in Δ and makes this choice known to the follower, who then chooses a strategy ω in Ω which will typically depend on δ and is denoted by $\omega(\delta)$. The payoffs to the players are those of the simultaneous game for the pair of strategies $(\delta, \omega(\delta))$. The strategy δ is executed without giving the leader an opportunity to reconsider it. This is the *commitment power* of the leader.

The simplest way to construct the leadership version is to start with the simultaneous game in extensive form. Player I moves first, and player II has a single information set since he is not informed about that move, and moves second. In the leadership game, that information set is dissolved into singletons, and the rest of the game, including payoffs, stays the same, as indicated in Figure 5.

The leadership game has perfect information because the follower is precisely informed about the announced strategy δ. He can therefore choose a best response $\omega(\delta)$ to δ. We assume that he always does this, even if δ is not a strategy announced by the leader in equilibrium. That is, we only consider *subgame perfect* equilibria. Any randomized strategy of the follower can be described as a *behavior strategy* defined by a 'local' probability distribution on the set Ω of choices of the follower for each announced strategy δ.

In a zero-sum game which has a value, leadership has no effect. This is the essence of the minmax theorem: each player can guarantee the value even if he announces his (mixed) strategy to his opponent and commits himself to playing it.

In a non-zero-sum game, leadership can sometimes serve as a coordination device and as a method for equilibrium selection. The game in Figure 6, for example, has two equilibria in pure strategies. There is no rule to select either equilibrium if the players are in symmetric positions. In the leadership game, if player I is made a leader, he will

Simultaneous game

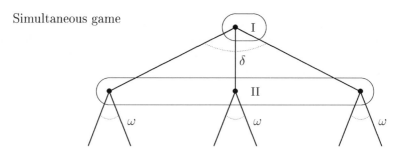

Leadership version

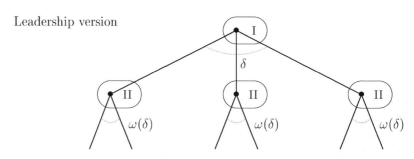

Figure 5. The simultaneous game and its leadership version in extensive form. The strategies δ of player I include all mixed strategies. In the simultaneous game, the choice ω of player II is the same irrespective of δ. In the leadership version, player I is the leader and moves first, and player II, the follower, may choose his strategy $\omega(\delta)$ depending on the announced leader strategy δ. The payoffs in both games are the same.

	L	R
T	8 / 9	0 / 0
B	0 / 0	9 / 8

Figure 6. Game where the unique equilibrium (T, L) of the leadership version is one of several equilibria in the simultaneous game.

select T, and player II will follow by choosing L. In fact, (T, L) is the *only* equilibrium of the leadership game. Player II may even consider it advantageous that player I is a leader and accept the payoff 8 in this equilibrium (T, L) instead of 9 in the other pure strategy equilibrium (B, R) as a price for avoiding the undesirable payoff 0.

However, a simultaneous game and its leadership version may have different equilibria. The simultaneous game in Figure 7 has a unique equilibrium in mixed strategies:

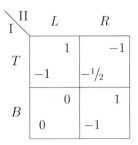

Figure 7. A simultaneous game with a unique mixed equilibrium. Its leadership version has a unique equilibrium where the leader announces the same mixed strategy, but the follower changes his strategy to the advantage of the leader.

player I plays T with probability $\frac{1}{3}$ and B with probability $\frac{2}{3}$, and player II plays L with probability $\frac{1}{3}$ and R with probability $\frac{2}{3}$. The resulting payoffs are $-\frac{2}{3}$, $\frac{1}{3}$.

In the leadership version, player I as the leader uses the *same* mixed strategy as in the simultaneous game, and player II gets the *same* payoff, but as a follower he will act to the *advantage* of the leader, for the following reason. Consider the possible announced mixed strategies, given by the probability p that the leader plays T. If player II responds with L or R, the payoffs to player I are $-p$ and $p/2 - 1$, respectively. When player II makes these choices, his own payoffs are p and $1 - 2p$, respectively. He therefore chooses R for $p < \frac{1}{3}$, L for $p > \frac{1}{3}$, and is indifferent for $p = \frac{1}{3}$. The resulting payoff to the leader as a function of p is shown in Figure 8. The leader tries to maximize this payoff, which he achieves only if the follower plays L. Thus, player I announces any p that is greater than or equal to $\frac{1}{3}$ but as small as possible. This means announcing

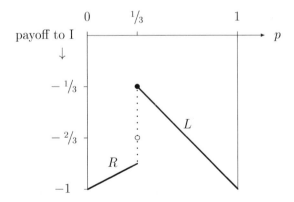

Figure 8. Payoff to player I in the game in Figure 7 as a function of the probability p that he plays T. It is assumed that player II uses his best response (L or R), which is unique except for $p^* = \frac{1}{3}$. Player I's equilibrium payoff in the simultaneous game is indicated by ○, that in the leadership version by ●.

exactly $\frac{1}{3}$, as pointed out by Avenhaus, Okada, and Zamir (1991): if the follower, who is then indifferent, were not to choose L with certainty, then it is easy to see that the leader could announce a slightly larger p, thus forcing the follower to play L and improve his payoff, in contradiction to the equilibrium property. Thus, the unique equilibrium of the leadership game is that player I announces $p^* = \frac{1}{3}$ and player II responds with L and deterministically, as described, for all other announcements of p, which are not materialized.

5.2. Announced inspection strategy

Inspection problems are a natural case for leadership games since the inspector can make his strategies public. The inspectee cannot announce that he intends to violate with some positive probability. The example in Figure 7 demonstrates the effect of leadership in an inspection game, with the inspector as player I and the inspectee as player II. This game is a special case of the recursive game in Figure 3 for two stages ($n = 2$), where the inspector has one inspection ($m = 1$). Inspecting at the first stage is the strategy T, and not inspecting, that is, inspecting at the second stage, is represented by B. For the inspectee, L and R refer to legal action and violation at the first stage, respectively. The losses to the players in the case of a caught violation in Figure 3 have the special form $a = \frac{1}{2}$ and $b = 1$.

In this leadership game, we have shown that the inspector announces his equilibrium strategy of the simultaneous game, but receives a better payoff. By the utility assumptions about the possible outcomes of an inspection, that better payoff is achieved by *legal* behavior of the inspectee. It can be shown that this applies not only to the simple game in Figure 7, but to the recursive game with general parameters n and m in Figure 3. However, the inspectee behaves legally only in the two-by-two games encountered in the recursive definition: if the inspections are used up ($n > 0$ but $m = 0$), then the inspectee can and will safely violate. Except for this case, the inspectee behaves legally if the inspector may announce his strategy.

The definition of this particular leadership game with general parameters n and m is due to Maschler (1966). However, he did not determine an equilibrium. In order to construct a solution, Maschler postulated that the inspectee acts to the advantage of the inspector when he is indifferent, and called this behavior 'Pareto-optimal'. Then he argued, as we did with the help of Figure 8, that the inspector can announce an inspection probability p that is slightly higher than p^*. In that way, the inspector is on the safe side, which should also be recommended in practice. Maschler (1967) uses leadership in a more general model involving several detectors that can help the inspector determine under which conditions to inspect.

Despite the advantage of leadership, announcing such a precisely calibrated inspection strategy looks risky. Above, the equilibrium strategy $p^* = \frac{1}{3}$ depends on the payoffs of the inspectee which might not be fully known to the inspector. Therefore, Avenhaus, Okada, and Zamir (1991) considered the leadership game for a simultaneous game with *incomplete information*. There, the gain to the inspectee for a successful violation is a payoff in some range with a certain probability distribution assumed by the inspector. In that leadership game, unlike in Figure 8, the inspector maximizes over a continuous payoff curve. He announces an inspection probability that just about forces the inspectee to act legally for *any* value of the unknown penalty b. This strategy has a higher equilibrium probability p^* and is safer than that used in the simultaneous game.

The simple game in Figure 7 and the argument using Figure 8 are prototypical for many leadership games. In the same vein, we consider now the inspection game in Figure 1 as a simultaneous game, and construct its leadership version. Recall that in this game, the inspector has collected some data and, using this data, has to decide whether the inspectee acted illegally or not. He uses a statistical test procedure which is designed to detect an illegal action. Then, he either calls an alarm, rejecting the null hypothesis H_0 of legal behavior of the inspectee in favor of the alternative hypothesis H_1 of a violation, or not. The first and second error probabilities α and β of a false rejection of H_0 or H_1, respectively, are the probabilities for a false alarm and an undetected violation.

As shown in Section 3.1, the only choice of the inspector is in effect the value for the false alarm probability α from $[0, 1]$. The non-detection probability β is then determined by the most powerful statistical test and defines the function $\beta(\alpha)$, which has the properties (3.7). Thereby, the convexity of β implies that the inspector has no advantage in choosing α randomly since this will not improve his non-detection probability. Hence, for constructing the leadership game as above in Section 5.1, the value of α can be announced deterministically. The possible actions of the inspectee are legal behavior H_0 and illegal behavior H_1. According to the payoff functions in (3.2), with $q = 0$ for H_0 and $q = 1$ for H_1, and the definition of $\beta(\alpha)$, this defines the game shown in Figure 9.

In an equilibrium of the simultaneous game, Theorem 3.1 shows that the inspector chooses α^* as defined by (3.8). Furthermore, the inspectee violates with positive probability q^* according to (3.4) and (3.9). This is different in the leadership version.

THEOREM 5.1. *The leadership version of the simultaneous game in Figure 9 has a unique equilibrium where the inspector announces α^* as defined by (3.8). In response to α, the inspectee violates if $\alpha < \alpha^*$, and acts legally if $\alpha \geq \alpha^*$.*

Inspector \ Inspectee	legal action H_0	violation H_1
$\alpha \in [0, 1]$	$-h\alpha$ $-e\alpha$	$-b + (1+b)\beta(\alpha)$ $-a - (1-a)\beta(\alpha)$

Figure 9. The game of Figure 1 with payoffs (3.2) depending on the false alarm probability α. The non-detection probability $\beta(\alpha)$ for the best test has the properties (3.7).

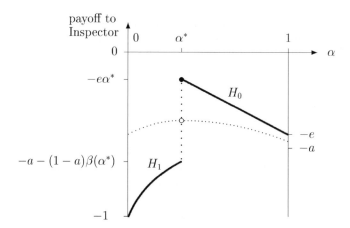

Figure 10. The inspector's payoff in Figure 9 as a function of α for the best responses of the inspectee. For $\alpha < \alpha^*$ the inspectee behaves illegally, hence the payoff is given by the curve H_1. For $\alpha > \alpha^*$ the inspectee behaves legally and the payoff is given by the line H_0. In the simultaneous game, the inspectee plays H_1 with probability q^* according to (3.9), so that the inspector's payoff, shown by the thin line, has its maximum for α^*; this is his equilibrium payoff, indicated by o. The inspector's equilibrium payoff in the leadership game is marked by •.

PROOF. The response of the inspectee is unique for any announcement of α except for α^*. Namely, the inspectee will violate if $\alpha < \alpha^*$, and act legally if $\alpha > \alpha^*$. The inspector's payoffs for these responses are shown in Figure 10. As argued with Figure 8, the only equilibrium of the leadership game is that in which the inspector announces α^*, and the inspectee acts legally in response.   □

Summarizing, we observe that in the simultaneous game, the inspectee violates with positive probability, whereas he acts legally in the leadership game. Inspector leadership serves as *deterrence* from illegal behavior. The optimal false alarm probability α^* of the inspector stays the same. His payoff is larger in the leadership game.

6. Conclusions

One of our claims is that inspection games constitute a real field of applications of game theory. Is this justified? Have these games actually been used? Did game theory provide more than a general insight, did it have an operational impact?

The decision to implement a particular verification procedure is usually *not* based on a game-theoretic analysis. Practical questions abound, and the allocation of inspection effort to various sites, for example, is usually based on rules of thumb. Most stratified sampling procedures of the IAEA are of this kind [IAEA (1989)]. However, they can often be justified by game-theoretic means. We mentioned that the subdivision of a plant into small areas with intermediate balances has no effect on the overall detection probability if the violator acts strategically – such a physical separation may only improve localizing a diversion of material. With all caution concerning the impact of a theoretical analysis, this observation may have influenced the design of some nuclear processing plants.

Another question concerns the proper scope of a game-theoretic model. For example, the course of second-level actions – after the inspector has raised an alarm – is often determined politically. In an inspection game, the effect of a detected violation is usually modeled by an unfavorable payoff to the inspectee. The particular magnitude of this penalty, as well as the inspectee's utility for a successful violation, is usually not known to the analyst. This is often used as an argument against game theory. As a counterargument, the signs of these payoffs often suffice, as we have illustrated with a rather general model in Theorem 3.1. Then, the part of the game where a violation is to be discovered can be reduced to a zero-sum game with the detection probability as payoff to the inspector, as first proposed by Bierlein (1968).

We believe that inspection models should be of this kind, where the merit of game theory as a technical tool becomes clearly visible. 'Political' parameters, like the choice of a false alarm probability, are exogenous to the model. Higher-level models describing the decisions of the states, like whether to cheat or not, and in the extreme whether 'to go to war' or not, should in our view not be blended with an inspection game. They are likely to be simplistic and will invalidate the analysis.

Which direction will or should research on inspection games take in the foreseeable future? The interesting concrete inspection games are stimulated by problems raised by practitioners. In that sense we expect continued progress from the application side, in particular in the area of environmental control where a fruitful interaction between theorists and environmental experts is still missing. Indeed there are open questions which shall be illustrated with the material accountancy example discussed in Section 3.2.

We have shown that intermediate inventories should be ignored in equilibrium. This means, however, that a decision about legal or illegal behavior can be made only at the end of the reference time. This may be too long if *detection time* becomes a criterion. If one models this appropriately, one enters the area of sequential games and sequential statistics. With a new 'non-detection game' similar to Theorem 3.1 and under reasonable assumptions one can show that then the probability of detection and the false alarm probability are replaced by the *average run lengths* under H_1 and H_0, respectively, that is, the expected times until an alarm is raised [see Avenhaus and Okada (1992)]. However, they need not exist and, more importantly, there is no equivalent to the Neyman–Pearson lemma which gives us constructive advice for the best sequential test. Thus, in general, sequential statistical inspection games represent a wide field for future research.

From a theoretical point of view, the leadership principle as applied to inspection games deserves more attention. In the case of sequential games, is legal behavior again an equilibrium strategy of the inspectee? How does the leadership principle work if more complicated payoff structures have to be considered? Does the inspector in general improve his equilibrium payoff by credibly announcing his inspection strategy? We think that for further research, a number of related ideas in economics, like principal–agent problems, should be compared more closely with leadership as presented here.

In this contribution, we have tried to demonstrate that inspection games have real applications and are useful tools for handling practical problems. Therefore, the major effort in the future, also in the interest of sound theoretical development, should be spent in deepening these applications, and – even more importantly – in trying to convince practitioners of the usefulness of appropriate game-theoretic models.

References

Altmann, J., et al. (eds.) (1992), Verification at Vienna: Monitoring Reductions of Conventional Armed Forces (Gordon and Breach, Philadelphia).

Anscombe, F.J., et al. (eds.) (1963), "Applications of statistical methodology to arms control and disarmament", Final Report to the U.S. Arms Control and Disarmament Agency under contract No. ACDA/ST-3 by Mathematica (Princeton, N.J.).

Anscombe, F.J., et al. (eds.) (1965), "Applications of statistical methodology to arms control and disarmament", Final Report to the U.S. Arms Control and Disarmament Agency under Contract No. ACDA/ST-37 by Mathematica (Princeton, N.J.).

Avenhaus, R. (1986), Safeguards Systems Analysis (Plenum, New York).

Avenhaus, R. (1994), "Decision theoretic analysis of pollutant emission monitoring procedures", Annals of Operations Research 54:23–38.

Avenhaus, R. (1997), "Entscheidungstheoretische Analyse der Fahrgast-Kontrollen", Der Nahverkehr 9/97 (Alba Fachverlag, Düsseldorf) 27–30.

Avenhaus, R., H.P. Battenberg and B.J. Falkowski (1991), "Optimal tests for data verification", Operations Research 39:341–348.

Avenhaus, R., and M.J. Canty (1996), Compliance Quantified (Cambridge University Press, Cambridge).

Avenhaus, R., M.J. Canty and B. von Stengel (1991), "Sequential aspects of nuclear safeguards: Interim inspections of direct use material", in: Proceedings of the 4th International Conference on Facility Operations-Safeguards Interface, American Nuclear Society, Albuquerque, New Mexico, 104–110.

Avenhaus, R., and A. Okada (1992), "Statistical criteria for sequential inspector-leadership games", Journal of the Operations Research Society of Japan 35:134–151.

Avenhaus, R., A. Okada and S. Zamir (1991), "Inspector leadership with incomplete information", in: R. Selten, ed., Game Equilibrium Models IV (Springer, Berlin) 319–361.

Avenhaus, R., and G. Piehlmeier (1994), "Recent developments in and present state of variable sampling", in: IAEA-SM-333/7, Proceedings of a Symposium on International Nuclear Safeguards 1994: Vision for the Future, Vol. I. IAEA, Vienna, 307–316.

Avenhaus, R., and B. von Stengel (1992), "Non-zero-sum Dresher inspection games", in: P. Gritzmann et al., eds., Operations Research '91 (Physica-Verlag, Heidelberg) 376–379.

Baiman, S. (1982), "Agency research in managerial accounting: A survey", Journal of Accounting Literature 1:154–213.

Baston, V.J., and F.A. Bostock (1991), "A generalized inspection game", Naval Research Logistics 38:171–182.

Battenberg, H.P., and B.J. Falkowski (1998), "On saddlepoints of two-person zero-sum games with applications to data verification tests", International Journal of Game Theory 27:561–576.

Bierlein, D. (1968), "Direkte Inspektionssysteme", Operations Research Verfahren 6:57–68.

Bierlein, D. (1969), "Auf Bilanzen und Inventuren basierende Safeguards-Systeme", Operations Research Verfahren 8:36–43.

Bird, C.G., and K.O. Kortanek (1974), "Game theoretic approaches to some air pollution regulation problems", Socio-Economic Planning Sciences 8:141–147.

Borch, K. (1982), "Insuring and auditing the auditor", in: M. Deistler, E. Fürst, and G. Schwödiauer, eds., Games, Economic Dynamics, Time Series Analysis (Physica-Verlag, Würzburg) 117–126. Reprinted in: K. Borch (1990), Economics of Insurance (North-Holland, Amsterdam) 350–362.

Borch, K. (1990), Economics of Insurance, Advanced Textbooks in Economics, Vol. 29 (North-Holland, Amsterdam).

Bowen, W.M., and C.A. Bennett (eds.) (1988), "Statistical methodology for nuclear material management", Report NUREG/CR-4604 PNL-5849, prepared for the U.S. Nuclear Regulatory Commission, Washington, DC.

Brams, S., and M.D. Davis (1987), "The verification problem in arms control: A game theoretic analysis", in: C. Ciotti-Revilla, R.L. Merritt and D.A. Zinnes, eds., Interaction and Communication in Global Politics (Sage, London) 141–161.

Brams, S., and D.M. Kilgour (1988), Game Theory and National Security (Basil Blackwell, New York) Chapter 8: Verification.

Cochran, W.G. (1963), Sampling Techniques, 2nd edn (Wiley, New York).

Derman, C. (1961), "On minimax surveillance schedules", Naval Research Logistics Quarterly 8:415–419.

Diamond, H. (1982), "Minimax policies for unobservable inspections", Mathematics of Operations Research 7:139–153.

Dresher, M. (1962), "A sampling inspection problem in arms control agreements: A game-theoretic analysis", Memorandum No. RM-2972-ARPA, The RAND Corporation, Santa Monica, CA.

Dye, R.A. (1986), "Optimal monitoring policies in agencies", RAND Journal of Economics 17:339–350.

Feichtinger, G. (1983), "A differential games solution to a model of competition between a thief and the police", Management Science 29:686–699.

Ferguson, T.S., and C. Melolidakis (1998), "On the inspection game", Naval Research Logistics 45:327–334.

Ferguson, T.S., and C. Melolidakis (2000), "Games with finite resources", International Journal of Game Theory 29:289–303.

Filar, J.A. (1985), "Player aggregation in the traveling inspector model", IEEE Transactions on Automatic Control AC-30:723–729.

Gal, S. (1983), Search Games (Academic Press, New York).

Gale, D. (1957), "Information in games with finite resources", in: M. Dresher, A.W. Tucker and P. Wolfe, eds., Contributions to the Theory of Games III, Annals of Mathematics Studies 39 (Princeton University Press, Princeton) 141–145.

Garnaev, A.Y. (1994), "A remark on the customs and smuggler game", Naval Research Logistics 41:287–293.

Goldman, A.J. (1984), "Strategic analysis for safeguards systems: A feasibility study, Appendix", Report NUREG/CR-3926, Vol. 2, prepared for the U.S. Nuclear Regulatory Commission, Washington, DC.

Goldman, A.J., and M.H. Pearl (1976), "The dependence of inspection-system performance on levels of penalties and inspection resources", Journal of Research of the National Bureau of Standards 80B:189–236.

Greenberg, J. (1984), "Avoiding tax avoidance: A (repeated) game-theoretic approach", Journal of Economic Theory 32:1–13.

Güth, W., and R. Pethig (1992), "Illegal pollution and monitoring of unknown quality – a signaling game approach", in: R. Pethig, ed., Conflicts and Cooperation in Managing Environmental Resources (Springer, Berlin) 276–332.

Höpfinger, E. (1971), "A game-theoretic analysis of an inspection problem", Unpublished manuscript, University of Karlsruhe.

Höpfinger, E. (1979), "Dynamic standard setting for carbon dioxide", in: S.J. Brams, A. Schotter and G. Schwödiauer, eds., Applied Game Theory (Physica-Verlag, Würzburg) 373–389.

Howson, J.T., Jr. (1972), "Equilibria of polymatrix games", Management Science 18:312–318.

IAEA (1989), IAEA Safeguards: Statistical Concepts and Techniques, 4th rev. edn. (IAEA, Vienna).

Jaech, J.L. (1973), "Statistical methods in nuclear material control", Technical Information Center, United States Atomic Energy Commission TID-26298, Washington, DC.

Kanodia, C.S. (1985), "Stochastic and moral hazard", Journal of Accounting Research 23:175–193.

Kaplan, R.S. (1973), "Statistical sampling in auditing with auxiliary information estimates", Journal of Accounting Research 11:238–258.

Karlin, S. (1959), Mathematical Methods and Theory in Games, Programming, and Economics, Vol. II: The Theory of Infinite Games (Addison-Wesley, Reading) (Dover reprint 1992).

Kilgour, D.M. (1992), "Site selection for on-site inspection in arms control", Arms Control 13:439–462.

Kilgour, D.M., N. Okada and A. Nishikori (1988), "Load control regulation of water pollution: An analysis using game theory", Journal of Environmental Management 27:179–194.

Klages, A. (1968), Spieltheorie und Wirtschaftsprüfung: Anwendung spieltheoretischer Modelle in der Wirtschaftsprüfung. Schriften des Europa-Kollegs Hamburg, Vol. 6 (Ludwig Appel Verlag, Hamburg).

Kuhn, H.W. (1963), "Recursive inspection games", in: F.J. Anscombe et al., eds., Applications of Statistical Methodology to Arms Control and Disarmament, Final report to the U.S. Arms Control and Disarmament Agency under contract No. ACDA/ST-3 by Mathematica, Part III (Princeton, NJ) 169–181.

Landsberger, M., and I. Meilijson (1982), "Incentive generating state dependent penalty system", Journal of Public Economics 19:333–352.

Lehmann, E.L. (1959), Testing Statistical Hypotheses (Wiley, New York).

Maschler, M. (1966), "A price leadership method for solving the inspector's non-constant-sum game", Naval Research Logistics Quarterly 13:11–33.

Maschler, M. (1967), "The inspector's non-constant-sum-game: Its dependence on a system of detectors", Naval Research Logistics Quarterly 14:275–290.

Mitzrotsky, E. (1993), "Game-theoretical approach to data verification" (in Hebrew), Master's Thesis, Department of Statistics, The Hebrew University of Jerusalem.

O'Neill, B. (1994), "Game theory models of peace and war", in: R.J. Aumann and S. Hart, eds., Handbook of Game Theory, Vol. 2 (North-Holland, Amsterdam) Chapter 29, 995–1053.

Piehlmeier, G. (1996), Spieltheoretische Untersuchung von Problemen der Datenverifikation (Kovač, Hamburg).

Reinganum, J.F., and L.L. Wilde (1986), "Equilibrium verification and reporting policies in a model of tax compliance", International Economic Review 27:739–760.

Rinderle, K. (1996), Mehrstufige sequentielle Inspektionsspiele mit statistischen Fehlern erster und zweiter Art (Kovač, Hamburg).

Rubinstein, A. (1979), "An optimal conviction policy for offenses that may have been committed by accident", in: S.J. Brams, A. Schotter and G. Schwödiauer, eds., Applied Game Theory (Physica-Verlag, Würzburg) 373–389.

Ruckle, W.H. (1992), "The upper risk of an inspection agreement", Operations Research 40:877–884.

Russell, G.S. (1990), "Game models for structuring monitoring and enforcement systems", Natural Resource Modeling 4:143–173.

Saaty, T.L. (1968), Mathematical Models of Arms Control and Disarmament: Application of Mathematical Structures in Politics (Wiley, New York).

Sakaguchi, M. (1977), "A sequential allocation game for targets with varying values", Journal of the Operations Research Society of Japan 20:182–193.

Sakaguchi, M. (1994), "A sequential game of multi-opportunity infiltration", Mathematica Japonicae 39:157–166.

Schleicher, H. (1971), "A recursive game for detecting tax law violations", Economies et Sociétés 5:1421–1440.

Thomas, M.U., and Y. Nisgav (1976), "An infiltration game with time dependent payoff", Naval Research Logistics Quarterly 23:297–302.

von Stackelberg, H. (1934), Marktform und Gleichgewicht (Springer, Berlin).

von Stengel, B. (1991), "Recursive inspection games", Technical Report S-9106, Universität der Bundeswehr München.

Weissing, F., and E. Ostrom (1991), "Irrigation institutions and the games irrigators play: Rule enforcement without guards", in: R. Selten, ed., Game Equilibrium Models II (Springer, Berlin).

Wölling, A. (2000), "Das Führerschaftsprinzip bei Inspektionsspielen", Universität der Bundeswehr München.

Chapter 52

ECONOMIC HISTORY AND GAME THEORY

AVNER GREIF*

Department of Economics, Stanford University, Stanford, CA, USA

Contents

*The research for this paper was supported by National Science Foundation Grants SES-9223974 and SBR-9602038. I am thankful to the editors, Professor Jacob Metzer, and an anonymous referee for very useful comments. First draft of this paper was written in 1996.

Handbook of Game Theory, Volume 3, Edited by R.J. Aumann and S. Hart

Abstract

This paper surveys the small, yet growing, literature that uses game theory for economic history analysis. It elaborates on the promise and challenge of applying game theory to economic history and presents the approaches taken in conducting such an application. Most of the essay, however, is devoted to studies in economic history that use game theory as their main analytical framework. Studies are presented in a way that highlights the range of potential topics in economic history that can be and have been enriched by a game-theoretical analysis.

Keywords

economic history, institutions

JEL classification: C7, N0

1. Introduction

Since the rise of cliometrics in the mid-60s, the main theoretical framework used by economic historians has been neo-classical theory. It is a powerful tool for examining non-strategic situations in which interactions are conducted within markets or shaped by them.[1] Game theory enables the analysis to go further by providing economic history with a theoretical framework suitable for analyzing strategic, economic, social, and political situations. Among such strategic situations are: economic exchange in which the number of participants is relatively small, political relationships within and between states, decision-making within regulatory and other governmental bodies, exchange in the absence of impartial, third-party enforcement, and intra-organizational relations. Such situations prevail even in modern market economies and were probably even more prevalent in pre-modern economies.

Furthermore, game-theoretic insights have provided support to the economic historians' long-held position that history matters. While neo-classical economics asserts that economies identical in their endowment, preferences, and technology will reach the same equilibrium, economic historians have long held that the historical experience of various economies indicates the limitations of this assertion. Game theory augmented this position by providing a theoretical framework whose insights reveal a role for history in economic systems. Game theory points, for example, to the potential sensitivity of outcomes to rules and hence to institutions, the possibility of multiple equilibria and hence the potential for distinct trajectories of institutional and economic changes, the crucial role of expectations and beliefs and hence the potential importance of the historical actors, and the possible role of evolutionary processes and change in equilibrium selection. In short, game theory indicates that within the framework of strategic rationality, different historical trajectories are possible in situations identical in terms of their endowment, preferences, and technology.[2]

Applying game theory to economic history can also potentially enrich game theory. History contains unique and, at times, detailed information regarding behavior in strategic situations, and thus it provides another laboratory in which to examine the relevance of the game-theoretic approach and its insights into positive economic analysis. Furthermore, historical analyses guided by game theory are likely to reveal theoretical issues that, if addressed, would contribute to the development of game theory and its ability to advance economic analysis.

This essay surveys the small, yet growing, literature that employs game theory in economic history. Section 2 briefly elaborates on the promise and challenge of applying game theory to economic history and presents the approaches taken in conducting such applications. Section 3 presents studies in economic history that either utilize

---

[1] On the Cliometric Revolution, see Williamson (1994). Hartwell (1973) surveys the methodological developments in economic history. For the many contributions generated by the neo-classical line of research in economic history, see McCloskey (1976). For recent discussion, see the session on "Cliometrics after Forty Years" in the AER 1997 (May). I further elaborate on these issues in Greif (1997a, 1998a).

[2] See Greif (1994a) for such a comparative game-theoretic analysis of two late medieval economies.

game theory as their main analytical framework or examine the empirical relevance of game-theoretical insights. This section contains two sub-sections. The first discusses economic history studies that use only general game-theoretical insights to guide the analysis. The second discusses economic history studies that use a context-specific game-theoretic model as their main theoretical framework. In either sub-section the studies are presented according to the issues they examine in economic history.

Clearly, it is impossible to elaborate on a myriad of papers in such a short essay, so a brief description of each paper is provided with only a few described in detail. Their (subjective) selection is influenced by their relative complexity, methodological contribution, or representativeness. Finally, since the goal of this essay is to survey applications of game-theoretical analysis to economic history, it does not systematically evaluate their arguments (although references to published comments on papers are provided).

2. Bringing game theory and economic history together

Economic history can benefit greatly from a theory enabling empirical analysis of strategic situations since issues central to economic history are inherently strategic. For example, economic history has always been concerned with the origin, impact, and path dependence of non-market economic, social, political, and legal institutions.[3] Indeed, this concern with non-market institutions logically follows from Adam Smith's legacy and the neo-classical economics which often identify the rise of the modern economic system with the expansion of the market system. This view implies, however, that an analysis of non-market situations is required to understand past economies, their functioning, and why some of them, but not others, transformed into market economies. Hence, a theoretical framework of strategic, non-market situations can expand our comprehension of issues central to economic history.

The ability of game theory – the existing theoretical framework for analyzing strategic situations to advance an empirical and historical study – should be judged empirically. Yet, certain conclusions of game-theoretical analysis make its application to economic history both challenging and promising. Game theory indicates that outcomes in strategic situations are potentially sensitive to details, that various equilibrium concepts are plausible, and (given an equilibrium concept) multiple equilibria may exist. Thus, applying game theory to economic history may be challenging since economic history is, first and foremost, an empirical field and economic historians are trying to understand what has actually transpired, why it transpired, and to what effect. One may argue that a theoretical analysis, whose conclusions regarding outcomes are non-robust

---

[3] For a discussion of the methodological differences between economic history and economics, see Backhouse (1985, pp. 216–221). For institutional studies during the nineteenth century in the German and English Historical Schools, see, for example, Weber (1987 [1927]); Cunningham (1882). On the general theory of path dependence see David (1988, 1992). See also Footnote 1.

and empirically inconclusive (in the sense that many outcomes are consistent with the theory), provides an inappropriate foundation for an empirical study.

Interestingly, however, the game-theoretical conclusions regarding non-robustness and inconclusiveness are in accordance with the conceptual foundations of historical analysis – namely, that outcomes depend on the details of the historical context, that "economic actors" can potentially matter, and that the non-economic aspects of the historical context, such as religious precedents or even chance, can influence economic outcomes. Game theory provides economic history with an explicit theoretical framework that does not lead to the ahistorical conclusion that the same preferences, technology, and endowments lead to a unique economic outcome in all historical episodes. The conclusions of game-theoretical analysis that challenge its empirical applicability make it a particularly promising theory for historical analysis since it can be used for analyzing strategic situations in a way that is sensitive to, and reveals the importance of, their historical dimensions.

Studies in economic history that use game theory have differed in their responses to the challenge and promise presented by the potential non-robustness and inconclusiveness of game-theoretical analysis. They all began with a historical study aimed at formulating the relevant issue to be examined. They used general game-theoretical insights – such as the possibility of coordination failure, the importance of credible commitment, or the problem of cooperation in a multi-person prisoners' game situation – to "frame" the analysis, that is to say, they explicitly specify the issues needing to be addressed, provide an organizing scheme for the historical evidence, or highlight the logic behind the historical evidence.[4] Some studies went so far as to undertake a game-theoretic analysis of various issues and they responded to the non-robustness and inconclusiveness problems in one of two (not mutually exclusive) ways.

Some studies responded to the non-robustness and inconclusiveness by basing the historical analysis only on those game-theoretic insights that are conclusive and robust.[5] The empirical investigation was thus guided by generic insights applicable in situations with such features. An example of such general insights would be that bargaining in the presence of asymmetric information can lead to negotiation failure. Reliance on general insights comes at the cost of limiting the ability to empirically substantiate a hypothesis. Without an explicit model it is difficult to enhance confidence in an argument by confronting the details of the historical episode with the details of the theoretical argument and its implications. In particular, without specifying the strategies employed by the players, it is difficult to empirically substantiate the analysis. Yet, the potential benefit of relying on general insights is the ability to discuss important situations without being constrained by the ability to explicitly model them.

---

[4] See, for example, North [(1981), Chapter 3] and Kantor (1991). "The power of game theory – and it's the way I've used it – is that it makes you structure the argument in formal terms, in precise terms..." [North (1993), p. 27].

[5] For a somewhat similar approach in the Industrial Organization literature, see Sutton (1992).

Other studies found it useful to confront non-robustness and inconclusiveness differently. A detailed empirical study of the historical episode under consideration was conducted and it provided the foundation for an interactive process of theoretical and historical examination aimed at formulating a context-specific model that captured an essence of the relevant strategic situation.[6] This interactive historical and theoretical analysis sufficiently constrained the model's specification, basing it on assumptions in which confidence can be gained independently from their predictive power, and ensured that the analysis did not impose the researcher's perception of a situation on the historical actors.[7] The resulting context-specific model provided the foundation for a game-theoretical analysis of the situation whose predictions could be compared with the historical evidence. At the same time, the model extended the examination of the extent to which its main conclusions are robust with respect to assumptions whose appropriateness is historically questionable. In short, these studies confronted non-robustness and inconclusiveness by utilizing a context-specific model. Such models enhance "general insight" analysis by generating falsifiable predictions, as well as the ability to check robustness and to gain a deeper understanding of the issues under consideration. Yet, despite these advantages, analysis based on a context-specific model is restricted to cases where such models can be formulated.

Economic history studies utilizing a game-theoretic, context-specific model mostly utilized two basic equilibrium concepts: Nash equilibrium and sub-game perfect equilibrium. These two concepts have the advantage of including most other equilibrium concepts as special cases, as well as having intuitive, common-sense interpretations.[8] Using these inclusive equilibrium concepts implies, however, that multiple equilibria are more likely to exist and the analysis is more likely to be inconclusive. In other words, it amplifies the two problems for empirical historical analysis associated with inconclusiveness, namely, identification and selection. Studies have differed in their responses to these problems.

Some studies, whose aim was to understand the logic beyond a particular behavior, considered the analysis complete when they revealed the existence of an equilibrium corresponding to that particular behavior. They did not aspire to substantiate that the behavior and expected behavior associated with that particular equilibrium were those that prevailed in the historical episode under consideration. If they attempted to account for cooperation, for example, they were satisfied with arguing that an equilibrium entailing cooperation existed. By and large, they neither tried to substantiate that this particular equilibrium – rather than another one entailing cooperation – indeed prevailed,

---

[6] Clearly, the essence of the issue that is captured should be both important and orthogonal to other relevant issues.

[7] For example, the King of England, Edward the First, noted in 1283 that insufficient protection to alien merchants' property rights deterred them from coming to trade. His remark enhances the confidence in the relevance of a model in which commitment to alien traders' property rights can foster their trade [Greif et al. (1994)].

[8] For an introduction to these concepts, see, for example, Fudenberg and Tirole (1991) or Rasmusen (1994).

nor did they examine how this equilibrium was selected or compare their analysis with a possible non-strategic account for cooperation. In other studies the need to identify a particular strategy was avoided by concentrating on the analysis of the set of equilibria.[9] This approach was adopted particularly in studies that examined the impact of changes in the rules of the game on outcomes.[10]

In other cases when the argument revolved around the empirical relevance of a particular strategy, the problem of identification was confronted by employing direct and indirect evidence to verify the use of this particular strategy (or some subset of the possible equilibrium strategies with particular features). Direct evidence of the historical relevance of a particular strategy is explicit documentary accounts reflecting the strategies that were used, or were intended to be used, by the decision-makers.[11] Such explicit documentary accounts are found in such diverse historical sources as business correspondence, private letters, legal procedures, the constitutions of guilds, the charters of firms, and records of public speeches.

Clearly, statements about intended courses of action can be just talk, but indirect evidence can enhance confidence in the empirical relevance of a particular strategy. Indirect evidence is an empirical confirmation of predictions generated under the assumption that a particular strategy was employed. The predictions generated by various economic history studies utilizing game theory cover a wide range of variables, such as price movements, contractual forms, dynamics of wealth distribution, exits, entries, price, and various responses to exogenous changes. In some studies it was possible to test these predictions econometrically but in others, because of the nature of their predictions, such tests could not be performed.[12] For example, the analysis of labor negotiation presented in Treble (1990), as discussed below, predicts that labor disputes should be a function of bargaining procedures. Because different procedures were in effect in the historical episode under consideration, this prediction could have been examined econometrically. The analysis in Greif (1989), which is also discussed below, predicts that traders' social structures should be a function of the strategy merchants used to punish an overseas agent who cheated a merchant. The analysis predicts in particular that a strategy of collective punishment would lead to a horizontal social structure in which traders would serve as both merchants and agents. While this prediction can be objectively verified, it cannot be econometrically tested. The advantage of confirming predictions based on an econometric test is that this test provides a significance level. Such analysis, nevertheless, is restricted only to issues that generate econometrically testable predictions. But one can increase the confidence in a hypothesis also by comparing its predictions – without using econometrics and hence having a significance level – with predictions generated under alternative hypotheses.

---

[9] See, for example, Greif et al. (1994), particularly Proposition 1.

[10] E.g., Milgrom et al. (1990); Greif et al. (1994).

[11] For an example of the extensive use of such evidence, see Greif (1989).

[12] For examples of econometric analyses see Porter (1983); Levenstein (1994). For non-econometric analyses see Greif (1989, 1993, 1994a); Rosenthal (1992).

So far, economic history has studied the problem of equilibrium selection entailed by multiple equilibria in a way that has been influenced more by the conceptual foundations of historical analysis rather than by game-theoretical literature dealing with refinements or evolutionary game theory. Most authors accounted for the selection of a particular equilibrium by invoking aspects of the historical context. One paper cited the public commitment of Winston Churchill to a particular strategy as being fundamental in the selection of a particular equilibrium.[13] Other papers pointed out that factors outside the game itself had influenced equilibrium selection. Among these were immigration that provided information networks, political changes that determined the initial set of players, and focal points provided by religious and social attitudes.[14] Some authors, particularly those interested in comparing two historical episodes, theoretically identified the range of parameters or variables that were required for one particular equilibrium to prevail rather than another. This theoretical prediction was compared with the empirical evidence regarding these variables in the historical episodes under consideration.[15]

3. Game-theoretical analyses in economic history

This section presents studies in economic history that use game theory. In line with the above discussion of the methodology employed by such studies, they are grouped according to those that use "generic insights" (Section 3.1) and those that use "context-specific models" (Section 3.2). In both sub-sections the presentation is organized by historical topics but it implicitly suggests the potential benefits of economic history studies using game theory to further empirical evaluation and to extend various aspects of the theory itself. I will return to this issue later in the section. Space limitation precludes a detailed examination of all the papers in either section. But to illustrate the methodological differences between the papers in the two sub-sections, I will provide a somewhat longer presentation of a study [Greif (1989, 1993, 1994a)] that uses a context-specific model.

3.1. The early years: Employing general game-theoretical insights

The first economic history papers that used general insights from game theory were published in the early 1980s. They examined such topics as regulations, market structure, and property-rights protection. As for their game-theoretic method, they used nested games, a situation in which the rules of one game are the equilibrium outcome of another game. Furthermore, they provided empirical evidence of problems encountered

---

[13] Maurer (1992).
[14] E.g., Greif (1994a).
[15] E.g., Rosenthal (1992); Greif (1994a); Baliga and Polak (1995).

in bargaining under incomplete information and suggested that off-the-path-of-play expected behavior might indeed influence economic outcomes, as formalized in the notion of sub-game-perfect equilibrium.

Regulations: Economic historians have for a long time emphasized the importance of the historical development of regulatory agencies and regulations in the US. Davis and North (1971), for example, argued that regulations were a welfare-enhancing process of institutional changes driven by the potential profit from regulating the economy. In contrast, using a game-theoretical analysis of endogenous regulations, Reiter and Hughes (1981) have argued that the process was not necessarily welfare-enhancing. In their formulation, economic agents and regulators are involved in a non-cooperative dynamic game with asymmetric information in which the regulators pursue their own agendas. To advance their agendas, they are also involved in a cooperative game with political agents in which they try to influence the political process through which the next period's legal and budgetary framework of the non-cooperative game is determined. While Reiter and Hughes did not attempt to explicitly solve the model, it provided them with a paradigm to discuss the emergence of the "modern regulated economy" as reflecting redistributive considerations, efficiency-enhancing motives, and political factors.

Interestingly, around the same time David (1982) combined cooperative and non-cooperative game theory in the opposite direction to examine "regulations" associated with the feudal system. His analysis used the Nash bargaining solution to examine transfers from peasants to lords of the manor. This analysis was incorporated within a broader game in which the feudal system itself reflects an equilibrium of the repeated game between peasants and a coalition of lords.

Eichengreen (1996) used game theory to examine the regulations that, according to his interpretation, were crucial to Europe's rapid post-World War II economic growth. The basic argument is inspired by the concept of sub-game perfection. Following the war there was a very high return on capital investment, but inducing investment required that investors were assured that *ex post*, after they had made a sunk investment, their workers would not hold them up and reap all the gains. At the same time, for workers not to find it optimal to act in this way they had to be assured that in the long run they would gain a share in the implied economic growth. Credible commitment by workers and investors alike was made through state intervention. The state acted as a third-party enforcer of labor relationship regulations and social welfare programs that ensured a sufficiently high rate of return to investors and workers alike.

Market structure: A market's structure is fundamental in determining an industry's conduct and performance. Traditionally, economic historians have not considered that the structure of a market can be influenced by strategic interactions. Yet Carlos and Hoffman (1986) argued that strategic considerations determined the structure of the fur industry in North America during the early nineteenth century. The two companies that operated in this industry from 1804 to 1821 (the Northwest Company and the Hudson's Bay Company) could have benefited from collusion or merger and there was no antitrust legislation to hinder either. Yet, both companies were engaged in an intense conflict that led to a depletion of the animal stock. Carlos and Hoffman argued that the persistence of this market structure reflects the difficulties of bargaining with incomplete information. General insights from bargaining models with incomplete information indicate that it is possible to fail to reach an *ex post* efficient agreement because each side attempts to misrepresent its type, players are likely to bargain over distributive mechanisms rather than allocations, and disagreement may result from a player's commitment to a tough strategy. Indeed, although the correspondence between the companies indicates that both recognized the gains from cooperation, each was trying to mislead the other. Furthermore, the two companies did not bargain over the allocation of joint profits but tried to reach a merger. After failing to merge, they bargained over a distribution of the territory that each would exploit as a monopolist. Also, negotiations prior to 1821 broke down partially because of the Hudson's Bay Company's commitment to a particular, very demanding strategy. The impetus for their final merger was the intervention of the government following a period of intense and ruinous competition. Hence, Carlos and Hoffman's analysis indicates that strategic considerations influenced the market's structure and provided "empirical evidence on the problems encountered in bargaining under incomplete information" (p. 968).[16]

Security of property rights: Financing the government by issuing public debt is one of the peculiar features of the pre-modern European economy. Arguably, this type of financing facilitated the rise of security markets [e.g., Neal (1990)] and provided the foundation for the modern welfare state. Yet, for a pre-modern ruler to gain access to credit he had to be able to commit to repay it despite the fact that he stood, roughly speaking, above the law. How could rulers commit to repay their debts? Why in some historical episodes did rulers renege on their obligations and in others respect them? Clearly, in a one-shot game between a ruler (who can request a loan and renege after receiving it) and a potential lender (who can decide whether or not to make a loan), the only sub-game perfect equilibrium entails no lending.

Veitch (1986) has argued, based on Telser's (1980) idea of self-enforcing agreements, that repetition and potential collective retaliation by lenders enlarged the equilibrium set and enabled rulers to commit to repay their debts, and hence to borrow. He has noted that in medieval Europe rulers often borrowed from members of a particular group, such as Jews, Templars, or the Italians, while debt repudiation was often carried out against the group as a whole rather than against particular members.[17] Veitch argued that this indicates that repudiation was curtailed by the threat of collective retaliation by the group. The threat was credible due to the ethnic, moral, or political relations among the lenders. The threat was effective as long as the ruler did not have any alternative group to borrow from, implying that the emergence of an alternative group would lead to

---

[16] The analysis is based on Myerson's (1984a, 1984b) work on a generalized Nash bargaining solution and Crawford's (1982a) model in which it is costly (for exogenous reasons) to change a strategy after committing to it. As Carlos and Hoffman (1988) later recognize in their response to Nye (1988), subsequent theoretical developments provided models better suited to capturing the essence of the historical situation.

[17] Or against a particular sub-group such as an Italian company.

repudiation against the previous group, as indeed was often the case.[18] Similarly, Root (1989) argued that during the 17th and 18th centuries corporate bodies, such as village communities, provincial estates, and guilds, enabled the King of France to commit to pay debts. They increased the opportunity cost of a breach, thereby restraining the king's ability to default and enabling him to borrow. Indeed, the rise of corporate bodies that loaned to the king in the eighteenth century is associated with a lower interest and bankruptcy rate relative to the seventeenth century.

North and Weingast (1989) and Weingast (1995) further expanded the study of the relations between credible commitment, property-rights security, and political power. If indeed security of property rights is a key to economic growth (as conjectured by North and Thomas (1973)), how was such security achieved in past societies governed by kings with military power superior to that of their subjects? North and Weingast argued that the Glorious Revolution of 1688 enabled the King of England to commit to such security, thereby providing institutional foundations for growth. During this revolution and the years of civil war prior to it, the Parliament established its ability and willingness to revolt against a king who abused property rights. This enabled the king to commit to the property rights of his subjects. Furthermore, to enhance the credibility of this threat and to limit the king's ability to renege, various measures were taken. The king's rights were clearly specified to foster coordination among members of the Parliament regarding which of the king's actions should trigger a reaction. The Parliament gained control over taxation and revenue allocation, an independent judiciary was established, and the king's prerogatives were curtailed. In support of the view that the Glorious Revolution enhanced property-rights security, North and Weingast pointed to the rise, during the eighteenth century, in sovereign debt and in the number and value of securities traded in England's private and public capital markets and the general decline in interest rates.[19]

3.2. Coming to maturity: Explicit models

3.2.1. Exchange and contract enforcement in the absence of a legal system

Neo-classical economics has long emphasized that an impartial legal system fosters exchange by providing contract enforcement. Yet even in contemporary developed and

---

[18] This analysis is insightful and novel but it is incomplete. For example, it is mistaken in arguing that a self-enforcing agreement among the Italian companies is a necessary condition for a sub-game perfect equilibrium in which the ruler does not repudiate.

[19] Carruthers (1990) criticized the claim that placing limits on the king enabled England to borrow, while Clark (1995) cast doubts on the claim that property rights were insecure prior to 1688. He examined the rate of return on private debt and land, and the price of land from 1540 to 1800, and was unable to detect any impact from the Glorious Revolution. Weingast (1995) applied the model of Greif et al. (1994) to further examine how constitutional changes during the Glorious Revolution enhanced the king's ability to borrow by increasing his ability to commit.

developing economies, much exchange is conducted without relying on contract enforcement provided by the state. (See further discussion in Greif (1997b).) This phenomenon has been even more prevalent in past economies where, in many cases, there was no state that could provide an impartial legal system. Yet, the institutional foundations of exchange in past societies were not studied in economic history before the introduction of game theory because there was no appropriate theoretical framework.

Some of the first economic history papers that utilized context-specific, explicit models were those that employed symmetric and asymmetric information repeated game models to examine institutions that provided informal contract enforcement in various historical episodes. They provided such enforcement by linking past conduct and future economic reward. Although such games usually have multiple equilibria and the equilibrium set is sensitive to their details, they were found to facilitate empirical examination when the analysis concentrated on the equilibrium set, and when the historical records were rich enough to constrain the model and enable identification of the equilibrium that prevailed. Apart from indicating the empirical relevance of repeated games (with perfect or imperfect monitoring), these studies show that game-theoretical analysis can highlight diverse aspects of a society, such as the inter-relations between economic institutions and social structures. They indicate how third-party enforcement can be made credible even in the absence of complete information, or a strategy that punishes someone who fails to punish a deviator. Finally, the studies indicate that it is misleading to view contract enforcement based on formal organizations and on repeated interaction as substitutes, since formal organizations may be required for the long hand of the future to sustain cooperation.

"Coalitions" and informal contract enforcement: Much of the spectacular European growth from the eleventh to the fourteenth centuries is attributed to the Commercial Revolution – the resurgence of Mediterranean and European long-distance trade. The actions and explicit statements of contemporaries indicate the important role played by overseas agents who managed the merchants' capital abroad. Operating through agents, however, required overcoming a commitment problem since agents who had control over others' capital could act opportunistically. To establish an efficient agency relationship, the agent had to commit *ex ante* to be honest *ex post* and not to embezzle the merchant's capital that was under his control (in the form of money, goods, and expensive packing materials). It is tempting to conclude, as Benson (1989) has argued, that an agent's concern about his reputation – namely, his concern about his ability to trade in the future – permitted such a commitment. Yet, this argument is unsatisfactory since it presents an incomplete theoretical analysis and, worse, since it implicitly claims that it is enough to comprehend a historical situation by examining only a theoretical possibility without examining any empirical evidence. Comprehending if and how the merchant-agent commitment problem was mitigated in a particular time and place requires detailed empirical research and a context-specific theoretical analysis.

A satisfactory empirical and theoretical analysis should address at least the following issues. If repetition enabled cooperation, should the model be of infinite or finite horizon? If an infinitely repeated game is appropriate, how was the unraveling problem mitigated (that is, why wouldn't an agent cheat in his old age)? Should the model be an incomplete-information model? Should it include a legal system? How was information acquired and transmitted? Should the set of traders and agents be considered exogenous? Could an agent begin operating as a merchant with goods he had embezzled? Who was to retaliate if an agent embezzled goods? Why was the threat of retaliation credible? What were the efficiency implications of the particular way in which the merchant-agent commitment problem was alleviated? Why did this particular way emerge?

Greif (1989, 1993, 1994a) examined these and related questions with respect to the Jewish Maghribi traders who operated during the eleventh century in the Muslim Mediterranean. The historical and theoretical evidence indicates that agency relations were not governed by the legal system and that the appropriate model is an infinitely repeated game with complete information. (Greif (1993) discusses why an incomplete-information model was ruled out and below I discuss how the unraveling problem was mitigated.) Specifically, it argues for the relevance of an efficiency-wage model with two particularly important features: matching is not completely random but is conditioned on the information available to the merchants, and sometimes a merchant has to cease operating through an honest agent.[20] A model incorporating these two features shows that a (sub-game perfect) equilibrium exists in which each merchant employs an agent from a particular sub-set of the potential agents and all merchants cease operating through any agent who ever cheated. This collective punishment is self-enforcing since the value of future relations with *all* the merchants keeps an agent honest. An agent who

---

[20] Specifically, the model is that of a One-Sided Prisoner's Dilemma game (OSPD) with perfect and complete information. There are M merchants and A agents in the economy, and it is assumed that $M < A$. Players live an infinite number of periods, agents have a time discount factor β, and an unemployed agent receives a per-period reservation utility of $\phi_u \geq 0$. In each period, an agent can be hired by only one merchant and a merchant can employ only one agent. Matching is random, but a merchant can restrict the matching to a sub-set of the unemployed agents that contains the agents who, according to the information available to the merchant, have previously taken particular sequences of actions. (The following assumes that the probability of re-matching with the same agent equals zero for all practical considerations.) A merchant who does not hire an agent receives a payoff of $\kappa > 0$. A merchant who hires an agent decides what wage ($W \geq 0$) to offer the agent. An employed agent can decide whether to be honest or to cheat. If he is honest, the merchant's payoff is $\gamma - W$, and the agent's payoff is W. Hence the gross gain from cooperation is γ, and it is assumed that cooperation is efficient, $\gamma > \kappa + \phi_u$. The merchant's wage offer is assumed credible, since in reality the agent held the goods and could determine the *ex post* allocation of gains. For that reason, if the agent cheats, the merchant's payoff is 0 and the agent's payoff is $\alpha > \phi_u$. Finally, a merchant prefers receiving κ to being cheated or paying $W = \alpha$, that is, $\kappa > \gamma - \alpha$. After the allocation of the payoffs, each merchant can decide whether to terminate his relations with his agent or not. There is a probability σ, however, that a merchant will be forced to terminate agency relations and this assumption captures merchants' limited ability to commit to future employment due to the need to shift commercial operations over places and goods and the high uncertainty of commerce and life during that period. For a similar reason, the merchants are assumed to be unable to condition wages on past conduct (indeed, merchants in neither group did so). Hence, attention is restricted to an equilibrium in which wages are constant over time. (For an efficiency wage model in which this result is derived endogenously, see MacLeod and Malcomson (1989). Their approach can be used for this analysis as well but is omitted to preserve simplicity.)

has cheated in the past and thus is not expected to be hired by merchants will not lose this value if caught cheating again. Thus, if a merchant nevertheless hires such an agent, the merchant has to pay a higher (efficiency) wage to keep the agent honest (relative to an agent who did not cheat in the past). Each merchant is thus induced to hire only honest agents – agents who are expected to be hired by others.

Acquiring and transmitting information during the late medieval period was costly, and hence the model should incorporate a merchant's decision to acquire information. Since merchants could gather information by forming an information-sharing network, suppose that each merchant can either "Invest" or "Not Invest" in "getting attached" to such a network before the game begins, and that his action is common knowledge. Investing entails a cost at each period in return for which the merchant learns the private histories of all the merchants who also invested; otherwise, he knows only his own history. Under the collectivist equilibrium, history has value and thus merchants are motivated to invest even though cheating does not occur on the equilibrium path.

The effectiveness of a collective punishment in inducing efficiency-enhancing agency relationships is undermined if an agent who cheated is not restricted from using the capital he embezzled as profitably as the merchant he cheated. If an agent can use the capital he embezzled as profitably as the merchant by becoming a merchant himself and hiring an agent, his lifetime expected utility is higher following cheating relative to a situation in which he cannot. This implies that to induce honesty, a merchant has to pay a higher wage to his agent. As a matter of fact, this wage has to be so high that it would be better for an individual to be an agent rather than a merchant. Agents would have to be paid more than half of the return on each business venture. The need to pay such a high wage, in turn, increases the set of situations in which merchants would not initiate efficiency-enhancing agency relationships, finding them to be unprofitable.

However, there is no historical evidence that agents were restricted exogenously from investing capital in trade. Such a restriction can be generated endogenously, however, under collective punishment. In particular, an agent who himself also acts as a merchant (and invests his capital through agents) can be endogenously deterred from embezzling under collective punishment (even when his share in a business venture is less than that of a merchant). When an agent also acts as a merchant, a strategy specifying non-punishment of agents who cheated a merchant who had cheated while acting as an agent is both self-enforcing and further reduces the agent's gain from cheating. It potentially enables hiring agents despite their ability to invest the capital they embezzled in trade.

When agency relations are governed by a "coalition" – by a group of merchants using the above strategy with respect to a particular group of agents – the collective punishment enables the employment of agents even when a specific merchant and agent are not expected to interact again. The gains from cooperation within the coalition (compared to hiring agents based on bilateral punishment), and the expectations concerning future hiring among the coalition's members, ensure the coalition's "closeness". Coalition members are motivated to hire and to be hired only by other members, while nonmembers are discouraged from hiring the coalition's members.[21] Similarly, since membership in the coalition is valuable, an overlapping generations version of the model in which sons inherit their fathers' membership and support them in their old age, shows how the unraveling problem can be avoided. Holding a son liable for his father's actions can motivate a father to be honest even in his old age.

The above theoretical discussion provides some of the conditions for and implications of governing agency relations by coalitions. Some of these implications are distinct from those generated by a bilateral efficiency-wage model [Shapiro and Stiglitz (1984)] or a model of incomplete information about agents' types. For example, within the coalition agency relations are likely to be flexible – merchants shift among agents and hire agents even for a short time in cases of need. Merchants also prefer hiring other merchants as their agents, perhaps through forms of business associations that require agents' capital investments. Further, members of the coalition are likely to be segregated from other merchants in the sense that they will not establish agency relations outside the coalition even if these relations – ignoring agency cost – are more profitable. Similarly, the sons of coalition members are likely to join the coalition.

Indeed, the *geniza* documents that reflect the operation of the Maghribis reveal the above conditions for, and implications of, governing agency relations by a coalition. They reflect a reciprocity based on a social and commercial information network with very flexible and non-bilateral agency relations. There was no merchant or agent "class" among the Maghribis while merchants hired other merchants as their agents and utilized forms of business associations that required agents' investments. Furthermore, the Maghribis did not establish agency relations with other (Jewish or non-Jewish) traders even when these relations were considered by them to be very profitable. To begin operating in a new trade center, some Maghribis immigrated to this center and began providing agency relations. Finally, traders' sons indeed supported their fathers in their old age and inherited membership in the coalition and family members were held morally (but not legally) responsible for each other. Notably, these observed features among the Maghribis did not prevail among the Italian traders who operated (particularly from the twelfth century) in the same area as the Maghribis, trading in the same goods, and using comparable naval technology. Bilateral, rather than collective, punishment was the norm that prevailed among Italian traders. (See further discussion below.)

In addition to this indirect evidence for the governance of agency relations among the Maghribis by a coalition, the *geniza* contains direct evidence on various aspects of the coalition. Explicit statements reflect the expectations for a multilateral punishment, the economic nature of the expected punishment, the linkage between past conduct and future employment, the interest that all coalition members took in the relations between a specific agent and merchant, and so forth. Further, the *geniza* reveals the existence of a set of cultural rules of behavior that alleviated the need for comprehensive agency contracts and coordinated responses by indicating what constituted "cheating".

---

[21] See, in particular, Greif (1994a).

The factors leading to the selection of this particular strategy are also reflected in the historical records, which suggest that multilateral punishment prevailed among the Maghribis because of a social process and cultural traits. The Maghribis were descendants of Jewish traders who left the increasingly politically insecure surroundings of Baghdad and emigrated to North Africa during the tenth century. Arguably, this emigration process, as well as their cultural background which emphasized collective responsibility, provided them with an initial social network for information transmission and made the collective punishment strategy a focal point. This particular social process and cultural background led to the governance of agency relations by a coalition, while the economic incentives generated by the coalition strengthened the Maghribis' distinct social identity. Indeed, the Maghribis retained their separate identity within the Jewish population until they were forced, for political reasons, to cease trading. This interrelation between social identity and the economic institution that governed agency relations suggests that the Maghribi traders' coalition did not necessarily have an efficient size. But because expectations regarding future employment and collective punishment were conditional on a particular ascribed social identity, there was no mechanism for the coalition to adjust to its economically optimal size.

Clay (1997) studied the informal contract enforcement institution that prevailed among long-distance American traders in Mexican California. Her evidence suggests that, in contrast to the situation among the Maghribis, the traders did not sever all their relations with an agent who cheated any of them. To understand this difference, Clay noted that an agent among these traders had a monopoly over credit transactions with members of a particular Mexican community. Since contract enforcement within each community was based on informal social sanctions, in order to sell on credit a trader had to settle in a community, marry locally, raise his children as Catholics, and speak Spanish at home. Furthermore, the small size of the Mexican communities implied that it was profitable for only one retailer to integrate in this way.

Clay incorporated this feature in an infinitely repeated game with imperfect monitoring, and found that the strategy of permanent and complete punishment of a trader who cheated in agency relations would have been Pareto-inferior. Such a strategy barred all the traders from operating in the community where the cheater had a monopoly over contract enforcement. A strategy entailing a partial boycott for a limited time following a first instance of cheating Pareto-dominates the complete boycott strategy. The boycott is partial in the sense that it does not preclude transactions requiring the use of the cheater's local enforcement ability. A complete boycott follows only after an act of cheating during a boycott. Direct and indirect evidence indicates that such a strategy was utilized by the traders. Hence, an environment different from that of the Maghribis led to a different Pareto-superior strategy.

Contract enforcement among "anonymous" individuals: Two studies examined contract enforcement over time and space among "anonymous" individuals.[22] Such contract

---

[22] More accurately, the exchange modeled in this line of research is "impersonal". See discussion in Greif (2000a).

enforcement over time was required at the Champagne Fairs in which, during the twelfth and the thirteenth centuries, much of the trade between northern and southern Europe was conducted. Milgrom, North and Weingast (1990) argued that in the large community of merchants who frequented the fairs, a reputation mechanism could not enable traders to commit to respect their obligations since such a large community lacked the social network required to make past actions known to all. Furthermore, the fairs' court could not directly impose its decision on traders after they left the fairs. Milgrom, North and Weingast suggested that a *Law Merchant* system, in which a court supplements a multilateral reputation mechanism, can ensure contract enforceability in such cases. Suppose that each pair of traders is matched only once and each trader knows only his own experience. Further assume that the court is capable only of verifying past actions and keeping records of traders who cheated in the past. Acquiring information and appealing to the court is costly for each merchant. Despite these costs, however, there exists a (symmetric sequential) equilibrium in this infinitely repeated, complete-information game in which cheating does not occur and merchants are induced to provide the court with the information required to support cooperation. It is the court's ability to activate the multilateral reputation mechanism by collecting and dispensing information that provides the appropriate incentives. Furthermore, there exists an equilibrium in which the traders' threat to withdraw from future trade is sufficient to deter the court from abusing its information to extort money from the traders. However, the paper does not go far enough in establishing the historical validity of its argument; it only points out that the fairs' authorities controlled who was permitted to enter the fairgrounds.

Court and other historical records from western and southern Europe dating back to the mid-twelfth century indicate the operation of another mechanism that enabled anonymous contracting over time and place. Traders applied a principle of community responsibility that linked the conduct of any trader with the obligations of every member of his community. For example, if a debtor from a specific community failed to appear at the place where he was supposed to meet his obligations, the lender could request the local court to confiscate the goods of any member of the debtor's community present at that locality. Those whose goods were confiscated could then seek a remedy from the original debtor. Traders used intra-community enforcement mechanisms to support inter-community exchange.

Historians have considered this "community responsibility system" for contract enforcement to be "barbaric" since it sometimes led to "retaliation phases" in which an accusation of cheating led to the end of trade between two communities for an extended time. Further, some other facts about the system – such as regulations aimed at increasing the cost of default to a lender, attempts of wealthy merchants from large communities to be exempt from the system, and its demise at the end of the thirteenth century – remain unexplained. Greif (1996a, 2000a) used a repeated, imperfect-monitoring game that captures the essence of the situation to explain these features as well as evaluate the pros and cons of the system.

The analysis indicates the rationale behind the costly retaliation phases as an on-the-equilibrium-path-of-play behavior required to maintain cooperation. They reflect asymmetric information between two local courts which, at times, reached different conclusions regarding the fulfillment or non-fulfillment of the contractual obligations. The regulations that increased the cost of default to the lender, and the attempts of wealthy merchants from large communities to be exempt from it, reflect the moral hazard problem generated by the system. Efficiency required that a lender verify the borrower's creditworthiness. But the system implied that he also considered the future possibility of obtaining compensation from the borrower's community, leading to misallocation of credit. Increasing the lender's cost of default mitigated this problem, while well-to-do members of wealthy communities were particularly interested in being exempt from the system since their community's wealth and size fostered the moral hazard problem. While they had the personal reputation required to borrow without community responsibility, their wealth enabled less creditworthy members of their community to borrow as well. These wealthy merchants thus had to pay the cost required to enable other members of their community to borrow and they gained less and paid more for the community responsibility system and therefore they wanted to be exempt from it. The model and historical evidence suggest that the decline of the system followed an increase in its cost in terms of retaliation phases and the moral hazard problem. The cost increase was due to the rising number of trading communities, the increased wealth of some communities, and social and political integration that enabled one to falsify his community's membership.

3.2.2. The state: Emergence, nature, and function

The European states were important economic decision-makers and the competition among them is often invoked to account for the rise of the Western World. Several studies used dynamic and repeated games with complete information and dynamic games with incomplete information to examine the relations between economic factors and the origin and nature of the European state. They indicate the importance of viewing a state as a self-enforcing institution, and the role of intra-state organizations in enhancing cooperation among various elements within the state, and they advance new interpretations of the parliamentary system. Finally, they reveal a reason why wars may occur even in a world with symmetric information and transferable utility.

The emergence and origin of the state: Among the most intriguing cases of state formation in medieval Europe is that of the Italian city states. They were formed through voluntary contracts and many of them experienced very rapid economic growth from the eleventh to the fourteenth centuries. Genoa is a case in point. It was established around 1096 for the explicit purpose of advancing the profits of its members, and indeed it emerged from obscurity to become a commercial empire that stretched from the Black Sea and beyond to northern Europe. Advancing the economic prosperity of Genoa required cooperation between Genoa's two dominant noble clans (and their political factions). They could militarily cooperate in raiding other political units or acquiring commercial rights from them, such as low customs or parts of ports which yielded rent every period after their acquisition. The acquisition of such rights was the key to the city's long-term economic prosperity. Yet, for cooperation to emerge, each clan had to expect to gain from it despite each clan's *ex post* ability to use its military power to challenge the other for its share in the gains. No clan made such an attempt from 1096 to 1164, but from 1164 to 1194 inter-clan warfare was frequent. Was inter-clan cooperation in acquiring rights prior to 1164 limited by the need to ensure the self-enforceability of the clans' contractual relations regarding the distribution of gains? Why did inter-clan warfare occur after 1169? Did the clans attempt to alter the rules of their game to enhance cooperation after 1164? These questions have been raised by historians but could not be addressed without an appropriate game-theoretical formulation.

Greif (1994a, 1998c) analyzed this situation as a dynamic game with complete information regarding the clans' military strength, but uncertainty regarding the outcome of a military conflict. The analysis indicates that self-enforceability may limit cooperation in the acquisition of rights. If the clans are at a mutual-deterrence equilibrium with less than the efficient number of rights, it may not be in a clan's interest to cooperate in acquiring additional rights. In a mutual-deterrence equilibrium with less than the efficient number of rights, neither clan challenges the other since the expected cost of the war and the cost implied by the possibility of defeat outweigh the expected gains from capturing the other clan's share. With additional rights, the increase in the expected benefit for a clan from challenging may exceed the increase in the expected cost of potential defeat, leading to either a military confrontation or a clan being forced to increase its investment in military ability. This "political cost" of acquiring rights implies that each clan may find it optimal to cooperate in acquiring less than the efficient number of rights.

The above analysis is incomplete since in the case of Genoa there is no justification for taking the clans' share in the gains as exogenous. Can the clans necessarily overcome the economic inefficiency implied by the political cost by re-allocating the gains? Suppose that one clan finds it beneficial to challenge the other given its share in the gains and military strength. Although the game is one of complete information, there still may be no other Pareto-superior equilibrium in which inter-clan war is prevented. It may not exist due to the uncertainty involved about military conflicts and the clans' ability to use its share in the gains to increase its military strength. Increasing the share in the gains of the clan that is about to challenge decreases the gains from a victory but increases the chance of winning. The other clan may thus be better off fighting while retaining its original allocation. Hence, the ability of Genoa's clans to cooperate could have been constrained by the extent to which their relations were self-enforcing.

But was this historically the case? Did the need for self-enforceability constrain the clans' economic cooperation? Under the assumption that it did, the model yields various predictions, such as the time-path of cooperation in raids and the acquisition of rights, investment in military strength, and responses (including inter-clan military confrontation) to various exogenous changes. These predictions are confirmed by the historical records. It was only in 1194 that a process of learning and severe external threat to Genoa motivated the clans to establish a self-enforcing organization which was known as a *podestà* (that is, a "power"). It altered the rules of the political game in Genoa to increase the set of parameters (including the number of rights) under which inter-clan cooperation could be achieved as a mutual deterrence equilibrium outcome. Furthermore, the podesteria coordinated on such an equilibrium. Essentially, the *podestà* was a non-Genoese hired for one year to govern Genoa and supported by his own military contingent. The *podesteria's* self-enforcing regulations were such that the *podestà* could commit to use his military power (only) against any clan attempting to militarily challenge another. It was under the *podesteria*, which formally lasted about 150 years, that Genoa reached its political and commercial apex. Understanding Genoa's commercial rise requires comprehension of its political foundations.[23]

Green (1993) analyzed the emergence of the parliamentary government of England during the thirteenth century, an event which arguably contributed to England's subsequent growth. His analysis supports the conjecture that a shift to parliamentary government alleviates the cost of communicating private information. The existing balance-of-power theory that views changes in governmental systems as indicative of changes in the technology of capturing or defending property, fails to explain a central provision in the Magna Carta (1215). Instead of requesting tax cuts, the English barons insisted that the king should request their consent for new taxes. Green argues that this request reflects the benefit of communication and exchange between the parties.

The loss of the English Crown's possessions in France shortly before 1215 and the growing complexity of European politics increased the threat of an external invasion of England and implied that the king had better information regarding such an invasion. To see why such an external threat and private information might make a political system based on communication Pareto-optimal, Green analyzed the following model. Consider a one-period game in which a ruler can always expropriate (at most) half of the subject's crop. There is some probability of an external threat to the whole crop which the ruler can successfully confront by (a) taking a costly action such as assembling an army, and then (b) confronting the threat (without any additional cost). If there is no threat the ruler prefers half the crop over taking the costly action. If there is a threat the ruler prefers taking the action if provided with two-thirds of the crop rather than not taking the action and getting no share at all. The subject can provide the ruler with two-thirds of the crop between the time of events (a) and (b). Whether the external threat is about to materialize or not is the ruler's private information. In this model there is a Bayesian Nash equilibrium in which the ruler communicates that the external threat has materialized by taking the costly action and the subject provides him with two-thirds of the crop. Despite the cost of the communication, this equilibrium Pareto-dominates the equilibria in which there is no communication. Hence, a shift to parliamentary government may reflect the benefit of such costly communication.

Committing to respect property rights: The five bankruptcies of the Spanish Crown (1557, 1575, 1596, 1607, and 1627) during the height of Spanish economic and political dominance are often used to demonstrate the limitations of a pre-modern public

---

[23] See also Rosenthal (1998), who examined how the game between kings and nobles in France and England influenced economic and political outcomes.

finance system [Cameron (1993), p. 137]. The rulers' inability to commit to the property rights of their lenders hindered their ability to borrow. In a detailed historical and game-theoretic analysis, Conklin (1998) advanced a different interpretation of these bankruptcies as reflecting a routine realignment of the king's finances. In other words, they do not reflect the failure of a system but its effective operation. These bankruptcies were not a wholesale repudiation of obligations to creditors. They were initiated by the Genoese, the king's foreign lenders, who ceased providing him with credit. In response, the king ceased paying the Genoese and negotiated a partial repayment of his debt to them with debt obligations directly linked to particular tax-generating assets the king had in Spain. The Genoese could sell these obligations to Spain's elite, namely, the king's local lenders.[24] After this realignment, the Genoese resumed lending.

For this analysis, Conklin used a repeated game with state variables, an important component of which is that the king "cares" about the welfare of his Spanish lenders. The justification for this specification was the king's dependency on these elite Spaniards for tax collection, administration, and military operations. Solving for a Pareto-optimal, sub-game-perfect equilibrium using a computer algorithm indicated that financial realignment should have occurred when the king reached the limit of his ability to commit to the Genoese. Shifting some of this debt to his elite to whom he could commit was a prerequisite for additional lending. This interpretation gains additional support, such as the particularities of the tax collection system and the king's willingness to prey on the wealth of particular Spaniards while continuing to pay his local debt.

3.2.3. Within states

Game theory facilitated the analyses of historical market structures (the number and relative size of firms within an industry), financial systems, legal systems, and development. At the same time, these analyses used historical data sets that enabled the examination of game-theoretical industrial organization models, confirmed game-theoretical predictions regarding the relations between rules and behavior, suggested the role of banks and the distribution of rent in initiating a move from one equilibrium to another, and inspired new game-theoretical models of financial and market structure.

Market structure and conduct: Business records from the period prior to and following the Sherman Act provide unique data sets for examining the relations between strategic behavior and market structures. Their analysis substantiated the importance of predation and reputation in influencing market structures, enabled empirical examination of models of tacit collusion, indicated limits of using only indirect (econometric) evidence in examining inter-firm interactions, and suggested that existing models of market structures are deficient in ignoring the multi-dimensionality of inter-firm interactions.

---

[24] In the process the Genoese had to bear some losses.

Burns (1986) examined the role that a reputation for predatory pricing played in the emergence of the tobacco trust between 1891 and 1906. An econometric analysis of the purchases of 43 rival firms by the old American Tobacco Company indicated that alleged predation significantly lowered the acquisition costs, both for asserted victims and, through reputation effects, for competitors that sold out peacefully. Similarly, Gorton (1996) examined the formation and implication of reputation in the Free Banking era (1838–60) during which new banks could enter the market and issue notes. Economic historians noticed, but found it difficult to explain, that "wildcat" ("fly by night") banking was not a pervasive problem during this period. The Diamond (1989) incomplete-information model suggests that if some banks were wildcats, some were not, and some could have chosen whether to be wildcats or not. A process of reputation formation may have deterred banks from choosing to become wildcats. Due to information asymmetry, all banks initially faced high discount rates, but following the default of the wildcats the discount rate declined due to the reputation acquired by the surviving firms. This decline, in turn, further motivated firms that could choose whether or not to be wildcats. Gorton conducted an econometric analysis of various aspects of these conjectures (particularly whether new banks' notes were discounted more and if this discount depended on the institutional environment), which confirmed the importance of reputation in preventing wildcat operations.[25]

Weiman and Levin (1994) combined direct and indirect (econometric) evidence to examine the development and implication of the strategy employed by the Southern Bell Telephone Company to acquire a monopoly position between 1894 and 1912. In contrast to the usual assumption in industrial organization, the strategic variable that enabled it to become a monopoly was not only price but also investment in toll lines ahead of demand, isolating independents in smaller areas, and influencing regulations by increasing the cost of competition to the users. Similarly, Gabel (1994) substantiated that between 1894 and 1910 AT&T acquired control over the telephone industry through predatory pricing. Despite its short-term cost, price reduction enabled AT&T to deter entry and to cheaply buy the property of independents. This strategy was facilitated by rate regulations and capital market imperfections that prevented independents from entering on a large scale.[26]

In a classic study, Porter (1983) examined collusion of a railroad cartel (the "Joint Executive Committee") established in 1789 to set prices for transport between Chicago and the East Coast. Using data from 1880 to 1886 the study tested and could not reject the relevance of Green and Porter's (1984) theory of collusion with imperfect monitoring and demand uncertainty. (The alternative was that price movements reflected exogenous shifts in demand and cost functions.) According to the Green and Porter theory, price wars occur on the equilibrium path due to the inability of firms to distinguish between

---

[25] See also Ramseyer's (1991) study of the relations between credible commitment and contractual relations in the prostitution industry in Imperial Japan.

[26] See also Nix and Gabel (1994).

shift in demand and a firm's defection. A sufficiently low price triggers a price war of some finite length, and although all firms realize that no deviation has occurred, punishment is required to retain collusion. Similar results were obtained by Ellison (1994), who compared the Green and Porter model with that of Rotemberg and Saloner (1986) in which price wars never transpire and on-the-path prices exhibit a counter-cyclical pattern.

The Green and Porter model also provides the theoretical framework used by Levenstein (1994, 1996a, 1996b) to examine collusion in the pre-WWI US bromine industry. An econometric analysis could not reject the Green and Porter model, indicating that price wars stabilized collusion. Yet, Levenstein (1996a) claimed that this conclusion is misleading. Using ample direct evidence, Levenstein substantiated that only a few wars stabilized collusion in the Green and Porter sense and these were short and mild. Long and severe price wars were either bargaining instruments aimed at influencing the distribution from collusion, or a profitable deviation from collusive behavior made possible by asymmetric technological changes. Finally, similar to the studies discussed above, her paper casts doubt on the empirical relevance of the strategic models of collusion which assume that price is the only strategic variable. In the bromine industry, collusion of firms was facilitated by altering the industry's information structure and marketing system [Levenstein (1996b)].

Financial systems and development: The nature and role of financial intermediaries that functioned prior to the rise of banks and securities markets have hardly been examined, limiting our understanding of pre-modern financial markets. Furthermore, the relations between development and distinct financial systems (differentiated by the nature and relative importance of banks and securities markets) have been examined by economic historians [Mokyr (1985), p. 37]. Only recently has the application of game theory to historical research facilitated the analysis of alternative intermediaries, enabled exploring the origins and implications of diverse financial systems, and suggested a heretofore unexamined role for banks in coordinating development.

Hoffman, Postel-Vinay, and Rosenthal (1998) used a repeated-game model and an unusual data set from Paris (1751) to theoretically and econometrically examine the operation of a credit system as an alternative to banks and security markets. In Old Regime France, notaries had property rights over the records of any transaction they registered. Hence, they had a monopoly over the information required for screening and matching potential borrowers and lenders. Their ability to provide credit market intermediation, however, could have been hindered by a "lock-in" effect. Unless they were able to commit not to exploit their monopoly power, potential participants in the credit market would have been deterred from approaching them. A game-theoretical formulation of this problem revealed that it could have been mitigated by an equilibrium in which notaries shared information with each other to reduce each notary's monopoly power over his clients. Indeed, the data confirms the behavior associated with this equilibrium.

There is a consensus among economic historians that different financial and industrial systems prevailed in the first and second major industrialized European countries

– namely, England and Germany.[27] English firms were relatively small and tended to be financed through tradeable bonds and arm's-length lending. The German firms, however, were large and tended to be financed by loans from particular banks which closely monitored them. Motivated by this difference, Baliga and Polak (1995) attempted to explore its origins and rationale using a dynamic game capturing the moral hazard problem inherent in industrial loans. Entrepreneurs would provide only second-best effort levels in the absence of monitoring, while costly monitoring induces more effort. Monitoring exhibits internal economies of scale while markets for tradeable loans exhibit external economies of scale.

The analysis provides the foundation for potentially beneficial future empirical analysis. It leads to comparative statics predictions regarding the relations between exogenous factors (such as base interest rates, firm size, and the lender's bargaining power) and the entrepreneur's choice of financial arrangements. Furthermore, it indicates the possible impact of the entrepreneurs' wealth and the market in government securities on the efficiency and selection of financial arrangements. When the analysis is further extended to an entry game in which entrepreneurs choose a firm's size and banks choose whether to acquire monitoring ability, multiple equilibria exist, indicating a possible rationale for the emergence and persistence of different systems.

The role of banks in coordinating development is suggested by the "big-push" theory of economic development [Murphy et al. (1989)]. When externalities make investment profitable only if enough firms invest at the same time, failure to coordinate on such a simultaneous investment may lead to an "underdevelopment trap". Inspired by this model and the positive historical correlation between rapid industrialization and large banks with some market power or with large equity holdings in industrial firms, Da Rin and Hellmann (1998) developed a model of the role of banks in coordinating a transition to an equilibrium in which firms invest. Utilizing a dynamic game with complete information they suggested that this positive correlation reflects the role of large banks in initiating a move from one equilibrium to another. A necessary condition for banks to coordinate industrialization is that at least one bank (or coordinated group of banks) is large enough to initiate a big push by subsidizing the investment of a critical mass of firms to induce them to invest. A bank's monopoly power or capital investment is required, however, to motivate that bank to coordinate by enabling it to benefit from the industrialization it triggered.

Law, development, and labor relations:[28] Game-theoretical models were used to evaluate the impact of legal rules and procedures on development and labor relations in various historical episodes. Rosenthal (1992) established that, despite the efficiency of several potential drainage and irrigation projects in France from 1700 to 1860, they were not carried out. In contrast, efficient projects were undertaken in England during this period, as well as in France after the French Revolution. The efficient projects were

---

[27] Some recent papers doubt this difference; see Fohlin (1994) and Kinghorn and Nye (1996).

[28] See also Milgrom et al. (1990) and Greif (1996a, 2000a).

not carried out since the village which had some *de facto* or *de jura* property rights over the land, and the lord who wanted to initiate the project, failed to reach an agreement regarding the distribution of the gains. Rosenthal conjectured that the distinct legal features of Old Regime France accounted for this failure.

To demonstrate that the legal features of the Old Regime could have inhibited reaching an agreement, Rosenthal used a dynamic incomplete-information game. Central to the model is asymmetric information regarding the legal validity of the village's rights over the land and the "burden of proof" rule, namely, whether the property rights will be assigned to the lord or the village if neither party is able to establish *de jura* rights over the land. The analysis indicates that legal prohibition on out-of-court settlements, the burden of proof rule that favored the village, and the high cost of using the legal system, increased the number of efficient projects that lords would find unprofitable to initiate. All these features of the legal system prevailed in Old Regime France but not in England or post-revolutionary France.[29]

Treble (1990) examined the impact of legal regulations on wage negotiations in the British coal industry from 1893–1914. These negotiations were conducted in "conciliation boards", and in cases of disagreement an arbitrator had to be used. The number of appeals made to arbitrators differed greatly among coal fields, and ranged from as low as 11 percent to as high as 56 percent (per number of negotiations). The economic historians' traditional explanation of these differences is that they reflected differences in the negotiators' personalities. Treble, however, modeled the bargaining process as a game which [unlike many other bargaining models, such as Farber (1980)] generated predictions regarding the frequency of appeals to arbitration. Treble's analysis predicted that because delay in reaching an agreement had a strategic value, the more the constitution of a particular conciliation board permitted delay without arbitration, the less arbitration would be used. When this hypothesis and the alternative, that appeal depended on personalities or reflected asymmetric information regarding the arbitrator's preference [Crawford (1982b)], was placed under econometric analysis, it could not be rejected.

Moriguchi (1998) has utilized a game-theoretic framework to conduct a comparative study of the evolution of labor relations in Japan and the US from 1900 to 1960. She modeled the relationships between potential employers and employees as a repeated game and examined the resulting set of equilibria for various parameter values. Furthermore, she examined the political ramifications of various equilibria and how, in turn, they influenced the endogenous political formation process of labor regulations. This framework enabled her to provide a novel interpretation of the evolution of labor relationships. The combined theoretical and historical analysis highlights, for example, the essence of employer paternalism as a mechanism to move from one equilibrium to another, and the importance of the Great Depression in causing a bifurcation of the equilibrium selected in each economy. Furthermore, the analysis indicates the importance

---

[29] See also Besley, Coate, and Guinnane (1993), who studied the evolution of laws governing provisions for the poor.

of various factors that led to the persistence of each equilibrium after the Great Depression, such as coordination cost and sunk investment in complementary regulations and technology.

3.2.4. Between states

Applying repeated and static games to historical analyses has indicated the importance of non-market institutions in influencing the historical process through which long-distance trade grew and the empirical relevance of the ideas behind renegotiation-proof equilibrium.[30] It has also lent support to the New International Trade Theory, by indicating the implications of the intra-firm incentive structure on inter-firm strategic interaction, how strategic international relations impact domestic economic policy, how cooperation can evolve, and what the relations are between equilibrium selection and credible, public communication.

International trade: Greif, Milgrom, and Weingast (1994) examined the operation and implications of an institution that enabled rulers during the late medieval period to commit to the security of alien traders' property rights. Having a local monopoly over coercive power, any medieval ruler faced the temptation to abuse the property of alien merchants who frequented his realm. Without an institution enabling a ruler to commit *ex ante* to secure alien merchants' rights, they would not have come to trade.

Since trade relationships were expected to be repeated, one may conjecture that a *bilateral reputation mechanism* in which a merchant whose rights were abused ceased trading, or an uncoordinated *multilateral reputation mechanism* in which a subgroup larger than the one that was abused ceased trading, could surmount this commitment problem. This conjecture, however, is wrong. Although each of these mechanisms can support some level of trade, neither can support the *efficient level of trade*. The bilateral reputation mechanism fails because, at the efficient level of trade, the value of future trade of the *"marginal"* traders to the ruler is zero, and hence the ruler is tempted to abuse their rights (irrespective of the distribution of the gains from trade and the ruler's discount factor). In a world fraught with information asymmetries, slow communication, and different possible interpretations of facts, the multilateral reputation mechanism is prone to fail for a similar reason. Hence, theoretically, a multilateral reputation mechanism can potentially overcome the commitment problem only when the merchants have some organization to coordinate their actions. Such a coordinating organization implies the existence of a Markov-perfect equilibrium in which traders come to trade (at the efficient level) as long as a boycott is not announced, but none of them come to trade if one is announced. The ruler respects the merchants' rights as long as a boycott is not announced, but abuses their rights otherwise. When a coordinating institution exists, trade may plausibly expand to its efficient level.

---

[30] The discussion above regarding agency relations and anonymous trade also relates to the institutions and long-distance trade. For an elaboration of the insights provided by game-theoretical analyses of long-distance trade in history, see Greif (1992).

Although the strategy just described leads to a perfect equilibrium, the theory in this form remains unconvincing considering the ideas behind renegotiation-proof equilibrium. According to the above equilibrium strategies, when a coordinating institution declares an embargo, merchants will pay attention to it because they expect that the ruler will abuse a violator's property rights. But are these expectations reasonable? Why would a ruler not *encourage* embargo-breakers rather than punish them? Encouragement is potentially credible since during an effective embargo the volume of trade shrinks and the value of the marginal trader increases; it is then possible for bilateral reputation mechanisms to become effective. This possibility limits the potential severity of an embargo and potentially hinders the ability of any coordinating organization to support efficient trade. In such cases, the efficient level of trade can be achieved when a multilateral reputation mechanism is supplemented by an organization with the ability to coordinate responses *and* ensure the traders' *compliance* with boycott decisions.

Direct and indirect historical evidence indicates that during the Commercial Revolution an institution with these attributes – the *merchant guild* – emerged and supported trade expansion and market integration. Merchant guilds exhibited a range of administrative forms – from a subdivision of a city administration, such as those of the Italian city-states, to an inter-city organization, such as the German Hansa. Yet, their functions were the same: to ensure the coordination and internal enforcement required to make the threat of collective action credible. The nature of these guilds and the dates of their emergence reflect historical as well as environmental factors. In Italy, for example, each city was large enough to ensure that its merchants were not "marginal" and its legal authority ensured their merchants' compliance with the guild's decisions. In contrast, the relatively small German cities had to organize themselves as one guild through a lengthy process to be able to inflict an effective boycott.

Irwin (1991) utilized a game-theoretical model to examine the competition between the English East India Company and the Dutch United East India Company during the early seventeenth century. The Dutch were able to achieve dominance in the trade in pepper brought from the Spice Islands of Indonesia, although both companies had similar costs and sold the pepper for the same price in the European market. To understand the sources of Dutch dominance, Irwin argued that the nature of the competition in the pepper market resembles Brander and Spencer's (1985) model of duopolistic competition in which two companies exporting a single good are engaged in a one-period Cournot (quantity) competition. The English and Dutch companies competed mainly in the market for pepper and both were state monopolies whose charters could have been revoked (*de jura* or *de facto*) in any period. If this was indeed the situation, Brander and Spencer's analysis indicates that any trade policy shifting one company's reaction function outward increases its Nash equilibrium profit while reducing that of the other. The policy that seems to have shifted the Dutch company's reaction function was instituted through its charter. It specified that its managers' wages should be a function of the company's profit *and* volume of trade, thereby shifting the company's reaction function outward [as in Fershtman and Judd (1987)].[31] The intra-firm incentive structure influenced inter-firm competition.

International relations: Maurer (1992) conducted a case study of the Anglo–German naval arms race from 1912 to 1914, providing an interesting example of actual equilibrium selection and the "evolution of cooperation" [Axelrod (1984)]. During this period the arms race resembled a repeated prisoners' dilemma game as both countries recognized the high cost it entailed. Some informal cooperation had been achieved when, in 1912, the First Lord of Britain's Admiralty, Winston Churchill, publicly announced a tit-for-tat strategy. Beyond the number of battleships already approved for building in Germany and Britain, Britain would build two battleships for each additional German battleship. The Germans adopted their best response of not producing additional battleships after testing the credibility of Churchill's announcement. Thus the arms race and public spending were diverted in other directions, such as construction of destroyers, the build-up of ground troops, and enhancing the battleships' features. Negotiation over a formal and broader arms-control agreement failed, however, particularly because both parties were concerned that the discussion would worsen their relations by raising the contentious security issues reflected in the naval competition and its link to the wider issue of the European balance of power.

3.2.5. Culture, institutions, and endogenous institutional dynamics

A long tradition in economic history argues that culture and institutions influence economic performance and growth.[32] The study of the inter-relations between institutions and culture, however, has been hindered by the lack of an appropriate theoretical framework. This has limited the ability to address questions that seem to be at the heart of developmental failures: Why do societies evolve along distinct institutional trajectories? Why do societies fail to adopt the institutional structures of more economically successful ones? How do past institutions influence the rate and direction of institutional change? Game theory provides a theoretical framework that facilitates addressing these questions and revealing why institutional dynamics is a historical process.

Greif (1994a, 1996b) integrated game-theoretical and sociological concepts to conduct a comparative historical analysis of the relations between culture and institutions.

---

[31] Irwin argues that this result supports the view of mercantilism as a strategic trade policy. Arcand and Brezis (1993) take a similar stand.

[32] E.g., North (1981). For an elaboration on various concepts of institutions in economics and economic history and the contribution of the game-theoretic perspective, see Greif (1997c, 1998b, 2000b). Similar to the above papers, institutions are defined here as non-technologically determined factors that direct and constrain behavior while being exogenous to each of the individuals whose behavior they influence but endogenous to the society as a whole. (Hence, an institution can reflect actions taken by all the individuals whose behavior the institution influences or actions taken by other individuals or organizations, such as the court or the legislature.)

The analysis considers the cultural factors that led two pre-modern traders' societies – the Maghribis from the Muslim world and the Genoese from the Latin world – to evolve along distinct institutional trajectories. It builds on a distinction between two elements of institutions – expectations and organizations – which is made possible by game theory. A player's expectations about the behavior of others are part of the non-technologically determined constraints that a player faces. Organizations such as the credit bureau, the court of law, or the firm, can potentially constrain behavior as well by changing the information available to players, changing payoffs associated with certain actions, or introducing another player (the organization itself). Clearly, organizations can be either exogenous or endogenous to the analysis (as discussed in Greif (1997c)). When they are endogenous their "introduction" means that they are transformed from being an off-the-path-of-play recognized or unrecognized possibility to an on-the-path-of-play reality. This framework enables the explicitation of a role of culture in institutional dynamics.

Specifically, as mentioned above, the different cultural heritages and social processes of the Maghribis and the Genoese seem to have led to a selection of distinct equilibria in the merchant-agent game. The Maghribis reached a "collectivist equilibrium" that entailed a collective punishment, while the Genoese reached an "individualist equilibrium" that entailed bilateral punishment. What is more surprising, however, is that a game-theoretical and empirical analysis indicates that once the distinct expectations associated with these strategies were formed with respect to agency relations they became "cultural beliefs" and transcended the original game in which they had been formed. They transcended it in the sense that they influenced subsequent responses to exogenous changes in the rules of the game and the endogenous process of organizational development, or, in other words, they became a cultural element that linked games.

Classical game theory does not say much about such inter-game linkages: actions to be taken following an expected change in the rules of the game are part of the (initial) equilibrium strategy combination, while an equilibrium that will be selected following an unexpected change in the rules of the game has no relation to the equilibrium that prevailed prior to the change. Yet, comparing the responses of both groups to exogenous changes indicates that the equilibria selected following unexpected changes in the rules of the game had a predictable relationship to the equilibria that prevailed prior to the change.[33] Cultural beliefs provided the initial conditions in a dynamic adjustment process through which the new equilibria were reached. Furthermore, the initial equilibria were related in a predictable manner to subsequently historical organizational innovations. Differences in organizational innovations among the two groups regarding organizations such as the guild, the court, the family firm, or the bill of lading, can be consistently accounted for as reflecting incentives generated by the expecta-

---

[33] Similar results emerged in experiments. See Camerer and Knez (1996).

tions that following an organizational change, the original cultural beliefs will still prevail.[34]

The analysis of institutions among the Maghribis and Genoese in the late medieval period suggests the historical importance of distinct cultures and initial organizational features in influencing institutional trajectories and economic development. Interestingly, the cultural and organizational distinctions among the Maghribi and Genoese societies resemble those found by social psychologists and development economists to differentiate contemporary developing and developed economies.

In any case, the analysis indicates how cultural traits and organizational features that crystallized in the past influence the direction of institutional change. Societies evolve along distinct institutional trajectories because past cultural beliefs and organizations transcend the particular situation in which they were formed and influence organizational innovations and responses to new situations. Furthermore, societies fail to adopt the institutional structures of more economically successful ones because they can only adopt organizational forms and formal rules of behavior. The behavioral implications of these, however, depend on the prevailing cultural beliefs, which may not change despite the introduction of new organizations and rules.

The analysis of agency relationships among the Maghribis and Genoese also sheds some light on the sources of endogenous institutional change. The rate of institutional changes depended on organizational innovations while their particular cultural beliefs provided members of each group with distinct incentives to invent new organizational forms.

Other studies that applied game theory to historical analysis indicate additional ways that past institutions influenced the rate of institutional change. As elaborated in Greif (2000b), existing institutions cause endogenous institutional change by directing processes through which "quasi-parameters" change. Quasi-parameters are institutional and other features – such as preferences, technological and other knowledge, and wealth distribution – that are either part of the rules of the game or can be taken as exogenous in a study of institutional statics because (or when) they change slowly and only their cumulative change causes institutions to change.[35] Existing institutions and their implications direct the process through which quasi-parameters change, and thereby they can, over time, cause existing institutions no longer to be an equilibrium – i.e., self-enforcing. In other words, processes caused by existing institutions can lead, over time, to endogenous institutional change.

---

[34] Yet, the theory cannot account for the timing of these organizational changes. Historically, it took a long time to introduce an organization despite the incentives and possibility of an earlier introduction.

[35] The main reasons for this discontinuous influence are that the mapping from quasi-parameters to an equilibrium is a set-to-point mapping and that institutions coordinate behavior. Because a particular equilibrium can prevail for a large set of quasi-parameters, there is a range in which they may change and equilibrium will still exist. At the same time, because changes in the cultural beliefs associated with the prevailing equilibrium – the shared expectations that coordinate behavior on it – require coordination, that equilibrium is likely to persist despite the changes in quasi-parameters. See further Greif (2000b).

To illustrate endogenous institutional change, consider again the above discussion of the community responsibility system that supported inter-community "anonymous" exchange during the late medieval period [Greif (2000a)]. During the late thirteenth century, attempts were made throughout Europe to abolish this system and to establish alternative contract enforcement institutions based, in particular, on a legal system administrated by state that held a trader, and not his community, liable for his contractual obligations. A game-theoretic and historical examination of the transition away from the community responsibility system suggests that, ironically, it reflects the system's own implications. The community responsibility system was eroding the quasi-parameters that made it self-enforcing. It enabled trade to expand and merchants' communities to grow in size, number, and economic and social heterogeneity. These changes implied that over time the community responsibility system was no longer an equilibrium. The increase in intra-communities' economic heterogeneity, for example, implied (as discussed previously) that some community members had to bear the cost of the community responsibility system without benefiting from it. Hence, they used their political influence within their community to abolish the system.

3.3. Conclusions

Although all the above studies integrate game-theoretical and economic history analyses, they differ in their objectives, their methodologies, and the weight placed on theory versus history. Yet, they forcefully indicate the potential contribution of combining economic history, game theory, and economics. Game theory has expanded the domain of economic history by permitting examination of important issues that could not be adequately addressed using a non-strategic framework. It enabled the examination of such diverse issues as contract enforcement in medieval trade, the economic implications of legal proceedings in Old Regime France, trade rivalry between the Dutch Republic and England, bargaining in England's coal mines, and the process through which the industrial structure emerged in the US. It provided, among other insights, new interpretations of the nature and economic implications of merchant guilds, the Glorious Revolution, the role banks play in development, the structures of industries, and the free-banking era.

More generally, these studies indicate the promise of applying game theory to economic history to advance our understanding of a variety of issues whose study requires strategic analysis. Among them are the nature and implications of the institutional foundations of markets, the legal system, the inter-relations between culture and institutions, the link between the potential use of violence and economic outcomes, the impact of strategic factors on market structures, and the economic implications of organizations for coordination and information transmission. Further, although all the above studies used equilibrium analysis, they illuminated the sources and implications of changes and path dependence. Incentives and expectations on and off the equilibrium path indicate the rationale behind the absence or occurrence of changes, while institutional path dependence was found, for example, to be due to acquired knowledge and information, economies of scale and scope associated with existing organizations or technology, coordination failure, distributional issues, capital market imperfections, and culture.[36]

Perhaps the most important contribution of analyses combining the game-theoretic and historical approaches is that they provide additional support for the importance of examining strategic situations and the empirical usefulness of game theory. Indeed, history provides an unusual laboratory in which to examine the empirical relevance of game theory since it contains unique data sets regarding strategic situations and the relationship between rules and outcomes. Interestingly, infinitely repeated games that have been considered by many as indicating the empirical irrelevance of game theory because they usually exhibit multiple equilibria, were found to be particularly useful for empirical analysis. Further, the studies discussed above also demonstrate the limitations of the theory and suggest directions for future development. For example, they indicate in the spirit of Schelling (1960) and Lewis (1969) that understanding equilibrium selection may require better comprehension of factors outside (the current formulation) of games, such as culture. Similarly, the study of organizations and organizational path dependence indicates the importance of considering the process through which the rules of the game are determined and the implications of organizations on the equilibrium set and equilibrium selection.

Finally, economic history analyses using game theory have enhanced our knowledge regarding issues central to economics, such as the nature and origin of institutions, the strategic determinants of industry structures and trade expansion, collusion, property rights, the economic implications of political institutions, labor relations, and the operation of capital markets. Hence, they provide an additional dimension to the long and productive collaboration between economic history and economics. Furthermore, these analyses indicate the need for and the benefit of combining theoretical and empirical research that transcends the boundaries of history, economics, political science, and sociology.

The application of game theory to historical analysis is still in its infancy. Yet, it seems to have already reaffirmed McCloskey's (1976) claims regarding the benefits of the interactions between economic history and economics in general. It has provided an improved set of evidence to evaluate current theories, suggested theoretical advances, and expanded our economic understanding of the nature, origin, and implications of various economic phenomena.

---

[36] On path dependence and institutions, see North (1990), who emphasizes economics of scale and scope, network externalities, and a subjective view of the world, and David (1994), who emphasizes conventions, information channels and codes as "sunk" organizational capital, interrelatedness, complementarities and precedents. E.g., knowledge: Hoffman et al. (1998); scale, scope, coordination, and culture: Greif (1994a), Weiman and Levin (1994); distribution: Rosenthal (1992). For recent discussions of institutional path dependence, see North (1990), David (1994), and Greif (1994a, 1997b, 1997c, 2000b).

References

Arcand, J.-L.L., and E. Brezis (1993), "Protectionism and power: A strategic interpretation of a mercantilism theory", Working Paper (Department of Economics and CRDE, University of Montreal).

Axelrod, R. (1984), The Evolution of Cooperation (Basic Books, New York).

Backhouse, R. (1985), A History of Modern Economic Analysis (Basil Blackwell, New York).

Baliga, S., and B. Polak (1995), "Banks versus bonds: A simple theory of comparative financial institutions", Cowles Foundation Discussion Paper no. 1100 (Yale University).

Brander, J.A., and B.J. Spencer (1985), "Export subsidies and international market share rivalry", Journal of International Economics 18:83–100.

Benson, B.L. (1989), "The spontaneous evolution of commercial law", Southern Economic Journal 55:644–661.

Besley, T., S. Coate and T.W. Guinnane (1993), "Why the workhouse test? Information and poor relief in nineteenth-century England", Working Paper (Economics Department, Princeton University).

Burns, M.R. (1986), "Predatory pricing and the acquisition cost of competitors", Journal of Political Economy 94:266–296.

Camerer, C., and M. Knez (1996), "Coordination in organizations: A game theoretic perspective", Working Paper (California Institute of Technology).

Cameron, R. (1993), A Concise Economic History of the World (Oxford University Press, Oxford).

Carlos, A.M., and E. Hoffman (1986), "The North American fur trade: Bargaining to a joint profit maximum under incomplete information, 1804–1821", Journal of Economic History 46:967–986.

Carlos, A.M., and E. Hoffman (1988), "Game theory and the North American fur trade: A reply", Journal of Economic History 48:801.

Carruthers, B.E. (1990), "Politics, popery and property: A comment on North and Weingast", Journal of Economic History 50:693–698.

Clark, G. (1995), "The political foundations of modern economic growth: England, 1540–1800", The Journal of Interdisciplinary History 26:563–588.

Clay, K. (1997), "Trade without law: Private-order institutions in Mexican California", The Journal of Law, Economics & Organization 13:202–231.

Conklin, J. (1998), "The theory of sovereign debt and Spain under Philip II", Journal of Political Economy 106:483–513.

Crawford, V.P. (1982a), "A theory of disagreement in bargaining", Econometrica 50:607–638.

Crawford, V.P. (1982b), "Compulsory arbitration, arbitral risk, and negotiated settlements: A case study in bargaining under asymmetric information", Review of Economic Studies 49:69–82.

Cunningham, W. (1882), The Growth of English Industry and Commerce (Cambridge University Press, Cambridge).

Da Rin, M., and T. Hellmann (1998), "Banks as catalysts for industrialization", Working Paper 103 (IGIER).

David, P.A. (1982), "Cooperative games for medieval warriors and peasants", mimeo (Department of Economics, Stanford University).

David, P.A. (1988), "Path dependence: Putting the past into the future of economics", Technical Report 533 (IMSSS, Stanford University).

David, P.A. (1992), "Path dependence and the predictability in dynamic systems with local network externalities: A paradigm for historical economics", in: C. Freeman and D. Foray, eds., Technology and the Wealth of Nations (Pinter Publishers, London).

David, P.A. (1994), "Why are institutions the "Carriers of History"?: Path-dependence and the evolution of conventions, organizations and institutions", Structural Change and Economic Dynamics 5:205–220.

Davis, E.L., and D.C. North (1971), Institutional Change and American Economic Growth (Cambridge University Press, Cambridge).

Diamond, D.W. (1989), "Reputation acquisition in debt market", Journal of Political Economy 97:828–862.

Eichengreen, B. (1996), "Institutions and economic growth: Europe after World War II", in: N. Crafts and G. Toniolo, eds., Economic Growth in Europe since 1945 (Cambridge University Press, Cambridge).

Ellison, G. (1994), "Theories of cartel and the joint executive committee", Rand Journal of Economics 25:37–57.

Farber, H.S. (1980), "An analysis of final-offer arbitration", Journal of Conflict Resolution 24:683–705.

Fershtman, C., and K.L. Judd (1987), "Equilibrium incentives in oligopoly", American Economic Review 77:927–940.

Fohlin, C.M. (1998), "Relationship banking, liquidity, and investment in the German industrialization", Journal of Finance 53:1737–1758.

Fudenberg, D., and J. Tirole (1991), Game Theory (MIT Press, Boston).

Gabel, D. (1994), "Competition in a network industry: The telephone industry, 1894–1910", The Journal of Economic History 54:543–572.

Gorton, G. (1996), "Reputation formation in early bank note markets", Journal of Political Economy 104:346–397.

Green, E.J. (1993), "On the emergence of parliamentary government. The role of private information", Quarterly Review: Federal Reserve Bank of Minneapolis, Winter:2–16.

Green, E.J., and R.H. Porter (1984), "Noncooperative collusion under imperfect price information", Econometrica 52:87–100.

Greif, A. (1989), "Reputation and coalitions in medieval trade: Evidence on the Maghribi traders", Journal of Economic History 49:857–882.

Greif, A. (1992), "Institutions and international trade: Lessons from the commercial revolution", American Economic Review 82:128–133.

Greif, A. (1993), "Contract enforceability and economic institutions in early trade: The Maghribi traders' coalition", American Economic Review 83:525–548.

Greif, A. (1994a), "Cultural beliefs and the organization of society: A historical and theoretical reflection on collectivist and individualist societies", Journal of Political Economy 102:912–950.

Greif, A. (1994b), "On the political foundations of the late medieval commercial revolution: Genoa during the twelfth and thirteenth centuries", Journal of Economic History 54:271–287.

Greif, A. (1996a), "Markets and legal systems: The development of markets in late medieval Europe and the transition from community responsibility to an individual responsibility legal doctrine", Manuscript (Stanford University).

Greif, A. (1996b), "On the study of organizations and evolving organizational forms through history: Reflection from the late medieval family firm", Industrial and Corporate Change 5:473–502.

Greif, A. (1997a), "Cliometrics after forty years: Microtheory and economic history", American Economic Review 87:400–403.

Greif, A. (1997b), "Contracting, enforcement and efficiency: Economics beyond the law", in: M. Bruno and B. Pleskovic, eds., Annual World Bank Conference on Development Economics (The World Bank, Washington, DC) 239–266.

Greif, A. (1997c), "Microtheory and recent developments in the study of economic institutions through economic history", in: D. Kreps and K. Wallis, eds., Advances in Economic Theory, Vol. II (Cambridge University Press, Cambridge) 79–113.

Greif, A. (1998a), "Historical institutional analysis: Game theory and non-market self-enforcing institutions during the late medieval period", Annales 53:597–633.

Greif, A. (1998b), "Historical and comparative institutional analysis", American Economic Review 88:80–84.

Greif, A. (1998c), "Self-enforcing political systems and economic growth: Late medieval Genoa", in: B. Bates, A. Greif, M. Levi, J.-L. Rosenthal and B. Weingast, eds., Analytic Narrative (Princeton University Press).

Greif, A. (2000a), "Impersonal exchange and the origin of markets: From the community responsibility system to individual legal responsibility in pre-modern Europe", in: M. Aoki and Y. Hayami, eds., Communities and Markets (Oxford University Press) Chapter 2.

Greif, A. (2000b), "Historical (and comparative) institutional analysis: Self-enforcing and self-reinforcing economic institutions and their dynamics" (Stanford University, Book MS).

Greif, A., P. Milgrom and B. Weingast (1994), "Coordination, commitment and enforcement: The case of the merchant guild", Journal of Political Economy 102:745–776.

Hartwell, R.M. (1973), "Good old economic history", Journal of Economic History 33:28–40.

Hoffman, P.T., G. Postel-Vinay and J.-L. Rosenthal (1998), "What do notaries do? Overcoming asymmetric information in financial markets: The case of Paris, 1751", Journal of Institutional and Theoretical Economics 154:499–530.

Irwin, D.A. (1991), "Mercantilism as strategic trade policy: The Anglo–Dutch rivalry for the East India trade", Journal of Political Economy 99:1296–1314.

Kantor, E.S. (1991), "Razorback, ticky cows and the closing of the Georgia open range: The dynamic of institutional change uncovered", Journal of Economic History 51:861–886.

Kinghorn, J.R., and J.V. Nye (1996), "The scale or production in Western economic development: A comparison of official industry statics in the United States, Britain, France and Germany, 1905–1913", Journal of Economic History 56:90–112.

Levenstein, M.C. (1994), "Price wars and the stability of collusion: Information and coordination in the nineteenth century chemical cartel", Working Paper (Economics Department, University of Michigan).

Levenstein, M.C. (1996a), "Do price wars facilitate collusion? A study of the bromine cartel before World War I", Exploration in Economic History 33:107–137.

Levenstein, M.C. (1996b), "Vertical restraints in the bromine cartel: The role of distributors in facilitating collusion", Working Paper (Department of Economics, University of Michigan).

Lewis, D. (1969), Convention: A Philosophical Study (Harvard University Press, Cambridge).

MacLeod, W.B., and J.M. Malcomson (1989), "Implicit contracts, incentive compatibility and involuntary unemployment", Econometrica 57:447–480.

Maurer, J.H. (1992), "The Anglo-German naval rivalry and informal arms control, 1912–1914", Conflict Resolution 36:284–308.

McCloskey, D.N. (1976), "Does the past have useful economics?", Journal of Economic Literature 14:434–461.

Milgrom, P.R., D. North and B.R. Weingast (1990), "The role of institutions in the revival of trade: The medieval law merchant, private judges and the Champagne fairs", Economics and Politics 2:1–23.

Mokyr, J., ed. (1985), The Economics of the Industrial Revolution (Rowman and Allanheld, Totowa, NJ).

Moriguchi, C. (1998), "The evolution of employment systems in the United States and Japan, 1900–1960: A comparative historical and institutional analysis", Ph.D. Thesis (Stanford University).

Murphy, K., A. Shleifer and R. Vishny (1989), "Industrialization and the big push", Journal of Economic Perspectives 97:1003–1026.

Myerson, R.B. (1984a), "Two-person bargaining problems with incomplete information", Econometrica 52:461–487.

Myerson, R.B. (1984b), "Cooperative games with incomplete information", International Journal of Game Theory 13:69–96.

Neal, L. (1990), The Rise of Financial Capitalism (Cambridge University Press, Cambridge).

Nix, J., and D. Gabel (1993), "AT&T strategic response to competition: Why not preempt entry?", Journal of Economic History 53:377–387.

North, D.C. (1981), Structure and Change in Economic History (Norton, New York).

North, D.C. (1990), Institutions, Institutional Change, and Economic Performance (Cambridge University Press, Cambridge).

North, D.C. (1993), "An interview with Douglass C. North", The Newsletter of the Cliometric Society 8:7–12, 24–28.

North, D.C., and R.P. Thomas (1973), The Rise of the Western World (Cambridge University Press, Cambridge).

North, D.C., and B. Weingast (1989), "Constitutions and commitment: Evolution of institutions governing public choice", Journal of Economic History 49:803–832.

Nye, J.V. (1988), "Game theory and the North American fur trade: A comment", Journal of Economic History 48:677–680.

Porter, R.H. (1983), "A study of cartel stability: The joint executive committee, 1880–1886", The Bell Journal of Economics 14:301–314.

Ramseyer, J.M. (1991), "Indentured prostitution in imperial Japan: Credible commitments in the commercial sex industry", Journal of Law Economic and Organization 7:89–115.

Rasmusen, E. (1994), Games and Information, Second Edition (Basil Blackwell, Oxford).

Reiter, S., and J. Hughes (1981), "A preface on modeling the regulated United States economy", Hofstra Law Review 9:1381–1421.

Root, H.L. (1989), "Tying the king's hands: Credible commitments and royal fiscal policy during the old regime", Rationality and Society 1:240–258.

Rosenthal, J.-L. (1992), The Fruits of Revolution: Property Rights, Litigation and French Agriculture, 1700–1860 (Cambridge University Press, Cambridge).

Rosenthal, J.-L. (1998), "The political economy of absolutism reconsidered", in: B. Bates, A. Greif, M. Levi, J.-L. Rosenthal and B. Weingast, eds., Analytic Narrative (Princeton University Press, Princeton).

Rotemberg, J.J., and G. Saloner (1986), "A supergame-theoretic model of price wars during booms", American Economic Review 76:390–407.

Schelling, T. (1960), The Strategy of Conflict (Harvard University Press, Cambridge).

Shapiro, C., and J.E. Stiglitz (1984), "Equilibrium unemployment as a worker discipline device", American Economic Review 74:433–444.

Sutton, J. (1992), Sunk Costs and Market Structure (MIT Press, Cambridge).

Telser, L. (1980), "A theory of self-enforcing agreements", Journal of Business 22:27–44.

Treble, J. (1990), "The pit and the pendulum: Arbitration in the British coal industry, 1893–1914", The Economic Journal 100:1095–1108.

Veitch, J.M. (1986), "Repudiations and confiscations by the medieval state", Journal of Economic History 46:31–36.

Weber, M. (1987 [1927]), General Economic History (Transaction Books, London). Translated by F.H. Knight.

Weiman, D.F., and R.C. Levin (1994), "Preying for monopoly? The case of Southern Bell Telephone Company, 1894–1912", Journal of Political Economy 102:103–126.

Weingast, B.R. (1995), "The political foundations of limited government: Parliament and sovereign debt in 17th and 18th centuries England", in: J.N. Drobak and J. Nye, eds., Frontiers of the New Institutional Economics, Volume in Honor of Douglass C. North (Academic Press, San Diego).

Williamson, S.H. (1994), "The history of cliometrics", in: Two Pioneers of Cliometrics (The Cliometrics Society, Miami University, Ohio).

Chapter 53

THE SHAPLEY VALUE

EYAL WINTER

Center for Rationality & Interactive Decision Theory, Hebrew University of Jerusalem, Jerusalem, Israel

Contents

Handbook of Game Theory, Volume 3, Edited by R.J. Aumann and S. Hart

Abstract

This chapter surveys some of the literature in game theory that has emerged from Shapley's seminal paper on the Value. The survey includes both contributions which offer different interpretations of the Shapley value as well as several different ways to characterize the value axiomatically. The chapter also surveys some of the literature that generalizes the notion of the value to situations in which *a priori* cooperation structure exists, as well as a different literature that discusses the relation between the Shapley value and models of non-cooperative bargaining. The chapter concludes with a discussion of the applied side of the Shapley value, primarily in the context of cost allocation and voting.

Keywords

Shapley value, cooperative games, coalitions, cooperation structures, voting

JEL classification: C71, D72

1. Introduction

To promote an understanding of the importance of Shapley's (1953) paper on the value, we shall start nine years earlier with the groundbreaking book by von Neumann and Morgenstern that laid the foundations of cooperative game theory. Unlike non-cooperative game theory, cooperative game theory does not specify a game through a minute description of the strategic environment, including the order of moves, the set of actions at each move, and the payoff consequences relative to all possible plays, but, instead, it reduces this collection of data to the coalitional form. The cooperative game theorist must base his prediction strictly on the payoff opportunities, conveyed by a single real number, available to each coalition: gone are the actions, the moves, and the individual payoffs. The chief advantage of this approach, at least in multiple-player environments, is its practical usefulness. A real-life situation fits more easily into a coalitional form game, whose structure has proved more tractable than that of a non-cooperative game, whether that be in normal or extensive form.

Prior to the appearance of the Shapley value, one solution concept alone ruled the kingdom of (cooperative) game theory: the von Neumann–Morgenstern solution. The core would not be defined until around the same time as the Shapley value. As set-valued solutions suggesting "reasonable" allocations of the resources of the grand coalition, the von Neumann–Morgenstern solution and the core both are based on the coalitional form game. However, no single-point solution concept existed as of yet by which to associate a single payoff vector to a coalitional form game. In fact the coalitional form game of those days had so little information in its black box that the creation of a single-point solution seemed untenable. It was in spite of these sharp limitations that Shapley came up with the solution. Using an axiomatic approach, Shapley constructed a solution remarkable not only for its attractive and intuitive definition but also for its unique characterization by a set of reasonable axioms. Section 2 of this chapter reviews Shapley's result, and in particular a version of it popularized in the wake of his original paper. Section 3 turns to a special case of the value for the class of voting games, the Shapley–Shubik index.

In addition to a model that attempts to predict the allocation of resources in multi-person interactions, Shapley also viewed the value as an index for measuring the power of players in a game. Like a price index or other market indices, the value uses averages (or weighted averages in some of its generalizations) to aggregate the power of players in their various cooperation opportunities. Alternatively, one can think of the Shapley value as a measure of the utility of players in a game. Alvin Roth took this interpretation a step further in formal terms by presenting the value as a von Neumann–Morgenstern utility function. Roth's result is discussed in Section 4.

Section 5 is devoted to alternative axiomatic characterizations of the Shapley value, with particular emphasis on Young's axiom of monotonicity, Hart and Mas-Colell's axiom of consistency, and the last-named pair's notion of potential (arguably a single-axiom characterization).

To be able to apply the Shapley value to concrete problems such as voting situations, it is important to be able to characterize it on sub-classes of games. Section 6 discusses several approaches in that direction. Section 7 surveys several attempts to generalize the Shapley value to a framework in which the environment is described by some *a priori* cooperation structure other than the coalitional form game (typically a partition of the set of players). Aumann and Drèze (1975) and Owen (1977) pioneered examples of such generalizations.

While the Shapley value is a classic cooperative solution concept, it has been shown to be sustained by some interesting strategic (bargaining) games. Section 8 looks at some of these results. Section 9 closes the chapter with a discussion of the practical importance of the Shapley value in general, and of its role as an estimate of parliamentary and voter power and as a rule for cost allocation in particular.

Space forbids discussion of the vast literature inspired by Shapley's paper. Certain of these topics, such as the various extensions of the value to NTU games, and the values of non-atomic games to emerge from the seminal book by Aumann and Shapley (1974), are treated in other chapters of this Handbook.

2. The framework

Recall that a *coalitional form game* (henceforth game) on a finite set of players $N = \{1, 2, 3, \ldots, n\}$ is a function v from the set of all coalitions 2^N to the set of real numbers $\mathbb{R}$ with $v(\emptyset) = 0$. $v(S)$ represents the total payoff or rent the coalition S can get in the game v.

A *value* is an operator ϕ that assigns to each game v a vector of payoffs $\phi(v) = (\phi_1, \phi_2, \ldots, \phi_n)$ in $\mathbb{R}^n$. $\phi_i(v)$ stands for i's payoff in the game, or alternatively for the measure of i's power in the game.

Shapley presented the value as an operator that assigns an expected marginal contribution to each player in the game with respect to a uniform distribution over the set of all permutations on the set of players. Specifically, let π be a permutation (or an order) on the set of players, i.e., a one-to-one function from N onto N, and let us imagine the players appearing one by one to collect their payoff according to the order π. For each player i we can denote by $p_\pi^i = \{j: \pi(i) > \pi(j)\}$ the set of players preceding player i in the order π. The marginal contribution of player i with respect to that order π is $v(p_\pi^i \cup i) - v(p_\pi^i)$. Now, if permutations are randomly chosen from the set Π of all permutations, with equal probability for each one of the $n!$ permutations, then the average marginal contribution of player i in the game v is

$$\phi_i(v) = 1/n! \sum_{\pi \in \Pi} \left[v\left(p_\pi^i \cup i\right) - v\left(p_\pi^i\right) \right], \tag{1}$$

which is Shapley's definition of the value.

While the intuitive definition of the value speaks for itself, Shapley supported it by an elegant axiomatic characterization. We now impose four axioms to be satisfied by a value:

The first axiom requires that players precisely distribute among themselves the resources available to the grand coalition. Namely,

EFFICIENCY. $\sum_{i \in N} \phi_i(v) = v(N)$.

The second axiom requires the following notion of symmetry:

Players $i, j \in N$ are said to be symmetric with respect to game v if they make the same marginal contribution to any coalition, i.e., for each $S \subset N$ with $i, j \notin S$, $v(S \cup i) = v(S \cup j)$. The symmetry axiom requires symmetric players to be paid equal shares.

SYMMETRY. *If players i and j are symmetric with respect to game v, then $\phi_i(v) = \phi_j(v)$.*

The third axiom requires that zero payoffs be assigned to players whose marginal contribution is null with respect to every coalition:

DUMMY. *If i is a dummy player, i.e., $v(S \cup i) - v(S) = 0$ for every $S \subset N$, then $\phi_i(v) = 0$.*

Finally, we require that the value be an additive operator on the space of all games, i.e.,

ADDITIVITY. $\phi(v + w) = \phi(v) + \phi(w)$, *where the game $v + w$ is defined by $(v + w)(S) = v(S) + w(S)$ for all S.*

Shapley's amazing result consisted in the fact that the four simple axioms defined above characterize a value uniquely:

THEOREM 1 [Shapley (1953)]. *There exists a unique value satisfying the efficiency, symmetry, dummy, and additivity axioms: it is the Shapley value given in Equation (1).*

The uniqueness result follows from the fact that the class of games with n players forms a 2^{n-1}-dimensional vector space in which the set of unanimity games constitutes a basis. A game u_R is said to be a unanimity game on the domain R if $u_R(S) = 1$, whenever $R \subset S$ and 0 otherwise. It is clear that the dummy and symmetry axioms together yield a value that is uniquely determined on unanimity games (each player in the domain should receive an equal share of 1 and the others 0). Combined with the additivity axiom and the fact that the unanimity games constitute a basis for the vector space of games, this yields the uniqueness result.

Here it should be noted that Shapley's original formulation was somewhat different from the one described above. Shapley was concerned with the space of all games that can be played by some large set of potential players U, called the universe. For every game v, which assigns a real number to every finite subset of U, a *carrier* N is a subset of U such that $v(S) = v(S \cap N)$ for every $S \subset U$. Hence, the set of players who actually participate in the game must be contained in any carrier of the game. If for some carrier N a player i is not in N, then i must be a dummy player because he does not affect the payoff of any coalition that he joins. Shapley imposed the *carrier axiom* onto this framework, which requires that within any carrier N of the game the players in N share the total payoff of $v(N)$ among themselves. Interestingly, this axiom bundles the efficiency axiom and the dummy axiom into one property.

3. Simple games

Some of the most exciting applications of the Shapley value involve the measurement of political power. The reason why the value lends itself so well to this domain of problems is that in many of these applications it is easy to identify the real-life environment with a specific coalitional form game. In politics, indeed in all voting situations, the power of a coalition comes down to the question of whether it can impose a certain collective decision, or, in a typical application, whether it possesses the necessary majority to pass a bill. Such situations can be represented by a collection of coalitions W (a subset of 2^N), where W stands for the set of "winning" coalitions, i.e., coalitions with enough power to impose a decision collectively. We call these situations "simple games". While simple games can get rather complex, their coalitional function v assumes only two values: 1 for winning coalitions and 0 otherwise (see Chapter 36 in this Handbook). If we assume monotonicity, i.e., that a superset of a winning coalition is likewise winning, then the players' marginal contributions to coalitions in such games also assume the values 0 and 1. Specifically, player i's marginal contribution to coalition S is 1 if by joining S player i can turn the coalition from a non-winning (or losing) to a winning coalition. In such cases, we can say that player i is *pivotal* to coalition S. Recalling the definition of the Shapley value, it is easy to see that in such games the value assigns to each player the probability of being pivotal with respect to his predecessors, where orders are sampled randomly and with equal probability. Specifically,

$$\phi_i(v) = \left| \left\{ \pi \in \Pi; \; p_\pi^i \cup i \in W \text{ and } p_\pi^i \notin W \right\} \right| / n!.$$

This special case of the Shapley value is known in the literature as the Shapley–Shubik index for simple games [Shapley and Shubik (1954)].

A very interesting interpretation of the Shapley–Shubik index in the context of voting was proposed by Straffin (1977). Consider a simple (voting) game with a set of winning coalitions W representing the distribution of power within a committee, say a parliament. Suppose that on the agenda are several bills on which players take positions. Let us take an *ex ante* point of view (before knowing which bill will be discussed) by assuming that the probability of each player voting in favor of the bill is p (independent over i). Suppose that a player is asking himself what the probability is that his vote will affect the outcome. Put differently, what is the probability that the bill will pass if and only if I support it? The answer to this question depends on p (as well as the distribution of power W). If p is 1 or 0, then I will have no effect unless I am a dictator. But because we do not know which bill is next on the agenda, it is reasonable to assume that p itself is a random variable. Specifically, let us assume that p is distributed uniformly on the interval $[0, 1]$. Straffin points out that with this model for random bills the probability that a player is effective is equivalent to his Shapley–Shubik index in the corresponding game (see Chapter 32 in this Handbook). We shall demonstrate this with an example.

EXAMPLE. Let $[3; 2, 1, 1]$ be a weighted majority game,[1] where the minimal winning coalitions are $\{1, 2\}$ and $\{1, 3\}$. Player 2 is effective only if player 1 votes for and player 3 votes against. For a given probability p of acceptance, this occurs with probability $p(1 - p)$. Since 2 and 3 are symmetric, the same holds for player 3. Now player 1's vote is ineffective only if 2 and 3 both vote against, which happens with probability $(1 - p)^2$. Thus player 1 is effective with probability $2p - p^2$. Integrating these functions between 0 and 1 yields $\phi_1 = 2/3$, $\phi_2 = \phi_3 = 1/6$.

It is interesting to note that with a different probability model for bills one can derive a different well-known power index, namely the Banzhaf index (see Chapter 32 in this Handbook). Specifically, if player k's probability of accepting the bill is p_k, and if $p_1, \ldots, p_n$ are chosen independently, each from a uniform distribution on $[0, 1]$, then player i's probability of being effective coincides with his Banzhaf index.

4. The Shapley value as a von Neumann–Morgenstern utility function

If we interpret the Shapley value as a measure of the benefit of "playing" the game (as was indeed suggested by Shapley himself in his original paper), then it is reasonable to think of different positions in a game as objects for which individuals have preferences. Such an interpretation immediately gives rise to the following question: What properties should these preferences possess so that the cardinal utility function that represents them coincides with the Shapley value? This question is answered by Roth (1977).

Roth defined a position in a game as a pair (i, v), where i is a player and v is a game. He then assumed that individuals have preferences defined on the mixture set M that contains all positions and lotteries whose outcomes are positions. Using "$\sim$" to denote indifference and "$\succ$" to denote strict preference, Roth imposed several properties on preferences. The first two properties are simple regularity conditions:

---

[1] In a weighted majority game $[q; w_1, \ldots, w_n]$, a coalition S is winning if and only if $\sum_{i \in S} w_i \geqslant q$.

A1. *Let v be a game in which i is a dummy. Then $(i, v) \sim (i, v_0)$, where v_0 is the null game in which every coalition earns zero. Furthermore, $(i, v_i) \succ (i, v_0)$, where v_i is the game in which i is a dictator, i.e., $v_i(S) = 1$ if $i \in S$ and $v_i(S) = 0$ otherwise.*

The second property, which relates to Shapley's symmetry axiom, requires that individual preferences not depend on the names of positions, i.e.,

A2. *For any game v and permutation π, $(i, v) \sim (\pi(i), \pi(v))$.*[2]

The two remaining properties are more substantial and deal with players' attitudes towards risk. The first of these properties requires that the certainty equivalent of a lottery that yields the position i in either game v or game w (with probabilities p and $1 - p$) be the position i in the game given by the expected value of the coalitional function with respect to the same probabilities. Specifically, for two positions (i, v) and (i, w), we denote by $[p(i, v); (1 - p)(i, w)]$ the lottery where (i, v) occurs with probability p and (i, w) occurs with probability $1 - p$.

A3. *Neutrality to Ordinary Risk*: $(i, (pw + (1 - p)v)) \sim [p(i, w); (1 - p)(i, v)]$.

Note that a weaker version of this property requires that $(i, v) \sim [(1/c)(i, cv); (1 - 1/c)(i, v_0)]$ for $c > 1$. It can be shown that this property implies that the utility function u, which represents the preferences over positions in a game, must satisfy $u(cv, i) = cu(v, i)$.

The last property asserts that in a unanimity game with a carrier of r players the utility of a non-dummy player is $1/r$ of the utility of a dictator. Specifically, let v_R be defined by $v_R(S) = 1$ if $R \subset S$ and 0 otherwise.

A4. *Neutrality to Strategic Risk*: $(i, v_R) \sim (i, (1/r)v_i)$.

An elegant result now asserts that:

THEOREM [Roth (1977)]. *Let u be a von Neumann–Morgenstern utility function over positions in games, which represents preferences satisfying the four axioms. Suppose that u is normalized so that $u(i, v_i) = 1$ and $u_i(i, v_0) = 0$. Then $u(i, v)$ is the Shapley value of player i in the game v.*

Roth's result can be viewed as an alternative axiomatization of the Shapley value. I will now survey several other characterizations of the value, which, unlike Roth's utility function, employ the standard concept of a payoff function.

---

[2] $\pi(v)$ is the game with $\pi(v)(S) = v(\pi(S))$, where $\pi(S) = \{j; \ j = \pi(i) \text{ for some } i \in S\}$.

5. Alternative axiomatizations of the value

One of the most elegant aspects of the axiomatic approach in game theory is that the same solution concept can be characterized by very different sets of axioms, sometimes to the point of seeming unrelated. But just as a sculpture seen from different angles is understood in greater depth, so is a solution concept by means of different axiomatizations, and in this respect the Shapley value is no exception. This section examines several alternative axiomatic treatments of the value.

Perhaps the most appealing property to result from Definition (1) of the Shapley value is that a player's payoff is only a function of the vector of his marginal contributions to the various coalitions. This raises an interesting question: Without forgoing the above property, how far can we depart from the Shapley value? "Not much", according to Young (1985), whose finding also yields an alternative axiomatization of the value.

For a game v, a coalition S, and a player $i \notin S$, we denote by $D_i(v, S)$ player i's marginal contribution to the coalition S with respect to the game v, i.e., $D_i(v, S) = v(S \cup i) - v(S)$. Young introduced the following axiom:

STRONG MONOTONICITY. *Suppose that v and w are two games such that for some $i \in N$ $D_i(v, S) \geqslant D_i(w, S)$. Then $\phi_i(v) \geqslant \phi_i(w)$.*

Young then showed that this property plays a central role in the characterization of the Shapley value. Specifically,

THEOREM [Young (1985)]. *There exists a unique value ϕ satisfying strong monotonicity, symmetry, and efficiency, namely the Shapley value.*

Note that Young's strong monotonicity axiom implies the marginality axiom, which asserts that the value of each player is only a function of that player's vector of marginal contributions. Young's axiomatic characterization of the value thus supports the claim that the Shapley value to some extent is a synonym for the principle of marginal contribution – a time-honored principle in economic theory. But we must be clear about what is meant by marginal contribution. In Young's framework, as in Shapley's definition of the value, players contribute by increasing the wealth of the coalition they join (or decreasing it if contributions are negative). This caused Hart and Mas-Colell (1989) to ask the following question: Can the Shapley value be derived by referring the players' marginal contributions to the wealth generated by the entire multilateral interaction, instead of tediously listing all the coalitions they can join? Offhand, the question sounds somewhat amorphous, for how is one to define an "entire interaction"? Absent a satisfactory definition, we shall proceed by way of supposition. Suppose each pair (N, v) is associated with a single real number $P(N, v)$ that sums up the wealth generated by the entire interaction in the game. We are already within an ace of defining marginal contributions. Specifically, player i's marginal contribution with respect to (N, v) is simply:

$$D^i P(N, v) = P(N, v) - P(N \setminus i, v),$$

where $(N \setminus i, v)$ is the game v restricted to the set of players $N \setminus i$. To be associated with a measure of power in the game, these marginal contributions need to add up to the total resources available to the grand coalitions, i.e., $v(N)$. So we will say that P is a potential function if

$$\sum_{i \in N} D^i P(N, v) = v(N). \tag{2}$$

Moreover, we normalize P to satisfy $P(\emptyset, v) = 0$. Given the mild requirement on p, there seems to be enough flexibility for many potential functions to exist. The remarkable result of Hart and Mas-Colell (1989) is that there is in fact only one potential function, and it is closely related to the Shapley value. Specifically,

THEOREM [Hart and Mas-Colell (1989)]. *There exists a unique potential function P. Moreover, the corresponding vector of marginal contributions $(D^1 P(N, v), \dots, D^n P(N, v))$ coincides with the Shapley value of the game (N, v).*

Let us note that by rewriting Equation (2) we obtain the following recursive equation:

$$P(N, v) = (1/|N|)\left[v(N) + \sum_{i \in N} P(N \setminus i, v) \right]. \tag{3}$$

Starting with $P(\emptyset, v) = 0$, Equation (3) determines p recursively. It is interesting to note that the potential is related to the notion of "dividends" used by Harsanyi (1963) to extend the Shapley value to games without transferable utility. Specifically, let $\sum_{T \subset N} a_T u_T$ be the (unique) representation of the game (N, v) as a linear combination of unanimity games. In any unanimity game u_T on the carrier T, the value of each player in T is the "dividend" $d_T = a_T/|T|$, and the Shapley value of player i in the game (N, v) is the sum of dividends that a player earns from all coalitions of which he is a member, i.e., $\sum_{\{T : i \in T\}} d_T$. Given the definition and uniqueness of the potential function, it follows that $P(N, v) = \sum_T d_T$, i.e., the total Harsanyi dividends across all coalitions in the game.

One can view Hart and Mas-Colell's result as a concise axiomatic characterization of the Shapley value – indeed, one that builds on a single axiom. In the same paper, Hart and Mas-Colell propose another axiomatization of the value by a different but related approach based on the notion of consistency.

Unlike in non-cooperative game theory, where the feasibility of a solution concept is judged according to strategic or optimal individual behavior, in cooperative game theory neither strategies nor individual payoffs are specified. A cooperative solution concept is considered attractive if it makes sense as an arbitration scheme for allocating costs or benefits. It comes as no surprise, then, that some of the popular properties used to support solution concepts in this field are normative in nature. The symmetry axiom is a case in point. It asserts that individuals indistinguishable in terms of their contributions to coalitions are to be treated equally. One of the most fundamental requirements of any legal system is that it be internally consistent. Consider some value (a single-point solution concept) ϕ, which we would like to use as a scheme for allocating payoffs in games. Suppose that we implement ϕ by first making payment to a group of players Z. Examining the environment subsequent to payment, we may realize that we are facing a different (reduced) game defined on the remaining players $N \setminus Z$ who are still waiting to be paid. The solution concept ϕ is said to be *consistent* if it yields the players in the reduced game exactly the same payoffs they are getting in the original game. Consistency properties play an important role in the literature of cooperative game theory. They were used in the context of the Shapley value by Hart and Mas-Colell (1989). The difference between Hart and Mas-Colell's notion of consistency and that of the related literature on other solution concepts lies in their definition of reduced game. For a given value ϕ, a game (N, v), and a coalition T, the reduced game (T, v_T) on the set of players T is defined as follows:

$$v_T(S) = v(S \cup T^c) - \sum_{i \in T^c} \phi_i(v|_{S \cup T^c}) \quad \text{for every } S \subset T,$$

where $v|_R$ denotes the game v restricted to the coalition R.

The worth of coalition S in the reduced game v_T is derived as follows. First, the players in S consider their options outside T, i.e., by playing the game only with the complementary coalition T^c. This restricts the game v to coalition $S \cup T^c$. In this game the total payoff of the members of T^c according to the solution ϕ is $\sum_{i \in T^c} \phi_i(v|_{S \cup T^c})$. Thus the resources left available for the members of S to allocate among themselves are precisely $v_T(S)$.

A value ϕ is now said to be *consistent* if for every game (N, v), every coalition T, and every $i \in T$, one has $\phi_i(T, v_T) = \phi_i(N, v)$.

Hart and Mas-Colell (1989) show that with two additional standard axioms the consistency property characterizes the Shapley value. Specifically,

THEOREM [Hart and Mas-Colell (1989)]. *There exists a unique value satisfying symmetry, efficiency and consistency, namely the Shapley value.*[3]

It is interesting to note that by replacing Hart and Mas-Colell's property with a different consistency property one obtains a characterization of a different solution concept, namely the pre-nucleolus. Sobolev's (1975) consistency property is based on the following definition of the reduced game:

$$v_T^*(S) = \max_{Q \subset N \setminus T} \left[v(Q \cup S) - \sum_{i \in Q} \phi_i(v) \right] \quad \text{if } S \neq T, \emptyset,$$

[3] Hart and Mas-Colell (1989) in fact showed that instead of the efficiency and symmetry axioms it is enough to require that the solution be "standard" for two-person games, i.e., that for such games $\phi_i(\{i, j\}, v) = v(i) + (1/2)[v(\{i, j\}) - v(i) - v(j)]$.

$$v_T^*(S) = \sum_{i \in T} \phi_i(v) \quad \text{if } S = T, \quad \text{and} \quad v_T^*(S) = 0 \quad \text{if } S = \emptyset.$$

Note that in Sobolev's definition the members of S evaluate their power in the reduced game by considering their *most attractive* options outside T. Furthermore, the collaborators of S outside T are paid according to their share in the original game v (and not according to the restricted game as in Hart and Mas-Colell's property). It is surprising that while the definitions of the Shapley value and the pre-nucleolus differ completely, their axiomatic characterizations differ only in terms of the reduced game on which the consistency property is based. This nicely demonstrates the strength of the axiomatic approach in cooperative game theory, which not only sheds light on individual solution concepts, but at the same time exposes their underlying relationships.

Hart and Mas-Colell's consistency property is also related to the "balanced contributions" property due to Myerson (1977). Roughly speaking, this property requires that player i contribute to player j's payoff what player j contributes to player i's payoff. Formally, the property can be written as follows:

BALANCED CONTRIBUTION. *For every two players i and j, $\phi_i(v) - \phi_i(v|_{N\setminus j}) = \phi_j(v) - \phi_j(v|_{N\setminus j})$.*

Myerson (1977) introduced a value that associates a payoff vector with each game v and graph g on the set N (representing communication channels between players). His result implies that the balanced contributions property, the efficiency axiom, and the symmetry axiom characterize the Shapley value.

We will close this section by discussing another axiomatization of the value, proposed by Chun (1989). It employs an interesting property which generalizes the strategic equivalence property traceable to von Neumann and Morgenstern (1944). Chun's *coalitional strategic equivalence* property requires that if we change the coalitional form game by adding a constant to every coalition that contains some (fixed) coalition $T \subset N$, then the payoffs to players outside S will not change. This means that the extra benefit (or loss if the added constant is negative) accrues only to the members of T. Formally:

COALITIONAL STRATEGIC EQUIVALENCE. *For all $T \subset N$ and $a \in \mathbb{R}$, let w_a^T be the game defined by $w_a^T(S) = a$ if $T \subset S$ and 0 otherwise. For all $T \subset N$ and $a \in \mathbb{R}$, if $v = w + w_a^T$, then $\phi_i(v) = \phi_i(w)$ for all $i \in N \setminus T$.*

Von Neumann and Morgenstern's strategic equivalence imposes the same requirement, but only for $|T| = 1$.

Another similar property proposed by Chun is *fair ranking*. It requires that if the underlying game changes in such a way that all coalitions but T maintain the same worth, then although the payoffs of members of T will vary, the ranking of the payoffs within T will be preserved. This directly reflects the idea that the ranking of players' payoffs within a coalition depends solely on the outside options of their members. Specifically:

FAIR RANKING. *Suppose that $v(S) = w(S)$ for every $S \neq T$; then for all $i, j \in T$, $\phi_i(v) > \phi_j(v)$ if and only if $\phi_i(w) > \phi_j(w)$.*

To characterize the Shapley value an additional harmless axiom is needed.

TRIVIALITY. *If v_0 is the constant zero game, i.e., $v_0(S) = 0$ for all $S \subset N$, then $\phi_i(v_0) = 0$ for all $i \in N$.*

The following characterization of the Shapley value can now be obtained:

THEOREM [Chun (1989)]. *The Shapley value is the unique value satisfying efficiency, triviality, coalitional strategic equivalence, and fair ranking.*

6. Sub-classes of games

Much of the Shapley value's attractiveness derives from its elegant axiomatic characterization. But while Shapley's axioms characterize the value uniquely on the class of all games, it is not clear whether they can be used to characterize the value on subspaces. It sounds puzzling, for what could go wrong? The fact that a value satisfies a certain set of axioms trivially implies that it satisfies those axioms on every subclass of games. However, the smaller the subclass, the less likely these axioms are to determine a *unique* value on it. To illustrate an extreme case of the problem, suppose that the subclass consists of all integer multiples of a single game v with no dummy players, and no two players are symmetric. On this subclass Shapley's symmetry and dummy axioms are vacuous: they impose no restriction on the payoff function. It is therefore easy to verify that any allocation of $v(N)$ can be supported as the payoff vector for v with respect to a value on this subclass that satisfies all Shapley's axioms. Another problem that can arise when considering subclasses of games is that they may not be closed under operations that are required by some of the axioms. A classic example of this is the class of all simple games. It is clear that the additivity axiom cannot be used on such a class, because the sum of two simple games is typically not a simple game anymore. One can revise the additivity axiom by requiring that it apply only when the sum of the two games falls within the subclass considered. Indeed, Dubey (1975) shows that with this amendment to the additivity[4] axiom, Shapley's original proof of uniqueness still applies to some subclasses of games, including the class of *all* simple games. However, even in conjunction with the other standard axioms, this axiom does not yield uniqueness in the class of all monotonic[5] simple games. To redress this problem, Dubey (1975) introduced an axiom that can replace additivity: for two simple games v and v', we define

---

[4] Specifically, one has to require the axiom only for games v_1, v_2 whose sum belongs to the underlying subclass.

[5] Recall that a simple game v is said to be monotonic if $v(S) = 1$ and $S \subset T$ implies $v(T) = 1$.

by min$\{v, v'\}$ the simple game in which S is winning if and only if it is winning in both v and v'. Similarly, we define by max$\{v, v'\}$ the game in which S is winning if and only if it is winning in at least one of the two games v and v'. Dubey imposed the property of

MODULARITY. $\phi(\min\{v, v'\}) + \phi(\max\{v, v'\}) = \phi(v) + \phi(v')$.

One can interpret this axiom within the framework of Roth's model in Section 3. Suppose that $\phi_i(v)$ stands for the utility of playing i's position in the game v, and player i is facing two lotteries. In the first lottery he will participate in either the game v or the game v' with equal probabilities. The other lottery involves two more "extreme" games: max$\{v, v'\}$, which makes winning easier than with v and v', and min$\{v, v'\}$, which makes winning harder. As before, each of these games will be realized with probability $1/2$. Modularity imposes that each player be "risk neutral" in the sense that he be indifferent between these two lotteries.

Note that min$\{v, v'\}$ and max$\{v, v'\}$ are monotonic simple games whenever v and v' are, so we do not need any conditioning in the formulation of the axiom. Dubey characterized the Shapley value on the class of monotonic simple games by means of the modularity axiom, showing that:

THEOREM [Dubey (1975)]. *There exists a unique value on the class of monotonic[6] simple games satisfying efficiency, symmetry, dummy, and modularity, and it is the Shapley–Shubik value.*

When trying to apply a solution concept to a specific allocation problem (or game), one may find it hard to justify the Shapley value on the basis of its axiomatic characterization. After all, an axiom like additivity deals with how the value varies as a result of changes in the game, taking into account games which may be completely irrelevant to the underlying problem. The story is different insofar as Shapley's other axioms are concerned, because the three of them impose requirements only on the specific game under consideration. Unfortunately, one cannot fully characterize the Shapley value by axioms of the second type only[7] (save perhaps by imposing the value formula as an axiom). Indeed, if we were able to do so, it would mean that we could axiomatically characterize the value on subclasses as small as a single game. While this is impossible, Neyman (1989) showed that Shapley's original axioms characterize the value on the additive class (group) spanned by a single game. Specifically, for a game v let $G(v)$ denote the class of all games obtained by a linear combination of restricted games of v, i.e., $G(v) = \{v'$ s.t. $v' = \sum_i k_i v|_{S_i}$, where k_i are integers and $v|_{S_i}$ is the game v restricted to the coalition $S_i\}$.

---

[6] The same result was shown by Dubey (1975) to hold for the class of superadditive simple games.

[7] A similar distinction can be made within the axiomatization of the Nash solution where the symmetry and efficiency axioms are "within games" while IIA and Invariance are "between games".

Note that by definition the class $G(v)$ is closed under summation of games, which makes the additivity axiom well defined on this class. Neyman shows that:

THEOREM [Neyman (1989)]. *For any v there exists a unique value on the subclass $G(v)$ satisfying efficiency, symmetry, dummy, and additivity, namely the Shapley value.*

It is worth noting that Hart and Mas-Colell's (1989) notion of Potential also characterizes the Shapley value on the subclass $G(v)$, since the Potential is defined on restricted games only.

7. Cooperation structures

One of Shapley's axioms which characterize the value is the symmetry axiom. It requires that a player's value in a game depend on nothing but his payoff opportunities as described by the coalitional form game, and in particular not on the his "name". Indeed, as we argued earlier, the possibility of constructing a unique measure of power axiomatically from very limited information about the interactive environment is doubtless one of the value's most appealing aspects. For some specific applications, however, we might possess more information about the environment than just the coalitional form game. Should we ignore this additional information when attempting to measure the relative power of players in a game? One source of asymmetry in the environment can follow simply from the fact that players differ in their bargaining skills or in their property rights. This interpretation led to the concept of the *weighted Shapley value* by Shapley (1953) and Kalai and Samet (1985) (see Chapter 54 in this Handbook). But asymmetry can derive from a completely different source. It can be due to the fact that the interaction between players is not symmetric, as happens when some players are organized into groups or when the communication structure between players is incomplete, thus making it difficult if not impossible for some players to negotiate with others. This interpretation has led to an interesting field of research on the Shapley value, which is mainly concerned with generalizations.

The earliest result in this field is probably due to Aumann and Drèze (1975), who consider situations in which there exists an exogenous coalition structure in addition to the coalitional form game. A coalition structure $B = (S_1, \ldots, S_m)$ is simply a partition of the set of players N, i.e., $\bigcup S_j = N$ and $S_i \cap S_j = \emptyset$ for $i \neq j$. In this context a value is an operator that assigns a vector of payoffs $\phi(B, v)$ to each pair (B, v), i.e., a coalition structure and a coalitional form game on N. Aumann and Drèze (1975) imposed the following axioms on such operators. First, the efficiency axiom is based on the idea that by forming a group, players can allocate to themselves only the resources available to their group. Specifically,

RELATIVE EFFICIENCY. *For every coalition structure $B = (S_1, \ldots, S_m)$ and $1 \leqslant k \leqslant m$, we have $\sum_{j \in S_k} \phi_j(B, v) = v(S_k)$.*

The remaining three axioms are straightforward extensions of their counterparts in the benchmark model.

SYMMETRY. *For every permutation π on N and every coalition structure B, we have* $\phi(\pi B, \pi v) = \pi \phi(B, v)$, *where πB is the coalition structure with $(\pi B)_i = \pi S_i$.*

DUMMY. *If i is a dummy player in v, then $\phi_i(B, v) = 0$ for all B.*

ADDITIVITY. $\phi(B, v + w) = \phi(B, v) + \phi(B, w)$ *for all B and any pair of games v and w.*

Aumann and Drèze showed that the above axioms characterize a value uniquely. This value assigns the Shapley value to each player in the game restricted to his group. Specifically, for each coalition S and game v, we define by $v|S$ the game on S given by $(v|_S)(T) = v(T)$ for all $T \subset S$. We now have:

THEOREM [Aumann and Drèze (1975)]. *For a given set of players N and a coalition structure B, there exists a unique B-value satisfying the four above axioms, namely* $\phi_i(B, v) = \phi_i(v|_{B(i)})$, *where $B(i)$ denotes the component of B that includes player i, and ϕ stands for the Shapley value.*

Aumann and Drèze then defined the coalition structure of all the major solution concepts in cooperative game theory, in addition to the Shapley value. In so doing they exposed a startling property of consistency satisfied by all but one of them. The Shapley value is the exception. Hart and Mas-Colell (1989) would later show that the Shapley value satisfies a version of this property (discussed earlier in Section 5).

In Aumann and Drèze's framework, the coalition structure can be thought of as representing contractual relationships that affect players' property rights over resources, i.e., players in $S \in B$ "own" a total resource $v(S)$, from which transfers to players outside S are forbidden. This is apparent from both the definition and the efficiency axiom that depends on the coalition structure. Roughly speaking, players from one component of the partition cannot share benefits with any player from another component. But in real life coalition formation often takes place merely for strategic reasons without affecting the fundamentals of the economy. A group may form in order to pressure another group, or to enhance the bargaining position of its members vis-à-vis other members without affecting the constraint that all players in N share the total pie $v(N)$. It is thus reasonable to ask whether it is possible to extend the Shapley value to this context as well. Owen (1977), Hart and Kurz (1983), and Winter (1989, 1991, 1992) take up this question. Owen proposed a value, like the Shapley value, in which each player receives his expected marginal contribution to coalitions with respect to a random order of players. But while the Shapley value assumes that all orders are of equal probability, the Owen value restricts orders according to the coalition structure. Specifically, for a given coalition structure B, let us consider a subset of the set of all orders, which includes only the orders in which players of the same component of B appear successively. Denote this set of orders $\Pi(B)$. The set of orders $\Pi(B)$ can be obtained by first ordering the components of B and then ordering players within each component of the partition. According to the Owen value, each player is assigned an expected marginal contribution to the coalition of preceding players with respect to a uniform distribution over the set of orders in $\Pi(B)$. More formally, let $B = (S_1, \ldots, S_m)$ be a coalition structure. Set $\Pi(B) = \{\pi \in \Pi;\ \text{if } i, j \in S_k \text{ and } \pi(i) < \pi(r) < \pi(j), \text{ then } r \in S_k\}$, which is the set of all permutations consistent with the coalition structure B. The Owen value of player i in the game v with the coalition structure B is now given by

$$\phi_i(B, v) = (1/|\Pi(B)|) \sum_{\pi \in \Pi(B)} \left[v\big(p_\pi^i \cup i\big) - v\big(p_\pi^i\big) \right].$$

Note that the Owen value satisfies efficiency with respect to the grand coalition, i.e., the total payoff across all players is $v(N)$. This is due to the fact that the total marginal contribution of all players with respect to a fixed order sums up to precisely $v(N)$.

Like the Shapley value, the Owen value can be characterized axiomatically in more than one way, including the way proposed by Owen himself and that of Hart and Kurz (1983). We will introduce here one version that was used in Winter (1989) for the characterization of level structure values, which generalize the Owen value.

First note that for a given coalition structure $B = (S_1, \ldots, S_m)$ and game v, we can define a "game between coalitions" in which each coalition S_i acts as a player. A coalition acting as a player will be denoted by $[S_i]$. Specifically, the worth of the coalition $\{[S_{i1}], \ldots, [S_{ik}]\}$ is $v(S_{i1} \cup S_{i2}, \ldots, \cup S_{ik})$. We will denote this game by v^*.

We now impose two symmetry axioms:

SYMMETRY ACROSS COALITIONS. *If players $[S_k]$ and $[S_l]$ are symmetric in the game v^*, then the total values for these coalitions are equal, i.e.,* $\sum_{S_l} \phi_j(B, v) = \sum_{S_k} \phi_j(B, v)$.

SYMMETRY WITHIN COALITIONS. *For any two players i and j who are symmetric in v and belong to the same coalition in B, i.e., $i, j \in S_k \in B$, we have $\phi_i(B, v) = \phi_j(B, v)$.*

We recall that the Owen value satisfies:

EFFICIENCY. $\sum_{i \in N} \phi_i(B, v) = v(N)$.

With the above axioms we now have:

THEOREM [Owen (1977)]. *For a given set of players N and a coalition structure B, the Owen value is the unique B-value satisfying symmetry across coalitions, symmetry within coalitions, dummy, additivity, and efficiency.*

Hart and Kurz (1983) formulated a different axiom which can be used to characterize the Owen value requiring that:

INESSENTIAL GAMES AMONG COALITIONS. *If $v(N) = \sum_k v(S_k)$, then for all k we have $\sum_{j \in S_k} \phi_j(B, v) = v(S_k)$.*

Hart and Kurz (1983) show that this axiom together with symmetry, additivity, and a carrier axiom combining dummy and efficiency yield a characterization of the Owen value. Hart and Kurz also used the Owen value to model processes of coalition formation.

Other papers take the Owen value as a starting-point either to suggest alternative axiomatizations or to develop further generalizations. Winter (1992) used the consistency and the potential approach to characterize the Owen value. Calvo, Lasaga, and Winter (1996) used Myerson's approach of balanced contributions for another axiomatization of the Owen value. In Owen and Winter (1992), the multilinear extension approach [see Owen (1972)] was used to propose an alternative interpretation of the Owen value based on a model of random coalitions. Finally, Winter (1991) proposed a general solution concept, special cases of which are the Harsanyi (1963) solution on standard NTU games and the Owen value for TU games with coalition structures (as well as other non-symmetric generalizations of the Shapley value).

As argued earlier, there is an intrinsic difference between Aumann and Drèze's interpretation of coalition structures and that of Owen. Thinking of coalition structures as unions or clubs, which define an asymmetric "closeness" relation between the various individuals, suggests several alternatives for introducing cooperation structures into the Shapley value framework. One such approach was proposed by Myerson (1977), who used graphs to represent cooperation structures between players. This important paper is discussed thoroughly in Monderer and Samet (see Chapter 54 in this Handbook). Another approach that is closer to Owen's (and indeed generalizes it) was analyzed by Winter (1989). Here cooperation is described by the notion of (hierarchical) level structures. Formally, a level structure is a finite sequence of partitions $B = (B_1, B_2, \ldots, B_m)$ such that B_i is a refinement of B_{i+1}. That is, if $S \in B_i$, then $S \subset T$ for some $T \in B_{i+1}$.

The idea behind level structure is that individuals inherit connections to other individuals by association to groups with various levels of commitment. In the context of international trade, one can think of B_m (the coarsest partition) as representing free trade agreements within, say, Nafta, the European Union, Mercusor, etc. Each bloc in this coalition structure is partitioned into countries, each country into federal states or regions, and so on down to the elementary units of households.

In order to define the extension of the Owen value within this framework, we adopt the interpretation that the partition B_j represents a stronger connection between players than that of the coarser B_k where $k > j$. Let us think of a permutation as the order by which players appear to collect their payoffs. To make payoffs dependent on the cooperation structure, we restrict ourselves to orders in which no player i follows player j if there is another player, say k, who is "closer" to player i and who has still not appeared. Formally, we can construct this set of orders inductively as follows:

For a given level structure $B = (B_1, \ldots, B_m)$, define

$$\Pi_m = \{\pi \in \Pi; \text{ for each } l, j \in S \in B_m \text{ and } i \in N, \ \pi(l) < \pi(i) < \pi(j) \\ \text{ implies } i \in S\}$$

and

$$\Pi_r = \{\pi \in \Pi_{r+1}; \text{ for each } l, j \in S \in B_r \text{ and } i \in N, \ \pi(l) < \pi(i) < \pi(j) \\ \text{ implies } i \in S\}.$$

The proposed value gives an expected marginal contribution to each player with respect to the uniform distribution over all the orders that are consistent with the level structure B, i.e., the orders in Π_1. Specifically,

$$\phi_i(B, v) = (1/|\Pi_1|) \sum_{\pi \in \Pi_1} \left[v(p_\pi^i \cup i) - v(p_\pi^i) \right]. \tag{4}$$

Note that when the level structure consists of only one partition (i.e., $m = 1$), we are back in Owen's framework. Moreover, in contrast to the special case of Owen, in which there is only one game between coalitions, this framework gives rise to m games between coalitions, one for each hierarchy level (partition). We denote these games by $v^1, v^2, \ldots, v^m$. The following axiom is an extension of the axiom of symmetry across coalitions that we defined earlier in relation to the Owen value.

COALITIONAL SYMMETRY. *Let* $B = (B_1, \ldots, B_m)$ *be a level structure. For each level* $1 \leqslant i \leqslant m$, *if* $S, T \in B_i$ *are symmetric as players* ([S] *and* [T]) *in the game* v^i *and if* S *and* T *are subsets of the same component in* B_j *for all* $j > i$, *then* $\sum_{r \in S} \phi_r(B, v) = \sum_{r \in T} \phi_r(B, v)$.

In order to axiomatize the level structure value, we need another symmetry axiom that requires equal treatment for symmetric players within the same coalition:

SYMMETRY WITHIN COALITIONS. *If* k *and* j *are two symmetric players with respect to the game* v, *where for every level* $1 \leqslant i \leqslant m$, *and any non-singleton coalition* $S \in B_1$, *then* $k \in S$ *iff* $j \in S$, *and* $\phi_i(B, v) = \phi_j(B, v)$.

Using straightforward generalizations of the rest of the axioms of the Owen value, it can be shown that:

THEOREM [Winter (1989)]. *There exists a unique level structure value satisfying coalitional symmetry, symmetry within coalitions, dummy, additivity, and efficiency. This value is given by* (4).

Several other approaches to cooperation structures have been proposed. We have already mentioned Myerson (1977), who uses a graph to represent bilateral connections (or communications) between individuals. An interesting application of Myerson's solution was proposed by Aumann and Myerson (1988). They considered an extensive form game in which players propose links to other players, sequentially. Using the Myerson value to represent the players' returns from each graph that forms, they analyze the endogenous graphs that form (given by the subgame perfect equilibria of the link formation game). Myerson (1980) discusses conference structures that are given formally by an arbitrary collection of coalitions representing a (possibly non-partitional) set of associations to which individuals belong. Derks and Peters (1993) consider a version of the Shapley value with restricted coalitions, representing a set of cooperation constraints. These are given by a mapping $\rho : 2^N \to 2^N$, such that (1) $\rho(S) \subset S$, (2) $S \subset T$ implies $\rho(S) \subset \rho(T)$, and (3) $\rho(\rho(S)) = \rho(S)$. This mapping can be interpreted as institutional constraints on the communications between players, i.e., $\rho(S)$ represents the most comprehensive agreement that can be attained within the set of players S. Van den Brink and Gilles propose Permission Structures, based on the idea that some interactions take place in hierarchical organizations in which cooperation between two individuals requires the consent of their supervisors. Permission structures are thus given by a mapping $p : N \to 2^N$, where $j \in p(i)$ stands for "j supervises i". The function p imposes exogenous restrictions on cooperation and allows for an extension of the Shapley value.

I will conclude this section by briefly discussing another interesting (asymmetric) generalization of the value. Unlike the others, this one was proposed by Shapley himself (see also Chapter 32 in this Handbook). Shapley (1977) examines the power of indices in political games, where players' political positions affect the prospects of coalition formation. Shapley's aim was to embed players' positions in an m-dimensional Euclidean space. The point $x_i \in \mathbb{R}^m$ of player i summarizes i's position (on a scale of support and opposition) on each of the relevant m "pure" issues. General issues faced by legislators typically involve a combination of several pure ones. For example, if the two pure issues are government spending (high or low) and national defense policy (aggressive or moderate), then the issue of whether or not to launch a new defense missile project is a combination of the two pure issues. Shapley's suggestion was to describe general issues as vectors of weights $w = (w_1, \ldots, w_m)$. Note that every vector w induces a natural order over the set of players. Specifically, j appears after i if $w \cdot x_i > w \cdot x_j$ (where $x \cdot y$ stands for the inner product of the vectors x and y). The main point to notice here is that different vectors induce different orders on the set of players. To measure the legislative power of each player in the game, one has to aggregate over all possible (general) issues. Let us therefore assume that issues occur randomly with respect to a uniform distribution over all issues (i.e., vectors w in $\mathbb{R}^m$). For each order of players π, let $\theta(\pi)$ be the probability that the random issue generates the order π. Thus the players' profile of political positions $(x_1, x_2, \ldots, x_n)$ is mapped into a probability distribution over the set of all permutations. Shapley's political value yields an expected marginal contribution to each player, where the random orders are given by the probability distribution θ. Note that the political value is in the class of Weber's random order values [see Weber (1988)]. A random order value is characterized by a probability distribution over the set of all permutations. According to this value, each player receives his expected marginal contribution to the players preceding him with respect to the underlying probability distribution on orders. The relation between the political value and the Owen value is also quite interesting. Suppose that the vector of positions is represented by m clusters of points in $\mathbb{R}^2$, where the cluster k consists of the players in S_k, whose positions are very close to each other but further away than those of other players (in the extreme case we could think of the members of S_k as having identical positions). It is pretty clear that the payoff vector that will emerge from Shapley's political value in this case will be very close to the Owen value for the coalition structure $B = (S_1, \ldots, S_m)$.

8. Sustaining the Shapley value via non-cooperative games

If the Shapley value is interpreted merely as a measure for evaluating players' power in a cooperative game, then its axiomatic foundation is strong enough to fully justify it. But the Shapley value is often interpreted (and indeed sometimes applied) as a scheme or a rule for allocating collective benefits or costs. The interpretation of the value in these situations implicitly assumes the existence of an outside authority – call it a planner – which determines individual payoffs based on the axioms that characterize the value. However, situations of benefit (or cost) allocation are, by their very nature, situations in which individuals have conflicting interests. Players who feel underpaid are therefore likely to dispute the fairness of the scheme by challenging one or more of its axioms. It would therefore be nice if the Shapley value could be supported as an outcome of some decentralized mechanism in which individuals behave strategically in the absence of a planner whose objectives, though benevolent, may be disputable. This objective has been pursued by several authors as part of a broader agenda that deals with the interface between cooperative and non-cooperative game theory. The concern of this literature is the construction of non-cooperative bargaining games that sustain various cooperative solution concepts as their equilibrium outcomes. This approach, often referred to in the literature as "the Nash Program", is attributed to Nash's (1950) groundbreaking work on the bargaining problem, which, in addition to laying the axiomatic foundation of the solution, constructs a non-cooperative game to sustain it.

Of all the solution concepts in cooperative game theory, the Shapley value is arguably the most "cooperative", undoubtedly more so than such concepts as the core and the bargaining set whose definitions include strategic interpretations. Yet, perhaps more than any other solution concept in cooperative game theory, the Shapley value emerges as the outcome of a variety of non-cooperative games quite different in structure and interpretation.

Harsanyi (1985) is probably the first to address the relationship between the Shapley value and non-cooperative games. Harsanyi's "dividend game" makes use of the relation between the Shapley value and the decomposition of games into unanimity games. To the more recent literature which uses sequential bargaining games to sustain cooperative solution concepts, Gul (1989) makes a pioneering contribution. In Gul's model, players meet randomly to conduct bilateral trades. When two players meet, one of them is chosen randomly (with equal probabilities $1/2-1/2$) to make a proposal. In a proposal by player i to player j at period t, player i offers to pay r_t to player j for purchasing j's resources in the game. If player j accepts the offer, he leaves the game with the proposed payoff, and the coalition $\{i, j\}$ becomes a single player in the new coalitional form game, implying that i now owns the property rights of player j. If j rejects the proposal by i, both players return to the pool of potential traders who meet through random matching in a subsequent period. Each pair's probability of being selected for trading is $2/(n_t(n_t - 1))$, where n_t is the number of players remaining at period t. The game ends when only a single player is left. For any given play path of the game, the payoff of player j is given by the current value of his stream of resources minus the payments he made to the other players. Thus, for a given strategy combination σ and a discount factor δ, we have

$$U^i(\sigma, \delta) = \sum_0^\infty (1 - \delta)[V(M_i^t) - r_i^t]\delta^t,$$

where M_i^t is the set of players whose resources are controlled by player i at time t and δ is a discount factor. Gul confines himself to the stationary subgame perfect equilibria (SSPE) of the game, i.e., equilibria in which players' actions at period t depend only upon the allocation of resources at time t. He argues that SSPE outcomes may not be efficient in the sense of maximizing the aggregate equilibrium payoffs of all the players in the economy, but he goes on to show that in any no-delay equilibrium (i.e., an equilibrium in which all pairwise meetings end with agreements) players' payoffs converge to the Shapley value of the underlying game when the discount factor approaches 1. Specifically,

THEOREM [Gul (1989)]. *Let $\sigma(\delta_k)$ be a sequence of SSPEs with respect to the discount factors $\{\delta_k\}$ which converge to 1 as k goes to infinity. If $\sigma(\delta_k)$ is a no-delay equilibrium for all k, then $U^i(\sigma(\delta_k), \delta_k)$ converges to i's Shapley value of V as k goes to infinity.*

It should be argued that in general the SSPE outcomes of Gul's game do not converge to the Shapley value as the discount factor approaches 1. Indeed, if delay occurs along the equilibrium path, the outcome may not be close to the Shapley value even for δ close to 1. Gul's original formulation of the above theorem required that $\sigma(\delta_k)$ be efficient equilibria (in terms of expected payoffs). Gul argues that the condition of efficiency is a sufficient guarantee that along the equilibrium path every bilateral matching terminates in an agreement. However, Hart and Levy (1999) show in an example that efficiency does not imply immediate agreement. Nevertheless, in a rejoinder to Hart and

Levy (1999), Gul (1999) points out that if the underlying coalitional game is strictly convex,[8] then in his model efficiency indeed implies no delay.

A different bargaining model to sustain the Shapley value through its consistency property was proposed by Hart and Mas-Colell (1996).[9] Unlike in Gul's model, which is based on bilateral agreements, in Hart and Mas-Colell's approach players submit proposals for payoff allocations to all the active players. Each round in the game is characterized by a set $S \subset N$ of "active players" and a player $i \in S$ who is designated to make a proposal after being randomly selected from the set S. A proposal is a feasible payoff vector x for the members in S, i.e., $\sum_{j \in S} x_j = v(S)$. Once the proposal is made, the players in S respond sequentially by either accepting or rejecting it. If all the members of S accept the proposal, the game ends and the players in S share payoffs according to the proposal. Inactive players receive a payoff of zero. If at least one player rejects the proposal, then the proposer i runs the risk of being dropped from the game. Specifically, the proposer leaves the game and joins the set of inactive players with a probability of $1 - p$, in which case the game continues into the next period with the set of active players being $S \setminus i$. Or the proposer remains active with probability p, and the game continues into the next period with the same set of active players. The game ends either when agreement is reached or when only one active player is left in the game. Hart and Mas-Colell analyzed the above (perfect information) game by means of its stationary subgame perfect equilibria, and concluded:

THEOREM [Hart and Mas-Colell (1996)]. *For every monotonic and non-negative*[10] *game v and for every $0 \leqslant p < 1$, the bargaining game described above has a unique stationary subgame perfect equilibrium (SSPE). Furthermore, if a_S is the SSPE payoff vector of a subgame starting with a period in which S is the active set of players, then $a_S = \phi(v|_S)$ (where ϕ stands for the Shapley value and $v|_S$ for the restricted game on S). In particular, the SSPE outcome of the whole game is the Shapley value of v.*

A rough summary of the argument for this result runs as follows. Let $a_{S,i}$ denote the equilibrium proposal when the set of active players is S and the proposer is $i \in S$. In equilibrium, player $j \in S$ with $j \neq i$ should be paid precisely what he expects to receive if agreement fails to be reached at the current payoff. As the protocol specifies, with probability p we remain with the same set of players and the next period's expected proposal will be $a_S = 1/|S| \sum_{i \in S} a_{S,i}$. With the remaining probability $1 - p$, players i will be ejected so that the next period's proposal is expected to be $a_{S \setminus i}$. We thus obtain the following two equations for $a_{S,i}$:

(1) $\sum_j a_{S,i}^j = v(S)$ (feasibility condition), and

---

[8] We recall that a game v is said to be strictly convex if $v(S \cup i) - V(S) > v(T \cup i) - v(T)$ whenever $T \subset S$ and $S \neq T$.

[9] In the original Hart and Mas-Colell (1996) paper, the bargaining game was based on an underlying non-transferable utility game.

[10] $v(S) \geqslant 0$ for all $S \subset N$, and $v(T) \leqslant v(S)$ for $T \subset S$.

(2) $a^j_{S,i} = pa^j_S + (1-p)a^j_{S\setminus i}$ (equilibrium condition).

Rewriting the second condition we notice that the two of them correspond precisely to the two properties of Myerson (1977), which we discussed in Section 5 and which together with efficiency characterize the value uniquely.

In a recent paper, Perez-Castrillo and Wettstein (1999) suggested a game that modifies that of Hart and Mas-Colell (1996) so as to allow the (random) order of proposals to be endogenized. The game runs as follows. Prior to making a proposal there is a bidding phase in which each player i commits to pay a payoff v^j_i to player j. These bids are made simultaneously. The identity of the proposal is determined by the bids. Specifically, the proposer is chosen to be the player i for which the difference between the bids made by i and the bids made to i is maximized, i.e., $i = \mathrm{argmax}_{k \in N}[\sum_j v^j_k - \sum_j v^k_j]$ (players' bids to themselves is always zero). If there is more than one player for which this maximum is attained, then the proposer is chosen from among these players with an equal probability for each candidate. Following player i's recognition to propose, the game proceeds according to Hart and Mas-Colell's protocol, with $p = 1$. Namely, upon rejection, player i leaves the game with probability 1. Perez-Castrillo and Wettstein (1999) show that this game implements the Shapley value in a unique subgame perfect equilibrium (since the game is of a finite horizon, no stationarity requirement is needed).

Almost all the bargaining games that have been proposed in the literature on the implementation of cooperative solutions via non-cooperative equilibria are based on the exchange of proposals and responses. A different approach to multilateral bargaining was adopted in Winter (1994). Rather than a model in which players make full proposals concerning payoff allocations and respond to such proposals, a more descriptive feature of bargaining situations is sought by assuming that players submit only demands, i.e., players announce the share they request in return for cooperation. A coalition emerges when the underlying resources are sufficient to satisfy the demands of all members. I will describe here a simple version of the Winter (1994) model and some of the results that follow.

Consider the order in which players move according to their name, i.e., player 1 followed by 2, etc. Each player i in his turn publicly announces a demand d_i (which should be interpreted as a statement by player i of agreeing to be a member of any coalition provided that he is paid at least d_i). Before player i makes his demand, we check whether there is a compatible coalition among the $i-1$ players who already made their demands. A coalition S is said to be compatible (to the underlying game v) if S can satisfy the demands of all its members, i.e., $\sum_{j \in S} d_j \leqslant v(S)$. If compatible coalitions exist, then the largest one (in terms of membership) leaves the game and each of its members receives his demand. The game then proceeds with the set of remaining players. If no such coalition exists, then player i moves ahead and makes his demand. The game ends when all players have made their demands. Those players who are not part of a compatible coalition receive their individually rational payoff.

Consider now a game that starts with a chance move that randomly selects an order with a uniform probability distribution over all orders and then proceeds in accordance with the above protocol. We call this game the demand commitment game. Winter (1994) shows that the demand commitment game implements the Shapley value if the underlying game is strictly convex.

THEOREM [Winter (1994)]. *For strictly convex (cooperative) games, the demand commitment game has a unique subgame perfect equilibrium, and each player's equilibrium payoff equals his Shapley value.*

Winter (1994) also considers a protocol that requires a second round of bidding in the event that the first round ends without a grand coalition that is compatible. It can be shown that with small delay costs the Shapley value emerges not only as the expected equilibrium outcome for each player, but that the actual demands made by the players in the first round coincide with the Shapley value of the underlying game.

Several other papers have followed the same approach in different contexts. Dasgupta and Chiu (1998) discuss a modified version of the Winter (1994) game, which allows for the implementation of the Shapley value in general games. Roughly, the idea is to allow outside transfers (or gifts) that will convexify a non-convex game. A balanced budget is guaranteed by a schedule of taxes dependent on the order of moves. Bag and Winter (1999) used a demand commitment-type mechanism to implement stable and efficient allocations in excludable public goods. Morelli (1999) modified Winter's (1994) model to describe legislative bargaining under various voting rules. Finally, Mutuswami and Winter (2000) used demand mechanisms of the same kind to study the formation of networks. They also noted that if the mechanism in Winter (1994) is amended to allow a compatible coalition to leave the game only when it is connected (i.e., only when it includes the last k players for some $1 \leqslant k \leqslant n$), then the resulting game implements the Shapley value not only in the case of convex games but in all games.

9. Practice

While game theory is thought of as "descriptive" in its attempt to explain social phenomena by means of formal modeling, cooperative game theory is primarily "prescriptive". It is not surprising that much of the literature on cooperative solution concepts finds its way not into economics journals but into journals of management science and operations research. Cooperative game theory does not set out to describe the way individuals behave. Rather, it recommends reasonable rules of allocation, or proposes indices to measure power. The prospect of using such a theory for practical applications is therefore quite attractive, the more so for its single-point solution and axiomatic foundation. In this section, I discuss two areas in which the Shapley value can be (and indeed has been) used as a practical tool: the measurement of voter power and cost allocation.

9.1. Measuring states' power in the U.S. presidential election

The procedure for electing a president in the United States consists of two stages. First, each state elects a group of representatives, or "Great Electors", who comprise the Electoral College. Second, the Electoral College elects the president by simple majority rule. It is assumed that each Great Elector votes for the candidate preferred by the majority of his/her state. Since the Electoral College of each state grows in proportion to its census count, a narrow majority in a densely populated state, like California, can affect an election's outcome more than wide majorities in several scarcely populated states. Mann and Shapley (1962) and Owen (1975) measured the voting power of voters from different states, using the Shapley value together with the interesting notion of compound simple games.

Let $M_1, M_2, \ldots, M_n$ be a sequence of n disjoint sets of players. Let $w_1, w_2, \ldots, w_n$ be a sequence of n simple games defined on the sets $M_1, \ldots, M_n$ respectively. And let v be another simple game defined on the set $N = \{1, 2, \ldots, n\}$. We will refer to the players in N as districts. The compound game $u = v[w_1, \ldots, w_n]$ is a simple game defined on the set $M = M_1 \cup M_2 \cup \cdots \cup M_n$ by $u(S) = v(\{j \mid w_j(S \cap M_j) = 1\})$. In words: We say that S wins in district j if S's members in that district form a winning coalition, i.e., if $w_j(S \cap M_j) = 1$. S is said to be winning in the game u if and only if the set of districts in which S is winning is itself a winning coalition in v. In the context of the presidential race, M_j is the set of voters in state j, w_j is the simple majority game in state j, and v is the electoral college game. Specifically, the electoral college game can be written as the following weighted majority game $[270; p_1, \ldots, p_{51}]$, where 51 stands for the number of states, and p_i is the number of electors nominated by a state i (e.g., 45 for California and 3 for the least populated states and the District of Columbia).

In general compound games, Owen has shown that the value of player i is the product of his value in the game within his district and the value of his district in the game v, i.e., $\phi_i(u) = \phi_j(v)\phi_i(w_j)$.

Since the districts' games are all symmetric simple majority games, the value of each player in the voting game in his state is simply 1 divided by the number of voters. To compute the value of the game v, Owen used the notion of a multilinear extension [see Owen (1972)]. Overall, he found that the power of a voter in a more populated state is substantially greater than that of a voter in a less populated state. For example, California voters enjoy more than three times the power of Washington D.C. voters.

Others have used the Shapley value (as well as other indices) to measure political power. Seidmann (1987) used it to compare the power of governments in Ireland following elections in the early and mid 80s. He argued that a government's durability greatly depends on the distribution of power across opposition parties, which can be estimated by means of the Shapley–Shubik index. Carreras, García-Jurado, and Pacios (1993) used the Shapley and the Owen value to evaluate the power of each of the parties in all the Spanish regional parliaments. Fedzhora (2000) uses the Shapley–Shubik index to study the voting power of the 27 regions (oblasts) in the Ukraine in the run-off stage of the presidential elections between 1994 and 1999. She also compares these indicators to the transfers that Ukrainian governments were making to the different regions. Another interesting application of the Shapley value to political science is due to Rapoport and Golan (1985). Immediately after the election of the tenth Israeli parliament, 21 students of political science, 24 Knesset members, and 7 parliamentary correspondents were invited to assess the political power ratios of the 10 parties represented in the Knesset. These assessments were then compared with various power indices, including the Shapley value. The value provided the best fit for 31% of the subjects, but the authors claimed the Banzhaf index performed better.

9.2. Allocation of costs

The problem of allocating the cost of constructing or maintaining a public facility among its various users is of great practical importance. Young (1994) (see Chapter 34 in this Handbook) offers a comprehensive survey of the relation between game theory and cost allocation. An interesting allocation rule for such problems, which is closely related to the Shapley value, emerges from the "Airport Game" of Littlechild and Owen (1973). Specifically, consider n planes of different size for which a runway needs to be built. Suppose that there are m types of planes and that the construction of a runway sufficient to service a type j plane is c_j with $c_1 < c_2 < \cdots < c_m$. Let N_j be the number of planes of type j so that $\bigcup N_j = N$ is the set of all planes which need to be serviced. A runway servicing a subgroup of planes $S \subset N$ will have to be long enough to allow the servicing of the largest plane in S. This gives rise to the following natural cost-sharing game (in coalitional form) defined on the set of planes: $c(S) = c_{j(S)}$ and $c(\emptyset) = 0$, where $j(S) = \max\{j \mid S \cap N_j \neq \emptyset\}$. Littlechild and Owen's (1973) suggestion was to use the game c to determine the allocation of cost by applying the Shapley value on the game.

Note that the game v can be decomposed into m unanimity games. Specifically, define $R_k = R_k \cup R_{k+1} \cup \cdots \cup R_m$ and consider the coalitional form games v_k with $v_k(S) = 0$ when $S \cap R_k = \emptyset$ and $v_k(S) = c_k - c_{k-1}$ otherwise (we set $c_0 = 0$). It is easy to verify that the sum of the games v_k is exactly the cost-sharing game, i.e., $v_1(S) + \cdots + v_m(S) = c(S)$ for every coalition of planes S. The additivity of the Shapley value implies that the value of the game c is $\phi(c) = \phi(v_1) + \cdots + \phi(v_m)$. But each v_k is a unanimity game with $\phi_i(v_k) = 0$ for $i \in N \setminus R_k$ and $\phi_i(v_k) = (c_k - c_{k-1})/|R_k|$ for $i \in R_k$. We therefore obtain that the Shapley value of the game c is given by $\phi_i(c) = (c_2 - c_1)/|R_1| + (c_3 - c_2)/|R_2| + \cdots + (c_j - c_{j-1})/|R_j|$ for a plane of type j.

The rule suggested by Littlechild and Owen (1973) has the following interesting interpretation. First, all players share equally the cost of a runway for type 1 planes. Then all players who need a larger facility share the marginal extra cost, i.e., $c_2 - c_1$. Next, all players who need yet a larger runway share equally the cost of upgrading to a runway large enough to service type 3 planes. We now continue in this manner until all the residual costs are allocated, which will ultimately allow for the acquisition of a runway long enough to service all planes.

A version of the Airport game was studied by Fragnelli et al. (1999). Their work was part of a research project funded by the European Commission with the aim of determining cost allocation rules for the railway infrastructure in Europe. Fragnelli et al. realized that the original Airport game is ill-suited to their problem since the maintenance cost of a railway infrastructure depends on the number of users. They constructed a new game which distinguishes construction costs (which do not depend on the number of users) from maintenance costs, and derived a simple formula for the Shapley value of the game. They also used real data concerning the maintenance of railway infrastructures to estimate the allocation of cost among different users.

References

Aumann, R.J., and J. Drèze (1975), "Cooperative games with coalition structures", International Journal of Game Theory 4:217–237.

Aumann, R.J., and R.B. Myerson (1988), "Endogenous formation of links between players and of coalitions: An application of the Shapley value", in: A.E. Roth, ed., The Shapley Value (Cambridge University Press, Cambridge), 175–191.

Aumann, R.J., and L.S. Shapley (1974), Values of Non-Atomic Games (Princeton University Press, Princeton).

Bag, P.K., and E. Winter (1999), "Simple subscription mechanisms for the production of public goods", Journal of Economic Theory 87:72–97.

Banzhaf, J.F. III (1965), "Weighted voting does not work: A mathematical analysis", Rutgers Law Review 19:317–343.

Calvo, E., J. Lasaga and E. Winter (1996), "On the principle of balanced contributions", Mathematical Social Sciences 31:171–182.

Carreras, F., I. García-Jurado and M.A. Pacios (1993), "Estudio coalicional de los parlamentos autonómicos españoles de régimen común", Revista de Estudios Políticos 82:152–176.

Chun, Y. (1989), "A new axiomatization of the Shapley value", Games and Economic Behavior 1:119–130.

Dasgupta, A., and Y.S. Chiu (1998), "On implementation via demand commitment games", International Journal of Game Theory 27:161–190.

Derks, J., and H. Peters (1993), "A Shapley value for games with restricted coalitions", International Journal of Game Theory 21:351–360.

Dubey, P. (1975), "On the uniqueness of the Shapley value", International Journal of Game Theory 4:131–139.

Fedzhora, L. (2000), "Voting power in the Ukrainian presidential elections", M.A. Thesis (Economic Education and Research Consortium, Kiev, Ukraine).

Fragnelli, V., I. García-Jurado, H. Norde, F. Patrone and S. Tijs (1999), "How to share railway infrastructure costs?", in: I. García-Jurado, F. Patrone and S. Tijs, eds., Game Practice: Contributions from Applied Game Theory (Kluwer, Dordrecht).

Gul, F. (1989), "Bargaining foundations of Shapley value", Econometrica 57:81–95.

Gul, F. (1999), "Efficiency and immediate agreement: A reply to Hart and Levy", Econometrica 67:913–918.

Harsanyi, J.C. (1963), "A simplified bargaining model for the n-person cooperative game", International Economic Review 4:194–220.

Harsanyi, J.C. (1985), "The Shapley value and the risk dominance solutions of two bargaining models for characteristic-function games", in: R.J. Aumann et al., eds., Essays in Game Theory and Mathematical Economics (Bibliographisches Institut, Mannheim), 43–68.

Hart, S., and M. Kurz (1983), "On the endogenous formation of coalitions", Econometrica 51:1295–1313.

Hart, S., and Z. Levy (1999), "Efficiency does not imply immediate agreement", Econometrica 67:909–912.

Hart, S., and A. Mas-Colell (1989), "Potential, value and consistency", Econometrica 57:589–614.

Hart, S., and A. Mas-Colell (1996), "Bargaining and value", Econometrica 64:357–380.

Kalai, E., and D. Samet (1985), "Monotonic solutions to general cooperative games", Econometrica 53:307–327.

Littlechild, S.C., and G. Owen (1973), "A simple expression for the Shapley value in a special case", Management Science 20:99–107.

Mann, I., and L.S. Shapley (1962), "The a-priori voting strength of the electoral college", American Political Science Review 72:70–79.

Monderer, D., and D. Samet (2002), "Variations on the Shapley value", in: R.J. Aumann and S. Hart, eds., Handbook of Game Theory, Vol. 3 (North-Holland, Amsterdam) Chapter 54, 2055–2076.

Morelli, M. (1999), "Demand competition and policy compromise in legislative bargaining", American Political Science Review 93:809–820.

Mutuswami, S., and E. Winter (2000), "Subscription mechanisms for network formation", Mimeo (The Center for Rationality and Interactive Decision-Making, The Hebrew University of Jerusalem).

Myerson, R.B. (1977), "Graphs and cooperation in games", Mathematics of Operations Research 2:225–229.

Myerson, R.B. (1980), "Conference structures and fair allocation rules", International Journal of Game Theory 9:169–182.

Nash, J. (1950), "The bargaining problem", Econometrica 18:155–162.

Neyman, A. (1989), "Uniqueness of the Shapley value", Games and Economic Behavior 1:116–118.

Owen, G. (1972), "Multilinear extensions of games", Management Science 18:64–79.

Owen, G. (1975), "Evaluation of a presidential election game", American Political Science Review 69:947–953.

Owen, G. (1977), "Values of games with a priori unions", in: R. Henn and O. Moeschlin, eds., Essays in Mathematical Economics and Game Theory (Springer-Verlag, Berlin), 76–88.

Owen, G., and E. Winter (1992), "The multilinear extension and the coalition value", Games and Economic Behavior 4:582–587.

Perez-Castrillo, D., and D. Wettstein (1999), "Bidding for the surplus: A non-cooperative approach to the Shapley value", DP 99-7 (Ben-Gurion University, Monaster Center for Economic Research).

Rapoport, A., and E. Golan (1985), "Assessment of political power in the Israeli Knesset", American Political Science Review 79:673–692.

Roth, A.E. (1977), "The Shapley value as a von Neumann–Morgenstern utility", Econometrica 45:657–664.

Seidmann, D. (1987), "The distribution of power in Da'il E'ireann", The Economic and Social Review 19:61–68.

Shapley, L.S. (1953), "A value for *n*-person games", in: H.W. Kuhn and A.W. Tucker, eds., Contributions to the Theory of Games II (Annals of Mathematics Studies 28) (Princeton University Press, Princeton), 307–317.

Shapley, L.S. (1977), "A comparison of power indices and a nonsymmetric generalization", P-5872 (The Rand Corporation, Santa Monica, CA).

Shapley, L.S., and M. Shubik (1954), "A method for evaluating the distribution of power in a committee system", American Political Science Review 48:787–792.

Sobolev, A.I. (1975), "The characterization of optimality principles in cooperative games by functional equations", Mathematical Methods in Social Sciences 6:94–151 (in Russian).

Straffin, P.D. (1977), "Homogeneity, independence and power indices", Public Choice 30:107–118.

Straffin, P.D. (1994), "Power and stability in politics", in: R.J. Aumann and S. Hart, eds., Handbook of Game Theory, Vol. 2 (North-Holland, Amsterdam) Chapter 32, 1127–1152.

Von Neumann, J., and O. Morgenstern (1944), Theory of Games and Economic Behavior (Princeton University Press, Princeton).

Weber, R. (1988), "Probabilistic values for games", in: A.E. Roth, ed., The Shapley Value (Cambridge University Press, Cambridge), 101–120.

Weber, R. (1994), "Games in coalitional form", in: R.J. Aumann and S. Hart, eds., Handbook of Game Theory, Vol. 2 (North-Holland, Amsterdam) Chapter 36, 1285–1304.

Winter, E. (1989), "A value for games with level structures", International Journal of Game Theory 18:227–242.

Winter, E. (1991), "A solution for non-transferable utility games with coalition structure", International Journal of Game Theory 20:53–63.

Winter, E. (1992), "The consistency and the potential for values with games with coalition structure", Games and Economic Behavior 4:132–144.

Winter, E. (1994), "The demand commitment bargaining and snowballing cooperation", Economic Theory 4:255–273.

Young, H.P. (1985), "Monotonic solutions of cooperative games", International Journal of Game Theory 14:65–72.

Young, H.P. (1994), "Cost allocation", in: R.J. Aumann and S. Hart, eds., Handbook of Game Theory, Vol. 2 (North-Holland, Amsterdam) Chapter 34, 1193–1235.

Chapter 54

VARIATIONS ON THE SHAPLEY VALUE

DOV MONDERER

The Technion, Haifa, Israel

DOV SAMET

Tel Aviv University, Tel Aviv, Israel

Contents

Handbook of Game Theory, Volume 3, Edited by R.J. Aumann and S. Hart

Abstract

This survey captures the main contributions in the area described by the title that were published up to 1997. (Unfortunately, it does not capture all of them.) The variations that are the subject of this chapter are those axiomatically characterized solutions which are obtained by varying either the set of axioms that define the Shapley value, or the domain over which this value is defined, or both.

In the first category, we deal mainly with probabilistic values. These are solutions that preserve one of the essential features of the Shapley value, namely, that they are given, for each player, by some averaging of the player's marginal contributions to coalitions, where the probabilistic weights depend on the coalitions only and not on the game. The Shapley value is the unique probabilistic value that is efficient and symmetric. We characterize and discuss two families of solutions: quasivalues, which are efficient probabilistic values, and semivalues, which are symmetric probabilistic values.

In the second category, we deal with solutions that generalize the Shapley value by changing the domain over which the solution is defined. In such generalizations the solution is defined on pairs, consisting of a game and some structure on the set of players. The Shapley value is a special case of such a generalization in the sense that it coincides with the solution on the restricted domain in which the second argument is fixed to be the "trivial" one. Under this category we survey mostly solutions in which the structure is a partition of the set of the players, and a solution in which the structure is a graph, the vertices of which are the players.

Keywords

Shapley value, quasivalue, semivalue, non-symmetric values, probabilistic values, weighted values

JEL classification: C71, D46, D63, D72, D74

1. Introduction

The variations that are the subject of this discussion are those axiomatically characterized solutions that are obtained by varying the set of axioms which define the Shapley
value, or the domain over which this value is defined, or both.

We first capture axiomatically, in Section 3, those solutions that preserve one of the
essential features of the Shapley value, namely, that they are given, for each player, by
some averaging of the player's marginal contributions to coalitions, where the probabilistic weights depend on the coalitions only and not on the game. We call the family
of all such solutions *probabilistic values*. The Shapley value is the unique probabilistic
value that is efficient and symmetric. In the following sections we characterize and discuss two families of solutions: *quasivalues*, which are efficient probabilistic values, and
semivalues, which are symmetric probabilistic values.

In Section 4 we discuss quasivalues. These solutions are related to the descriptions
of the Shapley value as a *random-order value*: it is the expected marginal contributions
of the players when they are randomly ordered. Indeed a solution is a quasivalue if and
only if it is a random-order value. The Shapley value is the special random-order value
in which the random orders are uniformly drawn.

Quasivalues can be equivalently described by choosing at random the "arrival" times
of the players, rather than their order, and taking the expected marginal contribution of a
player to the players who arrived before him. In particular, the Shapley value is obtained
when the arrival times are independent and each is uniformly distributed in the unit interval. When arrival times are independent, the quasivalue has another representation as
a *path value*: the value can be computed by integrating the derivative of the multilinear
extension of the game along a certain path.

In Section 5 we characterize and discuss a family of quasivalues – the *weighted Shapley values* – which was introduced by Shapley alongside the Shapley value.

Semivalues are discussed in Section 6. The symmetry axiom, which plays a central
role in the characterization of semivalues, is reflected in the probabilities that define
the value: they depend only on the size of coalitions. A semivalue does not, of course,
have to be efficient. We show that the specific deviation of a semivalue from efficiency
defines it uniquely. That is, if the magnitude of the deviation from efficiency is the same
for two semivalues in each game, then they coincide.

A probabilistic value defined on the set of all games is necessarily additive. When
the set of games is restricted to simple games (that is, games in which the worth of a
coalition is either 0 or 1), then additivity cannot be applied. Yet, all the generalizations
discussed so far can be axiomatically defined for this class of games when the *transfer*
axiom replaces that of additivity. This is done in Section 7.

Section 8 deals with solutions that generalize the Shapley value by changing the domain over which the solution is defined. In such generalizations the solution is defined
on pairs, consisting of a game and some structure on the set of players. The Shapley
value is a special case of such a generalization in the sense that it coincides with the solution on the restricted domain in which the second argument is fixed to be the "trivial"

one. Under this category we survey mostly solutions in which the structure is a partition of the set of the players, and a solution in which the structure is a graph the vertices of which are the players.

2. Preliminaries

Let N be a finite set with n elements, $n \geqslant 1$. Elements of N are called *players*. Any subset of N is called a *coalition*. For a coalition S, we denote $N \setminus S$ by S^c. We denote by $C = C(N)$ the set of all coalitions, and by $C_0 = C_0(N)$ the set of all nonempty coalitions.

A *game* on N is a set function $v : C \to R$, where R denotes the set of real numbers, with $v(\emptyset) = 0$. We will write $v(S \cup i)$ for $v(S \cup \{i\})$, and $v(S \setminus i)$ for $v(S \setminus \{i\})$. A game v is *additive* if for every pair of disjoint coalitions $S, T \in C$, $v(S \cup T) = v(S) + v(T)$, and it is *superadditive* if for each pair of such coalitions, $v(S \cup T) \geqslant v(S) + v(T)$. It is *monotone* if $v(S) \leqslant v(T)$ for every $S \subseteq T$. A game v is *simple* if $v(S) \in \{0, 1\}$ for every $S \in C$.

Player i is a *null* player in v if $v(S \cup i) - v(S) = 0$ for every $S \subseteq N \setminus i$. A coalition M is a *carrier* of v if M^c consists of null players. Hence, M is a carrier for v if and only if $v(S \cap M) = v(S)$ for all $S \subseteq N$. Player i is a *dummy* player in v if there exists an $a \in R$ such that $v(S \cup i) - v(S) = a$ for every $S \subseteq N \setminus i$.

The set of all games is denoted $G = G(N)$ and the set of all additive games is denoted $A = A(N)$. We naturally identify an additive game $v \in A$ with the vector $x \in R^N$ defined by $x_i = v(i)$ for every $i \in N$. For each such $x \in R^N$ and for every $S \in C$ we denote $x(S) = \sum_{i \in S} x_i$.

A *permutation* of N is a one-to-one function that maps N onto N. The set of all permutations on N is denoted by $\Theta = \Theta(N)$. The set of all one-to-one functions $\tau : N \to \{1, \ldots, n\}$ is denoted by $OR = OR(N)$. Each $\tau \in OR$ is identified with a complete order on N. In this order, i comes before j if $\tau(i) < \tau(j)$. For a game v and a permutation θ we define $\theta^* v \in G$ by $\theta^* v(S) = v(\theta(S))$ for all $S \in C$.

Let $H \subseteq G$. A *solution* on H is a function $\psi : H \to A$. We will use the terms *operator* and *solution* interchangeably. The domain of a solution is G whenever it is not explicitly specified otherwise. It is useful to single out a list of possible properties of solutions (axioms) that have been discussed in the literature.

- (A.1) ψ satisfies the *additivity axiom* or is *additive* if $\psi(u + v) = \psi(u) + \psi(v)$ for every $u, v \in G$.
- (A.2) ψ is *homogeneous* if $\psi(\alpha v) = \alpha \psi(v)$ for every $v \in G$ and $\alpha \in R$.
- (A.3) ψ satisfies the *linearity axiom*, or is *linear*, if it is additive and homogeneous.
- (A.4) ψ satisfies the *projection axiom*, or is a *projection*, if $\psi(\mu) = \mu$ for every $\mu \in A$.
- (A.5) ψ satisfies the *null player axiom* if $\psi v(i) = 0$ for every $v \in G$ and every null player i in v.
- (A.6) ψ satisfies the *dummy player axiom* if $\psi v(i) = a$ for $v \in G$ and for every dummy player i such that $v(S \cup i) - v(S) = a$ for every $S \subseteq N \setminus i$.

(A.7) ψ satisfies the *positivity axiom*, or ψ is *positive*, if $\psi v(i) \geqslant 0$ for every i whenever v is a monotone game.

(A.8) ψ satisfies the *Milnor axiom*, or is a *Milnor operator*, if for every v and $i \in N$,

$$\min_{S \subseteq N \setminus i} \big(v(S \cup i) - v(S)\big) \leqslant \psi v(i) \leqslant \max_{S \subseteq N \setminus i} \big(v(S \cup i) - v(S)\big).$$

(A.9) ψ satisfies the *symmetry axiom* or is *symmetric* if $\psi(\theta^* v) = \theta^* \psi v$ for every $v \in G$ and every permutation θ.

(A.10) ψ satisfies the *efficiency axiom* or is *efficient* if $\psi v(N) = v(N)$ for every $v \in G$.

(A.11) ψ satisfies the *carrier axiom* if $\psi v(M) = v(M)$ for every $v \in G$ and every carrier M of v.

We state the above axioms for solutions defined on G, but will sometimes use them for solutions defined on strict subsets of G. In such a case, an axiom is to be understood as holding whenever possible. Thus, for example, if ψ is a solution on $H \subset G$, the projection axiom should be read as $\psi \mu = \mu$ whenever $\mu \in H \cap A$.

3. Probabilistic values

For a finite set M we denote by $\Delta(M)$ the set of all probability measures on M. That is, $p \in \Delta(M)$ if $p : M \to [0, 1]$ and $\sum_{m \in M} p(m) = 1$. We denote by E_p the expectation operator with respect to p.

The set of all subsets of $N \setminus i$ is denoted by $C^i = C^i(N)$. For each game v, we denote by $D^i v$ the set function defined on C^i by $D^i v(S) = v(S \cup i) - v(S)$. We call $D^i v(S)$ the *marginal contribution* of i to S.

A solution ψ is called a *probabilistic value*, if for each player i there exists a probability $p^i \in \Delta(C^i)$, such that $\psi v(i)$ is the expected marginal contribution of i with respect to p^i. Namely,

$$\psi v(i) = E_{p^i}\big(D^i v\big) = \sum_{S \in C^i} p^i(S)\big(v(S \cup i) - v(S)\big). \tag{3.1}$$

The Shapley value, which inspired this definition, is the probabilistic value defined by the probabilities p^i defined by $p^i(S) = 1/\big(n\binom{n-1}{s}\big)$, where s denotes the number of elements in S. We denote the Shapley value by $\varphi = \varphi^N$. It can be easily verified that the probability measures p^i in (3.1) are uniquely determined by ψ. Probabilistic values can be axiomatically characterized as follows.

THEOREM 1. *Let ψ be a solution. Then the following three assertions are equivalent:*

(1) *ψ is a probabilistic value.*

(2) *ψ is a linear positive operator that satisfies the dummy axiom.*

(3) *ψ is a linear Milnor operator.*

Weber (1988) proved the equivalence of (1) and (2). Monderer (1988, 1990) proved the equivalence of (2) and (3).

Shapley (1953a) proved that the Shapley value is the unique solution that satisfies the additivity, symmetry and carrier axioms.[1] The following is one of many other characterizations of the Shapley value. We single out this particular characterization [which is proved in Weber (1988)] because it can serve as the basis for a unified approach to analyzing the variations on the Shapley value.

THEOREM 2. *The Shapley value is the unique solution which is a symmetric and efficient probabilistic value.*

The main variations on the Shapley value retain the condition of being a probabilistic value (i.e., that a player's value depends linearly and positively on his marginal contributions), while relaxing one of the other two conditions in Theorem 2. We call an efficient probabilistic value a *quasivalue*, and a symmetric probabilistic value a *semivalue*. Quasivalues are discussed in Sections 4 and 5, semivalues in Sections 6 and 7.

REMARKS.
  (1) As probabilistic values are characterized by axioms that do not mix values of different players, they can be defined and characterized for each player separately. Such an approach was taken by Weber (1988).
  (2) Weber (1988) proved that Theorem 1 holds for solutions defined on the class of monotone games. One can deduce from Monderer (1988) and from a remark in Weber that the equivalence of (1) and (3) in Theorem 1 continues to hold for solutions defined on the class of superadditive games.

4. Efficient probabilistic values (quasivalues)

We begin with an important characterization of quasivalues. For each $\tau \in OR$ we define the operator φ^τ by

$$\varphi^\tau v(i) = v\big(\{j \in N\colon \tau(j) \leqslant \tau(i)\}\big) - v\big(\{j \in N\colon \tau(j) < \tau(i)\}\big).$$

If we think of τ as the order in which players enter the "room" where the game takes place, then $\varphi^\tau(i)$ is the marginal contribution of i to the coalition of players that preceded him. For every $b \in \Delta(OR)$ we define a solution φ^b by

$$\varphi^b v(i) = \sum_{\tau \in OR} \varphi^\tau v(i) b(\tau).$$

---

[1] He proved this result for solutions defined on the set of superadditive games, with finite support, on a universal set of players; but his proof, as is well known, can also be used to characterize the Shapley value on the space $G(N)$ of all games on N.

Thus, $\varphi^b v(i)$ is the expected marginal contribution of i, where the order of entering the room is selected according to b.

A solution ψ is called a *random-order value* if there exists $b \in \Delta(OR)$ such that $\psi = \varphi^b$.

It is easy to verify that every random-order value is a quasivalue. Indeed, given $b \in \Delta(OR)$ we can define for each $i \in N$ and $S \in C^i$,

$$p^i(S) = b(\{\tau \in OR: \tau(j) < \tau(i) \text{ iff } j \in S\}). \tag{4.1}$$

If we insert the p^i's from (4.1) into (3.1), the resulting probabilistic value ψ coincides with φ^b. The following theorem of Weber (1988) shows that the converse is also true.

THEOREM 3. *A solution is a quasivalue if and only if it is a random-order value.*

Note that the Shapley value is a random-order value with $b(\tau) = 1/n!$ for every $\tau \in OR$.[2]

It can be easily verified that for every $b \in \Delta(OR)$ there exist n random variables $(z_i)_{i \in N}$ that are jointly distributed in the cube $[0, 1]^N$ with a density function and therefore with absolutely continuous marginal distribution functions $(\alpha_i)_{i \in N}$, such that for every $\tau \in OR$

$$b(\tau) = \mathrm{prob}(z_{\tau^{-1}(i)} < z_{\tau^{-1}(j)} \text{ for every } 1 \leqslant i < j \leqslant n). \tag{4.2}$$

This enables us to interpret φ^b in terms of random arrival times, as follows. We think of the value of the random variable z_i as the time of arrival of player i in the "room". The marginal contribution of i, when he arrives at time t, is his marginal contribution to the set of other players who arrived up to time t. Define

$$N(t) = \{j \in N: z_j \leqslant t\}. \tag{4.3}$$

Then $N(t)$ is a random coalition in C and it can be easily verified that

$$\varphi^b v(i) = \int_0^1 E\big(v\big(N(t) \cup i\big) - v\big(N(t) \setminus i\big)|z_i = t\big) \, d\alpha_i(t), \tag{4.4}$$

where the integral in (4.4) is the Riemann–Stieltjes integral. Thus, we have shown that a solution ψ is a random-order value, if and only if there exist n random variables

---

[2] Observe that for sufficiently large n the affine map $b \to \varphi^b$ is not one-to-one. Indeed, the affine dimension of the convex set $\Delta(OR)$ is $n! - 1$, while the affine dimension of the set of quasivalues is less than $n(2^{n-1} - 1)$, because each quasivalue ψ is uniquely determined by a vector $(p^1, p^2, \ldots, p^n)$ in $\times_{i \in N} \Delta(C^i)$ which, in addition, satisfies the necessary linear conditions for the efficiency of ψ. It is still to be checked whether the uniform distribution over OR is the only one that gives rise to the Shapley value.

$(z_i)_{i \in N}$, which are jointly distributed in the cube $[0, 1]^N$, and have a density function, such that $\psi v(i)$ is given by the right-hand side of (4.4), where $(\alpha_i)_{i \in N}$ are the marginal distribution functions.

Owen (1972) defined the multilinear extension of a game v. Each coalition S can be identified with an extreme point, e_S, of the cube $[0, 1]^N$, where e_S is the characteristic function of S. Therefore, each game v can be considered as a function on the extreme points of the cube and hence it can be uniquely extended to a multilinear function on the cube, which is denoted by $\bar{v}$. Owen showed that if the random variables that define b via (4.2) are independent, then for every i:

$$\varphi^b v(i) = \int_0^1 \frac{\partial \bar{v}}{\partial x_i} \left(\alpha(t) \right) d\alpha_i(t). \tag{4.5}$$

In particular, the Shapley value is determined by n i.i.d. random variables that are uniformly distributed in $[0, 1]$. Therefore,

$$\varphi v(i) = \int_0^1 \frac{\partial \bar{v}}{\partial x_i} (te) \, dt,$$

where $e = e_N$.

The presentation of the solution φ^b, in (4.5), gives rise to another natural solution. A solution ψ is a *path value* if there exists an absolutely continuous nondecreasing path $\alpha : [0, 1] \to [0, 1]^N$ with $\alpha(0) = 0$ and $\alpha(1) = e$ such that for every game v and every player i, $\psi v(i)$ is defined by the right-hand side of (4.5). Alternatively, ψ is a path value if there exists $b \in \Delta(OR)$ such that $\psi = \varphi^b$, and there exist n independent random variables that are distributed in $[0, 1]$ with absolutely continuous distribution functions such that (4.2) holds. It can be verified that not every quasivalue is a path value.

REMARKS.

(1) Gilboa and Monderer (1991) discussed quasivalues defined on subspaces of games. They gave necessary and sufficient conditions that ensure that such quasivalues can be extended to quasivalues on G.

(2) Monderer (1992) used the concept of quasivalue in order to provide a game-theoretic approach to the stochastic choice problem. He used Theorem 3 to confirm a conjecture of Block and Marschak (1960), which had already been proved before by Falmagne (1978). The relationship between the stochastic choice problem and quasivalues was further analyzed by Gilboa and Monderer (1992).

5. Weighted values

Weighted values comprise a special class of quasivalues, and more specifically of path values. We describe two forms of such values. Both were introduced by Shapley (1953a,

1953b), alongside the standard Shapley value, and were axiomatized by Kalai and Samet (1987).

Let $\omega \in R_{++}^N$ (that is, $\omega_i > 0$ for every $i \in N$). Recall that a unanimity game u_S, for $S \in C_0$, is defined by $u_S(T) = 1$ for $T \supseteq S$ and $u_S(T) = 0$ otherwise, and that the set of unanimity games, for all $S \in C_0$, is a basis for the linear space G of games. The *positively weighted value* φ^ω (which is called the ω-value) is the unique linear operator satisfying

$$\varphi^\omega u_S(i) = \begin{cases} \dfrac{\omega_i}{\omega(S)}, & \text{for } i \in S \\ 0, & \text{for } i \notin S \end{cases}$$

for each unanimity game u_S. If $\omega_i = \frac{1}{n}$ for every $i \in N$, then φ^ω is the Shapley value. Owen (1972) has shown that positively weighted values are path values, the path α being defined by

$$\alpha_i(t) = t^{\omega_i} \quad \text{for every } t \in [0, 1]. \tag{5.1}$$

Note that the mapping $\omega \to \varphi^\omega$ is homogeneous of degree zero, that is, $\varphi^{\lambda\omega} = \varphi^\omega$ for every $\lambda > 0$. Therefore, we can restrict our attention to vectors ω in the relative interior of $\Delta(N)$, which we call *weight vectors*. We turn now to an axiomatic characterization of positively weighted values. A solution ψ is *strictly positive* if for every monotone game v without null players, $\psi v(i) > 0$ for every $i \in N$. It is easily verified that a probabilistic value ψ is strictly positive, if and only if $p^i(S) > 0$ for every i and every $S \in C^i$, where $\{p^i : i \in N\}$ is the collection of probability measures that define ψ via (3.1). A coalition S is said to be a *coalition of partners* or a *p-type coalition*, in the game v, if, for every $T \subset S$ and each $R \subseteq S^c$, $v(R \cup T) = v(R)$. A solution ψ satisfies the *partnership axiom* if, whenever S is a p-type coalition,

$$\psi v(i) = \psi\big(\psi v(S) u_S\big)(i) \quad \text{for every } i \in S.$$

The proof of the following theorem is given in Kalai and Samet (1987):

THEOREM 4. *A solution is a positively weighted value, if and only if it is a strictly positive probabilistic value that satisfies the partnership axiom.*

The set of positively weighted values is not closed in the space of solutions. Obviously every solution which is the limit of positively weighted values is a probabilistic value that satisfies the partnership axiom. We turn now to characterize such solutions by introducing the concept of weight systems. A *weight system* is a vector $w = (w^S)_{S \in C_0}$ in $\times_{S \in C_0} \Delta(S)$ that satisfies

$$w_i^S = \frac{w_i^T}{w^T(S)} \tag{5.2}$$

whenever $i \in S \subseteq T$ and $w^T(S) > 0$. Note that any weight vector ω can be identified with the weight system w, where $w_i^S = \frac{\omega_i}{\omega(S)}$ for every $i \in S \in C_0$. The set of all weight systems is denoted by $\mathcal{W}$. For $w \in \mathcal{W}$ the *weighted value* φ^w (which is called the w-value) is defined as the unique linear operator which is defined for every unanimity game u_S by

$$\varphi^w u_S(i) = \begin{cases} w_i^S, & \text{for } i \in S, \\ 0, & \text{for } i \notin S. \end{cases} \tag{5.3}$$

The w-value of a unanimity game in (5.3) can be interpreted as follows. Every weight system w defines an ordered partition $(N_1, \ldots, N_k)$ of N (where k depends on w). The players in N_1 are those players i for whom $w_i^N > 0$. The coalition N_h, for $1 < h \leqslant k$, consists of all the players i in $(\bigcup_{l=1}^{h-1} N_l)^c$, for which $w_i^{(\bigcup_{l=1}^{h-1} N_l)^c} > 0$.

For $S \in C_0$, let $h = h(S)$ be the first index for which $S \cap N_h \neq \emptyset$. Then φ^w allocates among the players in $S \cap N_h$ their share, $u_S(N) = 1$, proportional to their weights in w^{N_h} while $\varphi^w u_S(i) = 0$ for all other players.

To show that every weighted value is a path value, Kalai and Samet (1987) defined independent random variables z_i, $i \in N$ with pairwise disjoint supports, such that if $i_h \in N_h$, for $h = 1, \ldots, k$, then $\text{prob}(z_{i_1} > z_{i_2} > \cdots > z_{i_k}) = 1$. The distribution of each z_i is defined on its support in $[0, 1]$ by a linear transformation of the distribution given in (5.1) with $\omega_i = w_i^{(\bigcap_{l=1}^{h-1} N_l)^c}$ when $i \in N_h$. Kalai and Samet (1987) proved:

THEOREM 5. *A solution is a weighted value if and only if it is a quasivalue that satisfies the partnership axiom.*

Hart and Mas-Colell (1989) presented two new approaches to value theory: the potential approach and the consistency approach. They showed that these approaches are naturally extended to the case of positively weighted values.

Here we present their theory using a slightly different notation. For a game v and a coalition S we denote the restriction of v to S by v^S. That is, $v^S \in G(S)$ and $v^S(T) = v(T)$ for every $T \in C(S)$. Let $AG = \bigcup_{S \subseteq N} G(S)$. A function $P : AG \to R$ is a ω-potential function, for a weight vector $\omega \in \Delta(N)$ (that is, $\omega_i > 0$ for every $i \in N$), if for every $v \in G$

$$P(v^\emptyset) = 0 \quad \text{and} \quad \sum_{i \in S}(P(v^S) - P(v^{S \backslash i})) = v(S), \quad \text{for every } S \neq \emptyset. \tag{5.4}$$

Note that if (5.4) holds, then it also holds for every $v \in G(T)$ for $S \subseteq T \subset N$.

Hart and Mas-Colell proved that there exists a unique ω-potential function, denoted by P_ω, and called the ω-potential function of G. It satisfies

$$P_\omega(v^S) = \int_0^1 \frac{\bar{v}(\alpha(t)e_S)}{t} \, dt,$$

where $\alpha(t)(i) = t^{\omega_i}$ for every $i \in N$, and where for $x, y \in R^N$, xy is the vector defined by $(xy)_i = x_i y_i$ for every $i \in N$. The e-potential function, P_e, is called the potential function (or simply the potential) of G. Note that $P_\omega v = E(\frac{1}{T} \int_0^1 v(N(t)) \, dt)$, where $N(t)$ is given by (4.3) with the random variables $(z_i)_{i \in N}$ defined by (5.1).

Hart and Mas-Colell also proved:

THEOREM 6. *Let* $\omega \in \Delta(N)$ *be a weight vector. Then for every game* v *and every* $i \in N$,

$$\varphi^\omega v(i) = \omega_i \left[P_\omega(v^N) - P_\omega(v^{N \setminus i}) \right].$$

We now present the consistency property of weighted values. A function $\psi : AG \to \bigcup_{S \in C} A(S)$ is called an *extended solution* if $\psi v^S \in A(S)$ for every $S \in C$. Thus, an extended solution is a collection of solutions ψ^S on $G(S)$ such that $\psi^S v^S = \psi v^S$ for every coalition S. Let ψ be an extended solution and let $T \subseteq S$. We define the reduced game $v_T^{(S, \psi)} \in G(T)$ as follows:

$$v_T^{(S, \psi)}(F) = v\big(F \cup (S \setminus T)\big) - \sum_{i \in S \setminus T} \psi v^{F \cup (S \setminus T)}(i), \quad \text{for all } F \subseteq T.$$

An extended solution ψ is *consistent* if for every game $v \in G$, coalition S, and coalition $T \subseteq S$,

$$\psi v_T^{(S, \psi)}(j) = \psi v^S(j) \quad \text{for every } j \in T.$$

Hart and Mas-Colell prove that for a fixed weight vector ω, the ω-weighted Shapley value, regarded in a natural way as an extended solution (that is, $(\varphi^\omega)^S = \varphi^{\omega^S}$, where ω^S is the restriction of ω to S), is consistent. Moreover, they proved that if an extended solution ψ is consistent and coincides with the ω-weighted extended value on two-person games, then it is the ω-weighted extended value on AG. They further used the consistency axiom to give an axiomatic characterization for positively weighted values.

THEOREM 7. *An extended solution is a* ω*-value, for some weight vector* ω, *if and only if it is consistent, and for two-person games it is efficient, TU-invariant*[3], *and monotonic.*

We end this discussion with a result concerning the weighted values of a game as a set-valued solution. Whatever the meaning of the weight vector w, it reflects certain properties or information about the players, not derived from the game, but rather given exogenously. If we lack this information, we can consider the less informative set-valued solution $\mathcal{W}(v) = \{\varphi^w v : w \in \mathcal{W}\}$. Let $C(v)$ denote the core of v and let $AC(v)$ be the anti-core of v, which consists of all $x \in A$ such that $x(S) \leqslant v(S)$ for every S and $x(N) = v(N)$. Monderer, Samet, and Shapley (1992) proved:

---

[3] That is, $\psi(av^{\{i,j\}} + \mu^{\{i,j\}}) = a\psi v^{\{i,j\}} + \mu^{\{i,j\}}$, for every $i \neq j$, $a > 0$ and $\mu \in A$.

THEOREM 8. *For every game v,*

$$C(v) \cup AC(v) \subseteq W(v).$$

That there is such a general relationship between the core and values is somewhat surprising in light of the difference in concept behind these solutions. Theorem 8 generalizes a theorem of Weber (1988), who proved that $C(v) \subseteq Q(v)$, where $Q(v)$ is the Weber set, i.e., the set of all $\psi(v)$ where ψ ranges over all quasivalues.

REMARKS.
  (1) Monderer, Samet and Shapley (1992) discussed geometrical and topological properties of the set of weight systems $\mathcal{W}$ (which is homeomorphic to the set of weighted values, because the map $w \to \varphi^w$ is one-to-one). $\mathcal{W}$ can easily be seen to be homeomorphic to the set of conditional systems that was discussed in the literature on non-cooperative game theory [e.g., see Myerson (1986) and McLennan (1989a, 1989b)]. The main theorem of McLennan (1989b) states that this set is homeomorphic to a ball. Monderer, Samet and Shapley (1992) proved that for a strictly convex game v, $\mathcal{W}$ is mapped homeomorphically onto the core of v by the mapping $w \to \varphi^w v$. This gives a new proof for McLennan's theorem. Moreover, the well-studied geometry of the core of strictly convex games [see Shapley (1971)] is reflected in the structure of $\mathcal{W}$.
  (2) Shapley (1981) proposed a family of weighted cost allocation schemes which, when translated into the language of cooperative game theory, yield the dual-weighted values. The dual game, v^*, of a game v is defined by: $v^*(S) = v(N) - v(S^c)$. For $w \in \mathcal{W}$, the dual w-weighted value is defined by: $\varphi^w_* v = \varphi^w v^*$. Shapley gave an axiomatization for solutions $\psi : G \times W \to A$ of the form $(v, \omega) \to \varphi^\omega_* v$. Kalai and Samet (1987) axiomatized the family of dual-weighted values, explored the relationship between weighted and dual-weighted values, and used this relationship in order to give an axiomatization of the Shapley value that does not use the symmetry axiom. Dual-weighted values are also discussed in Monderer, Samet and Shapley (1992), where it is shown that a theorem analogous to Theorem 8 can be stated for these solutions.

6. Symmetric probabilistic values (semivalues)

As semivalues are not efficient (except for the Shapley value), they cannot be considered as mechanisms for benefit or cost sharing. However, they can be interpreted as indices of power. The structure of power is usually described by a simple game, and therefore it is of interest to study semivalues as solutions defined on such games. But before doing so, we first analyze semivalues on all of G.

Recall that the probabilities $(p^i)_{i \in N}$, where $p^i \in \Delta(C^i)$, by which a probabilistic value ψ can be expressed, as in (3.1), are uniquely determined by ψ. If in addition ψ is symmetric, i.e., it is a semivalue, then it can easily be seen that $p^i(S)$ depends only on the number of elements of S. That is, there exists a function $\beta : \{0, 1, \ldots, n-1\} \to [0, 1]$ such that for every i,

$$p^i(S) = \beta(s), \tag{6.1}$$

where s denotes the number of elements in S. Obviously,

$$\sum_{s=0}^{n-1} \beta(s) \binom{n-1}{s} = 1. \tag{6.2}$$

Thus, we get the following characterization of semivalues given in Dubey, Neyman and Weber (1981):

THEOREM 9. *A solution ψ is a semivalue if and only if there exists a function*

$$\beta : \{0, 1, \ldots, n-1\} \to [0, 1],$$

that satisfies (6.2) such that $\psi = \varphi^\beta$, where for every game v and player i,

$$\varphi^\beta v(i) = \sum_{S \in C^i} \beta(s) \big(v(S \cup i) - v(S) \big), \quad i \in N. \tag{6.3}$$

Moreover, the correspondence $\beta \to \varphi^\beta$ is one-to-one.

Note that the Shapley value is φ^β for $\beta(s) = \dfrac{1}{n\binom{n-1}{s}}$ for every $0 \leqslant s \leqslant n-1$. The Banzhaf value[4] is the semivalue defined by $\beta(s) = \dfrac{1}{2^{n-1}}$.

By removing the efficiency axiom from the list of axioms defining the Shapley value, we obtain the family of semivalues. Thus, for a semivalue ψ, $\psi v(N)$ is not necessarily $v(N)$. It is interesting, then, to note that it is precisely what ψ gives to the grand coalition N, in the various games, which determines ψ. This is what we state and prove next.

THEOREM 10. *Let ψ_1 and ψ_2 be two semivalues[5] that satisfy for each game v,*

$$\psi_1 v(N) = \psi_2 v(N). \tag{6.4}$$

Then $\psi_1 = \psi_2$.

---

[4] See Section 7.

[5] This theorem can be proved for more general solutions, namely, for non-positive semivalues which are linear, symmetric solutions that satisfy the dummy axiom. Roth (1977) proved a special case of this stronger version of Theorem 10. He showed that the Banzhaf value is the only non-positive semivalue ψ that satisfies $\psi v(N) = \varphi^{bz} v(N)$ for every game v, where φ^{bz} is the Banzhaf value.

PROOF. Let $\mathcal{S}$ be the linear space of operators spanned by the set of semivalues. By Theorem 9, $\mathcal{S}$ is n-dimensional. For every game v we define a linear functional $f_v : \mathcal{S} \to R$, by $f_v(\psi) = \psi v(N)$. Fix an order τ on N, and for $k = 1, \ldots, n$, let v_k be the unanimity game on $\{\tau^{-1}(1), \ldots, \tau^{-1}(k)\}$. It is easily verified that $f_{v_1}, \ldots, f_{v_n}$ are linearly independent. Therefore, if $\psi \in \mathcal{S}$ and $f_{v_k}(\psi) = 0$ for $k = 1, \ldots, n$, then $\psi = 0$. Thus, if ψ_1 and ψ_2 satisfy (6.4), then $\psi_1 - \psi_2 = 0$. $\qquad\square$

Dubey, Neyman and Weber (1981) characterized semivalues defined on games with finite support on an infinite universe of players. Let U be an infinite set of players. Denote by $G_f = G_f(U)$ the space of all games on U with a finite carrier, and by $A_f = A_f(U)$ the space of all additive games in G_f. A solution, in this context, is a function $\psi : G_f \to A_f$. The symmetry of a solution is defined as before, with slight and obvious changes. Probabilistic solutions are defined as follows. For every finite subset of players M and for every $v \in G(M)$, denote by v_M the natural extension of v to $C(U)$, that is, $v_M(T) = v(T \cap M)$ for $T \subseteq U$. Every solution ψ on G_f defines a solution ψ^M on $G(M)$ by $\psi^M v(i) = \psi v_M(i)$ for every $v \in G(M)$ and every $i \in M$. We say that ψ is a probabilistic value on $G(U)$ if for every finite subset of players, M, ψ^M is a probabilistic value. Defining a semivalue as a probabilistic and symmetric solution, note that ψ is a semivalue if and only if ψ^M is a semivalue for every finite coalition M.

THEOREM 11. *Let U be an infinite set of players and let ψ be a solution on G_f. Then ψ is a semivalue if and only if there exists a Borel probability measure ξ on $[0, 1]$ such that for every finite coalition N with n elements, and for every $v \in G_f$ with carrier N, $\psi v = \psi_\xi v$, where,*

$$\psi_\xi v(i) = \sum_{S \subseteq N \setminus i} \beta^n(s)\big(v(S \cup i) - v(S)\big), \quad i \in N,$$

with

$$\beta^n(s) = \int_0^1 t^s (1-t)^{n-s-1} \, d\xi(t) \quad \text{for } 0 \leqslant s \leqslant n-1. \tag{6.5}$$

Moreover, the correspondence $\xi \to \psi_\xi$ is one-to-one.

Note that (6.5) can be rewritten in terms of random independent arrival times like (4.4),

$$\psi v(i) = \int_0^1 E\big(v(N(t) \cup i) - v(N(t) \setminus i) | z_i = t\big) \, d\xi(t),$$

where $N(t)$ is the random set defined in (4.5) with respect to n i.i.d. random variables $z_1, \ldots, z_n$, uniformly distributed on $[0, 1]$. Observe, however, that unlike (4.4) the integration here is with respect to ξ and not z_i.

REMARKS.

(1) Dubey, Neyman and Weber (1981) discussed continuity properties of semivalues on G_f. For every $v \in G_f$ the variation norm of v is: $\|v\| = \inf\{u(N) + w(N)\}$, where the supremum is over all monotone games $u, w \in G_f$ such that $u - w = v$. With this norm both G_f and A_f are normed spaces. The continuity of semivalues is defined with respect to this norm. It is proved that a semivalue ψ_ξ is continuous if and only if ξ is absolutely continuous with respect to the Lebesgue measure m on $[0, 1]$, and its Radon–Nikodym derivative with respect to m is bounded, when considered as a member of $L_\infty(0, 1)$, that is, $\|\frac{d\xi}{dm}\|_\infty < \infty$. Moreover, the mapping $g \to \psi_{\xi(g)}$, where $\xi(g)(A) = \int_A g\, dm$, is an isometry in the natural sense.

(2) Monderer (1988) discussed semivalues defined on symmetric subspaces of G. He proved that every such semivalue can be extended (not necessarily in a unique way) to a semivalue on G. Note that if ψ is a semivalue and G_ψ is the subspace of games v for which $\psi v(N) = v(N)$, then the restriction of ψ to any symmetric subspace of G_ψ is a Shapley value on this subspace. Monderer's theorem proves that every Shapley value on a symmetric subspace of games is obtained in this manner.

7. Indices of power

We denote the class of simple games by SG. A coalition S is a *winning coalition* in a simple game v if $v(S) = 1$, and it is a *losing coalition* if $v(S) = 0$. Of special interest are monotone simple games. In such games, a superset of a winning coalition is a winning coalition, and a subset of a losing coalition is a losing one.[6] This property reflects the natural idea, especially for voting games, that by adding members to a coalition it becomes stronger. The set of monotone simple games is denoted by MSG and $MSG \setminus \{0\}$ is denoted by MSG^+. The set of superadditive simple games is denoted by SSG, and SSG^+ is $SSG \setminus \{0\}$.

A solution defined on a subset of simple games is called an *index of power*. We refer the reader to Dubey and Shapley (1979), and Owen (1982) for a partial reference to articles about the use of indices of power in various areas, such as political science, law and electrical engineering.

Other than the linearity axiom, all axioms in Section 2 are applicable (and make sense with respect) to indices of power. We now present the *transfer* axiom, introduced by Dubey (1975),[7] which in some sense substitutes for the linearity axiom for indices of power. For any two games v and w we define the games $(v \vee w)$ and $(v \wedge w)$ by: $(v \vee w)(S) = \max(v(S), w(S))$ and $(v \wedge w)(S) = \min(v(S), w(S))$. Note that both games are simple whenever v and w are simple games.

---

[6] Many authors refer to a game $v \in SG$ as a 0–1 game, and reserve the notion "simple game" for what we called a monotone simple game, and some of those authors even exclude the game $v = 0$ from the set of simple games.

[7] The name was suggested by Weber (1988).

(A.12)   A solution ψ defined on a set M of simple games satisfies the *transfer* axiom
if for every $v, w \in M$ for which $v \vee w$ and $v \wedge w$ belong to M,

$$\psi(v) + \psi(w) = \psi(v \vee w) + \psi(v \wedge w).$$

The next theorem, which is proved in Weber (1988), characterizes probabilistic values, and hence quasivalues and semivalues on simple games.

THEOREM 12. *An index of power defined on the set of simple games, or on the set of monotone simple games, is a probabilistic value if and only if it satisfies the transfer, dummy, and positivity axioms. Consequently, an index of power is a quasivalue on these sets of games if and only if it satisfies the transfer, dummy, positivity and efficiency axioms, and it is a semivalue if and only if it satisfies the transfer, dummy, positivity and symmetry axioms.*

Note that every probabilistic value ψ that is defined on a set of simple games, which contains the unanimity games, can be uniquely extended to a probabilistic value on G. Moreover, if ψ is a quasivalue on such a set of simple games, then its extension is a quasivalue on G and if ψ is a semivalue, then its extension is a semivalue. This follows from the fact that the symmetry and efficiency axioms are "linear" axioms, in the sense that if they hold for a symmetric linear base of games then they hold for every game. The extension property of power indices implies, in particular, that every quasivalue on simple games is a random-order value (see Theorem 3).

Representations of indices of power as random-order values, or probabilistic values, have a special flavor. Let $v \neq 0$ be a simple monotone game, and let $\tau \in OR$. The *pivot* for τ in v is the unique player $j = j_\tau$ for whom the set R_j, of players preceding j in τ, is a losing coalition, while $R_j \cup j$ is winning. Thus, for a probability measure b on the set of orders $\Delta(OR)$, $\varphi^b v(i)$ is the b-probability that i is the pivot. In particular, the Shapley value [which is also called the Shapley–Shubik index of power – see Shapley and Shubik (1954)] of a player i in v is his probability of being the pivot when all orders have equal probability. We say that i is a *swing* for a coalition S in v, if $v(S) = 0$ and $v(S \cup i) = 1$. Let ψ be a probabilistic value defined by the probabilities $p^1, \ldots, p^n$ on $C^1, \ldots, C^n$, respectively. Then $\psi v(i)$ is the p^i-probability that i is a swing for a random coalition of the other voters.[8]

The Banzhaf index of power [Banzhaf (1965)] has been extensively discussed in the literature. Dubey and Shapley (1979) characterized the Banzhaf index on MSG^+ in the same way that Roth (1977) characterized the Banzhaf value on G (see Footnote 5). Lehrer (1988) gave another characterization, based on the behavior of an extended index of power when applied to games in which coalitions of the original game are amalgamated into one player, as follows. For every $\emptyset \subset T \subset N$ choose a player i_T in T. Let v

---

[8]   See Straffin (1988) for other probabilistic interpretations of the Shapley–Shubik and the Banzhaf indices of power.

be a simple game and let T be a nonempty coalition. Define two games $v_{[T]}$ and $v_{[T]}^m$ on the set of players $T^c \cup \{i_T\}$ as follows: $v_{[T]}(S) = v_{[T]}^m(S) = v(S)$ for every $S \subseteq T^c$, and for every such S,

$$v_{[T]}(S \cup i_T) = v(S \cup T) \quad \text{and} \quad v_{[T]}^m(S \cup i_T) = \max_{\emptyset \subset B \subseteq T} v(S \cup B).$$

THEOREM 13. *Let ψ be an extended semivalue on the sets of simple games on subsets of N,[9] and let $(i_T)_{\emptyset \subset T \subset N}$ be a collection of players such that $i_T \in T$. Then ψ is the Banzhaf index of power if and only if for every two-player coalition $T = \{i, j\}$,*

$$\psi v(i) + \psi v(j) \leqslant \psi v_{[T]}(i_T).$$

Moreover, ψ is the Banzhaf index if and only if for each such T,

$$\psi v(i) + \psi v(j) \leqslant \psi v_{[T]}^m(i_T).$$

REMARKS.
  (1) Einy (1987) characterized semivalues on $MSG(U)$, when U is an infinite universe set of players. Actually he proved the "semivalues" part of Theorem 12 in such a setup. In his papers (1987, 1988), he also provided a formula for the Banzhaf index of power of a monotone simple game v in terms of the minimal winning coalitions of this game.
  (2) For every positive weight vector $w = (w_1, \ldots, w_n)$, and $0 < c < 1$ we define the weighted majority game, $v_{w,c}$ with a quota c as follows:

$$v_{w,c}(S) = \begin{cases} 1, & \text{if } w(S) \geqslant c; \\ 0, & \text{if } w(S) < c. \end{cases}$$

  Dubey and Shapley (1979) have proved the following intriguing equality. For *any* probabilistic index of power ψ, $\int_0^1 \psi v_{w,c}(i) \, \mathrm{d}c = w_i$.

8. Values with a social structure

In all the variations on the Shapley value discussed in this section, there is some structure on the set of the players which is involved in determining the value of a game. All are extensions of the Shapley value, because when the structure on N is the "trivial" one, where triviality depends on the type of structure used in the solution, then the value is the Shapley value.

---

[9] That is, $\psi : \bigcup_{S \subseteq N} SG(S) \to \bigcup_{S \subseteq N} A(S)$, $\psi v \in A(S)$ for $v \in SG(S)$, and the restriction of ψ to $SG(S)$ is a semivalue for every $S \subseteq N$.

Aumann and Drèze (1975) were the first to introduce such a structure into the study of value, by changing the efficiency axiom. Instead of assuming that the grand coalition, N, forms and allocates its worth, $v(N)$, among its players, they assumed that there is a fixed partition $\pi = \{B_1, \ldots, B_m\}$ of N, which they called a *coalitional structure*, such that each coalition, B_k, forms and allocates its value, $v(B_k)$, among its members. Using the following axioms they characterized a unique solution called the π-value.

A solution ψ is π-*efficient* if for every game v, $\psi v(B_k) = v(B_k)$ for all $1 \leqslant k \leqslant m$. A permutation θ is π-*invariant* if $\theta(B_k) = B_k$ for every k, and ψ is π-*symmetric* if $\psi(\theta^* v) = \theta^* \psi v$ for every π-invariant permutation θ. Aumann and Drèze (1975) proved:

THEOREM 14. *There exists a unique probabilistic value φ^π which is π-symmetric and π-efficient, called the π-value, and it is given by:*

$$\varphi^\pi v(j) = \varphi v^{B_k}(j) \quad for \ j \in B_k. \tag{8.1}$$

Note that by (8.1), for $i \in B_k$, $\varphi^\pi v(i)$ depends only on the game v^{B_k}. Hence, the power of a player $i \in B_k$ to negotiate with players outside B_k is not taken into account when computing his share of $v(B_k)$.

Myerson (1977) studied a value, called the *fair allocation rule*, where the social structure is given by a graph with vertex set N. An arc (i, j) in graph g can be thought of as signifying some social relation, say a neighborhood or a channel of communication, between players i and j. For each coalition S, we denote by $g|_S$ the restriction of the graph g to the vertices in S. By $\pi_{g|_S}$ we denote the partition of S into the connected components of $g|_S$.

The axiomatization of the fair allocation rule is accomplished by fixing the game v, and varying the social structure. A function $\phi: GR \to A$, where GR is the set of all graphs on N, is a *fair allocation rule* if it satisfies two properties: it is *relatively efficient*, by which we mean that $\phi g(S) = v(S)$ for every connected component of g; and it is *fair* in the sense that for any two graphs g and h such that $h = g \cup \{(i, j)\}$,

$$\phi h(i) - \phi g(i) = \phi h(j) - \phi g(j).$$

Thus, ϕ is fair if, by adding an arc to a graph, the players forming this arc equally gain or lose.

Myerson (1977) proved:

THEOREM 15. *For any game $v \in G$, there exists a unique function $\phi^m: GR \to A$ which is a fair allocation, and it is given by,*

$$\phi_v^m(g) = \varphi(v|g),$$

where φ is the Shapley value, and $v|g$ is the game which satisfies for each coalition S,

$$(v|g)(S) = \sum_{T \in \pi_{g|S}} v(T).$$

Note that each of the coalitions in the partition of N to its connected components, i.e., the partition $\pi_{g|N}$, forms and allocates its worth among its members. Moreover, like the case of the π-value, the value of a player, in the fair allocation rule, is determined solely by the restriction of the game to the coalition in $\pi_{g|N}$, of which he is a member.

The following solution, introduced by Owen (1977), is based on a coalition structure, i.e., a partition $\pi = \{B_1, \ldots, B_m\}$ of N. But unlike the previous two solutions, it is a quasivalue, and as such it is efficient in the usual sense, that is, the players allocate the worth of the grand coalition N. By Theorem 3, a quasivalue is a random-order value, and the value proposed by Owen (1977) is the random-order value φ^{b_π}, defined by the probability distribution $b_\pi \in \Delta(OR)$, which we describe next.

We say that an order τ is consistent with π if for each $k = 1, \ldots, m$, all players in B_k arrive "together" in the following sense. If i arrives before j, and both are in B_k, then all the players arriving between i and j are also in B_k. Let $b_\pi \in \Delta(OR)$ be the probability measure that is uniformly distributed over all orders that are consistent with π. That is, it assigns the probability $\frac{1}{m!b_1!\cdots b_m!}$ to each such order, where b_k denotes the number of players in B_k.

This value, called the π-*coalitional structure* (cs) *value* was characterized in various setups by Owen (1977) and by Hart and Kurz (1983). We denote it by φ_π^{cs} and need some additional notations before we can present a modified version of their axiomatization. Let π be a partition of N. For each $v \in G$ we define a *partitional game* v_π, the players of which are the coalitions in π, as follows. For each $F \subseteq \pi$,

$$v_\pi(F) = v\left(\bigcup_{B \in F} B\right).$$

We say that B_1 and B_2 are symmetric players in v_π if for every $F \subseteq \pi$ that does not contain B_1 and B_2, $v_\pi(F \cup \{B_1\}) = v_\pi(F \cup \{B_2\})$. A solution ψ satisfies the π-*coalitional symmetry axiom* if $\psi v(B_1) = \psi v(B_2)$ whenever B_1 and B_2 are symmetric players in v_π. The proof of the following theorem can be derived from Owen (1977) and Hart and Kurz (1983):

THEOREM 16. *Let ψ be a solution and let π be a partition. Then ψ is the π coalitional structure value if and only if ψ is a random-order value that satisfies the π-symmetry and the π-coalitional symmetry axioms.*

Owen (1977) and Hart and Kurz (1983) explored some interesting properties of the Owen value. One such property is that the total value of a coalition B in π equals the

Shapley value of the player B in the game v_π. That is,

$$\varphi_\pi^{cs} v(B) = \varphi v_\pi(B) = \sum_{i \in B} \varphi v_\pi(i).$$

The next question is how do the players in B distribute the total amount available to them, $\varphi v_\pi(B)$. To answer this, we define for $B \in \pi$ a game $v_{(\pi,B)}$ in $G(B)$. Let $S \subseteq B$ and denote by $\pi_B | S$ the partition refining π by splitting B into S and $B \setminus S$. The game $v_{(\pi,B)}$ is defined by:

$$v_{(\pi,B)}(S) = \varphi^{b_{\pi_B | S}} v(S), \quad S \subseteq B.$$

THEOREM 17. *For each $i \in B$,*

$$\varphi^{b_\pi} v(i) = \varphi v_{(\pi,B)}(i).$$

Owen showed that Theorem 17 holds for $i \in S$ if $\pi_B | S$ is defined as the partition of $N \setminus (B \setminus S)$ obtained from π by replacing B with S (i.e., the players in $B \setminus S$ simply disappear). Hart and Kurz showed that Theorem 17 also holds when we define the game $v_{(\pi,B)}$ by defining $\pi_B | S$ as the partition obtained from π by replacing B with S and the singletons of $B \setminus S$.

REMARKS.
(1) Myerson (1980) extended the notion of the cooperation graph to a cooperation hypergraph, where a link can be formed among the members of a coalition with more than two players.
(2) Amer and Carreras (1995) extended the notion of the cooperation graph to the weighted cooperation graph, called a *cooperation index*. More formally, a cooperation index is a function $w : C \to [0, 1]$, where $w(S)$ may be interpreted as the probability of a communication link among the members of S. Naturally, it is required that $w(\{i\}) = 1$ for every player i. Amer and Carreras (1997) further discussed cooperation indices in the context of weighted Myerson values.
(3) Owen (1977) and Hart and Kurz (1983) gave an axiomatic characterization of the coalitional structure value as an operator which is defined on pairs (v, π), using axioms that consider changes in the values of such an operator when both the game and the partition are changed.
(4) McLean (1991) defined and analyzed coalitional structure random-order values, as follows. Let π be a partition of N and let $b \in \Delta(OR)$. b is *consistent with π* if $b(\tau) = 0$ for every order τ which is not consistent with π. A random order value is a *π random order coalitional structure value* if it is defined by some $b \in \Delta(OR)$ which is consistent with π. Levy and McLean (1989) defined and characterized weighted coalitional structure values.

(5) McLean and Ye (1996) developed the theory of coalitional structure semivalues for finite games.

(6) Winter (1992) characterized the Owen value by a consistency property which generalizes the one given in Hart and Mas-Colell (1989). He also generalized the potential approach and showed that the Owen value, Myerson value and A-D value can be developed through this approach.

(7) Winter (1989) developed the theory of values defined on pairs (v, p), where $v \in G$ and $p = (\pi_1, \ldots, \pi_m)$ is a decreasing sequence of partitions.[10]

(8) Lucas (1963) and Myerson (1976) dealt with values of generalized games, where the worth of a coalition also depends on the partition into coalitions generated by the players.

References

Amer, R., and F. Carreras (1995), "Games and cooperation indices", International Journal of Game Theory 3:239–258.

Amer, R., and F. Carreras (1997), "Cooperation indices and weighted Shapley values", Mathematics of Operations Research 22:955–968.

Aumann, R.J., and J.H. Drèze (1975), "Cooperative games with coalition structures", International Journal of Game Theory 4:217–237.

Banzhaf, J.F., III (1965), "Weighted voting does not work: A mathematical analysis", Rutgers Law Review 19:317–343.

Block, H.D., and J. Marschak (1960), "Random ordering and stochastic theories of responses", Cowles Foundation Paper, No. 147.

Dubey, P. (1975), "On the uniqueness of the Shapley value", International Journal of Game Theory 4:131–139.

Dubey, P., and L.S. Shapley (1979), "Mathematical properties of the Banzhaf power index", Mathematics of Operations Research 4:99–131.

Dubey, P., A. Neyman and R.J. Weber (1981), "Value theory without efficiency", Mathematics of Operations Research 6:122–128.

Einy, E. (1987), "Semivalues of simple games", Mathematics of Operations Research 12:185–192.

Einy, E. (1988), "The Shapley value on some lattices of monotonic games", Mathematical Social Sciences 15:1–10.

Falmagne, J.C. (1978), "A representation theorem for finite random scale systems", Journal of Mathematical Psychology 18:52–72.

Gilboa, I., and D. Monderer (1991), "Quasivalues on subspaces of games", International Journal of Game Theory 19:353–363.

Gilboa, I., and D. Monderer (1992), "A game theoretic approach to the binary stochastic choice problem", Journal of Mathematical Psychology 36:555–572.

Hart, S., and M. Kurz (1983), "Endogenous formation of coalitions", Econometrica 51:589–614.

Hart, S., and A. Mas-Colell (1989), "Potential, value and consistency", Econometrica 57:589–614.

Kalai, E., and D. Samet (1987), "On weighted Shapley values", International Journal of Game Theory 16:205–222.

Lehrer, E. (1988), "An axiomatization of the Banzhaf value", International Journal of Game Theory 17:89–99.

Levy, A., and R.P. McLean (1989), "Weighted coalition structure values", Games and Economic Behavior 1:234–249.

---

[10] The idea for dealing with such values appears in Owen (1977).

Lucas, W.F. (1963), "*n*-Person games in partition function form", Naval Research Logistics Quarterly 10:281–298.

McLean, R.P. (1991), "Random order coalition structure values", International Journal of Game Theory 20:109–127.

McLean, R.P., and T.X. Ye (1996), "Coalition structure semivalues of finite games", Rutgers University.

McLennan, A. (1989a), "Consistent systems in noncooperative game theory", International Journal of Game Theory 18:141–174.

McLennan, A. (1989b), "The space of conditional systems is a ball", International Journal of Game Theory 18:127–140.

Monderer, D. (1988), "Values and semivalues on subspaces of finite games", International Journal of Game Theory 17:321–338.

Monderer, D. (1990), "A Milnor condition for nonatomic Lipschitz games and its applications", Mathematics of Operations Research 15:714–723.

Monderer, D. (1992), "The stochastic choice problem: A game theoretic approach", Journal of Mathematical Psychology 36:547–554.

Monderer, D., D. Samet and L.S. Shapley (1992), "Weighted Shapley values and the core", International Journal of Game Theory 21:27–39.

Myerson, R.B. (1976), "Values of games in partition function form", International Journal of Game Theory 6:23–31.

Myerson, R.B. (1977), "Graphs and cooperation in games", Mathematics of Operations Research 2:225–229.

Myerson, R.B. (1980), "Conference structures and fair allocation rules", International Journal of Game Theory 9:169–182.

Myerson, R.B. (1986), "Multistage games with communication", Econometrica 54:323–358.

Owen, G. (1972), "Multilinear extension of games", Management Sciences 5:64–79.

Owen, G. (1977), "Values of games with a priori unions", in: R. Henn and O. Moeschlin, eds., Mathematical Economics and Game Theory (Springer, Berlin).

Owen, G. (1982), Game Theory, 2nd edn. (Academic Press, Orlando).

Roth, A.E. (1977), "Bargaining ability, the utility of playing a game, and models of coalition formation", Journal of Mathematical Psychology 16:153–160.

Roth, A.E., ed. (1988), "The Shapley Value (Cambridge University Press, Cambridge).

Shapley, L.S. (1953a), "A value for *n*-person games", Annals of Mathematics Studies 28:307–317.

Shapley, L.S. (1953b), "Additive and non-additive set functions", PhD Thesis (Dept. of Mathematics, Princeton University).

Shapley, L.S. (1971), "Cores of convex games", International Journal of Game Theory 1:11–26.

Shapley, L.S. (1981), "Discussant's comments: Equity considerations in traditional full cost allocation practices", in: Proceedings of the University of Oklahoma Conference on Cost Allocation, April 1981, 131–136.

Shapley, L.S., and M. Shubik (1954), "A method for evaluating the distribution of power in a committee system", American Political Science Review 48:787–792.

Straffin, P.D., Jr. (1988), "The Shapley–Shubik and Banzhaf power indices as probabilities", in: A.E. Roth, ed., The Shapley Value (Cambridge University Press, Cambridge) 71–81.

Weber, R.J. (1988), "Probabilistic values for games", in: A.E. Roth, ed., The Shapley Value (Cambridge University Press, Cambridge) 101–119.

Winter, E. (1989), "A value for games with level structures", International Journal of Game Theory 18:227–240.

Winter, E. (1992), "The consistency and potential for values of games with coalition structure", Games and Economic Behavior 4:132–144.

Chapter 55

VALUES OF NON-TRANSFERABLE UTILITY GAMES

RICHARD P. McLEAN

Department of Economics, NJ Hall, Rutgers University, New Brunswick, NJ, USA

Contents

Handbook of Game Theory, Volume 3, Edited by R.J. Aumann and S. Hart
© *2002 Elsevier Science B.V. All rights reserved*

Abstract

This chapter surveys a class of solution concepts for n-person games without transferable utility – NTU games for short – that are based on varying notions of "fair division". An NTU game is a specification of payoffs attainable by members of each coalition through some joint course of action. The players confront the problem of choosing a payoff or solution that is feasible for the group as a whole. This is a bargaining problem and its solution may be reasonably required to satisfy various criteria, and different sets of rules or axioms will characterize different solutions or classes of solutions. For transferable utility games, the main axiomatic solution is the *Shapley Value* and this chapter deals with *values of NTU games*, i.e., solutions for NTU games that coincide with the Shapley value in the transferable utility case. We survey axiomatic characterizations of the Shapley NTU value, the Harsanyi solution, the Egalitarian solution and the nonsymmetric generalizations of each. In addition, we discuss approaches to some of these solutions using the notions of potential and consistency for NTU games with finitely many as well as infinitely many players.

Keywords

non-transferable utility, Shapley value, Harsanyi solution, Egalitarian solution, consistency, potential, consistent value

JEL classification: C710, C780

1. Notation and definitions

1.1. General notation

Throughout this chapter, we work with a set of players $N = \{1, 2, \ldots, n\}$. If $S \subseteq N$, then $|S|$ denotes the cardinality of S and $\mathfrak{R}^S$ is the $|S|$-dimensional Euclidean space whose axes are labelled with the members of S. The non-negative (positive) orthant of $\mathfrak{R}^S$ will be denoted by $\mathfrak{R}^S_+$ ($\mathfrak{R}^S_{++}$). Similarly, the non-positive orthant $-\mathfrak{R}^S_+$ will be written $\mathfrak{R}^S_-$. If $x \in \mathfrak{R}^N$, then $x^S \in \mathfrak{R}^S$ is the "projection" of x onto $\mathfrak{R}^S$. The vector $(1, \ldots, 1)$ in $\mathfrak{R}^S$ will be denoted 1_S. The "indicator vector" $\chi_T \in \mathfrak{R}^N$ is defined by $\chi_T^i = 1$ if $i \in T$ and $\chi_T^i = 0$ if $i \notin T$. We write $x \geqslant y$ if $x^i \geqslant y^i$ for all i and $x > y$ if $x^i > y^i$ for all i. If x and u belong to $\mathfrak{R}^S$ and $t \in \mathfrak{R}$, then $x \cdot u$ denotes the inner product, tx denotes scalar multiplication, and $x * u \in \mathfrak{R}^S$ is defined by $(x * u)^i = x^i u^i$. If $A, B \subseteq \mathfrak{R}^N$, $y \in \mathfrak{R}^N$ and $t \in \mathfrak{R}$ then $A + B$ is the Minkowski sum of A and B, $A + y = A + \{y\}$, $y * A = \{y * x \mid x \in A\}$ and $tA = \{tx \mid x \in A\}$. The Pareto efficient boundary of A is denoted ∂A and the complement of A is written $\sim A$. If A is convex, then A is *smooth* if there exists a unique supporting hyperplane at each point of ∂A. A set A is *comprehensive* if $x \in A$ and $x \geqslant y$ imply that $y \in A$. To simplify the notation we will not distinguish between i and $\{i\}$. The symbol "\" denotes set theoretic subtraction.

1.2. Transferable utility games

A *TU game* is a set function $v: 2^N \to \mathfrak{R}$ satisfying $v(\emptyset) = 0$. (If X is a set, then 2^X denotes the power set of X.) The 2^{n-1}-dimensional vector space of TU games is denoted by G_N. If $\psi: G_N \to \mathfrak{R}^N$ is a function and $S \subseteq N$, define $(\psi v)(S) \equiv \sum_{i \in S} \psi^i(v)$. A game v is *monotonic* if $v(S) \leqslant v(T)$ whenever $S \subseteq T$. For each nonempty $T \subseteq N$, the *unanimity game* u_T is defined by $u_T(S) = 1$ if $T \subseteq S$ and $u_T(S) = 0$ otherwise. (It is well known that the collection $\{u_T\}_{\emptyset \neq T \subseteq N}$ is a basis for G_N.) Player $i \in N$ is a *null player* in v if $v(S \cup i) - v(S) = 0$ whenever $S \subseteq N \setminus i$, while player $i \in N$ is a *dummy player* in v if $v(S \cup i) - v(S) = v(i)$ whenever $S \subseteq N \setminus i$. Players i and j are *substitutes* in v if $v(S \cup i) = v(S \cup j)$ whenever $S \subseteq N \setminus \{i, j\}$. A coalition $S \subseteq N$ is *flat* in v if $v(S) = \sum_{i \in S} v(i)$.

The collection of $n!$ orderings of the elements of N will be denoted Ω_N. If $R = (i_1, \ldots, i_n)$ is an element of Ω_N and $i = i_k$, let $X^i(R) \equiv \{i_1, \ldots, i_{k-1}\}$ be the set of players that *precede* i in R. ($X^i(R) = \emptyset$ if $i = i_1$.) The *marginal contribution* of i in R in the game v is defined as $\Delta^i(R, v) \equiv v(X^i(R) \cup i) - v(X^i(R))$. Finally, let $M(\Omega_N)$ denote the set of probability measures on Ω_N.

1.3. Non-transferable utility games

An *NTU game* V is a correspondence that assigns to each nonempty $S \subseteq N$ a nonempty subset $V(S) \subseteq \mathfrak{R}^S$. If V is an NTU game, let $V^*(S) = V(S) \times \{0^{N \setminus S}\}$ for each $S \subseteq N$. (In many treatments, an NTU game is defined as V^* rather than V.) If V and W are NTU games, $t \in \Re$ and $\lambda \in \Re^N$, then $(V + W)(S) = V(S) + W(S)$, $(\lambda * V)(S) = \lambda^S * V(S)$, and $(tV)(S) = tV(S)$. Define Γ_N to be the set of NTU games V for which each set $V(S)$ is a proper, convex, closed, comprehensive subset of $\Re^S$.

In the discussion that follows, we will need to consider certain subclasses of Γ_N whose members satisfy some of the following additional restrictions:

(1) $V(N)$ is smooth;
(2) if $x, y \in \partial V(N)$ and $x \geqslant y$, then $x = y$;
(3) for each $S \subseteq N$, there exists an $x \in \Re^N$ such that $V^*(S) \subseteq V(N) + x$;
(4) if $\{x_m\}_{m=1}^{\infty}$ is a sequence in $V(S)$ with $x_{m+1} \geqslant x_m$ for each m, then there exists $y \in \Re^S$ such that $x_m \leqslant y$ for all m;
(5) for each $S \subseteq N$, the set of extreme points of $V(S)$ is bounded.

The main subclasses of Γ_N that we study are listed now:

Γ_N^1 consists of those games in Γ_N satisfying conditions (1), (2) and (3) above,
$\hat{\Gamma}_N^1$ consists of those games in Γ_N satisfying conditions (1), (2), (3) and (5) above,
Γ_N^2 consists of those games in Γ_N satisfying conditions (1) and (2) above,
Γ_N^3 consists of those games in Γ_N satisfying condition (4).

Condition (1) above is a regularity condition that guarantees that utility is locally transferable according to uniquely determined rates of transfer. Condition (2) is a "non-levelness" assumption and guarantees that a normal vector to a boundary point of $V(N)$ has positive components. Condition (3) is a weak version of monotonicity while (4) is a weak boundedness assumption. Condition (5) is important for the existence of some of the concepts we discuss below.

If $v \in G_N$, then the *NTU game corresponding* to v is defined by

$$V(S) = \left\{ x \in \Re^S \mid \sum_{i \in S} x^i \leqslant v(S) \right\}.$$

If V corresponds to v, we will sometimes say that v *generates* V. For each $T \subseteq N$, the NTU game corresponding to the unanimity game u_T will be denoted U_T. The collection of all NTU games corresponding to TU games is denoted Γ_N^{TU} and, obviously, $\Gamma_N^{TU} \subseteq \hat{\Gamma}_N^1$.

For any NTU game $V \in \Gamma_N$, let $d_V \in \Re^N$ be the vector with $d_V^i = \max\{x^i \mid x^i \in V(i)\}$ for each i. The collection of $V \in \Gamma_N$ with $d_V \in V(N)$ is denoted Γ_N^{IR}. A game $V \in \Gamma_N^{IR}$ is a *Pure Bargaining Problem* if

$$V(S) = \left\{ x \in \Re^S \mid x^i \leqslant d_V^i \text{ for each } i \in S \right\}$$

for each $S \neq N$, $d_V \notin \partial V(N)$ and the set $\{x \in V(N) \mid x \geqslant d_V\}$ is compact. The class of bargaining problems in Γ_N^{IR} is denoted Γ_N^B.

2. Values and bargaining solutions

We will record several solutions for TU games and bargaining problems whose NTU extensions are the focus of this chapter. We will also provide statements of characterization theorems since they will be referenced in the sketches of proofs of their NTU generalizations.

2.1. Random order values

For each $p \in M(\Omega_N)$, the *p-Shapley value* is the linear function $\varphi_p : G_N \to \Re^N$ defined as follows:

$$\varphi_p^i(v) = \sum_{R \in \Omega} p(R) \Delta_i (R, v).$$

A p-Shapley value is also called a *random order value* or a *quasivalue* and four axioms characterize the class of such values. In the statements of the axioms, $\psi : G_N \to \Re^N$ is a function.

S1: (*Linearity*) For every $v_1, v_2 \in G_N$ and real numbers α_1 and α_2,

$$\psi(\alpha_1 v_1 + \alpha_2 v_2) = \alpha_1 \psi(v_1) + \alpha_2 \psi(v_2).$$

S2: (*Efficiency*) For each $v \in G_N$, $(\psi v)(N) = v(N)$.

S3: (*Null Players*) If i is a null player in v, then $\psi^i(v) = 0$.

S4: (*Monotonicity*) If v is monotonic, then $\psi^i(v) \geqslant 0$ for all $i \in N$.

PROPOSITION 2.1.1 [Weber (1988)]. *An operator $\psi : G_N \to \Re^N$ satisfies S1–S4 if and only if there exists $p \in M(\Omega)$ such that $\psi = \varphi_p$.*

2.2. Weighted Shapley values

We now turn to a special class of p-Shapley values, the weighted Shapley values. If $w \in \Re_{++}^N$, let $p^w \in M(\Omega)$ be the probability measure defined as follows:

$$\text{for each } R = (i_i, i_2, \dots, i_n) \in \Omega_N, \; p^w(R) = \prod_{k=1}^{n} \left[w^{i_k} \Big/ \sum_{t=1}^{k} w^{i_t} \right].$$

Hence, orderings are constructed according to the following procedure. Player $i \in N$ is chosen to be at the end of the line with probability $w^i / \sum_{j \in N} w^j$. Given that the players in S are already in line, player $i \in N \backslash S$ is added to the front of the line with probability $w^i / \sum_{j \in N \backslash S} w^j$. Define the *weighted Shapley value with weight vector w* (or simply the *w-Shapley value*) as the operator $\varphi_{p^w} : G_N \to \Re^N$. For simplicity, we will denote φ_{p^w} as φ_w.

The w-Shapley value was defined in Shapley (1953a) and axiomatized in Kalai and Samet (1987) where it is called the "weighted Shapley value for simple weight systems". The w-Shapley value satisfies S1–S4 but two other axioms are required to characterize the class of w-Shapley values. First, S4 must be strengthened to guarantee that each $R \in \Omega_N$ occurs with positive probability.

S4': (*Positivity*) If $v \in G_N$ is monotonic and contains no null players, then $\psi^i(v) > 0$ for all $i \in N$.

The second axiom guarantees that the probability measure on orders has the special form p^w for some $w \in \Re^N_{++}$. A coalition $T \subseteq N$ is a *partnership* in v if $v(S \cup R) = v(A)$ whenever $S \subseteq T$, $S \neq T$ and $A \subseteq N \backslash T$. For example, T is a partnership in u_T.

S5: (*Partnership*) If T is a partnership in v, then $\psi^i(v) = \psi^i[(\psi v)(T)u_T]$ for each $i \in T$.

Informally, Axiom S5 states that the behavior of a partnership T in a game v is the same as the behavior of T in u_T in the sense that the fraction of $(\psi v)(T)$ received by $i \in T$ is the same as i's fraction of the one unit being allocated to T in u_T.

PROPOSITION 2.2.1 [Kalai and Samet (1987)]. *An operator $\psi : G_N \to \Re^N$ satisfies Axioms S1, S2, S3, S4' and S5 if and only if there exists $w \in \Re^N_{++}$ such that $\psi = \varphi_w$.*

REMARK 2.2.2. Axioms S1, S2, S3, S4 and S5 characterize the weighted Shapley values for "general weighted systems" [see Kalai and Samet (1987)].

2.3. The Shapley value

If $w \in \Re^N_{++}$ and $w^i = w^j$ for all $i, j \in N$, then $p^w(R) = \frac{1}{n!}$ for each $R \in \Omega$. In this case, we write φ_w simply as φ and refer to φ as the *Shapley value*. An additional axiom is required to guarantee equal weights.

S6: (*Substitution*) If i and j are substitutes in v, then $\psi^i(v) = \psi^j(v)$.

PROPOSITION 2.3.1 [Shapley (1953b)]. *An operator $\psi : G_N \to \Re^N$ satisfies Axioms S1, S2, S3 and S6 if and only if $\psi = \varphi$.*

2.4. Bargaining solutions on Γ^B_N

Let $w = (w^1, w^2, \dots, w^n) \in \Re^N_{++}$ be a vector of weights. For a game $V \in \Gamma^B_N$, define

$$\sigma_w(V) = \arg\max_{\substack{x \in V(N) \\ x \geqslant d_V}} \prod_{i \in N} (x^i - d^i_V)^{w^i}.$$

The set $\sigma_w(V)$ is a singleton because $V(N)$ is convex and the function

$$\sigma_w : \Gamma^B_N \to \Re^N$$

is the *weighted Nash Bargaining Solution* with the weight vector w or simply the w-NBS. One can characterize the w-NBS on Γ_N^B axiomatically and a proof may be found in Roth (1979) or Kalai (1977a).

If the components of the weight vector are identical, then we write σ_w simply as σ and refer to σ as the *Nash Bargaining Solution* or NBS for short. The NBS was defined and axiomatically characterized in Nash (1950) and a detailed discussion may be found in Chapter 24.

We will be interested in another solution for bargaining problems. Again, let $w \in \mathfrak{R}_{++}^N$, $V \in \Gamma_N^B$ and define

$$\tau_w(V) := d_V + t^*(V, w)w$$

where $t^*(V, w) = \max[t \mid d_V + tw \in V(N)]$. The function $\tau_w : \Gamma_N^B \to \mathfrak{R}^N$ is the w-proportional solution with weight vector w. If the components of w are all equal, then we write τ_w simply as τ and refer to τ as the (symmetric) proportional solution for V. The w-proportional solution was defined and axiomatized in Kalai (1977b).

3. The Shapley NTU value

3.1. The definition of the Shapley NTU value

Let $V \in \Gamma_N$ and for $\lambda \in \mathfrak{R}_{++}^N$, define

$$v_\lambda(S) = \sup[\lambda^S \cdot z \mid z \in V(S)].$$

The TU game v_λ is *defined for* V if $v_\lambda(S)$ is finite for each S.

DEFINITION 3.1.1. A vector $x \in \mathfrak{R}^N$ is a *Shapley NTU value payoff* for $V \in \Gamma_N$ if there exists $\lambda \in \mathfrak{R}_{++}^N$ such that:
   (1) $x \in V(N)$,
   (2) v_λ is defined for V,
   (3) $\lambda * x = \varphi(v_\lambda)$.

Condition (1) states that x is feasible for the grand coalition. Conditions (2) and (3) are best understood by first looking at a two-person game in Γ_N^B. In this case, conditions (1), (2) and (3) are equivalent to:

$$(x^1, x^2) \in \arg\max[\lambda^1 x^1 + \lambda^2 x^2 \mid x \in V(1, 2)]$$

and

$$\lambda^1(x^1 - d_V^1) = \lambda^2(x^2 - d_V^2).$$

These are precisely the two conditions that characterize (x^1, x^2) as the NBS of V. Intuitively, an arbitrator might desire to choose a payoff $(x^1, x^2) \in V(1, 2)$ that is "utilitarian" in the sense that $x^1 + x^2 = \max[z^1 + z^2 \mid z \in V(1, 2)]$ and "equitable" in the sense that $x^1 - d_V^1 = x^2 - d_V^2$ (i.e., net utility gains are equal). In general, these two criteria cannot be satisfied simultaneously. However, it may be possible to rescale the utility of the two players and achieve a balance of utilitarianism and equity in terms of the rescaled utilities. This is precisely what is accomplished by the NBS. Hence, conditions (1) and (2) of Definition 3.1.1 provide an extension of utilitarianism to the n-player problem while condition (3) is an extension of equity. (The indefinite article is used here because other extensions of equity and efficiency are possible as shown in Sections 4 and 6 below.)

An alternative definition of Shapley NTU payoff is possible using the idea of dividend in Harsanyi (1963).

DEFINITION 3.1.2. A vector $x \in \Re^N$ is a *Shapley NTU value payoff* for $V \in \Gamma_N$ if there exists $\lambda \in \Re_{++}^N$ and a collection of numbers $\{\alpha_T\}_{\emptyset \neq T \subseteq N}$ such that:
(1) $x \in V(N)$,
(2) v_λ is defined for V,
(3) $\lambda^i x^i = \sum_{T \subseteq N: \, i \in T} \alpha_T$ and $\sum_{i \in S} \sum_{T \subseteq S: \, i \in T} \alpha_T = v_\lambda(S)$ whenever $S \subseteq N$.

The equivalence of Definitions 3.1.1 and 3.1.2 follows from the fact that

$$\sum_{i \in S} \sum_{T \subseteq S: \, i \in T} \alpha_T = \sum_{T \subseteq S} |T| \alpha_T.$$

[See Aumann and Shapley (1974), Appendix A, Note 5.] The numbers α_T are interpreted as dividends that will be paid to the members of T. Suppose that the proper subcoalitions of S have all declared their dividends. Then each $i \in S$ collects $\sum_{T \subset S: \, S \neq T, i \in T} \alpha_T$ and condition (3) of Definition 3.1.2 requires that each member of S receive a dividend α_S equal to

$$\frac{1}{|S|} \left[v_\lambda(S) - \sum_{i \in S} \sum_{T \subseteq S: \, S \neq T, \, i \in T} \alpha_T \right].$$

Since each player in S receives the same dividend, condition (3) is an alternative view of the equity requirement of a Shapley value payoff.

From Definition 3.1.1, we see that, if V corresponds to $v \in G_N$, then λ must be a positive multiple of 1_N so that $\varphi(v)$ is the unique Shapley NTU value payoff for V. The Shapley NTU value is also an extension of the NBS to general NTU games in the sense that these solutions coincide for $V \in \Gamma_N^B$ whenever $\Phi(V) \neq \emptyset$.

PROPOSITION 3.1.3. *If* $V \in \Gamma_N^B$, *then* $\Phi(V) = \{\sigma(V)\}$.

PROOF. Using an argument from convex programming, it can be shown that $x = \sigma(V)$ if and only if there exists $\eta \in \Re^N_{++}$ such that (i) $\eta^i(x^i - d^i_V) = \eta^j(x^j - d^j_V)$ for all $i, j \in N$ and (ii) $\eta \cdot x = \max[\eta \cdot z \mid z \in V(N)]$. Let $V \in \Gamma^B_N$ and suppose $y \in \Phi(V)$ with associated comparison vector $\lambda \in \Re^N_{++}$. It follows that $y \in V(N)$ and that $v_\lambda(S)$ is flat for each $S \neq N$. In particular, $v_\lambda(S) = \sum_{i \in S} \lambda^i d^i_V$ for each $S \neq N$. Therefore,

$$\lambda^i\left(x^i - d^i_V\right) = \varphi^i(v_\lambda) - \lambda^i d^i_V = \frac{1}{n}\left[v_\lambda(N) - \sum_{j \in N} v_\lambda(j)\right]$$

for each i, (i) and (ii) are satisfied and it follows that $x = \sigma(V)$. Conversely, suppose that $x = \sigma(V)$. Then there exists $\lambda \in \Re^N_{++}$ such that (i) and (ii) are satisfied. Letting $v_\lambda(N) = \max[\lambda \cdot z \mid z \in V(N)]$ and $v_\lambda(S) = \sum_{i \in S} \lambda^i d^i_V$ for each $S \neq N$, we conclude that v_λ is defined for V. Since $\lambda^i x^i = \varphi^i(v_\lambda)$, it follows that x is a Shapley NTU value payoff for V. □

3.2. An axiomatic approach to the Shapley NTU value

In the previous section, it was shown that the set of Shapley NTU value payoffs coincides with the Shapley value for the TU games and the Nash Bargaining Solution for pure bargaining games. Thus, it is natural to synthesize the axiomatic characterization of these two solutions in an effort to construct a characterization for the Shapley NTU value. This approach is precisely the method of attack in Aumann (1985a) and Kern (1985).

Let $\Phi : \Gamma^1_N \to \Re^N$ be the correspondence that associates with each V in Γ^1_N the set of NTU value payoffs. The solution correspondence Φ will be referred to as the *Shapley correspondence*. Let $\tilde{\Gamma}^1_N$ be the set of NTU games $V \in \Gamma^1_N$ such that $\Phi(V) \neq \emptyset$. In the statement of the axioms, $\Psi : \tilde{\Gamma}^1_N \to \Re$ is a correspondence and U, V and W are games in $\tilde{\Gamma}^1_N$.

A0: *(Non-Emptiness)* $\Psi(V) \neq \emptyset$.

A1: *(Efficiency)* $\Psi(V) \subseteq \partial V(N)$.

A2: *(Conditional Additivity)* If $U = V + W$, then $[\Psi(V) + \Psi(W)] \cap \partial U(N) \subseteq \Psi(U)$.

A3: *(Unanimity)* If U_T corresponds to the unanimity game with carrier T, then $\Psi(U_T) = \{\chi_T / |T|\}$.

A4: *(Scale Covariance)* If $\alpha \in \Re^N_{++}$, then $\Psi(\alpha * V) = \alpha * \Psi(V)$.

A5: *(Independence of Irrelevant Alternatives)* If $V(N) \subseteq W(N)$ and $V(S) = W(S)$ for $S \neq N$, then $\Psi(W) \cap V(N) \subseteq \Psi(V)$.

Axioms A0 and A1 are self-explanatory. A2 states that whenever the sum of two solution payoffs is Pareto optimal in the sum game, then it must be a solution payoff for the sum game. A3 combines the symmetry and null player axioms of the TU theory. Mathematically, this axiom specifies a unique payoff for the NTU games corresponding to unanimity games. Axioms A4 and A5 are Nash's axioms extended to the NTU framework.

The next three lemmas [all found in Aumann (1985a)] provide the essential steps in the characterization of the Shapley correspondence.

LEMMA 3.2.1. *If* $\Psi : \tilde{\Gamma}_N^1 \to \Re$ *satisfies Axioms A0–A4 then* $\Psi(V) = \{\varphi(v)\}$ *whenever* V *corresponds to the TU game* v.

PROOF. For each real number α, let V^α correspond to the TU game αv. Using A0, A1, A2 and A3, we deduce that $\Psi(V^0) = \{0\}$. Using Axioms A1 and A2 again, it can be shown that $\Psi(V^{-1}) + \Psi(V) = \{0\}$ so A0 implies that $\Psi(V^{-1}) = -\Psi(V)$ and each is single-valued. Combining A4 with the results above, we deduce that $\Psi(\alpha V) = \alpha \Psi(V)$ for all scalars α. Using A3 it follows that $\Psi(U_T^\alpha) = \alpha \Psi(U_T) = \{\alpha \chi_T / |T|\} = \{\varphi(\alpha u_T)\}$. Since $\{u_T\}_{\emptyset \neq T \subseteq N}$ is a basis for G_N we can express each $v \in G_N$ as $v = \sum_{\emptyset \neq T \subseteq N} \alpha_T u_T$ so that $V = \sum_T U_T^{\alpha_T}$ for the NTU game V corresponding to v. Applying A1 and A2, we deduce that

$$\{\varphi(v)\} = \sum_T \{\varphi(\alpha_T u_T)\} = \sum_T \Psi(U_T^{\alpha_T}) \subseteq \Psi\left(\sum_T U_T^{\alpha_T}\right) = \Psi(V).$$

Since $\Psi(V)$ is a singleton set, $\Psi(V) = \{\varphi(v)\}$.   □

LEMMA 3.2.2. *If* Ψ *satisfies axioms A0–A4, then* $\Psi(V) \subseteq \Phi(V)$, *for each* $V \in \tilde{\Gamma}_N^1$.

PROOF. Choose $y \in \Psi(V)$. By Axiom A1, $y \in \partial V(N)$. Using Axiom A4 and the non-levelness assumption, we can assume without loss of generality that the unique normal to $V(N)$ at y is $\lambda = 1_N$. If V^0 corresponds to the TU game that is identically zero, then $y \in \partial(V(N) + V^0(N))$ and $V + V^0$ corresponds to v_λ. So using Lemma 3.2.1, A2 and Lemma 3.2.1 again, we get

$$\begin{aligned}
\lambda * y &\in \left[\Psi(V) + \{0\}\right] \cap \partial\left(V(N) + V^0(N)\right) \\
&= \left[\Psi(V) + \Psi(V^0)\right] \cap \partial\left(V(N) + V^0(N)\right) \\
&\subseteq \Psi\left(V + V^0\right) \\
&= \{\varphi(v_\lambda)\}.
\end{aligned}$$

Thus $\lambda * y = \varphi(v_\lambda)$ and we conclude that $y \in \Phi(V)$.   □

LEMMA 3.2.3. *If* Ψ *satisfies Axioms A0–A5, then* $\Phi(V) \subseteq \Psi(V)$, *for each* $V \in \tilde{\Gamma}_N^1$.

PROOF. Choose $y \in \Phi(V)$ and $\lambda \in \Re_{++}^N$ such that v_λ is defined and $\lambda * y = \varphi(v_\lambda)$. Let V_λ be the NTU game corresponding to v_λ and define a game W as follows:

$$W(S) = \begin{cases} V_\lambda(N) & \text{if } S = N, \\ \lambda^S * V(S) & \text{if } S \neq N. \end{cases}$$

Since $\lambda * y \in \Phi(W)$, it follows that $W \in \widetilde{\Gamma}_N^1$ and Axiom A0 implies that $\Psi(W) \neq \emptyset$. Again, let V^0 correspond to the TU game that is identically zero so that $V_\lambda = W + V^0$ and $\partial(W(N) + V^0(N)) = \partial W(N)$. Using Lemma 3.2.1 and A2 and A1, we get

$$
\begin{aligned}
\{\phi(v_\lambda)\} &= \Psi(V_\lambda) = \Psi(W + V^0) \\
&\supseteq \left[\Psi(W) + \{0\}\right] \cap \partial\left(W(N) + V^0(N)\right) \\
&= \Psi(W) \cap \partial W(N) = \Psi(W).
\end{aligned}
$$

Having already shown that $\Psi(W) \neq \emptyset$, we conclude that $\Psi(W) = \{\varphi(v_\lambda)\} = \{\lambda * y\}$. Since $\lambda * V(N) \subseteq W(N)$ and $\lambda^S * V(S) = W(S)$ for $S \neq N$, we can use A5 and A4 to deduce that $\{\lambda * y\} = \Psi(W) \cap [\lambda * V(N)] \subseteq \Psi(\lambda * V) = \lambda * \Psi(V)$ and, therefore, $y \in \Psi(V)$. $\qquad\square$

As the reader may verify, the Shapley correspondence Φ satisfies Axioms A0–A5 and the main characterization result for the Shapley NTU value can now be stated.

THEOREM 3.2.4 [Aumann (1985a)]. *A correspondence* $\Psi : \widetilde{\Gamma}_N^1 \to \mathfrak{R}^N$ *satisfies Axioms A0–A5 if and only if* $\Psi = \Phi$.

If A5 is dropped from the axiomatization, then one obtains the following result as a consequence of Lemma 3.2.2.

THEOREM 3.2.5 [Aumann (1985a)]. *The Shapley correspondence* Φ *satisfies Axioms A0–A4 and if* Ψ *is any other correspondence satisfying Axioms A0–A4, then* $\Psi(V) \subseteq \Phi(V)$.

In Kern (1985), an alternative version of the Independence of Irrelevant Alternatives (IIA) is offered.

A5′: (*Transfer Independence of Irrelevant Alternatives*) Let V and W be games and let $x \in \Psi(V)$. If there exists a $\lambda \in \mathfrak{R}_{++}^N$ and a collection $\{\xi_S\}_{S \subseteq N}$ with $\xi_N = x$ such that for each $S \subseteq N$, $\xi_S \in \partial V(S) \cap \partial W(S)$, λ^S is normal to $V(S)$ at ξ_S and λ^S is normal to $W(S)$ at ξ_S, then $x \in \Psi(W)$.

As a general rule, IIA-type axioms stipulate that, once a solution has been determined, we can alter the game "far away" from the solution and obtain a new game with the same solution. Indeed, this aspect of Nash's Bargaining Solution has been criticized on a variety of grounds and has resulted in alternatives to the NBS in which IIA has been replaced by other axioms. Axiom A5′ has a local flavor as well. If $\xi_S \in \partial V(S) \cap \partial W(S)$ and λ^S is normal to $V(S)$ and $W(S)$ at ξ_S, then $\partial V(S)$ and $\partial W(S)$ are "similar near ξ_S". This is certainly true for $S = N$ because of the smoothness assumption. When V and W are related in this way, the axiom requires that $x \in \Psi(W)$ if $x \in \Psi(V)$.

LEMMA 3.2.6. *If* $\Psi : \widetilde{\Gamma}_N^1 \to \mathfrak{R}^N$ *satisfies Axioms A0–A4 and A5′, then* $\Phi(V) \subseteq \Psi(V)$, *for all* $V \in \Gamma_1^*$.

PROOF. Let $y \in \Phi(V)$. Then $y \in \partial V(N)$ and there exists $\lambda \in \mathfrak{R}^N_{++}$ such that λ is a normal vector to $V(N)$ at y, v_λ is defined and $\lambda * y = \varphi(v_\lambda)$. Define the game W as in the proof of Lemma 3.2.3 and, using the same argument found there, conclude that $\lambda * y \in \Psi(W)$. Upon defining $\lambda^{-1} = (\frac{1}{\lambda^1}, \ldots, \frac{1}{\lambda^n})$, we deduce from A4 that $y \in \Psi(W_\lambda)$ where $W_\lambda = \lambda^{-1} * W$. Therefore, $y \in \partial W_\lambda(N)$ and from the definition of W it follows that λ is normal to $W_\lambda(N)$ at y. Since $W_\lambda(S) = V(S)$ for $S \neq N$, we conclude that there exists a collection $\{\xi_S\}_{S \subseteq N}$ with $\xi_N = y$ such that for each $S \subseteq N$, $\xi_S \in \partial V(S) \cap \partial W_\lambda(S)$ and λ^S is normal to $V(S)$ and $W_\lambda(S)$ at ξ_S. The conditions of A5' are therefore satisfied and it follows that $y \in \Psi(V)$. $\square$

It is easy to verify that the Shapley correspondence Φ satisfies A5' [see Kern (1985), Theorem 5.2] so Lemmas 3.2.2 and 3.2.6 combine to yield the following characterization.

THEOREM 3.2.7. *A correspondence* $\Psi : \widetilde{\Gamma}^1_N \to \mathfrak{R}^N$ *satisfies Axioms A0–A4 and A5' if and only if* $\Psi = \Phi$.

REMARK 3.2.8. Theorem 3.2.7 is essentially Theorem 5.5 of Kern (1985). However, Kern uses only Axioms A4 and A5' and a third axiom requiring that $\Psi(V) = \{\varphi(v)\}$ whenever V corresponds to the TU game v. Lemma 3.2.1 makes clear the relationship between these two axiom systems.

REMARK 3.2.9. In the statements of Theorems 3.2.4 and 3.2.7, the domain of the correspondence Ψ is $\widetilde{\Gamma}^1_N$ and some may find this an aesthetic drawback. However, there are smaller domains that one could work with. For example, the characterization theorems will still be true if one works with the smaller collection $\widehat{\Gamma}^1_N$. For $V \in \widehat{\Gamma}^1_N$, a fixed point argument can be used to verify that A0 holds. For a more detailed discussion see Aumann (1985a). The existence question is also treated at length in Kern (1985) where conditions different from smoothness and nonlevelness of $V(N)$ are identified that guarantee the existence of Shapley NTU value payoffs.

3.3. Non-symmetric generalizations of the Shapley NTU value

In the axiomatic characterization of the Shapley value, the Nash Bargaining Solution and the Shapley NTU value, some type of symmetry postulate is required. Basically, these symmetry axioms all guarantee that two players who affect the game in the same way will receive the same payoff. However, one can envision circumstances in which two players who are identical in terms of the data of the game might nonetheless be treated asymmetrically in the arbitrated solution. For example, two workers with the same productivity may be treated differently because one has attained seniority. Thus, an arbitrator might assign weights to the individuals that reflect the relative "importance" of each person in the resolution of the bargaining problem. Our next problem is to define and axiomatize a weighted Shapley NTU value that generalizes the symmetric

NTU value defined above and that coincides with the w-NBS on Γ_N^B and the w-Shapley value on Γ_N^{TU}.

DEFINITION 3.3.1. Let $w \in \Re_{++}^N$. A vector $x \in \Re^N$ is a w-*Shapley NTU value payoff* for $V \in \Gamma_N$ if there exists $\lambda \in \Re_{++}^N$ such that:
  (1) $x \in V(N)$,
  (2) v_λ is defined for V,
  (3) $\lambda * x = \varphi_w(v_\lambda)$.

The correspondence Φ_w that associates with each V the set of all w-Shapley value payoffs is called w-*Shapley correspondence*. Given Proposition 2.2.1, it is reasonable to conjecture that the axiomatic characterization of Φ_w weakens Axiom A3 (by dropping symmetry) and imposes NTU analogues of the positivity and partnership axioms (S4' and S5) used in the characterization of φ_w on G_N.

A3': (*Null Players*) For each $\emptyset \neq T \subseteq N$, there exists $q_T \in \Re^N$ such that $\Psi(U_T) = \{q_T\}$ and $q_T^i = 0$ if $i \notin T$.

This axiom generalizes the unanimity Axiom A3 in that it allows players within a carrier to receive different payoffs but it still requires Φ to be single-valued on the carrier games.

A6: (*Positivity*) If v is a monotonic TU game without null players and if V is the NTU game corresponding to v, then $x \in \Re_{++}^N$ for each $x \in \Psi(V)$.

A7: (*Partnership*) Let T be a partnership in a TU game v and let V correspond to v. If $x \in \Psi(V)$ then there exists $y \in \Psi((\sum_{j \in T} x^j) U_T)$ such that $x^i = y^i$ for each $i \in T$.

The next lemma generalizes Lemma 3.2.1 to the case where no symmetry requirement is imposed. Note that the domain of the correspondence Ψ is $\widehat{\Gamma}_N^1$.

LEMMA 3.3.2. *Suppose that a correspondence $\Psi : \widehat{\Gamma}_N^1 \to \Re^N$ satisfies Axioms A0, A1, A2, A3' and A4. Then there exists an operator $\psi : G_N \to \Re^N$ satisfying S1, S2 and S3 such that $\Psi(V) = \{\psi(v)\}$ whenever V corresponds to the TU game v.*

PROOF. Suppose Ψ satisfies the axioms and let $\{q_T\}_{\emptyset \neq T \subseteq N}$ be the collection specified by Axiom A3'. For each $\emptyset \neq T \subseteq N$, define $\psi(u_T) = q_T$ and linearly extend ψ to all of G_N. The resulting ψ is obviously linear and A1 implies that ψ is efficient. It can be verified using an induction argument that the null player axiom is satisfied. Using the same argument as that used in the proof of Lemma 3.2.1, we conclude that $\Psi(V)$ is a singleton whenever V corresponds to a TU game. Since

$$\{\psi(v)\} = \sum_T \{\alpha_T \psi(u_T)\} = \sum_T \Psi(U_T^{\alpha_T}) \subseteq \Psi\left(\sum_T U_T^{\alpha_T}\right) = \Psi(V),$$

it follows that $\Psi(V) = \{\psi(v)\}$. $\qquad\square$

LEMMA 3.3.3. *If* $\Psi : \widehat{\Gamma}^1_N \to \Re^N$ *satisfies Axioms A0, A1, A2, A3′, A4, A6 and A7, then there exists a* $w \in \Re^N_{++}$ *such that* $\Psi(V) = \{\varphi_w(v)\}$ *whenever V corresponds to the TU game v.*

PROOF. Suppose V corresponds to $v \in G_N$. Using Lemma 3.3.2, we conclude that $\Psi(V) = \{\psi(v)\}$ where $\psi : G_N \to \Re^N$ satisfies S1, S2 and S3. Since Ψ satisfies A6 and A7, it follows that ψ satisfies S4′ and S5 so applying Proposition 2.2.1 yields the result. $\square$

Using Lemma 3.3.3 above and the techniques of Lemmas 3.2.2 and 3.2.3, we can prove the following generalization of Theorem 3.2.4.

THEOREM 3.3.4 [Levy and McLean (1991)]. *A correspondence* $\Psi : \widehat{\Gamma}^1_N \to \Re^N$ *satisfies Axioms A0, A1, A2, A3′, A4, A5, A6 and A7 if and only if there exists a* $w \in \Re^N_{++}$ *such that* $\Psi(V) = \Phi_w$.

REMARK 3.3.5. Theorem 3.3.4 can be extended to the case of "general weight systems" of Kalai and Samet (1987). This extension is accomplished by weakening the positivity axiom A6 (to the monotonicity axiom A6′ below) and using Theorem 2 in Kalai and Samet (1987) instead of Proposition 2.2.1.

The w-Shapley correspondence obviously coincides with the w-Shapley value if V corresponds to a game $v \in G_N$. Furthermore, the w-Shapley correspondence is an extension of the w-NBS as well and we record a generalization of Proposition 3.1.3 that utilizes an essentially identical proof.

PROPOSITION 3.3.6. *If* $w \in \Re^N_{++}$ *and* $V \in \Gamma^B_N$, *then* $\Phi_w(V) = \{\sigma_w(V)\}$.

In Section 2.1, we presented the notion of random order value for TU games as a generalization of the w-Shapley value. In a similar way, Levy and McLean (1991) define and axiomatize a random order NTU value that generalizes the weighted Shapley NTU value discussed above and thus construct a taxonomy of NTU values with and without symmetry. Recall that $M(\Omega_N)$ is the set of probability measures on Ω_N and $\varphi_p : G_N \to \Re^N$ is the p-Shapley value corresponding to $p \in M(\Omega_N)$.

DEFINITION 3.3.7. Let $p \in M(\Omega_N)$. *A payoff vector* $x \in \Re^N$ *is a* p-*Shapley NTU value payoff if there exists* $\lambda \in \Re^N_{++}$ *such that:*
   (1) $x \in V(N)$,
   (2) v_λ is defined for V,
   (3) $\lambda * x = \varphi_p(v_\lambda)$.

The set of all p-Shapley NTU value payoffs of a game V is denoted $\Phi_p(V)$ and the correspondence that assigns to each V the set of p-Shapley NTU value payoffs is called the *p-Shapley correspondence*. Our characterization of Φ_p weakens the positivity Axiom A6 to A6′.

A6′: (*Monotonicity*) If v is a monotonic TU game and if V is the NTU game corresponding to v, then $x \in \mathfrak{R}_+^N$ whenever $x \in \Psi(V)$.

Suppose a correspondence Ψ satisfies A1, A2, A3′, A4 and A6′. By combining Lemma 3.3.2 and Proposition 2.1.1, we deduce that there exists $p \in M(\Omega)$ such that $\Psi(V) = \{\varphi_p(v)\}$ whenever V corresponds to $v \in G_N$. This observation and the techniques of Lemmas 3.2.2 and 3.2.3 yield the next characterization result.

THEOREM 3.3.8 [Levy and McLean (1991)]. *A correspondence* $\Psi : \widehat{\Gamma}_N^1 \to \mathfrak{R}^N$ *satisfies Axioms A0, A1, A2, A3′, A4, A5 and A6′ if and only if there exists a* $p \in M(\Omega)$ *such that* $\Psi = \Phi_p$.

4. The Harsanyi solution

4.1. The Harsanyi NTU value

A *payoff configuration* is a collection $\{x_S\}_{\emptyset \neq T \subseteq N}$ with $x_S \in \mathfrak{R}^S$ for each S. For simplicity, we will denote a payoff configuration by $\{x_S\}$. If $v \in G_N$ and $S \subseteq N$, then $v|_S$ represents the restriction of v to 2^S and $\varphi(v|_S)$ is the Shapley value of the $|S|$ player game $v|_S$. In Harsanyi (1963), a different extension of NBS and the Shapley value to general NTU games is given.

DEFINITION 4.1.1. A vector $x \in \mathfrak{R}^N$ is a *Harsanyi NTU value payoff* for $V \in \Gamma_N$ if there exists a payoff configuration $\{x_S\}$ with $x_N = x$ and $\lambda \in \mathfrak{R}_{++}^N$ such that:
  (1) $x_S \in \partial V(S)$ for each S,
  (2) $\lambda \cdot x = \max_{z \in V(N)} \lambda \cdot z$,
  (3) $\lambda^S * x_S = \varphi(\hat{v}_\lambda|_S)$ for each S where $\hat{v}_\lambda(S) = \lambda^S \cdot x_S$.

REMARK 4.1.2. Definition 3.1.1 for the Shapley NTU value can be restated so as to facilitate a comparison with the Harsanyi NTU value. In particular, a vector $x \in \mathfrak{R}^N$ is a Shapley NTU value payoff for $V \in \Gamma_N$ if there exists a payoff configuration $\{x_S\}$ with $x_N = x$ and $\lambda \in \mathfrak{R}_{++}^N$ such that:
  (1*) $x_S \in \partial V(S)$ for each S,
  (2*) $\lambda^S \cdot x_S = \max_{z \in V(S)} \lambda^S \cdot z$ for each S,
  (3*) $\lambda * x = \varphi(\hat{v}_\lambda)$ where $\hat{v}_\lambda(S) = \lambda^S \cdot x_S$.

Comparing the definitions of the Shapley and Harsanyi NTU values, we see that both solutions are defined in terms of a comparison vector $\lambda \in \mathfrak{R}_{++}^N$, a payoff configuration $\{x_S\}$ with $x_S \in \partial V(S)$ for each S, and a TU game $\hat{v}_\lambda$ where $\hat{v}_\lambda(S) = \lambda^S \cdot x_S$. The Shapley NTU value requires that $\hat{v}_\lambda(S) = \max_{z \in V(S)} \lambda^S \cdot z$ for every $S \subseteq N$ (so that $\hat{v}_\lambda = v_\lambda$) while the Harsanyi value requires this only for $S = N$. On the other hand, the Harsanyi

NTU value requires that $\lambda^S * x_S = \varphi(\hat{v}_\lambda|_S)$ for every $S \subseteq N$ while the Shapley NTU value requires this only for $S = N$. If $N = \{1, 2\}$, then the two definitions are identical and both coincide with the NBS. In light of the discussion in Section 3.1, conditions (2) and (3) in Definition 4.1.1 are, respectively, the extensions of the utilitarian and equity properties of the NBS. Comparing these conditions with 2^* and 3^* in the definition of the Shapley NTU value, we see that the Shapley NTU value emphasizes coalitional utilitarianism while the Harsanyi NTU value emphasizes coalitional equity.

The definition of Harsanyi NTU value can also be restated in terms of dividends in the spirit of Definition 3.1.2. For a proof of the equivalence of the two definitions, see Section 2.2 in Imai (1983).

DEFINITION 4.1.3. A vector $x \in \mathfrak{R}^N$ is a Harsanyi NTU value payoff for $V \in \Gamma_N$ if there exists a payoff configuration $\{x_S\}$ with $x_N = x$, a vector $\lambda \in \mathfrak{R}^N_{++}$ and a collection of numbers $\{\alpha_T\}_{\emptyset \neq T \subseteq N}$ such that:
  (1) $x_S \in \partial V(S)$ for each S,
  (2) $\lambda \cdot x = \max_{z \in V(N)} \lambda \cdot z$,
  (3) $\lambda^i x_S^i = \sum_{T \subseteq S: \, i \in T} \alpha_T$ for each $S \subseteq N$ and each $i \in S$.

As in Definition 3.1.2, the number α_T is interpreted as a dividend that each player in T receives as a member of T. If coalition S were to form, then each $i \in S$ would receive the payoff x_S^i. Condition (1) requires that x_S be efficient and condition (3) requires that i's rescaled payoff $\lambda^i x_S^i$ be equal to his accumulated dividends.

From Definition 4.1.1, it is easy to verify that, if V corresponds to $v \in G_N$, then $\varphi(v)$ is the unique Harsanyi NTU value payoff for V. To see this, note that if x is a Harsanyi value payoff, then the associated vector λ must be a positive scalar multiple of 1_N. If $\{x_S\}$ is the associated payoff configuration, then $1_S \cdot x_S = v(S)$ so v_λ is a positive multiple of v and, therefore, $x_N = x = \varphi(v)$. Conversely, define $x_S = \varphi(v|_S)$. Then $\{x_S\}$ satisfies the three conditions of the definition and $x_N = \varphi(v)$ is the Harsanyi NTU value payoff.

Like the Shapley NTU value, the Harsanyi NTU value is an extension of the NBS to general NTU games in the sense that the Harsanyi value coincides with the NBS for $V \in \Gamma_N^B$.

PROPOSITION 4.1.4. *Suppose that $V \in \Gamma_0^B$. If y is a Harsanyi NTU value payoff vector for V, then $y = \sigma(V)$.*

PROOF. Suppose y is a Harsanyi NTU value payoff for V. Then Definition 4.1.3 implies that there exist numbers $\{\alpha_T\}_{\emptyset \neq T \subseteq N}$ and a payoff configuration $\{x_S\}$ with $x_S \in \partial V(S)$ for all S such that $x_N = y$ and

$$\lambda^i x_S^i = \sum_{T \subseteq S: \, i \in T} \alpha_T$$

for each $i \in S \subseteq N$. Hence, $\lambda^i x^i_{\{i\}} = \alpha_{\{i\}}$ so $\alpha_{\{i\}} = \lambda^i d^i_V$. Let $S = \{i, j\}$. Then $\lambda^i x^i_S = \alpha_{\{i\}} + \alpha_S = \lambda^i d^i_V + \alpha_S$ so $x^i_S = d^i_V + \alpha_S / \lambda^i$. Since x_S and (d^i_V, d^j_V) are both in $\partial V(S)$ it follows that $\alpha_S = 0$. Using this argument inductively, we conclude that $\alpha_S = 0$ if $1 < |S| < n$. Hence, $x^i_S = d^i_V$ whenever $1 < |S| < n$. Therefore,

$$\lambda^i \left(x^i_N - d^i_V \right) = \alpha_N = \lambda^j \left(x^j_N - d^j_V \right)$$

for all $i, j \in N$. Combining this fact with condition (2) in Definition 4.1.3, we can use the argument in the proof of Proposition 3.1.3 and conclude that $x_N = y$ is the NBS for V. $\qquad\square$

4.2. The Harsanyi solution

Harsanyi's general approach for NTU games can be given an axiomatic foundation if we take the point of view that the entire payoff configuration should be treated as the outcome.

DEFINITION 4.2.1. A payoff configuration $\{x_S\}$ is a *Harsanyi solution* for $V \in \Gamma_N$ if there exists $\lambda \in \Re^N_{++}$ such that:
  (1) $x_S \in \partial V(S)$ for all $S \subseteq N$,
  (2) $\lambda \cdot x_N = \max_{z \in V(N)} \lambda \cdot z$,
  (3) $\lambda^S * x_S = \varphi(\hat{v}_\lambda |_S)$ for all $S \subseteq N$ where $\hat{v}_\lambda(S) = \lambda^S \cdot x_S$.

Note that we are distinguishing between a Harsanyi solution payoff configuration and a Harsanyi NTU value payoff vector. (In the literature, the term "Harsanyi solution" is ordinarily used for both of these concepts.) Obviously, it follows from the definition that if $\{x_S\}$ is a Harsanyi solution, then x_N is a Harsanyi NTU value payoff. The correspondence that associates with each $V \in \Gamma^2_N$ the set of Harsanyi solution payoff configurations for V is called the *Harsanyi solution correspondence* and is denoted H.

Let $\widetilde{\Gamma}^2_N$ be the set of $V \in \Gamma^2_N$ such that $H(V) \neq \emptyset$. In the axioms below, U, V and W are games and $\Psi : \widetilde{\Gamma}^2_N \to \prod_{S \subseteq N} R^S$ is a payoff configuration valued correspondence.

H0: (*Non-Emptiness*) $\Psi(V) \neq \emptyset$.
  H1: (*Efficiency*) If $\{x_S\} \in \Psi(V)$ then $x_S \in \partial V(S)$ for each $S \subseteq N$.
  H2: (*Conditional Additivity*) If $U = V + W$, $\{x_S\} \in \Psi(V)$, $\{y_S\} \in \Psi(W)$ and $x_S + y_S \in \partial U(S)$ for each S, then $\{x_S + y_S\} \in \Psi(V)$.
  H3: (*Unanimity Games*) If $T \subseteq N$ is nonempty, then $\Psi(U_T) = \{z_S\}$ where $z_S = \chi^S_T / |T|$ if $T \subseteq S$ and $z_S = 0$ otherwise.
  H4: (*Scale Covariance*) $\Psi(\alpha * V) = \{\{\alpha^S * x_S\} \mid \{x_S\} \in \Psi(V)\}$ for every $\alpha \in \Re^N_{++}$.
  H5: (*Independence of Irrelevant Alternatives*) If $\{x_S\} \in \Psi(W)$ and if $x_S \in V(S) \subseteq W(S)$ for each S, then $\{x_S\} \in \Psi(V)$.

The axioms for the Harsanyi solution correspondence on $\widetilde{\Gamma}^2_N$ can be readily compared with those for the Shapley correspondence on $\widetilde{\Gamma}^1_N$. Axioms H0–H5 are payoff configuration analogues of A0–A5 and they are used in Hart (1985a) to prove the following characterization theorem.

THEOREM 4.2.2 [Hart (1985a)]. *A correspondence* $\Psi : \tilde{\Gamma}_N^2 \to \prod_{S \subseteq N} R^S$ *satisfies Axioms H0–H5 if and only if* $\Psi = H$.

PROOF. For $v \in G_N$, let $h(v) = \{x_S\}$ where $x_S = \varphi(v|_S)$. If V corresponds to v, then $H(V) = \{h(v)\}$ so $\Gamma_N^{TU} \subseteq \tilde{\Gamma}_N^2$. (See the discussion preceding Proposition 4.1.4.) Now the proof proceeds in several steps:

Step 1: Since Ψ satisfies H0–H4, it follows that $\Phi(V) = \{h(v)\}$ whenever V corresponds to $v \in G_N$. This is the analogue of Lemma 3.2.1.

Step 2: Since Ψ satisfies H0–H4, one can show that $\Psi(V) \subseteq H(V)$ for each $V \in \tilde{\Gamma}_N^2$. This is the analogue of Lemma 3.2.2.

Step 3: Finally, it can be proved that $H(V) \subseteq \Psi(V)$ whenever Ψ satisfies H0–H5, thus providing a payoff configuration version of Lemma 3.2.3.

The proof is completed by verifying that H satisfies these axioms. □

The Harsanyi correspondence can be axiomatically characterized on Γ_N^2 if $H(V)$ is not required to be nonempty. To accomplish this, H3 must be strengthened and an additional axiom must be imposed.

H3$^+$: (*Unanimity Games*) For each real number c and nonempty $T \subseteq N$, let U_T^c be the game corresponding to the TU game cu_T. Then $\Psi(U_T^c) = \{z_S\}$ where $z_S = (c/|T|)\chi_T^S$ if $T \subseteq S$ and $z_S = 0$ otherwise.

H6: (*Zero Inessential Games*) If $0 \in \partial V(S)$ for each S, then $\{0\} \in \Psi(S)$.

THEOREM 4.2.3 [Hart (1985a)]. *A correspondence* $\Psi : \Gamma_N^2 \to \prod_{S \subseteq N} R^S$ *satisfies Axioms H1, H2, H3$^+$, H4, H5 and H6 if and only if* $\Psi = H$.

We conclude this section with a brief discussion of a weighted generalization of the Harsanyi solution analogous to the w-Shapley NTU value.

DEFINITION 4.2.4. Let $w \in \Re_{++}^N$. A payoff configuration $\{x_S\}$ is a w-*Harsanyi solution* for $V \in \Gamma_N$ if there exists $\lambda \in \Re_{++}^N$ and a collection of numbers $\{\alpha_T\}_{\emptyset \neq T \subseteq N}$ such that:

(1) $x_S \in \partial V(S)$,
(2) $\lambda \cdot x_N = \max_{z \in V(N)} \lambda \cdot z$,
(3) $\lambda^i * x_S^i = w^i \sum_{T \subseteq S: \, i \in T} \alpha_T$.

If the w^i are all equal then the w-Harsanyi solution reduces to the Harsanyi solution. Axiom H3 (or H3$^+$) imposes symmetry on the NTU value. If H3 is generalized in the same way that A3 is generalized to A3$'$, then the class of weighted Harsanyi solutions can be axiomatized. Other nonsymmetric generalizations of the Harsanyi solution [including NTU extensions of the TU coalition structure values studied in Owen (1977),

Hart and Kurz (1983) and Levy and McLean (1989)] are discussed in Winter (1991) and Winter (1992).

4.3. A comparison of the Shapley and Harsanyi NTU value

EXAMPLE 4.3.1. Let $N = \{1, 2, 3\}$ and for each $\varepsilon \in [0, 1/2]$, consider the game V_ε defined as:

$V_\varepsilon(i) = \{x \in R^{\{i\}} \mid x \leqslant 0\}$;
$V_\varepsilon(1, 2) = \{x \in R^{\{1,2\}} \mid (x^1, x^2) \leqslant (\frac{1}{2}, \frac{1}{2})\}$;
$V_\varepsilon(1, 3) = \{x \in R^{\{1,3\}} \mid (x^1, x^3) \leqslant (\varepsilon, 1 - \varepsilon)\}$;
$V_\varepsilon(2, 3) = \{x \in R^{\{2,3\}} \mid (x^2, x^3) \leqslant (\varepsilon, 1 - \varepsilon)\}$;
$V_\varepsilon(N) = \{x \in R^N \mid x \leqslant y \text{ for some } y \in C\}$;
where $C = \text{convex hull } \{(1/2, 1/2, 0), (\varepsilon, 0, 1 - \varepsilon), (0, \varepsilon, 1 - \varepsilon)\}$.

This game appears in Roth (1980) and has generated a great deal of discussion in the literature [see Aumann (1985b), Roth (1986), Aumann (1986) and Hart (1985b)] which we summarize here. The Shapley NTU value is $(\frac{1}{3}, \frac{1}{3}, \frac{1}{3})$ with comparison vector $\lambda = (1, 1, 1)$ and the Harsanyi NTU value is $(\frac{1}{2} - \frac{\varepsilon}{3}, \frac{1}{2} - \frac{\varepsilon}{3}, \frac{2\varepsilon}{3})$ with the same λ. Essentially, Roth argues that $(\frac{1}{2}, \frac{1}{2}, 0)$ is the only reasonable outcome in this game since $(\frac{1}{2}, \frac{1}{2})$ is preferred by $\{1, 2\}$ to any other attainable payoff and $\{1, 2\}$ can achieve $(\frac{1}{2}, \frac{1}{2})$ without the aid of player 3. In this example, $(\frac{1}{3}, \frac{1}{3}, \frac{1}{3})$ is the NTU value because the TU game v_λ corresponding to V_ε is symmetric although the NTU game V_ε is certainly not symmetric. Aumann counters that there are circumstances in which $(\frac{1}{3}, \frac{1}{3}, \frac{1}{3})$ is an expected outcome of bargaining. If coalitions form randomly, then 1 might be willing to strike a bargain with 3 in order to guarantee a positive payoff since, if he were to reject an offer by 3, then 2 might strike a bargain with 3 leaving 1 with nothing. A similar argument holds for 2. This argument is less compelling when ε is very small. Hence, one might argue that 3 should get something but that the payoff to 3 should decline as ε declines. In fact, it is precisely the Harsanyi NTU value that is close to $(\frac{1}{2}, \frac{1}{2}, 0)$ for small ε and, therefore, may better reflect the underlying structure.

EXAMPLE 4.3.2. Let $N = \{1, 2, 3\}$ and for each $\varepsilon \in [0, .2]$ consider a three-person, two-commodity pure exchange economy with initial endowments $a_1 = (1 - \varepsilon, 0)$, $a_2 = (0, 1 - \varepsilon)$ and $a_3 = (\varepsilon, \varepsilon)$ and utility functions $u_1(x, y) = \min\{x, y\} = u_2(x, y)$ and $u_3(x, y) = (1/2)(x + y)$. The NTU game V_ε associated with this economy is:

$V_\varepsilon(i) = \{x \in R^{\{i\}} \mid x^i \leqslant 0\}$ if $i = 1, 2$;
$V_\varepsilon(3) = \{x \in R^{\{3\}} \mid x^3 \leqslant \varepsilon\}$;
$V_\varepsilon(1, 2) = \{x \in R^{\{1,2\}} \mid x^1 + x^2 \leqslant 1 - \varepsilon, x^1 \leqslant 1 - \varepsilon, x^2 \leqslant 1 - \varepsilon\}$;
$V_\varepsilon(i, 3) = \{x \in R^{\{i,3\}} \mid (x^i, x^3) \leqslant \frac{1+\varepsilon}{2}, x^i \leqslant \varepsilon, x^3 \leqslant \frac{1+\varepsilon}{2}\}$ if $i = 1, 2$;
$V_\varepsilon(N) = \{x \in R^N \mid x^1 + x^2 + x^3 \leqslant 1, x^i \leqslant 1 \text{ for each } i\}$.

This example is a simplified version of one that appears in Shafer (1980) and is due to Hart and Kurz (1983). The payoffs at the unique Harsanyi NTU value (evaluated at the corresponding allocation) with comparison vector $(1, 1, 1)$ are: $u_1(\frac{1}{2} - \frac{5\varepsilon}{6}, \frac{1}{2} - \frac{5\varepsilon}{6}) = \frac{1}{2} - \frac{5\varepsilon}{6}$, $u_2(\frac{1}{2} - \frac{5\varepsilon}{6}, \frac{1}{2} - \frac{5\varepsilon}{6}) = \frac{1}{2} - \frac{5\varepsilon}{6}$, and $u_3(\frac{5\varepsilon}{3}, \frac{5\varepsilon}{3}) = \frac{5\varepsilon}{3}$. The payoffs at the unique Shapley NTU value (evaluated at the corresponding allocation) with comparison vector $(1, 1, 1)$ are: $u_1(\frac{5}{12} - \frac{5\varepsilon}{12}, \frac{5}{12} - \frac{5\varepsilon}{12}) = \frac{5}{12} - \frac{5\varepsilon}{12}$, $u_2(\frac{5}{12} - \frac{5\varepsilon}{12}, \frac{5}{12} - \frac{5\varepsilon}{12}) = \frac{5}{12} - \frac{5\varepsilon}{12}$, and $u_3(\frac{1}{6} + \frac{5\varepsilon}{6}, \frac{1}{6} + \frac{5\varepsilon}{6}) = \frac{1}{6} + \frac{5\varepsilon}{6}$.

Shafer argues that the Shapley NTU value is not a reasonable solution in this "economic" example. If $\varepsilon = 0$, then player 3 has no endowment yet he receives a nonzero allocation because of the "utility producing" properties of his utility function. The computation of the Shapley NTU value depends upon the assumption that utility is transferable and this is clearly at variance with the usual notions of ordinality in general equilibrium theory. However, one can use the same reasoning as found in the previous example and argue that if coalitions form randomly, then 3 might reasonably receive a nonzero payoff though such a payoff must be small if ε is small. This is precisely the property exhibited by the Harsanyi NTU value in this example.

What conclusion is to be drawn from these examples? In Hart (1985a), a case has been made that the Harsanyi NTU value is a good solution concept for NTU games with a small number of players while the Shapley NTU value is appropriate for "large" games. The latter statement seems justified in light of the equivalence principle discussed in Chapter 36 and Remark 7.2.7 below.

5. Multilinear extensions for NTU games

5.1. Multilinear extensions and Owen's approach to the NTU value

Another approach to the value of an NTU game is provided in Owen (1972b) and is based on the concept of multilinear extension. Let $v \in G_N$ and define $f_v : [0, 1]^n \to \Re$ as

$$f_v(x_1, \dots, x_n) = \sum_{S \subseteq N} \left[\prod_{i \in S} x_i \prod_{i \notin S} (1 - x_i) \right] v(S).$$

The function f_v is called the *multilinear extension* of v (MLE for short) and is defined in Owen (1972a). For each $i \in N$, it can be shown that

$$\frac{\partial f_v}{\partial x_i} = \sum_{S \subseteq N \setminus i} \prod_{j \in S} x_j \prod_{j \notin S \cup i} (1 - x_j) [v(S \cup i) - v(S)]$$

so that

$$\frac{\partial f_v}{\partial x_i}(t, \dots, t) = \sum_{S \subseteq N \setminus i} t^{|S|}(1 - t)^{n - |S| - 1} [v(S \cup i) - v(S)].$$

Upon integrating with respect to Lebesgue measure, one obtains

$$\int_0^1 \frac{\partial f_v}{\partial x_i}(t, \ldots, t)\, dt = \varphi^i(v).$$

In Owen (1972b), the MLE is defined in an analogous way for NTU games. For each $V \in \Gamma_N$, recall that $V^*(S) = V(S) \times \{0^{N\setminus S}\}$ and (abusing notation slightly) define

$$f_V(x_1, \ldots, x_n) = \sum_{S \subseteq N} \left[\prod_{i \in S} x_i \prod_{i \neq S} (1 - x_i) \right] V^*(S).$$

For each $x \in [0, 1]^n$, $f_V(x)$ is the sum of comprehensive, convex sets and is therefore comprehensive and convex. Suppose that there exists a differentiable function $F_V : \Re^N \times \Re^N \to \Re$ with the property that $\partial f_V(x) = \{u \in \Re^N \mid F_V(u, x) = 0\}$ whenever $x > 0$. That is, F_V implicitly defines the Pareto boundary of $f_V(x)$. To construct a value for V, Owen makes the following informal argument. Bargaining takes place over time. At time $t \in [0, 1]$, player i has accumulated a payoff of $u_i(t)$ as a result of a "commitment" of $x_i(t)$ up to time t. If the vector $u(t)$ is Pareto optimal given $x(t) > 0$, then $F_V(u(t), x(t)) = 0$. If the payoff and commitments of $j \in N\setminus i$ are fixed at time t, then a change in i's commitment from $x_i(t)$ to $x_i(t + \Delta t)$ should result in a corresponding change in i's payoff so as to maintain Pareto optimality. This means that

$$\frac{\partial F_V}{\partial u_i}\left[u_i(t + \Delta t) - u_i(t)\right] + \frac{\partial F_V}{\partial x_i}\left[x_i(t + \Delta t) - x_i(t)\right] \approx 0.$$

If r_i is the rate of change of commitment of player i, then (assuming x, u and F_V are differentiable) we are led to the following system of differential equations for $i \in N$:

$$\dot{u}_i(t) = -\left[\frac{\frac{\partial F_V}{\partial x_i}}{\frac{\partial F_V}{\partial u_i}}\right] r_i(t),$$

$$\dot{x}_i(t) = r_i(t),$$

$$u_i(0) = x_i(0) = 0.$$

The initial conditions are interpreted to mean that at $t = 0$, each player is committed and has accumulated a payoff of zero.

If $(\hat{x}, \hat{u})$ is a solution to this system with $\hat{x}(t) > 0$ for $0 < t \leqslant 1$, then for each such t, $F_V(\hat{u}(t), \hat{x}(t)) = 0$ so that $\hat{u}(t)$ is Pareto optimal for $f_V(\hat{x}(t))$. In particular, $\hat{u}(1)$ is Pareto optimal for $f_V(\hat{x}(1)) = f_V(1, \ldots, 1) = V(N)$. Owen refers to $\hat{u}(1)$ as a "generalized value" and has shown that both the Shapley value and the NBS are generalized values for the appropriate systems of differential equations.

PROPOSITION 5.1.1 [Owen (1972b)]. *Let V correspond to $v \in G_N$. Then for each $x > 0$, $F_V(u, x) = \sum_{i \in N} u_i - f_v(x)$. Furthermore, if $r_i(t) = 1$ for each i, then there is a solution $(\hat{x}, \hat{u})$ to the system of differential equations with $\hat{u}(1) = \varphi(v)$.*

PROOF. If f_v is the MLE of v, then $f_V(x) = \{u \in \Re^N \mid \sum_{i \in N} u_i \leqslant f_v(x)$ whenever $x > 0$ so that $F_V(u, x) = \sum_{i \in N} u_i - f_v(x)$. Hence, $\frac{\partial F_v}{\partial x_i} = -\frac{\partial f_v}{\partial x_i}$ and $\frac{\partial F_v}{\partial u_i} = 1$ so

$$\frac{\mathrm{d}\hat{u}_i}{\mathrm{d}t} = \frac{\partial f_v(\hat{x}_i(t))}{\partial x_i} = \frac{\partial f_v(t, \dots, t)}{\partial x_i}.$$

Upon integrating and using the fact that $\hat{u}(0) = 0$, we obtain

$$\hat{u}_i(1) = \int_0^1 \frac{\partial f_v}{\partial x_i}(t, \dots, t)\,\mathrm{d}t = \varphi^i(v). \qquad \square$$

PROPOSITION 5.1.2 [Owen (1972b)]. *Let $V \in \Gamma_N^B$ with $d_V = 0$. Then for each $x > 0$, $F_V(u, x) = g(u / \prod_{i \in N} x_i)$. Furthermore, if $r_i(t) = 1$ for each i, then there is a solution $(\hat{x}, \hat{u})$ to the system of differential equations with $\hat{u}(1) = \sigma(v)$.*

PROOF. Since $d_V = 0$, it follows that $f_V(x_1, \dots, x_n) = [\prod_{i \in N} x_i]V(N)$. If the Pareto optimal surface of $V(N)$ is defined by $g(u) = 0$, then $F_V(u, x) = g(u / \prod_{i \in N} x_i)$ whenever $x > 0$. Thus,

$$\frac{\partial F_V}{\partial u_i} = \frac{g_i(u / \prod_{i \in N} x_i)}{\prod_{j \in N} x_j}$$

where g_i is the partial derivative of g with respect to the ith argument. In addition,

$$\frac{\partial F_V}{\partial x_i} = -\left[\sum_k g_k\left(\frac{u}{\prod_{j \in N} x_j}\right)u_k\right]\left[\frac{1}{x_i \prod_{j \in N} x_j}\right].$$

Now let u^* be the NBS for V. Since $V(N)$ is convex, it follows that $g(u^*) = 0$ and $u_i^* g_i(u^*) = u_j^* g_j(u^*)$ for each $i, j \in N$. It is now easy to verify that $\hat{x}_i(t) = t$ and $\hat{u}_i(t) = t^n u_i^*$ solve the system of differential equations. Since $\hat{u}(1) = u^*$, we have the desired result. $\qquad \square$

REMARK 5.1.3. In the case of Proposition 5.1.1, it is not difficult to show that the initial value problem associated with F_V has a unique solution (i.e., the one given in the proof of the proposition) and, therefore, the Shapley value is the unique "generalized value" of an NTU game derived from a TU game. The uniqueness question is more complicated in the case of Proposition 5.1.2 due to the behavior of the function $F_V(u, x) = g(u / \prod_{i \in N} x_i)$ near $(u, x) = (0, 0)$. However, Hart and Mas-Colell (1991) have shown that there is a unique solution to the initial value problem associated with this F_V as well. Hence, there is a unique generalized value for games in Γ_N^B which is precisely Nash's Bargaining Solution.

5.2. A non-symmetric generalization

In the previous section, $r_i(t)$ was interpreted as the rate of change of commitment of i. In Owen (1972a), it also is interpreted as the density function of a random variable that describes the time at which i joins a coalition. In Propositions 5.1.1 and 5.1.2, this random variable is assumed to be uniformly distributed on $[0, 1]$ but there are other possibilities. Let $w = (w^1, \ldots, w^n) \in \Re_{++}^N$ and define $r_i(t) = w^i t^{w^i - 1}$. For this density, larger values of the weight parameter correspond to higher likelihoods of "late arrival".

PROPOSITION 5.2.1. *Let V correspond to $v \in G_N$. Then for each $x > 0$, $F_V(u, x) = \sum_{i \in N} u_i - f_v(x)$. Furthermore, if $r_i(t) = w^i t^{w^i - 1}$ for each i, then there is a solution $(\hat{x}, \hat{u})$ to the system of differential equations with $\hat{u}(1) = \varphi_w(v)$.*

PROOF. Following the proof of Proposition 5.1.1, we deduce that

$$\hat{u}_i(1) = \int_0^1 \nabla f_v\left(t^{w^1}, \ldots, t^{w^n}\right) w^i t^{w^i - 1} \, dt.$$

Since this integral is $\varphi_w^i(v)$ [Owen (1972a) or Kalai and Samet (1987)], the result follows. □

PROPOSITION 5.2.2. *Let $V \in \Gamma_N^B$ with $d_V = 0$. Then for each $x > 0$, $F_V(u, x) = g(u / \prod_{i \in N} x_i)$. Furthermore, if $r_i(t) = w^i t^{w^i - 1}$ for each i, then there is a solution $(\hat{x}, \hat{u})$ to the system of differential equations with $\hat{u}(1) = \sigma_w(V)$.*

PROOF. As in Proposition 5.1.2, let $d_V = 0$ and let the Pareto optimal surface of $V(N)$ be given by $g(u) = 0$. Then $\{u^*\} = \sigma_w(V)$ if and only if $g(u^*) = 0$ and

$$g_i(u^*) \frac{u_i^*}{w^i} = g_j(u^*) \frac{u_j^*}{w^j}$$

for each $i, j \in N$. Letting $W = \sum_{i \in N} w^i$, it is now easy to verify that $\hat{x}_i(t) = t^{w^i}$ and $\hat{u}_i(t) = t^W u_i^*$ solve the system of differential equations with $r_i(t) = w^i t^{w^i - 1}$ for each $i \in N$. Since $\hat{u}_i(1) = u_i^*$, we have the desired result. □

5.3. A comparison of the Harsanyi, Owen and Shapley NTU approaches

Although the solution derived from the MLE coincides with the Shapley value on TU games and the NBS on a class of pure bargaining games, it does not coincide with the

NTU value or the Harsanyi value for general NTU games. To see this, consider the following example from Owen (1972b).

EXAMPLE 5.3.1. Let $N = \{1, 2, 3\}$ and define V as follows:
$V(1, 2, 3) = \{u \in \Re^N \mid u_1 + u_2 + u_3 \leqslant 1, \ u_i \leqslant 1, \ u_i + u_j \leqslant 1 \text{ for all } i, j\}$;
$V(1, 2) = \{u \in \Re^{\{1,2\}} \mid u_1 + 4u_2 \leqslant 1, \ u_1 \leqslant 1, \ u_2 \leqslant 1/4\}$;
$V(S) = \{u \in \Re^S \mid u_i \leqslant 0 \text{ for all } i \in S\}$ otherwise.

The value for V computed using the MLE is $\hat{u}(1) = (0.5185, 0.4757, 0.0058)$. However, v_λ is defined for $\lambda = (1, 1, 1)$ and $v_\lambda(N) = 1$, $v_\lambda(1, 2) = 1$ and $v_\lambda(S) = 0$ otherwise, so the Shapley NTU value payoff is $\{(1/2, 1/2, 0)\}$. Furthermore, the Harsanyi NTU payoff is $(2/5, 2/5, 1/5)$ with $\lambda = (1, 1, 1)$ and the associated payoff configuration is $\{x_S\}$ where $x_N = (2/5, 2/5, 1/5)$, $x_{\{1,2\}} = (1/5, 1/5)$ and $x_S = 0$ otherwise.

6. The Egalitarian solution

6.1. Egalitarian values and Egalitarian solutions

Up to now, this chapter has been concerned with solutions for NTU games that yield the NBS and Shapley value on appropriate subclasses. However, one could take any pair of "related solutions" for pure bargaining games and TU games and look for a common extension to a larger class of NTU games. This program has been carried out in Kalai and Samet (1985) where they define the w-Egalitarian solution for NTU games which coincides with the w-Shapley value on Γ_N^{TU} and the w-proportional solution τ^w on Γ_N^B. There are actually two possible approaches to the Egalitarian solution, depending on whether one defines a "solution" to be a payoff vector in $\Re^N$ or a payoff configuration in $\prod_{S \subseteq N} \Re^S$.

DEFINITION 6.1.1. Let $w \in \Re_{++}^N$. A payoff configuration $\{x_S\}$ is a w-*Egalitarian solution* for $V \in \Gamma_N$ if there exists a collection of numbers $\{\alpha_T\}_{\emptyset \neq T \subseteq N}$ such that:
(1) $x_S \in \partial V(S)$ for each $S \subseteq N$,
(2) $x_S^i = w^i \sum_{T \subseteq S: \ i \in T} \alpha_T$ for all $i \in S$ and all $S \subseteq N$.

DEFINITION 6.1.2. A payoff vector $x \in \Re^N$ is a w-*Egalitarian value payoff* for $V \in \Gamma_N$ if there exists a w-Egalitarian solution payoff configuration $\{x_S\}$ for V with $x_N = x$.

REMARK 6.1.3. Comparing Definitions 4.2.4 and 6.1.1, we see that (at least structurally) a w-Egalitarian solution "looks like" a w-Harsanyi solution. As the reader may verify, the following more precise comparison can be made. If $\{x_S\}$ is a w-Egalitarian solution for V and $1_N \cdot x_N = \max[1_N \cdot z \mid z \in V(N)]$, then $\{x_S\}$ is a w-Harsanyi solution with comparison vector 1_N. On the other hand, every (symmetric) Harsanyi solution happens to be a w-Egalitarian solution. In particular, a payoff configuration $\{x_S\}$ is a (symmetric) Harsanyi solution for V if and only if there exists $\lambda \in \mathfrak{R}_{++}^N$ such that $\lambda \cdot x_N = \max[\lambda \cdot z \mid z \in V(N)]$ and $\{x_S\}$ is a w-Egalitarian solution with $w = (1/\lambda^1, \ldots, 1/\lambda^n)$.

The motivation for the definitions of Egalitarian solution and value is best understood by considering a two-person game $V \in \Gamma_N^B$. In this case, a payoff vector $x = (x^1, x^2)$ is a (symmetric) Egalitarian value payoff if:

$$x \in \partial V(1, 2) \quad \text{and} \quad x^1 - d_V^1 = x^2 - d_V^2.$$

Therefore, x is precisely the (symmetric) *proportional solution* $\tau(V)$ for the two-person bargaining problem as described in Section 2.4 [see Kalai (1977b)]. Geometrically, x is the intersection of the 45° line emanating from the point (d_V^1, d_V^2) and the boundary of $V(1, 2)$. In our discussion of the NBS, we stated that an arbitrator in ideal circumstances would like to choose a solution (x^1, x^2) that satisfies utilitarianism (i.e., $x^1 + x^2 = \max[z^1 + z^2 \mid z \in V(1, 2)]$) and equity (i.e., $x^1 - d_V^1 = x^2 - d_V^2$). Since these are generally incompatible, the NBS seeks a point (x^1, x^2) for which there exists some $\lambda \in \mathfrak{R}_{++}^{\{1,2\}}$ such that $\lambda * x$ will simultaneously satisfy these two criteria in the rescaled problem $\lambda * V$. By comparison, the proportional solution maintains the equity condition that $x^1 - d_V^1 = x^2 - d_V^2$ but dispenses with utilitarianism by requiring that x merely be efficient for N. Hence, conditions (1) and (2) of Definition 6.1.1 extend respectively the notions of efficiency and equity as they are interpreted in the proportional solution.

When V corresponds to a TU game v, it follows immediately from the definitions that the w-Egalitarian, w-Shapley and w-Harsanyi NTU value payoffs for V are identical and coincide with the w-Shapley value for V. Therefore, the w-Egalitarian value is a solution concept that extends the w-Shapley value and the w-proportional solution to the class of general NTU games.

REMARK 6.1.4. Definitions 6.1.1 and 6.1.2 translate into an algorithm for computing w-Egalitarian values and solutions. For each i, define $\alpha_{\{i\}} = d_V^i/w^i$. For each S with $|S| \geqslant 2$, inductively define α_S as $\max[t \mid z_S(t) \in V(S)]$ where $z_S^i(t) = w^i[t + \sum_{T \subseteq S, \ T \neq S: \ i \in T} \alpha_T]$. If this process is well defined for each $S \subseteq N$, then $\{z_S(\alpha_S)\}$ is a w-Egalitarian solution payoff configuration and $z_N(\alpha_N)$ is a w-Egalitarian value payoff.

6.2. An axiomatic approach to the Egalitarian value

The correspondence that associates with each V the set of w-Egalitarian value payoffs will be denoted E_w. In this section, we will provide an axiomatization of the w-Egalitarian value correspondence on the class Γ_N^3 (see Section 1.3). From Remark 6.1.4, it follows that $E_w(V)$ can consist of at most one element. Since $E_w(V) \neq \emptyset$ for each $V \in \Gamma_N^3$, we will treat $E_w : \Gamma_N^3 \to \mathfrak{R}^N$ as a function.

To state the axioms, we need a few more definitions and we follow the terminology in Hart (1985c). If $F : \Gamma_N^3 \to \Re^N$ is a function, then the vector $(F(V))^T \in \Re^T$ will be written $F^T(V)$. A coalition $T \subseteq N$ is a *carrier* for V if $V(S) = V(S \cap T) \times \Re_-^{S \setminus T}$ whenever $S \cap T \neq \emptyset$. A game is *individually rational monotonic* (IR monotonic for short) if $V_+(S) \times V_+(i) \subseteq V_+(S \cup i)$ where $V_+(T) = \{x \in V(T) \mid x^i \geq d_V^i,$ for each $i \in T\}$. If $\emptyset \neq T \subseteq N$ and $a \in \Re^N$ with $a^i = 0$ for $i \notin T$, define the NTU game $V_{T,a}$ as follows: $V_{T,a}(S) = \{a^S\} + \Re_-^S$ if $T \subseteq S$ and $V_{T,a}(S) = \Re_-^S$ otherwise. In the axioms below, $F : \Gamma_N^3 \to \Re^N$ is a function and V and W are NTU games.

E1: (*Individual Rationality*) If V is IR monotonic then $F^i(V) \geq d_V^i$ for each $i \in N$.

E2: (*Carrier*) If S is a carrier for V, then $F^S(V) \in \partial V(S)$.

E3: (*Additivity of Endowments*) If $a \in \Re^N$ with $a^i = 0$ for $i \notin T$ and if $F(V_{T,a}) = a$, then $F(V + V_{T,a}) = F(V) + a$.

E4: (*Egalitarian Monotonicity*) If $T \subseteq N$, $V(T) \subseteq W(T)$ and $V(S) = W(S)$ for all $S \neq T$, then $F^T(V) \leq F^T(W)$.

E5: (*Continuity*) F is continuous with respect to the Hausdorff topology on Γ_N^3.

E1 states that the payoff to player i must be individually rational if whenever i joins $S \subseteq N \setminus i$, such an arrangement does not eliminate individually rational payoffs that were available to i and S separately. E2 combines an efficiency property and a null player property. Since any superset T of a carrier S is also a carrier of V, it follows that $F^T(V) \in \partial V(T)$ whenever $S \subseteq T$. In particular, $F(V) \in \partial V(N)$. If S is a carrier of V and $i \notin S$, then E2 implies that $F^i(V) \leq 0$. E2 is an NTU extension of a similar axiom used in the characterization of the TU Shapley value [see Shapley (1953b)]. Axiom E3 has the following interpretation. Let $a \in \Re^N$ with $a^i = 0$ for $i \notin T$ and let V be an NTU game and define $W = V + V_{T,a}$. Then $W(S) = V(S) + \{a^S\}$ if $T \subseteq S$ and $W(S) = V(S)$ otherwise. Hence, W simply translates the payoffs of a superset S of T by a^S and leaves the attainable sets of other coalitions unchanged. If a is acceptable in the sense that $F(V_{T,a}) = a$, then E3 requires that the solution for W be precisely $F(V) + a$. Egalitarian monotonicity states that the members of a coalition T should be no worse off if their attainable set expands while those of all other coalitions remain unchanged. Finally, E5 is a regularity condition.

The main characterization theorem for the Egalitarian value may now be stated.

THEOREM 6.2.1 [Kalai and Samet (1985)]. *A function $F : \Gamma_N^3 \to \Re^N$ satisfies E1–E5 if and only if there exists $w \in \Re_{++}^N$ such that $F = E_w$.*

6.3. An axiomatic approach to the Egalitarian solution

The Shapley and Harsanyi NTU values do not satisfy E3 [see Hart (1985c), p. 318] or E4 [see Kalai and Samet (1985)]. In addition, the Shapley and Harsanyi correspondences have closed graphs but are not continuous. There is also another important distinction between these solutions (first indicated in Kalai (1977b) for σ_w vs τ_w) that lies in the way they respond to the rescaling of utilities. The Shapley NTU value and the Harsanyi solution are scale covariant in the sense of Axioms A4 and H4 while the Egalitarian value and the Egalitarian solution satisfy such a condition only for rescalings that put the same weight on *each* individual's utility.

Despite these differences, the Harsanyi solution and the Egalitarian solution are more closely related than one might believe at first glance. Recall that Theorem 4.2.3 states that a payoff configuration valued correspondence satisfies H1, H2, H3$^+$, H4, H5 and H6 on Γ_N^2 if and only if it is the Harsanyi solution. The next result shows that by dropping H4 (scale covariance) and enlarging the domain of the other axioms to all of Γ_N, we can actually characterize the Egalitarian solution in terms of the axioms used in the characterization of the Harsanyi solution. Let $\widehat{E}_w$ denote the payoff configuration valued correspondence that associates with each $V \in \Gamma_N$ the set of w-Egalitarian solution payoff configurations for V. The symmetric Egalitarian solution correspondence will be written simply as $\widehat{E}$.

THEOREM 6.3.1 [Samet (1985)]. *A correspondence* $\Psi : \Gamma_N \to \prod_{S \subseteq N} \Re^S$ *satisfies H1, H2, H3$^+$, H5 and H6 if and only if* $\Psi = \widehat{E}$.

We can characterize $\widehat{E}_w$ on Γ_N if we modify the unanimity games axiom H3$^+$ in an appropriate way.

H3^w: (*Unanimity Games*) For each real number c and nonempty $T \subseteq N$, let U_T^c be the game corresponding to the TU game cu_T. Then $\Psi(U_T^c) = \{z_S\}$ where $z_S^i = (cw^i / \sum_{j \in T} w^j)$ if $i \in T \subseteq S$ and $z_S^i = 0$ otherwise.

THEOREM 6.3.2 [Samet (1985)]. *A correspondence* $\Psi : \Gamma_N \to \prod_{S \subseteq N} \Re^S$ *satisfies H1, H2, H3^w, H5 and H6 if and only if* $\Psi = \widehat{E}_w$.

7. Potentials and values of NTU games

7.1. Potentials and the Egalitarian value of finite games

In Hart and Mas-Colell (1989), the notion of potential for TU games was introduced. These authors used the potential to provide a new axiomatization of the Shapley value for TU games using the concepts of reduced game and consistency and we briefly review their approach next.

In this section and in Section 8 below, let G denote the set of all pairs (N, v) where N is a finite set and v is a TU game in G_N. If (N, v) is a TU game and $S \subseteq N$, we will write $(S, v|_S)$ simply as (S, v). In a similar fashion, let Γ denote the set of all pairs (N, V) where N is a finite set and V is a NTU game in Γ_N satisfying the nonlevelness condition (3) in Section 1.3. In addition, let Γ^{TU} denote the games (N, V) where $V \in \Gamma_N^{TU}$.

Let $w = \{w^i\}$ be a collection of positive weights. A *w-potential* on G is a function $P_w : G \to \Re$ satisfying $P_w(\emptyset, v) = 0$ and

$$\sum_{i \in N} \left[P_w(N, v) - P_w(N \backslash i, v) \right] = v(N).$$

The following result provides a characterization of the weighted TU Shapley value in terms of potentials.

THEOREM 7.1.1 [Hart and Mas-Colell (1989)]. *There exists one and only one w-potential on G. Furthermore,*

$$w^i \left[P_w(N, v) - P_w(N \backslash i, v) \right] = \varphi_w^i(N, v)$$

for all $i \in N$.

The definition of potential can be extended to NTU games. A *w-potential* on Γ is a function $P_w : \Gamma \to \Re$ satisfying $P_w(\emptyset, V) = 0$ and

$$\left(w_i[P_w(N, V) - P_w(N \backslash i, V)] \right)_{i \in N} \in \partial V(N)$$

whenever $N \neq \emptyset$.

THEOREM 7.1.2 [Hart and Mas-Colell (1989)]. *There exists one and only one w-potential on Γ. Furthermore,*

$$w_i \left[P_w(N, V) - P_w(N \backslash i, V) \right] = E_w^i(N, V)$$

for all $i \in N$.

Recall (see Remark 6.1.3) that a payoff configuration $\{x_S\}$ is a (symmetric) Harsanyi solution for V if and only if there exists $\lambda \in \Re_{++}^N$ such that $\lambda \cdot x_N = \max[\lambda \cdot z \mid z \in V(N)]$ and $\{x_S\}$ is a *w*-Egalitarian solution with $w = (1/\lambda^1, \ldots, 1/\lambda^n)$. Hence, we have the following characterization of Harsanyi NTU value payoffs in terms of potentials. If $\lambda \in \Re_{++}^N$, let $\lambda^{-1} := (1/\lambda^1, \ldots, 1/\lambda^n)$.

THEOREM 7.1.3 [Hart and Mas-Colell (1989)]. *A payoff vector $x \in \Re^N$ is a Harsanyi NTU value payoff vector for the game (N, V) if and only if there exists a collection of positive weights $\{\lambda^i\}$ such that*

$$\lambda^i x^i = P_{\lambda^{-1}}(N, V) - P_{\lambda^{-1}}(N \backslash i, V) \quad \text{for each } i$$

and

$$\lambda \cdot x = \max_{z \in V(N)} \lambda \cdot z.$$

7.2. *Potentials and the Egalitarian solution of large games*

An *NTU game form* is a correspondence $V : \mathfrak{R}^n_+ \to \mathfrak{R}$ and a finite type NTU game with a continuum of players is defined as a pair (α, V) where $\alpha \in \mathfrak{R}^n_{++}$ and $V : \mathfrak{R}^n_+ \to \mathfrak{R}$ is a game form. A game form has the following interpretation. Each *profile* $x \in \mathfrak{R}_+$ is interpreted as a "coalition" and x^i represents the total "number" (i.e., mass) of players of type i in the coalition. If $a \in V(x)$, then a^i represents the per capita payoff to the players of type i in the coalition x. Hence, $a^i x^i$ is the total payoff to the type i players in coalition x.

A *solution function* for a game form V is a mapping $\xi : \mathfrak{R}^n_{++} \to \mathfrak{R}^n$ such that $\xi(x) \in \partial V(x)$ for each $x \in \mathfrak{R}^n_{++}$. In particular, $\xi^i(x)$ is interpreted as the payoff to each type i player and the total payoff to the players of type i is therefore $x^i \xi^i(x)$. The characterization of the Egalitarian solution for finite NTU games in terms of a potential function presented in Theorem 7.1.2 suggests an extension to the case of large games with finitely many types. If V is a game form, then a solution function $\xi : \mathfrak{R}^n_{++} \to \mathfrak{R}^n$ for V is Egalitarian if ξ admits a potential, i.e., if there exists a differentiable function $P : \mathfrak{R}^n_{++} \to \mathfrak{R}$ such that

$$\xi(x) = \nabla P(x)$$

for each $x \in \mathfrak{R}^n_{++}$. Thus, we are ideally interested in the existence of a differentiable function P such that $\nabla P(x) \in \partial V(x)$ for each $x \in \mathfrak{R}^n_{++}$. Unfortunately [as shown in Hart and Mas-Colell (1995a)], such a function may not exist, even for "well-behaved" game forms. This necessitates a weaker definition of solution and potential.

DEFINITION 7.2.1. Let V be a game form. A *generalized solution function* for V is a mapping $\xi : \mathfrak{R}^n_{++} \to \mathfrak{R}^n$ such that $\xi(x) \in \partial V(x)$ for almost all $x \in \mathfrak{R}^n_{++}$.

DEFINITION 7.2.2. A function $P : \mathfrak{R}^n_{++} \to \mathfrak{R}$ is a *Lipschitz potential* for V if P is Lipschitz continuous and $\nabla P(x) \in \partial V(x)$ for each $x \in \mathfrak{R}^n_{++}$ at which P is differentiable.

Every Lipschitz potential for V defines a generalized solution since (by the classical Rademacher theorem) a Lipschitz continuous function is differentiable almost everywhere. Even if we content ourselves with Lipschitz potentials (and their induced generalized solutions), we are still left with the problem of choosing a reasonable candidate from among the different Lipschitz potentials that may exist for a given game form. Taking their cue from certain problems in the theory of partial differential equations [see, e.g., Fleming (1969)], Hart and Mas-Colell (1995a) introduce the *variational potential* and show that, while not necessarily the unique Lipschitz potential, it does satisfy many desirable properties and is the "right" potential if one approaches the problem from an asymptotic viewpoint.

To construct the variational potential, let $v : \mathfrak{R}^n_{++} \times \mathfrak{R}^n \to \mathfrak{R} \cup \{\infty\}$ denote the support function of the set $V(x)$, i.e.,

$$v(x, p) = \sup_{a \in V(x)} p \cdot a.$$

Next, let $AC(x)$ denote the set of absolutely continuous functions $\gamma : [0, 1] \to \mathfrak{R}^n_{++} \cup \{0\}$ satisfying $\gamma(0) = 0$ and $\gamma(1) = x$. The *variational potential* for V is defined for each $x \in \mathfrak{R}^n_{++}$ as follows:

$$P^*(x) = \min_{\gamma(\cdot) \in AC(x)} \int_0^1 v(\gamma(t), \dot{\gamma}(t)) \, dt.$$

In order for $P^*(x)$ to be well defined (and have other desirable properties), the game form V must satisfy certain restrictions. Let $\mathcal{G}$ denote the set of game forms $V : \mathfrak{R}^n_+ \to \mathfrak{R}$ satisfying $(A.1)^*$–$(A.3)^*$ and $(A.4)$–$(A.9)$ in Hart and Mas-Colell (1995b). These assumptions include the familiar requirements that, for each $x \in \mathfrak{R}^n_+$, $V(x)$ be a nonempty, closed, convex, comprehensive, proper subset of $\mathfrak{R}^n$. However, these assumptions also guarantee that the support function of the set $V(x)$ satisfies certain regularity requirements, including differentiability in x.

THEOREM 7.2.3 [Hart and Mas-Colell (1995a)]. *Let $V \in \mathcal{G}$. For each $x \in \mathfrak{R}^n_{++}$, $P^*(x)$ is well defined and the resulting function $P^* : \mathfrak{R}^n_{++} \to \mathfrak{R}$ is Lipschitz continuous hence differentiable almost everywhere. Furthermore, $\nabla P^*(x) \in \partial V(x)$ whenever P^* is differentiable at x.*

To illustrate the variational potential, suppose that V is a *TU game form* defined by

$$V(x) = \left\{ a \in \mathfrak{R}^n \mid \sum_{i=1}^n a^i x^i \leqslant f(x) \right\}$$

where $f : \mathfrak{R}^n_{++} \to \mathfrak{R}$ is a (sufficiently well behaved) differentiable function. For each t, the support function v evaluated at $(x, p) = (\gamma(t), \dot{\gamma}(t))$ is finite only if for each i, $\dot{\gamma}^i(t) = \beta(t)\gamma^i(t)$ for some $\beta(t) \geqslant 0$. If we choose the normalization $\beta(t) = 1/t$, then $\gamma_x = tx$ is a solution to this system of differential equations satisfying the boundary conditions $\gamma_x(0) = 0$ and $\gamma_x(1) = x$. Therefore,

$$v(\gamma_x(t), \dot{\gamma}_x(t)) = \tfrac{1}{t} f(tx)$$

so that

$$P^*(x) = \int_0^1 v\big(\gamma_x(t), \dot{\gamma}_x(t)\big)\, dt = \int_0^1 \frac{1}{t} f(tx)\, dt$$

and we obtain

$$\frac{\partial P^*}{\partial x_i}(\alpha) = \int_0^1 \frac{\partial f}{\partial x_i}(t\alpha)\, dt$$

for each $\alpha \in \Re_{++}^n$. For finite NTU games generated by TU games, the Egalitarian solution coincides with the TU Shapley value. In the case of NTU game forms, the Egalitarian solution of a game form generated by a finite-type nonatomic TU game coincides with the Aumann–Shapley value and is obtained by the familiar diagonal formula.

In summary, each $V \in \mathcal{G}$ admits at least one Lipschitz potential, the variational potential P^*, and the variational potential can be used to define a generalized solution for the game (α, V). The variational potential may not be the unique Lipschitz potential even for well-behaved game forms [see Section V.8 of Hart and Mas-Colell (1995a)] so a given game form may have different generalized solutions corresponding to different Lipschitz potentials. However, the variational potential does exhibit several features that indicate that it is the "right" definition of Egalitarian solution, especially for applications. Some of these are summarized in the following result:

THEOREM 7.2.4 [Hart and Mas-Colell (1995a)]. *Suppose that $V \in \mathcal{G}$.*
  (i) *If Q is a Lipschitz potential for V, then $P^*(x) \geqslant Q(x)$ for all $x \in \Re_{++}^n$.*
  (ii) *If V admits a continuously differentiable potential P, then $P = P^*$.*
  (iii) *If V is a hyperplane game form, then the variational potential P^* is the unique Lipschitz potential for V.*

A complete specification of an Egalitarian solution requires that we define the solution for all profiles $x \in \Re_{++}^n$, even those at which P^* is not differentiable. If we are willing to allow the solution to be set-valued at points at which P^* is not differentiable, then the Lipschitz continuity of P^* suggests that Clarke's generalized gradient can be used to complete the description of an Egalitarian solution based on the variational potential.

DEFINITION 7.2.5. For each $\alpha \in \Re_{++}^n$ and each $V \in \mathcal{G}$, the *Egalitarian solution* $E^*(\alpha, V)$ is defined as

$$E^*(\alpha, V) = \big[\nabla^C P^*(\alpha) + \Re_+^n\big] \cap \partial V(\alpha)$$

where $\nabla^C P^*(x)$ denotes the Clarke generalized gradient at x [see Clarke (1983)].

Hart and Mas-Colell (1995a) prove that $\nabla^C P^*(x) = \{\nabla P^*(x)\}$ whenever P^* is differentiable at x. Since

$$\nabla^C P^*(x) = \operatorname{conv}\left\{y \mid y = \lim_k \nabla P^*(x_k),\ x_k \to x \text{ and } P^* \text{ is differentiable} \right.$$
$$\left. \text{at each } x_k \right\},$$

it follows that $[\nabla^C P^*(x) + \mathfrak{R}^n_+] \cap \partial V(x) \neq \emptyset$ if P^* is not differentiable at x.

Another strong justification for the variational Egalitarian solution is provided in Hart and Mas-Colell (1995b) where they develop an asymptotic approach to the definition of Egalitarian solution for a game form V. In the spirit of the asymptotic value of nonatomic TU games, the game form V is approximated by a sequence of finite NTU games $\{V_r\}$. If the associated sequence of Egalitarian solutions $\{E(V_r)\}$ has a limit, then that limit is defined as the "asymptotic Egalitarian solution" of the game form V.

Let $\alpha \in \mathfrak{R}^n_{++}$ be a profile in which each α_i is a positive integer. For each positive integer r, construct a finite NTU game (N_r, V_r) as follows. The player set N_r consists of $r\alpha_i$ players of type i and the total number of players is $|N_r| = r(\sum_i \alpha_i)$. For each $S \subseteq N_r$, let $m_i(S)$ denote the number of type i players in S and let $\mu^r(S) := (m_1(S)/r, \ldots, m_n(S)/r)$. Hence, $\mu^r(N_r) = \alpha$. The set $V_r(S)$ is defined as follows: $z \in V_r(S)$ if and only if there exists $(a_1, \ldots, a_n) \in V(\mu^r(S))$ such that $z_j = a_i$ whenever player $j \in N_r$ is of type i. As r increases the finite games (N_r, V_r) become better approximations of the continuum game (α, V). Since the potential preserves the symmetries of V_r, it is completely determined by the values of $\mu^r(S)$. To define the potential, let

$$L_r = \left\{(z_1, \ldots, z_n) \mid \text{for each } i,\ 0 \leqslant z_i \leqslant \alpha_i \text{ and } rz_i \text{ is an integer}\right\}.$$

The potential of (N_r, V_r) is a function $P_r : L_r \to \mathfrak{R}$ satisfying $P_r(0) = 0$ and for each $z \in L_r$,

$$DP_r(z) = \left(D_i P_r(z)\right)^n_{i=1} := \left(P_r(z) - P_r\left(z^{(i)}\right)\right)^n_{i=1} \in \partial V(z)$$

where

$$z^{(i)} = \left(z^1, \ldots, z^{i-1}, z^i - \tfrac{1}{r}, z^{i+1}, \ldots, z^n\right)$$

if $z^i \geqslant 1/r$ and $z^{(i)} = z$ otherwise.

THEOREM 7.2.6 [Hart and Mas-Colell (1995b)]. *Suppose that $V \in \mathcal{G}$. Then for each $\alpha \in \mathfrak{R}^N_{++}$:*
  (a) $\lim_{r \to \infty}(1/r) P_r(\alpha) = P^*(\alpha)$.
  (b) *If P^* is differentiable at α, then $\lim_{r \to \infty} DP_r(\alpha) = \nabla P^*(\alpha)$.*
  (c) $\lim_{r \to \infty} \operatorname{dist}[DP_r(\alpha), E^*(\alpha, V)] = 0$,
      where the $\operatorname{dist}[DP_r(\alpha), E^*(\alpha, V)] := \inf\{\|DP_r(\alpha) - \xi\| \mid \xi \in E^*(\alpha, V)\}$.

Property (a) states that the potentials of the approximating games converge to the variational potential P^* of (α, V) for every $\alpha \in \mathfrak{R}_{++}^N$. If P^* is differentiable at α, then $E^*(\alpha, V) = \{\nabla P^*(\alpha)\}$. Property (b) says that if P^* is differentiable at α, then the sequence of Egalitarian values for the approximating sequence of games converges to precisely the single Egalitarian value of (α, V) as defined by the variational potential. If P^* is not differentiable at α, then the sequence of Egalitarian values for the approximating sequence of games may not converge. However, every convergent subsequence of $\{DP_r(\alpha)\}_{r=1}^{\infty}$ will have a limit in $[\nabla^C P^*(\alpha) + \mathfrak{R}_+^n] \cap \partial V(\alpha)$.

REMARK 7.2.7. Given the relationship between the Harsanyi NTU value and the Egalitarian value for finite player games (see Remark 6.1.3), it is natural to propose an analogous notion of Harsanyi value for finite type NTU games with a continuum of players using the variational potential. This is accomplished in Hart and Mas-Colell (1996b) where the authors show that the Walras equilibria of a large economy need not be Harsanyi value allocations. This stands in contrast to the Value Equivalence Principle: in an economy with many agents and smooth preferences the Walras equilibria and the Shapley NTU value allocations coincide. For the special class of "diagonal Harsanyi value payoffs", Imai (1983) does provide an equivalence result for finite-type market games.

8. Consistency and values of NTU games

8.1. Consistency and TU games: The Shapley value

Many solutions of TU games have axiomatic characterizations that employ some notion of "reduced game" and "consistency". [For a survey see Driessen (1991).] Let (N, v) be a TU game, let S be a subset of N and let ψ be a solution. Loosely speaking, a reduced game v_S^{ψ} is a game with player set S derived from v by paying the players outside S their payoffs according to ψ. A solution ψ is consistent if the payoff to a player in S when ψ is applied to v_S^{ψ} is the same as that player's payoff when ψ is applied to the original game v. There are several possible constructions of a reduced game and Hart and Mas-Colell (1989) provide a definition that is appropriate for the study of the Shapley value. In particular, Hart and Mas-Colell show that the Shapley value is the unique solution that is "standard" for two-person games and is consistent with respect to their definition of reduced games.

Let G and Γ be defined as in Section 7.1 above. A *solution function* ψ on G is a mapping that associates with each $(N, v) \in G$ a payoff vector $\psi(N, v) = (\psi^i(N, v))_{i \in N} \in \mathfrak{R}^N$. A solution function ψ on Γ is a mapping that associates with each $(N, V) \in \Gamma$ a payoff vector $\psi(N, V) = (\psi^i(N, V))_{i \in N} \in \mathfrak{R}^N$. If (N, v) is a TU game, ψ is a solution function on G, and $T \subseteq N$, define the *reduced game* (T, v_T^{ψ}) as follows:

$$v_T^{\psi}(S) = \begin{cases} v(S \cup (N \backslash T)) - \sum_{i \in N \backslash T} \psi^i(S \cup (N \backslash T), v), & \text{if } \emptyset \neq S \subseteq T; \\ 0, & \text{if } S = \emptyset. \end{cases}$$

A solution function ψ on G is *consistent* if for each game (N, v) and each $T \subseteq N$,

$$\psi^i\left(T, v_T^\psi\right) = \psi^i(N, v) \quad \text{for all } i \in T.$$

THEOREM 8.1.1 [Hart and Mas-Colell (1989)]. *A solution function ψ on G is consistent and coincides with the Shapley value on two-person games if and only if $\psi(N, v) = \varphi(N, v)$ whenever $(N, v) \in G$.*

8.2. Consistency and NTU games: The Egalitarian value

The definition of reduced game and consistency can be extended to NTU games in a natural fashion. For each solution function ψ on Γ, each game $(N, V) \in \Gamma$ and each $T \subseteq N$, define the *reduced game* (T, V_T^ψ) as follows:

$$V_T^\psi(S) = \left\{ x \in \Re^S \mid \left(x, \left(\psi^j\left(S \cup (N \backslash T), V\right)_{j \in N \backslash T}\right) \in V\left(S \cup (N \backslash T)\right)\right) \right\}.$$

A solution function ψ on Γ is *consistent* if, for each (N, V) and $T \subseteq N$,

$$\psi^i\left(T, V_T^\psi\right) = \psi^i(N, V) \quad \text{for all } i \in T.$$

The definitions of reduced game and consistency coincide with those given in Section 8.1 if V corresponds to a TU game. The following result extends Theorem 8.1.1 to the NTU case.

THEOREM 8.2.1 [Hart and Mas-Colell (1989)]. *A solution function ψ on Γ is consistent and coincides with the proportional solution on two-person games if and only if $\psi(N, V) = E(N, V)$ whenever $(N, V) \in \Gamma$.*

8.3. The consistent value for hyperplane games

Since the NTU definition of consistency reduces to TU consistency, we can apply Theorem 8.1.1 and make a simple observation: any solution function on Γ that satisfies consistency and coincides with the TU Shapley value on two-player TU games must coincide with the TU Shapley value on the class of TU games. Theorem 8.2.1 illustrates this idea since any solution that is proportional on two-player games must coincide with the Shapley value on two-player TU games. A natural question now arises: does there exist a solution function on Γ that satisfies consistency and coincides with the Nash Bargaining Solution for two-player problems? If this question has an affirmative answer, then the resulting solution would be different from the Egalitarian solution function but would still coincide with the TU Shapley value on TU games. In fact, Maschler and Owen (1992) have shown that such a solution function does not exist. In particular, they have shown that consistency, symmetry, efficiency and scale covariance are incompatible. However, there does exist a solution function that extends the TU Shapley Value and the NBS as we will see in Section 8.4 below.

In order to construct this solution, we need to extend the concept of marginal contribution to NTU games. Let (N, V) be an NTU game. Suppose that $R = (i_1, \ldots, i_n)$ is an element of Ω and recall that $d_V^i = \max\{x^i \mid x^i \in V(i)\}$. The *marginal contribution* of i_k in R in the game V is defined inductively as follows:

$$\Delta^{i_1}(R, V) = d_V^{i_1}$$

and

$$\Delta^{i_k}(R, V) = \max\left\{x^{i_k} \mid \left(x^{i_k}, \left(\Delta^j(R, v)\right)_{j \in \{i_1, \ldots, i_{k-1}\}}\right) \in V(i_1, \ldots, i_{k-1}, i_k)\right\}$$

for each $k \geqslant 2$. Rearranging the numbers $\Delta^{i_1}(R, V), \ldots, \Delta^{i_n}(R, V)$ yields the marginal worth vector

$$\delta(R, V) := \left(\Delta^1(R, V), \ldots, \Delta^n(R, V)\right).$$

Since $\delta(R, V) \in \partial V(N)$ for each ordering R, a natural random order value for V is defined by $\sum_{R \in \Omega} \frac{1}{n!} \delta(R, V)$. However, this point is not guaranteed to be in $\partial V(N)$ unless $\partial V(N)$ is convex. If $V \in \Gamma^{TU}$, then $\partial V(N)$ is convex and $\Delta^i(R, V) = \Delta^i(R, v)$ (as defined in Section 1.2) where v generates V. In this case, $\sum_{R \in \Omega} \frac{1}{n!} \delta(R, V)$ is precisely the Shapley value payoff vector for the TU game that generates V. We will present a random order solution for general NTU games in the next section. First, we will consider a class of games introduced in Maschler and Owen (1989) on which they have defined a random order value that satisfies a weakened version of consistency.

An NTU game (N, V) is a *hyperplane game* if for each $S \subseteq N$, there exists a vector $a_S \in \mathfrak{R}_{++}^S$ and a real number $b(S)$ such that

$$V(S) = \left\{x \in \mathfrak{R}^S \mid a_S \cdot x \leqslant b(S)\right\}.$$

The set of all hyperplane games in Γ is denoted Γ^H. Hyperplane games are NTU games associated with "prize games" [see Hart (1994)]. The *consistent value* on Γ^H is the solution function $K : \Gamma^H \to \mathfrak{R}^n$ defined as

$$K(N, V) = \sum_{R \in \Omega} \frac{1}{n!} \delta(R, V).$$

Note that $K(N, V) \in \partial V(N)$ if $(N, V) \in \Gamma^H$ since every marginal worth vector $\delta(R, V) \in \partial V(N)$ and $\partial V(N)$ is convex. Furthermore, $K(N, V) = \varphi(N, v)$ if $V \in \Gamma^{TU}$ and v is the TU game that generates V.

The consistent value on Γ^H does not satisfy the Hart–Mas-Colell definition of consistency but it does satisfy two weaker notions of consistency.

DEFINITION 8.3.1. A solution function ψ on Γ is *completely consistent* if for each $(N, V) \in \Gamma$,

$$\sum_{S \subseteq N :\, |S|=k,\, i \in S} \psi^i \left(S, V_S^\psi \right) = \binom{|N| - 1}{k - 1} \psi^i(N, V) \quad \text{for all } i \in N$$

whenever $1 \leqslant k \leqslant |N|$.

DEFINITION 8.3.2. A solution function ψ on Γ is *globally consistent* if for each $(N, V) \in \Gamma$,

$$\sum_{\substack{S \subseteq N \\ S \neq \emptyset}} \psi^i \left(S, V_S^\psi \right) = 2^{n-1} \psi^i(N, V) \quad \text{for all } i \in N.$$

Clearly, consistency implies complete consistency and complete consistency implies global consistency. Complete consistency is defined in Maschler and Owen (1989) (where it is called consistency) and global consistency is defined in Orshan (1992). Global consistency can be used to axiomatize the consistent value on Γ^H. Players i and j are *substitutes* in V if for every $S \subseteq N \setminus \{i, j\}$, the following condition holds: for every $x_S \in \Re^S$ and for every $\xi \in \Re$,

$$(x_S, \xi) \in V(S \cup i) \quad \text{if and only if} \quad (x_S, \xi) \in V(S \cup j).$$

If V corresponds to a TU game v, then i and j are substitutes in V if and only if they are substitutes in v as defined in Section 1.2. If $a \in \Re^N$, let (N, V_a^{add}) be the game defined by

$$V_a^{add}(S) = \left\{ x \in \Re^S \mid \sum_{i \in S} x_i \leqslant \sum_{i \in S} a_i \right\}.$$

In the axioms below, ψ is a solution function on Γ.

C1: (*Efficiency*) $\psi(N, V) \in \partial V(N)$.
C2: (*Symmetry*) $\psi^i(N, V) = \psi^j(N, V)$ whenever i and j are substitutes in V.
C3: (*Scale Covariance*) For each $\lambda \in \Re_{++}^N$ and each $a \in \Re^N$,

$$\psi \left(N, \lambda * V + V_a^{add} \right) = \lambda * \psi(N, V) + \{a\}.$$

THEOREM 8.3.3 [Maschler and Owen (1989), Orshan (1992)]. *A solution function ψ on Γ^H satisfies C1, C2, C3 and global consistency if and only if $\psi = K$.*

Hart (1994) provides yet another axiomatization of K on Γ^H and to present that result, we need one more definition. For each i and each S containing i, define:

$$D^i(S, V) = \max\{x^i \mid (x^i, \psi(S\backslash i, V)) \in V(S)\}.$$

Note that $D^i(S, V)$ is uniquely determined by the condition $(D^i(S, V), \psi(S\backslash i, V)) \in \partial V(S)$.

C4: (*Marginality*) If (N, V) and (N, W) are games and if $D^i(S, V) = D^i(S, W)$ for each $S \subseteq N$, then $\psi^i(N, V) = \psi^j(N, W)$.

If ψ is efficient and if V corresponds to a TU game, then $D^i(S, V) = v(S) - v(S\backslash i)$ and C4 coincides with the marginality axiom for TU games found in Young (1988). In fact, the next result is an extension of Theorem 1 in Young (1988) to the hyperplane games.

THEOREM 8.3.4 [Hart (1994)]. *A solution function ψ on Γ^H satisfies C1, C2 and C4 if and only if $\psi = K$.*

8.4. The consistent value for general NTU games

We can now use the consistent value for hyperplane games as a tool to extend the consistent value to a class of general NTU games. Let (N, V) be an NTU game and let $\Lambda = \{\lambda_S\}_{S \subseteq N}$ be a collection with $\lambda_S \in \mathfrak{R}^S_{++}$. We will call such a collection a *comparison system*. If $\sup[\lambda_S \cdot z \mid z \in V(S)]$ is finite for each S, then we can define a hyperplane game V_Λ where, for each $S \subseteq N$,

$$V_\Lambda(S) = \{x \in \mathfrak{R}^S \mid \lambda_S \cdot x \leqslant \sup[\lambda_S \cdot z \mid z \in V(S)]\}.$$

In this case, we say that the hyperplane game V_Λ is *defined for* V. If V_Λ is defined for V, then the restriction (S, V_Λ) is a hyperplane game with player set S and it follows that $K(S, V_\Lambda) \in \partial V_\Lambda(S)$ for each $S \subseteq N$. As with the other solution concepts we have presented, we will distinguish between values as payoff vectors and solutions as payoff configurations.

DEFINITION 8.4.1. A payoff configuration $\{x_S\}$ is a *consistent solution payoff configuration* for (N, V) if there exists a comparison system $\Lambda = \{\lambda_S\}_{S \subseteq N}$ such that
  (i) $x_S \in V(S)$ for all S;
  (ii) V_Λ is defined for V;
  (iii) $x_S = K(S, V_\Lambda)$ for all S.

DEFINITION 8.4.2. A payoff vector $x \in \mathfrak{R}^n$ is a *consistent value payoff* for (N, V) if there exists a consistent solution payoff configuration $\{x_S\}$ such that $x_N = x$.

Every hyperplane game (hence every game generated by a TU game) has a unique consistent payoff configuration and a unique consistent value. If (N, V) is a hyperplane game and if $\{a_S\}_{S \subseteq N}$ is the associated collection of normals, then V_Λ is defined for V if and only if $\Lambda = \{ta_S\}_{S \subseteq N}$ for some $t > 0$. Hence, $V_\Lambda = V$ whenever V_Λ is defined, so the only possible consistent solution is the payoff configuration $\{K(S, V)\}_{S \subseteq N}$. Since V is a hyperplane game, it follows that $K(S, V) \in \partial V(S)$ for each S and we conclude that $\{K(S, V)\}_{S \subseteq N}$ is the unique consistent solution payoff configuration for V and that $K(N, V)$ is the unique consistent value payoff.

An alternative characterization of consistent solution payoff configurations is possible.

THEOREM 8.4.3 [Hart and Mas-Colell (1996a)]. *Let* (N, V) *be an NTU game. A payoff configuration* $\{x_S\}$ *is a consistent solution payoff configuration for V if and only if there exists a comparison system* $\Lambda = \{\lambda_S\}_{S \subseteq N}$ *such that:*

(i) $x_S \in \partial V(S)$ *for each* $S \subseteq N$;
(ii) V_Λ *is defined for V and* $\lambda_S \cdot x_S = \max[\lambda_S \cdot z \mid z \in V(S)]$ *for each* $S \subseteq N$;
(iii) *For each* $S \subseteq N$ *and for each* $i \in S$,

$$\sum_{j \in S \setminus i} \lambda_S^i \big(x_S^i - x_{S \setminus j}^i\big) = \sum_{j \in S \setminus i} \lambda_S^j \big(x_S^j - x_{S \setminus i}^j\big).$$

REMARK 8.4.4. Suppose that $V \in \Gamma_N^B$ and suppose that $\{x_S\}$ is a consistent solution payoff configuration for V with an associated comparison system $\{\lambda_S\}_{S \subseteq N}$. Conditions (i) and (ii) imply that $x_S = (d_V^i)_{i \in S}$ for each $S \subseteq N, S \neq N$ and condition (iii) implies that

$$\sum_{j \in N \setminus i} \lambda_N^i \big(x_N^i - d_V^i\big) = \sum_{j \in N \setminus i} \lambda_N^j \big(x_N^j - d_V^j\big).$$

Therefore, $\lambda_N^i(x_N^i - d_V^i) = \frac{1}{n} \sum_{j \in N} \lambda_N^j(x_N^j - d_V^j)$ for each $i \in N$. Applying condition (ii) to $S = N$, we conclude (as in the proof of Proposition 3.1.3) that $x_N = \sigma(V)$.

Hence, the consistent value coincides with the Shapley value on the TU games and the Nash Bargaining Solution on the pure bargaining games. Maschler and Owen (1992) provide an existence theorem for the consistent value using a fixed-point argument and Owen (1994) presents an example in which the consistent value is not a singleton set. The consistent value does not satisfy consistency on Γ (since the consistent value does not satisfy consistency on Γ^H). In addition, Owen (1994) shows that the consistent value does not satisfy complete consistency on Γ. A consistency based axiomatization on a suitably large class of NTU games that contains Γ^H and coincides with K on Γ^H remains an open problem.

8.5. Another comparison of NTU values

Recall that, in the game V_ε of Example 4.3.1, the Shapley NTU value payoff is $(\frac{1}{3}, \frac{1}{3}, \frac{1}{3})$ with comparison vector $\lambda = (1, 1, 1)$ and that the Harsanyi NTU value payoff is $(\frac{1}{2} - \frac{\varepsilon}{3}, \frac{1}{2} - \frac{\varepsilon}{3}, \frac{2\varepsilon}{3})$ with the same λ. Roth argues that $(\frac{1}{2}, \frac{1}{2}, 0)$ is the only reasonable outcome while Aumann counters that there are circumstances in which $(\frac{1}{3}, \frac{1}{3}, \frac{1}{3})$ is reasonable as an expected outcome, though this argument is less compelling when ε is very small. Hence, one might argue that 3 should get something but that the payoff to 3 should decline as ε declines, a feature captured by the Harsanyi NTU value in this example.

The unique consistent value payoff vector is $(\frac{1}{6} + \frac{\varepsilon}{3}, \frac{1}{6} + \frac{\varepsilon}{3}, \frac{2}{3}(1 - \varepsilon))$ with comparison system $\Lambda = \{1_S\}$. This solution can be justified using an argument based on a repeated bargaining model. Suppose that, with equal probability, one player is chosen randomly to propose a feasible payoff in $V_\varepsilon(N)$ and suppose that player i has been chosen. If i's first-round proposal is unanimously accepted, then the bargaining game ends and the players receive their specified payoffs. If at least one player rejects i's proposal, then i exits the game with probability ρ and stays in the game with probability $1 - \rho$. If S is the set of remaining players (so $S = N$ with probability $1 - \rho$ and $S = N \backslash i$ with probability ρ), then a member of S (possibly i if $S = N$) is chosen with equal probability to propose a feasible payoff in $V_\varepsilon(S)$ in the second round, and the process repeats itself. When a player exits, he receives a payoff of 0 and never reenters the bargaining process.

In order to obtain payoffs of $1/2$ in this bargaining game, players 1 and 2 need the agreement of player 3 (since $(\frac{1}{2}, \frac{1}{2}, 0) \in V_\varepsilon(N)$) or they need player 3 to have exited the game (since $(\frac{1}{2}, \frac{1}{2}) \in V_\varepsilon(1, 2)$). If 2 exits before 1 and 3, however, then 1 ends up with a payoff of ε (if he strikes a bargain with 3) or 0 (if he fails to strike a bargain with 3). Hence, 1 and 2 are concerned with striking a deal before 3 exits the bargaining process. In the extreme case of $\varepsilon = 0$, player 3 is in a powerful position. Players 1 and 2 may be willing to make a relatively large first-round concession to 3 (which 3 accepts) in order to avoid a payoff of 0 if 3 is not the first player to exit. This is reflected in the consistent value payoff of $(\frac{1}{6}, \frac{1}{6}, \frac{2}{3})$ in the game V_0. In the other extreme case in which $\varepsilon = 1$, players 1 and 2 can guarantee themselves a payoff of $1/2$ each, with or without the presence of player 3. Again, this situation is reflected in the consistent value payoff of $(\frac{1}{2}, \frac{1}{2}, 0)$ in the game V_1 and supports Roth's (static) argument in favor of $(\frac{1}{2}, \frac{1}{2}, 0)$. Finally, the game $V_{1/2}$ is completely symmetric and the consistent value of $(\frac{1}{3}, \frac{1}{3}, \frac{1}{3})$ reflects this symmetry. Note that, unlike the Shapley NTU value, the consistent value specifies symmetric payoffs only when the underlying NTU game itself is symmetric.

The justification for the consistent value in this example as the outcome of a dynamic bargaining model extends to general NTU games. In Hart and Mas-Colell (1996a), the authors present a dynamic bargaining model as described above and prove that, for a large class of NTU games (basically the class Γ_N^1), every limit point (as $\rho \to 0$) of a sequence of subgame perfect equilibrium payoff configurations is a consistent value payoff configuration. If the NTU game is a pure bargaining game, then every subgame perfect equibrium payoff vector of the associated dynamic game converges to the Nash Bargaining Solution. If V corresponds to a TU game v, then the associated dynamic game has a unique subgame perfect equibrium payoff vector that converges to the Shapley value of v.

9. Final remarks

9.1. Values and cores of convex NTU games

As we have repeatedly emphasized, the values and solutions presented in this chapter are all based on axiomatized notions of fairness. An important alternative concept is that of the core, which is defined in terms of coalitional stability. Let V be an NTU game. A vector $x \in \Re^N$ can be *improved upon* if there exists an $S \subseteq N$ and $y \in V(S)$ such that $y^i > x^i$, for each $i \in S$.

DEFINITION 9.1.1. A vector $x \in \Re^N$ is a core payoff for V if $x \in V(N)$ and x cannot be improved upon. The set of core payoffs of V is denoted core(V).

If V corresponds to $v \in G_N$, then $x \in$ core(V) if and only if $1_S \cdot x^S \geqslant v(S)$ for each $S \subseteq N$ and $1_N \cdot x \leqslant v(N)$. If $v \in G_N$, we will write core(v). In the TU case, it is not generally true that $\varphi(v) \in$ core(v) even if core(v) is not empty. However, $\varphi(v) \in$ core(v) if v is a convex game. A game $v \in G_N$ is *convex* (or *supermodular*) if $v(S) + v(T) \leqslant v(S \cup T) + v(S \cap T)$ whenever $S, T \subseteq N$.

Convex games were first studied in Shapley (1971), where it was shown that the value is an element of the core of a convex game. However, the actual result is more general than this and we describe it now. For $v \in G_N$ and $R \in \Omega$, define the marginal worth vector $\delta(R, v)$ as the point in $\Re^N$ whose ith component is $\Delta_i(R, v)$. In Theorem 3 of Shapley (1971), it is shown that the set of extreme points of the core of a convex game v is precisely the set $\{\delta(R, v)\}_{R \in \Omega}$ of marginal worth vectors and, therefore, the Shapley value belongs to the core. However, this also means that every random order value belongs to the core of a convex game, i.e., if v is convex, then

$$\bigcup_{p \in M(\Omega)} \varphi_p(v) = \text{core}(v).$$

Weber (1988) has shown that core(v) $\subseteq \bigcup_{p \in M(\Omega)} \varphi_p(v)$ for any game $v \in G_N$. Ichiishi (1981) proved that if the converse is true, then v is convex.

We are interested in the extent to which these results generalize to convex NTU games and the p-Shapley value correspondence Φ_p. A game $V \in \Gamma_N$ is cardinal convex if $V^*(S) + V^*(T) \subseteq V^*(S \cup T) + V^*(S \cap T)$ [see Sharkey (1981)]. It is easy to show that if v is a convex TU game and if V is the NTU game derived from v, then V is cardinal convex. It is also straightforward to show that if V is cardinal convex and if v_λ is defined V, then v_λ is a convex TU game. Define the inner core of an NTU game V to be

$$\text{incore}(V) = \left\{ x \in V(N) \mid v_\lambda \text{ is defined for some } \lambda \in \mathfrak{R}^N_{++} \text{ and } \lambda * x \in \text{core}(v_\lambda) \right\}.$$

By combining the definitions with the results for TU games mentioned above, we conclude that for any NTU game V, $\text{incore}(V) \subseteq \bigcup_{p \in M(\Omega)} \varphi_p(v)$ and in fact we have:

PROPOSITION 9.1.2 [Levy and McLean (1991)]. *If V is a cardinal convex game, then* $\bigcup_{p \in M(\Omega)} \Phi_p(V) = \text{incore}(V)$.

It is well known that $\text{incore}(V) \subseteq \text{core}(V)$ for any V so Proposition 9.1.2 generalizes results in Kern (1985, Theorem 4.4) and Ichiishi (1983, p. 148). Although $\text{incore}(V) \subseteq \text{core}(V)$, this containment can be strict if V does not correspond to a TU game. In fact, $\text{incore}(V)$ may be a nonempty, proper subset of $\text{core}(V)$ even if V is cardinal convex. For an example see Levy and McLean (1991).

9.2. Applications of values for NTU games with a continuum of players

Many political and economic situations are best modelled as games with infinitely many players. The value theory of such games comprises an extensive literature beginning with the fundamental study of Aumann and Shapley (1974). Values of NTU games with a continuum of players have been studied in several contexts. Value allocations are discussed extensively in Chapter 36 and, for economies with a continuum of agents (and satisfying a number of other regularity conditions), a "value equivalence" principle holds: the set of Walras allocations is equal to the set of value allocations. Another economic application of the NTU value arises in the theory of "Power and Taxes" as originally presented in Aumann and Kurz (1977a, 1977b). [See also Aumann, Kurz and Neyman (1987) and Osborne (1984) and the references cited therein.]

9.3. NTU values for games with incomplete information

In many bargaining problems, individuals possess private information that affects their evaluation or utility of different outcomes. In the presence of incomplete information, the resolution of bargaining is complicated by the fact that the arbitrator must request that the agents reveal their types before he can recommend an outcome. Of course, the agents will behave strategically and will "misrepresent" themselves to the arbitrator in a way favorable to themselves. For treatments that generalize the NBS to the incomplete information case, see Harsanyi and Selten (1972) and Myerson (1984a). In Myerson (1984b), the NTU value is extended to cooperative games with incomplete information in a way that generalizes the results of Myerson (1984a) for the pure bargaining case.

References

Aumann, R. (1985a), "An axiomatization of the non-transferable utility value", Econometrica 53:599–612.

Aumann, R. (1985b), "On the non-transferable utility value: A comment on the Roth–Shafer examples", Econometrica 53:667–677.

Aumann, R. (1986), "A rejoinder", Econometrica 54:985–989.

Aumann, R., and M. Kurz (1977a), "Power and taxes", Econometrica 45:1137–1161.

Aumann, R., and M. Kurz (1977b), "Power and taxes in a multi-commodity economy", Journal of Public Economics 9:139–161.

Aumann, R., M. Kurz and A. Neyman (1987), "Power and public goods", Journal of Economic Theory 42:108–127.

Aumann, R., and L.S. Shapley (1974), Values of Nonatomic Games (Princeton University Press, Princeton, NJ).

Clarke, F. (1983), Optimization and Nonsmooth Analysis (Wiley Interscience, New York).

Driessen, T. (1991), "A survey of consistency properties in cooperative game theory", SIAM Review 33:43–59.

Fleming, W. (1969), "The Cauchy problem for a nonlinear first order partial differential equation", Journal of Differential Equations 5:515–530.

Harsanyi, J. (1963), "A simplified bargaining model for the n-person game", International Economic Review 4:194–220.

Harsanyi, J., and R. Selten (1972), "A generalized Nash bargaining solution for two person bargaining games with incomplete information", Management Science 18:80–106.

Hart, S. (1985a), "An axiomatization of Harsanyi's non-transferable utility solution", Econometrica 53:1295–1313.

Hart, S. (1985b), "NTU games and markets: Some examples and the Harsanyi solution", Econometrica 53:1445–1450.

Hart, S. (1985c), "Axiomatic approaches to coalitional bargaining", in: A. Roth, ed., Game Theoretic Models of Bargaining (Cambridge University Press, Cambridge) 305–319.

Hart, S. (1994), "On prize games", in: N. Megiddo, ed., Essays in Game Theory in Honor of Michael Maschler (Springer, New York) 111–121.

Hart, S., and M. Kurz (1983), "On the endogenous formation of coalitions", Econometrica 51:1047–1064.

Hart, S., and A. Mas-Colell (1989), "Potential, value and consistency", Econometrica 57:589–614.

Hart, S., and A. Mas-Colell (1991), Personal Communication.

Hart, S., and A. Mas-Colell (1995a), "Egalitarian solutions of large games: A continuum of players", Mathematics of Operations Research 20:959–1002.

Hart, S., and A. Mas-Colell (1995b), "Egalitarian solutions of large games: The asymptotic approach", Mathematics of Operations Research 20:1003–1022.

Hart, S., and A. Mas-Colell (1996a), "Bargaining and value", Econometrica 64:357–380.

Hart, S., and A. Mas-Colell (1996b), "Harsanyi values of large economies: Nonequivalence to competitive equilibria", Games and Economic Behavior 13:74–99.

Ichiishi, T. (1981), "Supermodularity: Applications to convex games and the greedy algorithm for LP", Journal of Economic Theory 25:283–286.

Ichiishi, T. (1983), Game Theory for Economic Analysis (Academic Press, New York).

Imai, H. (1983), "On Harsanyi's solution", International Journal of Game Theory 12:161–179.

Kalai, E. (1977a), "Non-symmetric Nash solutions and replications of two person bargaining", International Journal of Game Theory 6:129–133.

Kalai, E. (1977b), "Proportional solutions to bargaining situations: Interpersonal utility comparisons", Econometrica 45:1623–1630.

Kalai, E., and D. Samet (1987), "On weighted Shapley values", International Journal of Game Theory 16:1623–1630.

Kalai, E., and D. Samet (1985), "Monotonic solutions to general cooperative games", Econometrica 53:307–327.

Kern, R. (1985), "The Shapley transfer value without zero weights", International Journal of Game Theory 14:73–92.

Kurz, M. (1994), "Game theory and public economics", in: R.J. Aumann and S. Hart, eds., Handbook of Game Theory, Vol. 2 (North-Holland, Amsterdam) Chapter 33, 1153–1192.

Levy, A., and R. McLean (1989), "Weighted coalition structure values", Games and Economic Behavior 1:234–249.

Levy, A., and R. McLean (1991), "An axiomatization of the non-symmetric NTU value", International Journal of Game Theory 19:109–127.

Maschler, M., and G. Owen (1989), "The consistent Shapley value for hyperplane games", International Journal of Game Theory 18:389–407.

Maschler, M., and G. Owen (1992), "The consistent Shapley value for games without side payments", in: R. Selten, ed., Rational Interaction: Essays in Honor of John Harsanyi (Springer, New York) 5–12.

Myerson, R. (1984a), "Two person bargaining problems with incomplete information", Econometrica 52:461–487.

Myerson, R. (1984b), "Cooperative games with incomplete information", International Journal of Game Theory 13:69–96.

Nash, J. (1950), "The bargaining problem", Econometrica 28:155–162.

Orshan, G. (1992), "The consistent Shapley value in hyperplane games from a global standpoint", International Journal of Game Theory 21:1–11.

Osborne, M. (1984), "Why do some goods bear higher taxes than others?", Journal of Economic Theory 32:301–316.

Owen, G. (1972a), "Multilinear extensions of games", Management Science 18:64–79.

Owen, G. (1972b), "Values of games without side payments", International Journal of Game Theory 1:94–109.

Owen, G. (1977), "Values of games with a priori unions", in: R. Henn and O. Moeschlin, eds., Essays in Mathematical Economics and Game Theory (Springer, New York) 76–88.

Owen, G. (1994), "The non-consistency and non-uniqueness of the consistent value", in: N. Megiddo, ed., Essays in Game Theory in Honor of Michael Maschler (Springer, New York) 155–162.

Peleg, B. (1986), "A proof that the core of an ordinal convex game is a von Neumann–Morgenstern solution", Mathematical Social Sciences 11:83–87.

Roth, A. (1979), Axiomatic Models of Bargaining, Lecture Notes in Economics and Mathematical Systems, No. 170 (Springer, New York).

Roth, A. (1980), "Values of games without side payments: Some difficulties with current concepts", Econometrica 48:457–465.

Roth, A. (1986), "On the NTU value: A reply to Aumann", Econometrica 54:981–984.

Samet, D. (1985), "An axiomatization of the Egalitarian solution", Mathematical Social Sciences 9:173–181.

Shafer, W. (1980), "On the existence and interpretation of value allocations", Econometrica 48:467–476.

Shapley, L. (1953a), "Additive and non-additive set functions", PhD Thesis (Department of Mathematics, Princeton University).

Shapley, L. (1953b), "A value for n-person games", in: H. Kuhn and A. Tucker, eds., Contributions to the Theory of Games II (Princeton University Press, Princeton, NJ) 307–317.

Shapley, L. (1969), "Utility comparisons and the theory of games", La Decision (Editions du Centre National de le Recherche Scientifique, Paris).

Shapley, L. (1971), "Cores of convex games", International Journal of Game Theory 1:11–26.

Sharkey, W. (1981), "Convex games without side payments", International Journal of Game Theory 11:101–106.

Thomson, W. (1994), "Cooperative models of bargaining", in: R.J. Aumann and S. Hart, eds., Handbook of Game Theory, Vol. 2 (North-Holland, Amsterdam) Chapter 35, 1237–1284.

Weber, R. (1988), "Probabilistic values for games", in: A. Roth, ed., The Shapley Value: Essays in Honor of Lloyd Shapley (Cambridge University Press, Cambridge) 101–119.

Winter, E. (1991), "On nontransferable utility games with coalition structure", International Journal of Game Theory 20:53–64.

Winter, E. (1992), "On bargaining position descriptions in NTU games: Symmetry versus asymmetry", International Journal of Game Theory 21:191–111.

Young, H.P. (1988), "Individual contribution and just compensation", in: A. Roth, ed., The Shapley Value: Essays in Honor of Lloyd Shapley (Cambridge University Press, Cambridge) 267–278.

Chapter 56

VALUES OF GAMES WITH INFINITELY MANY PLAYERS

ABRAHAM NEYMAN

*Institute of Mathematics and Center for Rationality and Interactive Decision Theory,
The Hebrew University of Jerusalem, Jerusalem, Israel*

Contents

Handbook of Game Theory, Volume 3, Edited by R.J. Aumann and S. Hart

Abstract

This chapter studies the theory of value of games with infinitely many players.

Games with infinitely many players are models of interactions with many players. Often most of the players are individually insignificant, and are effective in the game only via coalitions. At the same time there may exist big players who retain the power to wield single-handed influence. The interactions are modeled as cooperative games with a continuum of players. In general, the continuum consists of a non-atomic part (the "ocean"), along with (at most countably many) atoms. The continuum provides a convenient framework for mathematical analysis, and approximates the results for large finite games well. Also, it enables a unified view of games with finite, countable, or oceanic player-sets, or indeed any mixture of these.

The value is defined as a map from a space of cooperative games to payoffs that satisfies the classical value axioms: additivity (linearity), efficiency, symmetry and positivity. The chapter introduces many spaces for which there exists a unique value, as well as other spaces on which there is a value.

A game with infinitely many players can be considered as a limit of finite games with a large number of players. The chapter studies limiting values which are defined by means of the limits of the Shapley value of finite games that approximate the given game with infinitely many players.

Various formulas for the value which express the value as an average of marginal contribution are studied. These value formulas capture the idea that the value of a player is his expected marginal contribution to a perfect sample of size t of the set of all players where the size t is uniformly distributed on $[0,1]$. In the case of smooth games the value formula is a diagonal formula: an integral of marginal contributions which are expressed as partial derivatives and where the integral is over all perfect samples of the set of players. The domain of the formula is further extended by changing the order of integration and derivation and the introduction of a well-crafted infinitesimal perturbation of the perfect samples of the set of players provides a value formula that is applicable to many additional games with essential nondifferentiabilities.

Keywords

games, cooperative, coalitional form, transferable utility, value, continuum of players, non-atomic

JEL classification: D70, D71, D63, C71

1. Introduction

The Shapley value is one of the basic solution concepts of cooperative game theory. It can be viewed as a sort of average or expected outcome, or as an *a priori* evaluation of the players' expected payoffs.

The value has a very wide range of applications, particularly in economics and political science (see Chapters 32, 33 and 34 in this Handbook). In many of these applications it is necessary to consider games that involve a large number of players. Often most of the players are individually insignificant, and are effective in the game only via coalitions. At the same time there may exist big players who retain the power to wield single-handed influence. A typical example is provided by voting among stockholders of a corporation, with a few major stockholders and an "ocean" of minor stockholders. In economics, one considers an oligopolistic sector of firms embedded in a large population of "perfectly competitive" consumers. In all these cases, it is fruitful to model the game as one with a continuum of players. In general, the continuum consists of a non-atomic part (the "ocean"), along with (at most countably many) atoms. The continuum provides a convenient framework for mathematical analysis, and approximates the results for large finite games well. Also, it enables a unified view of games with finite, countable, or oceanic player-sets, or indeed any mixture of these.

1.1. An outline of the chapter

In Section 2 we highlight a sample of a few asymptotic results on the Shapley value of finite games. These results motivate the definition of games with infinitely many players and the corresponding value theory; in particular, Proposition 1 introduces a formula, called the diagonal formula of the value, that expresses the value (or an approximation thereof) by means of an integral along a specific path.

The Shapley value for finite games is defined as a linear map from the linear space of games to payoffs that satisfies a list of axioms: symmetry, efficiency and the null player axiom. Section 3 introduces the definitions of the space of games with a continuum of players, and defines the symmetry, positivity and efficiency of a map from games to payoffs (represented by finitely additive games) which enables us to define the value on a space of games as a linear, symmetric, positive and efficient map.

The value does not exist on the space of all games with a continuum of players. Therefore the theory studies spaces of games on which a value does exist and moreover looks for spaces for which there is a unique value. Section 4 introduces several spaces of games with a continuum of players including the two spaces of games, pNA and $bv'NA$, which played an important role in the development of value theory for non-atomic games. Theorem 2 asserts that there exists a unique value on each of the spaces pNA and $bv'NA$ as well as on pNA_∞. In addition there exists a (unique) value of norm 1 on the space $'NA$. Section 4 also includes a detailed outline of the proofs.

A game with infinitely many players can be considered as a limit of finite games with a large number of players. Section 5 studies limiting values which are defined by means of the limits of the Shapley value of finite games that approximate the given game with infinitely many players. Section 5 also introduces the asymptotic value. The asymptotic value of a game v is defined whenever all the sequences of Shapley values of finite games that "approximate" v have the same limit. The space of all games of bounded variation for which the asymptotic value exist is denoted $ASYMP$. Theorem 4 asserts that $ASYMP$ is a closed linear space of games that contains $bv'M$ and the map associating an asymptotic value with each game in $ASYMP$ is a value on $ASYMP$.

Section 6 introduces the mixing value which is defined on all games of bounded variation for which an analogous formula of the random order one for finite games is well defined. The space of all games having a mixing value is denoted MIX. Theorem 5 asserts that MIX is a closed linear space which contains pNA, and the map that associates a mixing value with each game v in MIX is a value on MIX.

Section 7 introduces value formulas. These value formulas capture the idea that the value of a player is his expected marginal contribution to a perfect sample of size t of the set of all players where the size t is uniformly distributed on $[0, 1]$. A value formula can be expressed as an integral of directional derivative whenever the game is a smooth function of finitely many non-atomic measures. More generally, by extending the game to be defined over all ideal coalitions – measurable $[0, 1]$-valued functions on the (measurable) space of players – this value formula provides a formula for the value of any game in pNA. Changing the order of derivation and integration results in a formula that is applicable to a wider class of games: all those games for which the formula yields a finitely additive game. In particular, this modified formula defines a value on $bv'NA$ or even on $bv'M$. When this formula does not yield a finitely additive game, applying the directional derivative to a carefully crafted small perturbation of perfect samples of the set of players yields a value formula on a much wider space of games (Theorem 7). These include games which are nondifferentiable functions (e.g., a piecewise linear function) of finitely many non-atomic probability measures.

Section 8 describes two spaces of games that are spanned by nondifferentiable functions of finitely many mutually singular non-atomic probability measures that have a unique value (Theorem 8). One is the space spanned by all piecewise linear functions of finitely many mutually singular non-atomic probability measures, and the other is the space spanned by all concave Lipschitz functions with increasing returns to scale of finitely many mutually singular non-atomic probability measures.

The value is defined as a map from a space of games to payoffs that satisfies a short list of plausible conditions: linearity, symmetry, positivity and efficiency. The Shapley value of finite games satisfies many additional desirable properties. It is continuous (of norm 1 with respect to the bounded variation norm), obeys the null player and the dummy axioms, and it is strongly positive. In addition, any value that is obtained from values of finite approximations obeys an additional property called diagonality. Section 9 introduces such additional desirable value properties as strong positivity and diagonality. Theorem 9 asserts that strong positivity can replace positivity and linearity in the characterization of the value on pNA_∞, and thus strong positivity and continuity can replace positivity and linearity in the characterization of a value on pNA. A strik-

ing property of the value on *pNA*, the asymptotic value, the mixing value, or any of
the values obtained by the formulas described in Section 7, is the diagonal property: if
two games coincide on all coalitions that are "almost" a perfect sample (within ε with
respect to finitely many non-atomic probability measures) of the set of players, their
values coincide. Theorem 11 asserts that any continuous value is diagonal.

Section 10 studies semivalues which are generalizations of the value that do not necessarily obey the efficiency property. Theorem 13 provides a characterization of all
semivalues on *pNA*. Theorems 14 and 15 provide results on asymptotic semivalues.

Section 11 studies partially symmetric values which are generalizations of the value
that do not necessarily obey the symmetry axiom. Theorems 15 and 16 characterize in
particular all the partially symmetric values on *pNA* and *pM* that are symmetric with
respect to those symmetries that preserve a fixed finite partition of the set of players.
A corollary of this characterization on *pM* is the characterization of all values on *pM*.
A partially symmetric value, called a μ-value, is symmetric with respect to all symmetries that preserve a fixed non-atomic population measure μ and (in addition to linearity,
positivity and efficiency), obeys the dummy axiom. Such a partially symmetric value is
called a μ-value. Theorem 18 asserts the uniqueness of a μ-value on *pNA*(μ). Theorem 19 asserts that an important class of nondifferentiable markets have an asymptotic
μ-value.

Games that arise from the study of non-atomic markets, called market games, obey
two specific conditions: concavity and homogeneity of degree 1. The core of such games
is nonempty. Section 12 provides results on the relation between the value and the core
of market games. Theorem 20 asserts the coincidence of the core and the value in the
differentiable case. Theorem 21 relates the existence of an asymptotic value of nondifferentiable market games to the existence of a center of symmetry of its core. Theorems 22 and 23 describe the asymptotic μ-value and the value of market games with a
finite-dimensional core as an average of the extreme points of the core.

2. Prelude: Values of large finite games

This section introduces several asymptotic results on the Shapley value ψ of finite
games. These results can serve as an introduction to the theory of values of games
with infinitely many players, by motivating the definitions of games with a continuum
of players and the corresponding value theory.

We start with the introduction of a representation of games with finitely many play-
ers. Let ℓ be a fixed positive integer and $f : \mathbb{R}^\ell \to \mathbb{R}$ a function with $f(0) = 0$. Given
$w_1, \ldots, w_\ell \in \mathbb{R}^\ell_+$, define the *n*-person game $v = [f; w_1, \ldots, w_\ell]$ by

$$v(S) = f\left(\sum_{i \in S} w_i\right).$$

A special subclass of these games is weighted majority games where $\ell = 1$, $0 < q <$
$\sum_{i=1}^n w_i$ and $f(x) = 1$ if $x \geqslant q$ and 0 otherwise. The asymptotic results relate to se-

quences of finite games where the function f is fixed and the vector of weights varies. If the function f is differentiable at x, the directional derivative at x of the function f in the direction $y \in \mathbb{R}^\ell$ is $f_y(x) := \lim_{\varepsilon \to 0+} (f(x + \varepsilon y) - f(x))/\varepsilon$.

The following result follows from the proofs of results on the asymptotic value of non-atomic games. It can be derived from the proof in Kannai (1966) or from the proof in Dubey (1980).

PROPOSITION 1. *Assume that f is a continuously differentiable function on $[0, 1]^\ell$. For every $\varepsilon > 0$ there is $\delta > 0$ such that if $w_1, \ldots, w_n \in \mathbb{R}_+^\ell$ with (the normalization) $\sum_{i=1}^n w_i = (1, \ldots, 1)$ and $\max_{i=1}^n \|w_i\| < \delta$, then*

$$\left| \psi v(i) - \int_0^1 f_{w_i}(t, \ldots, t)\, dt \right| \leqslant \varepsilon \|w_i\|,$$

where $v = [f; w_1, \ldots, w_n]$. Thus, for every coalition S of players,

$$\left| \psi v(S) - \int_0^1 f_{w(S)}(t, \ldots, t)\, dt \right| < \varepsilon,$$

where $\psi v(S) = \sum_{i \in S} \psi v(i)$ and $w(S) = \sum_{i \in S} w_i$.

The limit of a sequence of games of the form $[f; w_1^k, \ldots, w_{n_k}^k]$, where f is a fixed function and $w_1^k, \ldots, w_{n_k}^k$ is a sequence of vectors in $\mathbb{R}_+^\ell$ with $\sum_{i=1}^{n_k} w_i^k = (1, \ldots, 1)$ and $\max_{i=1}^{n_k} \|w_i\| \to_{k \to \infty} 0$ can be modeled by a 'game' of the form $f \circ (\mu_1, \ldots, \mu_\ell)$ where $\mu = (\mu_1, \ldots, \mu_\ell)$ is a vector of non-atomic probability measures defined on a measurable space $(I, \mathcal{C})$. If f is continuously differentiable, the limiting value formula thus corresponds to:

$$\varphi(f \circ \mu)(S) = \int_0^1 f_{\mu(S)}(t, \ldots, t)\, dt.$$

The next results address the asymptotics of the value of weighted majority games. The n-person weighted majority game $v = [q; w_1, \ldots, w_n]$ with $w_1, \ldots, w_n \in \mathbb{R}_+$ and $0 < q < \sum_{i=1}^m w_i$, is defined by $v(S) = 1$ if $\sum_{i \in S} w_i \geqslant q$ and 0 otherwise.

PROPOSITION 2 [Neyman (1981a)]. *For every $\varepsilon > 0$ there is $K > 0$ such that if $v = [q; w_1, \ldots, w_n]$ is a weighted majority game with $w_1, \ldots, w_n \in \mathbb{R}^+$, $\sum_{i=1}^n w_i = 1$ and $K \max_{i=1}^n w_i < q < 1 - K \max_{i=1}^n w_i$,*

$$\left| \psi v(S) - \sum_{i \in S} w_i \right| < \varepsilon$$

for every coalition $S \subset \{1, \ldots, n\}$.

In particular, if $v_k = [q_k; w_1^k, \ldots, w_{n_k}^k]$ is a sequence of weighted majority games with $w_i^k \geqslant 0$, $\sum_{i=1}^{n_k} w_i^k \to_{k \to \infty} 1$, and $q_k \to_{k \to \infty} q$ with $0 < q < 1$, then for any sequence of coalitions $S_k \subset \{1, \ldots, n_k\}$, $\psi v_k(S_k) - \sum_{i \in S_k} w_i^k \to_{k \to \infty} 0$. The limit of such a sequence of weighted majority games is modeled as a game of the form $v = f_q \circ \mu$ where μ is a non-atomic probability measure and $f_q(x) = 1$ if $x \geqslant q$ and 0 otherwise. The non-atomic probability measure μ describes the value of this limiting game:

$$\varphi v(S) = \mu(S).$$

The next result follows easily from the result of Berbee (1981).

PROPOSITION 3. *Let* $w_1, \ldots, w_n, \ldots$ *be a sequence of positive numbers with*

$$\sum_{i=1}^{\infty} w_i = 1$$

and $0 < q < 1$. *There exists a sequence of nonnegative numbers* ψ_i, $i \geqslant 1$, *with* $\sum_{i=1}^{\infty} \psi_i = 1$, *such that for every* $\varepsilon > 0$ *there is a positive constant* $\delta > 0$ *and a positive integer* n_0 *such that if* $v = [q'; u_1, \ldots, u_n]$ *is a weighted majority game with* $n \geqslant n_0$, $|q' - q| < \delta$ *and* $\sum_{i=1}^{n} |w_i - u_i| < \delta$, *then*

$$\left| \psi v(i) - \psi_i \right| < \varepsilon.$$

Equivalently, let $v_k = [q_k; w_1^k, \ldots, w_{n_k}^k]$ be a sequence of weighted majority games with $\sum_{i=1}^{n_k} w_i^k \to_{k \to \infty} 1$ and $w_j^k \to_{k \to \infty} w_j$ and $q_k \to_{k \to \infty} q$. Then the Shapley value of player i in the game v_k converges as $k \to \infty$ to a limit ψ_i and $\sum_{i=1}^{\infty} \psi_i = 1$. The limit of such a sequence of weighted majority games can be modeled as a game of the form $v = f_q \circ \mu$ where μ is a purely atomic measure with atoms $i = 1, 2, \ldots$ and $\mu(i) = w_i$. The sequence ψ_i describes the value of this limiting game:

$$\varphi v(S) = \sum_{i \in S} \psi_i.$$

3. Definitions

The space of players is represented by a measurable space $(I, \mathcal{C})$. The members of set I are called *players*, those of $\mathcal{C}$ – *coalitions*. A game in coalitional form is a real-valued function v on $\mathcal{C}$ such that $v(\emptyset) = 0$. For each coalition S in $\mathcal{C}$, the number $v(S)$ is interpreted as the total payoff that the coalition S, if it forms, can obtain for its members; it is called the *worth* of S. The set of all games forms a linear space over the field of real numbers, which is denoted G. We assume that $(I, \mathcal{C})$ is isomorphic to

([0, 1], $\mathcal{B}$), where $\mathcal{B}$ stands for the Borel subsets of [0, 1]. A game v is *finitely additive* if $v(S \cup T) = v(S) + v(T)$ whenever S and T are two disjoint coalitions. A distribution of payoffs is represented by a finitely additive game. A value is a mapping from games to distributions of payoffs, i.e., to finitely additive games that satisfy several plausible conditions: linearity, symmetry, positivity and efficiency. There are several additional desirable conditions: e.g., continuity, strong positivity, a null player axiom, a dummy player axiom, and so on. It is remarkable, however, that a short list of plausible conditions that define the value suffices to determine the value in many cases of interest, and that in these cases the unique value satisfies many additional desirable properties. We start with the notations and definitions needed formally to define the value, and set forth four spaces of games on which there is a unique value.

A game v is *monotonic* if $v(S) \geqslant v(T)$ whenever $S \supset T$; it is of *bounded variation* if it is the difference of two monotonic games. The set of all games of bounded variation (over a fixed measurable space $(I, \mathcal{C})$) is a vector space. The *variation* of a game $v \in G$, $\|v\|$, is the supremum of the variation of v over all increasing chains $S_1 \subset S_2 \subset \cdots \subset S_n$ in $\mathcal{C}$. Equivalently, $\|v\| = \inf\{u(I) + w(I) : u, w$ are monotonic games with $v = u - w\}$ ($\inf \emptyset = \infty$). The variation defines a topology on G; a basis for the open neighborhood of a game v in G consists of the sets $\{u \in G : \|u - v\| < \varepsilon\}$ where ε varies over all positive numbers. Addition and multiplication of two games are continuous and so is the multiplication of a game and a real number. A game v has bounded variation if $\|v\| < \infty$. The space of all games of bounded variation, BV, is a Banach algebra with respect to the variation norm $\| \ \|$, i.e., it is complete with respect to the distance defined by the norm (and thus it is a Banach Space) and it is closed under multiplication which is continuous. Let $\mathcal{G}$ denote the group of automorphisms (i.e., one-to-one measurable mappings Θ from I onto I with Θ^{-1} measurable) of the underlying space $(I, \mathcal{C})$. Each Θ in $\mathcal{G}$ induces a linear mapping Θ_* of BV (and of the space of all games) onto itself, defined by $(\Theta_* v)(S) = v(\Theta S)$. A set of games Q is called *symmetric* if $\Theta_* Q = Q$ for all Θ in $\mathcal{G}$. The space of all finitely additive and bounded games is denoted FA; the subspace of all measures (i.e., countably additive games) is denoted M and its subspace consisting of all non-atomic measures is denoted NA. The space of all non-atomic elements of FA is denoted AN. Obviously, $NA \subset M \subset FA \subset BV$, and each of the spaces NA, AN, M, FA and BV is a symmetric space. Given a set of games Q, we denote by Q^+ all monotonic games in Q, and by Q^1 the set of all games v in Q^+ with $v(I) = 1$. A map $\varphi : Q \to BV$ is called *positive* if $\varphi(Q^+) \subset BV^+$; *symmetric* if for every $\Theta \in \mathcal{G}$ and v in Q, $\Theta_* v \in Q$ implies that $\varphi(\Theta_* v) = \Theta_*(\varphi v)$; and *efficient* if for every v in Q, $(\varphi v)(I) = v(I)$.

DEFINITION 1 [Aumann and Shapley (1974)]. Let Q be a symmetric linear subspace of BV. A *value* on Q is a linear map $\varphi : Q \to FA$ that is symmetric, positive and efficient.

The definition of a value is naturally extended also to include spaces of games that are not necessarily included in BV. Thus let Q be a linear and symmetric space of games.

DEFINITION 2. A *value* on Q is a linear map from Q into the space of finitely additive games that is symmetric, positive (i.e., $\varphi(Q^+) \subset BV^+$), efficient, and $\varphi(Q \cap BV) \subset FA$.

There are several spaces of games that have a unique value. One of them is the space of all games with a finite support, where a subset T of I is called a *support* of a game v if $v(S) = v(S \cap T)$ for every S in C. The space of all games with finite support is denoted *FG*.

THEOREM 1 [Shapley (1953a)]. *There is a unique value on FG.*

The unique value ψ on *FG* is given by the following formula. Let T be a finite support of the game v (in *FG*). Then, $\psi v(S) = 0$ for every S in C with $S \cap T = \emptyset$, and for $t \in T$,

$$\psi v(\{t\}) = \frac{1}{n!} \sum_{\mathcal{R}} [v(\mathcal{P}_t^{\mathcal{R}} \cup \{t\}) - v(\mathcal{P}_t^{\mathcal{R}})],$$

where n is the number of elements of the support T, the sum runs over all $n!$ orders of T, and $\mathcal{P}_t^{\mathcal{R}}$ is the set of all elements of T preceding t in the order $\mathcal{R}$.

Finite games can also be identified with a function v defined on a finite subfield π of C: given a real-valued function v defined on a finite subfield π of C with $v(\emptyset) = 0$, the Shapley value of the game v is defined as the additive function $\psi : \pi \to \mathbb{R}$ defined by its values on the atoms of π; for every atom a of π,

$$\psi v(a) = \frac{1}{n!} \sum \mathcal{R}[v(\mathcal{P}_a^{\mathcal{R}} \cup a) - v(\mathcal{P}_a^{\mathcal{R}})],$$

where n is the number of atoms of π, the sum runs over all $n!$ orders of the atoms of π, and $\mathcal{P}_a^{\mathcal{R}}$ is the union of all atoms of π preceding a in the order $\mathcal{R}$.

4. The value on *pNA* and *bv'NA*

Let *pNA* (*pM*, *pFA*) be the closed (in the bounded variation norm) algebra generated by NA (*M*, *FA*, respectively). Equivalently, *pNA* is the closed linear subspace that is generated by powers of measure in NA^1 [Aumann and Shapley (1974), Lemma 7.2]. Similarly, *pM* (*pFA*) is the closed algebra that is generated by powers of elements in *M* (*FA*). Let $\| \ \|_\infty$ be defined on the space of all games G by

$$\|v\|_\infty = \inf\{\mu(I) \colon \mu \in NA^+, \ \mu - v \in BV^+, \ \mu + v \in BV^+\} + \sup\{|v(S)| \colon S \in C\}.$$

In the definition of $\|v\|_\infty$, we use the common convention that $\inf \emptyset = \infty$. The definition of $\| \ \|_\infty$ is an analog of the C^1 norm of continuously differentiable functions; e.g., if f is continuously differentiable on $[0, 1]$ and μ is a non-atomic probability measure, $\|f \circ \mu\|_\infty = \max\{|f'(t)|:\ 0 \leqslant t \leqslant 1\} + \max\{|f(t)|:\ 0 \leqslant t \leqslant 1\}$. Note that $\|uv\|_\infty \leqslant \|u\|_\infty \|v\|_\infty$ for two games u and v. The function $v \mapsto \|v\|_\infty$ defines a topology on G; a basis for the open neighborhood of a game v in G consists of the sets $\{u \in G:\ \|u - v\|_\infty < \varepsilon\}$ where ε varies over all positive numbers. The space of games pNA_∞ is defined as the $\|\ \|_\infty$ closed linear subspace that is generated by powers of measures in NA^1. Obviously, $pNA_\infty \subset pNA$ ($\|\ \|_\infty \geqslant \|\ \|$). Both spaces contain many games of interest. If μ is a measure in NA^1 and $f : [0, 1] \to \mathbb{R}$ is a function with $f(0) = 0$, then $f \circ \mu \in pNA$ if and only if f is absolutely continuous [Aumann and Shapley (1974), Theorem C] and in that case $\|f \circ \mu\| = \int_0^1 |f'(t)|\,dt$ where f' is the (Radon–Nikodym) derivative of the function f, and $f \circ \mu \in pNA_\infty$ if and only if f is continuously differentiable on $[0, 1]$. Also, if $\mu = (\mu_1, \ldots, \mu_n)$ is a vector of NA measures with range $\{\mu(S):\ S \in \mathcal{C}\} =: \mathcal{R}(\mu)$, and f is a C^1 function on $\mathcal{R}(\mu)$ with $f(\mu(\emptyset)) = 0$, then $f \circ \mu \in pNA_\infty$ [Aumann and Shapley (1974), Proposition 7.1].

Let $bv'NA$ ($bv'M$, $bv'FA$) be the closed linear space generated by games of the form $f \circ \mu$ where $f \in bv' =: \{f : [0, 1] \to \mathbb{R} \mid f$ is of bounded variation and continuous at 0 and at 1 with $f(0) = 0\}$, and $\mu \in NA^1$ ($\mu \in M^1$, $\mu \in FA^1$). Obviously, $pNA_\infty \subset pNA \subset bv'NA \subset bv'M \subset bv'FA$.

THEOREM 2 [Aumann and Shapley (1974), Theorems A, B; Monderer (1990), Theorem 2.1]. *Each of the spaces pNA, bv'NA, pNA_∞ has a unique value.*

The proof of the uniqueness of the value on these spaces relies on two basic arguments. Let $\mu \in NA^1$ and let $\mathcal{G}(\mu)$ be the subgroup of $\mathcal{G}$ of all automorphisms Θ that are μ-measure-preserving, i.e., $\mu(\Theta S) = \mu(S)$ for every coalition S in $\mathcal{C}$. Then the set of all finitely additive games u that are fixed points of the action of $\mathcal{G}(\mu)$ (i.e., $\Theta_* u = u$ for every Θ in $\mathcal{G}(\mu)$) are games of the form $\alpha\mu$, $\alpha \in \mathbb{R}$ (here we use the standardness assumption that $(I, \mathcal{C})$ is isomorphic to $([0, 1], \mathcal{B})$). Therefore, if Q is a symmetric space of games and $\varphi : Q \to FA$ is symmetric, $\mu \in NA^1$, and $f : [0, 1] \to \mathbb{R}$ is such that $f \circ \mu \in Q$, then $\Theta_*(f \circ \mu) = f \circ \mu$ for every $\Theta \in \mathcal{G}(\mu)$. It follows that the finitely additive game $\varphi(f \circ \mu)$ is a fixed point of $\mathcal{G}(\mu)$, which implies that $\varphi(f \circ \mu) = \alpha\mu$. If, moreover, φ is also efficient, then $\varphi(f \circ \mu) = f(1)\mu$. The spaces pNA and $bv'NA$ are closed linear spans of games of the form $f \circ \mu$. The space pNA_∞ is in the closed linear span of games of the form $f \circ \mu$ with $f \circ \mu$ in pNA_∞. Therefore, each of the spaces pNA_∞, pNA and $bv'NA$ has a dense subspace on which there is at most one value.

To complete the uniqueness proof, it suffices to show that any value on these spaces is indeed continuous. For that we apply the concept of internality. A linear subspace Q of BV is said to be *reproducing* if $Q = Q^+ - Q^+$. It is said to be *internal* if for all v in Q,

$$\|v\| = \inf\{u(I) + w(I):\ u, w \in Q^+ \text{ and } v = u - w\}.$$

Clearly, every internal space is reproducing. Also any linear positive mapping φ from a closed reproducing space Q into BV is continuous, and any linear positive and efficient mapping φ from an internal space Q into BV has norm 1 [Aumann and Shapley (1974), Propositions 4.15 and 4.7]. Therefore, to complete the proof of uniqueness, it is sufficient to show that each of the spaces pNA, $bv'NA$ and pNA_∞ is internal. The closure of an internal space is internal [Aumann and Shapley (1974), Proposition 4.12] and any linear space of games that is the union of internal spaces is internal. Therefore one demonstrates that each of these spaces is the union of spaces that contain a dense internal subspace. There are several dense internal subspaces of pNA [Aumann and Shapley (1974), Lemma 7.18; Reichert and Tauman (1985); Monderer and Neyman (1988), Section 5]. The proof of the internality of $bv'NA$ is more involved [Aumann and Shapley (1974), Section 8].

Note that if Q and Q' are two linear symmetric spaces of games and $Q \supset Q'$, the existence of a value on Q implies the existence of a value on Q'. However, the uniqueness of a value on one of them does not imply uniqueness on the other. There are, for instance, subspaces of pNA which have more than one value [Hart and Neyman (1988), Example 4]. There are however results that provide sufficient conditions for subspaces of pNA to have a unique value. For a coalition S we define the *restriction* of v to S, v_S, by $v_S(T) = v(S \cap T)$ for every $T \in \mathcal{C}$. A set of games M is *restrictable* if $v_S \in M$ for every $v \in M$ and for every $S \in \mathcal{C}$. Hart and Monderer [(1997), Theorem 3.6] prove the uniqueness of the value on restrictable subspaces of pNA_∞ that contain NA. This is the non-atomic version of the analogous result for finite games proved in Hart and Mas-Colell (1989) and Neyman (1989).

Existence of a value on pNA and $bv'NA$ is proved in Aumann and Shapley, Chapter I. We outline below a proof that is based on the proof of Mertens and Neyman (2001). Assume that $v = \sum_{i=1}^{n} f_i \circ \mu_i$ with $f_i \in bv'$ and $\mu_i \in NA^1$. Define $\varphi v(S) = \sum_{i=1}^{n} f_i(1)\mu_i(S)$. It follows from the continuity at 0 and 1 of each of the functions f_i that

$$\sum_{i=1}^{n} f_i(1)\mu_i(S) = \lim_{\varepsilon \to 0+} \frac{1}{\varepsilon} \int_0^{1-\varepsilon} \left[\sum_{i=1}^{n} f_i\big(t + \varepsilon\mu_i(S)\big) - \sum_{i=1}^{n} f_i(t) \right] dt.$$

By Lyapunov's Theorem, applied to the vector measure $(\mu_1, \ldots, \mu_n)$, for every $0 \leqslant t \leqslant 1 - \varepsilon$ there are coalitions $T \subset R$ with $\mu_i(T) = t$ and $\mu_i(R) = t + \varepsilon\mu_i(S)$. Therefore, if v is monotonic,

$$v(R) - v(T) = \sum_{i=1}^{n} f_i\big(t + \varepsilon\mu_i(S)\big) - \sum_{i=1}^{n} f_i(t) \geqslant 0,$$

and therefore $\varphi v(S) \geqslant 0$. In particular, if $0 = \sum_{i=1}^{n} f_i \circ \mu_i - \sum_{j=1}^{k} g_j \circ \nu_j$ where $f_i, g_j \in bv'$ and $\mu_i, \nu_j \in NA^1$, $\sum_{i=1}^{n} f_i(1)\mu_i = \sum_{j=1}^{k} g_j(1)\nu_j$, and thus also φv is well defined, i.e., it is independent of the presentation. Obviously, $\varphi v \in NA \subset FA$ and $\varphi v(I) = v(I)$ and φ is symmetric and linear. Therefore φ defines a value on the linear and symmetric subspace of $bv'NA$ of all games of the form $\sum_{i=1}^{n} f_i \circ \mu_i$ with n

a positive integer, $f_i \in bv'$ and $\mu_i \in NA^1$. We next show that φ is of norm 1 and thus has a (unique) extension to a continuous value (of norm 1) on $bv'NA$. For each $u \in FA$, $\|u\| = \sup_{S \in \mathcal{C}} |u(S)| + |u(S^c)|$. Therefore, it is sufficient to prove that for every coalition $S \in \mathcal{C}$, $|\varphi v(S)| + |\varphi v(S^c)| \leqslant \|v\|$. For each positive integer m let $S_0 \subset S_1 \subset S_2 \subset \cdots \subset S_m$ and $S_0^c \subset S_1^c \subset \cdots \subset S_m^c$ be measurable subsets of S and $S^c = I \setminus S$ respectively with $\mu_i(S_j) = \frac{j}{m+1}\mu_i(S)$ and $\mu_i(S_j^c) = \frac{j}{m+1}\mu_i(S^c)$. For every $0 \leqslant t \leqslant \frac{1}{m+1}$ let I_t be a measurable subset of $I \setminus (S_m \cup S_m^c)$ with $\mu_i(I_t) = t$. Define the increasing sequence of coalitions $T_0 \subset T_1 \subset \cdots \subset T_{2m}$ by $T_0 = I_t$, $T_{2j-1} = I_t \cup S_j \cup S_{j-1}^c$ and $T_{2j} = I_t \cup S_j \cup S_j^c$, $j = 1, \ldots, m$. Obviously,

$$\|v\| \geqslant \sum_{j=1}^{2m} |v(T_j) - v(T_{j-1})|$$

$$\geqslant \left| \sum_{j=0}^{m-1} v(T_{2j+1}) - v(T_{2j}) \right| + \left| \sum_{j=1}^{m} v(T_{2j}) - v(T_{2j-1}) \right|.$$

Set $\varepsilon = \frac{1}{m+1}$. Note that

$$\frac{1}{\varepsilon} \int_0^\varepsilon \sum_{j=0}^{m-1} \left[\sum_{i=1}^{n} f_i(t + \varepsilon j + \varepsilon \mu_i(S)) - \sum_{i=1}^{n} f_i(t + j\varepsilon) \right] dt$$

$$= \frac{1}{\varepsilon} \int_0^{1-\varepsilon} \left[\sum_{i=1}^{n} f_i(t + \varepsilon \mu_i(S)) - \sum_{i=1}^{n} f_i(t) \right] dt \to_{m \to \infty} \varphi v(S),$$

and similarly

$$\frac{1}{\varepsilon} \int_0^\varepsilon \sum_{j=1}^{m} \left[\sum_{i=1}^{n} f_i(t + \varepsilon j) - \sum_{i=1}^{n} f_i(t + j\varepsilon - \varepsilon \mu_i(S^c)) \right] dt$$

$$= \frac{1}{\varepsilon} \int_\varepsilon^1 \left[\sum_{i=1}^{n} f_i(t) - \sum_{i=1}^{n} f_i(t - \varepsilon \mu_i(S^c)) \right] dt \to_{m \to \infty} \varphi v(S^c).$$

As

$$v(T_{2j+1}) - v(T_{2j}) = \sum_{i=1}^{n} f_i(t + \varepsilon j + \varepsilon \mu_i(S)) - \sum_{i=1}^{n} f_i(t + j\varepsilon),$$

and

$$v(T_{2j}) - v(T_{2j-1}) = \sum_{i=1}^{n} f_i(t + 2j\varepsilon) - \sum_{i=1}^{n} f_i(t + 2j\varepsilon - \varepsilon \mu_i(S^c)),$$

we deduce that for each fixed $0 \leqslant t \leqslant \varepsilon$,

$$\left| \sum_{j=0}^{m-1} \left[\sum_{i=1}^{n} f_i \big(t + \varepsilon j + \varepsilon \mu_i(S) \big) - \sum_{i=1}^{n} f_i(t + j\varepsilon) \right] \right|$$
$$+ \left| \sum_{j=1}^{m} \left[\sum_{i=1}^{n} f_i(t + \varepsilon j) - \sum_{i=1}^{n} f_i \big(t + j\varepsilon - \varepsilon \mu_i(S^c) \big) \right] \right| \leqslant \|v\|$$

and therefore $|\varphi v(S)| + |\varphi v(S^c)| \leqslant \|v\|$. It follows that the map φ, which is defined on a dense subspace of $bv'NA$ and with values in NA, is linear, symmetric, efficient, and of norm 1, and therefore φ has a unique extension to a continuous value on $bv'NA$.

The bounded variation of the functions f_i is used in the above proof to show that the integrals $\int_a^b f(x)\,dx$, $0 \leqslant a < b \leqslant 1$, are well defined. Therefore, the proof also demonstrates the result of Tauman (1979), i.e., that there is a unique value of norm 1 on $In'NA$: the closed linear space of games that is generated by games of the form $f \circ \mu$ where $\mu \in NA^1$ and $f \in In' = \{ f : [0, 1] \to \mathbb{R} \mid f(0) = 0,\ f$ integrable and continuous at 0 and $1 \}$. The integrability requirement can also be dropped. Mertens and Neyman (2001) prove the existence of a value (of norm 1) on the spaces $'AN \supset 'NA$: the closure in the variation distance of the linear space spanned by all games $f \circ \mu$ where $\mu \in AN^1 \supset NA^1$ and f is a real-valued function defined on $[0, 1]$ with $f(0) = 0$ and continuous at 0 and 1. Moreover, the continuity of the function f at 0 and 1 can be further weakened by requiring the upper and lower averages of f on the intervals $[0, \varepsilon]$ and $[1 - \varepsilon, 1]$ to converge to $f(0) = 0$ and $f(1)$, respectively.

The existence of a value (of norm 1) on pNA and $bv'NA$ also follows from various constructive approaches to the value, such as the limiting approach discussed immediately below and the value formulas approach (Section 7). Other constructive approaches include the mixing approach (Section 6) that provides a value on pNA. Each constructive approach describes a way of "computing" the value for a specific game.

The space of all games that are in the (feasible) domain of each approach will form a subspace and the associated map will be a value. Each of the approaches has its advantages and each sheds additional interpretive light on the value obtained. We start with limiting value approaches.

5. Limiting values

Limiting values are defined by means of the limits of the Shapley value of the finite games that approximate the given game.

Given a game v on $(I, \mathcal{C})$ and a finite measurable field π, we denote by v_π the restriction of the game v to the finite field π. The real-valued function v_π (defined on π)

can be viewed as a finite game. The Shapley value for finite games ψv_π on π is given by

$$(\psi v_\pi)(a) = \frac{1}{n!} \sum_{\mathcal{R}} \left[v\left(\mathcal{P}_a^{\mathcal{R}} \cup a\right) - v\left(\mathcal{P}_a^{\mathcal{R}}\right) \right],$$

where n is the number of atoms of π, the sum runs over all orders $\mathcal{R}$ of the atoms of π, and $\mathcal{P}_a^{\mathcal{R}}$ is the union of all atoms preceding a in the order $\mathcal{R}$.

5.1. The weak asymptotic value

The family of measurable finite fields (that contain a given coalition T) is ordered by inclusion, and given two measurable finite fields π and π' there is a measurable finite field π'' that contains both π and π'. This enables us to define limits of functions of finite measurable fields. The weak asymptotic value of the game v is defined as $\lim_\pi \psi v_\pi$ whenever the limit of $\psi v_\pi(T)$ exists where π ranges over the family of measurable finite fields that include T. Equivalently,

DEFINITION 3. Let v be a game. A game φv is said to be the *weak asymptotic value* of v if for every $S \in \mathcal{C}$ and every $\varepsilon > 0$, there is a finite subfield π of $\mathcal{C}$ with $S \in \pi$ such that for any finite subfield π' with $\pi' \supset \pi$, $|\psi v_{\pi'}(S) - \varphi v(S)| < \varepsilon$.

REMARKS. (1) Let v be a game. The game v has a weak asymptotic value if and only if for every S in $\mathcal{C}$ and $\varepsilon > 0$ there exists a finite subfield π with $S \in \pi$ such that for any finite subfield π' with $\pi' \supset \pi$, $|\psi v_{\pi'}(S) - \psi v_\pi(S)| < \varepsilon$.

(2) A game v has at most one weak asymptotic value.

(3) The weak asymptotic value φv of a game v is a finitely additive game, and $\|\varphi v\| \leqslant \|v\|$ whenever $v \in BV$.

THEOREM 3 [Neyman (1988), p. 559]. *The set of all games having a weak asymptotic value is a linear symmetric space of games, and the operator mapping each game to its weak asymptotic value is a value on that space. If $ASYMP^*$ denotes all games with bounded variation having a weak asymptotic value, then $ASYMP^*$ is a closed subspace of BV with $bv'FA \subset ASYMP^*$.*

REMARKS. The linearity, symmetry, positivity and efficiency of ψ (the Shapley value for games with finitely many players) and of the map $v \mapsto v_\pi$ imply the linearity, symmetry, positivity and efficiency of the weak asymptotic value map and also imply the linearity and symmetry of the space of games having a weak asymptotic value. The closedness of $ASYMP^*$ follows from the linearity of ψ and from the inequalities $\|\psi v_\pi\| \leqslant \|v_\pi\| \leqslant \|v\|$. The difficult part of the theorem is the inclusion $bv'FA \subset ASYMP^*$, on which we will comment later. The inclusion $bv'FA \subset ASYMP^*$ implies in particular the existence of a value on $bv'FA$.

5.2. The asymptotic value

The asymptotic value of a game v is defined whenever all the sequences of the Shapley values of finite games that "approximate" v have the same limit. Formally: given T in C, a T-admissible sequence is an increasing sequence $(\pi_1, \pi_2, \ldots)$ of finite subfields of C such that $T \in \pi_1$ and $\bigcup_i \pi_i$ generates C.

DEFINITION 4. A game φv is said to be the *asymptotic value* of v, if $\lim_{k \to \infty} \psi v_{\pi_k}(T)$ exists and $= \varphi v(T)$ for every $T \in C$ and every T-admissible sequence $(\pi_i)_{i=1}^{\infty}$.

REMARKS. (1) A game v has an asymptotic value if and only if $\lim_{k \to \infty} \psi v_{\pi_k}(T)$ exists for every T in C and every T-admissible sequence $\mathcal{P} = (\pi_k)_{k=1}^{\infty}$, and the limit is independent of the choice of $\mathcal{P}$.

(2) For any given v, the asymptotic value, if it exists, is clearly unique.

(3) The asymptotic value φv of a game v is finitely additive, and $\|\varphi v\| \leqslant \|v\|$ whenever $v \in BV$.

(4) If v has an asymptotic value φv then v has a weak asymptotic value, which $= \varphi v$.

(5) The *dual* of a game v is defined as the game v^* given by $v^*(S) = v(I) - v(I \setminus S)$. Then $\psi v_{\pi}^* = \psi v_{\pi} = \psi(\frac{v + v^*}{2})_{\pi}$, and therefore v has a (weak) asymptotic value if and only if v^* has a (weak) asymptotic value if and only if $(\frac{v + v^*}{2})$ has a (weak) asymptotic value, and the values coincide.

The space of all games of bounded variation that have an asymptotic value is denoted *ASYMP*.

THEOREM 4 [Aumann and Shapley (1974), Theorem F; Neyman (1981a); and Neyman (1988), Theorem A]. *The set of all games having an asymptotic value is a linear symmetric space of games, and the operator mapping each game to its asymptotic value is a value on that space. ASYMP is a closed linear symmetric subspace of BV which contains $bv'M$.*

REMARKS. (1) The linearity and symmetry of ψ (the Shapley value for games with finitely many players), and of the map $v \to v_{\pi}$, imply the linearity and symmetry of the asymptotic value map and its domain. The efficiency and positivity of the asymptotic value follows from the efficiency and positivity of the maps $v \to v_{\pi}$ and $v_{\pi} \to \psi v_{\pi}$. The closedness of *ASYMP* follows from both the linearity of ψ and the inequalities $\|\psi v_{\pi}\| \leqslant \|v\|_{\pi} \leqslant \|v\|$.

(2) Several authors have contributed to proving the inclusion $bv'M \subset ASYMP$. Let FL denote the set of measures with at most finitely many atoms and M_a all purely atomic measures. Each of the spaces, $bv'NA$, $bv'FL$ and $bv'M_a$ is defined as the closed subspace of BV that is generated by games of the form $f \circ \mu$ where $f \in bv'$ and μ is a probability measure in NA^1, FL^1, or M_a^1 respectively.

Kannai (1966) and Aumann and Shapley (1974) show that $pNA \subset ASYMP$. Artstein (1971) proves essentially that $pM_a \subset ASYMP$. Fogelman and Quinzii (1980) show that $pFL \subset ASYMP$. Neyman (1981a, 1979) shows that $bv'NA \subset ASYMP$ and $bv'FL \subset ASYMP$ respectively. Berbee (1981), together with Shapley (1962) and the proof of Neyman [(1981a), Lemma 8 and Theorem A], imply that $bv'M_a \subset ASYMP$. It was further announced in Neyman (1979) that Berbee's result implies the existence of a partition value on $bv'M$. Neyman (1988) shows that $bv'M \subset ASYMP$.

(3) The space of games, $bv'M$, is generated by scalar measure games, i.e., by games that are real-valued functions of scalar measures. Therefore, the essential part of the inclusion $bv'M \subset ASYMP$ is that whenever $f \in bv'$ and μ is a probability measure, $f \circ \mu$ has an asymptotic value. The tightness of the assumptions on f and μ for $f \circ \mu$ to have an asymptotic value follows from the next three remarks.

(4) $pFA \notin ASYMP$ [Neyman (1988), p. 578]. For example, $\mu^3 \notin ASYMP$ whenever μ is a positive non-atomic finitely additive measure which is not countably additive. This illustrates the essentiality of the countable additivity of μ for $f \circ \mu$ to have an asymptotic value when f is a polynomial.

(5) For $0 < q < 1$, we denote by f_q the function given by $f_q(x) = 1$ if $x \geqslant q$ and $f_q(x) = 0$ otherwise. Let μ be a non-atomic measure with total mass 1. Then $f_q \circ \mu$ has an asymptotic value if and only if μ is positive [Neyman (1988), Theorem 5.1]. There are games of the form $f_q \circ \mu$, where μ is a purely atomic signed measure with $\mu(I) = 1$, for which $f_{\frac{1}{2}} \circ \mu$ does not have an asymptotic value [Neyman (1988), Example 5.2]. These two comments illustrate the essentiality of the positivity of the measure μ.

(6) There are games of the form $f \circ \mu$, where $\mu \in NA^1$ and f is continuous at 0 and 1 and vanishes outside a countable set of points, which do not have an asymptotic value [Neyman (1981a), p. 210]. Thus, the bounded variation of the function f is essential for $f \circ \mu$ to have an asymptotic value.

(7) The set of games $DIAG$ is defined [Aumann and Shapley (1974), p. 252] as the set of all games v in BV for which there is a positive integer k, measures $\mu_1, \ldots, \mu_k \in NA^1$, and $\varepsilon > 0$, such that for any coalition $S \in \mathcal{C}$ with $\mu_i(S) - \mu_j(S) \leqslant \varepsilon$ for all $1 \leqslant i, j \leqslant k$, $v(S) = 0$. $DIAG$ is a symmetric linear subspace of $ASYMP$ [Aumann and Shapley (1974)].

(8) If μ_1, μ_2, μ_3, are three non-atomic probabilities that are linearly independent, then the game v defined by $v(S) = \min[\mu_1(S), \mu_2(S), \mu_3(S)]$ is not in $ASYMP$ [Aumann and Shapley (1974), Example 19.2].

(9) Games of the form $f_q \circ \mu$ where $\mu \in M^1$ and $0 < q < 1$ are called *weighted majority games*; a coalition S is winning, i.e., $v(S) = 1$, if and only if its total μ-weight is at least q. The measure μ is called the *weight measure* and the number q is called the *quota*. The result $bv'M \subset ASYMP$ implies in particular that weighted majority games have an asymptotic value.

(10) A *two-house weighted majority game* v is the product of two weighted majority games, i.e., $v = (f_q \circ \mu)(f_c \circ \nu)$, where $\mu, \nu \in M^1, 0 < c < 1$, and $0 < q < 1$. If $\mu, \nu \in$

NA^1 with $\mu \neq \nu$ and $q \neq 1/2$, ν has an asymptotic value if and only if $q \neq c$. Indeed, if $q < c$, $(f_q \circ \mu)(f_c \circ \nu) - (f_c \circ \nu) \in DIAG$ and therefore

$$(f_q \circ \mu)(f_c \circ \nu) \in bv'NA + DIAG \subset ASYMP,$$

and if $q = c \neq 1/2$, $(f_q \circ \mu)(f_c \circ \nu) \notin ASYMP$ [Neyman and Tauman (1979), proof of Proposition 5.42]. If μ has infinitely many atoms and the set of atoms of ν is disjoint to the set of atoms of μ, $(f_q \circ \mu)(f_c \circ \nu)$ has an asymptotic value [Neyman and Smorodinski (1993)].

(11) Let $\mathcal{A}$ denote the closed algebra that is generated by games of the form $f \circ \mu$ where $\mu \in NA^1$ and $f \in bv'$ is continuous. It follows from the proof of Proposition 5.5 in Neyman and Tauman (1979) together with Neyman (1981a) that $\mathcal{A} \subset ASYMP$ and that for every $u \in \mathcal{A}$ and $v \in bv'NA$, $uv \in ASYMP$.

The following should shed light on the proofs regarding asymptotic and limiting values.

Let π be a finite subfield of $\mathcal{C}$ and let v be a game. We will recall formulas for the Shapley value of finite games by applying these formulas to the game v_π. This will enable us to illustrate proofs regarding the limiting properties of the Shapley value ψv_π of the finite game v_π. Let $A(\pi)$, or A for short, denote the set of atoms of π. The Shapley value of the game v_π to an atom a in A is given by

$$\psi v_\pi(a) = \frac{1}{|A|!} \sum_{\mathcal{R}} [v(P_a^{\mathcal{R}} \cup a) - v(P_a^{\mathcal{R}})],$$

where $|A|$ stands for the number of elements of A, the summation ranges over all orders $\mathcal{R}$ of the atoms in A, and $P_a^{\mathcal{R}}$ is the union of all atoms preceding a in the order R. An alternative formula for ψv_π could be given by means of a family X_a, $a \in A(\pi)$, of i.i.d. real-valued random variables that are uniformly distributed on $(0, 1)$. The values of X_a, $a \in A(\pi)$, induce with probability 1 an order $\mathcal{R}$ on A; a precedes b if and only if $X_a < X_b$. As X_a, $a \in A$, are i.i.d., all orders are equally likely. Thus, identifying sets with their indicator functions, and letting $I(X_a \leqslant X_b)$ denote the indicator of the event $X_a \leqslant X_b$, the value formula is

$$\psi v_\pi(a) = E\left[v\left(\sum_{b \in A} bI(X_b \leqslant X_a)\right) - v\left(\sum_{b \in A} bI(X_b < X_a)\right)\right].$$

Thus, by conditioning on the value of X_a, and letting $S(t)$, $0 \leqslant t \leqslant 1$, denote the union of all atoms b of π for which $X_b \leqslant t$, we find that

$$\psi v_\pi(a) = \int_0^1 E\big(v(S(t) \cup a) - v(S(t) \setminus a)\big) \, dt.$$

Assume now that the game v is a continuously differentiable function of finitely many non-atomic probability measures; i.e., $v = f \circ \mu$, where $\mu = (\mu_1, \ldots, \mu_n) \in (NA^1)^n$ and $f : \mathcal{R}(\mu) \to \mathcal{R}$ is continuously differentiable. Then

$$v\big(S(t) \cup a\big) - v(S(t) \setminus a) = \sum_{i=1}^{n} \frac{\partial f}{\partial \xi_i}(y) \mu_i(a),$$

where y is a point in the interval $[\mu(S(t) \setminus a), \mu(S(t) \cup a)]$. Thus, in particular, $\sum_{a \in \pi} |v(S(t) \cup a) - v(S(t) \setminus a)|$ is bounded.

For any scalar measure v, $v(S(t))$ is the sum $\sum_{a \in A(\pi)} v(a) I(X_a \leqslant t)$, i.e., it is a sum of $|A(\pi)|$ independent random variables with expectation $tv(a)$ and variance $t(1 - t)v^2(a)$. Therefore, $E(v(S(t))) = tv(I)$ and

$$\mathrm{Var}\big(S(t)\big) = t(1 - t) \sum_{a \in A(\pi)} v^2(a) \leqslant t(1 - t)v(I) \max\{v(a) : a \in A(\pi)\}.$$

Therefore, if $\mu = (\mu_1, \ldots, \mu_n)$ is a vector of measures, and $(\pi_k)_{k=1}^{\infty}$ is a sequence of finite fields with $\max\{|\mu_j(a)| : 1 \leqslant j \leqslant n, a \in A(\pi_k)\} \to_{k \to \infty} 0$, $\mu(S(t))$ converges in distribution to $t\mu(I)$; so if T is a coalition with $T \in \pi_k$ for every k, then

$$\sum_{\substack{a \subset T \\ a \in A(\pi_k)}} \big[v\big(S_k(t) \cup a\big) - v\big(S_k(t) \setminus a\big)\big] \to_{k \to \infty} \sum_{i=1}^{n} \frac{\partial f}{\partial x_i}\big(t\mu(I)\big)\mu_i(T),$$

and therefore $\lim_{k \to \infty} \psi v_{\pi_k}(T)$ exists and

$$\lim_{k \to \infty} \psi v_{\pi_k}(T) = \int_0^1 f_{\mu(T)}\big(t\mu(I)\big) \, dt.$$

For any T-admissible sequence $(\pi_k)_{k=1}^{\infty}$ and non-atomic measure v, $\max\{v(a) : a \in A(\pi_k)\} \to_{k \to \infty} 0$. Therefore any game $v = f \circ \mu$, where f is a continuously differentiable function defined on the range of μ, $\{\mu(S) : S \in \mathcal{C}\}$, has an asymptotic value φv whenever $\mu = (\mu_1, \ldots, \mu_n)$ is a vector of non-atomic measures.

Also, given any $\varepsilon > 0$ and a non-atomic element v of FA^+, there exists a partition π such that $\max\{v(a) : a \in A(\pi)\} < \varepsilon$. Therefore $v = f \circ \mu$ has a weak asymptotic value φv whenever $\mu = (\mu_1, \ldots, \mu_n)$ is a vector of non-atomic bounded finitely additive measures. In both cases the value is given by

$$\varphi v(S) = \int_0^1 f_{\mu(S)}\big(t\mu(I)\big) \, dt.$$

The following tool which is of independent interest will be used later on to show how to reduce the proof of $bv'NA \subset ASYMP$ (or $bv'M \subset ASYMP$) to the proof that $f_q \circ \mu \in ASYMP$ where $f_q(x) = I(x \geqslant q)$, $0 < q < 1$, and $\mu \in NA^1$ (or $\mu \in M^1$).

Let v be a monotonic game. For every $\alpha > 0$, let v^α denote the indicator of $v \geqslant \alpha$, i.e., $v^\alpha(S) = I(v(S) \geqslant \alpha)$, i.e., $v^\alpha(S) = 1$ if $v(S) \geqslant \alpha$ and 0 otherwise. Note that for every coalition S, $v(S) = \int_0^\infty v^\alpha(S)\,d\alpha$. Then for every finite field π, $\psi v_\pi = \int_0^\infty \psi(v^\alpha)_\pi\,d\alpha = \int_0^{v(I)} \psi(v^\alpha)_\pi\,d\alpha$. Therefore if $(\pi_k)_{k=1}^\infty$ is a T-admissible sequence such that for every $\alpha > 0$, $\lim_{k\to\infty}\psi(v^\alpha)_{\pi_k}(T)$ exists, then from the Lebesgue's bounded convergence theorem it follows that $\lim_{k\to\infty}\psi v_{\pi_k}(T)$ exists and

$$\lim_{k\to\infty}\psi v_{\pi_k}(T) = \int_0^{v(I)} \left(\lim_{k\to\infty}\psi v_{\pi_k}^\alpha(T)\right)d\alpha.$$

Thus, if for each $\alpha > 0$, $v^\alpha = I(v \geqslant \alpha)$ has an asymptotic value φv^α, then v also has an asymptotic value φv, which is given by $\int_0^\infty \varphi v^\alpha(T)\,d\alpha = \varphi v(T)$.

The spaces *ASYMP* and *ASYMP** are closed and each of the spaces $bv'NA$, $bv'M$ and $bv'FA$ is the closed linear space that is generated by scalar measure games of the form $f \circ \mu$ with $f \in bv'$ monotonic and $\mu \in NA^1$, $\mu \in M^1$, $\mu \in FA^1$ respectively. Also, each monotonic f in bv' is the sum of two monotonic functions in bv', one right continuous and the other left continuous. If $f \in bv'$, with f monotonic and left continuous, then $(f \circ \mu)^* = g \circ \mu$, with $g \in bv'$ monotonic and right continuous.

Therefore, to show that $bv'NA \subset ASYMP$ ($bv'M \subset ASYMP$ or $bv'FA \subset ASYMP^*$), it suffices to show that $f \circ \mu \in ASYMP$ (or $\in ASYMP^*$) for any monotonic and right continuous function f in bv' and $\mu \in NA^1$ ($\mu \in M^1$ or $\mu \in FA^1$). Note that

$$(f \circ \mu)^\alpha(S) = I\big(f(\mu(S)) \geqslant \alpha\big) = I\big(\mu(S) \geqslant \inf\{x: f(x) \geqslant \alpha\}\big) = f_q\big(\mu(S)\big),$$

where $q = \inf\{x: f(x) \geqslant \alpha\}$. Thus, in view of the above remarks, to show that $bv'NA \subset ASYMP$ (or $bv'M \subset ASYMP$), it suffices to prove that $f_q \circ \mu \in ASYMP$ for any $0 < q < 1$ and $\mu \in NA^1$ (or $\mu \in M^1$), where $f_q(x) = 1$ if $x \geqslant q$ and $= 0$ otherwise.

The proofs of the relations $f_q \circ \mu \in ASYMP$ for $0 < q < 1$ and $\mu \in NA^1$ (or $\mu \in M_a^1$ or $\mu \in M^1$) rely on delicate probabilistic results, which are of independent mathematical interest. These results can be formulated in various equivalent forms. We introduce them as properties of Poisson bridges. Let $X_1, X_2, \ldots$ be a sequence of i.i.d. random variables that are uniformly distributed on the open unit interval $(0, 1)$. With any finite or infinite sequence $w = (w_i)_{i=1}^n$ or $w = (w_i)_{i=1}^\infty$ of positive numbers, we associate a stochastic process, $S_w(\cdot)$, or $S(\cdot)$ for short, given by $S(t) = \sum_i w_i I(X_i \leqslant t)$ for $0 \leqslant t \leqslant 1$ (such a process is called a pure jump Poisson bridge). The proof of $f_q \circ \mu \in ASYMP$ for $\mu \in NA^1$ uses the following result, which is closely related to renewal theory.

PROPOSITION 4 [Neyman (1981a)]. *For every $\varepsilon > 0$ there exists a constant $K > 0$, such that if $w_1, \ldots, w_n$ are positive numbers with $\sum_{i=1}^n w_i = 1$, and $K \max_{i=1}^n w_i < q < 1 - K \max_{i=1}^n w_i$ then*

$$\sum_{i=1}^n \left| w_i - \mathrm{Prob}\big(S(X_i) \in [q, q+w_i)\big)\right| < \varepsilon.$$

The proof of $f_q \circ \mu \in ASYMP$ for $\mu \in M_a$ uses the following result, due to Berbee (1981).

PROPOSITION 5 [Berbee (1981)]. *For every sequence* $(w_i)_{i=1}^{\infty}$ *with* $w_i > 0$, *and every* $0 < q < \sum_{i=1}^{\infty} w_i$, *the probability that there exists* $0 \leqslant t \leqslant 1$ *with* $S(t) = q$ *is* 0, *i.e.*,

$$\text{Prob}(\exists t, \ S(t) = q) = 0.$$

6. The mixing value

An *order* on the underlying space I is a relation $\mathcal{R}$ on I that is transitive, irreflexive, and complete. For each order $\mathcal{R}$ and each s in I, define the initial segment $I(s; \mathcal{R})$ by $I(s; \mathcal{R}) = \{t \in I : s\mathcal{R}t\}$. An order $\mathcal{R}$ is measurable if the σ-field generated by all the initial segments $I(s; \mathcal{R})$ equals $\mathcal{C}$. Let ORD be the linear space of all games v in BV such that for each measurable $\mathcal{R}$, there exists a unique measure $\varphi(v; \mathcal{R})$ satisfying

$$\varphi(v; \mathcal{R})\big(I(s; \mathcal{R})\big) = v\big(I(s; \mathcal{R})\big) \quad \text{for all } s \in I.$$

Let μ be a probability measure on the underlying space $(I, \mathcal{C})$. A μ-mixing sequence is a sequence $(\Theta_1, \Theta_2, \ldots)$ of μ-measure-preserving automorphisms of $(I, \mathcal{C})$, such that for all S, T in $\mathcal{C}$, $\lim_{k \to \infty} \mu(S \cap \Theta_k T) = \mu(S)\mu(T)$. For each order $\mathcal{R}$ on I and automorphism Θ in $\mathcal{G}$ we denote by $\Theta\mathcal{R}$ the order on I defined by $\Theta s \Theta \mathcal{R} \Theta t \Leftrightarrow s\mathcal{R}t$. If v is absolutely continuous with respect to the measure μ we write $v \ll \mu$.

DEFINITION 5. Let $v \in ORD$. A game φv is said to be the *mixing value* of v if there is a measure μ_v in NA^1 such that for all μ in NA^1 with $\mu_v \ll \mu$ and all μ-mixing sequences $(\Theta_1, \Theta_2, \ldots)$, for all measurable orders $\mathcal{R}$, and all coalitions S, $\lim_{k \to \infty} \varphi(v; \Theta_k \mathcal{R})(S)$ exists and $= (\varphi v)(S)$.

The set of all games in ORD that have a mixing value is denoted MIX.

THEOREM 5 [Aumann and Shapley (1974), Theorem E]. *MIX is a closed symmetric linear subspace of BV which contains pNA, and the map* φ *that associates a mixing value* φv *to each* v *is a value on MIX with* $\|\varphi\| \leqslant 1$.

7. Formulas for values

Let v be a vector measure game, i.e., a game that is a function of finitely many measures. Such games have a representation of the form $v = f \circ \mu$, where $\mu = (\mu_1, \ldots, \mu_n)$ is a vector of (probability) measures, and f is a real-valued function defined on the range $\mathcal{R}(\mu)$ of the vector measure μ with $f(0) = 0$.

When f is continuously differentiable, and $\mu = (\mu_1, \ldots, \mu_n)$ is a vector of non-atomic measures with $\mu_i(I) \neq 0$, then $v = f \circ \mu$ is in $pNA_\infty (\subset pNA \subset ASYMP)$ and the value (asymptotic, mixing, or the unique value on pNA) is given by Aumann and Shapley (1974, Theorem B),

$$\varphi v(S) \equiv \varphi(f \circ \mu)(S) = \int_0^1 f_{\mu(S)}\big(t\mu(I)\big)\,dt$$

(where $f_{\mu(S)}$ is the derivative of f in the direction of $\mu(S)$); i.e.,

$$\varphi(f \circ \mu)(S) = \sum_{i=1}^n \mu_i(S) \int_0^1 \frac{\partial f}{\partial x_i}\big(t\mu_1(I), \ldots, t\mu_n(I)\big)\,dt.$$

The above formula expresses the value as an average of derivatives, that is, of marginal contributions. The derivatives are taken along the interval $[0, \mu(I)]$, i.e., the line segment connecting the vector 0 and the vector $\mu(I)$. This interval, which is contained in the range of the non-atomic vector measure μ, is called the *diagonal* of the range of μ, and the above formula is called the *diagonal formula*. The formula depends on the differentiability of f along the diagonal and on the game being a non-atomic vector measure game.

Extensions of the diagonal formula (due to Mertens) enable one to apply (variations of) the diagonal formula to games with discontinuities and with essential nondifferentiabilities. In particular, the generalized diagonal formula defines a value of norm 1 on a closed subspace of BV that includes all finite games, $bv'FA$, the closed algebra generated by $bv'NA$, all games generated by a finite number of algebraic and lattice operations[1] from a finite number of measures, and all market games that are functions of finitely many measures.

The value is defined as a composition of positive linear symmetric mappings of norm 1. One of these mappings is an extension operator which is an operator from a space of games into the space of ideal games, i.e., functions that are defined on ideal coalitions, which are measurable functions from $(I, \mathcal{C})$ into $([0, 1], \mathcal{B})$.

The second one is a "derivative" operator, which is essentially the diagonal formula obtained by changing the order of integration and differentiation. The third one is an averaged derivative; it replaces the derivatives on this diagonal with an average of derivatives evaluated in a small invariant perturbation of the diagonal. The invariance of the perturbed distribution is with respect to all automorphisms of $(I, \mathcal{C})$.

First we illustrate the action and basic properties of the "derivative" and the "averaged derivative" mappings in the context of non-atomic vector measure games.

The derivative operator [Mertens (1980)] provides a diagonal formula for the value of vector measure games in $bv'NA$; let $\mu = (\mu_1, \ldots, \mu_n) \in (NA^+)^n$ and let $f : \mathcal{R}(\mu) \to \mathbb{R}$

---

[1] Max and min.

with $f(0) = 0$ and $f \circ \mu$ in $bv'NA$. Then f is continuous at $\mu(I)$ and $\mu(\emptyset) = 0$, and the value of $f \circ \mu$ is given by

$$
\begin{aligned}
\varphi(f \circ \mu)(S) &= \lim_{\varepsilon \to 0+} \frac{1}{\varepsilon} \int_0^{1-\varepsilon} \left[f\big(t\mu(I) + \varepsilon\mu(S)\big) - f\big(t\mu(I)\big) \right] dt \\
&= \lim_{\varepsilon \to 0+} \frac{1}{2\varepsilon} \int_\varepsilon^{1-\varepsilon} \left[f\big(t\mu(I) + \varepsilon\mu(S)\big) - f\big(t\mu(I) - \varepsilon\mu(S)\big) \right] dt.
\end{aligned}
$$

The limits of integration in these formulas, $1 - \varepsilon$ and ε, could be replaced by 1 and 0 respectively whenever f is extended to $\mathbb{R}^n$ in such a way that f remains continuous at $\mu(I)$ and $\mu(\emptyset)$.

Let $\mu = (\mu_1, \dots, \mu_n) \in (NA^+)^n$ be a vector of measures. Consider the class $\mathcal{F}(\mu)$ of all functions $f : \mathcal{R}(\mu) \to \mathbb{R}$ with $f(0) = 0$, f continuous at 0 and $\mu(I)$, $f \circ \mu$ of bounded variation, and for which the limit

$$
\lim_{\varepsilon \to 0} \frac{1}{2\varepsilon} \int_\varepsilon^{1-\varepsilon} \left[f\big(t\mu(I) + \varepsilon\mu(S)\big) - f\big(t\mu(I) - \varepsilon\mu(S)\big) \right] dt
$$

exists for all S in $\mathcal{C}$. Note that $f \in \mathcal{F}(\mu)$ whenever f is continuously differentiable with $f(0) = 0$. An example of a Lipschitz function that is not in $\mathcal{F}(\mu)$ where μ is a vector of 2 linearly independent probability measures is $f(x_1, x_2) = (x_1 - x_2) \sin \log |x_1 - x_2|$. Denote this limit by $Dv(S)$ and note that Dv is a function of the measures $\mu_1, \dots, \mu_n$; i.e., $Dv = \bar{f} \circ \mu$ for some function $\bar{f}$ that is defined on the range of μ. The derivative operator D acts on all games of the form $f \circ \mu$, $\mu \in (NA^+)^n$, and $f \in \mathcal{F}(\mu)$; it is a value of norm 1 on the space of all these games for which $D(f \circ \mu) \in FA$.

Let $v = f \circ \mu$, where $\mu \in (NA^+)^n$ and $f \in \mathcal{F}(\mu)$. Then $D(f \circ \mu)$ is of the form $\bar{f} \circ \mu$, where $\bar{f}$ is linear on every plane containing the diagonal $[0, \mu(I)]$, $\bar{f}(\mu(I)) = f(\mu(I))$ and $\|\bar{f} \circ \mu\| \leqslant \|f \circ \mu\|$. There is no loss of generality in assuming that the measures $\mu_1, \dots, \mu_n$ are independent (i.e., that the range of the vector measure $\mathcal{R}(\mu)$ is full-dimensional), and then we identify $\bar{f}$ with its unique extension to a function on $\mathbb{R}^n$ that is linear on every plane containing the interval $[0, \mu(I)]$.

The averaging operator expresses the value of the game $\bar{f} \circ \mu$ ($= D(f \circ \mu)$ for f in $\mathcal{F}(\mu)$) as an average, $\mathrm{E}(\sum_{i=1}^n \frac{\partial \bar{f}}{\partial x_i}(z)\mu_i)$, where z is an n-dimensional random variable whose distribution, P_μ, is strictly stable of index 1 and has a Fourier transform ξ to be described below. When $\mu_1, \dots, \mu_n$ are mutually singular, $z = (z_1, \dots, z_n)$ is a vector of independent "centered" Cauchy distributions, and then

$$
\xi(y) = \exp\left(-\sum_{i=1}^n \mu_i(I) \right) = \exp\left(-\sum_{i=1}^n \|\mu_i\| \right).
$$

Otherwise, let $v \in NA^+$ be such that $\mu_1, \ldots, \mu_n$ are absolutely continuous with respect to v; e.g., $v = \sum \mu_i$. Define $N_\mu : \mathbb{R}^n \to \mathbb{R}^n$ by

$$N_\mu(y) = \int_I \left| \sum_{i=1}^n (d\mu_i/dv) y_i \right| dv.$$

Then z is an n-dimensional random variable whose distribution has a Fourier transform $\xi_z(y) = \exp(-N_\mu(y))$.

Let Q be the linear space of games generated by games of the form $v = f \circ \mu$ where $\mu = (\mu_1, \ldots, \mu_n) \in (NA^+)^n$ is a vector of linearly independent non-atomic measures and $f \in \mathcal{F}(\mu)$. The space Q is symmetric and the map $\varphi : Q \to FA$ given by

$$\varphi(f \circ \mu) = E_{P_\mu}\big(\bar{f}_{\mu(S)}(z)\big) = E_{P_\mu}\left(\sum_{i=1}^n \frac{\partial \bar{f}_i}{\partial x_i}(z)\mu_i(S) \right)$$

is a value of norm 1 on Q and therefore has an extension to a value of norm 1 on the closure $\bar{Q}$ of Q.

The space $\bar{Q}$ contains $bv'NA$, the closed algebra generated by $bv'NA$, all games generated by a finite number of algebraic and lattice operations from a finite number of non-atomic measures and all games of the form $f \circ \mu$ where $\mu \in (NA^+)^n$ and f is a continuous concave function on the range of μ.

To show that $E_{P_\mu}(\bar{f}_{\mu(S)}(z))$ is well defined for any f in $\mathcal{F}(\mu)$, one shows that $\bar{f}$ is Lipschitz and therefore on the one hand $[\bar{f}(z + \delta\mu(S)) - \bar{f}(z)]/\delta$ is bounded and continuous, and on the other hand, given S in $\mathcal{C}$, for almost every z (with respect to the Lebesgue measure on $\mathbb{R}^n$) $\bar{f}_{\mu(S)}(z)$ exists and is bounded (as a function of z). The distribution P_μ is absolutely continuous with respect to Lebesgue measure and therefore $E_{P_\mu}(\bar{f}_{\mu(S)}(z))$ is well defined and equals $\lim_{\delta \to 0} E_{P_\mu}([\bar{f}(z + \delta\mu(S)) - \bar{f}(z)]/\delta)$ (using the Lebesgue bounded convergence theorem).

To show that $E_{P_\mu}(\bar{f}_{\mu(S)}(z))$ is finitely additive in S, assume that S, T are two disjoint coalitions. For any $\delta > 0$, let P_μ^δ be the translation of the measure P_μ by $-\delta\mu(S)$, i.e.,

$$P_\mu^\delta(B) = P_\mu\big(B + \delta\mu(S)\big).$$

Since P_μ^δ converges in norm to P_μ as $\delta \to 0$ (i.e., $\|P_\mu^\delta - P_\mu\| \to_{\delta \to 0} 0$), and $[\bar{f}(z + \delta\mu(T)) - \bar{f}(z)]/\delta$ is bounded, it follows that

$$\lim_{\delta \to 0} E_{P_\mu}\big([\bar{f}(z + \delta\mu(S) + \delta\mu(T)) - \bar{f}(z + \delta\mu(S))]/\delta\big)$$
$$= \lim_{\delta \to 0} E_{P_\mu^\delta}\big([\bar{f}(z + \delta\mu(T)) - \bar{f}(z)]/\delta\big)$$
$$= \lim_{\delta \to 0} E_{P_\mu}\big([\bar{f}(z + \delta\mu(T)) - \bar{f}(z)]/\delta\big) = E_{P_\mu}\big(\bar{f}_{\mu(T)}(z)\big).$$

Therefore

$$
\begin{aligned}
\mathrm{E}_{P_\mu}\big(\bar{f}_{\mu(S\cup T)}(z)\big) &= \lim_{\delta\to 0}\mathrm{E}_{P_\mu}\big(\big[\bar{f}\big(z+\delta\mu(S)+\delta\mu(T)\big) \\
&\quad - \bar{f}\big(z+\delta\mu(S)\big)+\bar{f}\big(z+\delta\mu(S)\big)-\bar{f}(z)\big]/\delta\big) \\
&= \mathrm{E}_{P_\mu}\big(\bar{f}_{\mu(S)}(z)\big)+\mathrm{E}_{P_\mu}\big(\bar{f}_{\mu(T)}(z)\big).
\end{aligned}
$$

We now turn to a detailed description of the mappings used in the construction of a value due to Mertens on a large space of games.

Let $B(I,\mathcal{C})$ be the space of bounded measurable real-valued functions on $(I,\mathcal{C})$, and $B_1^+(I,\mathcal{C}) = \{f\in B(I,\mathcal{C})\colon 0\leqslant f\leqslant 1\}$ the space of "ideal coalitions" (measurable fuzzy sets). A function $\bar{v}$ on $B_1^+(I,\mathcal{C})$ is *constant sum* if $\bar{v}(f)+\bar{v}(1-f)=\bar{v}(1)$ for every $f\in B_1^+(I,\mathcal{C})$. It is *monotonic* if for every $f,g\in B_1^+(I,\mathcal{C})$ with $f\geqslant g$, $\bar{v}(f)\geqslant\bar{v}(g)$. It is *finitely additive* if $\bar{v}(f+g)=\bar{v}(f)+\bar{v}(g)$ whenever $f,g\in B_1^+(I,\mathcal{C})$ with $f+g\leqslant 1$. It is *of bounded variation* if it is the difference of two monotonic functions, and the variation norm $\|\bar{v}\|_{IBV}$ (or $\|\bar{v}\|$ for short) is the supremum of the variation of $\bar{v}$ over all increasing sequences $0\leqslant f_1\leqslant f_2\leqslant\cdots\leqslant 1$ in $B_1^+(I,\mathcal{C})$. The symmetry group $\mathcal{G}$ acts on $B_1^+(I,\mathcal{C})$ by $(\Theta_* f)(s)=f(\Theta s)$.

DEFINITION 6. An *extension operator* is a linear and symmetric map ψ from a linear symmetric set of games to real-valued functions on $B_1^+(I,\mathcal{C})$, with $(\psi v)(0)=0$, so that $(\psi v)(1)=v(I)$, $\|\psi v\|_{IBV}\leqslant\|v\|$, and ψv is finitely additive whenever v is finitely additive, and ψv is constant sum whenever v is constant sum.

It follows that an extension operator is necessarily positive (i.e., if v is monotonic so is ψv) and that every extension operator ψ on a linear subspace Q of BV has a unique extension to an extension operator $\bar{\psi}$ on $\bar{Q}$ (the closure of Q).

One example of an extension operator is Owen's (1972) multilinear extension for finite games. For another example, consider all games that are functions of finitely many non-atomic probability measures, i.e., $Q=\{f\circ(\mu_1\ldots,\mu_n)\colon n$ a positive integer, $\mu_i\in NA^1$ and $f\colon\mathcal{R}(\mu_1,\ldots,\mu_n)\to\mathbb{R}$ with $f(0)=0\}$, where $\mathcal{R}(\mu_1,\ldots,\mu_n)=\{(\mu_1(S),\ldots,\mu_n(S))\colon S\in\mathcal{C}\}$ is the range of the vector measure $(\mu_1,\ldots,\mu_n)$. Then Q is a linear and symmetric space of games. An extension operator on Q is defined as follows: for $\mu=(\mu_1,\ldots,\mu_n)\in(NA^1)^n$ and $g\in B_1^+(I,\mathcal{C})$, let $\mu(g)=(\int_I g\,d\mu_1,\ldots,\int_I g\,d\mu_n)$ and define $\psi(f\circ\mu)(g)=f(\mu(g))$. Then it is easy to verify that ψ is well defined, linear, and symmetric, that $\psi v(1)=v(I)$, ψv is finitely additive whenever v is, and ψv is constant sum whenever v is. The inequality $\|\psi v\|_{IBV}\leqslant\|v\|$ follows from the result of Dvoretsky, Wald, and Wolfowitz [(1951), p. 66, Theorem 4], which asserts that for every $\mu\in(NA^1)^n$ and every sequence $0\leqslant f_1\leqslant f_2\leqslant\cdots\leqslant f_k\leqslant 1$ in $B_1^+(I,\mathcal{C})$, there exists a sequence $\emptyset\subset S_1\subset S_2\subset\cdots\subset S_k\subset I$ with $\mu(f_i)=\mu(S_i)$. This extension operator satisfies additional properties: $(\psi v)(S)=v(S)$, $\psi(vu)=(\psi v)(\psi u)$, $\psi(f\circ v)=f\circ\psi v$ whenever $f\colon\mathbb{R}\to\mathbb{R}$ with $f(0)=0$, $\psi(v\wedge u)=(\psi v)\wedge(\psi u)$ and $\psi(v\vee u)=(\psi v)\vee(\psi u)$. The NA-topology on $B(I,\mathcal{C})$ is the minimal topology for which the functions $f \mapsto \mu(f)$ are continuous for each $\mu \in NA$. For every game $v \in pNA$, ψv is *NA*-continuous. Moreover, as the characteristic functions $I_S \in B^1_+(I, \mathcal{C})$, $S \in \mathcal{C}$, are *NA*-dense in $B^1_+(I, \mathcal{C})$, ψv is the unique *NA*-continuous function on $B^1_+(I, \mathcal{C})$ with $\psi v(S) = v(S)$. Similarly, for every game v in pNA' – the closure in the supremum norm of pNA – there is a unique function $\bar{v} : B^1_+(I, \mathcal{C}) \to \mathbb{R}$ which is *NA*-continuous and $\bar{v}(S) = v(S)$. Moreover, the map $v \mapsto \bar{v}$ is an extension operator on pNA' and it satisfies: $\overline{vu} = \bar{v}\bar{u}$, $\overline{v \wedge u} = \bar{v} \wedge \bar{u}$ and $\overline{v \vee u} = \bar{v} \vee \bar{u}$ for all $v, u \in pNA'$ and $\overline{(f \circ v)} = f \circ \bar{v}$ whenever $f : \mathbb{R} \to \mathbb{R}$ is continuous with $f(0) = 0$.

Another example is the following natural extension map defined on $bv'M$. Assume that $\mu \in M^1$ with a set of atoms $A = A(\mu)$. Let $v = f \circ \mu$ with $f \in bv'$ and $\mu \in NA^1$. For $g \in B^1_+(I, \mathcal{C})$ define $\psi v(g) = \mathrm{E}(f(\mu(gA^c + \sum_{a \in A} X_a a)))$ where $A^c = I \setminus A$ and $(X_a)_{a \in A}$ are independent $\{0, 1\}$-valued random variables with $\mathrm{E}(X_a) = g(a)$. If $v = \sum_{i=1}^{k} v_i$, $v_i = f_i \circ \mu_i$ with $\mu_i \in M^1$ and $f_i \in bv'$ we define $\psi v(g) = \sum_{i=1}^{k} \bar{v}_i(g)$. It is possible to show that $\|\psi v\|_{IBV} \leqslant \|v\|$ and thus in particular $\bar{v}$ is well defined (i.e., it is independent of the representation of v) and can be extended to the closure of all these finite sums, i.e., it defines an extension map on $bv'M$.

The space $bv'NA$ is a subset of both $bv'M$ and $\overline{Q}$ and the above two extensions that were defined on $\overline{Q}$ and on $bv'M$ coincide on $bv'NA$, and thus induce (by restriction) an extension operator ψ on $bv'NA$ and on pNA. The extension on $bv'NA$ can be used to provide formulas for the values on $bv'NA$ and on pNA.

Given v in $\overline{Q}$, we denote by $\bar{v}$ the extension (ideal game) ψv. We also identify S with the indicator of the set S, χ_S, and the scalar t also stands for the constant function in $B^+_1(I, \mathcal{C})$ with value t. Then, if $v \in pNA$, for almost all $0 < t < 1$, for every coalition S, the derivative $\partial \bar{v}(t, S) =: \lim_{\delta \to 0}[\bar{v}(t) + \delta_S) - \bar{v}(t)]/\delta$ exists [Aumann and Shapley (1974), Proposition 24.1] and the value of the game is given by

$$\varphi v(S) = \int_0^1 \partial \bar{v}(t, S) \, \mathrm{d}t.$$

For any game v in $bv'NA$, the value of v is given by

$$\begin{aligned}
\varphi v(S) &= \lim_{\delta \to 0+} \frac{1}{\delta} \int_0^{1-\delta} \left[\bar{v}(t + \delta S) - \bar{v}(t)\right] \mathrm{d}t \\
&= \lim_{\delta \to 0} \frac{1}{2\delta} \int_{|\delta|}^{1-|\delta|} \left[\bar{v}(t + \delta S) - \bar{v}(t - \delta S)\right] \mathrm{d}t.
\end{aligned}$$

More generally, given an extension operator ψ on a linear symmetric space of games Q, we extend $\bar{v} = \psi v$, for v in Q, to be defined on all of $B(I, \mathcal{C})$ by $\bar{v}(f) = \bar{v}(\max[0, \min(1, f)])$ (and then we can replace the limits $|\delta|$ and $1 - |\delta|$ in the above integrals with 0 and 1 respectively, and define the derivative operator ψ_D by

$$\psi_D \bar{v}(\chi) = \lim_{\delta \to 0} \int_0^1 \frac{\bar{v}(t + \delta\chi) - \bar{v}(t - \delta\chi)}{2\delta} \, \mathrm{d}t$$

whenever the integral and the limit exists for all χ in $B(I, \mathcal{C})$. The integral exists whenever $\bar{v}$ has bounded variation. The derivative operator is positive, linear and symmetric. Assuming continuity in the NA-topology of $\bar{v}$ at 1 and 0, ψ_D also satisfies $\psi_D \bar{v}(\alpha + \beta \chi) = \alpha \bar{v}(1) + \beta [\psi_D \bar{v}](\chi)$ for every $\alpha, \beta \in \mathbb{R}$, so that $\psi_D \bar{v}$ is linear on every plan in $B(I, \mathcal{C})$ containing the constants and ψ_D is efficient, i.e., $\psi_D \bar{v}(1) = \bar{v}(1)$.

The following [due to Mertens (1988a)] will (essentially) extend the previously mentioned extension operators to a large class of games that includes $bv'FA$ and all games of the form $f \circ \mu$ where $\mu = (\mu_1, \ldots, \mu_n) \in NA$. Intuitively we would like to define $\psi_E v(f)$ for f in $B_1^+(I, \mathcal{C})$ as the "limit" of the expectation $v(S)$ with respect to a distribution P of a random set that is very similar to the ideal set f.

Let $\mathcal{F}$ be the set of all triplets $\alpha = (\pi, \nu, \varepsilon)$ where π is a finite subfield of C, ν is a finite set of non-atomic elements of FA and $\varepsilon > 0$. $\mathcal{F}$ is ordered by $\alpha = (\pi, \nu, \varepsilon) \leqslant \alpha' = (\pi', \nu', \varepsilon')$ if $\pi' \supset \pi$, $\nu' \supset \nu$ and $\varepsilon' \leqslant \varepsilon$. $(\mathcal{F}, \leqslant)$ is filtering-increasing with respect to this order, i.e., given $\alpha = (\pi, \nu, \varepsilon)$ and $\alpha' = (\pi', \nu', \varepsilon')$ in $\mathcal{F}$ there is $\alpha'' = (\pi'', \nu'', \varepsilon'')$ with $\alpha'' \geqslant \alpha$ and $\alpha'' \geqslant \alpha'$.

For any $\alpha = (\pi, \nu, \varepsilon)$ in $\mathcal{F}$ and f in $B_1^+(I, \mathcal{C})$, $P_{\alpha, f}$ is the set of all probabilities P with finite support on $\mathcal{C}$ such that $|\sum_{S \in \mathcal{C}} SP(S) - f| \equiv |E_P(S) - f| < \varepsilon$ uniformly on I (i.e., for every $t \in I$, $|\sum_{S \in \mathcal{C}} P(S)I(t \in S) - f(t)| < \varepsilon$), and such that for any T_1, T_2 in π with $T_1 \cap T_2 = \emptyset$, $T_1 \cap S$ is independent of $T_2 \cap S$, and for all μ in $\nu \mu(S \cap T_1) = \int_{T_1} f \, d\mu$. That $P_{\alpha, f} \neq \emptyset$ follows from Lyapunov's convexity theorem on the range of a vector of non-atomic elements of FA.

For any game v, and any $f \in B_1^+(I, \mathcal{C})$, let

$$\bar{v}(f) = \lim_{\alpha \in \mathcal{F}} \sup_{P \in P_{\alpha, f}} E_P\big(v(S)\big) \equiv \lim_{\alpha \in \mathcal{F}} \sup_{P \in P_{\alpha, f}} \sum_{S \in \mathcal{C}} v(S)P(S)$$

and

$$\underline{v}(f) = \lim_{\alpha \in \mathcal{F}} \inf_{P \in P_{\alpha, f}} E_P\big(v(S)\big).$$

The definition of $\bar{v}$ is extended to all of $B(I, \mathcal{C})$ by $\bar{v}(f) = \bar{v}(\max(0, \min(1, f)))$.

Note that if v is a game with finite support, then for any f in $B_1^+(I, \mathcal{C})$, $\bar{v}(f) = \underline{v}(f)$ coincides with Owen's multilinear extension, i.e., letting T be the finite support of v,

$$\bar{v}(f) = \sum_{S \subset T} \left[\prod_{s \in S} f(s) \prod_{s \in T \setminus S} (1 - f(s)) \right] v(S)$$

and whenever v is a function of finitely many non-atomic elements of FA, $v_1, \ldots, v_n$, i.e., $v = g \circ (v_1, \ldots, v_n)$ where g is a real-valued function defined on the range of $(v_1, \ldots, v_n)$, then for any $f \in B_1^+(I, \mathcal{C})$

$$\bar{v}(f) = \underline{v}(f) = g\left(\int f \, dv_1, \ldots, \int f \, dv_n \right).$$

Let $V_\varepsilon = \{\chi \in B_1^+(I, C) : \sup \chi - \inf \chi \leqslant \varepsilon\}$. Let D be the space of all games v in BV for which $\frac{1}{\varepsilon} \sup\{\bar{v}(\chi) - \underline{v}(\chi) : \chi \in V_\varepsilon\} \to 0$ as $\varepsilon \to 0+$. Obviously, $D \supset D_\varepsilon$ where D_ε is the set of all games v for which $\bar{v}(\chi) = \underline{v}(\chi)$ for any $\chi \in V_\varepsilon$. Next we define the derivative operators φ_D; its domain is the set of all games v in D for which the integrals (for sufficiently small $\varepsilon > 0$) and the limit

$$[\varphi_D v](\chi) = \lim_{\tau \to 0} \int_0^1 \frac{\bar{v}(t + \tau \chi) - \bar{v}(t - \tau \chi)}{2\tau} \, dt$$

exist for every χ in $B(I, C)$.

The integrals always exist for games in BV. The limit, on the other hand, may not exist even for games that are Lipschitz functions of a non-atomic signed measure. However, the limit exists for many games of interest, including the concave functions of finitely many non-atomic measures, and the algebra generated by $bv'FA$. Assuming, in addition, some continuity assumption on $\bar{v}$ at 1 and 0 (e.g., $\bar{v}(f) \to \bar{v}(1)$ as $\inf(f) \to 1$ and $\bar{v}(f) \to \bar{v}(0)$ as $\sup(f) \to 0$) we obtain that $\varphi_D v$ obeys the following additional properties: $\varphi_D v(\chi) + \varphi_D v(1 - \chi) = v(1)$ whenever v is a constant sum; $[\varphi_D v](a + bx) = av(1) + b[\varphi_D v](x)$ for every $a, b \in \mathbb{R}$.

THEOREM 6 [Mertens (1988a), Section 1]. *Let $Q = \{v \in Dom\,\varphi_D : \varphi_D v \in FA\}$. Then Q is a closed symmetric space that contains DIFF, DIAG and $bv'FA$ and $\varphi_D : Q \to FA$ is a value on Q.*

For every function $w : B(I, C) \to \mathbb{R}$ in the range of φ_D, and every two elements χ, h in $B(I, C)$, we define $w_h(\chi)$ to be the (two-sided) directional derivative of w at χ in the direction h, i.e.,

$$w_h(\chi) = \lim_{\delta \to 0} \left[w(\chi + \delta h) - w(\chi - \delta h) \right] / (2\delta)$$

whenever the limit exists. Obviously, when w is finitely additive then $w_h(\chi) = w(h)$. For a function w in the range of φ_D and of the form $w = f \circ \mu$ where $\mu = (\mu_1, \ldots, \mu_n)$ is a vector of measures in NA, f is Lipschitz, satisfying $f(a\mu(I) + bx) = af(\mu(I)) + bf(x)$ and for every y in $L(\mathcal{R}(\mu))$ – the linear span of the range of the vector measure μ – for almost every x in $L(\mathcal{R}(\mu))$ the directional derivative of f at x in the direction y, $f_y(x)$ exists, and then obviously $w_h(\chi) = f_{\mu(h)}(\mu(\chi))$.

The following theorem will provide an existence of a value on a large space of games as well as a formula for the value as an average of the derivatives $(\varphi_D \circ \varphi_E v)_h(\chi) \equiv w_h(\chi)$ where χ is distributed according to a cylinder probability measure on $B(I, C)$ which is invariant under all automorphisms of (I, C). We recall now the concept of a cylinder measure on $B(I, C)$.

The algebra of cylinder sets in $B(I, C)$ is the algebra generated by the sets of the firm $\mu^{-1}(B)$ where B is any Borel subset of $\mathbb{R}^n$ and $\mu = (\mu_1, \ldots, \mu_n)$ is any vector of measures. A cylinder probability is a finitely additive probability P on the algebra of cylinder sets such that for every vector measure $\mu = (\mu_1, \ldots, \mu_n)$, $P \circ \mu^{-1}$ is a countably additive probability measure on the Borel subsets of $\mathbb{R}^n$. Any cylinder probability P on $B(I, \mathcal{C})$ is uniquely characterized by its Fourier transform, a function on the dual that is defined by $F(\mu) = E_P(\exp(i\mu(\chi))$. Let P be the cylinder probability measure on $B(I, \mathcal{C})$ whose Fourier transform F_P is given by $F_P(\mu) = \exp(-\|\mu\|)$. This cylinder measure is invariant under all automorphisms of $(I, \mathcal{C})$. Recall that $(I, \mathcal{C})$ is isomorphic to $[0, 1]$ with the Borel sets, and for a vector of measures $\mu = (\mu_1, \ldots, \mu_n)$, the Fourier transform of $P \circ \mu^{-1}$, $F_{P \circ \mu^{-1}}$ is given by $F_{P \circ \mu^{-1}}(y) = \exp(-N_\mu(y))$, and $P \circ \mu^{-1}$ is absolutely continuous with respect to the Lebesgue measure, with moreover, continuous Radon–Nikodym derivatives.

Let Q_M be the closed symmetric space generated by all games v in the domain of φ_D such that either $\varphi_D v \in FA$, or $\varphi_D(v)$ is a function of finitely many non-atomic measures.

THEOREM 7 [Mertens (1988a), Section 2]. *Let $v \in Q_M$. Then for every h in $B(I, \mathcal{C})$, $(\varphi_D(v))_h(\chi)$ exists for P almost every χ and is P integrable in χ, and the mapping of each game v in Q_M to the game φv given by*

$$\varphi v(S) = \int \big(\varphi_D(v)\big)_S(\chi) \, dP(\chi)$$

is a value of norm 1 on Q_M.

REMARKS. The extreme points of the set of invariant cylinder probabilities on $B(I, \mathcal{C})$ have a Fourier transform $F_{m,\sigma}(\mu) = \exp(im\mu(1) - \sigma\|\mu\|)$ where $m \in \mathbb{R}$, $\sigma \geqslant 0$. More precisely, there is a one-to-one and onto map between countably additive measures Q on $\mathbb{R} \times \mathbb{R}_+$ and invariant cylinder measures on $B(I, \mathcal{C})$ where every measure Q on $\mathbb{R} \times \mathbb{R}_+$ is associated with the cylinder measure whose Fourier transform is given by

$$F_Q(\mu) = \int_{\mathbb{R} \times \mathbb{R}_+} F_{m,\sigma}(\mu) \, dQ(m, \sigma).$$

The associated cylinder measure is nondegenerate if $Q(r = 0) = 0$, and in the above value formula and theorem, P can be replaced with any invariant nondegenerate cylinder measure of total mass 1 on $B(I, \mathcal{C})$.

Neyman (2001) provides an alternative approach to define a value on the closed space spanned by all bounded variation games of the form $f \circ \mu$ where μ is a vector of non-atomic probability measures and f is continuous at $\mu(\emptyset)$ and $\mu(I)$, and Q_M. This alternative approach stems from the ideas in Mertens (1988a) and expresses the value as a limit of averaged marginal contributions where the average is with respect to a distribution which is strictly stable of index 1.

For any $\mathbb{R}^n$-valued non-atomic vector measure μ we define a map φ_μ^δ from $Q(\mu)$ – the space of all games of bounded variation that are functions of the vector measure μ and are continuous at $\mu(\emptyset)$ and at $\mu(I)$ – to BV. The map φ_μ^δ depends on a small positive constant $\delta > 0$ and the vector measure $\mu = (\mu_1, \ldots, \mu_n)$.

For $\delta > 0$ let $I_\delta(t) = I \ (3\delta \leqslant t < 1 - 3\delta)$ where I stands for the indicator function. The essential role of the function I_δ is to make the integrands that appear in the integrals used in the definition of the value well defined.

Let $\mu = (\mu_1, \ldots, \mu_n)$ be a vector of non-atomic probability measures and $f : \mathcal{R}(\mu) \rightarrow \mathbb{R}$ be continuous at $\mu(\emptyset)$ and $\mu(I)$ and with $f \circ \mu$ of bounded variation.

It follows that for every $x \in 2\mathcal{R}(\mu) - \mu(I)$, $S \in \mathcal{C}$, and t with $I_\delta(t) = 1$, $t\mu(I) + \delta x$ and $t\mu(I) + \delta x + \delta\mu(S)$ are in $\mathcal{R}(\mu)$ and therefore the functions $t \mapsto I_\delta(t)f(t\mu(I) + \delta x)$ and $t \mapsto I_\delta(t)f(t\mu(I) + \delta x + \delta\mu(S))$ are of bounded variation on $[0, 1]$ and thus in particular they are integrable functions. Therefore, given a game $f \circ \mu \in Q(\mu)$, the function $F_{f,\mu}$, defined on all triples (δ, x, S) with $\delta > 0$ sufficiently small (e.g., $\delta < 1/9$), $x \in \mathbb{R}^n$ with $\delta x \in 2\mathcal{R}(\mu) - \mu(I)$, and $S \in \mathcal{C}$ by

$$F_{f,\mu}(\delta, x, S) = \int_0^1 I_\delta(t) \frac{f(t\mu(I) + \delta^2 x + \delta^3\mu(S)) - f(t\mu(I) + \delta^2 x)}{\delta^3} \, dt,$$

is well defined.

Let P_μ^δ be the restriction of P_μ to the set of all points in $\{x \in \mathbb{R}^n : \delta x \in 2\mathcal{R}(\mu) - \mu(I)\}$. The function $x \mapsto F_{f,\mu}(\delta, x, S)$ is continuous and bounded and therefore the function φ_μ^δ defined on $Q(\mu) \times \mathcal{C}$ by

$$\varphi_\mu^\delta(f \circ \mu, S) = \int_{AF(\mu)} F_{f,\mu}(\delta, x, S) \, dP_\mu^\delta(x),$$

where $AF(\mu)$ stands for the affine space spanned by $\mathcal{R}(\mu)$, is well defined.

The linear space of games $Q(\mu)$ is not a symmetric space. Moreover, the map φ_μ^δ violates all value axioms. It does not map $Q(\mu)$ into FA, it is not efficient, and it is not symmetric. In addition, given two non-atomic vector measures, μ and ν, the operators φ_μ^δ and φ_ν^δ differ on the intersection $Q(\mu) \cap Q(\nu)$. However, it turns out that the violation of the value axioms by φ_μ^δ diminishes as δ goes to 0, and the difference $\varphi_\mu^\delta(f \circ \mu) - \varphi_\nu^\delta(g \circ \nu)$ goes to 0 as $\delta \rightarrow 0$ whenever $f \circ \mu = g \circ \nu$. Therefore, an appropriate limiting argument enables us to generate a value on the union of all the spaces $Q(\mu)$.

Consider the partially ordered linear space $\mathcal{L}$ of all bounded functions defined on the open interval $(0, 1/9)$ with the partial order $h \succ g$ if and only if $h(\delta) \geqslant g(\delta)$ for all sufficiently small values of $\delta > 0$. Let $L : \mathcal{L} \rightarrow \mathbb{R}$ be a monotonic (i.e., $L(h) \geqslant L(g)$ whenever $h \succ g$) linear functional with $L(\mathbf{1}) = 1$.

Let Q denote the union of all spaces $Q(\mu)$ where μ ranges over all vectors of finitely many non-atomic probability measures. Define the map $\varphi : Q \rightarrow \mathbb{R}^\mathcal{C}$ by $\varphi v(S) = L(\varphi_\mu^\delta(v, S))$ whenever $v \in Q(\mu)$. It turns out that φ is well defined, φv is in FA (whenever $v \in Q$) and φ is a value of norm 1 on Q. As φ is a value of norm 1 on Q and the continuous extension of any value of norm 1 defines a value (of norm 1) on the closure, we have:

PROPOSITION 6. *φ is a value of norm 1 on $\overline{Q}$.*

8. Uniqueness of the value of nondifferentiable games

The symmetry axiom of a value implies [see, e.g., Aumann and Shapley (1974), p. 139] that if $\mu = (\mu_1, \ldots, \mu_n)$ is a vector of mutually singular measures in NA^1, $f : [0, 1]^n \to \mathbb{R}$, and $v = f \circ \mu$ is a game which is a member of a space Q on which a value φ is defined, then

$$\varphi v = \sum_{i=1}^{n} a_i(f)\mu_i.$$

The efficiency of φ implies that $\sum_{i=1}^{n} a_i(f) = f(1, \ldots, 1)$, and if in addition f is symmetric in the n coordinates all the coefficients $a_i(f)$ are the same, i.e., $a_i(f) = \frac{1}{n} f(1, \ldots, 1)$. The results below will specify spaces of games that are generated by vector measure games $v = f \circ \mu$ where μ is a vector of mutually singular non-atomic probability measures and $f : [0, 1]^n \to \mathbb{R}$ and on which a unique value exists.

Denote by $\overline{M}_+^k$ the linear space of function on $[0, 1]^k$ spanned by all concave Lipschitz functions $f : [0, 1]^k \to \mathbb{R}$ with $f(0) = 0$ and $f_y(x) \leqslant f_y(tx)$ whenever x is in the interior of $[0, 1]^k$, $y \in \mathbb{R}_+^k$ and $0 < t \leqslant 1$. $\overline{M}_\ell^k$ is the linear space of all continuous piecewise linear functions f on $[0, 1]^k$ with $f(0) = 0$.

DEFINITION 7 [Haimanko (2000d)]. *CON_s* (respectively, *LIN_s*) is the linear span of games of the form $f \circ \mu$ where for some positive integer k, $f \in \overline{M}_+^k$ (respectively, $f \in \overline{M}_+^k$) and $\mu = (\mu_1, \ldots, \mu_k)$ is a vector of mutually singular measures in NA^1.

The spaces *CON_s* and *LIN_s* are in the domain of the Mertens value; therefore there is a value on *CON_s* and on *LIN_s*. If $f \in \overline{M}_+^k$ or $f \in \overline{M}_\ell^k$, the directional derivative $f_y(x)$ exists whenever x is in the interior of $[0, 1]^k$ and $y \in \mathbb{R}^k$, and for each fixed x, the directional derivative of the function $y \mapsto f_y(x)$ in the direction $z \in \mathbb{R}^k$,

$$\lim_{\varepsilon \to 0+} \frac{f_{y+\varepsilon z}(x) - f_y(x)}{\varepsilon},$$

exists and is denoted $\partial f(x; y; z)$. If $\mu = (\mu_1, \ldots, \mu_k)$ is a vector of mutually singular measures in NA^1 and φ_M is the Mertens value,

$$\varphi_M(f \circ \mu)(S) = \int_0^1 \partial f\big(t\mathbf{1}_k; X; \mu(S)\big) \, dt$$

where $\mathbf{1}_k = (1, \ldots, 1) \in \mathbb{R}^k$ and $X = (X_1, \ldots, X_k)$ is a vector of independent random variables, each with the standard Cauchy distribution.

THEOREM 8 [Haimanko (2000d) and (2001b)]. *The Mertens value is the unique value on CON_s and on LIN_s.*

9. Desired properties of values

The short list of conditions that define the value, linearity, positivity, symmetry and efficiency, suffices to determine the value on many spaces of games, like pNA_∞, pNA and $bv'NA$. Values were defined in other cases by means of constructive approaches. All these values satisfy many additional desirable properties like continuity, the projection axiom, i.e., $\varphi v = v$ whenever $v \in FA$ is in the domain of φ, the dummy axiom, i.e., $\varphi v(S^c) = 0$ whenever S is a carrier of v, and stronger versions of positivity.

9.1. Strong positivity

One of the stronger versions of positivity is called strong positivity: Let $Q \subseteq BV$. A map $\varphi : Q \to FA$ is *strongly positive* if $\varphi v(S) \geqslant \varphi w(S)$ whenever $v, w \in Q$ and $v(T \cup S') - v(T) \geqslant w(T \cup S') - w(T)$ for all $S' \subseteq S$ and T in $\mathcal{C}$.

It turns out that strong positivity can replace positivity and linearity in the characterization of the value on spaces of "smooth" games.

THEOREM 9 [Monderer and Neyman (1988), Theorems 8 and 5].
  (a) *Any symmetric, efficient, strongly positive map from pNA_∞ to FA is the Aumann–Shapley value.*
  (b) *Any symmetric, efficient, strongly positive and continuous map from pNA to FA is the Aumann–Shapley value.*

9.2. The partition value

Given a value φ on a space Q, it is of interest to specify whether it could be interpreted as a limiting value. This leads to the definition of a partition value.

A value φ on a space Q is a *partition value* [Neyman and Tauman (1979)] if for every coalition S there exists a T-admissible sequence $(\pi_k)_{k=1}^\infty$ such that $\varphi v(T) = \lim_{k \to \infty} \psi v_{\pi_k}(T)$.

It follows that any partition value is continuous (with norm $\leqslant 1$) and coincides with the asymptotic value on all games that have an asymptotic value [Neyman and Tauman (1979)]. There are however spaces of games that include *ASYMP* and on which a partition value exists. Given two sets of games, Q and R, we denote by $Q * R$ the closed linear and symmetric space generated by all games of the form uv where $u \in Q$ and $v \in R$.

THEOREM 10 [Neyman and Tauman (1979)]. *There exists a partition value on the closed linear space spanned by ASYMP, $bv'NA * bv'NA$ and $\mathcal{A} * bv'NA * bv'NA$.*

9.3. The diagonal property

A remarkable implication of the diagonal formula is that the value of a vector measure game $f \circ \mu$ in pNA or in $bv'NA$ is completely determined by the behavior of f near the diagonal of the range of μ. We proceed to define a diagonal value as a value that depends only on the behavior of the game in a diagonal neighborhood. Formally, given a vector of non-atomic probability measures $\mu = (\mu_1, \ldots, \mu_n)$ and $\varepsilon > 0$, the $\varepsilon - \mu$ diagonal neighborhood, $\mathcal{D}(\mu, \varepsilon)$, is defined as the set of all coalitions S with $|\mu_i(S) - \mu_j(S)| < \varepsilon$ for all $1 \leqslant i, j \leqslant n$. A map φ from a set Q of games is diagonal if for every $\varepsilon - \mu$ diagonal neighborhood $\mathcal{D}(\mu, \varepsilon)$ and every two games v, w in $Q \cap BV$ that coincide on $\mathcal{D}(\mu, \varepsilon)$ (i.e., $v(S) = w(S)$ for any S in $\mathcal{D}(\mu, \varepsilon)$), $\varphi v = \varphi w$.

There are examples of non-diagonal values [Neyman and Tauman (1976)] even on reproducing spaces [Tauman (1977)]. However,

THEOREM 11 [Neyman (1977a)]. *Any continuous value* (*in the variation norm*) *is diagonal.*

Thus, in particular, the values on (*pNA* and) *bv'NA* [Aumann and Shapley (1974), Proposition 43.1], the weak asymptotic value, the asymptotic value [Aumann and Shapley (1974), Proposition 43.11], the mixing value [Aumann and Shapley (1974), Proposition 43.2], any partition value [Neyman and Tauman (1979)] and any value on a closed reproducing space, are all diagonal.

Let *DIAG* denote the linear subspace of BV that is generated by the games that vanish on some diagonal neighborhood. Then $DIAG \subset ASYMP$ and any continuous (and in particular any asymptotic or partition) value vanishes on DIAG.

10. Semivalues

Semivalues are generalizations and/or analogies of the Shapley value that do not (necessarily) obey the efficiency, or Pareto optimality, property. This has stemmed partly from the search for value function that describes the prospects of playing different roles in a game.

DEFINITION 8 [Dubey, Neyman and Weber (1981)]. A *semivalue* on a linear and symmetric space of games Q is a linear, symmetric and positive map $\psi : Q \to FA$ that satisfies the projection axiom (i.e., $\psi v = v$ for every v in $Q \cap FA$).

The following result characterizes all semivalues on the space *FG* of all games having a finite support.

THEOREM 12 [Dubey, Neyman and Weber (1981), Theorem 1].
  (a) *For each probability measure ξ on* $[0, 1]$, *the map ψ_ξ that is given by*

$$\psi_\xi v(S) = \int_0^1 \partial \bar{v}(t, S) \, d\xi(t),$$

where $\bar{v}$ is Owen's multilinear extension of the game v, defines a semivalue on
FG. Moreover, every semivalue on FG is of this form and the mapping $\xi \to \psi_\xi$
is 1–1.

(b) ψ_ξ is continuous (in the variation norm) on G if and only if ξ is absolutely
continuous w.r.t. the Lebesgue measure λ on $[0, 1]$ and $d\xi/d\lambda \in L_\infty[0, 1]$. In
that case, $\|\psi_\xi\|_{BV} = \|d\xi/d\lambda\|_\infty$.

Let W be the subset of all elements g in $L_\infty[0, 1]$ such that $\int_0^1 g(t)\, dt = 1$.

THEOREM 13 [Dubey, Neyman and Weber (1981), Theorem 2]. *For every g in W, the
map* $\psi_g : pNA \to FA$ *defined by*

$$\psi_g v(S) = \int_0^1 \partial \bar{v}(t, S) g(t)\, dt$$

*is a semivalue on pNA. Moreover, every semivalue on pNA is of this form and the map
$g \to \psi_g$ maps W onto the family of semivalues on pNA and is a linear isometry.*

Let W_c be the subset of all elements g in W which are (represented by) continuous
functions on $[0, 1]$. We identify W (W_c) with the set of all probability measures ξ on
$[0, 1]$ which are absolutely continuous w.r.t. the Lebesgue measure λ on $[0, 1]$, and
their Radon–Nikodym derivative $d\xi/d\lambda \in W$ ($\in W_c$).

DEFINITION 9. Let v be a game and let ξ be a probability measure on $[0, 1]$. A game
φv is said to be the *weak asymptotic ξ-semivalue* of v if, for every $S \in \mathcal{C}$ and every
$\varepsilon > 0$, there is a finite subfield π of $\mathcal{C}$ with $S \in \pi$ such that for any finite subfield π'
with $\pi' \supset \pi$,

$$\left| \psi_\xi v_{\pi'}(S) - \varphi v(S) \right| < \varepsilon.$$

REMARKS. (1) A game v has a weak asymptotic ξ-semivalue if and only if, for every
S in $\mathcal{C}$ and every $\varepsilon > 0$, there is a finite subfield π of $\mathcal{C}$ with $S \in \pi$ such that for any
subfield π' with $\pi' \supset \pi$, $|\psi_\xi v_{\pi'}(S) - \psi_\xi v_\pi(S)| < \varepsilon$.
   (2) A game v has at most one weak asymptotic ξ-semivalue.
   (3) The weak asymptotic ξ-semivalue φv of a game v is a finitely additive game and
$\|\varphi v\| \leqslant \|v\| \|d\xi/d\lambda\|_\infty$ whenever $\xi \in W$.

Let ξ be a probability measure on $[0, 1]$. The set of all games of bounded variation
and having a weak asymptotic ξ-semivalue is denoted $ASYMP^*(\xi)$.

THEOREM 14. *Let ξ be a probability measure on $[0, 1]$.*
   (a) *The set of all games having a weak asymptotic ξ-semivalue is a linear symmetric
       space of games and the operator that maps each game to its weak asymptotic
       ξ-semivalue is a semivalue on that space.*

(b) *ASYMP*$^*(\xi)$ *is closed* $\Leftrightarrow \xi \in W \Leftrightarrow pFA \subset ASYMP^*(\xi) \Leftrightarrow pNA \subset ASYMP^*(\xi)$.
(c) $\xi \in W_c \Rightarrow bv'NA \subset bv'FA \subset ASYMP^*(\xi)$.

The asymptotic ξ-semivalue of a game v is defined whenever all the sequences of the ξ-semivalues of finite games that "approximate" v have the same limit.

DEFINITION 10. Let ξ be a probability measure on $[0, 1]$. A game φv is said to be the *asymptotic ξ-semivalue* of v if, for every $T \in \mathcal{C}$ and every T-admissible sequence $(\pi_i)_{i=1}^{\infty}$, the following limit and equality exists:

$$\lim_{k \to \infty} \psi_\xi v_{\pi_k}(T) = \varphi v(T).$$

REMARKS. (1) A game v has an asymptotic ξ-semivalue if and only if for every S in C and every S-admissible sequence $\mathcal{P} = (\pi_i)_{i=1}^{\infty}$ the limit, $\lim_{k \to \infty} \psi_\xi v_{\pi_i}(S)$, exists and is independent of the choice of $\mathcal{P}$.
(2) For any given v, the asymptotic ξ-semivalue, if it exists, is clearly unique.
(3) The asymptotic ξ-semivalue φv of a game v is finitely additive and $\|\varphi v\|_{BV} \leqslant \|v\|_{BV} \|d\xi/d\lambda\|_\infty$ (when $\xi \in W$).
(4) If v has an asymptotic ξ-semivalue φv, then v has a weak asymptotic ξ-semivalue $(=\varphi v)$.

The space of all games of bounded variation that have an asymptotic ξ-semivalue is denoted *ASYMP*(ξ).

THEOREM 15.
(a) [Dubey (1980)]. $pNA_\infty \subset ASYMP(\xi)$ *and if* φv *denotes the asymptotic ξ-semivalue of* $v \in pNA_\infty$, *then* $\|\varphi v\|_\infty \leqslant 2\|v\|_\infty$.
(b) $pNA \subset pM \subset ASYMP(\xi)$ *whenever* $\xi \in W$.
(c) $bv'NA \subset bv'M \subset ASYMP(\xi)$ *whenever* $\xi \in W_c$.

11. Partially symmetric values

Non-symmetric values (quasivalues) and partially symmetric values are generalizations and/or analogies of the Shapley value that do not necessarily obey the symmetry axiom. A (symmetric) value is covariant with respect to all automorphisms of the space of players. A partially symmetric value is covariant with respect to a specified subgroup of automorphisms. Of particular importance are the trivial group, the subgroups that preserve a finite (or countable) partition of the set of players and the subgroups that preserve a fixed population measure.

The group of all automorphisms that preserve a measurable partition Π, i.e., all automorphisms θ such that $\theta S = S$ for any $S \in \Pi$, is denoted $\mathcal{G}(\Pi)$. Similarly, if Π is a σ-algebra of coalitions $(\Pi \subset \mathcal{C})$ we denote by $\mathcal{G}(\Pi)$ the set of all automorphisms θ

such that $\theta S = S$ for any $S \in \Pi$. The group of automorphisms that preserve a fixed measure μ on the space of players is denoted $\mathcal{G}(\mu)$.

11.1. Non-symmetric and partially symmetric values

DEFINITION 11. Let Q be a linear space of games. A *non-symmetric value (quasi-value)* on Q is a linear, efficient and positive map $\psi: Q \to FA$ that satisfies the projection axiom. Given a subgroup of automorphisms $\mathcal{H}$, an *$\mathcal{H}$-symmetric value* on Q is a non-symmetric value ψ on Q such that for every $\theta \in \mathcal{H}$ and $v \in Q$, $\psi(\theta_* v) = \theta_*(\psi v)$. Given a σ-algebra Π, a *Π-symmetric value* on Q is a $\mathcal{G}(\Pi)$-symmetric value. Given a measure μ on the space of players, a *μ-value* is a $\mathcal{G}(\mu)$-symmetric value.

A *coalition structure* is a measurable, finite or countable, partition Π of I such that every atom of Π is an infinite set. The results below characterize all the $\mathcal{G}(\Pi)$-symmetric values, where Π is a fixed coalition structure, on finite games and on smooth games with a continuum of players. The characterizations employ path values which are defined by means of monotonic increasing paths in $B^1_+(I, \mathcal{C})$. A *path* is a monotonic function $\gamma: [0, 1] \to B^1_+(I, \mathcal{C})$ such that for each $s \in I$, $\gamma(0)(s) = 0$, $\gamma(1)(s) = 1$, and the function $t \mapsto \gamma(t)(s)$ is continuous at 0 and 1. A path γ is called *continuous* if for every fixed $s \in I$ the function $t \mapsto \gamma(t)(s)$ is continuous; *NA-continuous* if for every $\mu \in NA$ the function $t \to \mu(\gamma(t))$ is continuous; and *Π-symmetric* (where Π is a partition of I) if for every $0 \leqslant t \leqslant 1$, the function $s \mapsto \gamma(t)(s)$ is Π-measurable. Given an extension operator ψ on a linear space of games Q (for $v \in Q$ we use the notation $\bar{v} = \psi v$) and a path γ, the *γ-path extension* of v, $\bar{v}_\gamma$, is defined [Haimanko (2000c)] by

$$\bar{v}_\gamma(f) = \bar{v}(\gamma^*(f))$$

where $\gamma^*: B^1_+(I, \mathcal{C}) \to B^1_+(I, \mathcal{C})$ is defined by $\gamma^*(f)(s) = \gamma(f(s))(s)$. For a coalition $S \in \mathcal{C}$ and $v \in Q$, define

$$\varphi^\varepsilon_\gamma(v)(S) = \frac{1}{\varepsilon} \int_\varepsilon^{1-\varepsilon} \left[\bar{v}_\gamma(tI + \varepsilon S) - \bar{v}_\gamma(tI) \right] dt$$

and $DIFF(\gamma)$ is defined as the set of all $v \in Q$ such that for every $S \in \mathcal{C}$ the limit $\varphi_\gamma(v)(S) = \lim_{\varepsilon \to 0+} \varphi^\varepsilon_\gamma(v)(S)$ exists and the game $\varphi_\gamma(v)$ is in FA.

PROPOSITION 7 [Haimanko (2000c)]. *$DIFF(\gamma)$ is a closed linear subspace of Q and the map φ_γ is a non-symmetric value on $DIFF(\gamma)$. If Π is a measurable partition (or a σ-algebra) and γ is Π-symmetric then φ_γ is a Π-symmetric value.*

In what follows we assume that $Q = bv'NA$ with the natural extension operator and Π is a fixed coalition structure.

THEOREM 16 [Haimanko (2000c)]. (a) γ *is NA-continuous* $\Leftrightarrow pNA \subset DIFF(\gamma)$.

(b) *Every non-symmetric value on* $pNA(\lambda)$, $\lambda \in NA^1$, *is a mixture of path values.*

(c) *Every* Π*-symmetric value (where* Π *is a coalition structure) on pNA (on FG) is a mixture of* Π*-symmetric path values.*

REMARKS [Haimanko (2000c)]. (1) There are non-symmetric values on *pNA* which are not mixtures of path values.

(2) Every linear and efficient map $\psi : pNA \to FA$ which is Π-symmetric with respect to a coalition structure Π obeys the projection axiom. Therefore, the projection axiom of a non-symmetric value is not used in (c) of Theorem 16.

The above characterization of all Π-symmetric values on $pNA(\lambda)$ as mixtures of path values relies on the non-atomicity of the games in *pNA*. A more delicate construction, to be described below, leads to the characterization of all Π-symmetric values on *pM*.

For every game v in $bv'M$ there is a measure $\lambda \in M^1$ such that $v \in bv'M(\lambda)$. The set of atoms of λ is denoted by $A(\lambda)$. Given two paths γ, γ', define for $v \in bv'M(\lambda)$, $f \in B^1_+(I, \mathcal{C})$

$$\bar{v}_{\gamma,\gamma'}(f) = \bar{v}_{\gamma''}$$

where $\gamma'' = \gamma[I - A(\lambda)] + \gamma'[A(\lambda)]$ and $v \to \bar{v}$ is the natural extension on $bv'M$. It can be verified that $\bar{v}_{\gamma,\gamma'}$ is well defined on $bv'M$, i.e., the definition is independent of the measure λ. For a coalition $S \in \mathcal{C}$ and $v \in bv'M$, define

$$\varphi^\varepsilon_{\gamma,\gamma'}(v)(S) = \frac{1}{\varepsilon} \int_\varepsilon^{1-\varepsilon} \left[\bar{v}_{\gamma,\gamma'}(tI + \varepsilon S) - \bar{v}_{\gamma,\gamma'}(tI) \right] dt.$$

We associate with $bv'M$ and the pair of paths γ and γ' the set $DIFF(\gamma, \gamma')$ of all games $v \in bv'M$ such that for every $S \in \mathcal{C}$ the limit $\varphi_{\gamma,\gamma'}(v)(S) = \lim_{\varepsilon \to 0+} \varphi^\varepsilon_{\gamma''}(v)(S)$ exists and the game $\varphi_{\gamma,\gamma'}(v)$ is in *FA*.

THEOREM 17 [Haimanko (2000c)]. (a) $DIFF(\gamma, \gamma')$ *is a closed linear subspace of* $bv'M$ *and the map* $\varphi_{\gamma,\gamma'} : DIFF(\gamma, \gamma') \to FA$ *is a non-symmetric value.*

(b) *Given a measurable partition (or a* σ*-algebra)* Π, $\varphi_{\gamma,\gamma'}$ *is* Π*-symmetric whenever* $s \mapsto \gamma(t)(s)$ *and* $s \mapsto \gamma'(t)(s)$ *are* Π *measurable for every* t.

(c) *If* Π *is a coalition structure, any* Π*-symmetric value on pM is a mixture of maps* $\varphi_{\gamma,\gamma'}$ *where the paths* γ *and* γ' *are continuous and* Π*-symmetric.*

REMARKS [Haimanko (2000c)]. (1) If γ is *NA*-continuous and the discontinuities of the functions $t \mapsto \gamma'(t)(s)$, $s \in I$, are mutually disjoint, $pM \subset DIFF(\gamma, \gamma')$. In particular if γ and γ' are continuous, $pM \subset DIFF(\gamma, \gamma')$ and therefore if in addition $\gamma(t)$ and $\gamma'(t)$ are constant functions for each fixed t, γ and γ' are $\mathcal{G}$-symmetric and thus $\varphi_{\gamma,\gamma'}$

defines a value of norm 1 on *pM*. These infinitely many values on *pM* were discovered by Hart (1973) and are called the Hart values.

(2) A surprising outcome of this systematic study is the characterization of all values on *pM*. Any value on *pM* is a mixture of the Hart values.

(3) If Π is a coalition structure, any Π-symmetric linear and efficient map $\psi : bv'M \to FA$ obeys the projection axiom. Therefore, the projection axiom of a non-symmetric value is not used in the characterization (c) of Theorem 17.

Weighted values are non-symmetric values of the form φ_γ where γ is a path described by means of a measurable weight function $w : I \to \mathbb{R}_{++}$ (which is bounded away from 0): for every $\gamma(s)(t) = t^{w(s)}$. Hart and Monderer (1997) have successfully applied the potential approach of Hart and Mas-Colell (1989) to weighted values of smooth (continuously differentiable) games: to every game $v \in pNA_\infty$ the game $(P_w v)(S) = \int_0^1 \frac{\bar{v}(t^w S)}{t} \, dt$ is in pNA_∞ and $\varphi_\gamma v(S) = (\overline{P_w} v)_{wS}(I)$. In addition they define an asymptotic w-weighted value and prove that every game in pNA_∞ has an asymptotic w-weighted value.

Another important subgroup of automorphisms is the one that preserves a fixed (population) non-atomic measure μ.

11.2. Measure-based values

Measure-based values are generalizations and/or analogues of the value that take into account the coalition worth function and a fixed (population) measure μ. This has stemmed partly from applications of value theory to economic models in which a given population measure μ is part of the models.

Let μ be a fixed non-atomic measure on $(I, \mathcal{C})$. A subset of games, Q, is μ-symmetric if for every $\Theta \in \mathcal{G}(\mu)$ (the group of μ-measure-preserving automorphisms), $\Theta_* Q = Q$. A map φ from a μ-symmetric set of games Q into a space of games is μ-symmetric if $\varphi\Theta_* v = \Theta_* \varphi v$ for every Θ_* in $\mathcal{G}(\mu)$ and every game v in Q.

DEFINITION 12. Let Q be a μ-symmetric space of games. A μ-*value* on Q is a map from Q into finitely additive games that is linear, μ-symmetric, positive, efficient, and obeys the dummy axiom (i.e., $\varphi v(S) = 0$ whenever the complement of S, S^c, is a carrier of v).

REMARKS. The definition of a μ-value differs from the definition of a value in two aspects. First, there is a weakening of the symmetry axiom. A value is required to be covariant with the actions of all automorphisms, while a μ-value is required to be covariant only with the actions of the μ-measure-preserving automorphism. Second, there is an additional axiom in the definition of a μ-value, the dummy axiom. In view of the other axioms (that φv is finitely additive and efficiency), this additional axiom could be viewed as a strengthening of the efficiency axiom; the μ-value is required to obey $\varphi v(S) = v(S)$ for any carrier S of the game v, while the efficiency axiom requires it only for the carrier $S = I$. In many cases of interest, the dummy axiom follows from the other axioms of the value [Aumann and Shapley (1974), Note 4, p. 18]. Obviously, the value on *pNA* obeys the dummy axiom and the restriction of any value that obeys the dummy axiom to a μ-symmetric subspace is a μ-value. In particular, the restriction of the unique value on *pNA* to $pNA(\mu)$, the closed (in the bounded variation norm) algebra generated by non-atomic probability measures that are absolutely continuous with respect to μ, is a μ-value. It turns out to be the unique μ-value on $pNA(\mu)$.

THEOREM 18 [Monderer (1986), Theorem A]. *There exists a unique μ-value on $pNA(\mu)$; it is the restriction to $pNA(\mu)$ of the unique value on pNA.*

REMARKS. In the above characterization, the dummy axiom could be replaced by the projection axiom [Monderer (1986), Theorem D]: there exists a unique linear, μ-symmetric, positive and efficient map $\varphi : pNA(\mu) \to FA$ that obeys the projection axiom; it is the unique value on $pNA(\mu)$. The characterizations of the μ-value on $pNA(\mu)$ is derived from Monderer [(1986), Main Theorem], which characterizes all μ-symmetric, continuous linear maps from $pNA(\mu)$ into *FA*.

Similar to the constructive approaches to the value, one can develop corresponding constructive approaches to μ-values. We start with the asymptotic approach.

Given T in $\mathcal{C}$ with $\mu(T) > 0$, a $\mu-T$-*admissible sequence* is a T-admissible sequence $(\pi_1, \pi_2, \ldots)$ for which $\lim_{k \to \infty}(\min\{\mu(A): A$ is an atom of $\pi_k\}/\max\{\mu(A): A$ is an atom of $\pi_k\}) = 1$.

DEFINITION 13. A game φv is said to be the μ-*asymptotic value* of v if, for every T in $\mathcal{C}$ with $\mu(T) > 0$ and every $\mu-T$-admissible sequence $(\pi_i)_{i=1}^{\infty}$, the following limit and equality exists

$$\lim_{k \to \infty} \psi v_{\pi_k}(T) = \varphi v(T).$$

REMARKS. (1) A game v has a μ-asymptotic value if and only if for every T in $\mathcal{C}$ with $\mu(T) > 0$ and every $\mu-T$-admissible sequence $\mathcal{P} = (\pi_i)_{i=1}^{\infty}$, the limit, $\lim_{k \to \infty} \psi v_{\pi_k}(T)$ exists and is independent of the choice of $\mathcal{P}$.

(2) For a given game v, the μ-asymptotic value, if it exists, is clearly unique.

(3) The μ-asymptotic value φv of a game v is finitely additive and $\|\varphi v\| \leqslant \|v\|$ whenever v has bounded variation.

The space of all games of bounded variation that have a μ-asymptotic value is denoted by *ASYMP*$[\mu]$.

THEOREM 19 [Hart (1980), Theorem 9.2]. *The set of all games having a μ-asymptotic value is a linear μ-symmetric space of games and the operator mapping each game to its μ-asymptotic value is a μ-value on that space. ASYMP$[\mu]$ is a closed μ-symmetric* *linear subspace of BV which contains all functions of the form $f \circ (\mu_1, \ldots, \mu_n)$ where $\mu_1, \ldots, \mu_n$ are absolutely continuous w.r.t. μ and with Radon–Nikodym derivatives, $d\mu_i/d\mu$, in $L_2(\mu)$ and f is concave and homogeneous of degree 1.*

12. The value and the core

The Banach space of all bounded set functions with the supremum norm is denoted BS, and its closed subspace spanned by pNA is denoted pNA'. The set of all games in pNA' that are homogeneous of degree 1 (i.e., $\bar{v}(\alpha f) = \alpha\bar{v}(f)$ for all f in $B_1^+(I, \mathcal{C})$ and all α in $[0, 1]$), and superadditive (i.e., $v(S \cup T) \geqslant v(S) + v(T)$ for all S, T in $\mathcal{C}$), is denoted H'. The subset of all games in H' which also satisfy $v \geqslant 0$ is denoted H'_+. The core of a game v in H' is non-empty and for every S in $\mathcal{C}$, the minimum of $\mu(S)$ when μ runs over all elements μ in core (v) is attained. The exact cover of a game v in H' is the game $\min\{\mu(S): \mu \in \text{Core}(v)\}$ and its core coincides with the core of v. A game $v \in H'$ that equals its exact cover is called exact. The closed (in the bounded variation norm) subspace of BV that is generated by pNA and $DIAG$ is denoted by $pNAD$.

THEOREM 20 [Aumann and Shapley (1974), Theorem I]. *The core of a game v in $pNAD \cap H'$ (which obviously contains $pNA \cap H'$) has a unique element which coincides with the (asymptotic) value of v.*

THEOREM 21 [Hart (1977a)]. *Let $v \in H'_+$. If v has an asymptotic value φv, then φv is the center of symmetry of the core of v. The game v has an asymptotic value φv when either the core of v contains a single element ν (and then $\varphi v = \nu$), or v is exact and the core of v has a center of symmetry ν (and then $\varphi v = \nu$).*

REMARKS. (1) The second theorem provides a necessary and sufficient condition for the asymptotic value of an exact game v in H'_+ to exist: the symmetry of its core. There are (non-exact) games in H'_+ whose cores have a center of symmetry which do not have an asymptotic value [Hart (1977a), Section 8].

(2) A game v in H'_+ has a unique element in the core if and only if $v \in pNAD$.

(3) The conditions of superadditivity and homogeneity of degree 1 are satisfied by all games arising from non-atomic (i.e., perfectly competitive) economic markets (see Chapter 35 in this Handbook). However, the games arising from these markets have a finite-dimensional core while games in H'_+ may have an infinite-dimensional core.

The above result shows that the asymptotic value fails to exist for games in H'_+ whenever the core of the game does not have a center of symmetry. Nevertheless, value theory can be applied to such games, and when it is, the value turns out to be a "center" of the core. It will be described as an average of the extreme points of the core of the game $v \in H'_+$.

Let $v \in H'_+$ have a finite-dimensional core. Then the core of v is a convex closed subset of a vector space generated by finitely many non-atomic probability measures, $\nu_1, \ldots, \nu_n$. Therefore the core of v is identified with a convex subset K of $\mathbb{R}^n$, by $a = (a_1, \ldots, a_n) \in K$ if and only if $\sum a_i \mu_i \in \text{Core}(v)$. For almost every z in $\mathbb{R}^n$ there exists a unique $a(z) = (a_1(z), \ldots, a_n(z))$ in K with $\sum_{i=1}^n a_i(z) z_i = \min\{\sum_{i=1}^n a_i z_i : (a_1, \ldots, a_n) \in K\}$.

THEOREM 22 [Hart (1980)]. *If $\nu_1, \ldots, \nu_n$ are absolutely continuous with respect to μ and $d\nu_i/d\mu \in L_2(\mu)$, then $v \in ASYMP[\mu]$ and its μ-asymptotic value $\varphi_\mu v$ is given by*

$$\varphi_\mu v = \int_{\mathbb{R}^n} \sum a_i(z) \nu_i \, dN(z)$$

where N is a normal distribution on $\mathbb{R}^n$ with 0 expectation and a covariance matrix that is given by

$$E_N(z_i z_j) = \int_I \left(\frac{d\nu_i}{d\mu}\right)\left(\frac{d\nu_j}{d\mu}\right) d\mu.$$

Equivalently, N is the distribution on $\mathbb{R}^n$ whose Fourier transform φ_N is given by

$$\varphi_N(y) = \exp\left(-\int_I \left(\sum_{i=1}^n y_i \left(\frac{d\nu_i}{d\mu}\right)\right)^2 d\mu\right).$$

THEOREM 23 [Mertens (1988b)]. *The Mertens value of the game $v \in H'_+$ with a finite-dimensional core, φv, is given by*

$$\varphi v = \int_{\mathbb{R}^n} \left(\sum a_i(z) \nu_i\right) dG(z)$$

where G is the distribution on $\mathbb{R}^n$ whose Fourier transform φ_G is given by $\varphi_G(y) = \exp(-N_v(y))$, $v = (\nu_1, \ldots, \nu_n)$ and N_v is defined as in the section regarding formulas for the value.

It is not known whether there is a unique continuous value on the space spanned by all games in H'_+ which have a finite-dimensional core. We will state however a result that proves uniqueness for a natural subspace. Let M_F be the closed space of games spanned by vector measure games $f \circ \mu \in H'_+$ where $\mu = (\mu_1, \ldots, \mu_k)$ is a vector of mutually singular measures in NA^1.

THEOREM 24 [Haimanko (2001a)]. *The Mertens value is the only continuous value on M_F.*

13. Comments on some classes of games

13.1. Absolutely continuous non-atomic games

A game v is *absolutely continuous with respect to a non-atomic measure* μ if for every $\varepsilon > 0$ there is $\delta > 0$ such that for every increasing sequence of coalitions $S_1 \subset S_2 \subset \cdots \subset S_{2k}$ with $\sum_{i=1}^{k} \mu(S_{2i}) - \mu(S_{2i-1}) < \delta$, $\sum_{i=1}^{k} |v(S_{2i}) - v(S_{2i-1})| < \varepsilon$. A game v is *absolutely continuous* if there is a measure $\mu \in NA^+$ such that v is absolutely continuous with respect to μ. A game of the form $f \circ \mu$ where $\mu \in NA^1$ is absolutely continuous (with respect to μ) if and only if $f : [0, \mu(I)] \to \mathbb{R}$ is absolutely continuous. An absolutely continuous game has bounded variation and the set of all absolutely continuous games, AC, is a closed subspace of BV [Aumann and Shapley (1974)]. If v and u are absolutely continuous with respect to $\mu \in NA^+$ and $v \in NA^+$ respectively, then vu is absolutely continuous with respect to $\mu + v$. Thus AC is a closed subalgebra of BV.

It is not known whether there is a value on AC.

13.2. Games in pNA

The space *pNA* played a central role in the development of values of non-atomic games. Games arising in applications are often either vector measure games or approximated by vector measure games. Researchers have thus looked for conditions on vector (or scalar) measure games to be in *pNA*. Tauman (1982) provides a characterization of vector measure games in *pNA*. A vector measure game $v = f \circ \mu$ with $\mu \in (NA^1)^n$ is in *pNA* if and only if the function that maps a vector measure v with the same range as μ to the game $f \circ v$, is continuous at μ, i.e., for every $\varepsilon > 0$ there is $\delta > 0$ such that if $v \in (NA^1)^n$ has the same range as the vector measure μ and $\sum_{i=1}^{n} \|\mu_i - v_i\| < \delta$ then $\|f \circ \mu - f \circ v\| < \varepsilon$. Proposition 10.17 of [Aumann and Shapley (1974)] provides a sufficient condition, expressed as a condition on a continuous non-decreasing real-valued function $f : \mathbb{R}^n \to \mathbb{R}$, for a game of the form $f \circ \mu$ where μ is a vector of non-atomic measures to be in *pNA*. Theorem C of Aumann and Shapley (1974) asserts that the scalar measure game $f \circ \mu$ ($\mu \in NA^1$) is in *pNA* if and only if the real-valued function $f : [0, 1] \to \mathbb{R}$ is absolutely continuous. Given a signed scalar measure μ with range $[-a, b]$ ($a, b > 0$), Kohlberg (1973) provides a sufficient and necessary condition on the real-valued function $f : [-a, b] \to \mathbb{R}$ for the scalar measure game $f \circ \mu$ to be in *pNA*.

13.3. Games in bv′NA

Any function $f \in bv'$ has a unique representation as a sum of an absolutely continuous function f^{ac} that vanishes at 0 and a singular function f^s in bv', i.e., a function whose variation is on a set of measure zero. The subspace of all singular functions in bv' is denoted s'. The closed linear space generated by all games of the form $f \circ \mu$ where

$f \in s'$ and $\mu \in NA^1$ is denoted $s'NA$. Aumann and Shapley (1974) show that $bv'NA$ is the algebraic direct sum of the two spaces pNA and $s'NA$, and that for any two games, $u \in pNA$ and $v \in s'NA$, $\|u + v\| = \|u\| + \|v\|$. The norm of a function $f \in bv'$, $\|f\|$, is defined as the variation of f on $[0, 1]$. If $\mu \in NA^1$ and $f \in bv'$ then $\|f \circ \mu\| = \|f\|$. Therefore, if $(f_i)_{i=1}^{\infty}$ is a sequence of functions in bv' with $\sum_{i=1}^{\infty} \|f_i\| < \infty$ and $(\mu_i)_{i=1}^{\infty}$ is a sequence of non-atomic probability measures, the series $\sum_{i=1}^{\infty} f_i \circ \mu_i$ converges to a game $v \in bv'NA$. If the functions f_i are absolutely continuous the series converges to a game in pNA. However, not every game in pNA has a representation as a sum of scalar measure games. If the functions f_i are singular, the series $\sum_{i=1}^{\infty} f_i \circ \mu_i$ converges to a game in $s'NA$, and every game $v \in s'NA$ is a countable (possibly finite) sum, $v = \sum_{i=1}^{\infty} f_i \circ \mu_i$, where $f_i \in s'$ with $\|v\| = \sum_{i=1}^{\infty} \|f_i\|$ and $\mu_i \in NA^1$.

A simple game is a game v where $v(S) \in \{0, 1\}$. A weighted majority game is a (simple) game of the form

$$v(S) = \begin{cases} 1 & \text{if } \mu(S) \geqslant q \\ 0 & \text{if } \mu(S) < q \end{cases}$$

or

$$v(S) = \begin{cases} 1 & \text{if } \mu(S) > q \\ 0 & \text{if } \mu(S) \leqslant q \end{cases}$$

where $\mu \in FA^+$ and $0 < q < \mu(I)$. The finitely additive measure μ is called the *weight measure* and q is called the *quota*. Every simple monotonic game in $bv'NA$ is a weighted majority game, and the weighted measure is non-atomic.

References

Artstein, Z. (1971), "Values of games with denumerably many players", International Journal of Game Theory 1:27–37.

Aubin, J. (1981), "Cooperative fuzzy games", Mathematics of Operations Research 6:1–13.

Aumann, R.J. (1975), "Values of markets with a continuum of traders", Econometrica 43:611–646.

Aumann, R.J. (1978), "Recent developments in the theory of the Shapley value", in: Proceedings of the International Congress of Mathematicians, Helsinki, 994–1003.

Aumann, R.J., R.J. Gardner and R. Rosenthal (1977), "Core and value for a public goods economy: An example", Journal of Economic Theory 15:363–365.

Aumann, R.J., and M. Kurz (1977a), "Power and taxes", Econometrica 45:1137–1161.

Aumann, R.J., and M. Kurz (1977b), "Power and taxes in a multi-commodity economy", Israel Journal of Mathematics 27:185–234.

Aumann, R.J., and M. Kurz (1978), "Power and taxes in a multi-commodity economy (updated)", Journal of Public Economics 9:139–161.

Aumann, R.J., M. Kurz and A. Neyman (1987), "Power and public goods", Journal of Economic Theory 42:108–127.

Aumann, R.J., M. Kurz and A. Neyman (1983), "Voting for public goods", Review of Economic Studies 50:677–693.

Aumann, R.J., and M. Perles (1965), "A variational problem arising in economics", Journal of Mathematical Analysis and Applications 11:488–503.

Aumann, R.J., and U. Rothblum (1977), "Orderable set functions and continuity III: Orderability and absolute continuity", SIAM Journal on Control and Optimization 15:156–162.

Aumann, R.J., and L.S. Shapley (1974), Values of Non-atomic Games (Princeton University Press, Princeton, NJ).

Berbee, H. (1981), "On covering single points by randomly ordered intervals", Annals of Probability 9:520–528.

Billera, L.J. (1976), "Values of non-atomic games", Mathematical Reviews 51:2093–2094.

Billera, L.J., and D.C. Heath (1982), "Allocation of shared costs: A set of axioms yielding a unique procedure", Mathematics of Operations Research 7:32–39.

Billera, L.J., D.C. Heath and J. Raanan (1978), "Internal telephone billing rates: A novel application of nonatomic game theory", Operations Research 26:956–965.

Billera, L.J., D.C. Heath and J. Raanan (1981), "A unique procedure for allocating common costs from a production process", Journal of Accounting Research 19:185–196.

Billera, L.J., and J. Raanan (1981), "Cores of non-atomic linear production games", Mathematics of Operations Research 6:420–423.

Brown, D., and P.A. Loeb (1977), "The values of non-standard exchange economies", Israel Journal of Mathematics 25:71–86.

Butnariu, D. (1983), "Decompositions and ranges for additive fuzzy measures", Fuzzy Sets and Systems 3:135–155.

Butnariu, D. (1987), "Values and cores of fuzzy games with infinitely many players", International Journal of Game Theory 16:43–68.

Champsaur, P. (1975), "Cooperation versus competition", Journal of Economic Theory 11:394–417.

Cheng, H. (1981), "On dual regularity and value convergence theorems", Journal of Mathematical Economics 6:37–57.

Dubey, P. (1975), "Some results on values of finite and infinite games", Ph.D. Thesis (Cornell University).

Dubey, P. (1980), "Asymptotic semivalues and a short proof of Kannai's theorem", Mathematics of Operations Research 5:267–270.

Dubey, P., and A. Neyman (1984), "Payoffs in nonatomic economies: An axiomatic approach", Econometrica 52:1129–1150.

Dubey, P., and A. Neyman (1997), "An equivalence principle for perfectly competitive economies", Journal of Economic Theory 75:314–344.

Dubey, P., A. Neyman and R.J. Weber (1981), "Value theory without efficiency", Mathematics of Operations Research 6:122–128.

Dvoretsky A., A. Wald and J. Wolfowitz (1951), "Relations among certain ranges of vector measures", Pacific Journal of Mathematics 1:59–74.

Fogelman, F., and M. Quinzii (1980), "Asymptotic values of mixed games", Mathematics of Operations Research 5:86–93.

Gardner, R.J. (1975), "Studies in the theory of social institutions", Ph.D. Thesis (Cornell University).

Gardner, R.J. (1977), "Shapley value and disadvantageous monopolies", Journal of Economic Theory 16:513–517.

Gardner, R.J. (1981), "Wealth and power in a collegial polity", Journal of Economic Theory 25:353–366.

Gilboa, I. (1989), "Additivization of nonadditive measures", Mathematics of Operations Research 14:1–17.

Guesnerie, R. (1977), "Monopoly, syndicate, and Shapley value: About some conjectures", Journal of Economic Theory 15:235–251.

Haimanko, O. (2000a), "Value theory without symmetry", International Journal of Game Theory 29:451–468.

Haimanko, O. (2000b), "Partially symmetric values", Mathematics of Operations Research 25:573–590.

Haimanko, O. (2000c), "Nonsymmetric values of nonatomic and mixed games", Mathematics of Operations Research 25:591–605.

Haimanko, O. (2000d), "Values of games with a continuum of players", Ph.D. Thesis (The Hebrew University of Jerusalem).

Haimanko, O. (2001a), "Payoffs in non-differentiable perfectly competitive TU economies", Journal of Economic Theory (forthcoming).

Haimanko, O. (2001b), "Cost sharing: The non-differentiable case", Journal of Mathematical Economics 35:445–462.

Hart, S. (1973), "Values of mixed games", International Journal of Game Theory 2:69–85.

Hart, S. (1977a), "Asymptotic values of games with a continuum of players", Journal of Mathematical Economics 4:57–80.

Hart, S. (1977b), "Values of non-differentiable markets with a continuum of traders", Journal of Mathematical Economics 4:103–116.

Hart, S. (1979), "Values of large market games", in: S. Brams, A. Schotter and G. Schwodiauer, eds., Applied Game Theory (Physica-Verlag, Heidelberg), 187–197.

Hart, S. (1980), "Measure-based values of market games", Mathematics of Operations Research 5:197–228.

Hart, S. (1982), "The number of commodities required to represent a market game", Journal of Economic Theory 27:163–169.

Hart, S., and A. Mas-Colell (1989), "Potential, value and consistency", Econometrica 57:589–614.

Hart, S., and A. Mas-Colell (1995a), "Egalitarian solutions of large games. I. A continuum of players", Mathematics of Operations Research 20:959–1002.

Hart, S., and A. Mas-Colell (1995b), "Egalitarian solutions of large games. II. The asymptotic approach", Mathematics of Operations Research 20:1003–1022.

Hart, S., and A. Mas-Colell (1996), "Harsanyi values of large economies: Nonequivalence to competitive equilibria", Games and Economic Behavior 13:74–99.

Hart, S., and D. Monderer (1997), "Potential and weighted values of nonatomic games", Mathematics of Operations Research 22:619–630.

Hart, S., and A. Neyman (1988), "Values of non-atomic vector measure games: Are they linear combinations of the measures?", Journal of Mathematical Economics 17:31–40.

Imai, H. (1983a), "Voting, bargaining, and factor income distribution", Journal of Mathematical Economics 11:211–233.

Imai, H. (1983b), "On Harsanyi's solution", International Journal of Game Theory 12:161–179.

Isbell, J. (1975), "Values of non-atomic games", Bulletin of the American Mathematical Society 83:539–546.

Kannai, Y. (1966), "Values of games with a continuum of players", Israel Journal of Mathematics 4:54–58.

Kemperman, J.H.B. (1985), "Asymptotic expansions for the Shapley index of power in a weighted majority game", mimeo.

Kohlberg, E. (1973), "On non-atomic games: Conditions for $f \circ \mu \subset pNA$", International Journal of Game Theory 2:87–98.

Kuhn, H.W., and A.W. Tucker (1950), "Editors' preface to Contributions to the Theory of Games (Vol. I), in: Annals of Mathematics Study 24 (Princeton University Press, Princeton, NJ).

Lefvre, F. (1994), "The Shapley value of a perfectly competitive market may not exist", in: J.-F. Mertens and S. Sorin, eds., Game-Theoretic Methods in General Equilibrium Analysis (Kluwer Academic, Dordrecht), 95–104.

Lyapunov, A. (1940), "Sur les fonction-vecteurs completement additions", Bull. Acad. Soc. URRS Ser. Math. 4:465–478.

Mas-Colell, A. (1977), "Competitive and value allocations of large exchange economies", Journal of Economic Theory 14:419–438.

Mas-Colell, A. (1980), "Remarks on the game-theoretic analysis of a simple distribution of surplus problem", International Journal of Game Theory 9:125–140.

McLean, R. (1999), "Coalition structure values of mixed games", in: M.H. Wooders, ed., Topics in Mathematical Economics and Game Theory: Essays in Honor of Robert J. Aumann (American Mathematical Society, Providence, RI), 105–120.

McLean, R.P., and W.W. Sharkey (1998), "Weighted Aumann–Shapley pricing", International Journal of Game Theory 27:511–523.

Mertens, J.-F. (1980), "Values and derivatives", Mathematics of Operations Research 5:523–552.

Mertens, J.-F. (1988a), "The Shapley value in the non-differentiable case", International Journal of Game Theory 17:1–65.

Mertens, J.-F. (1988b), "Nondifferentiable TU markets: The value", in: A.E. Roth, ed., The Shapley Value, Essays in Honor of Loyd S. Shapley (Cambridge University Press, Cambridge), 235–264.

Mertens, J.-F. (1990), "Extensions of games, purification of strategies, and Lyapunov's theorem", in: J.J. Gabszewicz, J.-F. Richard and L. Wolsey, eds., Economic Decision-Making: Games, Econometrics and Optimisation. Contributions in Honour of J.H. Drèze (Elsevier Science Publishers, Amsterdam), 233–279.

Mertens, J.-F., and A. Neyman (2001), "A value on $'AN$", Center for Rationality DP-276 (The Hebrew University of Jerusalem).

Milchtaich, I. (1997), "Vector measure games based on measures with values in an infinite dimensional vector space", Games and Economic Behavior 24:25–46.

Milchtaich, I. (1995), "The value of nonatomic games arising from certain noncooperative congestion games", Center for Rationality DP-87 (The Hebrew University of Jerusalem).

Milnor, J.W., and L.S. Shapley (1978), "Values of large games. II: Oceanic games", Mathematics of Operations Research 3:290–307.

Mirman, L.J., and A. Neyman (1983), "Prices for homogeneous cost functions", Journal of Mathematical Economics 12:257–273.

Mirman, L.J., and A. Neyman (1984), "Diagonality of cost allocation prices", Mathematics of Operations Research 9:66–74.

Mirman, L.J., J. Raanan and Y. Tauman (1982), "A sufficient condition on f for $f \circ \mu$ to be in *pNAD*", Journal of Mathematical Economics 9:251–257.

Mirman, L.J., D. Samet and Y. Tauman (1983), "An axiomatic approach to the allocation of fixed cost through prices", Bell Journal of Economics 14:139–151.

Mirman, L.J., and Y. Tauman (1981), "Valeur de Shapley et repartition equitable des couts de production", Cahiers du Seminaire d'Econometrie 23:121–151.

Mirman, L.J., and Y. Tauman (1982a), "The continuity of the Aumann–Shapley price mechanism", Journal of Mathematical Economics 9:235–249.

Mirman, L.J., and Y. Tauman (1982b), "Demand compatible equitable cost sharing prices", Mathematics of Operations Research 7:40–56.

Monderer, D. (1986), "Measure-based values of non-atomic games", Mathematics of Operations Research 11:321–335.

Monderer, D. (1988), "A probabilistic problem arising in economics", Journal of Mathematical Economics 17:321–338.

Monderer, D. (1989a), "Asymptotic measure-based values of nonatomic games", Mathematics of Operations Research 14:737–744.

Monderer, D. (1989b), "Weighted majority games have many μ-values", International Journal of Game Theory 18:321–325.

Monderer, D. (1990), "A Milnor condition for non-atomic Lipschitz games and its applications", Mathematics of Operations Research 15:714–723.

Monderer, D., and A. Neyman (1988), "Values of smooth nonatomic games: The method of multilinear approximation", in: A.E. Roth, ed., The Shapley Value: Essays in Honor of Lloyd S. Shapley (Cambridge University Press, Cambridge), 217–234 (Chapter 15).

Monderer, D., and W.H. Ruckle (1990), "On the symmetry axiom for values of non-atomic games", International Journal of Mathematics and Mathematical Sciences 13:165–170.

Neyman, A. (1977a), "Continuous values are diagonal", Mathematics of Operations Research 2:338–342.

Neyman, A. (1977b), "Values for non-transferable utility games with a continuum of players", TR 351, S.O.R.I.E. (Cornell University).

Neyman, A. (1979), "Asymptotic values of mixed games", in: O. Moeschlin and D. Pallaschke, eds., Game Theory and Related Topics (North-Holland, Amsterdam), 71–81.

Neyman, A. (1981a), "Singular games have asymptotic values", Mathematics of Operations Research 6:205–212.

Neyman, A. (1981b), "Decomposition of ranges of vector measures", Israel Journal of Mathematics 40:54–65.

Neyman, A. (1982), "Renewal theory for sampling without replacement", Annals of Probability 10:464–481.

Neyman, A. (1985), "Semi-values of political economic games", Mathematics of Operations Research 10:390–402.

Neyman, A. (1988), "Weighted majority games have asymptotic value", Mathematics of Operations Research 13:556–580.

Neyman, A. (1989), "Uniqueness of the Shapley value", Games and Economic Behavior 1:116–118.

Neyman, A. (1994), "Values of games with a continuum of players", in: J.-F. Mertens and S. Sorin, eds., Game-Theoretic Methods in General Equilibrium Analysis, Chapter IV (Kluwer Academic Publishers, Dordrecht), 67–79.

Neyman, A. (2001), "Values of non-atomic vector measure games", Israel Journal of Mathematics 124:1–27.

Neyman, A., and R. Smorodinski (1993), "The asymptotic value of the two-house weighted majority games", mimeo.

Neyman, A., and Y. Tauman (1976), "The existence of nondiagonal axiomatic values", Mathematics of Operations Research 1:246–250.

Neyman, A., and Y. Tauman (1979), "The partition value", Mathematics of Operations Research 4:236–264.

Osborne, M.J. (1984a), "A reformulation of the marginal productivity theory of distribution", Econometrica 52:599–630.

Osborne, M.J. (1984b), "Why do some goods bear higher taxes than others?", Journal of Economic Theory 32:301–316.

Owen, G. (1972), "Multilinear extension of games", Management Science 18:64–79.

Owen, G. (1982), Game Theory, 2nd edn. (Academic Press, New York), Chapter 12.

Pallaschke, D., and J. Rosenmüller (1997), "The Shapley value for countably many players", Optimization 40:351–384.

Pazgal, A. (1991), "Potential and consistency in TU games with a continuum of players and in cost allocation problems", M.Sc. Thesis (Tel Aviv University).

Raut, L.K. (1997), "Construction of a Haar measure on the projective limit group and random order values of nonatomic games", Journal of Mathematical Economics 27:229–250.

Raut, L.K. (1999), "Aumann–Shapley random order values of non-atomic games," in: M.H. Wooders, ed., Topics in Mathematical Economics and Game Theory: Essays in Honor of Robert J. Aumann (American Mathematical Society, Providence, RI), 121–136.

Reichert, J., and Y. Tauman (1985), "The space of polynomials in measures is internal", Mathematics of Operations Research 10:1–6.

Rosenmüller, J. (1971), "On core and value", in: R. Henn, ed., Operations Research-Verfahren (Anton Hein, Meisenheim), 84–104.

Rosenthal, R.W. (1976), "Lindahl solutions and values for a public goods example", Journal of Mathematical Economics 3:37–41.

Rothblum, U. (1973), "On orderable set functions and continuity I", Israel Journal of Mathematics 16:375–397.

Rothblum, U. (1977), "Orderable set functions and continuity II: Set functions with infinitely many null points", SIAM Journal on Control and Optimization 15:144–155.

Ruckle, W.H. (1982a), "Projections in certain spaces of set functions", Mathematics of Operations Research 7:314–318.

Ruckle, W.H. (1982b), "An extension of the Aumann–Shapley value concept to functions on arbitrary Banach spaces", International Journal of Game Theory 11:105–116.

Ruckle, W.H. (1988), "The linearizing projection: Global theories", International Journal of Game Theory 17:67–87.

Samet, D. (1981), Syndication in Market Games with a Continuum of Traders, DP 497 (KGSM, Northwestern University).

Samet, D. (1984), "Vector measures are open maps", Mathematics of Operations Research 9:471–474.

Samet, D. (1987), "Continuous selections for vector measures", Mathematics of Operations Research 12:536–543.

Samet, D., and Y. Tauman (1982), "The determination of marginal cost prices under a set of axioms", Econometrica 50:895–909.

Samet, D., Y. Tauman and I. Zang (1984), "An application of the Aumann–Shapley prices to the transportation problems", Mathematics of Operations Research 9:25–42.

Santos, J.C., and J.M. Zarzuelo (1998), "Mixing weighted values for non-atomic games", International Journal of Game Theory 27:331–342.

Santos, J.C., and J.M. Zarzuelo (1999), "Weighted values for non-atomic games: An axiomatic approach", Mathematical Methods of Operations Research 50:53–63.

Shapiro, N.Z., and L.S. Shapley (1978), "Values of large games I: A limit theorem", Mathematics of Operations Research 3:1–9.

Shapley, L.S. (1953a), "A value for *n*-person games", in: H.W. Kuhn and A.W. Tucker, eds., Contributions to the Theory of Games, Vol. II (Princeton University Press, Princeton, NJ), 307–317.

Shapley, L.S. (1953b), "Additive and non-additive set functions", Ph.D. Thesis (Princeton University).

Shapley, L.S. (1962), "Values of games with infinitely many players", in: M. Maschler, ed., Recent Advances in Game Theory (Ivy Curtis Press, Philadelphia, PA), 113–118.

Shapley, L.S. (1964), "Values of large games VII: A general exchange economy with money", RM 4248 (Rand Corporation, Santa Monica).

Shapley, L.S., and M. Shubik (1969), "Pure competition, coalitional power and fair division", International Economic Review 10:337–362.

Sroka, J.J. (1993), "The value of certain *pNA* games through infinite-dimensional Banach spaces", International Journal of Game Theory 22:123–139.

Tauman, Y. (1977), "A nondiagonal value on a reproducing space", Mathematics of Operations Research 2:331–337.

Tauman, Y. (1979), "Value on the space of all scalar integrable games", in: O. Moeschlin and D. Pallaschke, eds., Game Theory and Related Topics (North-Holland, Amsterdam), 107–115.

Tauman, Y. (1981a), "Value on a class of non-differentiable market games", International Journal of Game Theory 10:155–162.

Tauman, Y. (1981b), "Values of market games with a majority rule", in: O. Moeschlin and D. Pallaschke, eds., Game Theory and Mathematical Economics (North-Holland, Amsterdam), 103–122.

Tauman, Y. (1982), "A characterization of vector measure games in *pNA*", Israel Journal of Mathematics 43:75–96.

Weber, R.J. (1979), "Subjectivity in the valuation of games", in: O. Moeschlin and D. Pallaschke, eds., Game Theory and Related Topics (North-Holland, Amsterdam), 129–136.

Weber, R.J. (1988), "Probabilistic values for games", in: A.E. Roth, ed., The Shapley Value, Essays in Honor of Lloyd S. Shapley (Cambridge University Press, Cambridge), 101–119.

Wooders, M.H., and W.R. Zame (1988), "Values of large finite games", in: A.E. Roth, ed., The Shapley Value, Essays in Honor of Lloyd S. Shapley (Cambridge University Press, Cambridge), 195–206.

Young, H.P. (1985), "Monotonic solutions of cooperative games", International Journal of Game Theory 14:65–72.

Young, H.P. (1985), "Producer incentives in cost allocation", Econometrica 53:757–765.

Chapter 57

VALUES OF PERFECTLY COMPETITIVE ECONOMIES*

SERGIU HART

Center for Rationality and Interactive Decision Theory, Department of Economics, and Department of Mathematics, The Hebrew University of Jerusalem, Jerusalem, Israel

Contents

*The author thanks Robert J. Aumann and Andreu Mas-Colell for their comments.

Handbook of Game Theory, Volume 3, Edited by R.J. Aumann and S. Hart

Abstract

Perfectly competitive economies are economic models with many agents, each of whom is relatively insignificant. This chapter studies the relations between the basic economic concept of *competitive* (or *Walrasian*) *equilibrium,* and the game-theoretic solution concept of *value*. It includes the classical *Value Equivalence Theorem* together with its many extensions and generalizations, as well as recent results.

Keywords

perfect competition, value, Shapley value, competitive equilibrium, value equivalence principle

JEL classification: C7, D5

1. Introduction

This chapter is devoted to the study of economic models with many agents, each of whom is relatively insignificant. These are referred to as *perfectly competitive* models. The basic economic concept for such models is the *competitive* (or *Walrasian*) *equilibrium*, which prescribes prices that make the total demand equal to the total supply, i.e., under which the markets clear. The fact that each agent is negligible implies that he cannot singly affect the prices, and so he takes them as given when finding his optimal consumption, or "demand".

Game-theoretic concepts, particularly "cooperative game-theoretic" solutions (like core and value), are supported by very different considerations from the above. First, such solutions are based on the "coalitional game" which, in assigning to each coalition the set of possible allocations to its members, abstracts away from specific mechanisms (like prices). And second, they prescribe outcomes that satisfy certain desirable criteria or postulates (like efficiency, stability, or fairness).

Nevertheless, it turns out that game-theoretic solutions may yield the competitive outcomes. The prime example is the Core Equivalence Theorem (see the chapter of Anderson (1992) in this *Handbook*) which states that, in perfectly competitive economies, the core coincides with the set of competitive allocations.

Here we will study another central solution concept, the (Shapley) *value* (see the chapter of Winter (2002) in this *Handbook*). The value, originally introduced by Shapley (1953) on the basis of a number of simple axioms,[1] turns out to measure the "average marginal contributions" of the players. It is perhaps best viewed as an "expected outcome" or an "a priori utility" of playing the game.

The first work on values of perfectly competitive economies is due to Shapley (1964) (see also Shapley and Shubik (1969)), who considered differentiable exchange economies (or markets) with transferable utilities (TU), and showed that, as the number of players increases by replication, the Shapley values converge to the competitive equilibrium.

To extend this result, two developments were needed. One was to develop the theory of value for games with a continuum of players (the "limit game") and, in particular, to obtain the Shapley (1964) result in this framework; see Aumann and Shapley (1974) (and the chapter of Neyman (2002) in this *Handbook*). The other was to define the concept of value for general non-transferable utility (NTU) games; see Harsanyi (1963) and Shapley (1969) (and the chapter of McLean (2002) in this *Handbook*). This culminated in the general *Value Equivalence Theorem* of Aumann (1975): In a differentiable economy with a continuum of traders, the set of Shapley (1969) value allocations coincides with the set of competitive allocations.

Next, the non-differentiable case was studied by Champsaur (1975) (limit of replicas) and by Hart (1977b) (a continuum of traders), who showed that one direction always

---

[1] Efficiency, symmetry, dummy player and additivity.

Table 1
The Value Principle

		Limit of replicas	Continuum of agents
Differentiable	**TU**	Shapley (1964)	Aumann and Shapley (1974)
(Equivalence)	**NTU**	Mas-Colell (1977)	Aumann (1975)
General *(Inclusion)*	**TU** **NTU**	Champsaur (1975)	Hart (1977b)

holds – all value allocations are competitive – but not necessarily the converse. Together with the work of Mas-Colell (1977) (limit of replicas in the NTU case), the picture was completed, resulting in the *Value Principle*: *In perfectly competitive economies, every Shapley value allocation is competitive, and the converse holds in the differentiable case*; see Table 1.

In the TU-case, the concept of value is clear and undisputed (of course, its definition in the case of a continuum of players entails many difficulties). This is not so in the NTU-case. There are different ways of extending the Shapley TU-value, and which one is more appropriate is a matter of some controversy (see the references in Subsection 5.4). The Value Principle stated above was exclusively established for the Shapley (1969) NTU-value (also known as the "λ-transfer value"). Recently, Hart and Mas-Colell (1996a) studied the Harsanyi (1963) NTU-value in perfectly competitive economies, and showed that these value allocations may be different from the competitive ones, even in the differentiable case (and that, moreover, this is a robust phenomenon). Since the Harsanyi NTU-value is no less appropriate than the Shapley NTU-value (and possibly even more so), this casts doubt on the general validity of the Value Principle, in particular when the economy exhibits substantial NTU features; see Subsection 5.4.

The chapter is organized as follows: Section 2 presents the basic model of an exchange economy with a continuum of agents, together with the definitions of the appropriate concepts. The Value Principle results are stated in Section 3. An informal (and hopefully instructive) proof of the Value Equivalence Theorem is provided in Section 4. Section 5 is devoted to additional material, generalizations, extensions and alternative approaches.[2]

2. The model

As we saw in the Introduction, there are various ways to model "perfectly competitive" economies. One is to consider sequences of economies with finitely many agents, where the number of agents increases to infinity; see Subsection 5.5. Another is to study directly the "limit" economy with a continuum of agents. Indeed, to model precisely the

---

[2] Precise definitions and statements are given for one main result (the Value Equivalence Theorem with a continuum of traders); the rest of the material is presented in a less formal manner.

intuitive notion that an individual's influence on the whole economy is negligible, one needs infinitely many participants. As Aumann (1964, p. 39) says, "... the most natural model for this purpose contains a *continuum* of participants, similar to the continuum of points on a line or the continuum of particles in a fluid".

Our basic model is that of a *pure exchange economy* (or *market*) with a continuum of traders, and consists of the following:

E1 The *commodities* (or *goods*): $\ell = 1, 2, \ldots, L$; the consumption set of each agent is thus[3] $\mathbb{R}_+^L$.

E2 The *space of traders*: A measure space $(T, \mathfrak{C}, \mu)$, where: T is the set of *traders* (or *agents*); $\mathfrak{C}$, a σ-field of subsets of T, is the set of *coalitions*; and μ, a probability measure[4] on $\mathfrak{C}$, is the *population measure* (i.e., for every coalition $S \in \mathfrak{C}$, the number $\mu(S) \in [0, 1]$ is the "mass" of S, relative to the whole space T).

E3 The *initial endowments*: For every trader $t \in T$, a commodity vector $\mathbf{e}(t) = (\mathbf{e}^1(t), \mathbf{e}^2(t), \ldots, \mathbf{e}^L(t)) \in \mathbb{R}_+^L$, where $\mathbf{e}^\ell(t)$ is the quantity of good ℓ initially owned by trader t.

E4 The *utility functions*: For every trader $t \in T$, a function $u_t : \mathbb{R}_+^L \to \mathbb{R}$, where $u_t(x)$ is the utility of t from the commodity bundle $x \in \mathbb{R}_+^L$.

We make the following assumptions:

G1 (Measurability) $\mathbf{e}(t)$ is measurable in t and $u_t(x)$ is jointly measurable in (t, x) (relative to the Borel σ-field on $\mathbb{R}_+^L$ and the σ-field $\mathfrak{C}$ on T).

G2 (Endowments) Every commodity is present in the market, i.e.,[5] $\int_T \mathbf{e}\, d\mu \gg 0$.

G3 (Utilities) The utility functions u_t are all strictly monotonic (i.e., $x \gneq y$ implies $u_t(x) > u_t(y)$), continuous and uniformly bounded (i.e., there exists M such that $|u_t(x)| \leqslant M$ for all x and t).

G4 (Standardness) $(T, \mathfrak{C}, \mu)$ is isomorphic to[6] $([0, 1], \mathfrak{B}, \nu)$, the unit interval $[0, 1]$ with the Borel σ-field $\mathfrak{B}$ and the Lebesgue measure ν; in particular, μ is a non-atomic measure.

All of these assumptions are standard; we will refer to G1–G4 (in addition, of course, to E1–E4) as the *general case*. Some assumptions (like uniform boundedness of the utility functions) are made for simplicity of presentation only; the reader should consult the relevant papers for more general setups.

Finally, we introduce an additional set of assumptions known as the *differentiable case*:

---

[3] $\mathbb{R}$ is the real line, and $\mathbb{R}_+^L$ is the non-negative orthant of the L-dimensional Euclidean space.

[4] I.e., μ is non-negative and $\mu(T) = 1$.

[5] For vectors $x, y \in \mathbb{R}^L$, we take $x \geqslant y$ and $x \gg y$ to mean, respectively, $x^\ell \geqslant y^\ell$ and $x^\ell > y^\ell$ for all ℓ. Next, $x \gneq y$ means $x \geqslant y$ and $x \neq y$ (i.e., $x^\ell \geqslant y^\ell$ for all ℓ, with at least one strict inequality).

[6] There is little loss of generality here; see Aumann and Shapley (1974, Chapter VIII).

D1 The utility functions u_t are all concave and continuously differentiable on[7] $\mathbb{R}_+^L$, and their gradients ∇u_t are uniformly bounded and uniformly positive in bounded subsets of[8] $\mathbb{R}_+^L$.

D2 The initial endowments are uniformly bounded (i.e., there exists M such that $\mathbf{e}^\ell(t) \leqslant M$ for all ℓ and t).

An *allocation* $\mathbf{x}$ is a feasible outcome of this exchange economy, i.e., a redistribution of the total initial endowment. That is, $\mathbf{x}: T \to \mathbb{R}_+^L$ is an integrable function, where $\mathbf{x}(t) \in \mathbb{R}_+^L$ is agent t's final commodity bundle, which satisfies:

$$\int_T \mathbf{x}\,d\mu = \int_T \mathbf{e}\,d\mu. \tag{2.1}$$

We now define the two concepts that will be the concern of this chapter: competitive equilibrium and value. An allocation $\mathbf{x}$ is *competitive* (or *Walrasian*) if there exists a *price vector* $p \in \mathbb{R}_+^L$, $p \neq 0$ such that, for every agent $t \in T$, the bundle $\mathbf{x}(t)$ is maximal with respect to u_t in t's budget set $B_t(p) := \{x \in \mathbb{R}_+^L: p \cdot x \leqslant p \cdot \mathbf{e}(t)\}$ (i.e., $u_t(\mathbf{x}(t)) \geqslant u_t(y)$ for all $y \in B_t(p)$). We refer to $\mathbf{e}(t)$ as t's *supply*, and to $\mathbf{x}(t)$ as t's *demand*; equality (2.1) then states that total demand equals total supply.

An allocation $\mathbf{x}$ is a *Shapley* (1969) (*NTU-*)*value* allocation (see the chapter of McLean (2002) in this *Handbook*) if there exists a collection $\lambda = (\lambda_t)_{t \in T}$ of nonnegative weights (formally, a measurable function $\lambda: T \to \mathbb{R}_+$) such that[9]

$$\int_S \lambda_t u_t(\mathbf{x}(t))\,d\mu(t) = \varphi v_\lambda(S), \quad \text{for every } S \in \mathfrak{C}, \tag{2.2}$$

where v_λ is the TU-game defined by

$$v_\lambda(S) := \max\left\{\int_S \lambda_t u_t(\mathbf{y}(t))\,d\mu(t): \mathbf{y}(t) \in \mathbb{R}_+^L \text{ for all } t \in S\right.$$

$$\left. \text{and } \int_S \mathbf{y}\,d\mu = \int_S \mathbf{e}\,d\mu\right\} \tag{2.3}$$

for all coalitions $S \in \mathfrak{C}$, and φv_λ is the "asymptotic value" (see the chapter of Neyman (2002) in this *Handbook*) of the game v_λ. That is, if utilities were transferable between

---

[7] Including the boundary of $\mathbb{R}_+^L$; that is, u_t is continuously differentiable in the interior of $\mathbb{R}_+^L$, and all its partial derivatives can be continuously extended to the boundary of $\mathbb{R}_+^L$.

[8] I.e., for every $M < \infty$ there exist $c > 0$ and $C < \infty$ such that $c < \partial u_t(x)/\partial x^\ell < C$ for all t, ℓ and x with $\|x\| \leqslant M$.

[9] Equivalently, in "density" terms, $\lambda_t u_t(\mathbf{x}(t)) = (d(\varphi v_\lambda)/d\mu)(t)$ for μ-almost every t. The right-hand side (which is the Radon–Nikodym derivative of φv_λ with respect to μ) may be viewed as the "value of t"; in our informal arguments, we will abuse notation and write it as $\varphi v_\lambda(t)$, so (2.2) becomes $\lambda_t u_t(\mathbf{x}(t)) = \varphi v_\lambda(t)$ for (almost) every t (just like the case when there are only finitely many agents).

the traders at the rates λ, then the value of the resulting TU-game (2.3) would in fact be achievable without transfers (2.2).

3. The Value Principle

In this section we will state the main results on the relations between value and competitive equilibria. The following two theorems comprise what is known as the *Value Principle*.

THEOREM 3.1 (Value Equivalence). *In the differentiable case*: *The set of Shapley value allocations coincides with the set of competitive allocations.*

THEOREM 3.2 (Value Inclusion). *In the general case*: *The set of Shapley value allocations is included in the set of competitive allocations* (*and there are cases where the inclusion is strict*).

In other words: First, value allocations are always competitive. And second, competitive allocations are value allocations provided the economy is appropriately differentiable (or smooth).

In the framework of the previous section (markets with a continuum of agents), the "general case" refers to E1–E4 together with G1–G4; the "differentiable case", to E1–E4, G1–G4 and D1–D2. Value Equivalence has been proved by Aumann (1975). We will provide an informal – and hopefully instructive – proof in the next section. Value Inclusion has been proved by Hart (1977b); see Subsections 5.1 and 5.2 below. As stated in the Introduction, parallel results have been obtained in other frameworks (like limits of sequences, transferable utility, etc.; recall Table 1 and see Section 5).

All of these results use the concept of NTU-value due to Shapley (1969) (defined in the previous section). However, there are other notions of an NTU-value. A prominent one is the *Harsanyi* (1963) *NTU-value*. We have:

PROPOSITION 3.3. *In the differentiable case*: *The set of Harsanyi value allocations may be disjoint from the set of competitive allocations.*

This is shown by Hart and Mas-Colell (1996a), who provide a robust example with a unique Harsanyi value which is different from the unique competitive equilibrium. This clearly raises doubts about either the Harsanyi value or the Value Principle. In view of the additional literature, Hart and Mas-Colell (1996a) suggest that, since the Harsanyi concept is no less appropriate than the Shapley one (and perhaps even more so), it follows that the Value Principle may not be valid in general – particularly when the economy is far removed from the transferable utility case. See Subsection 5.4.

4. Proof of Value Equivalence

We now provide an informal proof for the Value Equivalence Theorem 3.1, in the frame-work of Section 2 (under all the assumptions there, including differentiability); in particular, "value" stands for "Shapley NTU-value".

4.1. Value allocations are competitive allocations

Let $\mathbf{x}$ be a value allocation with corresponding weights λ. We ignore technical compli-cations and assume that $\mathbf{x}(t) \gg 0$ and $\lambda_t > 0$ for all t. Also, we will write "for all t" where in fact it should say "for μ-almost every t", and we regard a point t as if it were a "real agent".[10]

First, we have

$$\int_T \lambda_t u_t\big(\mathbf{x}(t)\big)\, d\mu(t) = v_\lambda(T) \tag{4.1}$$

by (2.2) for T together with $\varphi v_\lambda(T) = v_\lambda(T)$ (since the TU-value is efficient). Thus the maximum in (2.3) for T is achieved at $\mathbf{x}$. Therefore[11]

$$\lambda_t \nabla u_t\big(\mathbf{x}(t)\big) = \lambda_{t'} \nabla u_{t'}\big(\mathbf{x}(t')\big) \quad \text{for all } t, t' \in T \tag{4.2}$$

(this is standard: if not, then a reallocation of goods between t and t' would increase the total utility $v_\lambda(T)$). Let p be the common value of[12] $\lambda_t \nabla u_t(\mathbf{x}(t))$ for all t:

$$\nabla u_t\big(\mathbf{x}(t)\big) = (1/\lambda_t)p \quad \text{for all } t \in T.$$

The utility functions u_t are concave; therefore $u_t(y) \leqslant u_t(\mathbf{x}(t)) + \nabla u_t(\mathbf{x}(t)) \cdot (y - \mathbf{x}(t))$, from which it follows that

$$\text{if } p \cdot y \leqslant p \cdot \mathbf{x}(t) \text{ then } u_t(y) \leqslant u_t\big(\mathbf{x}(t)\big). \tag{4.3}$$

Next, we claim that the asymptotic value φv_λ of v_λ satisfies

$$\varphi v_\lambda(t) = \lambda_t u_t\big(\mathbf{x}(t)\big) + p \cdot (\mathbf{e}(t) - \mathbf{x}(t)). \tag{4.4}$$

The intuition for this is as follows: The value of t is t's marginal contribution to a ran-dom coalition Q (this holds in the case of finitely many players by Shapley's formula,

---

[10] As in the case of finitely many agents. More appropriately, one should view dt as the "agent"; we prefer to use t since it may appear less intimidating for some readers.

[11] We write $\nabla u(x)$ for the gradient of u at x, i.e., the vector of partial derivatives $(\partial u(x)/\partial x^\ell)_{\ell=1,\ldots,L}$.

[12] Letting $p = (p^\ell)_{\ell=1,\ldots,L}$, one may interpret p^ℓ as the Lagrange multiplier (or "shadow price") of the constraint $\int_T \mathbf{x}^\ell = \int_T \mathbf{e}^\ell$.

and it extends to the general case since the asymptotic value is the limit of values of finite approximations). Now a "random coalition" Q, when there is a continuum of players, is a perfect sample of the grand coalition T. (Assume for instance that we have a large finite population, consisting of two types of players: $1/3$ of them are of type 1, and $2/3$ of them of type 2. The coalition of the first half of the players in a random order will most probably contain, by the Law of Large Numbers, about $1/3$ players of type 1 and $2/3$ of type 2. This holds for "one half" as well as for any other proportion strictly between 0 and 1.) Therefore the average marginal contribution of t to Q is essentially the same as t's marginal contribution to the grand coalition[13] T, which we can write suggestively as

$$\varphi v_\lambda(t) = v_\lambda(T) - v_\lambda\big(T\backslash\{t\}\big).$$

Since $v_\lambda(T)$ is the result of a maximization problem, its derivative ("in the direction t") is obtained by keeping the optimal allocation fixed and multiplying the change in the constraints by the corresponding Lagrange multipliers.[14] The optimal allocation is $\mathbf{x}$ (by (4.1)); rewriting $\int_T \mathbf{x}\,d\mu = \int_T \mathbf{e}\,d\mu$ as $\int_{T\backslash\{t\}} \mathbf{x}\,d\mu = \int_{T\backslash\{t\}} \mathbf{e}\,d\mu + (\mathbf{e}(t) - \mathbf{x}(t))\,d\mu(t)$ shows that the change in the constraints is[15] $\mathbf{e}(t) - \mathbf{x}(t)$; and the Lagrange multipliers are p. Therefore indeed

$$\varphi v_\lambda(t) = \lambda_t u_t\big(\mathbf{x}(t)\big) + p \cdot \big(\mathbf{e}(t) - \mathbf{x}(t)\big).$$

In other words, the marginal contribution of t to the total utility $v_\lambda(T)$ consists of t's weighted utility contribution $\lambda_t u_t(\mathbf{x}(t))$, together with his net contribution of commodities $\mathbf{e}(t) - \mathbf{x}(t)$, evaluated according to the "shadow prices" p (i.e., the marginal weighted utilities, which are common to everyone by (4.2)). We have thus shown (4.4).

Comparing (2.2) (see Footnote 9) and (4.4) implies that

$$p \cdot \big(\mathbf{e}(t) - \mathbf{x}(t)\big) = 0 \quad \text{for all } t,$$

which together with (4.3) shows that $\mathbf{x}$ is a competitive allocation relative to the price vector p.

4.2. Competitive allocations are value allocations

Let $\mathbf{x}$ be a competitive allocation with associated price vector p. From the fact that $\mathbf{x}(t)$ is the demand of t at the prices p (i.e., $p \cdot y \leqslant p \cdot \mathbf{e}(t)$ implies $u_t(y) \leqslant u_t(\mathbf{x}(t))$), it

---

[13] Technically, one uses here the homogeneity of degree 1 of v, hence the homogeneity of degree 0 of its derivatives.

[14] Since the first order condition (that of vanishing derivatives) is satisfied at optimal allocations, the change in the optimal allocation is, in the first order, zero. This is the so-called "Envelope Theorem"; see, e.g., Mas-Colell, Whinston and Green (1995, Section M.L).

[15] In "density" terms.

follows that the gradient $\nabla u_t(\mathbf{x}(t))$ is proportional to p (again, we assume for simplicity that $\mathbf{x}(t) \gg 0$ for all t). That is, there exist $\lambda_t > 0$ such that

$$\lambda_t \nabla u_t(\mathbf{x}(t)) = p \quad \text{for all } t.$$

This implies that the maximum in the definition of $v_\lambda(T)$ (for this collection $\lambda = (\lambda_t)_t$) is attained at $\mathbf{x}$.

As we saw in the previous subsection (recall (4.4)), the asymptotic value φv_λ of v_λ satisfies

$$\varphi v_\lambda(t) = \lambda_t u_t(\mathbf{x}(t)) + p \cdot (\mathbf{e}(t) - \mathbf{x}(t)).$$

But $\mathbf{x}(t)$ is t's demand at p, so $p \cdot \mathbf{x}(t) = p \cdot \mathbf{e}(t)$ (by monotonicity), and therefore $\varphi v_\lambda(t) = \lambda_t u_t(\mathbf{x}(t))$, thus completing the proof that $\mathbf{x}$ is indeed a value allocation (by (2.2)).

5. Generalizations and extensions

5.1. The non-differentiable case

In the general case (i.e., without the differentiability assumptions), the Value Inclusion Theorem says that every Shapley value allocation is competitive, and that the converse need no longer hold. In fact, in the non-differentiable case the asymptotic value of v_λ may not exist (since different finite approximations lead to different limits),[16] and so there may be no value allocations at all. Moreover, even if value allocations do exist, they correspond only to some of the competitive allocations; roughly speaking, values "select" certain "centrally located" equilibria.

The proof that every Shapley value is competitive is based on generalizing the value formula (4.4); see Hart (1977b) and the next subsection.

5.2. The geometry of non-atomic market games

A *(TU-)market game* is a game that arises from an economy as in Section 2 (see (2.3)).

To illustrate the geometrical structure that underlies the results of this chapter, suppose for simplicity that there are finitely many types of agents, say n, with a continuum of mass $1/n$ of each type. A coalition S is then characterized by its composition, or *profile*, $s = (s_1, s_2, \ldots, s_n) \in \mathbb{R}_+^n$, where s_i is the mass of agents of type i in S. Formula (2.3) then defines a function $v = v_\lambda$ over profiles s; that is,[17] $v : \mathbb{R}_+^n \to \mathbb{R}$. Such a

---

[16] A necessary condition for the existence of the asymptotic value is that the core possess a center of symmetry; see Hart (1977a, 1977b). Note that the convex hull of k points in general position does not have a center of symmetry when $k \geqslant 3$ (it is a triangle, a tetrahedron, etc.).

[17] In fact, only $s \leqslant (1/n, 1/n, \ldots, 1/n)$ matter.

function is clearly super-additive (i.e., $v(s + s') \geq v(s) + v(s')$ for all $s, s' \in \mathbb{R}^n_+$) and homogeneous of degree 1 (i.e., $v(\alpha s) = \alpha v(s)$ for every $s \in \mathbb{R}^n_+$ and $\alpha \geq 0$), and hence concave.[18]

Let $\bar{s} = (1/n, 1/n, \ldots, 1/n)$ denote (the profile of) the grand coalition. A payoff vector[19] $\pi = (\pi_1, \pi_2, \ldots, \pi_n)$, where π_i is the payoff of each agent of type i, is in the *core* of v if $\sum_{i=1}^n \pi_i \bar{s}_i = v(\bar{s})$ and $\sum_{i=1}^n \pi_i s_i \geq v(s)$ for every $s \leq \bar{s}$. Thus π is nothing other than a normal vector, or "supergradient" (recall that v is concave) to v at[20] $\bar{s}$. In particular, if v is differentiable, then the core has a unique element: the gradient $\nabla v(\bar{s})$ of v at $\bar{s}$.

As for the TU-value of v, it is obtained as in Subsection 4.1 (see the considerations leading to (4.4)) as follows. Assume first that v is continuously differentiable. The profile q of a random coalition Q is (approximately) proportional to that of the grand coalition, i.e., $q \approx \alpha \bar{s}$ for some $\alpha \in (0, 1)$ (by the Law of Large Numbers). The marginal contribution of an agent of type i to such a coalition Q is thus given by the partial derivative $(\partial v/\partial s_i)(q) \approx (\partial v/\partial s_i)(\alpha \bar{s})$, which equals $(\partial v/\partial s_i)(\bar{s})$ by homogeneity. Therefore the value of type i is $(\partial v/\partial s_i)(\bar{s})$, and the value payoff vector is $\nabla v(\bar{s})$ – identical to the core.[21]

If v is not differentiable, then the partial derivatives are replaced by directional derivatives, which correspond to supergradients, and are again independent of α by homogeneity. The set of supergradients is convex, therefore averaging (over random coalitions) implies that the value vector is also a supergradient of v at $\bar{s}$ – and so in this case too the value belongs to the core.

Summarizing: *In a non-atomic TU-market game, the value belongs to the core, and is the unique element in the core in the differentiable case.* But the core and the set of competitive allocations coincide – this is the Core Equivalence Theorem (see the chapter of Anderson (1992) in this *Handbook*, and note that differentiability is not required) – which yields the Value Principle in the TU-case.

Moving now to the NTU-case, note that if $\pi = (\pi_1, \pi_2, \ldots, \pi_n)$ is a Shapley NTU-value with weights[22] $\lambda = (\lambda_1, \lambda_2, \ldots, \lambda_n)$, then the vector $(\lambda_1 \pi_1, \lambda_2 \pi_2, \ldots, \lambda_n \pi_n)$ is the TU-value of v_λ (see (2.2)) and so, as we have seen above, it is a supergradient of v_λ at $\bar{s}$. Therefore $\sum_{i=1}^n \lambda_i \pi_i s_i \geq v_\lambda(s)$ for every s, implying that π cannot be improved upon by a coalition of profile s (recall the definition of v_λ as a maximum). In other words,

---

[18] Such a function may be called "conical": its *subgraph* $\{(s, \xi) \in \mathbb{R}^n_+ \times \mathbb{R} : \xi \leq v(s)\}$ is a convex cone.

[19] We only consider type-symmetric payoffs, where agents of the same type receive the same payoffs.

[20] I.e., the graph of the linear function $h(s) := \pi \cdot s$ is a supporting hyperplane to the subgraph of v (which is a convex cone; see Footnote 18) at the point $(\bar{s}, v(\bar{s}))$ (and thus also on the whole "diagonal" $\{(\alpha \bar{s}, v(\alpha \bar{s})):$ $\alpha \geq 0\}$).

[21] Another proof of this statement can be obtained using the *potential approach* of Hart and Mas-Colell (1989): Given v, there exists a unique function $P \equiv P(v) : \mathbb{R}^n_+ \to \mathbb{R}$ with $P(0) = 0$ and $s \cdot \nabla P(s) = v(s)$ for all s – the *potential of* v – and moreover $\nabla P(\bar{s})$ is the value payoff vector. When v is homogeneous of degree 1, the Euler equation implies $s \cdot \nabla v(s) = v(s)$, and so $P = v$ and $\nabla P(\bar{s}) = \nabla v(\bar{s})$ (or: value = core).

[22] λ_i is the weight corresponding to agents of type i (recall that, for simplicity, we assume type-symmetry).

π belongs to the (NTU) core of the economy, and so it is a competitive allocation by the Core Equivalence Theorem. This establishes the Value Inclusion result of Theorem 3.2.

For precise presentations of these topics, the reader is referred to Shapley (1964), Hart (1977a, 1977b) and Hart and Mas-Colell (1996a, Section VII).

5.3. Other non-atomic TU-values

In the non-differentiable case, many of the relevant TU-games may not have an asymptotic value.[23] This happens when different sequences of finite games that approximate the given non-atomic game have values that converge to different limits. (The simplest such case is the so-called "3-gloves market"; see Aumann and Shapley (1974, Section 19) and recall Footnote 16.) Therefore one looks for other TU-value concepts.

One approach (first used by Aumann and Kurz (1977)) modifies the definition of the asymptotic value by considering only those finite approximations with players that are (approximately) equal in size, where "size" is determined by the underlying population measure μ. The resulting *measure-based value* (or μ-value) is shown by Hart (1980) to exist under wide conditions and, moreover, to yield a competitive allocation (i.e., the Value Inclusion result of Theorem 3.2 continues to hold for this value). An explicit formula, involving an appropriate normal distribution, is obtained for the resulting competitive price; this price may then be viewed as an "expected equilibrium price", where the expectation is taken over random samples of the agents.

Another approach, due to Mertens (1988a, 1988b), defines a value concept on a large space of non-atomic games, which includes all markets (whether differentiable or not); again, the Value Inclusion Theorem 3.2 holds for the Mertens value as well.

5.4. Other NTU-values

Extending the concept of the Shapley value from the class of TU-games to the class of NTU-games is a conceptually challenging problem. One looks for an NTU-solution concept that, in particular: (i) extends the Shapley (1953) value for TU-games; (ii) extends the Nash (1950) bargaining solution for pure bargaining problems; (iii) is covariant with independent rescalings of the utility functions;[24] and (iv) satisfies postulates like efficiency and symmetry. However, all this does not yet determine the solution, and additional constructs are needed.

The two major *NTU-value* concepts are due to *Harsanyi* (1963) and to *Shapley* (1969). In fact, Shapley introduced his NTU-value originally as a simplification of the Harsanyi NTU-value. It then turned out to be of interest in its own right, and has been

---

[23] See however Hart (1977b, Theorem C).

[24] In the TU-case, where there is a common medium of utility transfer, one may multiply all utilities by the *same* constant without changing the problem. In contrast, in the NTU-case, each player's utility may be multiplied by a *different* constant.

used in various models, economic and otherwise. However, further research has indicated that the Harsanyi concept may well be the more appropriate one, at least in some cases; see Aumann (1985a, 1985b, 1986), Hart (1985a, 1985b), Roth (1980, 1986) and Shafer (1980) for some of the issues related to the interpretation and appropriateness of these concepts.

The Harsanyi NTU-values of large differentiable economies[25] were studied by Hart and Mas-Colell (1996a). It is shown, first, that the Value Inclusion result holds in some cases;[26] and, second, that in general the two (non-empty) sets of allocations – the Harsanyi values and the competitive equilibria – may be disjoint. Moreover, this "non-equivalence" is a robust phenomenon, and a result of being far removed from the transferable utility case (i.e., the lack of substitutability among the agents' utilities).

Another NTU-value that has recently been analyzed is the *consistent NTU-value* (see Maschler and Owen (1989, 1992), and also Hart and Mas-Colell (1996b)). For a first study on the connections between this value and the core, see Leviatan (1998).

5.5. Limits of sequences

Rather than analyze the limit economy with a continuum of agents, one may instead consider sequences of finite economies. The simplest such approach, originally used in the study of the Core Equivalence Theorem, is that of "replicas": Each agent is replaced by r identical agents, and one looks at the limit as r increases to infinity.

The results here are the same as in Theorems 3.1 and 3.2: Limits of Shapley value allocations are competitive, and the converse holds in the differentiable case. See Shapley (1964) (differentiable, TU); Champsaur (1975) (non-differentiable, TU and NTU); Mas-Colell (1977) and Cheng (1981) (differentiable, NTU); and also Wooders and Zame (1987a, 1987b) for a "space of characteristics" framework (non-differentiable, TU and NTU).

5.6. Other approaches

Among the other approaches to the Value Equivalence result, one should mention the axiomatic characterization of solutions of large economies due to Dubey and Neyman (1984, 1997), which captures many of the interesting concepts – competitive equilibrium, core, value – and thus implies their equivalence. Other works use non-standard analysis [Brown and Loeb (1977)] or fuzzy games [Butnariu (1987)].

---

[25] The framework is that of a continuum of agents of finitely many types, as in Subsection 5.2. Moreover, as it is shown in Hart and Mas-Colell (1995a, 1995b), the analysis is borne out by limits of finite approximations.

[26] Specifically, when the Harsanyi value is "tight".

5.7. Ordinal vs. cardinal

The competitive allocations and the core are clearly "ordinal" concepts: They depend only on the preference relations of the agents and not on the particular utility representations. How about the value?

The construction in Section 2 uses given utility functions u_t. However, the relative weights λ_t are obtained *endogenously* (as a "fixed-point" where (2.2) holds). Thus, if we were to apply *linear* transformations to the functions u_t (with different coefficients for different t) the NTU-value allocations would not change. Moreover, a careful look at the proof of the Value Equivalence result in Section 3 shows that only "local" information is used (e.g., gradients or marginal rates of substitution), and so applying *differentiable monotonic* transformations will not matter either. Therefore, the NTU-value depends only on the preference orders, and is indeed an "ordinal" concept. In fact, Aumann (1975) (and others) works directly with collections of utility representations (rather than with one utility representation and weights λ, as we do here[27]).

5.8. Imperfect competition

Perfect competition corresponds to all agents being individually insignificant. Imperfect competition results when there are "large" agents. This is modelled by population measures that are no longer non-atomic; the atoms are precisely the large agents. An interesting question is whether it is worthwhile for a coalition to become an atom, i.e., a "monopoly". For when this is measured by the value, see Guesnerie (1977) and Gardner (1977) (compare this with the results for the core; see the chapter of Gabszewicz and Shitovitz (1992) in this *Handbook*).

For other economic applications, the reader is referred to the chapter of Mertens (2002) in this *Handbook*.

References

Anderson, R.M. (1992), "The core in perfectly competitive economies", in: R.J. Aumann and S. Hart, eds., Handbook of Game Theory, Vol. 1 (North-Holland, Amsterdam) Chapter 14, 413–457.

Aumann, R.J. (1964), "Markets with a continuum of traders", Econometrica 32:39–50.

Aumann, R.J. (1975), "Values of markets with a continuum of traders", Econometrica 43:611–646.

Aumann, R.J. (1985a), "An axiomatization of the non-transferable utility value", Econometrica 53:599–612.

Aumann, R.J. (1985b), "On the non-transferable utility value: A comment on the Roth–Shafer examples", Econometrica 53:667–678.

Aumann, R.J. (1986), "Rejoinder", Econometrica 54:985–989.

Aumann, R.J., and M. Kurz (1977), "Power and taxes in a multi-commodity economy", Israel Journal of Mathematics 27:185–234.

Aumann, R.J., and L.S. Shapley (1974), Values of Non-Atomic Games (Princeton University Press).

---

[27] This "λ-approach" is commonly used, and we hope it will thus be easier on the reader.

Brown, D., and P.A. Loeb (1977), "The values of non-standard exchange economies", Israel Journal of Mathematics 25:71–86.

Butnariu, D. (1987), "Values and cores of fuzzy games with infinitely many players", International Journal of Game Theory 16:43–68.

Champsaur, P. (1975), "Cooperation vs. competition", Journal of Economic Theory 11:394–417.

Cheng, H. (1981), "On dual regularity and value convergence theorems", Journal of Mathematical Economics 6:37–57.

Dubey, P., and A. Neyman (1984), "Payoffs in nonatomic economies: An axiomatic approach", Econometrica 52:1129–1150.

Dubey, P., and A. Neyman (1997), "An equivalence principle for perfectly competitive economies", Journal of Economic Theory 75:314–344.

Gabszewicz, J.J., and B. Shitovitz (1992), "The core in imperfectly competitive economies", in: R.J. Aumann and S. Hart, eds., Handbook of Game Theory, Vol. 1 (North-Holland, Amsterdam) Chapter 15, 459–483.

Gardner, R.J. (1977), "Shapley value and disadvantageous monopolies", Journal of Economic Theory 16:513–517.

Guesnerie, R. (1977), "Monopoly, syndicate, and Shapley value: About some conjectures", Journal of Economic Theory 15:235–251.

Harsanyi, J.C. (1963), "A simplified bargaining model for the n-person cooperative game", International Economic Review 4:194–220.

Hart, S. (1977a), "Asymptotic value of games with a continuum of players", Journal of Mathematical Economics 4:57–80.

Hart, S. (1977b), "Values of non-differentiable markets with a continuum of traders", Journal of Mathematical Economics 4:103–116.

Hart, S. (1980), "Measure-based values of market games", Mathematics of Operations Research 5:197–228.

Hart, S. (1985a), "An axiomatization of Harsanyi's non-transferable utility solution", Econometrica 53:1295–1313.

Hart, S. (1985b), "Non-transferable utility games and markets: Some examples and the Harsanyi solution", Econometrica 53:1445–1450.

Hart, S., and A. Mas-Colell (1989), "Potential, value and consistency", Econometrica 57:589–614.

Hart, S., and A. Mas-Colell (1995a), "Egalitarian solutions of large games: I. A continuum of players", Mathematics of Operations Research 20:959–1002.

Hart, S., and A. Mas-Colell (1995b), "Egalitarian solutions of large games: II. The asymptotic approach", Mathematics of Operations Research 20:1003–1022.

Hart, S., and A. Mas-Colell (1996a), "Harsanyi values of large economies: Non-equivalence to competitive equilibria", Games and Economic Behavior 13:74–99.

Hart, S., and A. Mas-Colell (1996b), "Bargaining and value", Econometrica 64:357–380.

Leviatan, S. (1998), "Consistent values and the core in continuum market games with two types", The Hebrew University of Jerusalem, Center for Rationality DP-171.

Maschler, M., and G. Owen (1989), "The consistent Shapley value for hyperplane games", International Journal of Game Theory 18:389–407.

Maschler, M., and G. Owen (1992), "The consistent Shapley value for games without side payments", in: R. Selten, ed., Rational Interaction (Springer) 5–12.

Mas-Colell, A. (1977), "Competitive and value allocations of large exchange economies", Journal of Economic Theory 14:419–438.

Mas-Colell, A., M.D. Whinston and J.R. Green (1995), Microeconomic Theory (Oxford University Press).

McLean, R.P. (2002), "Values of non-transferable utility games", in: R.J. Aumann and S. Hart, eds., Handbook of Game Theory, Vol. 3 (North-Holland, Amsterdam) Chapter 55, 2077–2120.

Mertens, J.-F. (1988a), "The Shapley value in the non-differentiable case", International Journal of Game Theory 17:1–65.

Mertens, J.-F. (1988b), "Nondifferentiable TU markets: The value", in: A.E. Roth, ed., The Shapley Value (Cambridge University Press) 235–264.

Mertens, J.-F. (2002), "Some other applications of the value", in: R.J. Aumann and S. Hart, eds., Handbook of Game Theory, Vol. 3 (North-Holland, Amsterdam) Chapter 58, 2185–2201.

Nash, J. (1950), "The bargaining problem", Econometrica 18:155–162.

Neyman, A. (2002), "Values of games with infinitely many players", in: R.J. Aumann and S. Hart, eds., Handbook of Game Theory, Vol. 3 (North-Holland, Amsterdam) Chapter 56, 2121–2167.

Roth, A.E. (1980), "Values of games without side payments: Some difficulties with current concepts", Econometrica 48:457–465.

Roth, A.E. (1986), "On the non-transferable utility value: A reply to Aumann", Econometrica 54:981–984.

Shafer, W. (1980), "On the existence and interpretation of value allocations", Econometrica 48:467–477.

Shapley, L.S. (1953), "A value for n-person games", in: H.W. Kuhn and A.W. Tucker, eds., Contributions to the Theory of Games II, Annals of Mathematics Studies 28 (Princeton University Press) 307–317.

Shapley, L.S. (1964), "Values of large games VII: A general exchange economy with money", RM-4248-PR (The Rand Corporation, Santa Monica, CA).

Shapley, L.S. (1969), "Utility comparison and the theory of games", La Décision (Editions du CNRS, Paris) 251–263.

Shapley, L.S., and M. Shubik (1969), "Pure competition, coalitional power, and fair division", International Economic Review 24:1–39.

Winter, E. (2002), "The Shapley value", in: R.J. Aumann and S. Hart, eds., Handbook of Game Theory, Vol. 3 (North-Holland, Amsterdam) Chapter 53, 2025–2054.

Wooders, M., and W.R. Zame (1987a), "Large games: Fair and stable outcomes", Journal of Economic Theory 42:59–93.

Wooders, M., and W.R. Zame (1987b), "NTU values of large games", Stanford University, I.M.S.S.S. TR-503 (mimeo).

Chapter 58

SOME OTHER ECONOMIC APPLICATIONS OF THE VALUE*

JEAN-FRANÇOIS MERTENS

*Center for Operations Research and Econometrics (CORE), Université Catholique de Louvain,
Louvain-la-Neuve, Belgium*

Contents

*Based on Chapter 10 of *Game-Theoretic Methods in General Equilibrium Analysis*, J.-F. Mertens and S. Sorin (editors), Vol. 77 of *NATO ASI Series D*, Kluwer Academic Publishers, 1994. Permission to use this material is gratefully acknowledged.

Handbook of Game Theory, Volume 3, Edited by R.J. Aumann and S. Hart

Abstract

Selected applications of the NTU Shapley value to settings other than that of general equilibrium in perfect competition. Three models are considered: Taxation and Redistribution, Public Goods, and Fixed Prices. In each case, the value leads to significant conceptual insights which, while compelling once understood, are far from obvious at the outset.

Keywords

NTU Shapley value, taxation, redistribution, public goods, fixed prices

JEL classification: H23, D45, D72, C71

1. Introduction

The concept of value has proved to be a powerful tool in analyzing economic models. Indeed, since the Shapley value can be interpreted in terms of "marginal worth", it is closely related to traditional economic ideas (see Chapter 57 of this *Handbook,* which presents the Value Equivalence Theorem). Though other important applications exist, we focus here on three applications of values to economic models outside the general equilibrium model; each describes a different departure from the market environment. The first two are economic-political models dealing with taxation. Taxation has (at least) two purposes: redistribution and the financing of public goods. The classical literature assumes that a benevolent government aims to maximize some social utility function. The game approach is quite different: it assumes that the policy of the government is determined by the interests of those who elect it. Both with redistribution and with public goods, the value is a natural tool to analyze the corresponding games. The last section deals with fixed-price economies.

Throughout this chapter, we will provide intuitions on why one would expect the results to hold (or not), rather than detailed proofs. For the latter, the reader may consult the original papers.

2. The basic model

Define a competitive economy as $(T, \ell, \mathbf{e}, (u_t)_{t \in T})$ where $T = [0, 1]$ stands for the set of traders, endowed with Lebesgue measure μ; $\mathbb{R}_+^\ell$ is the commodity space; $\mathbf{e} : T \to \mathbb{R}_+^\ell$, integrable, is the initial allocation; $u_t : \mathbb{R}_+^\ell \to \mathbb{R}$ is the utility function of t. An *allocation* is a (measurable) map $\mathbf{x} : T \to \mathbb{R}_+^\ell$ such that $\int_T \mathbf{x}_t \mu(\mathrm{d}t) = \int_T \mathbf{e}_t \mu(\mathrm{d}t)$.

An individual trader is best viewed as an "infinitesimal subset" $\mathrm{d}t$ of T. Thus $\mathbf{e}_t \mu(\mathrm{d}t)$ is trader $\mathrm{d}t$'s endowment, $\mathbf{x}_t \mu(\mathrm{d}t)$ denotes what he gets under an allocation $\mathbf{x}$, and $u_t(\mathbf{x}_t)\mu(\mathrm{d}t)$ is the utility he derives from it. We assume $u_t(0) = 0$.

Value allocations. Consider a positive "weight" function $\lambda : T \to \mathbb{R}_+$ (integrable). The worth $v_\lambda(S)$ of a coalition $S \subset T$ with respect to λ is the maximum "weighted total utility" it can get on its own, i.e., when the members of S reallocate their total endowment among each other. In the above setting,

$$v_\lambda(S) = \max \left\{ \int_S \lambda_t u_t(\mathbf{x}_t)\mu(\mathrm{d}t) \text{ such that } \int_S \mathbf{x}_t \mu(\mathrm{d}t) = \int_S \mathbf{e}_t \mu(\mathrm{d}t) \right\}.$$

The value of a trader is his average marginal contribution to the coalitions to which he belongs, where "average" means expectation with respect to a distribution induced by a random order of the players. Thus, the value of trader $\mathrm{d}t$ is

$$(\varphi v_\lambda)(\mathrm{d}t) = E\left[v_\lambda(S_{\mathrm{d}t} \cup \mathrm{d}t) - v_\lambda(S_{\mathrm{d}t}) \right], \tag{1}$$

where S_{dt} is the set of players "before dt" in a random order, and E stands for "expectation".

An allocation $\mathbf{x}$ is called a *value allocation* if there exists a weight function λ such that

$$(\varphi v_\lambda)(dt) = \lambda_t u_t(\mathbf{x}_t)\mu(dt)$$

for all t; in words, if it assigns to each trader – and hence to each coalition – precisely its (weighted) value.

Value equivalence.   It is useful for the sequel to review here the argument showing that a value allocation $\mathbf{x}$ is also Walrasian. Note first that it is *efficient* – achieves the worth $v_\lambda(T)$ of T (i.e., $\int_T \lambda_t u_t(\mathbf{x}_t)\mu(dt) = v_\lambda(T)$); this follows from the efficiency of the value (i.e., $v_\lambda(T) = (\varphi v_\lambda)(T)$) and the definition of value allocation. Denoting the gradient of u_t at x by $u'_t(x)$, we have

$$\forall t_1, t_2 \in T, \ \lambda_{t_1} u'_{t_1}(\mathbf{x}_{t_1}) = \lambda_{t_2} u'_{t_2}(\mathbf{x}_{t_2})$$

(otherwise, society could always profitably reallocate its endowment, contradicting $\mathbf{x}$'s efficiency). Call this common value p.

Next, for any coalition S, define an *S-allocation* as a (measurable) map $\mathbf{x}^S : S \to \mathbb{R}^\ell_+$ with $\int_S \mathbf{x}^S_t \mu(dt) = \int_S \mathbf{e}_t \mu(dt)$; it is just like an allocation, except that it reallocates within S only. As with allocations, if such an $\mathbf{x}^S$ achieves $v_\lambda(S)$, then

$$\forall t_1, t_2 \in S, \ \lambda_{t_1} u'_{t_1}\left(\mathbf{x}^S_{t_1}\right) = \lambda_{t_2} u'_{t_2}\left(\mathbf{x}^S_{t_2}\right);$$

in particular, this applies to S_{dt}. Now S_{dt} is a "perfect sample" of T, in the sense that in S_{dt} the distribution of players is the same as in T. From this it may be seen that the common value of the gradients for S_{dt} is also p. Hence, dt's contribution to S_{dt} is twofold: the weighted utility of dt itself, and the change in the other traders' aggregate utility due to dt's joining the coalition. Under the new optimal allocation of the endowment of $S_{dt} \cup dt$, dt gets $\mathbf{x}_t \mu(dt)$ and therefore its weighted utility is $\lambda_t u_t(\mathbf{x}_t)\mu(dt)$. So $\mathbf{e}_t \mu(dt) - \mathbf{x}_t \mu(dt)$ must be distributed among the traders in S_{dt}, so their increase in utility is $p \cdot [\mathbf{e}_t - \mathbf{x}_t]\mu(dt)$. So

$$(\varphi v_\lambda)(dt) = p \cdot [\mathbf{e}_t - \mathbf{x}_t]\mu(dt) + \lambda_t u_t(\mathbf{x}_t)\mu(dt),$$

so from the definition of "value allocation" it follows that

$$p \cdot [\mathbf{e}_t - \mathbf{x}_t] = 0.$$

Now $(\mathbf{x}_t, u_t(\mathbf{x}_t))$ is on the boundary of the convex hull of the graph of u_t. Otherwise, it would be possible to split trader dt (i.e., the bundle $\mathbf{x}_t$ dt allocated to him) into several players so that the weighted sum of their utilities would be greater than dt's utility, contradicting **x**'s efficiency. Together with the equality of the gradients, this yields

$$\forall x \in \mathbb{R}_+^\ell: \lambda_t \big[u_t(\mathbf{x}_t) - u_t(x) \big] \geqslant p \cdot (\mathbf{x}_t - x).$$

Hence, $\mathbf{x}_t$ maximizes $u_t(x)$ under the constraint $p \cdot x \leqslant p \cdot \mathbf{x}_t$, and hence under the constraint $p \cdot x \leqslant p \cdot \mathbf{e}_t$. So **x** is a Walrasian allocation corresponding to the price system p.

3. Taxation and redistribution

Much of public economics regards the government as an exogenous benevolent economic agent that aims to maximize some social utility, usually the sum of individual utilities [see Arrow and Kurz (1970)]. This ignores the workings of a democracy, in which government actions are determined by the pressures of competing interests, whose strength depends on political power (voting) and economic power (resources). This section, based on Aumann and Kurz (1977a, 1977b), applies this idea to the issues of taxation and redistribution. In our model, each agent has an endowment and a utility function, taxation and redistribution policy is decided by majority voting, and each agent can destroy part or all of his endowment, which we may think of, e.g., as labor. Thus while any majority can levy taxes even at 100% and redistribute the proceeds among its own members, anyone can decide not to work, so that the majority gets nothing by taxing the income from his labor. Though he will not feel better (no utility of leisure is assumed), he can use this as a threat to make the majority compromise. This influences the composition of the majority coalition and the tax policy it adopts.

We start from the previous model but with a single commodity. Since we want to allow for threats and non-transferable utilities, we use the Harsanyi–Shapley NTU value. Suppose given a weight function λ. Then the worth of the all-player coalition T is

$$v_\lambda(T) = \max\left\{ \int_T \lambda_t u_t(\mathbf{x}_t) \mu(\mathrm{d}t) \text{ such that } \int_T \mathbf{x}_t \mu(\mathrm{d}t) = \int_T \mathbf{e}_t \mu(\mathrm{d}t) \right\}.$$

Suppose now that two complementary coalitions S and $T \setminus S$ form. Think of $v_\lambda(S)$ as the aggregate utility of S if it bargains against $T \setminus S$. As in Nash (1953), suppose that the two parties can commit themselves to carrying out *threat strategies* if no satisfactory agreement is reached. If under these strategies S and $T \setminus S$ get f and g respectively, then they are bargaining for $v_\lambda(T) - f - g$, and, under the symmetry assumption, this is split evenly. Hence, S gets $\frac{1}{2}(v_\lambda(T) + f - g)$ and $T \setminus S$ gets $\frac{1}{2}(v_\lambda(T) + g - f)$, so that the derived game between S and $T \setminus S$ is a constant-sum game.

The optimal threat strategy for the majority coalition is to tax at 100%, while the optimal threat strategy for the minority is to destroy all of its endowment; indeed, in this way the majority ensures its own endowment, and the minority ensures getting nothing, while each side is prevented from getting more. Thus the reduced game value is

$$q_\lambda(S) = \begin{cases} \max\{\int_S \lambda_t u_t(\mathbf{x}_t)\mu(dt) \text{ s.t. } \int_S \mathbf{x}_t\mu(dt) = \int_S e_t\mu(dt)\}, & \text{if } \mu(S) > \tfrac{1}{2}; \\ 0, & \text{if } \mu(S) < \tfrac{1}{2}; \end{cases}$$

and we have

$$v_\lambda(S) = \tfrac{1}{2}\big[q_\lambda(T) + q_\lambda(S) - q_\lambda(T\backslash S)\big];$$

it follows that

$$(\varphi v_\lambda)(dt) = (\varphi q_\lambda)(dt) = E\big[q_\lambda(S_{dt} \cup dt) - q_\lambda(S_{dt})\big].$$

As in the previous section, S_{dt} is almost certainly a perfect sample of T. We can then use the self-explanatory notation $S_{dt} = \theta T$, where $\theta \in [0, 1]$ is the size of the sample. If we think of the players as entering a room at a uniform rate but in a random order, then θ can be viewed as the random time when dt enters the room, and thus is uniformly distributed. Hence

$$(\varphi q_\lambda)(dt) = \int_0^1 \big[q_\lambda(\theta T \cup dt) - q_\lambda(\theta T)\big]\,d\theta.$$

Now there are three different cases:
  (1) both θT and $\theta T \cup dt$ are majorities;
  (2) both are minorities;
  (3) θT is a minority, but $\theta T \cup dt$ is a majority; i.e., dt is pivotal.
  Suppose $q_\lambda(T)$ is achieved at the allocation $\mathbf{x}$. Then dt's expected contribution in the three cases is as follows, case 1 being as in the last section:
  (1) $\tfrac{1}{2}\lambda_t[u_t(\mathbf{x}_t) - u_t'(\mathbf{x}_t) \cdot (\mathbf{x}_t - e_t)]\mu(dt)$ $[= \int_{\frac{1}{2}}^1 [q_\lambda(\theta T \cup dt) - q_\lambda(\theta T)]\,d\theta]$;
  (2) $\tfrac{1}{2}(0 - 0)[= \int_0^{\frac{1}{2}-\mu(dt)}(0)\,d\theta]$;
  (3) $\tfrac{1}{2}q_\lambda(T)\mu(dt)$ $[= \int_{\frac{1}{2}-\mu(dt)}^{\frac{1}{2}}[q_\lambda(\tfrac{1}{2}T) - 0]\,dt]$.
All in all, therefore,

$$\varphi v_\lambda(dt) = \varphi q_\lambda(dt) = \frac{1}{2}\big(\lambda_t[u_t(\mathbf{x}_t) - u_t'(\mathbf{x}_t) \cdot (\mathbf{x}_t - e_t)] + q_\lambda(T)\big)\mu(dt).$$

By the definition of value allocations, $\varphi v_\lambda(dt) = \lambda_t u_t(\mathbf{x}_t)\mu(dt)$ and, as in the previous section, $\lambda_t u_t'(\mathbf{x}_t)$ is constant in t, say $= p$. It follows that

$$\lambda_t u_t(\mathbf{x}_t) = q_\lambda(T) - p \cdot (\mathbf{x}_t - e_t),$$

μ-a.e. In the case of a single commodity ($\ell = 1$) this comes to

$$\mathbf{e}_t - \mathbf{x}_t = \frac{u_t(\mathbf{x}_t)}{u'_t(\mathbf{x}_t)} - C \tag{2}$$

for all t, where $C = q_\lambda(T)/p$ is a constant. Defining *value allocation* as in Section 2, we find that (2) is necessary and sufficient for $\mathbf{x}$ to be a value allocation. Moreover, (2) has a single solution $(\mathbf{x}, C)$, and C is positive. This constitutes the main result in Aumann and Kurz (1977a).

Now to the economic interpretation. The left side of (2) is the (signed) tax on t. Note that when the utility functions u_t are increasing and concave, $\mathbf{e}_t$ is increasing in $\mathbf{x}_t$ or, more intuitively, $\mathbf{x}_t$ is increasing in $\mathbf{e}_t$ – but with slope $< \frac{1}{2}$. That is to say, marginal tax rates are between 50% and 100%.

Though there is no explicit uncertainty in the model, define the *fear of ruin* as the ratio $u_t(\mathbf{x}_t)/u'_t(\mathbf{x}_t)$. To explain why, consider its reciprocal u'/u. Suppose that some player is ready to play a game in which with probability $1 - p$ his initial fortune x is increased by a small amount ε and with probability p he is ruined and his fortune is 0. If he is indifferent between playing the game or not, p/ε is a measure of his boldness (at x) – and so, its reciprocal ε/p a measure of his fear (of ruin). Indifference means that

$$u(x) = p \cdot 0 + (1 - p) \cdot u(x + \varepsilon).$$

Hence, when ε goes to zero, ε/p goes to $u(x)/u'(x)$. Thus, the tax equals the fear of ruin at the net income, less a constant tax credit.

EXAMPLE. We end this section with an example of an explicit calculation of tax policy. Let all agents t have the same utilities $u_t(x) = u(x) = x^\alpha$, where $0 < \alpha \leqslant 1$. The fear of ruin is then

$$\frac{u(x)}{u'(x)} = \frac{x^\alpha}{\alpha x^{\alpha - 1}} = \frac{x}{\alpha}.$$

Integrating, we get

$$C = \int_T \frac{u(\mathbf{x})}{u'(\mathbf{x})} + 0 = \int_T \frac{x}{\alpha} = \frac{1}{\alpha} \int_T \mathbf{e},$$

since $\int_T \mathbf{e} = \int_T \mathbf{x}$. Hence the tax $\mathbf{e}_t - \mathbf{x}_t$ is given by

$$\mathbf{e}_t - \mathbf{x}_t = \frac{1}{1 + \alpha} \left(\mathbf{e}_t - \int_T \mathbf{e} \right);$$

this may be thought of as a uniform tax at the rate of $1/(1 + \alpha)$, followed by an equal distribution of the proceeds among the whole population. Recalling that the exponent α

is an inverse measure of risk aversion, we conclude that the more risk averse the agents are, the larger the tax rate is.

As a comment, recall that all estimates of coefficients of relative risk aversion are strongly negative – say ranging from $\alpha = -2$ to $\alpha = -15$ (!) (cf., e.g., Drèze (1981)), with $u_t(x) = \frac{x^{\alpha_t} - e^{\alpha_t}}{\alpha_t}$, and note the limiting result for $\alpha = 0$. This is clearly due to the enormous power attributed here to the majority: for $\alpha \leqslant 0$, it can – and will under the above "optimal threats" – put the minority effectively outside its consumption set.

It might be interesting to investigate in this context (say α's distributed between -15 and $+1$) the implications of the traditional requirement that laws should be anonymous. For example, formulations of the sort: t's net tax $T(y_t, r)$ is a function only of his reported (i.e., not destroyed) endowment y_t ($0 \leqslant y_t \leqslant e_t$) and of the distribution r (in duality with the space of all Lipschitz functions of y) of those y_t's such that:
   (a) budget balance always holds,
   (b) T is jointly continuous, monotone and with slope $\leqslant 1$ in y,
   (c) an increase in r in the sense of first-order stochastic dominance decreases T'_y for all y and $T(0, \cdot)$, and
   (d) absence of "monetary illusion" – since the model satisfies it, – i.e., for $\lambda > 0$, denoting by r^λ the image of r under $y \mapsto \lambda y$, one has $T(\lambda y, r^\lambda) = \lambda T(y, r)$.
One might even want to add progressivity:
   (e) $T''_y \geqslant 0$, or equivalently: consumption is concave in y; or still equivalently: tax is convex in consumption.
And then the normal effect of an increased dispersion of r (second-order stochastic dominance) would also be to decrease T'_y for all y and $T(0, \cdot)$: more tax receipts becoming available, everyone and every marginal dollar should benefit from it. (Observe that the above tax policy, $T(y, r) = \frac{1}{1+\alpha}(y - \bar{y})$, satisfies all those requirements.) The aim here is clearly not to have to play explicitly with ideas like a higher utility for society of individuals with a lower risk aversion, and still to exhibit some similar effect. And hopefully less than all the above requirements would be sufficient to determine the solution.

4. Public goods without exclusion

The second purpose of taxation is the financing of public goods. As with redistribution, considering government as subject to the influence of its electors, rather than benevolently maximizing some social welfare function, sheds new light on the subject. As in Auman, Kurz and Neyman (1983, 1987), a *Public Goods Economy* is modeled by a continuum of agents, endowed with resources and voting rights, who take part in the production of non-exclusive public goods. More precisely, when a coalition forms, it chooses one strategy amongst the available, which together with the complementary coalition's choice determines the bundle of public goods that is produced. A natural question is the dependence of the outcome of the game on the distribution of voting rights, which should at first sight exert a major influence. But the Harsanyi–Shapley NTU Value leads to the surprising new and pessimistic result that the distribution of voting rights has (little or) nothing to do with the choice. Aside from its proof, this result is reinforced by an economic argument based on the implicit price of a vote.

A nonatomic *public goods economy* is modeled by the set of agents $T = [0, 1]$, the space $\mathbb{R}_+^\ell$ of resources, and the *public goods space* $\mathbb{R}_+^m$. There is a correspondence G from $\mathbb{R}_+^\ell$ to $\mathbb{R}_+^m$, representing the production function, which takes compact and nonempty values. $u_t : \mathbb{R}_+^m \to \mathbb{R}$ is t's utility function. v is a nonatomic voting measure with $v(T) = 1$.

Every agent has resources that can be used to produce public goods. The voting measure v is not necessarily identical to the population measure μ. For example, it is possible that noncitizens do not have the right to vote.

A vote takes place and the majority decides which public good(s) will be produced. The minority is not entitled to produce public goods but, as in the previous model, it has the right to destroy its resources. The important point is that once the public good has been produced, the minority may not be excluded from consuming it.

THEOREM 1. *In the voting game, the value outcomes are independent of the voting measure* v.

Consider an example with two public goods, TV and libraries. There are two kinds of agents in two equally weighted sets: one kind is fond of TV programs, the other of books. Assume further that TV fans possess all the voting rights. Then one might expect that after the voting, TV programs will be the only available leisure. But as it happens, both TV and books will be available in equal amounts! Whatever the voting rights, the same bundle of public goods will be chosen.

SKETCH OF PROOF. We first describe the Harsanyi–Shapley NTU value of a game, also used in Section 3. For a fixed weight function λ, every coalition S announces a threat strategy z_S that it will carry out if negotiations with the complementary coalition $T \setminus S$ fail. Together with $T \setminus S$'s strategy, z_S yields S a total payoff of $V(S) = U(z_S, z_{T \setminus S})(S)$. After these announcements, the players are in a fixed-threat TU game, and its solution is taken to be the Shapley value. So each player wants the threats of the various coalitions of which he is a member to be chosen so as to maximize his own final payoff according to this Shapley value. This can be shown to imply that all members of any given coalition S unanimously want to maximize $H(S) = V(S) - V(T \setminus S)$. Since the same holds for $T \setminus S$, the *optimal threat strategies* z_S^* and $z_{T \setminus S}^*$ are the saddle points of the two-person zero-sum game $H(S)$. Let $q_\lambda(S) = U(z_S^*, z_{T \setminus S}^*)(S)$ be the total payoff to S when both S and $T \setminus S$ carry out their optimal threats. Define $w_\lambda(S) = q_\lambda(S) - q_\lambda(T \setminus S) = H(S)(z_S^*, z_{T \setminus S}^*)$; thus $w_\lambda(S)$ measures the "bargaining power" of S – its ability to threaten. Define also $v_\lambda(S) = \frac{1}{2}(w_\lambda(S) + w_\lambda(T))$. As argued in the previous section, $v_\lambda(S)$ is the total utility S can expect to result from an efficient compromise with $T \setminus S$. Observe that $\varphi v_\lambda = \varphi q_\lambda$ (since the difference is a game where every coalition gets the same as its complement), but whereas q_λ might depend on the particular choices of optimal threats z_S^*, that is not so for v_λ. The function v_λ is the

Harsanyi coalitional form of the game with weight function λ, and as before, we define a *value allocation* as a bundle $\mathbf{y}$ achieving the Harsanyi–Shapley NTU value, i.e., such that

$$\varphi v_\lambda(S) = \int_S \lambda_t u_t(\mathbf{y}_t)\mu(dt) \quad \text{for every } S.$$

We know

$$(\varphi v)(dt) = \int_0^1 \left[v(\theta T \cup dt) - v(\theta T)\right]d\theta.$$

We want to compare this expression for two different voting measures ν and ξ:

$$(\varphi v^\nu - \varphi v^\xi)(dt) = \int_0^1 \left[(v^\nu - v^\xi)(\theta T \cup dt) - (v^\nu - v^\xi)(\theta T)\right]d\theta.$$

We call a coalition S *even* if S is either a majority under both ν and ξ or a minority under both ν and ξ. If S is even, so is $T \backslash S$ and their strategic options are the same under ν and ξ, i.e., $v^\nu(S) - v^\xi(S) = 0$. Every perfect sample of the whole population θT is even – it is determined by its size. For $\theta T \cup dt$, it is even if $\theta > \frac{1}{2}$ or $\theta < \frac{1}{2} - \delta$ [with $\delta = \max(\nu(dt), \xi(dt))$]. The previous difference thus amounts to:

$$(\varphi v^\nu - \varphi v^\xi)(dt) = \int_{\frac{1}{2}-\delta}^{\frac{1}{2}} \left[(v^\nu - v^\xi)(\theta T \cup dt)\right]d\theta$$

$$= \frac{1}{2}\int_{\frac{1}{2}-\delta}^{\frac{1}{2}} \left[(w^\nu - w^\xi)(\theta T \cup dt)\right]d\theta.$$

If $w^\nu(\theta T \cup dt)$ is achieved at the outcome $\mathbf{y}$, then, by additivity of the integral, and homogeneity:

$$w^\nu(\theta T \cup dt) = H^\mathbf{y}(\theta T \cup dt)$$
$$= U^\mathbf{y}(\theta T) + 2U^\mathbf{y}(dt) - U^\mathbf{y}((1-\theta)T)$$
$$= (2\theta - 1)U^\mathbf{y}(T) + 2\lambda_t u_t(\mathbf{y}_t)\mu(dt).$$

Now $2\theta - 1$ as well as $\mu(dt)$ are infinitesimal. Independently of the voting measure ν, we have shown that in the relevant range ($\theta \in [\frac{1}{2} - \delta, \frac{1}{2}]$), $w^\nu(\theta T \cup dt)$ is infinitesimal: the idea is that, under those circumstances, both the coalition and its complement resemble $\frac{1}{2}T$, thus are close to each other, and whatever the outcome, they enjoy the same utility derived from the *common* consumption of the same public good. Nobody enjoys a real bargaining advantage, and the efficient compromise induced by the Shapley value leads to equal treatment.

Going back to more technical arguments, assume $u_t(y) \leqslant K$ for all feasible y and all t. Then

$$U^y(T) \leqslant K \int \lambda_t \mu(\mathrm{d}t).$$

Given any coalition S, and $\delta > 0$, partition S into $S_1 \cup \cdots \cup S_n$, with $(v + \xi)(S_i) \leqslant \delta$, and $n \leqslant 2/\delta$. Then we obtain

$$\varphi v^v(S) - \varphi v^\xi(S)$$

$$= \frac{1}{2} \sum_{i=1}^{n} \int_{\frac{1}{2}-\delta}^{\frac{1}{2}} \left[(2\theta - 1)\left(U^{y_i^v}(T) - U^{y_i^\xi}(T)\right) + 2 \int_{S_i} \lambda_t \left(u_t\left(y_i^v\right) - u_t\left(y_i^\xi\right)\right) \mu(\mathrm{d}t) \right] \mathrm{d}\theta$$

$$\leqslant \frac{1}{2} \cdot n \cdot \left| \int_{\frac{1}{2}-\delta}^{\frac{1}{2}} (2\theta - 1) \, \mathrm{d}\theta \right| \cdot (2K) \int \lambda_t \mu(\mathrm{d}t) + \frac{1}{2} \cdot 4K\delta \int_S \lambda_t \mu(\mathrm{d}t)$$

$$\leqslant 2K\delta \left[\int_T \lambda_t \mu(\mathrm{d}t) + \int_S \lambda_t \mu(\mathrm{d}t) \right] \leqslant 4K\delta \int_T \lambda_t \mu(\mathrm{d}t).$$

This being true for all δ, we obtain $\varphi v^v(S) = \varphi v^\xi(S)$.

This being true for any weight function λ, the result follows. □

Though counterintuitive, this result might have been guessed from a similar analysis conducted in a transferable-utility (TU) context: we allow agents to trade their votes for money. The outcome of such a TU game consists of a public goods vector and a side-payment vector. In the book vs. TV game, it can easily be derived from the following conditions:

(1) It is Pareto-optimal.
(2) Under sensible assumptions, the only Pareto-optimal situation involves a production of $(\frac{1}{2}, \frac{1}{2})$ with the accompanying schedule of side-payments from book fans to TV fans.
(3) To have this outcome approved, as book fans need only 50% of the vote and thus can play TV fans off against each other, they drive the price of a vote down to zero and can achieve the desired outcome without effective payments.

In the case of non-exclusive public goods, every agent endowed with a voting right is potentially a free-rider: he believes his personal vote not to influence the final outcome, which he may like or dislike, and is thus ready to sell it even at a low price. This is wrong in the case of redistribution or more generally of exclusive public goods, where, as here, voting is assumed not to be secret, and the identity of the voters is crucial in eventually determining everyone's payoff.

To conclude this section, we stress the connection between the TU and NTU situations: the TU games that we obtain corresponding to the latter are similarly public good economies, but in addition have a single desirable private good available for transfers.

5. Economies with fixed prices

5.1. Introduction

In economies with fixed prices all trading must take place at exogenously given prices, which determine (together with the positive orthant) a new consumption set – the net trade set – for each trader. This model has been used to describe market failures such as unemployment. In general, price rigidities prevent market clearance: a trader's consumption set is exogenous and, under the standard assumptions, his utility function has an absolute maximum, a satiation point, generally in the interior of the consumption set; it may be the case that for any price vector, at least one of the traders' utility functions is satiated; this trader uses less than the maximum budget available to him, creating a total budget excess.

EXAMPLE. Consider a fixed-price exchange economy with one commodity and two traders. Let their net trade sets be $X_1 = X_2 = [-5, +5]$ and their utility functions $u_1(x) = -(x-1)^2$ and $u_2(x) = -(x+2)^2$. Since the satiation points are respectively 1 and -2, if $p > 0$ then $x_1^* = 0$ and $x_2^* = -2$. Hence, the market does not clear. Similarly if $p < 0$ or if $p = 0$.

This suggests a generalization of the equilibrium concept in the general class of markets with satiation: the total budget excess is divided among all the traders, as dividends, so that supply matches demand. However, Drèze and Müller (1980) extended the First Theorem of Welfare Economics to this equilibrium concept, proving it to be too broad: with appropriate dividends, one can obtain any Pareto-optimum.

In this respect, the Shapley value leads to more specific results (Aumann and Drèze, 1986): the income allocated to a trader depends only and monotonically on his trading opportunities and not on his utility function! This will be formally stated in Section 5.4, which includes a sketch of the proof. A formulation in the particular context of fixed-price economies will then be presented.

5.2. Dividend equilibria

Define a market with satiation as $M^1 = (T; \ell; (X_t)_{t \in T}; (u_t)_{t \in T})$, where $T = \{1, \ldots, k\}$ is a finite set of traders, $\mathbb{R}^\ell$ is the space of commodities, $X_t \subset \mathbb{R}^\ell$ is trader t's net trade set, supposed to be compact, convex, with nonempty interior and containing 0, and u_t is trader t's utility function, assumed concave and continuous on X_t.

A price vector is any element in $\mathbb{R}^\ell$.

Let $B_t = \{x \in X_t \mid u_t(x) = \max_{y \in X_t} u_t(y)\}$ be the set of satiation points of trader t. B_t is nonempty, i.e., every trader has at least one satiation point. For simplicity, traders such that $0 \in B_t$ may be taken out of the economy. They are fully satisfied with their initial endowment. Thus we will suppose that $\forall t, 0 \notin B_t$.

An *allocation* is a vector $\mathbf{x} \in \prod_{t \in T} X_t$ such that $\sum_{t \in T} \mathbf{x}_t = 0$.

As noted above, competitive equilibria may fail to exist since, whatever the price vector, a trader may well refuse to make use of his entire budget, thus preventing the market from clearing. The idea of dividends is to let the other traders use the excess budget.

A *dividend* is a vector $c \in \mathbb{R}^k$. A *dividend equilibrium* is a triple constituted of a price q, a dividend c and an allocation $\mathbf{x}$ such that, for all t, $\mathbf{x}_t$ maximizes $u_t(x)$ on X_t subject to $q \cdot x \leqslant c_t$.

5.3. Value allocations

This is the "finite" version of the definition in Section 2.

A *comparison vector* is a non-zero vector $\lambda \in \mathbb{R}_+^k$. For each λ and each coalition $S \subset T$, the *worth of S according to λ* is

$$v_\lambda(S) = \max \left\{ \sum_{t \in S} \lambda_t u_t(\mathbf{x}_t) \text{ s.t. } \sum_{t \in S} \mathbf{x}_t = 0 \text{ and } \forall t \in S, \ \mathbf{x}_t \in X_t \right\};$$

$v_\lambda(S)$ is the maximum total utility that coalition S can get by internal redistribution when its members have weights λ_t.

An allocation is called a *value allocation* if there exists a comparison vector λ such that $\lambda_t u_t(\mathbf{x}_t) = \varphi v_\lambda(t)$ for all t, where φv_λ is the Shapley value of the game v_λ.

5.4. The main result

M^n, the n-fold replica of market M^1, is the market with satiation where every agent of M^1 has n twins. Formally stated:
- $T^n = \bigcup_{i \in T} T_i^n$ (the set of nk traders).
- $\forall i \in T$, $|T_i^n| = n$ (there are n traders of type i).
- $\forall t \in T_i^n$, $u_t = u_i$ and $X_t = X_i$.

ELBOW ROOM ASSUMPTION. For all $J \subset \{1, \ldots, k\}$,

$$0 \notin \mathrm{bd} \left[\sum_{i \in J} B_i + \sum_{i \notin J} X_i \right].$$

In words, if it is possible to satiate all traders in some J simultaneously, then it is also possible to do so when they are restricted to the relative interior of their satiation sets, and the others to that of their net trade sets. Note that since the right-hand side is the boundary of a convex subset of $\mathbb{R}^\ell$, its dimension is at most $(\ell - 1)$. Since the possible J are finite in number, the assumption holds for all but an $(\ell - 1)$-dimensional set of total endowments. In that respect, it is generic.

An allocation $\hat{\mathbf{x}}$ of M^n is an *equal treatment* allocation if traders of the same type are assigned the same net trade. Trivially, there is then an associated allocation $\mathbf{x}$ of M^1.

THEOREM 2. *Consider a sequence* $(\mathbf{x}^n)_{n \in \mathbb{N}}$ *where* $\mathbf{x}^n$ *is an allocation corresponding to an equal treatment value allocation in* M^n. *Let* $\mathbf{x}^\infty$ *be a limit of a subsequence of* $(\mathbf{x}^n)_{n \in \mathbb{N}}$. *Then, there is a dividend* (*vector*) c *and a price vector* q *such that* $(q, c, \mathbf{x}^\infty)$ *is a dividend equilibrium where*
- *c is nonnegative, i.e.,* $\forall i$, $c_i \geqslant 0$
- *c is monotonic, i.e.,* $\forall i$, $X_i \subset X_j \Rightarrow c_i \leqslant c_j$.

What gives substance to the theorem is the following existence result.

PROPOSITION 3. *There exists an equal treatment value allocation for every* M^n.

SKETCH OF PROOF OF THEOREM 2. This sketch is very informal. To make things simpler, suppose that:
- $\sum_i \lambda_i^n = 1$ (normalization);
- $\forall i$, $\mathbf{x}_i^n$ and λ_i^n converge;
- $\forall i$, $\mathbf{x}_i^n$ and $\mathbf{x}_i^\infty$ are in $\text{int}(X_{it})$;
- $\forall i$, u_i is strictly concave and continuously differentiable on X_i.

Call those types i for which $\lim_{n \to \infty} \lambda_i^n = 0$ *lightweight*, the rest *heavyweight*. Suppose that the weights of all the lightweight types converge to 0 at the same speed. We have

$$\forall (i, j), \ \forall n, \ \lambda_i^n u_i'\left(\mathbf{x}_i^n\right) = \lambda_j^n u_j'\left(\mathbf{x}_j^n\right) \ (=: q^n).$$

In the limit,

$$\forall (i, j), \ \lambda_i^\infty u_i'\left(\mathbf{x}_i^\infty\right) = \lambda_j^\infty u_j'\left(\mathbf{x}_j^\infty\right) \ (=: q^\infty).$$

Now consider the contribution of a trader to a coalition S.

If S is "large enough", it is very likely to be a good sample of the population T^n. Thus an optimal allocation for S is approximately the optimal $\mathbf{x}^n$ for T^n. The first term of the new trader's contribution is what he gets for himself. Since he does not change the optimal allocation by much, this is $\lambda_i^n u_i(\mathbf{x}_i^n)$. The second term is his influence on the other traders' utilities. Since the net trade must equal zero and all gradients are equal, this is approximately $-q^n \cdot \mathbf{x}_i^n$. Thus, the contribution is approximately $\Delta = \lambda_i^n u_i(\mathbf{x}_i^n) - q^n \cdot \mathbf{x}_i^n$.

If S is "small", the previous considerations do not hold. However, a new trader's contribution to a (small) coalition is uniformly bounded. This follows, e.g., from the continuous differentiability of the utilities on the compact net trade sets. Moreover, the probability P^n of S being a small coalition goes to zero as n goes to infinity. Denote by δ_i^n the expected contribution of a trader t of type i conditional on the coalition being small.

Now we have (very roughly):

$$\varphi v_\lambda^n(t) \approx \left(1 - P^n\right)\Delta + P^n \delta_i^n.$$

Since $\varphi v_\lambda^n(t) = \lambda_i^n u_i(\mathbf{x}_i^n)$ we have:

$$\lambda_i^n u_i(\mathbf{x}_i^n) \approx (1 - P^n)\Delta + P^n \delta_i^n.$$

Hence,

$$q^n \cdot \mathbf{x}_i^n \approx \frac{P^n}{1 - P^n}\left(\delta_i^n - \lambda_i^n u_i(\mathbf{x}_i^n)\right). \tag{3}$$

Note that $P^n/(1-P^n) \to 0$.

Suppose there is no lightweight type. If simultaneous satiation of all traders is possible, then this is the Shapley value for all sufficiently large n, since then $\lambda_i^n > 0$ for all i. It is clear that the theorem holds then, for any price system q, and $c_i = C$ sufficiently large. Otherwise, $u_i'(\mathbf{x}_i^\infty) \neq 0$ for at least some i. Hence, because of the equality of the gradients, $u_j'(\mathbf{x}_j^\infty) \neq 0$ for all j. Hence $q^\infty \neq 0$ and for all i, the gradient of u_i at $\mathbf{x}_i^\infty$ is in the direction of q^∞. Going to the limit in (3) gives $q^\infty \cdot \mathbf{x}_i^\infty = 0$. Hence, $\mathbf{x}_i^\infty$ maximizes $u_i(x)$ on X_i subject to $q^\infty \cdot x \leqslant 0$.

Hence, it is an ordinary competitive equilibrium and trivially a dividend equilibrium.

Suppose now that type i is lightweight. We have $q^\infty = 0$. Hence $u_j'(\mathbf{x}_j^\infty) = 0$ for any heavyweight type j; that is to say, $\mathbf{x}_j^\infty$ satiates j. Before letting n go to infinity, divide equality (3) by $\|q^n\|$. We shall see that δ's order of magnitude is greater than that of $\lambda_i^n u_i(\mathbf{x}_i^n)$. Assume, for simplicity, that the sequence $q^n/\|q^n\|$ converges to some point q. We get:

$$q \cdot \mathbf{x}_i^\infty = \lim_{n \to \infty} \frac{P^n}{(1-P^n)\|q^n\|}\left(\delta_i^n - \lambda_i^n u_i(\mathbf{x}_i^n)\right).$$

Denote this quantity c_i.

If $u_i'(\mathbf{x}_i^\infty) = 0$, then $\mathbf{x}_i^\infty$ maximizes u_i over X_i. Hence it maximizes u_i over $\{x \in X_i, q \cdot x \leqslant c_i\}$. The same holds if $u_i'(\mathbf{x}_i^\infty) \neq 0$, because then q is in the direction of $u_i'(\mathbf{x}_i^\infty)$.

We claim that c_i is nonnegative, depends monotonically on X_i, and not at all on u_i.

A lightweight trader t's joining a "small" coalition S contributes to three utilities: (i) his own, (ii) that of other lightweight traders, and (iii) that of heavyweight traders.

Roughly (again), since we are considering a small coalition, the heavyweight traders are probably not all simultaneously satiated. Since the weights in (i) and (ii) tend to zero when n becomes large, when trader t joins the coalition, his resources are best used if distributed to unsatiated heavyweight traders.

His ability to give his resources depends only and monotonically on X_t, and not on u_t.

Though the optimal redistribution of t's resources may involve several heavyweight traders, giving it all to only one trader gives a lower bound to (iii): δ_i^n is of larger order than λ_i^n.

This establishes our claim about c_i.

If trader t is heavyweight, contributions (i) and (ii) in δ_i^n are at least cancelled out by $\lambda_i^n u_i(\mathbf{x}_i^n)$, since $\lambda_i^\infty u_i(\mathbf{x}_i^\infty)$ is the most t can get. Thus, the rest of the argument is as before with

$$q.x_i^\infty \leqslant \lim_{n\to\infty}\left(\frac{\varepsilon^n}{\|q^n\|}\right) \text{ [component (iii) of δ]}$$

and c_i defined as the right side of this inequality.

Since i is heavyweight, $\mathbf{x}_i^\infty$ satiates. By the above inequality, it also satisfies the budget constraint.   □

5.5. Concluding remarks

A question eluded until now is: "What about the core in an economy that allows satiation?" Remember that an allocation is in the core if it cannot be improved upon by any coalition. This can be understood either in a strong or a weak way. The strong version requires that no "weak improvement" is possible. That is to say that it is not possible that some members are strictly better off while others are not worse off. With this definition the Core Equivalence Theorem for replica economies applies. And since the set of competitive equilibria may be empty, the same holds for the core. On the other hand, the "weak" core may be too large and not even enjoy the equal treatment property. Either way, the core does not yield much insight into these economies.

References

Arrow, K.J., and M. Kurz (1970), Public Investment, the Rate of Return, and Optimal Fiscal Policy (Johns Hopkins Press, Baltimore, MD).

Aumann, R.J. (1964), "Markets with a continuum of traders", Econometrica 32:39–50.

Aumann, R.J. (1975), "Values of markets with a continuum of traders", Econometrica 43:611–646.

Aumann, R.J., and J.H. Drèze (1986), "Values of markets with satiation or fixed prices", Econometrica 54:1271–1318.

Aumann, R.J., and M. Kurz (1977a), "Power and taxes", Econometrica 45:1137–1161.

Aumann, R.J., and M. Kurz (1977b), "Power and taxes in a multi-commodity economy", Israel Journal of Mathematics 27:186–234.

Aumann, R.J., M. Kurz and A. Neyman (1983), "Voting for public goods", Review of Economic Studies 50:677–693.

Aumann, R.J., M. Kurz and A. Neyman (1987), "Power and public goods", Journal of Economic Theory 42:108–127.

Champsaur, P. (1975), "Cooperation versus competition", Journal of Economic Theory 11:394–417.

Drèze, J.H. (1981), "Inferring risk tolerance from deductibles in insurance contracts", The Geneva Papers on Risk and Insurance 20:48–52.

Drèze, J.H., and H. Müller (1980), "Optimality properties of rationing schemes", Journal of Economic Theory 23:150–159.

Harsanyi, J.C. (1959), "A bargaining model for the cooperative n-person game", in: A.W. Tucker and R.D. Luce, eds., Contributions to the Theory of Games IV, Ann. of Math. Studies 40 (Princeton University Press, Princeton, NJ) 325–355.

Mertens, J.-F., and S. Sorin (eds.) (1994), Game-Theoretic Methods in General Equilibrium Analysis, Vol. 77 of NATO ASI Series – Series D: Behavioural and Social Sciences (Kluwer Academic Publishers, Dordrecht).

Nash, J.F. (1953), "Two-person cooperative games", Econometrica 21:128–140.

Shapley, L.S. (1953), "A Value for n-person games", in: H.W. Kuhn and A.W. Tucker, eds., Contributions to the Theory of Games II, Ann. of Math. Studies 28 (Princeton University Press, Princeton, NJ) 307–317.

Shapley, L.S. (1969), "Utility comparisons and the theory of games", in: La Décision (Editions du Centre National de la Recherche Scientifique (CNRS), Paris) 251–263.

Chapter 59

STRATEGIC ASPECTS OF POLITICAL SYSTEMS

JEFFREY S. BANKS*

*Division of Humanities and Social Sciences, California Institute of Technology,
Pasadena, CA, USA*

Contents

*Deceased December 21, 2000. The editors are grateful to David Austen-Smith and Richard McKelvey, who saw the manuscript through the final stages of the production process. It was evident from Jeff's marginal notes that he had intended to add material to the introductory section. However, since the paper seemed otherwise complete, it was decided not to alter the manuscript from the state that Jeff left it in.

Handbook of Game Theory, Volume 3, Edited by R.J. Aumann and S. Hart

Abstract

Early results on the emptiness of the core and the majority-rule-chaos results led to the recognition of the importance of modeling institutional details in political processes. A sample of the literature on game-theoretic models of political phenomena that ensued is presented. In the case of sophisticated voting over certain kinds of binary agendas, such as might occur in a legislative setting, equilibria exist and can be nicely characterized. Endogenous choice of the agenda can sometimes yield "sophisticated sincerity", where equilibrium voting behavior is indistinguishable from sincere voting. Under some conditions there exist agenda-independent outcomes. Various kinds of "structure-induced equilibria" are also discussed. Finally, the effect of various types of incomplete information is considered. Incomplete information of how the voters will behave leads to probabilistic voting models that typically yield utilitarian outcomes. Uncertainty among the voters over which is the preferred outcome yields the pivotal voting phenomenon, in which voters can glean information from the fact that they are pivotal. The implications of this phenomenon are illustrated by results on the Condorcet Jury problem, where voters have common interests but different information.

Keywords

voting, political theory, social choice, asymmetric information

JEL classification: D71, D72, D82, C70

1. Introduction

In this paper I review a small, selective sample of the existing literature on game-theoretic models of political phenomena.

Let X denote a set of outcomes, and $N = \{1, \dots, n\}$ a finite set of voters, with $n \geqslant 2$ and with each $i \in N$ having a binary preference relation R_i on X. In what follows we consider two separate environments: (1) the finite environment, where X is finite and, for all $i \in N$, R_i is a weak order; and (2) the spatial environment, where $X \subset \mathfrak{R}^k$ is compact and convex and each R_i is a continuous and strictly convex weak order.[1] In either case let $\mathcal{R}^n$ denote the set of admissible preference profiles, with common element ρ.

2. The cooperative approach

One popular approach to collective decision-making is to suppress any explicit behavioral theory and model the interaction as a simple cooperative game with ordinal preferences. Thus let $\mathcal{D} \subset 2^N$ denote the set of decisive or winning coalitions, required to be non-empty, monotonic ($C \in \mathcal{D}$ and $C \subset C'$ implies $C' \in \mathcal{D}$) and proper ($C \in \mathcal{D}$ implies $N \backslash C \notin \mathcal{D}$). The (strict) social preference relation of $\mathcal{D}$ at ρ is defined as

$$x P_{\mathcal{D}}(\rho) y \quad \text{iff} \quad \exists C \in \mathcal{D} \text{ s.t. } x P_i y \ \forall i \in C,$$

and the core of $\mathcal{D}$ at ρ is

$$\mathcal{C}(\mathcal{D}, \rho) = \{x \in X : \nexists y \in X \text{ s.t. } y P_{\mathcal{D}}(\rho) x\}.$$

Results on the possible emptiness of the majority-rule core have been known for some time. For instance, in the finite environment Condorcet (1785) constructed his famous three-person, three-alternative "paradox": $x P_1 y P_1 z$, $y P_2 z P_2 x$, $z P_3 x P_3 y$. In the spatial environment, as convex preferences are single-peaked we know from Black (1958) that the majority-rule core will be non-empty in one dimension; however as Figure 1 demonstrates it is easy to construct three-person, two-dimensional examples where the core is empty.

In this example individuals' preferences are "Euclidean": for all $x, y \in X$, $x R_i y$ if and only if $\|x - x^i\| \leqslant \|y - x^i\|$, where x^i is i's ideal point. These results are generalized to arbitrary simple rules by reference to the rule's Nakamura number, $\eta(\mathcal{D})$, which is equal to ∞ if the rule is collegial, i.e., $\bigcap_{C \in \mathcal{D}} C \neq \emptyset$, and is otherwise equal to

$$\min \left\{ |\mathcal{D}'| : \mathcal{D}' \subseteq \mathcal{D} \ \& \ \bigcap_{C \in \mathcal{D}'} C = \emptyset \right\}.$$

---

[1] These assumptions on preferences are stronger than necessary for some of the results below; however they facilitate comparisons across a variety of models.

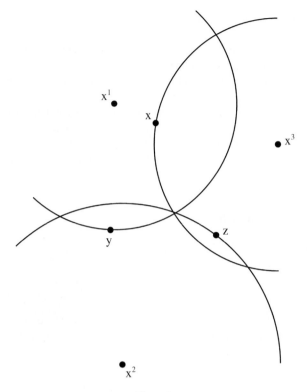

Figure 1.

In words, the Nakamura number of a non-collegial simple rule $\mathcal{D}$ is the number of coalitions in the smallest non-collegial subfamily of $\mathcal{D}$. For non-collegial rules this number
ranges from a low of three (e.g., majority rule with n odd) to a high of n (e.g., a quota
rule with $q = n - 1$).

Part (a) of the following is due to Nakamura (1979), while part (b) is due to
Schofield (1983) and Strnad (1985):

PROPOSITION 1.
  (a) *In the finite environment, $\mathcal{C}(\mathcal{D}, \rho) \neq \emptyset$ for all $\rho \in \mathcal{R}^n$ if and only if $|X| < \eta(\mathcal{D})$.*
  (b) *In the spatial environment, $\mathcal{C}(\mathcal{D}, \rho) \neq \emptyset$ for all $\rho \in \mathcal{R}^n$ if and only if
      $k < \eta(\mathcal{D}) - 1$.*

Thus, for any simple rule one can identify, for both the finite and the spatial environment, a critical number such that core non-emptiness is only guaranteed when the
number of alternatives or the dimension of the policy space is below this number. As a
corollary, we see that if the number of alternatives is at least n, or the dimension of the

policy space at least $n - 1$, then for *any* non-collegial rule the core will be empty for some profile of preferences.

Furthermore, generalizing the logic of the Plott (1967) conditions for a majority-rule core point, one can identify a second critical number (necessarily larger than the first) for non-collegial simple rules in the spatial model with the following property: assuming individual preferences are in addition representable by smooth utility functions, if the dimension of the policy space is above this second number the core will "almost always" be empty. Specifically, the set of utility profiles for which the core is empty will be open and dense in the Whitney C^∞ topology on the space of smooth utility profiles [Schofield (1984), Banks (1995), Saari (1997)]. Therefore core non-emptiness for non-collegial simple rules will be quite tenuous in high-dimensional policy spaces. Finally, McKelvey (1979) showed that when the core of a strong ($C \notin \mathcal{D}$ implies $N \backslash C \in \mathcal{D}$) simple rule is empty, the strict social preference relation $P_\mathcal{D}(\rho)$ is extremely badly behaved, in the following sense: given any two alternatives $x, y \in X$ there exists a finite set of alternatives $\{a_1, \ldots, a_m\}$ such that $x P_\mathcal{D}(\rho) a_1 P_\mathcal{D}(\rho) a_2 P_\mathcal{D}(\rho) \ldots P_\mathcal{D}(\rho) a_m P_\mathcal{D}(\rho) y$. That is, one can get from any one alternative to any other, and back again, via the strict social preference relation.[2]

It is difficult to overstate the impact these "chaos" theorems have had on the formal modeling of politics. Most importantly for this paper, the theorems gave rise to a newfound appreciation for political institutions such as committee systems, restrictive voting rules, etc. The theorems implied that additional structure was required on the collective decision-making process so as to generate a well-posed, i.e., equilibrium, model. These institutional features then became the foundation of or the motivation for various non-cooperative game forms used to study specific political phenomena. In what follows we will examine a set of these game forms in depth, maintaining for the most part a focus on majority rule. Further, given the intrinsic appeal of majority-rule core points when they exist, we will inquire as to whether a game form is "Condorcet consistent"; that is, do the equilibrium outcomes correspond to majority-rule core points when the latter exist?

3. Sophisticated voting and agendas

Consider the finite environment, where we make the additional assumptions that each individual's preference relation is a linear order, thereby ruling out individual indifference; and that n is odd. Together these imply that the strict majority preference relation is complete (although of course not necessarily transitive), and therefore that any majority-rule core alternative is unique and is strictly majority-preferred to any other alternative, i.e., it is a "Condorcet winner".

---

[2] An analogous result holds for non-strong rules when the strict social preference relation is replaced with the weak social preference relation [Austen-Smith and Banks (1999)]. One implication of this is that, when the core is empty, the transitive closure of the weak social preference relation exhibits universal indifference.

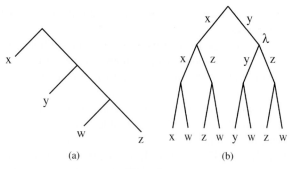

Figure 2.

We consider sequential voting methods for selecting a unique outcome from X. Specifically, the voting procedures we examine can be characterized as occurring on a *binary voting tree* $\Gamma = (\Lambda, Q, \theta)$ having many features in common with an extensive form game of perfect information: Λ is a finite set of nodes, and Q is a precedence relation on Λ generating a unique initial node λ_1 as well as a unique path between all $\lambda, \lambda' \in \Lambda$ for which $\lambda Q \lambda'$. The precedence relation Q partitions Λ into decision nodes Λ^d and terminal nodes Λ^t, where $\Lambda^t = \{\lambda \in \Lambda : \lambda' Q \lambda \text{ for no } \lambda' \in \Lambda\}$ and $\Lambda^d = \Lambda \backslash \Lambda^t$. The qualifier "binary" implies that each decision node $\lambda \in \Lambda^d$ has precisely two nodes immediately following it, $l(\lambda)$ and $r(\lambda)$ (for "left" and "right", respectively), and decision nodes characterize instances when a majority vote is taken over which of these two nodes to move. Finally, the function $\theta : \Lambda^t \to X$ assigns to each terminal node exactly one of the alternatives; we require θ to be onto, so that each element of X is the selected outcome for some sequence of majority decisions. Figure 2 gives two examples of voting trees.

Given a voting tree Γ, a strategy for any $i \in N$ is a decision rule $\sigma_i : \Lambda^d \to \{l, r\}$ describing how i votes at each decision node. A profile of strategies $\sigma = (\sigma_1, \ldots, \sigma_n)$ then determines a unique sequence of majority-rule decisions through the voting tree, ultimately leading to a particular terminal node $\lambda(\sigma) \in \Lambda^t$, and hence an outcome $\theta(\lambda(\sigma)) \in X$. Thus for any profile of preferences ρ we have a well-defined (ordinal) non-cooperative game. "Sophisticated" voting strategies are generated by iteratively deleting weakly dominated strategies [Farquharson (1969)], which can be characterized as follows [McKelvey and Niemi (1978), Gretlein (1983)]: (1) at all final decision nodes (i.e., those only followed by terminal nodes) individual i votes for her preferred alternative among those associated with the subsequent two terminal nodes (note that this constitutes the unique undominated Nash equilibrium in the subgame); (2) at penultimate decision nodes i votes for her preferred alternative from those derived from (1), given common knowledge of preferences; and so on, back up the voting tree. This process, which is equivalent to simply applying the majority preference relation $P(\rho)$ via backwards induction to Γ, can be characterized formally by associating with each node $\lambda \in \Lambda$ its *sophisticated equivalent* $s(\lambda) \in X$, which is the outcome that will ultimately be selected if node λ is reached. The mapping $s(\cdot)$ is defined inductively by: (a) for all $\lambda \in \Lambda^t$, $s(\lambda) = \theta(\lambda)$, (b) for all $\lambda \in \Lambda^d$, $s(\lambda) = s(l(\lambda))$ if $s(l(\lambda))P(\rho)s(r(\lambda))$ and $s(\lambda) = s(r(\lambda))$ otherwise. This process leads to a unique outcome $s(\lambda_1)$, referred to as the *sophisticated voting outcome* (although the strategies generating $s(\lambda_1)$ need not be unique since for some λ it may be that $l(\lambda) = r(\lambda)$). The outcome $s(\lambda_1)$ obviously depends on the voting tree Γ; however it depends on the preference profile ρ only through the majority preference relation, $P(\rho)$, and so we denote the sophisticated voting outcome associated with Γ and P as $s(\Gamma, P)$. Finally, we know from McGarvey (1953) that for any complete and asymmetric relation P on X there exists a value of n and a profile of linear orders ρ such that P is the majority preference relation associated with (n, ρ). Since these parameters are here unrestricted, spanning over all complete and asymmetric relations P on X is equivalent to spanning over all (n, ρ).

The qualifier "sophisticated" is meant to contrast such behavior with "sincere" voting, in which branches are labeled with alternatives and at each decision node voters simply select the branch associated with the more preferred alternative. For instance, let $X = \{w, x, y, z\}$ and $n = 3$, with preferences given by $z P_1 x P_1 y P_1 w$, $w P_2 z P_2 x P_2 y$, $y P_3 w P_3 z P_3 x$, and consider the subgame in Figure 2(b) beginning at node λ. Sincere voting would require 1 to vote for z over y; however, since w will majority-defeat z if z is selected at λ, y will defeat w if y is selected at λ, and $y P_1 w$, 1's sophisticated strategy is to vote for y at λ. On the other hand, both 2's and 3's sophisticated strategy prescribes voting sincerely at λ.

It is evident that the specifics of the voting tree play an important role in the determination of the sophisticated voting outcome; for instance, x is the outcome in Figure 2(a) but y is the outcome in Figure 2(b) given the above preference profile. For any majority preference relation P define $V(P) = \{x \in X : \exists \Gamma \text{ s.t. } x = s(\Gamma, P)\}$ as the set of alternatives which are the sophisticated voting outcome for some binary voting game, and consider the following definition of the top cycle set:

$$T(P) = \Big\{ x \in X : \forall y \neq x \in X, \ \exists \{a_1, \dots, a_r\} \subseteq X \text{ s.t.}$$
$$a_1 = x, \ a_r = y, \text{ and } \forall t \in \{1, \dots, r-1\}, \ a_t P a_{t+1} \Big\}.$$

In words, x is in the top cycle set if and only if we can get from x to *every* other alternative via the majority preference relation P. If x is a Condorcet winner then it is easily seen that $T(P) = \{x\}$. Otherwise $T(P)$ contains at least three elements, and can in general include Pareto-dominated alternatives; for instance, in the above example $T(P) = X$, but $z P_i x$ for all $i \in N$.

McKelvey and Niemi (1978) prove that for all P, $V(P) \subseteq T(P)$, while Moulin (1986) proves that for all P, $T(P) \subseteq V(P)$. Taken together we therefore have,

PROPOSITION 2. *For all P, $V(P) = T(P)$.*

Thus all binary voting procedures are Condorcet consistent, and so if P admits a Condorcet winner then the equilibrium outcome will be independent of the voting procedure in place. Conversely, in the absence of a Condorcet winner the voting procedure itself will play some role in determining the collective outcome, where at times these outcomes can be inefficient.

Much of the work on voting procedures and agendas has focused on the *amendment* voting procedure of which the game in Figure 2(b) is an example. Thus two alternatives are put to a vote; the winner is then paired with a new alternative, and so on. This procedure is attractive for a number of reasons, not the least of which is that it is consistent with the rules of the US Congress governing the perfecting of a bill through a series of amendments to the bill (i.e., Roberts' Rules of Order). An amendment procedure is characterized by an ordering of X or *agenda* $\alpha : \{1, \ldots, m\} \rightarrow X$, where the first vote is between $\alpha(1)$ and $\alpha(2)$, the winner then faces $\alpha(3)$, etc. Let Γ_α denote the voting game associated with the agenda α, and let $A(P) = \{x \in X : \exists \alpha \text{ s.t. } x = s(\Gamma_\alpha, P)\}$ denote the set of sophisticated voting outcomes when restricting attention to amendment procedures.

Building on the work of Miller (1980) and Shepsle and Weingast (1984), Banks (1985) provides the following characterization of the set $A(P)$: let $\mathcal{T}(P) = \{Y \subseteq X : P \text{ is transitive on } Y\}$ be those subsets of alternatives for which the majority preference relation is transitive, and $\mathcal{E}(P) = \{Y \subseteq X : \forall z \notin Y \; \exists y \in Y \text{ s.t. } yPz\}$ be those subsets of alternatives for which every non-member of the set is majority-defeated by some member of the set. If P is transitive on the set Y then there will exist a unique P-maximal element of Y; label this alternative $m(Y)$ and define $B(P) = \{x \in X : x = m(Y) \text{ for some } Y \in \mathcal{T}(P) \cap \mathcal{E}(P)\}$. Note that (i) for all P, $B(P) \subseteq \mathcal{T}(P)$; (ii) as with $T(P)$, if a Condorcet winner exists $B(P)$ will coincide with this alternative, and otherwise consist of at least three elements; and (iii) in contrast to $T(P)$ any $x \in B(P)$ cannot be Pareto-dominated by any other alternative.[3] For instance, in the above example $B(P) = \{y, z, w\}$, excluding the Pareto-dominated alternative x.

PROPOSITION 3. *For all P, $A(P) = B(P)$.*

Therefore the amendment procedure always generates Pareto-efficient outcomes. On the other hand, in the absence of a Condorcet winner there remains a sensitivity of this outcome to the particular agenda employed.

Two theoretical conclusions naturally arise at this level of the analysis. The first is that we should at times observe individuals voting against their "myopic" best interests, as voter 1 does in the above example. The second is that in the absence of a Condorcet winner there will be a diversity of induced preferences over the set of possible agendas; therefore to the extent the agenda is itself endogenously determined we have merely pushed the collective decision problem back one step (albeit limiting the relevant set of alternatives to $A(P)$). We consider each of these conclusions in turn.

---

[3] If zP_ix for all $i \in N$ then zPy for all y such that xPy by the transitivity of individuals' preferences; so if $x = m(Y)$ for some $Y \in \mathcal{T}(P)$ then $Y \notin \mathcal{E}(P)$.

With respect to "insincere" behavior, consider the spatial environment, with preference profile ρ for which the majority-rule core is empty, and suppose that $m \leqslant n$ alternatives from X are selected to make up an amendment procedure in the following manner: initially voter m selects an alternative $x \in X$ to be $\alpha(m)$, i.e., the last alternative to be considered; then voter $m - 1$, with knowledge of $\alpha(m)$, selects $\alpha(m - 1)$, etc., with voter 1 selecting $\alpha(1)$. A proposal strategy for $i \leqslant m$ is thus of the form $\pi_i : X^{(m-i)} \to X$; given a profile of proposal strategies $\pi = (\pi_1, \ldots, \pi_m)$ let $\mathbf{x}(\pi) = (\alpha(1; \pi), \ldots, \alpha(m; \pi))$ denote the resulting amendment agenda. An equilibrium is then defined as (a) sophisticated voting strategies over any set of m alternatives from X, and (b) subgame perfect proposal strategies for all $i \leqslant m$. Finally, recall that for amendment procedures each node $\lambda \in \Lambda^d$ is a vote between alternatives $\alpha(j)$ and $\alpha(k)$, for some $j, k \leqslant m$, and hence sincere voting is well-defined. Austen-Smith (1987) proves the following.

PROPOSITION 4. *In any equilibrium the proposal strategies π^* are such that all individuals vote sincerely on $\mathbf{x}(\pi^*)$.*

Therefore when the alternatives on the agenda are endogenously chosen, we have "sophisticated sincerity" as the equilibrium voting behavior of the individuals will be indistinguishable from sincere voting. In particular, observing sincere voting is not inconsistent with sophisticated voting; rather, the evidence of sophisticated voting will be much more indirect, with its influence felt through the proposals that make up the agenda.

With respect to the endogenous choice of agenda for a fixed set of alternatives, suppose there exists a "status quo" alternative which must be considered last in the agenda. For instance, after various amendments to a bill have been considered and either adopted or rejected, the bill in its final form is voted on against the status quo of "no change" in the policy. Let $x_\circ \in X$ denote this status quo policy, and let $\mathcal{A}(x_\circ)$ denote the set of agendas with $\alpha(m) = x_\circ$. Finally, say that $x \in X$ is an agenda-independent outcome under the amendment procedure if $x = s(\Gamma_\alpha, P)$ for all $\alpha \in \mathcal{A}(x_\circ)$. The presence of an agenda-independent outcome would obviously extinguish any conflict over the choice of agenda.

One possibility of course is that $x_\circ$ itself is an agenda-independent outcome, which only occurs when $x_\circ$ is a Condorcet winner. Alternatively, consider the following condition:

Condition C: $\exists x \in X$ s.t.   (i) $x P x_\circ$, and

                              (ii) $x P y \; \forall y \neq x$ s.t. $y P x_\circ$.

To see that this is sufficient for agenda independence, note that an amendment procedure generates a particular form of assignment θ of terminal nodes to alternatives; for instance, in Figure 2(b) each of x, y, and z are paired with the final alternative w at least once, and all final pairings involve w. This is a general feature of amendment procedures, so consider the voting game formed by replacing each final decision node $\lambda \in \Gamma_\alpha$ with the majority-preferred alternative among its two immediate successors, $s(\lambda)$. Label this voting game $\Gamma'_\alpha(P)$, and note that the alternatives in this game, denoted X', are given by any $y \in X$ for which $y P x_o$, along with possibly x_o itself (if $x_o P z$ for some $z \in X$). Now if there exists an alternative $x^* \in X'$ which is majority-preferred to all others in X' (i.e., x^* is a Condorcet winner in X'), then applying Proposition 2 we have that $x^* = s(\Gamma'_\alpha(P), P)$ regardless of the structure of the voting game $\Gamma_\alpha(P)$, in particular regardless of the ordering $\alpha \in \mathcal{A}(x_o)$. But then x^* is an agenda-independent outcome.

As an example of an environment where Condition C holds, consider the following classic model of distributive politics, which we label DP [Ferejohn et al. (1987)]: each $i \in N$ is the sole representative from district i in a legislature, and district i has associated with it a project that would generate political benefits and costs of $b_i > 0$ and $c_i > 0$, respectively. Assume $c_i \neq c_j$ for all $i, j \in N$, and assume without loss of generality that $c_1 < c_2 < \cdots < c_n$. The problem before the legislature is to decide which (if any) of the projects to fund; thus an outcome is a vector $x = (x_1, \ldots, x_n) \in \{0, 1\}^n = X$, where $x_i = 1$ denotes funding the ith project and $x_i = 0$ denotes rejection. For any $x \in X$ let $F(x) \subseteq N$ denote the projects that are funded.

If project i is funded the benefits accrue solely to legislator i, whereas the costs are divided equally among all the legislators; if project i is rejected, legislator i receives zero benefits but still must bear her share of the costs of any funded projects. Thus the utility for legislator i derived from any outcome $x \in X$ can be written

$$u_i(x) = b_i x_i - \frac{\sum_{j=1}^n c_j x_j}{n}.$$

Define $x^a \in X$ by $F(x^a) = \{i \in N: i \leqslant (n+1)/2\}$ and assume that for all $i \in F(x^a)$, $u_i(x^a) > 0$; that is, the least-cost majority of projects generates a strictly positive utility for those legislators receiving projects. Finally, let the status quo alternative be $x_o = (0, \ldots, 0)$, i.e., no project is funded.

PROPOSITION 5. *x^a is the agenda-independent outcome in DP under the amendment procedure.*

To see this, note first that for an alternative x to be majority-preferred to x_o it must be that $F(x) > n/2$, for otherwise a majority of legislators are receiving zero benefits while bearing positive costs, thereby generating a utility strictly less than zero. Second, among those alternatives with $F(x) > n/2$ there is one alternative that is majority-preferred to all others, namely x^a. This follows since for any $x \neq x^a$ such that $F(x) > n/2$, all $i \in \{1, \ldots, (n+1)/2\}$ must be bearing a strictly higher cost (since project cost is increasing in the index) while receiving no higher benefit (either $x_i = 1$ and i receives the same benefit or else $x_i = 0$ and i receives a strictly lower benefit). Therefore the majority coalition of $\{1, \ldots, (n+1)/2\}$ prefers x^a to any other outcome which has a

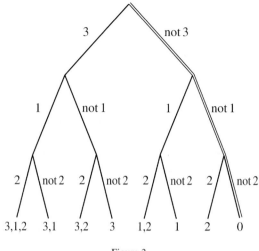

Figure 3.

majority of projects funded, and since $u_i(x^a) > 0$ for all members of this coalition x^a is majority-preferred to x_0 as well. Therefore Condition C holds, with x^a the agenda-independent outcome.

Notice that for this problem there exists another voting procedure which might actually seem more natural, namely, where each project is considered one at a time, and accepted or rejected according to a majority vote. Such a procedure would again generate an outcome $x \in \{0, 1\}^n$, although the procedure would look quite different from the amendment procedure. For instance, suppose $n = 3$, and the projects are considered in the order $3, 1, 2$; then the binary voting tree would look as in Figure 3 (where for simplicity we list $F(x)$ rather than x). Using the above utilities we can compute the sophisticated voting strategies: at each of the four final decision nodes the coalition $\{1, 3\}$ prefers not to fund project 2, so this project is defeated regardless of the previous votes. Given this, the coalition $\{2, 3\}$ prefers not to fund project 1 at the two penultimate nodes, and similarly the coalition $\{1, 2\}$ prefers not to fund project 3 at the initial node. Therefore the sophisticated voting outcome is that *none* of the three projects is funded. Furthermore, this logic is independent of the order in which the projects are considered, and immediately extends to the n-project environment. Therefore, the outcome under the "project-by-project" voting procedure is invariably the status quo $x_0 = (0, \ldots, 0)$.

We can generalize from this example and create a different form of agenda independence for the special case where the outcome set X is of the form $X = \{0, 1\}^k$, with the interpretation being that $x_i = 1$ signifies acceptance of issue i and $x_i = 0$ rejection. Equivalently, an outcome can be thought of as a (possibly empty) subset $J \subseteq K = \{1, \ldots, k\}$ of accepted issues, with individual preference and hence majority preference then defined over these subsets. An issue-by-issue procedure considers each of the k issues one at a time, and is characterized by an agenda $\iota : K \to K$ which orders the k issues. Let Γ_ι be the issue-by-issue voting game associated with the agenda ι, and say that $x \in X$ is an agenda-independent outcome under the issue-by-issue procedure if $x = s(\Gamma_\iota, P)$ for all ι.

Consider the following restriction on majority preference:

Condition S: For all $J \subseteq \{1, \ldots, k\}$ and all $j \in J$, $J \setminus \{j\} P J$.

In words, slightly smaller (by inclusion) subsets of accepted issues are preferred by a majority to larger. To see that S is sufficient to guarantee agenda independence, note that any final decision node corresponds to a vote of the form J vs. $J \setminus \{j\}$ (where $\iota(j) = k$). If S holds then issue j will be rejected regardless of the earlier votes, i.e., regardless of the contents of J. By backwards induction, then, the penultimate votes are of the form J' vs. $J' \setminus \{d\}$ (where $\iota(d) = k - 1$) and again d is defeated regardless of J'. Continuing this logic, we see that if S holds then all issues are rejected for any ordering of the issues. Thus we have not only agenda independence but we have identified the sophisticated voting outcome as $(0, \ldots, 0)$, all issues rejected.

Returning to the above example, we have,

PROPOSITION 6. $x_\circ$ *is the agenda-independent outcome in DP under the issue-by-issue procedure.*

Therefore, as with the amendment procedure, if attention is restricted to issue-by-issue procedures there will not exist any conflict over the agenda to use. Finally, note that if it were put to a vote whether an amendment procedure or an issue-by-issue procedure should be adopted in DP, a majority (namely, the members of $F(x^a)$) would prefer the former to the latter regardless of the agenda subsequently employed.

4. The spatial environment

It is evident that what drives the previous result is the decomposability of the collective choice problem into n smaller problems ("fund project i", "don't fund project i"), as well as the separability of legislators' preferences across these n problems. This decomposability has a natural analogue in the spatial environment, when we think of each dimension of the policy space as a separate issue to be decided. To keep matters simple we make the simplifying assumption that X is rectangular, $X = [\underline{x}_1, \overline{x}_1] \times \cdots \times [\underline{x}_k, \overline{x}_k]$, so that the feasible choices along any one dimension are independent of the choices along the others. We return to the notion of separable preferences below.

As in the previous section, with n odd and each R_i strictly convex the majority-rule core, if non-empty, will consist of a single element, x^c, where x^c is strictly majority-preferred to all other alternatives. On the other hand, Figure 1 demonstrated that, no matter how nice individual preferences are, in two or more dimensions a majority-rule core point rarely exists, i.e., for every alternative there is another that is majority-preferred to it. From this perspective then no alternative is stable, in that for any alternative there exists a majority coalition with the willingness and ability to change the collective decision to some other alternative. Now suppose that we constrain the influence of majority coalitions by requiring that any such change from one alternative to another must be accomplished via a sequence of one-dimension-at-a-time changes. For any $x, y \in X$, define $I(x, y) = \{j \in \{1, \dots, k\}: x_j \neq y_j\}$ as the set of issues along which x and y disagree. Then since not all of these changes will necessarily be accepted, in considering a move from x to y the set of possible outcomes are those points in X of the form $x + (y - x)^J$, where $J \subseteq I(x, y)$ and for any vector $z \in \Re^d$ and subset $J \subseteq \{1, \dots, k\}$, $z_i^J = z_i$ if $i \in J$ and $z_i^J = 0$ otherwise. For instance, in two dimensions the set of possible outcomes is given by $\{(x_1, x_2), (x_1, y_2), (y_1, x_2), (y_1, y_2)\}$.

Thus for any ordered pair of alternatives $(x, y) \in X^2$ and any ordering ι of the elements of $I(x, y)$ we have a well-defined issue-by-issue voting procedure $\Gamma_\iota(x, y)$; and given individual and hence majority preferences P on X we can compute the sophisticated voting outcome $s(\Gamma_\iota(x, y), P)$.[4] Say that an outcome $x^* \in X$ is a *sophisticated voting equilibrium* if for any $y \neq x^* \in X$ and any ordering ι of $I(x^*, y)$, $s(\Gamma_\iota(x^*, y), P) = x^*$. That is, an alternative is a sophisticated voting equilibrium if no dimension-by-dimension change can be produced when all individuals vote sophisticatedly. Finally, say that individual i's preferences are *separable* if for all $(x, y) \in X^2$, $J \subseteq I(x, y)$ and $j \in J$, $[x + (y - x)^J]R_i[x + (y - x)^{J \setminus \{j\}}]$ if and only if $[x + (y - x)^{\{j\}}]R_i[x]$. Kramer (1972) then proves the following:

PROPOSITION 7. *If, for all $i \in N$, R_i is separable, then there exists a sophisticated voting equilibrium.*

Since preferences are strictly convex, continuous and separable, and n is odd, along issue dimension j there will exist a unique point x_j^m given by the median of the individuals' ideal points. Letting $x^* = (x_1^m, \dots, x_d^m)$, separability of preferences then implies that for any $y \in X$ Condition S above is satisfied: for any subset of changes $J \subseteq I(x^*, y)$ an individual's preferences over J and $J \setminus \{j\}$ (where $j \in J$) are completely determined by their preferences along dimension j, and since $x_j^* = x_j^m$ a majority prefers $J \setminus \{j\}$ to J.

Kramer (1972) actually proves the existence of an issue-by-issue median, i.e., an alternative for which no one-dimensional change could attract a majority, without requiring separability. However he also shows by way of example that without separability one can construct situations in which two one-dimensional changes away from x^m, together with sophisticated voting, will occur. Note that if a Condorcet winner x^c exists, it must be that $x^c = x^m$, and so issue-by-issue voting is Condorcet consistent. On the

---

[4] Since individuals may be indifferent between accepting and rejecting a change on an issue we add the behavioral assumption that when indifferent an individual votes against the change.

other hand, even when a Condorcet winner does not exist a sophisticated voting equilibrium *will* exist (with separable preferences), due to the constraints placed on movements through the policy space.

Kramer (1972) is the first example of what became known as the *structure-induced equilibrium* approach to collective decision-making (as opposed to preference-induced equilibrium, i.e., the core). A second example is given by Shepsle (1979) where, rather than directly impeding a majority's influence through the policy space, the choice of collective outcome is decentralized into a set of k one-dimensional decisions by subgroups of individuals. Thus define a *committee system* as an onto function $\kappa : N \to K = \{1, \ldots, k\}$, with the interpretation that $\kappa(i)$ is the single issue upon which $i \in N$ is influential. For all $j \in K$, let $N^{\kappa}(j) = \{i \in N : \kappa(i) = j\}$ denote the set of individuals, or "committee" assigned to issue j, and for simplicity assume $|N^{\kappa}(j)|$ is odd for all $j \in K$. The committees play "Nash" against one another, taking the decisions on the other issues as given, and use majority rule to determine the outcome on their issue. Thus let $F(x; j) = \{y \in X : y_l = x_l \text{ for all } l \neq j\}$ denote the set of feasible alternatives the committee $N^{\kappa}(j)$ can generate, given that x summarizes the choices by the other committees, and similarly let $C^{\kappa}(x; j) \subset F(x; j)$ denote the set of majority-rule core points for the committee. An outcome $x^* \in X$ is a κ-*committee equilibrium* if, for all $j \in K$, $x^* \in C^{\kappa}(x^*; j)$; that is, no majority of any committee can implement a preferred outcome.

PROPOSITION 8. *For all κ, a κ-committee equilibrium exists.*

Since individual preferences are strictly convex and continuous on X, for each $(x, j) \in X \times K$ the preferences of each $i \in N^{\kappa}(j)$ will be single-peaked on $F(x; j)$ and hence $C^{\kappa}(x; j)$ will be equal to the unique median of the ideal points of members of $N^{\kappa}(j)$ along $F(x; j)$. And since preferences are continuous and the median function is continuous, we have that $C^{\kappa}(\cdot; j) : X \to X$ is a continuous function, as is the function $\mathcal{B} : X \to X$ defined by $\mathcal{B}(x) = (C_1^{\kappa}(x; 1), \ldots, C_k^{\kappa}(x; k))$. By Brouwer's fixed-point theorem there exists $x^* \in X$ such that $x^* = \mathcal{B}(x^*)$.

A few comments on this model are in order. The first is that if committees were to move sequentially as opposed to simultaneously, equilibria need not exist since the induced preferences for members of the committee moving first need not be single-peaked along their issue when the latter committees' responses are taken into account [Denzau and Mackay (1981)]. On the other hand, if individual preferences are assumed to be separable it is easily seen that this will not be a concern, and in fact the equilibrium will be independent of the order in which the committees report their choices.

Second, although as stated this model is a combination of cooperative (within committee) and non-cooperative (between committee) elements, we can easily make the entire process non-cooperative: let the strategy space for each $i \in N^{\kappa}(j)$ be given by $[\underline{x}_j, \overline{x}^j]$, with the outcome function being the Cartesian product of the median choices along each dimension. If individuals' preferences are separable then $i \in N^{\kappa}(j)$ would have a dominant strategy to choose her ideal outcome from $[\underline{x}_j, \overline{x}^j]$, thereby implementing the (now unique) κ-committee equilibrium. If preferences are non-separable this ideal outcome may depend on the median choices along the other dimensions. However, taking these other choices as fixed, i's best response is to choose her ideal outcome from $[\underline{x}_j, \overline{x}^j]$, and therefore any κ-committee equilibrium will constitute a Nash equilibrium of the above game (although there may be others).

Finally, in contrast to Kramer (1972) here one does not necessarily get Condorcet consistency, even with separable preferences. This is so because the "core" voter can only be on a single committee, and even then there is no assurance she gets her way along the relevant dimension. To see this let $k = 3$ and $n = 9$, and let all individuals have Euclidean preferences, with ideal points $x^1 = (0, 0, 0)$, $x^2 = x^3 = (1, 0, 0)$, $x^4 = x^5 = (-1, 0, 0)$, $x^6 = (0, 1, 0)$, $x^7 = (0, -1, 0)$, $x^8 = (0, 0, 1)$, and $x^9 = (0, 0, -1)$. Then the majority-rule core is at $(0, 0, 0)$; but if the committee for the first dimension is given by $\{1, 2, 3\}$ the first coordinate in any equilibrium is 1. On the other hand, if κ is such that $N^\kappa(1) = \{1, 2, 4\}$, $N^\kappa(2) = \{3, 6, 7\}$, and $N^\kappa(3) = \{5, 8, 9\}$, then in fact the κ-committee equilibrium *is* at the core.

The above model was motivated by the committee system employed in the US Congress. Parliamentary democracies also typically display a certain structure in their collective decision-making, a structure which might again be leveraged to produce equilibrium predictions [Laver and Shepsle (1990), Austen-Smith and Banks (1990)]. In parliaments the locus of control is much more at the party level as opposed to the individual level; however to avoid modeling intraparty behavior we imagine an n-party world with perfect party discipline so as to remain consistent with our n-individual analysis. As with the committee model each dimension of the policy space is associated with a different "ministry", the differences being that here a single party controls a ministry, and a party can control more than one ministry. Thus let $\mathcal{P}(K)$ denote the power set of K, and define an issue allocation as a mapping $\omega: N \to \mathcal{P}(K)$ such that

$$\bigcup_{i \in N} \omega(i) = K \quad \text{and,} \quad \text{for all } i, j \in N, \ \omega(i) \cap \omega(j) = \emptyset.$$

That is, each issue is assigned to precisely one party, with $\omega(i)$ denoting the issues under the influence of party i. Let Ω denote the (finite) set of such mappings. Parties for which $\omega(\cdot) \neq \emptyset$ constitute the governing coalition, and play Nash against one another to determine an outcome from X; thus the strategy sets are of the form $S_i^\omega = \prod_{j \in \omega(i)} [\underline{x}_j, \overline{x}_j]$, and the outcome function is simply the admixture of the various choices. Let $E(\omega) \subset X$ denote the set of Nash equilibrium outcomes for the allocation ω. Given our stated assumptions on individual/party preferences, a standard argument establishes that $E(\omega)$ is non-empty for each allocation ω in Ω. Further, as in the committee model, if preferences are separable then each member of the governing coalition will have a dominant strategy, implying for each $\omega \in \Omega$ the Nash equilibrium will be unique. Hereafter we assume this uniqueness holds regardless of the separability of preferences.

In the above model of Shepsle (1979) the committee system κ was exogenous, thereby leaving the puzzle of collective choice somewhat unfinished. In contrast, here we endogenize the choice of allocation: let $E = \bigcup_{\omega \in \Omega} E(\omega) \subset X$ denote the (finite) set of outcomes implementable by some allocation, and note that each $i \in N$ has well-defined preferences defined over E and hence (by uniqueness) well-defined induced preferences over the set of allocations Ω. Rather than modeling the choice of allocation and hence the government formation process explicitly [as in, e.g., Austen-Smith and Banks (1988) or Baron (1991)], Laver and Shepsle (1990) and Austen-Smith and Banks (1990) explore various core concepts associated with the set E of implementable outcomes. The most straightforward concept, although by no means the most reasonable, is to identify the weighted majority core on the restricted set of outcomes, E. That is, each party has a certain number of seats and hence "weight" in the legislature; an alternative $x \in E$ is an "allocation" core if there does not exist $y \in E$ which a weighted majority prefers. Note that any such prediction gives not only a policy from X, but also the identity of the government and the distribution of policy influence within the government. Furthermore, since the set of implementable outcomes E is much smaller than the set of all possible outcomes X, these allocation core points can exist even when a core point in X does not. In fact, since E is finite the former need not lead the precarious existence the latter often do in multiple dimensions with respect to small changes in preferences. Finally, since $\omega(i) = K$, i.e., allocating all issues to a single party, is allowed, we have that the profile of ideal points $\{x^1, \ldots, x^n\}$ is necessarily a subset of E. Therefore, if we add the assumptions that the party weights are such that no coalition has precisely $1/2$ the weight (i.e., weighted majority rule is strong) and that each R_i is representable by a continuously differentiable utility function $u_i : X \to \Re$, then when x^c exists (and is interior to X) it must be that $x^c = x^i$ for some $i \in N$, and hence we have a weighted version of Condorcet consistency holding. This would give an instance of one-party government where that party does not itself hold a majority of the seats (i.e., a one-party, minority government).

Austen-Smith and Banks (1990) prove the following:

PROPOSITION 9. *If $n = 3$ and, for all $i \in N$, R_i is Euclidean, then the allocation core is non-empty.*

In fact, with $n = 3$ the allocation core is either equal to one of the ideal points, or else is equal to the issue-by-issue median (cf. Figure 1). Even without a general existence theorem, Laver and Shepsle (1996) take the model's predictions to the data on post-World War II coalitional governments in parliamentary systems, and find an admirable degree of consistency.

Each of the above three models employs some element of "real world" legislative decision-making to provide the analytical traction necessary to generate equilibrium predictions. The final model we consider in this section is relatively more generic, and is meant to capture individual and collective behavior in more unstructured situations. Specifically, we look at a discrete time, infinite horizon model of bargaining where, in contrast to Rubinstein-type bargaining with a deterministic proposer sequence, the proposer in each period is randomly recognized. Thus in period $t = 1, 2, \ldots$, individual $i \in N$ is recognized with probability $\mu_i \geqslant 0$ as the current proposer, where $\sum_{j \in N} \mu_j = 1$. Upon being recognized she selects some outcome $x \in X$, and upon observing x each individual votes either to accept or reject. If the coalition of individuals voting to accept are in $\mathcal{D}$, where $\mathcal{D}$ is some simple rule, the process ends and the outcome is (x, t); otherwise the process moves to period $t + 1$. Individual i's preferences are represented by a continuous and strictly concave utility function $u_i : X \to \mathfrak{R}_{++}$ and discount factor $\delta_i \in [0, 1]$, with the utility to i from the outcome (x, t) then being $\delta_i^{t-1} u_i(x)$. This model, due to Banks and Duggan (2000), extends to the spatial environment the earlier work of Baron and Ferejohn (1989) in which this bargaining protocol was studied in a "divide-the-dollar" environment with "selfish" linear preferences, i.e., $X = \{x \in [0, 1]^n : \sum_{i \in N} x_i = 1\}$ and for all $i \in N$ and $x \in X$, $u_i(x) = x_i$, and where the focus was on majority rule.

Banks and Duggan (2000) restrict attention to equilibria in stationary strategies, consisting of a proposal $p_i \in X$ offered anytime i is recognized; and a voting rule $v_i : X \to \{accept, reject\}$ independent both of history and of the identity of the proposer. To prove existence they need to allow randomization in the proposal-making; thus each $i \in N$ selects a probability distribution $\pi_i \in \mathcal{P}(X)$ (with the latter endowed with the topology of weak convergence). However, each individual observes the realized policy proposal prior to voting. Finally, they characterize only "no-delay" equilibria in which the first proposal is accepted with probability one. When $\delta_i < 1$ for all $i \in N$ it is easily seen that all stationary equilibria will involve no delay; however, since individuals here are allowed to be perfectly patient (i.e., $\delta_i = 1$) this amounts to a selection from the set of all stationary equilibria.

Let $\mathbf{u} = (u_1, \dots, u_n)$, and as before let $\mathcal{C}(\mathcal{D}, \mathbf{u}) \subseteq X$ denote the core of the simple rule $\mathcal{D}$ at the profile $\mathbf{u}$.

PROPOSITION 10.
   (a) *There exist stationary no-delay equilibria;*
   (b) *If $\delta_i = 1$ for all $i \in N$ and either $\mathcal{D}$ is collegial or X is one-dimensional, then $(\pi_1^*, \dots, \pi_n^*)$ is a stationary no-delay equilibrium if and only if $\pi_i^*(\{x^*\}) = 1$ for all $i \in N$ and for some $x^* \in \mathcal{C}(\mathcal{D}, \mathbf{u})$.*

Comparing this result to Proposition 1, and cognizant of the fact that here we are assuming strict concavity rather than strict quasi-concavity (i.e., strictly convex preferences), we see first that equilibria exist for all profiles $\mathbf{u}$, all simple rules $\mathcal{D}$, and all dimensions of the policy space k. Further, when individuals are perfectly patient we get a core equivalence result precisely when the core is non-empty for all profiles $\mathbf{u}$ and simple rules $\mathcal{D}$ (i.e., $k = 1$) and when the core is non-empty for all profiles $\mathbf{u}$ and all dimensions of the policy space k (i.e., $\mathcal{D}$ collegial). Banks and Duggan (2000) also show that the set of stationary no-delay equilibria is upper-hemicontinuous in (among other parameters) the discount factors, and hence if individuals are "almost" perfectly patient all equilibrium outcomes will be "close" to the core.

Another interesting comparison is with the equilibrium predictions of the issue-by-issue model and the parliamentary model described above. In Figure 1 both of these models predict the issue-by-issue median x^m as the unique equilibrium outcome. On the other hand, it is readily apparent that in the bargaining model under majority rule any equilibrium proposal lies on the contract curve between the proposer and one of the remaining individuals, with the exact location of the proposals (and the probability distribution over them) depending on the underlying parameters. Thus, the former two models predict a much more centrist outcome than does the bargaining model.

5. Incomplete information

All of the models surveyed in the previous two sections have been examples of political processes in which the set of individuals N directly chooses the collective outcome from X. We begin this section with the classic model of *representative* democracy, namely, the Downsian model of electoral competition, and analyze it in the context of Proposition 1. In the basic model two candidates, A and B ($\notin N$) simultaneously choose policies $x_a, x_b \in X$; upon observing these policies individuals simultaneously vote for either A or B (so no abstentions), with the majority-preferred candidate winning the election and implementing her announced policy. The candidates have no policy interests themselves, and only care about winning the election: the winning candidate receives a utility payoff of 1, while the loser receives -1. Given any (x_a, x_b)-pair each $i \in N$ has a weakly dominant strategy to vote for the candidate offering the preferred policy according to u_i, with i voting for each with probability 0.5 when $u_i(x_a) = u_i(x_b)$. Thus, given knowledge of voter preferences the two candidates are engaged in a symmetric, zero-sum game with common strategy space X, and so in any equilibrium each must receive an expected payoff of zero. It is easily seen that a pure strategy equilibrium to this game exists if and only if the majority-rule core is non-empty: if the core is empty then given any x_b there exists a policy which majority-defeats it, and hence if adopted by A would generate payoffs $(1, -1)$. Conversely, if x_C is in the core candidate C guarantees herself an expected payoff of zero, and so if x_a and x_b are core policies (x_a, x_b) constitutes an equilibrium.

From Section 2 we know that the majority-rule core in the spatial environment is equal to the median voter's ideal point when the dimension of the policy space is one, thus giving the original prediction of Downs (1957) that both candidates would locate there. On the other hand, we also know that when this dimension is greater than one the core is almost always empty, and so pure strategy equilibria typically fail to exist. One possible escape route of course is to consider mixed strategies, as was done by Banks and Duggan (2000) in the bargaining context. However, while existence of mixed strategy equilibria is assured in the finite environment,[5] in the spatial environ-

---

[5] See [Laffond et al. (1993)] for a characterization of the support of the mixed strategy equilibria in the finite environment.

ment it has proven to be somewhat more elusive given the inherent discontinuities in the majority preference relation and hence in the above zero-sum game [however see Kramer (1978)]. Alternatively, one can smooth over these discontinuities by positing a sufficient amount of uncertainty, from the candidates' perspective, about the intended behavior of the voters. To this end let the probability $i \in N$ votes for candidate A, given announced policy positions (x_a, x_b), be $p_i(u_i(x_a), u_i(x_b))$, with $1 - p_i$ being the probability of voting for B (so again no abstentions). Candidates are assumed to maximize expected vote, which, in the absence of abstentions, is equivalent to maximizing expected plurality. Thus A solves

$$\max_{x \in X} \sum_{i \in N} p_i\big(u_i(x), u_i(x_b)\big),$$

with a similar expression for B. As in the bargaining model, we assume that, for all $i \in N$, u_i is strictly concave and maps into the positive real line.

PROPOSITION 11.
  (a) *If, for all $i \in N$, $p_i(\cdot)$ is concave increasing in its first argument and convex decreasing in its second, then a pure strategy equilibrium exists.*
  (b) *If $p_i(\cdot)$ can be written as*

$$p_i\big(u_i(x_a), u_i(x_b)\big) = \hat{p}\big(u_i(x_a) - u_i(x_b)\big),$$

  where $\hat{p}$ is increasing and differentiable, then in equilibrium $x_a = x_b = x^u$, where x^u solves

$$\max \sum_{i \in N} u_i(x).$$

  (c) *If $p_i(\cdot)$ can be written as*

$$p_i\big(u_i(x_a), u_i(x_b)\big) = \tilde{p}\big(u_i(x_a)/u_i(x_b)\big),$$

  where $\tilde{p}$ is increasing and differentiable, then in equilibrium $x_a = x_b = x^n$, where x^n solves

$$\max \sum_{i \in N} \ln\big(u_i(x)\big).$$

Thus, while (a) [due to Hinich et al. (1972)] provides sufficient conditions for the existence of a pure strategy equilibrium in the candidate game, (b) and (c) demonstrate that, in certain situations, these equilibrium policies will coincide with either utilitarian or Nash social welfare optima. Beyond any normative significance, this implies that the candidates' equilibrium policies are actually independent of their expected plurality functions, and, therefore, independent of the structure of the incomplete information in the model (other than the requirement that $\hat{p}$ and $\tilde{p}$ only depend on i through u_i). Part (b) extends the result of Lindbeck and Weibull (1987) for a model of income redistribution, while (c) is an extension of a result for the familiar binary Luce model employed by Coughlin and Nitzan (1981) and others, where

$$p_i\big(u_i(x_a), u_i(x_b)\big) = \frac{u_i(x_a)}{u_i(x_a) + u_i(x_b)}.$$

One weakness of the probabilistic voting model is that the analysis starts "in the middle", by positing some voter probability functions without generating them from primitives. This can be remedied by assuming that, from the candidates' perspective, there exist some additional considerations beyond a comparison of policies which influence the voters' decisions and which are unknown to the candidates. We can think of these as non-policy characteristics of the candidates, or (equivalently) as the candidates' fixed positions on some other policy dimensions. For example, if $i \in N$ votes for candidate A if and only if $u_i(x_a) > u_i(x_b) + \beta$, where β is distributed according to a smooth function G, then $p_i(u_i(x_a), u_i(x_b))$ is simply equal to $G(u_i(x_a) - u_i(x_b))$ and (b) is satisfied. Additionally, if G is uniform on an interval $[\underline{\beta}, \overline{\beta}]$ sufficiently large (and containing zero), then

$$p_i\big(u_i(x_a), u_i(x_b)\big) = \frac{u_i(x_a) - u_i(x_b) - \underline{\beta}}{\overline{\beta} - \underline{\beta}},$$

and hence p will be linearly increasing in $u_i(x_a)$ and linearly decreasing in $u_i(x_b)$, and therefore (a) is satisfied. Alternatively, the "bias" term β could enter in multiplicatively rather than additively, thereby generating a form as in (c). One immediate consequence of moving back to this more primitive stage is to qualify the welfare statements above: it is not the sum of the individual utilities that is actually being maximized, but rather the sum of the *known* components of their utilities.[6]

In the previous model incomplete information concerning voter behavior was present, in that the candidates could not perfectly predict how policy positions would be translated into votes. However, the decision problem on the part of the voters was straightforward: given only two alternatives, and with complete information about their own preferences, each voter's weakly dominant strategy was simply to vote for her preferred alternative. Suppose now that there exists some uncertainty among the voters about which is the preferred alternative [Austen-Smith and Banks (1996)]. Specifically, let $X = \{A, B\}$, and let there be two possible states of the world, $S = \{A, B\}$, with $\pi \in (0, 1)$ the prior probability that the state is A. Individual i's preferences over outcome-state pairs (x, s) are represented by $u_i : X \times S \to \{0, 1\}$, with $u_i(x, s) = 1$ if

---

[6] See [Banks and Duggan (1999)] for these and other extensions of results in probabilistic voting.

and only if $x = s$. Hence, in contrast to the heterogeneous preferences assumed in all of the above models, here the individuals have a common interest in selecting the outcome to match the state. Individuals simultaneously vote for either A or B, with B being the outcome if and only if h individuals vote for it, where $h \in \{1, \ldots, n\}$. Thus $h = (n + 1)/2$ means that majority rule is being employed, whereas $h = n$ means that B is the outcome only if the individuals unanimously vote for B.

While individuals share a common interest, they possess potentially different information concerning the true state, and hence the optimal choice, in that prior to voting each $i \in N$ receives a private signal $t_i \in T_i$ correlated with the true state. Thus let $v_i : T_i \to X$ denote a voting strategy for i, $v = (v_1, \ldots, v_n)$ a profile of voting strategies, $t = (t_1, \ldots, t_n) \in \prod_{i \in N} T_i$ a profile of signals, and $d = (d_1, \ldots, d_n) \in X^n$ a profile of individual decisions. Given a common prior $p(\cdot)$ over profiles of signals, then, we have a Bayesian game among the voters, with the expected utility of a pair (d, t) equal to the probability the true state is equal to the collective decision conditional on t. Defining $c(d) \in X$ by $c(d) = B$ if and only if $|\{i \in N: d_i = B\}| \geqslant h$, we can write this as $\Pr[s = c(d); t]$, which (via Bayes' Rule) is equal to $\Pr[s = c(d) \, \& \, t]/p(t)$. Moreover, since $p(t_{-i}; t_i) = p(t)/\sum_{T_{-i}} p(t_i, t_{-i})$, we can simplify the expression for the expected utility of i choosing d_i, conditional on t_i and v, from

$$EU(d_i; t_i, v) = \sum_{T_{-i}} p(t_{-i}; t_i) \frac{\Pr[s = c(d_i, v(t_{-i})) \, \& \, t]}{p(t)}$$

to

$$EU(d_i; t_i, v) = \sum_{T_{-i}} \frac{\Pr[s = c(d_i, v(t_{-i})) \, \& \, t]}{\sum_{T_{-i}} p(t_i, t_{-i})}.$$

Now in comparing the difference in expected utility from voting for A or B, we can obviously ignore the denominator above; further, this difference will only be non-zero when i is pivotal in determining the outcome. Therefore let

$$T_{-i}^p(v) = \left\{ t_{-i} \in T_{-i}: \left| \{j \in N \backslash \{i\}: v_j(t_j) = B\} \right| = h - 1 \right\}$$

be the set of others' signals for which, according to v, i is pivotal. Finally, when i is pivotal her decision is equal to the collective decision, and hence we get that

$$EU(A; t_i, v) - EU(B; t_i, v) \geqslant 0 \quad \text{if and only if}$$
$$\sum_{T_{-i}^p(v)} \left[\Pr[s = A \, \& \, t] - \Pr[s = B \, \& \, t] \right] \geqslant 0.$$

Thus, in identifying which is the better decision each individual conditions on being pivotal, which (through v) limits the set of possible inferences she can have about others' information.

The presence of this "pivotal" information can have dramatic effects on individuals' voting behavior. The first thing to note is that, unlike when information is complete, sincere voting is no longer a weakly dominant strategy, where v_i is sincere if, for all $t_i \in T_i$, $v_i(t_i) = A$ if and only if $E[u_i(A, \cdot); t_i] > E[u_i(B, \cdot); t_i]$. As a simple example, let $n = 5$, $\pi = 0.5$, $h = (n + 1)/2$, and for all $i \in N$ let $T_i = \{a, b\}$ with $\Pr[t_i = a; s = A] = \Pr[t_i = b; s = B] = q > 0.5$. Thus there are four pure strategies: "always vote A", "always vote B", "vote A if and only if $t_i = a$", and "vote A if and only if $t_i = b$", with the third strategy being sincere. Suppose voters 1 and 2 play "always A" while 3 and 4 play the sincere strategy; then 5 should play "always B" since the only time 5 is pivotal is when $t_3 = t_4 = b$, in which case even if $t_5 = a$ Bayes' Rule implies the more likely state is B. Therefore sincere voting is not a weakly dominant strategy. On the other hand, sincere voting *does* constitute an equilibrium: if the others are behaving sincerely and i is pivotal, she infers that two of the others have observed a and two have observed b. Given the symmetry of the situation these signals cancel out when applying Bayes' Rule, and so it is optimal for i to behave sincerely as well.

The reason sincere voting constitutes an equilibrium is that majority rule is the "right" rule to use in symmetric situations such as this, in the sense that if all the signals were known choosing A if and only if a majority of the signals were a would maximize the likelihood of making the correct decision [Austen-Smith and Banks (1996)]. Conversely, suppose we keep all the underlying parameters the same but use unanimity rule, $h = n$ [Feddersen and Pesendorfer (1998)]. Then it is easily seen that sincere voting does *not* constitute an equilibrium: under unanimity voter i is pivotal only when all others are voting for B. But if the others vote for B if and only if their signals are b, i prefers to vote for B when pivotal even if $t_i = a$, as the (inferred) collection of $n - 1$ b's overwhelms her one a. On the other hand, everyone adopting the "always B" strategy is not an equilibrium either, since in this case i is always pivotal and hence infers nothing about the others' signals. Thus when $t_i = a$ i should vote for A. (Of course, everyone adopting the "always A" strategy is an equilibrium, since nobody is ever pivotal.) Therefore, to find interesting symmetric equilibria Feddersen and Pesendorfer (1998) look to mixed strategies, letting $\sigma(t_i) \in [0, 1]$ be the probability a type-t_i individual votes for B. Beyond symmetry, they require the strategies to be "responsive" to the private signals, so that $\sigma(a) \neq \sigma(b)$. It is readily apparent that, given the symmetry of the environment, any such equilibrium under unanimity will be of the form $0 < \sigma(a) < 1 = \sigma(b)$, i.e., vote for B upon observing the signal b, and randomize upon observing a.

The goal is to compare unanimity rule to majority rule, as well as any other rule, in terms of the equilibrium probability of the collective making an incorrect decision. As they let the population size n vary, we can parametrize the rules by $\gamma \in (0, 1)$ and set $h = \lceil \gamma n \rceil$ where for any $r \in \Re_+$, $\lceil r \rceil$ is the smallest integer greater than or equal to r (e.g., $\gamma = 0.5$ is majority rule).

PROPOSITION 12.

(a) *Under unanimity rule there exists a unique responsive symmetric equilibrium; as $n \to \infty$ $\sigma(a) \to 1$, and the probabilities of choosing A when the state is B and choosing B when the state is A are bounded away from zero.*

(b) *For any $\gamma \in (0, 1)$ there exists $n(\gamma)$ such that for all $n \geqslant n(\gamma)$ there exists a responsive symmetric equilibrium; as $n \to \infty$ any sequence of such equilibria has the probabilities of choosing A when the state is B and choosing B when the state is A converging to zero.*

Therefore from the perspective of error probabilities unanimity rule is unambiguously the worst rule for large collectives. Furthermore, Feddersen and Pesendorfer (1998) show by way of a twelve-person example that "large" need not be very large. They interpret this result as casting doubt on the appropriateness of using unanimity rule in, for instance, legal proceedings.[7]

The preceding result showed how a model that Proposition 1(a) would suggest is trivial under complete information, namely, when there are only two alternatives, becomes not so trivial with the judicious introduction of incomplete information. Our last model shows the same phenomenon with respect to Proposition 1(b), and is due to Gilligan and Krehbiel (1987). Consider a one-dimensional policy space $X = \Re$ within which two players, labelled the committee C and the floor F, interact to select a final policy.[8] Two different procedures are considered, yielding two different game forms: under a "closed" rule C makes a take-it-or-leave-it proposal $p \in \Re$, where if F rejects the proposal the status quo policy $s \in \Re$ is implemented. In contrast, under the open rule F can select any policy following C's proposal, thereby rendering the latter "cheap talk".

As with Feddersen and Pesendorfer (1998), Gilligan and Krehbiel (1987) are interested in a comparison of these two rules with respect to the equilibrium outcomes they generate, with the motivation here being that F has the ability to choose *ex ante* which procedure to adopt. With complete information F would surely prefer the open rule, given its lack of constraints. However, suppose that under either rule, if $z \in \Re$ is the chosen policy the final outcome is not z but is rather $x = z + t$, where t is drawn uniformly from [0, 1] and whose value is known to C but unknown to F. Thus a strategy for C under either rule is a mapping from [0, 1] to $\Re$; however, to distinguish the two, let $p(t)$ denote C's proposal under the closed rule and $m(t)$ C's "message" under the open rule. Similarly, let $a(p)$ denote the probability F accepts a proposal of $p \in \Re$ from C under the closed rule, and $p(m)$ the policy F selects following the message $m \in \Re$ under the open rule. Players' utilities over final outcomes are quadratic: $u_c(x) = -(x - x_c)^2$, $x_c > 0$, and $u_f(x) = -x^2$. Thus x_c measures the heterogeneity in the players' preferences. Given this specification, if F's updated belief about the value of t has mean $\bar{t}$ her best response under the closed rule is to accept p if $\|p - \bar{t}\| \leqslant \|s - \bar{t}\|$ and reject otherwise, i.e., select p or s depending on which is closer to $\bar{t}$. Under the open rule, given mean $\bar{t}$ F's best response is simply to select $p = -\bar{t}$.

As with most signaling games there exist multiple equilibria under either rule. Under the open rule Gilligan and Krehbiel select the "most informative" equilibria, which have

---

[7] However see [Coughlan (2000)] for a rebuttal.
[8] We can think of F as the median member of the legislature, and C as the median member of a committee.

the following partition structure [Crawford and Sobel (1982)]: let $t_0 = 0$ and $t_{N(x_c)} = 1$, where $N(x_c)$ is the largest integer satisfying $|2N(1 - N)x_c| < 1$. For $1 < i < N(x_c)$ define t_i by

$$t_i = t_1 i + 2i(1 - i)x_c;$$

thus the equilibria will be parametrized by the choice of t_1. Given a distinct set $\{m_1, \ldots, m_N\} = M$ of messages C's strategy is of the form $m^*(t) = m_j$ if and only if $t \in [t_{j-1}, t_j)$. Given F's updated belief her optimal response to any $m_j \in M$ is $p^*(m_j) = -(t_j - t_{j-1})/2$. Faced with this strategy C's strategy above is optimal if and only if for all $i = 1, \ldots, N(x_c)$, t_i is indifferent between sending $m^*(t_i - \varepsilon)$ and $m^*(t_i + \varepsilon)$, which holds only when

$$t_{i+1} = 2t_i - t_{i-1} - 4x_c,$$

which is a second-order difference equation whose solution is given above. Of particular relevance is the fact that $N(x_c)$ is decreasing in x_c, and tends to infinity as x_c tends to zero.

The equilibrium Gilligan and Krehbiel select under the closed rule has the following path of play: (1) for $t \in [0, s_1] \cup [s_3, 1]$, $p^*(t) = x_c - t$, and $a^*(p^*(t)) = 1$; (2) for $t \in (s_1, s_2]$, $p^*(t) = 4x_c + s$, and $a^*(p^*(t)) = 1$; and (3) for $t \in (s_2, s_3)$, $p^*(t) = \overline{p} \in (s, s + 4x_c)$ and $a^*(p^*(t)) = 0$. Thus, when the shift parameter t is sufficiently small or sufficiently large, C is able to separate and implement her optimal outcome. A third region of types pools together and has their proposal accepted, while a fourth has their proposal rejected, with the logic of the latter being that $s + t$, the outcome upon rejection, is already close to x_c (in fact, $s_3 = x_c - s$).

PROPOSITION 13. *There exists $\overline{x} > 0$ such that if $x_c < \overline{x}$ then the floor's ex ante equilibrium utility under the closed rule is greater than that under the open rule.*

Therefore, as long as the preferences of the committee and floor are not too dissimilar, the floor prefers to grant the committee a certain amount of agenda control. The intuition behind this result is the following: under both the open and the closed rule the equilibrium expected utility of F is decreasing in x_c, so that as their preferences become less diverse F is better off under either rule. For the open rule this is true since the more homogeneous the preferences the greater is C's ability to transmit information, leading to a more informed decision by F. Under the closed rule the set of types that separate and subsequently implement their ideal outcome is itself increasing as x_c decreases, and as x_c decreases C's ideal outcome becomes more favorable from F's perspective. Now in selecting the closed rule over the open rule F faces a distributional loss, due to C's being able to bias the resulting outcome in her favor, as well as an informational gain, due to the increased amount of information transmitted and the assumed risk aversion of F. As x_c decreases this distributional loss tends to be outweighed by the informational gain, implying the closed rule is preferable.

References

Austen-Smith, D. (1987), "Sophisticated sincerity: voting over endogenous agendas", American Political Science Review 81:1323–1330.

Austen-Smith, D., and J. Banks (1988), "Elections, coalitions, and legislative outcomes", American Political Science Review 82:409–422.

Austen-Smith, D., and J. Banks (1990), "Stable governments and the allocation of policy portfolios", American Political Science Review 84:891–906.

Austen-Smith, D., and J. Banks (1996), "Information, rationality, and the Condorcet jury theorem", American Political Science Review 90:34–45.

Austen-Smith, D., and J. Banks (1999), "Cycling of simple rules in the spatial model", Social Choice and Welfare 16:663–672.

Banks, J. (1985), "Sophisticated voting outcomes and agenda control", Social Choice and Welfare 1:295–306.

Banks, J. (1995), "Singularity theory and core existence in the spatial model", Journal of Mathematical Economics 24:523–536.

Banks, J., and J. Duggan (1999), "The theory of probabilistic voting in the spatial model of elections", Working paper.

Banks, J., and J. Duggan (2000), "A bargaining model of collective choice", American Political Science Review 94:73–88.

Baron, D. (1991), "A spatial bargaining theory of government formation in parliamentary systems", American Political Science Review 85:137–164.

Baron, D., and J. Ferejohn (1989), "Bargaining in legislatures", American Political Science Review 83:1181–1206.

Black, D. (1958), The Theory of Committees and Elections (Cambridge University Press, Cambridge).

Condorcet, M. (1785/1994), Foundations of Social Choice and Political Theory, Transl. I. McLean and F. Hewitt, eds. (Edward Elgar, Brookfield, VT).

Coughlan, P. (2000), "In defense of unanimous jury verdicts: mistrials, communication, and strategic voting", American Political Science Review 94:375–393.

Coughlin, P., and S. Nitzan (1981), "Electoral outcomes with probabilistic voting and Nash social welfare maxima", Journal of Public Economics 15:113–122.

Coughlin, P. (1992), Probabilistic Voting Theory (Cambridge University Press, New York).

Crawford, V., and J. Sobel (1982), "Strategic information transmission", Econometrica 50:1431–1451.

Denzau, A., and R. Mackay (1981), "Structure-induced equilibria and perfect foresight expectations", American Journal of Political Science 25:762–779.

Downs, A. (1957), An Economic Theory of Democracy (Harper and Row, New York).

Farquharson, R. (1969), The Theory of Voting (Yale University Press, New Haven).

Feddersen, T., and W. Pesendorfer (1998), "Convicting the innocent: the inferiority of unanimous jury verdicts", American Political Science Review 92:23–36.

Ferejohn, J., M. Fiorina and R. McKelvey (1987), "Sophisticated voting and agenda independence in the distributive politics setting", American Journal of Political Science 31:169–193.

Gilligan, T., and K. Krehbiel (1987), "Collective decision making and standing committees: an informational rationale for restrictive amendment procedures", Journal of Law, Economics and Organization 3:287–335.

Gretlein, R. (1983), "Dominance elimination procedures on finite alternative games", International Journal of Game Theory 12:107–113.

Hinich, M., J. Ledyard and P. Ordeshook (1972), "Nonvoting and the existence of equilibrium under majority rule", Journal of Economic Theory 4:144-153.

Kramer, G. (1972), "Sophisticated voting over multidimensional choice spaces", Journal of Mathematical Sociology 2:165–180.

Kramer, G. (1978), "Existence of electoral equilibrium", in: P. Ordeshook, ed., Game Theory and Political Science (NYU Press, New York).

Laffond, G., J. Laslier and M. Le Breton (1993), "The bipartisan set of a tournament", Games and Economic Behavior 5:182–201.

Laver, M., and K. Shepsle (1990), "Coalitions and cabinet government", American Political Science Review 84:873–890.

Laver, M., and K. Shepsle (1996), Making and Breaking Governments (Cambridge University Press, New York).

Lindbeck, A., and J. Weibull (1987), "Balanced-budget redistributions as the outcome of political competition", Public Choice 52:273–297.

McGarvey, D. (1953), "A theorem on the construction of voting paradoxes", Econometrica 21:608–610.

McKelvey, R. (1979), "Generalized conditions for global intransitivities in formal voting models", Journal of Economic Theory 47:1085–1112.

McKelvey, R., and R. Niemi (1978), "A multistage game representation of sophisticated voting for binary procedures", Journal of Economic Theory 18:1–22.

Miller, N. (1980), "A new solution set for tournaments and majority voting: further graph-theoretical approaches to the theory of voting", American Journal of Political Science 24:68–96.

Moulin, H. (1986), "Choosing from a tournament", Social Choice and Welfare 3:271–291.

Nakamura, K. (1979), "The vetoers in a simple game with ordinal preferences", International Journal of Game Theory 8:55–61.

Plott, C. (1967), "A notion of equilibrium and its possibility under majority rule", American Economic Review 57:787–806.

Saari, D. (1997), "The generic existence of a core for q-rules", Economic Theory 9:219–260.

Schofield, N. (1983), "Generic instability of majority rule", Review of Economic Studies 50:695–705.

Schofield, N. (1984), "Social equilibrium cycles on compact sets", Journal of Economic Theory 33:59–71.

Shepsle, K. (1979), "Institutional arrangements and equilibrium in multidimensional voting models", American Journal of Political Science 23:37–59.

Shepsle, K., and B. Weingast (1984), "Uncovered sets and sophisticated voting outcomes with implications for agenda institutions", American Journal of Political Science 28:49–74.

Strnad, J. (1985), "The structure of continuous-valued neutral monotonic social functions", Social Choice and Welfare 2:181–195.

Chapter 60

GAME-THEORETIC ANALYSIS OF LEGAL RULES AND INSTITUTIONS

JEAN-PIERRE BENOîT* and LEWIS A. KORNHAUSER[†]

Department of Economics and School of Law, New York University, New York, USA

Contents

*The support of the C.V. Starr Center for Applied Economics is acknowledged.

[†] Alfred and Gail Engelberg Professor of Law, New York University. The support of the Filomen D'Agostino and Max E. Greenberg Research Fund of the NYU School of Law is acknowledged.

We thank John Ferejohn, Mark Geistfeld, Marcel Kahan, Martin Osborne and William Thomson for comments.

Handbook of Game Theory, Volume 3, Edited by R.J. Aumann and S. Hart

Abstract

We offer a selective survey of the uses of cooperative and non-cooperative game theory in the analysis of legal rules and institutions. In so doing, we illustrate some of the ways in which law influences behavior, analyze the mechanism design aspect of legal rules and institutions, and examine some of the difficulties in the use of game-theoretic concepts to clarify legal doctrine.

Keywords

economic analysis of law, Coase Theorem, accidents, positive political theory, voting power

JEL classification: C7, D7, K0

1. Introduction

Game-theoretic ideas and analyses have appeared explicitly in legal commentaries and judicial opinions for over thirty years. They first emerged in the immediate aftermath of early reapportionment cases that redrew districts in federal and state elections in the United States. As the U.S. courts struggled to articulate the meaning of equal representation and to give content to the slogan "one person, one vote", they considered, and for a time endorsed, a concept of voting power that has game-theoretic roots. At approximately the same time, the publication of Calabresi (1961) and Coase (1960) provoked an outpouring of economic analyses of legal rules and institutions that continues to grow. In the late 1960s, the first game-theoretic models of contract doctrine [Birmingham (1969a, 1969b)] appeared. Models of accident law in Brown (1973) and Diamond (1974a, 1974b) followed a few years later. In the mid-1980s, the explicit use of game theory in the economic analysis of law burgeoned.

The early uses of game theory encompass the two different ways in which game theory has been applied to legal problems. On the one hand, as in the reapportionment cases, courts and analysts have adopted game-theoretic tools to further the goals of traditional Anglo-American legal scholarship. In this tradition, both judge and commentator seek to rationalize a set of related judicial decisions by articulating the underlying normative framework that unifies and justifies the judicial doctrine. In this use, which we shall call *doctrinal analysis*, game-theoretic concepts elaborate the meaning of key, doctrinal ideas that define what the law seeks to achieve.

On the other hand, as in the analyses of contract and tort, game theory has provided a theory of how individuals respond to legal rules. The analyst considers the games defined by a class of legal rules and studies how the equilibrium behavior of individuals varies with the choice of legal rule. In legal cultures that, as in the United States, regard law as an instrument for the guidance and regulation of social behavior, this analysis is a necessary step in the task facing the legal policymaker. It is worth emphasizing that nothing in this analytic structure ties the analyst, or the policymaker, to an "economic" objective such as efficiency or maximization of social welfare.

In this essay, we offer a selective survey of the uses of game theory in the analysis of law. The use of game theory as an adjunct of doctrinal analysis has been relatively limited, and our survey of this area is correspondingly reasonably exhaustive. In contrast, the use of game theory in understanding how legal rules affect individual behavior has been pervasive. Game theory has been explicitly applied to analyses of contract,[1] torts between strangers,[2] bankruptcy and corporate reorganization,[3] corpo-

---

[1] E.g., Birmingham (1969a, 1969b), Shavell (1980), Kornhauser (1983), Rogerson (1984), and Hermalin and Katz (1993).

[2] E.g., Brown (1973), Diamond (1974a, 1974b), Diamond and Mirrlees (1975), and Shavell (1987).

[3] E.g., Baird and Picker (1991), Bebchuk and Chang (1992), Gertner and Scharfstein (1991), Schwartz (1988, 1993), and White (1980, 1994).

rate takeovers and other corporate problems,[4] product liability,[5] and the decision to settle rather than litigate[6] and the consequent effects on the likelihood that plaintiff will prevail at trial.[7]

Kornhauser (1989) argues that virtually all of the economic analysis of law can be understood as an exercise in mechanism design in which a policymaker chooses among legal rules that define games played by citizens. Moreover, certain traditional subdisciplines of economics and finance, such as industrial organization, public finance, and environmental economics, are inextricably tied to legal rules. Game-theoretic models in those fields invariably shed light on legal rules and institutions.

Given the size and diversity of this latter literature, we cannot provide a comprehensive survey of game-theoretic models of legal rules and hence make no attempt to do so. Furthermore, in some areas at least, such literature surveys already exist: see, for example, Shavell (1987) on torts, Cooter and Rubinfeld (1989) on dispute resolution, Kornhauser (1986) on contract remedies, Pyle (1991) on criminal law, Hart and Holmstrom (1987) and Holmstrom and Tirole (1989) on the incomplete contracting literature, Symposium (1991) on corporate law generally and Harris and Raviv (1992) on the capital structure of firms. In addition, Baird, Gertner and Picker (1994) is an introduction to game theory that presents the concepts through a wide variety of legal applications.

Having forsworn comprehensive coverage, we have three aims in this survey. Our first is to suggest the wide variety of legal rules and institutions on which game theory as a behavioral theory sheds light. In particular, we illustrate the ways in which legal rules influence individual behavior. Our second is to emphasize the mechanism design aspect that underlies many game-theoretic analyses of the law. Our third is to examine some of the difficulties in the use of game-theoretic concepts to clarify legal doctrine.

In Section 2, we examine the simple two-person game that underlies much of the economic analysis of accidents as well as the analysis of contract remedies and tort. In this model, the legal rule serves as a direct incentive mechanism.

In Section 3, we examine the insights that game-theoretic models have provided into the "Coase Theorem" which often serves as an analytic starting point for economic analyses of law. In Section 4 we introduce a literature that extends the use of game theory as a behavioral predictor from private law to public law and the analysis of constitutions. Section 5 discusses the small literature that has used game theory in doctrinal analysis. Section 6 offers some concluding remarks.

---

[4] E.g., Bebchuk (1994), Bulow and Klemperer (1996), Cramton and Schwartz (1991), Grossman and Hart (1982), Kahan and Tuckman (1993) and Sercu and van Hulle (1995).
[5] E.g., Oi (1973), Kambhu (1982), Polinsky and Rogerson (1983), and Geistfeld (1995).
[6] E.g., Bebchuk (1984), Daughety and Reinganum (1993), Hay (1995), Png (1982), Reinganum (1988), Reinganum and Wilde (1986), Shavell (1982), Spier (1992, 1994).
[7] E.g., Priest and Klein (1984), Eisenberg (1990), Wittman (1988).

2. Economic analysis of the common law when contracting costs are infinite

2.1. The basic model

Brown (1973) offers a simple model of accident law. We discuss it extensively here for two reasons. First, it underlies many, if not most, economic analyses of law that examine how legal rules can solve externality problems. Second, and more importantly, this simple model of torts illustrates clearly the use of game theory as a theory of the behavior of agents in response to legal rules.

There are three decisionmakers: the legal policymaker, generally thought of as the Court, and two agents, called the (potential) injurer I and the (potential) victim V. Agents I and V are each independently engaged in an activity that may result in harm to V. They simultaneously choose an amount x and y, respectively, to expend on care. These amounts determine the probability $p(x, y)$ with which V may experience a loss and the value $L(x, y)$ of such a loss. The functions $p(x, y)$ and $L(x, y)$ are common knowledge. The policymaker seeks to regulate the (risk-neutral) agents' choices of care levels.

The policymaker selects a legal rule from a family of permissible rules of the form $R(x, y; X, Y)$, where $R(x, y; X, Y)$ is the proportion of the loss $L(x, y)$ legally imposed on I, and the parameters $X \geqslant 0$ and $Y \geqslant 0$ will be interpreted as I's and V's standard of care, respectively.

Although this framework is simple, it enables a comparison of many legal rules. Thus, a general regime of *negligence with contributory negligence* is characterized by a loss assignment function of the form:

$$R(x, y; X, Y) = \begin{cases} 1 & \text{if } x < X \text{ and } y \geqslant Y, \\ 0 & \text{otherwise.} \end{cases} \tag{1a}$$

Under a *pure negligence* rule the injurer is responsible whenever she fails to take an "adequate" level of care; this corresponds to $0 < X < \infty$, $Y = 0$. With *strict liability* the injurer is responsible for damage regardless of the level of care she takes; this corresponds to $X = \infty$, and $Y = 0$.[8] Under strict liability with contributory negligence the injurer is responsible whenever the victim takes an adequate level of care; thus, $X = \infty$, $0 < Y < \infty$.

---

[8] The legal literature generally distinguishes between "strict liability rules" and "negligence rules". As the text indicates, a strict liability rule (without contributory negligence) can be understood as a member of the general class of negligence with contributory negligence rules. A different distinction between strict liability and negligence rules is suggested by Equation (1b) below; namely, that under strict liability the default bearer of liability, i.e., the party who bears the loss when each party meets her standard of care, is the injurer rather than the victim.

Strict liability with dual contributory negligence has a different pattern of liability defined by:

$$R(x, y; X, Y) = \begin{cases} 1 & \text{if } x < X \text{ or } y \geqslant Y, \\ 0 & \text{otherwise.}^9 \end{cases} \tag{1b}$$

Under a rule of *comparative negligence*, R may assume values strictly between 0 and 1 when each party fails to meet its standard of care.

Any given rule $R(x, y; X, Y)$ defines a two-person non-zero-sum game between I and V. Each agent seeks a level of care that minimizes her expected personal costs. Thus, I chooses x to minimize:

$$x + p(x, y)L(x, y)R(x, y; X, Y) \tag{2}$$

while V chooses y in order to minimize:

$$y + p(x, y)L(x, y)\big[1 - R(x, y; X, Y)\big]. \tag{3}$$

In principle, an economic analysis studies the equilibria of these various games. Once these equilibria have been identified, the policymaker can choose the rule that induces the best equilibrium, given the social objective function. The analysis is thus not inextricably linked to a search for an efficient or welfare maximizing rule. One can as easily search for the rule that minimizes the accident rate or equalizes care expenditures or for any other social objective that the policymaker seeks to implement. One way to proceed would be to analyze the case law to identify the Court's objective function and then choose the legal rule that induces the equilibrium that is best in these terms.

In practice, however, and perhaps unfortunately, the literature has to a large extent chosen as objective function the minimization of the expected social costs of accidents.[10] That is, the minimization of:

$$x + y + p(x, y)L(x, y). \tag{4}$$

Let (x^*, y^*) be the (assumed) unique minimizing care levels. Consider the common law regime of negligence with contributory negligence as characterized by (1a). Clearly choosing standards $X = x^*$ and $Y = y^*$ induces the agents to take the optimal levels

---

[9] In fact, if we rename the parties, the pattern of liability under this rule is identical to the pattern under negligence with contributory negligence. Put differently, in Brown's model, the loss is perfectly transferable between parties and the legal rule simply assigns the loss to one party or another. This is not an adequate model of *personal injury*, since a personal injury may alter V's utility *function*, and some injuries may not be compensable. See, for example, Arlen (1992) for a modification that addresses the problem of personal injury.

[10] Diamond (1974b), Kornhauser and Schotter (1992), and Endres and Querner (1995) study how the equilibria change as the standards of care change and are thus examples of exceptions to this general practice.

of care. However, these are not the only such standards. In fact, an infinite number of rules of the form $\{X = x^*, Y \in [0, y^*]\}$ and of the form $\{X \in [M, \infty]$ for sufficiently large M, $Y = y^*\}$ induce an equilibrium in which the agents adopt care levels that are socially optimal.

Consider first a rule with $X = x^*$ and $Y \leqslant y^*$. When I chooses x^*, V is responsible for the total loss, i.e., $R = 0$. From (3) V chooses y to minimize:

$$y + p(x^*, y)L(x^*, y).$$

But this is just (4) shifted down by x^* (given that I chooses x^*) so that (3) is also minimized at y^*. Likewise, given V's choice of y^*, I minimizes (2) by choosing x^*.[11] A similar argument shows that (x^*, y^*) is the equilibrium for those games in which $Y = y^*$ and X is sufficiently large.

Thus, efficiency can be obtained even if the court can only observe the level of care being taken by one of the parties. In particular, if the court can only observe x, then an appropriate pure negligence rule is efficient – the court sets $X = x^*$, $Y = 0$ and lets V, whose action it cannot observe, optimize against I's choice. Similarly, when the court only observes y, the strict liability with contributory negligence rule ($X = \infty$, $Y = y^*$) is efficient.

Notice that these payoff functions, which accurately capture the structure of the negligence rule, are discontinuous at the standard of care.[12] This discontinuity plays an important role in generating the efficient outcome, because it allows both agents to "see" the full cost of their decisions. For example with the rule $X = x^*$ and $Y = 0$, I bears the full cost of the accident in the event he chooses $x < x^*$, and none of the cost at $x = x^*$. Therefore, at a choice of $x = x^*$, I's marginal cost of lowering x is the entire accident cost. At the same time, when I chooses $x = x^*$, V bears the entire loss and sees the full cost of her actions.

2.2. Extensions within accident law

Several modifications of this basic framework have been studied. Shavell (1987) analyzes a model in which each party chooses both a level of care and a level of activity, but the allocation of the loss still depends only on the parties' choice of care levels. Unsurprisingly, no legal rule can induce both parties to choose efficient activity levels and efficient care levels; after all, the court has only one instrument (for each party) to

---

[11] (x^*, y^*) is the only pure strategy equilibrium for games with the standards of care as indicated. Suppose I chooses $x' < x^*$. If V chooses $y' < Y$, then V's cost is $y' + p(x', y')L(x', y') > x^* - x' + y^* + p(x^*, y^*)L(x^*, y^*) > y^* > Y$. Therefore V would choose Y and I's best response to Y is $X = x^* \neq x'$. If I chooses $x > x^*$, she makes unnecessary expenditures on care.

[12] Kahan (1989), however, argues that these payoff functions do not correctly model the measure of damages. He contends that, in an appropriate model, the payoffs are continuous functions. These functions still show the agent the full marginal cost of her decisions.

control the two decisions that each party makes. One can still identify the rule that is best given the social objective function. Shavell (1987) also investigates the impact of an insurance market on care decisions. (See also Winter (1991) for a discussion of the dynamics of the market for liability insurance.)

A number of other variants of this model have also been analyzed. Green (1976) and Emons and Sobel (1991), for example, study the equilibria of the basic model of negligence with contributory negligence when injurers differ in the cost of care. Beard (1990) and Kornhauser and Revesz (1990) consider the effects of the potential insolvency of the parties on care choice. Shavell (1983) and Wittman (1981) analyze games with a sequential structure. Leong (1989) and Arlen (1990) study negligence with contributory negligence rules when both the victim and the injurer suffer damage. Arlen (1992) studies these same rules when the agents suffer personal (or non-pecuniary) injury.

Several authors have integrated the analysis of the incentives to take care into a fuller description of the litigation process that enforces liability rules. Ordover (1978) shows that when litigation costs are positive some injurers will be negligent in equilibrium. Craswell and Calfee (1986) studies how uncertainty concerning legal standards affects the equilibria. Polinsky and Shavell (1989) studies the effects of errors in fact-finding on the probability that a victim will bring suit and hence on the incentives to comply with the law. Hylton (1990, 1995) introduces costly litigation and legal error. Png (1987) and Polinsky and Rubinfeld (1988) study the effect of the rate of settlement on compliance. Png (1987) and Hylton (1993) consider how the rule allocating the costs of litigation between the parties affects the incentives to take care. Cooter, Kornhauser, and Lane (1979) assumes that the court only observes "local" information about the private costs of care and the costs of accidents. In their model the courts adjust the standards of care over time and converge to the optimum. Polinsky (1987b) compares strict liability to negligence when the injurer is uncertain about the size of the loss the victim will suffer.

Many authors have compared the efficiency of negligence with contributory negligence rules to comparative negligence rules. Under a rule of comparative negligence injurer and victim share the loss when both fail to meet the standard of care. That is, the injurer pays the share

$$R(x, y; X, Y) = \begin{cases} 1 & \text{for } x < X \text{ and } y \geqslant Y, \\ f(x, y; X, Y) & \text{for } x < X \text{ and } y < Y, \\ 0 & \text{otherwise } (x \geqslant 0), \end{cases}$$

where $0 \leqslant f(x, y; X, Y) \leqslant 1$. Landes and Posner (1980) argues that both rules can induce efficient caretaking; Haddock and Curran (1985) and Cooter and Ulen (1986) argue for the superiority of comparative negligence when actual care levels are observed with error; Edlin (1994) argues that with the appropriate choice of standards of care the two rules perform equally well even when such error is present. Rubinfeld (1987) considers the performance of comparative negligence when the court cannot observe the private costs of care.

2.3. Other interpretations of the model

This model can be recast to yield insights into other areas of law. Indeed, some [Cooter (1985)] have suggested that the pervasiveness of this model shows the inherent unity of the common law. While this claim seems too strong – common law rules govern a wider set of interactions than those described here – the model does illuminate a wide variety of legal institutions.

To begin, we reinterpret the problem in terms of remedies for breach of contract. Consider, for example, two parties, a buyer B and a seller S, contemplating an exchange.[13] B requires a part produced by S. The value to B of the part depends upon the level of investment y that B makes in advance of delivery. Denote this value by $v(y)$. y is known as the "reliance" level, as it indicates the extent to which B is relying upon receiving the part. S agrees to deliver the part and chooses an investment level x. This results in a probability $[1 - p(x)]$ that S will successfully deliver the part and a probability $p(x)$ that he will not perform the contract. B pays S an amount k up front for the part. A legal rule $R(x, y)$ determines the amount $R(x, y)v(y)$ that S must pay B if S fails to deliver the part. S then chooses x to maximize:

$$k - x - p(x)v(y)R(x, y), \tag{5}$$

while B simultaneously chooses y to maximize:

$$-k - y + \big[1 - p(x)\big]v(y) + p(x)R(x, y)v(y). \tag{6}$$

This model is perfectly congruent to Brown's basic model of Section 2.1. To see this, first subtract a term $g(y)$ from the victim's objective function (3). We interpret $g(y)$ as V's (added) payoff from her investment of y, should no accident occur.

In Brown's basic model $g(y)$ is zero. This corresponds to a potential victim who, by taking care, can reduce the value of her assets that are at risk, say by putting some of them in a fireproof safe, but whose actions have no effect on the overall value of the assets. On the other hand, consider a potential victim whose investment determines the value of some assets should no accident occur, but which assets will be completely lost in the event of an accident. Then $g(y) = L(y) (= L(x, y))$ in (3). If this investment affects only the value of the assets, and not the probability that they are lost, then $p(x, y) = p(x)$. Now the accident model of Equations (2) and (3) is identical to the breach of contract model (5) and (6) (with $v(y) \equiv L(y)$ and $k = 0$).

---

[13] The model in the text is similar to ones presented in Shavell (1980), Kornhauser (1983) and Rogerson (1984).

Thus, the accident interpretation and the contract interpretation differ largely in the characterization of the net benefits of the victim's action. Despite this congruence, courts, and analysts, have treated these situations differently. Rather than rules setting standards of care X and Y to determine liability for breach of contract, courts have commonly used a "reliance measure", $R(x, y) = R(y) = y/v(y)$ (so that $R(y)v(y) = y$),[14] and an expectation measure, $R(x, y) = 1$. These measures of contract damages are rules of "strict liability" as the amount of the seller's liability is not contingent on its own decisions; they are also "actual damage" measures as they depend on the actual, as opposed to any hypothetical, choice of the Buyer.

It is easy to see that neither of these two damage measures will generally induce an efficient level of care and reliance. The total expected social welfare of the contract is

$$[1 - p(x)]v(y) - x - y.$$

Let (x^*, y^*) maximize this social welfare function.[15] Thus,

$$p'(x^*) = -1/v(y^*), \text{ or } p(x^*) = 0 \text{ and } p'(x^*) < 1/v(y^*),$$
$$v'(y^*) = 1/(1 - p(x^*)).$$

When $R(y) = 1$, B's maximizing choice, y_e, is independent of S's action and is implicitly defined by

$$v'(y_e) = 1.$$

Given this, S's equilibrium action is defined by:

$$p'(x_e) = -1/v(y_e).$$

Note that $(x_e, y_e) = (x^*, y^*)$ only when $p(x^*) = 0$. When $p(x^*) \neq 0$, this expectation measure leads to too much reliance on the part of B; S, however, adopts the optimal amount of care given B's reliance.

Under the reliance measure $R(y) = y/v(y)$, B again maximizes independently of S by setting:

$$v'(y_r) = 1.$$

---

[14] $R(y)v(y) = y$ is one interpretation of a reliance measure. Some legal commentators understand reliance as a measure that places the promisee in the same position he would have been in had the contract not been made. Reliance then includes the opportunity cost of entering the contract. On this interpretation, reliance damages would equal expectation damages in a competitive market.

[15] We assume sufficient conditions for a unique nonzero solution, and that all our functions are differentiable.

S's equilibrium action is now defined by:

$$p'(x_r) = -1/y_r.$$

Again y_r is efficient only if $p(x_r) = 0$. In this case, S adopts the efficient level of care too. Otherwise, B under-relies and S breaches too frequently given B's reliance.

We may also use this example to suggest how non-efficiency concerns might enter into the legal analysis. Consider a technology in which, by taking a reasonable amount of care, S can get p close to 0; further reductions in p are very costly without much gain. Indeed, there may be a residual uncertainty which S cannot avoid. Then the solutions to the three programs above may well all result in a very small probability of nonperformance. The solutions will all be close to each other, so that fairness considerations may easily outweigh efficiency considerations. Under the reliance measure, for instance, when S fails to deliver despite his "best efforts" he must only repay B for B's out-of-pocket (or reliance) losses, rather than the loss of B's potential gain, which may be quite high.[16]

Brown's model of accident law is easily converted to a model of the taking of property. The Fifth Amendment to the Constitution of the United States requires that "property not be taken for a public use without compensation". How does the level of required compensation affect the extent to which the landowner develops her land and the nature of the public takings?

We follow the model in Hermalin (1995). Let the landowner choose an amount y to expend on land development, resulting in a property value $v(y)$. $v(y)$ is an increasing function of her investment. The value of the land to the government is a random value whose realization is unknown at the time the landowner chooses his investment level. Let $q(a)$ denote the probability that the value of the land in public hands will be greater than a, and let $p(y) = q(v(y))$. Assume that the state appropriates the land if and only if it is worth more in public hands than in private hands. As before, we

---

[16] One might convert this fairness argument into an efficiency argument by extending the analysis to consider why B does not choose to integrate vertically with S.

A number of other aspects of contract law have also been modeled. Hadfield (1994) considers complications created by the inability of courts to observe actions. Hermalin and Katz (1993) examines why contracts do not cover all contingencies. Katz (1990a, 1990b, 1993) has studied doctrines surrounding offer and acceptance, which determine when a contract has been formed. Various authors – e.g., Posner and Rosenfield (1977), White (1988), and Sykes (1992) – have considered the related doctrines of impossibility and impracticability which are, in a sense, the converse of breach as they state conditions under which contractual obligation is discharged. Rasmusen and Ayres (1993) examine the doctrines governing unilateral and bilateral mistake. Polinsky (1983, 1987a) addresses some more general questions of the risk-bearing features of contract that are raised by the mistake and impossibility doctrines.

The rule of Hadley v. Baxendale has also received much attention as in Perloff (1981), Ayres and Gertner (1989, 1992), Bebchuk and Shavell (1991), and Johnston (1990). The latter four articles are particularly concerned with the role of the legal rule in inducing parties to reveal information.

consider compensation rules of the form $R(y)v(y)$. Then the landowner chooses y to maximize:

$$-y + \big(1 - p(y)\big)v(y) + p(y)R(y)v(y). \tag{7}$$

Note that (7) is identical to (6), when $k = 0$ and p is written more generally as $p(x, y)$. Suppose that the state is benevolent so that its objective function is net social value:

$$-y + \big[1 - p(y)\big]v(y) + \mathrm{E}\big[a|a > v(y)\big]. \tag{8}$$

The government chooses $R(y)$ to maximize (8), given the landowner's program (7).[17] As a simple example, consider a rule $R(y) = \alpha$ and suppose that the land is either worth more to the state than $\max v(y)$ or worth less than $v(0)$ (perhaps a new rail line must be situated). Thus $p'(y) \equiv 0$, $\mathrm{E}'(v(y)) = 0$ and efficiency demands that:

$$v'(y) = 1/\big[1 - p(y)\big].$$

The landowner maximizes by setting:

$$v'(y) = 1/\big[1 - p(y)(1 - \alpha)\big].$$

Efficiency calls for no compensation, i.e., $\alpha = 0$. However, if $p(y)$ is small, then setting $\alpha = 1$ may be fairer while causing only a small loss in total welfare.

It is striking that radically different legal institutions are captured by models that are structurally so similar. Why do legal institutions vary so much across what seem to be structurally similar strategic situations? The literature does not offer a clear answer; indeed, it has not explicitly addressed the issue. Three suggestions follow.

First, the nature of the externality differs slightly across legal institutions. In the accident case, the likelihood of an adverse outcome depends on the care choices of each party. In the contract case, the seller determines the likelihood of breach while the buyer determines the size of the loss. In the takings context, the landowner determines the market value of the land. Perhaps the shifting judicial approaches reflect important differences in the court's ability to monitor the differing transactions. Second, the differences in institutions may reflect different values that the courts seek to further in the differing contexts. Finally, the judges who developed these common law solutions may not have

---

[17] Analyses of takings as in Blume et al. (1984) and Fischel and Perry (1989) are more complex because they are set in a broader framework in which the government must also set taxes in order to finance the takings. In this context one is able to consider a non-benevolent government as well. This government acts in the interest of the majority; see Fischel and Perry (1989) and Hermalin (1995). Hermalin, whose analysis is explicitly normative, examines a set of non-standard compensation rules as well. We have ignored these complexities because we are interested in showing the doctrinal reach of the simple model.

perceived the common strategic features of these situations and thus may have generated a multiplicity of responses.[18]

3. Is law important? Understanding the Coase Theorem

The analysis of accident law in the prior section assumed that the parties could not negotiate or enforce agreements concerning the level of care that each should adopt. The interpretation of the model as one of accidents between strangers presents a situation in which such a bar to negotiation is plausible; the pedestrian contemplating a stroll rarely has the opportunity to negotiate with all drivers who might strike her during her ramble. Under other interpretations, however, the barriers to negotiation are less. In these contexts, one must confront a seductive, though ill-specified, argument for the irrelevance of certain legal rules.

This irrelevance argument is generally attributed to Ronald Coase (1960) who argued through two examples that the assignment of liability would not, in the absence of *transaction costs*, affect the economic efficiency of the behavior of agents because an "incorrect" assignment of liability could always be corrected through private, enforceable agreements. This insight has come to be known as the Coase Theorem. In the context of contract remedies, for example, the parties could negotiate a contingent claims contract specifying the level of reliance that the buyer should undertake. Coase also emphasized that the assignment of liability might indeed matter when transaction costs were present. Coase never actually stated a "theorem" and subsequent scholars have had difficulty formulating a precise statement. Precision is difficult because Coase argued from example and did not clearly define several key concepts in the argument. In particular, "transaction costs" are never defined and the legal background in which the parties act is underspecified.

Game theorists have generally treated the theorem as a claim about bargaining and, as we shall see, have used both the tools of cooperative and non-cooperative game theory to illuminate our understanding of Coase's examples. The Coase Theorem, however, is also a claim about the importance of legal rules to bargaining outcomes. At the outset, we must thus understand precisely which legal rules are in question so that they may be modeled appropriately.

The Coase Theorem implicates two sets of legal rules: the rules of contract and the assignment of entitlements. The rules of contract determine which agreements are legally enforceable and what the consequences of non-performance of a contract are. For the Coase Theorem to hold, the courts should be willing to enforce every contract and the consequences of non-performance should be sufficiently grave to deter non-performance. Absent this possibility, repetition and reputation effects might ensure compliance with contracts in certain situations.

---

[18] We thank a referee for suggesting this third possibility.

The entitlement defines the initial package of legal relations over which the parties negotiate.[19] This package identifies the set of actions which the entitlement holder is free to undertake and the consequences to others for interfering with the entitlement holder's enjoyment of her entitlement. The law and economics literature, drawing on Calabresi and Melamed (1972), has generally focused on the two forms of entitlement protection called property rules and liability rules. Under a property rule, all transfers must be consensual; consequently the transfer price is set by the entitlement holder herself. Under a liability rule, an agent may force the transfer of the entitlement; if the parties fail to agree on a price, a court sets the price at which the entitlement transfers.

Consider, for example, two adjoining landowners A and B. B wishes to erect a building which will obstruct A's coastal view. Suppose A has an entitlement to her view. Property rule protection of A's entitlement means that A may enjoin B's interfering use of his land and thus subject B to severe fines or imprisonment for contempt should B erect the building. Put differently, A may set any price, including infinity, for the sale of his view, without regard to the "reasonableness" of this price. Under a liability rule, A's recourse against B is limited to an action for damages. Put differently, should B erect a building against A's wishes, the courts will determine a reasonable "sale" price for the loss of A's view.

The Coase Theorem is commonly understood as a claim that, in a world of "zero transaction costs", the nature and assignment of the entitlement does not affect the final allocation of real resources; in this sense, then, legal rules are irrelevant. One might ask two very different questions about the asserted irrelevance of legal rules. First, one might assume a perfect law of contract and specify more precisely what "zero transaction costs" means in order to assure the irrelevance of the assignment of entitlements. Alternatively, one might ask how "imperfect" contract law could be without disturbing the irrelevance of the assignment of the entitlements. Following most of the literature, we investigate the first question.

Much controversy revolves around the concept of transaction costs. At the most basic level, transaction costs involve such things as the time and money needed to bring the various parties together and the legal fees incurred in signing a contract. One may

---

[19] An entitlement is actually a complex bundle of various types of legal relations. A classification of legal relations offered by Hohfeld (1913) is useful. He identified eight fundamental legal relations which can be presented as four pairs of jural opposites:

right	privilege	power	immunity
duty	no-right	disability	liability

When an individual has a right to do X, she may invoke state power to prevent interference with her doing X. By contrast, if she has the privilege to do X, no one else may invoke the state to interfere (though they may be able to "interfere" in other ways). An individual may hold a privilege without holding a right and conversely. When an individual has a power, she has the authority to alter the rights or privileges of another; if the individual holds an immunity (with respect to a particular right or privilege), no one has a power to alter it. Kennedy and Michelman (1980) apply this classification to illuminate different regimes of property and contract.

go further, however, and, somewhat tautologically, consider a transaction cost to constitute anything which prevents the consummation of efficient trades. Game theory has clarified the different impediments to trade by focusing on two types of barriers: those that arise from strategic interaction per se and those that arise from incomplete information.

We might then understand the Coase Theorem as one of two claims about bargaining.[20] First, we might understand the Coase Theorem as a claim that bargains are efficient regardless of the assignment and nature of the entitlement. Call this the Coase Strong Efficient Bargaining Claim. Second, we might understand the Coase Theorem as a claim that, although bargaining is not always efficient, the degree of efficiency is unaffected by the nature and assignment of the entitlement. Call this the Coase Weak Efficient Bargaining Claim.

The game-theoretic investigations do provide some support for Coase's valuable insight.[21] However, the bulk of the analysis suggests that both Coase Claims are, in general, false. One set of arguments focuses on the strategic structure of games of complete information. A second set considers the problems created by bargaining under incomplete information. A third set circumvents the bargaining problem using cooperative game theory. The next three subsections discuss each set of arguments in turn.

First, it is worthwhile emphasizing one point. Discussions of the Coase Theorem often characterize it as a claim that, under appropriate conditions, the legal regime is economically irrelevant. This loose characterization obscures an important fact. At most, the Coase Theorem asserts that *efficiency* may be unaffected by the assignment of the entitlements. There are, of course, other economic considerations. In particular, Coase never asserted that the final *distribution* of assets would be unaffected by the legal regime. On the contrary, every "proof" of the Coase Theorem relies on the parties negotiating transfers, the direction of which depends on the assignment of the entitlement. Thus, even in circumstances in which the Coase Theorem holds, the choice of a legal rule will certainly matter to the parties affected by its announcement and it may also matter to a policymaker.

3.1. The Coase Theorem and complete information

Reconsider the injurer/victim model of Section 2.1. Suppose that I engages in an activity that may cause a fixed amount of damage, $L(x, y) \equiv d$, to agent V. Agent I can avoid causing this damage at a cost of c. That is, $p(x, y) = 0$ if $x \geqslant c$, $p(x, y) = 1$ if $x < c$. Both c and d lie in the interval $(0, 1)$. We assume that the benefit to I of the

---

[20] The literature on the Coase Theorem is vast and offers a variety of characterizations of the claim. For overviews of the literature that offer accounts consistent with that in the text see Cooter (1987) and Veljanovski (1982).

[21] Hoffman and Spitzer (1982, 1985, 1986) and Harrison and McAfee (1986) provide experimental evidence in support of the Coase Claims.

activity is greater than 1, so that it is always socially desirable for I to undertake the potentially harmful activity.

We can view V as being entitled to be free from damage. As previously noted, an entitlement can be protected with a property rule or a liability rule. Consider a compensation rule that specifies a lump sum M that I must pay V for damage. We interpret $M \geqslant 1$ as a property rule, since I will only fail to take care if V consents.[22] V's entitlement is protected by a liability rule when $0 < M < 1$. Clearly, the larger M the greater the protection afforded V. When $M = 0$, V has no protection; this can be interpreted as giving I an entitlement to cause damage that is protected by a property rule.

A surface analysis suggests that if $M = 0$, I will not take care since the damage I causes is external to I, whereas if $M = 1$, I will take care since $c < 1$. In particular, when $M = 0$ and $c < d$, I will cause harm to V, and when $M = 1$ and $c > d$ I will take care, even though both these outcomes are inefficient.

The Coase Theorem holds that this conclusion is unduly negative. For instance, when $M = 0$ and $c < d$, V could induce I to take care by offering I a side payment of, say, $c + (d - c)/2$. This is possible provided that "transaction costs" are not too high. If making the side payment involved a separate cost greater than $(d - c)$, no mutually beneficial payment could be made.

When transaction costs are negligible then, for any values of c and d, and for any value of M, we might expect the firms to negotiate a mutually beneficial transfer which yields the efficient level of care. Note that even if this expectation is correct, it does *not* imply that the legal regime M is irrelevant, as M affects the direction and level of the transfer.

While the insight that firms will have an incentive to "bargain around" any legal rule is important, the blunt assertion that they will successfully do so is hasty. Exactly how are the firms to settle on an appropriate transfer? What if, with $c < d$, V offers to pay c to I, while I insists upon receiving d? Negotiations could break down, resulting in the inefficient outcome of no care being taken. In any case, Coase did not specify the bargaining game which would permit us to rule out this possibility.

Suppose that firms always seek to reach mutually beneficial arrangements and that they can costlessly implement any contract upon which they agree. Consider the polar cases of $M = 0$, and $M = 1$. A simple dominance argument shows that in the former case I will never take care when $c > d$, since it is dominated for V to pay I more than d to take care. Thus, there will never be more than an efficient level of care taken. Similarly, when $M(x, d) = 1$ there will never be less than an efficient level of care. Without additional assumptions no more than this can be guaranteed. In particular, the legal regime may affect the level of care taken and the Coase Theorem fails in both its

---

[22] This characterization of a property rule is not wholly satisfactory because the court might enforce a property rule by injunction (and by contempt proceedings should the injunction be violated). Thus this characterization might differ from the operation of such a legal rule should I act irrationally.

forms. To establish a Coase Theorem one requires additional structure. Notice that the material facts do not in themselves determine this additional structure. To evaluate the Coase Theorem we must posit a bargaining game and examine how its solution changes with legal regimes.

Consider any legal rule M and a simple game in which V makes a take-it-or-leave-it offer to I. Then, in the unique subgame perfect equilibrium, when $c < d$ V offers I a side-payment of $\max[c - M, 0]$ to take care and I takes care. Hence I takes care regardless of the legal rule. When $c > d$, V offers to accept a side-payment of $\min[c, M]$ from I if I does not take care, and under every legal rule I does not take care. In both cases, the efficient outcome is reached regardless of the legal rule $M(c, d)$.

This is a particularly simple model of bargaining. Rubinstein (1982) offers a more sophisticated model in which two parties alternatingly propose divisions of a surplus until one of the parties accepts the other's offer. In our present situation, if $c < d$ there is a surplus when care is taken and if $c > d$ there is a surplus when care is not taken, regardless of the rule $M(c, d)$. Suppose that when $c < d$, in period 1 V offers I a payment to take care. If I rejects the payment, in period 2 I makes a counterdemand of a payment. If V rejects I's proposal he makes a counteroffer in period 3 and so forth. When $c > d$ the game is the same except that the payments are for not taking care. Each party attaches a subjective time cost to delay and the game ends when an offer is accepted or when I decides to act unilaterally. Rubinstein's results imply that for all $M(c, d)$ the parties instantly agree upon an efficient transfer.

Thus, we have some support for the Coase Strong Efficient Bargaining Claim. However, the result that bargaining will be efficient is delicate. Shaked has shown that with three or more agents bargaining in Rubinstein's model may be inefficient.[23] Avery and Zemsky (1994) present a unifying discussion of a variety of bargaining models that result in inefficient bargaining.[24] Perhaps most importantly, we have been assuming complete information. We consider incomplete information in the next section.

3.2. The Coase Theorem and incomplete information

Generally, we might expect a firm to possess more information about its costs than outside agents possess. We now modify the model of the previous section in this direction. Specifically, following Kaplow and Shavell (1996), suppose that while I knows the cost c of taking care, V knows only that c is distributed with continuous positive density $f(c)$ on the unit interval. Similarly, while V knows the level of damage d which I would cause, I knows only that d is distributed with density $g(d)$ on the unit interval.

---

[23] Shaked's model is discussed in Osborne and Rubinstein (1990).

[24] These models, which include Fernandez and Glazer (1991) and Haller and Holden (1990), all have complete information.

We focus on the two rules $M = 0$ and $M = 1$. Call these rules M_0 and M_1, respectively. Notice that to implement M_0 and M_1 the court only needs to know whether or not damage has occurred.

Efficiency requires that I take care whenever $c < d$. Let $S_i = \{(c,d)|I$ takes care under rule $M_i\}$. The strong efficient bargaining claim asserts that $S_0 = S_1 = \{(c,d)|c \leqslant d\}$.[25] Call this set E. The weak efficient bargaining claim could be taken to mean one of two things; that $S_0 = S_1 \neq E$ or that, although $S_0 \neq S_1$, M_0 and M_1 are equally inefficient (given a measure of the degree of inefficiency of rules).

Again consider the simple bargaining procedure in which V makes a take-it-or-leave-it offer to I. Consider first the situation when M_0 prevails. Then V may seek to bribe I to take care. We calculate how V's offer to buy I's right to cause harm varies with V's damage cost d. Suppose V of type d makes an offer of $o(d)$ to I. I accepts if $c \leqslant o(d)$. Thus, when V makes an offer of $o(d)$, he incurs expected costs of $F(o(d))o(d) + (1 - F(o(d))d$. These costs are minimized by that $o(d)$ which satisfies

$$o(d) = d - F\big(o(d)\big)/f\big(o(d)\big).$$

Call this value $o_0(d)$. Thus:

$$S_0 = \big\{(c,d) \mid c \leqslant o_0(d)\big\}.$$

Since $o_0(d) < d$ efficiency fails in the presence of incomplete information, the strong efficiency claim is false.[26] To check the Weak Efficiency Claim we examine the effect of a change in the legal regime. Consider the situation when M_1 prevails. V may now offer to sell I the right to cause the harm. Suppose that V offers to forego his damage payment of 1 at a price $o(d)$. I accepts if $c > o(d)$ as it would be cheaper for I to buy the right than to avoid the harm. Thus, when V makes an offer of $o(d)$, he expects profits of $(1 - F(o(d)))(o(d) - d)$. These profits are maximized by that $o(d)$ which satisfies

$$o(d) = d + \big[1 - F\big(o(d)\big)\big]/f\big(o(d)\big).$$

Call this value $o_1(d)$. Thus:

$$S_1 = \big\{(c,d) \mid c \leqslant o_1(d)\big\}.$$

Since $o_1(d) > d$, again efficiency fails. There is an important difference in regimes M_0 and M_1, however. Under M_0, there are instances when efficiency demands

---

[25] Of course, if $c = d$ efficiency permits that care either be taken or not taken. Throughout, we will be a little loose as to what happens at degenerate "equality" points.

[26] Indeed, Myerson and Satterthwaite (1983) show that, quite generally, bargaining under incomplete information will be inefficient.

that I take care but he does not. In no case does I take care when he should not. In contrast, under M_1, there may be too much care taken, but never too little.[27]

From this comparison, we may conclude that the Weak Efficiency Claim is false in both its forms. For instance, if F places more weight in the interval $[o_0(d), d]$ than in the interval $[d, o_1(d)]$ then M_0 results in an inefficient level of care more often than M_1 does.[28]

3.3. A cooperative game-theoretic approach to the Coase Theorem

In the example of the previous section the firms were in a situation of asymmetric information. One might interpret the difficulties imposed by this asymmetric information as an *informational* transaction cost, thereby rescuing the Coase Theorem. Similarly, one might interpret the inefficiencies that arise with three or more agents in a Rubinstein-type bargaining model as resulting from *strategic* transaction costs. On this view, cooperative game theory would seem to provide the best hope for the Coase Strong Efficiency Claim. After all, the characteristic function form of a game simply posits that firms can reach any agreement they choose, effectively assuming away any "bargaining costs".

Consider the classic externality problem in which air pollution generated by two factories interferes with the operation of a nearby laundry. There are two possible assignments of the entitlement: (1) the factories pay for damage caused to the laundry and (2) the factories do not pay for such damage. Each of these assignments yields a game in characteristic function form that derives from the underlying technological interaction between factories and laundry. In this context, one might formulate the Coase Theorem as a claim either that the (relevant) solution to each of the games is the same or that each solution contains only equally efficient allocations. Although there is no general theory that studies (or even formulates) this problem, it is easy to provide examples that contradict each claim. We present here a modification of the examples underlying Aivazian and Callen (1981) and Aivazian, Callen, and Lipnowski (1987).

There are three firms. Firms 1 and 2 are industrial factories, and firm 3 is a laundry. Each firm can operate at one level or shut down. Firm 3 operating on its own earns a profit of 9. If either firm 1 or 2 operates, however, a pollution cost of 2 is imposed on firm 3. If both firms 1 and 2 operate, independently or jointly, a cost of 8 is imposed on firm 3. Firms 1 and 2 operating independently each earn a profit of 1, regardless of the actions of other firms. Firms 1 and 2 operating together earn joint profits of 7.

---

[27] This is in keeping with a claim from the previous section.

[28] Several authors have considered the differential effects of legal rules on the revelation of private information. Ayres and Gertner (1989) argues that this consideration should inform judicial choice of contract default rules. The response to their claim [Johnston (1990)], their reply [Ayres and Gertner (1992)], and a related analysis [Bebchuk and Shavell (1991)] provide insight into the force of Coase Theorem arguments in contract law. See also Ayres and Talley (1995) and Kaplow and Shavell (1996). Johnston's (1995) study of the effects of blurriness in the entitlement in situations of incomplete information provides related insights.

We must define two characteristic function form games: one in which the laundry legally bears the pollution cost, and one in which the factories legally bear the cost. In general, defining a characteristic function game from underlying data is not altogether obvious. For instance, in defining the value of a firm operating on its own, should it be assumed that the other firms operate independently or jointly? Should it be assumed that the other firms maximize their profits or that they minimize the profits of the firm under consideration? Our example has been chosen to circumvent these difficulties. Here, the same game is defined regardless of how these questions are answered.

We first derive the characteristic function of the game in which the laundry bears the pollution costs. We have $v(3) = 1$; this value of 1 is obtained by taking the 9 that firm 1 earns by operating and subtracting the cost of 8 which is imposed upon it by firms 1 and 2 operating separately or jointly. $v(23) = 7$ since the best that these firms can do jointly is have firm 2 shut down and firm 3 operate. Their profits are then 9 minus the cost of 2 imposed by firm 1. $v(123) = 9$; firm 3 alone operates, since having one of the other firms also operate would yield the grand coalition a total of 8 and having all three firms operate would yield it 3. Similar reasoning yields the following characteristic function:

$$v(1) = v(2) = v(3) = 1; \quad v(12) = v(13) = v(23) = 7; \quad v(123) = 9.$$

Consider now a liability regime in which the court awards damages to the laundry based upon the maximal possible return to any coalition containing the laundry. Under such a regime we have:

$$w(1) = w(2) = w(12) = 0; \quad w(3) = w(13) = w(23) = w(123) = 9.$$

The Coase Theorem relates the solutions of the two games v and w. Possibly the most important solution concept for cooperative games is the core. The core of the liability game w consists of the unique point $(0, 0, 9)$. On the other hand, the no-liability game v has an empty core.

Informal Coase-type reasoning would hold that in the absence of a liability rule, firm 3 will simply bribe firms 1 and 2 not to produce, say by offering each of them 3.5. However, such an arrangement is unstable. Firm 1, for instance, could suggest a separate agreement just between itself and firm 3. Since the core is empty, there is in fact no arrangement among the three firms which cannot be blocked by a subset of them. Although there are no (obvious) transaction costs, it is quite possible that the firms will fail to settle on an efficient outcome. On the other hand, in the liability regime of game w, the efficient outcome in which only firm 3 produces is the obvious stable outcome.

Given that the core of v is empty we might want to consider other solution concepts. The Shapley value and nucleolus are both the point $(0, 0, 9)$ for w and the point $(3, 3, 3)$ for v. These solution concepts always exist and are always efficient for superadditive games so that this Coasean result is not terribly revealing about the validity of the Strong Efficiency Claim.[29]

Now consider the bargaining set.[30] For the game w the bargaining set still consists of the unique point $(0, 0, 9)$. For v, however, while the efficient outcome $(3, 3, 3)$ with coalitional structure $\{(123)\}$ is an element of the bargaining set, the inefficient point $(1, 3.5, 3.5)$ with coalitional structure $\{1, (23)\}$ is also an element of the bargaining set.[31] Note that this inefficient outcome in which firms 2 and 3 form a coalition is better for firms 2 and 3 than the efficient element of the bargaining set (and better than the Shapley value and nucleolus).

The complete information, no transaction cost world of cooperative game theory might seem to be the perfect setting for the Coase Theorem. However, even here the theorem is problematic; the choice of entitlement might very well matter to a legal policymaker concerned solely with efficiency. Clearly, a policymaker with other concerns or faced with a more recalcitrant environment or a policymaker designing the contract regime rather than assigning the entitlement will also need to evaluate the solutions of games defined by different legal rules.

4. Game-theoretic analyses of legal rulemaking

Legal controversies arise over broad questions concerning the structure of adjudication and the consequences of specific legal doctrines. Game theory has played an increasingly important role in illuminating these structural questions. Schwartz (1992), Cameron (1993), Kornhauser (1992, 1996) and Rasmusen (1994), for instance, have analyzed the practice of precedent both within a court and with respect to lower court obedience to higher court decisions. Others have addressed the claim that judicial review of legislative action is countermajoritarian (see, e.g., Ferejohn and Shipan (1990) and the effects of specific decisions of the United States Supreme Court [e.g., Cohen and Spitzer (1995)].

Eskridge and Ferejohn (1992) employs a now standard model to illuminate difficult questions of the interrelation of Congress, administrative agencies and the courts. In principle, under the U.S. Constitution policy is promulgated through Congress, with the policy views of the President accommodated to some extent because of the existence of the presidential veto power. The recent rise of lawmaking promulgated by administrative agencies under the control of the President has shifted additional power

---

[29] The difference in the solutions for the two games, however, may be of great significance to a legal policymaker whose social objective extends beyond efficiency to fairness or other concerns.

[30] See Aumann and Maschler (1964) for a definition of the bargaining set. We are using the definition of M^{**}.

[31] Other elements of the bargaining set are $(1, 1, 1)$ for coalitional structure $\{(1), (2), (3)\}$, $(3.5, 3.5, 1)$ for coalitional structure $\{(1, 2), (3)\}$, and $(3.5, 1, 3.5)$ for coalitional structure $\{(1, 3)\}$. See Aivasian, Callen, and Lipnowski (1987) for a derivation (in an equivalent game).

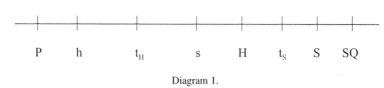

Diagram 1.

to the President. In response, Congress has attempted to redress the balance by assuming the power to veto administrative action. In *Immigration and Nationalization Service v. Chadha*, the Supreme Court ruled such legislative vetoes unconstitutional. Eskridge and Ferejohn analyze the consequences of these decisions in the following simple model.

There are four actors: The House of Representatives, the Senate, the President, and an executive Agency. They play a sequential game which results in some policy which can be represented as a point along a one-dimensional continuum. Diagram 1 presents a possible distribution of preferences. P is the ideal point of the President; H is the ideal point of the median voter in the House; h is the ideal point of the "2/3rds" house voter;[32] S is the ideal point of the median voter in the Senate; s is the ideal point of the 2/3rds Senate voter; SQ is the status quo. Preferences are spatial so that a person with ideal policy X_i prefers x to y if $|x - X_i| < |y - X_i|$.

Consider the *Legislation Game* which involves the House, the Senate, and the President. This game determines whether the United States will shift away from the prevailing status quo policy, SQ. In stage 1 of the Legislation Game, the House chooses some point x in the policy space. If $x = SQ$ the game ends and the policy SQ is maintained. If $x \neq SQ$ the game continues. In stage 2, the Senate approves or rejects the proposed point x. If the Senate rejects x, the game ends with SQ. If the Senate approves x, the game proceeds to stage 3. In stage 3, the President either signs the bill x or vetoes it. If the President signs the bill the game ends and the bill x becomes law. If the President vetoes the bill, the game proceeds to stage 4, in which the Senate and the House move simultaneously. If two thirds of each house approves x, the bill is passed and becomes law. If x lacks a two-thirds majority in either house, then SQ prevails.

For any specified distribution of ideal points, this game is easily solved. Consider the pattern displayed in Diagram 1. Since the status quo is to the right of all ideal policies, clearly some new policy will be enacted. t_S in Diagram 1 is a point such that $|t_S - S| = |S - SQ|$. The new policy will be no further to the left of S than t_S, since the Senate would reject any such proposal in favor of SQ. In fact, in the unique subgame perfect equilibrium for the configuration of Diagram 1, the House proposes t_S and this is approved by both the Senate and the President.

Next consider the Delegation Game in which Congress does not formulate policy, but instead delegates this task to an executive agency (such as the Environmental Protection

[32] The 2/3rds voter is the voter such that she and all voters to her right just constitute a 2/3 majority.

Agency). In the *Delegation Game*, the Agency chooses a policy y in stage 0 and then the above Legislation Game is played with y serving as the new status quo. Since the executive agency is staffed by the President we will assume that its ideal point is that of the President (i.e., P). With the configuration of Diagram 1, the unique subgame perfect equilibrium policy outcome is now h.

The delegation of power to the Agency dramatically shifts the equilibrium policy from t_S to h. How might Congress moderate Agency's exercise of its discretion? In the *Two-House Veto Game* and the *One-House Veto Game*, Congress intervenes directly. These two games model the effects of *Chadha*, where the Supreme Court ruled such supervision unconstitutional.

In the Two-House Veto Game, Congress delegates decision-making authority to an executive agency but reserves the right to overturn any policy proposal by a majority vote of each house. Specifically, in stage 1 of the Two-House Veto Game, the Agency chooses a policy y. In stage 2, the House votes on the policy. If a majority approves the policy it is adopted. Otherwise, the game proceeds to stage 3. If a majority of the Senate also approves y, it is adopted. Otherwise, the game ends with the original status quo SQ remaining in place.

Reconsider the ideal point configuration Diagram 1. The point t_H is such that $|t_H - H| = |H - SQ|$. Clearly, t_H is the equilibrium of the Two-House Veto Game as any point to the left of t_H will be vetoed by both Houses. The Two-House Veto Game thus moderates the action of the Agency by bringing the equilibrium from h to t_H, which is closer to the policy t_S that would be enacted directly by Congress (with consent of the President).

In the One-House Veto Game a specific house of Congress has the power to approve or disapprove of the Agency's policy choice. Clearly the outcome will be t_H if the House has the veto power and the policy t_S if the Senate has the veto power.

Thus, *Chadha*'s announcement of a constitutional prohibition on the power of Congress to veto agency decisions dramatically restricts Congress' ability to control power that it statutorily delegates to an executive agency. The previous analysis of the Legislation, Delegation and Veto games suggests how one might analyze the way that different institutional structures determine which legal policies prevail. Of course, a complete analysis would compare equilibria across all possible distributions of preferences of the relevant institutional actors.

5. Cooperative game theory as a tool for analysis of legal concepts

The prior discussion has examined the use of game theory to understand the behavioral consequences of legal rules. As we have seen, the structure of much legal doctrine can be illuminated by models of simple games. These models have had an important impact on the legal commentary of cases and legislation. Moreover, courts have occasionally cited such game-theoretic analyses in support of particular decisions.[33]

Game-theoretic tools have also played another role in legal analysis, a role that may have had a more direct impact on legal development. Game theory, particularly cooperative game theory, permits a conceptual analysis of ideas central to some areas of legal doctrine. This doctrinal approach may explicate the social objective underlying legal rules.[34]

In this section, we discuss two doctrinal analyses. We first examine the treatment of an index of voting power by the United States' judicial system. We then consider a case law problem of fair division.

5.1. Measures of voting power

In 1962, the Supreme Court of the United States began a voting rights revolution when it held in *Baker v. Carr* (1962) that the apportionment of state legislatures is subject to

[33] The Supreme Court of the United States has referred to game-theoretic analyses of settlement in at least three of its opinions. Two of these cases concern settlement under joint and several liability. *McDermott, Inc. v. Amclyde et al.* (1994) cites Easterbrook, Landes and Posner (1981), Polinsky and Shavell (1980) and Kornhauser and Revesz (1993); *Texas Industries v. Radcliff Materials* (1981) cites Cirace (1980), Easterbrook, Landes and Posner (1981) and Polinsky and Shavell (1980). The dissent in *Marek v. Chesny* (1985) cites Shavell (1982) and Katz (1987) in a discussion of fee-shifting.

The intermediate appellate courts in the federal system and the state courts of the United States also occasionally refer to game-theoretic analyses of settlement or of joint and several liability (or of both). Judge Easterbrook, himself a noted scholar in the field of economic analysis of law, has cited Shavell (1982) and Katz (1987) in *Packer Engineering v. Kratville* (1992), Easterbrook, Landes and Posner (1981), Polinsky and Shavell (1980) and Kornhauser and Revesz (1989) in *In re Oil Spill of Amoco Cadiz*, 954 F.2d 1279 (7th Cir. 1991) (per curiam), Kornhauser and Revesz (1989) in *Akzo Coatings v. Aigner Corp.* (1994), and Shavell (1982) in *Premier Electric Construction Company v. National Electrical Contractors Assn.* (1987). In addition, Rosenberg and Shavell (1985) were cited by the First Circuit in *Weinberger v. Great Northern Nekoosa Corp.* (1991) and Easterbrook, Landes and Posner (1981), Polinsky and Shavell (1980) and Kornhauser and Revesz (1993) were cited by the District of Columbia Court of Appeals in *Berg v. Footer* (1996). *In re Larsen* (1992) cites Bebchuk (1988) and Rosenberg and Shavell (1985).

Again, two economically sophisticated judges, Judge Calabresi and Judge Easterbrook, have cited Ayres and Gertner (1989) on the problems of information revelation in *American National Fire Insurance Co. v. Kenealy and Kenealy* (1995) and *Harnischfeger Corp. v. Harbor Insurance Co.* (1991), respectively. Similarly, both Judges Easterbrook (in *United States v. Herrera* (1995)) and Posner (in *In re Hoskins* (1996)) have cited Baird, Gertner and Picker (1994) on game theory generally.

It is worth noting that Coase (1960), the most cited article in the legal academic literature, is only cited 42 times by state and federal courts of the United States. Fifteen of these citations are from United States Court of Appeals for the Seventh Circuit. (LEXIS search done on 24 April 2000.)

[34] The distinction between the use of game theory to predict behavior and game theory to explicate doctrine is sometimes fuzzy. For example, one might rationalize the structure of product liability doctrine as a set of legal rules that induce efficient behavior on the part of manufacturers and consumers. This rationalization might use game theory both as a predictive tool (to show that some particular rule in fact induces efficient conduct) and as a conceptual tool (to illuminate the meaning of efficiency, e.g., to distinguish between ex ante, interim, and ex post efficiency).

judicial review under the 14th amendment of the U.S. Constitution. In the cases that immediately followed, the Court announced that the U.S. Constitution guarantees each citizen "fair and effective representation" (*Reynolds v. Sims* (1964)). It enunciated a maxim of "one person, one vote" to guide states in legislative reapportionment. In the context of legislative elections from single-member districts, this slogan had a clear meaning and an immediate impact: it required redistricting when the population discrepancy between the largest and smallest districts was too great.[35] As the Court widened the reach of jurisdictions and governmental bodies subject to judicial oversight it gradually confronted a wider and wider range of election practices which required it to elaborate a more complete theory of fair representation. In particular, the Court encountered systems that weight the votes of legislators to counterbalance population disparities as well as systems in which larger districts elect proportionately more legislators. Federal and state courts then confronted the question: When do these systems satisfy the requirements of fair and effective representation?

In 1954, Shapley and Shubik (1954) offered an interpretation of the Shapley value as an index of voting power. They interpreted the Shapley value as measuring the chance a voter would be critical to the passage of a bill. These ideas influenced John Banzhaf III who, as a law student, proposed a related index in Banzhaf (1965).[36] Banzhaf also interpreted his index as a measure of the chance a voter would be critical. Banzhaf (1965) used this index to analyze weighted voting schemes and Banzhaf (1968) applied the index to the problem of multi-member districts. In each article, Banzhaf acknowledged the influence of the Shapley value on his thinking.[37]

Banzhaf (1965) had a substantial impact on litigation; even while in draft, it was introduced into the deliberations of the state and federal courts considering a plan for the temporary reapportionment of the New York Legislature in which weighted voting would be used. (For a discussion of the plan see *Glinski v. Lomenzo* (1965) and *WMCA, Inc. v. Lomenzo* (1965).) After he published his articles, Banzhaf continued to intervene in reapportionment litigation; the courts thus frequently considered his view. Before we examine Banzhaf's influence on the case law, we define the two power indices mentioned and briefly discuss the relation between them. Straffin (1994) provides a comprehensive survey of the technical literature.

Consider a simple, proper voting game (N, v). ($v(S) = 0$ or 1 for all coalitions S; $v(S) = 1$ implies $v(N/S) = 0$.) Let θ_i be the number of coalitions S in which i is a

---

[35] Of course, as the Court was soon forced to learn, there are many ways to divide a population into roughly equal districts.

[36] The Banzhaf index is equivalent to measures independently proposed by Coleman (1973), Dahl (1957), and Rae (1969). On this equivalence, see Dubey and Shapley (1979).

[37] Footnote 32 to Banzhaf (1965) begins, "Although the definition presented in this article is based in part on the ideas of Shapley–Shubik, their definition is rejected because... [it] has not been shown that the order in which votes are cast is significant; ... Likewise it seems unreasonable to credit a legislator with different amounts of voting power depending on when or for what reasons he joins a particular voting coalition". Footnote 55 of Banzhaf (1968) thanks Shapley.

swing voter; that is, the number of coalitions S such that $v(S) = 1$ and $v(S - \{i\}) = 0$. The unnormalized Banzhaf power index β_i is given by:

$$\beta_i = \theta_i / 2^{n-1}$$

while the normalized Banzhaf index is:

$$\beta'_i = \theta_i / \sum_i \theta_i.$$

The Shapley–Shubik power index is defined as:

$$\sigma_i = \sum_{S \text{ is a swing for } i} (s - 1)!(n - s)!/n! \quad \text{where } s = |S|.$$

An individual's (unnormalized) Banzhaf index gives the probability that she will be a swing voter under the assumption that all coalitions are equally likely to form. An individual's Shapley–Shubik index gives the probability that she will be a swing voter under the assumption that it is equally likely that a coalition of any size will form.[38] Although these two indices often give similar measures of power, they need not. Indeed, they may differ radically both in the relative weights they assign and in the rank order they place players. (See Straffin (1988) for examples.)

Straffin (1988) derives the indices within a probabilistic model in which there are n voters and each voter i has probability p_i of voting "yes" on a given issue, and a probability $(1 - p_i)$ of voting "no" on the issue. Voter i's power is given by the probability that her vote will affect the outcome of the election. Thus, a voter's power depends on the joint distribution from which the p_i's are drawn. Straffin shows that (a) σ_i measures i's power if $p_i = p$ for all i, and p is drawn from the uniform distribution on $[0, 1]$ while (b) β_i measures i's power if, for each i, p_i is drawn independently from the uniform distribution on $[0, 1]$. Equivalently, β_i measures i's power on the assumption that each voter independently has a 50% chance of voting "yes" on the issue.[39]

We turn now to the judicial reception of these ideas. The maxim of "one person, one vote" was articulated in the context of a political structure in which legislators are elected from single-member districts and each legislator has one vote within the legislature. The early reapportionment cases therefore identified one political structure – equal-sized, single-member districts – that satisfies the constitutional requirement that

---

[38] More precisely, the Shapley–Shubik index assumes that it is equally likely that a coalition of any size will form, and that all coalitions of a given size are equally like to form. An equivalent characterization is that the Shapley–Shubik index gives the probability that as an individual joins a coalition she will be a swing under the assumption that all orders of coalition formation are equally likely.

[39] For an axiomatic characterization of the two indices see Dubey and Shapley (1979).

each citizen's vote have equal influence on political outcomes.[40] Other political structures, however, were soon challenged. These challenges required the judiciary to elaborate a conception of "equal voting power".

Two political structures were of particular interest. Suppose that districts vary in size. In systems of weighted voting, each district elects one legislator but the elected representatives of larger districts are given more votes within the legislature. What set of voting weights, if any, yield equal influence of each citizen's vote? In the second political structure, larger districts elect more representatives. How should representatives be allocated among the districts to satisfy the constitutional mandate?

The Banzhaf index (and Shapley–Shubik index) suggest answers to these questions. Citizens in a representative democracy play a compound voting game. They elect representatives, and these representatives then pass legislation. There are three ways that the Banzhaf index could be used in this context. First, it could be used to measure the power of the elected representatives voting in the legislature. Second, it could be used to measure the power of citizens when voting for representatives. Finally, it could be used to measure the net power of citizens in the compound voting game. Since the probability that a citizen is a swing voter for the final passage of legislation equals the probability that her vote is crucial to getting her legislator elected times the probability that the legislator's vote is crucial, this last measure of Banzhaf power is simply the product of the first two.

From roughly 1967 through 1990 the Banzhaf index served as a judicially accepted criterion for the determination of the power of representatives at the legislative level in weighted voting schemes. The Banzhaf index had particular importance in New York State where many county boards elected supervisors from towns with widely varying populations.[41]

Indeed, Banzhaf's first article analyzed the striking case of the Nassau County Board of Supervisors in 1958 and 1964. Voters in different municipalities elected a total of six representatives. Since the municipalities differed in size, the elected representatives were given different voting weights, presumably to equalize the power of the citizens in the different municipalities. Incredibly, the weights were distributed among the representatives in such a way that three representatives had no swings – their votes could never influence the outcome (see Table 1). The weighting schemes masked the fact that in both 1958 and 1964 the six representatives were actually playing a game of three-player majority rule; three of the representatives had power $1/3$ and three had power 0 under both the Shapley and the normalized Banzhaf index.

---

[40] The case law focuses on the formulation of the conception of right articulated in *Reynolds v. Sims* (1964). There the U.S. Supreme Court stated that "each and every citizen has an inalienable right to full and effective participation in the political processes of his State's legislative bodies" (at 565). Moreover, "full and effective participation by all citizens in state government requires ... that each citizen have an equally effective voice in the election of members of his state legislature" (at 565).

[41] Imrie (1973) and Lucas (1983) provide brief accounts of the use of the Banzhaf index in New York State.

Table 1
Board of Supervisors – Nassau County

		1958		
Municipality	Population	Weight of representative	Unnormalized power	Normalized power
Hempstead1	618,065	9	1/2	1/3
Hempstead2		9	1/2	1/3
N. Hempstead	184,060	7	1/2	1/3
Oyster Bay	164,716	3	0	0
Glen Cove	19,296	1	0	0
Long Beach	17,999	1	0	0
		1964		
Hempstead1	728,625	31	1/2	1/3
Hempstead2		31	1/2	1/3
N. Hempstead	213,225	21	0	0
Oyster Bay	285,545	28	1/2	1/3
Glen Cove	22,752	2	0	0
Long Beach	25,654	2	0	0

Subsequently, in *Iannucci v. Board of Supervisors of Washington County* (1967) the New York Court of Appeals, the highest court in New York State, endorsed the Banzhaf index as the appropriate standard for determining the weights accorded to representatives of districts of varying sizes. The Court stated that the principle of "one person, one vote" was satisfied as long as the "[Banzhaf] power of a representative to affect the passage of legislation by his vote ... roughly corresponds to the proportion of the population of his constituency" (at 252). Interestingly, the Court explicitly endorsed the abstract nature of the power calculus, ruling that "the sole criterion [of the constitutionality of weighted voting schemes] is the mathematical voting power which each legislator possesses *in theory* (emphasis added) ... and not the actual voting power he possesses in fact..." (at 252). In *Franklin v. Kraus* (1973), the New York Court of Appeals again approved a weighted voting system that made the Banzhaf power index of each representative proportional to his district's size.

In both these cases the court used the Banzhaf index to determine the voting power of the legislators. The courts, however, did not carry the Banzhaf reasoning through to the citizen level. It can be shown that the probability that an individual is a swing voter in the election of her legislator is (approximately) inversely proportional to the *square root* of the population of her district. Thus, equalizing each citizen's net Banzhaf power in the compound voting game would require that a legislator be given power proportional to this square root. Nonetheless, the courts decided that a legislator's power should be proportional to the population itself.

How do we account for the failure of the courts to consider the citizen's power in the compound game? Apparently, the courts adopted a different view of the relation between representative and citizen than that implicit in Banzhaf.[42] The swing voter analysis assumes that citizens disagree about which candidate would best represent the jurisdiction and that the prevailing candidate acts in the interests of those citizens *who voted for her.* On a different conception of representation, however, once elected a candidate represents the interests of *all* citizens within her constituency. Each citizen in a district then exerts an equal influence upon the district representative, so that a citizen's influence is inversely proportional to the district population. Hence, under this conception, a representative's power should be proportional to the number of constituents in her district, as the courts ruled.

We now consider judicial treatment of multi-member districts. In *Abate v. Mundt* (1969), the New York Court of Appeals upheld a system of multi-member districts that allocated supervisors to election districts in proportion to district populations. The court again ignored the compound game analysis. It should be noted, however, that unlike Banzhaf's analysis of weighted voting schemes, his analysis of multi-member districts is seriously flawed. His assertion that voters in districts should be given representatives in proportion to the square root of the size of the population is unwarranted.

First of all, it is clear that no recommendation on the number of representatives in a multi-member district can be made independently of the voting method being used to elect these representatives. Is each voter being given one vote or the same number of votes as there are representatives or is some other method in use? Second, if individuals are given many votes, then what is the proper assumption to make on the way these votes are cast? Clearly, the assumption that an individual is casting her own votes independently of each other is absurd. Finally, what assumption is being made about the correlation of the voting behavior of representatives voted for by a single individual? If the correlation is perfect, then, in effect, the multi-member districting scheme results in a weighted voting game in the legislature. Of course, the ratios of voting powers in such a game are not necessarily equal to the ratios of voting weights. On the other hand, if these representatives are voting differently from each other, in what sense are they all representing the same individual? There may well be such a sense, but it is ill-captured by the Banzhaf conception, which considers the probability that a voter will be decisive on an issue.

A few years after *Abate*, the United States Supreme Court (in *Whitcomb v. Chavis* (1973)) rejected the Banzhaf index in the judicial evaluation of multi-member districts. The Court argued that the index offered only a theoretical measure of the voter's power while constitutional considerations had to rest on disparities in actual voting power.[43] Thus, the theoretical character of the power index which had pleased the New York Court of Appeals, moved the United States Supreme Court to reject the index.

---

[42] Banzhaf (1965) is noncommittal about the relationship between a citizen and his representative. Banzhaf (1968) explicitly endorses the compound voting computation suggested above.

[43] Banzhaf was an expert witness for plaintiffs at trial and signed the plaintiff-appellee's brief to the United States Supreme Court. Interestingly, plaintiffs sought single-member districts as a remedy in *Whitcomb v. Chavis*.

In 1989 the U.S. Supreme Court explicitly carried out a compound voting game analysis when considering the weighted voting scheme used by the Board of Estimate of New York City.[44] The court rejected the Banzhaf index, again citing its "theoretical" nature. On the basis of this decision, one federal district court struck down as unconstitutional the weighted voting scheme that then prevailed in Nassau County.[45] Two federal district courts have approved a weighted voting system in the New York counties of Roxbury and Schoharie respectively which assigned *weights* proportionally to population.[46] Thus, for the time being at least, the Banzhaf index appears to have been discredited for all United States legislative apportionment purposes.

This brief history illustrates both the potential significance and difficulties of judicial use of game-theoretic (or other formal) concepts in the articulation of legal rules. The Banzhaf index governed the shape of weighted voting systems for roughly twenty years and partially shaped the judicial treatment of voting systems that consisted of unequal sized districts. Unfortunately, however, the courts and the community of political scientists and game theorists failed to engage in a constructive dialogue that would have enabled the courts to formulate a coherent conception of equal and effective representation for each citizen. The courts never clearly articulated or resolved three important questions. First, as we remarked earlier, the courts had to choose whether to focus attention on the game constituted by voters electing representatives, the game played by the elected representatives in the legislature, or the compound game. The role of citizens in the compound game seems normatively the correct one for the courts to consider, but they were never clear on this point. Second, and as a consequence of their failure to consider the compound game explicitly, the courts did not articulate a theory of representation that would allow analysis of the compound game. Finally, the courts did not question whether power was the appropriate concept for expressing equality of representation. The indices of voting power measure the probability that an individual will be a swing voter under various assumptions; the related concept of satisfaction measures the probability that the outcome will be the one that the voter desires. In a town with like-minded citizens each one will have zero voting power since decisions will always be unanimous and no individual voter will be crucial. On the other hand, each citizen will have a satisfaction of one. The courts should have at least considered whether satisfaction was a more appropriate concept than power on which to construct the right to fair and effective representation.

The analytic community on the other hand bears some responsibility for the judicial failure to confront these three questions. Neither Banzhaf himself nor other commentators on the judicial efforts clearly framed the question in terms of the compound game.

---

[44] *Morris v. Board of Estimate of New York City* (1989).

[45] *Jackson v. Nassau County Board of Supervisors* (1993). The weights at issue in the litigation had been allocated so that the Banzhaf power of each representative was proportional to the population of the representative's district.

[46] *Roxbury Taxpayers Alliance v. Delaware County Board of Supervisors* (1995) and *Reform of Schoharie County v. Schoharie County Bd. of Supervisors* (1997).

Moreover, discussions of the importance of the concept of representation to the analysis of the game were sparse. Finally, although a voter's satisfaction is arguably more important than a voter's power, the former concept has received comparatively little attention from game theorists.[47]

5.2. Fair allocation of losses and benefits

Adjudication resolves disputes among parties. Courts are thus often more concerned, at least superficially, with the fair resolution of the dispute between the parties before the court than with the ex ante behavior that may avoid similar disputes in the future (or reshape them). Many classes of disputes, in fact, seem explicitly directed at the fair division of property among disputing claimants. The most obvious area of law in which questions of this type arise is bankruptcy where the court must allocate the assets of the bankrupt among its creditors in a way that distributes the loss – i.e., the difference between the (valid) claims of creditors and the value of the estate – fairly.[48] Other areas include: the interpretations of wills that divide more assets among the heirs than are actually available; the rules of contribution when the share of an insolvent tortfeasor must be allocated among the remaining joint tortfeasors; rate proceedings for common carriers such as telecommunications or electricity providers when the common costs of service provision must be allocated among different classes of customers; and the determination of fee awards to class action attorneys when the common costs of representation must be allocated among different classes of plaintiffs.

One might reasonably ask what conception of fairness is embodied in legal rules governing these allocations? What explains variation, if any, in these rules? In the case of the common law, a judge confronted with a dispute consults the decisions in analogous, previously decided cases and attempts to articulate a principle or set of principles that rationalizes these outcomes. Here, answering the above questions is particularly important and difficult.

Talmudic law, like Anglo-American law, is a form of common law adjudication. Aumann and Maschler (1985) and O'Neill (1982) have deployed game-theoretic tools to explicate in part the rules implicit in two areas of Talmudic law that concern the fair allocation of property among disputing claimants.

Aumann and Maschler consider the following bankruptcy problem from the Talmud. There are three creditors on an estate. Creditor 1 is owed 100; creditor 2 is owed 200, and creditor 3 is owed 300. The estate is worth less than the 600 total claim. The Talmud states that when the estate is valued at 100, the creditors should each receive $33\frac{1}{3}$; when

---

[47] On satisfaction see Brams and Lake (1978) and Straffin, Davis, and Brams (1982).

[48] In fact bankruptcy law must balance other concerns with fairness. In particular, the rules of bankruptcy may affect the pre-bankruptcy behavior of both debtor and creditors. Moreover, even after bankruptcy, the value of a debtor corporation is generally greater as an ongoing concern than liquidated; capturing this surplus may entail creating incentives to some parties such as managers or specific creditors that otherwise seem unfair. Finally, we have ignored concerns about priority of claims of different types of creditors.

the estate is valued at 200, the creditors should receive, respectively, 50, 75, and 75; when the estate is valued at 300, the allocation should be (50, 100, 150). What principle underlies the prescribed allocations? How should these three examples be extended to other bankruptcy cases?

As a preliminary, consider the "contested-garment problem" from the Mishnah: one individual claims an entire garment; a second individual claims one-half of the same garment. The Talmud allocates the garment $(3/4, 1/4)$ to the two claimants respectively. Here, the underlying principle is well understood by Talmudic scholars. The second claimant implicitly concedes half the garment to the first, while the first claimant concedes nothing. Each claimant is given the amount that is conceded to him, and the remainder is divided equally. However, this principle is articulated only for the case of two claimants. How can this principle for the two-claimant case be extended to the n-claimant case?

Define a general claim problem $P = (A; c_1, c_2, \ldots, c_n)$, where A is the total amount of assets to be distributed and c_i is i's claim on A. Aumann and Maschler call a distribution x of the amount A "contested-garment-consistent" if (x_i, x_j) is the contested-garment solution to the claim problem $(x_i + x_j; c_i, c_j)$ for all claimants i and j. They show that each claim problem has a unique contested-garment-consistent solution. Moreover, this solution gives the allocations prescribed by the Talmud for the above three bankruptcy problems.

Note that while the notion of contested-garment consistency does not appear elsewhere in the game theory literature, this "consistency" approach is certainly game-theoretic in spirit. For instance, Harsanyi has characterized the product solution to the n-person Nash bargaining problem as the unique solution which is Nash-consistent for all pairs of bargainers.

As in O'Neill (1982), Aumann and Maschler turn the claim problem into a cooperative game by associating the following characteristic function game with it:

$$ v(S) = \max\left(A - \sum_{i \in N-S} c_i, 0 \right) \quad \text{for } S \subseteq N. $$

This characteristic function is constructed on the assumption that claims are satisfied in full as they arrive and that a coalition can only *guarantee* itself what it would receive if it were the last to make a claim on the assets A. Alternatively, a coalition's worth is what it obtains when it concedes the claims of the other claimants. Aumann and Maschler demonstrate that the contested-garment-consistent solution (and hence the Talmudic division to the three bankruptcy problems) is the nucleolus of this coalitional game.[49]

While this result is remarkable and Aumann and Maschler's discussion is insightful and revealing, various objections can be raised. These objections pertain to the limitations of the game-theoretic approach in general.

---

[49] See Benoît (1998) for a simple proof of this result.

A difficulty in applying game theory is that "real world" situations are not presented as games. Rather a game must be defined. While the characteristic function Aumann and Maschler associate with the claim problem is a reasonable one, other specifications are also reasonable. For instance, we could define a coalition's worth to be its expected payment assuming that the other claimants do not form coalitions, that all claimants are paid off as they arrive, and that all orders of arrival are equally likely.

Similarly, while consistency notions are interesting, several can be defined. Thus, call a solution x to a claim problem "CG2-consistent" if for all $S \subset N$, $(\sum_S x_i, \sum_{N/S} x_i)$ is the contested-garment solution to $(A; \sum_S c_i, \sum_{N/S} c_i)$. As the claim problem $(500; 100, 200, 300)$ shows, there is no CG2-consistent solution.[50] This non-existence might be taken as evidence that CG2-consistency is the wrong notion of consistency, but it might just as well indicate that the contested-garment example should not be extended to more than two people.[51]

As noted earlier, two questions arise in the analysis of a body of common law cases. What is the general rule that is to be derived from the disposition of particular cases, and what is the conception of fairness embodied in this law? These two questions are related because a rule that rationalizes a set of cases must both "fit" those cases and offer an attractive rationale for them. After all, many rules will be consistent with the decisions in the set of cases.

Aumann and Maschler explicitly address the first question. Implicitly, they address the second as well. The nucleolus of a bankruptcy problem minimizes the maximum loss suffered by any coalition. Thus, it has an egalitarian spirit, but one in which coalitions and not just individuals are taken as objects of concern.

O'Neill considers a problem in which a testator bequeaths shares in his estate, valued at 120, that total to more than one. The entire estate is bequeathed to heir 1, one-half the estate to heir 2, one-third the estate to heir 3 and one-quarter the estate to heir 4. According to O'Neill, the 12th-century scholar Ibn Ezra claimed that the division $(97/144, 25/144, 13/144, 9/144)$ is the appropriate one.

Ibn Ezra reasoned that heir 4 lays claim to a fourth of the estate, but that all four heirs make equal claim to this quarter and so this quarter is split evenly. Next heir 3 claims one third of the estate. He has already received his share of $1/4$ of the estate, so now he and the three larger claimants receive an additional $1/3(1/3-1/4)$ for a total of $13/144$. Proceeding in this manner yields the above numbers.

In contrast to the earlier bankruptcy problem, here the reasoning underlying the division is given explicitly. O'Neill is interested in both extending this reasoning and proposing alternative, albeit related, solutions. Notice that while each heir only makes a claim for himself, this claim can be taken as implicitly imputing a division of the remainder of the assets to the other heirs. Indeed, given a specific heir's claim and an allocation

---

[50] Solving for x_i by taking $S = \{i\}$ for each claimant one at a time yields $(50, 150, 250)$ as the putative CG2-solution, but this does not sum to 500.

[51] Aumann and Maschler consider a principle similar to CG2-consistency. They show how a modification of this principle can be used to again derive the CG-consistent solution.

rule, use the rule to impute a division of the remainder of the assets to each of the other heirs. Since no heir has any greater legitimacy than any other, compute the allocation that results from averaging the various divisions obtained in this manner. O'Neill calls a rule consistent if the allocation thus obtained is the same as the allocation the rule prescribed for the initial problem. Ibn Ezra's solution is not consistent in this sense.

Again associate the characteristic function $v(S) = \max(A - \sum_{i \in N-S} c_i, 0)$ with this claim problem. The Shapley value yields an allocation which is consistent in O'Neill's sense. Furthermore, the Shapley allocation affords an interesting interpretation. Suppose that each heir is simply paid off in full as he presents his claim until the estate is exhausted. The allocation which results from averaging over all possible orders in which the claims could be presented is the Shapley value.

When making allocations, courts might want to balance practical considerations with ethical ones. For instance, courts may not want to give claimants the incentive to combine or split their claims. Thus, they might require that $(x_1, \ldots, x_{i,}, \ldots, x_j, \ldots, x_n)$ is the solution to $(A; c_1, \ldots, c_i, \ldots, c_j, \ldots, c_n)$ if and only if $(x_1, \ldots, x_i + x_j, \ldots, x_n)$ is the solution to $(A; c_1, \ldots, c_i + c_j, \ldots, c_n)$. The only such rule is proportional allocation.[52]

One should note that U.S. law generally does not resolve claims problems similarly to the Talmudic solutions that inspired O'Neill and Aumann and Maschler. In bankruptcy law, the assets A are, in fact, allocated among claimants proportionally to their claims. U.S. law presumably adopts this approach because creditors may freely transfer claims on the debtor-in-bankruptcy. With inheritances the problem is considered one of interpreting the intent of the testator. If no clear intent can be inferred, the will is invalid, the individual dies intestate, and the state's default rules of inheritance govern. Determination of intent depends critically on the precise wording of the document. See, e.g., *Estate of Heisserer v. Loos* (1985), *In re Vismar's Estate* (1921), and *Rodisch v. Moore* (1913).

The shares of insolvent defendants among joint tortfeasors are generally allocated using one of two rules. Under legal contribution, the tortfeasor against whom the plaintiff chooses to execute the judgment must bear the entire share of the insolvent parties. Under equitable contribution, these insolvent shares are divided among the solvent tortfeasors either equally or proportionally to their own shares. (Kornhauser and Revesz (1990) summarizes the law in this area.)

6. Concluding remarks

Law pervades daily life. Legal rules structure relations among individuals, institutions, and government. The theory of games provides a framework within which to study the effects these legal rules have on the behavior of these agents. In a legal culture that considers law a means to guide behavior to specified goals, this framework will interest

---

[52] See Chun (1988), de Frutos (1994) and O'Neill (1982). Thomson (1997) provides a survey of axiomatic models of bankruptcy.

not only legal scholars but also lawmakers directly as they attempt to design rules that forward their goals. In the last decade, analysts have deployed increasingly sophisticated models to an ever-widening range of legal rules. Our discussion has merely scratched the surface of this literature. Many analyses of legal rules focus not on the behavioral effects of these rules but on their conceptual and normative foundations. The tools of game theory have also offered insights into these areas.

References

Articles

Aivazian V.A., and J.L. Callen (1981), "The Coase Theorem and the empty core", Journal of Law & Economics 24:175–181.

Aivazian, V.A., J.L. Callen and I. Lipnowski (1987), "The Coase Theorem and coalitional stability", Economica 54:517–520.

Arlen, J. (1990), "Re-examining liability rules when injurers as well as victims suffer losses", International Review of Law and Economics 10:233–239.

Arlen, J. (1992), "Liability for physical injury when injurers as well as victims suffer losses", Journal of Law, Economics & Organization 8:411–426.

Aumann, R., and M. Maschler (1964), "The bargaining set for cooperative games", in: M. Dresher, L.S. Shapley and A.W. Tucker, eds., Advances in Game Theory (Princeton University Press, Princeton), 443–476.

Aumann, R., and M. Maschler (1985), "Game-theoretic analysis of a bankruptcy problem from the Talmud", Journal of Economic Theory 36:195–213.

Avery, C., and P.B. Zemsky (1994), "Money burning and multiple equilibria in bargaining", Games and Economic Behavior 7:154–168.

Ayres, I. and R. Gertner (1989), "Filling gaps in incomplete contracts: An economic theory of default rules", Yale Law Journal 99:87–130.

Ayres, I., and R. Gertner (1992), "Strategic contractual inefficiency and the optimal choice of legal rules", Yale Law Journal 101:729–772.

Ayres, I., and E. Talley (1995), "Solomonic bargaining: Dividing a legal entitlement to facilitate Coasean trade", Yale Law Journal 104:1027–1117.

Baird, D., R. Gertner and R. Picker (1994), Game Theory and the Law (Harvard University Press, Cambridge, MA).

Baird, D., and R. Picker (1991), "A simple non-cooperative bargaining model of corporate reorganization", Journal of Legal Studies 20:311–350.

Banzhaf, J.F. III (1965), "Weighted voting doesn't work", Rutgers Law Review 19:317–343.

Banzhaf, J.F. III (1966), "Multi-member electoral districts", Yale Law Journal 75:1309–1338.

Barton, J. (1972), "The economic basis of damages for breach of contract", Journal of Legal Studies 1:277–304.

Beard, T.R. (1990), "Bankruptcy and care choice", Rand Journal of Economics 12:626–634.

Bebchuk, L.A. (1984), "Litigation and settlement under imperfect information", Rand Journal of Economics 15:404–415.

Bebchuk, L.A. (1988), "Suing solely to extract a settlement offer", Journal of Legal Studies 17:437–450.

Bebchuk, L.A. (1994), "Efficient and inefficient sales of corporate control", Quarterly Journal of Economics 109:957–993.

Bebchuk, L.A., and S. Shavell (1991), "Information and the scope of liability for breach of contract: The rule of Hadley v. Baxendale", Journal of Law, Economics & Organization 7:284–312.

Becker, G. (1968), "Crime and punishment: An economic approach", Journal of Political Economy 76:169–217.

Benoit, J.-P. (1998), "The nucleolus is contested-garment-consistent: A direct proof", Journal of Economic Theory 77:192–196.

Birmingham, R.L. (1969a), "Damage measures and economic rationality: The geometry of contract law", Duke Law Journal 1969:49–71.

Birmingham, R.L. (1969b), "Legal and moral duty in game theory: Common law contract and Chinese analogies", Buffalo Law Review 18:99–117.

Blume, L., D. Rubinfeld and P. Shapiro (1984), "The taking of land: When should compensation be paid?", Quarterly Journal of Economics 99:71–92.

Brams, S.J., and M. Lake (1978), "Power and satisfaction in a representative democracy", in: P.C. Ordeshhook, ed., Game Theory and Political Science, 529–562.

Brown, J.P. (1973), "Toward an economic theory of liability", Journal of Legal Studies 2:323–349.

Bulow, J., and P. Klemperer (1996), "Auctions versus negotiations", American Economic Review 86:180–194.

Calabresi, G. (1961), "Some thoughts on risk distribution and the law of torts", Yale Law Journal 70:499–553.

Calabresi, G., and A.D. Melamed (1972), "Property rules, liability rules and inalienability: One view of the cathedral", Harvard Law Review 85:1089–1128.

Cameron, C. (1993), "New avenues for modeling judicial politics", University of Rochester Wallis Institute of Political Economy Working Paper.

Chun, Y. (1988), "The proportional solution for rights problems", Mathematical Social Sciences 15:231–246.

Cirace, J. (1980), "A game-theoretic analysis of contribution and claim reduction in antitrust treble damages suits", St. John's Law Review 55:42–62.

Coase, R. (1960), "The problem of social cost", Journal of Law & Economics 3:1–44.

Cohen, L.R., and M.L. Spitzer (1995), "Judicial deference to agency action: A rational choice theory and an empirical test", mimeo.

Coleman, J.C. (1973), "Loss of power", American Sociological Review 38:1–17.

Cooter, R.D. (1985), "The unity of tort, property and contract", California Law Review 73:1–51.

Cooter, R.D. (1987), "Coase Theorem", in: J. Eatwell, M. Milgate and P. Newman, eds., The New Palgrave Dictionary of Economics, Vol. 1 (Macmillan, London) 457–460.

Cooter, R.D., L.A. Kornhauser and D. Lane (1979), "Liability rules, limited information, and the role of precedent", Bell Journal of Economics and Management Science 10:366–373.

Cooter, R.D., and D.L. Rubinfeld (1989), "Economic analysis of legal disputes and their resolution", Journal of Economic Literature 27:1067–1097.

Cooter, R.D., and T. Ulen (1986), "An economic case for comparative negligence", New York University Law Review 61:1067–1110.

Cooter, R.D., and T. Ulen (1988), Law and Economics (Scott, Foresman and Co., London).

Cramton P., and A. Schwartz (1991), "Using auction theory to inform takeover regulation", Journal of Law, Economics & Organization 7:27–54.

Craswell, R. (1988), "Precontractual investigation as an optimal precaution problem", Journal of Legal Studies 17:401–436.

Craswell, R., and J.E. Calfee (1986), "Deterrence and uncertain legal standards", Journal of Law, Economics & Organization 2:279–303.

Dahl, R.A. (1957), "The concept of power", Behavioral Science 2:201–215.

Daughety, A., and J. Reinganum (1993), "Endogenous sequencing in models of settlement and litigation", Journal of Law, Economics & Organization 9:314–348.

De Frutos, M.A. (1994), "Coalitional manipulation in a bankruptcy problem", mimeo.

Diamond, P.A. (1974a), "Single-activity accidents", Journal of Legal Studies 3:107–162.

Diamond, P.A. (1974b), "Accident law and resource allocation", Bell Journal of Economics and Management Science 5:366–405.

Diamond, P.A. and E. Maskin (1979), "An equilibrium analysis of search and breach of contract I: Steady states", Bell Journal of Economics and Management Science 10:282–316.

Diamond, P.A., and J.A. Mirrlees (1975), "On the assignment of liability: The uniform case", Bell Journal of Economics and Management Science 6:487–516.

Dubey, P., and L.S. Shapley (1979), "Mathematical properties of the Banzhaf power index", Mathematics of Operations Research 4:99–131.

Edlin, A.S. (1994), "Efficient standards of due care: Should courts find more parties negligent under comparative negligence?", International Review of Law and Economics 15:21–34.

Eisenberg, T. (1990), "Testing the selection effect: A new theoretical framework with empirical tests", Journal of Legal Studies 19:337–358.

Emons, W. and J. Sobel (1991), "On the effectiveness of liability rules when agents are not identical", Review of Economic Studies 58:375–390.

Endres, A., and I. Querner (1995), "On the existence of care equilibria under tort law", Journal of Institutional and Theoretical Economics 151:348–357.

Eskridge, W.N., and J. Ferejohn (1992), "Making the deal stick: Enforcing the original constitutional structure of lawmaking in the modern regulatory state", Journal of Law, Economics & Organization 8:165–189.

Farrell, J. (1987), "Information and the Coase Theorem", Journal of Economic Perspectives 1:113–129.

Ferejohn, J., and C. Shipan (1990), "Congressional influence on bureaucracy", Journal of Law, Economics & Organization 6:1–20.

Fernandez, R., and J. Glazer (1991), "Striking for a bargain between two completely informed agents", American Economic Review 81:240–252.

Fischel, W.A., and P. Shapiro (1989), "A constitutional choice model of compensation for takings", International Review of Law and Economics 9:115–128.

Geistfeld, M. (1995), "Manufacturer moral hazard and the tort-contract issue in products liability", International Review of Law and Economics 15:241–257.

Glazer, J., and A. Ching-to Ma (1989), "Efficient allocation of a "prize": King Solomon's dilemma", Games and Economic Behavior 1:222–233.

Green, J. (1976), "On the optimal structure of liability rules", Bell Journal of Economics and Management Science 7:553–574.

Grossman, S., and O. Hart (1980), "Disclosure laws and takeover bids", Journal of Finance 35:323–334.

Haddock, D., and C. Curran (1985), "An economic theory of comparative negligence", Journal of Legal Studies 14:49–72.

Hadfield, G. (1994), "Judicial competence and the interpretation of incomplete contracts", Journal of Legal Studies 23:159–184.

Haller, H., and S. Holden (1990), "A letter to the editor on wage bargaining", Journal of Economic Theory 52:232–236.

Harris, M., and A. Raviv (1992), "Financial contracting theory", in: J.-J. Laffont, ed., Advances in Economic Theory: Sixth World Congress, 64–150.

Harrison, G.W., and M. McKee (1985), "Experimental evaluation of the Coase Theorem", Journal of Law & Economics 28:653–670.

Hart, O., and B. Holmstrom (1987), "The theory of contracts", in: T. Bewley, ed., Advances in Economic Theory: Fifth World Congress, 71–155.

Hay, B.L. (1995), "Effort, information, settlement, trial", Journal of Legal Studies 29:29–62.

Hermalin, B. (1995), "An economic analysis of takings", Journal of Law, Economics & Organization 11:64–86.

Hermalin, B. and M. Katz (1993), "Judicial modification of contracts between sophisticated parties: A more complete view of incomplete contracts and their breach", Journal of Law, Economics & Organization 9:230–255.

Hoffman, E. and M.L. Spitzer (1982), "The Coase Theorem: Some experimental tests", Journal of Law & Economics 25:73–98.

Hoffman, E. and M.L. Spitzer (1985), "Entitlements, rights, and fairness: An experimental examination of subjects' concepts of distributive justice", Journal of Legal Studies 14:259–298.

Hoffman, E., and M.L. Spitzer (1986), "Experimental tests of the Coase Theorem with large bargaining groups", Journal of Legal Studies 15:149–171.

Hohfeld, W.N. (1913), "Some fundamental legal conceptions as applied in judicial reasoning", Yale Law Journal 23:16–59.

Holmstrom, B., and J. Tirole (1989), "The theory of the firm", in: R. Schmalensee and R.D. Willig, eds., Handbook of Industrial Organization 1:61–133.

Hurwicz, L. (1995), "What is the Coase Theorem?", Japan and the World Economy 7:49–74.

Hylton, K.N. (1990), "Costly litigation and legal error under negligence", Journal of Law, Economics & Organization 6:433–452.

Hylton, K.N. (1993), "Litigation cost allocation rules and compliance with the negligence standard", Journal of Legal Studies 22:457–476.

Hylton, K.N. (1995), "Compliance with the law and the trial selection process: A theoretical framework", manuscript.

Imrie, R.W. (1973), "The impact of the weighted vote on representation in municipal governing bodies in New York State", Annals of the New York Academy of Sciences 219:192–199.

Johnston, J.S. (1990), "Strategic bargaining and the economic theory of contract default rules", Yale Law Journal 100:615–664.

Johnston, J.S. (1995), "Bargaining under rules vs. standards", Journal of Law, Economics & Organization 11:256–281.

Kahan, M. (1989), "Causation and incentives to take care under the negligence rule", Journal of Legal Studies 18:427–447.

Kahan, M., and B. Tuchman (1993), "Do bondholders lose from junk bond covenant changes?", Journal of Business 66:499–516.

Kambhu, J. (1982), "Optimal product quality under asymmetric information and moral hazard", Bell Journal of Economics and Management Science 13:483–492.

Kaplow, L., and S. Shavell (1996), "Property rules versus liability rules: An economic analysis", Harvard Law Review 109:713–789.

Katz, A. (1987), "Measuring the demand for litigation", Journal of Law, Economics & Organization 3:143–176.

Katz, A. (1990a), "The strategic structure of offer and acceptance: Game theory and the law of contract formation", Michigan Law Review 89:216–295.

Katz, A. (1990b), "Your terms or mine? The duty to read the fine print in contracts", Rand Journal of Economics 21:518–537.

Katz, A. (1993), "Transaction costs and the legal mechanics of exchange: When should silence in the face of an offer be construed as acceptance?", Journal of Law, Economics & Organization 9:77–97.

Kennedy, D., and F. Michelman (1980), "Are property and contract efficient?", Hofstra Law Review 8:711–770.

Kornhauser, L.A. (1983), "Reliance, reputation and breach of contract", Journal of Law and Economics 26:691–706.

Kornhauser, L.A. (1986), "An introduction to the economic analysis of contract remedies", University of Colorado Law Review 57:683–725.

Kornhauser, L.A. (1989), "The new economic analysis of law: Legal rules as incentives", in: N. Mercuro, ed., Developments in Law and Economics, 27–55.

Kornhauser, L.A. (1992), "Modeling collegial courts. II. Legal doctrine", Journal of Law, Economics & Organization 8:441–470.

Kornhauser, L.A. (1996), "Adjudication by a resource-constrained team: Hierarchy and precedent in a judicial system", Southern California Law Review 68:1605–1629.

Kornhauser, L.A. and R.L. Revesz (1990), "Apportioning damages among potentially insolvent actors", Journal of Legal Studies 19:617–652.

Kornhauser, L.A. and R.L. Revesz (1993), "Settlements under joint and several liability", New York University Law Review 68:427–493.

Kornhauser, L.A., and A. Schotter (1992), "An experimental study of two-actor accidents", New York University C.V. Starr Center for Applied Economics Working Paper #92-57.

Landes, W., and R. Posner (1980), "Joint and multiple tortfeasors: An economic analysis", Journal of Legal Studies 9:517–555.

Leong, A. (1989), "Liability rules when injurers as well as victims suffer losses", International Review of Law and Economics 9:105–111.

Lucas, W.F. (1983), "Measuring power in weighted voting systems", in: S. Brams, W. Lucas and P. Straffin, eds., Political and Related Models, 183–238.

Mailath, G.J., and A. Postlewaite (1990), "Asymmetric information bargaining problems with many agents", Review of Economic Studies 57:351–367.

Moulin, H. (1988), Axioms of Cooperative Decision Making (Cambridge University Press, Cambridge).

Myerson, R.B., and M.A. Satterthwaite (1983), "Efficient mechanisms for bilateral trading", Journal of Economic Theory 29:265–281.

Oi, W. (1973), "The economics of product safety", Bell Journal of Economics and Management Science 4:3–28.

O'Neill, B. (1982), "A problem of rights arbitration from the Talmud", Mathematical Social Sciences 2:345–371.

Ordover, J.A. (1978), "Costly litigation in the model of single-activity accidents", Journal of Legal Studies 7:243–262.

Osborne, M.J., and A. Rubinstein (1990), Bargaining and Markets (Academic Press, San Diego, CA).

Perloff, J. (1981), "Breach of contract and the foreseeability doctrine of Hadley v. Baxendale", Journal of Legal Studies 10:39–64.

Png, I.P.L. (1983), "Strategic behavior in suit, settlement, and trial", Bell Journal of Economics and Management Science 14:539–550.

Png, I.P.L. (1987), "Litigation, liability and incentives for care", Journal of Public Economics 34:61–85.

Polinsky, A.M. (1983), "Risk sharing through breach of contract remedies", Journal of Legal Studies 12:427–444.

Polinsky, A.M. (1987a), "Fixed price vs. spot price contracts: A study in risk allocation", Journal of Law, Economics & Organization 3:27–46.

Polinsky, A.M. (1987b), "Optimal liability when the injurer's information about the victim's loss is imperfect", International Review of Law and Economics 7:139–147.

Polinsky, A.M., and W. Rogerson (1983), "Product liability, consumer misperception, and market power", Bell Journal of Economics and Management Science 14:581–589.

Polinsky, A.M., and D.L. Rubinfeld (1988), "The deterrent effects of settlements and trials", International Review of Law and Economics 8:109–116.

Polinsky, A.M., and S. Shavell (1989), "Legal error, litigation, and the incentive to obey the law", Journal of Law, Economics & Organization 5:99–108.

Posner, R., and A.M. Rosenfield (1977), "Impossibility and related doctrines in contract law: An economic analysis", Journal of Legal Studies 6:83–118.

Priest, G.L., and B. Klein (1984), "The selection of disputes for litigation", Journal of Legal Studies 13:1–56.

Rae, D.W. (1969), "Decision rules and individual values in constitutional choice", American Political Science Review 63:40–56.

Rasmusen, E. (1994), "Judicial legitimacy as a repeated game", Journal of Law, Economics & Organization 10:63–83.

Rasmusen, E., and I. Ayres (1993), "Mutual and unilateral mistake in contract law", Journal of Legal Studies 22:309–344.

Reinganum, J. (1988), "Plea bargaining and prosecutorial discretion", American Economic Review 78:713–728.

Reinganum, J., and L. Wilde (1986), "Settlement and litigation and the allocation of litigation costs", RAND Journal of Economics 17:557–566.

Rogerson, W. (1984), "Efficient reliance and damage measures for breach of contract", Bell Journal of Economics and Management Science 15:39–53.

Rosenberg, D., and S. Shavell (1985), "A model in which suits are brought for their nuisance value", International Review of Law and Economics 5:3–13.

Rubinfeld, D. (1987), "The efficiency of comparative negligence", Journal of Legal Studies 16:373–394.

Rubinstein, A. (1982), "Perfect equilibrium in a bargaining model", Econometrica 50:1151–1172.

Samuelson, W. (1985), "A comment on the Coase Theorem", in: A.E. Roth, ed., Game-Theoretic Models of Bargaining, 321–339.

Schwartz, A. (1989), "A theory of loan priorities", Journal of Legal Studies 18:209–261.

Schwartz, A. (1993), "Bankruptcy workouts and debt contracts", Journal of Law and Economics 36:595–632.

Schwartz, E.P. (1992), "Policy, precedent, and power: A positive theory of supreme court decision-making", Journal of Law, Economics & Organization 8:219–252.

Sercu, P., and C. van Hulle (1994), "Financing instruments, security design, and the efficiency of takeovers: A note", International Review of Law and Economics 15:373–393.

Shapley, L.S., and M. Shubik (1954), "A method for evaluating the distribution of power in a committee system", American Political Science Review 48:787–792.

Shavell, S. (1980), "Damage measures for breach of contract", Bell Journal of Economics and Management Science 11:466–490.

Shavell, S. (1982), "Suit settlement and trial: A theoretical analysis under alternative methods for the allocation of legal costs", Journal of Legal Studies 11:55–82.

Shavell, S. (1983), "Torts in which victims and injurers move sequentially", Journal of Law & Economics 26:589–612.

Shavell, S. (1987), Economic Analysis of Accident Law (Harvard University Press, Cambridge, MA).

Shavell, S. (1995), "The appeals process as a means of error correction", Journal of Legal Studies 24:379–426.

Sobel, J. (1985), "Disclosure of evidence and resolution of disputes: Who should bear the burden of proof?", in: A.E. Roth, ed., Game-Theoretic Models of Bargaining, 341–361.

Spier, K. (1992), "The dynamics of pretrial negotiation", Review of Economic Studies 59:93–108.

Spier, K. (1994), "Pretrial bargaining and the design of fee shifting rules", RAND Journal of Economics 25:197–214.

Straffin, P.D., Jr. (1988), "The Shapley–Shubik and Banzhaf power indices as probabilities", in: A.E. Roth, ed., The Shapley Value: Essays in Honor of Lloyd Shapley, 71–81.

Straffin, P.D., Jr. (1994), "Power and stability in politics", in: R.J. Aumann and S. Hart, eds., Handbook of Game Theory, Vol. 2 (North-Holland, Amsterdam) Chapter 32, 1127–1152.

Straffin, P.D., Jr., M.D. Davis and S.J. Brams (1982), "Power and satisfaction in an ideologically divided voting body", in: M.J. Holler, ed., Power, Voting, and Voting Power, 239–255.

Sykes, A.O. (1990), "The doctrine of commercial impracticability in a second-best world", Journal of Legal Studies 19:43–94.

Symposium (1991), "Just winners and losers: The application of game theory to corporate law and practice", University of Cincinnati Law Review 60:2–448.

Thomson, W. (1997), "Axiomatic analyses of bankruptcy and taxation problems: A survey", mimeo.

Veljanovski, C.J. (1982), "The Coase theorems and the economic theory of markets and law", Kyklos 35:53–74.

Wang, G.H. (1994), "Litigation and pretrial negotiations under incomplete information", Journal of Law, Economics & Organization 10:187–200.

Weingast, B., and M. Moran (1983), "Bureaucratic discretion of congressional control? Regulatory policy-making by the Federal Trade Commission", Journal of Political Economy 91:756–800.

White, M.J. (1980), "Public policy toward bankruptcy: Me-first and other priority rules", Bell Journal of Economics and Management Science 11:550–564.

White, M.J. (1988), "Contract breach and contract discharge due to impossibility: A unified theory", Journal of Legal Studies 17:353–376.

White, M.J. (1994), "Corporate bankruptcy as a filtering device: Chapter 11. Reorganization and out-of-court debt restructurings", Journal of Law, Economics & Organization 10:268–295.

Winter, R. (1991), "Liability insurance", Journal of Economic Perspectives 5:115–136.

Wittman, D. (1981), "Optimal pricing of sequential inputs: Last clear chance, mitigation of damages, and related doctrines in the law", Journal of Legal Studies 10:65–92.

Wittman, D. (1988), "Dispute resolution, bargaining, and the selection of cases for trial: A study of the generation of biased and unbiased data", Journal of Legal Studies 17:313–352.

Young, H.P. (1994), Equity: In Theory and Practice (Princeton University Press, Princeton, NJ).

Cases

Abate v. Mundt, 25 N.Y.2d 309 (1969)

Abate v. Mundt, 403 U.S. 182 (1971)

Akzo Coatings, Inc v. Aigner Corp., 30 F.3d 761 (7th Cir. 1994)

American National Fire Insurance Co. v. Kenealy and Kenealy, 72 F.3d 264 (2d Cir. 1995)

Baker v. Carr, 369 U.S. 186 (1962)

Bechtle v. Board of Supervisors of County of Nassau, 441 N.Y.S. 2d 403 (2d Dept 1981)

Berg v. Footer, 673 A. 2d 1244 (D.C. App., 1996)

Board of Estimate of New York City v. Morris, 489 U.S. 688 (1989)

Chapman v. Meier, 420 U.S. 1 (1975)

Chevron, U.S.A., Inc v. Natural Resources Defense Council Inc, 467 U.S. 837 (1984)

Estate of Heisserer v. Loos, 698 S.S. 2d 6 (Mo. Ct. of Appeals 1985)

Franklin v. Kraus, 72 Misc.2d (Sup.Ct., Nassau Cty 1972) reversed 32 NY 2d 234 (1973) reargument denied, motion to amend remittitur granted 33 N.Y. 2d 646 (1973) appeal dismissed for want of a substantial federal question 415 U.S. 904 (1974)

Franklin v. Mandeville, 57 Misc.2d 1072 (Sup.Ct. Nassau Cty.1968) aff'd 299 N.Y.S.2d 953 (2d Dept 1969) modified 26 NY2d 65 (1970)

Glinski v. Lomenzo, 16 N.Y.2d 27 (1965)

Harnischefeger Corp. v. Harbor Insurance Co., 927 F.2d 974 (7th Cir 1991)

Holder v. Hall, 114 S.Ct. 2581 (1994)

Iannucci v. Board of Supervisors of Washington County, 20 N.Y.2d 244 (1967)

Immigration and Nationalization Service v. Chadha, 462 U.S. 919 (1983)

In re Hoskins, 102 F.3d 311 (7th Cir. 1996)

In re Larsen, 616 A.2nd 529 (Pa. S. Ct 1992)

In re Oil Spill of Amoco Cadiz, 954 F.2d 1279 (7th Cir 1991)

In re Vismar's Estate, 191 NYS 752 (Surrogates Ct 1921)

Jackson v. Nassau County Board of Supervisors, 818 F. Supp. 509 (EDNY 1993)

Kilgarlin v. Hill, 386 U.S. 120 (1966)

League of Women Voters v. Nassau County Board of Supervisors, 737 F.2d 155 (2d Cir. 1984) cert denied sub nom. *Schmertz v. Nassau County Board of Supervisors*, 469 U.S. 1108 (1985)

Marek v. Chesny, 473 U.S. 1 (1985)

McDermott Inc. v. Amclyde et al., 511 S. Ct. 202 at notes 14, 20, and 24 (1994)

Packer Engineering v. Kratville, 965 F.2d 174 (7th Cir. 1992)

Premier Electric Construction Company v. National Electrical Contractors Assn., 814 F.d 358 (7th Cir. 1987)

Reynolds v. Sims, 377 U.S. 533 (1964)

Rodisch v. Moore, 101 N.E. 206 (Ill. Sup. Ct 1913)

Roxbury Taxpayers Alliance v. Delaware County Board of Supervisors, 886 F. Supp. 242 (NDNY 1995) aff'd, 80 F.3d 42 (2d Cir.), cert. denied sub nom. *MacDonald v. Delaware County Board of Supervisors*, 519 U.S. 872 (1996)

Reform of Schoharie County v. Schoharie County Board of Supervisors, 975 F. Supp. 191 (NDNY,1997)

Texas Industries v. Radcliff Materials, 451 U.S. 630 (1981)

United States v. Herrera, 70 F.3d 444 (7th Cir, 1995)

Weinberger v. Great Northern Nekoosa Corp., 925 F.2d 518 (1st Cir., 1991)

Whitcomb v. Chavis, 403 U.S. 124 (1971)

WMCA Inc. v. Lomenzo, 246 F. Supp. 953 (S.D.N.Y. 1965) affirmed per curiam sub nom *Travia v. Lomenzo*, 382 U.S. 287 (1965)

Chapter 61

IMPLEMENTATION THEORY

THOMAS R. PALFREY*

Division of Humanities and Social Sciences, California Institute of Technology, Pasadena, CA, USA

Contents

*I thank John Duggan, Matthew Jackson, and Sanjay Srivastava for helpful comments and many enlightening discussions on the subject of implementation theory. Suggestions from Robert Aumann, Sergiu Hart and two anonymous readers are also gratefully acknowledged. The financial support of the National Science Foundation is gratefully acknowledged.

Handbook of Game Theory, Volume 3, Edited by R.J. Aumann and S. Hart

Abstract

This chapter surveys the branch of implementation theory initiated by Maskin (1999). Results for both complete and incomplete information environments are covered.

Keywords

implementation theory, mechanism design, game theory, social choice

JEL classification: 025, 026

1. Introduction

Implementation theory is an area of research in economic theory that rigorously investigates the correspondence between normative goals and institutions designed to achieve (implement) those goals. More precisely, given a normative goal or welfare criterion for a particular class of allocation problems, or domain of environments, it formally characterizes organizational mechanisms that will guarantee outcomes consistent with that goal, assuming the outcomes of any such mechanism arise from some specification of equilibrium behavior. The approaches to this problem to date lie in the general domain of game theory because, as a matter of definition in the implementation theory literature, an institution is modelled as a mechanism, which is essentially a non-cooperative game form. Moreover, the specific models of equilibrium behavior are usually borrowed from game theory. Consequently, many of the same issues that intrigue game theorists are the focus of attention in implementation theory: How do results change if informational assumptions change? How do results depend on the equilibrium concept governing stable behavior in the game? How do results depend on the distribution of preferences of the players or the number of players? Also, many of the same issues that arise in social choice theory and welfare economics are also at the heart of implementation theory: Are some first-best welfare criteria unachievable? What is the constrained second-best solution? What is the general correspondence between normative axioms on social choice functions and the possibility of strategic manipulation, and how does this correspondence depend on the domain of environments? What is the correspondence between social choice functions and voting rules?

In order to limit this chapter to manageable proportions, attention will be mainly focused on the part of implementation theory using non-cooperative equilibrium concepts that follows the seminal paper by Maskin (1999).[1] This body of research has its roots in the foundational work on decentralization and economic design of Hayek, Koopmans, Hurwicz, Reiter, Marschak, Radner, Vickrey and others that dates back more than half a century.

The limitation to this somewhat restricted subset of the enormous literature of implementation theory and mechanism design excludes four basic categories of results. The first category is implementation via dominant strategy equilibrium, perhaps more familiarly known as the characterization of strategyproof mechanisms. This research, mainly following the seminal work of Gibbard (1973) and Satterthwaite (1975), has close connections with social choice theory, and for that reason has already been treated in some depth in Chapter 31 of this Handbook (Vol. II). The second category excluded is implementation via solution concepts that allow for coalitional behavior, most notably, strong equilibrium [Dutta and Sen (1991c)] and coalition-proof equilibrium [Bernheim and Whinston (1987)]. The third category involves practical issues in the design of mechanisms. There is a vast literature identifying specific classes of mechanisms such

---

[1] The published article is a revision of a paper that was circulated in 1977.

as divide-and-choose, sequential majority voting, auctions, and so forth, and studying/characterizing the range of social choice rules that are implementable using such mechanisms. The fourth category is the many applications of the revelation principle to design mechanisms in Bayesian environments.[2] This includes much work in auction theory, bargaining, contracting, and regulation. These other categories are already covered in some detail in several other chapters of this Handbook.

Later in this chapter we discuss practical aspects of implementation, as it relates to the specific mechanisms used in the constructive proofs of the main theorems. But a few introductory remarks about this are probably in order. In contrast to the literature devoted to studying classes of "natural" mechanisms, the most general characterizations of implementable social choice rules often resort to highly abstract mechanisms, with little or no concern for practical application. The reason for this is that the mechanisms constructed in these theorems are supposed to apply to arbitrary implementable social choice rules. A typical characterization result in implementation theory takes as given an equilibrium concept (say, subgame perfect Nash equilibrium) and tries to identify necessary and sufficient conditions for a social choice rule to be implementable, under minimal domain restrictions on the environment. The method of proof (at least for sufficiency) is to construct a mechanism that will work for *any* implementable social choice rule. That is, a *standard game form* is identified which will implement all such rules with only minor variation across rules. It should come as no surprise that this often involves the construction of highly abstract mechanisms.

A premise of the research covered in this chapter is that to create a general foundation for implementation theory it makes sense to begin by identifying general conditions on social choice rules (and domains and equilibrium concepts) under which there *exists* some implementing mechanism. In this sense it is appropriate to view the highly abstract mechanisms more as vehicles for proving existence theorems than as specific suggestions for the nitty gritty details of organizational design. This is not to say that they provide no insights into the principles of such design, but rather to say that for most practical applications one could (hopefully) strip away a lot of the complexity of the abstract mechanisms. Thus a natural next direction to pursue, particularly for those interested in specific applications, is the identification of practical restrictions on mechanisms and the characterization of social choice rules that can be implemented by mechanisms satisfying such restrictions. Some research in implementation theory is beginning to move in this direction, and, by doing so, is beginning to bridge the gap between this literature and the aforementioned research which focuses on implementation using very special classes of procedures such as binary voting trees and bargaining rules.

---

[2] The approach based on the revelation principle studies the *truthful implementation* [Dasgupta, Hammond and Maskin (1979)] of allocation rules [referred to as *weak implementation* in Palfrey and Srivastava (1993, p. 15)]. That is, an allocation rule is truthfully implementable if there is some equilibrium of some mechanism that leads to the desired allocation rule. This chapter only deals with *full* implementation, where the implementing mechanism must have the property that *all* equilibria of the mechanism lead to the desired allocation rule.

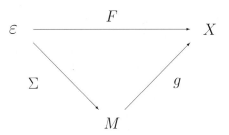

Figure 1. Mount–Reiter diagram.

Finally, there are other surveys that interested readers would find useful. These include Allen (1997), Jackson (2002), Maskin (1985), Moore (1992), Palfrey (1992), Palfrey and Srivastava (1993), and Postlewaite (1985).

1.1. The basic structure of the implementation problem

The simple structure of the general implementation problem provides a convenient way to organize the research that has been conducted to date in this area, and at the same time organize the possibilities for future research. Before presenting this classification scheme, it is useful to present one of the most concise representations of the implementation problem, called a *Mount–Reiter diagram*, a version of which originally appeared in Mount and Reiter (1974).

Figure 1 contains the basic elements of an implementation problem. The notation is as follows:

$\mathcal{E}$: The domain of environments. Each *environment*, $e \in \mathcal{E}$, consists of a set of *feasible outcomes*, $A(e)$, a set of *individuals*, $I(e)$, and a *preference profile* $R(e)$, where $R_i(e)$ is a weak order on $A(e)$.[3]

X: The *outcome space*. Note: $A(e) \subseteq X$ for all $e \in \mathcal{E}$.

$F \subseteq \{ f : \mathcal{E} \to X \}$: The *welfare criterion* (or *social choice set*) which specifies the set of acceptable mappings from environments to outcomes. An element, f, of F is called a *social choice function*.

$M = M_1 \times \cdots \times M_N$: The *message space*.

$g : M \to X$: The *outcome function*.

$\mu = \langle M, g \rangle$: The *mechanism*.

$\Sigma = $ The *equilibrium concept* that maps each μ into $\Sigma_\mu \subseteq \{ \sigma : \mathcal{E} \to M \}$.

As an example to illustrate what these different abstract concepts might be, consider the domain of pure exchange economies. Each element of the domain would consist of a

---

[3] Except where noted, we will assume that the set of feasible outcomes is a constant A that is independent of e, and the planner knows A. Furthermore, we will typically take the set of individuals $I = \{1, 2, \ldots, N\}$ as fixed.

set of traders, each of whom has an initial endowment, and the set of feasible outcomes would just be the set of all reallocations of the initial endowment. Many implementation results rely on domain restrictions, which in this illustration would involve standard assumptions such as strictly increasing, convex preferences, and so forth. The welfare criterion, or social choice set, might consist of all social choice functions satisfying a list of conditions such as individual rationality, Pareto optimality, interiority, envy-freeness, and so forth. One common social choice set is the set of all selections from the Walrasian equilibrium correspondence. For this case the message space might be either an agent's entire preference mapping or perhaps his demand correspondence, and the "planner" would take on the role of the auctioneer. An example of an outcome function would be the allocations implied by a deterministic pricing rule (such as some selection from the set of market clearing prices), given the reported demands. Common equilibrium concepts employed in these settings are Nash equilibrium or dominant strategy equilibrium.

The arrows of the Mount–Reiter diagram indicate that the diagram has the commutative property that, under the equilibrium concept Σ, the set of desirable social choice functions defined by F correspond exactly to the outcomes that arise under the mechanism μ. That is, $F = g \circ \Sigma_\mu$. When this happens, we say that "μ implements F via Σ in $\mathcal{E}$". Whenever there exists some mechanism such that that statement is true, we say "F is *implementable* via Σ in $\mathcal{E}$". Implementation theory then looks at the relationship between domains, equilibrium concepts, welfare criteria, and implementing mechanisms, and the various questions that may arise about this relationship. The remainder of this chapter summarizes a small part of what is known about this. Because this has been the main focus of the literature, the discussion here will concentrate primarily on *the existence question*: Under what conditions on F, Σ, and $\mathcal{E}$ does there exist a mechanism μ such that μ implements F via Σ in $\mathcal{E}$?

1.2. Informational issues

To this point, nothing has been said about what information can be used in the construction of a mechanism nor about what information the individuals have about $\mathcal{E}$. A common interpretation given to the implementation problem is that there is a mythical agent, called "the planner", who has normative goals and very limited information about the environment (typically one assumes that the planner only knows $\mathcal{E}$ and X). The planner then must elicit enough information from the individuals to implement outcomes in a manner consistent with those normative goals. This requires creating incentives for such information to be voluntarily and accurately provided by the individuals. Nearly all of the literature in implementation theory assumes that the details of the environment are not directly verifiable by the planner, even *ex post*.[4] Thus, implementation theory characterizes the limits of a planner's power in a society with decentralized information.

---

[4] There are some results on auditing and other *ex post* verification procedures. See, for example, Townsend (1979), Chander and Wilde (1998) and the references they cite.

The information that the individuals have about $\mathcal{E}$ is also an important consideration. This information is best thought of as part of the description of the domain. The main information distinction that is usually made in the literature is between *complete information* and *incomplete information*. Complete information models assume that e is common knowledge among the individuals (but, of course, unknown to the planner). Incomplete information assumes that individuals have some private information. This is usually modeled in the Harsanyi (1967, 1968) tradition, by defining an environment as a profile of types, one for each individual, where a type indexes an individual's information about other individuals' types. In this manner, an environment (preference profile, set of feasible allocations, etc.) is uniquely defined for each possible type profile.

One branch of implementation theory addresses a somewhat different informational issue. Given a social choice correspondence and a domain of environments, how much information about the environment is minimally needed to determine which outcome should be selected, and how close to this level of minimal information gathering (informationally efficient) do different "natural" mechanisms come? In particular, this question has been asked in the domain of neoclassical pure exchange environments, where the answer is that the Walrasian market mechanism is informationally efficient [Hurwicz (1977); Mount and Reiter (1974)]. With few exceptions that branch of implementation theory does not directly address questions of incentive compatibility of mechanisms. This chapter will not cover the contributions in that area.

1.3. Equilibrium: Incentive compatibility and uniqueness

In principle, the equilibrium concept Σ could be almost anything. It simply defines a systematic rule for mapping environments into messages for arbitrary mechanisms. However, nearly all work in implementation theory and mechanism design restricts attention to equilibrium concepts borrowed from noncooperative game theory, all of which require the *rational response property* in one form or another. That is, each individual, given their information, preferences, and a set of assumptions about how other individuals are behaving, and given a set of rules $\langle M, g \rangle$, adopts a rational response, where rationality is usually based on maximization of expected utility.[5]

The requirement of implementation can be broken down into two components. The first component is *incentive compatibility*:[6] This is most transparent for the special case of social choice functions (i.e., F is a singleton). If a mechanism $\langle M, g \rangle$ implements a social choice function f, it must be the case that there is an equilibrium strategy profile

---

[5] There are some exceptions, notably the use of maximin strategies [Thomson (1979)] and undominated strategies [Jackson (1992)], which do not require players to adopt expected utility maximizing responses to the strategies of the other players.

[6] This is sometimes referred to as *Truthful Implementability* [Dasgupta, Hammond, and Maskin (1979)] because, in the framework where individual preferences and information are represented as "types", if the mechanism is *direct* in the sense that each individual is required to report their type (i.e., $M = T$), then the truthful strategy $\sigma(t) = t$ is an equilibrium of the direct mechanism $\mu = \langle T, f \rangle$.

$\sigma : \mathcal{E} \to M$ such that $g \circ \sigma = f$. The second component is *uniqueness*: If a mechanism $\langle M, g \rangle$ implements a social choice function f, it must be the case, for all social choice functions $h \neq f$, that there is *not* an equilibrium strategy profile σ' such that $g \circ \sigma' = h$.

For the more general case of the implementation of a social choice set, F, these two components extend in the natural way. If a mechanism $\langle M, g \rangle$ implements a social choice set F, it must be the case that, for each $f \in F$ there is an equilibrium strategy profile σ such that $g \circ \sigma = f$. If a mechanism $\langle M, g \rangle$ implements a social choice set F, it must be the case, for all social choice functions $h \notin F$, that there is *not* an equilibrium strategy profile σ such that $g \circ \sigma = h$.

1.4. The organization of this chapter

The remainder of this paper is divided into three sections. Section 2 presents characterizations of implementation under conditions of complete information for several different equilibrium concepts. In particular, the relatively comprehensive characterization for *Nash implementation* (i.e., implementation in complete information domains via Nash equilibrium) is set out in considerable detail. The partial characterizations for *refined Nash implementation* (subgame perfect equilibrium, undominated Nash equilibrium, dominance solvable equilibrium, etc.) are then discussed. This part also describes the problem of implementation by games of perfect information, and a few results in the area, particularly with regard to "voting trees", are briefly discussed. Finally results for *virtual implementation* (both in Nash equilibrium and using refinements) are described, where social choice functions are only approximated as the equilibrium outcomes of mechanisms.

Section 3 explains how the results for complete information are extended to incomplete information "Bayesian" domains, where environments are represented as collections of Harsanyi type-profiles, and players are assumed to have well-defined common knowledge priors about the distribution of types.

Section 4 discusses some "difficult" problems in the area of implementation theory that have either been ignored or studied in only the simplest settings. This includes dynamic issues such as renegotiation of mechanisms and dynamic allocation problems, considerations of simplicity, robustness, and bounded rationality, and issues of incomplete control by the planner over the mechanism (side games played by the agents, or preplay communication).

2. Implementation under conditions of complete information

By complete information, we mean that individual preferences and feasible alternatives are common knowledge among all the individuals. This does *not* mean that the planner knows these preferences. The planner is assumed to know only the set of possible individual preference profiles and the set of feasible allocations.[7] For this reason, we simplify the notation considerably for this section of the chapter. First, we represent the domain by $\boldsymbol{R}$, the set of possible preference profiles, with typical element $R = (R_1, R_2, \ldots, R_N)$, and the set of feasible alternatives is A. A social choice set can, without loss of generality, be represented as a correspondence F mapping $\boldsymbol{R}$ into subsets of A. We denote the image of F at R by $F(R)$.

2.1. Nash implementation

Consider a mechanism $\mu = \langle M, g \rangle$ and a profile R. The pair (μ, R) defines an N-player noncooperative game.

DEFINITION 1. A message profile $m^* \in M$ is called a *Nash equilibrium of μ at R* if, for all $i \in I$, and for all $m_i \in M_i$

$$g(m^*) R_i g(m_i, m^*_{-i}).$$

Therefore, the definition of Nash implementability is:

DEFINITION 2. A social choice correspondence F is *Nash implementable in $\boldsymbol{R}$* if there exists a mechanism $\mu = \langle M, g \rangle$ such that:
   (a) For every $R \in \boldsymbol{R}$ and for every $x \in F(\boldsymbol{R})$, there exists $m^* \in M$ such that m^* is a Nash equilibrium of μ at R and $g(m^*) = x$.
   (b) For every $R \in \boldsymbol{R}$ and for every $y \notin F(R)$, there does not exist $m^* \in M$ such that m^* is a Nash equilibrium of μ at R and $g(m^*) = y$.

Alternatively, writing $\sigma^*(R)$ as the set of Nash equilibria of μ at R, we can state (a) and (b) as:
   (a') $F(R) \subseteq g \circ \sigma^*(R)$ for all $R \in \boldsymbol{R}$,
   (b') $g \circ \sigma^*(R) \subseteq F(R)$ for all $R \in \boldsymbol{R}$.
   Condition (a) corresponds to what we have referred to as incentive compatibility and condition (b) is what we have referred to as uniqueness. We proceed from here by characterizing the implications of (a) and (b).

2.1.1. Incentive compatibility

Suppose a social choice function $f : \boldsymbol{R} \rightarrow A$ is implementable via Nash equilibrium. Then there exists a mechanism μ that implements f via Σ in $\mathcal{E}$. What exactly does this

---

[7] Nearly always, the set of feasible allocations is taken as fixed in implementation theory. A notable exception to this is Hurwicz, Maskin, Postlewaite (1995). In this paper, the mechanism has individuals report endowments as well as preferences to the planner, but it is assumed that it is impossible for an individual to overstate his endowment (although understatements are possible). See Hong (1996, 1998) and Tian (1994) for extensions to Bayesian environments.

mean? First, it means that there is a Nash equilibrium of μ that yields f as the equilibrium outcome function. With complete information this turns out to have very little bite. That is, examples of social choice functions that are not "incentive compatible" when the individuals have complete information are rather special. How special, you might ask. First, the examples must have only two individuals. This fact is quickly established below as Proposition 3.

PROPOSITION 3. *In complete information domains with $N > 2$, every social choice function f is incentive compatible (i.e., there exists a mechanism such that (a) is satisfied).*

PROOF. Consider the following mechanism, which we call the *agreement mechanism*. Let $M_i = \boldsymbol{R}$ for all $i \in I$. That is, individuals report profiles.[8] Arbitrarily pick some $a_0 \in A$. Partition the message space into two parts, M_a (called the *agreement region*) and M_d (called the *disagreement region*).

$$M_a = \{m \in M \mid \exists j \in I, \ R \in \boldsymbol{R} \text{ such that } m_i = R \text{ for all } i \neq j\};$$
$$M_d = \{m \in M \mid m \notin M_a\}.$$

In other words, the agreement region consists of all message profiles where either every individual reports the same preference profile, or all except one individual report the same preference profile. The outcome function is then defined as follows.

$$g(m) = \begin{cases} f(R) & \text{if } m \in M_a, \\ a_o & \text{if } m \in M_d. \end{cases}$$

It is easy to see that if the profile of preferences is R, then it is a Nash equilibrium for all individuals to report $m_i = R$, since unilateral deviations from unanimous agreement will not affect the outcome. Therefore, (a) is satisfied. □

Therefore, incentive compatibility[9] is not an issue when information is complete and $N > 2$. Of course the problem with this mechanism is that *any unanimous message profile is a Nash equilibrium at any preference profile*. Therefore, this mechanism does not satisfy the uniqueness requirement (b) of implementation. We return to this problem shortly, after addressing the question of incentive compatibility for the $N = 2$ case.

---

[8] This mechanism could be simplified further by having each agent report a feasible outcome.

[9] The reader should not confuse this with a number of negative results on implementation of social choice functions via Nash equilibrium when mechanisms are restricted to being "direct" mechanisms ($M_i = R_j$). If individuals do not report profiles, but only report their own component of the profile (sometimes called *privacy-preserving mechanisms*), then clearly incentive compatibility can be a problem. This kind of incentive compatibility is usually called strategyproofness, and is closely related to the problem of implementation under the much stronger equilibrium concept of dominant strategy equilibrium. See Dasgupta, Hammond, and Maskin (1979).

When $N = 2$ the outcome function of the mechanism used in Proposition 3 is not well defined, since a unilateral deviation from unanimous agreement is not well defined. If $m_1 = R$ and $m_2 = R'$, then it is unclear whether $g(m) = f(R)$ or $g(m) = f(R')$. There are some simple cases where incentive compatibility is assured when $N = 2$. First, if there exists a uniformly bad outcome, w, with the property that, for all $a \in A$, and for all $R \in \mathbf{R}$, $a R_i w$, $i = 1, 2$. In that case, the mechanism above can be modified so that M_a requires unanimous agreement, and $a_o = w$. Clearly any unanimous report of a profile is a Nash equilibrium regardless of the actual preferences of the individuals, so this modified mechanism satisfies (a) but fails to satisfy (b).

A considerably weaker assumption, called *nonempty lower intersection*, is due to Dutta and Sen (1991b) and Moore and Repullo (1990). We state a slightly weaker version below, which is sufficient for the incentive compatibility requirement (a) when $N = 2$. They define a slightly stronger version that is needed to satisfy the uniqueness requirement (b).

DEFINITION 4. A social choice function f satisfies *Weak Nonempty Lower Intersection*[10] if, for all $R, R' \in \mathbf{R}$, such that $R \neq R'$, $\exists c \in A$ such that $f(R) R_1 c$ and $f(R') R_2' c$.

The definition of social choice correspondences is similar:

DEFINITION 5. A social choice correspondence F satisfies *Weak Nonempty Lower Intersection* if for all $R, R' \in \mathbf{R}$, such that $R \neq R'$, and for all $a \in f(R)$ and $b \in f(R')$, $\exists c \in A$ such that $a R_1 c$ and $b R_2' c$.

To see that this is a sufficient condition for (a), consider implementing the social choice function f. From Definition 4, we can define a function $c(R, R')$ for $R \neq R'$ with the property that $f(R) R_1 c(R, R')$ and $f(R') R_2' c(R, R')$. We can then modify the mechanism above by:

$$ g(m) = \begin{cases} f(R) & \text{if } m_1 = m_2 = R, \\ c(R', R) & \text{if } m_1 = R \text{ and } m_2 = R'. \end{cases} $$

This mechanism is illustrated in Figure 2. It is easy to check that weak nonempty lower intersection guarantees $m = (R, R)$ is a Nash equilibrium when the actual profile is R.

There are two interesting special cases where Nonempty Lower Intersection holds. The first is when there exists a universally "bad" outcome [Moore and Repullo (1990)] with the property that it is strictly less preferred than all outcomes in the range of the social choice rule, for all agents, at all profiles in the domain.[11] This is satisfied by any

---

[10] The stronger version, called *Nonempty Lower Intersection*, requires $f(R) P_1 c$ and $f(R') P_2' c$.

[11] Notice that this is a joint restriction on the domain and the social choice function.

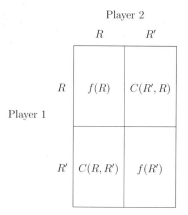

Figure 2. (Weak) Nonempty Lower Intersection: $f(R)R_1C(R, R')$, $f(R')R_2'C(R, R')$ and $f(R')R_1'C(R', R)$, $f(R)R_2C(R', R) \Rightarrow f(R) \in NE(R), f(R') \in NE(R')$.

nonwasteful social choice rule in exchange economies with free disposal and strictly increasing preferences, since destruction of the endowment is a bad outcome. The second special case is any Pareto efficient and individually rational interior social choice correspondence in exchange economies (with or without free disposal) with strictly convex and strictly increasing preferences and fixed initial endowments [Dutta and Sen (1991b); Moore and Repullo (1990)].

2.1.2. Uniqueness

Clearly, incentive compatibility places few restrictions on Nash implementable social choice functions (and correspondences) with complete information. The second requirement of uniqueness is more difficult, and the major breakthrough in characterizing this was the classic paper of Maskin (1999). In that paper, he introduces two conditions, which are jointly sufficient for Nash implementation when $N \geqslant 3$. These conditions are called *Monotonicity* and *No Veto Power* (*NVP*).

DEFINITION 6. A social choice correspondence F is *Monotonic* if, for all $R, R' \in R$

$$\left(x \in F(R), x \notin F(R')\right) \Rightarrow \exists i \in I, \ a \in A \text{ such that } x R_i a P_i' x.$$

The agent i and the alternative a are called, respectively, the *test agent* and the *test alternative*. Stated in the contrapositive, this says simply that if x is a socially desired alternative at R, and x does not strictly fall in preference for anyone when the profile is changed to R', then x must be a socially desired alternative at R'. Thus monotonic social choice correspondences must satisfy a version of a nonnegative responsiveness criterion with respect to individual preferences. In fact, this is a remarkably strong requirement for a social choice correspondence. For example, it rules out nearly any scoring rule, such as the Borda count or Plurality voting. Several other examples of nonmonotonic social choice functions in applications to bilateral contracting are given in Moore and Repullo (1988). One very nice illustration of a nonmonotonic social choice correspondence is given in the following example. It is a variation on the "King Solomon's Dilemma" example of Glazer and Ma (1989) and Moore (1992), where the problem is to allocate a baby to its true mother.

2.1.3. Example 1

There are two individuals in the game (Ms. α and Ms. β), and four possible alternatives:

> $a =$ give the baby to Ms. α,
>
> $b =$ give the baby to Ms. β,
>
> $c =$ divide the baby into two equal halves and give each mother one half,
>
> $d =$ execute both mothers and the child.

Assume the domain consists of only two possible preference profiles depending on whether α or β is the real mother, and we will call these profiles R and R' respectively. They are given below:

$$R^\alpha = a \succ b \succ c \succ d, \qquad R'_\alpha = a \succ c \succ b \succ d,$$
$$R_\beta = b \succ c \succ a \succ d, \qquad R'_\beta = b \succ a \succ c \succ d.$$

The social choice function King Solomon wishes to implement is $f(R) = a$ and $f(R') = b$. This is not monotonic. Consider the change from R to R'. Alternative a does not fall in either player's preference order as a result of this change. However, $f(R') = b \neq a$, a contradiction of monotonicity. Notice however that this social choice function is incentive compatible since there is a universally bad outcome, d, which is ranked last by both players in both of their preference orders.

A second example, from a neoclassical 2-person pure exchange environment illustrates the geometry of monotonicity.

2.1.4. Example 2

Consider allocation x in Figure 3.

Suppose $x \in f(R)$ where the indifference curves through x of the two individuals are labelled R_1 and R_2 respectively in that figure. Now consider some other profile R' where $R_2 = R'_2$, and R'_1 is such that the lower contour set of x for individual 1 has expanded. Monotonicity would require $x \in f(R')$. Put another way (formally stated in the definition), if f is monotonic and $x \notin f(R')$ for some $R' \neq R$, then one of the two individuals must have an indifference curve through x that either crosses the R-indifference

Agent 2

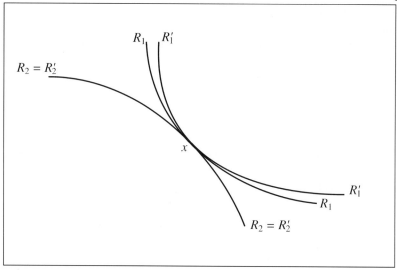

Agent 1

Figure 3. Illustration of monotonicity: $x \in F(R) \Rightarrow x \in F(R')$.

curve through x or bounds a strictly smaller lower contour set. Figure 4 illustrates the (generic) case in which the R'-indifference curve of one of the individuals (individual 1, in the figure) crosses the R-indifference curve through x. Thus, in this example agent 1 is the test agent. One possible test alternative $a \in A$ (an alternative required in the definition of monotonicity which has the property that $x R_i a P'_i x$) is marked in that figure.

2.1.5. Necessary and sufficient conditions

Maskin (1999) proved that monotonicity is a necessary condition for Nash implementation.

THEOREM 7. *If F is Nash implementable then F is monotonic.*

PROOF. Consider any mechanism μ that Nash implements F and consider some $x \in F(R)$ and some Nash equilibrium message, m^*, at profile R, such that $g(m^*) = x$. Define the "option set"[12] for i at m^* as

$$O_i(m^*; \mu) = \{a \in A \mid \exists m'_i \in M_i \text{ such that } g(m'_i, m^*_{-i}) = a\}.$$

---

[12] This is similar to the role of option sets in the strategyproofness literature.

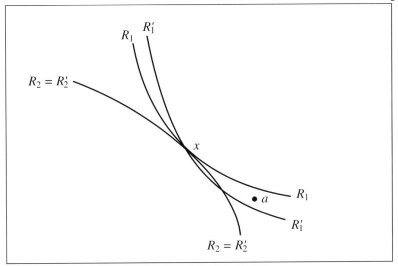

Figure 4. Illustration of tests agent and test allocation: $x R_1 a P_1' x$.

That is, fixing the messages of the other players at m^*_{-i}, the range of possible outcomes that can result for some message by i under the mechanism μ is $O_i(m^*; \mu)$.
By the definition of Nash equilibrium, $x_i^* R_i a$ for all i and for all $a \in O_i(m^*; \mu)$. Now
consider some new profile R' where $x \notin F(R')$. Since μ Nash implements F, it must
be that m is not a Nash equilibrium at R'. Thus there exist some i and some alternative
$a \in O_i(m^*; \mu)$ such that $a P_i' x$. Thus a is the test alternative and i is the test agent as
required in Definition 6, with the property that $x R_i a P_i' x$.   □

The second theorem in Maskin (1999) was proved in papers by Williams (1984,
1986), Saijo (1988), McKelvey (1989), and Repullo (1987).[13] It provides an elegant
and simple sufficient condition for Nash implementation for the case of three or more
agents. This is a condition of near unanimity, called No Veto Power (NVP).

DEFINITION 8.   A social choice correspondence F satisfies *No Veto Power* (NVP) if,
for all $R \in \mathbf{R}$ and for all $x \in A$, and $i \in I$,

$$[x R_j y \text{ for all } j \neq i, \text{ for all } y \in Y] \Rightarrow x \in F(R).$$

[13] Groves (1979) credits Karl Vind for some of the ideas underlying the now-standard constructive proofs
of sufficiency, and writes down a version of the theorem that uses these ideas in the proof. That version
of the theorem, as well as the version of the theorem proved by Williams (1984, 1986), imposes stronger
assumptions than in Maskin (1999).

THEOREM 9. *If $N \geqslant 3$ and F is Monotonic and satisfies NVP, then F is Nash implementable.*

PROOF. The proof given here is constructive and is similar to one in Repullo (1987). A very general mechanism is defined, and then the rest of the proof consists of demonstrating that the mechanism implements any social choice function that satisfies the hypotheses of the theorem. This is usually how characterization theorems are proved in implementation theory. Consider the following generic mechanism, which we call the *agreement/integer mechanism*:

$$M_i = \boldsymbol{R} \times A \times \{0, 1, 2, \ldots\}.$$

That is, each individual reports a **profile**, an **allocation**, and an **integer**. The outcome function is similar to the agreement mechanism, except the disagreement region is a bit more complicated, and agreement must be with respect to an allocation and a profile.

$$M_a = \big\{m \in M \mid \exists j, \ R \in \boldsymbol{R}, \ a \in F(R) \text{ such that } m_i = (R, a, z_i),$$
$$\text{where } z_i = 0 \text{ for each } i \neq j\big\};$$
$$M_d = \{m \in M \mid m \notin M_a\}.$$

The outcome function is defined as follows. The outcome function is constructed so that, if the message profile is in M_a, then the outcome is either a or the allocation announced by individual j, which we will denote a_j. If the outcome is in M_d, then the outcome is a_k where k is the individual announcing the highest integer (ties broken by a predetermined rule). This feature of the mechanism has become commonly known as an *integer game* (although in actuality, it is only a piece of the original game form). Formally,

$$g(m) = \begin{cases} a & \text{if } m \in M_a \text{ and } a_j P_j a, \\ a_j & \text{if } m \in M_a \text{ and } a R_j a_j, \\ a_k & \text{if } m \in M_d \text{ and } k = \max\{i \in I \mid z_i \geqslant z_j \text{ for all } j \in I\}. \end{cases}$$

Recall that we must show that $F(R) \subseteq NE_\mu(R)$ and $NE_\mu(R) \subseteq F(R)$ for all R, where $NE_\mu(R) = \{a \in A \mid \exists m \in M \text{ such that } a = g(m) R_i g(m_i', m_{-i}) \text{ for all } i \in I, m_i' \in M_i\}$ is the set of Nash equilibrium outcomes to μ at R.

(1) $F(R) \subseteq NE_\mu(R)$.

At any R, and for any $a \in F(R)$, there is a Nash equilibrium in which all individuals report $m_i = (R, a, 0)$. Such a message lies in M_a and any unilateral deviation also lies in M_a. The only unilateral deviation that could change the outcome is a deviation in which some player j reports an alternative a_j such that $a R_j a_j$. Therefore, a is a R_j-maximal element of $O_j(m; \mu)$ for all $j \in I$, so $m = (R, a, 0)$ is a Nash equilibrium.

(2) $NE_\mu(R) \subseteq F(R)$.

This is the more delicate part of the proof, and is the part that exploits Monotonicity and NVP. (Notice that part (1) of the proof above exploited only the assumption that $N \geqslant 3$.) Suppose that $m \in NE_\mu(R)$ and $g(m) = a \notin F(R)$. First notice that it cannot be the case that all individuals are reporting $(R', a, 0)$ where $a \in F(R')$ for some $R' \in \boldsymbol{R}$. This would put the outcome in M_a and Monotonicity guarantees the existence of some $j \in I, b \in A$ such that a $R'_j b P_j a$, so player j is better off changing to a message $(\cdot, b, \cdot)$ which changes the outcome from a to b. Thus $m_i \neq m_j$ for some i, j. Whenever this is the case, the option set for at least $N - 1$ of the agents is the entire alternative space, A. Since $a \notin F(R)$ and F satisfies NVP, it must be that there is at least one of these $N - 1$ agents, k, and some element $c \in A$ such that $c P_k a$. Since the option set for k is the entire alternative space, A, individual k is better off changing his message to $(\cdot, c, z_k) \neq m_j$ where $z_k > z_j$, $j \neq k$, which will change the outcome from a to c. This contradicts the hypothesis that m is a Nash equilibrium. $\square$

Since these two results, improvements have been developed to make the characterization of Nash implementation complete and/or to reduce the size of the message space of the general implementing mechanism. These improvements are in Moore and Repullo (1990), Dutta and Sen (1991b), Danilov (1992),[14] Saijo (1988), Sjöström (1991) and McKelvey (1989) and the references they cite.

The last part of this section on Nash implementation is devoted to a simple application to pure exchange economies. It turns out the Walrasian correspondence satisfies both Monotonicity and NVP under some mild domain restrictions. First notice that in private good economies with strictly increasing preferences and three or more agents, NVP is satisfied vacuously. Next suppose that indifference curves are strictly quasi-concave and twice continuously differentiable, endowments for all individuals are strictly positive in every good, and indifference curves never touch the axis. It is well known that these conditions are sufficient to guarantee the existence of a Walrasian equilibrium and to further guarantee that all Walrasian equilibrium allocations in these environments are interior points, with every individual consuming a positive amount of every good in every competitive equilibrium. Finally, assume that "the planner" knows everyone's endowment.[15]

---

[14] Of these, Danilov (1992) establishes a particularly elegant necessary and sufficient condition (with three or more players), which is a generalization of the notion of monotonicity, called *essential monotonicity*. However, these results are limited somewhat by this assumption of universal domain. Nash implementable social choice correspondences need not satisfy essential monotonicity under domain restrictions. Yamato (1992) gives necessary and sufficient conditions on restricted domains.

[15] This assumption can be relaxed. See Hurwicz, Maskin, and Postlewaite (1995). The Walrasian correspondence can also be modified somewhat to the "constrained Walrasian correspondence" which constrains individual demands in a particular way. This modified competitive equilibrium can be shown to be implementable in more general economic domains in which Walrasian equilibrium allocations are not guaranteed to be locally unique and interior. See the survey by Postlewaite (1985), or Hurwicz (1986).

Since the planner knows the endowments, a different mechanism can be constructed for each endowment profile. Thus, to check for monotonicity it suffices to show that the Walrasian correspondence, with endowments fixed and only preferences changing, is monotonic. If a is a Walrasian equilibrium allocation at R and not a Walrasian equilibrium allocation at R', then there exists some individual for whom the supporting price line for the equilibrium at R is not tangent to the R'_i indifference curve through a. But this is just the same as the illustration in Figure 4, and we have labelled allocation b as a "test allocation" as required by the monotonicity definition. The key is that for a to be a Walrasian equilibrium allocation at R and not a Walrasian equilibrium allocation at R' implies that the indifference curves through x at R and R' cross at x.

As mentioned briefly above, there are many environments and "nice" (from a normative standpoint) allocation rules that violate Monotonicity, and in the $N = 2$ case ("bilateral contracting" environments) NVP is simply too strong a condition to impose on a social choice function. There are two possible responses to this problem. One possibility, and the main direction implementation theory has pursued, is that Nash equilibrium places insufficient restrictions on the behavior of individuals.[16] This leads to consideration of implementation using refinements of Nash equilibrium, or *refined Nash implementation*. A second possibility is that implementability places very strong restrictions on what kinds of social choice functions a planner can hope to enforce in a decentralized way. If not all social choice functions can be implemented, then we need to ask "how close" we can get to implementing a desired social choice function. This has led to the work in *virtual implementation*. These two directions are discussed next.

2.2. Refined Nash implementation

More social choice correspondences can be implemented using refinements of Nash equilibrium. The reason for this is straightforward, and is easiest to grasp in the case of $N \geqslant 3$. In that case, the incentive compatibility problem does not arise (Proposition 3), so the only issue is (ii) uniqueness. Thus the problem with Nash implementation is that Nash equilibrium is too permissive an equilibrium concept. A nonmonotonic social choice function fails to be implementable simply because there are too many Nash equilibria. It is impossible to have $f(R)$ be a Nash equilibrium outcome at R and at the same time avoid having $a \neq f(R)$ also be a Nash equilibrium outcome at R. But of

---

[16] One might argue to the contrary that in other ways Nash equilibrium places too strong a restriction on individual behavior. Both directions are undoubtedly true. Experimental evidence has shown that both of these are defensible. On the one hand, some refinements of Nash equilibrium have received experimental support indicating that additional restrictions beyond mutual best response have predictive value [Banks, Camerer, and Porter (1994)]. On the other hand, many experiments indicate that players are at best imperfectly rational, and even violate simple basic axioms such as transitivity and dominance. Thus, from a practical standpoint, it is very important to explore the implementation question under assumptions other than the simple mutual best response criterion of Nash equilibrium.

course this is exactly the kind of problem that refinements of Nash equilibrium can be used for. The trick in implementation theory with refinements is to exploit the refinement by constructing a mechanism so that precisely the "bad" equilibria (the equilibria whose outcomes lie outside of F) are refined away, while the other equilibria survive the refinement.[17]

2.2.1. Subgame perfect implementation

The first systematic approach to extending the Maskin characterization beyond Nash equilibrium in complete information environments was to look at implementation via subgame perfect Nash equilibrium [Moore and Repullo (1988); Abreu and Sen (1990)]. They find that more social choice functions can be implemented via subgame perfect Nash equilibrium than via Nash equilibrium. The idea is that sequential rationality can be exploited to eliminate certain bad equilibria. The following simple example in the "voting/social choice" tradition illustrates the point.

2.2.2. Example 3

There are three players on a committee who are to decide between three alternatives, $A = \{a, b, c\}$. There are two profiles in the domain, denoted R and R'. Individuals 1 and 2 have the same preferences in both profiles. Only player 3 has a different preference order under R than under R'. These are listed below:

$$R_1 = R_1' = a \succ b \succ c, \qquad R_2 = R_2' = b \succ c \succ a,$$
$$R_3 = c \succ a \succ b, \qquad R_3' = a \succ c \succ b.$$

The following social choice function is Pareto optimal and satisfies the Condorcet criterion that an alternative should be selected if it is preferred by a majority to any other alternative:

$$f(R) = b, \qquad f(R') = a.$$

This social choice function violates monotonicity since b does not go down in player 3's rankings when moving from profile R to R' (and no one else's preferences change). Therefore it is not Nash implementable. However, the following trivial mechanism (extensive form game form) implements it in subgame perfect Nash equilibrium:

---

[17] Earlier work by Farquharson (1957/1969), Moulin (1979), Crawford (1979) and Demange (1984) in specific applications of multistage games to voting theory, bargaining theory, and exchange economies foreshadows the more abstract formulation in the relatively more recent work in implementation theory with refinements.

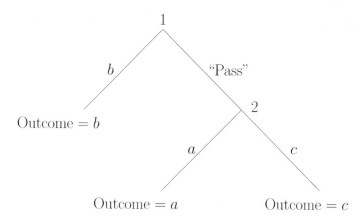

Figure 5. Game tree for Example 2.

Stage 1: Player 1 either chooses alternative b, or passes. The game ends if b is chosen. The game proceeds to stage 2 if player 1 passes.

Stage 2: Player 3 chooses between a and c. The game ends at this point.

The voting tree is illustrated in Figure 5.

To see that this game implements f, work back from the final stage. In stage 2, player 3 would choose c in profile R and a in profile R'. Therefore, player 1's best response is to choose b in profile R and to pass in profile R'. Notice that there is another Nash equilibrium under profile R', where player 2 adopts the strategy of choosing c if player 1 passes, and thus player 1 chooses b in stage 1. But of course this is not sequentially rational and is therefore ruled out by subgame perfection.

2.2.3. Characterization results

Abreu and Sen (1990) provide a nearly complete characterization of social choice correspondences that are implementable via subgame perfect Nash equilibrium, by giving a general necessary condition, which is also sufficient if $N \geqslant 3$ for social choice functions satisfying NVP. This condition is strictly weaker than Monotonicity, in the following way. Recall that monotonicity requires, for any R, R' and $a \in A$, with $a = f(R)$, $a \neq f(R')$, the existence of a test agent i and a test allocation b such that $aR_ibR_i'a$. That there is some player and some allocation that produces a preference switch with $f(R)$ when moving from R to R'. The weakening of this resulting from the sequential rationality refinement is that the preference switch does not have to involve $f(R)$ directly. Any preference switch between two alternatives, say b and c, will do, as long as these alternatives can be indirectly linked to $f(R)$ in a particular fashion. We formally state this necessary condition and call it *indirect monotonicity*,[18] to contrast it with the

---

[18] Abreu and Sen (1990) call it *Condition α*.

direct linkage to $f(R)$ of the test alternative in the original definition of monotonicity.

DEFINITION 10. A social choice correspondence F satisfies *indirect monotonicity* if there $\exists B \subseteq A$ such that $F(R) \subseteq B$ for all $R \in \mathcal{R}$, and if for all R, R' and $a \in A$, with $a \in F(R), a \notin F(R'), \exists L < \infty$, and $\exists$ a sequence of agents $\{j_0, \ldots, j_L\}$ and $\exists$ sequences of alternatives $\{a_0, \ldots, a_{L+1}\}$, $\{b_0, \ldots, b_L\}$ belonging to B such that:

    (i) $a_k R_{jk} a_{k+1}, k = 0, 1, \ldots, L$;

    (ii) $a_{L+1} P'_{jL} a_L$;

    (iii) $b_k P'_{jk} a_k, k = 0, 1, \ldots, L$;

    (iv) $(a_{L+1} R'_{jL} b \ \forall b \in B) \Rightarrow (L = 0 \text{ or } j_{L-1} = j_L)$.

The key parts of the definition of indirect monotonicity are (i) and (ii). A less restrictive version of indirect monotonicity consisting of only parts (i) and (ii) was used first by Moore and Repullo (1988) as a weaker necessary condition for implementation in subgame perfect equilibrium with multistage games.

The main two general theorems about implementation via subgame perfect implementation are the following. The proofs [Abreu and Sen (1990)] are long and tedious and are omitted, although an outline of the proof for the sufficiency result is given. Similar results, but slightly less general, can be found in Moore and Repullo (1988).

THEOREM 11 (Necessity). *If a social choice correspondence F is implementable via subgame perfect Nash equilibrium, then F satisfies indirect monotonicity.*

THEOREM 12 (Sufficiency). *If $N \geqslant 3$ and F satisfies NVP and indirect monotonicity, then F is implementable via subgame perfect Nash equilibrium.*

PROOF. This is an outline of the sufficiency proof. Since F satisfies indirect monotonicity, there exists the required set B and for any (R, R', a) such that $a \in F(R)$ and $a \notin F(R')$ there exists an integer L and the required sequences $\{j_k(R, R', a)\}_{k=0,1,\ldots,L}$ and $\{a_k(R, R', a)\}_{k=0,1,\ldots,L+1}$ that satisfy (i)–(iv) of Definition 6. In the first stage of the mechanism, all agents announce a triple of the form (m_{i1}, m_{i2}, m_{i3}) where $m_{i1} \in R$, $m_{i2} \in A$, and $m_{i3} \in \{0, 1, \ldots\}$. The first stage of the game then conforms fairly closely to the agreement/integer mechanism, with a minor exception. If there is too much disagreement (there exist three or more agents whose reports are different) the outcome is determined by m_{i2} of the agent who announced the largest integer. If there is unanimous agreement in the first two components of the message, so all agents send some (R, a, z_i) and $a \in F(R)$, then the game is over and the outcome is a. The same is true if there is only one disagreeing report in the first two components, unless the dissenting report is sent by $i_0(R, m_{i01}, a)$, in which case the first of a sequence of at most L "binary" agreement/integer games is triggered in which either some agent gets to choose his most preferred element of B or the next in the sequence of binary agreement/integer games is triggered. If the game ever gets to the $(L + 1)$st stage, then the outcome is a_{L+1} and the game ends.

The rest of the proof follows the usual order. First, one shows that for all $R \in \mathcal{R}$ and for all $a \in F(R)$ there is a subgame perfect Nash equilibrium at R with a as the equilibrium outcome. Second, one shows that for all $R \in \mathcal{R}$ and for all $a \notin F(R)$ there is no subgame perfect Nash equilibrium at R with a as the equilibrium outcome. $\square$

In spite of its formidable complexity, some progress has been made tracing out the implications of indirect monotonicity for two well-known classes of implementation problems: exchange economies and voting. Moore and Repullo (1988) show that any selection from the Walrasian equilibrium correspondence satisfies indirect monotonicity, in spite of being nonmonotonic. There are also some results for the $N = 2$ case that can be found in Moore (1992) and Moore and Repullo (1988) which rely on sidepayments of a divisible private good. The case of voting-based social choice rules contrasts sharply with this. Abreu and Sen (1990), Palfrey and Srivastava (1991a) and Sen (1987) show that many voting rules fail to satisfy indirect monotonicity, as do most runoff procedures and "scoring rules" (such as the famous Borda rule). However, a class of voting-based social choice correspondences, including the Copeland rule, is implementable via subgame perfect Nash equilibrium [Sen (1987)]. Some related findings are in Moulin (1979), Dutta and Sen (1993), and the references they cite.

There are a number of applications that exploit the combined power of sidepayments and sequential mechanisms. See Glazer and Ma (1989), Varian (1994), and Jackson and Moulin (1990). Moore (1992) also gives some additional examples.

2.2.4. Implementation by backward induction and voting trees

In general, it is not possible to implement a social choice function via subgame perfect Nash equilibrium without resorting to games of imperfect information. At some point, it is necessary to have a stage with simultaneous moves. Others have investigated the implementation question when mechanisms are restricted to games of perfect information. In that case, the refinement implied by solving the game in its last stage and working back to earlier moves, generates similar behavior as subgame perfect equilibrium.[19] Example 3 above illustrates how it is possible for nonmonotonic social choice functions to be implemented via backward induction. The work of Glazer and Ma (1989) illustrates how economic contracting and allocation problems similar in structure to Example 1 (King Solomon's Dilemma) can be solved with backward induction implementation if sidepayments are possible. Crawford's work (1977, 1979) on bargaining

---

[19] In fact, it is exactly the same if players are assumed to have strict preferences. Much of the work in this area has evolved as a branch of social choice theory, where it is common to work with environments where A is finite and preferences are linear orders on A (i.e., strict).

mechanisms[20] proves in fairly general bargaining settings that games of perfect information can be used to implement nonmonotonic social choice functions that are fair. The general problem of implementation by backward induction has been studied by Hererro and Srivastava (1992) and Trick and Srivastava (1996). The characterizations, unfortunately, are quite cumbersome to deal with, and the necessary conditions for implementation via backward induction are virtually impossible to check in most settings. But some useful results have been found for certain domains.

Closely related to the problem of implementation by backward induction is implementation by voting trees, using the solution concept of *sophisticated voting* as developed by Farquharson (1957/1969). Sophisticated voting works in the following way. First, a binary voting tree is defined, which consists of an initial pair of alternatives, which the individuals vote between. Depending on which outcome wins a majority of votes,[21] the process either ends or moves on to another, predetermined pair and another vote is taken. Usually one of the alternatives in this new vote is the winner of the previous vote, but this is not a requirement of voting trees. The tree is finite, so at some point the process ends regardless of which alternative wins. Sophisticated voting means that one starts at the end of the voting tree, and, for each "final" vote, determines who will win if everyone votes sincerely at that last stage. Then one moves back one step to examine every penultimate vote, and voters vote taking account of how the final votes will be determined. Thus, as in subgame perfect Nash equilibrium, voters have perfect foresight about the outcomes of future votes, and vote accordingly.

The problem of implementation by voting trees was first studied in depth by Moulin (1979), using the concept of dominance solvability, which reduces to sophisticated voting [McKelvey and Niemi (1978)] in binary voting trees. There are two distinct types of sequential voting procedures that have been investigated in detail. The first type consists of *binary amendment procedures*. In a binary amendment procedure, all the alternatives (assumed to be finite) are placed in a fixed order, say, $(a_1, a_2, \ldots, a_{|A|})$. At stage 1, the first two alternatives are voted between. Then the winner goes against the next alternative in the list, and so forth. A major question in social choice theory, and for that matter, in implementation theory, is how to characterize the set of social choice functions that are implementable by binary amendment procedures via sophisticated voting. This work is closely related to work by Miller (1977), Banks (1985), and others, which explores general properties of the majority-rule dominance relation, and following in the footsteps of Condorcet, looks at the implementability of social choice correspondences that satisfy certain normative properties. Several results appear in Moulin (1986), who identifies an "adjacency condition" that is necessary for implementation via binary voting trees. For more details, the reader is referred to the chapter on Social Choice Theory by Moulin (1994) in Volume II of this Handbook.

---

[20] This includes the divide-and-choose method and generalizations of it.

[21] It is common to assume an odd number of voters for obvious reasons. Extensions to permit even numbers of voters are usually possible, but the occurrence of ties clutters up the analysis.

More recent results on implementation via binary voting trees are found in Dutta and Sen (1993). First, they show that implementability by sophisticated voting in binary voting trees implies implementability in backward induction using games of perfect information. They also show that several well-known selections from the top-cycle set[22] are implementable, but that certain selections that have appealing normative properties are not implementable.

2.2.5. Normal form refinements

There are some other refinements of Nash equilibrium that have been investigated for mechanisms in the normal form. These fall into two categories. The first category relies on dominance (either strict or weak) to eliminate outcomes that are unwanted Nash equilibria. This was first explored in Palfrey and Srivastava (1991a) where implementation via *undominated Nash equilibrium* is characterized. Subsequent work that explores this and other variations of dominance-based implementation in the normal form includes Jackson (1992), Jackson, Palfrey, and Srivastava (1994), Sjöström (1991), Tatamitani (1993) and Yamato (1993). Using a different approach, Abreu and Matsushima (1990) obtain results for implementation in iteratively weakly undominated strategies, if randomized mechanisms can be used and small fines can be imposed out of equilibrium. The work by Abreu and Matsushima (1992a, 1992b), Glazer and Rosenthal (1992), and Duggan (1993) extends this line of exploiting strategic dominance relations to refine equilibria by looking at iterated elimination of strictly dominated strategies and also investigating the use of these dominance arguments to design mechanisms that *virtually* implement (see Section 2.3 below) social choice functions.

The second category of refinements looks at implementation via trembling hand perfect Nash equilibrium. The main contribution here is the work of Sjöström (1993).

The central finding of the work in implementation theory using normal form refinements is that essentially anything can be implemented. In particular, it is the case that dominance-based refinements are more powerful than refinements based on sequential rationality, at least in the context of implementation theory. A simple result is in Palfrey and Srivastava (1991a), for the case of undominated Nash equilibrium.

DEFINITION 13. Consider a mechanism $\mu = \langle M, g \rangle$. A message profile $m^* \in M$ is called an *undominated Nash equilibrium of μ at R* if, for all $i \in I$, for all $m_i \in M_i$, $g(m^*) R_i g(m_i, m^*_{-i})$ and there does not exist $i \in I$ and $m_i \in M$ such that

$$g(m_i, m_{-i}) R_i g(m^*_i, m_{-i}) \quad \text{for all } m_{-i} \in M_{-i} \text{ and}$$
$$g(m_i, m_{-i}) P_i g(m^*_i, m_{-i}) \quad \text{for some } m_{-i} \in M_{-i}.$$

---

[22] The top-cycle set at R is the minimal subset, TC, of A, with the property that for all a, b such that $a \in TC$ and $b \notin TC$, a majority strictly prefers a to b. This set has a very prominent role in the theory of voting and committees.

In other words, m^* is an undominated Nash equilibrium at R if it is a Nash equilibrium and, for all i, m_i^* is not weakly dominated.

THEOREM 14. *Suppose R contains no profile where some agent is indifferent between all elements of A. If $N \geqslant 3$ and F satisfies NVP, then F is implementable in undominated Nash equilibrium.*

PROOF. The proof of this theorem is quite involved. It uses a variation on the agreement/integer game, but the general construction of the mechanism uses an unusual technique, called tailchasing. Consider the standard agreement/integer game, μ, used in the proof of Theorem 9. If m^* is a Nash equilibrium of μ at R, but $g(m^*) \notin f(R)$, then one can make m^* dominated at R by amending the game in a simple way. Take some player i and two alternatives x, y such that $x P_i y$. Add a message m_i' for a player i and a message for each of the other players $j \neq i$, m_j', such that

$$g(m_i^*, m_{-i}) = g(m_i', m_{-i}) \quad \text{for all } m_{-i} \neq m_{-i}',$$
$$g(m_i^*, m_{-i}') = y, \qquad g(m_i', m_{-i}') = x.$$

Now strategy m^* is dominated at R. Of course, this is not the end of the story, since it is now possible that (m_i', m_{-i}^*) is a new undominated Nash equilibrium which still produces the undesired outcome $a \notin f(R)$. To avoid this, we can add another message for i, m_i'' and another message for the other players $j \neq i$, m_j'' and do the same thing again. If we repeat this an infinite number of times, we have created an infinite sequence of strategies for i, each one of which is dominated by the next one in the sequence. The complication in the proof is to show that in the process of doing this, we have not disturbed the "good" undominated Nash equilibria at R and have not inadvertently added some new undominated Nash equilibria. $\square$

This kind of construction is illustrated in the following example.

2.2.6. Example 4

This example is from Palfrey and Srivastava (1991a, pp. 488–489).

$A = \{a, b, c, d\}$, $N = 2$, $\boldsymbol{R} = \{R, R'\}$:

$$R_1 = R_2 \qquad \underline{R_1'} = R_2'$$

a	a
b	b
cd	c
	d

$F(R) = \{a, b\}$, $F(R') = \{a\}$.

It is easy to show that there is no implementation with a finite mechanism, and any implementation must involve an infinite chain of dominated strategies for one player in profile R'. One such mechanism is:

	Player 2				
	M_2^1	M_2^2	M_2^3	M_2^4	$\cdots$
m_1^1	a	c	c	c	$\cdots$
m_1^2	c	b	d	d	$\cdots$
m_1^3	c	b	c	d	$\cdots$
m_1^4	c	b	c	c	$\cdots$
$\cdot$	$\cdot$	$\cdot$	$\cdot$	$\cdot$	
$\cdot$	$\cdot$	$\cdot$	$\cdot$	$\cdot$	
$\cdot$	$\cdot$	$\cdot$	$\cdot$	$\cdot$	

(Player 1 labels rows.)

2.2.7. Constraints on mechanisms

Few would argue that mechanisms of the sort used in the previous example solve the implementation problem in a satisfactory manner.[23] This concern motivated the work of Jackson (1992) who raised the issue of *bounded* mechanisms.

DEFINITION 15. A mechanism is *bounded relative to* $\mathbf{R}$ if, for all $R \in \mathbf{R}$, and $m_i \in M_i$, if m_i is weakly dominated at R, then there exists an undominated (at R) message $m_i' \in M_i$ that weakly dominates m_i at R.

In other words, mechanisms that exploit infinite chains of dominated strategies, as occurs in tailchasing constructions, are ruled out. Note that, like the best response criterion, it is not just a property of the mechanism, but a joint restriction on the mechanism *and the domain*. Jackson (1992)[24] shows that a weaker equilibrium notion than Nash equilibrium, called "undominated strategies", has a similar property to undominated Nash implementation, namely that essentially all social choice correspondences are implementable. He shows that if mechanisms are required to be bounded, then very restrictive results reminiscent of the Gibbard–Satterthwaite theorem hold, so that almost no social choice function is implementable via undominated strategies with bounded mechanisms. However, these negative results do not carry over to undominated Nash implementation.

---

[23] In fact, few would argue that *any* of the mechanisms used in the most general sufficiency theorems are particularly appealing.

[24] That is also the first paper to seriously raise the issue of mixed strategies. All of the results that have been described so far in this paper are for *pure strategy implementation*. Only very recently have results been appearing that explicitly address the mixed strategy problem. See for example the work by Abreu and Matsushima (1992a) on virtual implementation.

Following the work of Jackson (1992), Jackson, Palfrey, and Srivastava (1994) provide a characterization of undominated Nash implementation using bounded mechanisms and requiring the best response property. They find that the boundedness restriction, while ruling out some social choice correspondences, is actually quite permissive. First of all, all social choice correspondences that are Nash implementable are implementable by bounded mechanisms [see also Tatamitani (1993) and Yamato (1993)]. Second, in economic environments with free disposal, any interior allocation rule is implementable. Furthermore, there are many allocation rules that fail to be subgame perfect Nash implementable that are implementable via undominated Nash equilibrium using bounded mechanisms.

Boundedness is not the only special property of mechanisms that has been investigated. Hurwicz (1960) suggests a number of criteria for judging the adequacy of a mechanism. Saijo (1988), McKelvey (1989), Chakravorti (1991), Dutta, Sen, and Vohra (1994), Reichelstein and Reiter (1988), Tian (1990) and others have argued that message spaces should be as small as possible and have given results about how small the message spaces of implementing mechanisms can be. Related work focuses on "natural" mechanisms, such as restrictions to price-quantity mechanisms in economic environments [Saijo et al. (1996); Sjöström (1995); Corchon and Wilkie (1995)]. Abreu and Sen (1990) argue that mechanisms should have a best response property relative to the domain for which they are designed. Continuity of outcome functions as a property of implementing mechanisms has been the focus of several papers [Reichelstein (1984); Postlewaite and Wettstein (1989); Tian (1989); Wettstein (1990, 1992); Peleg (1996a, 1996b) and Mas-Colell and Vives (1993)].

2.3. Virtual implementation

2.3.1. Virtual Nash implementation

A mechanism virtually implements a social choice function[25] if it can (exactly) implement arbitrarily close approximations of that social choice function. The concept was first introduced by Matsushima (1988). It is immediately obvious that, regardless of the domain and regardless of the equilibrium concept, the set of virtually implementable social choice functions contains the set of all implementable social choice functions. What is less obvious is how much more is virtually implementable compared with what is exactly implementable. It turns out that it makes a big difference.

One way to see why so much more is virtually implementable can be seen by referring back to Figure 3. That figure shows how the preferences R and R' must line up in order for monotonicity to have any bite in pure exchange economies. As can readily

---

[25] The work on virtual implementation limits attention to single-valued social choice correspondences. Since the results in this area are so permissive (i.e., few social choice functions fail to be virtually implementable), this does not seem to be an important restriction.

be seen, this is not a generic picture. Rather, Figure 4 shows the generic case, where monotonicity places no restrictions on the social choice at R' if $a = f(R)$. Virtual implementation exploits the nongenericity of situations where monotonicity is binding.[26] It does so by implementing lotteries that produce, in equilibrium at R, $f(R)$ with very high probability, and some other outcomes with very low probability.

In finite or countable economic environments, every social choice function is virtually implementable if individuals have preferences over lotteries that admit a von Neumann–Morgenstern representation and if there are at least three agents.[27] The result is proved in Abreu and Sen (1991) for the case of strict preferences and under a domain restriction that excludes unanimity among the preferences of the agents over pure alternatives. They also address the 2-agent case, where a nonempty lower intersection property is needed.

A key difference between the virtual implementation construction and the Nash implementation construction has to do with the use of lotteries instead of pure alternatives in the test pairs. In particular, virtual implementation allows test pairs involving lotteries in the neighborhood (in lottery space) of $f(R)$ rather than requiring the test pairs to *exactly involve* $f(R)$. It turns out that by expanding the first allocation of the test pair to any neighborhood of $f(R)$, one can always find a test pair of the sort required in the definition of monotonicity.

There are several ways to illustrate why this is so. Perhaps the simplest is to consider the case of von Neumann–Morgenstern preferences for lotteries. If an individual maximizes expected utility, then his indifference surfaces in lottery space are parallel hyperplanes. For the case of three pure alternatives, this is illustrated in Figure 6.

For this three alternative case, consider two preference profiles, R and R', which differ in some individual's von Neumann–Morgenstern utility function. This means that the slope of the indifference lines for this individual has changed. Accordingly, in *every* neighborhood of *every* interior lottery in Figure 6, there exists a test pair of lotteries such that this agent has a preference switch over the test pair of lotteries. Now consider a social choice function that assigns a pure allocation to each preference profile, but that fails to satisfy monotonicity. In other words, the social choice function assigns one of the vertices of the triangle in Figure 6 to each profile. We can perturb this social choice function ever so slightly so that instead of assigning a vertex, it assigns an interior lottery, x, arbitrarily close to the vertex. This "approximation" of the social choice function satisfies monotonicity because there exists agent i (whose von Neumann–Morgenstern utilities have changed), and a lottery y such that $x R_i y R'_i x$. In this way, every (interior)

---

[26] This fact that monotonicity holds generically is proved formally in Bergin and Sen (1996). They show for classical pure exchange environments with continuous, strictly monotone (but not necessarily convex) preferences there exists a dense subset of utility functions that always "cross" (i.e., there are never tangencies of the sort depicted in Figure 2).

[27] In fact, more general lottery preferences can be used, as long as they satisfy a condition that guarantees individuals prefer lotteries that place more probability weight on more-preferred pure alternatives.

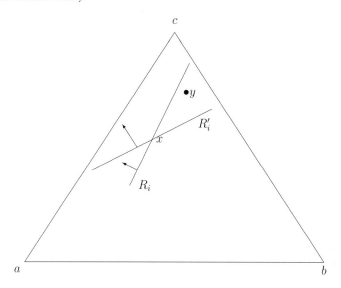

Figure 6. Vertices a, b, and c represent pure alternatives, with all other points representing lotteries over those alternatives. The indifference curves passing through lottery x for agent i under two von Neumann–Morgenstern utility functions are labelled R_i and R'_i, with the direction of preference marked with arrows. Lottery y satisfies $x R_i y R'_i x$.

approximation of every pure social choice function in this simple example is monotonic and hence (if veto power problems are avoided) implementable.

Abreu and Sen (1991) prove that this simple construction outlined above for the case of von Neumann–Morgenstern preferences and $|A| = 3$ is very general. The upshot of this is that moving from exact to virtual implementation completely eliminates the necessity of monotonicity.

2.3.2. Virtual implementation in iterated removal of dominated strategies

An even more powerful result is established in Abreu and Matsushima (1992a). They show that if only virtual implementation is required, then in finite-profile environments one can find mechanisms such that not only is there a unique Nash equilibrium that approximately yields the social choice function, but the Nash equilibrium is strictly dominance solvable. They exploit the fact that for each agent there is a function h from his possible preferences to lottery space, h, such that if $R_i \neq R'_i$, then $h(R_i) R_i h(R'_i)$ and $h(R'_i) R'_i h(R_i)$. The message space of agent i consists of a single report of i's own preferences and multiple (ordered) reports of the entire preference profile, with the final outcome patching together pieces of a lottery, each piece of which is determined by some portion of the reported profiles. The payoff function is then constructed so that falsely reporting one's own type as R' at R will lead to individual i receiving the $g(R')$ lottery instead of the $g(R)$ lottery with some probability, so this false report is a strictly dominated strategy. Incentives are provided so that subsequent[28] reports of the profile will "agree" with earlier reports in a particular way. The first defection from agreement is punished severely enough to make it unprofitable to defect from the truthful self-reports of the first component of the message space. The degree of approximation can then be made as fine as one wishes simply by requiring a very large number of reported profiles.

Formally, a message for i is a $K + 1$ vector $m_i = (m_i^0, m_i^1, \ldots, m_i^K)$, where the first component is an element of i's set of possible preferences and the other K components are each elements of the set of possible preference profiles. The outcome function is then pieced together in the following way. Let ε be some small positive number. With probability ε/I (where I is the number of players), the outcome is based only on m_i^0, and equals $h(m_i^0)$, so i is strictly better off reporting m_i^0 honestly. With probability ε^2/I agent i is rewarded if, for all $k = 1, \ldots, K, m_i^k = m^0$ whenever $m_j^h = m^0$ for all $j \int = i$ for all $h < k$. That is, i gets a small reward (in expected terms) for honestly revealing his preference, and then gets an order of magnitude *smaller* reward for always agreeing with the vector of first reports (including his own). These are the only pieces of the outcome function that are affected by m^0. Clearly, for ε small enough the first order loss overwhelms any possible second order gain from falsely reporting m_i^0. Thus messages involving false reports of m_i^0 are strictly dominated.

The remaining pieces of the outcome function (each of which is used with probability $(1 - \varepsilon - \varepsilon^2)/K$) correspond to the final K components of the messages, where each agent is reporting a preference profile. If everyone agrees on the kth profile, then that kth piece of the outcome function is simply the social choice function at that commonly reported profile. For K large enough, the gain one can obtain from deviating and reporting $m_i^k \neq m^0$ in the kth piece can be made arbitrarily small. But the penalty for being the first to report $m_i^k \neq m^0$ is constant with respect to K, so this penalty will exceed any gain from deviating when K is large. Thus deviating at $h = k + 1$ can be shown to be dominated once all strategies involving deviations at $h < k + 1$ have been eliminated. Variations on this "piecewise" approximation technique also appear in Abreu and Matsushima (1990) where the results are extended to incomplete information (see below) and Abreu and Matsushima (1994) where a similar technique is applied to exact implementation via iterated elimination of *weakly* dominated strategies.[29]

This kind of construction is quite a bit different from the usual Maskin type of construction used elsewhere in the proofs of implementation theorems. It has a number of attractive features, one of which is the avoidance of any mixed strategy equilibria.

---

[28] The term "subsequent" should not be interpreted as meaning that the profiles are reported sequentially, since the game is simultaneous-move. Rather, the vector of reported profiles is *ordered*, so subsequent refers to the reported profile with the next index number. Glazer and Rubinstein (1996) show that there is a similar sequential game that can be constructed which is dominance solvable following similar logic.

[29] Glazer and Perry (1996) show that this mechanism can be reconstructed as a multistage mechanism which can be solved by backward induction. Glazer and Rubinstein (1996) propose that this reduces the computational burden on the players.

In other constructions, mixed strategies are usually just ignored. This can be problematic as an example of Jackson (1992) shows that there are some Nash implementable social choice correspondences that are impossible to implement by a finite mechanism without introducing other mixed strategy equilibria. A second feature is that in finite domains one can implement using finite message spaces. While this is also true for Nash implementation when the environment is finite, there are several examples that illustrate the impossibility of finite implementation in other settings. Palfrey and Srivastava (1991a) show that sometimes infinite constructions are needed for undominated Nash implementation, and Dutta and Sen (1994b) show that Bayesian Nash implementation in finite environments can require infinite message spaces.

Glazer and Rosenthal (1992) raise the issue that in spite of the obvious virtues of the implementing mechanism used in the Abreu and Matsushima (1992a) proof, there are other drawbacks. In particular, Glazer and Rosenthal (1992) argue that the kind of game that is implied by the mechanism is precisely the same kind of game that game theorists have argued as being fragile, in the sense that the predictions of Nash equilibrium are not *a priori* plausible. Abreu and Matsushima (1992b) respond that they believe iterated strict dominance is a good solution concept for predictive[30] purposes, especially in the context of their construction. However, preliminary experimental findings [Sefton and Yavas (1996)] indicate that in some environments the Abreu–Matsushima mechanisms perform poorly.

This is part of an ongoing debate in implementation theory about the "desirability" of mechanisms and/or solution concepts in the constructive existence proofs that are used to establish implementation results. The arguments by critics are based on two premises: (1) equilibrium concepts, or at least the ones that have been explored, do not predict equally well for all mechanisms; and (2) the quality of the existence result is diminished if the construction uses a mechanism that seems unattractive. Both premises suggest interesting avenues of future research.

An initial response to (1) is that these are empirical issues that require serious study, not mere introspection. The obvious implication is that experimental[31] work in game theory will be crucial to generating useful predictive models of behavior in games. This in turn may require a redirection of effort in implementation theory. For example, from the game theory experiments that have been conducted to date, it is clear that limited rationality considerations will need to be incorporated into the equilibrium concepts, as will statistical (as opposed to deterministic) theories of behavior.[32]

---

[30] In implementation theory, it is the predictive value of the solution concept that matters. One can think of the solution concept as the planner's model for predicting outcomes that will arise under different mechanisms and in different environments. If the model predicts inaccurately, then a mechanism will fail to implement the planner's targeted social choice function.

[31] The use of controlled experimentation in settling these empirical questions is urged in Abreu and Matsushima's (1992b) response to Glazer and Rosenthal (1992).

[32] See, for example, McKelvey and Palfrey (1995, 1996, 1998).

Possible responses to (2) are more complicated. The cheap response is that the implementing mechanisms used in the proofs are not meant to be representative of mechanisms that would actually be used in "real" situations that have a lot of structure. These are merely mathematical techniques, and any mechanism used in a "real" situation should exploit the special structure of the situation. Since the class of environments to which the theorems apply is usually very broad, the implementing mechanisms used in the constructive proofs must work for almost any imaginable setting. The question this response begs is: *for a specific problem of interest, can a "reasonable" mechanism be found?* The existence theorems do not answer this question, nor are they intended to. That is a question of specific application. So far, even with the alternative mechanisms of Abreu–Matsushima, the mechanisms used in general constructive existence theorems are impractical. However, some nice results for familiar environments exist [e.g., Crawford (1979); Moulin (1984); Jackson and Moulin (1990)] that suggest we can be optimistic about finding practical mechanisms for implementation in some common economic settings.

3. Implementation with incomplete information

This section looks at the extension of the results of Section 2 to the case of incomplete information. Just as most of the results above are organized around Nash equilibrium and refinements, the same is done in this section, except the baseline equilibrium concept is Harsanyi's (1967, 1968) Bayesian equilibrium.[33]

3.1. The type representation

The main difference in the model structure with incomplete information is that a domain specifies not only the set of possible preference profiles, but also the information each agent has about the preference profile and about the other agents' information. We adopt the "type representation" that is familiar to literature on Bayesian mechanism design [see, e.g., Myerson (1985)].

An *incomplete information domain*[34] consists of a set, I, of n agents, a set, A, of feasible alternatives, a set of types, T_i, for each agent $i \in I$, a von Neumann–Morgenstern utility function for each agent, $u_i : T \times A \to \mathcal{R}$, and a collection of conditional probability distributions $\{q_i(t_{-i}|t_i)\}$, for each $i \in I$ and for each $t_i \in T_i$. There are a variety of familiar domain restrictions that will be referred to, when necessary, as follows:

(1) *Finite types*: $|T_i| < \infty$.
(2) *Diffuse priors*: $q_i(t_{-i}|t_i) > 0$ for all $i \in I$, for all $t_i \in T_i$, and for all $t_{-i} \in T_{-i}$.

---

[33] This should come as no surprise to the reader, since Bayesian equilibrium is simply a version of Nash equilibrium, adapted to deal with asymmetries of information.

[34] Myerson (1985) calls this a Bayesian Collective Decision Problem.

(3) *Private values:*[35] $u_i(t_i, t_{-i}, a) = u_i(t_i, t'_{-i}, a)$ for all $i, t_i, t_{-i}, t'_{-i}, a$.

(4) *Independent types:* $q_i(t_{-i}|t_i) = q_i(t_{-i}|t'_i)$ for all $i, t_i, t'_{-i}, t_{-i}, a$.

(5) *Value-distinguished types:* For all $i, t_i, t'_i \in T_i, t_i \neq t'_i, \exists a, b$ such that $u_i(t_i, t_{-i}, a) > u_i(t_i, t_{-i}, b)$ and $u_i(t'_i, t_{-i}, b) > u_i(t'_i, t_{-i}, a)$ for all $t_{-i} \in T_{-i}$.

A *social choice function* (or *allocation rule*) $f : T \to A$ assigns a unique outcome to each type profile. A *social choice correspondence*, F, is a collection of social choice functions. The set of all allocation rules in the domain is denoted by X, so in general, we have $f \in F \subseteq X$. A mechanism $\mu = \langle M, g \rangle$ is defined as before. A *strategy* for i is a function mapping T_i into M_i, denoted $\sigma_i : T_i \to M_i$. We also denote type t_i of player i's interim utility of an allocation rule $x \in X$ by:

$$U_i(x, t_i) = E_t\{u_i(x(t), t)|t_i\},$$

where E_t is the expectation over t. Similarly, given a strategy profile σ in a mechanism μ, we denote type t_i of player i's *interim utility of strategy σ in μ* by:

$$U_i(\sigma, t_i) = E_t\{u_i(g(\sigma(t)), t)|t_i\}.$$

3.2. Bayesian Nash implementation

The Bayesian approach to full implementation with incomplete information was initiated by Postlewaite and Schmeidler (1986). Bayesian Nash implementation, like Nash implementation, has two components, *incentive compatibility* and *uniqueness*. The main difference is that incentive compatibility imposes genuine restrictions on social choice functions, unlike the case of complete information. When players have private information, the planner must provide the individual with incentives to reveal that information, in contrast to the complete information case, where an individual's report of his information could be checked against another individual's report of that information. Thus, while the constructions with complete information rely heavily on mutual auditing schemes that we called "agreement mechanisms", the constructions with incomplete information do not.[36]

DEFINITION 16. A strategy σ is a *Bayesian equilibrium* of μ if, for all i and for all $t_i \in T_i$

$$U_i(\sigma, t_i) \geqslant U_i(\sigma'_i, \sigma_{-i}, t_i) \quad \text{for all } \sigma'_i : T_i \to M_i.$$

---

[35] In this case, we simply write $u_i(t_i, a)$, since i's utility depends only on his own type.

[36] There are special exceptions where mutual auditing schemes can be used, which include domains in which there is enough redundancy of information in the group so that an individual's report of the state may be checked against the joint report of the other individuals. This requires a condition called *Non-Exclusive Information* (NEI). See Postlewaite and Schmeidler (1986) or Palfrey and Srivastava (1986). Complete information is the extreme form of NEI.

DEFINITION 17. A social choice function $f : T \to A$ (or allocation rule $x : T \to A$) is *Bayesian implementable* if there is a mechanism $\mu = \langle M, g \rangle$ such that there exists a Bayesian equilibrium of μ and, for every Bayesian equilibrium, σ, of M, $f(t) = g(\sigma(t))$ for all $t \in T$.

Implementable social choice sets are defined analogously.

For the rest of this section, we restrict attention to the simpler case of *diffuse types*, defined above. Later in the chapter, the extension of these results to more general information structures will be explained.

3.2.1. Incentive compatibility and the Bayesian revelation principle

Paralleling the definition for complete information, a social choice function (or allocation rule) is called (Bayesian) incentive compatible if and only if it can arise as a Bayesian equilibrium of some mechanism. The *revelation principle* [Myerson (1979); Harris and Townsend (1981)] is the simple proposition that an allocation rule x can arise as the Bayesian equilibrium to some mechanism if and only if truth is a Bayesian equilibrium of the direct[37] mechanism, $\mu = \langle T, x \rangle$. Thus, we state the following.

DEFINITION 18. An allocation rule x is *incentive compatible* if, for all i and for all $t_i, t_i' \in T_i$

$$U_i(x, t_i) \geqslant E_t \{ u_i(x(t_i', t_{-i}), t) | t_i \}.$$

3.2.2. Uniqueness

Just as the multiple equilibrium problem can arise with complete information, the same can happen with incomplete information. In particular, direct mechanisms often have this problem (as was the case with the "agreement" direct mechanism in the complete information case). Consider the following example.

3.2.3. Example 5

This is based on an allocation rule investigated in Holmstrom and Myerson (1983).

There are two agents, each of whom has two types. Types are equally likely and statistically independent and individuals have private values. The alternative set is $A = \{a, b, c\}$. Utility functions are given by (u_{ij} denotes the utility to type j of player i):

$$u_{11}(a) = 2u_{11}(b) = 1u_{11}(c) = 0, \qquad u_{12}(a) = 0u_{12}(b) = 4u_{12}(c) = 9,$$
$$u_{21}(a) = 2u_{21}(b) = 1u_{21}(c) = 0, \qquad u_{22}(a) = 2u_{22}(b) = 1u_{22}(c) = -8.$$

---

[37] A mechanism is *direct* if $M_i = T_I$ for all $i \in I$.

The following social choice function, f, is incentive compatible and efficient (where f_{ij} denotes the outcome when player 1 is type i and player 2 is type j):

$$f_{11} = a, \quad f_{12} = b, \quad f_{21} = c, \quad f_{22} = b.$$

It is easy to check that for the direct revelation mechanism $\langle T, f \rangle$, there is a "truthful" Bayesian equilibrium where both players adopt strategies of reporting their actual type, i.e., f is incentive compatible. However, there is another equilibrium of $\langle T, f \rangle$, where both players always report type 2 and the outcome is always b. We call such strategies in the direct mechanism *deceptions*, since such strategies involve falsely reported types. Denoting this deceptive strategy profile as α, it defines a new social choice function, which we call f_α, defined by $f_\alpha(t) \equiv f(\alpha(t))t$. This illustrates that this particular allocation rule is not Bayesian Nash implementable by the direct mechanism. However, it turns out to be possible to add messages, augmenting[38] the direct mechanism into an "indirect" mechanism that implements f. One way to do this is by giving player 1 another pair of messages, call them "truth" and "lie", one of which must be sent along with the report of his type. The outcome function is then defined so that $g(m) = f(t)$ if the vector of reported types is t and player one says "truth". If player 1 says "lie", then $g(m) = f(t_1, t_2')$ where t_1 is player 1's reported type and t_2' is the opposite of player 2's reported type. This is illustrated in Figure 7.

It is easy to check that if the players use the α deception above, then player 1 will announce "lie", which is not an equilibrium since player 2 would be better off always responding by announcing type 1. In fact, simple inspection shows that there are no longer any Bayesian equilibria that lead to social choice functions different from f, and (truth, "truth") is a Bayesian equilibrium[39] that leads to f.

3.2.4. Bayesian monotonicity

Given that the incentive compatibility condition holds, the implementation problem boils down to determining for which social choice functions it is possible to augment the direct mechanism as in the example above, to eliminate unwanted Bayesian equilibria. This is the so-called *method of selective elimination* [Mookherjee and Reichelstein (1990)] that is used in most of the constructive sufficiency proofs in implementation theory. Again paralleling the complete information case, there is a simple necessary condition for this to be possible, which is an "interim" version of Maskin's monotonicity condition (Definition 6), called *Bayesian monotonicity*.

DEFINITION 19. A social choice correspondence F is *Bayesian monotonic* if, for every $f \in F$ and for every joint deception $\alpha: T \to T$ such that $f_\alpha \notin F$, $\exists i \in I, t_i \in T_i$,

---

[38] The terminology "augmented" mechanism is due to Mookherjee and Reichelstein (1990).

[39] There is another Bayesian equilibrium that also leads to f. See Palfrey and Srivastava (1993).

Player 2

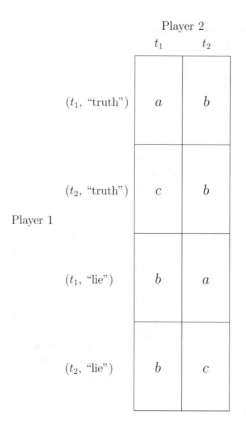

Figure 7. Implementing mechanism for Holmstrom–Myerson example.

and an allocation rule $y : T \to A$ such that $U_i(f_\alpha, t_i) < U_i(y_\alpha, t_i)$ and, for all $t_i' \in T_i$, $U_i(f, t_i') \geqslant U_i(y, t_i')$.

The intuition behind this condition is simpler than it looks. In particular, think of the relationship between f and f_α being roughly the same in the above definition as the relationship between R and R', the difference being that with asymmetric information, we need to consider changes in the entire social choice function f, rather than limiting attention to the particular change in type profile from R to R' (or t to t', in the type notation). So, if $f_\alpha \notin F$ (analogous to $a \notin F(R')$ in the complete information formulation), we need a test agent, i, and a test allocation *rule* y (analogous to test allocation, in the monotonicity definition), such that i's (interim) preference between f and y is the reverse of his preference between f_α and y_α (with the appropriate quantifiers and qualifiers included). Thus the basic idea is the same, and involves a test agent and a test allocation rule.

3.2.5. Necessary and sufficient conditions

We are now ready to state the main result regarding necessary conditions[40] for Bayesian implementation.

THEOREM 20. *If F is Bayesian Nash implementable, then F is Bayesian monotonic, and every f $\in$ F is incentive compatible.*

PROOF. The necessity of incentive compatibility is obvious. The proof for necessity of Bayesian monotonicity follows the same logic as the proof for necessity of monotonicity with complete information (see Theorem 7). □

As with complete information, sufficient conditions generally require an allocation rule to be in the social choice correspondence if there is nearly unanimous agreement among the individuals about the "best" allocation rule. This is the role of NVP in the original Maskin sufficiency theorem. There are two ways to guarantee this. The first way (and by far the simplest) is to make a domain assumption that avoids the problem by ruling out preference profiles where there is nearly unanimous agreement. The prototypical example of such a domain is a pure exchange economy. In that case, there is a great deal of conflict across agents, as each agent's most preferred outcome is to be allocated the entire societal endowment, and this most preferred outcome is the same regardless of the agent's type. Another related example includes the class of environments with side-payments using a divisible private good that everyone values positively, the best-known case being quasi-linear utility. We present two sufficiency results, one for the case of pure exchange economies, and the second for a generalization. We assume throughout that information is diffuse and $n \geqslant 3$.

Consider a pure exchange economy with asymmetric information, E, with L goods and n individuals, where the societal endowment[41] is given by $w = (w_1, \ldots, w_L)$. The alternative set, A, is the set of all nonnegative allocations of w across the n agents.[42] The set of feasible allocation rules mapping T into A are denoted X.

THEOREM 21. *Assume $n \geqslant 3$ and information is diffuse. A social choice function $x \in X$ is Bayesian Nash implementable if and only if it satisfies incentive compatibility and Bayesian monotonicity.*

---

[40] There are other necessary conditions. For example, F must be closed with respect to common knowledge concatenations. See Postlewaite and Schmeidler (1986) or Palfrey and Srivastava (1993) for details.

[41] We will not be addressing questions of individual rationality, so the initial allocation of the endowment is left unspecified.

[42] One could permit free disposal as well, but this is not needed for the implementation result. The constructions by Postlewaite and Schmeidler (1986) and Palfrey and Srivastava (1989a) assume free disposal. We do not assume it here, but do assume diffuse information. Free disposability simplifies the constructions when information is not diffuse, by permitting destruction of the entire endowment (i.e., all agents receive 0) when the joint reported type profile is not consistent with any type profile in T.

PROOF. Only if follows from Theorem 20. It is only slightly more difficult. Once again, we use a variation on the agreement/integer game, adapted to the incomplete information framework. Notice, however, that there is always "agreement" with diffuse types, since each player is submitting a different component of the type profile, and all reported type profiles are possible. Each player is asked to report a type and either an allocation rule that is *constant* in his own type (but can depend on other players' types) or a nonnegative integer. Thus:

$$M_i = T_i x \{X_{-i} \cup \{0, 1, \ldots\}\}$$

where X_{-i} denotes the set of allocation rules that are constant with respect to t_i. The *agreement region* is the set of message profiles where each player sends a reported type and "0". The *unilateral disagreement region* is the set of message profiles where exactly one agent reports a type and something other than "0". Finally, the *disagreement region* is the set of all message profiles with at least two agents failing to report "0". In the agreement region, the outcome is just $x(t)$, where t is the reported type profile. In the unilateral disagreement region the outcome is also just $x(t)$, unless the disagreeing agent, i, sends $y \in X_{-i}$ with the property that $U_i(x, t_i') \geqslant U_i(y, t_i')$ for all $t_i' \in T_i$. In that case, the outcome is $y(t)$. In the disagreement region, the agent who submits the highest integer[43] is allocated w and everyone else is allocated 0.

Notice how the mechanism parallels very closely the complete information mechanism. The structure of the unilateral disagreement region is such that if all individuals are reporting truthfully, no player can unilaterally falsely report and/or disagree and be better off. By incentive compatibility it does not pay to announce a false type. The fact that y does not depend on the disagreer's types implies that it doesn't pay to report y and a false (or true) type. Therefore, there is a Bayesian equilibrium in the agreement region, where all players truthfully report their types. There can be no equilibrium outside the agreement region, because there would be at least two agents each of whom could unilaterally change his message and receive w. Thus the only possible other equilibria that might arise would be in the agreement region, where agents are using a joint deception α. But the Bayesian monotonicity condition (which f satisfies by assumption) says that either $x_\alpha = x$ or there exists a y, and i, and a t_i, such that $U_i(x, t_i') \geqslant U_i(y, t_i')$ for all $t_i' \in T_i$ but $U_i(y_\alpha, t_i) > U_i(x_\alpha, t_i)$. Since it is easy to project y onto X_{-i} [see Palfrey and Srivastava (1993)] and preserve these inequalities, it follows that i is better off deviating unilaterally and reporting y instead of "0".   □

The extension of the above result to more general environments is simple, as long as individuals have different "best elements" that do not depend on their type. For each i, suppose that there exists an alternative b_i such that $U_i(b_i, t) \geqslant U_i(a, t)$ for all $a \in A$

---

[43] In this region, if a player sends an allocation instead of an integer, this is counted as "0". Ties are broken in favor of the agent with the lowest index.

and $t \in T$, and further suppose that for all i, j it is the case that $U_i(b_i, t) > U_i(b_j, t)$ for all $t \in T$. If this condition holds, we say players have distinct best elements.

THEOREM 22. *If $n \geqslant 3$, information is diffuse and players have distinct best elements, then f is Bayesian implementable if and only if f is incentive compatible and Bayesian monotonic.*

PROOF. Identical to the proof of Theorem 21, except in the disagreement region the outcome is b_i, where i is winner of the integer game. □

Jackson (1991) shows that the condition of distinct best elements can be further weakened, and their result is summarized in Palfrey and Srivastava (1993, p. 35). An even more general version that considers nondiffuse as well as diffuse information structures is in Jackson (1991). That paper identifies a condition that is a hybrid between Bayesian monotonicity and an interim version of NVP, called *monotonicity-no-veto* (MNV). The earlier papers by Postlewaite and Schmeidler (1986) and Palfrey and Srivastava (1987, 1989a) also consider nondiffuse information structures.

Dutta and Sen (1991b) provide a sufficient condition for Bayesian implementation, when $n \geqslant 3$ and information is diffuse, that is even weaker than the MNV condition of Jackson (1991). They call this condition *extended unanimity*, and it, like MNV, incorporates Bayesian monotonicity. They also prove that when this condition holds and T is finite, then any incentive compatible social choice function can be implemented using a finite mechanism. They do this using a variation on the integer game, called a *modulo game*,[44] which accomplishes the same thing as an integer game but only requires using the first n positive integers.

Dutta and Sen (1994b) raise an interesting point about the size of the message space that may be required for implementation of a social choice function. They present an example of a social choice function that fails their sufficiency condition (it violates unanimity and there are only two agents), but is nonetheless implementable via Bayesian equilibrium. But they are able to show that any implementing mechanisms must use an infinite message space, in spite of the fact that both A and T are finite.

Dutta and Sen (1994a) extend their general characterization of Bayesian implementable social choice correspondences when $n \geqslant 3$ to the $n = 2$ case, using an interim version of the nonempty lower intersection property that they used in their $n = 2$ characterization with complete information [Dutta and Sen (1991a)]. This complements some earlier work on characterizing implementable social choice functions by Mookherjee and Reichelstein (1990). Dutta and Sen (1994a) extend these results to characterize

---

[44] The modulo game is due to Saijo (1988) and is also used in McKelvey (1989) and elsewhere. A potential weakness of a modulo game is that it typically introduces unwanted mixed strategy equilibria that could be avoided by the familiar greatest-integer game.

Bayesian implementable social choice correspondences for the $n = 2$ case, for "economic environments".[45]

All of the results described above are restricted (either explicitly or implicitly) to finite sets of player types. Obviously for many applications in economics this is a strong requirement. Duggan (1994b) provides a rigorous treatment of the many difficult technical problems that can arise when the space of types is uncountable. He extends the results of Jackson (1991) to very general environments, and identifies some new, more inclusive conditions that replace previous assumptions about best elements, private values, and economic environments. The key assumption he uses is called *interiority*, which is satisfied in most applications.

3.3. Implementation using refinements of Bayesian equilibrium

Just as in the case of complete information, refinements permit a wider class of social choice functions to be implemented. These fall into two classes: dominance-based refinements using simultaneous-move mechanisms, and sequential rationality refinements using sequential mechanisms. In both cases, the results and proof techniques have similarities to the complete information case.

3.3.1. Undominated Bayesian equilibrium

The results for implementation using dominance refinements in the normal form are limited to undominated Bayesian implementation, where a nearly complete characterization is given in Palfrey and Srivastava (1989b). An undominated Bayesian equilibrium is a Bayesian equilibrium where no player is using a weakly dominated strategy. There are several results, some positive and some negative. First, in private value environments with diffuse types and value-distinguished types, any incentive compatible allocation rule satisfying no veto power is implementable via undominated Bayesian equilibrium. The proof assumes the existence of best and worst elements[46] for each type of each agent, but does not require No Veto Power. They also show that with non-private values, some additional very strong restrictions are needed, and, moreover, the assumption of value distinction is critical.[47]

---

[45] The term *economic* is vague. "Informally speaking, an economic environment is one where it is possible to make some individual strictly better off from any given allocation in a manner which is independent of her type. This hypothesis, while strong, will be satisfied if there is a transferable private good in which the utilities of both individuals are strictly increasing". [Dutta and Sen (1994a, p. 52).]

[46] Notice that if A is finite or more generally if A is compact and preferences are continuous, then best and worst elements exist. The proof can be extended to cover some special environments where best elements do not exist, such as the quasi-linear utility case.

[47] The assumption of value distinction is stronger than might appear. It rules out environments where two types of an agent differ only in their beliefs about the other agents. One can imagine some natural environments where value distinction might be violated, such as financial trading environments, where a key feature of the information structure involves what agents know about what other agents know.

Two simple voting public goods examples illustrate both the power and the limitations (with common values) of the undominated refinement.

3.3.2. Example 6

There are three agents, two feasible outcomes, $A = \{a, b\}$, private values, independent types, and each player can be one of two types. Type α strictly prefers a to b and type β strictly prefers b to a, and the probability of being type α is q, with $q^2 \geqslant \frac{1}{2}$. The "best solution" according to almost any set of reasonable normative criteria is to choose a if and only if at least two agents are type α. Surprisingly, this "majoritarian" solution, while incentive compatible, is not implementable via Bayesian equilibrium. It is fairly easy to show that for any mechanism that produces the majoritarian solution as a Bayesian equilibrium that mechanism will have another equilibrium in which outcome b is produced at every type profile. However, it is easy to see that the majoritarian solution is implementable via *undominated* Bayesian equilibrium, since it is the unique undominated Bayesian equilibrium[48] outcome in the direct mechanism.

3.3.3. Example 7

This is the same as Example 6 (two feasible outcomes, three agents, two independent types and type α occurs with probability $q^2 > \frac{1}{2}$), except there are common values.[49] The common preferences are such that if a majority of agents are type α, then everyone prefers a to b, and if a majority of agents are type β, then everyone prefers b to a. We call these "majoritarian preferences". Obviously, there is a unique best social choice function for essentially any non-malevolent welfare criterion, which is the majoritarian (and *unanimous*, as well!) solution: choose a if and only if at least two agents are type α.

First observe that because of the common values feature of this example, players no longer have a dominant strategy in the direct game for agents to honestly report their true type. (Of course, truth is still a Bayesian equilibrium of the direct game.) One can show [Palfrey and Srivastava (1989b)] that this social choice function is not even implementable in undominated Bayesian equilibrium. In particular, any mechanism which produces the majoritarian solution as an undominated Bayesian equilibrium always has another undominated Bayesian equilibrium where the outcome is always b.

The point of Example 7 is to illustrate that with common values, using refinements may have only limited usefulness in a Bayesian framework. We know from the work in complete information that implementation requires the existence of test agents and test

---

[48] Notice that it is actually a *dominant strategy equilibrium* of the direct mechanism. This example illustrates how it is possible for an allocation rule to be dominant strategy implementable (and strategy-proof), but not Bayesian Nash implementable.

[49] By common values we mean that every agent has the same type-contingent preferences. A related mechanism design problem is explored in more depth by Glazer and Rubinstein (1998).

pairs of allocations that involve (often delicate) patterns of preference reversal between preference profiles. Analogously, in the Bayesian setting such preference reversals must occur across *type* profiles. With private values, such preference reversals are easy to find. With common values and/or non-value-distinguished types, such preference reversals often simply do not exist, even in very natural examples of social choice functions that satisfy No Veto Power and $N > 2$.

3.3.4. Virtual implementation with incomplete information

For virtual implementation, Abreu and Matsushima have an extension of their complete information paper on the use of iterated elimination of strictly dominated strategies [Abreu and Matsushima (1990)] for implementation in incomplete information environments. They show that, under a condition they call *measurability* and some additional minor restrictions on the domain, any incentive compatible social choice function defined on finite domains can be virtually implemented by iterated elimination of strictly dominated strategies. Abreu and Matsushima (1994) conjecture that with some additional assumptions (such as the ability to use small monetary transfers) one may be able to obtain exact implementation via iterated elimination of weakly dominated strategies in finite incomplete information domains.

Duggan (1997) looks at the related issue of virtual implementation in Bayesian equilibrium (rather than iterated elimination of dominated strategies). He shows that the measurability of Abreu and Matsushima (1990) is not necessary for virtual Bayesian implementation, and uses a condition called *incentive consistency*.

Serrano and Vohra (1999) have recently shown that both measurability and incentive consistency are stronger conditions than originally implied in these papers. Results and examples in that paper provide insight into the kinds of domain restrictions that are required for the permissive virtual implementation results. In a sequel to that paper, Serrano and Vohra (2000) identify a useful and easy-to-check domain restriction, called *type diversity*. They show that any incentive-compatible social choice function is virtually Bayesian implementable as long as the environments satisfy this condition. Type diversity requires that different types of an agent have different interim preferences over pure alternatives. In private-values models, it reduces to the familiar condition of value-distinguished types [Palfrey and Srivastava (1989b)]. We turn next to the question of implementation using sequential rationality refinements, where results parallel (to an extent) the results for subgame perfect implementation.

3.3.5. Implementation via sequential equilibrium

There are some papers that partially characterize the set of implementable social choice functions for incomplete information environments using the equilibrium refinement of sequential equilibrium. The main idea behind these characterizations is the same as the ideas behind the results for subgame perfect equilibrium implementation under conditions of complete information. Instead of requiring a test pair involving the social choice function, x, as is required in Bayesian monotonicity, all that is needed is *some* (interim) preference reversal between *some* pair of allocation rules, plus an appropriate sequence of allocation rules that indirectly connect x with the test pair of allocation rules.

The details of the conditions analogous to indirect monotonicity for incomplete information are messy to state, because of quantifiers and qualifiers that relate to the posterior beliefs an agent could have at different stages of an extensive form game in which different players are adopting different deceptions. However, the intuition behind the condition is similar to the intuition behind Condition α in Abreu and Sen (1990).

As with the necessary and sufficient results for Bayesian implementation, results are easiest to state and prove for the special case of economic environments, where No Veto Power problems are assumed away and where there are at least three agents.

Bergin and Sen (1998) have some results for this case. They identify a condition which is sufficient for implementation by sequential equilibrium in a two-stage game. That paper also makes the point that with incomplete information there exist social choice functions that are implementable sequentially, but are not implementable via undominated Bayesian equilibrium (or via iterated dominance), and are not even virtually implementable. This contrasts with the complete information case, where any social choice function in economic environments that is implementable via subgame perfect Nash equilibrium is also implementable via undominated Nash equilibrium. They are able to obtain these very strong results by showing that the "consistent beliefs" condition of sequential equilibrium can be exploited to place restrictions on equilibrium play in the second stage of the mechanism.

Baliga (1999) also looks at implementation via sequential equilibrium, limited to finite stage extensive games. His paper makes additional restrictions of private values and independent types, which lead to a significant simplification of the analysis.

A more general approach is taken in Brusco (1995) who does not limit himself to stage games nor to economic environments. He looks at implementation via perfect Bayesian equilibrium and obtains the incomplete information equivalent to indirect monotonicity which he calls "Condition $\beta+$". This condition is then combined with No Veto Power in a manner similar to Jackson's (1991) monotonicity-no-veto condition to produce *sequential monotonicity-no-veto*. His main theorem[50] is that any incentive-compatible social choice function satisfying SMNV is implementable in perfect Bayesian equilibrium. He also identifies a weaker condition than $\beta+$ (called Condition β), which he proves is necessary for implementation in perfect Bayesian equilibrium. However, Brusco's (1995) results are weaker than Bergin and Sen (1998) because his conditions on the requisite sequence of test allocations include a universal quantifier on beliefs that makes it much more difficult to guarantee existence of the sequence. Bergin and Sen show that the very tight condition of belief consistency can replace the universal quantifier. Loosely speaking, Brusco's results exploit *only* the sequential

---

[50] The main theorem is stated more generally. In particular he allows for social choice correspondences, which means that the additional restriction of closure under the common knowledge concatenation is required.

rationality part of sequential equilibrium, while Bergin and Sen exploit both sequential rationality *and* belief consistency. This seemingly minor distinction actually makes quite a difference in proving what can be implemented.

Duggan (1995) focuses on sequentially rational implementation[51] in quasi-linear environments where the outcomes are lotteries over a finite set of public projects (and transfers). He shows that any incentive-compatible social choice function is implementable in private-values environments with diffuse priors over a finite type space if there are three or more agents. He also shows that these results can be extended, with some modifications, in a number of directions: two agents; the domain of exchange economies; infinite type spaces; bounded transfers; nondiffuse priors; and social choice correspondences. He also shows how a "belief revelation mechanism" can be used if the planner does not know the agents' prior beliefs, as long as these prior beliefs are common knowledge among the agents.

4. Open issues

Open problems in implementation theory abound. Several issues have been explored at only a superficial level, and others have not been studied at all. Some of these have been mentioned in passing in the previous sections. There are also numerous untied loose ends having to do with completing full characterizations of implementability under the solution concepts discussed above.

4.1. Implementation using perfect information extensive forms and voting trees

Among these uncompleted problems is implementation via backward induction and the closely related problem of implementing using voting trees. If this class of implementation problems has a shortcoming, it is that extensions to the more challenging problem of implementation in incomplete information environments are limited. The structure of the arguments for backward induction implementation fails to extend nicely to incomplete information environments, as we know from the literature on sophisticated voting with incomplete information [e.g., Ordeshook and Palfrey (1988)].

4.2. Renegotiation and information leakage

Many of the above constructions have the feature that undesirable (e.g., Pareto inefficient, grossly inequitable, or individually irrational) allocations are used in the mechanism to break unwanted equilibria. The simplest examples arise when there exists a universally bad outcome that is reverted to in the event of disagreement. This is not necessarily a problem in some settings, where the planner's objective function may conflict

---

[51] Duggan (1995) defines *sequentially rational implementation* as implementation simultaneously in Perfect Bayesian Equilibrium and Sequential Equilibrium.

with Pareto optimality from the point of view of the agents (as in many principal-agent problems). However, in some settings, most obviously exchange environments, one usually thinks of the mechanism as being something that the players themselves construct in order to achieve efficient allocations. In this case, one would expect agents to renegotiate outcomes that are commonly known among themselves to be Pareto dominated.[52] Maskin and Moore (1999) examine the implications of requiring that the outcomes always be Pareto optimal. This approach has the virtue of avoiding mixed strategy equilibria and also avoiding implausibly bad outcomes off the equilibrium path. It has the defect of implicitly permitting the outcome function of the mechanism to depend on the preference profile, which makes the specification of renegotiation somewhat arbitrary. Ideally, one would wish to specify a bargaining game that would arise in the event that an outcome is reached which is inefficient.[53] But then the bargaining game itself should be considered part of the mechanism, which leads to an infinite regress problem.

More generally, the planner may be viewed directly as a player, who has state-contingent preferences over outcomes, just as the other players do. The planner has prior beliefs over the states or preference profiles (even if the players themselves have complete information). Given these priors, the planner has induced preferences over the allocations. This presents a commitment problem for the planner, since the outcome function must adhere to these preferences. This places restrictions both on the mechanism that can be used and also on the social choice functions that can be implemented. Several papers have been written recently on this subject, which vary in the assumptions they make about the extent to which the planner can commit to a mechanism, the extent to which the planner may update his priors after observing the reported messages, and the extent to which the planner participates directly in the mechanism. The first paper on this subject is by Chakravorti, Corchon, and Wilkie (1994) and assumes that the social choice function must be consistent with some prior the planner might have over the states, which implies that the outcome function is restricted to the range of the social choice function. The planner is not an active participant and does not update beliefs based on the messages of the players, nor does he choose an outcome that is optimal given his beliefs. Baliga, Corchon, and Sjöström (1997) obtain results with an actively participating planner,[54] who acts optimally, given the messages of the players and cannot commit ex ante to an outcome function. Thus, the mechanism consists of a message space for the players and a planner's strategy of how to assign messages to outcomes. This strategy replaces the familiar outcome function, but is required to be sequentially rational. Thus, the mechanism is really a two-stage (signalling) game, and an equilibrium of the mechanism must satisfy the conditions of Perfect Bayesian Equilibrium. Baliga and Sjöström (1999) obtain further results on interactive implementation with an

---

[52] One doesn't have to look very hard to find counterexamples to this in the real world. Institutional structures (such as courts) are widely used to enforce ex post inefficient allocations in order to provide salient incentives. Some forms of criminal punishments, such as incarceration and physical abuse, fall into this category.

[53] See, for example, Aghion, Dewatripont, and Rey (1994) or Rubinstein and Wolinsky (1992).

[54] That is, the planner also submits messages. They call this "interactive implementation".

uninformed planner who can commit to an outcome function, and who also participates in the message-sending stage.

Another sort of renegotiation arises in incomplete information settings if the social choice function calls for allocations that are known by at least some of the players to be inefficient. In particular, for certain type realizations, some players may be able to propose to replace the mechanism with a new one that all other types of all other players would unanimously prefer to the outcome of the social choice function. This problem of lack of *durability*[55] [Holmstrom and Myerson (1983)] opens up another kind of renegotiation problem, which may involve the potential leakage of information between agents in the renegotiation process. Related issues are also addressed in Sjöström (1996), and in principal-agent settings by Maskin and Tirole (1992) and in exchange economies by Palfrey and Srivastava (1993).

Also related to this kind of renegotiation is the problem of preplay communication among the agents. It is well known that preplay communication can expand the set of equilibria of a game, and a similar thing can happen in mechanism design. This can occur because information can be transmitted by preplay communication and because communication opens up possibilities for coordination that were impossible to achieve with independent actions. In nearly all of the implementation theory research, it is assumed that preplay communication is impossible (i.e., the message space of the mechanism specifies all possible communication). An exception is Palfrey and Srivastava (1991b), which explicitly looks at designing mechanisms that are "communication-proof", in the sense that the equilibria with arbitrary kinds of preplay communication are interim-payoff-equivalent to the equilibria without preplay communication. They show that for a wide range of economic environments one can construct communication-proof mechanisms to implement any interim-efficient, incentive-compatible allocation rule. Chakrovorti (1993) also looks at implementation with preplay communication.

4.3. Individual rationality and participation constraints

A close relative to the renegotiation problem is individual rationality. Participation constraints pose similar restrictions on feasible mechanisms as do renegotiation constraints, and have similar consequences for implementation theory. In particular, individual rationality gives every player a right to veto the final allocation, which restricts the use of unreasonable outcomes off the equilibrium path, which are often used in constructive proofs to knock out unwanted equilibria. Jackson and Palfrey (2001) show that individual rationality and renegotiation can be treated in a unified structure, which they call *voluntary implementation*.

Ma, Moore, and Turnbull (1988) investigate the effect of participation constraints on the implementability of efficient allocations in the context of an agency problem of

---

[55] Closely related to this are the notions of *ratifiability* and *secure allocations* [Cramton and Palfrey (1995)] and *stable allocations* [Legros (1990)].

Demski and Sappington (1984).[56] In the mechanism originally proposed by Demski and Sappington (1984), there are multiple equilibria. Ma, Moore, and Turnbull (1988) show that a more complicated mechanism eliminates the unwanted equilibria. Glover (1994) provides a simplification of their result.

4.4. Implementation in dynamic environments

Many mechanism design problems and allocation problems involve intertemporal allocations. One obvious example is bargaining when delay is costly. In that case both the split of the pie and the time of agreement are economically important components of the final allocation. Recently, Rubinstein and Wolinsky (1992) looked at the renegotiation-proof problem in implementation theory by appending an infinite horizon bargaining game with discounting to the end of each inefficient terminal node. This is an alternative approach to the same renegotiation problem that Maskin and Moore (1999) were concerned about. However, like the rest of implementation theory, their interest is in implementing static allocation rules (i.e., no delay) in environments that are (except for the final bargaining stages) static. This is true for all the other sequential game constructions in implementation theory: time stands still while the mechanism is played out.

Intertemporal implementation raises additional issues. Consider, for example, a setting in which every day the same set of agents is confronted with the next in a series of connected allocation problems, and there is discounting. A preference profile is not an infinite sequence of "one shot" profiles corresponding with each time period. A social choice function is a mapping from the set of these profile sequences into allocation sequences. Renegotiation-proofness would impose a natural time consistency constraint that the social choice function would have to satisfy from time t onward, for each t. With this kind of structure one could begin to look at a broader set of economic issues related to growth, savings, intertemporal consumption, and so forth.

Jackson and Palfrey (1998) follow such an approach in the context of a dynamic bargaining problem with randomly matched buyers and sellers. Many buyers and sellers are randomly matched in each period, and bargain under conditions of complete information. Those pairs who reach agreements leave the market, and the unsuccessful pairs are rematched in the following period. This continues for a finite number of periods. The matching process is taken as exogenous, and the implementation problem is to design a single bargaining mechanism to implement the efficiently dynamic allocation rule (which pairs of agents should trade, as a function of the time period), subject to the constraints of the matching process and individual rationality. They identify necessary and sufficient conditions for an allocation rule to be implemented and show that it is often the case that constrained efficient allocations are not implementable under *any* bargaining game.

---

[56] There is a large and growing literature on implementation theoretic applications to agency problems. The techniques used there apply many of the ideas surveyed in this chapter. See, for example, Arya and Glover (1995), Arya, Glover, and Rajan (2000), Ma (1988), and Duggan (1998).

There are some very simple intertemporal allocation problems that could be investigated as a first step. One example is the one-sector growth model of Boylan et al. (1996) which compares different political mechanisms for deciding on investments. As a second example, Bliss and Nalebuff (1984) look at an intertemporal public goods problem. There is a single indivisible public good which can be produced once and for all at any date $t = 1, 2, 3, \ldots$, and preferences are quasilinear with discounting. The production technology requires a unit of private good for the public good to be provided. Thus, an allocation is a time at which the public good is produced and an infinite stream of taxes for each individual, as a function of the profile of preferences for the public good. Bliss and Nalebuff (1984) look at the equilibrium of a specific mechanism, the voluntary contribution mechanism. At each point in time an individual must decide whether or not to privately pay for the public good, depending on their type. The unique equilibrium is for types that prefer the public good more strongly to pay earlier. Thus the public good is always produced by having the individual with the strongest preference for the public good pay for it, and the time of delay before production depends on what the highest valuation is and on the distribution of types. One could generalize this as a dynamic implementation problem, which would raise some interesting questions: What other allocation rules are implementable in this setting? Is the Bliss and Nalebuff (1984) equilibrium allocation rule interim incentive efficient?[57]

4.5. Robustness of the mechanism

The approach of double implementation looks at robustness with respect to the equilibrium concept. For example, Tatamitani (1993) and Yamato (1993) both consider implementation simultaneously in Nash equilibrium and undominated Nash equilibrium. The idea is that if we are not fully confident about which equilibrium concept is more reasonable for a certain mechanism, then it is better if the mechanism implements a social choice function via two different equilibrium concepts (say, one weak and one strong concept). Other papers adopting this approach include Peleg (1996a, 1996b) and Corchon and Wilkie (1995). The constructive proofs used by Jackson, Palfrey, and Srivastava (1994) and Sjöström (1994) also have double implementation properties.

A second approach to robustness is to require continuity of the implementing mechanism. This was discussed briefly earlier in the chapter. Continuity will protect against *local* mis-specification of behavior by the agents. Duggan and Roberts (1997) provide a formal justification along these lines.[58] On the other hand, the preponderance of discontinuous mechanisms in real applications (e.g., auctions) suggests that the issue of mechanism continuity is of little practical concern in many settings. Related to the continuity requirement is dynamical stability, under various assumptions of the adjustment

---

[57] Notice that it is not ex post efficient since there is always delay in producing the public good.

[58] Corchon and Ortuño-Ortin (1995) consider the question of robustness with respect to the agents' prior beliefs.

path. See Hurwicz (1994), Cabrales (1999), Kim (1993). Chen (2002) and Chen and Tang (1998) have experimental results demonstrating that stability properties of mechanisms can have significant consequences.

There is also the issue of robustness with respect to coalitional deviations. Implementation via strong Nash equilibrium [Maskin (1979); Dutta and Sen (1991c); Suh (1997)] and coalition-proof equilibrium [Boylan (1998)] are one way to address these issues. The issue of collusion among the agents is related to this [Baliga and Brusco (2000); Sjöström (1999)]. These approaches look at robustness with respect to profitable deviations by coalitions. Eliaz (1999) has looked at implementation via mechanisms where the outcome function is robust to *arbitrary* deviations by subsets of players, a much stronger requirement.

An alternative approach to robustness, which is related to virtual implementation, involves considerations of equilibrium models based on stochastic choice,[59] so that social choice functions may be approximated "on average" rather than precisely.[60] Implementation theory (even the relaxed problem of virtual implementation) so far has investigated special deterministic models of individual behavior. The key assumption for obtaining results is that the equilibrium model that is assumed to govern individual behavior under any mechanism *is exactly correct*. Many of the mechanisms have no room for error. One would generally think of such fragile mechanisms as being nonrobust. Similarly (especially in the Bayesian environments), the details of the environment, such as the common knowledge priors of the players and the distribution of types, are known to the planner *precisely*. Often mechanisms rely on this exact knowledge. It should be the case that if the model of behavior or the model of the environment is not completely accurate, the equilibrium behavior of the agents does not lead to outcomes too far from the social choice function one is trying to implement.

This problem suggests a need to investigate mechanisms that either do not make special use of detailed information about the environment (such as the distribution of types) or else look at models that permit statistical deviation from the behavior that is predicted under the equilibrium model. In the latter case, it may be more natural to think of social choice functions as type-contingent random variables rather than as *deterministic* functions of the type profile. Related to the problem of robustness of the mechanisms and the possible use of statistical notions of equilibrium is bounded rationality. The usual defense for assuming in economic models that all actors are fully rational, is that it is a good first cut on the problem and often captures much of the reality of a given economic situation. In any case, most economists regard the fully rational equilibrium as an appropriate benchmark in most situations. Unfortunately, since implementation theorists and mechanism designers get to *choose* the economic game in very special ways, this

---

[59] One such approach is quantal response equilibrium [McKelvey and Palfrey (1995, 1996, 1998)].

[60] Note the subtle difference between this and virtual implementation. In the virtual implementation approach, something is virtually implementable if a nearby social choice function can be exactly implemented. The players do not make choices stochastically, although the outcome functions of the mechanisms may use lotteries.

rationale loses much of its punch. It may well be that the games where rational models predict poorly are precisely those games that implementation theorists are prone to designing. Integer games, modulo games, "grand lottery" games like those used in virtual implementation proofs, and the enormous message spaces endemic to all the general constructions would seem for the most part to be games that would challenge the limits of even the most brilliant and experienced game player. If such constructions are unavoidable we really need to start to look beyond models of perfectly rational behavior. Even if such constructions are avoidable, we have to be asking more questions about the match (or mismatch) between equilibrium concepts as predictive tools and limitations on the rationality of the players.

References

Abreu, D., and H. Matsushima (1990), "Virtual implementation in iteratively undominated strategies: Incomplete information", mimeo.

Abreu, D., and H. Matsushima (1992a), "Virtual implementation in iteratively undominated strategies: Complete information", Econometrica 60:993–1008.

Abreu, D., and H. Matsushima (1992b), "A response to Glazer and Rosenthal", Econometrica 60:1439–1442.

Abreu, D., and H. Matsushima (1994), "Exact implementation", Journal of Economic Theory 64:1–20.

Abreu, D., and A. Sen (1990), "Subgame perfect implementation: A necessary and almost sufficient condition", Journal of Economic Theory 50:285–299.

Abreu, D., and A. Sen (1991), "Virtual implementation in Nash equilibrium", Econometrica 59:997–1021.

Aghion, P., M. Dewatripont and P. Rey (1994), "Renegotiation design with unverifiable information", Econometrica 62:257–282.

Allen, B. (1997), "Implementation theory with incomplete information", in: S. Hart and A. Mas-Colell, eds., Cooperation: Game Theoretic Approaches (Springer, Heidelberg).

Arya, A., and J. Glover (1995), "A simple forecasting mechanism for moral hazard settings", Journal of Economic Theory 66:507–521.

Arya, A., J. Glover and U. Rajan (2000), "Implementation in principal-agent models of adverse selection", Journal of Economic Theory 93:87–109.

Baliga, S. (1999), "Implementation in economic environments with incomplete information: The use of multistage games", Games and Economic Behavior 27:173–183.

Baliga, S., and S. Brusco (2000), "Collusion, renegotiation, and implementation", Social Choice and Welfare 17:69–83.

Baliga, S., L. Corchon and T. Sjöström (1997), "The theory of implementation when the planner is a player", Journal of Economic Theory 77:15–33.

Baliga, S., and T. Sjöström (1999), "Interactive implementation", Games and Economic Behavior 27:38–63.

Banks, J. (1985), "Sophisticated voting outcomes and agenda control", Social Choice and Welfare 1:295–306.

Banks, J., C. Camerer and D. Porter (1994), "An experimental analysis of Nash refinements in signalling games", Games and Economic Behavior 6:1–31.

Bergin, J., and A. Sen (1996), "Implementation in generic environments", Social Choice and Welfare 13:467–448.

Bergin, J., and A. Sen (1998), "Extensive form implementation in incomplete information environments", Journal of Economic Theory 80:222–256.

Bernheim, B.D., and M. Whinston (1987), "Coalition-proof Nash equilibrium, II: Applications", Journal of Economic Theory 42:13–29.

Bliss, C., and B. Nalebuff (1984), "Dragon-slaying and ballroom dancing: The private supply of a public good", Journal of Public Economics 25:1–12.

Boylan, R. (1998), "Coalition-proof implementation", Journal of Economic Theory 82:132–143.

Boylan, R., J. Ledyard and R. McKelvey (1996), "Political competition in a model of economic growth: Some theoretical results", Economic Theory 7:191–205.

Brusco, S. (1995), "Perfect Bayesian implementation", Economic Theory 5:419–444.

Cabrales, A. (1999), "Adaptive dynamics and the implementation problem with complete information", Journal of Economic Theory 86:159–184.

Chakravorti, B. (1991), "Strategy space reductions for feasible implementation of Walrasian performance", Social Choice and Welfare 8:235–246.

Chakravorti, B. (1993), "Sequential rationality, implementation and pre-play communication", Journal of Mathematical Economics 22:265–294.

Chakravorti, B., L. Corchon and S. Wilkie (1994), "Credible implementation", mimeo.

Chander, P., and L. Wilde (1998) "A general characterization of optimal income taxation and enforcement", Review of Economic Studies 65:165–183.

Chen, Y. (2002), "Dynamic stability of Nash-efficient public goods mechanisms: Reconciling theory and experiments", Games and Economic Behavior, forthcoming.

Chen, Y., and F.-F. Tang (1998), "Learning and incentive-compatible mechanisms for public goods provision: An experimental study", Journal of Political Economy 106:633–662.

Corchon, L., and I. Ortuño-Ortin (1995), "Robust implementation under alternative information structures", Economic Design 1:157–172.

Corchon, L., and S. Wilkie (1995), "Double implementation of the ratio correspondence by a market mechanism", Economic Design 2:325–337.

Cramton, P., and T. Palfrey (1995), "Ratifiable mechanisms: Learning from disagreement", Games and Economic Behavior 10:255–283.

Crawford, V. (1977), "A game of fair division", Review of Economic Studies 44:235–247.

Crawford, V. (1979), "A procedure for generating Pareto-efficient egalitarian equivalent allocations", Econometrica 47:49–60.

Danilov, V. (1992), "Implementation via Nash equilibrium", Econometrica 60:43–56.

Dasgupta, P., P. Hammond and E. Maskin (1979), "The implementation of social choice rules: Some general results on incentive compatibility", Review of Economic Studies 46:185–216.

Demange, G. (1984), "Implementing efficient egalitarian equivalent allocations", Econometrica 52:1167–1177.

Demski, J., and D. Sappington (1984), "Optimal incentive contracts with multiple agents", Journal of Economic Theory 33:152–171.

Duggan, J. (1993), "Bayesian implementation with infinite types", mimeo.

Duggan, J. (1994a), "Virtual implementation in Bayesian equilibrium with infinite sets of types", mimeo (California Institute of Technology, CA).

Duggan, J. (1994b), "Bayesian implementation", Ph.D. dissertation (California Institute of Technology, CA).

Duggan, J. (1995), "Sequentially rational implementation with incomplete information", mimeo (Queens University).

Duggan, J. (1997), "Virtual Bayesian implementation", Econometrica 65:1175–1199.

Duggan, J. (1998), "An extensive form solution to the adverse selection problem in principal/multi-agent environments", Review of Economic Design 3:167–191.

Duggan, J., and J. Roberts (1997), "Robust implementation", Working paper (University of Rochester, Rochester).

Dutta, B., and A. Sen (1991a), "A necessary and sufficient condition for two-person Nash implementation", Review of Economic Studies 58:121–128.

Dutta, B., and A. Sen (1991b), "Further results on Bayesian implementation", mimeo.

Dutta, B., and A. Sen (1991c), "Implementation under strong equilibrium: A complete characterization", Journal of Mathematical Economics 20:49–67.

Dutta, B., and A. Sen (1993), "Implementing generalized Condorcet social choice functions via backward induction", Social Choice and Welfare 10:149–160.

Dutta, B., and A. Sen (1994a), "Two-person Bayesian implementation", Economic Design 1:41–54.

Dutta, B., and A. Sen (1994b), "Bayesian implementation: The necessity of infinite mechanisms", Journal of Economic Theory 64:130–141.

Dutta, B., A. Sen and R. Vohra (1994), "Nash implementation through elementary mechanisms in exchange economies", Economic Design 1:173–203.

Eliaz, K. (1999), "Fault tolerant implementation", Working paper (Tel Aviv University, Tel Aviv).

Farquharson, R. (1957/1969), Theory of Voting (Yale University Press, New Haven).

Gibbard, A. (1973), "Manipulation of voting schemes", Econometrica 41:587–601.

Glazer, J., and C.-T. Ma (1989), "Efficient allocation of a 'prize' — King Solomon's dilemma", Games and Economic Behavior 1:222–233.

Glazer, J., and M. Perry (1996), "Virtual implementation by backwards induction", Games and Economic Behavior 15:27–32.

Glazer, J., and R. Rosenthal (1992), "A note on the Abreu–Matsushima mechanism", Econometrica 60:1435–1438.

Glazer, J., and A. Rubinstein (1996), "An extensive game as a guide for solving a normal game", Journal of Economic Theory 70:32–42.

Glazer, J., and A. Rubinstein (1998), "Motives and implementation: On the design of mechanisms to elicit opinions", Journal of Economic Theory 79:157–173.

Glover, J. (1994), "A simpler mechanism that stops agents from cheating", Journal of Economic Theory 62:221–229.

Groves, T. (1979), "Efficient collective choice with compensation", in: J.-J. Laffont, ed., Aggregation and Revelation of Preferences (North-Holland, Amsterdam) 37–59.

Harris, M., and R. Townsend (1981), "Resource allocation with asymmetric information", Econometrica 49:33–64.

Harsanyi, J. (1967, 1968), "Games with incomplete information played by Bayesian players", Management Science 14:159–182, 320–334, 486–502.

Hererro, M., and S. Srivastava (1992), "Implementation via backward induction", Journal of Economic Theory 56:70–88.

Holmstrom, B., and R. Myerson (1983), "Efficient and durable decision rules with incomplete information", Econometrica 51:1799–1819.

Hong, L. (1996), "Bayesian implementation in exchange economies with state dependent feasible sets and private information", Social Choice and Welfare 13:433–444.

Hong, L. (1998), "Feasible Bayesian implementation with state dependent feasible sets", Journal of Economic Theory 80:201–221.

Hong, L., and S. Page (1994), "Reducing informational costs in endowment mechanisms", Economic Design 1:103–117.

Hurwicz, L. (1960), "Optimality and information efficiency in resource allocation processes", in: K. Arrow, S. Karlin and P. Suppes, eds., Mathematical Methods in the Social Sciences (Stanford University Press, Stanford) 27–46.

Hurwicz, L. (1972), "On informationally decentralized systems", in: C.B. McGuire and R. Radner, eds., Decision and Organization (North-Holland, Amsterdam).

Hurwicz, L. (1973), "The design of mechanisms for resource allocation", American Economic Review 61:1–30.

Hurwicz, L. (1977), "Optimality and informational efficiency in resource allocation problems", in: K. Arrow, S. Karlin and P. Suppes, eds., Mathematical Methods in the Social Sciences (Stanford University Press, Stanford) 27–48.

Hurwicz, L. (1986), "Incentive aspects of decentralization", in: K. Arrow and M. Intriligator, eds., Handbook of Mathematical Economics, Vol. III (North-Holland, Amsterdam) 1441–1482.

Hurwicz, L. (1994), "Economic design, adjustment processes, mechanisms, and institutions", Economic Design 1:1–14.

Hurwicz, L., E. Maskin and A. Postlewaite (1995), "Feasible Nash implementation of social choice corre-spondences when the designer does not know endowments or production sets", in: J. Ledyard, ed., The Economics of Information Decentralization: Complexity, Efficiency, and Stability (Kluwer, Amsterdam).

Jackson, M. (1991), "Bayesian implementation", Econometrica 59:461–477.

Jackson, M. (1992), "Implementation in undominated strategies: A look at bounded mechanisms", Review of Economic Studies 59:757–775.

Jackson, M. (2002), "A crash course in implementation theory", in: W. Thomson, ed., The Axiomatic Method: Principles and Applications to Game Theory and Resource Allocation, forthcoming.

Jackson, M., and H. Moulin (1990), "Implementing a public project and distributing its cost", Journal of Economic Theory 57:124–140.

Jackson, M., and T. Palfrey (1998), "Efficiency and voluntary implementation in markets with repeated pair-wise bargaining", Econometrica 66:1353–1388.

Jackson, M., and T. Palfrey (2001), "Voluntary implementation", Journal of Economic Theory 98:1–25.

Jackson, M., T. Palfrey and S. Srivastava (1994), "Undominated Nash implementation in bounded mecha-nisms", Games and Economic Behavior 6:474–501.

Kalai, E., and J. Ledyard (1998), "Repeated implementation", Journal of Economic Theory 83:308–317.

Kim, T. (1993), "A stable Nash mechanism implementing Lindahl allocations for quasi-linear environments", Journal of Mathematical Economics 22:359–371.

Legros, P. (1990), "Strongly durable allocations", CAE Working Paper No. 90-05 (Cornell University).

Ma, C. (1988), "Unique implementation of incentive contracts with many agents", Review of Economic Stud-ies 55:555–572.

Ma, C., J. Moore and S. Turnbull (1988), "Stopping agents from cheating", Journal of Economic Theory 46:355–372.

Mas-Colell, A., and X. Vives (1993), "Implementation in economies with a continuum of agents", Review of Economic Studies 10:613–629.

Maskin, E. (1999), "Nash implementation and welfare optimality", Review of Economic Studies 66:23–38.

Maskin, E. (1979), "Implementation and strong Nash equilibrium", in: J.-J. Laffont, ed., Aggregation and Revelation of Preferences (North-Holland, Amsterdam).

Maskin, E. (1985) "The theory of implementation in Nash equilibrium", in: L. Hurwicz, D. Schmeidler and H. Sonnenschein, eds., Social Organization: Essays in Memory of Elisha Pazner (Cambridge University Press, Cambridge).

Maskin, E., and J. Moore (1999), "Implementation with renegotiation", Review of Economic Studies 66:39–56.

Maskin, E., and J. Tirole (1992), "The principal-agent relationship with an informed principal, II: Common values", Econometrica 60:1–42.

Matsushima, H. (1988), "A new approach to the implementation problem", Journal of Economic Theory 45:128–144.

Matsushima, H. (1993), "Bayesian monotonicity with side payments", Journal of Economic Theory 59:107–121.

McKelvey, R. (1989), "Game forms for Nash implementation of general social choice correspondences", Social Choice and Welfare 6:139–156.

McKelvey, R., and R. Niemi (1978), "A multistage game representation of sophisticated voting for binary procedures", Journal of Economic Theory 81:1–22.

McKelvey, R., and T. Palfrey (1995), "Quantal response equilibria for normal form games", Games and Eco-nomic Behavior 10:6–38.

McKelvey, R., and T. Palfrey (1996), "A statistical theory of equilibrium in games", Japanese Economic Review 47:186–209.

McKelvey, R., and T. Palfrey (1998), "Quantal response equilibria for extensive form games", Experimental Economics 1:9–41.

Miller, N. (1977), "Graph-theoretic approaches to the theory of voting", American Journal of Political Science 21:769–803.

Mookherjee, D., and S. Reichelstein (1990), "Implementation via augmented revelation mechanisms", Review of Economic Studies 57:453–475.

Moore, J. (1992), "Implementation, contracts, and renegotiation in environments with complete information", in: J.-J. Laffont, ed., Advances in Economic Theory, Vol. 1 (Cambridge University Press, Cambridge).

Moore, J., and R. Repullo (1988), "Subgame perfect implementation", Econometrica 46:1191–220.

Moore, J., and R. Repullo (1990), "Nash implementation: A full characterization", Econometrica 58:1083–1099.

Moulin, H. (1979), "Dominance-solvable voting schemes", Econometrica 47:1337–1351.

Moulin, H. (1984), "Implementing the Kalai–Smorodinsky bargaining solution", Journal of Economic Theory 33:32–45.

Moulin, H. (1986) "Choosing from a tournament", Social Choice and Welfare 3:271–291.

Moulin, H. (1994), "Social choice", in: R.J. Aumann and S. Hart, eds., Handbook of Game Theory, Vol. 2 (North-Holland, Amsterdam) Chapter 31, 1091–1125.

Mount, K., and S. Reiter (1974), "The informational size of message spaces", Journal of Economic Theory 8:161–192.

Myerson, R. (1979), "Incentive compatibility and the bargaining problem", Econometrica 47:61–74.

Myerson, R. (1985), "Bayesian equilibrium and incentive compatibility: An introduction", in: L. Hurwicz, D. Schmeidler and H. Sonnenschein, eds., Social Goals and Organization: Essays in Memory of Elisha Pazner (Cambridge University Press, Cambridge) 229–259.

Ordeshook, P., and T. Palfrey (1988), "Agendas, strategic voting, and signaling with incomplete information", American Journal of Political Science 32:441–466.

Palfrey, T. (1992), "Implementation in Bayesian equilibrium: The multiple equilibrium problem", in: J.-J. Laffont, ed., Advances in Economic Theory, Vol. 1 (Cambridge University Press, Cambridge).

Palfrey, T., and S. Srivastava (1986), "Private information in large economies", Journal of Economic Theory 39:34–58.

Palfrey, T., and S. Srivastava (1987), "On Bayesian implementable allocations", Review of Economic Studies 54:193–208.

Palfrey, T., and S. Srivastava (1989a), "Implementation with incomplete information in exchange economies", Econometrica 57:115–134.

Palfrey, T., and S. Srivastava (1989b), "Mechanism design with incomplete information: A solution to the implementation problem", Journal of Political Economy 97:668–691.

Palfrey, T., and S. Srivastava (1991a), "Nash implementation using undominated strategies", Econometrica 59:479–501.

Palfrey, T., and S. Srivastava (1991b), "Efficient trading mechanisms with preplay communication", Journal of Economic Theory 55:17–40.

Palfrey, T., and S. Srivastava (1993), Bayesian Implementation (Harwood Academic Publishers, Reading).

Peleg, B. (1996a), "A continuous double implementation of the constrained Walrasian equilibrium", Economic Design 2:89–97.

Peleg, B. (1996b), "Double implementation of the Lindahl equilibrium by a continuous mechanism", Economic Design 2:311–324.

Postlewaite, A. (1985), "Implementation via Nash equilibria in economic environments", in: L. Hurwicz, D. Schmeidler and H. Sonnenschein, eds., Social Goals and Organization: Essays in Memory of Elisha Pazner (Cambridge University Press, Cambridge) 205–228.

Postlewaite, A., and D. Schmeidler (1986), "Implementation in differential information economies", Journal of Economic Theory 39:14–33.

Postlewaite, A., and D. Wettstein (1989), "Feasible and continuous implementation", Review of Economic Studies 56:603–611.

Reichelstein, S. (1984), "Incentive compatibility and informational requirements", Journal of Economic Theory 34:32–51.

Reichelstein, S., and S. Reiter (1988), "Game forms with minimal strategy spaces", Econometrica 56:661–692.

Repullo, R. (1987), "A simple proof of Maskin's theorem on Nash implementation", Social Choice and Welfare 4:39–41.

Rubinstein, A., and A. Wolinsky (1992), "Renegotiation-proof implementation and time preferences", American Economic Review 82:600–614.

Saijo, T. (1988), "Strategy space reductions in Maskin's theorem: Sufficient conditions for Nash implementation", Econometrica 56:693–700.

Saijo, T., Y. Tatamitani and T. Yamato (1996), "Toward natural implementation", International Economic Review 37:949–980.

Satterthwaite, M. (1975), "Strategy-proofness and Arrow's conditions: Existence and correspondence theorems for voting procedure and social welfare functions", Journal of Economic Theory 10:187–217.

Sefton, M., and A. Yavas (1996), "Abreu–Matsushima mechanisms: Experimental evidence", Games and Economic Behavior 16:280–302.

Sen, A. (1987), "Two essays on the theory of implementation", Ph.D. dissertation (Princeton University).

Serrano, R., and R. Vohra (1999), "On the impossibility of implementation under incomplete information", Working Paper #99-10 (Brown University, Department of Economics, Providence) (forthcoming in Econometrica).

Serrano, R., and R. Vohra (2000), "Type diversity and virtual Bayesian implementation", Working Paper #00-16 (Brown University, Department of Economics, Providence).

Sjöström, T. (1991), "On the necessary and sufficient conditions for Nash implementation", Social Choice and Welfare 8:333–340.

Sjöström, T. (1993), "Implementation in perfect equilibria", Social Choice and Welfare 10:97–106.

Sjöström, T. (1994), "Implementation in undominated Nash equilibrium without integer games", Games and Economic Behavior 6:502–511.

Sjöström, T. (1995), "Implementation by demand mechanisms", Economic Design 1:343–354.

Sjöström, T. (1996), "Credibility and renegotiation of outcome functions in implementation", Japanese Economic Review 47:157–169.

Sjöström, T. (1999), "Undominated Nash implementation with collusion and renegotiation", Games and Economic Behavior 26:337–352.

Suh, S.-C. (1997), "An algorithm checking strong Nash implementability", Journal of Mathematical Economics 25:109–122.

Tatamitani, Y. (1993), "Double implementation in Nash and undominated Nash equilibrium in social choice environments", Economic Theory 3:109–117.

Thomson, W. (1979), "Maximin strategies and elicitation of preferences", in: J.-J. Laffont, ed., Aggregation and Revelation of Preferences (North-Holland, Amsterdam) 245–268.

Tian, G. (1989), "Implementation of the Lindahl correspondence by a single-valued, feasible, and continuous mechanism", Review of Economic Studies 56:613–621.

Tian, G. (1990), "Completely feasible continuous implementation of the Lindahl correspondence with a message space of minimal dimension", Journal of Economic Theory 51:443–452.

Tian, G. (1994), "Bayesian implementation in exchange economies with state dependent preferences and feasible sets", Social Choice and Welfare 16:99–120.

Townsend, R. (1979), "Optimal contracts and competitive markets with costly state verification", Journal of Economic Theory 21:265–293.

Trick, M., and S. Srivastava (1996), "Sophisticated voting rules: The two tournaments case", Social Choice and Welfare 13:275–289.

Varian, H. (1994), "A solution to the problem of externalities and public goods when agents are well-informed", American Economic Review 84:1278–1293.

Wettstein, D. (1990), "Continuous implementation of constrained rational expectations equilibria", Journal of Economic Theory 52:208–222.

Wettstein, D. (1992), "Continuous implementation in economies with incomplete information", Games and Economic Behavior 4:463–483.

Williams, S. (1984), "Sufficient conditions for Nash implementation", mimeo (University of Minnesota).

Williams, S. (1986), "Realization and Nash implementation: Two aspects of mechanism design", Econometrica 54:139–151.

Yamato, T. (1992), "On Nash implementation of social choice correspondences", Games and Economic Behavior 4:484–492.

Yamato, T. (1993), "Double implementation in Nash and undominated Nash equilibria", Journal of Economic Theory 59:311–323.

Chapter 62

GAME THEORY AND EXPERIMENTAL GAMING

MARTIN SHUBIK*

Cowles Foundation, Yale University, New Haven, CT, USA

Contents

*I would like to thank the Pew Charitable Trust for its generous support.

Handbook of Game Theory, Volume 3, Edited by R.J. Aumann and S. Hart

Abstract

This is a survey and discussion of work covering both formal game theory and experimental gaming prior to 1991. It is a useful preliminary introduction to the considerable change and emphasis which has taken place since that time where dynamics, learning, and local optimization have challenged the concept of noncooperative equilibria.

Keywords

experimental gaming, game theory, context, minimax, and coalitional form

JEL classification: C71, C72, C73, C90

1. Scope

This article deals with experimental games as they pertain to game theory. As such there is a natural distinction between experimentation with abstract games devoted to testing a specific hypothesis in game theory and games with a scenario from a discipline such as economics or political science where the game is presented in the context of some particular activity, even though the same hypothesis might be tested.[1]

John Kennedy, professor of psychology at Princeton and one of the earliest experimenters with games, suggested[2] that if you could tell him the results you wanted and give him control of the briefing he could get you the results.

Context appears to be critical in the study of behavior in games. R. Simon, in his Ph.D. thesis (1967), controlled for context. He used the same two-person zero-sum game with three different scenarios. One was abstract, one described the game in a business context and the other in a military context. The players were selected from majors in business and in military science. As the game was a relatively simple two-person zero-sum game a reasonably strong case could be made out for the maxmin solution concept. The performance of all students was best in the abstract scenario. Some of the military science students complained about the simplicity of the military scenario and the business students complained about the simplicity of the business scenario.

Many experiments in game theory with abstract games are to "see what happens" and then possibly to look for consistency of the outcomes with various solution concepts.

In teaching the basic concepts of game theory, for several years I have used several simple games. In the first lecture when most of the students have had no exposure to game theory I present them with three matrices, the Prisoner's Dilemma, Chicken and the Battle of the Sexes (see Figures 1a, b and c). The students are asked if they have had any experience with these games. If not, they are asked to associate the name with the game. Table 1 shows the data for 1988.

There appears to be no significant ability to match the words with the matrices.[3]

There are several thousand articles on experimental matrix games and several hundred on experimental economics alone, most of which can be interpreted as pertaining to game theory as well as to economics. Many of the business games are multipurpose and contain a component which involves game theory. War gaming, especially at the tactical level and to some extent at the strategic level, has been influenced by game theory [see Brewer and Shubik (1979)]. No attempt is made here to provide an exhaustive review of the vast and differentiated body of literature pertaining to blends of operational, experimental and didactic gaming. Instead a selection is made concentrating on

---

[1] For example, one might hypothesize that the percentage of points selected in an abstract game which lie in the core will be greater than the points selected in an experiment repeated with the only difference being a scenario supplied to the game.

[2] Personal communication, 1957.

[3] The rows and columns do not all add to the same number because of some omissions among 30 students.

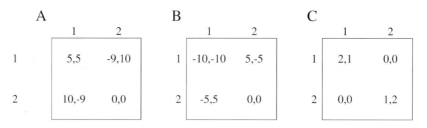

Figure 1.

Table 1

Guessed name	Actual game		
	Chicken	Battle of the Sexes	Prisoner's Dilemma
Chicken	7	11	8
Battle of the Sexes	9	12	8
Prisoner's Dilemma	10	7	9

the modeling problems, formal structures and selection of solution concepts in game theory.

Roth (1986), in a provocative discussion of laboratory experimentation in economics, has suggested three kinds of activities for those engaged in economic experimental gaming. They are "speaking to theorists", "searching for facts" and "whispering in the ears of princes". The third of these activities, from the viewpoint of society, is possibly the most immediately useful. In the United States the princes tend to be corporate executives, politicians and generals or admirals. However, the topic of operational gaming merits a separate treatment, and is not treated here.

2. Game theory and gaming

Before discussing experimental games which might lead to the confirmation or rejection of theory, or to the discovery of "facts" or regularities in human behavior, it is desirable to consider those aspects of game theory which might benefit from experimental gaming.

Perhaps the most important item to keep in mind is the fundamental distinction between the game theory and the approach of social psychology to the same problems. Underlying a considerable amount of game theory is the concept of *external symmetry*. Unless stated otherwise, all players in game theory are treated as though they are intrinsically symmetric in all nonspecified attributes. They have the same perceptions, the same abilities to calculate and the same personalities. Much of game theory is devoted to deriving normative or behavioral solution concepts to games played by personality-free players in context-free environments. In particular, one of the impressive achievements of formal game theory has been to produce a menu of different solution concepts which suggest ingeniously reasoned sets of outcomes to nonsymmetric games played by externally symmetric players.

In bold contrast, the approach of the social psychologists has for the most part been to take symmetric games and observe the broad differences in play which can be attributed to nonsymmetric players. Personality, passion and perception count and the social psychologist is interested to see how they count.

A completely different view from either of the above underlies much of the approach of the economists interested in studying mass markets. The power of the mass market is that it turns the social science problem of competition or cooperation among the few into a nonsocial science problem. The message of the mass market is not that personality differences do not exist, but that in the impersonal mass market they do not matter. The large market has the individual player pitted against an aggregate of other players. When the large market becomes still larger, all concern for the others is attenuated and the individual can regard himself as playing against a given inanimate object known as the market.

2.1. The testable elements of game theory

Before asking who has been testing what, a question which needs to be asked is what is there to be tested concerning game theory. A brief listing is given and discussed in the subsequent sections.

Context: Does context matter? Or do we believe that strategic structure *in vitro* will be treated the same way by rational players regardless of context? There has been much abstract experimental gaming with no scenarios supplied to the players. But there also is a growing body of experimental literature where the context is explicitly economic, political, military or other.

External symmetry and limited rationality: The social psychologists are concerned with studying decision-makers as they are. A central simplification of game theory has been to build upon the abstract simplification of rational man. Not only are the players in game theory devoid of personality, they are endowed with unlimited rationality, unless specified otherwise. What is learned from Maschler's schoolchildren, Smith's or Roth's undergraduates or Battaglio's rats [Maschler (1978), Smith (1986), Roth (1983), Battaglio et al. (1985)] requires considerable interpretation to relate the experimental results with the game abstraction.

Rules of the game and the role of time: In much game theory time plays little role. In the extensive form the sequencing of moves is frequently critical, but the cardinal aspects of the length of elapsed time between moves is rarely important (exceptions are in dueling and search and evasion models). In particular, one of the paradoxes in attempting to experiment with games in coalitional form is that the key representation – the characteristic functions (and variants thereof) – implicitly assumes that the bargaining, communication, deal-making and tentative formation of coalitions is costless and timeless. Whereas in actual experiments the amount of time taken in bargaining and discussion is often a critical limiting factor in determining the outcome.

Utility and preference: Do we carry around utility functions? The utility function is an extremely handy fiction for the mathematical economist and the game theorist applying game theory to the study of mass markets, but in many of the experiments with matrix games or market games the payoffs have tended to be money (on a performance or hourly basis or both), points or grades [see Rapoport, Guyer and Gordon (1976)]. For example, Mosteller and Nogee (1951) attempted to measure the utility function of individuals with little success. In most experimental work little attempt is made to obtain individual measures before the experiment. The design difficulties and the costs make this type of pre-testing prohibitively difficult to perform. Depending on the specifics of the experiment it can be argued that in some experiments all one needs to know is that more money is preferred to less; in others the existence of a linear measure of utility is important; in still others there are questions concerning interpersonal comparisons. See however Roth and Malouf (1979) for an experimental design which attempts to account for these difficulties.[4]

Even with individual testing there is the possibility that the goals of the isolated individual change when placed in a multiperson context. The maximization of relative score or status [Shubik (1971b)] may dominate the seeking of absolute point score.

Solution concepts: Given that we have an adequate characterization of the game and the individuals about to play, what do we want to test for? Over the course of the years various game theorists have proposed many solution concepts offering both normative and behavioral justifications for the solutions. Much of the activity in experimental gaming has been devoted to considering different solution concepts.

2.2. *Experimentation and representation of the game*

The two fundamental representations of games which have been used for experimentation have been the strategic and cooperative form. Rarely has the formal extensive form been used in experimentation. This appears to be due to several fundamental reasons. In the study of bargaining and negotiation in general we have no easy simple description of the game in extensive form. It is difficult to describe talk as moves. There is a lack of structure and a flexibility in sequencing which weakens the analogy with games such as chess where the extensive form is clearly defined.

In experimenting with the game in matrix form, frequently the experimenter provides the same matrix to be played several times; but clearly this is different from giving the individual a new and considerably larger matrix representing the strategic form of a multistage game. Shubik, Wolf and Poon (1974) experimented with a game in matrix form played twice and a much larger matrix representing the strategic form of the two-stage game to be played once. Strikingly different results were obtained with considerable emphasis given in the first version to the behavior of the opponent on his first trial.

---

[4] Unfortunately, given the virtual duality between probability and utility, they have to introduce the assumption that the subjective probabilities are the same as those presented in the experiment. Thus one difficulty may have been traded for another.

How big a matrix can an experimental subject handle? The answer appears to be a 2×2 unless there is some form of special structure on the entries; thus for example Fouraker, Shubik and Siegel (1961) used a 57×25 matrix in the study of triopoly, but this matrix had considerable regularities (being generated from continuous payoff functions from a Cournot triopoly model).

3. Abstract matrix games

3.1. Matrix games

The 2×2 matrix game has been a major source for the provision of didactic examples and experimental games for social scientists with interests in game theory. Limiting ourselves to strict orderings there are $4! \times 4! = 576$ ways in which two payoff entries can be placed in each of four cells of a 2×2 matrix. When symmetries are accounted for there are 78 strategically different games which remain [see Rapoport, Guyer and Gordon (1976)]. If ties are considered in the payoffs then the number of strategically different games becomes considerably larger. Guyer and Hamburger (1968) counted the strategically different weakly ordinal games and Powers (1986) made some corrections and provided a taxonomy for these games. There are 726 strategically different games. O'Neill, in unpublished notes dating from 1987, has estimated that the lower bound on the number of different $2 \times 2 \times 2$ games is $(8!)^3/(8 \cdot 6) = 1,365,590,016,000$. For the 3×3 game we have $(9!)^2/((3!)^2 \cdot 2) = 1,828,915,200$. Thus we see that the reason why the $2 \times 2 \times 2$ and the 3×3 games have not been studied, classified and analyzed in any generality is that the number of different cases is astronomical.

Before considering the general nonconstant-sum game, the first and most elementary question to ask concerns the playing of two-person zero- and constant-sum games. This splits into two questions. The first is how good a predictor is the saddlepoint in two-person zero-sum games with a pure strategy saddlepoint. The second is how good a predictor is the maxmin mixed strategy of a player's behavior. Possibly the two earliest published works on the zero-sum game were those of Morin (1960) and Lieberman (1960) who considered a 3×3 game with a saddlepoint. He has 15 pairs of subjects play a 3×3 for 200 times each at a penny a point. He used the last 10 trials as his statistic. In these trials approximately 90% selected their optimal strategy. One may query the approach of using the first 190 trials for learning. Lieberman (1962) also considered a zero-sum 2×2 game played against a stooge using a minimax mixed strategy. Rather than approach an optimal strategy the live player appeared to follow or imitate the stooge.

Rapoport, Guyer and Gordon [(1976, Chapter 21)] note three ordinally different games of pure opposition. Using 4 to stand for the most desired outcome and 1 for the least desired, Game 1 below has a saddlepoint with dominated strategies for each player, Game 2 has a saddlepoint but only one player has a dominated strategy and Game 3 has no saddlepoint.

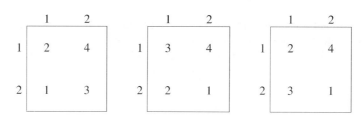

Figure 2.

Column Player

	1	2	3	J
1	0	1	1	0
2	1	0	1	0
3	1	1	0	0
J	0	0	0	1

Row Player

Figure 3.

They report on one-shot experiments by Frenkel (1973) with a large population of players with reasonably good support for the saddlepoints but not for the mixed strategy. The saddlepoint for small games appears to provide a good prediction of behavior. Shubik has used a 2×2 zero-sum game with a saddlepoint as a "rationality test" prior to having students play in nonzero-sum games and one can bet with safety that over 90% of the students will select their saddlepoint strategy. The mixed strategy results are by no means that clear.

More recently, O'Neill (1987) has considered a repeated zero-sum game with a mixed strategy solution. The O'Neill experiment has 25 pairs of subjects play a card-matching game 105 times. Figure 3 shows the matrix. The + is a win for Row and the − a loss. There is a unique equilibrium with both players randomizing using probabilities of 0.2, 0.2, 0.2, 0.4, each. The value of the game is 0.4.

The students were initially given $2.50 each and at each stage the winner received 5 cents from the loser. The aggregate data involving 2625 joint moves showed high consistency with the predictions of minimax theory. The proportion of wins for the row player was 40.9% rather than 40% as predicted by theory. Brown and Rosenthal (1987) have challenged O'Neill's analysis but his results are of note [see also O'Neill (1991)].

One might argue that minimax is a normative theory which is an extension of the concept of rational behavior and that there is little to actually test beyond observing that naive players do not necessarily do the right thing. In saddlepoint games there is reasonable evidence that they tend to do the right thing [see, for example, Lieberman (1960)] and as the results depend only on ordinal properties of the matrices this is doubly comforting. The mixed strategies however in general require the full philosophical mysteries of the measurement of utility and choice under uncertainty. One of the attractive features of the O'Neill design is that there are only two payoff values and hence the utility structure is kept at its simplest. It would be of interest to see the O'Neill experiment replicated and possibly also performed with a larger matrix (say a 6 × 6) with the same structure.

It should be noted, taking a cue from Miller's (1956) classical article, that unless a matrix has special structure it may not be reasonable to experiment with any size larger than 2 × 2. A 4 × 4 with general entries would require too much memory, cognition and calculation for nontrained subjects.

One of the most comprehensive research projects devoted to studying matrix games has been that of Rapoport, Guyer and Gordon (1976). In their book entitled *The 2 × 2 Game*, not only do they count all of the different games, but they develop a taxonomy for all 78 of them. They begin by observing that there are 12 ordinally symmetric games and 66 nonsymmetric games. They classify them into games of complete, partial and no opposition. Games of complete opposition are the ordinal variants of two-person constant-sum games. Games of no opposition are classified by them as those which have the best outcome to each in the same cell. Space limitations do not permit a full discussion of this taxonomy, yet it is my belief that the problem of constructing an adequate taxonomy for matrix games is valuable both from the viewpoint of theory and experimentation. In particular, although much stress has been laid on the noncooperative equilibrium for games in strategic form, there appears to be a host of considerations which can all be present simultaneously and contribute towards driving a solution to a predictable outcome. These may be reflected in a taxonomy. A partial list is noted here: (1) uniqueness of equilibria; (2) symmetry of the game; (3) safety levels of equilibria; (4) existence of dominant strategies; (5) Pareto optimality; (6) interpersonal comparison and sidepayment conditions; (7) information conditions.[5]

Of all of the 78 strategically different 2 × 2 games I suggest that the following one is structurally the most conducive to the selection of its equilibrium point solution.[6] The cell (1, 1) will be selected not merely because it is an equilibrium point but also because it is unique, Pareto-optimal, results from the selection of dominant strategies,

	1	2
1	4,4	3,2
2	2,3	1,1

---

[5] In an experiment with several 2 × 2 games played repeatedly with low information where each of the players was given only his own payoff matrix and information about the move selected by his competitor after each of a sequence of plays, the ability to predict the outcome depended not merely on the noncooperative equilibrium but on the coincidence of an outcome having several other properties [see Shubik (1962)].

[6] Where the convention is that 4 is high and 1 low.

is symmetric and coincides with maxmin ($P_1 - P_2$). Rapoport et al. (1976) have this tied for first place with a closely related no-conflict game, with the off-diagonal entries reversed. This new game does not have the coincidence of the maxmin difference solution with the equilibrium.

The work of Rapoport, Guyer and Gordon (1976) to this date is the most exhaustive attempt to ring the changes on conditions of play as well as structure of the 2×2 game.

3.2. Games in coalitional form

If one wishes to be a purist, a game in coalitional or characteristic function form cannot actually be played. It is an abstraction with playing time and all other transaction costs set at zero. Thus any attempt to experiment with games in characteristic function form must take into account the distinction between the representation of the game and the differences introduced by the control and the time utilized in play.

With allowances made for the difficulties in linking the actual play with the characteristic function, there have been several experiments utilizing the characteristic function especially for the three-person game and to some extent for larger games. In particular, the invention of the bargaining set by Aumann and Maschler (1964) was motivated by Maschler's desire to explain experimental evidence [see Maschler (1963, 1978)].

The book of Kahan and Amnon Rapoport (1984) has possibly the most exhaustive discussion and report on games in coalitional form. They have eleven chapters on the theory development pertaining to games in coalitional form, a chapter on experimental results with three-person games, a chapter[7] on games for $n \geqslant 4$, followed by concluding remarks which support the general thesis that the experimental results are not independent of normalization and of other aspects of context; that no normative game-theoretic solution emerges as *generally* verifiable (often this does not rule out their use in special context-rich domains).

It is my belief that a fruitful way to utilize a game in characteristic function form for experimental purposes is to explore beliefs or norms of students and others concerning how the proceeds from cooperation in a three-person game with sidepayments *should* be split. Any play by three individuals of a three-person game in characteristic function form is no longer represented by the form but has to be adjusted for the time and resource costs of the process of play as well as controlled for personality.

In a normative inquiry Shubik (1975a, 1978) has used three characteristic functions for many years. They were selected so that the first game has a rather large core, the second a single-point core that is highly nonsymmetric and the third no core. For the most part the subjects were students attending lectures on game theory who gave their opinions on at least the first game without knowledge of any formal cooperative solution.

$$\text{Game 1:} \quad v(1) = v(2) = v(3) = 0; \quad v(12) = 1, v(13) = 2, v(23) = 3;$$
$$v(123) = 4.$$

[7] These two chapters provide the most comprehensive summary of experimental results with games in characteristic form up until the early 1980s. These include Apex games, simple games, Selten's work on the equal division core and market games.

Table 2

	Percentage in core		Percentage even split
	Game 1	Game 2	Game 3
1980	86	28.3	36
1981	86.8	5	32.5
1983	87	21.8	7.5
1984	89.5	19	–
1985	93	20	–
1988	92	11	5

Game 2: $v(1) = v(2) = v(3) = 0;$ $v(12) = 0, v(13) = 0, v(23) = 4;$
$v(123) = 4.$

Game 3: $v(1) = v(2) = v(3) = 0;$ $v(12) = 2.5, v(13) = 3, v(23) = 3.5;$
$v(123) = 4.$

Table 2 shows the percentage of individuals selecting a point in the core for games 1 and 2. The smallest sample size was 17 and the largest was 50. The last column shows the choice of an even split in the game without a core.

There was no discernible pattern for the game without a core except for a selection of the even split, possibly as a focal point. There was little evidence supporting the value or nucleolus.

When the core is "flat" and not too nonsymmetric (game 1), it is a good predictor. When it is one point but highly skewed, its attractiveness is highly diminished (game 2).

Three-person games have been of considerable interest to social psychologists. Caplow (1956) suggested a theory of coalitions which was experimented with by Gamson (1961) and later the three-person simple majority game was used in experiments by Byron Roth (1975). He introduced a scenario involving a task performed by a scout troop with three variables: the size of the groups considered as a single player, the effort expended, and the economic value added to any coalition of players. He compared both questionnaire response and actual play. His experimental results showed an important role for equity considerations.

Selten, since 1972, has been concerned with equity considerations in coalitional bargaining. His views are summarized in a detailed survey [Selten (1987)]. In essence, he suggests that limited rationality must be taken into consideration and that normalization of payoffs and then equity considerations in division provide important structure to bargaining. He considers and compares both the work by Maschler (1978) on the bargaining set and his theory of equal division payoff bounds [Selten (1982)]. Selten defines "satisficing" as the player obtaining his aspiration level which in the context of a characteristic function game is a lower bound on what a player is willing to accept in a genuine coalition. The equal division payoff bounds provide an easy way to obtain aspiration levels, especially in a zero-normalized three-person game. The specifics for the calculation of the bounds are lengthy and are not reproduced here. There are however three points to be made. (1) They are chosen specifically to reflect limited rationality. (2) The criterion of success suggested is an "area computation", i.e., no point prediction is used as a test of success, but a range or area is considered. (3) The examination of data generated from four sets of experiments done by others "strongly suggests that equity considerations have an important influence on the behavior of experimental subjects in zero-normalized three-person games". This also comes through in the work of W. Güth, R. Schmittberger and B. Schwartz (1982) who considered one-period ultimatum bargaining games where one side proposes the division of a sum of money M into two parts, $M - x$ and x. The other side then has the choice of accepting or rejecting the split with both sides obtaining 0. The perfect equilibrium is unique and has the side that moves first take all. This is not what happened experimentally.

The approach in these experiments is towards developing a descriptive theory. As such it appears to this writer that the structure of the game and both psychological and socio-psychological factors must be taken into account explicitly. The only way that *process* considerations can be avoided is by not actually playing the game but asking individuals to state how they think they should or they would play. The characteristic function is an idealization. The coalitions are not coalitions, they are costless coalitions-in-being. Attempts to experiment with playing games in characteristic function form go against the original conception of the characteristic form. As the results of Byron Roth indicated *both* the questionnaire and actual play of the game are worth utilizing and they give different results. Much is still to be learned in comparing both approaches.

3.3. Other games

One important but somewhat overlooked aspect of experimental gaming is the construction and use of simple paradoxical games to illustrate problems in game theory and to find out what happens when these games are played. Three examples are given. Hausner, Nash, Shapley and Shubik (1964) in the early 1950s constructed a game called "so long sucker" in which the winner is the survivor; in order to win it is necessary, but not sufficient, to form coalitions. At some point it will pay someone to "doublecross" his partner. However, to a game theorist the definition of doublecross is difficult to make precise. When John McCarthy decided to get revenge for being doublecrossed by Nash, Nash pointed out that McCarthy was "being irrational" as he could have easily calculated that Nash would have to do what he did given the opportunity.

The dollar auction game [see Shubik (1971a)] is another example of an extremely simple yet paradoxical game. The highest bidder pays to win a dollar, but the second-highest bidder must also pay but receives nothing in return. Experiments with this game belong primarily to the domain of social psychology and a recent book by Tegler (1980) covers the experimental work. O'Neill (1987) however has suggested a game-theoretic solution to the auction with limited resources.

The third game is a simple game attributed to Rosenthal. The subjects are confronted with a game in extensive form as indicated in the game tree in Figure 4. They are told that they are playing an unknown opponent and that they are player 2 and have been given the move. They are required to select their move and justify the choice.

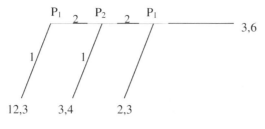

Figure 4.

The paradox comes in the nature of the expectations that player 2 must form about player 1's behavior. If player 1 were "rational" he would have ended the game: hence the move should never have come to player 2.

In an informal[8] classroom exercise in 1985 17 out of 21 acting as player 2 chose to continue the game and 4 terminated giving reasons for the most part involving the stupidity or the erratic behavior of the competition.

4. Experimental economics

The first adequately documented running of an experimental game to study competition among firms was that of Chamberlain (1948). It was a report of a relatively informal oligopolistic market run in a class in economic theory. This was a relatively isolated event [occurring before the publication of the Nash equilibrium (1951)] and the general idea of experimentation in the economics of competition did not spread for around another decade. Perhaps the main impetus to the idea of experimentation on economic competition came via the construction of the first computer-based business game by Bellman et al. (1957). Within a few years business games were widely used both in business schools and in many corporations. Hoggatt (1959, 1969) was possibly the earliest researcher to recognize the value of the computer-based game as an experimental tool.

In the past two decades there has been considerable work by Vernon Smith primarily using a computer-based laboratory to study price formation in various market structures. Many of the results appear to offer comforting support for the virtues of a competitive price system. However, it is important to consider that possibly a key feature in the functioning of an anonymous mass market is that it is designed implicitly or explicitly to *minimize both* the socio-psychological and the more complex game-theoretic aspects of strategic human behavior. The crowning joy of economic theory as a social science is its theory of markets and competition. But paradoxically markets appear to work the best

[8] The meaning of "informal" here is that the students were not paid for their choice. I supplied them with the game as shown in Figure 4 and asked them to write down what they would do, if as player 2 they were called on to move. They were asked to explain their choice.

when the individuals are converted into nonsocial anonymous entities and competition is attenuated, leaving a set of one-person maximization problems.[9]

This observation is meant not to detract from the value of the experimental work of Smith and many others who have been concerned with showing the efficiency of markets but to contrast the experimental program of Smith, Plott and others concerned with the design of economic mechanisms with the more general problems of game theory.[10]

Are economic and game-theoretic principles sufficiently basic that one might have cleaner experiments using rats and pigeons rather than humans? Hopefully, animal passions and neuroses are simpler than those of humans. Kagel et al. (1975) began to investigate consumer demand in rats. The investigations have been primarily aimed at isolated individual economic behavior in animals, including an investigation of risky choice [Battaglio et al. (1985)]. Kagel (1987) offers a justification of the use of animals, noting problems of species extrapolation and cognition. He points out that (he believes that) animals unlike humans are not selected from a background of political, economic and cultural contexts, which at least to the game theorist concerned with external symmetry is helpful.

The Kagel paper is noted (although it is concerned only with individual decision-making) to call attention to the relationship between individual decision-making and game-playing behavior. In few, if any, game experiments is there extensive pre-testing of one-person decision-making under exogenous uncertainty. Yet there is an implicit assumption made that the subjects satisfy the theory for one-person behavior. Is the jump from one- to two-person behavior the same for humans, sharks, wolves and rats?

Especially given the growth of interest in game theory applications to animal behavior, it would be possible and of interest to extend experimentation with animals to two-person zero- and nonconstant-sum games. In much of the experimental work in game theory anonymity is a key control factor. It should be reasonably straightforward to have animals (not even of the same species) in separate cages, each with two choices leading to four prizes from a 2×2 matrix. There is the problem as to how or if the animals are ever able to deduce that they are in a two-animal competitive situation. Furthermore, at least if the analogy with economics is to be carried further, can animals be taught the meaning and value of money, as money figures so prominently in business games, much of experimental economics, experimental game theory and in the actual economy? I suspect that the answer is no. But how far can animal analogies be made seems to be a reasonable question.

Another approach to the study of competition has been via the use of robots in computerized oligopoly games. Hoggatt (1969) in an imaginative design had a live player play two artificial players separately in the same type of game. One artificial player

---

[9] But even here the dynamics of social psychology appears in mass markets. Shubik (1970) managed to run a simulated stock market with over 100 individuals using their own money to buy and sell shares in four corporations involved in a business game. Trading was at four professional broker-operated trading posts. Both a boom and a panic were encountered.

[10] See the research series in experimental economics starting with Smith (1979).

was designed to be cooperative and the other competitive. The live players tended to cooperate with the former and to compete with the latter.

Shubik, Wolf and Eisenberg (1972), Shubik and Riese (1972), Liu (1973), and others have over the years run a series of experiments with an oligopolistic market game model with a real player versus either other real players or artificial players [Shubik, Wolf and Lockhart (1971)]. In the Systems Development Corporation experiments markets with 1, 2, 3, 5, 7, and 10 live players were run to examine the predictive value of the noncooperative equilibrium and its relationship with the competitive equilibrium. The average of each of the last few periods tended to lie between the beat-the-average and noncooperative solution.

A money prize was offered in the Shubik and Riese (1972) experiment with 10 duopoly pairs to the firm whose profit relative to its opponent was the highest. This converted a nonzero-sum game into a zero-sum game and the predicted outcome was reasonably good.

In a series of experiments run with students in game theory courses the students each first played in a monopoly game and then played in a duopoly game against an artificial player. The monopoly game posed a problem in simple maximization. There was no significant correlation in performance (measured in terms of profit ranking) between those who performed well in monopoly and those who performed well in duopoly.

Business games have been used for the most part in business school education and corporate training courses with little or no experimental concern. Business game models have tended to be large, complex and *ad hoc*, but games to study oligopoly have been kept relatively simple in order to maintain experimental control as can be seen by the work of Sauermann and Selten (1959); Hoggatt (1959); Fouraker, Shubik and Siegel (1961); Fouraker and Siegel (1963); Friedman (1967, 1969); and Friedman and Hoggatt (1980). A difficulty with the simplification is that it becomes so radical that the words (such as firm production cost, advertising, etc.) bear so little resemblance to the complex phenomena they are meant to evoke that experience without considerable ability to abstract may actually hamper the players.[11]

The first work in cooperative game experimental economics was the prize-winning book of Siegel and Fouraker (1960) which contained a series of elegant simple experiments in bilateral monopoly exploring the theories of Edgeworth (1881), Bowley (1928) and Zeuthen (1930). This was interdisciplinary work at its best with the senior author being a quantitatively oriented psychologist. The book begins with an explicit statement of the nature of the economic context, the forms of negotiation and the information conditions. In game-theoretic terms the Edgeworth model leads to the no-sidepayment core, the Zeuthen model calls for a Nash–Harsanyi value and the Bowley solution is the outcome of a two-stage sequential game with perfect information concerning the news.

---

[11] Two examples where the same abstract game can be given different scenarios are the Ph.D. thesis of R. Simon, already noted, and a student paper by Bulow where the same business game was run calling the same control variable advertising in one run and product development in another.

In the Siegel–Fouraker experiments subjects were paid, instructed to maximize profits and anonymous to each other. Their moves were relayed through an intermediary. In their first experiments they obtained evidence that the selection of a Pareto-optimal outcome increased with increasing information about each others' profits.

The simple economic context of the Siegel–Fouraker experiments combined with care to maintain anonymity gave interesting results consistent with economic theory. In contrast the imaginative and rich teaching games of Raiffa (1983) involving face-to-face bargaining and complex information conditions illustrate how far we are from a broadly acceptable theory of bargaining backed with experimental evidence. Roth (1983), recognizing especially the difficulties in observing bargainers' preferences, has devised several insightful experiments on bargaining. In particular, he has concentrated on two-person bargaining where failure to agree gives the players nothing. This is a constant-threat game and the various cooperative fair division theories such as those of Nash, Shapley, Harsanyi, Maschler and Perles, Raiffa, and Zeuthen can all be considered.

Both in his book and in a perceptive article Roth (1979, 1983) has stressed the problems of the game-theoretic approach to bargaining and the importance of and difficulties in linking theory and experimentation. The value and fair division theories noted are predicated on the two individuals knowing their own and each others' preferences. Abstractly, a bargaining game is defined by a convex compact set of outcomes S and a point t in that set which represents the disagreement payoff. A solution to a bargaining game (S, t) is a rule applied to (S, t) which selects a point in S. The game theorist, by assuming external symmetry concerning the personalities and other particular features of the players and by making the solution depend just upon the payoff set and the no-agreement point, has thrown away most if not all of the social psychology.

Roth's experiments have been designed to allow the expected utility of the participants to be determined. Participants bargained over the probability that they would receive a monetary prize. If they did not agree within a specified time both received nothing. The experiment of Roth and Malouf (1979) was designed to see if the outcome of a binary lottery game[12] would depend on whether the players are explicitly informed of their opponent's prize. They experimented with a partial and a complete information condition and found that under incomplete information there was a tendency towards equal division of the lottery tickets, whereas with complete information the tendency was towards equal expected payoffs.

The experiment of Roth and Murnighan (1982) had one player with a $20 prize and the other with a $5 prize. A 4×2 factorial design of information conditions and common knowledge was used: (1) neither player knows his competitor's prize; (2) and (3) one does; (4) both do. In one set of instances the state of information is common knowledge, in the other it is not common knowledge.

The results suggested that the effect of information is primarily a function of whether the player with the smaller prize is informed about both. If this is common knowledge

---

[12] If a player obtained 40% of the lottery tickets he would have a 40% chance of winning his personal prize and a 60% chance of obtaining nothing.

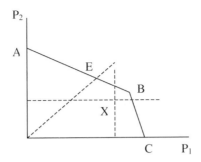

Figure 5.

there is less disagreement than if it is not. Roth and Schoumaker (1983) considered the proposition that if the players were Bayesian utility maximizers then the previous results could indicate that the effect of information was to change players' subjective beliefs.

The efforts of Roth and colleagues complement the observations of Raiffa (1983) on the complexities faced in experimenting with bargaining situations. They clearly stress the importance of information. They avoid face-to-face complications and they bring new emphasis to the distinction that must be made between behavioral and normative approaches. Perhaps one of the key overlooked factors in the understanding of competition and bargaining is that they are almost always viewed by economists and game theorists as resolution mechanisms for individuals who know what things are worth to themselves and others. An alternative view is that the mechanisms are devices which help to clarify and to force individuals to value items where previously they had no clear scheme of evaluation. In many of the experiments money is used for payoffs possibly because it is just about the only item which is clearly man-made and artificial that exists in the economy in such a form that it is normatively expected that most individuals will have a conscious value for it even though they have hardly worked out valuations for most other items.

Schelling (1961) suggested the possible importance of "natural" focal points in games. Stone (1958) performed a simple experiment in which two players were required to select a vertical and a horizontal line respectively on a payoff diagram as shown in Figure 5. If the two lines intersect within or on the boundary (indicated by *ABC*) the players receive payoffs determined by the intersection of the lines (*X* provides an example). If the lines intersect outside of *ABC* then both players receive nothing.

Stone's results show a bimodal distribution with the two points selected being *B*, a "prominent point" and *E*, an equal-split point. In class I have run a one-sided version of the Stone experiment six times. The results are shown in Table 3 and are bimodal.

A more detailed discussion of some of the more recent work[13] on experimental gaming in economics is to be found in the survey by Roth (1988), which includes the consid-

---

[13] A bibliography on earlier work together with commentary can be found in Shubik (1975b). A highly useful source for game experiment literature is the volumes edited by Sauermann starting with Sauermann (1967–78).

	E	B	Other offers*
1980	9	16	13
1981	10	11	11
1983	12	8	4
1984	8	6	5
1985	8	10	7
1988	24	15	7

Table 3

* These are scattered on several other points.

erable work on bidding and auction mechanisms [a good example of which is provided by Radner and Schotter (1989)].

5. Experimentation and operational advice

Possibly the closest that experimental economics comes to military gaming is in the testing of specific mechanisms for economic allocation and control of items involving some aspects of public goods. Plott (1987) discusses policy applications of experimental methods. This is akin to Roth's category of "whispering into the ears of princes", or at least bureaucrats. He notes agenda manipulation for a flying club [Levine and Plott (1977)], rate-filing policies for inland water transportation [Hong and Plott (1982)] and several other examples of institution design and experimentation. Alger (1986) considers the use of oligopoly games for policy purposes in the control of industry. Although this work is related to the concerns of the game theorist and indicates how much behavior may be guided by structure, it is more directly in the domain of economic application than aimed at answering the basic questions of game theory.

6. Experimental social psychology, political science and law

6.1. Social psychology

It must be stressed that the same experimental game can be approached from highly different viewpoints. Although interdisciplinary work is often highly desirable, much of experimental gaming has been carried out with emphasis on a single discipline. In particular, the vast literature on the Prisoner's Dilemma contains many experiments strictly in social psychology where the questions investigated include how play is influenced by sex differences or differences in nationality.

6.2. Political science

We may divide much of gaming experiments in political science into two parts. One is concerned with simple matrix experiments used for the most part for analogies to situations involving threats and international bargaining. The second is devoted to experimental work on voting and committees.

The gaming section of *Conflict Resolution* contains a large selection of articles heavily oriented towards 2×2 matrix games with the emphasis on the social psychology and political science interpretations. Much of the political science gaming activities are not of prime concern to the game theorist.

The experiments dealing with voting and committees are more directly related to strictly game-theoretic concerns. An up-to-date survey is provided by McKelvey and Ordeshook (1987). They note two different ways of viewing much of the work. One is a test of basic propositions and the other an attempt to gain insights into poorly understood aspects of political process. In the development of game theory at the different levels of theory, application and experimentation, a recognition of the two different aspects of gaming is critical. The cooperative form is a convenient fiction. In the actual playing of games, mechanisms and institutions cannot be avoided. Frequently the act of constructing the playable game is a means of making relevant case distinctions and clarifying ill-specified models used in theory.

6.3. Law

There is little of methodological interest to the game theorist in experimental law and economics. A recent lengthy survey article by Hoffman and Spitzer (1985) provides an extensive bibliography on auctions, sealed bids, and experiments with the provision of public goods (a topic on which Plott and colleagues have done considerable work). The work on both auctions and public goods is of substantive interest to those interested in mechanism design. But essentially as yet there is no indication that the intersection between law, game theory and experimentation is more than some applied industrial organization where the legal content is negligible beyond a relatively simplistic misinterpretation of the contextual setting of threats.

The cultural and academic gap between game theorists and the laissez-faire school of law and economics is typified by virtually totemistic references to "The Coase Theorem". This theorem, to the best of this writer's understanding, amounts to a casual comment by Coase (1960) to the effect that two parties who can harm each other, but who can negotiate, will arrive at an efficient outcome. Hoffman and Spitzer (1982) offer some experimental tests of the Coase so-called theorem, but as there is no well-defined context-free theorem, it is somewhat difficult to know what they are testing beyond the proposition that if people can threaten each other but also can make mutually beneficial deals they may do so.

7. Game theory and military gaming

The earliest use of formal gaming appears to have been by the military. A history and discussion of war gaming has been given elsewhere [Brewer and Shubik (1979)]. The expenditures of the military on games and simulations for training and planning purposes are orders of magnitude larger than the expenditures of all of the social sciences on experimental gaming. Although the games devoted to doctrine, i.e., the actual testing of overall planning procedures, can in some sense be regarded as research on the relevance and viability of overall strategies, at least in the open literature there is surprisingly little connection between the considerable activity in war gaming and the academic research community concerned with both game theory and gaming.

War games such as the Global War Game at the U.S. Naval War College may involve up to around 800 players and staff and take many weeks spread over several years to play. The compression of game time as compared to clock time is not far from one to one. In the course of play threats and counterthreats are made. Real fears are manifested as to whether the simulated war will go nuclear. Yet there is little if any connection between these activities and the scholarly game-theoretic literature on threats.

8. Where to with experimental gaming?

Is there such a thing as context-free experimental gaming? At the most basic level the answer must be no. The act of running an experimental game involves process. Games that are run take time and require the specification of process rules.

(1) In spite of the popularity of the 2×2 matrix game and the profusion of cases for the $2 \times 2 \times 2$ and 3×3, several basic theoretical and experimental questions remain to be properly formulated and answered. In particular with the 2×2 game there are many special cases, games with names which have attracted considerable attention. What is the class of special games for the larger examples? Are there basic new phenomena which appear with larger matrix games? What, if any, are the special games for the $2 \times 2 \times 2$ and the 3×3?

(2) Setting aside game theory, there are basic problems with the theory of choice under risk. The critique of Kahneman and Tversky (1979, 1984) and various explanations of the Allais paradox indicate that our understanding of one-person reaction to risk is still open to new observations, theory and experimentation. It is probable that different professions and societies adopt different ways to cope with individual risk. This suggests that studies of fighter pilots, scuba divers, high rise construction workers, demolition squads, mountaineers and other vocations or avocations with high-risk characteristics is called for.

The social psychologists indicate that group pressures may influence corporate risk-taking [Janis (1982)]. In the economics literature it seems to have been overlooked that probably over 80% of the economic decisions made in a society are *fiduciary* decisions with someone acting with someone else's money or life at stake. Yet no experimentation and little theory exists to account for fiduciary risk behavior.

A socialized individual has a far better perception of the value of money than the value of thousands of goods. Yet little direct consideration has been given to the important role of money as an intermediary between goods and their valuation.

(3) The experimental evidence is overwhelming that one-point predictions are rarely if ever confirmed in few-player games. Thus it appears that solutions such as the value

are best considered as normative or as benchmarks for the behavior of abstract players. In mass markets or voting much (but not all) of the social psychology may be wiped out: hence chances for prediction are improved. For few-person games more explicitly interdisciplinary work is called for to consider how much of the observations are being explained by game theory, personal or social factors.

(4) Game-theoretic solutions proposed as normative procedures can be viewed as what we should teach the public, or as reflecting what we believe public norms to be. Solutions proposed as reflecting actual behavior (such as the noncooperative theory) require a different treatment and concern for experimental validation. The statement that individuals should select the Pareto-optimal noncooperative equilibrium if there is one is representative of a blend of behavioral and normative considerations. There is a need to clarify the relationship between behavioral and normative assumptions.

(5) Little attention appears to have been paid the effects of time. Roth, Murnighan and Schoumaker (1988) have evidence concerning the importance of the end effect in bargaining. But both at the level of theory and experimentation clock time seems to have been of little concern, even though in military operational gaming questions concerning how time compression or expansion influences the game are of considerable concern.

(6) The explicit recognition that increasing numbers of players change the nature of communication and information[14] requires more exploration of numbers between 3 and say 20. How many is many has game-theoretical, psychological and socio-psychological dimensions. One can study how game-theoretic solutions change with numbers, but at the same time psychological and socio-psychological possibilities change.

(7) Game-theoretic investigations have cast light on phenomena such as bluff, threat and false revelation of preference. All of these items at one time might have been regarded as being almost completely in the domain of psychology or social psychology, yet they were susceptible to game-theoretic analysis. In human conflict and cooperation words such as faith, hope, charity, envy, rage, revenge, hate, fear, trust, honor and rectitude all appear to play a role. As yet they do not appear to have been susceptible to game-theoretic analysis and perhaps they may remain so if we maintain the usual model of rational man. It might be different if a viable theory can be devised to consider *context-rational man* who, although he may calculate and try to optimize, is limited in terms of capacity constraints on perception, memory, calculation and communication. It is possible that the instincts and emotions are devices that enable our capacity-constrained rational man to produce an aggregated cognitive map of the context of his environment simple enough that he can act reasonably well with his limited facilities. Furthermore, actions based on emotion such as rage may be powerful aggregated signals.

The development of game theory represented an enormous step forward in the realization of the potential for rational calculation. Yet paradoxically it showed how enormous the size of calculation could become. Computation, information and communication

---

[14] Around thirty years ago at MIT there was considerable interest in the structure of communication networks. None of this material seems to have made any mark on game-theoretic thought.

capacity considerations suggest that at best in situations of any complexity we have common context rather than common knowledge.

There is constant feedback between theory, observation and experimentation. Experimental game theory is only at its beginning but already some of the messages are clear. Context matters. The game-theoretic model of rational man is not an idealization but an approximation of a far more complex creature which performs under severe constraints which appear in the concerns of the psychologists and social psychologists more than in the game-theoretic literature.

References

Alger, D. (1986), Investigating Oligopolies within the Laboratory (Bureau of Economics, Federal Trade Commission).

Aumann, R.J., and M. Maschler (1964), "The bargaining set for cooperative games", in: M. Dresher, L.S. Shapley and A.W. Tucker, eds., Advances in Game Theory (Princeton University Press, Princeton, NJ), 443–476.

Battaglio, R.C., J.H. Kagel and D. McDonald (1985), "Animals' choices over uncertain outcomes: Some initial experimental results", American Economic Review 75:597–613.

Bellman, R., C.E. Clark, D.G. Malcolm, C.J. Craft and F.M. Ricciardi (1957), "On the construction of a multistage multiperson business game", Journal of Operations Research 5:469–503.

Bowley, A.L. (1928), "On bilateral monopoly", The Economic Journal 38:651–659.

Brewer, G., and M. Shubik (1979), The War Game (Harvard University Press, Cambridge, MA).

Brown, J.N., and R.W. Rosenthal (1987), Testing the Minimax Hypothesis: A Re-Examination of O'Neill's Game Experiment (Department of Economics, SUNY, Stony Brook, NY).

Caplow, T. (1956), "A theory of coalitions in the triad", American Sociological Review 21:489–493.

Chamberlain, E.H. (1948), "An experimental imperfect market", Journal of Political Economy 56:95–108.

Coase, R. (1960), "The problem of social cost", The Journal of Law and Economics 3:1–44.

Edgeworth, F.Y. (1881), Mathematical Psychics (Kegan Paul, London).

Fouraker, L.E., M. Shubik and S. Siegel (1961), "Oligopoly bargaining: The quantity adjuster models", Research Bulletin 20 (Department of Psychology, Pennsylvania State University). Partially reported in: L.E. Fouraker and S. Siegel (1963), Bargaining Behavior (McGraw Hill, Hightstown, NJ).

Frenkel, O. (1973), "A study of 78 non-iterated ordinal 2×2 games", University of Toronto (unpublished).

Friedman, J. (1967), "An experimental study of cooperative duopoly", Econometrica 33:379–397.

Friedman, J. (1969), "On experimental research on oligopoly", Review of Economic Studies 36:399–415.

Friedman, J., and A. Hoggatt (1980), An Experiment in Noncooperative Oligopoly: Research in Experimental Economics, Vol. 1, Supplement 1 (JAI Press, Greenwich CT).

Gamson, W.A. (1961), "An experimental test of a theory of coalition formation", American Sociological Review 26:565–573.

Goldstick, D., and B. O'Neill (1988), "Truer", Philosophy of Science 55:583–597.

Güth, W., R. Schmittberger and B. Schwartz (1982), "An experimental analysis of ultimatum bargaining", Journal of Economic Behavior and Organization 3:367–388.

Guyer, M., and H. Hamburger (1968), "A note on a taxonomy of 2×2 games", General Systems 13:205–219.

Hausner, M., J. Nash, L.S. Shapley and M. Shubik (1964), "So long sucker: A four-person game", in: M. Shubik, ed., Game Theory and Related Approaches to Social Behavior (Wiley, New York), 359–361.

Hoffman, E., and M.L. Spitzer (1982), "The Coase Theorem: Some experimental tests", Journal of Law and Economics 25:73–98.

Hoffman, E., and M.L. Spitzer (1985), "Experimental law and economics: An introduction", Columbia Law Review 85:991–1031.

Hoggatt, A.C. (1959), "An experimental business game", Behavioral Science 4:192–203.

Hoggatt, A.C. (1969), "Response of paid subjects to differential behavior of robots in bifurcated duopoly games", Review of Economic Studies 36:417–432.

Hong, J.T., and C.R. Plott (1982), "Rate-filing policies for Inland water transportation: An experimental approach", Bell Journal of Economics 13:1–19.

Kagel, J.H. (1987), "Economics according to the rats (and pigeons too): What have we learned and what can we hope to learn?" in: A.E. Roth, ed., Laboratory Experimentation in Economics: Six Points of View (Cambridge University Press, Cambridge), 155–192.

Kagel, J.H., R.C. Battaglio, H. Rachlin, L. Green, R.L. Basmann and W.R. Klemm (1975), "Experimental studies of consumer demand behavior using laboratory animals", Economic Inquiry 13:22–38.

Kagel, J.H., and A.E. Roth (1985), Handbook of Experimental Economics (Princeton University Press, Princeton, NJ).

Kahan, J.P., and A. Rapoport (1984), Theories of Coalition Formation (L. Erlbaum Associates, Hillsdale, NJ).

Kahneman, D., and A. Tversky (1979), "A prospect theory: An analysis of decisions under risk", Econometrica 47:263–291.

Kahneman, D., and A. Tversky (1984), "Choices, values and frames", American Psychologist 39:341–350.

Kalish, G., J.W. Milnor, J. Nash and E.D. Nering (1954), "Some experimental N-person games", in: R.H. Thrall, C.H. Coombs and R.L. Davis, eds., Decision Processes (John Wiley, New York), 301–327.

Levine, M.E., and C.R. Plott (1977), "Agenda influence and its implications", Virginia Law Review 63:561–604.

Lieberman, B. (1960), "Human behavior in a strictly determined 3×3 matrix game", Behavioral Science 5:317–322.

Lieberman, B. (1962), "Experimental studies of conflict in some two-person and three-person games", in: J.H. Criswell et al., eds., Mathematical Methods in Small Group Processes (Stanford University Press, Stanford, CA), 203–220.

Liu, T.C. (1973), "Gaming and oligopolistic competition", Ph.D. Thesis (Department of Administrative Sciences, Yale University, New Haven, CT).

Maschler, M. (1963), "The power of a coalition", Management Science 10:8–29.

Maschler, M. (1978), "Playing an N-person game: An experiment", in: H. Sauermann, ed., Coalition-Forming Behavior: Contributions to Experimental Economics, Vol. 8 (Mohr, Tübingen), 283–328.

McKelvey, R.D., and P.C. Ordeshook (1987), "A decade of experimental research on spatial models of elections and committees", SSWP 657 (California Institute of Technology, Pasadena, CA).

Miller, G.A. (1956), "The magic number seven, plus or minus two: Some limits on our capacity for processing information", Psychological Review 63:81–97.

Morin, R.E. (1960), "Strategies in games with saddlepoints", Psychological Reports 7:479–485.

Mosteller, F., and P. Nogee (1951), "An experimental measurement of utility", Journal of Political Economy 59:371–404.

Murnighan, J.K., A.E. Roth and F. Schoumaker (1988), "Risk aversion in bargaining: An experimental study", Journal of Risk and Uncertainty 1:95–118.

Nash, J.F., Jr. (1951), "Noncooperative games", Annals of Mathematics 54:289–295.

O'Neill, B. (1986), "International escalation and the dollar auction", Conflict Resolution 30:33–50.

O'Neill, B. (1987), "Nonmetric test of the minimax theory of two-person zero-sum games", Proceedings of the National Academy of Sciences, USA 84:2106–2109.

O'Neill, B. (1991), "Comments on Brown and Rosenthal's reexamination", Econometrica 59:503–507.

Plott, C.R. (1987), "Dimensions of parallelism: Some policy applications of experimental methods", in: A.E. Roth, ed., Laboratory Experimentation in Economics: Six Points of View (Cambridge University Press, Cambridge) 193–219.

Powers, I.Y. (1986), "Three essays in game theory", Ph.D. Thesis (Yale University, New Haven, CT).

Radner, R., and A. Schotter (1989), "The sealed bid mechanism: An experimental study", Journal of Economic Theory 48:179–220.

Raiffa, H. (1983), The Art and Science of Negotiation (Harvard University Press, Cambridge, MA).

Rapoport, A., M.J. Guyer and D.G. Gordon (1976), The 2 × 2 Game (University of Michigan Press, Ann Arbor).

Roth, B. (1975), "Coalition formation in the triad", Ph.D. Thesis (New School of Social Research, New York, NY).

Roth, A.E. (1979), Axiomatic Models of Bargaining, Lecture Notes in Economics and Mathematical Systems, Vol. 170 (Springer-Verlag, Berlin).

Roth, A.E. (1983), "Toward a theory of bargaining: An experimental study in economics", Science 220:687–691.

Roth, A.E. (1986), "Laboratory experiments in economics", in: T. Bewely, ed., Advances in Economic Theory (Cambridge University Press, Cambridge), 245–273.

Roth, A.E. (1987), Laboratory Experimentation in Economics: Six Points of View (Cambridge University Press, Cambridge).

Roth, A.E. (1988), "Laboratory experimentation in economics: A methodological overview", The Economic Journal 98:974–1031.

Roth, A.E., and M. Malouf (1979), "Game-theoretic models and the role of information in bargaining", Psychological Review 86:574–594.

Roth, A.E., and J.K. Murnighan (1982), "The role of information in bargaining: An experimental study", Econometrica 50:1123–1142.

Roth, A.E., J.K. Murnighan and F. Schoumaker (1988), "The deadline effect in bargaining: Some experimental evidence", American Economic Review 78:806–823.

Roth, A.E., and F. Schoumaker (1983), "Expectations and reputations in bargaining: An experimental study", American Economic Review 73:362–372.

Sauermann, H. (ed.) (1967–78), Contributions to Experimental Economics, Vols. 1–8 (Mohr, Tübingen).

Sauermann, H., and R. Selten (1959), "Ein Oligopolexperiment", Zeitschrift für die Gesamte Staatswissenschaft 115:427–471.

Schelling, J.C. (1961), Strategy and Conflict (Harvard University Press, Cambridge, MA).

Selten, R. (1982), "Equal division payoff bounds for three-person characteristic function experiments", in: R. Tietz, ed., Aspiration Levels in Bargaining and Economic Decision-Making, Lecture Notes in Economics and Mathematical Systems, Vol. 213 (Springer-Verlag, Berlin), 265–275.

Selten, R. (1987), "Equity and coalition bargaining in experimental three-person games", in: A.E. Roth, ed., Laboratory Experimentation in Economics: Six Points of View (Harvard University Press, Cambridge, MA), 42–98.

Shubik, M. (1962), "Some experimental non-zero-sum games with lack of information about the rules", Management Science 8:215–234.

Shubik, M. (1970), "A note on a simulated stock market", Decision Sciences 1:129–141.

Shubik, M. (1971a), "The dollar auction game: A paradox in noncooperative behavior and escalation", The Journal of Conflict Resolution 15:109–111.

Shubik, M. (1971b), "Games of status", Behavioral Sciences 16:117–129.

Shubik, M. (1975a), Games for Society, Business and War (Elsevier, Amsterdam).

Shubik, M. (1975b), The Uses and Methods of Gaming (Elsevier, Amsterdam).

Shubik, M. (1978), "Opinions on how one should play a three-person nonconstant-sum game", Simulation and Games 9:301–308.

Shubik, M. (1986), "Cooperative game solutions: Australian, Indian and U.S. opinions", Journal of Conflict Resolution 30:63–76.

Shubik, M., and M. Reise (1972), "An experiment with ten duopoly games and beat-the-average behavior", in: H. Sauermann, ed., Contributions to Experimental Economics (Mohr, Tübingen), 656–689.

Shubik, M., and G. Wolf (1977), "Beliefs about coalition formation in multiple-resource three-person situations", Behavioral Science 22:99–106.

Shubik, M., G. Wolf and H.B. Eisenberg (1972), "Some experiences with an experimental oligopoly business game", General Systems 17:61–75.

Shubik, M., G. Wolf and S. Lockhart (1971), "An artificial player for a business market game", Simulation and Games 2:27–43.

Shubik, M., G. Wolf and B. Poon (1974), "Perception of payoff structure and opponent's behavior in repeated matrix games", Journal of Conflict Resolution 18:646–656.

Siegel, S., and L.E. Fouraker (1960), Bargaining and Group Decision-Making: Experiments in Bilateral Monopoly (Macmillan, New York).

Simon, R.I. (1967), "The effects of different encodings on complex problem solving", Ph.D. Dissertation (Yale University, New Haven, CT).

Smith, V.L. (1965), "Experimental auction markets and the Walrasian hypothesis", Journal of Political Economy 73:387–393.

Smith, V.L. (1979), Research in Experimental Economics, Vol. 1 (JAI Press, Greenwich, CT).

Smith, V.L. (1986), "Experimental methods in the political economy of exchange", Science 234:167–172.

Stone, J.J. (1958), "An experiment in bargaining games", Econometrica 26:286–296.

Tegler, A.I. (1980), Too Much Invested to Quit (Pergamon, New York).

Zeuthen, F. (1930), Problems of Monopoly and Economic Warfare (Routledge, London).

AUTHOR INDEX

n indicates citation in footnote.

SUBJECT INDEX

HANDBOOKS IN ECONOMICS

1. HANDBOOK OF MATHEMATICAL ECONOMICS (in 4 volumes)
   Volumes 1, 2 and 3 edited by Kenneth J. Arrow and Michael D. Intriligator
   Volume 4 edited by Werner Hildenbrand and Hugo Sonnenschein

2. HANDBOOK OF ECONOMETRICS (in 6 volumes)
   Volumes 1, 2 and 3 edited by Zvi Griliches and Michael D. Intriligator
   Volume 4 edited by Robert F. Engle and Daniel L. McFadden
   Volume 5 edited by James J. Heckman and Edward Leamer
   Volume 6 is in preparation (editors James J. Heckman and Edward Leamer)

3. HANDBOOK OF INTERNATIONAL ECONOMICS (in 3 volumes)
   Volumes 1 and 2 edited by Ronald W. Jones and Peter B. Kenen
   Volume 3 edited by Gene M. Grossman and Kenneth Rogoff

4. HANDBOOK OF PUBLIC ECONOMICS (in 4 volumes)
   Edited by Alan J. Auerbach and Martin Feldstein

5. HANDBOOK OF LABOR ECONOMICS (in 5 volumes)
   Volumes 1 and 2 edited by Orley C. Ashenfelter and Richard Layard
   Volumes 3A, 3B and 3C edited by Orley C. Ashenfelter and David Card

6. HANDBOOK OF NATURAL RESOURCE AND ENERGY ECONOMICS
   (in 3 volumes). Edited by Allen V. Kneese and James L. Sweeney

7. HANDBOOK OF REGIONAL AND URBAN ECONOMICS (in 3 volumes)
   Volume 1 edited by Peter Nijkamp
   Volume 2 edited by Edwin S. Mills
   Volume 3 edited by Paul C. Cheshire and Edwin S. Mills

8. HANDBOOK OF MONETARY ECONOMICS (in 2 volumes)
   Edited by Benjamin Friedman and Frank Hahn

9. HANDBOOK OF DEVELOPMENT ECONOMICS (in 4 volumes)
   Volumes 1 and 2 edited by Hollis B. Chenery and T.N. Srinivasan
   Volumes 3A and 3B edited by Jere Behrman and T.N. Srinivasan

10. HANDBOOK OF INDUSTRIAL ORGANIZATION (in 3 volumes)
    Volumes 1 and 2 edited by Richard Schmalensee and Robert R. Willig
    Volume 3 is in preparation (editors Mark Armstrong and Robert H. Porter)

11. HANDBOOK OF GAME THEORY with Economic Applications (in 3 volumes)
    Edited by Robert J. Aumann and Sergiu Hart

12. HANDBOOK OF DEFENSE ECONOMICS (in 1 volume)
Edited by Keith Hartley and Todd Sandler

13. HANDBOOK OF COMPUTATIONAL ECONOMICS (in 1 volume)
Edited by Hans M. Amman, David A. Kendrick and John Rust

14. HANDBOOK OF POPULATION AND FAMILY ECONOMICS (in 2 volumes)
Edited by Mark R. Rosenzweig and Oded Stark

15. HANDBOOK OF MACROECONOMICS (in 3 volumes)
Edited by John B. Taylor and Michael Woodford

16. HANDBOOK OF INCOME DISTRIBUTION (in 1 volume)
Edited by Anthony B. Atkinson and François Bourguignon

17. HANDBOOK OF HEALTH ECONOMICS (in 2 volumes)
Edited by Anthony J. Culyer and Joseph P. Newhouse

18. HANDBOOK OF AGRICULTURAL ECONOMICS (in 4 volumes)
Edited by Bruce L. Gardner and Gordon C. Rausser

19. HANDBOOK OF SOCIAL CHOICE AND WELFARE (in 2 volumes)
Volume 1 edited by Kenneth J. Arrow, Amartya K. Sen and Kotaro Suzumura
Volume 2 is in preparation (editors Kenneth J. Arrow, Amartya K. Sen and Kotaro Suzumura)

FORTHCOMING TITLES

HANDBOOK OF RESULTS IN EXPERIMENTAL ECONOMICS
Editors Charles Plott and Vernon L. Smith

HANDBOOK OF ENVIRONMENTAL ECONOMICS
Editors Karl-Goran Mäler and Jeff Vincent

HANDBOOK OF THE ECONOMICS OF FINANCE
Editors George M. Constantinides, Milton Harris and René M. Stulz

HANDBOOK ON THE ECONOMICS OF GIVING, RECIPROCITY AND ALTRUISM
Editors Serge-Christophe Kolm and Jean Mercier Ythier

HANDBOOK ON THE ECONOMICS OF ART AND CULTURE
Editors Victor Ginsburgh and David Throsby

HANDBOOK OF ECONOMIC GROWTH
Editors Philippe Aghion and Steven N. Durlauf

All published volumes available